Advances in Intelligent Systems and Computing

Volume 1304

AF885854

The series "Advances in Intelligent Systems and Computing" contains publications on theory, applications, and design methods of Intelligent Systems and Intelligent Computing. Virtually all disciplines such as engineering, natural sciences, computer and information science, ICT, economics, business, e-commerce, environment, healthcare, life science are covered. The list of topics spans all the areas of modern intelligent systems and computing such as: computational intelligence, soft computing including neural networks, fuzzy systems, evolutionary computing and the fusion of these paradigms, social intelligence, ambient intelligence, computational neuroscience, artificial life, virtual worlds and society, cognitive science and systems, Perception and Vision, DNA and immune based systems, self-organizing and adaptive systems, e-Learning and teaching, human-centered and human-centric computing, recommender systems, intelligent control, robotics and mechatronics including human-machine teaming, knowledge-based paradigms, learning paradigms, machine ethics, intelligent data analysis, knowledge management, intelligent agents, intelligent decision making and support, intelligent network security, trust management, interactive entertainment, Web intelligence and multimedia.

The publications within "Advances in Intelligent Systems and Computing" are primarily proceedings of important conferences, symposia and congresses. They cover significant recent developments in the field, both of a foundational and applicable character. An important characteristic feature of the series is the short publication time and world-wide distribution. This permits a rapid and broad dissemination of research results.

Indexed by SCOPUS, DBLP, EI Compendex, INSPEC, WTI Frankfurt eG, zbMATH, Japanese Science and Technology Agency (JST), SCImago.

All books published in the series are submitted for consideration in Web of Science.

More information about this series at http://www.springer.com/series/11156

Madjid Tavana · Nadia Nedjah ·
Reda Alhajj
Editors

Emerging Trends in Intelligent and Interactive Systems and Applications

Proceedings of the 5th International Conference on Intelligent, Interactive Systems and Applications (IISA2020)

Springer

Editors
Madjid Tavana
The Business Systems and Analytics Department
La Salle University
Philadelphia, PA, USA

Nadia Nedjah
Departamento de Engenharia Eletrônica e Telecomunicações, Faculdade de Engenharia
Universidade do Estado do Rio de Janeiro
Rio de Janeiro, Brazil

Reda Alhajj
Department of Computer Science
University of Calgary
Calgary, AB, Canada

ISSN 2194-5357 ISSN 2194-5365 (electronic)
Advances in Intelligent Systems and Computing
ISBN 978-3-030-63783-5 ISBN 978-3-030-63784-2 (eBook)
https://doi.org/10.1007/978-3-030-63784-2

This Springer imprint is published by the registered company Springer Nature Switzerland AG
The registered company address is: Gewerbestrasse 11, 6330 Cham, Switzerland

Preface

IISA 2020, the 5th International Conference on Intelligent, Interactive Systems and Applications, provides a forum for presenting new results and developments, but—maybe even more—a forum for vivid discussions among international specialists from diverse areas and backgrounds. All submissions have been subjected to a rigorous and tough peer-review process. The program committee has selected over 100 full papers. The proceedings are organized into nine topics on different systems' orientations, such as analytical systems, database management systems, electronics systems, energy systems, intelligent systems, network systems, optimization systems, and pattern recognition systems and applications. It is noteworthy to point out that several special issues will be published based on extended versions of some of the best accepted and presented papers.

Research works on analytics systems investigate utility predominant to systems in daily routines. They apply to many real-world applications, such as health care systems and business-oriented systems. Analytic systems come at the intersections of many computer science fields, such as information infrastructures, data analytics, decision support systems, text analytics, and persuasive technologies. These systems are based mainly on analytic databases, which are at the heart of business intelligence and big data analytics, and usually function as part of more extensive data warehouses. They are popular because they offer faster query times, simpler maintenance, and easier scalability due to their less volatile nature. Analytic systems usually consider metrics such as effectiveness, efficiency, usability, portability, interpretability to determine the system's impact. Nowadays, these systems collect measurements from daily practices instead of only computational experiments. This requires a significant software engineering effort before determining the analytic system's utility. Strategic research in the development of analytic systems consists of improving transparency in the decision-making process. This proceedings reports 29 research works related to analytic systems.

Database management systems are essential to the development of efficient information systems. Data management is the process of preparing, storing, organizing, and maintaining the data collected by an organization. Effective data management is fundamental to run business applications and provide analytical information that assets in guiding executive decision-making and strategic planning business managers. Databases are the most common platform used to store corporate data. They contain a collection of data that is structured so that it can be managed, updated, and maintained effectively, yet efficiently. They are used in transaction processing systems that create operational data, such as customer records and sales orders, and data warehouses, which store consolidated data sets from business systems for data analytics. Three research papers related to analytic systems are included in this proceedings.

Electronics systems can be defined as a set of physical components that are connected in a way that implements the prescribed functionality. In general, the system relies on some input devices such as sensors to obtain information about the situation, processes the collected data via electronic gates, to yield results in the form of signals to be used to trigger some output actuators, such motors, to achieve the intended function. Electronic control systems are also viewed as a process that transforms one signal into another in order to give the desired system response. Electronic systems may be analog using continuous signals in time or digital handling only discrete signals in time. Of course, it is possible to have a hybrid electronic system based on both continuous and digital signals and components. Many real-world devices rely on an electronic control system. Related to electronic systems, the reader may enjoy eleven papers reported in this proceedings.

Efficient solutions to energy-related problems form a significant challenge for humanity in this century worldwide. Energy systems usually encompass power systems and their optimization, power generation, power trading, electricity risk management, among many other related aspects. Modeling is the best tool for designing effective yet efficient energy systems, including issues such as process optimization, synthesis, design, and operation. Nowadays, most of the research efforts regarding energy systems focus on new and clean energy, such as photovoltaic and wind systems. Such power systems are engineered to supply usable solar power by means of photovoltaics. They are composed of several such components, including solar panels to absorb and convert sunlight into electricity, a solar inverter to convert the output from direct to alternating current, as well as mounting, cabling, and other electrical accessories to set up an overall system. There are also research initiatives regarding electric car designs. This proceedings includes seven research papers related to efficient energy solutions.

Engineering concentrates mainly on finding and developing a new efficient solution to real-world problems. Engineering intelligent systems are one way to find these solutions. It is based on computing and artificial intelligence. In most solutions, an intelligent system senses and reacts to their environments. This type of system usually includes a processing unit, such a general-purpose processor and/or a dedicated one, to infer from the sensed data the right actions that should be applied to the environment in order to achieve what the system is supposed to do. Most intelligent systems are powered by a computational intelligence technique to solve the hard problem at hand. There are many such techniques, such as artificial neural networks, fuzzy logic, evolutionary computation, swarm intelligence, among others. This proceedings reports 19 research papers on intelligent systems.

Network systems are all about automating the transfer of information from one computer to another or from one part of the system to another. A fundamental shift in technology and concepts made networking systems possible. They explore a wide range of technology, specifically regarding software and hardware supporting different kinds of network topology configurations, such as LANs, WANs, and Wi-Fi networks. It is also noteworthy to point out the wireless sensor networks that allow developing solutions for many engineering problems that are distributed by their nature. Networked systems can also be regarded as high-performance computing systems, wherein many cores and multicores complex architectures form the foundation of the proposed solutions. This proceedings reports 16 research papers on network systems.

Optimization is a well-established discipline. Many of the real-world engineering problems can be rendered as optimization problems of a cost function that models all the behavior of the problem variables. In this case, one or several optimal or pseudo-optimal solutions of the defined cost function are sought. The problems considered are usually hard problems and cannot be solved using well-known analytical techniques. Thus, researchers usually exploit iterative exploratory processes to reach a good feasible solution to the problem at hand. Evolutionary and swarm-based computationally intelligent methods are used for each goal. Also, there is a lot of research work oriented toward finding good strategies to form the core of the techniques that help in the optimization process. This proceedings includes seven research works related to optimization systems.

Pattern recognition is the process of identifying patterns mostly by using machine learning. Pattern recognition can be defined as the classification of data based on prior knowledge about patterns that have their classes known. It is based on recognizing some regularities in the presented data and attempting to configure a model to recognize new patterns that have not been seen by the model. Pattern recognition is a technique at the heart of many smart systems. Pattern recognition systems are usually trained from known training data aiming at a general model that is able to recognize unknown data. Machine learning is a fundamental tool for pattern recognition. Computational vision is one of the fields that profits a large spectrum from pattern recognition to advance the state-of-the-art regarding smart system engineering. This proceedings reports four research papers on pattern recognition systems.

All the aforementioned research topics are applied to engineers leading to better and more efficient systems to advance the state-of-the-art toward a smarter human life. Thus, this proceedings also bring forth many reports on successful applications, ranging from abstract ones to more practical systems. This proceedings includes 24 applied research papers.

Besides the presentations of all the included diverse research works, the conference program also features some interesting cutting-edges keynote presentations. The plenary talks will focus on crucial topics that have recently attracted much attention and will be presented by outstanding experts from the area. We believe all participants will enjoy and benefit from the rich program.

It is a privilege and honor for us to participate in this edition of IISA 2020, and we hope we will be with you once again next year for IISA 2021. It has also been a rewarding experience, and we express our sincere thanks to all program committee members for their cooperation in shaping the conference and running the refereeing process. We highly appreciate the hard work and timely feedback from the reviewers who did an excellent job. Last but not least, we would like to thank all the authors, organizers, and participants for supporting IISA 2020. Thank you for your time and effort in preparing and submitting your papers and your patience through the long process. Your work is the backbone of the conference.

We wish you all a pleasant and fruitful meeting, and we look forward to your submission to and participation in IISA 2021.

Madjid Tavana
Nadia Nedjah
Reda Alhajj

Contents

Electronics Systems

Energy Systems

Network Systems

Pattern Recognition Systems

Analytical Systems

Consumer's Optimal Decision and the Role of Insurance in the Risk State—An Analytical Perspective Based on the Expected Utility Function

Wenting Cao[1,2](✉)

[1] School of Economics and Management, Yunnan Normal University, Kunming 650500, China
caowt@163.com
[2] School of Economics, Sichuan University, Chengdu 610065, China

Abstract. This paper starting from the hypothesis of "rational economic man", establishes a simple insurance decision model and combines the expected utility function to analyze the function and social value of insurance. The conclusion is as follows: Since the utility of definite income is greater than the utility of expected income under uncertain conditions, the purchase of insurance can enhance the welfare of the individual. Risk averse consumer purchases insurance is a rational economic behavior, there is a promotion in utility for individuals, and an improvement in welfare for society. Insurance companies should fully diversify risks and reduce the moral hazard of insurance contracts, so as to promote the effective operation of the insurance market.

Keywords: Expected utility function · Risk averse · Insurers' behavior · Policyholders' behavior · Insurance decision

1 Introduction

Insurance is a financial system arrangement based on economic security. Currently, the insurance industry manages risk by using engineering models and actuarial experience (Weimin Dong [1]). The purpose of insurance is to diversify and reduce risk. (Kunreuther H [2]). Insurance is sometimes compared to gambling, but from an economic point of view, the two functions are completely opposite. Gambling creates new risks, and insurance eliminates or greatly reduces existing risks. Just as architects can play an important role in helping society become more resilient (Roaf, S., Crichton, D. & Nicol, F [3]), so do insurance companies (Crichton, D [4]). The insurance industry has an increasingly important role in helping society to adapt and become more resilient (Crichton, D [5]). Currently, some ways in which insurers can help are: Assistance with identifying areas at risk; Catastrophe modelling; Economic incentives to discourage construction in the flood plain etc. (Crichton D [6]). From an individual's point of view, consumers replace large uncertain losses with small costs (insurance fees), and such losses will exist without insurance. In general, Insurance has the macro social utility and the micro utility of economic compensation. Based on the assumption of a

M. Tavana et al. (Eds.): IISA 2020, AISC 1304, pp. 3–10, 2021.
https://doi.org/10.1007/978-3-030-63784-2_1

rational economic person, this paper establishes a simple insurance decision-making model, combines the expected utility function to analyze the function and social value of insurance, and obtains relevant inspirations on how to promote the efficient operation of the insurance market.

2 Overview of Expected Utility Functions

2.1 Expression of Expected Utility Function

The economy consists of a group of economic participants (individual, family, enterprise, government, etc.). A participant's economic needs are described by his preference for different consumption plans (The consumption plan refers to the possible consumption choices of consumers. It depends on the future state of the economy and thus contains different realization values), which is the participant's ranking of all possible consumption plans. Another way of expressing preferences is the utility function. The utility function U depends on the probability of future states and the preference for consumption in different states. The expected utility function is the expected value of utility of different consumption paths (Consumption path refers to a specific realization value of a consumption plan, such as (c_0, c_{1w}) called a consumption path), that is, the utility of uncertain consumption plans. Since each consumption path corresponds to a possible state, and each state is mutually exclusive, the occurrence of one state means that the other states have not occurred. Therefore, the expected utility function has a very intuitive explanation: the utility obtained by consumption in an uncertain state is the weighted average of the utility obtained by the consumption path in each possible state, and the weight is the probability of the occurrence of the corresponding state. The expected utility function can be simply expressed as: $E[U(W)] = U(\pi;\quad W_1, W_2) = \pi U(W_1) + (1-\pi)U(W_2)$. If the uncertain event has n uncertain results: $W_1, W_2, \ldots W_n$, And there are n corresponding probabilities: $\pi_1, \pi_2, \ldots \pi_n$, The expected utility function can be written as: $E[U(W)] = U(\pi_1, \pi_2, \ldots \pi_n;\quad W_1, W_2, \ldots W_n) = \sum_{i=1}^{n} \pi_i \times U(W_i)$.

2.2 Expected Utility Functions for Different Risk Preferences

In reality, people have different attitudes towards risk. Some people are very cautious in their work and will take some risk management measures (such as purchasing insurance) for future uncertain events. Some people have a strong tolerance for risk and are willing to participate in various forms of gambling. Still others have an indifferent attitude towards risk, neither have a preference nor aversion.

(1) The expected utility function of a risk averse buyer. If a person is a risk consumer in an uncertain event, the utility of his expected value is greater than his expected utility, that is: $U[\pi W_1 + (1-\pi)W_2] > \pi U(W_1) + (1-\pi)U(W_2)$. The expected utility function at this time is strictly concave, that is, $d^2U/dW^2 < 0$, The consumer is a risk averse person.

(2) The expected utility function of a risk preference person, if a person is a risk preference consumer in an uncertain event, the utility of his expected value is less than his expected utility, that is: $U[\pi W_1 + (1-\pi)W_2] < \pi U(W_1) + (1-\pi) U(W_2)$. The expected utility function at this time is strictly convex. that is $d^2U/dW^2 > 0$, The consumer is a risk preference person.

(3) The expected utility function of a risk neutral person. For a risk neutral consumer, the utility of his expected value is equal to his expected utility, that is: $U[\pi W_1 + (1-\pi)W_2] = \pi U(W_1) + (1-\pi)U(W_2)$. The utility function at this time is a linear function, that is $d^2U/dW^2 = 0$.

3 Analysis of Insurance Decision Based on Expected Utility Function

In economics, risk averse buyer is considered a typical feature of human beings. A normative model of choice, such as expected utility theory, implies that risk averse consumers should value insurance, as it protects them against large losses relative to their wealth (Kunreuther H [2]). Does buying insurance improve the level of welfare for individuals? First, establish a simple insurance decision model, analyze how the premium rate and insurance amount are determined from the perspective of the insurance company (insurer) and consumer (policyholder), and then place it in the expected utility function of risk averse buyer, comparing the utility brought by different decision results.

3.1 Simple Insurance Decision Model

First, define the meaning of each variable in the simple insurance decision model. Suppose someone has a family wealth of W. If there is a risk (such as fire, theft, car accident, etc.), his wealth will suffer losses. Assume that the loss of wealth due to a risk is L, and the probability of such loss (risk) occurring is π. If the consumer has not purchased insurance in advance, his expected utility is $E(U) = (1-\pi) \times U(W) + \pi \times U(W-L)$. If the consumer has purchased insurance in advance (the insurance amount is K, the premium rate is γ, and the premium paid is γK), the consumer will get two results: No risk encountered (no loss occurred), his property is $W_1 = W - \gamma K$; Risk occurred (loss suffered) and his property is $W_2 = W - L + K - \gamma K$.

(1) Insurer Decision: How to Determine Premium Rates γ

As a rational economic entity, insurance company still follows the principle of profit maximization when making decisions. The profit of insurance company mainly comes from premium income, and the pricing of insurance products is a key issue in the decision-making process of insurance company. No matter whether the policyholder is out of insurance or not, the insurance company can always get the premium income of γK. Suppose there are a large number of consumers insured in the society, and the risks faced by each consumer are independent of each other. The insurance company's expected profit from each consumer is $\pi \times (\gamma K - K) + (1-\pi) \times \gamma K = \gamma K - \pi K$. If there are n consumers insured, the profit of the insurance company is $n \times (\gamma K - \pi K)$. Since there are many insurance companies in the insurance market, assuming that there is no threshold to restrict the entry and exit of the insurance industry, the insurance market can be approximated as a perfect competition market. Competition transfers capital from the department with low profit to the department with high profit. Finally, the economic profit of each enterprise in the perfect competition market is zero. The profit of the insurance company is zero, that is $n \times (\gamma K - \pi K) = 0$, And then introduce $\gamma = \pi$, That is, the insurance premium rate is equal to the probability that the insured person will suffer a total loss. This means that insurance company provides customers with a completely "fair" insurance premium rate under intense competitive pressure. In fact, this hypothesis has also been verified in reality. Many insurance companies with strong strengths, wide business and cross-regional spread their risks sufficiently, and the insurance premium rates they provide are very close to "fair" rates.

(2) Policyholder Decision: How to Determine the Amount of Insurance K

As a rational economic person, the policyholder still follows the principle of profit maximization when making decisions. The policyholder's expected utility function is $E(U) = (1 - \pi) \times U(W - \gamma K) + \pi \times U(W - L + K - \gamma K)$. In order to maximize the expected utility of the policyholder, let $\partial E(U)/\partial K = 0$. get $-\gamma(1 - \pi)U'(W - \gamma K) + (1 - \gamma)\pi U'(W - L + K - \gamma K) = 0$. Since insurance company uses "fair" insurance premium rate ($\gamma = \pi$), get $U'(W - \gamma K) = U'(W - L + K - \gamma K)$. As a risk averse person, his expected utility function is concave, the second derivative is negative, and the expected utility function is monotone with diminishing marginal utility. Since it is a monotonic form of function, equal marginal utility means that the amount of property on both sides of the equation is also equal, that is $W - \gamma K = W - L + K - \gamma K$, Thus introduce $K = L$. This means that when insurance company is pricing at "fair" insurance premium rate, risk averse buyer fully insures the property that may suffer losses (insurance amount = loss amount).

3.2 Insurance Decision Analysis Based on Expected Utility Function

A large number of statistics show that most people are risk averse consumers in most cases. Suppose a consumer is a risk averse buyer, and his expected utility function is shown in Fig. 1. Suppose the consumer has two different income W_1 and W_2, his expected value is $E(W) = \pi W_1 + (1 - \pi)W_2$, and the corresponding utility value is $U[E(W)] = C$, this is a level of utility at a risk-free income. At the same time, the expected utility corresponding to the two incomes W_1 and W_2 is $U(W) = \pi U(W_1) + (1 - \pi)U(W_2)$, this is a level of utility at risk, and the expected utility is some linear combination of $U(W_1)$ and $U(W_2)$. The value of the linear combination (the value of the expected utility) must be located at a point on the line AB (such as the T point). No matter where it is located, the corresponding utility value (S point) is lower than $U[E(W)]$ (the utility value corresponding to the C point). Figure 1 clearly shows that line AB is located below the utility function. The expected utility of W_1 and W_2 is $U(W) = \pi U(W_1) + (1 - \pi)U(W_2)$, which always corresponds to a certain income CE. Simultaneously the income of $E(W) = \pi W_1 + (1 - \pi)W_2$ can also be regarded as a certain value. The risk premium is regarded as the difference between $E(W)$ and CE ($P = E(W) - CE$), which can be seen as the price paid by consumer for risk.

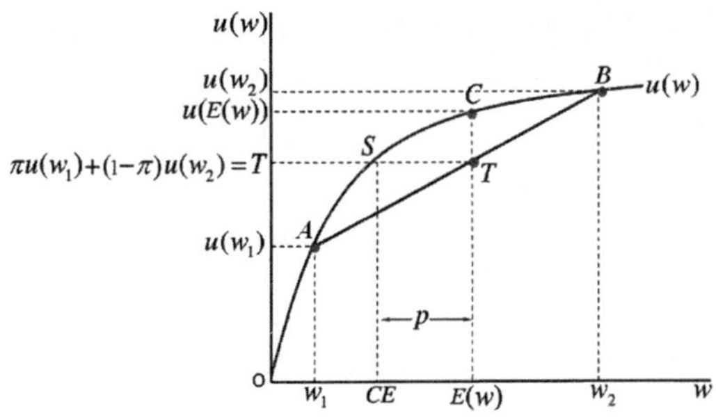

Fig. 1. Risk premium for risk averse buyer

For a consumer, if he does not encounter any risk after the insurance application, the premium already paid is γK and his property is $W - \gamma K$. If he encounters risks after he insures, his property is $W - L + K - \gamma K = W - \gamma K$ (because $K = L$). Therefore, regardless of whether or not the consumer encounters risks after insurance, his property is $W - \gamma K$. If the consumer refuses to buy insurance, the loss due to risk is L, and the probability of risk occurring is $\pi(\pi = \gamma)$. The expected income of the person who is not willing to buy insurance is $W - \pi L = W - \gamma K$. Although the number is the same as the consumer who bought the insurance, $W - \gamma K$ is the expected value under an uncertain condition. For a risk averse person, the expected utility function is strictly concave. The utility corresponding to $W - \gamma K$ under certain conditions is greater than the utility under uncertain conditions, that is $U(W - \gamma K) > \pi U(W - L) + (1 - \pi)U(W)$. So for individuals, purchasing insurance can increase their benefits (Fig. 2).

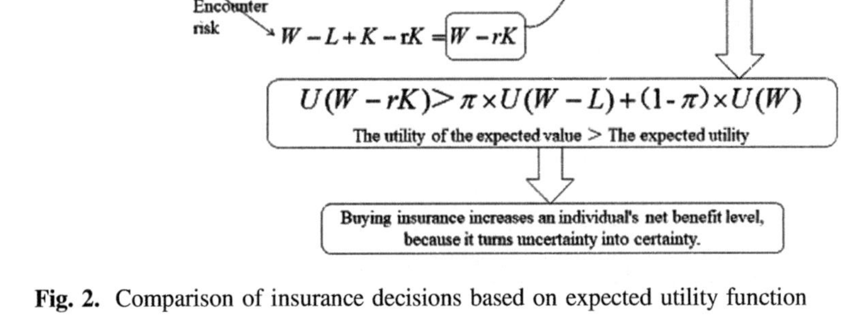

Fig. 2. Comparison of insurance decisions based on expected utility function

4 Practical Problems in the Application of Insurance: Will it Reduce the Level of Social Welfare?

4.1 Impact on Social Welfare of Insurer's Failure to Implement "Fair" Rate

In fact, in insurance practice, the insurance company's premium rate γ is not equal to (usually higher than) the probability of the total loss of the policyholders. Because, in reality, the premiums of insurance company are composed of pure premiums and additional premiums. The additional premiums must take into account the various operating costs and financial costs of the insurance company. Generally, the pricing actuarial assumptions of insurance company is conservative, and the premium rate is relatively high in order to generate more distributable surplus in practice. At this time, the price of insurance will be higher than the "fair" insurance price, and insurance company shares some of the net benefits with consumers. In general, it does not affect the improvement of social welfare, because turning uncertainty into certainty is an improvement in utility. Therefore, the premium rate γ is higher than the probability π of the insured persons' overall loss, which does not affect the establishment of the above conclusion.

4.2 Impact on Personal Utility of Policyholder's Failure to Adopt Full Insurance

Another important fact is that in the course of insurance practice, the policyholder does not always take the form of full insurance when applying for insurance. Because the policyholder cannot assess the size of future losses in advance (also refer to the market value of the insurance subject at the time of encountering risk), the determination of insurance amount not only considers the probability of the occurrence of the risk and the value of the insured object, but also considers the policyholder's own economic status. After the insurance amount is determined, in the actual process of claim, the insurance assessor must also evaluate the actual value of the insurance subject to obtain the degree of protection (Degree of protection = Sum insured/Insurance value. Compensation amount = Degree of protection × Amount of loss). The amount of compensation is determined based on the degree of protection. When the degree of protection is equal to 1 (full insurance) or greater than 1 (excess insurance), the compensation amount is equal to the loss amount, up to the insurance amount; When the degree of protection is less than 1 (insufficient insurance), the compensation obtained by the policyholder is part of the sum insured, at most it does not exceed the sum insured. Therefore, no matter whether it is full insurance, excess insurance or insufficient insurance, the maximum compensation amount does not exceed the sum insured. Therefore, the analysis of the assumption of full insurance ($K = L$) above does not violate the actual situation of insurance claims. It can represent the principle of loss compensation in insurance, that is, the purpose of insurance is to stop losses rather than to make profits. In general, policyholders diversify their risks through insurance, transform uncertain risks into definite results, and obtain loss compensation and economic payment when risks occur. This is an improvement of welfare for individuals and society.

4.3 Analyzing the Role of Insurance from the Perspective of Expected Utility Function

The concavity of the risk averse buyer's expected utility function suggests that utility under definite income is higher than the utility of the expected value under uncertainty. The role of insurance is also reflected in this. By purchasing insurance, each person spreads the risk to all relevant policyholders, converts the uncertainty of the future into certainty, and thus minimizes his own risk. While improving personal utility, it also improves the welfare of the entire society. The most immediate benefit of the insurance function (diversification risk, loss compensation and economic payments) is the Pareto improvement of the whole society.

5 Conclusion

Based on the analysis above, it can be concluded that the purchase of insurance by risk averse buyer is a rational economic behavior. There is an increase in utility for individuals and an improvement in welfare levels for society. How to promote the effective operation of the insurance market and make insurance effectively play the role of risk dispersion is worth thinking about. The prerequisite for the effective operation of the insurance market is that the insurance company can continue to operate. In the analysis of the insurer's decision mentioned above, an important assumption is mentioned that there are a large number of consumers insured in the society and the risks faced by consumers are independent of each other. When risk is not sufficiently independent, insurance cannot fully play the function of risk diversification, which is why some natural disasters with a wide range of coverage cannot be covered (exclusions for insurance contracts). Because some natural disasters (such as earthquakes, typhoons, etc.) occur once, the policyholders in the region cannot be spared, at this time, insurance companies cannot make up for losses by spreading risks. In the face of such large-scale risks, it is worth thinking about how to effectively diversify the risks. Some mature practices have been tried, such as establishing policy-based insurance institutions, developing insurance options, and issuing bonds to catastrophe reinsurance, etc. Another factor in ensuring that insurance markets work efficiently is to minimize moral hazard. If there is moral hazard, the premium rate calculated by the insurance company based on the probability in the original risk state will cause the insurance company to fall into a financial crisis and affect the ability of the insurance company to operate sustainably. For this reason, insurance companies will also make some restrictions when signing insurance contracts with policyholders, such as the statement of guarantee clauses in insurance contracts. The effective operation of the insurance market provides a more favorable market environment for various insurance products. The abundant insurance products provide consumers with the possibility to fully diversify their risks, which is conducive to expanding the coverage of insurance and ultimately achieving Pareto improvement of the whole social welfare.

References

1. Dong, W.: Paying the price: the status and role of insurance against natural disasters in the United States. Earthq. Spectra **15**(3), 599–600 (1999)
2. Kunreuther, H.: The role of insurance in reducing losses from extreme events: the need for public–private partnerships. Geneva Pap. Risk Insur. Issues Pract. **40**(4), 741–762 (2015)
3. Roaf, S., Crichton, D., Nicol, F.: Adapting Buildings and Cities for Climate Change, pp. 32–50. Architectural Press and Elsevier Press, Oxford (2009)
4. Crichton, D.: The role of private insurance companies in managing flood risks in the UK and Europe. In: Urban Flood Risk Management: Introduction-1st International Expert Meeting on Urban Flood Management, pp. 159–160 (2005)
5. Crichton, D.: What can cities do to increase resilience? Philos. Trans. Math. Phys. Eng. Sci. **365**(1860), 2731–2739 (2007)
6. Crichton, D.: Role of insurance in reducing flood risk. Geneva Pap. Risk Insur. Issues Pract. **33**(1), 117–132 (2008)

Correlation Analysis of Children's Stroller Information

Liu Xia[1], Chen Qianwen[1], Qiao Feng[1], and Pei Fei[2(✉)]

[1] China National Institute of Standardization, Beijing 100191, China
[2] China Standardization Press Co. Ltd., Beijing 102218, China
liuxia1010@163.com

Abstract. The concerns of the respondents on the stroller mainly focus on the appropriate age groups, materials, stability, and conformity certification marks, reaching 81.64%, 79.91%, 76.46%, and 74.95%, respectively. The purchase channels are mainly through the large stores or malls, specialty stores, online shopping, and large chain supermarkets, reaching 86.18%, 80.13%, 65.01%, and 56.16%, respectively. The consideration of the consumers on safety indicators focus on the stability, the rationality of the design, and irritating odors of related devices when purchasing a stroller. The correlation is analyzed among the ages, education levels, occupations, purchase channels and the related information attention degrees of the respondents. There are significant statistical differences among consumers' concerns when purchasing children's strollers. The correlation between the age characteristics and the channel choice for consumers to purchase the stroller is not obvious. The correlation of relevant chi-square test between the genders and consumer purchase channels is not obvious.

Keywords: Correlation · Children's stroller · Consumers · Endowments · Information

1 Introduction

Children's strollers are favored by parents as a standing item for babies to go out. However, there are also many blind spots with respect to the safety problems in the use of children's stroller. While bringing great convenience to parents, the safety accidents caused by the children's stroller are not common. Relevant research shows that 10.6% of the surveyed households have experienced injuries or dangerous situations in the use of children's stroller. The situation of the safe consumption of stroller products is not optimistic, for which 47.6% of parents do not know the provision that "the products in the catalog of children's stroller must be subjected to compulsory product certification (3C Certification) before leaving the factory, being sold, imported or used in other business activities. Only 44.75% of the parents know that the State has promulgated safety standards for children's bicycles. The awareness rate of the safety standards for children's tricycles, strollers, baby walkers, and electric strollers is very low, at 11.8%, 17.2%, 15.0%, and 11.3%, respectively. More than 85% of parents do not know to check the safety items that whether there are small parts, sharp tips, dangerous gaps on the stroller, whether the braking device, folding or locking mechanism is effective, whether the frame parts are firm, or whether the driving is stable.

M. Tavana et al. (Eds.): IISA 2020, AISC 1304, pp. 11–19, 2021.
https://doi.org/10.1007/978-3-030-63784-2_2

From the perspective of purchasing principles, different types of strollers should be selected and purchased according to the age groups and growth needs of children. Try best to buy a stroller product with a single function, and resolutely refuse to buy the stroller without the 3C mark. At the same time, attention shall be paid to the product recall information and consumption warning information released by the General Administration of Quality Supervision, Inspection and Quarantine, and stay away from the stroller products with hidden safety hazards in time. With our country has been paying more and more attention to the development of the domestic market of Children's stroller, the endowment characteristics for consumers to purchase children's stroller may have a certain influence on the selected products while the consumers with different endowments may have different concerns about the quality and safety of the Children's stroller.

2 Survey Conditions

There are 463 questionnaires distributed and returned. They are all valid questionnaires. The contents of the questionnaire include the basic information, the attention degree to the basic information on children's stroller, the purchase channels and so on.

2.1 Personal Information

In this survey, 25–35-year-old respondents accounted for the highest proportion, more than half of the sample size. It is accounted for 53.35% of the total number of the respondents. Then the following ages are 18–25 years old, accounting for 23.54%. The age range from 35–50 years old is accounted for 20.52%. There also is a small number of 50–65 years old (2.38%). The specific distributions are shown in Fig. 1.

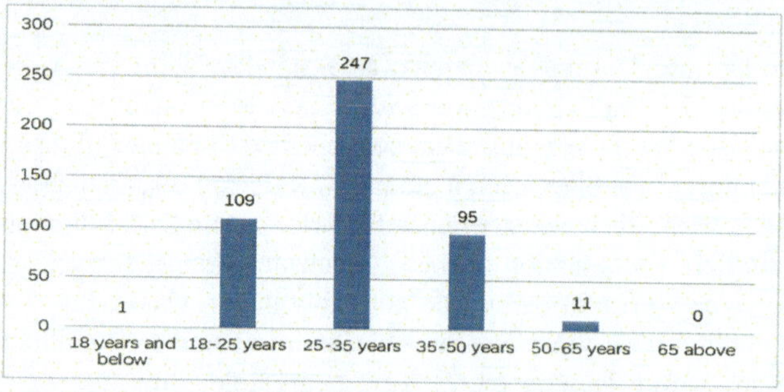

Fig. 1. Age distribution of the respondents

In this survey, the proportion of female respondents is nearly twice that of male respondents. There is a big difference between the sex distribution ratio (38.88:61.12) and the actual sex distribution ratio (104.90:100) in China. This deviation may lead to the distortion of the influence of the gender characteristics of consumers on the purchase channels, information concerns and judgment of harmful scenes in the data analysis stage. The ratio is that the male accounts for 38.88% and the female is 61.12%, which is shown in Fig. 2.

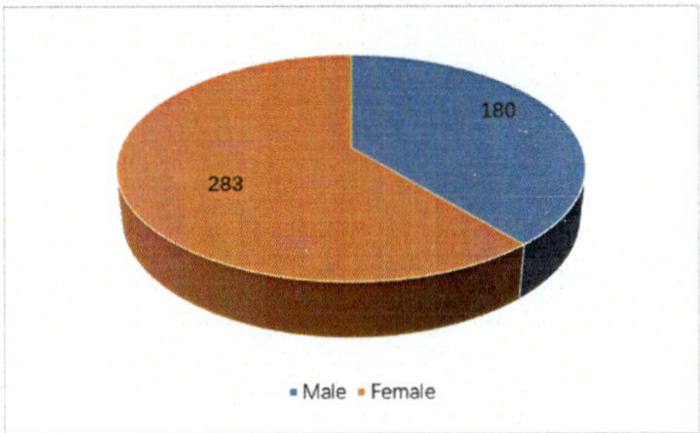

Fig. 2. Gender distribution of the respondents

In this survey, the overall education level of consumers is relatively high, accounting for more than 80% of the total of undergraduates, masters and above, of which undergraduates account for the largest proportion. This education distribution may have some impact on the data of consumers' purchase channels, information, safety indicators, and judgment of injurious scenes, resulting in the deviation between the collection results and the actual situation. The specific distributions are shown in Fig. 3.

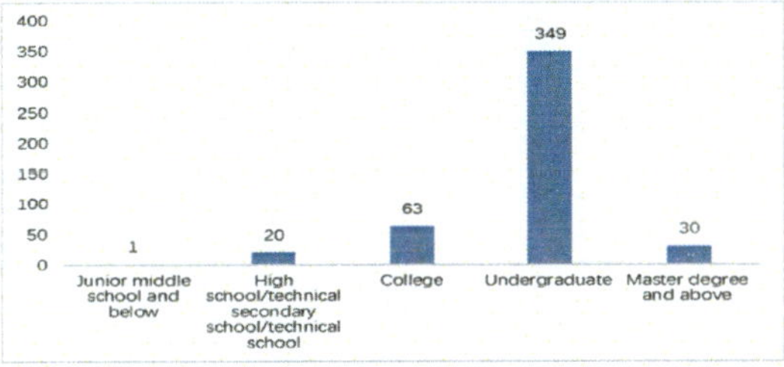

Fig. 3. Education level distribution of the of the respondents

Their occupation is mostly the company employees (49.89%), the professionals (teachers/doctors/lawyer, etc.) (17.28%), the public institution/civil servants/ government staff (12.74%) and students (8.86%), indicating that the cultural level of the respondents during this survey is relatively high, the specific distributions are shown in Fig. 4.

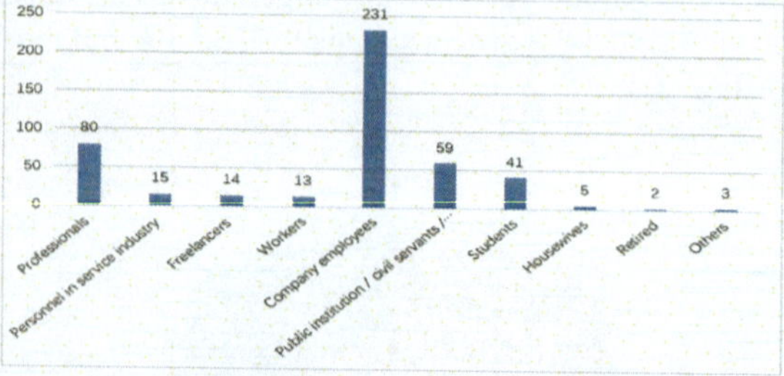

Fig. 4. Occupation distribution of respondents

2.2 Respondents Attention

The product related information is listed in Fig. 5. The ratio for appropriate age group is 81.64%. The ratios for texture of material, stability and price are 79.91%, 76.46% and 61.99%, respectively. The ratios for cautions, weight and user guide are 43 2%, 39.74% and 34.56%, respectively. The other factors include factory name, factory address, and brand (18.79%), place of origin-domestic (13.17%), place of origin-abroad (9.5%) and others (1.14%). Consumers pay special attention to the appropriate age group, the texture of material and the stability of the Children's stroller.

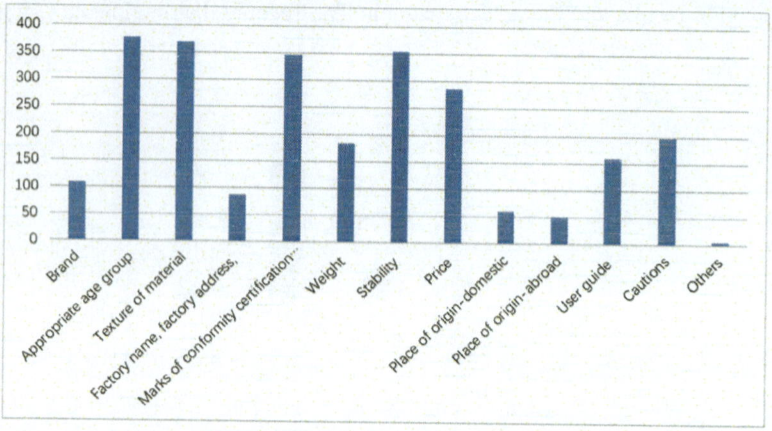

Fig. 5. Attention of respondents to children's stroller information

2.3 Purchase Channels of Respondents

The purchase channels to buy children's stroller include specialty stores, large store etc. The distributions are shown in Fig. 6. The respondents are mainly purchasing Children's stroller in large store/shopping malls (86.18%). The ratio in specialty stores is 80.13%. The ratio in on-line shopping is 65.01%. The ratio in large supermarket chains is 56.16%.

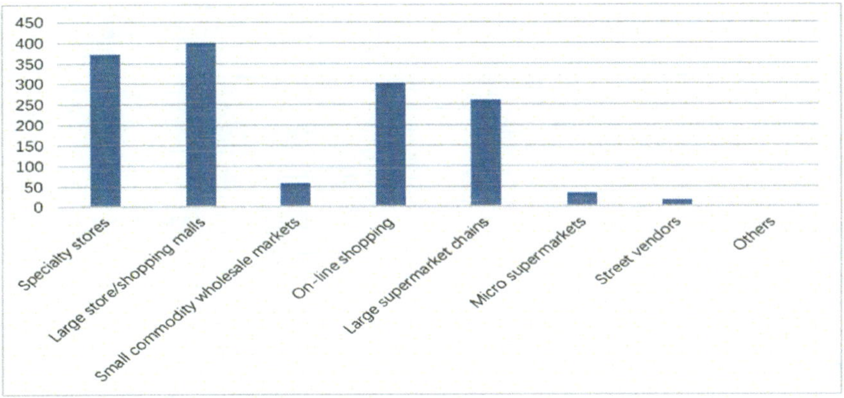

Fig. 6. Purchase channels of respondents

2.4 Considerations of Respondents

The safety indicators when buying Children's stroller mainly include the stability, the rationality of the design, the odor, etc. of the related devices. There are differences in the focus on different types of children's stroller, taking children's bicycles and babe strollers as examples, see Table 1 and Table 2 for specific indicators. Consumers' considerations on the safety indicators of children's bicycles focus on the braking device (85.96%), the protection of various exposed parts (81.86%), the rationality of functional protrusion (67.6%), too wide brake lever (56.16%), and chain cover (52.27%), etc. Consumers' considerations on the safety indicators of babe stroller focus on the puncture wound (81.43%), seat belt firmness (78.19%), vehicle body firmness (75.59%), unpleasant odor (63.07%), braking device (60.69%), etc.

Table 1. Proportion of the respondents considering safety indicators of children's bicycles

Options	Subtotals	Proportions
Braking device	398	85.96%
Chain cover	242	52.27%
Too wide brake lever	260	56.16%
Protection of exposed parts	379	81.86%
Rationality of protrusion	313	67.6%
Others	8	1.73%
Total	463	

Table 2. Proportion of the respondents considering safety indicators of babe strollers

Options	Subtotals	Proportions
Unpleasant odor	292	63.07%
Braking device	281	60.69%
Puncture wounds	377	81.43%
Vehicle body firmness	350	75.59%
Seat belt firmness	362	78.19%
Reliability of folding device	254	54.86%
Flame cloth	155	33.48%
Others	4	0.86%
Total	463	

3 Analysis on Correlation

3.1 Characteristics of Consumers and Purchase Channels

It can be seen from the cluster chart above that consumers of all ages take large stores or shopping malls, exclusive stores and online shopping as the three main purchasing channels, which is shown in Fig. 7. As can be seen from the cluster bar diagram above, consumers of all ages take large stores/malls, specialty stores, online shopping, and large chain supermarkets as the main purchasing channels, while relatively fewer consumers chose other channels. Since the chi-square test is the degree of deviation between the actual observed value and the theoretically inferred value of the statistical samples, such a degree of deviation between the actual observed value and the theoretically inferred value determines the size of the chi-square value. The larger the chi-square value is, the greater the deviation degree between the two is, and vice versa; if the two values are equal, the chi-square value will be zero, indicating that the theoretical values are completely consistent.

A chi-square test is performed on the correlation between age characteristics and the channels consumers choose to purchase a stroller with the test results shown in Table 3. It can get to know through the chi-square test that the correlation of the age characteristics to consumers' choice of channels for purchasing the stroller is not obvious. The correlative chi-square test between gender and consumers' choice of purchase channels is shown in Table 4. According to the chi-square test, it is known that the correlation of the gender characteristics to consumers' choice of channels for purchasing the stroller is not obvious.

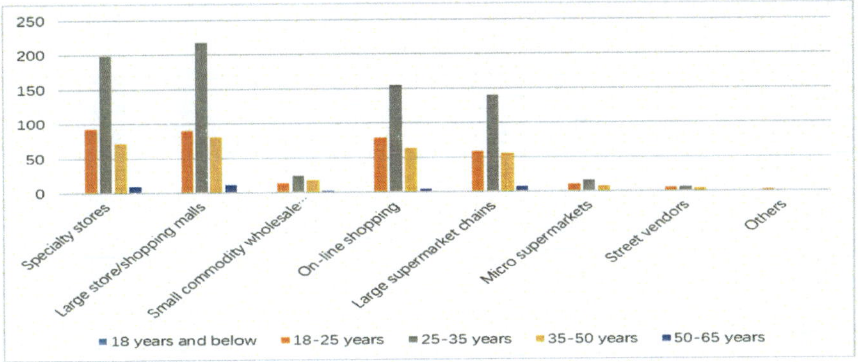

Fig. 7. Correlation between the characteristics of consumers and their purchase channels

Table 3. Chi-square test of age characteristics

	Value	df	Asymptotic Sig
Pearson's chi-squared test	18.509a	28	.913
Likelihood ratio	19.559	28	.880
Linear and linear combinations	.048	1	.827
N in valid cases	1437		

Table 4. Chi-square test of gender characteristics

	Value	df	Asymptotic Sig
Pearson's chi-squared test	2.319a	7	.940
Likelihood ratio	3.022	7	.883
Linear and linear combinations	.087	1	.768
N in valid cases	1298		

3.2 Consumers and Their Attention to the Stroller Information

As for the question of "What product-related information will you pay attention to when purchasing a children's stroller?" The collected data are the categorical variable, as can be seen from Table 5, the concerns of consumers focused on "Appropriate age group", "Texture of materials", "Stability", "Conformity certificate and national standards", and what is the secondary is the place for origin, factory name, and brand. A chi-square test is performed for such a categorical variable, from which it can be known that there is a significant statistical difference ($p < 0.05$) between the concerns of consumers when purchasing a stroller. The value of Chi-square is 1093.789. The value of df is 12, and the Asymptotic Sig is 0.000.

Table 5. Concerns to the product information

	Observed number	Expected number	Residual error
Brand	109	199.0	−90.0
Appropriate age group	378	199.0	179.0
Texture of materials	370	199.0	171.0
Factory name and address	87	199.0	−112.0
Certificate and national standards	347	199.0	148.0
Weight	184	199.0	−15.0
Stability	354	199.0	155.0
Price	287	199.0	88.0
Domestic place of origin	61	199.0	−138.0
Foreign place of origin	44	199.0	−155.0
User guide	160	199.0	−39.0
Cautions	200	199.0	1.0
Others	6	199.0	−193.0
Total	2587		

4 Conclusion

Consumers' concerns can be divided into two categories when they buy a stroller. Both categories of consumers are more concerned about material, stability, certification, whether national standards are marked, applicable age and other information, and pay less attention to information such as place of origin, address and brand. The difference between the two groups is that one group of consumers also pay attention to precautions, instructions and product weight. The difference between the two groups may be related to the education level of consumers. The results show that the degree of attention of the respondents on the concerns of the stroller is mainly appropriate age group, the texture of material, stability, and conformity certification and marks, reaching 81.64%, 79.91%, 76.46%, and 74.95%, respectively. The purchase channels are mainly through large stores/malls (86.18%), specialty stores (80.13%), online shopping (65.01%), and large chain supermarkets (56.16%) to buy children's stroller. The consideration of consumers on safety indicators focuses on the stability, the rationality of design, odors of related devices. The considerations of consumers on the safety indicators for children's bicycles are focused on the braking device (85.96%), the protection of various exposed parts (81.86%), the rationality of the functional protrusion (67.6%), and too wide brake lever (56.16%), chain cover (52.27%), etc. The considerations of consumers on the safety indicators of babe strollers focus on puncture wounds (81.43%), seat belt firmness (78.19%), vehicle body firmness (75.59%), unpleasant odor (63.07%), and braking devices (60.69%) and so on.

Acknowledgements. This paper has been supported by the national project "Research on key technical standards for quality and safety control of consumer goods" (2016YFF02022600), and "Research on common technology for integrative services by internet plus" (2017YFF0209604),

Central basic scientific research project "Research on Consumer Goods Safety Hazard Identification and risk Assessment based on scenario Simulation" (552018Y-5928), and "Research on the construction of consumer goods safety injury scenarios and risk assessment methods of intelligent service robots based on VR experiment" (552020Y-7462).

References

1. Ma, S.: Research on the quality and safety risk early warning technology of labor protection supplies based on social tolerance. University of Science and Technology Beijing (2015)
2. Zunxiong, Liu: Efficacy simulation of class data goodness-of-fit test. Stat. Decis. **34**(24), 86–87 (2018)
3. Xiaoling, Lv: Analysis on the investment risk tolerance of banking clients. J. Appl. Stat. Manag. **06**, 1062–1065 (2007)
4. Jinkins, D.: Conspicuous consumption in the United States and China. J. Econ. Behav. Organ. **127**, 115–132 (2016)
5. Wu, Z.B., Xu, J.P.: Managing consistency and consensus in group decision making with hesitant fuzzy linguistic preference relations. Omega Int. J. Manage. Sci. **65**, 28–40 (2016)
6. Papetti, P., Costa, C., Antonucci, F., et al.: A RFID web-based info tracing system for the artisanal Italian cheese quality traceability. Food Control **27**(1), 234–241 (2012)
7. Liu, X,, Luo, H.-Q.: Research on the methods of safety hazard identification for the consumer products1 based on product life cycle. In: The International Conference on Computer and Information Science, Safety Engineering (CAISSE), pp. 156–161 (2012)
8. Dhar, R., Wertenbroch, K.: Consumer choice between hedonism and utilitarian goods. J. Mark. Res. **37**(1), 60–71 (2000)
9. Heffetz, O.: A test of conspicuous consumption: visibility and income elasticities. Rev. Econ. Stat. **93**(4), 1101–1117 (2010)
10. Xie, G., Jiang, Y., Shi, C.: The influence of E - WOM dispersion on purchase intention: the moderating role of endowment effect. Chin. J. Manag. **16**(3), 425–438,455 (2019)

Game on the Tacit Collusion of the Telecom Market Under the Full-Business Operating Environment

Qiming Tang[1], Yongyao Chen[2], and Meijuan Li[1](✉)

[1] Yunnan Normal University, Kunming, China
limeijuan927@163.com
[2] Yunnan University, Kunming, China

Abstract. This paper constructs the single-stage game model and the infinite repeated game model to analyze the conditions of the tacit collusion of the telecommunication market under the full-business operation environment. The results show that the Nash equilibrium of single-stage game is: (betrayal, betrayal); in the repeated game, when the discount factor $\delta > \frac{73}{81}$, the profit of the tacit collusion among telecom operators is greater than the profit that they can obtain by betraying each other, so the telecom operators will choose the tacit collusion to obtain high monopoly profits. Therefore, the regulatory agencies should choose the appropriate regulatory methods according to the operator's production cost differences and discount factor, which prevents telecom operators from colluding and raising price, urges operators to strengthen technological innovation, improves the quality of telecom products, and creates an orderly competitive telecom market.

Keywords: Full-business operation · Telecom market · Tacit collusion

1 Introduction

The full-business operation means that telecom operators operate mobile, data, and fixed networks at the same time, and carry out operation modes of access services, communication services, value-added services, and content utilization, and comprehensively operate communications service services. The tacit collusion is the non-cooperative collusion formed by the participants based on a common interest. The participants maintain a conspiracy relationship for a long period of time because of common interests. The participants do not have obvious negotiation and negotiation behavior, nor do they sign contracts and agreements. It is the activity which the tacit collaborators voluntarily participate to jointly increase price or limit production while maintaining their independence. In 2008, China Telecom purchased China Unicom's CDMA Network. China Satellite's basic telecom business merged into China Telecom to form new China Telecom, China Tietong merged into China Mobile, China Netcom and China Unicom merged into new China Unicom, Finally, China Mobile, China Unicom and China Telecom form an oligarch monopoly in China's mobile communications market, and enter a new era of full-business operations, which accompany

M. Tavana et al. (Eds.): IISA 2020, AISC 1304, pp. 20–26, 2021.
https://doi.org/10.1007/978-3-030-63784-2_3

multi-oligopolistic competition games. The formation of a three-oligopolistic competitive landscape means that collusion between telecom operators can result in higher profits. Telecom operators have strong dependence and constraints each other. There are competing contradictions and the potential of mutual cooperation. They will use the tacit collusion to ease the fierce competition in the telecom market and obtain monopoly profits.

2 Model Assumptions

This section analyzes the single-stage game model and the infinite-repeat game model to find out the conditions of the tacit collusion under a full-business operating environment.

It is a hypothesis that there are two telecom operators in the telecom market, and both operators conduct full-business operations. Assume that the output of telecom operator 1 is Q1 and the output of telecom operator 2 is Q2. At this time, the total output of the telecom market is: Q = Q1 + Q2, P is the market price, which is the function of the total production levelQ, and the market demand function is P = (Q) = a −bQ, (a > 0, b > 0). Assuming that the marginal costs of the two telecom operators are all constant.

3 Model Analysis

3.1 Single Stage Game Model

The Tacit Collusion of Two Telecom Operators. Assume that telecom operator 1 and telecom operator 2 conduct tacit collusion, the total collaborated production is Qthe marginal cost is c, and total revenue function is: TR = PQ = (a−bQ)Q = aQ−bQ2.

The total revenue function performs a first-order derivation with yield Q to obtain the marginal revenue MR-a-2bQ. Since the monopolist follows the principle of maximizing profits by MR = MC to determine the output and price, the following equation is obtained:

$$Q^M = \frac{a-c}{2b}, P^M = \frac{a+c}{2} \tag{1}$$

The output of telecom operators 1 and 2 is:

$$Q1^M = Q1^M = \frac{1}{2}Q = \frac{a-c}{4b} \tag{2}$$

The monopoly profits of each telecom operator is:

$$\pi 1 = \pi 2 = \frac{(a - c)^2}{8b} \tag{3}$$

The Betrayal of a Telecom Operator

Both Parties Determine Their Output and Prices by Maximizing Their Profits. Under the Cournot model, each of the two telecom operators decides his own production to maximize his profits. Both parties do not know each other's production information before making the decision.

The operator 1 profit function is:

$$\pi 1 = (a - bQ1 - bQ2 - c)Q1 \tag{4}$$

The operator 2 profit function is:

$$\pi 2 = (a - bQ1 - bQ2 - c)Q2 \tag{5}$$

From the first-order conditions of profit maximization, the optimal output of each operator is obtained:

$$Q1^N = \frac{a - c}{3b}, Q2^N = \frac{a - c}{3b} \tag{6}$$

It can be seen that the two telecom operators have the same output. At this time, the two telecom operators have the same maximum profits. The total output of the telecom market is: $Q^N = Q1^N + Q2^N = \frac{2(a-c)}{3b}$, and the respective maximum profits are:

$$\pi 1^N = \pi 2^N = \frac{(a - c)^2}{9b} \tag{7}$$

An Operator Chooses to Betray. Assuming that telecom operator 1 is betrayed, it determines output based on its own profit maximization principle. From the first-order condition $\frac{\partial \pi 1}{\partial Q1} = 0$, the yield under monopoly conditions can be obtained:

$$Q1 = \frac{a - c}{2b} - \frac{Q2}{2} \tag{8}$$

In the current period when the telecom operator was betrayed, the telecom operator 2 still choses to collude without their knowledge. From formula (2), it can be seen that the output of the telecom operator 2 is: $Q2 = \frac{1}{2}Q = \frac{a-c}{4b}$, Substitute (8) for the following:

$$Q1^B = \frac{3(a - c)}{8b} \tag{9}$$

Substituting Eqs. (8) and (9) into Eqs. (4) and (5), the profit of the two operators can be calculated as:

$$\pi 1^B = \frac{9(a-c)^2}{64b}, \pi 2^B = \frac{6(a-c)^2}{64b}$$

When telecom operator 2 is betrayed, it can be obtained by the same method:

$$Q1^B = \frac{a-c}{4b}, Q2^B = \frac{3(a-c)}{8b}, \pi 1^B = \frac{6(a-c)^2}{64b}, \pi 2^B = \frac{9(a-c)^2}{64b}$$

Based on the above three kinds of game profit results, the following matrix game representation can be made:

From Table 1, we can see that in the single-stage game, whether the telecom operator 1 choose to tacit collusion or betrayal, the dominant strategies of the telecom operators are: betrayal. Regardless of whether the telecom operator 2 choose to tacit collusion or betrayal, the dominant strategies of telecom operators are: betrayal, Nash equilibrium is: (betrayal, betrayal). Therefore, in a single-stage game, when operators only have one cooperation, tacit collusion is difficult to form.

Table 1. Matrix representation of the size of telecom operators' profit in the case of tacit collusion and betrayal

	Operator2	
Operator 1	Tacit Collusion	Betrayal
Tacit Collusion	$\frac{(a-c)^2}{8b}, \frac{(a-c)^2}{8b}$	$\frac{6(a-c)^2}{64b}, \frac{9(a-c)^2}{64b}$
Betrayal	$\frac{9(a-c)^2}{64b}, \frac{6(a-c)^2}{64b}$	$\frac{(a-c)^2}{9b}, \frac{(a-c)^2}{9b}$

3.2 Infinite Repeated Game Model

Assuming that telecom operators follow the grim strategy in an infinitely repeated game. The ruthless strategy means that before the operator 1 betrays the operator 2, the operator 2 has been maintaining cooperation with the operator 1. but because of the betrayal of the operator 1, the operator 2 has chosen to betray the tactic collusion in the remaining games to punish. The operator 1 will no longer cooperate with the operator 1.

In the repeated game, the discount factor is assumed to be δ and $\delta \in (0,1)$. When two telecom operators have chosen to tactic collusion with each other, the profits of operator i in each period are:

$$\pi i^C = \pi i^M + \pi i^M \delta + \pi i^M \delta^2 + \cdots = \frac{\pi i^M}{1-\delta} = \frac{(a-c)^2}{8b(1-\delta)} \quad (10)$$

When operator i betrays, its profit in each period is:

$$\pi i^B = \pi i^B + \pi i^N \delta + \pi i^N \delta^2 + \cdots = \frac{9(a-c)^2}{64b} + \frac{(a-c)^2}{9b(1-\delta)} \tag{11}$$

Only $\pi i^C > \pi i^B$, there is $\delta > \frac{73}{81}$. It can be seen that when δ is bigger, the higher the collusion's profit, the greater the possibility of a tacit understanding between the operators. when $\delta > \frac{73}{81}$, the profit of the telecom operators in the tacit agreement is greater than that of the mutual betrayals. Therefore, the telecom operators will choose to conspire to obtain a high monopoly profit, and the treachery will trigger both sides, the price war will reduce the monopoly profits of both parties. Therefore, in infinitely repeated games, telecom operators often make tacit collusion.

4 Conclusions and Countermeasures

4.1 Conclusions

This paper analyzes the tacit collusion behavior of the telecom market by constructing a single-stage game and a repeated game model. The results show that in a single-stage game, tacit collusion is difficult to form. Although telecom operators choose to obtain the profit of the tacit conspiracy is greater than the profits that they choose to betray. If an operator unilaterally chooses to betray, the profits can bring to the telecom operator are higher than the tacit complicity, so the tacit collusion is unstable, the Nash equilibrium of the telecom operators is: (betrayal, betrayal). When operators only have one cooperation, the incentive for betrayal is obvious, the tacit collusion is difficult to form, telecom operators will experience a price war. However, in a multi-stage iterative game, the situation will accordingly change.

In the repeated game, the strength of the telecom operators' tacit collusion motivation depends on the discount factor δ. Under the condition that the discount factor $\delta > \frac{73}{81}$, the telecom operator's profits of tacit collusion is greater than the profits that can be obtained by betraying each other. Significantly, the possibility of tacit collusion is large, and operators can achieve the purpose of monopolizing the market through tacit collusion.

4.2 Suggestion

Preventing Sequential Price Increases of Telecom Operators. The price tacit collusion of the telecom operators makes the profits of all conspiracy operators maximise, and the incentive of tacit collusion is striking. The three oligopolistic operators in the industry have a strong ability to manipulate prices, so operators will conspire to raise prices and obtain excess monopoly profits. When a certain operator posts price increasing information, it causes other operators in the industry to use the same price increasing rate and violates the market price increasing behavior, which indicates that there is a tacit collusion in the telecom market. The sequential price increasing behavior of telecom operators may not be based on changes in market conditions, but rather the result of tacit collusion between operators. The regulatory agencies should pay attention to the strengths among operators, the sequence of price increase, the time of price increase, the magnitude of the price increase and the change in profit after the price increase, and govern the price information inconsistent with the market rules, and regulate the operators who engage in tacit collusion.

Improving Product Quality. Under the full-business operating environment, Stronger alternatives to telecom services operated by telecom operators, more selective and easier to collude. Horizontal and vertical product differences are conducive to the existence of tacit collusion. When there are quality differences in products, operators with high product quality have incentives to betray collusion, because the profits by betraying collusion are higher than those under collusion conditions. Therefore, the telecom operators have the incentive to increase product quality in order to obtain higher monopoly profits. The telecom operators will optimize the allocation of their own advantageous resources, deepen business models, strengthen technological innovation and research and development, develop new products, increase product added value, innovate product portfolio, expand service content, segment consumer markets, and improve the quality of telecom products for consumers. The differentiated and refined services will carried out, and new products in the areas of data traffic, voice, Internet, and other services will continuously researched, and strengthen the brand image of operators.

Improving the Market Supervision Mechanism. It is important to create a market environment that encourages all telecom operators to compete actively and strictly supervise and administer the telecom market in accordance with laws and regulations. When formulating regulatory policies, regulatory agencies should select appropriate regulatory methods based on operators' differences in production costs and discount factors, continue to improve the supervision mechanism of the telecom industry, guide the orderly competition among telecom operators. The more similar the production costs are among telecom operators, the greater the possibility of achieving tacit collusion; the greater the difference in production costs, the less possibility it is to achieve tacit collusion. High-cost telecom operators are more inclined to produce lower total output and set higher prices, which will help high-cost incumbents obtain high monopoly profits, and at the same time prevent new entrants from entering the telecom market to compete with them. The telecom supervisory department must supervise the prices and production of product, and follow up and investigate whether there is tacit collusion among the operators, and whether the price and output of each operator are in accordance with the price and output of the tacit agreement. When discounting factors is high, the telecom operators can gain higher monopoly profits, The motivation for tacit collusion is significant. When there is a tacit collusion among the operators, strict regulation is implemented in accordance with the law, which increase consumer surplus and improve the overall market competition efficiency and social welfare.

References

1. Fen, H.: Game analysis of the competition pattern of mobile communication market before and after telecom reorganization. China Secur. Futur. **2**, 35–38 (2011)
2. Liu, J.: Research on the analysis and complexity of the competition game process in telecommunication business market, pp. 54–68. Doctor Thesis, Jilin University (2016)
3. Wang, L.: Game of the impact of number portability on competition of telecom operators in mobile internet environment, pp. 44–60. Master Thesis, Nanchang University (2015)
4. Li, M.: Vertical market delineation and access regulation based on price behavior of telecom companies. Econ. Rev. **2**, 48–54 (2011)
5. Li, M.: Review of the theoretical research on access pricing and telecommunication network competition. Prod. Res. **7**, 254–256 (2012)
6. Li, M., Zhang, L.: Study on the influence of the number portability policy on the collusion of telecom operators. Ind. Technol. Econ. **36**(11), 104–109 (2017)
7. Zhang, Q., Liu, G.: Analysis of the existence and stability of tacit collusion in the telecommunications industry. Tech. Econ. Manag. Res. **7**, 111–114 (2011)
8. Zhang, Q., Liu, G.: Vertical differentiation and collusion analysis of telecom operators. Bus. Res. **1**, 41–46 (2013)
9. Wang, R.: Analysis of game behavior of oligopoly monopoly market-taking China telecom industry as an example. Hebei Enterp. **6**, 68–69 (2014)
10. Zheng, S.: Analysis of the pro-collusion effect of horizontal mergers-take the restructuring of China's telecom industry in 2008 as an example, pp. 37–41. Master Thesis, Tianjin University of Business (2010)
11. Yang, S., et al.: The analysis of competitive game of telecom enterprises based on hybrid strategy. Explor. Econ. Issues **8**, 179–183 (2014)
12. Zhou, Y.: Research on the regulation of interconnection and interconnection of China's telecom industry under the full-service operating environment, pp. 42–46. Master Thesis, Yunnan Normal University (2015)

Discussion on the Training Mode of Data Capability for Economics and Management Majors

Guihua Han(✉), Mintyu Lin, and Cuilin Li

Hubei Business College,
No. 634, Xiongchu Avenue, Hongshan District, Wuhan, Hubei, China
hgh415@126.com

Abstract. In order to adapt to the development of new engineering education, colleges and universities pay more attention to students' data capability to adapt to the era of big data, this article starts with the content of the concept of data capability, discusses on the training mode of data capability for economics and management majors, which is expound from three aspects like the data thinking ability and data analysis ability and data application ability, for further deepening the talent cultivation reforms of economics and management majors, it provides a certain reference value.

Keywords: Big data · Data capability · Data thinking · Data analysis

1 Introduction

Almost all personal Internet activity is now digitized. With the continuous application of big data technology, each of our actions is recorded in digital form. When these data are applied to all walks of life, we also change from being a recorder to being a data user. In order to better understand the world, adapt to the world development, solve the future works, study and even life problems, we must learn to deal with all kinds of information, especially digital information.

With the popularity of smart phones, mobile network, Internet of things, cloud computing and big data, data has been involved in all aspects of human life. The ability which is to build models based on massive data and extract valuable information in order to form individual and Integrated knowledge is considered as one of the basic production capacities in the digital age.

From E-learning (Electronic learning) to the M-learning (Mobile learning), U-learning (Ubiquitous learning) [1], until wisely study now, a digital learning environment based on the information technology application not only changes our learning environment in space and time, but also offers limitless possibilities for the mobile, intelligentization and individuation of learning process and learning resources. The wisdom which extracts information from data, transforms information into knowledge, and then solves the problem is a new understanding of "learning".

Under the background of the current era of big data, for meeting the need of the times, various universities are actively promoting the development of new engineering

M. Tavana et al. (Eds.): IISA 2020, AISC 1304, pp. 27–32, 2021.
https://doi.org/10.1007/978-3-030-63784-2_4

education, paying more attention to cultivate the students' data capability for adapting to the big data era. For the students majoring in economics and management, under the background of big data era, they should have higher data-sensitive, therefore, the reform of professional talent cultivation of economics and management becomes the general trend of the times, making the students to have the capability to collect, collate and analyze data has become a hot topic in the reform direction.

2 The Connotation of the Data Capability Concept

The European Digital Agenda, officially released by the European union on May 19, 2010, states that "Learning to use ICT and Digital media is an essential skill for modern citizens" and emphasize that Digital capabilities are "the ability that using information society technologies confidently and critically for work, leisure and communication". Specifically, the conceptual model of data capability [1] includes three levels: data thinking ability, data analysis ability and data application ability. The core of data thinking ability is data sensitivity, and the core of data analysis ability is data processing, the core of data application is data evaluation, as shown in Fig. 1.

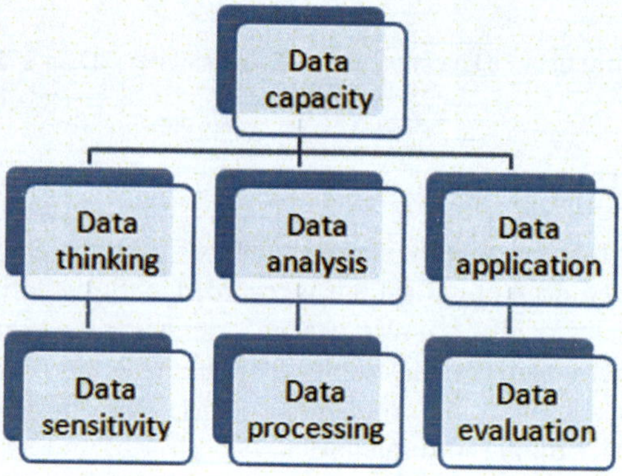

Fig. 1. The conceptual model of data capability

3 Data Capability Training

This paper mainly expounds the data capability training of economics and management majors from three aspects: data thinking ability, data analysis ability and data application ability.

3.1 Training the Students' Data Thinking Ability

Data thinking ability [2] requires people keep sensitive to data. The first thing is to be able to obtain data. Colleges and universities could offer relevant courses, also by means of lectures, open classes and online communication and so on to guide students to study independently, so as to deepen the students' understanding the today's social changes, enrich their own knowledge, which enable the students to have the abilities of systematic and critical thinking. And they could collect a large number of data, and extract the essence, from which to analyze and sort out a variety of useful data by using of the existing knowledge and modern information means, according to the task requirements of the relevant courses study. In the past, sampling has always been the main means of data acquisition, which is the helpless choice for human beings when they cannot obtain the overall data information. In the era of big data, people can obtain and analyze more data, even all the data related to them, and no longer rely on sampling, which can bring a more comprehensive understanding, and can more clearly find the details that the sample cannot reveal. which can be represented by lists or graphs.

3.2 Training the Students' Data Analysis Ability

Data analysis ability [3] is a high-level thinking activity. In the courses of economics and management, data analysis is based on the principle of statistical analysis, which can help enterprises make correct judgment and make reasonable marketing plans. With the rapid increase of data in modern society, the thinking mode of statistics and probability is a necessary skill for modern people, which has an important influence on people's work, life and study, the details are as follows:

The mean, median, and modal are the measuring units that describe the central tendency of the data sets. Students should apply the theoretical knowledge in probability and statistics to data analysis to develop their thinking flexibility.

The data analysis of economics and management students focuses on economic activities, so the projects of their data analysis should be close to real life, so that it makes the students really realize that the data is around us and closely related to our lives. On this basis, it is necessary to consciously train students to think statistically, access to the project data information, so as to do further data processing for the next step of data application preparation.

Students establish the association between a large number of data from the acquisition of data information, according to the master degree of data analysis, choose the appropriate effective data analysis tools or methods, extract useful information.

3.3 Training the Students' Data Application Ability

Data application ability [4, 5] refers to the ability of solving problem by using of the existing resources or acquired data. In specific applications, the ultimate goal of data analysis is to solve problems, which is a higher level requirement. For this reason, the students majoring in economics and management should be familiar with industry knowledge and business process, determine the way to solve problems according to the results of data analysis, select appropriate means, and apply them to planning schemes such as marketing and management, so as to optimize business process and meet the needs of customers.

Take the sales data of a cake shop as an example, the data of the 10 submission records before a certain day is shown in Table 1. Data analysis is conducted on this basis, and the statistical results are shown in Table 2. The data in Table 2 are further analyzed and expanded to get Fig. 2. In Fig. 2, we get a more intuitive conclusion: the top sellers are meat floss buns and tiramisu, followed by sandwiches and garlic cheese bread, but raisin toast and egg tarts are not so popular.

In order to ensure the diversity of goods sold in cake shops and also to ensure the interests maximization, the corresponding adjustment measures can be formulated according the above conclusion. For meat floss buns and tiramisu with better sales, the production scale should be appropriately increased according to a certain proportion, while for goods with poor sales, the production should be less.

4 Establishing an Effective Training Mode of "Data Ability" for Economics and Management Students

4.1 Adjustment of Curriculum System

In combination with the major characteristics, courses such as market survey, probability theory, statistics and statistical software are properly set to make the training of data ability run through the whole course learning process [6], which is conducive to the training of students' data capabilities. Compiling the relevant courses textbooks suitable for economics and management majors, which add more practical training cases to improve students' ability of data thinking, data analysis and data application.

Table 1. The first 10 submission records data of a cake shop

ID	Goods	Amount
1	Raisin toast	1
1	Meat floss buns	3
2	Tiramisu	3
2	Sandwiches	3
3	Garlic cheese bread	3
3	Meat floss buns	2
4	Tiramisu	3
4	Egg tarts	1
4	Meat floss buns	1

Table 2. Commodity quantity

Goods	Raisin toast	Tiramisu	Sandwiches	Garlic cheese bread	Meat floss buns	Egg tarts
Number	1	6	3	3	6	1

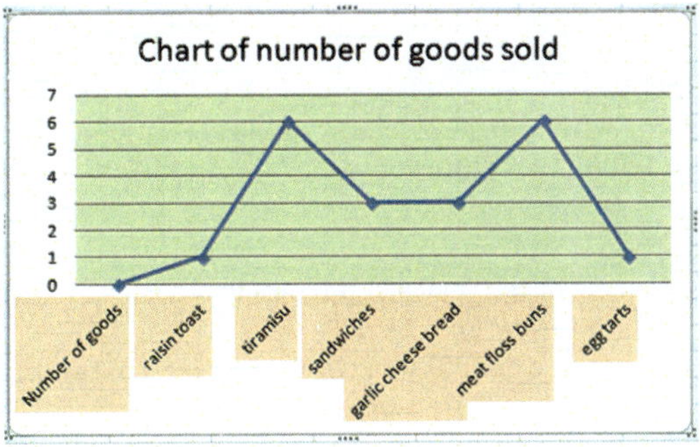

Fig. 2. Commodity quantity diagram

4.2 Reform of Teaching Methods

The combination of theory teaching and practice teaching focuses more on the cultivation of students' ability to solve problems comprehensively by using data. The basic methods of data analysis mastered by students should be combined with practical training and professional needs [7]. They should complete the case project in a group, submit detailed reports and participate in the defense, and finally take the completion quality of the project as one of the assessment contents of the course.

5 The Cultivation Effect of Data Capability

Through the adjustment of the curriculum system and the reform of teaching methods, the students majoring in economics and management have achieved good results in the cultivation of data capability. Students' ability to obtain, analyze and apply data has been significantly improved [8]. Through the study of relevant courses, the students of economics and management will establish the concept of data processing and be able to consciously think about the real problems from the perspective of data, that is, abstract the real problems to obtain data, and use the relevant statistical knowledge to analyze and apply the data.

6 Conclusion

Under the background of big data, this paper analyzes the data capability of economic management major, expounds the ability training of data thinking, data analysis and data application, establishes the training mode of curriculum system adjustment and teaching method reform, and cultivates the economic management professional with data ability.

References

1. Hu, W.: Digital ability development of students based on adaptive learning platform. China Educ. Tech. Equip. **10**(19), 3–5 (2017)
2. Wang, F.: Training of data analysis ability of college students in the new era. J. Heilongjiang Coll. Educ. **33**(5), 22–23 (2014)
3. Liu, Y.: The cultivation of college students' practical ability under the background of big data era. Labor Secur. World (Theory Ed.) **24**, 134–135 (2013)
4. Liu, P.: A brief discussion on data ability cultivation in the era of big data. Telecom World **22**, 46–47 (2017)
5. Cheng, G., Li, M.: Cultivation mechanism of big data capability of enterprise. J. Mod. Inf. **34** (3), 7–11 (2014)
6. Deng, W.: Research on data analysis ability of economics and management major. Digit. Commun. **40**(2), 93–95 (2013)
7. Yang, N.: Exploration on the cultivation mode of computing thinking in economics and management majors under the background of big data. J. High. Learn. Res. **22**, 77–78 (2016)
8. Chen, S., Shi, X.: Comprehensive evaluation of students' "data" ability in Liaoning University of Technology. J. Liaoning Univ. Technol. (Soc. Sci. Ed.) **20**(5), 104–106 (2018)

Research on Service Quality Evaluation System of Automobile Maintenance Enterprises Based on AHP

Xiao Juan Yang(✉), Shu Quan Xv, Fu Jia Liu, and Guo Fang Wu

Research Institute of Highway Ministry of Transport, Beijing, China
xj.y@rioh.cn

Abstract. With the development of automobile industry, automobile maintenance industry is closely related to people's production and life, and plays an increasingly important role in the development of China's market economy. The level of service quality, only through scientific evaluation can be found shortcomings and problems, for the future improvement and improvement of service quality direction. How to develop and improve the service quality evaluation system of automobile maintenance enterprises in the new situation, so that it can adapt to the construction of socialist market economy and meet the needs of people's living standards, has become an important topic. This paper mainly expounds the theoretical basis of automobile maintenance quality evaluation, analyzes the current situation of service quality evaluation of automobile maintenance industry in China and the existing problems in automobile maintenance industry. Based on the research of advanced technology and technology of automobile maintenance in China, a comprehensive and perfect service quality evaluation index system of automobile maintenance enterprises is proposed based on the theory of AHP, which is for further regulation Model China's maintenance service quality evaluation work, lead enterprises to achieve technological progress, and promote industry transformation and upgrading to provide theoretical support.

Keywords: Automobile maintenance · Service quality · Analytic hierarchy process

1 Introduction

With the increase of car ownership in China, the domestic demand for car maintenance business continues to increase, and the industry's core profit gradually turns to the after-sales service link [1]. Automobile maintenance service has a mature market environment in foreign countries, but it is relatively backward in China. Good service quality can not only enhance the brand image of the enterprise, establish a good reputation, but also win a larger market share and more profits for the enterprise. The evaluation and management of service quality of automobile maintenance enterprises is related to the control and supervision of automobile manufacturers to their 4S stores, the improvement of their own ability and competitiveness. However, in the process of evaluation, if the assessment scope is too large to meet the actual business needs, the

M. Tavana et al. (Eds.): IISA 2020, AISC 1304, pp. 33–38, 2021.
https://doi.org/10.1007/978-3-030-63784-2_5

assessment method is not appropriate, which makes the assessment of service quality become a mere formality. If the assessment scope is too small, it can not accurately cover all aspects of automobile maintenance service, which is easy to generalize. Therefore, a good service quality evaluation should not only cover all the links involved, but also adjust the assessment indicators at any time to meet the changing needs.

In the past, the research on service quality evaluation of maintenance enterprises focused on the level of automobile maintenance business, which did not fully cover the overall service quality of enterprises. However, with the transformation and upgrading of maintenance industry, in the context of paying more attention to service quality, comprehensive and three-dimensional enterprise service quality evaluation indicators should be established [2–4]. On the other hand, the main way to evaluate the enterprise is to survey the customers and score the traditional experts, which limits the evaluation factors, because the content that the customers can really perceive is rather one-sided, and the enterprise management, operation and other deep-seated reasons that affect the service quality cannot be deeply explored. This paper constructs a service quality evaluation index system of automobile maintenance enterprises. By selecting scientific and reasonable evaluation methods, in order to promote competitiveness, and help departments to formulate scientific and reasonable management systems and measures Support.

2 Related Research

Service quality is the sum of the characteristics and characteristics of the service or service industry of product production to meet the requirements and potential needs [5]. PZB put forward the theoretical gap analysis model of service quality in 1985, and analyzed five kinds of gaps related to service quality between service providers and service consumers [6], namely, the perception gap of management, quality standard, service delivery, market communication and quality service. These gaps can be used to analyze the inconsistencies and the lack of necessary communication between enterprises and customers, service managers, service employees and service communication. The proposed gap analysis model of service quality lays the theoretical foundation for service quality evaluation. The SERVQUAL is completely based on the customer perception, that is to say, taking the customer's subjective consciousness as the key point, first measuring the customer's desire for service, so as to calculate the difference between the two and take it as the basis of judging the service quality level. However, it is worth noting that the research object of PZB is only limited to the some industries, so many conclusions are not necessarily universal applicability, while the limitation in the application of service quality evaluation of automobile repair is that it does not combine the characteristics of automobile products and services [7].

3 Influencing Factors and Main Evaluation Indexes of Service Quality of Maintenance Enterprises

The service content of automobile maintenance enterprises mainly involves service concept, service system, personnel allocation, site environment, facilities and equipment, maintenance process, maintenance parts, customer tracking, etc. This paper mainly considers the relevant evaluation level of service quality, and establishes a multi index and multi-level service quality evaluation index system of maintenance enterprises based on the AHP and the specific situation of the automobile maintenance industry, as shown in Fig. 1 below.

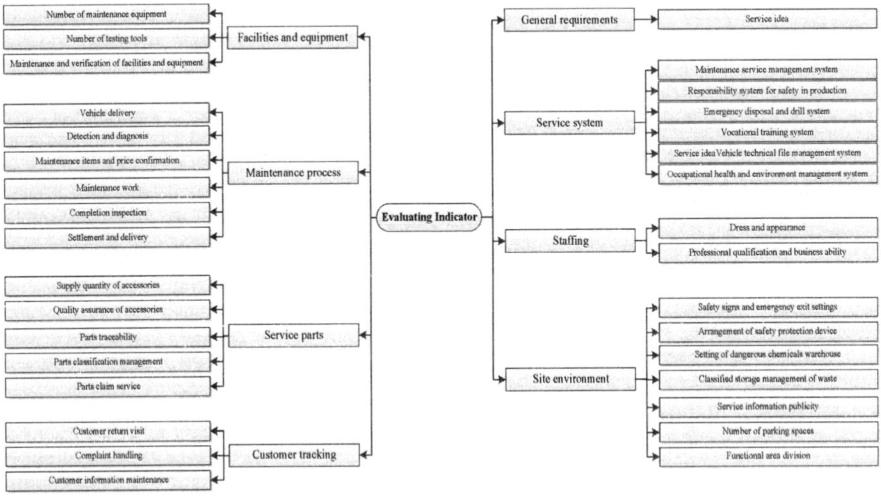

Fig. 1. Service quality evaluation index of automobile maintenance enterprises

4 Service Quality Evaluation of Automobile Maintenance Enterprises Based on AHP

4.1 Basic Concepts

Analytic hierarchy process (AHP) has strong operability and is widely used in the field of satisfaction evaluation. AHP was put forward by T.L. Saaty in the 1970s [8]. The analytic hierarchy process obtains several judgment matrices of the level, and then obtains the relatively important weight of the index through solution analysis, which is very suitable for solving the problem that is not easy to be completely quantitative. The general model of the hierarchical structure model is shown in Fig. 2.

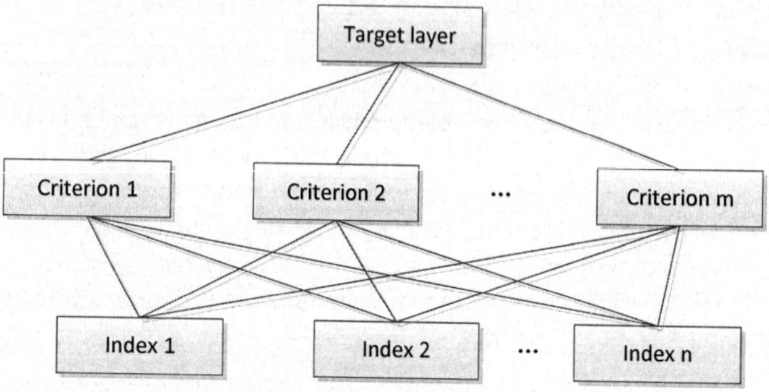

Fig. 2. Hierarchy model

4.2 Analysis Steps

(1) Build a hierarchy model.
(2) Construct judgment matrix.
(3) Calculate the relative weight of single sorting and do consistency test.
(4) Calculate the total sorting weight and consistency test.

5 Establishment of Evaluation Model

In this analysis, a three-level evaluation index system for service quality of automobile maintenance enterprises will be established. After the industry experts and maintenance enterprise personnel are invited to score, the AHP method will be used to calculate and finally determine the weight of indicators at all levels. The overall evaluation index and its weight are shown in Table 1 below.

Table 1. Evaluation index and weight

First index	Second index	Third index	Weight	Second index	Third index	Weight
Service quality of automobile maintenance enterprises 1.0	General requirements 0.03	Service idea	0.03	Facilities and equipment 0.15	Number of maintenance equipment	0.06
	Service system 0.10	Maintenance service management system	0.02		Number of testing tools	0.06
		Responsibility system for safety in production	0.02		Maintenance and verification of facilities and equipment	0.03
		Emergency disposal and drill system	0.02	Maintenance process 0.38	Vehicle delivery	0.03
		Vocational training system	0.02		Detection and diagnosis	0.06
		Service ideaVehicle technical file management system	0.01		Maintenance items and price confirmation	0.04
		Occupational health and environment management system	0.01		Maintenance work	0.15
	Staffing 0.05	Dress and appearance	0.02		Completion inspection	0.06
		Professional qualification and business ability	0.03		Settlement and delivery	0.04
	Site environment 0.10	Safety signs and emergency exit settings	0.01	Service parts 0.12	Supply quantity of accessories	0.02
		Arrangement of safety protection device	0.02		Quality assurance of accessories	0.03
		Setting of dangerous chemicals warehouse	0.01		Parts traceability	0.03
		Classified storage management of waste	0.02		Parts classification management	0.01
		Service information publicity	0.01		Parts claim service	0.03
		Number of parking spaces	0.01	Customer tracking 0.07	Customer return visit	0.02
		Functional area division	0.02		Complaint handling	0.03
					Customer information maintenance	0.02

6 Model Application

This evaluation is conducted by the combination of expert document review and on-site inspection. On site inspection includes the open evaluation of experts of the audit team, secret visit of mysterious customers, questionnaire survey of previous services, etc., which makes the evaluation methods more rich and diverse. Through the use of the service quality evaluation index system of automobile maintenance enterprises established in this paper, brand 4S stores, comprehensive maintenance enterprises, small repair shops and other enterprise types are selected in Beijing for classified evaluation. The evaluation results and grade standards are shown in Table 2 below.

Table 2. Evaluation results and grade standards

Fraction	<0.6	0.6–0.75	0.75–0.9	>0.9
Grade	Unqualified	Pass	Good	Excellent
4S shop of a brand				0.91
A comprehensive maintenance enterprise			0.78	
A small repair shop in a community		0.65		

7 Conclusion

Using the theory of AHP, we establishes the service quality evaluation system of automobile maintenance enterprises, specifically constructs a comprehensive evaluation index model, puts forward practical evaluation methods, and solves the problem that the service quality evaluation of automobile maintenance enterprises in China is not comprehensive and feasible at present by realizing more comprehensive, detailed, scientific and reasonable evaluation indexes It provides technical support for maintenance enterprises to improve service mode, enhance core competitiveness and promote the overall improvement of service quality of the industry.

References

1. Development Research Centre of the State Council: Growth potential and external development environment of China's automobile industry (2004)
2. Liang, Y.: Research on service quality evaluation of automobile maintenance. Heilongjiang Transp. Technol. (1), 181–182 (2015)
3. Zhu, J., Hu, W.: Research on evaluation index analysis and evaluation system of automobile maintenance service quality. J. Hubei Univ. Technol. **22**(5), 88–91 (2007)
4. Chou, C.: Research on service quality evaluation of automobile maintenance. Wuhan University of Technology, Wuhan (2009)
5. Parasuraman, A., Zeithaml, V.A., Leonard, L.: A conceptual model of service quality and its implications for future research. J. Mark. **49**(9), 44 (1985)
6. Lieb, C., Bentz, A.: The use of third-party logistics services by large American manufacturers. Transp. J. **43**(3), 23–24 (2004)
7. Gao, K.: Automobile evaluation based on fuzzy analytic hierarchy process. J. Shenzhen Polytech. (4), 23–25 (2006)
8. Du, D., Pang, Q., Wu, Y.: Modern comprehensive evaluation method and case selection. Tsinghua University Press, Beijing (2008)

A Method of Determining Membership Function in Fuzzy Comprehensive Evaluation

Guoliang Dong(✉), Chaozhou Chen, and Guofang Wu

Research Institute of Highway Ministry of Transport, Beijing 100088, China
gl.dong@rioh.cn

Abstract. Choosing different membership functions in fuzzy comprehensive evaluation will affect the evaluation results. The use of an inappropriate membership function may lead to large deviations in the evaluation results. Discuss the methods and steps of building the membership function for a practical application of risk assessment. Select some sample points and calculate the difference between the original input value and the deblurred calculated value. Take appropriate measures to gradually reduce or eliminate the deviation, and ultimately make the deviation within an appropriate range. Conclusions are drawn by calculating sample points: The anti-fuzzy evaluation results are consistent with the original input values, and the final membership degree reached the expected requirements.

Keywords: Fuzzy comprehensive evaluation method · Risk evaluation · Membership function · Deblurred calculated value · Normalized membership value · Unnormalized membership value

1 Introduction

During the risk assessment, different evaluation methods will be choosen according to application scenarios [1–3]. The fuzzy comprehensive evaluation (FCE) method is widely used in different occasions [4–6]. In the FCE method, the determination of the membership function is important [7, 8]. Inappropriate membership will lead to large deviation. Select some sample points and calculate the subset of membership. The anti-fuzzification calculation is performed, and the devia-tion between the deblurred calculated value (DCV) and the original input value (OIV) is used for testing whether the membership degree is appropriate. Take appropriate measures to gradually reduce or eliminate the deviation between the DCV and the OIV, and the membership function finally obtained meets the expected requirements.

2 Membership Function

The membership function should objectively reflect the relationship between the input and the output. The following principles should be followed while building the membership function [9].

M. Tavana et al. (Eds.): IISA 2020, AISC 1304, pp. 39–46, 2021.
https://doi.org/10.1007/978-3-030-63784-2_6

1) The membership function should have a unimodal characteristic.
2) The membership function should be symmetrical.
3) Each element belongs to at least one region but should not exceed two regions.
4) The same element should not take the maximum value in two regions.
5) The overlap should not cross the maximum of the two membership functions.

3 Methods and Steps for Determining Membership Function

The membership reflects the degree to which the evaluation index belongs to each evaluation level [10, 11]. Use the membership function to convert the evaluation score into a subset of membership. Taking a certain risk evaluation as an example, the evaluation level is divided into 5 levels as shown in Table 1.

Table 1. Evaluation levels

Risk classification,V	v1	v2	v3	v4	v5
Evaluation score,x	0–20	20–40	40–60	60–80	80–100

3.1 Preliminary Determination of Membership

For each evaluation level, the Cauchy symmetric distribution is selected [12]. The membership function of evaluation level v1 is:

$$u_{v1}(x) = \frac{1}{1+\alpha(x-c)^2} \tag{1}$$

Taking the midpoint as the maximum point of membership, then $c = 10$. In order to make the membership value outside the 0–20 as small as possible, take u_{v1} (30) = 0.1, and the calculation is $\alpha = 0.0225$. The membership function of the level v1 is:

$$u_{v1}(x) = \frac{1}{1+0.0225 * (x-10)^2} \tag{2}$$

Using the same method, membership functions of other evaluation levels can be obtained. The membership curves are shown in Fig. 1.

To verify the rationality of the selected membership function, some sample points are selected. The defuzzification calculation is performed on the calculated membership degree subset. Compare the difference between the DCV and the OIV. The smaller the difference, the more reasonable the membership function. For areas with large deviations, appropriate methods and measures need to be taken to reduce the deviation value to an appropriate range.

Assuming a certain evaluation index, the original evaluation score is 2 points. The original membership value for each evaluation level can be obtained according to the

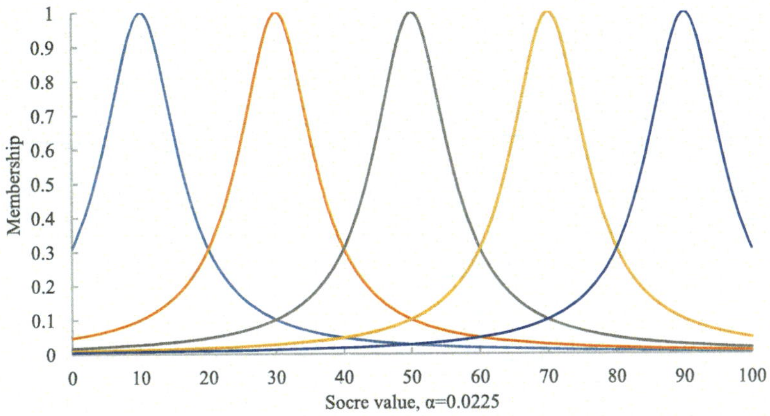

Fig. 1. Membership function graph 1

formula. Normalize the original calculated membership value and get the normalized membership subset *R = (0.824 0.108 0.038 0.019 0.011)*. The anti-fuzzy direction calculation is performed by the following rank parameter column vector of each rank.

$$C = (10\ 30\ 50\ 70\ 90)^T$$

Then the evaluation score is $k = R.C = \sum r_j.c_j = 15.7$. There is a large deviation between the DCV (15.7) and the OIV (2). The same method is used to other sample points to obtain the normalized membership degree subsets, the DCV, and the deviation between the DCV and the OIV, as shown in Table 2.

Table 2. The OIV, normalized subset of membership, the DCV, deviation between the DCV and the OIV

No.	OIV,x	Uv1(x)	Uv2(x)	Uv3(x)	Uv4(x)	Uv5(x)	DCV,x1	x1 − x
1	2	0.825	0.107	0.038	0.019	0.011	15.7	13.7
2	8	0.881	0.079	0.023	0.011	0.006	13.7	5.7
3	15	0.744	0.190	0.040	0.017	0.009	17.1	2.2
4	20	0.447	0.447	0.068	0.025	0.013	24.2	4.2
5	30	0.079	0.810	0.079	0.021	0.010	31.4	1.5
6	80	0.013	0.025	0.068	0.447	0.447	75.8	−4.2
7	85	0.009	0.017	0.040	0.190	0.744	82.9	−2.2
8	95	0.008	0.014	0.029	0.088	0.861	85.6	−9.4
9	100	0.014	0.023	0.045	0.121	0.797	83.3	−16.8

It can be seen from Table 2 that there is a large deviation between the DCV and the OIV. It can be seen from the curve in Fig. 1 that the curve's slope in each interval is steep, and there are more intersections between the curves. The defects need to be corrected.

3.2 Reduce the Slope of the Curve

In order to reduce the slope of the curve, take u_{v1} (20) = 0.9 and the calculation is $\alpha = 0.0011$. The membership function of the evaluation level v1 is:

$$u_{v1}(x) = \frac{1}{1 + 0.0011 * (x - 10)^2} \tag{3}$$

Using the same method, other membership functions can be obtained. The membership curves are shown in Fig. 2.

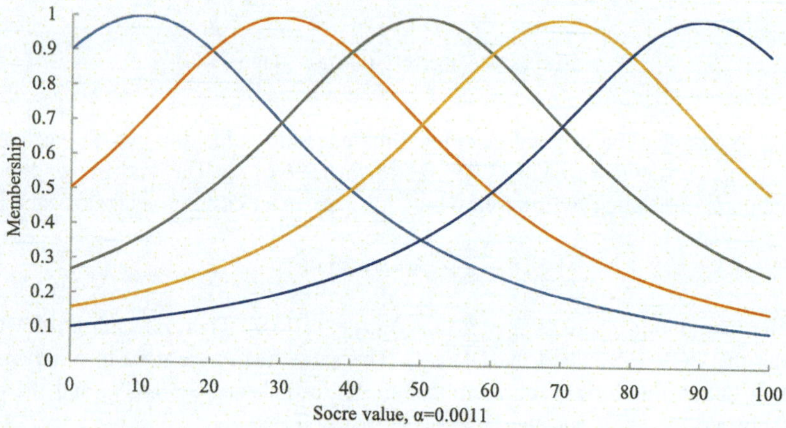

Fig. 2. Membership function graph 2

For the sample points, the normalized membership degree subsets, the DCV, and the deviation between the DCV and the OIV are calculated, as shown in Table 3.

It can be seen from Table 3 that the deviation value is larger than in Table 2. Compare the curves in Fig. 1 and Fig. 2. The curve in Fig. 2 is smoother than in Fig. 1, but the intersection area in Fig. 2 is more than that in Fig. 1.

3.3 Reduce the Intersection Area

In order to reduce the mutual interference between the intersection regions of the membership curve, different membership curves should be used in different intervals. In the evaluation interval, use a gentler curve and the sample points in the evaluation interval get a higher degree of membership. Outside the interval, a curve with a steep slope is adopted and the membership value in the marginal interval drops rapidly. Ensure that there are no more than 2 intersections of the curve. Therefore, formula 3 is used within the interval, and straight lines are used for smoothing outside the interval.

Table 3. The OIV, normalized subset of membership, the DCV, deviation between the DCV and the OIV

No.	OIV,x	Uv1(x)	Uv2(x)	Uv3(x)	Uv4(x)	Uv5(x)	DCV,x1	x1 − x
1	2	0.462	0.265	0.140	0.081	0.052	29.9	27.9
2	8	0.433	0.284	0.148	0.083	0.052	30.7	22.7
3	15	0.379	0.312	0.166	0.090	0.054	32.6	17.6
4	20	0.330	0.330	0.184	0.098	0.057	34.4	14.4
5	30	0.235	0.339	0.235	0.123	0.068	39.0	9.0
6	80	0.057	0.098	0.184	0.330	0.330	65.6	−14.4
7	85	0.054	0.090	0.166	0.312	0.379	67.4	−17.6
8	95	0.000	0.000	0.000	0.000	1.000	90.0	−5.0
9	100	0.000	0.000	0.000	0.000	1.000	90.0	−10.0

When $0 \leq x \leq 20$, formula 3 is used. When $x > 30$, $u_{v1}(x) = 0$. Correct the membership value between 20 and 30 to a straight line. The membership function of evaluation level v1 is:

$$u_{v1}(x) = \begin{cases} 1/(1+0.0011*(x-10)^2) & 0 \leq x \leq 20 \\ (30-x)*0.09 & 20 < x \leq 30 \\ 0 & 30 < x \end{cases} \tag{4}$$

Using the same method, other membership functions can be obtained. The membership curves are shown in Fig. 3.

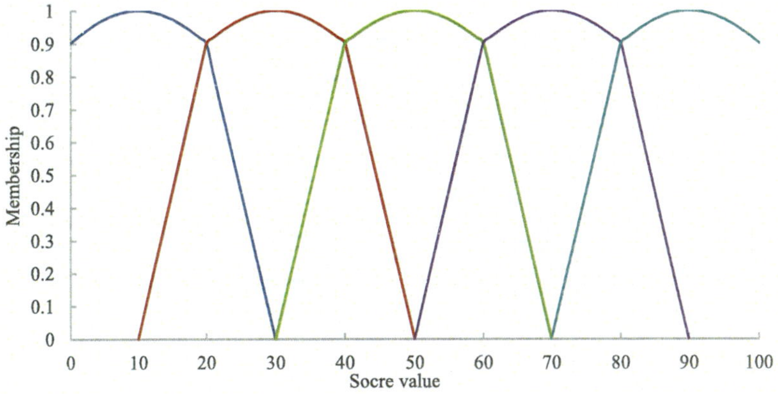

Fig. 3. Membership function graph 3

For the sample points, the normalized membership degree subsets, the DCV, and the deviation between the DCV and the OIV are calculated, as shown in Table 4.

Table 4. The OIV, normalized subset of membership, the DCV, deviation between the DCV and the OIV

No.	OIV,x	Uv1(x)	Uv2(x)	Uv3(x)	Uv4(x)	Uv5(x)	DCV,x1	x1 − x
1	2	1.000	0.000	0.000	0.000	0.000	10.0	8.0
2	8	1.000	0.000	0.000	0.000	0.000	10.0	2.0
3	15	0.684	0.360	0.000	0.000	0.000	16.3	1.3
4	20	0.500	0.500	0.000	0.000	0.000	20.0	0.0
5	30	0.000	1.000	0.000	0.000	0.000	30.0	0.0
6	80	0.000	0.000	0.000	0.500	0.500	80.0	0.0
7	85	0.000	0.000	0.000	0.316	0.684	83.7	−1.3
8	95	0.000	0.000	0.000	0.000	1.000	90.0	−5.0
9	100	0.000	0.000	0.000	0.000	1.000	90.0	−10.0

From the data in Table 4 and other calculated data, it can be seen that when $10 < x < 90$ and $10 < x < 90$, the deviation between the DCV and the OIV is within a reasonable range. When $0 \leq x \leq 10$, the DCV is a fixed value of 10. When $90 \leq x \leq 100$, the DCV is 90. The membership function with large deviation (when $x \leq 10$ or $x \geq 90$) should be corrected.

3.4 Determine the Final Membership Function

There is a large deviation for the membership function of the first and the last interval, which needs to be corrected. The membership functions while $x \leq 10$ or $x \geq 90$ are compensated by straight lines and the original membership value is used as the resulst of anti-fuzzy calculation which can eliminate the deviation. For other intervals, the normalized membership subset is used as the result of anti-fuzzification calculation. The final membership function is as follows:

$$u_{v1}(x) = \begin{cases} 0.1 * x & 0 \leq x \leq 10 \\ 1/(1 + 0.0011 * (x - 10)^2) & 10 < x \leq 20 \\ (30 - x) * 0.09 & 20 < x \leq 30 \\ 0 & x > 30 \end{cases} \tag{5}$$

$$u_{v2}(x) = \begin{cases} (x - 10) * 0.09 & 10 < x \leq 20 \\ 1/(1 + 0.0011 * (x - 30)^2) & 20 < x \leq 40 \\ (50 - x) * 0.09 & 40 < x \leq 50 \\ 0 & x \leq 10, x > 50 \end{cases} \tag{6}$$

$$u_{v3}(x) = \begin{cases} (x - 30) * 0.09 & 30 < x \leq 40 \\ 1/(1 + 0.0011 * (x - 50)^2) & 40 < x \leq 60 \\ (70 - x) * 0.09 & 60 < x \leq 70 \\ 0 & x \leq 30, x > 70 \end{cases} \tag{7}$$

$$u_{v4}(x) = \begin{cases} (x-50)*0.09 & 50 < x \leq 60 \\ 1/(1+0.0011*(x-70)^2) & 60 < x \leq 80 \\ (90-x)*0.09 & 80 < x \leq 90 \\ 0 & x \leq 50, x > 90 \end{cases} \quad (8)$$

$$u_{v5}(x) = \begin{cases} 0 & x \leq 70 \\ (x-70)*0.09 & 70 < x \leq 80 \\ 1/(1+0.0011*(x-90)^2) & 80 < x \leq 90 \\ 1+1/90*(x-90) & x > 90 \end{cases} \quad (9)$$

Table 5. The OIV, membership degree subset, the DCV, deviation between the DCV and the OIV

No.	OIV,x	Uv1(x)	Uv2(x)	Uv3(x)	Uv4(x)	Uv5(x)	DCV,x1	x1 − x
1*	2	0.200	0.000	0.000	0.000	0.000	2.0	0.0
2*	8	0.800	0.000	0.000	0.000	0.000	8.0	0.0
3	15	0.684	0.360	0.000	0.000	0.000	16.3	1.3
4	20	0.500	0.500	0.000	0.000	0.000	20.0	0.0
5	30	0.000	1.000	0.000	0.000	0.000	30.0	0.0
6	80	0.000	0.000	0.000	0.500	0.500	80.0	0.0
7	85	0.000	0.000	0.000	0.316	0.684	83.7	−1.3
8*	95	0.000	0.000	0.000	0.000	1.056	95.0	0.0
9*	100	0.000	0.000	0.000	0.000	1.111	100.0	0.0

Notes: No. 1, 2, 8, and 9 are unnormalized values, and others are normalized values.

The membership degree subsets, the DCV, and the deviation between the DCV and the OIV are calculated, as shown in Table 5.

It can be seen from Table 5 that the deviation between DCV and the OIV is within a reasonable range. The membership function finally obtained meets the predetermined requirements.

The finally determined subsets of membership levels for each evaluation level are calculated according to formulas 5–9. If $x \leq 10$ or $x \geq 90$, the unnor-malized membership values are used to construct the membership subset. If $10 < x < 90$, the normalized membership values are used.

4 Conclusion

Apply the membership evaluation principle, select sample points, and calculate the membership subset. Perform the anti-fuzzification calculation on the membership subset, and compare the deviation between the reverse calculation result and the input value. By correcting the slope of the membership curve, calculating the sub-area of the membership, and reducing the intersection area of the membership curve, the problem of excessive deviation between the calculated value and the input value of anti-fuzzification is gradually solved. The membership function finally obtained meets the expected requirements.

Acknowledgments. This paper is funded by the Special Fund Project for Science and Technology Innovation (2019-C507) of the Institute of Highway Science, Ministry of Transport.

References

1. Wu, G., Li, C.: Application of analytic hierarchy process in safety risk assessment of CFST arch bridge construction. W. Transp. Sci. Technol. **132**(07), 123–127 (2018)
2. Wang, S., Fu, D., Chen, J., et al.: Method for determining the weight of subjective evaluation index of automobile dynamic performance based on entropy method. J. Highw. Transp. Res. Dev. **32**(7), 153–158 (2015)
3. Zhou, D., Wang, H., Li, W.: Research on the risk of liquid ammonia leakage based on AHP and fuzzy theory. J. Wuhan Univ. Technol. **03**, 81–87 (2017)
4. Guo, F., Meng, Y.: Research on coal quality grading evaluation based on fuzzy evaluation method. Math. Pract. Cogn. **7**, 1–8 (2019)
5. Zhu, W., Mei, S.: Application of fuzzy comprehensive evaluation method in hospital internal performance appraisal. Sci. Educ. Lit. (First Issue) **454**(04), 115–118 (2019)
6. Wu, J., Fan, W.: Research on risk assessment system for dangerous goods road transportation. J. Highw. Transp. Res. Dev. **32**(12), 6–11 (2015)
7. Qian, K., Qin, M., Song, Y.: Multi-objective optimal dispatch of charging stations based on fuzzy membership. Electr. Power Constr. **41**(2), 118–124 (2020)
8. He, Y.: Fuzzy reconstruction algorithm and its application based on label semantics and prototype theory (2017)
9. Wang, J., Lu, Z.: Method for determining membership function in fuzzy control. Henan Sci. **18**(4), 348–351 (2000)
10. Wang, Y., Huang, J., Liu, K., et al.: Evaluation and spatial variability of paddy soil fertility in typical counties in Northeast China. J. Plant Nutr. Fertil. **26**(2), 256–266 (2020)
11. Yang, M., Lu, H., Zhang, Y.: National comprehensive transportation system safety evaluation and countermeasures based on the comprehensive index model. In: World Transportation Conference (2019)
12. Xiao, W.: Research on the evaluation of the transportation safety of automobile units. Nanjing University of Science and Technology (2009)

Analysis of Computer Software Technology under Big Data Environment

Lei Guiping(✉)

Guangdong University of Science and Technology, Dongguan 523083, China
122432141@qq.com

Abstract. With the rapid development of modern social information technology and the interaction of economic and social information more frequently, all industries are in the background of the era of big data. In this case. The development level of computer software technology has become an important factor that affects and even determines the speed of social development. Facing the increasingly complex Internet network structure and traffic data information, how computer software technology is facing new and emerging challenges, strengthening the computing power and data processing capacity of the big data application center has become a decisive factor affecting the development of the contemporary computer industry. Based on the analysis of the problems in the field of computer software development in the era of big data, this paper gives a brief introduction to the software technology, hoping to help readers understand the relevant technology in the computer software industry.

Keywords: Big data era · Computer software technology · Information technology

1 Social Problems in Computer Software Technology

1.1 The Accuracy of the Information Cannot be Guaranteed

In the era of big data, various information technologies are showing a blowout development. The wide application of self-media, cloud computing, file sharing and other technologies has increased people's interaction channels. At the same time, networked electronic products have become popular, and people have realized real-time communication through network connection [1]. Under this circumstances, massive amounts of information data circulate on the Internet, and everyone is the sender and receiver of information. Due to the increase of online traffic and information, it is difficult for the public to judge the authenticity of massive information, At the same time, the cost of making rumors is reduced, and stakeholders may disseminate false information for their own purposes and spread the false information through some technical means. In this way, the false information spreads from one to the next in the Internet, confusing the public and forming the effect of "three people become a tiger". In the current era, everyone lives in the torrent of the Internet, and inevitably we will receive all kinds of information from the Internet. In addition to being misled by false information, it is easy for us to become the disseminator of rumors unknowingly.

M. Tavana et al. (Eds.): IISA 2020, AISC 1304, pp. 47–52, 2021.
https://doi.org/10.1007/978-3-030-63784-2_7

The existence of false information is no doubt not conducive to the construction of a healthy network environment [2].

1.2 Reliability of Data Transmission

In the network information communication mechanism, it is divided into five layers of network technology architecture, which are application layer, transmission layer, network layer, middle layer and physical layer. Among them, data transmission activities in communication between networks are mainly in the transmission layer. In the process of network data transmission, packet loss may occur due to the queuing delay of data nodes, namely routers, congestion mechanism or time delay in network links. This situation will cause data transmission problems in the communication mechanism [3]. After a packet is lost, there are generally two kinds of processing situations. One is that the receiving party does not receive the data and sends the abnormal information back to the sending party through the network. The other is that the receiver does not receive the data, but does not report abnormality. The loss caused by the latter is often much larger than the former, in some specific communication mechanisms, if the parties to the communication have some important business secrets, data loss occurred in the process of information transmission, such as product delivery, and there was no abnormal information feedback from the network, it is likely to lead to failure of cooperation, resulting in economic losses that are unaffordable for companies or individuals. At present, two protocols, UDP and TCP, are generally used in the transport layer. Among them, TCP is more widely used because the mechanism of this protocol can ensure more reliable information transmission between the receiver and the sender. However, data transmission in the network cannot completely prevent the occurrence of packet loss.

1.3 Information Leakage

Nowadays, the situation of personal information leakage is very serious, I believe that many people are suffering from spam SMS and unknown phone harassment. The existence of these situations more or less reflects the information leakage of people in contemporary society [4]. What's more, some unscrupulous companies distribute and sell the collected user information at will for profit-making purposes. These behaviors pose a threat to citizens' personal privacy and personal safety. In addition to company actions, some people lack awareness of protecting others' personal information. For example, it is unreasonable to dig into the personal information of celebrities on the Internet. In such a society where personal information is sold at will and released in a malicious way, people's sense of security is insufficient, and citizens whose personal information is leaked will have troubles in life.

2 Example Analysis of Computer Software Technology in the Context of Big Data

2.1 Software Information Collection and Collation

In the era of big data, there are abundant data sources, and there is a lot of real and false information. The collection and processing of massive information requires a powerful data processing center. However, a single host cannot undertake a large number of computing tasks and provide the computing power required by the Internet. Therefore, nowadays, virtual servers are generally built into a large data center by a collection of multiple computers to process the data, as shown in Fig. 1. In the process of data collection and processing in the era of big data, it is required to increase the calculation speed. Currently, MapReduce parallel processing technology is generally used. This technology can simultaneously perform a large amount of data to complete the cutting, process the data in parallel, organize the processing results and feedback these three tasks. The application of parallel processing technology in this technology improves the computing speed and increases the utilization of computers in data application centers.

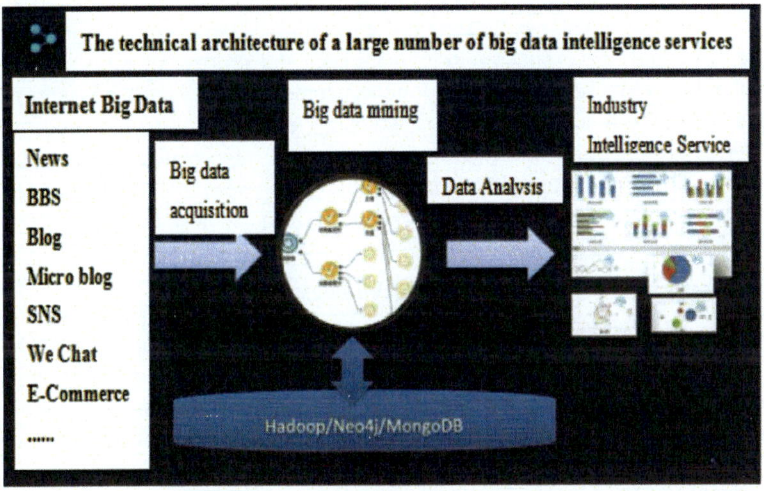

Fig. 1. Schematic diagram of big data service technology architecture

In the development of the modern era, the collection and processing technology of big data information has developed rapidly, and new technologies have been continuously developed and applied. Different techniques are used in different information processing processes. The following techniques are more common. The first is system log collection. Some distributed architectures implement system log collection. The collection process has the characteristics of fast transmission speed. There is also a network data collection method. This method is widely used in unstructured data collection. Through web crawlers and website open APIs, unstructured files in the network are converted into structured files and stored on the local computer [5]. Other data collection methods are not as widely used as the above two, so I will not repeat them here.

2.2 Software Information Storage Technology

In information technology processing, There are various forms of information, text, pictures, audio and video are becoming more and more diverse in one form on the network. The content of information is becoming more and more quantified, and the trend of more and more complicated source methods is developing on the Internet. These conditions make it more difficult to store information. First of all, the space requirement for information storage is greater. In addition, the process of searching, modifying and transferring information is more demanding. In the modern development process, the processing of information often needs to consider factors such as storage cost, processing time, and accuracy. Based on the above considerations, the concept of cloud computing has emerged. Under this technology, a large amount of information can be stored and accessed by powerful servers, which improves the effect of information storage.

In the process of building a modern society, the application of cloud storage services has been quite extensive, through the application of cloud storage services, it can solve the time and space constraints of people in the process of transmitting information and data. In modern society, as long as people have electronic products that can be connected to the Internet, they can download, view, and upload resources in cloud storage. Massive data and information further form a huge network database, which further strengthens the integration efficiency of various resources and provides more convenient information services for the operation of the society [6].

2.3 Software Information Security Technology

The development of information technology not only facilitates people's lives, but also brings trouble to the development of society. Based on the above information processing problems to consider, people increasingly recognize the importance of information security technology, people are becoming more and more aware of the network security mechanism in the process of Internet information transmission. Just like the lock shown in Fig. 2, it is safe to lock. In the daily use of the Internet, ordinary netizens need to strengthen their security awareness, not randomly click unknown links, or browse websites that are not protected by security protocols. In addition, industry practitioners pay attention to the encryption and decryption of important documents, strengthen the self-binding ethical standards of the industry, and do not use their own technology and relevant knowledge to do things that violate ethics and laws [7].

At present, more and more mature virtualization technology has improved the collection of data in the computer software industry. In the actual application process, virtualization technology can make the computer's data processing out of the limitations of hardware facilities, thereby improving the computer's ability to process big data, shorten the operation time, and optimize information processing performance.

Fig. 2. An example of information security issues

Virtualization technology is now one of the more mature computer software technologies, through data analysis and image processing, greatly shorten the cost of similar information processing. However, at the same time, the wide application of virtualization technology also aggravates the severity of network information security problem to some extent. Hackers can attack end-to-end security vulnerabilities in the application of virtualization technology, and implant Trojan horses, viruses, worms and other network insecurity factors on the client. In practical application, the scientific application and further development of virtualization technology require industry practitioners to abide by industry ethics, and network security engineers to find network security vulnerabilities and repair them. Meanwhile, network firewall is constantly strengthened to ensure network security [8].

3 Conclusion

The rapid development of computer software technology in the era of big data. The development of information technology has promoted the progress of human society in the fields of finance, science and technology, and has facilitated people's lives, but it has also brought certain social problems. Under the development trend of the current era, computer software technology will continue to develop for a period of time in the future. The emergence of problems requires the joint efforts of everyone in all walks of life in society to explore and solve them. The development of software technology bears the burden of leading the development trend of the times and providing technical support for social development. The complexity of massive information and information processing in today's society makes it increasingly difficult for information processing. Industry practitioners should follow the development trend of the times and update their own ideas concept, promote the continuous development of computer information processing technology in the era of big data, so as to meet the needs of social development.

References

1. Yong, W.: Analysis of computer software technology in the era of big data. Commun. World **27**(04), 71–72 (2020)
2. Zongwei, G.: Computer software technology analysis under big data environment. Comput. Prod. Circ. **04**, 31 (2020)
3. Jigang, Z.: Computer software technology analysis under big data environment. Inf. Recording Mater. **21**(01), 137–138 (2020)
4. Hongqiang, H.: Analysis of computer software technology in the era of big data. Mod. Ind. Eco. Inform. **9**(11), 58–59+108 (2019)
5. Huanzheng, S.: Application research of computer software technology in big data environment. Wirel. Internet Technol. **16**(14), 126–127 (2019)
6. Tonghan, Z.: Application research of computer software technology under big data environment. Electr. World (03), 42+44 (2019)
7. Heo, M., Toomey, N.: Learning with multimedia: the effects of gender, type of multimedia learning resources, and spatial ability. Comput. Educ. **146**, 103747 (2020)
8. Greenbaum, L.E., Ukomadu, C., Tchorz, J.S.: Clinical translation of liver regeneration therapies: A conceptual road map. Biochem. Pharmacol. **175**, 113847 (2020)

Research on ShadowsocksR Traffic Identification Based on Xgboost Algorithm

Ji Qingbing[1], Deng Xiaoyan[2(✉)], Ni Lulin[1], and Lei Haijun[3]

[1] China Electronics Technology Group Corporation, 30 Group, Beijing, China
[2] Chengdu University of Information Technology, Chengdu, China
dxy@cuit.edu.cn
[3] Chengdu Normal University Deyang High School, Chengdu, China

Abstract. It is very difficult to identify Shadowsocks (SS) traffic, most of which stay in the laboratory environment, and there are very few published research results in this field at home and abroad. ShadowsocksR (SSR) is an enhanced version of SS. It can disguise the traffic of SS as that of conventional protocol, such as HTTP traffic, TLS traffic, etc., which makes it more difficult to identify SSR traffic. Based on Xgboost algorithm, this paper proposes a method to identify SSR traffic for the first time. The experimental results show that this method has a good recognition effect on SSR traffic, and the precision, the recall, the accuracy is all above 95.3%.

Keywords: ShadowsocksR · Xgboost · K-means

1 Introduction

With the rapid development of the Internet, the number of network users is growing explosively. According to the survey and statistics of Miniwatts [1], a well-known Internet business company, by March 3, 2020, the global Internet users have reached 4.57 billion, accounting for 58.6% of the current global population. With the increase of the number of network users, the explosive growth of traffic information is followed, which brings great challenges to network management. In order to manage and plan network traffic, network operators often use traffic identification technology to monitor network traffic.

With the improvement of people's attention to the security of network communication, various kinds of tunneling technology, proxy technology, traffic camouflaging technology are also proposed. The encrypted traffic generated by the communication tools developed by these technologies is difficult to be identified. SS is a new anonymous communication tool based on Socks5, which has the features of high efficiency, fast speed and simple deployment. It is one of the most popular tools to turn over the wall. Its operation principle is similar to that of anonymous communication software based on proxy technology, but it is different from Tor and SSH and other common anonymous networks which use some common protocols to communicate. SS has a set of its own protocol for encrypted communication, so it is more difficult to identify the encrypted traffic generated by SS. At present, there are only a few literatures on identifying SS, and most of them stay in the laboratory environment. SSR is an

M. Tavana et al. (Eds.): IISA 2020, AISC 1304, pp. 53–61, 2021.
https://doi.org/10.1007/978-3-030-63784-2_8

enhanced version of SS. It can disguise the traffic of SS as that of conventional protocol, such as HTTP traffic, TLS traffic, etc., which makes it more difficult to identify SSR traffic. Here, aiming at HTTP camouflaging traffic of SSR, this paper proposes a method to identify SSR traffic for the first time. This method uses the statistical features of flow load, the entropy features of load information of the first four packets of a single flow and the new entropy features obtained from the analysis of the entropy features of load information by K-means clustering to supervised learn the HTTP traffic of SSR. The identification precision can reach 95.3%, the recall can reach 99.9%, and the accuracy can reach 99.7%.

The structure of this paper is as follows: in Sect. 2, we introduce some works related to our research. In Sect. 3, we state the communication principle of SS and SSR. In Sect. 4, we briefly introduce the principle of machine learning algorithm and related parameter setting methods that will be used in our identification model. Then in Sect. 5, we introduce the data set, feature selection and other methods that we will use to build our identification model. Next, in Sect. 6, we use the machine learning algorithm in Sect. 4, the data and the relevant features of the flow load in Sect. 5, to carry out the experiment on identifying SSR traffic, and analyze the experimental results. Finally, the work of this paper is summarized in Sect. 7.

2 Related Work

Literature [2] introduces the recent development of traffic identification technology, most of which are for non-encrypted traffic. Currently, the main technologies of traffic identification are based on port, payload, machine learning.

The identification technology based on port is simple to deploy and suitable for traditional network environment. By identifying the destination port number in the TCP/UDP packet header, it finds the network application corresponding to the port and completes the traffic classification [3]. The disadvantage is also obvious: application traffic that does not use the default port is not recognized.

In addition to detecting the IP header and TCP/UDP header of traffic, the identification technology based on payload also detects part or all of the payload content of data packets. Therefore, this method is also called deep packet inspection (DPI) method. Compared with the identification method based on port number, this method has a more in-depth analysis of traffic, directly reviews the content of traffic, so it has a higher classification accuracy. However, DPI is ineffective in identifying the traffic generated by many applications is encrypted.

In order to solve the above problems, researchers have introduced the concept of machine learning into encrypted traffic identification. The main idea is to train an identification model by defining a set of statistical features related to traffic, such as packet length, packet arrival time, etc., and then use machine learning algorithm to achieve the identification of network traffic [4–6].In reference [7], the Random Forest (RF) Algorithm, a semi-supervised learning algorithm, is applied into the traffic identification of SS with an accuracy of 85%. Zeng Xuemei et al. [8] proposed an identification method for SS based on flow context and host behavior with an accuracy of 93.43% in the laboratory environment. In reference [9], the identification of SS traffic based on Xgboost algorithm is realized, and the accuracy is 94.63%. SSR is an enhanced version of SS, which uses traffic camouflage technology, which makes it more difficult to identify SSR. Up to now, there is no published literature on identifying SSR.

3 Principle of Shadowsocks

3.1 Shadowsocks

Shadowsocks (SS) is a software of over the wall based on Socks5 protocol. Like most principles of proxy software, SS first establishes an SSH based encryption channel, which includes servers outside the firewall; secondly, SS proxy by using established channels, which means that the real server is requested through SSH server. Finally, the SSH server sends back the response data using the established channel. Because SSH itself is based on RSA encryption technology, the firewall cannot analyze the key words of encrypted data in the transmission process, so as to prevent reconnection. SSH has targeted interference problems, so SS splits Socks5 protocol into two parts: server side and client side. The communication principle of SS is shown in Fig. 1.

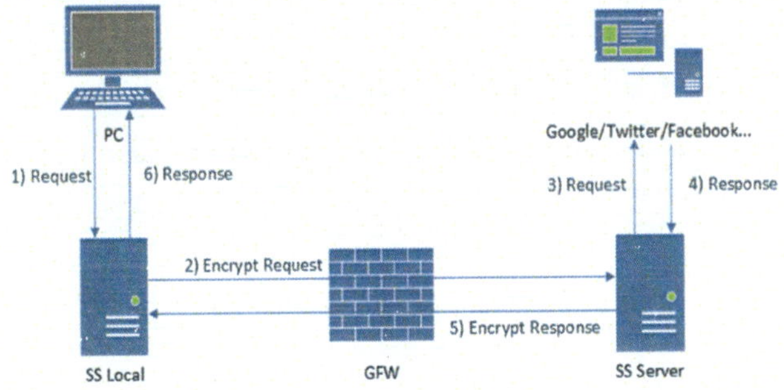

Fig. 1. The communication principle of SS

As we can see in Fig. 1, every running step of SS bypasses the detection of the firewall and will not be interfered by the firewall. That's why it's hard to identify SS.

3.2 ShadowsocksR

ShadowsocksR (SSR) is the upgraded version and enhanced version of SS. The whole operation principle of SSR is basically the same as that of SS, just to ensure the security and anti-identification of data, SSR adds the options of "Protocol" and "obfuscation" plug-in on the basis of "encryption" of SS; the main function of "Protocol" plug-in is to encapsulate the data in a certain format before "encryption" of network application data, so as to increase the security of data; The main function of the "obfuscation" plug-in is to disguise the network application data as a standard network protocol including HTTP and HTTPS after "encryption", so as to increase the anti- identification of data.

Currently, the latest version of SSR is 5.15, in which the "Protocol" plug-in supports 13 options, the "obfuscation" supports 6 options, and the "encryption" supports more than a dozen encryption algorithms, such as AES, idea, RC4, etc.

4 Clustering Algorithm and Ensemble Learning

4.1 K-Means Clustering Algorithm

K-means clustering algorithm is a simple iterative clustering algorithm, using distance [10] as the similarity measurement index, so as to find K classes in a given data set, and the center of each class is obtained according to the mean value of all the values in the class, and the center of each class is described by the clustering center. Procedure is as follows:

1) Randomly select k samples as the initial clustering center.
2) Calculate the distance between each sample and cluster center.
3) Regress each sample to the nearest cluster center.
4) Solving the mean value of each class as a new clustering center.
5) If the clustering center remains unchanged or the number of iterations is reached, the algorithm ends, otherwise it returns to 2).

4.2 Xgboost Ensemble Learning Algorithm

Xgboost (extreme gradient boosting) [11] is an improved algorithm of traditional Gradient Boosting Decision Tree algorithm, which is improved based on AdaBoost and GBDT. Xgboost has the features of high accuracy, flexible and difficult to overfit and so on.

The objective function after Taylor's second order expansion is Eq. (1).

$$Obj^{(t)} \cong \sum\nolimits_{i=1}^{n} \left[l\left(y_i, \hat{y}_i^{(t-1)} + g_i f_t(x_i) + \frac{1}{2} h_i f_t^2(x_i) \right) \right] + \Omega(f_t) + constant \quad (1)$$

Definition g_i and h_i as follows:

$$g_i = \partial_{\hat{y}(t-1)} l\left(y_i, \hat{y}_i^{(t-1)} \right) \text{ and } h_i = \partial^2_{\hat{y}(t-1)} l\left(y_i, \hat{y}_i^{(t-1)} \right)$$

After the constant term *constant* is removed, a concise objective function is obtained, as shown in formula (2):

$$\sum\nolimits_{i=1}^{n} \left[g_i f_t(x_i) + \frac{1}{2} h_i f_t^2(x_i) \right] + \Omega(f_t) \quad (2)$$

This objective function only depends on the first derivative and the second derivative of each sample point on the loss function. Next, we define the complexity of the tree, which includes the number of leaf nodes of a decision tree and the sum of the square of the output scores of each leaf node. The more leaf nodes, the weaker generalization ability. The complexity calculation formula of the tree is as follows:

$$\Omega(f_t) = \gamma T + \frac{1}{2}\lambda \sum\nolimits_{j=1}^{T} \omega_j^2 \tag{3}$$

After the definition of decision tree complexity, the objective function can be rewritten as Eq. (4).

$$Obj^{(t)} = \sum\nolimits_{i=1}^{n} \left[\left(\sum\nolimits_{i \in I_j} g_i \right) \omega_j + \frac{1}{2} \left(\sum\nolimits_{i \in I_j} h_i + \lambda \right) \omega_j^2 \right] + \gamma T \tag{4}$$

Where, the sample set on each leaf $I_j = \{i | q(x_i) = j\}$.

The objective loss function contains T independent quadratic functions, and then defines $G_j = \sum_{i \in I_j} g_i$, $H_j = \sum\limits_{i \in I_j} h_i$. The objective function can be rewritten as Eqs. (5).

$$Obj^{(t)} = \sum\nolimits_{j=1}^{T} \left[G_j \omega_j + \frac{1}{2} \left(H_j + \lambda \right) \omega_j^2 \right] + \gamma T \tag{5}$$

The training process is to minimize the target loss function, and when $\omega_j^* = \frac{G_j}{H_j + \lambda}$, the objective loss function is the smallest. $Obj = -\frac{1}{2} \sum_{j=1}^{T} \frac{G_j^2}{H_j + \lambda} + \gamma T$, and Obj is a loss function for scoring tree structure, so the smaller the value, the better the tree structure (Table 1).

Table 1. Dataset distribution information

Data set	Pattern	Data size	Number of flows	Collect time
Training set	HTTP traffic	11.8G	5000	<3 months
	Non-HTTP traffic	17.3G	5000	<4 months
Test set 1	HTTP traffic	39.7G	659	>12 months
	Non-HTTP traffic		21430	
Test set 2	HTTP traffic	426 M	142	<2 months
	Non-HTTP traffic		1238	
Test set 3	HTTP traffic	20.4G	3500	<5 days
	Non-HTTP traffic		395	

5 The Method of Identifying ShadowsocksR

5.1 Data Set

The HTTP traffic in the above table is camouflaged by the encrypted traffic when using HTTP mode to visit foreign website resources in SSR. Non-HTTP traffic is the ordinary traffic generated by visiting domestic public websites, including some real HTTP protocol traffic. In order to ensure the accuracy of the training model, we average the proportion of positive samples (HTTP traffic) and negative samples (Non-HTTP traffic) in the training set. The collection time in the above table is relative to the time calculated on January 1, 2020.

5.2 Choose of Features

Because the SSR traffic has some features of HTTP protocol after camouflaged as HTTP traffic, and still retains some features of encrypted traffic, we extract the statistical features of flow payload, the information entropy of the first four packets' payload of a single flow and other new entropy features obtained by clustering analysis, and take these features as the classification feature to measure the randomness of SSR traffic.

Table 2. Table captions should be placed above the tables.

Feature name	Description
upp_n	Total number of uplink payload packets
dpp_n	Total number of downlink payload packets
ul_t	Total number of uplink payload
dl_t	Total number of downlink payload
fd	Flow duration time
ul_a	The average value of uplink payload
dl_a	The average value of downlink payload
ul_min	The minimum number of uplink payload
dl_min	The minimum number of downlink payload
ul_max	The maximum number of uplink payload
dl_max	The maximum number of downlink payload
ul_v	The variance of uplink payload
dl_v	The variance of downlink payload
fir_e	The information entropy of the first package' payload
fir_d	Direction of the first package
sec_e	The information entropy of the second package' payload
ec_d	Direction of the second package
thi_e	The information entropy of the third package' payload
thi_d	Direction of the third package
fou_e	The information entropy of the fourth package' payload
fou_d	Direction of the fourth package
other_e	other new entropy features obtained by clustering analysis

When calculating the information entropy of payload, in order to average the differences between HTTP protocol features and encrypted traffic features, we add fixed printable characters to average the non-printable characters and control characters in the traffic, so as to reduce the possibility that the information entropy of payload inverse difference between the same type of traffic is too large, which is beneficial to the classification results.

5.3 Cluster Analysis

We use two analysis methods: one is equal width analysis, the other is K-means clustering analysis.

In the left figure, the information entropy of the first packet' payload of HTTP traffic in training set is analyzed by equal width analysis and K-means clustering analysis, and in the right figure, the information entropy of the second packet' payload of HTTP traffic in training set is analyzed by equal width analysis and K-means clustering analysis. The abscissa represents the classification generated by different methods, while the ordinate represents the proportion of each category in the whole. The reason for adopting the above drawing expression is to find the significant outliers of the data set and estimate the value range of the data; the value range of the information entropy of the first package' and the second package' payload is determined through the analysis of the outliers excluded and the feedback value of the accuracy (Fig. 2).

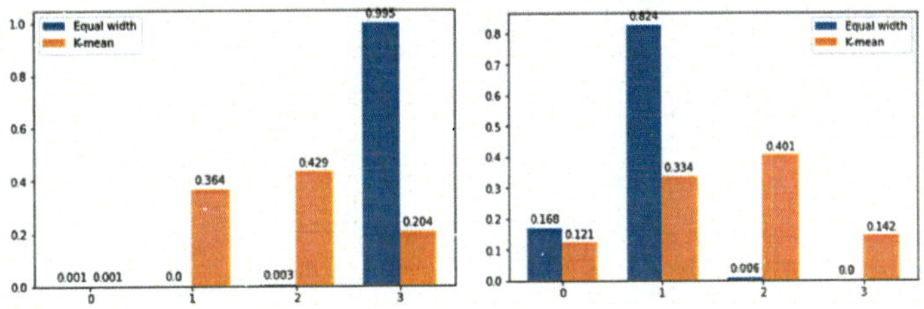

Fig. 2. Classification of data set 4 by equal width method and K-means clustering analysis method

6 Experiment and Evaluation

The data of training set is imported into Xgboost training model, and the appropriate parameters are adjusted to perform the task of two classification of training set. The following figure describes the importance of each feature in the training process, the abscissa is the importance of the feature, and the ordinate describes the feature dimension of the training data. For example, the importance of the first dimension (see Fig. 3) is 191.

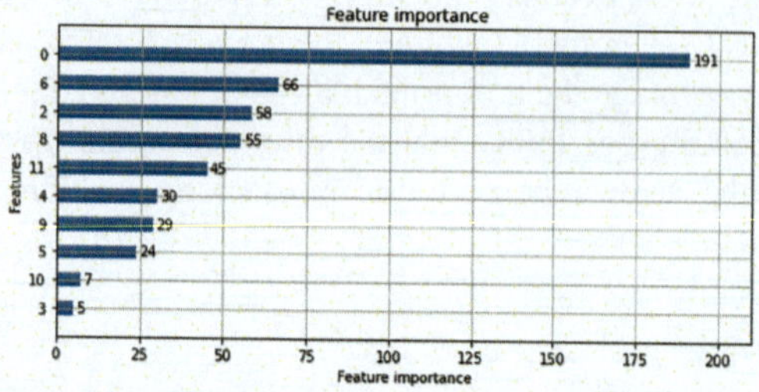

Fig. 3. Importance arrangement of training model features

Table 3 describes the predicted results of the test set in the training model.

According to the prediction results in Table 2 and Table 3, through clustering analysis, we can clear the outlier data in the training set, improve the robustness of the training model, and have high accuracy, recall and accuracy for both the long-term data (test set 1) and the short-term data (test set 2). And it also has high precision and recall when the data set with small proportion of positive samples (3.0% (test set 1), 10.3% (test set 2)) and large proportion (89.8% (test set 3)) are tested.

Table 3. Test results of test set

Test set	Precision	Recall	Accuracy
Test set1	95.3%	99.9%	99.7%
Test set2	98.1%	100%	99.9%
Test set3	99.6%	99.5%	99.3%

7 Conclusion

In this paper, we focused on the identification of traffic camouflaged with HTTP protocol in SSR. According to the features of HTTP traffic, we use the equal width method and K-means clustering analysis method to analyze, extract the hidden features of HTTP traffic, clean the discrete point traffic of training set, and classify HTTP traffic and non-HTTP traffic based on Xgboost algorithm. The experimental results show that the training model has high robustness and ideal precision, recall and accuracy.

References

1. Miniwatts Marketing Group: Internet usage statistics-world internet users and population stats, June 2020. https://www.internetworldstats.com/stats.htm
2. Chen, L., Gao, S., Liu, B., Lu, Z.: Research status and development trends on network encrypted traffic identification. Netinfo Secur. (03), 19–25(2019)
3. Wang, Y., Xiang, Y., Yu, S.Z.: An automatic application signature construction system for unknown traffic. Concurrency Comput. Pract. Experience **22**(13), 1927–1944 (2010)

4. Perera, P., Tian, Y.C., Fidge, C., et al.: A comparison of supervised machine learning algorithms for classification of communications network traffic. In: International Conference on Neural Information Processing, pp. 445–454. Springer, Cham (2017)
5. Andrew, et al.: Internet traffic classification using bayesian analysis techniques. ACM SIGMETRICS Perform. Eval. Rev. **33**(5), 50–60 (2005)
6. Zuev, D., Moore, A.W.: Traffic classification using a statistical approach. Passive and Active Network Measurement. 6th International Workshop, pp. 32–45. Springer-Verlag, USA (2005)
7. Deng, Z.Y., Liu, Z.H., Chen, Z.G., et al.: The random forest based detection of Shadowsock's traffic. In: 9th International Conference on Intelligent Human-Machine Systems and Cybernetics (IHMSC), pp. 75–77. IEEE, Hangzhou (2017)
8. Xuemei, Z.E.N.G., Xingshu, C.H.E.N., Guolin, S.H.A.O., et al.: Flow context and host behavior based shadowsocks's traffic identification. Spec. Sect. Secur. Priv. Cloud IoT **7**, 41017–41032 (2019)
9. Cheng, J., Li, Y., Huang, C., Yu, A., Zhang, T.: ACER: detecting Shadowsocks server based on active probe technology. J. Comput. Virol. Hacking Tech. **16**(3), 217–227 (2020). https://doi.org/10.1007/s11416-020-00353-z
10. Liu, Y., Li, W., Li, Y.: Network traffic classification using k-means clustering. In: Second International Multi-Symposiums on Computer and Computational Sciences, pp. 360–365. IEEE, Hangzhou (2007)
11. Xuhang, L.: SSH Traffic Identification Based on XGBoost. Harbin University of Science and Technology, Harbin (2019)

Research on the Relevance and VaR of GEM Market Based on Vine Copula

Zhang Xin(✉) and Zou Yumei

Shandong University of Science and Technology, Qingdao 266000, Shandong Province, China
1428741044@qq.com

Abstract. This article selects the Standard & Poor's GEM index (SPHKGEM), GEM Strategic Emerging Industries Index (SZSCSEII), the Growth Enterprise Market 50 Index (SZEXT50), the Growth Enterprise Market Fundamental Market Index (SZSCPMI) as the research objects from January 2, 2018 to December 2, 2019. Establish the C-Vine structure for 2018 and 2019 for the four indexes, and then establish the C-Vine structure for the overall yield, and measure the VaR of the asset portfolio that conforms to the C-Vine structure in 2018 and 2019. The results show that the GEM Strategic Emerging Industries Index is in a key node position among the four GEM indices, and it shows a high correlation with other indexes under the C-Vine structure. The GEM Strategic Emerging Industries Index.The VaR in 2019 combined with the SPHKGEM, the SZEXT50, and the SZSCPMI decreased compared with 2018. This is related to China's continued implementation of a stable monetary policy, continuous promotion of the opening of the financial industry, the government's encouragement of technology companies, and the development of high-tech industries.

Keywords: C-Vine · Non-linear correlation coefficient · GEM · VaR value at Risk

1 Introduction

The correlation between financial markets has always attracted the attention of scholars. When a financial crisis occurs, huge fluctuations in a financial market spread to other financial markets through certain channels. The government can formulate appropriate financial policies and conduct macro-control based on the correlation between this market and the crisis market. Chinext is the second stock exchange market, which is different from the main board market. It is designed to provide financing channels and growth space for entrepreneurial enterprises, minor enterprises and high-tech industrial enterprises that need capitalizing and development. The growth enterprise market is an important replenish to the main board market and occupies an significant place in the capital market.

At present, Vine Copula and VaR have been widely used in financial market measurement. Huang [1] utilized the Copula-GARCH model to estimate the VaR of the stock market portfolio. Zhou [2] grouped various institutions in the financial market

M. Tavana et al. (Eds.): IISA 2020, AISC 1304, pp. 62–69, 2021.
https://doi.org/10.1007/978-3-030-63784-2_9

according to their industry and used Copula to measure the overall VaR. Lin Bo [3] introduced the concept, characteristics and calculation method of VaR in the application of VaR method based on the GARCH model in the Shanghai Stock Market. The GARCH model was used to estimate and predict the VaR value of the stock market, and an empirical study was conducted on the risk measurement of the Shanghai Composite Index. Based on GARCH family model, Li Lin and Zhang Qinglong [4] calculated the VaR value of liquidity gap under two different distribution assumptions, and compared and analyzed the liquidity risk of large and small commercial banks in China. The results show that the VaR of China's large commercial banks is increasing year by year, while the VaR of small and medium-sized banks shows a steady decline, but on the whole, the VaR of small and medium-sized commercial banks is significantly higher. Zhao Zhao [5] and others based on the GARCH model to measure and compare the overall risk of China's GEM market. Wu Jilin and Zhang Erhua [6] built a markov regime switching Student-t Copula tudied the structural changes of correlation between the China's stock market and US and Japanese stocks, and found that US, Japanese and Hong Kong stocks have obvious risk spillover effect on the continent market. Liu Zhenya [7] used this model to study the dynamic relationship between the New York stock market and the China's stock market, and found that American stock market has influence on China's stock market. Yan Taihua and Chen Mingyu [8] studied the return series of the sse composite Index, and the research showed that this model could well describe the stage characteristics of stock market volatility. Vine Copula is a model of complex multivariate correlation structure proposed by Bedford and Cooke [9, 10] based on Joe [11]. Zhang Bangzheng et al. [12] used the R-Vine Copula method to study the correlation between the Shanghai Industrial Index and found that the Vine structure can better characterize the correlation between the industrial indexes. Han Chao [13] used the high-dimensional dynamic C-Vine and D-Vine Copula to model the sequence and then used the Monte Carlo method to calculate the combined risk VaR. It provides a new model and method for quantitative risk indicators and risk supervision. Hu Zhiqiang and Zhao Meijuan [14]. Research on IPO underpricing system and IPO underpricing under the multi-partial t-Copula model: Using partial t-Copula function to analyze the underpricing of IPO underpricing t-Copula function in the empirical analysis based on the main board, Small and Medium board and The Growth Enterprise Market. As a result, it was found that the degree of underpricing of The Growth Enterprise Market is lower than that of the main board market and the Small and Medium board market. Bai Zhengwu et al. [15] discovered a long-term high dependence between the small and medium board and the tail of the GEM on the basis of the correlation between the dynamic t-Copula model and my country's capital market, and the fluctuation range between the GEM and the main board is relatively strong, so it is necessary to guard against risks.

Based on the above research, no one has combined GEM with Vine to study the correlation between GEM stocks. This paper innovatively uses two Vine Copula to study the correlations between different structures that may exist in the GEM stock market. Finally, this paper measures the liquidity risk between 2018 and 2019 by VaR.

2 Model Building

2.1 Vine Copula

Common Vine Copula models are divided into two categories, and only the C-Vine Copula model will be explained below. The corresponding four-dimensional C-Vine structure is shown in Fig. 1.

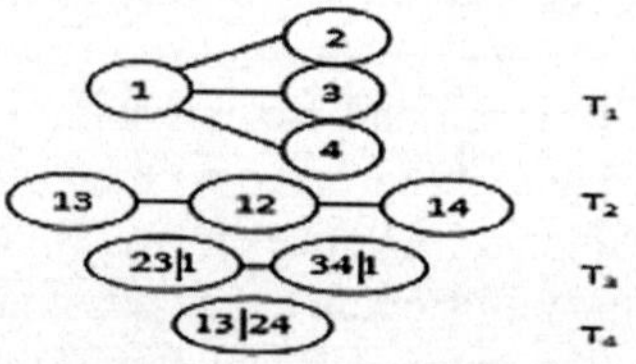

Fig. 1. C-Vine structure

Bedford and Cooke (2001) gave the density function expression C-Vine of n-dimensional C-Vine:

$$f(x_1,\ldots,x_n) = \Pi_{k=1}^{n} f(X_k)\Pi_{j=1}^{n-1}\Pi_{i=1}^{n-j} c_{j,j+i|1,\ldots,j-1}\left(F(X_j|X_1,\ldots,X_{j-1}), F(X_j|X_1,\ldots,X_{j-1})|\theta_{j,j+i|1,\ldots,j-1}\right)$$

Among them, i and j represent the variable serial number, $\theta_{j,j+i|1,\ldots,j-1}$ and $\theta_{i,j+1|j+1|i+1,\ldots,i+j-1}$ are the correlation parameters of the C-Vine Copula function, $c_{j,j+i|1,\ldots,j-1}$ and $c_{i,j+1|i+1,\ldots,i+j+1}$ represent density functions, $F(\cdot|\cdot)$ represents the edge distribution function.

2.2 VaR

Value at Risk, or VaR. It refers to a certain period of time, a certain level of confidence, a portfolio of the largest losses faced.

$$\mathrm{Prob}(\Delta p \leq \mathrm{VaR}) = 1 - \alpha$$

Δp is the alteration in the market value of the portfolio during the period. This formula indicates that within a certain holding period Δt, at a given confidence level $1 - \alpha$, the maximum loss of the portfolio will not exceed VaR.

Combining VaR:

$$VaR = \sqrt{VaR(\alpha)^2 + VaR(\beta)^2 + 2 * VaR(\alpha) * VaR(\beta) * Rho}$$

Rho is the correlation coefficient of assets α and β.

2.3 Parameter Estimation

The non-parametric kernel density estimation method is served to obtain the logarithmic marginal distribution function. The expression of kernel density estimation is as follows:

$$f(x) = \frac{1}{Th_T}\sum_{t=1}^{N} K\left(\frac{x - r_t}{h_T}\right)$$

Where $\{r_t\}_{t=1}^{T}$ is the logarithmic rate series. The quality of the kernel estimation result mainly depends on the kernel function and the window width. When the sample capacity is large enough, the selection of kernel function has rarely influence on the estimation results. In this paper, the normal kernel function with good smoothness is selected. Because financial time series usually show the characteristics of peaks and thick tails. In this paper, the optimal window width $\hat{h}$ is obtained by using the optimal window width selection principle proposed by Bowman [19]:

$$\hat{h} = 1.06\hat{\sigma}N^{-1/5}$$

$$\hat{\sigma} = \frac{\text{median}\left(\{|x_t - \hat{\mu}|\}_{t=1}^{N}\right)}{0.6745}$$

Where $\hat{\sigma}$ is the estimated value of the sample standard deviation and $\hat{\mu}$ is the median of the financial sample sequence $\{x_t\}_{t=1}^{N}$. The sequence Copula function will be obtained by kernel density estimation, and the correlation coefficient will be estimated by maximum likelihood.

3 Empirical Research

3.1 Descriptive Statistics

Standard & Poor's GEM index, the GEM Strategic Emerging Industries Index, the Growth Enterprise Market 50 Index, and the Growth Enterprise Market Basic Market Index were selected as the research objects. The sample interval was from January 2, 2018 to December 2, 2019, excluding trading days. After the transaction or non-corresponding data, there are 465 sets of data. The logarithmic rate of return expression is as follows:

$$R_t = (\text{Ln}P_t - \text{Ln}P_{t-1}) \times 100$$

Table 1 gives the values of the basic statistics of the four index returns .It can be seen from Table 1 that the logarithmic returns of the four indexes are negative during the sample period, and all four indexes have left-biased characteristics, indicating that each data has an obvious left tail, and all four indexes have spikes. The characteristics of thick tail. From the JB normal test and the ADF stationarity test, all data refuse the original hypothesis of normal distribution, and are all stationary series.

Table 1. Basic statistics of the logarithmic returns of the four indexes

	mean	std	skewness	kurtosis	JB	ADF
SPHKGEM	−0.263	1.111	−1.135	8.321	648.354***	−19.869***
SZSCSEII	−0.026	1.847	−0.145	4.910	72.342***	−21.447***
SZEXT50	−0.024	1.894	−0.040	4.625	51.280***	−21.497***
SZSCPMI	−0.029	5.309	−0.289	4.927	78.375***	−21.485***

Note: *, ** and *** indicate significant at 10%, 5% and 1% levels, respectively.

3.2 Risk Study Based on Vine

The kernel density estimation method is used to convert the four index returns to a sequence of (0,1) uniform distribution, and the Kendall-rank correlation coefficient table between the four indexes is obtained (Table 2).

Table 2. Rank correlation coefficients of the four indices in 2018 and 2019

2019	2018			
	SPHKGEM	SZSCSEII	SZEXT50	SZSCPMI
SPHKGEM		0.247	0.245	0.237
SZSCSEII	0.105		0.880	0.919
SZEXT50	0.112	0.915		0.855
SZSCPMI	0.092	0.0876	0.825	

The S&P GEM Index is numbered 1 in the C-Vine structure, the GEM Strategic Emerging Industries Index is 2, the GEM 50 Index is 3, and the GEM Basic Market Index is 4. As can be make out from Table 3, the kendall correlation coefficient of the GEM Strategic Emerging Industries Index and the GEM Basic Market Index is the largest, at 0.896, followed by the GEM Strategic Emerging Industries Index and the GEM 50 Index at 0.895, and again The GEM basic market index and the GEM 50 index were 0.838. According to the principle of the largest sum of absolute values of rank correlation coefficients, the GEM strategic emerging industry index can be determined as the key node of the first tree of C-Vine. The structure of C-Vine is shown in Fig. 2.

Table 4 is the optimal Copula and parameter estimation results of the edges of the first-level tree and the second-level tree in the total C-Vine structure fitted with four indices. It can be obtain from Fig. 3 that emerging industries have the strongest relevance and deepest influence on the GEM with other sectors. Strategic emerging industries are built on the basis of major frontier technological breakthroughs, representing the new direction of future technological and industrial development, and reflecting the current development trend of the world's knowledge economy, circular

Table 3. Rank correlation coefficients of total returns of the four indexes

	SPHKGEM	SZSCSEII	SZEXT50	SZSCPMI	Sum
SPHKGEM		0.180	0.183	0.168	0.531
SZSCSEII	0.180		0.895	0.896	1.971
SZEXT50	0.183	0.895		0.838	1.916
SZSCPMI	0.168	0.896	0.838		1.902

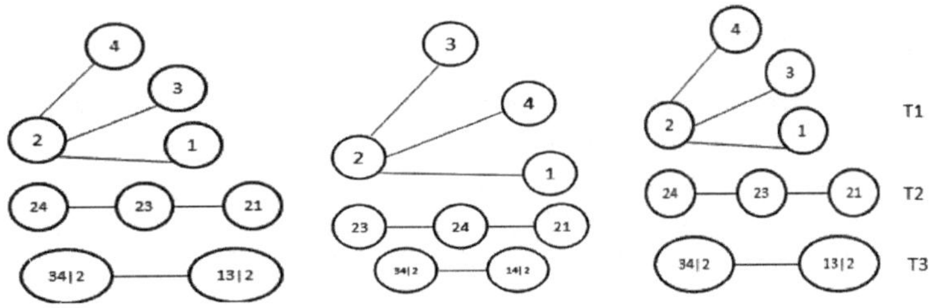

(a)2018 C-Vine structure **(b)** 2019 C-Vine structure **(c)** Total C-Vine structure

Fig. 2. Vine structure of four indexes

economy, and low-carbon economy. Society has an industry that has overall driving and a major leading role. From the three C-Vine structures, we can see the driving and leading role of the emerging industries of ChiNext to the overall economic and social environment of ChiNext.

Table 4. Optimal Copula and parameter estimation of T1 and T2 edges in the total C-Vine structure

T1		24	23	21
	Copula type	Joe	Survival Clayton	Survival Gumbel
	Parameter estimation	15.94***	14.35***	1.24***
T2		43|2	31|2	
	Copula type	t	N	
	Parameter estimation	−0.09***/8.70***	0.04*	

3.3 VaR

Most of the companies listed on the ChiNext market are engaged in high-tech business and have a high growth rate. But they are often set up for a relatively short period of time with a small scale and less outstanding performance but a plenty room for growth. Therefore, selecting the value of the combined VaR determined according to the C-Vine structure, a comparative analysis of the VaR between 2018 and 2019 is carried out with the outcomes shown in Table 5. It is found that VaR, the combination of the GEM Strategic Emerging Industries Index and the remaining three indexes, is decreased compared with 2018. Among them, the GEM Strategic Emerging Industries

Index–S&P GEM Index portfolio has the smallest change in liquidity risk, from 3.680 to 3.634, and the GEM Strategic Emerging Industries Index–GEM 50 Index portfolio has the largest change in VaR, from 6.579 to 5.379, followed by the GEM Strategic Emerging Industries Index–the GEM Basic Market Index, fell from 6.188 to 5.381. At this stage, the factors affecting the Hong Kong stock market are still overseas markets. Under the environment of the Fed's contraction and interest rate hikes and liquidity tightening, Hong Kong stocks face an increased risk of valuation callbacks. Therefore, the GEM Strategic Emerging Industries Index and the S&P GEM Index combination flow Sexual risk changes are minimal. The more open A-shares, the greater the influence on Hong Kong stocks, and the future trend of Hong Kong stocks and A-shares will become stronger and stronger. While China continues to implement a prudent monetary policy and continues to promote the opening up of the financial industry, the government encourages technology companies to develop high-tech industries to reduce the risk of the mainland's GEM market.

Table 5. VaR results in 2018 and 2019

Year	2018	2019
SZSCSEII - SPHKGEM	3.680	3.634
SZSCSEII - SZEXT50	6.579	5.379
asSZSCSEII - SZSCPMI	6.188	5.381

4 Conclusion

This article uses the S&P GEM Index, the GEM Strategic Emerging Industries Index, the GEM 50 Index, and the GEM Basic Market Index to construct the C-Vine and D-Vine structures, the result shows that the GEM Strategic Emerging Industries Index has the strongest correlation with other GEM indexes. This is related to China's a series of policies of encouraging the integrated development of digitalization, networking and intelligence, giving priority to cultivating and vigorously developing a number of strategic emerging industrial clusters, and promoting the deep integration of the Internet, big data, AI and the substantial economy. In the comparison of VaR in 2018 and 2019, it is found that the combined risk of the GEM Strategic Emerging Industries Index-Standard & Poor's GEM Index is decreased but the decline is relatively small, while the combined risks of the GEM Strategic Emerging Industries Index-the GEM 50 Index and the GEM Strategic Emerging Industries Index–the GEM Basic Market Index are decreased significantly compared to the same period in 2018, which is related to the policies of implementing the strategic deployments such as building a cyber power and a digital China, and actively promoting the deep integration of capital markets and technological innovation. Various domestic policy support will help GEM to perform well in the future.

References

1. Huang, J.J., et al.: Estimating value at risk of portfolio by conditional Copula-GARCH method. Insur. Math. Econ. **45**(3), 315–324 (2009)
2. Zhou, Q., Chen, Z., Ming, R.: Copula-based grouped risk aggregation under mixed operation. Appl. Math. **61**(1), 103–120 (2016). https://doi.org/10.1007/s10492-016-0124-z
3. Bo, L.: Application of VaR method based on GARCH model in Shanghai Stock Market. Times Finance (27), 250–251+253 (2013)
4. Lin, L., Qinglong, Z.: A comparative study of liquidity risk of different types of commercial banks–an empirical analysis based on the GARCH-VaR model. China Price **03**, 50–53 (2020)
5. Zhao, Z., Li, B.: Comparison of Risk Measurement of my country's Growth Enterprise Market Based on GARCH Models. Econ. Manage (3), 58–64 (2016)
6. Jilin, W., Erhua, Z.: Subprime mortgage crisis, market risk and the dependence of the stock market. World Econ. **33**(03), 95–108 (2010)
7. Zhenya, L.: The impact of the New York stock market on China's A-share market. Nankai Econ. Res. (03), 13–26 + 53 (2006)
8. Taihua, Y., Mingyu, C.: Time series analysis of shanghai stock index weekly return rate based on markov switching model. China Manage. Sci. **17**(06), 33–38 (2009)
9. Bedford, T., Cooke, R.M.: Probability density decomposition for conditionally dependent random variables modeled by vines. Ann. Math. Artif. Intell. **32**, 245–268 (2001)
10. Bedford, T., Vines, C.R.M.: A new graphical model for dependent random variable. Ann. Stat. **30**(3), 1031–1068 (2002)
11. Joe, H.: Families of m-Variate distributions with given margins and m(m−1)/2 BIvariate dependence parameters. Lect. Notes-Monogr. Ser. **28**, 120–141 (1996)
12. Bangzheng, Z., Yu, W.: Research on the correlation of Shanghai Stock Exchange Index based on R_Vine Copula method. J. Beijing Inst. Technol. (Soc. Sci. Edn.) **17**(03), 100–108 (2015)
13. Chao, H., Taihua, Y.: Risk analysis of exchange rate portfolio based on high-dimensional dynamic vine Copula. China Manage. Sci. **25**(02), 10–20 (2017)
14. Zhiqiang, H., Meijuan, Z.: Research on the IPO underpricing system and IPO underpricing under the multi-partial t-Copula model: an empirical analysis based on the main board, small and medium-sized board and the GEM. Econ. Rev. **199**(03), 150–162 (2016)
15. Zhengwu, B., Shuzhen, Z., Xuerui, Z.: Research on the correlation of my country's capital market based on dynamic t-Copula model. Manage. Modernization **39**(06), 5–7 (2019)

Analysis of Internet Public Opinion on "Delayed Retirement Age"–Based on Latent Dirichlet Allocation

Luo Yuting(✉) and Zhao Mingqing

Shandong University of Science and Technology, Qingdao 266590, China
luoyuting0824@163.com

Abstract. Information sharing in the age of the Internet makes it impossible for the online public opinion to be neglected in the policy-making, analyzing and guiding the online public opinion reasonably and effectively have great significance to improving government administrative ability. "Delayed retirement age" has always been one of the hot topics in the public. Therefore, this article uses python to obtain Weibo texts, combines with the Baidu index to select the hot topics related to "delayed retirement age", identifies three hot topics: "pension increase", "pension insurance reform" and "ageing population". Based on Latent Dirichlet Allocation, this article segments the Internet public opinion and selects relevant keywords under each hot topic, using text tendency analysis to analyze the public opinion from January to May in 2020. The empirical analysis of the online public opinion on the topic of "delayed retirement age" shows that the overall public opinion tendency from January to May is pessimistic. This article explains this result and gives corresponding suggestions.

Keywords: Retirement age · Internet public opinion · Text preference · Latent Dirichlet Allocation

1 Introduction

With the advent of the Internet era, network tools are gradually becoming one of the key means for the spread and acquisition of public opinion. On April 28, 2020, the China Internet Network Information Center released the 45th "Statistical Report on the Development Status of China's Internet". Until March 2020, the number of Internet users in China was 904 million, and the Internet penetration rate reached 64.5%. It's obviously that online public opinion has become a major component of social public opinion, and its analysis has important practical significance. This helps the government to guide policy implementation, grasps the psychological sentiment of netizens, and creates a benign social atmosphere for policy implementation. Thus, it promotes the improvement of government image and credibility, and promotes the reform of management methods.

As the aging of population in China continues to intensify, the pension insurance reform is accelerating, delaying the retirement age as a livelihood policy. And raising the legal retirement age has become an inevitable choice. The spread of the Internet has

M. Tavana et al. (Eds.): IISA 2020, AISC 1304, pp. 70–76, 2021.
https://doi.org/10.1007/978-3-030-63784-2_10

expanded the reach and influence of delayed retirement age reforms, and it has brought unprecedented challenges to policy credibility. Therefore, the analysis of online public opinion on the policy of "delayed retirement age" is helpful for the government to understand the people's attitude towards it and decide whether to continue to implement or revise the policy, and allow the public to have certain policies in advance.

The aging population has become a social problem faced by many countries, and public opinion also affects the government's decision-making to a certain extent. Therefore, in recent years, scholars have carried out many studies around the delayed retirement age and analysis of public opinion.

In China, Yang Vana [1] used an actuarial model shows that the current pension balance and the accumulated balance of the current contribution rate would be short in 2050, and the pension account cannot be maintained until 2050. Zhou Xinyi et al. [2] found that the implementation of delayed retirement can significantly reduce the contribution rate that has to be increased to ensure the target replacement rate by constructing a cross-period iterative model that includes a "consolidated account" pension system. Through collating online public opinion surveys, Lin Yuming et al. [3] found that netizens' persistent opposition to the policy of postponement of retreat has emerged as a one-sided phenomenon of standing public opinion. Based on social network analysis of the discussion of "postponement of reform", Shi Penghui [4] found that the role of Internet public opinion leaders in the development of the situation. Xu Xin et al. [5] tried to combine statistical ideas and semantic-based text tendency analysis methods, and proposed a tendency classification method based on pattern extraction and matching. The technical realization of Internet corpus tendency analysis is expounded. Based on the research of text tendency recognition method, Wang Lancheng et al. [6] adopted machine learning method to analyze the tendency of the subject network public opinion text. The analysis results showed that the combination of the two can improve the comprehensiveness and domain pertinence of concepts in emotional ontology.

Abroad, Galasso [7] speculated that the retirement age of all countries will be postponed by fitting the political and economic scenarios of 2050. According to the German pension insurance data, the results in [8] showed that pension reform will cause delays in retirement and employment. Increases and changes in the unemployment rate. Li [9] proposed ways to extend the legal retirement age, reduce the pension replacement rate, and expand the coverage of pension insurance. Dong et al. [10] found 57 hot spots attributed to public policies or incidents that were more likely to produce confrontations and contradictions, and stimulate people's enthusiasm for discussion by studying 57 online public events. To achieve democratization of the public. Savigny [12] showed that the Internet as an expansion of a new communication method, the emergence of online public opinion poses new challenges to the management of traditional media and political parties. Zheng [13] investigated Weibo accounts registered by government agencies, Now it has become a new channel for the government to provide public services and interact with citizens. Qiu et al. [14] pointed out that the behavior of netizens driven by online public opinion will help restore social fairness, morality and political rule. Under China's fragmented public domain, the government as an agent of public resources, it plays the role of participant and inspector. Internet public opinion plays an important role in cultivating participatory democracy.

To sum up, there is currently little research on the Internet public opinion about "delayed retirement age". Considering that China will implement the delayed retirement age policy, and public sentiment is an important factor that must be considered when making policy, this article is based on the Latent Dirichlet Allocation(LDA), combined with Baidu index, through the text tendency analysis of "delayed retirement age" for online public opinion analysis.

2 Model Building

2.1 Hot Topic Selection and Word Segmentation Processing

In order to achieve a good analysis effect, the selection of hot topics related to "delayed retirement age" is very important. Only by finding a suitable hot topic set can we make an accurate and comprehensive explanation of people's emotions. Therefore, this article first combines the Baidu index to search, and selects the relevant search term of "delayed retirement age". Use the relevant search term Baidu index as the search quantity index, enter each relevant search term in the Baidu index, obtain the time series data of its search volume, and select the top three related search terms as the hot topic. The octopus collector crawls the microblog data related to the hot topic, and uses the LDA [15] for word segmentation processing.

2.2 Text Tendency Analysis

Use the R software to match the microblog text after word segmentation with the loaded dictionary package, and use the adaptive adjustment tendency method to adjust the weight of the emotion words, so as to obtain the quantification index of each public opinion's tendency. Among them, if the index result is less than 0, it indicates that the public opinion tends to be pessimistic; if the index result is greater than 0, it indicates that the public opinion tends to be optimistic. In addition, the larger the absolute value of the indicator, the higher the degree of pessimism or optimism in the public opinion tendency. Due to the large amount of public opinion data, the public sentiment tendency after word segmentation by combining Chinese dictionaries is relatively clear, so this article will not discuss the situation when the index result is equal to 0. The symbols used in the text are shown in Table 1.

Because of the obvious professionalism of the "delayed retirement age" online public opinion corpus, the content of the public opinion corpus under different topics is obviously different, so this article uses adaptive adjustment of the propensity method [16] to calculate the second assignment of the propensity of keywords. So as to reflect the difference of different vocabulary's propensity in different topics of public opinion. The topic sentiment lexicon used in this method is obtained by subject sampling public opinion after keyword tagging and keyword extraction and part-of-speech screening. The specific calculation process of adaptively adjusting the index of propensity is as follows:

Table 1. Symbol description

Symbol	Description	Symbol	Description
Ω_H	Optimistic thesaurus	H_i	The number of times the $i-th$ vocabulary appears in optimistic public opinion
Ω_D	Pessimistic Thesaurus	D_i	The number of times the $i-th$ vocabulary appears in pessimistic public opinion
$Hnum_k$	The number of occurrences of optimistic words in the $k-th$ public opinion	H_{ki}	The number of times the $i-th$ optimistic vocabulary appears in the $k-th$ public opinion
$Dnum_k$	The number of occurrences of pessimistic words in the $k-th$ public opinion	D_{ki}	The number of times the $i-th$ pessimistic word appears in the $k-th$ public opinion
WH_i	The optimistic weight of the $i-th$ optimistic vocabulary	HD_k	Propensity of the $k-th$ opinion
WD_i	The optimistic weight of the $i-th$ pessimistic vocabulary	M	The total of public oponion

(1) According to the difference between the number of occurrences of optimistic vocabulary and pessimistic vocabulary in each public opinion, determine the category of the public opinion: define that the corresponding public opinion tendency is optimistic when $Hnum_k - \text{Dnum}_k > 0$; it is pessimistic public opinion when $Hnum_k - \text{Dnum}_k < 0$.

(2) Adjust the optimistic weight of the vocabulary according to the number of times the optimistic vocabulary appears in the optimistic public opinion and pessimistic public opinion; adjust the lexical pessimistic weight according to the number of times the pessimistic vocabulary appears in the optimistic public opinion and the pessimistic public opinion; the weight of the words in the vocabulary that does not appear in the news Is 1. The calculation formula is as follows:

$$WH_i = \frac{H_i}{H_i + D_i}, i \in \Omega_H WD_i = \frac{D_i}{H_i + D_i}, i \in \Omega_D$$

(3) Recalculate the attributes of public opinion based on optimistic and pessimistic lexical weights, the calculation formula is as follows:

$$HD_k = \sum_{i \in \Omega_H} H_{ki} WH_i - \sum_{i \in \Omega_D} D_{ki} WD_i$$

(4) Calculation $\frac{\sum_k HD_k}{M}$, from this, the self-adjusting propensity index for each period is obtained, and then the trend of public opinion is determined.

3 An Empirical Analysis of Online Public Opinion

Because the evolution of online public opinion on hot topics has a periodical nature and the delay in retirement age is closely related to the interests of the people, it also coincides with the convening of the "NPC & CPPCC". It triggered the public's re-discussion of "delayed retirement age". And because of the participants, the increase increases the breadth and depth of the issue, forming a double discussion between cyberspace and the real community. Therefore, this article uses the Baidu index to select "pension increase", "pension insurance reform" and "population aging" as the hot topics related to "delayed retirement age", use the octopus collector to crawl Weibo on 20201-May public opinion on "delayed retirement age", apply the LDA to process the relevant public opinion of hot topics, and get the relevant keywords corresponding to each hot topic and the probability that each keyword appears in the corresponding hot topic, and record it for. 10 keywords with a high probability of appearing in each hot topic was selected in this paper. The results are shown in Table 2.

Table 2. Probability value of each keyword under hot topics.

Pension increase		Pension insurance reform		Population aging	
Key words	*P*	Key words	*P*	Key words	*P*
Increase	0.0224	Retirement	0.0982	Retirement	0.0413
Need	0.0187	Delay	0.0611	Delay	0.0349
Retirement	0.0151	Plan	0.0138	Pension	0.0287
Delay	0.0114	Pension	0.0105	Population structure	0.0255
Result	0.0114	Basic	0.0071	Age	0.0192
Pension	0.0114	Foundation	0.0071	Progressive	0.0161
Low	0.0114	Change	0.0071	Social security	0.0129
Adjust	0.0077	Target	0.0071	Reform	0.0129
Receive	0.0077	Well	0.0071	Demographic dividend	0.0129
Money	0.0077	Employment	0.0071	Increase	0.0098

In terms of pension increase, words such as "up-regulation", "need" and "retirement" have a large value, so the public opinion has a high degree of attention and has a strong judgment significance on the tendency of this topic, "low", "Cause" and other expressions express the negative emotions of the people; in terms of pension insurance reform, the words "retirement", "delay" and "basic" are relatively large, and the distribution and score of the words generally show the positive emotions of the people; In terms of population aging, words such as "retirement" and "pension" have a large value, so public opinion containing such keywords has a strong sense of judgment. Through analysis, it is found that the words with "retired" and "delayed" appear in the vocabularies with higher scores among the three hot topics, indicating that the hot topics have a better ability to interpret the delayed public opinion.

By adaptively adjusting the propensity analysis method, the public opinion tendencies of the three hot topics such as pension increase, pension insurance reform and population aging are determined, and the public opinion index is calculated, and the weighted average of the public opinion indexes of the three hot topics is shown in Fig. 1.

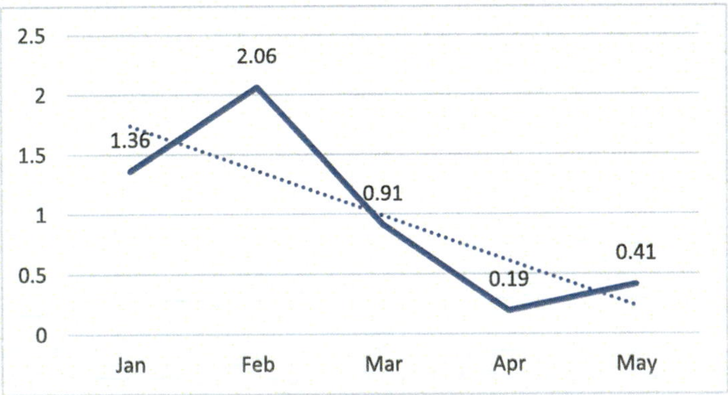

Fig. 1. The average tendency of delayed public opinion from January to May 2020.

From January to May 2020, public opinion tends to be pessimistic as a whole. The relevant policies introduced at the end of 2019 have restored the public's attention to delayed retirement and once improved. With the advent of the epidemic, people's emotions are more prone to extremes. And Under the tense atmosphere at that time, the public was more inclined to express their emotions on the Internet. The later economic downturn caused the public to have heated discussions about the delayed retirement and showed pessimism about it; on the eve of the two sessions in 2020, related proposals for delayed retirement appeared one by one in front of the people, the government work report gave people confidence, and the public's public opinion about the delay in retirement showed a positive trend.

4 Conclusion

In this paper, three hot topics: "pension increase", "pension insurance reform", and "ageing population" are developed under the theme of "delayed retirement age". Combining with the LDA and Baidu index, words and keywords was selectd. The adaptive adjustment tendency analysis method analyzed the public opinion of the delayed retirement age from January to May 2020, and the results showed that the overall public opinion tendency from January to May was pessimistic. Although the public opinion expressed by netizens cannot be regarded as the general attitudes and opinions of the people, it is to a certain extent a reflection of the real public opinion in the cyberspace. Therefore, the government should fully consider the opinions and views of all sectors of society and take into account the interests of all walks of life to reasonably guide the development of online public opinion and actively and steadily advance the step-by-step retirement age policy.

In this paper, since that the selected time span is not long enough and the selected platform is relatively single, the time span and platform can be further expanded in the later period to obtain more convincing and more comprehensive system results.

Acknowledgments. This article is one of the research results of the project of the Shandong University of Science and Technology's graduate tutor's guidance ability improvement plan (project number: KDYC17018) and the Shandong University of Science and Technology professional degree graduate teaching case database construction project (project number: SKDYAL17010).

References

1. Yang, F.: Study on the impact of delayed retirement on pension sustainability. Macroecon. Res. **5**, 91–101 (2016)
2. Zhou, X., Deng, L., Gong, F.: Population aging, pension insurance contribution rate and target replacement rate of basic pension. Finance Trade Res. **31**(02), 57–67 (2020)
3. Lin, Y., Liu, J.: Public policy analysis of the evolution of network public opinion. J. Intell. **35**(8), 151–155 (2016)
4. Shi, P.: Empirical studies of network public opinion based on social network analysis. J. Mod. Inf. **33**(02), 27–31 (2013)
5. Xu, X., Yu, F., Zhang, L.: A method and its application of text semantic orientation. Mod. Libr. Inf. Technol. **10**, 54–62 (2011)
6. Wang, L., Xu, Z.: Tendency analysis of thematic networks public opinion based on sentiment ontology. Inf. Control **42**(01), 46–52 (2013)
7. Vincenzo, G.: Postponing retirement: the political effect of aging. J. Pub. Econ. **92**(10), 2157–2169 (2008)
8. Staubli, S., Zweimüller, J.: Does raising the early retirement age increase employment of older workers? J. Pub. Economics **108**, 18–32 (2013)
9. Yue, L.: Paradoxical effects of increasing the normal retirement age: a prospective evaluation. Eur. Econ. Rev. **101**, 512–527 (2018)
10. Tiance, D., Chen, X., Liang, X.: Social media and internet public events. Telematics Inform. **34**(3), 726–739 (2017)
11. Juliane, U., Kristin, B.: Tracing public opinion online-an example of use for social network analysis in communication research. Procedia-Soc. Behav. Sci. **100**, 108–126 (2013)
12. Heather, S.: Public opinion, political communication and the internet. Politics **22**(1), 1–8 (2002)
13. Lei, Z.: Social media in chinese government: drivers, challenges and capabilities. Gov. Inf. Q. **30**(4), 369–376 (2013)
14. Lin, Q., Han, L., Chi, Y., Chiu, P.: Online collective behaviors in china: dimensions and motivations. Anal. Soc. Issues Pub. Policy **15**(1), 44–68 (2015)
15. Blei, D., Ng, A., Jordan, M., et al.: Latent Dirichlet allocation. J. Mach. Learn. Res. 993–1022 (2003)
16. Zhu, J., Xie, B., Luo, X., et al.: Net-mediated public opinion analysis of China's real estate. J. Appl. Stat. Manage. **35**(04), 722–741 (2016)

Research on Service Evaluation of Automobile Inspection and Testing Based on AHP - Fuzzy Comprehensive Evaluation Method

Chen Chaozhou(✉), Xu Shuquan, and Liu Fujia

Research Institute of Highway Ministry of Transport, Beijing 100088, China
cz.chen@rioh.cn

Abstract. In order to grasp the current status, problems and service weaknesses of automobile inspection and testing services, and to propose service improvement and optimization measures. Five criteria-level indicators and corresponding 14 subdivision indexes of automobile inspection and testing service evaluation index system were established. Through the AHP-fuzzy comprehensive evaluation method of automobile inspection and testing service evaluation model, the index weights are determined and the current status of the automotive inspection and testing service is comprehensively evaluated. The weight ranking of the service indicators is professionalism(B_3)>economy(B_2) >convenience(B_1)>transparency(B_5)>comfort(B_4), the top three items in the order of the weight of all items are technical interpretation, cost of delay and dress & language. According to the comprehensive evaluation, the total score of the automotive inspection and testing service is 90.65, and the index evaluation score of the criterion layer is ranked as B_1>B_4>B_3>B_2>B_5, according to 5 levels of service satisfaction reviews and evaluations The total score shows that the service evaluation is generally at a satisfactory level, and the scores of convenience, comfort, economy and other indicators are also above the satisfaction line (>85), this shows that in recent years, the reform of inspection and testing services aimed at reducing costs and increasing efficiency and improving testing services has achieved certain results. For specific indicators, it is proposed that the professional capacity building of inspection and testing services should be continuously strengthened, and the technical interpretation to customers and the level of standardization and standardization of external services should be emphasized, and the information management of inspection and testing services should be strengthened as well.

Keywords: Road transportation · Service evaluation · AHP-fuzzy comprehensive evaluation · Automobile inspection and testing · Index system

1 Introduction

Standardizing and strengthening the management of automobile inspection and testing services has a positive significance for serving road transportation, ensuring the safe operation of transportation, and promoting social and economic development [1]. Under the new situation, the service management and development of automobile

M. Tavana et al. (Eds.): IISA 2020, AISC 1304, pp. 77–85, 2021.
https://doi.org/10.1007/978-3-030-63784-2_11

inspection and testing agencies face new issues [2], China is vigorously advancing the reform of inspection and testing services. At the national level, it is required to improve the inspection and testing services of in-use vehicles from the aspects of booking services, optimizing processes, improving efficiency. Therefore, it is necessary to strengthen the systematic research on automobile inspection and testing services. Analytic Hierarchy Process (AHP) and fuzzy comprehensive evaluation methods can usually be used for service evaluation analysis. AHP is a quantitative analysis method that can treat complex problems as a unified system and comb through the orderly hierarchy of interactions among various factors in the system to comprehensively analyze multiple factors in the system. It can achieve quantitative analysis of qualitative factors, and to a certain extent test and reduce the influence of subjective factors. [3], it has the characteristics of simple operation, high efficiency, easy to understand, comprehensive and practical [4, 5], usually used to determine the weight of each indicator in the evaluation process [6]. Fuzzy comprehensive evaluation is to regard the fuzzy object under investigation and the fuzzy concept reflecting the fuzzy object as a fuzzy set. By establishing the relevant operations and transformations of fuzzy membership functions and fuzzy sets, a structured comprehensive evaluation matrix is obtained, and the evaluation level of the evaluation object and its degree of belonging to each level are further obtained to achieve the purpose of quantitative analysis of the fuzzy object [3, 4]. There are a few literatures that have studied automobile inspection and testing services [7], however, in general, there is relatively little research in this area. The existing literature is mainly analyzed from the perspective of capacity building of inspection and testing institutions, but there are few studies from the perspective of customer perception and the combination of new policy requirements. In order to quantify the assessment of service status and pinpoint service failure points, this paper uses AHP-fuzzy comprehensive evaluation method to quantify the current automotive inspection and testing service work. Different from the existing related research, the service demand is analyzed from the customer's perspective, combined with the new policy requirements, an automobile inspection and testing service evaluation system is established, and the fuzzy comprehensive evaluation method is used to carry out empirical evaluation and analysis.

2 Evaluation Index System

According to the analytic hierarchy process, combined with reform, industry practice and expert demonstration, an automobile inspection and testing service system was established (Table 1).

Table 1. Automobile inspection and testing service evaluation index system

Target layer A	Criterion layer B	Index layer C
Satisfaction of automotive inspection and testing services	Convenience (B_1)	Service process (C_1)
		Appointment test (C_2)
		Off-site testing, annual review(C_3)
		Detection efficiency (C_4)
	Economy(B_2)	Charge standard (C_5)
		Cost of delay (C_6)
	Professionalism (B_3)	Opinion tracking (C_7)
		Technical interpretation (C_8)
		Dress & language (C_9)
	Comfort (B_4)	Service attitude (C_{10})
		Rest area experience (C_{11})
		Environmental sanitation (C_{12})
	Transparency (B_5)	Information disclosure (C_{13})
		Logo slogan settings (C_{14})

3 Evaluation Index Weight

3.1 Method

Combining the expert index scoring method and the 1-9 scale method, the factors at the same level are compared and compared with the common upper level factors to determine a ratio and construct a judgment matrix $\boldsymbol{B}$. Through matrix calculation, each index weight set is obtained and According to the consistency test, the index weights at different levels are obtained [8, 9]. The specific steps are as follows:

Construct a Judgment Matrix. Based on the scale method, a judgment matrix $\boldsymbol{B} = (b_{ij})_{m\times n}$ is constructed for evaluation indicators at different levels to complete the transition from qualitative to quantitative analysis.

Single Level Sorting. Calculate the eigenvalue and eigenvector of judgment matrix $\boldsymbol{B}$, and normalize them to obtain the weight values of the same level factors relative to the common previous level.

Consistency Test. In order to test the consistency of the judgment matrix $\boldsymbol{B}$, the index **C.I** which measures the deviation of the judgment matrix, the average random consistency index **R.I** and the consistency ratio **C.R** are introduced. The conversion formula between the three parameters is as follows:

$$\mathrm{C.I} = \frac{\lambda_{\max} - n}{n - 1} \quad (1)$$

In the formula,$\lambda_{\max}$ is the maximum eigenvalue of judgment matrix $\boldsymbol{B}$, **R.I** takes the value according to Table 2. When **C.R**<0.1, it is considered that the judgment matrix has satisfactory consistency, otherwise it is necessary to adjust and reconstruct the judgment matrix until it meets the consistency test.

Table 2. Average random consistency indexes

Matrix order	3	4	5	6	7
R.I	0.52	0.89	1.12	1.26	1.36

Multi-level Total Sorting. Calculate the synthetic weight of the factors of each layer to the target layer, get the weight of all indicators, and complete the total ranking of the layers.

3.2 Actual Examples

Taking the determination of the corresponding index weights based on "convenience B_1" as an example, an analysis and comparison of C_1, C_2, C_3, and C_4 yields a judgment matrix $\boldsymbol{B}_1$ based on "convenience B_1",Then normalize the matrix $\boldsymbol{B}$ by columns and add the rows to obtain the index weight value (K_i) as shown in Table 3. Consistency test on the judgment matrix: in the case of matrix order n = 4, **R.I** = 0.89, then **C. I** = 0.048, **C.R** = 0.0539 < 0.1, therefore, $\boldsymbol{B}_1$ meet the consistency requirements.

Table 3. Index weights based on convenience

B_1	C_1	C_2	C_3	C_3	K_i
C_1	1	1/5	1/8	3	0.0876
C_2	5	1	1/2	5	0.3177
C_3	8	2	1	8	0.5468
C_4	1/3	1/5	1/8	1	0.0479

Similarly, the index weights based on the index layers under B_2, B_3, B_4, and B_5 and the index weights based on the criteria layer under A can be further obtained, and then the full index weights can be obtained (Table 4).

Table 4. Weight of all indexes

	B_1	B_2	B_3	B_4	B_5	All index weights
	0.1203	0.3204	0.4837	0.0305	0.0451	
C_1	0.0876	–	–	–	–	0.0105
C_2	0.3177	–	–	–	–	0.0382
C_3	0.5468	–	–	–	–	0.0658
C_4	0.0479	–	–	–	–	0.0058
C_5	–	0.2500	–	–	–	0.0801
C_6	–	0.7500	–	–	–	0.2403
C_7	–	–	0.0938	–	–	0.0454
C_8	–	–	0.7396	–	–	0.3577
C_9	–	–	0.1666	–	–	0.0806
C_{10}	–	–	–	0.7838	–	0.0239
C_{11}	–	–	–	0.1349	–	0.0041
C_{12}	–	–	–	0.0813	–	0.0025
C_{13}	–	–	–	–	0.800	0.0361
C_{14}	–	–	–	–	0.200	0.0090

4 Fuzzy Comprehensive Evaluation

4.1 Evaluation Steps

Constructing Factor Sets and Weight Sets (step 1). According to Table 1, the number of indicators in the criterion layer based on the target layer or the number of indicators in the indicator layer based on the criterion layer is counted as a factor set $\boldsymbol{U} = \{u_1, u_2, \ldots, u_i\}$, and each element is recorded as u_i (i = 1, 2, …, m). According to Table 4, the obtained index weights are recorded as a weight set $\boldsymbol{K} = \{k_1, k_2, \ldots, k_i\}$, and the weights corresponding to each index are recorded as k_i (i = 1, 2, …, m), $k_i \geq 0$ and $\sum k_i = 1$.

Create Comment Sets and Value Sets (step 2). The comment set is a set of total judgment results that the evaluator may make to the judged object [10]. Since the factors corresponding to the criteria layer and the indicator layer are the same, the 5-level division method is used here to construct a review set $\boldsymbol{V} = \{v_1, v_2, v_3, v_4, v_5\}$ for the service satisfaction of the inspection and testing agency, which is defined as five levels (very satisfied, satisfied, average, dissatisfied and very dissatisfied) in turn, and let the value set corresponding to the review set be $\boldsymbol{M} = \{95, 85, 75, 65, 55\}$.

Construct the Membership Matrix (step 3). Select a specific area for the questionnaire survey. The content of the questionnaire is set according to the indicators of each layer given in Table 1, and the 5-level evaluation form scale option proposed in step 2 is used. After completing the statistics of the questionnaire, you can obtain the number of comments of different ratings corresponding to each evaluation index, and then further construct a membership matrix (denoted as $\boldsymbol{R}$), namely:

$$\boldsymbol{R}=\begin{pmatrix}\boldsymbol{R}_1\\ \boldsymbol{R}_2\\ \vdots\\ \boldsymbol{R}_m\end{pmatrix}=\begin{bmatrix}r_{11} & r_{12} & \cdots & r_{1n}\\ r_{21} & r_{22} & \cdots & r_{2n}\\ \vdots & \vdots & \ddots & \vdots\\ r_{m1} & r_{m2} & \cdots & r_{mn}\end{bmatrix} \tag{2}$$

In the formula, $\boldsymbol{R}_m$ represents the evaluation level fuzzy vector of the m-th index, which can be regarded as a fuzzy subset on the comment set $\boldsymbol{V}$; r_{mn} represents the membership of the m-th index to the n-th level, which is numerically equal to the m-th index Choose the ratio of the number of people in the corresponding level to the total number of people participating in the evaluation.

Obtain Comprehensive Evaluation Judgment Matrix (step 4). Establish a comprehensive evaluation model and calculate the evaluation results, make $\boldsymbol{S}$ as:

$$\boldsymbol{S}=\boldsymbol{K}\circ\boldsymbol{R}=[k_1,k_2,\cdots,k_m]\circ\begin{bmatrix}r_{11} & r_{12} & \cdots & r_{1n}\\ r_{21} & r_{22} & \cdots & r_{2n}\\ \vdots & \vdots & \ddots & \vdots\\ r_{m1} & r_{m2} & \cdots & r_{mn}\end{bmatrix}=[s_1,s_2,\cdots,s_n] \tag{3}$$

In the formula, $\boldsymbol{K}$ is the weight vector obtained by step 1, $\boldsymbol{R}$ is the fuzzy judgment matrix (membership matrix) obtained by step 3, and "$\circ$" is the fuzzy composition operator. The corresponding evaluation value s_i in $\boldsymbol{S}$ corresponds to the membership degree of level i, $s_i \geq 0$ and $\sum s_i = 1$.

Determine the Evaluation Score (step 5). Calculate the evaluation score according to following formula:

$$\boldsymbol{J}=\boldsymbol{S}*\boldsymbol{M}^{-1} \tag{4}$$

In the formula, $\boldsymbol{J}$ represents the evaluation score and $\boldsymbol{S}$ is the comprehensive evaluation judgment matrix obtained by step 4. $\boldsymbol{M}$ is the value set vector corresponding to the comment set obtained in step 2.

4.2 Actual Example

Set the questionnaire content against the service indicators constructed in Table 1, and conduct a questionnaire survey on the service situation of the inspection and testing institutions (a total of 495 questionnaires were recovered). Combined with the index weights of the index system and the results of the questionnaire survey, the fuzzy comprehensive evaluation is as follows:

According to Table 1, the factor set $\boldsymbol{U}_A = \{u_{B_1}, u_{B_2}, \ldots, u_{B_5}\}$ based on the target layer is established, and the factor set based on the criteria layer for automobile inspection and testing service evaluation is:

$$U_{B_1} = \{u_{C1}, u_{C2}, u_{C3}, u_{C4}\},$$

$$U_{B_2} = \{u_{C5}, u_{C6}\},$$

$$U_{B_3} = \{u_{C7}, u_{C8}, u_{C9}\},$$

$$U_{B_4} = \{u_{C10}, u_{C11}, u_{C12}\},$$

$$U_{B_5} = \{u_{C13}, u_{C14}\}.$$

The weight of each evaluation factor corresponds to the weight value obtained in the index system. The corresponding set is the weight set, and the weight set is $\boldsymbol{K_0}, \boldsymbol{K_1}, \boldsymbol{K_2}, \boldsymbol{K_3}, \boldsymbol{K_4}, \boldsymbol{K_5}$, where $\boldsymbol{K_0}$ represents the weight set of the criterion layer to the target layer, and $\boldsymbol{K_1}, \boldsymbol{K_2}, \boldsymbol{K_3}, \boldsymbol{K_4}, \boldsymbol{K_5}$ represents the weight set of the index layer to the criterion layer, respectively.

According to the evaluation process method, construct the corresponding comment set and value set.

Taking the criterion level index "convenience B_1" as an example, let the sub-sets based on B_1 be: $\boldsymbol{R}_{C_1}$, $\boldsymbol{R}_{C_2}$, $\boldsymbol{R}_{C_3}$, $\boldsymbol{R}_{C_4}$, then:

$\boldsymbol{R}_{C_1}$={0.8384, 0.1111, 0.0283, 0.0081, 0.0141}, $\boldsymbol{R}_{C_2}$= {0.9192, 0.0606, 0.0121, 0.0040, 0.0040}, $\boldsymbol{R}_{C_3}$={0.9354, 0.0485, 0.0081, 0.0040, 0.0040}, $\boldsymbol{R}_{C_4}$= {0.7596, 0.1677, 0.0424, 0.0081, 0.0222}.

Then the membership matrix based on convenience is:

$$\boldsymbol{R}_{B_1} = \begin{pmatrix} \boldsymbol{R}_{C_1} \\ \boldsymbol{R}_{C_2} \\ \boldsymbol{R}_{C_3} \\ \boldsymbol{R}_{C_4} \end{pmatrix} = \begin{bmatrix} 0.8384 & 0.1111 & 0.0283 & 0.0081 & 0.0141 \\ 0.9192 & 0.0606 & 0.0121 & 0.0040 & 0.0040 \\ 0.9354 & 0.0485 & 0.0081 & 0.0040 & 0.0040 \\ 0.7596 & 0.1677 & 0.0424 & 0.0081 & 0.0222 \end{bmatrix}$$

Similarly, other membership matrices can be obtained.

From the evaluation step 4 in Sect. 3.1, we can know that through the evaluation model, fuzzy operations on $\boldsymbol{K_1}$ and $\boldsymbol{R}_{B_1}$ can obtain the evaluation results of the membership degree based on convenience as: $\boldsymbol{S}_{B_1}$ = (0.9133,0.0635,0.0128,0.0046, 0.0058).

Similarly, the evaluation results of membership degree based on economy, professionalism, comfort and transparency can be obtained. Construct the first-level fuzzy evaluation judgment matrix of customer service evaluation of automobile inspection and testing agencies as $\boldsymbol{R_A}$:

$$\boldsymbol{R_A} = \begin{pmatrix} \boldsymbol{S}_{B_1} \\ \boldsymbol{S}_{B_2} \\ \boldsymbol{S}_{B_3} \\ \boldsymbol{S}_{B_4} \\ \boldsymbol{S}_{B_5} \end{pmatrix} = \begin{bmatrix} 0.9133 & 0.0635 & 0.0128 & 0.0046 & 0.0058 \\ 0.5949 & 0.1843 & 0.1008 & 0.0892 & 0.0308 \\ 0.7863 & 0.1519 & 0.0381 & 0.0057 & 0.0180 \\ 0.7990 & 0.1388 & 0.0405 & 0.0058 & 0.0159 \\ 0.7868 & 0.0146 & 0.0432 & 0.0117 & 0.0117 \end{bmatrix}$$

Let $\boldsymbol{S_A}$ be the final evaluation result based on the target layer, then through fuzzy operation $\boldsymbol{K_0} \circ \boldsymbol{R_A}$, we get: $\boldsymbol{S_A}$=(0.7458,0.1461,0.0558,0.0328,0.0194).

Let $\boldsymbol{M} = (95, 85, 75, 65, 55)$ be the score matrix corresponding to the review set and $\boldsymbol{J}_{con}$ be the service evaluation score based on convenience, then: $\boldsymbol{J}_{con}=\boldsymbol{S_{B_1}} * \boldsymbol{M}^{-1}=93.73$.

Similarly, the scores based on economy, professionalism, comfort and transparency are 87.23, 91.71, 91.99 and 80.64, respectively, and the total score of the evaluation of automobile inspection and testing services is 90.65.

5 Conclusion

Established an evaluation index system for automobile inspection and testing services based on 14 subdivision indexes of 5 criteria layer elements. Through the AHP method, the overall weight ranking of each index was obtained. The importance of the indicators at the criterion level is ranked as professionalism>economy>convenience>transparency>comfort. It can be seen that service professionalism is particularly important. This index is the first prerequisite for good inspection and testing services. Inspection and testing service institutions should strengthen professional capacity building. In the ranking of all weights, the top three indicators are C_8, C_6 and C_9, which should be properly paid attention to.

A fuzzy comprehensive evaluation model for automobile inspection and testing services was established, and a quantitative evaluation score for automobile inspection and testing services was obtained. It can be seen from the scoring results that according to the statistics of the five-level comment set, the customer satisfaction with the services of the automobile inspection and testing agency is: 74.6% are very satisfied, 14.6% are satisfied, 5.6% are generally, 3.3% are dissatisfied, and the very dissatisfied accounted for 1.9%. The total score of automotive inspection and service evaluation is 90.65, the overall result of service evaluation is at a satisfactory level, the reform of automotive inspection and detection has achieved remarkable results, and is basically recognized by the society. It can be seen from the ranking of the scores obtained by the five criteria layer indicators that transparency is the weakest link in the current automotive inspection and testing service, and the service transparency needs to be further strengthened.

The established vehicle inspection and testing service evaluation model can realize the quantitative evaluation of vehicle inspection and testing services, and is operable. The evaluation result is in line with objective reality and has guiding significance. It provides an effective scientific method for improving automobile inspection and testing services and has application value.

Acknowledgments. This paper is funded by the research project (2019-05-074) of the Transportation Standards (Quota) of the Ministry of Transport of China.

References

1. Ye, L.: An analogy study on the relationship between highway transportation and socio-economic development. Highway **1**(11), 138–143 (2014)
2. Wang, L., Li, X., He, J.: Research on the development strategy of China's inspection and testing service industry during the "13th Five-Year Plan" period. Economics 11–15 (2015)
3. Liu, Y., Zhou, Y., Bao, W.: Comprehensive performance evaluation of concrete beam bridge based on fuzzy evaluation theory and AHP. Highw. Transp. Sci. Technol. **29**(12), 96–100 (2012)
4. Huang, H.: Research on Passenger Satisfaction Evaluation of Changsha City Based on AHP-Fuzzy Comprehensive Evaluation Method. Central South University of Forestry and Technology (2014)
5. Yang, H., Wang, W.: Research on customer satisfaction evaluation system of third-party logistics enterprises. Marketing **1**(27), 181–193 (2015)
6. Zou, L., Yin, X., Feng, Y.: Research on E-commerce logistics service evaluation model based on AHP. Logistics Technol. **34**(8), 97–99 (2015)
7. Zhang, J., Qi, L., Li, G.: Research on the evaluation of the third-party inspection and testing service model under "Internet+". Sci. Technol. Ind. **17**(3), 107–114 (2017)
8. Zhang, C., Meng, Q., Wang, W.: Comprehensive evaluation of high-viscosity modified asphalt based on multi-level fuzzy evaluation. Highw. Transp. Sci. Technol. **32**(5), 36–55 (2015)
9. Zhang, B.: AHP and its Application Cases, 1st edn. Electronic Industry Press, Beijing (2014)
10. Zhou, W., Xie, Y., Zhang, Z.: Selection of accelerated loading equipment based on AHP. Highw. Transp. Sci. Technol. **29**(7), 38–44 (2012)

Analysis of Acoustic Features of Mongolian Long Tone

Shuzhen Ma, Gegen Tana, and Axu Hu(✉)

Key Laboratory of China's Ethnic Languages and Information Technology of Ministry of Education, Northwest Minzu Univereity, Lanzhou 730030, Gansu, China
1607650039@qq.com

Abstract. As an ancient Mongolian art, Mongolian long tone not only contains the essence of Mongolian culture in content, but also has unique characteristics in terms of singing methods and vocal skills. In this paper, using the method of experimental phonetics, the Mongolian long-toned voice signal is collected by the entire singing and speech energy of the four long tones, the marked signal parameters are extracted, and then the fundamental frequency and energy parameters are analyzed to obtain the basic rhythm features of long notes.

Keywords: Mongolian long tone · Acoustic analysis · Singing formants

1 Introduction

The vocal art of Mongolian music includes two types of singing, the long tone and the short tone. The long tone has become an intangible cultural heritage. The Long tone folk songs are a genre of folk songs created by the Mongolians who have long lived in the Mongolian grasslands. The Mongolian long tone can be said to be the most beautiful and typical cultural style and artistic survival of the Mongolian people. At the same time, the long tone, as a typical representative of the nomadic civilization of the grassland, contains a unique and profound concept of human beings in the pursuit of survival, made important contributions to human culture, and shined with wisdom on the grassland. In this paper, by collecting the Mongolian long tone speech signal, using the experimental speech acoustic analysis method, the marked signal parameters are extracted, and then the fundamental frequency and energy parameters are analyzed in order to obtain the long tone basic prosody features.

2 Overall Energy Distribution

Energy represents the level of volume, and in singing represents the level of air flow and the amplitude of tonal vibration of the singer during singing. Mass extraction of energy values is done with Praat software, using SPSS19.0 as the statistical distribution, and four long tones energy overall distribution maps are obtained, in which the abscissa is energy and the ordinate is the number of times the energy value appears. In order to more objectively and meticulously analyze the art form of Mongolian long

M. Tavana et al. (Eds.): IISA 2020, AISC 1304, pp. 86–94, 2021.
https://doi.org/10.1007/978-3-030-63784-2_12

tone, we extracts the vocal and verbal energy of the four long tones*("Rich and Vast Alxa" (m1, f1), "Petite Chestnut Horse" (m2), "Hang Gai in the North" (f2))*, using horizontal and vertical comparative analysis methods, analyzes the style of the long tone in a comprehensive way (Figures 1, 2, 3, and 4).

2.1 Overall Singing Energy

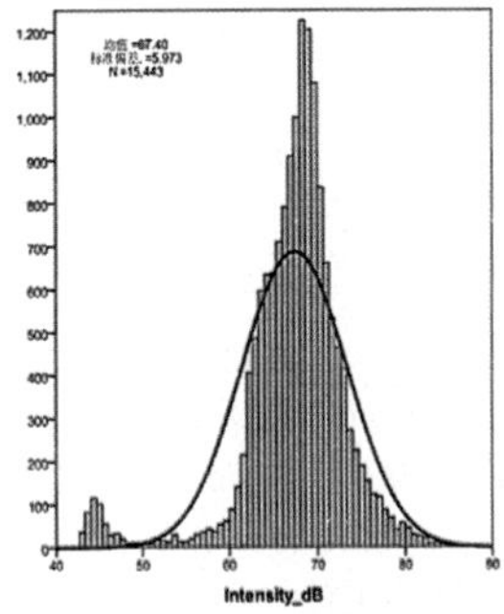

Fig. 1. m1 overall singing energy distribution

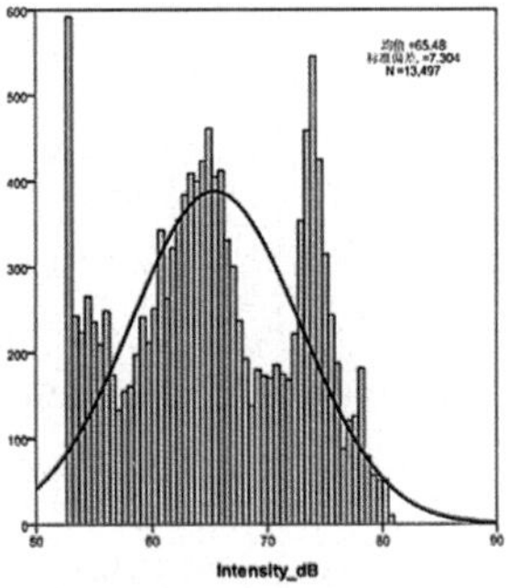

Fig. 2. f1 overall singing energy distribution

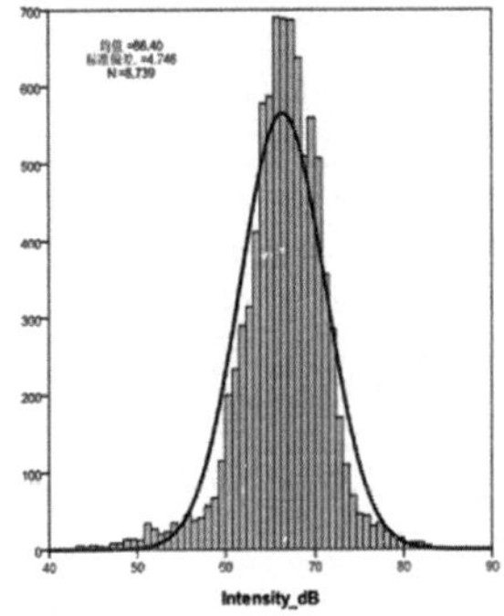

Fig. 3. m2 overall singing energy distribution diagram

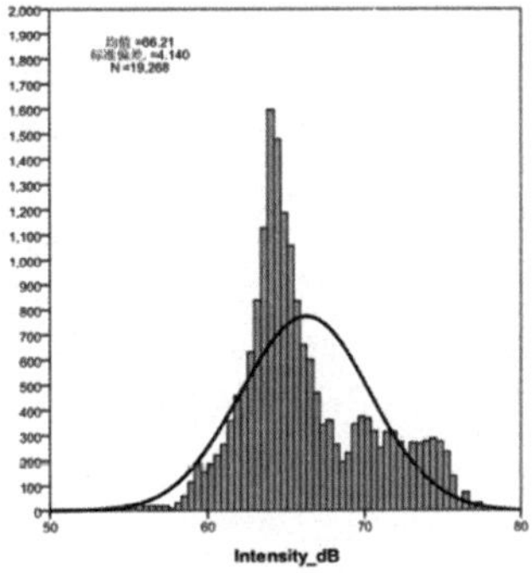

Fig. 4. f2 overall singing energy distribution diagram

By observing the above four pictures, we found the average of the singing energy of these four long tones is about 65 dB–67 dB, and the maximum difference is 1.92 dB. The main concentrated areas of the singing energy of each long tone are different. "Singing energy" is not too high, and when singing, male voice is very concentrated, female voice is more dispersed (Table 1).

Table 1. Overall singing energy data of different tracks

Gender	Male		Female	
Repertoire	*"Rich and Vast Alxa"*	*"Petite Chestnut Horse"*	*"Rich and Vast Alxa"*	*"Hang Gai in the North"*
Duration (s)	164.8	93.3	144	205.6
Energy max (dB)	86.7	86.7	81	77.6
Energy min (dB)	42.6	43.3	52.6	54.3
Energy mean (dB)	67.4	66.4	65.5	66.2

Observing the above table, we can find:

(1) When men and women sing the same long tone "Rich and Vast Alxa", their respective singing treatment methods are different. The duration of male singing is 20.8S longer than that of female singing, which means that different singers have different pauses in the lyrics of each sentence and the pronunciation of each sound is different. Different singers will perform interpretations based on their own understanding and experience of the track. The highest energy of the male voice is 5.7 dB higher than the highest energy of the female voice, and the average energy of male voices is 1.9 dB higher than female voices.

(2) Judging from the performance of the four tones singing, the highest energy reflects the characteristic that the highest energy of male voices is higher than the highest energy of female voices, and the lowest energy is lower than female voices. which reflects the intensity span is greater than the female voice when male voices are singing.

(3) The average value of energy when singing four long tones is between 65 dB–68 dB (Figures 5, 6, 7 and 8).

2.2 Overall Reading Energy

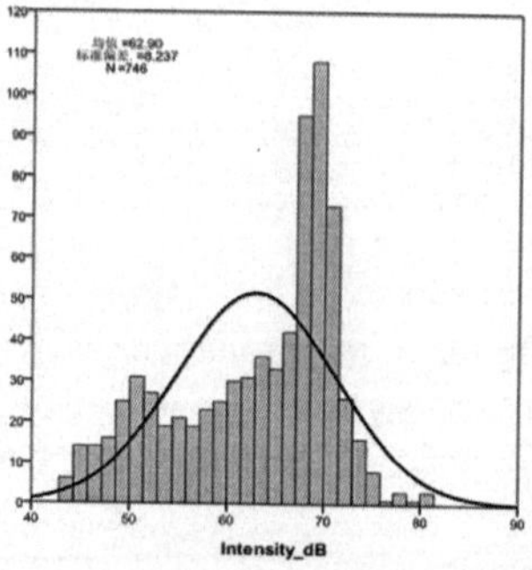

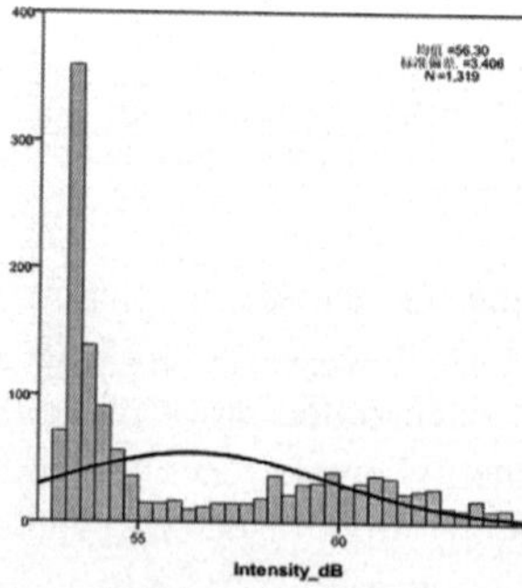

Fig. 5. m1 overall energy distribution diagram **Fig. 6.** f1 overall energy distribution diagram

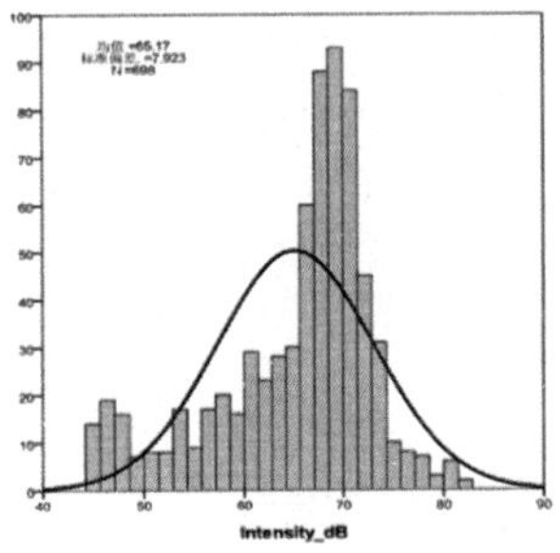

Fig. 7. m2 overall energy distribution diagram

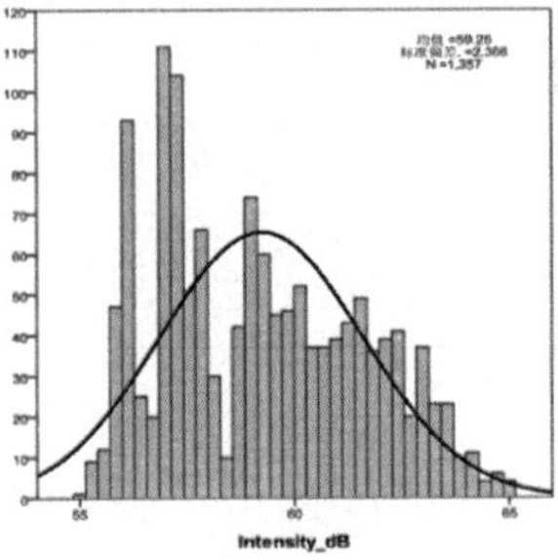

Fig. 8. f2 overall energy distribution diagram

It is not difficult to find through the above four pictures: The average energy of read these four long tone is about 56 dB–66 dB, and the maximum difference is 8.87 dB. The main concentrated areas of energy presented by each long tone are also different. "read energy" is not too high, and when reading aloud, male voices are concentrated, female voices are more dispersed (Table 2).

Table 2. Overall energy data table for different tracks

Gender	Male		Female	
Repertoire	*"Rich and Vast Alxa"*	*"Petite Chestnut Horse"*	*"Rich and Vast Alxa"*	*"Hang Gai in the North"*
Duration (s)	8.0	7.5	14.1	14.5
Energy max(dB)	81.2	82.3	64.8	65.1
Energy max (dB)	43.9	44.5	52.9	54.9
Energy max (dB)	62.9	65.17	56.3	59.25

Observing the above table, we can find:

(1) Men and women are reading the same long tone "Rich and Vast Alxa". Their reading methods are different. Contrary to singing, female voices spend more time than male voices when reading, which is 6.1 s more. The highest energy of male voice is 16.4 dB higher than that of female voice, and the lowest energy is 1 dB lower than female voice, the average energy of male voice is 6.6 dB higher than female voice.
(2) Judging from the situation when the four chiefs adjusted the reading, it showed the same situation as when singing, that is, the vocal intensity of male voices when singing and reading is larger than that of female voices.
(3) The average value of energy when reading the four long tones is between 56 Db–66 dB.

Singing is a combination of language and music. Although "singing is an extension of speaking", the change of singing energy greatly exceeds the change of speech energy. Singing is greater than the fluctuation of speech. The specific performance is:

the difference between the maximum and minimum values when singing is between 23 dB and 45 dB, and the difference between the maximum and minimum values when reading is between 10 dB and 38 dB, which means that the span of the sound intensity when singing is larger than the span when speaking, which reflects that the state of the vibration amplitude change of the vocal cords when singing is more flexible than the state when speaking.

3 Energy Analysis of the Music Paragraph

Period–Music paragraph is the smallest structure that constitutes an independent passage. The structure of the long tone is more common with two passages, the lyrics of the upper and lower sections are the same, and they are repeated with the same melody. To this end, we divides the entire long tone and analyzes the energy of the long tone from the perspective of the same long note (Figures 9, 10, 11 and 12).

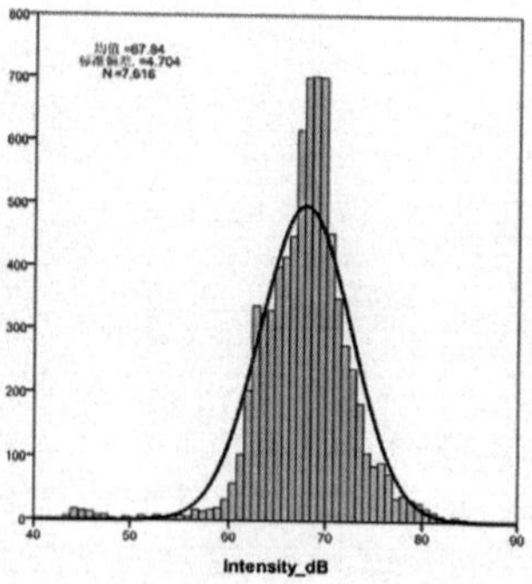

Fig. 9. Singing energy distribution of m1 first section

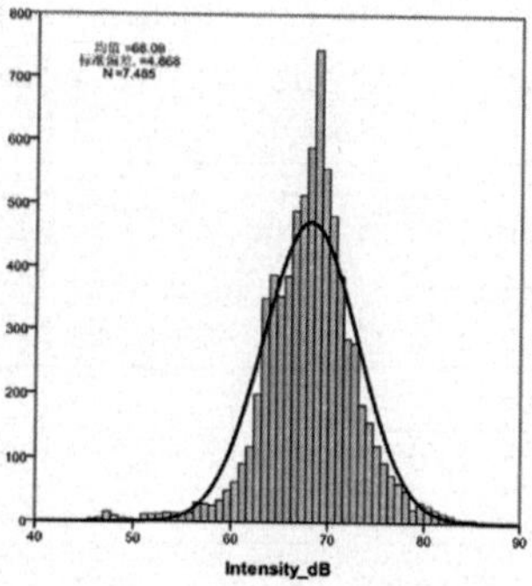

Fig. 10. Singing energy distribution of m1 second section

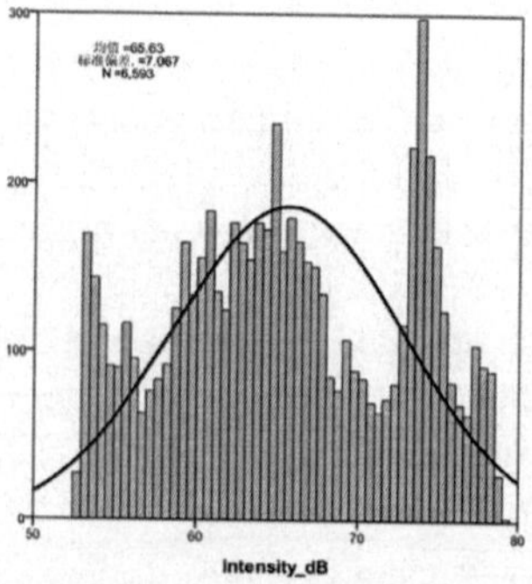

Fig. 11. f1 sing energy distribution diagram of the first section

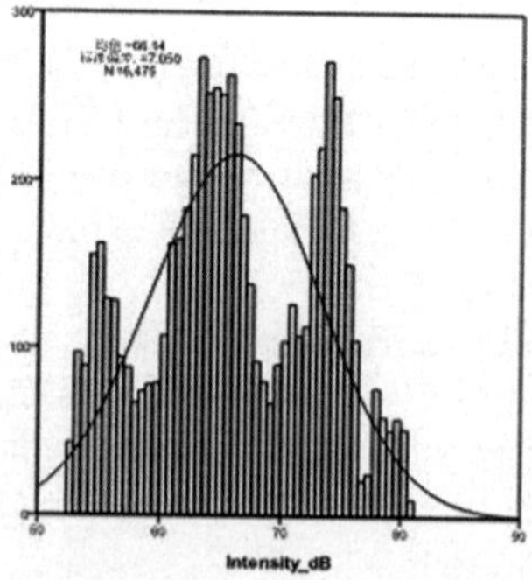

Fig. 12. f1 sing energy distribution of the second section

It can be seen from the figure above: the average sing energy of male voices in the upper and lower sections of "Rich and Vast Alxa" is around 68 dB, and that of female voices is around 66 dB. This shows that the long tone singers have a balanced energy distribution when singing, so that the energy values of the upper and lower sections show a good fit. The specific performance is shown in the table below (Table 3):

Table 3. Singing energy data of upper and lower sections of the same song

Repertoire	m1 (upper)	m2 (lower)	f1 (upper)	f2 (lower)
Duration (s)	81.3	79.9	70.4	69.1
Energy max(dB)	85	86.7	79.1	81
Energy max (dB)	43.2	46	52.8	52.6
Energy max (dB)	67.8	68.1	65.6	66.1

Observing the table above, you can find:

(1) When singing long tone, the lyrics in the upper and lower sections are the same, but the singers are not allocated the same singing time for the upper and lower sections. When the male sings, the upper section is 81.3 s, the lower is 79.9 s, but only 1.4 s apart. When the female sings, the upper and lower passages are only 1.3 s apart.
(2) The highest energy value of the upper period is 85 dB when the male sings, and the highest energy value is 86.7 dB in the lower period. The highest energy value of the latter period is only 1.7 dB higher than that of the previous. The highest energy value of the upper is 79.1 dB, the highest energy value of the lower passage is 81 dB, and the highest energy value of the latter is only 1 dB higher than that of the previous.

According to the sing energy data of the upper and lower period of the long tone, the following is a scatter overlap distribution diagram. As shown below, it is found that the energy of male singing is concentrated between 60 dB and 80 dB, and the energy distribution of female singing is scattered, reflecting when the male singing, the voice is always high, and the female voice is high and low, and it is frustrated (Figures 13 and 14).

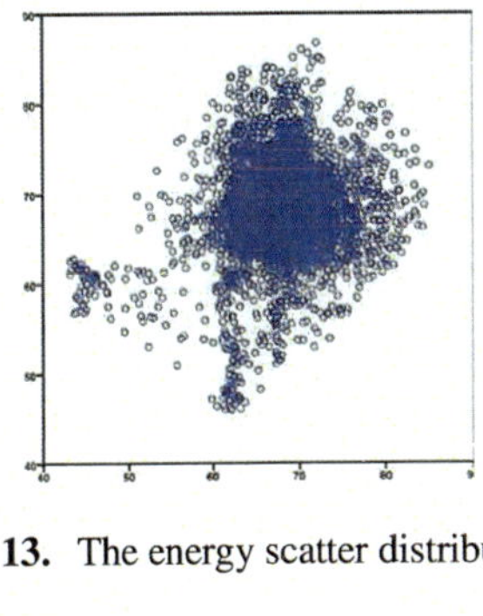

Fig. 13. The energy scatter distribution of m1

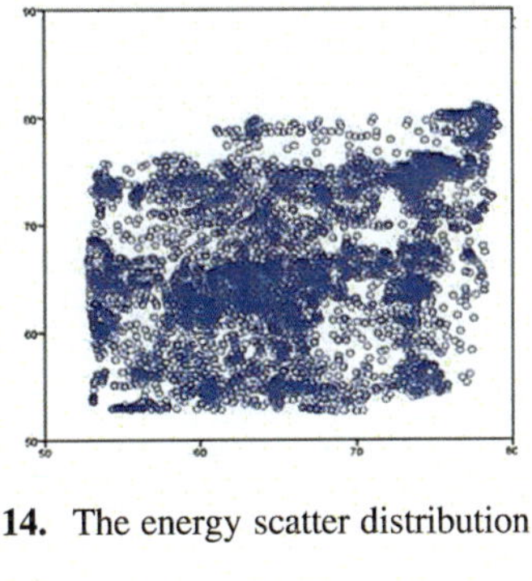

Fig. 14. The energy scatter distribution of f1

4 Singing Formant

The formant refers to the impulse response of the sound channel. If the sound channel is regarded as a resonant cavity, the formant is the resonant frequency of the cavity. The international music acoustics standard that defines the quality of the voices of singers is whether there is a formant at the sound spectrum of about 2200–2800 Hz, which is the most sensitive area of the human ear. This formant is called the "singing formant". That is to say, as long as the singer masters the correct pronunciation method, then he can increase the formant energy density of each harmonic, thereby producing a bright and penetrating singing effect.

Good resonance is very important for a singer. It is an important means of expression in the singing process. It can make people's voices more crisp, bright, and have a strong penetration. The Mongolian long tone has this bright, brilliant and penetrating sound quality. Such a voice can be spread farther in the vast prairie. In terms of the sampling content, this article collected the "singing formants" of male and female voices during singing, and the male and female voices formed the sonogram of singing formants during recitation and singing. The following six pictures show singing formant and spectral envelope when male and female singing and reading at the same time (Figures 15, 16, 17, 18, 19, 20, 21 and 22).

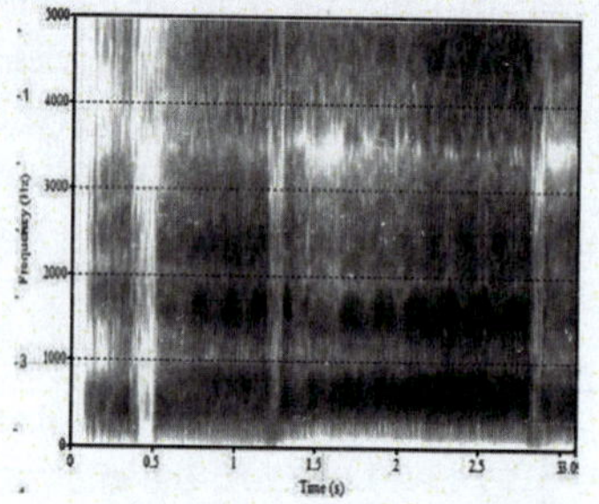

Fig. 15. m2 singing formant phonogram

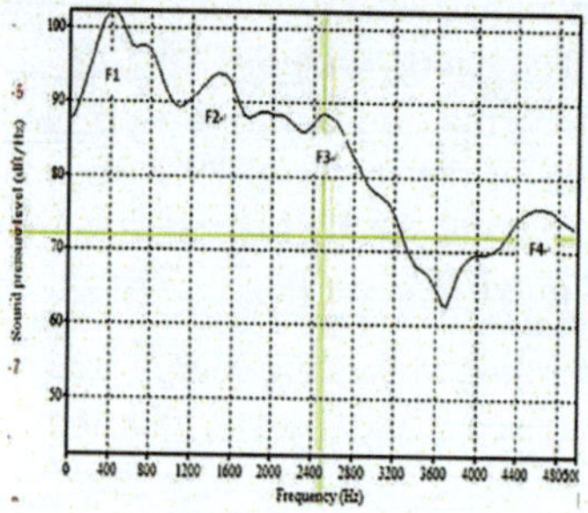

Fig. 16. m2 singing syllable spectrum envelope

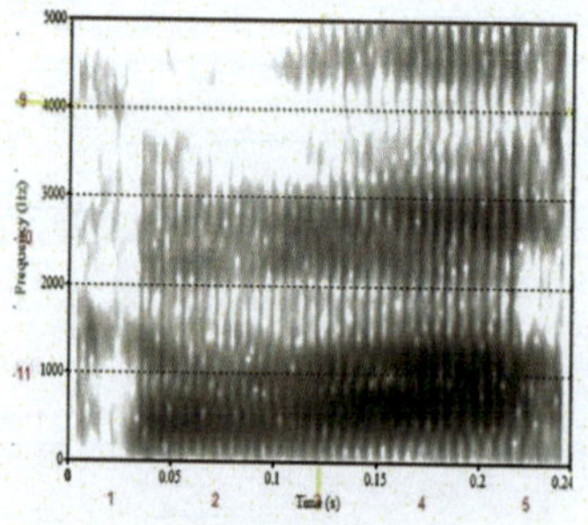

Fig. 17. m2 reading aloud

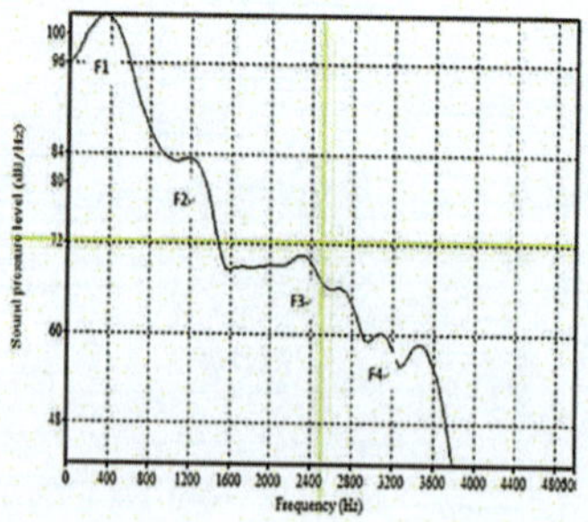

Fig. 18. m2 reading first syllable spectrum envelope

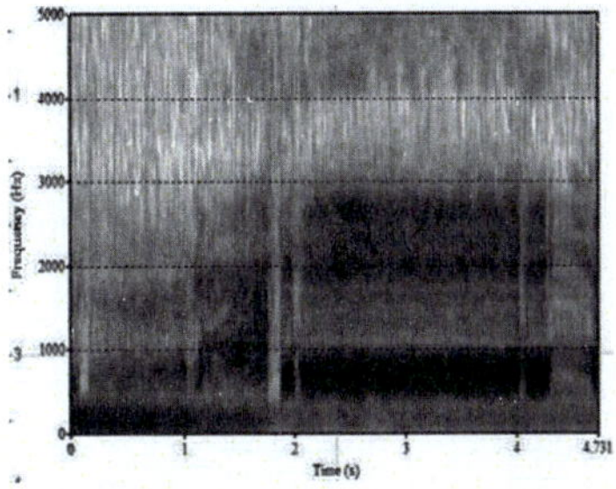

Fig. 19. F3 singing formant speech

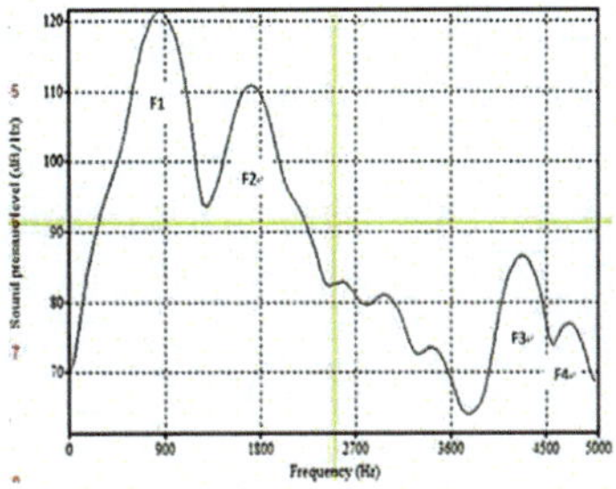

Fig. 20. f3 singing first syllable spectrum envelope

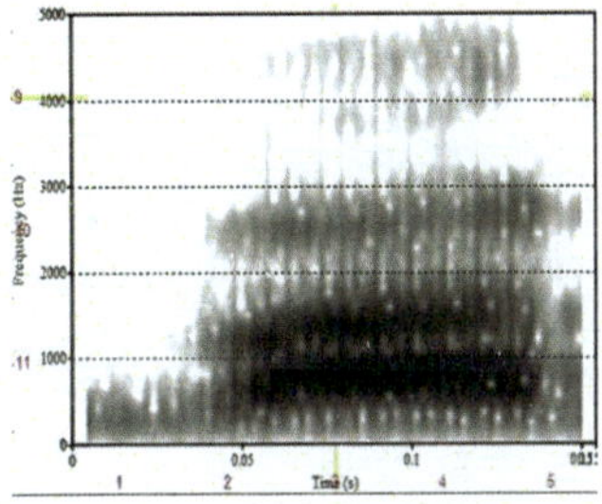

Fig. 21. f3 reading aloud graph

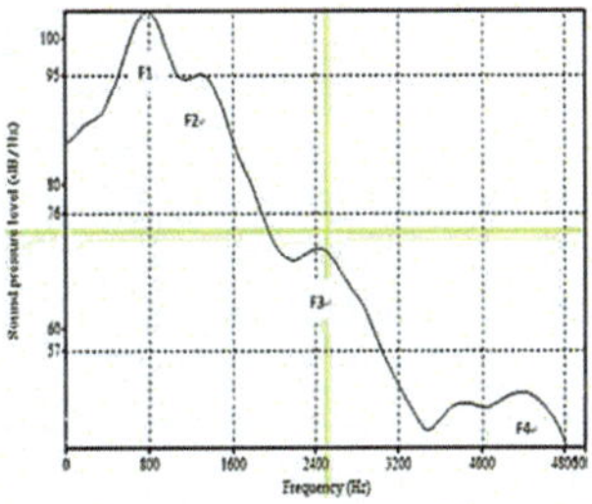

Fig. 22. f3 reading aloud first syllable spectrum envelop

By comparing the sonogram and spectrum envelope of the above two long tones when singing, a comparative analysis of the formants of men and women when singing long tones and reading lyrics. It can be concluded that:

Observation of the m2 singing diagram shows that the darkening bars are formed at 1000 Hz, 2000 Hz and 2500 Hz, respectively, but the strong energy black bars are not formed in other regions; there is almost no energy distribution in the regions above 4800 Hz. Observing the f3 singing phonogram, it can be seen that the phonogram forms thick black bars near 1000 Hz, 2000 Hz, and 2800 Hz, respectively, but fails to form stronger black bars in other areas. Combined with the spectral envelope, it can be seen that during the singing process, the distribution range of low-frequency formants (F1, F2) is mainly in the range of 400 Hz–1800 Hz, while the third (F3), fourth (F4) and other high-frequency formants are probably around 2400 Hz–4800 Hz, which is the "singing formant" mentioned above.

The air flow rushes out during the vocalization, and this time makes the vocal cords vibrate and the fundamental frequency is generated. Different singing types have different singing characteristics. The larynx expands more fully in long-tune singing, that is, the low frequency is rich and the sound is loud; the high frequency is extremely penetrating and the tone is brilliant. When the speaker sings to a high note, the throat knot will run up, which has a great relationship with the voice mechanism of the speaker when singing. There are a lot of vocal skills and techniques when singing long tones. These are not created when the long tones are produced, but are constantly generated in the process of being passed on from generation to generation. These singing skills are inherited by generations of long tones. It is absorbed and based on inheriting the achievements of the predecessors, and in the long-term practice, it is obtained by adding some of the singing skills that you have mastered when singing. Therefore, the intangible cultural heritage of Mongolian long tone should be protected and passed on by modern science and technology.

Acknowledgments. This work was fifinancially supported by Central University Fundamental Research Fund (No. 31920200028) and Northwest Minzu Univereity 2020 Graduate Research Innovation Project (No. Yxm2020102).

References

1. He, H., Jiatai, C., Yuling, Z.: Mongolian Prosodic Feature Acoustic Parameter Database. Inner Mongolia University Press, Inner Mongolia (2001)
2. Bei, W., Yufang, Y., Shinan, L.: Standard Chinese sentences stressed syllable pitch changing mode study. Acta Acustica (3)(2002)
3. Ertai, Q.: Mongolian Grammar. Inner Mongolia people's Publishing House, Hohhot (1991)
4. Halazhen. Ordos Wedding Big Songs. Inner Mongolia People's Publishing House, Hohhot (2011)
5. Shanying, Y.: Different types of singer resonance peak features singing tone and the mechanism of formation research. Vocal Music Study **4** (2010)

Research on the Application of Fine Execution of Big Data Empowering Court

Jiejing Yao[1(✉)] and Peng Hui[2]

[1] Shanghai University of Political Science and Law, Shanghai, China
penghui@sass.org.cn
[2] Shanghai Adcademcy of Social Sciences, Shanghai, China

Abstract. At the current stage, the implementation of innovative big data has achieved rich results. (1) With the help of big data execution thinking, we can determine the current residential address, business address, frequent access place, real-time location of the executed person, and solve the problem of finding the executed person through the way of big data electronic delivery; (2) According to the superposition of the subject database and the subject database, the hidden property of the executor can be found by deep big data mining technology, so as to solve the problem of finding property difficulty. (3) The electronic inspection of the data property of the person subjected to execution can be controlled, and a more effective and convenient data-based method can be adopted. At the same time, the problem of avoiding execution difficulty can be solved by promoting the face recognition Limited high-tech, using social new media innovation to persuade execution, and continuously enriching the Ecological Internet execution.

Keywords: Fine execution · Big data · Empowering court

1 Introduction

1.1 Learning of Domain Knowledge

According to the characteristics of multiple data sources, different data structures, and sparse data in court execution, it is necessary to solve the problem of completing, correcting, and reasoning the noisy and incomplete knowledge map containing multiple relationships. The data source of the executive business has the characteristics of fragmentation, strong personal privacy, high noise ratio, and even more deficiencies, which can quickly bring messy information to the process of building a knowledge map automatically [1]. Based on this, it is necessary to integrate the multiple complex relationships contained in the ontology knowledge of internal judicial data and external social network data, master and use multi-dimensional digital representation, feature extraction, semantic analysis, and other means, so as to refine the knowledge organization mode of the court implementation ontology. Then, the resource base of big data of court execution is gradually improved, and a multi-level ontology knowledge map base of court execution and its spatiotemporal evolution model is constructed.

M. Tavana et al. (Eds.): IISA 2020, AISC 1304, pp. 95–100, 2021.
https://doi.org/10.1007/978-3-030-63784-2_13

1.2 Cognitive Psychological Calculation

It is necessary to establish a unified quantitative model and intention calculation system of hidden property game behavior on the basis of coherent psychology. For example, when facing the financial transaction behavior in the virtual network, the mining technology of analyzing hidden property clues can focus on the psychological characteristics, cognitive bias, reference dependence, decreasing sensitivity, overestimating low probability events, underestimating high probability events, loss aversion and other attributes of the executee, through which the behavior process from cognition to decision-making can be depicted [2]. To explore the motive, purpose, and cognition of the executed person's deep inner will, explain the influence mechanism of cognitive psychology and external environment on the hidden property behavior of the executed person, and establish a unified quantitative model of the action mechanism. In the model or data-driven statistical analysis, the potential intention of the executee is integrated, and a new method of calculating the execution intention based on the combination of data-driven and cognitive guidance is proposed to predict the behavior track of the executee.

1.3 Generalization of the Small Sample Set

Under the limited condition of external data sharing for the court's execution business, it is necessary to study the generalization of a small sample set based on the probability graph model and multi-task learning. In the process of implementing the complete process control, the complexity of the case or the lack of valuable data, as well as the inadequacy of the implementation data sharing, actually caused the difficulty of fine implementation [3]. For example, in the judicial auction process of property disposal, there may be irregularities in the evaluation of value and the determination of the reserve price of the auction, as well as irregularities of auction companies and malicious bidding of bidders. Therefore, on the basis of comprehensive judicial auction data, third-party evaluation institutions, and other fundamental data, historical evaluation, and auction appraisal data, it is necessary to conduct in-depth discussion on the rational early warning technology of auction differences and auction violations based on the comprehensive analysis of auction data.

1.4 Human-machine Cooperation

As far as the current artificial intelligence system is concerned, it can't achieve the human intelligence with high autonomous learning ability, especially for the essential application fields with high professional technology content such as court execution decision-making [4]. Therefore, the critical problem for the future success of the application of big data and artificial intelligence in refined execution is to be able to better integrate human situation awareness, cognitive ability, machine computing, and storage ability, and integrate the experience of judicial experts and machine learning analysis and prediction ability, so as to establish a hybrid enhanced intelligence and interpretable decision-making theory of human-computer cooperation.

2 Clue Searching of Fusion Track

2.1 Activity Tracking of the Executed Person

Because the data sources of the executee in the network are complex and diverse, it is necessary to calculate based on wide-area distribution, collect and integrate multi-source heterogeneous data, including distributed crawling of spatial network data, distributed real-time access of private network data, etc. On this basis, a series of complete processes, including data cleaning, data integration, data transformation, and data induction, are formed with high-efficiency preprocessing technology [5].

The key points are: (1) Extract entities of different granularity from the massive activity track data, and analyze the relationship between entities, so as to build a large-scale entity network that can reflect the behaviors, attributes, and relationships between the executor or entities offline. It needs to support the description of behavior relationships across multiple nodes, which is a description method based on the domain knowledge base to study the characteristics of key personnel and group rows. (2) With the help of big data, we can also analyze the behavior habits and preferences of the executor behind the messy data so as to find the behaviors and tracks more in line with the interests and habits of the parties and conduct targeted in-depth learning and modeling. Finally, by combining the entity network of the executee, we can realize the all-round characterization of the behavior characteristics of the executee, and construct the knowledge map of the activity tracking of the multi-source heterogeneous executee. (3) Aiming at the effective spatiotemporal index technology of mobile objects, the track data in complex application scenarios are mined and analyzed. With the help of credit card, third-party payment, consumption records and other relevant location data of the parties' previous dining places, we can analyze their eating and consumption habits, and provide efficient solutions for the court to locate the executee. (4) As far as the existing early-warning integral model is concerned, it depends too much on human experience, and then it will be challenging to carry out accurate early-warning. Therefore, it is necessary to build a large-scale distributed model training platform, discuss the automatic learning and improvement of early warning model machine learning algorithm based on knowledge and experience, and based on the large-scale entity network, build the behavior characteristics and location trajectory analysis model for the executed person, so as to realize the intelligent trajectory behavior early warning of the executed person based on model analysis.

2.2 The Construction of the Clue Mining System

At the technical level, the first crack points are the completion, error correction, and reasoning of large noise and incomplete execution knowledge maps containing multiple relations. In the court execution business, the heterogeneous multi-source or partial lack of data sources is easy to introduce noise into the process of automatic knowledge mapping [6]. In the integration of internal judicial data and external social network data, the ontology knowledge contains complex and multiple relationships, which require the use of multi-dimensional digital representation, feature extraction, and semantic analysis, and other means to refine the ontology knowledge organization mode of court enforcement. Integrate the court execution big data resource library and the third party massive executed person online and offline data information, construct a multi-level court execution ontology knowledge graph and its Spatio-temporal evolution model, and accurately correct, complete reasoning.

On this basis, for the data acquired from the internal and external data interaction, it is necessary to deeply mine the spatial structure relationship between the sequential data and the data on the cross-section, and propose the analysis and processing of time and space coupling and the calculation framework of multimodal features, as well as the design of efficient algorithm of multimodal information fusion, so as to build the features of high-quality fusion of multimodal data and prior knowledge Engineering. Specifically, it shows how to deeply mine and builds large-scale network behavior patterns of the executed entities, analyzes the virtual network behavior characteristics of the executed based on social big data, and integrates the court big data resource base to implement multimodal features system.

Next, we need to build an adaptive trajectory prediction model based on behavior portrait. This model introduces the concept of the executee's portrait into the field of the court's executee's behavior analysis, analyzes the behavior characteristics of the executee from a multi-dimensional perspective, so as to solve the effective collection and organization of the executee's information in the big data environment, and enhances the single variable risk differentiation ability through the method of variable derivation. In terms of implementation technology, the random forest model and Logistic regression model in machine learning theory can be integrated to construct an RFML + LG model to generate a series of sub-models with portraits of the executed person. Before the statistical modeling, the decision tree in a random forest is used to analyze, and binary decision tree variables are generated. Then, the output of the random forest model is imported into the logistic regression model for statistical modeling, which lays the algorithm foundation for the big data information of the executed person to be transformed into their behavior portrait.

3 The Clue Mining of Network Financial Behavior

3.1 Hidden Property Discovery Mechanism

The portrait of the executor is a comprehensive description and integration of the financial distribution, personality, psychological behavior characteristics, and other existing data of the executor. The dimension perspective of multi-dimensional portrait includes: the growth process, financing distribution, physical property distribution, social relationship network, personality, hobbies and psychological behavior, financial transaction network, etc. Based on this idea, it is necessary to establish in-depth cooperation between the court and government agencies, financial institutions, etc., so as to obtain the real estate, vehicle, industrial and commercial tax, online consumption, financial asset transaction data of the executed person, as well as the dynamic data of online consumption records, banks, funds, bonds, stock transactions and other assets involved in the case [7]. In this way, a distributed data storage, query and processing system is constructed to mine the characteristics of users' social relations and transaction behaviors, and display them visually on the big data platform. In fact, due to the large number of cases of historical property concealment, the complexity of clues of property concealment, and the variety of concealment methods and channels, there are not many high-value information that can be timely and accurately retrieved and extracted from historical cases in the process of implementation. Therefore, it is necessary to start with the collection and analysis of hidden cases of historical property, to realize incremental learning through big data analysis technology and case reasoning technology, to establish a hidden property case database including retrieval, reuse, correction, preservation and answer functions, to achieve Quick and accurate solution.

3.2 Mining Model of Hidden Property

For this model, the first mock exam is built by combining the technology of big data analysis and integration with the case reasoning technology in AI. Based on case representation, case attributes, case retrieval and case self-learning process, a case database of hidden property is constructed for a large number of cases of historical hidden property. Then, use the feature retrieval dimension of the case for correlation analysis, and then use hierarchical clustering and other data mining techniques to find the law of hidden property in historical cases, extract the tacit knowledge in hidden property cases, and form mathematics that includes multiple hidden property modes model. On this basis, based on cognitive psychology, a unified quantitative model is established to determine hidden property and other game behaviors. Among them, the focus is to put the psychological characteristics, cognitive bias, reference basis, loss avoidance and other characteristics of the executee into the framework of game analysis, depict the behavior process of the executee from cognition to decision-making, and reflect the influence mechanism of psychological factors and external environmental factors on the behavior of the executee hiding property. Then, the parameter weights are determined through the hidden property game analysis framework, and then the quantitative probability values of various hidden property behaviors are calculated. Through the probability model constructed by this method, the probability between the multi-dimensional static information and dynamic characteristics of the executed person is described, and the information of the executed person is visualized simply and intuitively. Then, through Bayesian grid, hidden Markov model, conditional random field, the correlation dependence between different dimensions of the executed person is analyzed, and the input characteristics of the executed person and the property involved in the case are discussed. The relationship between behavior recognition.

Acknowledgments. This article obtained the national key R & D plan "Research on Fine Execution Technology and Equipment of Full Process Management" (2018YFC0830403)

References

1. Granger, M.P., Irion, K.: The court of justice and the data retention directive in digital rights Ireland: telling off the EU legislator and teaching a lesson in privacy and data protection. Eur. Law Rev. **39**(4), 835–850 (2014)
2. Cate, F.H., Mayer-Schönberger, V.: Notice and consent in a world of big data. Int. Data Priv. Law **3**(2), 67–73 (2013)
3. Hipgrave, S.: Smarter fraud investigations with big data analytics. Netw. Secur. **12**, 7–9 (2013)
4. Koops, B.J.: The trouble with European data protection law. Int. Data Priv. Law **4**(4), 250–261 (2014)
5. Kemp, R.: Legal aspects of managing big data. Comput. Law Secur. Rev. **30**(5), 482–491 (2014)
6. Frické, M.: Big data and its epistemology. J. Assoc. Inf. Sci. Technol. **66**(4), 651–661 (2015)
7. Camacho, J.: Visualizing big data with compressed score plots: approach and research challenges. Chemom. Intell. Lab. Syst. **135**(11), 110–125 (2014)

A Dynamic Correlation Method of Fragmented Web Resources

Haibo Hou[1], Qiurong Zhu[1(✉)], Yu Zhang[2], and Jiangbing Yang[1]

[1] Beijing Lanxum New Technology Co. Ltd., Beijing, China
zhuqiurong@lanxum.com
[2] Cyberspace Institute of Advanced Technology, Guangzhou University, Guangzhou 510006, China

Abstract. In a large-scale network environment, the traffic component of the gateway is complex and changeable, and the number of web pages is huge and unevenly distributed, which poses a challenge to the malicious web pages detection. Among them, the most challenge comes from the dynamic correlation of web resources. Due to Internet advertising, CDN acceleration mechanisms, cloud services, and high compression coding, these responses are usually multi-source links and are transmitted in fragmented form. The web page resources such as text, LOGO, pictures, audio and video received after an access request may come from different servers, and the information is out of order, cluttered and fragmented. How to obtain web page source such as web pages source code, audio and video, file and picture in a high-speed network environment and associated assembled into a complete web page efficiently so that they can make detection, which is of great significance for malicious web page homology detection, application network usage situation analysis, and traffic perception management. This paper focuses on the problem of resource association in the situation of large-scale traffic, and proposes a multi-domain feature association method to associate multimedia resources as much as possible to obtain a complete web page relatively. The experimental results show that the multi-domain association success rate can reach more than 87%, which has improved over 60% compared with the existing association methods based on a single feature.

Keywords: Dynamic correlation · Fragmented web resources · CDN · Passive traffic filtering

1 Introduction

With the continuous development of the HTTP protocol and the widespread application of network acceleration technologies such as CDN, the Internet topology is becoming more complex. “Computer is the network” marks the Web site as an interactive platform for application delivery and traffic-intensive real-time multimedia content. In the period of Web 2.0, the complexity of the page structure has increased significantly compared to Web 1.0. According to the statistics results of HTTP Archive [1], the average transmission size of all web pages counted on January 1, 2019 was 1966 KB, and the average number of requests generated by each webpage reached 75 times.

M. Tavana et al. (Eds.): IISA 2020, AISC 1304, pp. 101–110, 2021.
https://doi.org/10.1007/978-3-030-63784-2_14

Therefore, the traffic behavior characteristics of web services under high-speed network environment are changing rapidly. The response information of massive websites presents the characteristics of multi-source links and fragmentation. The website's LOGO, pictures, and content may come from different servers. In this way, the return information of the target website cannot be obtained quickly and accurately, which brings difficulties to the identification and detection of malicious web pages. Therefore, it is necessary to find a method that can associate heterogeneous resources and analyze the relationship between resources. Existing research methods on webpage resource association by scholars domestic and overseas are mainly based on traffic characteristics, and the research direction is mainly used for traffic source tracing, anomaly detection, user tracking, and billing management.

Butkiewicz et al. [2] proposed a browser-based measurement framework for studying users' online behavior. The measurement framework correlation analysis all HTTP requests and HTTP responses generated when the web page is rendered. The association method is a browser-based active webpage association measurement. Through the browser plug-in, all local webpage loading events can be accurately recorded, and then the HTTP stream association results are obtained based on the log information analysis.

Goldberg et al. [3] proposed a browser fingerprint recognition method based on the user-agent field in the HTTP session by analyzing the content of the HTML web page, and realized the association of HTTP requests to different users.

Neasbitt et al. [4] automatically reconstructed the user-browser interaction process through network tracking, and designed a browser plug-in ClickMiner to discriminate user access behavior. Experimental results show that ClickMiner can correctly reconstruct 82% to 90% of user-browser interactions have false positive rates between 0.74% and 1.16%.

Franziska et al. [5] proposed a client-side method that analyzes and associates the data traffic obtained by deploying browser plug-ins, and detects five kinds of third-party trackers based on the way of operating the browser's status. But the browser plug-in-based method is not suitable for passive traffic environments.

Gunes/Acar et al. [6] designed a browser-based fingerprint recognition framework FPDetective, and through semi-automatic analysis of JS scripts and Flash source code, generated browser fingerprints to achieve large-scale analysis of TopList websites. However, this method requires code-level analysis of web pages and is not suitable for large-scale real-time analysis.

Existing methods or client-based methods, actively associate by collecting browser state characteristics, which cannot adapt to passive measurement of large-scale traffic on the gateway; or in another situation of using Referer, Cookie, User-agent and other fields of HTTP requests, but do not consider each of these associated fields' applicable scenarios, and there are limitations such as difficulty in guaranteeing the association success rate when used alone.

2 Dynamic Correlation Process

Based on a long-term observation of Internet gateway access logs, this chapter focus on the problem of multi-source fragmented transmission of web resources, and proposes an association method based on multi-feature fusion, which can filter part of invalid traffic while achieving the association of fragmented web resources.

2.1 Multi-domain Association Ideas

This paper proposes a multi-feature association method RLT (Compromise of Referer and Location and TCP) by mining attribute fields such as Referer and their timing relationships in HTTP requests. The RLT model combines three resource association methods: Referer, redirection and TCP long connection and they are shown as follows:

1. The fields used in the three association methods may have poor or even unavailable association effects due to service provider or protocol settings. Fusion and complementation can alleviate this problem and improve the association effect;
2. There are differences in the applicable scene of correlation field. The Referer and redirection methods can perform cross-domain resource association, while TCP long connection is more suitable for resource association in the same domain.

The three resource association methods of Referer, redirection, and TCP long connection are described in detail below.

1. Based on Referer association method

The Referer-based association method is mainly based on the fields in the HTTP protocol. According to RFC7231, the HTTP referer is one of the HTTP header fields and is an important part of an HTTP request. Resources usually carry their own Referer information in order to notify the server of its origin. Under normal circumstances, the HTTP request header sent by the browser has Referer. The Referer request header contains the address of the previous page, and follows the link of the current request page from this page.

Therefore, by matching the Referer with the URL of the source page, you can retroactively locate the resource pre-resource node from the traffic, so as to achieve the association between different HTTP streams.

2. Relevance-based association method

According to RFC2616, the HTTP protocol allows the server to redirect client requests to other locations. In the HTTP protocol, the three most common types of redirection are: 301 Redirect, 302 Redirect, and Meta Fresh. 301 Redirect stands for Permanently Moved, 302 Redirect stands for Temporarily Moved, and Meta Fresh stands for redirecting to a new webpage after a certain time. The server triggers a redirect by sending a special response (redirect) with a status code of 3xx. When the browser receives a redirect response, it will use the new URL provided by the response and load it immediately. When a resource is redirected to another resource, the response side often carries the location field, so it can use the transmission information when the resource is redirected to another location to associate the resource.

3. Association method based on TCP long connection

As we all know TCP is a connection-oriented, reliable, byte stream-based transport layer communication protocol. HTTP requests and HTTP responses are transmitted through the TCP connection channel to transmit request and response data. TCP connections can usually be divided into long connections and short connections, of which long connections are also called persistent connections.

HTTP 1.0 uses short connections by default, that is, each time the browser and server perform an HTTP operation, a connection is established, and the connection will be terminated when the task ends. If an HTML or other type of Web page accessed by the client browser contains other Web resources, such as JavaScript files, image files, CSS files and so on; Each time the browser encounters such a Web resource, it will create a HTTP session. As of HTTP 1.1, long connections are used by default to maintain connection characteristics. Using the long-connected HTTP protocol, code will be added to the response header: Connection: keep-alive, and keep-alive has a hold time, which can be set in the server software.

HTTP long connection is used for negotiation to reuse TCP connection, which can save resources and time consumption. In the situation of using a long connection, the TCP connection used to transfer HTTP data between the client and the server is not closed. When a web page requests resources, it usually establishes several TCP connections with the server to request resources. Several HTTP messages transmitted in one TCP session are usually from the same page. As shown in Fig. 1, if there are several HTTP messages multiplexing the same TCP session information within a set period of time, it can be said that the information comes from the same page.

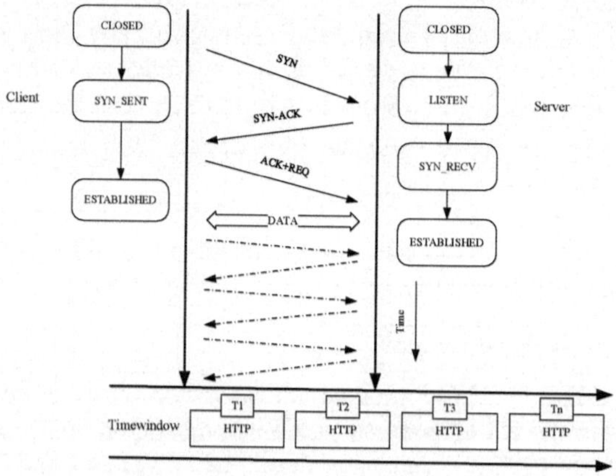

Fig. 1. Association based on TCP long connection

Whether it is flow measurement, flow correlation, or resource correlation, the time dimension is an important feature. In the large-scale traffic measurement scenario, setting a time window is help to analyze large-scale traffic in real time under the condition of limited resources; when the stream is associated, the request response time of user browsing web pages and choosing to click the hyperlink, and the request response time of browser sending embedded resource object are exist obvious differences. Selecting an appropriate time threshold can distinguish the two access behaviors to achieve the homologous flow association. At the same time, the user's browsing time for the web page is usually limited. Setting a reasonable time window can help analyze resources in web pages more accurately, so as to achieve resource association.

According to the definition of the web standards organization W3.ORG, in general, accessing and loading a web page requires operations such as redirection, cache detection, DNS resolution, three-way handshake, request and response, and so on. The specific model [7] is shown in Fig. 2.

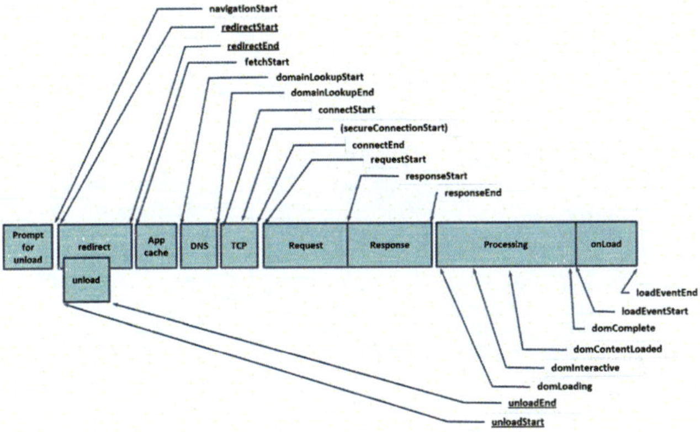

Fig. 2. Webpage loading model

Google Analytics carried out website speed analysis on this basis, and proposed data indicators to measure page load speed [8]: page load time, redirect time, domain name query time, server connection time, server response time, and web pages download time, including the time required for the browser to parse HTML, load CSS, JS, pictures and other files and render display page content, but without browsing time, so the web page load time is largely equal to the sum of the redirect time, domain name query time, server connection, server response time, and web page download time.

Based on the existing statistics and the experimental situation in this paper, the time window and settings proposed in this paper are as follows:

1. Page loading time: Indicates the time window when the user starts to visit the first page, waits for the page to load, and then selects one of the hyperlinks. In order to distinguish two related web page traffic effectively, this article chooses to set the time value to 3 s.
2. Embedded resource loading time: The maximum time interval between the time when the embedded resource is requested and the time when the first page is loaded. Based on the analysis of HTTP traffic on Alexa's China's top 100 traffic website, this article sets the embedded resource loading time to 950 ms.

3. Resource correlation time window: This article sets the resource correlation time window to 5 s, which not only ensures that the correlation time window can cover the loading time of most web pages, but also considers setting a smaller time window. It is help to achieve the actual needs of efficiency analyze in the background of large-scale traffic. By setting the correlation time window, it is possible to ensure that relatively complete web resources are collected and acquired for most pages, and the validity and accuracy of real-time correlation in a high-traffic environment is guaranteed.

2.2 System Structure and Workflow

The multi-feature association method RLT system structure is shown in Fig. 3. The workflow of RLT is divided into four processes: gateway traffic collection, association feature extraction, association feature fusion, and association verification.

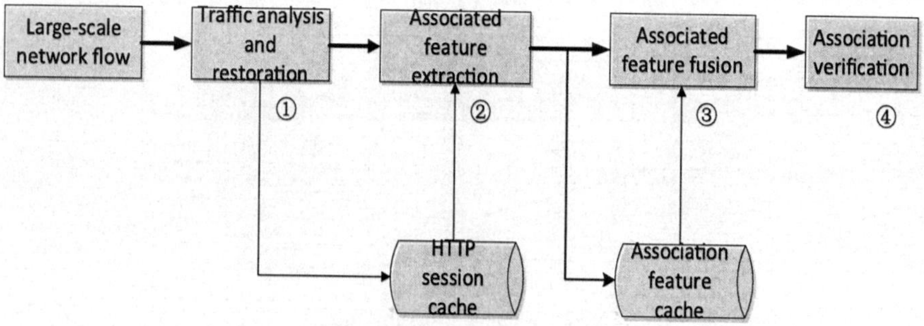

Fig. 3. Web resource association process

According to filtering rules, the large-scale network traffic analysis and restoration process caches valid traffic, parses required HTTP fields according to the HTTP protocol format, and identifies web resource HTTP field as association objects. Due to the complexity, high concurrency, and high traffic of gateway traffic, Efficient network flow processing technology must be used to capture network packets. For this, RLT uses the network flow security platform developed by the National Engineering Laboratory of Information Content Security Technology. The network flow security platform is a high-performance network security development platform for large-scale network flow processing. It has full-stack protocol analysis and restoration capabilities, improves the flexible deployment capability by shielding details below the transport layer through abstract plane.

In the link of association feature extraction, through analyzing the HTTP flow features corresponding to the extracted resources, a series of HTTP sessions between the user browser and the web service site when the user clicks the URL link to visit the web resources are mined out in the massive HTTP field data cached, and the data structure with high search efficiency is cached in the server memory.

At the moment of multi-domain feature associations, features of Referer, Location, and TCP long connections are fused. Within a certain time window, multiple HTTP streams that access the same web page are associated to obtain a web page resource as complete as possible.

The webpage URL obtained by the system association is used to actively access the webpage resource corresponding to the URL, and the consistency between the associated webpage and the benchmark webpage is compared to verify the accuracy of the RLT association model.

3 Experimental Evaluation

3.1 Data Acquisition

In this paper, we use manual and automatic methods to collect experimental data. The correlation research and experimental verification of gateway traffic need the marked HTTP traffic. Part of the marked traffic is collected manually. We visiting the top domestic websites of CNZZ, the URL of the visited websites is recorded manually, and the traffic generated by visiting these websites is captured. At the same time, in order to improve the efficiency of large-scale data collection, this paper uses Selenium automation framework [9] and Chrome browser to build a data collection system. When collecting data, it calls chrome interface through Selenium automation framework to simulate user behavior and click the URL in these websites to open a new page. When executing the test script, the browser automatically performs click, input, open and verification operations according to the script, which is consistent with the real user operations.

This time, we visited the top 5000 domestic websites of CNZZ industry, and collected 7 groups of data by day. In order to evaluate the efficiency of different association methods as much as possible, the data retained except the websites due to visit timeout, visit error or incomplete log record. The basic information of association data set is shown in Table 1.

Table 1. Correlation Data Set

Serial number	Number of websites accessed successfully	TCP connection number	HTTP sessions number
1	4386	79755	392388
2	4521	83225	405852
3	4531	82637	405998
4	4464	81358	396201
5	4507	81686	404113
6	4473	81523	399952

3.2 Experimental Results

RLT model integrates three association methods: Referer, redirection and TCP long connection. The fusion association enables the above three association methods to learn from each other in different scenarios and improve the overall association success rate through complementarity. It can be seen from Fig. 4 that in this paper, the RLT model can be associated with data by an average of 97%, which is higher than the success rate of the above three methods, and the association effect is better.

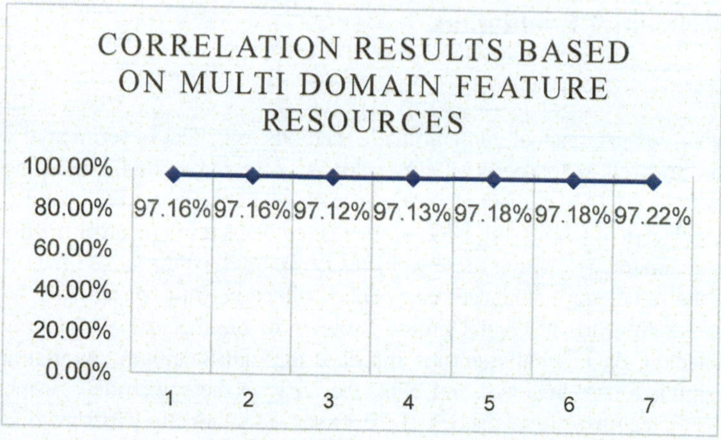

Fig. 4. RLT mix correlation result

Figure 5 is a comparison of the association results obtained by Location, Referer, TCP long connection, Referer & Location and the fusion method RLT based on the data set in this paper.

It can be seen from the figure that the correlation rate based on Location is the lowest, and the correlation rate based on the fusion method RLT in this paper is the highest, reaching an average of over 97%. Among them, the Referer-based Resurf method has an association rate of about 66%, the RLT association rate has increased by approximately 21%, and the association rate of the Referer & Location-based TNG method is 70%, which has increased by approximately 17%, Compared with the TCP-based long connection, the 85% correlation rate has also increased by about 12%.

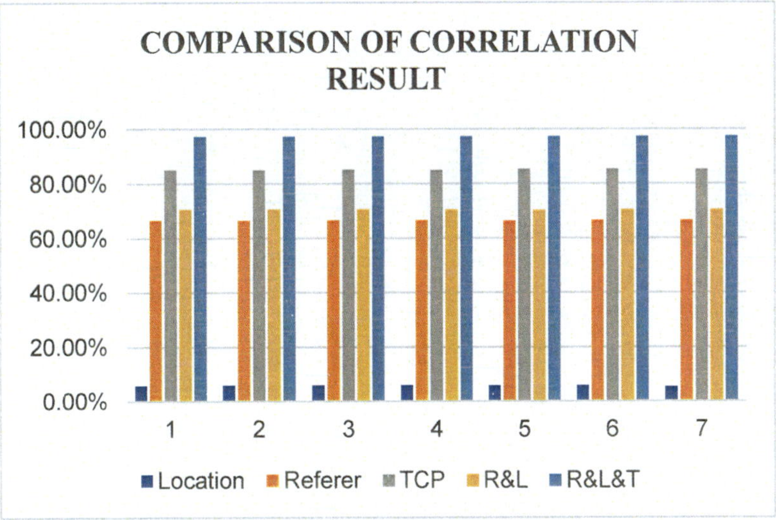

Fig. 5. Comparison of correlation result

4 Conclusion

Due to the large number of requests and responses in a large-scale high-speed network streaming environment, webpage data presents the characteristics of multi-source links and fragmented transmission. This paper mainly solves this problem, correlates fragmented data, and stitches it into a complete webpage content.

The experimental results show that the association rate of the RLT method proposed in this paper can reach more than 97%. Compared with the traditional method, RLT can associate 12% more web resources at least, indicating that the method in this article is better at associating web resources and overcome the effects of fragmentation transmission.

Acknowledgments. This work is supported by National Key R&D Program of China (No. 2018YFB0803900).

References

1. https://httparchive.org/reports/state-of-the-web
2. Butkiewicz, M., Madhyastha, H.V., Sekar, V.: Understanding website complexity: measurements, metrics, and implications. In: Proceeding of the 2011 AMC SIGCOMM Conference on Internet Measurement Conference, pp. 313–328. ACM (2011)
3. Maier, G., Schneider, F., Feldmann, A.: NAT usage in residential broadband networks. In: Passive and Active Measurement, pp. 32–41. Springer Berlin Heidelberg (2011)
4. Neasbitt, C., Perdisci, R., Li, K., et al.: ClickMiner: towards forensic reconstruction of user-browser interactions from network traces. In: Proceedings of the 2014 ACM SIGSAC Conference on Computer and Communications Security, pp. 1244–1255. ACM (2014)

5. Roesner, F., Kohno, T., Wetherall, D.: Detecting and defending against third-party tracking on the web. In: Proceedings of the 9th USENIX Conference on Networked Systems Design and Implementation, pp. 12–15. USENIX Association (2012)
6. Acar, G., Juarez, M., Nikiforakis, N., et al.: Fpdetective: dusting the web for fingerprinters. In: Proceedings of the 2013 ACM SIGSAC Conference on Computer & Communications Security, pp. 1129–1140. ACM (2013)
7. https://www.w3.org/TR/navigation-timing/#processing-model
8. https://analytics.google.com/analytics/web/
9. Wang, X., Xu, P.: Build and auto testing framework based on selenium and fitnesse. In: Proceedings of the International Conference on Information Technology and Computer Science, ITCS, pp. 436–439 (2009)

A Robust Predictive Current Control for SPMSM Based on Internal Model Disturbance Observer

Xiaoning Mu[1,2(✉)], Fanquan Zeng[1,2], Yang Zhou[1,2], Yebing Cui[1,2], and Yao Yao[1,2]

[1] Shanghai Institute of Spaceflight Control Technology, Shanghai 201109, China
2277498961@qq.com
[2] Shanghai Engineering Research Center of Servo System, Shanghai 201109, China

Abstract. The control accuracy of traditional deadbeat predictive control of surface-mount permanent magnet synchronous motors (SPMSM) is susceptible to model parameters and environmental disturbances. This paper introduces an internal model disturbance observer (IMDO) into the current loop predictive control, which can real-time compensation of *d* and *q* axis voltages to improve current loop control accuracy. First, according to the control model of SPMSM, calculate the stability domain of the predictive control algorithm; then, the observer is designed, and the system disturbance obtained by the internal model control equation is multiplied by the compensation coefficient to the *d* and *q* axis voltages. Finally, the simulation results verify that the proposed algorithm can improve the robustness of predictive control and improve the control performance of predictive control when the parameters are disturbed.

Keywords: Internal Model Disturbance Observer · Robust control · Current prediction · SPMSM

1 Introduction

SPMSM has the characteristics of simple mechanical structure, high power density and small torque ripple. It has been widely used in places where high control accuracy and high reliability are required, such as the aerospace field, high-precision CNC machining machine tools, industrial robots, etc. [1, 2].

Deadbeat Predictive Current Control (DPCC) can obtain better current dynamic and static performance [3]. However, deadbeat predictive control relies on accurate motor model parameters, and changes in motor parameters will cause the predicted voltage to deviate from the ideal value, thereby reducing motor control performance [4]. In [5, 6], a current stabilization algorithm is proposed to reduce the torque ripple, but this method ignores the effects of permanent magnet resistance and flux linkage mismatch.

The internal model disturbance observer has the advantages of simple structure, convenient design, strong robustness, etc., and is wildly used in the field of motor control [7–10]. In [7], the internal model control is introduced into the design of current

M. Tavana et al. (Eds.): IISA 2020, AISC 1304, pp. 111–119, 2021.
https://doi.org/10.1007/978-3-030-63784-2_15

loop to make the current steady-state fluctuations smaller. In [8], the adaptive control method combined with the internal model disturbance observer is applied to AC permanent magnet linear motors. In [9], an internal model observer is introduced in DPCC, which can predicts disturbance. In [10], an internal model control with extended state observer is proposed, which can reduce the external torque disturbance.

Based on the discrete mathematical model of SPMSM, a new DPCC based on the internal model disturbance observer is proposed. The advantage of IMDO in disturbance observation can be used to solve the parameter sensitivity problem of DPCC. Firstly, the control method and stability domain of DPCC are analyzed; according to the principle of internal model control, an observer is constructed to estimate the systems disturbance of the d and q axis, then the disturbance is added to the d and q axis control voltages through the compensation coefficient to improve the immunity of the current loop performance. Finally, the effectiveness of the observer is verified through simulation.

2 Control Strategy of DPCC

In the control strategy of DPCC, d and q axis current command in the next beat ($\mathbf{i}^*_{d-axis}(n+1)$, $\mathbf{i}^*_{q-axis}(n+1)$), and d and q axis current feedback of the current beat ($\mathbf{i}_{d-axis}(n)$, $\mathbf{i}_{q-axis}(n)$) can obtain the control voltages ($u_d(n)$, $u_q(n)$) through the discrete current model [4]. The simplified control strategy of DPCC is shown in Fig. 1.

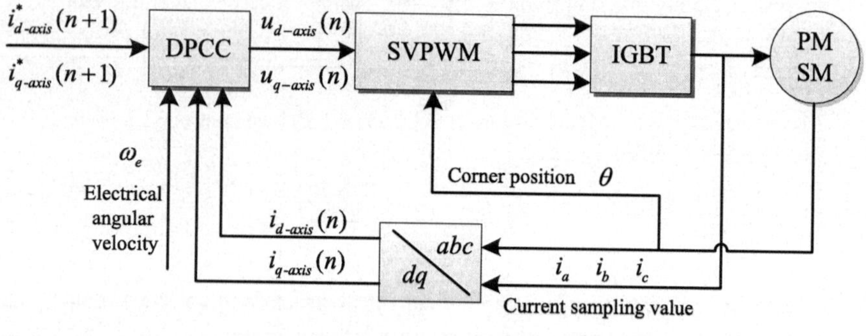

Fig. 1. Simplified control strategy of DPCC

3 The Design of IMDO

When considering the parameter error, the current equation of PMSM can be expressed as:

$$\dot{i} = A_o i + B(u_S - f) \tag{1}$$

where $\dot{i} = [di_{d-axis}/dt \quad di_{q-axis}/dt]^T$; $i = [i_{d-axis} \quad i_{q-axis}]^T$; $u_s = u - D'$; $u_s = [u_{sd-axis} \quad u_{sq--axis}]^T$; $u = [u_{d-axis} \quad u_{q-axis}]^T$; $\mathbf{A}_o = [-R/L \quad 0; \ 0 \quad -R/L]$;

$\boldsymbol{D}' = [-L\omega_e i_{q-axis} \quad -L\omega_e i_{d-axis} + \omega_e \psi_f]^T$; $\mathbf{f} = [f_{d-axis} \quad f_{q-axis}]^T$; $\boldsymbol{B}$=$[1/L \quad 0; \quad 0 \quad 1/L]$. In the above matrix, $u_{sd-axis}$ and $u_{sq--axis}$ are stator voltage; i_{d-axis} and i_{q-axis} are stator current; R and L are stator resistance and inductance respectively; ψ_f stands for magnetic flux; f_{d-axis} and f_{q-axis} are the designed disturbance.

The total disturbances f_{d-axis} and f_{q-axis} can be expressed as:

$$\begin{cases} f_{d-axis} = \varepsilon_{d-axis} + \Delta L \frac{\mathrm{d}i_{d-axis}}{\mathrm{d}_t} + \Delta R i_{d-axis} - \Delta L i_{q-axis}\omega_e \\ f_{q-axis} = \varepsilon_{q-axis} + \Delta L \frac{\mathrm{d}i_{q-axis}}{\mathrm{d}_t} + \Delta R i_{q-axis} + \Delta L \omega_e i_{d-axis} + \omega_e \Delta\psi_f \end{cases} \tag{2}$$

where ε_{d-axis} and ε_{q-axis} are unmeasured interference; R_0, L_0, ψ_{f0} represent the actual motor stator resistance, actual inductance and actual magnetic flux respectively; ΔL, ΔR and $\Delta\psi_f$ are the inductance deviation, resistance deviation and flux linkage deviation respectively, $\Delta L = L_0 - L$, $\Delta R = R_0 - R$, and $\Delta\psi_f = \psi_{f0} - \psi_f$.

Assuming that the coupling term and back EMF can be fully compensated, the state estimation equation of Eq. (1) is established as:

$$\dot{\hat{i}} = A_o \hat{i} + B(u_S - \hat{f}) \tag{3}$$

in the equation, $\hat{\boldsymbol{i}}$ and $\hat{\boldsymbol{f}}$ are the estimated variables of $\boldsymbol{i}$ and $\boldsymbol{f}$ respectively.

Define the error variables $\tilde{i} = i - \hat{i}$, $\tilde{f} = \hat{f} - f$, then obtain the new equation of current state by subtracting Eq. (1) and Eq. (3):

$$\begin{cases} \dot{\tilde{i}} = A_o \tilde{i} + B\tilde{f} \\ Y = \tilde{i} \end{cases} \tag{4}$$

Equation (4) represents the error state equation between the estimated and actual state of current. Using the design principle of the internal model disturbance observer, taking the q-axis observer as an example, assuming that the reference input is 0, and $\hat{f}_{q-axis}$ is regarded as the control variable, then the tracking error e_{q-axis} can be expressed as:

$$\begin{cases} e_{q-axis} = 0 - Y_2 \\ \dot{e} = -\dot{Y}_2 = -\dot{\tilde{i}}_{q-axis} \end{cases} \tag{5}$$

Introducing two new state variables z and m, it can get a new equation as:

$$\begin{cases} z = \dot{\tilde{i}}_{q-axis} \\ m = \dot{\tilde{f}}_{q-axis} \\ \dot{z} = \ddot{\tilde{i}}_{q-axis} = az + bm \end{cases} \tag{6}$$

Equation (5) and Eq. (6) can constitute a new augmented system equation:

$$\dot{z} = A'z + B'm \tag{7}$$

where $z = [e_{q-axis} \quad z]^T$; $A' = [0 \quad -1; \quad 0 \quad a]$; $B' = [0 \quad b]^T$.

The rank of controllable discriminant matrix of the above formula is that: rank $[\boldsymbol{B}' \quad \boldsymbol{A}'\boldsymbol{B}']$= 2, it means that formula (7) is fully controllable. According to the principle of state feedback, it is easy to realize state feedback control, by setting $m = -kz$ and $k = [k_1 \quad k_2]$, make the controlled system gradually stable, and get the control amount m as follow:

$$\dot{\hat{f}}_{q-axis} = m = -k_1 e_{q-axis} - k_2 z = -k_1 \tilde{i}_{q-axis} - k_2 \dot{\tilde{i}}_{q-axis} \tag{8}$$

where k_1 and k_2 are undetermined coefficients.

In a sampling period of the control system, when $\dot{f}_{q-axis} = 0$ is unchanged, then Eq. (8) can be further expressed as:

$$\dot{\hat{f}}_{q-axis} = -k_1 \tilde{i}_{q-axis} - k_2 \dot{\tilde{i}}_{q-axis} \tag{9}$$

The characteristic root of the system realized by Eq. (7) is $SI - (A' - B'k)$, and the characteristic root distribution of the standard second-order oscillation system is used to optimize k_1 and k_2, then it can get that:

$$[k_1 \quad k_2]^T = \left[-\omega_n^2/b \quad (a + 2\zeta\omega_n)/b\right]^T \tag{10}$$

Then, the discretized internal model disturbance observer can be obtained, and the control structure is shown in Fig. 2.

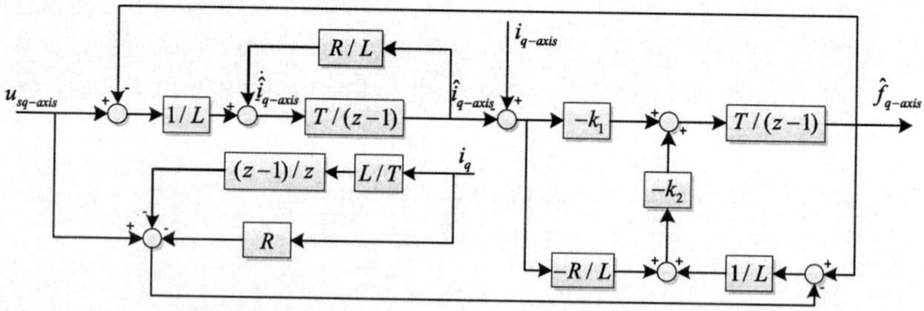

Fig. 2. Control structure of q-axis IMDO

4 The New Robust DPCC

The new robust-DPCC (RPCC) uses $\hat{f}_{d-axis}$ and $\hat{f}_{q-axis}$ as the reference voltage for system disturbance compensation, then the final output voltage reference after compensation is given by:

$$\begin{bmatrix} u^*_{dref-axis}(n) \\ u^*_{qref-axis}(n) \end{bmatrix} = \begin{bmatrix} u^*_{d-axis}(n) \\ u^*_{q-axis}(n) \end{bmatrix} - \xi \begin{bmatrix} \hat{f}_{d-axis}(n+1) \\ \hat{f}_{q-axis}(n+1) \end{bmatrix} \tag{11}$$

where ξ is the compensation coefficient.

The control structure of RPCC is shown below (Fig. 3):

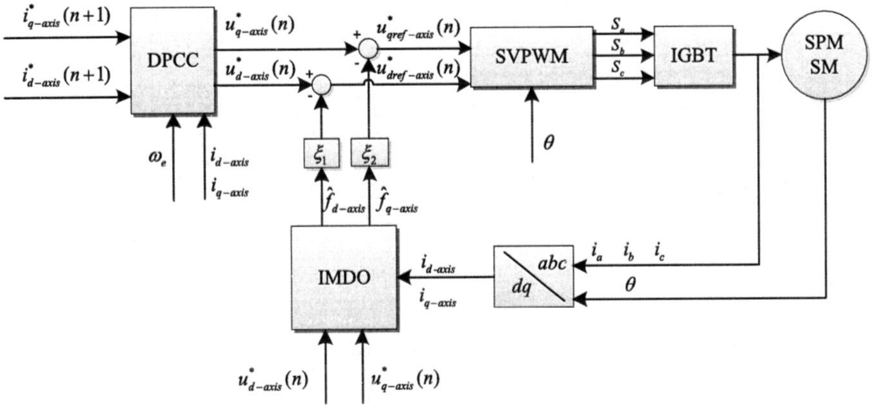

Fig. 3. Control structure of RPCC

5 Simulation Results and Analysis

5.1 Simulation Conditions

Table 1 is the motor simulation parameters. The algorithm was performed through Matlab/Simulink.

Table 1. Parameters of SPMSM

Parameters	Values
Stator resistance R/Ω	0.958
Stator inductance L/mH	5.25
Flux linkage ψ_f/Wb	0.1827
Pole pairs p_n	4
$J/(\text{kg}\cdot\text{m}^2)$	0.003
B/(N·m·s)	0.008
Sampling time T/s	0.0001

The simulation conditions are set as follows: the DC bus voltage is 311 V; the speed step is given from 0 to 1000 r/min at 0.01 s; the specified load is suddenly applied at 0.08 s; then the load is suddenly unloaded at 0.13 s; set k_1 as −0.00001; set k_2 as 0.2.

5.2 Simulation of RPCC

Figure 4 to Fig. 5 are the waveform comparisons between the traditional DPCC and the RPCC in the case of inductance parameter mismatch. The simulation selects *d*-axis current feedback and *d*-axis current error in two cases of $L_0 = 0.5L$ and $L_0 = 1.5L$ respectively for comparison.

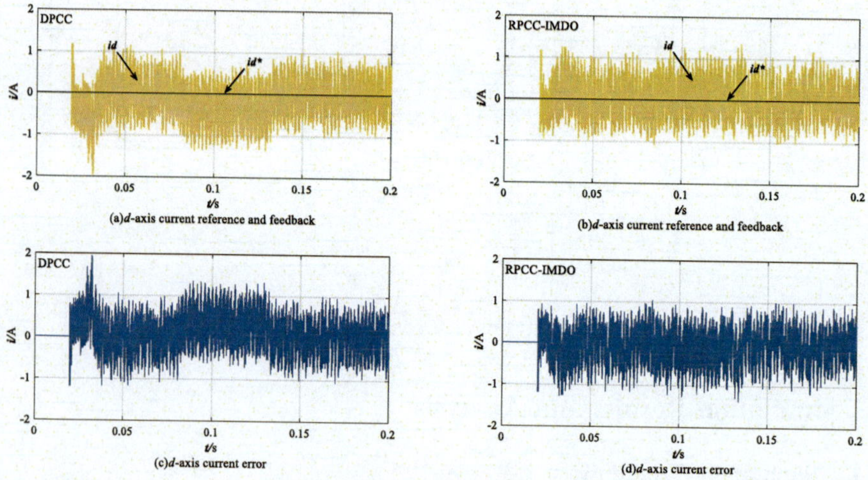

Fig. 4. Current feedback and current error on *d*-axis ($L_0 = 0.5L$)

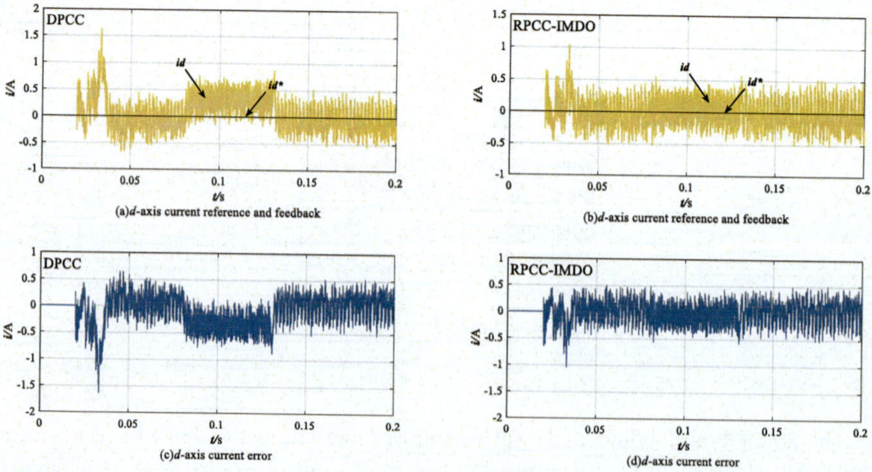

Fig. 5. Current feedback and current error on *d*-axis ($L_0 = 1.5L$)

According to Fig. 4(c)–(d) and Fig. 5(c)–(d), the method of RPCC can make the static error less on d-axis and the system more stable. Regardless of whether the inductance value is too large or too small, after adding IMDO, the d-axis current feedback fluctuation can be reduced, and the command can be closely followed, unlike the large static error generated in traditional predictive control.

Similarly, Fig. 6 to Fig. 7 are the waveform comparisons in the case of flux linkage mismatch. The simulation selects q-axis current feedback and q-axis current error in two cases of $\psi_{f0} = 0.5\psi_f$ and $\psi_{f0} = 1.5\psi_f$ respectively for comparison.

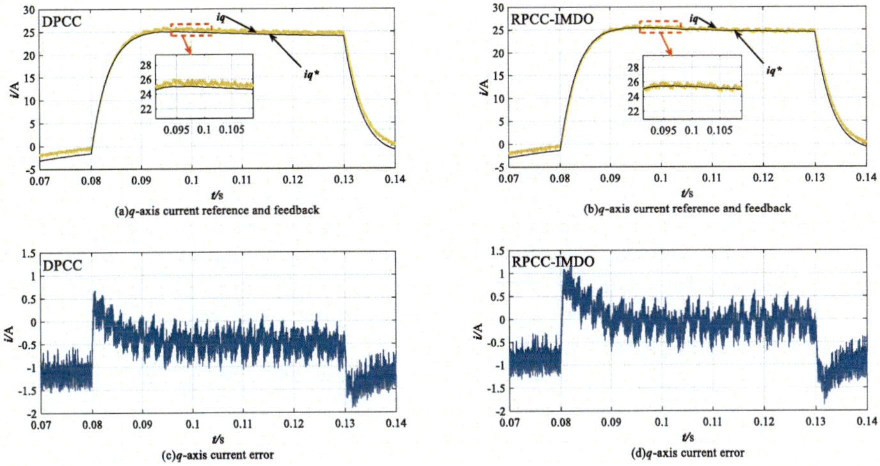

Fig. 6. Current feedback and current error on q-axis ($\psi_{f0} = 0.5\psi_f$)

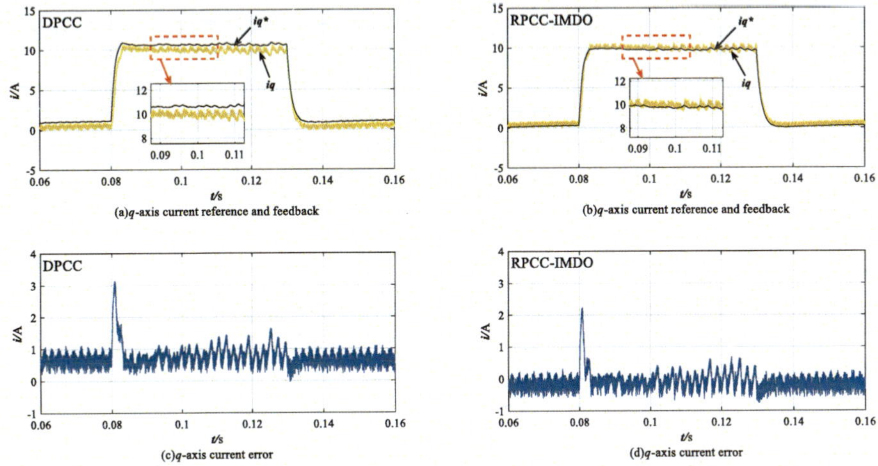

Fig. 7. Current feedback and current error on q-axis ($\psi_{f0} = 1.5\psi_f$)

Comparing Fig. 6(c)–(d), when the load changes suddenly, adding a disturbance observation compensation controller will react more quickly, improving the dynamic response capability of the current loop. At the same time, the static error will be smaller after stabilization, and the torque output will be more stable. It can be seen from Fig. 7 (c)–(d) that when the flux linkage is too large, there will always be a static difference in the q-axis current of the DPCC, especially when a sudden load torque is applied, the instantaneous static difference will be very large. In contrast, under the new method, the static error can be stabilized near 0 with less fluctuation.

6 Conclusion

According to the mathematical model of SPMSM, DPCC based on the internal model disturbance observer has been studied. The IMDO can effectively predict the disturbance, and use the observed disturbance at the next moment to perform real-time disturbance compensation for $u_{d\text{-}axis}$ and $u_{q\text{-}axis}$. Meanwhile, IMDO will not weaken the stability of traditional DPCC, it improves the robustness of the system. Simulation results verify the practicability of the proposed algorithm. In short, the addition of IMDO solves the problem of parameter sensitivity in traditional DPCC, it proposes feasible solutions for predictive control and enables DPCC to better play its unique advantages in motor control. Furthermore, IMDO will be applicable to various control systems that require disturbance observation.

References

1. Wang, G., Chen, X., Zeng, F.: Permanent magnetic servo motor's space vector control system and MATLAB Simulation. Flight Control Detect. **1**(3), 059–062 (2018)
2. You, L., Wang, Y., Wu, J.: Controlling of BLDC motor based on gray prediction. Flight Control Detect. **2**(3), 103–108 (2019)
3. Holtz, J.: Advanced PWM and predictive control-an overview. IEEE Trans. Ind. Electron. **6** (63), 3837–3844 (2016)
4. Luo, Y., Zhang, C.: A comparative experimental analysis of PMSM between deadbeat prediction current control and field-oriented control. In: 10th International Conference on Applied Energy (ICAE2018), Hong Kong, China, pp. 2488–2493 (2019)
5. Siami, M., Khaburi, D.A., Rodriguez, J.: Torque ripple reduction of predictive torque control for PMSM drives with parameter mismatch. IEEE Tran. Power Electron. **9**(32), 7160–7168 (2017)
6. Zhang, Z., Li, Z., Kazmierkowski, M.P., Rodriguez, J., Kennel, R.: Robust predictive control of three-level NPC back-to-back power converter PMSG wind turbine systems with revised predictions. IEEE Trans. Power Electron. **11**(33), 9588–9598 (2018)
7. Huan, Y., Xiong, S.: An internal model control-based observer for current loops in permanent magnet synchronous motor. Proc. CSEE **36**(11), 3070–3075 (2016)
8. Wang, L., Zhao, J., Dong, F., He, Z.: High-bandwidth and strong robust predictive current control strategy research for permanent-magnet synchronous linear motor based on adaptive internal model observer. Proc. CSEE **39**(10), 3098–3106 (2019)

9. Yin, Z., Bai, C., Du, C., Liu, J.: Deadbeat predictive current control for permanent magnet linear synchronous motor based on internal model disturbance observer. Trans. China Electro Tech. Soc. **33**(24), 5741–5750 (2018)
10. Liu, C., Luo, G., Xue, Z., Zhou, Z., Chen, Z.: A PMSM speed servo system based on internal model control with extended state observer. In: IECON 2017 - 43rd Annual Conference of the IEEE Industrial Electronics Society, Beijing, pp. 1729–1734 (2017)

Mathematical Model of Ship Collision Avoidance in Narrow Channel Overtaking Situation

Keyin Miao, Renqiang Wang(✉), Jianming Sun, and Hua Deng

Navigation College, Jiangsu Maritime Institute, Nanjing 2111709, China
wangrenqiang2009@126.com

Abstract. In order to solve the collision avoidance problem in the case of ship overtaking in a narrow water area, a ship collision avoidance model was proposed based on ship domain. On the basis of the channel planning and design specifications, the qua-ternary ship area of the narrow channel was reshaped, and a model of the minimum ship area of a ship in the waterway is constructed to further determine the safe passing distance of the ship when overtaking. Then, a mathematical model of ship steering motion based on the combination of the ship's steering collision avoidance geometric model, the collision avoidance action target was determined, that is, the closest encounter distance (DCPA) and the closest encounter time (TCPA), so that the above two collision avoidance action indicators met the actual collision avoidance (DCPA is greater than ship domain, TCPA is as large as possible). Simulation experiments showed that when using this model for overtaking and collision avoidance, the overtaking ship's trajectory of motion was in line with the actual navigation. The minimum lateral distance to the overtaken ship was 80 m and the minimum lateral distance to shallow water was 55 m.

Keywords: Ship · Collision avoidance · Overtaking · Narrow channel · Mathematical model

1 Introduction

The frequent navigation of ships in narrow waterways and the narrow channel width make it more difficult for ships to steer, resulting in collisions of ships from time to time. To solve the problem, various methods were used to carry out theoretical application research [1–3]. The article starts from the actual meeting of ships sailing in narrow waterways, and conducts research from the perspective of waterway planning. First, the safe passage distance between ships in the case of overtaking is determined according to the waterway design planning. Then, the appropriate collision avoidance model is constructed with the help of the classic collision avoidance geometric model. Last, the direction-changing collision avoidance action of the overtaking ship is solved.

M. Tavana et al. (Eds.): IISA 2020, AISC 1304, pp. 120–127, 2021.
https://doi.org/10.1007/978-3-030-63784-2_16

2 Safe Distance in Ship Overtaking

In order to ensure safety when a ship is steering in a narrow waterway, it can keep a safe distance from other sailing vessels in vicinity. When ship is sailing in restricted waters with limited width and depth, it is necessary to maintain the safe distance from the channel boundary to avoid stranding accidents. In actual navigation, the size of the safety distance is generally determined according to the navigation experience of the ship's pilot.

In the process of studying the "safe encounter distance", the concept of the ship domain [4] appeared. Technically, the boundary of the ship domain is essentially outside the radius of the Safety Distance at Closest Point of Approach (SDCPA), where SDCPA equals ship domain. This paper draws on the ideas of the quaternary ship domain [5] and uses the channel design codes to reshape the dynamic domain of ships in restricted waters. It can be obtained from the channel design specifications that the plane composition of navigation width.

The width of a single lane is:

$$W = A + 2c \tag{1}$$

The width of the double lane is:

$$W = 2A + b + 2c \tag{2}$$

The width of the track strip is:

$$A = n(L\sin\gamma + B) \tag{3}$$

where, n is the ship drift multiple, L is the design captain, γ is the wind pressure declination, c is the surplus width between the ship and the bottom edge of the channel, mainly to avoid the ship's wall effect, b is the surplus width between ships. In order to affect the interaction of ships, B is the design ship width.

Therefore, for a two-way channel, the minimum channel width on one side is: $W = A + b/2 + c$. Therefore, it can be deduced that the two-way course situation is the largest design ship, and the best position of the ship should be reserved $W_{left} = (A + b)/2$ on the left and $W_{right} = A/2 + c$ on the right. Further, according to the ship domain, the length of the right half axis of the designed ship type ship in the ship area of the channel is $W_{ss} = W_{right}$, and the length of the left half axis is $W_{ps} = W_{left}$.

Similarly, for ship sailing in the channel, there should be a channel corresponding to its ship type, the width of which can be expressed as:

$$W' = A' + c' + b' \tag{4}$$

$$A' = n(L'\sin\gamma' + B') \tag{5}$$

where, A' is the width of the track strip of any ship, c' is the safety margin after considering the effect of any ship and the wall at the edge of the channel, and b' is the inter-ship effect between any ship and the ship on the left side. After the safety margin.

Then the left and right horizontal semi-axes of the ship field are:

$$\begin{cases} R_s = A'/2 + c' \\ R_p = (A' + b')/2 \end{cases} \tag{6}$$

The stern half-axis of the ship area in the case of narrow-channel overtaking is temporarily not considered. The value of the first half axis of the ship field should mainly consider the ship's steering performance, that is, the size of the new course distance.

Considering the small steering angle of the ship in restricted waters, the speed drop is not large, the traverse distance is very small, and the curved motion in steering is ignored. In this paper, it is treated as a straight line until it turns to a new course [6]. For this reason, the new course distance can be expressed as:

$$R_{NCD} = vg\left(T + \frac{t}{2} + \frac{57.3}{K \cdot \delta}\tan\frac{\theta}{2}\right) \tag{7}$$

where, t is the steering time of the ship, δ is the rudder angle (°), θ is the heading change angle, T is the ship's follow-ability index, and K is the ship's cycle-ability index.

When actually avoiding collision, the above values should be corrected. In view of the fact that most commercial ships are sterns, the captain's correction is the maximum value and the total length of the ship is L. In addition, it is necessary to consider the shallow water situation, that is, the minimum distance r_0 between the limit position of the rich water depth and the position of the obstacle. Therefore, the revised formula for the bow half axis is:

$$R_{fore} = R_{NCD} + L + r_0 \tag{8}$$

3 Steering Geometric Model of Collision Avoidance

Steering avoidance is the most commonly used avoidance method by pilots. According to practical experience, as long as the surrounding navigable waters allow, pilots generally tend to use steer avoidance methods [8]. Suppose the steering angle adopted by the chasing ship when the target ship enters the dynamic boundary is θ. If the relative positional relationship between the two ships is unchanged, the own speed component on the x and y axes becomes:

$$\begin{cases} v'_{xo} = v'_o \sin(\varphi_o + \theta) \\ v'_{yo} = v'_o \cos(\varphi_o + \theta) \end{cases} \tag{9}$$

After the ship was redirected, the relative speed and course between the two ships became:

$$\begin{cases} v'_{xR} = v_{xT} - v'_{xo} \\ v'_{yR} = v_{yT} - v'_{yo} \end{cases} \tag{10}$$

The relative speed v'_R is:

$$v'_R = \sqrt{v'^2_{xR} + v'^2_{yR}} \tag{11}$$

The relative course φ'_R is:

$$\varphi'_R = \arctan \frac{v'_{xR}}{v'_{yR}} + \alpha \tag{12}$$

where, $\alpha = \begin{cases} 000°, v'_{xo} \geq 0, v'_{yo} \geq 0 \\ 180°, v'_{xo} < 0, v'_{yo} < 0 \\ 180°, v'_{xo} \geq 0, v'_{yo} < 0 \\ 360°, v'_{xo} < 0, v'_{yo} \geq 0 \end{cases}$.

Relative bearing of target ship:

$$\theta'_T = \alpha'_T - (\varphi_o + \Delta\varphi) \tag{13}$$

After changing the direction, the relative motion line equation becomes:

$$y = x \cdot \cot\left(\varphi'_R - \varphi_o - \Delta\varphi\right) + \left(y_{t2} - x_{t2} \cot\left(\varphi'_R - \varphi_o - \Delta\varphi\right)\right) \tag{14}$$

The equation of the relative motion line of own ship relative to other ships after redirection can also be obtained by the method.

4 Ship Steering Motion Equation

The rules for collision avoidance stipulate that the ship giving way should take appropriate steering or variable speed avoidance action to make way for the direct-going ship, and should satisfy such action so that the two ships can pass the safe distance. Regardless of whether the steering is combined with variable speed actions, only steering or only slowing down, this belongs to the maneuvering action of the ship

[9]. The equation of the three-degree-of-freedom maneuvering motion of a separate ship is:

$$\begin{cases} m \cdot (\dot{u} - vr) = X_H + X_P + X_R \\ m \cdot (\dot{v} + ur) = Y_H + Y_P + Y_R \\ I_z \cdot \dot{r} = N_H + N_P + N_R \\ \dot{\phi} = r \\ \dot{x}_{G0} = u \cdot \cos \phi - v \cdot \sin \phi \\ \dot{y}_{G0} = u \cdot \sin \phi + v \cdot \cos \phi \end{cases} \tag{15}$$

where, m represents the mass of the ship, I_z represents the moment of inertia of the ship, H represents the hull, P represents the propeller, R represents the rudder, φ represents the course of the ship, u represents the longitudinal partial velocity of the ship, v represents the lateral partial velocity of the ship, and r represents the angular velocity of the ship's bow. Because the ship not only moves forward, but also moves left and right, the speed of the ship is $V = \sqrt{u^2 + v^2}$.

So, combined with the above-mentioned ship steering collision avoidance geometric model, the target parameter values (DCPA and TCPA) of the steering collision avoidance action can be further obtained.

5 Computer Experiment

The ship steering and collision avoidance model constructed in the narrow watercourse overtaking situation is used to conduct simulation experiments [10]. The initial conditions of the simulation are as follows: the width of the narrow channel is 400 m, the effective width of the channel is 350 m, the two ships travel along the channel, the initial position of the overtaking ship is at the axis −30 m, and the initial position of the overtaken ship is at the axis 250 m. After 35 s, the overtaking ship took steering action, and after 180 s, the steering was stable, and the lateral distance between the two ships remained above 80 m. The trajectories of the two ships in Fig. 1 are in line with the actual situation of collision avoidance maneuvering in the case of narrow-channel navigation vessels overtaking; at the same time, in the whole process, the overtaking vessel can maintain a distance of more than 55 m from the shore shallow water. The curves of the control rudder angle and course of the overtaking ship are shown in Fig. 2 and Fig. 3. It can be seen from the figure that the real-time curve of the ship's control rudder angle and heading is in line with the actual evasive maneuvering situation of the ship.

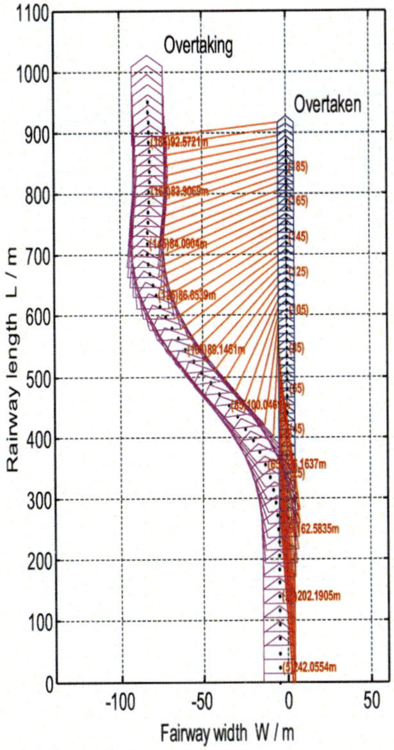

Fig. 1. The trajectory of the two ships in overtaking

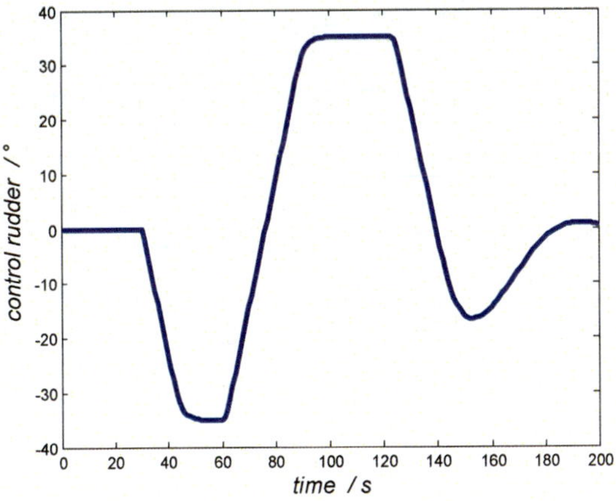

Fig. 2. Rudder angle of the overtaking ship

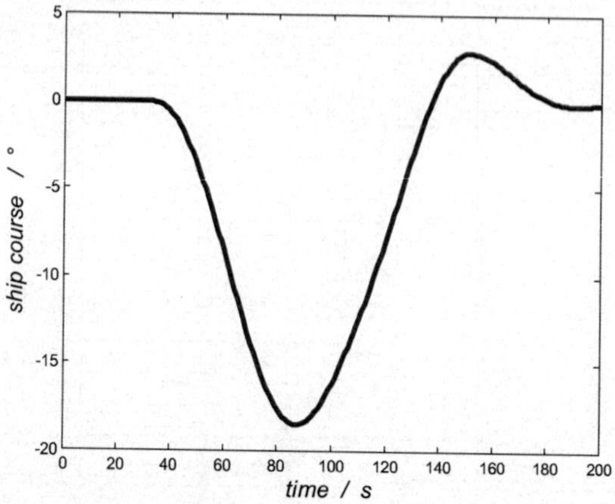

Fig. 3. The course of the overtaking ship

6 Conclusion

Reshape the quaternary of ship field in narrow waterway on the basis of the channel planning and design specifications, and construct a model of the minimum ship field that conforms to the ship sailing in the channel; thereafter, based on the mathematical model of ship maneuvering motion, combined with the geometric model of ship steering collision avoidance The goal of collision avoidance action is determined, and a ship collision avoidance model under narrow water channel overtaking situation is constructed. This model can solve the problem of actual collision avoidance maneuvering in the case of ships overtaking in narrow waters. Simulation experiments show that when the model is used for pursuit and collision avoidance, the trajectory of the chasing ship's chasing movement is in line with the actual sailing. The minimum lateral distance to be kept with the chased ship is 80 m, and the minimum lateral distance to the shallow water is 55 m.

Acknowledgments. This work was supported by Natural Science Research Project of Universities in Jiangsu Province under Grant No. 19KJD580001, 19KJA150005 and 18KJB580003.

References

1. Cao, J.-F., Ling, Z.-H., Gao, C., et al.: Collision avoidance and formation control for multi-agent based on swarming. J. Syst. Simul. **26**(3), 562–566 (2014)
2. Zhao, Y., Jiao, L., Zhou, R., et al.: UAV formation control with collision avoidance using improved artificial potential fields. In: Proceedings of the 36th Chinese Control Conference, pp. 6219–6224 (2017)
3. Huayan, P.U., Ding, F., Li, X., et al.: Maritime autonomous collision avoidance in a dynamic environment based on collision cone of ellipse. Chin. J. Sci. Instrum. **38**(7), 1756–1762 (2017)

4. Wang, R.: Research on the decision-making of ship collision avoidance based on the field of ships. Dalian Maritime University (2012)
5. Wang, N.: A novel analytical framework for dynamic quaternion ship domains. J. Navig. **62** (4), 265–281 (2013)
6. Xu, H.: Research on the decision-making of collision avoidance in narrow channels based on shipbuilding in mutual seeing. Dalian Maritime University (2017)
7. Zhao, Y., Wang, R., Zhang, X., et al.: Modeling of Chinese east China sea merchant ship steering and avoidance fishing vessels based on ship maneuvering. China Navig. **38**(4), 64–67 (2015)
8. Wang, R., Zhao, Y., Xie, B.: Mathematical model of ship dynamic steering collision avoidance action. J. Dalian Marit. Univ. **40**(1), 17–20 (2014)
9. Jia, X., Yang, Y.: Mathematical Model of Ship Motion. Dalian Maritime University Press, Dalian (1999)
10. Zhao, Y., Wang, R.: Sigmoid function based path modelling of obstacle avoidance for ship in restricted water. MATEC Web Conf. **232**, 25 (2018)

Prediction and Analysis of Air Ticket Based on ARIMA Model

Qingyun Chi(✉), Menglin Liu, and Bin Yang

School of Information Science and Engineering, Zaozhuang University, Zaozhuang, China
cqysd@163.com

Abstract. Due to the high volatility, high randomness and vulnerability to many factors, the prediction of air ticket price has become a very challenging problem. Considering the characteristics of ticket prices, this paper utilizes Python to crawl the information of all flights from Shanghai to Beijing in one year on an e-commerce platform through a web crawler. The rules of ticket prices are analyzed based on the ARIMA model, so as to make reference to the prediction of future ticket prices.

Keywords: ARIMA model · Crawler · Time series analysis · Ticket price

1 Introduction

Most travellers usually choose to take the high-speed rail. Of course, if this is the most cost-effective for the closer journey, the price of buying air tickets in advance may be cheaper than the high-speed rail ticket for the long-distance travellers [1, 2]. To crawl, analyze and predict the ticket data in advance is more convenient and cost-effective to travel [3, 4].

In recent years, with the liberalization of China's civil aviation industry's independent pricing policy, airlines have become freer to price their flight products, which has made ticket price setting strategies more flexible and diversified. Airlines will formulate complex pricing strategy to ensure that its revenue is maximized [5].

Airlines often adjust ticket prices dynamically according to the actual flight attendance and passenger demand, and this information is not available externally [6–8]. The information of influencing factors that can be obtained is very limited, and the air ticket price itself has great volatility and randomness. Many factors make the air ticket price prediction become a very challenging problem. When the travel demand of passengers matches the capacity of airlines, it is very meaningful to study the ticket price [9–11].

The time series is a historical value that records a thing at the same time interval within a certain period of time, usually based on the year, month, and day [12–14]. By discovering the inherent laws and trends of these data, the trend of the data will be analyzed. Using the collected real ticket price data set of an e-commerce platform, through data analysis, it is found that the ticket price data has the characteristics of no

M. Tavana et al. (Eds.): IISA 2020, AISC 1304, pp. 128–135, 2021.
https://doi.org/10.1007/978-3-030-63784-2_17

single cycle of a single sequence and the regularity of multiple consecutive flight sequences.

For non-stationary time series, several times of scoring are used to turn it into a stationary series. Then use the ARIMA model to model the stationary sequence, then inversely transform it into the original sequence, and determine which model the random sequence is suitable for according to the statistical characteristics of the autocorrelation function and partial autocorrelation function, and then determine the model order.

2 The Process of Establishing ARIMA Model

2.1 The Modeling Process is Divided into the Following Three Steps

Calculate the autocorrelation function and partial autocorrelation function of the sample sequence check its censoring, tailing and periodicity, and determine which model form the sequence is more suitable for.

After determining the applicable model for the sequence, the parameters of the sample are estimated. First make a preliminary estimate, then make an accurate estimate under a certain indicator.

In order to test the model, it is necessary to test the model. The residual sequence is tested to determine whether it is a white noise sequence. If the residual is not a white noise sequence, it is necessary to return to the first or second step for reconstruction or parameter selection.

2.2 Model Recognition and Order Determination

ARIMA (p, d, q) is one of the time series prediction analysis methods. The full name of ARIM is Autoregressive Integrated Moving Average Model, which involves three parameters. Parameter autoregressive terms are represented by p; The number of moving average terms is represented by q, while the number of differences when the time series becomes stationary is represented by d.

The problem of identifying and determining the order of the model is mainly to determine the three parameters p, d and q. The order d of the difference is usually obtained by observing the illustration, the first or second order. Here we mainly introduce the determination of p and q. Let's start with two functions. The property of these two functions can be used to determine the type of model.

2.3 Autocorrelation Function

The autocorrelation function (ACF) describes the linear correlation between the observed values of time series and the observed values in the past. The calculation formula is as follows:

$$ACF(k) = \rho_k = \frac{cov(y_{t,}y_{t-k})}{Var(y_t)} \quad (1)$$

Where, k represents the number of lagging periods. If $k = 2$, it represents y_t and y_{t-k}.

2.4 Partial Autocorrelation Function

The partial autocorrelation function (PACF) describes the linear correlation between the expected past observations of time series under the condition of given intermediate observations.

Describe the correlation between y_t and y_{t-3}, When k is 3. But y_{t-1} and y_{t-2} is also affect the correlation. The impact is eliminated by PACF and included in ACF.

2.5 Parameter Estimation

When using the truncation of ACF and PACF to judge the ARMA model, the order of P and Q cannot be determined. In order to determine the order of P and Q more accurately, it must be combined with the commonly used order determination criteria. The most widely used are AIC and BIC.

2.6 Akaike Information Criterion (AIC)

The AIC criterion is a weighting function of the fitting accuracy and the number of parameters. The model that minimizes AIC function is considered as the optimal model.

Generally, AIC is defined as:

$$AIC = 2\mathrm{K} - 2\ln(\mathrm{L}) \tag{2}$$

In the above formula (2), parameter K represents the number of parameters of ARIMA model, and parameter L represents the likelihood function. In a group of models, the minimum AIC value is usually chosen as the best model.

The likelihood function is usually used to reflect the difference between the two models. The greater the difference, the greater the likelihood function. However, if the likelihood function difference is not too obvious, the first parameter of the model plays a decisive role. In this sense, the fewer model parameters, the better.

In general, as the model complexity increases, the value of L increases with the value of K. That is, when the value of parameter K increases, the value of likelihood function L also increases, leading to a decrease in AIC. However, when the value of K reaches a certain value, the growth of likelihood function will slow down, eventually leading to the increase of AIC value. In this case, the model will be too complex, resulting in over-fitting of the model. Therefore, AIC can not only improve the model fit, but also reduce overfitting by introducing penalty terms to minimize model parameters.

2.7 Bayesian Information Criterion (BIC)

There are some shortcomings in AIC criterion. When the sample size is large, the information provided by the fitting error in THE AIC criterion will be amplified, and the penalty factor of the number of parameters is always 2, independent of the sample size. Therefore, when the sample size is large, the model selected by AIC criterion will not converge, which is the shortcoming of AIC criterion. BIC is chosen to make up the deficiency of AIC.

The penalty term involved in BIC model is related to the number of model parameters, but the penalty term is larger than AIC model. Because the BIC model fully considers the number of samples, if we select a large enough sample, we can avoid the model being too complex due to high accuracy. BIC standard definition is as shown in Formula (3):

$$BIC = \mathrm{kln(n)} - 2\mathrm{ln(L)} \tag{3}$$

In Formula 3, parameter k represents the number of model parameters, and sample number and likelihood function are respectively represented by parameter n and l.

The expression Kln(n) represents the penalty term. If the dimension is too large and the training sample is small, the excessively high dimension can be effectively avoided.

3 Empirical Research

3.1 Data Source and Description

Based on the monthly air ticket sales data of a certain e-commerce platform in recent years, this paper establishes a time series model, which is recorded as the original data in the appendix (Fig. 1).

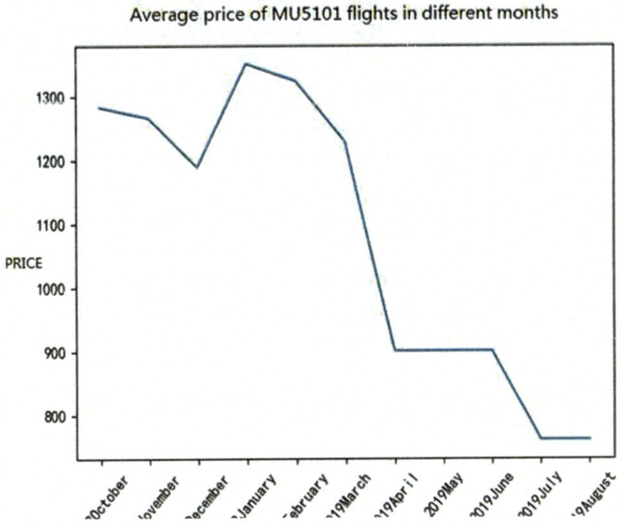

Fig. 1. Average price of MU5101 flights in different months

The price fluctuation of each flight is different. Some flights have obvious price fluctuation, while others do not. The research result makes the model result in the actual situation is not significant. Some flight prices always have obvious fluctuations and the data characteristics are relatively obvious. In this case, the results of the test model are meaningful and more generally applicable. Taking MU5101 flight as an example, the change trend of ticket price is shown in the graph, and it can be roughly seen that a certain ticket price has obvious non-stationarity through the graph, showing a certain slope downward trend.

According to the price changes can be judged out of the off-season peak season, the big festival near will almost rise, winter vacation, summer vacation, Chinese New year during the price is also higher.

3.2 Modeling and Prediction

In order to determine the model form and lag order, ACF and PACF were used to complete the identification. For price, the first-order difference, combined with ACF and PACF graphs, suggests that the p value of autoregressive order is 0, and the q value of moving average order is 0 (Figs. 2 and 3).

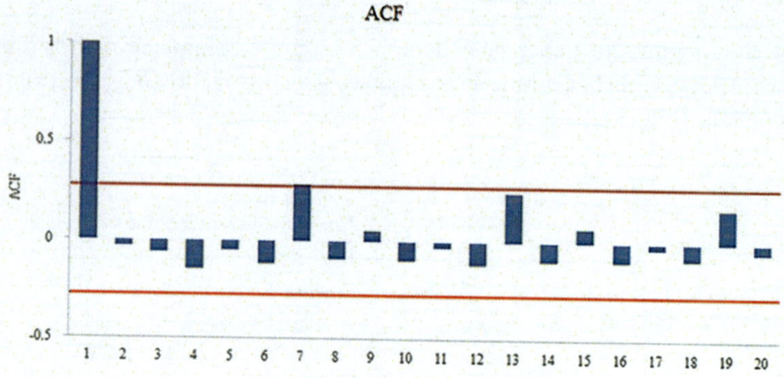

Fig. 2. ACF of air ticket

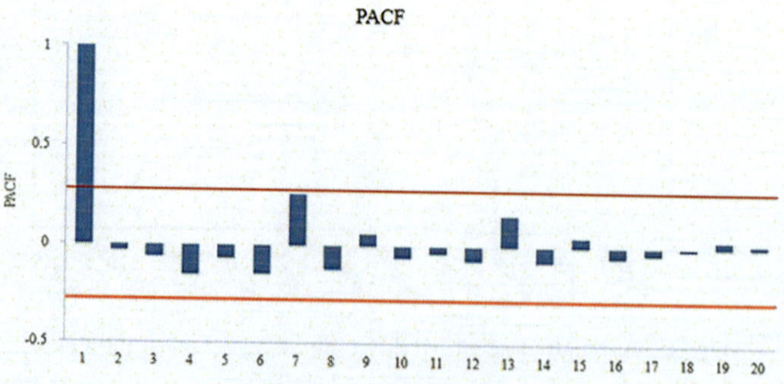

Fig. 3. PACF of air ticket

Model recognition refers to the process of determining whether the time series analysis is applied to AR(p), MA(q) or ARMA (p, q), In order to determine the order of P and Q more accurately, it is combined with the commonly used order determination criteria.

The following table shows the results of the model construction, including model parameters, Q statistics and information criteria:

Table 1. ARMA (2,1,3) model parameter table.

Item	Symbols	Values
Constant item	c	−3.349
AR	$\alpha 1$	−0.991
	$\alpha 2$	0.006
	$\alpha 3$	0.793
MA	$\beta 1$	−0.982
	$\beta 2$	−0.811
Q	Q6(p)	2.068(0.150)
	Q12(p)	8.611(0.197)
	Q18(p)	10.918(0.536)
	Q24(p)	11.693(0.863)
	Q30(p)	12.967(0.967)
AIC	AIC	688.886
	BIC	702.129

* $p < 0.05$ ** $p < 0.01$

The MRIMA model requires the residual of the model to be white noise, that is, the residual has no autocorrelation. White noise can be tested by Q statistics.

For example, in Table 1, Q6 is used to test whether the first 6 order autocorrelation coefficient of residual error satisfies white noise. If the corresponding P value is greater than 0.1, it indicates that the test satisfies white noise, and Q6 can be directly decomposed under normal circumstances.

The AIC and BIC values of information criterion were used to compare multiple analysis models. The lower the two values, the better. If analyzed for many times, the change of the two values can be compared to comprehensively explain the optimization process of model construction.

The Price, combination AIC, by modeling and comparing multiple potential alternative models, the optimal model is finally found as follows: $y(t) = y(t) = -3.349 - 0.991*y(t-1) + 0.006*y(t-2) + 0.793*\varepsilon(t-1) - 0.982*\varepsilon(t-2) - 0.811*\varepsilon(t-3)$.

From the results of Q statistics, if the p value of Q6 is greater than 0.1, the original hypothesis can not be rejected at the level of significance of 0.1. The residual error of the model is white noise, and the model basically meets the requirements (Fig. 4).

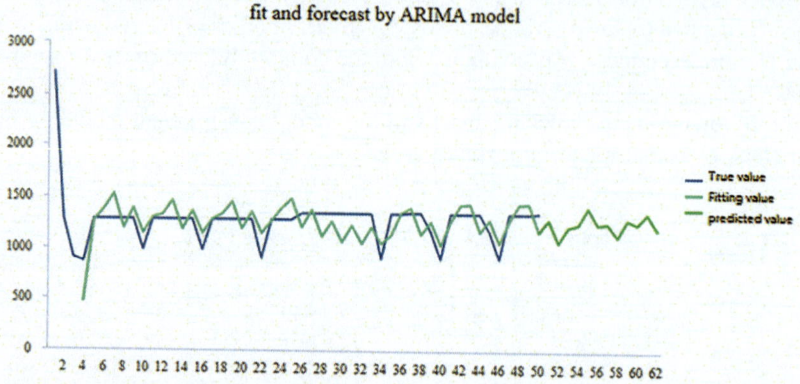

Fig. 4. Fit and forest by ARIMA model

First, the figure above shows the prediction of the latest 12 data periods of the model.

Second, the actual values and fitting values of the last 50 phases of the original time series are only shown in the figure (if the original time series is less than 50 phases, the actual sequence data is displayed).

4 Summary

The airline's fare strategy can be found regularly, which is easier to find when a certain amount of data is crawled and visualized. Based on the time series analysis of the characteristics of the air ticket price of an e-commerce platform, this paper obtains the ARIMA (p, D, q) model with a high degree of fit. It is found that the air ticket price of this airline fluctuates around an average level, including a certain long-term trend.

References

1. Tziridis, K., Kalampokas, T., Papakostas, G., Diamantaras, K.I.: Airfare prices prediction using machine learning techniques. In: 25th European Signal Processing Conference (EUSIPCO), pp. 1036–1039 (2017)
2. Zheng, X., Niu, K., Ma, J., Zhang, Z., Li, X., Li, Q.: A prediction algorithm for airfare based on time series. In: 7th International Conference on Computer Engineering and Networks, pp. 1–7 (2017)
3. Udapure, T.V., Kale, R.D., Dharmik, R.C.: Study of web crawler and its different types. IOSR J. Org. **16**(1), 01–05 (2014)
4. Yu, J., Li, M., Zhang, D.: A distributed web crawler model based on cloud computing. In: Proceedings of the 2nd Information Technology and Mechatronics Engineering Conference, pp. 276–279 (2016)
5. Huang, Q., Li, Q., Yan, Z., Fu, H.: A novel incremental parallel web crawler based on focused crawling. J. Comput. Inf. Syst. **9**(6), 2461–2469 (2013)

6. Kumar, M., Bhatia, R.: Design of a mobile web crawler for hidden web. In: International Conference on Recent Advances in Information Technology, pp. 186–190 (2016)
7. Liu, C.H., Wu, C.J., Chen, H.M.: Testing of AJAX-based web applications using hierarchical state model. In: IEEE 13th International Conference on E-Business Engineering, pp. 250–256 (2016)
8. Cao, S., Yuan, Z., Zhang, C., Zhao, L.: LOS classification for urban rail transit passages based on passenger perceptions. J. Transp. Syst. Eng. Inf. Technol. **9**(2), 99–104 (2009)
9. Xu, X., Han, F., Dai, F.: Design and implementation of air traffic flow statistics and prediction system. J. Civ. Aviat. Univ. China **23**(4), 1–5 (2005)
10. Chen, Y., Deng, S., Liu, Y.: Design and implementation of film ticket booking system based on J2ME. Comput. Mod. **1**(7), 133–135 (2009)
11. Gu, Z., Wang, S., Zhao, Y.: Prediction model of air ticket price based on time series. J. Civ. Aviat. Univ. China **31**(2), 80–84 (2013)
12. Zhang, W.: Design and implementation of air ticket booking system based on web. Wirel. Internet Technol. **24**, 60–61 (2015)
13. Hang, B.: Design and implementation of cinema online booking system. In: 2011 International Symposium on Computer Science and Society, pp. 196–199 (2011)
14. Groves, W., Gini, M.: On optimizing airline ticket purchase timing. ACM Trans. Intell. Syst. Technol. **7**(1), 1–28 (2015)

A General Model for Publicly Verifiable Secret Sharing

Wei Zhao(✉) and Feng Li

Science and Technology on Communication Security Laboratory,
Chengdu 610041, People's Republic of China
zhaowei9801@163.com

Abstract. Publicly verifiable secret sharing (PVSS) is of remarkable significance for confidential data processing and storage in complex interactive systems. In this correspondence subtly deviating properties of preceding PVSS models are studied and a general PVSS model is proposed to satisfy possible requirements without ambiguity. The new model comprises a secret sharing algorithm, a secret recovering algorithm, cryptosystems for shareholders and relative zero-knowledge protocols. Two kinds of security for PVSS are formally defined: all-or-nothing security and indistinguishability security. It is also justified that the zero-knowledge protocol in the distribution phase does not influence security for the new PVSS model.

Keywords: Publicly Verifiable Secret Sharing · Zero-knowledge protocol · Indistinguishable

1 Introduction

Secret sharing and its extensions play an important role in a number of scenarios of big data and modern communication network, e.g., secure multiparty computation, electronic voting [Sch99], electronic cash, software key escrow [Mic92], distributed key generation [FS01, GJKR07, ZI03], confidential data storage [Tra19]. Particularly, because of its information-theoretic security secret sharing is a promising candidate means to protect confidentiality and integrity of outsourcing digital data [BMV06], considering the menace of quantum computing due to the fact that Shor's algorithm breaks popular number-theory-based public-key cryptography [Sho97]. Secret sharing, independently invented by Shamir [Sha79] and Blakley [Blk79], is a mechanism which distributes a secret among a group of shareholders. The shares in some pre-defined access set are combined together to recover the secret.

Verifiable secret sharing (VSS) was first proposed by [CGMA] to deal with a dishonest dealer who may send incorrect shares, or dishonest shareholders who are free to release false shares at the reconstruction phase. Auxiliary information ensures that even if the dealer is malicious, the shareholders can later collectively find a unique secret. Feldman [Fel87] and Pedersen [Ped92] respectively constructed practical and effective VSS schemes. Actually, their VSS schemes also implicitly detect dishonest behavior of shareholders. Anyhow, the conventional VSS schemes employ a complain-and-reply phase, with a burden of communication cost to check dishonest behavior.

M. Tavana et al. (Eds.): IISA 2020, AISC 1304, pp. 136–143, 2021.
https://doi.org/10.1007/978-3-030-63784-2_18

Further, Goldreich [Gldr04] formally defined VSS as a secure multiparty computation protocol to compute random functions depending on some secret sharing scheme.

Stadler [Std96] explicitly introduced the concept of publicly verifiable secret sharing (PVSS) and publicly verifiable encryption (PVE), in which any one can check that a secret is correctly distributed. Actually, the scheme by [CGMA] implicitly has this special property. Stadler [Std96] gave two specific PVSS schemes in which dishonest shareholders are also detected, one for discrete logarithm and the other for RSA root problem. The schemes in [Std96] use Σ-protocols for public verification and is non-interactive à la Fiat-Shamir.

Fujisaki and Okamoto [FO98] gave a formal model of (t, n)-threshold PVSS with all-or-nothing security, though the model did not detect malicious shareholders. They also constructed a specific PVSS scheme using commitment schemes and Σ-protocols. This scheme is secure based on two assumptions. One is the modified RSA assumption, and the other conjectures that without public verification an probabilistic polynomial-time adversary cannot determine whether transmitted data is correct distribution of a unique secret.

Schoenmakers [Sch99] described a model for non-interactive PVSS, which explicitly required that a shareholder should prove that her released share is correct. A specific PVSS scheme is given based on the computational Diffie-Hellman (CDH) assumption and the random oracle assumption, and it is stated to leak no partial information of secrets based on the decisional Diffie-Hellman (DDH) assumption and the random oracle assumption.

Young and Yung [YY01] gave a specific PVSS scheme for discrete logarithm where shareholders can use different cryptosystems.

Boudot and Traoré [BT99] gave PVSS schemes for discrete logarithm and integer factoring respectively with fast and delayed recovery, and announced to have improved efficiency of previous schemes in [BG97, FO98, Mao98, Std96, YY99].

The rest of this paper is as follows. In Sect. 2 we review previous PVSS definitions and PVSS schemes, analyze their subtle differences and give a formal mathematical definition for PVSS. Security for our new PVSS model is also characterized, including all-or-nothing security and indistinguishability security. It is also proved that our security definitions are irrelevant to the zero-knowledge protocol in the distribution phase. In Sect. 3 we analyze the advantages of the new PVSS model. Finally, in Sect. 4 we give a summary.

2 A PVSS Model

The communication channels are public and authenticated, e.g. by public key infrastructure. Each shareholder announces her public key cryptosystem, and each cryptosystem has a zero-knowledge protocol to prove correct decryption. The dealer assigns a secret sharing scheme and a zero-knowledge protocol which ensures him to prove correct distribution of a secret.

A polynomial-time recognized binary relation $R \subseteq \{0,1\}^* \times \{0,1\}^*$ is used to define restriction on secrets. For $(s, w) \in R$, s is a secret and its image w is its constraint. There exists a polynomial $p(\cdot)$ such that $|w| \leq p(|s|)$ for all $(s, w) \in R$. It is efficient to compute an image w for secret s in some preceding PVSS schemes, but the relation R is not necessarily functional. The relation R is said to be non-traceable if for every probabilistic polynomial-time algorithm ADV, every positive polynomial poly$(\cdot)$, and sufficiently large $|s|$,

$$Pr[\mathrm{ADV}(w, |s|) = s : (s, w) \in R] < 1/\mathrm{poly}(|s|).$$

The conditional probability takes over internal coin tosses of ADV and random variables s, w. Some relations are shown below.

Discrete Logarithm. Let G be a cyclic group of prime order and let g be one of its generators. $R_{\mathrm{DL}} = \{(s, w) \in \mathbb{Z} \times G : g^s = w\}$.

Integer Factoring. $R_{\mathrm{IF}} = \{(s, w) \in \mathbb{N} \times \mathbb{N} : s \,\mathrm{divides}\, w\}$.

Null. R_{Null} is an empty set, i.e. there is no restriction on secrets.

Let $I = \{1, 2, \cdots, n\}$ be the set of shareholders and $i \in I$ represents the i-th shareholder. Further, A is an access structure, i.e. A is constituted by subsets of I and every subset in A is qualified to recover the secret. Additionally, A is monotone, i.e. $A_1 \in A$ and $A_1 \subseteq A_2$ imply $A_2 \in A$.

SHA is a probabilistic polynomial-time algorithm generating secret shares. The input is a secret s and the output is a public share psh and secret shares s_i, $i \in I$.

$$SHA(s) = \left(psh, \{s_i\}_{i \in I}\right). \tag{1}$$

REC is a deterministic polynomial-time algorithm reconstructing the secret. Any qualified set of shareholders are able to obtain the secret. If Eq. (1) holds, then

$$REC\left(psh, \{s_i\}_{i \in T}\right) = s, \forall T \in A. \tag{2}$$

Each shareholder $i \in I$ registers a public key cryptosystem. For simplicity we denote by E_i the public key and the encryption algorithm, and by D_i the secret key and the decryption algorithm. The dealer encrypts a share using its corresponding cryptosystem,

$$E_i(s_i) = c_i, \forall i \in I. \tag{3}$$

The cryptosystem of the i-th shareholder defines a language $L_i = \{(x, c) : D_i(c) = x\}$. Let ZK_i be a zero-knowledge protocol for the language L_i. The prover is the i -th shareholder and anyone can be the verifier. A correct common input is a cipher with its decryption like $(D_i(c), c)$, and the auxiliary input for the prover is D_i.

When distributing a secret s, the dealer operates as (1) and (3). For convenience we denote the composite algorithm of (1) and (3) by

$$DIS(s, r) = \left(psh, \{c_i\}_{i \in I}\right), \tag{4}$$

where r is concatenation of random inputs used in algorithm SHA and in each encryption (3).

Considering together the secret sharing scheme, the restriction relation R and cryptosystems for shareholders, we define a language L characterizing data distributed by an honest dealer:

$$L = \left\{ \left(x, w, \{x_i\}_{i \in I}\right) : \exists s, \text{s.t.} \forall T \in A, \text{REC}\left(x, \{D_i(x_i)\}_{i \in T}\right) = s \, \text{and} \, (s, w) \in R \right\}. \tag{5}$$

Let ZK_L is a zero-knowledge protocol for the language L. In our PVSS model the prover is the dealer and everyone can be the verifier. A correct common input is data outputted from the algorithm DIS and the auxiliary input for the prover is s and r in Eq. (4).

Let $ppar$ denote all public parameters of the PVSS scheme, including public keys $E_i, i \in I$ and parameters of the secret sharing scheme to operate SHA and REC. The algorithm DIS depends on $ppar$. The sizes of all parameters are related, and hence we use a common security parameter k, e.g. k is security parameter for all cryptosystems and the secret size is no less than k.

Then $\left(R, \text{SHA}, \text{REC}, ZK_L, \{E_i, D_i, ZK_i\}_{i \in I}\right)$ constitute a PVSS scheme with access structure A under the relation R. It works as below.

Setup. Choose a large enough security parameter k. The restriction relation R and the secret sharing scheme are defined. Every shareholder sets up her public key cryptosystem. We denote this phase by $\text{GEN}(k) = \left(ppar, \{D_i\}_{i \in I}\right)$.

Distribution. To distribute a secret s under its restriction w, the dealer operates the algorithm DIS as (4). Then he publishes w and the public share psh, and also sends c_i to the i-th shareholder. If the dealer has to prove a correct distribution, then he runs the zero-knowledge protocol ZK_L.

Reconstruction. Every shareholder decrypts the cipher she received, i.e. $D_i(c_i) = \tilde{s}_i, i \in I$. Then the i-th shareholder releases sei and proves $(\tilde{s}_i, c_i) \in L_i$ by the zero-knowledge protocol ZK_i. If succeeding shareholders constitute a qualified set T in A, they pool their shares and recover the secret as Eq. (2).

If all zero-knowledge protocols, i.e. ZK_L and $\{ZK_i\}_{i \in I}$, are non-interactive, the PVSS scheme is called non-interactive.

3 PVSS Security

In a convectional VSS model private channels are available and hence information security is feasible. However, communication channels for PVSS are public, thus an adversary with unbounded computational power is always able to access to secrets. Therefore, it is adequate that an adversary against PVSS is modeled by a probabilistic polynomial-time Turing machine.

No shareholder succeeds in cheating in the reconstruction phase since every released share is checked by $ZK_i, i \in I$. Malicious shareholders should be detected and punished. Thus, we focus on security of the distribution phase. As for distribution, it is only in the zero-knowledge protocol ZK_L that there possibly exists interactive communication. Additionally, the zero-knowledge protocol ZK_L is of no help to an adversary since the verifier can be anyone, including every shareholder. Then it is rational to only consider a passive static adversary and define security without influence by ZK_L.

Security of PVSS is related to all components: the restricting relation R, the secret sharing scheme and cryptosystems for shareholders. However, even if every component is secure, the PVSS may be vulnerable. An extreme example is that the image of a secret under the restriction relation is a cipher of the secret by cryptosystem of a shareholder. In that case, the adversary has only to corrupt one shareholder to gain the secret.

Definition 3.1. A PVSS scheme is **all-or-nothing secure** if for every probabilistic polynomial-time algorithm ADV, every subset $T \notin A$, every positive polynomial $\text{poly}(\cdot)$, and all sufficiently large k,

$$Pr\left[\text{ADV}\left(\text{DIS}(s,r), w, \{D_i\}_{i\in T}\right) = s : \text{GEN}(k) = \left(ppar, \{D_i\}_{i\in I}\right), (s,w) \in R\right] < 1/\text{poly}(k),$$

where the probability takes over the internal coin tosses of ADV and random variables s, w, r.

Furthermore, indistinguishability is defined by the following game between an adversary and a challenger.

(1) The challenger generates parameters: $\text{GEN}(k) = \left(ppar, \{D_i\}_{i\in I}\right)$. The adversary chooses a group of shareholders $T \notin A$. The challenger gives $ppar$, $\{D_i\}_{i\in T}$ to the adversary and retains other secret parameters $\{D_i\}_{i\notin T}$.
(2) After performing some operations, e.g. encryption, the adversary submits to the challenger two distinct secrets s_0, s_1 and w satisfying $(s_0, w) \in R$ and $(s_1, w) \in R$.
(3) The challenger uniformly selects at random a bit $\sigma \in \{0, 1\}$ and a random bit string r, operates the algorithm DIS, and sends $\text{DIS}(s_\sigma, r)$ to the adversary.
(4) The adversary is free to perform any number of additional computation and finally outputs a guess for the value of σ.

A PVSS scheme is indistinguishable if any probabilistic polynomial-time adversary has only a negligible advantage to win the game.

In a formal way, indistinguishability is defined as below.

Definition 3.2. A PVSS scheme is **indistinguishable** if for every probabilistic polynomial-time algorithm ADV, every polynomial-time algorithm FIND, every subset $T \notin A$, every positive polynomial $\text{poly}(\cdot)$, and sufficiently large k,

$$\left| Pr\left[\text{ADV}\left(\text{DIS}(s_\sigma, r), w, \{D_i\}_{i\in A}, y\right) = \sigma : \begin{array}{c} \text{GEN}(k) = \left(ppar, \{D_i\}_{i\in I}\right); \\ \text{FIND}\left(ppar, \{D_i\}_{i\in T}, y\right) = (s_0, s_1, w) \\ s.t. (s_0, w) \in R, (s_1, w) \in R; \\ \text{A random bit } \sigma \in \{0,1\}; \end{array}\right] - \frac{1}{2} \right| < 1/\text{poly}(k),$$

where input y is random bits for the algorithm FIND. The conditional probability takes over internal coin tosses of ADV and random variable r.

Remark: The security definitions only make sense for PVSS under a non-traceable relation. Otherwise, too much information of the secret is leaked by the relation R. In addition, if under the restriction relation an image determines a unique secret, e.g., the discrete logarithm relation, then indistinguishability is not feasible. It is also noticed that an indistinguishable PVSS scheme is necessarily all-or-nothing secure.

4 Advantages

The new model given in this article generally proceeds involved ingredients of available PVSS. The goal of PVSS is to distribute a secret and recover it with provable correctness. Since Stadler [Std96] introduced PVSS and PVE, a variety of schemes followed. However, they have underlying subtle differences and hence meet different goals. Our PVSS model definitely and clearly combine those aspects in a formal way, avoiding ambiguity and meeting requirements in different applications.

Controllability. A PVSS scheme is said to be controllable if the dealer is able to legally distribute an assigned secret, i.e. the dealer distributes a given secret and honest shareholders can recover the same secret. A universal PVSS model should be controllable. Actually, PVSS defined by [Std96] is implicitly controllable. Anyhow, in some other scenarios like [Sch99], the dealer randomly generates a secret by themselves and fails to distribute an assigned secret. Thus, non-controllable PVSS, in which the dealer has no complete control on the secret to be distributed, may work in environments where the secret comes only from a pseudorandom generator of the dealer, or may be used to obtain forward security in special cryptographic applications.

Restriction. In the scheme by [Sch99] the secret is free, while secrets are subject to various constraints in other schemes. For example, the secret may be a discrete logarithm in a cyclic group, a factor of a large RSA modulus, a signature of a known message, or the secret key with respect to a known public key of a given cryptosystem. Furthermore, restriction over secrets exerts influence on PVSS security.

Verifiable Recovery. In the reconstruction phase, shareholders may submit false shares. The abstract PVSS model defined by [FO98] do not detect dishonest shareholders while the specific scheme in [FO98] uses RSA encryption and its decryption can be checked. In most other PVSS schemes, e.g. [Std96, YY01, BT99], secret restriction or public verification for the distribution phase can implicitly help check false shares in the reconstruction phase. The PVSS scheme by [Sch99] explicitly demands each shareholder to prove that the share she released is exactly decrypted from the cipher she received from the dealer.

PVSS Security. Most previous PVSS schemes emphasize zero-knowledgeness of public verification. The formal model by [FO98] defined all-or-nothing security. In fact, PVSS security depends on several aspects: public key cryptosystems for shareholders, public verification protocols and assigned restriction over the secret.

A universal model should explain how all aspects influence PVSS security. By the new model one sees that the PVSS scheme by [Sch99] is not indistinguishable.

Independent Verification. Separate responsibility is clear in a universal PVSS model. The different components of a scheme are expected to be independent: data transmission, public verification for distribution, and public verification for reconstruction. However, definitions may differ from each other in special scenarios. For example, verifiable encryption defined by [CD00] needs both the cipher and information from public verification to recover the secret.

Simple Setup. To initialize a PVSS scheme, we usually assign a secret sharing scheme and register public key cryptosystems for shareholders. Some schemes need more special parameters. For example, the scheme by [Sch99] needs a random element in a discrete group whose discrete logarithm is known neither by the dealer nor by any shareholder, and the schemes by [FO98, BT99] need an RSA modulus whose factoring is known neither by the dealer nor by any shareholder. Hence an external trusted data source is needed, or the dealer and shareholders participate in a secure multiparty computation protocol to compute public parameters. In most cases a PVSS scheme is expected to be simply initialized without extra communication resources.

5 Conclusion

Previous PVSS schemes deviate in several aspects for various applications. In this paper a general PVSS model is constructed with mathematical characterization of PVSS security, including all-or-nothing security and indistinguishability security. Our PVSS model, like other known models, use zero-knowledge proof systems. PVSS employs public key infrastructure (PKI) and has no private channel. Compared with ordinary VSS, PVSS has lower communication cost: it probably uses fewer interaction rounds. The new general model defines all aspects of a PVSS system.

References

[BG97] Bellare, M., Goldwasser, S.: Verifiable partial key escrow. In: Proceedings of the 4th ACM Conference on Computer and Communications Security, pp. 78–91 (1997)

[Blk79] Blakley, B.: Safeguarding cryptographic keys. In: Proceedings of AFIPS 1979 National Computer Conference, vol. 48, pp. 313–317 (1979)

[BT99] Boudot, F., Traoré, J.: Efficient publicly verifiable secret sharing schemes with fast or delayed recovery. In: Proceedings of the Second International Conference on Information and Communication Security, LNCS, vol. 1726, pp. 87–102 (1999)

[BMV06] Buchmann, J., May, A., Vollmer, U.: Perspectives for cryptographic longterm security. Commun. ACM **49**(9), 50–55 (2006)

[CD00] Camenish, J., Damgard, I.: Verifiable encryption, group encryption, and their applications to separable group signatures and signature sharing schemes. In: Advances in Cryptography - ASIACRYPT 2000, LNCS, vol. 1976, pp. 331–345 (2000)

[CGMA] Chor, B., Goldwasser, S., Micali, S., Awerbuch, B.: Verifiable secret sharing and achieving simultaneity in the presence of faults. In: 26th IEEE Symposium on Foundations of Computer Science, pp. 383–395 (1985)

[Fel87] Feldman, P.: A practical scheme for non-interactive verifiable secret sharing. In: Proceedings of 28th IEEE Symposium on Foundations of Computer Science, pp. 427–437 (1987)

[FO98] Fujisaki, E., Okamoto, T.: A practical and provably secure scheme for publicly verifiable secret sharing and its applications. In: Advances in Cryptography – EUROCRYPT 1998, LNCS, vol. 1403, pp. 32–46 (1998)

[FS01] Fouque, P.A., Stern, J.: One round threshold discrete-log key generation without private channels. In: The International Conference on Theory and Practice of Public-Key Cryptography 2001, LNCS, vol. 1992, pp. 301–316 (2001)

[Gldr04] Goldreich, O.: The Foundations of Cryptography - Basic Applications. Now Publishers Inc. (2004)

[GJKR07] Gennaro, R., Jarecki, S., Krawczyk, H., Rabin, T.: Secure distributed key generation for discrete-log based cryptosystems. J. Crypt. **20**, 51–83 (2007)

[Mao98] Mao, W.: Guaranteed correct sharing of integer factorization with off-line shareholders. In: Public Key Cryptography, LNCS, vol. 1431, pp. 27–42 (1998)

[Mic92] Micali, S.: Fair public-key cryptosystem. In: Advances in Cryptology – Crypto 1992, LNCS, vol. 740, pp. 113–138. Springer, Heidelberg (1992)

[Ped92] Pedersen, T.: Non-interactive and information-theoretic secure verifiable secret sharing. In: Advances in Cryptology – CRYPTO 1991, LNCS, vol. 576, pp. 129–140 (1992)

[Sha79] Shamir, A.: How to share a secret. Commun. ACM **22**(11), 612–613 (1979)

[Sho97] Shor, P.W.: Polynomial-time algorithms for prime factorization and discrete logarithms on a quantum computer. SIAM J. Comp. **26**(5), 1484–1509 (1997)

[Std96] Stadler, M.: Publicly verifiable secret sharing. In: Advances in Cryptography - EUROCRYPT'96, LNCS, vol. 1070, pp. 190–199 (1996)

[Sch99] Schoenmakers, B.: A simple publicly verifiable secret sharing scheme and its application to electronic voting. In: Advances in Cryptography – CRYPTO 1999, LNCS, vol. 1666, pp. 148–164 (1999)

[Tra19] Traverso, G.G.: Long-term confidential secret sharing-based distributed storage systems [Dissertation]. Darmstadt, Technische Universität, FG Informatik (2019)

[YY99] Young, A., Yung, M.: Auto-recoverable auto-certifiable cryptosystems. In: EUROCRYPT 1998, LNCS, vol. 1403, pp. 17–31. Springer, Heidelberg (1998)

[YY01] Young, A., Yung, M.: A PVSS as hard as discrete log and shareholder separability. In: PKC 2001, LNCS, vol. 1992, pp. 287–299 (2001)

[ZI03] Zhang, R., Imai, H.: Round optimal distributed key generation of threshold cryptosystem based on discrete logarithm problem. In: ACNS 2003, LNCS, vol. 2846, pp. 96–110 (2003)

Popularity Prediction of Food Safety Internet Public Opinion Using LSTM and Attention Mechanism

Bo Song[1], Xiaofen Gu[2], Junliang He[1](✉), Wei Yan[1], and Tianjiao Zhang[3]

[1] China Institute of FTZ Supply Chain, Shanghai Maritime University, Shanghai 201306, People's Republic of China
jlhe@shmtu.edu.cn

[2] School of Logistics Engineering, Shanghai Maritime University, Shanghai 201306, People's Republic of China

[3] College of Information Technology, Shanghai Ocean University, Shanghai 201306, People's Republic of China

Abstract. Food safety is of the top priority and always causes the country's high attention. To predict the popularity of public opinion on food safety is helpful for dealing with public events related with food safety. This paper proposes a popularity prediction model based on LSTM and attention mechanism. Firstly, an index system for evaluating the popularity of Internet public opinion is proposed, where the index weight is determined by information entropy. Different from existing prediction methods, the proposed model uses multi-index public opinion collected from multiple platforms to calculate the total popularity, and adopts an attention mechanism to capture the interaction between different indexes and platforms. The case of "LaTourangelle walnut oil" was selected for analysis. The result shows that the proposed prediction model out performes other two classical models and can predict the development trend of food safety events more accurately.

Keywords: Popularity prediction · LSTM · Attention mechanism · Food safety

1 Introduction

With the rapid development of Internet, we-media has become an important place for people to express and disseminate opinions. Recently, a lot of food safety incidents have been exposed on Internet via we-media, some of which have attracted huge public attention and triggered government actions. Food safety incidents may arouse different types of Internet public opinion, the degree of attention the public pay to the incident is generally measured by the popularity of public opinion [1]. To accurately predict the popularity of Internet public opinion on various food safety incidents helps us to recognize major incidents and take countermeasures in advance.

Although the frequent occurrence of food safety problems has caused more and more scholars' attention, there are relatively few studies focusing on the online public

M. Tavana et al. (Eds.): IISA 2020, AISC 1304, pp. 144–152, 2021.
https://doi.org/10.1007/978-3-030-63784-2_19

opinion of food safety. At this stage, Internet public opinion researches on food safety are mainly carried out from four aspects: Firstly, the process, characteristics and influencing factors of the evolution and spread of food safety Internet public opinion are studied through game, infectious diseases and other models [2]. The second is to construct a public opinion early warning and monitoring model based on identifying food safety hot spots. Geng et al. [3] proposes a novel early warning modeling method based on the deep radial basis function neural network that integrates an analytic hierarchy process. The third is to carry out sentiment analysis on food safety Internet public opinion. Li [4] comprehensively uses LSTM and CNN deep learning technology to propose a C-LSTM classification algorithm for the sentiment of Internet public opinion. Fourthly, based on questionnaire surveys or case studies, explore the strategies of Internet public opinion management for food safety [5]. Inspired by these researches, this paper comprehensively considers multiple quantitative indexes from multiple platforms and proposes a prediction model based on the LSTM and attention mechanism. It is possible to predict the popularity of an event in the entire network using this model.

2 Quantitative Index of Internet Public Opinion Popularity

The premise of predicting the popularity of Internet public opinion is to quantitatively define and calculate its popularity. The current calculation of the popularity of Internet public opinion can be divided into two categories: one is to directly use a single index such as the amount of information published as the popularity. The other is to include a variety of features such as time, amount of comments and user identity into the evaluation index system [6]. The basic principles of scientific, systematic and hierarchical should be considered when selecting the indexes, and the convenience of data acquisition and the simplicity of calculation also need consideration. Therefore, referring to the thesis by Kemal [7], this article selects four indexes to measure the Internet public opinion popularity of a platform: *information release*, *likes*, *comments* and *views*.

The above four indexes are used to measure the popularity of the public opinion. However, there is a big difference in quantity between different types of indexes, which makes it difficult to compare and intuitively reflect the specific popularity value of events, so the data needs to be standardized. The following normalization function is used for conversion.

$$y = (2 - \frac{2}{a^{-x}}) \times 100 \quad (1)$$

In the above normalization function, the value of parameter a needs to be determined, and the corresponding parameter value can be calculated by setting the quantity value x when the popularity value y reaches 100 according to historical experience. The popularity formulas for different indexes from different platforms are calculated according to the above method, the difference is that the quantity value that causes the popularity value to reach 100 depends on the platform and the characteristic of the indexes.

For the public opinion popularity value of a food safety incident on a certain platform, the linear synthesis method is used to synthesize the four indexes' popularity values of the platform. The synthesized coefficient or weight is determined by using information entropy [8]. Similarly, the overall popularity of the event is also obtained by linear synthesis.

3 Prediction Model of Internet Public Opinion Popularity

In previous researches, most scholars focused on one platform, such as Sina Weibo, Twitter or WeChat, when predicting the popularity of Internet public opinion. However, the outbreak of public opinion on any major event in cyberspace is not only reflected in one platform, but also affects each other among different platforms. Therefore, it is necessary to consider multiple platforms and their interactions.

3.1 Notations and Problem Description

Assume M platforms are selected for studying the popularity of food safety events and each platform has N indexes. Choosing one of the indexes from a platform as the target sequence for prediction and other time series are used as features. Given a time window length of T, this article uses $\boldsymbol{X} = (\boldsymbol{X}^1, \boldsymbol{X}^2, \ldots \boldsymbol{X}^M) \in R^{N \times T \times M}$ to represent the time series formed by all platforms in the past T periods, where $X_t^i = (x_t^{i,1}, x_t^{i,2}, \ldots x_t^{i,N})^{\mathrm{T}} \in R^N$ represents the sequence of N indexes of the $i-th$ platform at time t. In this paper, $\boldsymbol{Y} = (y^1, y^2, \ldots y^M) \in R^{T \times M}$ is used to represent the time series composed of the data from various platforms with the same index in the past T periods. Therefore, the problem is to predict the sequence value of an index from target platform in the future $T + \tau$ periods based on the data from the past T periods.

3.2 Prediction Model Based on LSTM and Attention Mechanism

The architecture diagram of the prediction model proposed is shown in Fig. 1, which can be divided into three modules: the data module is mainly used to obtain and sort out the data from multiple platforms to constitute a multi-platform-and-dimensional time series as the original input. In the prediction module, it follows the encoder-decoder architecture. The encoder encodes the original input sequence, and the decoder predicts the output sequence of a dimension from target platform, so as to serve as the input in the popularity calculation formula. After obtaining the prediction sequence of all indexes, the popularity calculation module will obtain the popularity value of the whole network according to the formulas proposed in Sect. 2.

Index Attention Mechanism. For any platform, there is a complex correlation between time series of different indexes. In order to understand the correlation between the target sequence and the time series composed of other indexes in the same platform, this paper introduces the following index attention mechanism:

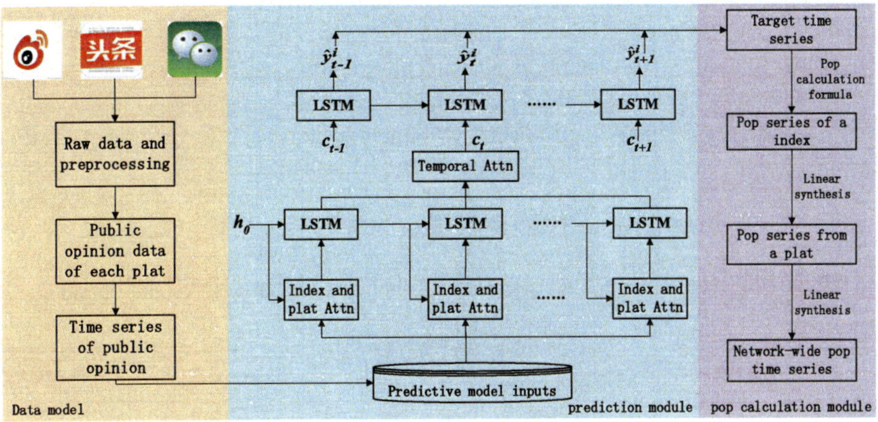

Fig. 1. The framework of prediction model

$$e_t^k = \boldsymbol{v}_n^{\mathrm{T}} tanh(\boldsymbol{W}_n[\boldsymbol{h}_{t-1}; \boldsymbol{s}_{t-1}] + \boldsymbol{U}_n X^{i,k} + \boldsymbol{b}_n) \tag{2}$$

$$\alpha_t^k = \frac{exp(e_t^k)}{\sum_{j=1}^{N} exp(e_t^j)} \tag{3}$$

Among them, the parameters to be learned are $\boldsymbol{v}_n, \boldsymbol{b}_n \in R^T, \boldsymbol{W}_n \in R^{T \times 2l}$ and $\boldsymbol{U}_n \in R^{T \times T}$, l is the number of neurons in the LSTM unit under the coding layer. The index attention weight is determined by the input and the historical state (such as $\boldsymbol{h}_{t-1}$ and $\boldsymbol{s}_{t-1}$) in the coding layer. This paper uses the softmax function to normalize e_t^k. Attention weight a_t^k refers to the degree to which the time series of the $k-th$ index affects the target time series at time t. After calculating the attention weight, for time t, the output vector of its index attention mechanism is:

$$\tilde{\boldsymbol{X}}_t^{index} = (\alpha_t^1 x_t^{i,1}, \alpha_t^2 x_t^{i,2}, \ldots, \alpha_t^N x_t^{i,N})^{\mathrm{T}} \tag{4}$$

Platform Attention Mechanism. The outbreak of public opinion of any major event in cyberspace is not only reflected in one platform, but also affects other platforms. Since directly taking all time series from all platforms as the input of the encoder to capture the correlation between different platforms will lead to higher computing cost and lower performance, the following platform attention mechanism is adopted:

$$P_t^i = \boldsymbol{v}_m^{\mathrm{T}} tanh(\boldsymbol{W}_m[\boldsymbol{h}_{t-1}; \boldsymbol{s}_{t-1}] + \boldsymbol{U}_m \boldsymbol{y}^i + \boldsymbol{Z}_m \boldsymbol{X}^i \boldsymbol{u}_m + \boldsymbol{b}_m) \tag{5}$$

$$\beta_t^i = \frac{exp(p_t^i)}{\sum_{j=1}^{M} exp(p_t^j)} \tag{6}$$

Among them, the parameters to be learned are $\boldsymbol{v}_m, \boldsymbol{u}_m, \boldsymbol{b}_m \in R^T, \boldsymbol{W}_m \in R^{T\times 2l}$, $\boldsymbol{U}_{\mathrm{m}} \in R^{T\times T}$ and $\boldsymbol{Z}_m \in R^{T\times N}$, this attention mechanism adaptively selects the relevant platform for prediction by referring to the historical state (such as $\boldsymbol{h}_{t-1}$ and $\boldsymbol{s}_{t-1}$) in the coding layer, the target sequence, and the corresponding time series of other platforms. At time t, the output vector of its platform attention mechanism is:

$$\tilde{X}_t^{plat} = (\beta_t^1 y_t^1, \beta_t^2 y_t^2, \ldots, \beta_t^M y_t^M)^{\mathrm{T}} \tag{7}$$

Aggregate the output under two attention mechanism as the input of the encoder:

$$\tilde{\boldsymbol{X}}_t = [\tilde{\boldsymbol{X}}_t^{index}; \tilde{\boldsymbol{X}}_t^{plat}] \tag{8}$$

Where, $\tilde{\mathbf{X}}_t \in R^{M+N}$. Then the hidden layer state of the LSTM unit at time t is:

$$\boldsymbol{h}_t = f_e(\boldsymbol{h}_{t-1}, \tilde{\boldsymbol{X}}_t) \tag{9}$$

Temporal Attention Mechanism. Since the performance of the encoder-decoder architecture will gradually decrease with the increase of the length of the encoder [9], this paper introduces a temporal attention mechanism, which can automatically capture the trend and periodicity of the target sequence.

$$\mathrm{u}_{t'}^o = \boldsymbol{v}_d^{\mathrm{T}} tanh(\boldsymbol{W}_{\mathrm{d}}[\boldsymbol{d}_{t'-1}; \boldsymbol{s}'_{t'-1}] + \boldsymbol{Z}_{\mathrm{d}}\boldsymbol{h}_o + \boldsymbol{b}_d) \tag{10}$$

$$\chi_{t'}^o = \frac{\exp(u_{t'}^o)}{\sum_{j=1}^T \exp(u_{t'}^j)} \tag{11}$$

Where, $\boldsymbol{Z}_d \in R^{l\times l}$, $\boldsymbol{W}_{\mathrm{d}} \in R^{l\times 2g}$, $\boldsymbol{v}_d$ and $\boldsymbol{b}_{\mathrm{d}}$ are parameters to be learned, $\boldsymbol{d}_{t'-1}$ and $\boldsymbol{s}'_{t'-1}$ are historical states of the decoding layer, and g is the number of neurons of the LSTM unit in the decoding layer. Softmax function is used to normalize $u_{t'}^o$, and $\chi_{t'}^o$ is the attention weight. At time t', the output vector of the temporal attention mechanism is:

$$\boldsymbol{c}_{t'} = \sum_{o=1}^T \chi_{t'}^o \boldsymbol{h}_o \tag{12}$$

In the decoder, after calculating $c_{t'}$, the hidden layer state of the decoder is updated by combining the output $\hat{y}_{t'-1}^i$ of the decoder at the previous moment:

$$\boldsymbol{d}_{t'} = f_{\mathrm{d}}(\boldsymbol{d}_{t'-1}, [\hat{y}_{t'-1}^i; \boldsymbol{c}_{t'}]) \tag{13}$$

f_{d} is an update function. $d_{t'}$ is the new hidden layer state of the final prediction.

$$\hat{y}_{t'}^i = \boldsymbol{v}_y^{\mathrm{T}}(\boldsymbol{W}_l[\boldsymbol{c}_{t'}; \boldsymbol{d}_{t'}] + \boldsymbol{b}_l) + b_y \tag{14}$$

Where, $W_l \in R^{g\times(l+g)}$ and $b_l \in R^g$ map the stitching vector $[c_{t'}; d_{t'}] \in R^{l+g}$ to the size of decoder hidden layer. Finally, a linear transformation is used to generate the final predicted value.

This model adopts the back-propagation algorithm for training. During the model training process, Adam's optimization algorithm is used to train the model by minimizing the mean square error between the prediction and the actual value vector.

4 Experiment and Result

4.1 Raw Data and Preprocessing

In this paper, the "LaTourangelle Walnut Oil Incident" is selected as the research object. There are many types of propagation platforms in cyberspace, so this article chooses three platforms with greater influence and different types: Sina Weibo, WeChat and Toutiao to study. "LaTourangelle" as the search keyword, relevant data was collected from these platforms with the help of Python.

The collected data was screened and sorted to eliminate duplicate or useless information. A total of 2666 pieces of valid data were obtained by three plats. In order to construct the time series required, each platform takes half an hour as a unit to calculate the total amount of each index and the total number of samples is $4 \times 434 = 1302$. From the Fig. 2, which is the data from one plat, it can be seen that the data of each index fluctuates greatly, and the main changes are gathered in the first 100 moments. This phenomenon revealed that the generation and dissemination of Internet public opinion is huge, which leads to the shortening of the whole life cycle of it. The purpose of popularity prediction is to be able to perceive the future development direction of the event before the major food safety event erupts, so this article chooses to predict the budding or climax period of the event, and the selected period is 2019/7/15 23:00-2019/7/18 8:30 (a total of 116 time periods) as the training set, and 2019/7/18 9:00-2019/7/19 8:30 (a total of 48 time periods) as the test set.

After the dataset is divided, the data is normalized to $[0,1]$ according to Min-Max. During the model training phase, set to predict the data of the next 12 moments. This paper uses the grid search method to adjust the main parameters, which include time window length, learning rate, bath size and etc.

4.2 Comparison of Popularity Prediction Models

To verify the validity of the model, this paper compares the proposed model with two common models, ARIMA and LSTM. In the ARIMA model considering seasonal factors, the data of the previous 48 moments are inputs, and the grid search is used to select the optimal parameters for prediction. The LSTM model is similar to the prediction model proposed without attention mechanism. The result is as follows (Table 1):

After predicting the future time series of each index, the inverse normalization process is carried out first. The popularity value of a platform can be calculated according to formulas proposed in Sect. 2. According to the popularity algorithm

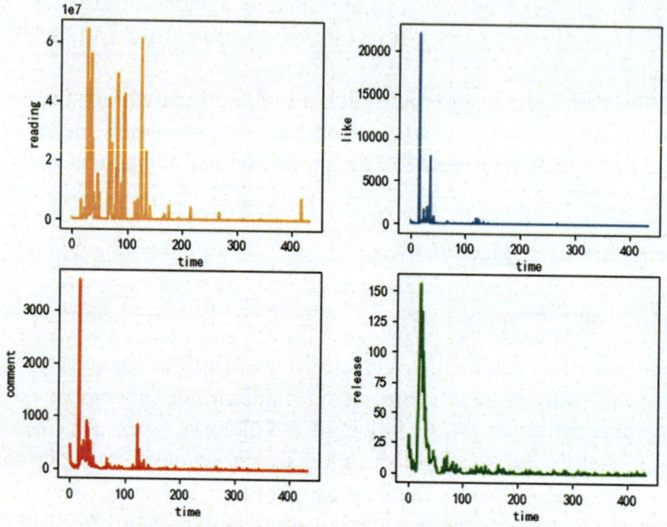

Fig. 2. Public opinion from Sina Weibo

Table 1. Performance comparison among different models

Indicator	Model	Release	Like	Comment	View
MSE	LSTM+Attn	0.008586	0.001783	0.001873	0.002870
	ARIMA	0.015010	0.002149	0.002371	0.004466
	LSTM	0.013229	0.001339	0.002071	0.007829
RMSE	LSTM+Attn	0.076788	0.032316	0.028678	0.037309
	ARIMA	0.116232	0.038964	0.040167	0.060516
	LSTM	0.103685	0.032387	0.044203	0.075807
MAE	LSTM+Attn	0.053883	0.015579	0.014905	0.008169
	ARIMA	0.097611	0.025640	0.039937	0.038814
	LSTM	0.101769	0.027967	0.038607	0.069994

introduced by Jin *et al.* [10], this article assigns weight values of 0.5, 0.25, and 0.25 to Sina Weibo, WeChat, and Toutiao to calculate the popularity of the whole network. Figure 3 is a comparison chart of the predicted and actual popularity.

Compared with the actual popularity curve, the predicted popularity is relatively stable and higher than the actual one. In the ever-changing era, the generation and dissemination of Internet public opinion is rapid, which leads to the shortening of the whole life cycle of it. An effective prediction model can give an alert before a major food safety event outbreaks, so it is reasonable and necessary to keep the predicted curve higher than the actual one, and the trends of the two curves are basically the same, indicating that the popularity prediction model is effective.

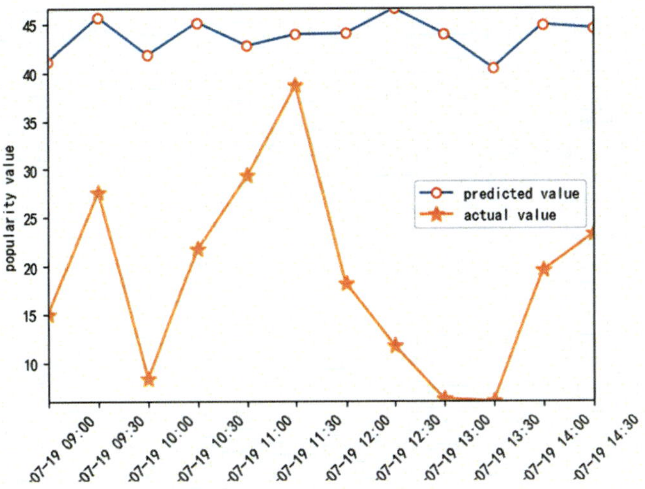

Fig. 3. Comparison of popularity

5 Conclusion and Prospect

This paper firstly determine the amount of information release, likes, comments, and views as indexes to measure the popularity of public opinion, and then propose a popularity prediction model based on the LSTM and attention mechanism, which comprehensively considering the mutual interaction of multiple quantitative indexes from different platforms. The result with "LaTourangelle Walnut Oil" data from Sina Weibo, WeChat and Toutiao shows that the prediction accuracy of the proposed model is higher than that of ARIMA and LSTM, and can accurately give the popularity curve in line with the actual situation.

However, this article also has some limitations: based on the model proposed in this article, although the predicted popularity curve and the actual one are basically the same, there is still a gap between the two curves. Internet public opinion is subject to multi-party supervision and restrictions, and it has certain complexity and particularity. Using the above numerical indexes can only capture the general trend of public opinion. The influence of text features such as topic and content on public opinion popularity is ignored. This will be studied in the future research.

Acknowledgments. This work is sponsored by National Natural Science Foundation of China (71601113, 71602114), Shanghai Science & Technology Committee Research Project (17040501700, 18DZ1206802) and Shanghai Sailing Program (17YF1407700).

References

1. Chen, G., Kong, Q., Xu, N., Mao, W.: NPP: a neural popularity prediction model for social media content. Neurocomputing **333**, 221–230 (2019)

2. Elder, L., Greene, S., Lizotte, M.K.: The gender gap on public opinion towards genetically modified food. Soc. Sci. J. **55**(4), 500–509 (2018)
3. Geng, Z., Shang, D., Han, Y., Zhong, Y.: Early warning modeling and analysis based on a deep radial basis function neural network integrating an analytic hierarchy process: a case study for food safety. Food Control **96**, 329–342 (2019)
4. Feng, Y., Li, Y.: A document clustering model based on convolutional autoencoder. Mod. Inf. Technol. **2**(2), 12–15 (2018)
5. Lassen, J.: Listened to, but not heard! The failure to represent the public in genetically modified food policies. Public Underst. Sci. **27**(8), 923–936 (2018)
6. Bao, Z., Liu, Y., Zhang, Z., Liu, H., Cheng, J.: Predicting popularity via a generative model with adaptive peeking window. Phys. A **522**, 54–68 (2019)
7. Akyol, K., Sen, B.: Modeling and predicting of news popularity in social media sources. CMC Comput. Mater. Contin. **61**(1), 69–80 (2019)
8. Gao, J., Liu, F., Zhang, J., Hu, J., Cao, Y.: Information entropy as a basic building block of complexity theory. Entropy **15**(9), 3396–3418 (2013)
9. Xiao, X., Wang, L., Ding, K., Xiang, S., Pan, C.: Deep hierarchical encoder–decoder network for image captioning. IEEE Trans. Multimedia **21**(11), 2942–2956 (2019)
10. Jin, J., Chen, A.: The popularity of network events and topics: an operational measurement design based on the propagation effect. Mod. Commun. (J. Commun. Univ. China) **39**(5), 71–75 (2017)

A Monitoring Method of Power Wireless Private Network Based on Distributed Big Data Stream Processing

Weijun Zheng[1], Junyu Liu[2], Zhe Liu[2(✉)], Jinghui Fang[1], Weiwu Qi[2], and Quan Xiao[2]

[1] Jiaxing Power Supply Station, State Grid Zhejiang Electric Power Co., Ltd., Jiaxing 314033, Zhejiang, China

[2] Beijing Fibrlink Communications Co., Ltd., Sihezhuang Road, Beijing 100070, Fengtai, China

liuzhe@sgitg.sgcc.com.cn

Abstract. Aiming at the problems of slow monitoring speed and low accuracy in the process of power wireless private network monitoring and automated operation and maintenance, a network monitoring method based on distributed big data stream processing is proposed. The monitoring data source of the system node to be tested is collected by sensors to grasp the operation status of the power grid and collect data in real time. At the same time, the Storm computing framework is used to process real-time detection of equipment faults in real time to meet the needs of rapid processing such as power grid status monitoring abnormal detection and fault analysis. Simulation experiment results show that the distributed big data stream processing proposed in this paper has a great advantage in the processing of data of electric power wireless private network, which greatly improves the data efficiency of the power grid.

Keywords: Power wireless private network · Big data · Monitoring method · Smart grid

1 Introduction

With the large-scale promotion and construction of electric power wireless private networks, the problem of weak network monitoring and operation and maintenance methods is becoming increasingly serious. In the course of operation, the power wireless private network will inevitably be affected by external or internal factors and cause abnormalities. At the same time, the complexity of the grid structure and the real-time nature of the grid data make it extremely difficult to monitor and locate abnormal events. How to develop a method for quickly processing the data stream of the power wireless private network and effectively monitoring the operation status and network status of the power grid is a huge challenge facing the smart grid [1–3].

M. Tavana et al. (Eds.): IISA 2020, AISC 1304, pp. 153–160, 2021.
https://doi.org/10.1007/978-3-030-63784-2_20

In recent years, big data has been applied in smart grids. Big data technology can be used to collect, store and perform correlation analysis from massive grid data to obtain new and valuable information. However, research on the direction of big data streaming relatively few, the application of smart grid big data processing technology at home and abroad is still in its infancy. State Grid has developed a big data batch processing system using the Hadoop platform, which greatly improves the data processing efficiency, but it cannot process power data in real time and reliably, and the real-time processing capacity is insufficient [4]. The limitations of the real-time, volatile, and disorderly data flow of the power wireless private network and the reliability and timeliness of big data processing make traditional processing methods no longer applicable. Wang et al. used Hive's database to realize rapid analysis and query of multi-dimensional data, but they still lacked in reliability [5]. Zhao et al. used the sliding window technology of data flow to analyze online real-time data, but there is no specific system implementation scheme [6]. Therefore, this article provides a new idea for the rapid processing of big data in the power wireless private network.

Based on an in-depth understanding of the processing requirements of big data for power wireless private networks, this paper proposes a distributed big data stream processing framework and introduces Storm streaming computing framework to process big data of electricity [7–10]. By collecting and integrating real-time data of power grid condition monitoring equipment, it gradually becomes an effective data stream. The system uses distributed storage to quickly process the monitoring data stream, which meets the needs of the state monitoring and abnormal analysis business for rapid analysis of the data stream. The system has powerful automatic analysis capabilities and brand-new operation interactions, and redefines the network operation and maintenance analysis products and network operation and maintenance work modes of the power wireless private network.

2 Anomaly Detection Architecture of Power Wireless Private Network

The network monitoring system of the power wireless private network collects network planning data, network optimization data, network monitoring data, base station measurement data and other data of the power wireless private network. The internal monitoring system of the system is used to discover network abnormalities, diagnose the causes of abnormalities, and provide solutions. The proposed scheme is for reference for decision-making of operation and maintenance workers and network optimization personnel, as shown in Fig. 1 below.

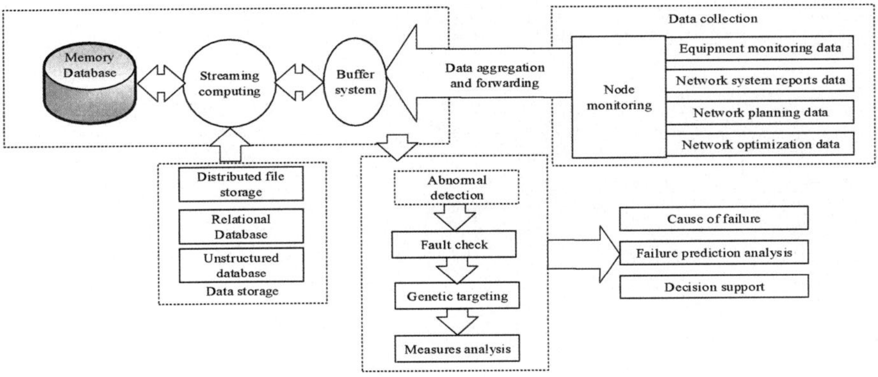

Fig. 1. System flow schematic

2.1 Data Collection

The streaming data in the power wireless private network mainly comes from remote monitoring equipment, network system report data, and various network planning data. These power data are large in scale and many types. In order to adapt to the harsh working environment, the remote monitoring equipment has good reliability and high integration, and the key equipment is: business test module, data transmission MODEM, global positioning GPS (configurable), storage equipment, power supply equipment, central processing unit. These devices send real-time network status information to the power grid every 1 min. When a node fails, the data file will be transmitted to other nodes without loss, ensuring the integrity of the data.

2.2 Data Transmission

In order to ensure that the speed of the system collecting power data and the speed of processing data are synchronized, the power data needs to be buffered. This article uses a distributed storage method to avoid problems caused by long task wait time and low resource utilization, which is beneficial to control the speed of data transmission to the system and ensure the integrity of data.

2.3 Intelligent Operation and Maintenance

The operation and maintenance mechanism of the power wireless private network combined with the comprehensive analysis and evaluation of the collected various power data, finds network anomalies and automatically diagnoses the causes of the anomalies, and adopts the advanced interactive mode design to show the intelligent analysis results. Its work is very close to the mode of technical experts manually diagnosing network problems. The work of manpower mainly lies in the import of original data and decision-making of analysis results. The intermediate analysis process is automatically completed by the system without manual intervention. The ability of intelligent analysis greatly reduces manual analysis work and can greatly improve the work efficiency of personnel (Fig. 2).

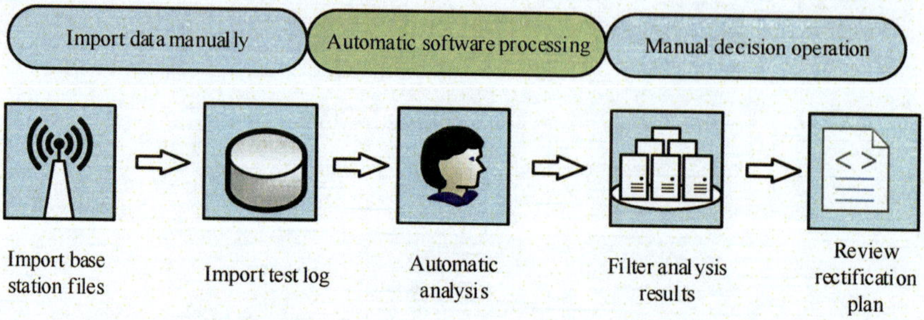

Fig. 2. Schematic diagram of intelligent analysis.

3 Network State Monitoring Based on Streaming Computing

According to the characteristics of real-time, volatile and disordered data flow of the smart grid, the stream processing system is real-time compared with batch processing. It will continuously calculate the data that enters the system at any time in the form of a stream, and there is no need to operate on the entire data set, and the response results can be quickly obtained through stream processing. It has a good application prospect in scenes with real-time requirements. It has been widely used in many neighborhoods such as bank transactions, stock markets, websites and real-time traffic flow analysis.

3.1 Storm-Based Streaming Computing Framework

Hadoop can handle massive amounts of data, but the system has large delays, slow responses, and complicated operation and maintenance. The Storm streaming computing framework adopted in this article is an open source distributed real-time computing framework. It is open sourced by Twitter and has simple operation and maintenance, which makes up for the shortcomings of the Hadoop system and can meet more data processing needs. Storm's streaming framework, as shown in Fig. 3.

The Storm model mainly has two components, the main control node Nimbus and the working node Supervisor. Nimbus is responsible for sending code to the machine to distribute the work and monitor the running status. Supervisor is responsible for monitoring the work of the machine, and chooses to start or shut down the worker according to needs. Zookeeper is an external resource required by Storm. Storm can be abstracted as a topology, and a topology is a graph composed of data sources (Spouts) and data stream processing components (Bolts). Spout is responsible for continuously reading data from the queue and sending it to the downstream Bolt to form a data stream. Bolt completes the logical realization of the function, and the grouping strategy in topology can flexibly formulate the realization of the architectural function.

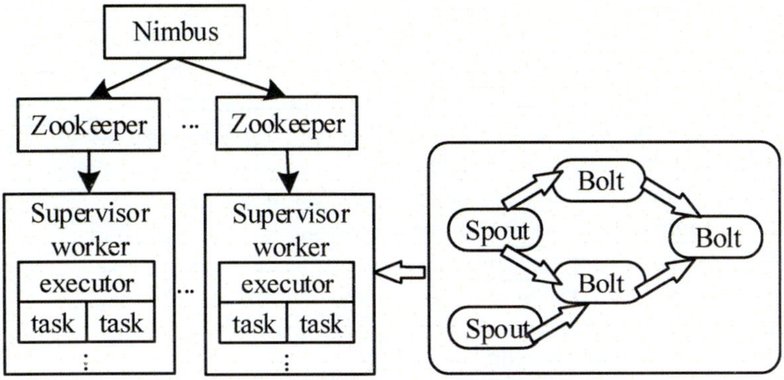

Fig. 3. Storm streaming framework.

3.2 Storm-Based Streaming Computing Framework

This paper will use the streaming computing system Storm to monitor the power wireless private network, and design the sliding window to detect abnormally the big data flow of the power grid state. The monitoring framework of power wireless private network based on streaming data processing is divided into four modules: real-time data collection, data conversion, streaming computing and data storage module, which meets the goal of meeting the needs of power wireless private network business monitoring. As shown in Fig. 4.

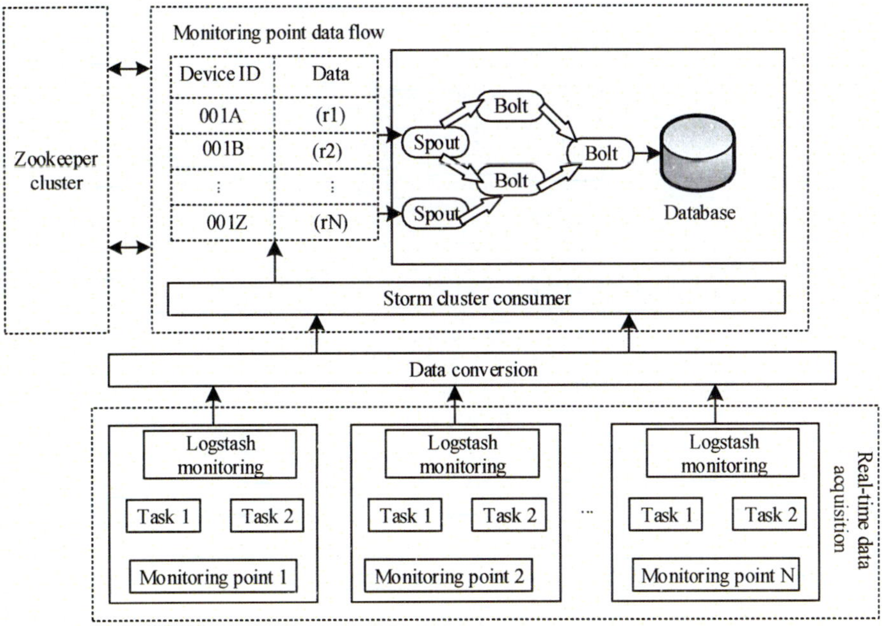

Fig. 4. Smart grid big data real-time stream processing framework.

This article uses the Logstash software deployed on the monitoring node to perform real-time data collection. Logstash is a pipeline with real-time data transmission capabilities. It is responsible for transmitting data information from the input end of the pipeline to the output end of the pipeline. It is commonly used as the log collection equipment in the log relationship system. After the monitoring is completed, the test results are returned to the data conversion module through the backhaul module, and the data transmitted from Logstash is cached to coordinate the data collection rate and the system data processing rate to ensure that the data can be processed stably during processing. Zookeeper constitutes a coordinated control module, which realizes the coordinated control of business logic topology and Storm cluster mapping, so that data can be processed quickly and efficiently.

4 Experiments and Results

4.1 Experimental Test Environment

This article establishes a Storm cluster consisting of a master control node and 7 working nodes. Eight PCs with the same configuration are selected to build a Storm cluster environment. A Linux operating system is deployed on each PC, one of which is a PC Run the Nimbus and Supervisor daemons, and the remaining 7 units only run Supervisor. It is assumed that the experiment is conducted under the distributed Storm cluster without other tasks. Configure the software environment on all hosts in the cluster as shown in Table 1.

Table 1. Software environment configuration table

Name of software	Version
Operating system	Linux
Storm	0.9.0
JDK	Jdk.1.7.0_07
Zookeeper	Zookeeper3.4.5
Python	Python-2.7.3
MySQL	5.7

The hardware of the system built in this article uses industrial-grade devices, the processor uses TI's SitaraTM series ARM® Cortex®-A8 AM3354, the main frequency is 800 MHz, and the configuration is 512 MB DDR3 memory and 1 GB NandFlash memory. The system runs an embedded Linux operating system, properly trims the Linux kernel and develops special drivers to meet the actual functional requirements of the system. The Linux operating system has the advantages of open source and secondary development, and the operating system runs stably and reliably.

4.2 Experimental Data

In order to verify the performance of the distributed big data stream processing framework of the electric power wireless private network constructed in this paper, the data used in this experiment comes from the real-time data of a provincial grid big data platform in March, and a total of 1,000 sets of remote Monitoring equipment, each equipment collects and calculates data every 1 min, and generates about 500,000 pieces of data every day. During the experiment, each piece of data was processed according to the position of the monitoring point, and one piece was sent to the system within 1 s, and then 60 pieces of data were simultaneously input into the power wireless data processing framework based on flow calculation.

4.3 Analysis of Results

In order to more intuitively see the actual application effect of the distributed data stream processing technology proposed in this paper, the number of working nodes of the Storm cluster is set to 8, and the number of experimental data source components and logical processing components are set to 6 and 7, respectively. The collected data streams are run in the cluster and stand-alone respectively, and the experimental comparison and analysis are performed with the data processing time as the standard, and abnormal monitoring of the data of the power wireless private network is performed. The results of the time comparison experiment are shown in Fig. 5.

As can be seen from the figure, with the increase of data scale, the processing efficiency of the streaming computing framework based on Storm cluster proposed in this paper is higher than that of the stand-alone processing. When the data size is small, the streaming computing method used in this paper takes longer than the stand-alone operating environment. This is because when the test data is running in the Storm cluster, additional work such as task distribution requires some time.

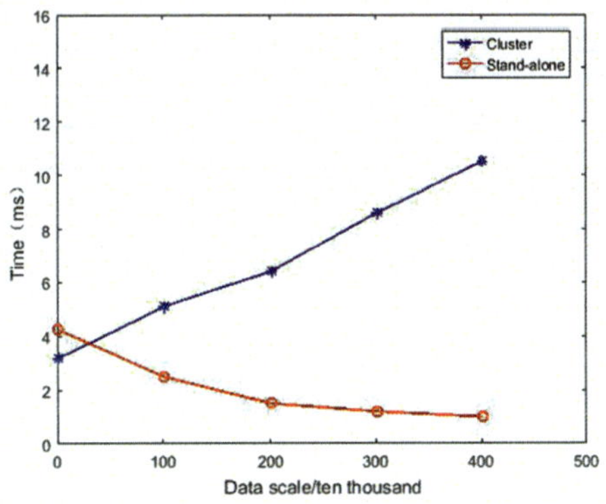

Fig. 5. Comparison of cluster and stand-alone execution time.

According to the experimental results, the distributed data stream processing technology under the big data analysis proposed in this paper has a greater advantage in the efficiency of data stream processing compared to traditional technologies, which can greatly save processing time.

5 Summary

This paper studies the network monitoring technology under the power wireless private network, and proposes a network monitoring method based on distributed big data stream processing, and analyzes in detail from the modules of real-time data collection, data conversion, streaming computing and data storage. In the Storm cluster environment, the big data flow of electricity is taken as the research object, and the data is used to simulate the network monitoring experiment. The simulation experiment proves that the method proposed in this paper is extremely effective, which further improves the real-time processing of the monitoring data stream of the electric power private network, and can provide a theoretical basis for the processing of the data stream under the analysis of subsequent power big data.

Acknowledgments. This research was funded by the Scientific Project Subsidized by STATE GRID Corporation of China, grant number 5700-201919233A-0-0-00.

References

1. Wu, R.J., Wei, W.: Research on radio monitoring data quality and cleaning technology. J. China Radio. 36–39 (2019)
2. Xu, Y., Zhou, S.W., Ding, F.: Research on evaluation and optimization methods of radio monitoring station deployment efficiency. J. Chin. J. Sens. Actuators 1389–1392 (2018)
3. Ying, Y.Z., Xie, J.W.: Research on signal monitoring technology based on sensor cooperative network. J. Radio Eng. 15–18 (2014)
4. Liu, S.R., Song, Y.Q., Zhu, Y.L., Wang, D.W.: Research on storage of smart grid state monitoring data based on Hadoop. J. Comput. Sci. **40**(1), 81–84 (2013)
5. Wang, D.W., Xiao, K., Xiao, L.: Hive-based power equipment status information data warehouse. J. Power Syst. Prot. Control **41**(9), 125–130 (2013)
6. Zhao, J.K., Yang, G.F., Mu, L.S.: Application research of data stream technology in power grid automation. J. Power Syst. Technol. **35**(8), 6–11 (2011)
7. Sun Q., Jiang S.Y. (2019) Streaming big data real-time processing technology, platform and application. J. China Collective Economy., 156–157
8. Yao, J., Zhang, H., Wei, L., Guo, J.: Research on planning and evaluation method of spectrum monitoring system for electric power wireless private network. Electr. Power Eng. Technol. **38**(04), 62–67 (2019)
9. Wang, H., Gu, S., Wan, Y.S., Yu, J.: Tracking and locating method of interference source for power wireless private network. Guangdong Electr. Power **32**(06), 86–93 (2019)
10. Zhang, Y., Cao, W., Shao, S.: Hybrid monitoring sleep method for low-power wireless sensor network nodes. Comput. Technol. Dev. **26**(10), 196–199 (2016)

Research on Online Reputation of Goods Based on Emotional Analysis

Xiaotong Yan, Zhijie Zhao(✉), Xiaowei Han, Zhipeng Fan, and Jialin Zhang

Harbin University of Commerce, Harbin 15000, HL, China
zhijiezzz@aliyun.com

Abstract. Online reputation of goods plays an important role in improving market efficiency and creating an orderly competitive environment in the network market. Based on the detailed possibility theory, this paper proposes an improved online reputation measurement model of goods, extracts two thinking paths in the process of forming consumers' online reputation attitude, and analyzes the factors that influence the change of goods reputation. In this paper, the text emotion analysis technology is used to deeply mine the text information of e-commerce comments to reflect the attributes and attitudes of consumers, the usefulness of a single comment is determined by other comment features. And combined with the search index on the search engine to determine the online reputation of goods. Grab the mobile phone review data on JD.com platform for experimental verification to show the current online reputation ranking of goods, consumers' attention to various attributes of goods, emotional attitudes and so on. This paper expands the evaluation perspective of online commodity reputation, and from the perspective of cognition and emotion. It can better discover the true wishes of consumers and provide targeted countermeasures and suggestions for enterprises to improve goods and services.

Keywords: Online reputation of Goods · Online reviews · Emotional analysis · ELM

1 Introduction

E-commerce market has the characteristics of low information search cost and fierce market competition. However, there are still serious problems of consumer information asymmetry and "lemon market" [1]. The reputation of commodity network refers to the evaluation of the reputation and popularity of products put on the market by stakeholders in e-commerce activities in terms of quality, brand and style [2]. It can verify the commodity signal sent by the seller afterwards, help users better understand and buy goods [3], gradually transform into trust in enterprises and commodities [4, 5], and have a certain impact on commodity premium [6–8]. The after-sales comment mechanism set up by the e-commerce platform can effectively collect consumers' subjective evaluation opinions on specific goods, and these user feedback will affect the impression of goods in the minds of potential consumers and eventually form a commodity reputation [9]. The platform network market has a competitive market

M. Tavana et al. (Eds.): IISA 2020, AISC 1304, pp. 161–170, 2021.
https://doi.org/10.1007/978-3-030-63784-2_21

structure, and the cost of consumers to "exit" is low. Therefore, commodity reputation plays a stronger role in the platform network market [10], and it is more necessary to evaluate the reputation of goods accurately and objectively [11, 12].

As an important mechanism of e-commerce, online reputation evaluation needs to clarify the evaluation objects, stakeholders involved and usage methods [13–15]. The paper [16, 17] uses the questionnaire method to divide different dimensions to measure the online reputation of goods. Chai adopts the comment level with decreasing weight and limited memory reputation score to highlight the feedback generated by the recent trade [18]. Wang uses the method of literature research to think that commodity reputation includes commodity reviews, impressions, product scores and trading records [19]. Lv adopts the method of game theory and believes that the more cooperative behaviors, the higher the reputation value of goods [20]. Zhang establishes a reputation model and improves the marketing strategy by identifying the key patterns, relationships, indicators and parameters of customer ratings and comments [21]. Li selects the ten most representative comments according to the characteristics of the comments. Using the method of emotional analysis to gain commodity reputation, and the efficiency of text summary is high [22]. Hung constructs a hybrid multi-criteria decision-making model, which believes that online reputation of goods is affected by user participation and collaboration [23]. There are also some scholars analyzes other reputation objects from the cognitive field. Reyes uses questionnaire method to collect data and uses ELM model to analyze the influencing factors of hotel reputation credibility [24]. Lina uses ELM model to analyze the impact of consumers' reputation evaluation on travel agencies as a peripheral path [25]. To sum up, we find that most of the measurement of commodity reputation adopts questionnaire survey, simulation or tends to pay attention to the single characteristic dimension of score, which can not truly reflect the multi-dimensional preference of consumers to a large extent. The commodity review path of e-commerce platform can be traced back to the source, which has longer continuity and faster response speed than traditional marketing activities. The content of consumer comments can express commodity opinions and information more accurately, eliminating the impact of a single evaluation level, which usually varies from person to person. At present, there are also some scholars mining the review text, but rarely analyze which thinking paths consumers use to judge commodity reputation from the perspective of consumer cognition.

The purpose of this paper is to explore the factors that affect the online reputation of consumers based on the detailed possibility theory, and to construct a commodity online reputation measurement model based on the emotional analysis of the comment text. First of all, by analyzing the two thinking paths of consumer persuasion and attitude change, it is determined that the online reputation of goods is affected by consumers' emotional satisfaction tendency, evaluator's trust, trading time window, negative evaluation sensitivity and product popularity. Then build a measurement model. Finally, collect JD.com website 102 mobile phone comment data and search engine data for empirical analysis, get reputation ranking, commodity attribute improvement suggestions, and put forward marketing suggestions.

2 Model Construction

2.1 Analysis of Influencing Factors

The online reputation of goods is mainly implied in consumers' online reviews and is related to their popularity. The exhaustive possibility model (Elaboration Likelihood Model, ELM) was put forward by social psychologists Richard E Petty and John T Cacioppo in the 1980s [26]. In the previous research on the change of people's behavior and attitude towards persuasion based on the exhaustive possibility model, it is generally believed that the judgment based on the content of the message text and comments with attitude tendency is the central path through deep thinking. In addition, the credibility of information sources and media are simple clues, consumers are less likely to think through self-cognition, and the psychological process of attitude change is usually a marginal path [27]. R Filieri and F McLeay found that central paths related to information quality and marginal paths related to rankings affect consumers' acceptance of the information contained in comments [28]. Cheung believes that when people get comment information, most of them only read the comments on the first few pages, which are relatively recent or at the top of the list, and then go through the edge path [29]. According to Gao, online reputation expresses consumers' comprehensive evaluation of a commodity. In addition to this comment information, the popularity of one of the dimensions of reputation measurement can be obtained by consumers' simple recall of the product, and the impact of attitude change is carried out through the marginal path [30]. This paper analyzes the factors that influence the change of consumers' attitude towards online reputation of goods. The marginal path involves the credibility of online reviews, as well as the characteristics of commentators, useful votes, the timeliness of comments and the popularity of products. At this time, users are more likely to think only shallow and use simple clues and reasoning to judge. If potential consumers want to judge the content quality of online comments, consumers need to spend more energy on self-cognitive thinking and evaluate information through the central path in the detailed possibility theory.

2.2 Construction of Online Reputation Model of Commodity Based on Emotional Analysis

How to make better use of the online word-of-mouth information hidden in massive online reviews, make a more accurate and comprehensive evaluation of goods, and help potential consumers to complete purchase decisions with confidence is the primary problem to be solved in this study. At present, most qualitative studies believe that commodity reputation can be divided into two dimensions: reputation and popularity. In this paper, when building the model, we first calculate the commodity word-of-mouth. According to the factors that affect the usefulness of commodity comments, the weight coefficient of the opinions of each comment publisher is calculated. Combined with the text mining and emotional semantic analysis of the specific content of the comment text, the consumers' satisfaction with the emotional tendency of each attribute of the commodity is calculated. Then calculate the commodity popularity and replace it with Baidu search index. According to the results of the above steps, the online reputation evaluation model is obtained by linear regression.

Data Acquisition and Preprocessing

This paper takes the e-commerce review data and Baidu search index of goods as the research object. The dimensions of comment collection include commodity name, comment publisher ID, whether it is a member, comment star, comment text content, comment release time, comment argument praise, and so on. Baidu search index crawls data under commodity keywords from the date of listing to the date of comment crawling.

First of all, the spam comments are eliminated. Then, the online comment text data used for emotion analysis is further preprocessed, including word segmentation, part-of-speech tagging, low-frequency word filtering and syntactic dependency analysis. In this paper, Jieba software is used for word segmentation and part of speech tagging; TF-IDF method is used to filter low-frequency words, in order to take only a certain number of attribute words as commodity attribute evaluation matrix in the subsequent model establishment process, $w(A) = (w(a_1), w(a_2), w(a_n))^T$, to avoid the problem that the matrix is too sparse, and take the TF-IDF value as the weight of the attribute.

Calculate the Opinion Weighting Coefficient of Each Comment Publish

$$w(r_i) = \frac{1 + HV_{s(i)}}{p + \Sigma_{i=1}^{P} HV_{s(i)}} \times e^{(-\Delta\lambda t)} \times \alpha(\mu - St) \tag{1}$$

Among them, $w(r_i)$ in formula (1) indicates the weight of comments in Article I $HV_{s(i)}$ indicates that comments in Article I are praised for their usefulness , $(1 + HV_{s(i)})$ is the total number of comments given in Article I, p indicates the total number of comments. $e^{(-\Delta\lambda t)}$ indicates that the time decline factor is used to weaken the impact of early online reviews on reputation, t is the difference between the current time and the comment time, λ is used to denote the decline rate of e, the range of values is $0 \leq e^{(-\Delta\lambda t)} \leq 1$. St represents the comment list star $0 \leq St \leq 5, 0 \leq \alpha \leq 1$ is used to enhance the weight of negative evaluation. The root formula (3) can get the weight of each comment, thus the comment weight matrix $w(R) = (w(r_1), w(r_2), \ldots, w(r_p))$.

Calculate Consumers' Emotional Tendency Towards Goods

Then, this paper analyzes the emotional tendency implied in the comment text, and calculates the weight of commodity attributes and the emotional satisfaction of commodity attributes. This model uses the method of emotion analysis in natural language processing, and specifically uses Prost model (dependency analysis, triple, word2vec, support vector machine, ranking and other methods for emotion analysis). It can collect as many attribute words and emotion words of comment triple as possible, obtain emotion words as accurately as possible, and identify the emotional tendency of e-commerce comments with high accuracy.

Triplet Extraction of Comment Text

The triplet for the evaluation of individual attributes of goods is extracted through syntactic dependency relations, which is composed of "attribute words", "emotion words" and "degree words". For example, for the comment unit "the screen is very great", the attribute word is "screen", the emotion word is "great", and the degree word is "very", so the triple is <s creen, great, very>.

Feature Vectorization

The extracted triplet is trained by word2vec neural network to obtain the vectorized representation of the space. Because word2vec can only transform words into vectors, but can't directly deal with sentences, we first obtain each word vector in the triple after the reduction of the previous clause, and then linearly express the attribute comment vector.

SVM Classifier Classifies the Emotion of the Specific Attributes of the Comment Text

After the triple is vectorized, it carries on the emotion classification operation. If an attribute of a commodity is not mentioned in a comment, the missing value NA is used to express the emotional tendency of the comment to evaluate this attribute, and the evaluation matrix of commodity attribute in the above measurement model is obtained.

Extract the Weight of Commodity Attribute Evaluation and Improvement Suggestions Keywords

The ultimate goal is to get a specific analysis of the emotional tendency of each attribute of commodity evaluation, so it is necessary to classify the attributes separately and summarize the results as the weight of commodity attribute evaluation. For the attributes with low praise rate of emotional analysis, it is necessary to extract keywords for improvement. Using the TF-IDF method in the second step above, the importance of each word to the attribute is obtained, and the first ten most important words are taken as the keywords for the improvement of the attribute.

Calculate the Popularity of Goods

$$AR = Ln(Awareness) \tag{2}$$

In addition to online reviews, commodity popularity is also a dimension that constitutes the online reputation of goods. In this study, the popularity will be replaced by the average value of Baidu search index from the date of listing to the date of search. However, the final calculation results related to online reviews range from −1 to 1. There are small differences and great differences in popularity between them. If the two are multiplied directly, the influence of online reviews will be covered. In order to solve this problem, this study uses the most commonly used method to solve the problem of excessive heteroscedasticity to take the logarithm of the popularity of goods.

$$PR = w(R) \times M \times w(A) \tag{3}$$

Obtaining the three parameters in the above formula (3) is the key to calculating the reputation of goods. Where, $w(R) = (w(r_1), w(r_2), \ldots, w(r_p))$ represents the vector composed of the weight of each comment on the product, and p is the total number of comments. $M = (a_{ij})_{p \times n}$ represents the commodity attribute evaluation matrix obtained after the emotional analysis of the commodity online review, and the a_{ij} element represents the j attribute evaluation emotional tendency of the article I review of the commodity. $w(A) = (w(a_1), w(a_2), w(a_n))^T$ represents the vector composed of the weight of each attribute of a commodity, where n is the number of attributes.

Commodity online reputation model, as shown in formula (4).

$$CR = PR * AR \tag{4}$$

3 Experimental and Data Analysis

This paper selects the mobile comments on JD.com platform and Baidu search index as empirical data. Write a web crawler in python to capture online reviews of JD.com Mall's top 102 smartphones. Baidu search index crawls data under commodity keywords from the date of listing to the date of comment crawling.

According to the commodity online reputation model, the experimental results are as follows. The attribute weight of the mobile phone is shown in Table 1. Indicating how much consumers attach importance to the attribute.

Table 1. Mobile phone attribute weight.

Attribute	Weight	Attribute	Weight	Attribute	Weight
Battery	0.176502901	Fever	0.063518881	Processor	0.065806737
Camera	0.143112239	Appearance	0.061192556	Parts	0.052481729
Price	0.077086824	Logistics	0.058975611	Feel	0.064518881
Screen	0.071363264	Function	0.055744076	Other	0.047845969
System	0.070033603	After sales	0.055335611	…	…

It can be seen that consumers mentioned the most mobile phone battery and photo taking function in the comments, with weights as high as 0.176502901 and 0.143112239. Price, screen and system are followed, but they are basically the same as other attributes. In advertising promotion and e-commerce detail page design, manufacturers can focus on battery performance, camera photography effect, cost performance, screen display effect and system fluency.

In this paper, based on the emotional analysis of mobile phone comments text, it is found that the attributes with low praise rate are battery, camera, screen, price and system. In addition, we use the TF-IDF method to extract the first six most important attributes under these attributes as suggestions for improvement. For example, the

Table 2. Battery improvement triple.

Negative comment	Favorable comment	Neutral comment
battery, life, short	battery, in general, durable	battery, problem, big
battery, durable, not	battery, life, one day	battery, hot, very
battery, average	battery, life, promotion	battery, heats up, often
battery, ok	battery, fine, very	battery, heating
battery, durable, very	battery, nothing, wrong	battery, doesn't work
battery, makes, do	battery, performance, good	battery, power down, fast

battery improvement feature triple <" attribute words", "emotional words", "degree words"> the results are shown in Table 2.

According to the most critical negative evaluation attribute of the lower consumer satisfaction, we found that what mobile phone manufacturers urgently need to improve are battery life, battery heating, camera layout, camera entering dust, screen color, screen quality, price stability, system fluency and so on. Improving the quality of products and services targeted by enterprises can help potential consumers quickly establish trust in goods and promote the formation of purchase decisions.

According to the calculation results of attribute evaluation matrix, comment weight matrix and mobile phone attribute weight matrix. We can calculate the online word-of-mouth of 102 mobile phones. The popularity of goods can be calculated according to Baidu search index. The specific results are shown in Table 3.

Table 3. Mobile phone reputation and popularity.

Num	Reputation	Popularity	Num	Reputation	Popularity
1	0.8755995	8.8094	2	0.8366226	7.1098
3	0.7222304	7.2742	4	0.5615953	8.2031
5	0.6790824	4.1541	...	...	...

As can be seen in the table, goods with high reputation are often well-known. Businesses should pay attention to the promotion of mobile phone quality at the same time. There are other cases, such as mobile phone 5, although the reputation is not high, only 0.5615953, but a large number of users need to search, and the popularity is as high as 8.2031. The emergence of this situation is often due to the fact that the Internet further expands the effect of brand or commodity crisis, resulting in serious damage to reputation.

In order to more intuitively show the reputation ranking of mobile phones, we select the mobile phone with the best online reputation (8.08325405) as the benchmark, set its online reputation as 100. Then calculate the relative reputation of other mobile phones. The data obtained from this processing is shown in Table 4.

Table 4. Relative reputation of mobile phone.

Num	Online reputation	Relative reputation	Num	Online reputation	Relative reputation
1	7.71350623	95.62	2	5.94821936	73.58
3	5.25364838	64.86	4	4.60682241	56.86
…	…	…	…	…	…
99	6.62979213	82.51	100	8.08325405	100
101	4.48960243	55.54	102	7.37841793	91.52

As a result, we can get a ranking of mobile phone online reputation. Reputation ranking represents the status in the minds of consumers. By scoring and ranking, manufacturers can understand the general situation of products. For example, the score of No. 101 mobile phone is only 55.54. For the low score mobile phone, we can refer to the top mobile phone to comprehensively consider improving our product design.

4 Conclusion

With the application of detailed possibility theory, this paper expands the influencing factors of commodity online reputation on e-commerce platform from the aspect of consumer cognition, and improves the commodity online reputation measurement model based on emotional analysis. Climbing the comments and related search data of 102 mobile phones on the JD.com platform, it is found that among all the attributes of mobile phones, battery and camera consumers pay the highest attention, which is often mentioned in the comments. At the same time, the consumer satisfaction of battery, camera, price and screen is relatively low. Merchants can refer to the results of emotional analysis to improve the phone in terms of battery life, camera sealing and screen material selection. Reputation represents the will of consumers and can be converted into trust over time. Businesses should focus on the gap between products put on the market and products with high reputation and strive to narrow the gap. The model established in this paper has strong applicability in the field of electronic consumption of e-commerce platform. Business interaction mechanism and reputation early warning mechanism can be established in the future. Timely feedback to the merchant on which aspects of the product need to be improved, and remind the merchant to take coping strategies when the reputation is below the threshold.

Acknowledgments. This work is supported in part by the project of postdoctoral research start-up fund in 2019 (No.LBH-Q19028).

References

1. Matherly, T.: A panel for lemons? Positivity bias, reputation systems and data quality on MTurk. Eur. J. Mark. **53**(2), 195–223 (2019)

2. Díaz, M. R., Rodríguez, T. F. E.: Determining the reliability and validity of online reputation databases for lodging : Booking.com, TripAdvisor, and HolidayCheck. J. Vacation Mark.**24**(3), 261–274 (2017).
3. Hossain, M.A., Rahman, S., Chowdhury, T.A., Chan, C., Yang, X., Su, Q.: How signaling mechanisms reduce "lemons" from online group buying (OGB) markets? a study of China. Int. J. Phys. Distrib. Logistics Manag. **48**(7), 658–681 (2018)
4. Zablocki, A., Schlegelmilch, B., Houston, M.J.: How valence, volume and variance of online reviews influence brand attitudes. AMS Rev. **3**(6) (2018)
5. Shad Manaman, H., Jamali, S., AleAhmad, A.: Online reputation measurement of companies based on user-generated content in online social networks. Comput. Hum. Behav. **54**, 94–100 (2016)
6. Seraina, C., Buhalis, D., Kountouri.L, Eleftherios, G., Andrianos, E.: The impact of online reputation on hotel profitability. Int. J. Contemp. Hosp. **32**(1), 20–39 (2020)
7. Foroudi, P.: Influence of brand signature, brand awareness, brand attitude, brand reputation on hotel industry's brand performance. Int. J. Hospitality Manag. **10**, 1–5 (2018)
8. Sami, Y., Berri, J., Qurishi, A., Mohammed, A.: measuring reputation and influence in online social networks: a systematic literature review. IEEE Access, 8(6), 24–50 (2020)
9. Al-Obeidat, F., Spencer, B., Al Taei, M.: Identifying major tasks and minor tasks within online reviews. Future Gener.Comput. Syst. **110**, 413–421 (2020)
10. Xiaobo, T., Xinrui, Z., Jiankun, Y.: Research on the relationship between online review, perceived usefulness and New Product Diffusion. China Soft Sci. **7**, 162–171 (2017)
11. Ghose, A., Ipeirotis, P.G., Li, B.: Designing ranking systems for hotels on travel search engines by mining user-generated and crowdsourced content. Mark. Sci. **31**(3), 493–520 (2012)
12. Manes, E., Tchetchik, A.: The role of electronic word of mouth in reducing information asymmetry: an empirical investigation of online hotel booking. J. Bus. Res. **85**, 185–196 (2018)
13. Ghiasi, H., Brojeny, M.F., Gholamian, M.R.: A reputation system for e-marketplaces based on pairwise comparison. Knowl. Inf. Syst. **56**(2), 613–636 (2017)
14. Walker, K., Dyck, B.: The primary importance of corporate social responsibility and ethicality in corporate reputation an empirical study. Bus. Soc. Rev. **119**(1), 147–174 (2014)
15. Cioppi, M., Curina, I., Forlani, F., Pencarelli, T.: Online presence, visibility and reputation: A systematic literature review in management studies. J. Res. Int. Mark. **13**(4), 547–577 (2019)
16. Dutot, V., Castellano, S.: Designing a measurement scale for E-Reputation. Corp. Reputation Rev. **18**(4), 294–313 (2015)
17. Yinghui, Z.: Research on credit evaluation model of C2C e-commerce based on reputation. Shopping mall Modernization **8**, 41–43 (2016)
18. Chai, Y., Li, D., Wu, Y.: Recommendation algorithm for mobile E-commerce based on reputation. In: Proceedings of the 2019 11th International Conference on Machine Learning and Computing-ICMLC2019, pp. 217–223 (2019)
19. Xuhui, W., Qilin, Z.: Formation mechanism and governance mechanism of "lemon problem" in platform network market-based on Alibaba's case study. China soft Sci. **10**, 31–52 (2017)
20. Lv, J., Wang, T., Wang, H., Yu, J., Wang, Y.: A SECPG model for purchase behavior analysis in social e-commerce environment. Int. J. Commun. Syst. **4**, 41–49 (2020)
21. Lemin, Z., Xin, W., Ningyu, W., Qingxin, L.: A model of online goods reputation estimation and sales strategy formulation based on user feedback, Electron. Technol. Softw. Eng. **4**, 67–70 (2020)

22. Li, X., Zhu, S., Yu, J., Yang, A., Zhang, M.: A commodity evaluation algorithm based on commodity review abstracts. In: 2018 International Conference on Artificial Intelligence and Big Data (ICAIBD), pp.1–5 (2018)
23. Hung, Y.H., Huang, T.L., Hsieh, J.C., Tsuei, H.J., Cheng, C.C., Tzeng, G.H.: Online reputation management for improving marketing by using a hybrid MCDM model. Knowl. Based Syst. **35**, 87–93 (2012)
24. Reyes-Menendez, A., Saura, J.R., Martinez-Navalon, J.G.: The impact of e-WOM on hotels management reputation: exploring TripAdvisorreview credibility with the ELM model. IEEE Access **7**, 68–77 (2019)
25. Wang, L.: Effect of perceived quality information of Chinese online travel agency service on customer satisfaction, trust and purchase intention: based on elaboration likelihood model. J. Hospitality Tourism Stud.**20**(1), 19–37 (2018)
26. Petty, R.E., Cacioppo, J.T.: The elaboration likelihood model of persuasion. Commun. Persuasion **1**(24), 123–205 (1986)
27. Pierro, A., Mannetti, L., Kruglanski, A.W.: Relevance override: on the reduced impact of "cues" under high-motivation conditions of persuasion studies. J. Pers. Soc. Psychol. **86**(2), 251–264 (2004)
28. Filieri, R., Mcleay, F.: E-WOM and accommodation: an analysis of the factors that influence travelers' adoption of information from online reviews. J. Travel Res. **53**(1), 44–57 (2013)
29. Cheung, C.M.K., Lee, M.K.O., Rabjohn, N.: The impact of electronic word-of-mouth: the adoption of online opinions in online customer communities. Internet Res. **18**(3), 229–247 (2008)
30. Gao, Q., Tian, Y., Tu, M.: Exploring factors influencing Chinese user's perceived credibility of health and safety information on Weibo. Comput. Hum. Behav. **9**(45), 21–31 (2015)

Feature Selection Method Based on Chi-Square Test and Minimum Redundancy

Yuxian Wang(✉) and Changyin Zhou

Shandong University of Science and Technology, Qingdao 266590, China
YuxianW123@163.com

Abstract. This paper studies the feature selection problem of high-dimensional classification sample data, and proposes a feature selection method based on the combination of chi-square test and minimum redundancy. Firstly, the chi-square test is used to select the sample data highly related to the classes and reduce the data scale quickly and effectively. On this basis, the minimum redundancy algorithm is used to further remove the redundancy and realize feature selection. In order to verify the feasibility of the method proposed in this paper, the method is applied to the feature selection of ALT-ALB-AML, Breast-A and Stomach data sets. By using support vector machine, K-nearest neighbor classification and geometric barycenter classification to predict and classify high-dimensional classification sample data, it is found that the feature selection method presented in our paper can effectively classify and improve classification accuracy compared with chi-square test or minimum redundancy-maximum relevance or simple combination of the two.

Keywords: Feature Selection · Chi-square Test · Minimum Redundancy-Maximum Relevance · Gene Expression Data

1 Introduction

Facing the problems of large amount of data, many missing values and large noise in high-dimensional classification sample data, dimensionality reduction of data has become one of the chief tasks of data mining. Feature selection is a method to reduce data dimension, which is an essential step in data mining and application. Feature selection can reduce data dimension and remove redundant and irrelevant features from the original data effectively [1].

The methods of feature selection are divided into three categories generally [2], including embedded, wrapper and filter. No matter which method is adopted, the key point of feature selection is to select the feature that is most relevant to the classes and has the least redundancy. The method named minimum redundancy-maximum relevance (MRMR) [3], proposed by Ding and Peng, measures the correlation between classes and features and the redundancy of features through mutual information. The feature subsets selected by this method are characterized by strong discrimination and low redundancy. This method can effectively reduce the size of the feature subsets. On

M. Tavana et al. (Eds.): IISA 2020, AISC 1304, pp. 171–178, 2021.
https://doi.org/10.1007/978-3-030-63784-2_22

this basis, Sakar [4] proposed a method for selecting features based on a combination of kernel correlation analysis with MRMR. Kernel canonical correlation analysis is used to measure the relevance between classes and features, measure the redundancy of features. After that, Mandal [5] introduced non-dominated feature subsets to the minimum redundancy and maximum correlation feature selection method, which improved the classification accuracy. Yao Minghai [6] improved the redundancy evaluation function of MRMR. Chen Jiangpeng [7] proposed a naive Bayesian classifier based on MRMR. Radovic [8] proposed a feature selection method of TMRMR, which can process multivariate time data. Chen Chen [9] combined chi-square test and random forest Gini index to find the best feature subset, and classified it with support vector machine, and got good results.

Although the MRMR feature selection method can select the feature subset which is highly correlated with the classes and has low redundancy between features, the calculation cost is quite high. In this paper, chi-square test and minimum redundancy method are combined to select the features highly related to the classes through chi-square test, and further select the feature subset with low redundancy. For better classification, this paper proposes a geometric center of gravity classification. First, this paper finds the geometric center of gravity of different types of data in the training set. Then, the distance between the data in the test set and the center of gravity coordinates is calculated in sequence, and the class of the data is the class of the nearest center of gravity coordinates. Finally, the conclusion is obtained by comparing with the classification accuracy results of support vector machine (SVM) and K-nearest neighbor classification (KNN).

The structure of the paper is as follows: In Sect. 2, the basic concepts of chi-square test and the MRMR method are introduced, and an algorithm based on the combination of chi-square test and minimum redundancy is proposed. In Sect. 3, the proposed algorithm is applied to three data sets to illustrate its effectiveness. The conclusion is drawn in the last part.

2 Model Introduction and Algorithmic Process

The high-dimensional classification sample data to be studied in this paper has the following data formats:

$$D = (X, C), \tag{1}$$

in which, $X_{n \times m} = \left(x_{ij}\right)_{n \times m}$, $C = (c_1, \cdots, c_n)^T$, $c_i = 1, -1$.

In the high-dimensional classification sample data, the feature dimension m is much larger than the sample number n, but some features are not important for sample classification, so it is necessary to pretreatment the data before using the high-dimensional classification sample data, such as feature selection.

2.1 Chi-Square Test

Chi-square test is used to compare the goodness of fit between actual frequency and theoretical frequency. It is often used to analyze the relevance between two classification variables [10]. The original hypothesis of the χ^2 test is that two variables are independent. The larger the calculated χ^2 value is, the greater the deviation between the observed value and the theoretical value is, which tends to reject the null hypothesis.

In this paper, the χ^2 test is applied to the feature data and the class data, and the features that are more relevant to the classes are selected. Assuming that a feature has p values and q classes, a contingency table is established (see Table 1).

Table 1. Frequency contingency table of the values of the features and classes

Classes features	*1*	2	...	q	sum
1	k_{11}	k_{12}	...	k_{1q}	k_1
2	k_{21}	k_{22}	...	k_{2q}	k_2
...	...	...	...	...	...
p	k_{p1}	k_{p2}	...	k_{pq}	k_p
sum	n_1	n_2	...	n_q	n

χ^2 is calculated as follows:

$$\chi^2 = \sum_{i=1}^{p}\sum_{j=1}^{q}\frac{\left(k_{ij} - E_{ij}\right)^2}{E_{ij}}, \tag{2}$$

$$E_{ij} = k_i \times \frac{n_j}{n}\; i = 1, 2, \cdots, p,\; j = 1, 2, \cdots, q. \tag{3}$$

In Formula (2), k_{ij} is the actual frequency of the feature i and the class j, E_{ij} is the expected frequency of the feature i and the class j. In the Formula (3), k_i is the frequency of the feature i, n_j is the frequency of the class j and n is the total number of samples.

Features and classes are used as two attribute indicators. Chi-square goodness of fit test is used to determine the correlation between the two. If the correlation is large, it will enter the feature subset to be retained, otherwise the data will be deleted. χ^2 test can be calculated in parallel, but the correlation between the selected feature and the feature to be selected is not considered. Therefore, the optimal feature subset cannot be obtained.

2.2 Minimum Redundancy-Maximum Relevance

The minimum redundancy-maximum relevance method (MRMR) is based on mutual information. The purpose of this method is to select feature subsets, in which features have strong correlation with classes and low feature redundancy. In order to make the mutual information between classes and features larger and the mutual information between features smaller, this paper uses the mutual information difference criterion to search feature subsets.

Minimum redundancy refers to minimizing the relevance between features,

$$\min W, W = \frac{1}{|F|^2} \sum_{i,j \in F} I(x_i, x_j). \tag{4}$$

In the Formula (4), F is the feature subset to be selected, $|F|$ is the number of features in the subset F, and x_i, x_j is the feature data of columns i and j respectively. $I(x_i, x_j)$ is the mutual information between feature i and feature j [11],

$$I(x_i, x_j) = \sum_{i,j} p(x_i, x_j) \log \frac{p(x_i, x_j)}{p(x_i) p(x_j)}, i \neq j. \tag{5}$$

Maximum relevance refers to maximizing the relevance between features and classes,

$$\max V, V = \frac{1}{|F|} \sum_{i \in F} I(C, x_i). \tag{6}$$

In the Formula (6), C is the class variable, $C = \{c_1, c_2, \cdots, c_n\}$, $I(C, x_i)$ is the mutual information between class C and feature i.

$$MRMR = \max \left[\sum_{i \in F} I(C, x_i) - \frac{1}{|F|} \sum_{i,j \in F} I(x_i, x_j) \right]. \tag{7}$$

As shown in Formula (7), the mutual information difference criterion is to maximize $V - W$.

2.3 Algorithm based on χ^2 Test and Minimum Redundancy.

Considering that, if the χ^2 test is simply combined with the minimum redundant maximum relevance, firstly, the χ^2 test is used to select the feature data highly relevant to the classes, and then in the minimum redundant maximum relevance algorithm, the mutual information is used to select the data highly relevant to the classes, resulting in effective data loss. Based on this, this paper proposes a feature selection algorithm (χ^2-MR) combining χ^2 test and minimum redundancy. The specific algorithm steps are as follows.

Step 1: Input the feature data D, class C, the threshold value P of χ^2 test and the feature number k of output.
Step 2: Set feature subset F as empty.

Step 3: Use χ^2 test to calculate the correlation between feature x_i, $i = 1, 2, \cdots, N$, and class C, calculate χ^2 value according to Formula (2), and calculate corresponding p_i.
Step 4: If $p_i < P$, add feature x_i to feature subset F_1.
Step 5: According to Formula (4), calculate the mean of mutual information between each feature and other features in feature subset F_1 in turn.
Step 6: Sort the mean of the mutual information for all the features from smallest to largest, the first k features with the smallest value enter the final feature subset F.
Step 7: Output the feature subset F.

In Step 2, set the feature subset F to an empty set. In Step 3, χ^2 test is performed on all feature data and class data to select feature data that are highly relevant to classes. When the χ^2 value is larger, the correlation is larger and the corresponding probability p_i is smaller. When the probability p_i is less than the threshold P, the feature is added to the feature subset F_1.

The next steps belong to the minimum redundancy algorithm. In order to minimize the redundancy of selected features, in Step 5, the mean of mutual information between each feature and other features in feature subset F_1 is calculated successively. The mean of mutual information is sorted from small to large. The smaller the mean value is, the less the redundancy of this feature and other features is. Select the first k features with the smallest mean to enter the final feature subset F. Thus, after the selecting method χ^2 - MR, features that are more relevant to the classes and less redundant from other features are picked out.

3 Numerical Experiments

In this paper, three data sets of gene expression are used in the experiment, and the detailed description of the data sets is shown in the Table 2. Three data sets are normalized and discretized to reduce the influence of dimension on experimental results.

Table 2. Experimental data sets

Data sets	Features	Samples and distribution
ALT-ALB-AML [12]	999	38(19 + 8 + 11)
Breast-A [12]	1213	98(11 + 51 + 36)
Stomach [13]	18556	129(20 + 109)

In the experiment, support vector machine, k-nearest neighbor method and geometric barycenter method are used to classify the data sets. Among them, the kernel function of SVM is linear kernel, K = 4 in KNN classifier, and the distance is set to Manhattan distance. The threshold value of χ^2 test P = 0.01.

To reduce the effect of over fitting and other factors, the data sets are cross-checked five times, and then the average values are determined. The model algorithm is tested in MATLAB R2014a. The classification accuracy obtained from the experiment is shown in Tables 3, 4 and 5.

Table 3. Classification accuracy of ALT-ALB-AML(%)

Classifier	Algorithm	Number of features in *F*							
		1	2	3	4	5	6	7	8
SVM	χ^2	68.6	69.4	71.1	**98.9**	**97.4**	97.3	97.4	97.3
	MRMR	71.8	76.8	73.9	79.3	83.9	76.4	82.1	92.5
	χ^2-MRMR	68.4	**77.4**	74.3	85.4	92.2	92.0	94.9	94.7
	χ^2-MR	**75.7**	72.9	94.6	**98.9**	92.1	**97.5**	**97.5**	92.1
KNN	χ^2	66.6	66.4	67.6	94.5	95.4	95.3	96.8	97.2
	MRMR	74.7	74.5	67.4	72.8	71.0	68.6	81.6	85.5
	χ^2-MRMR	56.2	61.4	76.9	95.9	93.7	86.8	90.5	94.4
	χ^2-MR	69.3	68.2	70.7	67.9	76.1	79.3	75.0	86.8
Geometric barycenter method	χ^2-MRMR	50.4	52.2	53.0	98.2	97.2	97.4	97.4	97.1
	MRMR	52.7	63.1	60.8	61.6	67.9	84.4	84.6	88.9
		73.9	57.4	80.5	97.5	97.4	97.4	94.6	**97.5**
	χ^2-MR	57.4	73.6	**97.4**	93.9	93.3	93.3	95.4	94.9

Table 4. Classification accuracy of Breast-A (%)

Classifier	Algorithm	Number of features in *F*							
		1	2	3	4	5	6	7	8
SVM	χ^2	83.6	82.0	84.9	83.7	83.4	84.3	83.9	84.7
	MRMR	77.4	70.7	82.4	83.0	78.2	84.7	85.7	88.0
	χ^2-MRMR	62.2	68.4	67.2	71.5	71.9	73.3	77.5	80.6
	χ^2-MR	89.0	81.8	**85.7**	**89.6**	**96.0**	87.8	91.0	88.9
KNN	χ^2	81.4	76.1	73.6	82.5	81.3	78.6	82.0	82.9
	MRMR	74.5	71.9	75.7	78.8	69.6	71.0	68.4	76.8
	χ^2-MRMR	54.2	66.5	73.6	78.4	77.4	80.5	82.5	84.7
	χ^2-MR	85.0	72.1	78.6	78.3	82.7	89.6	86.8	85.8
Geometric barycenter method	χ^2	59.2	74.5	76.4	76.6	73.5	76.5	79.6	74.4
	MRMR	73.7	73.0	80.9	74.5	63.3	75.5	78.6	77.6
	χ^2-MRMR	63.3	60.3	59.1	71.6	81.6	79.8	83.6	87.7
	χ^2-MR	**89.6**	**89.0**	84.9	87.6	87.9	**96.0**	**96.0**	**94.9**

Table 5. Classification accuracy of Stomach (%)

Classifier	Algorithm	Number of features in *F*							
		1	2	3	4	5	6	7	8
SVM	χ^2	92.3	92.4	92.4	92.7	92.7	93.4	**93.4**	92.7
	MRMR	92.3	92.3	92.2	92.3	92.3	92.2	92.3	92.3
	χ^2-MRMR	92.2	92.3	92.0	**93.3**	92.7	93.0	92.3	91.7
	χ^2-MR	93.3	**93.3**	**92.7**	93.2	93.0	92.6	92.9	**93.5**
KNN	χ^2	92.4	92.3	92.6	91.9	90.2	92.9	92.7	92.4
	MRMR	91.0	91.2	89.9	87.6	92.3	92.0	92.0	91.2
	χ^2-MRMR	92.3	92.2	92.3	91.8	92.2	92.3	92.2	91.5
	χ^2-MR	92.4	92.7	92.4	92.4	**93.5**	**93.6**	93.2	93.0
Geometric barycenter method	χ^2	71.3	72.1	86.8	83.8	83.8	88.3	91.5	93.0
	MRMR	77.5	65.1	53.4	69.0	66.7	65.1	66.6	69.8
	χ^2-MRMR	81.5	72.9	76.7	73.6	80.6	79.8	85.3	83.7
	χ^2-MR	**94.3**	86.1	84.5	90.7	91.3	89.9	87.6	92.3

The bold font in the table is the optimal correct classification rate predicted by different feature selection methods and classifiers when the feature subset size is the same. From the experimental results, we can see that the feature selection algorithm χ^2-MR has better performance than the single χ^2 test, the MRMR algorithm, or the combination of the two simplex feature selection algorithm χ^2-MRMR. Among the three classification methods, SVM achieves better classification accuracy in ALT-ALB-AML and Stomach data sets, while geometric classification is more suitable for Breast-A. It shows that the most suitable classifiers for different data sets are different. When the size of feature subset is below 8, the accuracy of disease prediction and classification can be very good, which further proves the high redundancy of gene expression data for disease prediction and classification.

4 Conclusion

This paper presents a feature selection algorithm based on chi-square test and minimum redundancy, and a classification method based on geometric center of gravity. Through several classifiers on ALT-ALB-AML, Breast-A and Stomach datasets, it is found that the feature selection algorithm combined with chi-square test and minimum redundancy is better than chi-square test or minimum redundancy and maximum relevance algorithm or simply the two, and the effectiveness of the algorithm is verified. For gene expression data, the algorithm combining chi-square test and minimum redundancy effectively selects the features of strong correlation with classes and low redundancy, and geometric barycenter classification effectively improves the correct classification rate. Although the algorithm effectively removes redundant genes and improves the correct classification rate, some experimental data sets are unbalanced, the number of samples is small, there is a certain contingency, which needs to be further solved in the future.

References

1. Avrim, L.B.: Pat L: Selection of relevant features and examples in machine learning. Artif. Intell. **97**(1), 245–271 (1997)
2. Saeys, Y., Inza, I., Pedro, L.: A review of feature selection techniques in bioinformatics. Bioinformatics **23**(19), 2507–2517 (2007)
3. Ding, C., Peng, H.: Minimum redundancy feature selection from microarray gene expression data. J. Bioinform. Comput. Biol.**3**(2), 185–205 (2005)
4. Sakar, C.O., Kursun. O., Gurgen, F.: A feature selection method based on kernel canonical correlation analysis and the minimum redundancy-maximum Relevance filter method. Expert Syst. Appl. **39**, 3432–3437 (2012)
5. Mandal, M., Mukhopadhyay, A.: An improved minimum redundancy maximum relevance approach for feature selection in gene expression data. Procedia Technol. **10**, 20–27 (2013)
6. Yao, H., Wang, N., Qi, M., Li, Y.: Research on the improved maximum relevance minimum redundancy feature selection method. Comput. Eng. Appl. **50**(9), 116–122 (2014)
7. Chen, J., Hu, B.: Application of naive Bayesian classifier based on maximum relevance and minimum redundancy. Chin. J. Health Stat. **32**(6), 932–934 (2015)
8. Radovic M, Ghalwash M, Filipovic N: Obradovic: minimum redundancy maximum relevance feature selection approach for temporal gene expression data. BMC Bioinform. **18** (9) (2017). https://doi.org/10.1186/s12859-016-1423-9
9. Chen, C., Liang, X.: Feature selection method based on Gini index and chi-square test. Comput. Eng. Des. **40**(8), 2342–2345 (2019)
10. Guo, Y., Li, G.: A filtering feature selection framework for multi marker data. CAAI Trans. Intell. Syst. **9**(3), 292–297 (2014)
11. Gurban, M., Thiran, J.P.: Information theoretic feature extraction for audio-visual speech recognition. IEEE Trans. Signal Process. **57**(12), 4765–4776 (2009)
12. Cancer Program Legacy Publication Resources. https://portals.broadinstitute.org/cgi-bin/cancer/datasets.cgi. Accessed 16 Apr 2018
13. The Cancer Genome Atlas. https://portal.gdc.cancer.gov.Accessed 06 Dec 2019

An Approach to Stock Price Prediction Based on News Sentiment Analysis

Xiao Huang(✉)

Beijing University of Technology, Beijing 100101, China
872215409@qq.com

Abstract. The sentiment of the stock market can reflect the behavior of investors to a certain extent and influence their investment decisions. As a kind of unstructured data, market news can reflect and guide the overall environmental sentiment of the market. Together with stock prices, it can become a crucial market reference data, which can effectively help investors' investment decisions. This paper proposes a vectoring method that can accurately and quickly establish a multiple-bit emotional feature for massive news data. It uses a support Victor Machine (SVM) model to predict the impact of financial news on the stock market and uses bootstrap to alleviate the problem of over-fitting. The results of the experiments conducted on the Shanghai and Shenzhen stock indexes indicate that the proposed method can improve the prediction accuracy by about 8% compared with the traditional model, and obtained an excess return of 6.52% in the three-month backtest, which proves its effectiveness.

Keywords: Stock market forecasting · Financial sentiment-driven · News · Text features · Trading signals · Artificial intelligence

1 Introduction

The stock market is an extremely important financial market. As an important participant in the stock market, investors' emotional changes will be quickly reflected in the market. With the development of information technology, the channels and speeds for investors to obtain information have changed. The rich information contained in social media and news sites has gradually become another important factor that affects investors' investment expectations and even investment decisions. In other words, in addition to the impact of structural change data such as stock price historical data, the stock market also receives some indirect effects of unstructured data, such as the impact of fermentation of news events on investor behavior [1]. Financial news inducement magnifies investors' attitude towards the stock market [2]. With the proliferation of news, blogs, forums, and social network speech, the text on the Internet provides information that reflects the psychology of investors [3].

However, although these data are important, they are difficult to apply directly to pricing and the prediction results are heavily dependent on subjective experience. Such a method not only has omissions but also tends to be unbalanced in emotional understanding and accuracy, never causing serious deviations in results. Therefore, to ensure the accuracy and timeliness of the analysis, this paper builds a prediction model

M. Tavana et al. (Eds.): IISA 2020, AISC 1304, pp. 179–185, 2021.
https://doi.org/10.1007/978-3-030-63784-2_23

of financial news characteristics on the stock market for online financial news and historical stock price data. The model is based on support vector machines, builds the training set based on stock historical data, and reduces over-fitting by bootstrap to improve the model generalization ability.

The main contributions of this paper are as follows: 1) Use multiple dimensional features to determine the emotional polarity of financial news, and give different feature weights to the feature entries according to the obviousness of the features, to improve the analytical ability of text data; 2) Based on the stock time series data. Construction, to avoid the risk of emotional bias of human judgment; 3) The weight of the optimization model must be super-parametric, and bootstrap is used to randomly sample to avoid over-fitting. The empirical results verify that the model in this paper is indeed important for financial news processing.

2 Financial News Sentiment Mining Model

2.1 Model Structure

The traditional news sentiment mining model has some defects, mainly concentrated in the following aspects. 1) Inaccurate description of characteristic words. This model judges the special vocabulary of financial news, but it only uses a general dictionary, which lacks the corresponding accuracy. 2) The lack of emotion in financial news. Financial news is usually expressed as a terminology; vector and traditional models rarely use emotion to analyze news behavior at the semantic level. 3) The financial news mark is not comprehensive. The traditional model only marks each financial news and assigns a label (such as positive or negative), and then in the classification model, the news fragments are sorted according to the time-stamp and labeled with the label value. In addition to this, some model methods only classify the tag values for financial news tags.

Given this, this article improves from the following aspects: First, in the text processing of financial news, the polarity of emotions is judged, a feature lexicon is constructed, and the news is analyzed using multiple features to ensure prediction accuracy and reduce judgment errors; Secondly, for the labeling of news, it is combined with minute-level stock data, and the program is used to automatically label, which reduces the time cost of manually labeling a large amount of news based on ensuring the accuracy of prediction, and also verifies the large feasibility of rapid labeling of batch news; again. According to bootstrap random sampling in all samples to reduce over-fitting; besides, the training set and the test set are strictly separated to ensure the reliability and accuracy of the results. The processing model proposed in this paper is shown in Fig. 1.

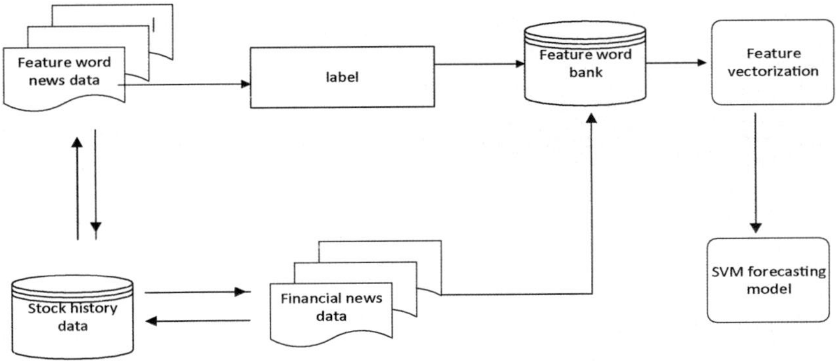

Fig. 1. Structure of text feature mining model

2.2 Label Generation

To exclude the impact of the stock's fluctuations, this article uses the changes in the stock price's abnormal return rate before and after the news release to quantify its impact on the stock market fluctuations.

Normal rate of return. Normal return refers to the expected return of the stock market in the absence of financial news. [4] Usually, the fixed average return model is used to estimate the normal return. Before calculating the normal rate of return, the rate of return needs to be calculated. The rate of return index is calculated using the following formula:

$$Ri,t = (pi,t + 10 - pi,t)/Pi,t \tag{1}$$

where Ri,t means financial news i at the time of publication £ after 10 min, the Shanghai and Shenzhen stocks refer to the changing rate of return, Pi,t means the index of financial news i at the time of publication, and Pi,t + 10 means the index of financial news i after the publication time of £10 min.

For financial news i, the stock index changes of T0 $\sim$ T10 the stock index changes as a sliding window, through regression analysis of the fixed average return model, [5] the estimated value of u is:

$$\mu = \mathrm{Ri,t} + \varepsilon i,t \tag{2}$$

where u represents the unpublished normal rate of return of financial news,

e i,t represents the estimated value of the deviation between Ri,t.

Abnormal rate of return. Abnormal rate of return refers to the difference between the actual rate of return and the normal rate of return after 10 min at the time t of financial news i:

$$\mathrm{AR}_{i,t} + 10 = R_{i,t+10} - \mu \tag{3}$$

According to the literature, the abnormal rate of return can be regarded as the impact of financial news on the market and is divided into positive and negative. The formula is as follows:

$$label = \begin{cases} 1, & if \quad ARi, t \leq 0 \\ -1, & if \quad ARi, t > 0 \end{cases} \tag{4}$$

Where the label shows the emotional label for news i.

2.3 Feature Word Mining

There are many feature words in the content of financial news, which cannot be directly used for data processing like structured data. Therefore, it needs to be improved from the feature transformation of the news itself and understanding. First, convert the text data to the corresponding feature space. Secondly, a feature vocabulary is established to store representative feature words. Special Characteristics, the thesaurus is used to judge the similarity of the news features of the training set to improve the accuracy of the news feature segmentation. This article refers to the feature word segmentation method used by Schumaker et al. in segmenting feature words.

In the experimental process, for the selection of related word features, standard chi-square system measurement is used to achieve. The chi-square test can better reflect the relevance of a word vector in different categories, and the words that reflect the text characteristics in financial news. The form of the chi-square test is shown in Eq. (5):

$$\chi^2(t, c) = \frac{N*(AD - CB)^2}{(A + C)*(\mathrm{B} + \mathrm{D})*(\mathrm{A} + \mathrm{B})*(\mathrm{C} + \mathrm{D})} \tag{5}$$

where N represents the total number of documents in the training data set; A represents the number of documents containing both entries belonging to category c; B represents the documents containing entries £, but not belonging to category c number of documents; C represents the number of documents that belong to category c, but do not contain the entry £; D indicates the number of documents that do not belong to category c and do not contain the entry £; X indicates the final calculated chi-square value.

For the commonly used tf-idf method, the higher the word frequency, the higher the tf-idf value of the document [6]. Obviously, it does not need to extract high-frequency words in the document, but needs to extract words with obvious features. In the chi-square test, it is more suitable for extracting the characteristic terms of financial news. So in particular, the chi-square test can calculate the chi-square value of different feature entries. Therefore, the chi-square value can be used to obtain the relevant feature words in the order of feature discrimination. To make these words more accurate in the judgment of news, in the subsequent experiments, the different feature words are given corresponding feature weights.

2.4 Model Training

After completing the construction of the feature vocabulary, the training text needs to be preprocessed, except for removing unnecessary stop words and punctuation marks. And it is necessary to compare it with the feature words in the vocabulary sex comparison, finally obtaining the feature vector mark. Then it is significant to use SVM for model classification training [7]. The experiment iterated the training process, cross-validated each iteration, and used bootstrap for random sampling, randomly assigned training set (80%) and test set (20%). The pseudo-code of the training process is as follows.

Algorithm 1

Input: a training set $S = \{(x_{11}, x_{21},\ldots, x_{1m}, y_1).(x_{21,}x_{22} \ldots, x_{2m}, y_2), \ldots, (x_{n1}, x_{n2},\ldots, x_{nm}, y_n)\}$

Output:

1. while accuracy < 75 do
2. Strain, Stest ← S where Strain is the sampled training set and Stest is the remaining test set. This step mainly uses the bootstrap method to randomly sample from the data set with replacement.
3. Perform cross-validation to obtain accuracy.
4. end while.

3 Test Results and Evaluation

3.1 Experiment Preparation

This article captures a total of 15,968 financial news about the Shanghai and Shenzhen macro stock markets in 2016–2018 from Sina Finance, Flush, and China Securities. The financial news is all in simplified Chinese, and the experiment is based on the SSE 500 index in the same time period. Market reference. The experimental environment uses Python 3. 7. The compilation platform is Pycharm.

3.2 Experimental Results and Comparison

This article will refer to the BP neural network (BP-NN) proposed in the reference, Bernoulli's Naive Bayes Model (B-NB), which is widely used in text classification, and the sentiment discrimination model proposed in this paper for comparative experiments [8]. The accuracy of each model is as listed in Tables 1 and 2.

Table 1. Accuracy of each model in teat set (positive)

Positive prediction	Accuracy rate	Recall rate
BP-NN	0.53	0.54
SVM	0.65	0.68
B-NB	0.62	0.55

Table 2. Accuracy of each model in teat set (negative)

Negative prediction	Accuracy rate	Recall rate
BP-NN	0.54	0.59
SVM	0.63	0.62
B-NB	0.68	0.55

It can be seen that in the judgment of positive news, SVM has better accuracy than the other two models. But in the classification of Negative news, Bernoulli's Naive Bayes has certain advantages over SVM. This may be because the market strives to create an upward atmosphere, and positive news dominates, so negative news is generally more credible and has a stronger linear correlation with stock price fluctuations, and it is more suitable for the Bernoulli Naive Bayes algorithm. [9] According to the previous analysis, it is obvious that SVM with better news prediction is more advantageous. The effect of BP-NN is not very good, because the total amount of corpus is limited, and the neural network does not handle the small data set very well [10]. Figure 2 shows the results of applying the model of this paper to the Shanghai 500 for a 3-month return test. It can be seen that excess returns have been increasing steadily, proving the effectiveness of the proposed method.

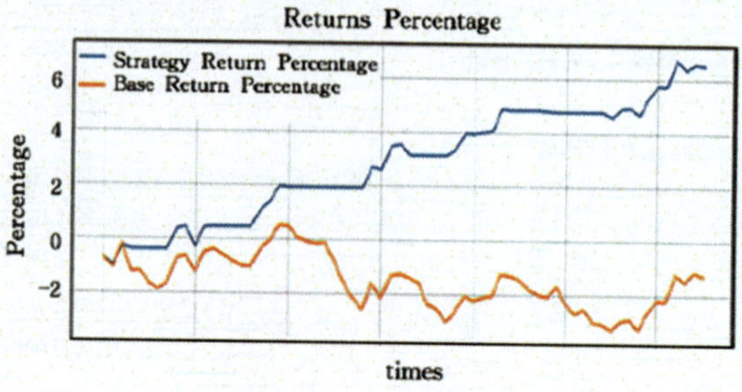

Fig. 2. Income comparison

4 Conclusion

This article proposes an SVM model that predicts changes in the market through the change in the emotional polarity of financial news and is verified through experiments. The goal of the next step is to further expand the source of financial texts, not limited to news and announcements, and to combine its polarity and influence factors to further explore the breadth and depth of financial text mining. In the future, if the SVM model can be adopted in most financial areas, the phenomenon of information asymmetry will be greatly reduced.

References

1. Oliveria, N., Cortez, P., Areal, N.: Stock market sentiment lexicon acquisition using microblogging data and satistical measures. Decis. Support Syst. **85**, 62–73 (2016)
2. Long, W., Tang, Y., Tian, Y.: Investor sentiment identification based on the Universum SVM. Neural Comput. Appl. **30**(2), 661–667 (2018)
3. Perikos, I., Hatzil, Y.I.: Recognizing emotions in the text using ensemble of classifiers. Eng. Appl. Artif. Intell. **51**, 191–201 (2016)
4. Hajek, P.: Combining bag-of-words and sentiment features of annual reports to predict abnormal stock returns. Neural Comput. Appl. **29**(7), 343–358 (2018)
5. Yao, W.D., Wang, R.J.: An empirical study of the relationship between stock market volatility and policy events from the perspective of structural decomposition- based on EEMD Algorithm. Shanghai Econ. Res. **1**, 71–80 (2016)
6. Yan, H., Hongbing, O.: Financial time series prediction based on deep learning. Wirel. Personal Commun. **102**(2), 683–700 (2017). https://doi.org/10.1007/s11277-017-5086-2.
7. Sim, H.S., Hae, I.K., Jae, J.A.: Is deep learning for image recognition applicable to stock market prediction? Complexity, 1–10 (2019), 19 February 2019. https://doi.org/10.1155/2019/4324878.
8. Li, X., Xie, H., Wang, R., Cai, Y., Cao, J., Wang, F., Min, H., Deng, X.: Empirical analysis: stock market prediction via extreme learning machine. Neural Comput. Appl. **27**(1), 67–78 (2014). https://doi.org/10.1007/s00521-014-1550-z
9. Golmohammadi, K., Zaiane, O.R., Diaz, D.: Detecting stock market manipulation using supervised learning algorithms. pp. 435–441. IEEE (2014). https://doi.org/10.1109/DSAA.2014.7058109
10. Chong, E., Han, C., Park, F.C.: Deep learning networks for stock market analysis and prediction: methodology, data representations, and case studies. Expert Syst. Appl. **83**, 187–205 (2017). https://doi.org/10.1016/j.eswa.2017.04.030https://doi.org/10.1016/j.eswa.2017.04.030

A Stock Prediction Method Based on LSTM

Sijie Zhou(✉)

School of Electrical Engineering, Sichuan University, Chengdu 610065, China
zzz_1527@163.com

Abstract. Length for the current cycle when the neural network in stock forecasting widespread lag problem, put forward a kind of improved stock prediction method based on the LSTM, first by multidimensional vector input, select other companies with stock prices higher correlation coefficient of the daily closing price of stocks, combination forecasting stock price data as input vector of the model itself; Secondly, different eigenvectors are selected as input vectors by feature engineering, and eigenvector combinations that can significantly reduce the predicted hysteresis can be obtained through repeated training. Finally, through the emotional analysis of news texts related to stock companies, the emotional score obtained is taken as the model input vector. The prediction results of Tencent stock show that this method not only improves the prediction accuracy, but also significantly improves the prediction hysteresis.

Keywords: Stock prediction · LSTM · CNN

1 Introduction

The stock market is a highly complex nonlinear dynamic system. There are many internal and external factors affecting its fluctuation, and most of them are difficult to quantify. Meanwhile, the information and data sets related to stocks are very large, which makes the traditional non-artificial intelligence methods often unsatisfactory in stock price prediction. With the development of artificial intelligence and big data technology, more and more machine learning algorithms emerge, such as decision tree genetic algorithm support vector machine logistic regression and deep learning network model (such as convolutional neural network (CNN)) cyclic neural network (RNN) and long-duration cyclic neural network (LSTM). Among them,LSTM overcomes the shortcoming that RNN will forget the previous state information over time, and has the characteristics suitable for processing and predicting important events with long intervals and delays in time series, In recent years, LSTM has been widely used in many fields, including speech recognition, document recognition, handwriting recognition and image analysis. Literature [10] compares the prediction results of Shanghai Composite Index volatility with 18 classical models, and the experiment shows that LSTM has obvious advantages in prediction. However, as one of the time series prediction models, THE LSTM is still dominated by the input of historical data in stock prediction. Therefore, there is a general lag problem. Aiming at this lag, this paper improves the LSTM prediction method through three steps of multi-dimensional vector input feature engineering and emotion analysis, Filter input variables and features,

M. Tavana et al. (Eds.): IISA 2020, AISC 1304, pp. 186–193, 2021.
https://doi.org/10.1007/978-3-030-63784-2_24

optimize model input, reduce lag caused by input historical data, and improve model prediction efficiency and accuracy.

2 Structure and Calculation Principle of LSTM

LSTM is a specific form of RNN, and RNN will encounter the problem of gradient disappearance or gradient expansion when dealing with long-term dependence and long-distance nodes in time series. LSTM adopts the mechanism of control gate, by adding memory cells, input gate, forgetting gate and output gate, the weight of self circulation is not fixed, so that the integration at different times can be achieved when the model parameters are fixed The dynamic change of scale avoids the disappearance or expansion of gradient.

(1) Cellular state
In addition to propagating the hidden state ht forward, another hidden state, cell state, is added to LSTM at each sequence index position t, which is recorded as ct.
(2) Calculation principle of input gate, forgetting gate and output gate

The gate of LSTM at each sequence index position t includes forgetting gate, input gate and output gate. The input gate is responsible for processing the input of the current sequence position, which consists of two parts: the first part uses sigmoid activation function, and the output is it; the second part uses tanh activation function, and the output is at.

$$i_t = \sigma(W_i * h_{t-1} + U_i * X_t + b_i), \tag{1}$$

$$a_t = tanh(W_o * h_{t-1} + U_o * X_t + b_o), \tag{2}$$

Where: W_i, U_i, b_i, W_o, U_o and b_o are the coefficient and bias of the linear relationship; σ is sigmoid activation function.

The role of forgetfulness gate in LSTM model is to control the activation value f of forgetfulness gate at time t of hidden cell state in the upper layer through a certain probability. The formula is as follows

$$f_t = \sigma\left(W_f * h_{t-1} + U_f * X_t + b_f\right), \tag{3}$$

Where: W_f, U_f, b_f are the coefficient and bias of the linear relationship.

According to the calculation results of the input gate and forgetting gate, the cell state update value C_t at time t can be obtained, and the formula is as follows

$$C_t = C_{t-1} * f_t + i_t * a_t, \tag{4}$$

The value of the output gate can be calculated from the cell state update value as follows

$$O_t = \sigma(W_o * h_{t-1} + U_o * X_t + b_o), \tag{5}$$

$$h_t = O_t * \tanh(C_t), \tag{6}$$

Where, W_o, U_o, b_o are the coefficients and biases of the linear relationship.

With the above calculation, the LSTM can effectively utilize the input to enable its long-term memory function.

3 Stock Price Forecasting Method Based on LSTM

3.1 Forecasting Process

To predict daily closing price of the stock company tencent, for example, is given based on the prediction of stock price process and realization of the LSTM, stock price forecasting method based on the LSTM by learning history data, find the stock daily closing price and input variables of the nonlinear relationship, and the nonlinear relationship between generations into used to predict the input data, get the closing price of the future, forecasting process is as follows:

1) Collect the historical data of Tencent stock closing price, confirm the data of training set and test set;
2) The previous day's closing price of the stock in the training set was used as the input vector, and the next day's closing price was used as the target value to enter the LSTM model;
3) Train the model and predict the test data.

In this paper, the daily closing price of Tencent solstice on March 29, 2016 (except the rest day) on March 29, 2018 is used for training and prediction. The data of the first 400 days is used as the training set, and the remaining data is used as the evaluation index of the predicted results of the test set:

1) Prediction deviation

   $\text{acc} = \frac{1}{L}\left|\sum_{i=1}^{L}\frac{(y_i - y_i')}{y_i}\right|$, where L is the forecast days of the test set, y_i is the predicted value, y_i' represents the true value of the test set. The smaller the acc, the higher the prediction accuracy.
2) Prediction lag

   Expressed as $\text{lag} = \sum_{i=1}^{M} t_i - \sum_{i=1}^{N} t_k$. Where, M, N respectively, represent the sum of the maximum and minimum values in the test set prediction value and the real value sequence, t_i and t_k represents the ordering values of the extreme values of the i and k of the predicted value and the real value sequence of the test set, respectively. The smaller the lag, the smaller the lag.

3.2 Prediction Results and Analysis

The predicted value and the true value of the closing price of the test set are shown in Fig. 1, where the blue curve is the true value and the red curve is the predicted value and the output result is acc = 0.027, lag = 51.

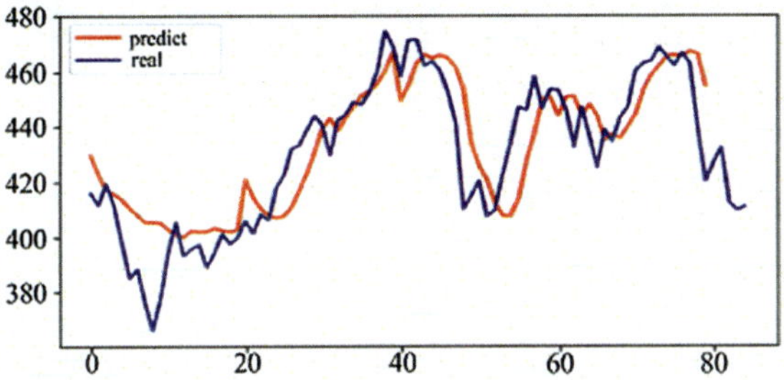

Fig. 1. The predicted value and the true value of the closing price of the test set

As can be seen from Fig. 1, in terms of accuracy, the predicted value of some dates (0–20 days) has a large deviation from the true value. In terms of hysteresis, the predicted value curve and the real value curve have obvious delayed translation, that is, the change of the predicted result is often behind the actual change, and the hysteresis is obvious.

4 Improved Stock Forecasting Method Based on LSTM

By above knowable, the stock price prediction based on LSTM has large improvement in accuracy and hysteresis, the main reason is the company's stock's closing price is not only related to its own historical data, also may be associated with other factors from the internal factors, the company's internal management structure of company policy change main shareholder stake on company earnings, etc. can all influence investor's investment strategy, so as to affect the stock price; From the perspective of external factors, the behavior of competitors and partners related to macroeconomic performance, news and public opinions will also affect the stock price forecast based on LSTM, which requires the input of digital data in the form of days. As for many influencing factors mentioned above, The paper mainly considers the company with investment or cooperative relationship and news public opinion and other factors to improve the prediction method.

4.1 Dimensional Vector Input

Taking the stock price forecast of Tencent as an example, the correlation coefficient r of the statistical index is used to determine which company's stock price is highly correlated with Tencent's stock price. The value range of r is -1–1. The closer to 1, the higher the positive correlation is.

$$r = \frac{\sum (x-\bar{x})(y-\bar{y})}{\sqrt{\sum (x-\bar{x})^2}\sqrt{\sum (y-\bar{y})^2}}.$$

There are several kinds of data that need to be calculated: the companies that invest in Tencent mainly come from JPMorgan Chase; the companies that Tencent invests mainly come from 58 Tongcheng, Ali pictures, Jingdong, etc.; Tencent platform concept stocks refer to the companies that have cooperative relations with Tencent, mainly including Changshan Beiming, information development, Annie shares, etc. According to the date, the closing price of the above company's historical stock and Tencent's historical closing price are corresponded one by one, forming two series respectively. SPSS statistical analysis software is used to calculate the correlation coefficient, and the significance level p is given. When p is not more than 0.05, the correlation coefficient is significant. The top three positive correlations were: JPMorgan, 0898; Jingdong, 0.887; Huanan City, 0.734. Table 1 shows the correlation calculation results of some companies. The corresponding companies are JPMorgan Chase, Jingdong, South China City, Ali film, information development, Changshan Beiming, 58 Tongcheng, Annie Co., Ltd., zhangqu technology, zhongqingbao and Leju.

Take jpmorgan Chase as an example, and give a schematic diagram of its changes with the historical stock price of Tencent. 4. In order to show its correlation more conveniently, logarithmic coordinate is used in the vertical axis.

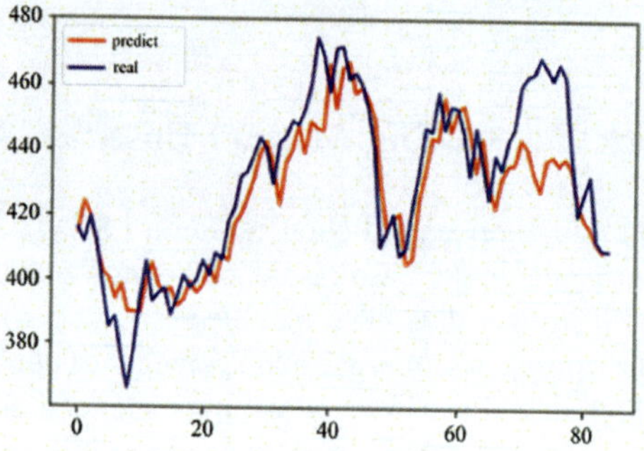

Fig. 2. The predicted value of the closing price based on LSTM

As shown in Fig. 2, The blue curve represents the change in the logarithm of Tencent's closing price, while the red curve represents the change in the logarithm of jpmorgan's closing price. The two logarithm curves show a consistent trend.

The above three companies from the solstice 29 March 2016 on March 29, 2018 daily closing price of shares with tencent closing itself as a 4 d input vector, and the input LSTM model for 400 days before the training and prediction data as the training set, the data as a test set, to possess the ability to improve the prediction accuracy and hysteresis. The output result is acc = 0.023, lag = 12.

4.2 Feature Engineering

RLT model integrates three association methods: Referer, redirection and TCP long connection. The fusion association enables the above three association methods to learn from each other in different scenarios and improve the overall association success rate through complementarity. It can be seen from Fig. 5 that in this paper, the RLT model can be associated with data by an average of 97%, which is higher than the success rate of the above three methods, and the association effect is better.

Feature is extracted from the data to predict the results of useful information, the characteristics of the project's main purpose is to find the better task implementation characteristics, based on the characteristics of choice, eliminate irrelevant or redundant features, reduce the feature dimension, improve the accuracy of the algorithm, at the same time reduces the time required to program is running.

Stock prediction based on LSTM application lag exists, the reason mainly lies in the input data are based on the previous day's stock price data found in the process of model training, when dealing with the data of neuron weights reach a high level, the training error reduced to a small level, cause the trained neural network algorithm of past data actually turned into translation, the paper consider the stock of the daily closing price the day opening price high There is also a correlation between the lowest price, so the data of the day is also taken as the input feature vector, which makes the corresponding weight of the data of the day before the input become smaller, thus reducing the lag of the prediction.

The closing price of jpmorgan's shares in Beijing Donghua South City the previous day and Tencent's opening price of the same day were taken as 6-dimensional input vectors, the data of the first 400 days were taken as training set, and the rest data were taken as test set. LSTM was used to conduct training and prediction, and the accuracy and hysteresis were further improved, and the output result was acc = 0.016, lag = -108.

4.3 Emotional Analysis of News

News public opinion is one of the important factors that affect the share price, which affects investor's decision, so it should be one of the characteristics of the quantitative analysis of public opinion to sentiment analysis technical paper USES web crawler to crawl sina finance and economics website related to tencent news text, and use of the existing text analysis tools - tencent Wen Zhi to crawl news emotional analysis, get daily news emotional score value Q. Q = positive scores - negative scores, the closer to 100, and Q show emotions tend to be more positive, the closer the Q - 100, shows that emotions tend to be more negative for using tencent Wen Zhi Carry out the emotional analysis chart of news.

Jpmorgan jingdong south China city stock's closing share price the day before, and tencent day opening price high emotional score value of the lowest price is the day before tencent etc. as 7 d input vector, the first 400 days of data as a training set, the data as a test set, using the LSTM training and forecasting, again to improve prediction accuracy and the output of lag. The output result is acc = 0.010, lag = -189. As shown in Fig. 3.

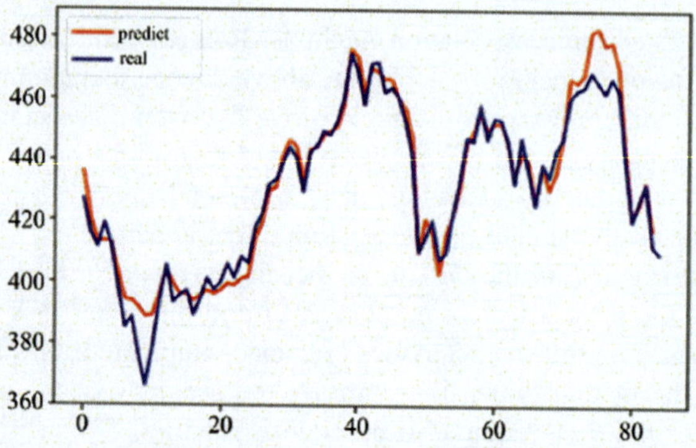

Fig. 3. The predicted value of the closing price based on LSTM with emotional analysis

The numerical comparison between the predicted and actual values of some test sets is shown in Table 1.

Table 1. The predicted value of the test set is compared with the actual value

Sequence number	Predicted value	Actual value	Sequence number	Predicted value	Actual value
1	435.7	426.8	2	421.3	415.8
3	412.9	411.4	4	413.4	419.2
5	412.7	411.6	6	402.9	398.0
7	394.2	385.0	8	393.8	388.4
9	391.4	376.0	10	388.5	366.0
11	388.9	378.0	12	396.0	394.0
13	404.1	405.4	14	399.5	393.2
15	395.9	395.6	16	396.0	397.2
17	393.7	389.0	18	395.5	394.2
19	397.1	401.2	20	396.1	397.4
21	397.2	399.8	22	399.4	405.8
23	398.8	401.2	24	401.1	408.0

5 Conclusion

Paper is common stock prediction based on LSTM lag problem, through the cause analysis, adopt multidimensional vector input characteristics of the engineering and news sentiment analysis three steps, evaluate the model input variables and the characteristics and selection, optimization model of input, improve efficiency of tencent company stock prediction results show that the method in terms of hysteresis and accuracy are improved significantly.

References

1. Chang, P.C., Wu, J.L.: A takagi-sugeno fuzzy model combined with a support vector regression for stock trading forecasting[J]. Applied Soft Computing **38**, 831–842 (2016)
2. Huo, J., Zheng, Y., Chen, X.: Implementation of transaction trend prediction model based on regression analysis[J]. Journal of Baoshan Teachers' College **117**(1), 19–23 (2009)
3. Yu, Y., Si, X., Hu, C., Zhang, J.: A review of recurrent neural networks: LSTM cells and network architectures[J]. Neural Comput. **31**(1), 1–36 (2019)
4. K Rollner, Vanstone B, Finnie G. Financial time series forecasting with machine learning techniques: a survey. In: Proceedings of 18th European Symposium on Artificial Neural Networks Computational Intelligence and Machine Learning, Bruges(Belgium), 2010: 28–30.

Research on DTC Fuzzy PID Control of Permanent Magnet Synchronous Motor Based on SVPWM

Menglin Ma(✉) and Mengda Li

Shanghai Dianji University, Shanghai 201306, People's Republic of China
1227597514@qq.com

Abstract. Permanent magnet synchronous motor has been widely used because of its simple structure, high efficiency and low loss. It has important research value to improve its control level. Direct torque control (DTC) technology directly controls the torque by looking up the table. Because of its fast dynamic response and strong robustness, it has been widely used in motor control. However, the traditional direct torque control has some problems, such as large torque ripple and serious current distortion, which is not conducive to motor control. By analyzing the shortcomings of the traditional direct torque control (DTC) and realizing its shortcomings, a Direct Torque Fuzzy PID control of PMSM Based on SVPWM is designed. SVPWM is used to replace search module and fuzzy PID controller is used to replace PI regulator. The simulation model of the system is established in Matlab/Simulink. Simulation results show the effectiveness of the optimization strategy.

Keywords: PMSM · DTC · Fuzzy PID · SVPWM

1 Introduction

Most of the research on PMSM is focused on the control. As one of the classical control strategies, direct torque control is more intuitive and concise than other control methods [4].

It mainly observes the stator flux linkage by detecting the stator resistance, which saves a lot of complicated calculation [1]. Compared with vector control, it reduces the need for motor models. The electromagnetic torque has a simple structure and fast response speed, which improves the robustness of the system [12].

In view of the shortcomings of traditional direct torque control, such as serious stator phase current distortion and large torque ripple, fuzzy PID controller is used to replace PI governor, and SVPWM technology and direct torque control technology are combined to replace traditional direct torque. The control technology improves the control performance of torque and flux linkage.

M. Tavana et al. (Eds.): IISA 2020, AISC 1304, pp. 194–203, 2021.
https://doi.org/10.1007/978-3-030-63784-2_25

2 PMSM Mathematical Model and Traditional DTC Control Scheme

2.1 Mathematical Model of PMSM in Coordinate System

The DQ rotor flux rotation coordinate system is used to control the PMSM. The coordinate system relationship is shown in Fig. 1.

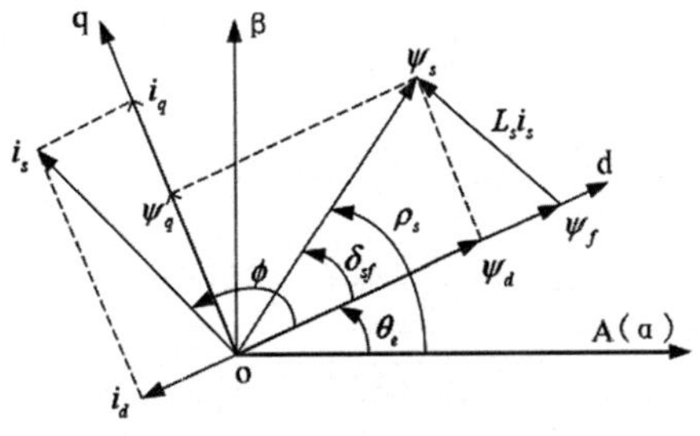

Fig. 1. Relationship between different coordinate systems

After coordinate transformation, the mathematical model of PMSM in DQ rotor flux rotation coordinate system is as follows:

Voltage equation:

$$\begin{cases} u_d = R_s i_d + \dfrac{d\psi_d}{dt} - \omega_s \psi_q \\ u_q = R_s i_q + \dfrac{d\psi_q}{dt} - \omega_s \psi_d \end{cases} \tag{1}$$

In formula (1), the meaning of each symbol is as follows (Fig. 2):

Nane	Meaning
U_d, U_q	the components of stator voltage in dq coordinate system
I_d, I_q	the components of stator current
Rs	the stator resistance
ψ_d, ψ_q	the components of stator flux in dq coordinate system

Fig. 2. Formula (1) symbolic interpretation

Flux linkage equation:

$$\begin{cases} \psi_d = L_d i_d + \psi_f \\ \psi_q = L_q i_q \end{cases} \tag{2}$$

In formula (2): L_d, L_q are synchronous inductors of stator vertical axis and direct axis respectively; ψ_f is permanent magnetic flux linkage, with value of 0.183 wb.

Torque equation:

$$T_e = \frac{3}{2}P(\psi_d i_d - \psi_d i_d) \tag{3}$$

In formula (3): P is the motor pole pair.

2.2 Principle of Direct Torque Control

DTC control deals with formula (3), and the final torque equation is:

$$T_e = \frac{3p}{4L_d L_q}|\psi_s|(2\psi_f L_q \sin\delta + |\psi_s|(L_d - L_q)\sin 2\delta_S) \tag{4}$$

In formula (4), the meaning of each function is as follows (Fig. 3):

Name of name	Note
$\lvert\psi_s\rvert$	The amplitude of the stator flux
δ_s	The torque angle
ψ_f	Permanent magnet flux and fixed value

Fig. 3. Formula (2) symbolic interpretation

When the amplitude of stator flux linkage is controlled to a fixed value, then the electromagnetic torque is only related to the load angle. DTC controls torque through load angle [2].

2.3 PMSM-DTC System Composition

The pmsm-dtc system diagram is shown in Fig. 4.

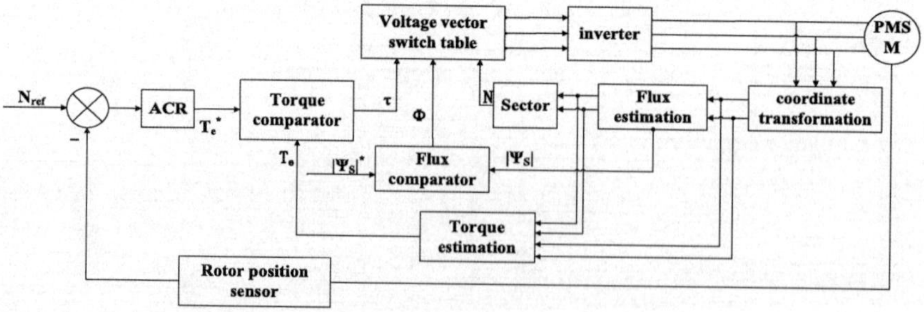

Fig. 4. PMSM-DTC overall structure chart

From above, PMSM-DTC system consists of speed regulator, hysteresis controller (including torque hysteresis control and flux hysteresis control), switch table, stator flux and electromagnetic torque estimation module, PMSM, inverter, etc. [11].

2.4 PMSM-DTC System of Simulation

2.4.1 Analysis

Matlab software is used to model the traditional DTC mode. The parameters of the control system are set as follows: the simulation time is set as 0.4s, the load is added at 0.2S, the inertia set by the permanent magnet motor is 0.003 J (kg. M^2), the given voltage is 311 V, the speed of the control motor is 1000 r/min, and the stator resistance is 0.956 Ω. Observe the output speed, torque and stator flux track of the system after adding load.

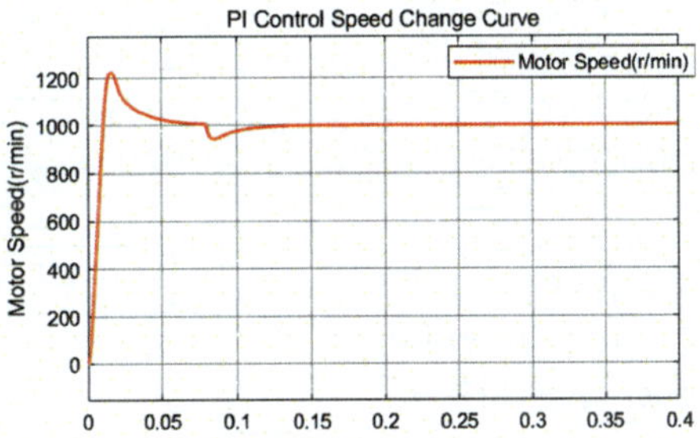

Fig. 5. Speed change curve

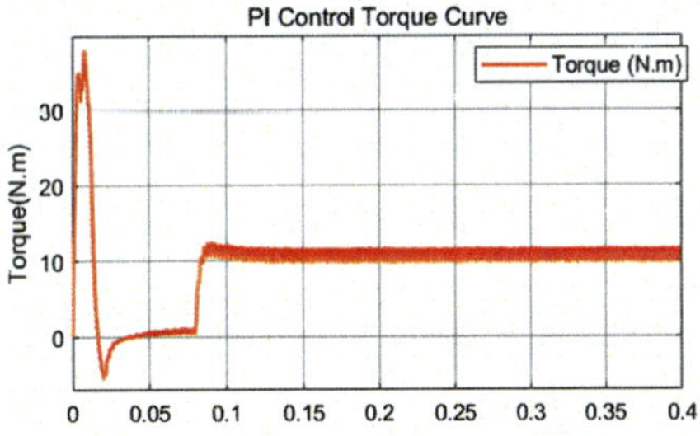

Fig. 6. Change curve of electromagnetic torque

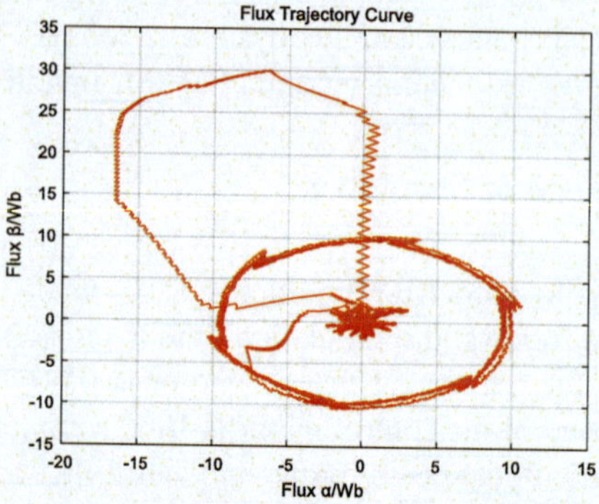

Fig. 7. Track of flux linkage

By analyzing the simulation results, we can find that when the motor speed in Fig. 5 rises from zero to rated speed, it has very fast response speed. However, it is not difficult to see that the torque and flux linkage in Figs. 6 and 7 fluctuate greatly, which is difficult to meet the requirements of the actual speed control system.

3 Improved PMSM-DTC

3.1 PMSM-DTC Based on SVPWM

The vertex of the trajectory of the eight voltage vectors generated by SVPWM is a circle, and then the circle is divided into six parts, called sectors. The vector position is shown in figure a below:

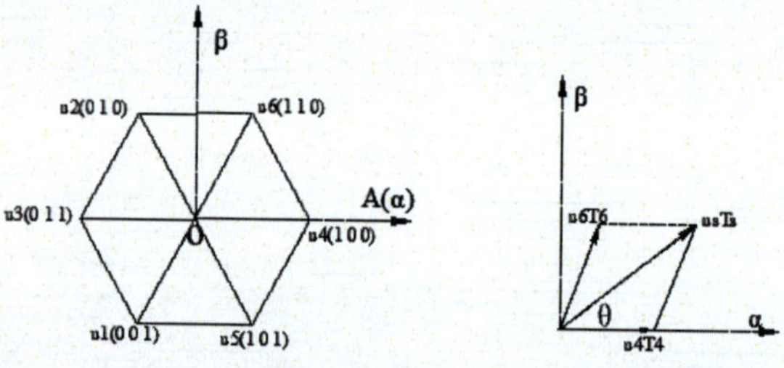

a b

Fig. 8. Space sector region and voltage vector synthesis method

Suppose we need to output a space voltage us. Suppose it is in a sector, we first take out a sector separately, as shown in Fig. 8b, and then use two adjacent voltage space vectors to represent it, which is equivalent to:

$$\begin{cases} Tu_s = T_4 u_4 + T_6 u_6 \\ T = T_4 + T_6 + T_0 \end{cases} \quad (5)$$

In formula (5): T is the PWM interrupt period; T_4 and T_6 are the effective action time of switching signals on U_4 and U_6; T_0 is the effective action time of zero voltage vector. The derived method is also suitable for calculating the space vector voltage of other sectors.

3.2 Fuzzy PID Controller

Fuzzy control PID is used to replace the original PI speed regulator.

Input of fuzzy controller: speed error E and error change rate E_C. According to the designed fuzzy rules and membership functions, fuzzy reasoning is carried out to obtain the variation of K_P, K_i and K_d, ΔK_P, ΔK_i and ΔK_d to adjust PID parameters. The automatic adjustment of three parameters of the controller is realized according to the following formula.

$$\begin{cases} k_p^{\circ} = \Delta k_p + k_p \\ k_i^{\circ} = \Delta k_i + k_i \\ k_d^{\circ} = \Delta k_d + k_d \end{cases} \quad (6)$$

In formula (6): K_P, K_i and K_d correspond to proportional parameters, integral parameters and differential parameters of PID respectively. And ΔK_P, ΔK_i and ΔK_d represent the variation of proportion, integral and differential.

3.2.1 Fuzzy Rule Base

The universe of fuzzy controller with two inputs is defined as {− 3, − 2, − 1, 0, 1, 2, 3}. {Nb, nm, NS, Z, PS, PM, Pb} are corresponding fuzzy subsets. Here, triangular membership function is used to solve the fuzzy problem.

The fuzzy rule table of fuzzy PID controller is as follows (Table 1):

Table 1. ΔK_P, Δk_i, Δk_d fuzzy PID control rules.

		EC						
		NB	NM	NS	ZE	PS	PM	PB
E	NB	PB	PB	PM	PM	PS	ZE	NS
	NM	PB	PB	PM	PS	PS	ZE	NS
	NS	PM	PM	PM	PS	ZE	NS	NS
	ZE	PM	PM	PS	ZE	NS	NM	NM
	PS	PS	PS	ZE	NS	NS	NM	NM
	PM	PS	ZE	NS	NM	NM	PS	PS
	PB	ZO	ZO	NM	NM	NM	NB	NB

3.3 Improved System and Simulation

The improved overall structure is shown in Fig. 9:

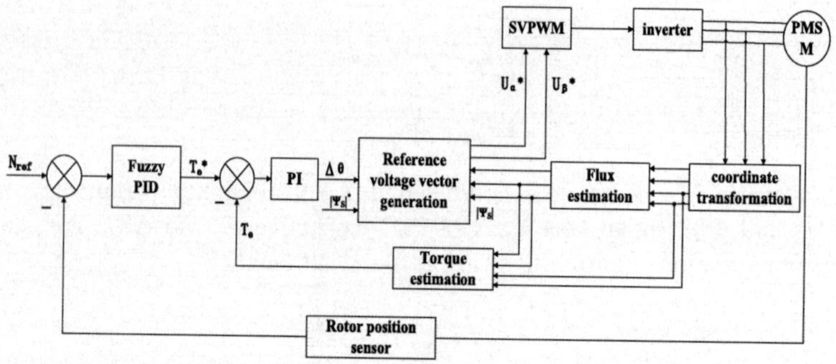

Fig. 9. Improved pmsm-dtc system block diagram.

In order to facilitate the comparison with pmsm-dtc system, the simulation data setting of the improved system is exactly the same as the traditional system simulation in the previous section.

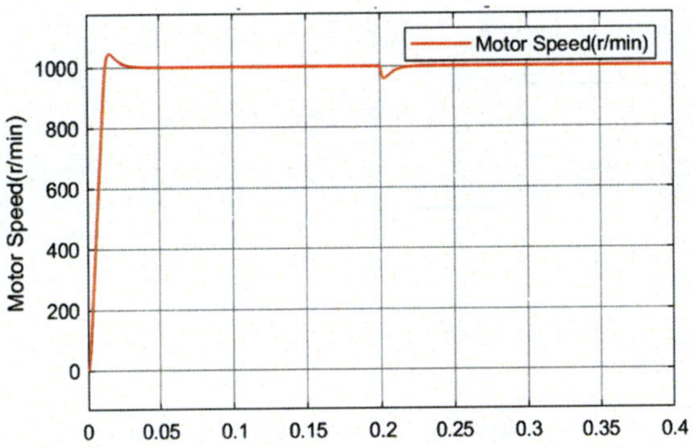

Fig. 10. Speed change curve of improved system.

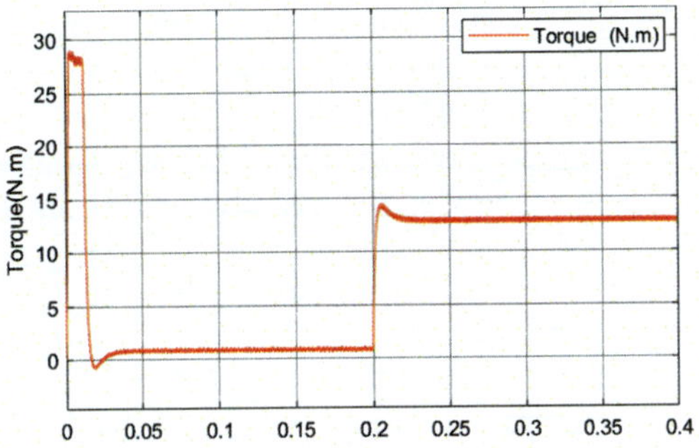

Fig. 11. Improved electromagnetic torque curve.

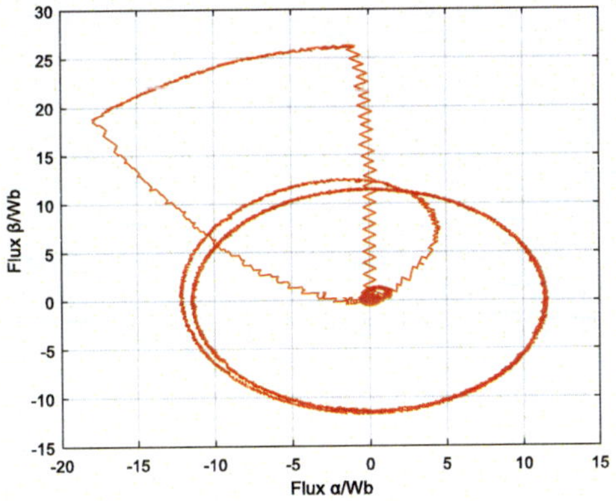

Fig. 12. Improved stator flux trajectory curve.

It is not difficult to see from Fig. 10 that when the motor is rapidly increased from 0 to 1000 r/min, the speed decreases slightly after loading within 0.2S, but it can still quickly recover to the set value, with fast response speed and small overshoot; in Fig. 11, the torque ripple is greatly improved, it is found that the response speed increases a lot, and the stator flux linkage track in Fig. 12 is also consistent with the theoretical circle.

4 Conclusion

Through the analysis of the first simulation, many shortcomings of the traditional system are found. This paper uses the combination of fuzzy PID and SVPWM algorithm to improve the overall performance of the system. Through the analysis and comparison of software simulation models, the results show that the improved control method can reduce the output torque ripple, which verifies the correctness and performance superiority of the improved system theory.

However, due to the limited installation of position sensor in many occasions, it is impossible to directly install the sensor to obtain the rotor speed and position of the motor, so the research on sensorless vector control should be increased in the future.

References

1. Yin, X., Yuan, D., Zhen, Y., Shao, Z., Xiangjie, L.V.: DTC technology of permanent magnet synchronous motor based on SVPWM. J. Syst. Simul. **31**(11), 2535–2542 (2019)
2. Zhao, S., Xu, Y.: Comparative study on vector control and direct torque control characteristics of permanent magnet synchronous motor. Electron. Q. (10), 99–106 (2019)
3. Bin, W., Kun, Z., Xiaoyan, S.: Permanent magnet synchronous motor direct torque control based on sliding mode variable structure. J. Anhui Univ. Sci. and Technol. (Natural Science Edition) **39**(05), 19–24 (2019)
4. Zhang, L., Gui, X., Yan, L.: Research on direct torque control of permanent magnet synchronous motor based on sliding mode observer. Micromotor **47**(07), 61–64 (2019)
5. Ye, F.: Deadbeat direct torque control of permanent magnet synchronous motor based on SVPWM. Shenyang University of technology (2019)
6. Wu, J.: Research on PMSM direct torque control based on space vector modulation. Xiangtan University (2019)
7. Zhang, R.: Research on direct torque control of permanent magnet synchronous motor. Liaoning University of Petroleum and Chemical Technology (2019)
8. Ding, S.: Research on direct torque control of permanent magnet synchronous motor based on sliding mode control. Northeast Petroleum University (2019)
9. Guo, Z.: Research on SVM-DTC control strategy of permanent magnet synchronous motor. Dalian University of technology (2018)
10. Deng, J.: Simulation experiment research on direct torque control of permanent magnet synchronous motor. Shanghai Institute of Electrical Engineering (2018)
11. Qin, D.: Research and implementation of permanent magnet synchronous motor direct torque control system. Anhui University of Technology (2018)

12. Hu, Y., Tian, C., Gu, Y.: In-depth research on direct torque control of permanent magnet synchronousmotor. In: 28th Annual Conference of the IEEE Industrial Electronics Society. IEEE Industrial ElectronicsSociety, Sevilla 2002(3), pp. 1060–1065 (2002)
13. Xiaolei, W., Keyou, Z., Peiliang, C.: Permanent magnet synchronous motor stator flux linkage twelve-section direct torque control. J. Qingdao Univ. **21**(4), 66–69 (2006)
14. Dong, S., Hu, Y., Yan, W., et al.: An improveddirect torque control method for permanent magnet synchronous motor. J. Beijing Univ. Chem. Technol. (Natural Science) **46**(3), 105–111 (2019)
15. Wu, C., Cheng, Y., Liu, Z., Hu, W.: Study on efficiency optimization control strategy of permanent magnet synchronous motor system. Electron.Measur. Technol. **43**(10), 36-41 (2020)
16. Zhong, Z., You, J., Zhou, S.: Optimization of direct torque control algorithm for permanent magnet synchronous motor based on spatial analytical model observer. Motor Control Appl. **47**(05), 22–27 (2020)
17. Yaohua, Li., Haohao, S., Xiangzhen, M.: Simplified predictive control strategy for direct torque control system of surface permanent magnet synchronous motor. J. Elect. Mach. Control **24**(04), 96–103 (2020)
18. Xuxia, S., Lin, M., Kai, W., Yidong, L.: Research on speed sensorless direct torque control system. Power Electron. Technol. **54**(03), 44–47 (2020)
19. AI, X., Wang, W., Wang, H.: 12 sector direct torque control method for internal permanent magnet synchronous motor. Acta Solar Energy Sinica **41**(01), 325–332 (2020)
20. Li, L., Lin, G.: improved model predictive direct torque control of permanent magnet synchronous motor . J. Motor Control **24**(01), 10–17 (2020)

The Research on Stock Price Prediction Based on Machine Learning Model

Shulue Xu(✉)

Beijing University of Post and Telecommunication, Beijing 100876, China
xushulve@bupt.edu.cn

Abstract. Increasing number of people has participated in the stock market, with the developments of economic and security system in recent years. However, because of the complexity of stock market, how to adopt the most appropriate trading strategy according to the trend of share prices, has become a problem concerned to both scholars and investors. Therefore, how to effectively predict the goings of stock price has become a heated topic in the research field. In this paper, the researchers analyzed the stocking data and its variation of quoted companies by four machine learning models, which indicated that these types of models are able to help improving the accuracy of predictions. What's more, the models built by Tensorflow had a better performance in indexes considered, compared to the model built by K-nearest neighbor model, logistic regression model and support vector machine model. The accuracy of different models has improved when they are adjusted for financial characteristics. The researchers have evaluated various kinds of machine learning models' performance in predicting the stocking prices.

Keywords: Machine leaning · Stock prediction · KNN

1 Introduction

Increasing number of people has participated in the stock market, with the development of economic and security system in recent years. However, because of the complexity of stock market, how to adopt the most appropriate trading strategy according to the trend of share prices, has become a problem concerned to both scholars and investors.

Although stock market is always regarded as a non-linear and unstable system, it is still possible to follow the pattern of its changes under a relevant stable environment. With the integration of the relevant fields- mathematical statistics, probability theory and neurosciences- in recent years, the theory of machine learning experienced a developing stage. The possibility of utilizing various types of machine learning models in stock prediction field has been concerned. Therefore, more and more people has tried to predict stock price by machine learning methods. In this paper, the researchers analyzed the cleaned stock price histories, gained usable models through training, made comparisons between the predictions by several indexes, then made improvements on the structures and parameters according to the weakness exploded. Finally, the results before and after improvements compared and analyzed, to give a better explanation fork the works of modifications.

M. Tavana et al. (Eds.): IISA 2020, AISC 1304, pp. 204–210, 2021.
https://doi.org/10.1007/978-3-030-63784-2_26

Before we go through the process and results of modelling, some of the essential technologies and background knowledge are introduced below:

2 Fundamental Methods in Machine Learning

Machine learning is a subject which allows machines to acquire new knowledge and skills, and meanwhile, be able to reconstruct existing knowledge structures, by continuously improving their own skills in a way that is similar to the learning behavior of human beings. The relationships between experiences (E), task (T) and measurements of certain performance (P) are as in Fig. 1, where the performance for a certain task will improve as the experiences (E) accumulates, which indicates that machine can acquire learning abilities through the process of getting more experiences (E).

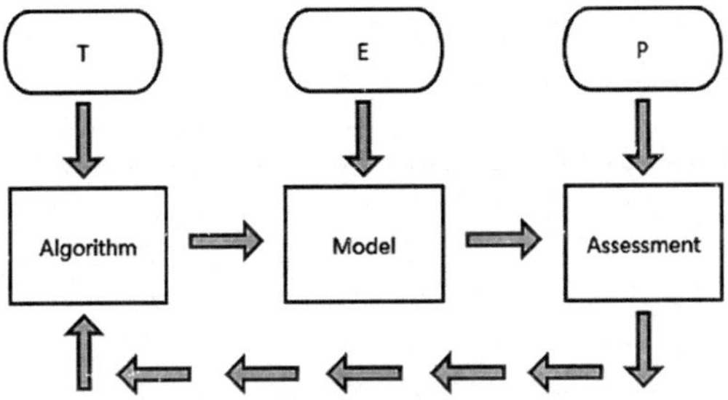

Fig. 1. Relationships between T, E and P

2.1 Stock Price Prediction Model

1) K-Nearest Neighbor

K-Nearest Neighbor (K-NN) means to calculate the distance between the testing objects and training objects by the method shows in Formula 1, then to find the object with highest frequency of occurrence in the K-nearest objects accordingly, and finally to defined the category of testing objects by the category of the found objects.

$$d_{12} = \sqrt{(x_1 - x_2)^2 + (y_1 - y_2)^2}$$
$$d_{12} = |x_1 - x_2| + |y_1 - y_2|$$

2) Logical Regression and Support Vector Machine

Logical Regression (LP) model uses sigmoid function as the logic regression unit to do the fittings, as well as constraining the domain of targets. The model can be well explained by formula 2:

$$h_e(x) = g(\theta^T x) g(z) = \frac{1}{1+e^{-2}}$$

For the purpose of making the figure as smooth and as similar to unary quadratic function as possible, other than using root mean square error, we can use the function shows in Formula 3 to define the global optimization point, i.e. to turn the figure into optimizing cross entropy function and to minimize the losses, as shows in Formula 4:

$$\cos t[h_e(x), y] = \begin{cases} -\log[h_e(x)], y = 1 \\ -\log[1 - h_e(x)], = 0 \end{cases}$$

Support Vector Machine (SVM) means to use dichotomous method on vectors, the essence is to try to create largest distance between two separated categories, which will make the separation more credible. SVM is able to project the linear inseparability data, such as stock indexes, into perfect linear separable or basically linear separable spaces, thus to turn the "ups or downs" of stock price, into a classification problem. The hinge loss function with regularization terms is shown in Formula 5:

$$j(\theta) = -\frac{1}{m}\left[\sum_{i=1}^{m} y^{(i)} \log h_e(x)^{(i)} + (1 - y^{(i)}) \log(1 - h_e(x)^{(i)})\right]$$

3) TensorFlow and Keras
Multi-Layer Perceptron (MLP) is what we mainly used in building neural networks, in which the inputs, outputs and hidden layers are included. The stock price will be input into the model and then feed forward through the hidden layers to produce the output by specific weights. Loss functions are also used to evaluate and optimize, as shows in Formula 6:

$$L(\hat{y}, y) = -[y \log\ y + (1 - y) \log(1 - \hat{y})]$$

3 Stock Price Prediction Method

3.1 Assessment Indexes

1) Partial Regression Indexes:
In regression indexes, y is used to represent the true value, ŷ represents the predicted value, thus in the process of simulation, small y and ŷ value indicates a good

predicted result. The formulas for calculating mean square error (MSE), root mean square error (RMSE) and mean absolute error (MAE) are shown as in Formula 7:

$$MSE(y,\hat{y}) = \frac{1}{m}\sum\nolimits_{i=1}^{m}(y_i - \hat{y}_i)^2$$
$$RMSE(y,\hat{y}) = \frac{1}{m}\sum\nolimits_{i=1}^{m}(y_i - \hat{y}_i)^2$$

2) Classification Evaluation Indexes
Following indexes are used in the the research, in which "TP" represents "True Positive", means that the real value is true and the predicted value is positive. "FP" represents "False Positive", means that the real value is false but the predicted value is positive. "FN" represents "False Negative", means that the real value is false and the predicted value is negative:

$$\begin{aligned} \text{Precision} &= \text{TP} \div (\text{TP} + \text{FP}) \\ \text{Recall Rate} &= \text{TP} \div (\text{TP} + \text{FN}) \\ \text{Accuracy(Correct Rate)} &= (\text{TP} + \text{TN}) \div (\text{TP} + \text{FP} + \text{TN} + \text{FN}) \end{aligned}$$

4 Building Prediction Model for Stock Price

We used web crawler to acquire the trading indexes and S&P 500 index of 20 domestic quoted companies, since they are able to provided vast and representative data. The process of data pre-processing includes 5 parts in total: 1) Fill in the missing values, detect and remove the abnormal values and noises. 2) Reduce the scale of data used for data mining. 3) Cleaning the text data by removing the embedded characters that can make the data dislocated. 4) Data discretization 5) Data normalization.

Use the theory part of supervised learning model strategy to model the trading index of multiple stocks and therefore to predict the price trend. Apart from that, synchronous and asynchronous prediction methods are also utilized to analyze the price trend in neural networks.

5 Experiments

We will use moving average technique on the data set. The first step is to create a data box that contains only the date and closing price columns. Then breaking it down into training sets and validation sets to validate our predictions. When splitting data into training sets and validation sets, we cannot use random segmentation because it would break the time order. So, this paper uses the data from last year as a validation set, and the data from the previous four years as a training set.

The RMSE value of the forecast result is 20.198824562125374, which indicates that the forecast result of the model is good. However, judging the value of the RMSE alone does not help us evaluate the predicted results of the model. The predicted values and actual values are shown in the same figure, which is Fig. 2.

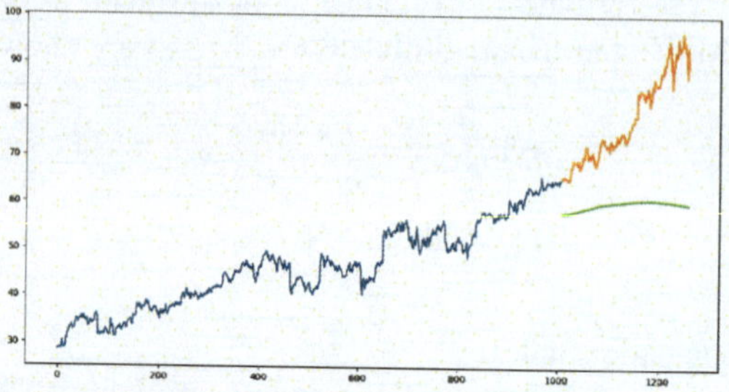

Fig. 2. Prediction results of the moving average model

It can be seen from RMSE value and Fig. 4 that RMSE value is close to 105, but the predicted result of the closing price is poor. The predicted value has the same change trend as the observed value of the training set, and they both begin to show an upward trend and then gradually decline, but there is a big difference in value.

This paper sorts the data set in ascending order firstly, and then creates a separate data set so that any new features created do not affect the original data. Furthermore, we assumed that the first and last days of the week have a far greater impact on the closing price of stocks than the rest of the week. Therefore, we have created a feature to identify whether a given day is Monday/Friday or Tuesday/Wednesday/Thursday.

Finally, the data are divided into training set and validation set to check the performance of the model. The RMSE value predicted by the linear regression model is 15.076923947374754, which is higher than that of the moving average model, which clearly indicates that the prediction results of the linear regression model are poor. In this paper, the predicted value and the actual value are shown in Fig. 3.

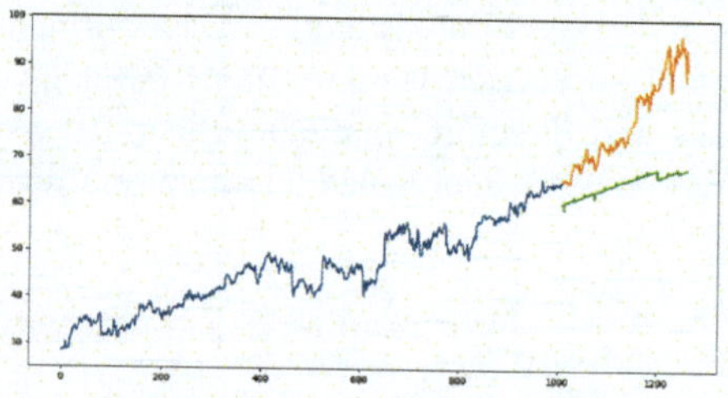

Fig. 3. Prediction results of the linear regression model

Linear regression is a simple technique that is easy to explain, but it has some obvious drawbacks. One problem of using regression algorithms is that the model overfits dates and months. The model will consider the value of the same date of a month ago or a year ago, rather than the previous value from a predictive perspective.

As you can see from the chart above, stock shares fell in January 2016 and January 2017. The model predicts the same for January 2018. Linear regression techniques are good for solving problems such as hypermarket sales, where independent features are useful for determining target values.

The actual stock price as well as the predicted price gotten through neural networks based on the stock price historical data are shown in Fig. 4:

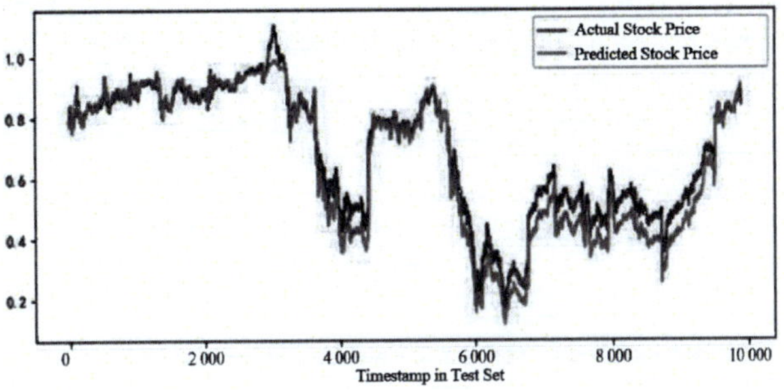

Fig. 4. Actual Stock Price and Predicted Stock Price

The mean square error for the result is (0.10 ± 0.05), thus, the value gotten from testing set is 0.0015, meaning that the fitting degree is relatively high. Moreover, this method is more appropriate for practical uses, since it is based on the historical prices of stocks.

6 Conclusion

In this paper, it is found that the machine learning algorithm can help to find the an appropriate trading strategy by evaluating a certain range of indexes and values. By selecting the appropriate parameters and evaluation indexes in combination with certain scenarios, the transform from the prediction of future prices to the improvement of actual benefits can be realized. However, practically speaking, the accuracy of prediction also depends on the data we use, so it is impossible to train a model to be applicable to all scenarios. Given the reason that the sample data chosen in this paper is still not perfectly comprehensive, the reality of the stock market, political and economic environment can be complex, it can be hard to produce a generalized model for predictions. Therefore, for building a more generalized model, we will need to investigate further into the trading scenarios, make more specific classifications and more comprehensively analyze data and modifying models accordingly, so that the thoughts of "machine learning" can be better combined with the realistic.

References

1. Arora, N., Parimala, M.: Financial analysis: stock market prediction using deep learning algorithms. Ssrn Electron. J.(2019)
2. Sim, H.S., Kim, H.I., Ahn, J.J., Silva, T.C.: Is deep learning for image recognition applicable to stock market prediction? Hindawi (2019)

3. Das, S., Mokashi, K., Culkin, R.: Are markets truly efficient? experiments using deep learning algorithms for market movement prediction. Algorithms **11**(9), 138 (2018)
4. Yan, H., Ouyang, H.: Financial time series prediction based on deep learning. Wirel. Personal Commun. **102**, 1–18 (2017)
5. Stock Time Series Prediction Based on Deep Learning
6. Golmohammadi, K., Zaiane, O.R., Díaz, D.: Detecting stock market manipulation using supervised learning algorithms. In: International Conference on Data Science & Advanced Analytics. IEEE (2015)
7. Li, X., Xie, H., Wang, R., et al.: Empirical analysis: stock market prediction via extreme learning machine. Neural Comput. Appl. **27**(1), 67–78 (2016)

Research on Evaluation Model and Method of Comprehensive Benefits for Multi-station Integration

Chen Jing[1], Zhang Yuyuan[1], Jin Qiang[2(✉)], Cui Kai[2], Yuan Fusheng[3], and Zou Ying[3]

[1] Beijing Electric Power Economic Research Institute Co., Ltd., Beijing 100037, China
[2] State Grid Economic and Technological Research Institute Co., Ltd., Beijing 102209, China
jinq1980@126.com
[3] State Grid Information & Telecommunication Co., Ltd., Beijing Branch, Beijing 100031, China

Abstract. The power system integration station based on resource integration and facilities sharing has entered the pilot and promotion stage. This paper make a research in the benefit evaluation of the integration station. Firstly, three typical integration scenarios are analyzed, including energy hub, data hub and business hub. Then based on the principles of consistency, measurability, comparability, mutual independence and integrity between evaluating indicator and evaluating purpose, the comprehensive evaluation system of multi-station integration (MSI) is constructed, and the evaluation indexes are analyzed systematically. After that, the comprehensive benefit evaluation method of integration station is proposed based on ideal point method and the relative nearness degree as well as evaluation result is calculated by index weigh through analytic hierarchy process (AHP). Finally, the feasibility and effectiveness of the proposed strategy are verified by three practical integration station cases. This method provides technical support for the construction and evaluation of the MSI.

Keywords: Multi-station integration · Typical scenarios · Comprehensive benefit evaluation

1 Introduction

With the in-depth development of the energy revolution marked by the technological revolution, the network hub role of power grid connecting energy production and energy consumption has become increasingly prominent, and the efficient, intelligent, comprehensive construction and operation of power grid resources has become an inevitable technology development trend [1, 2]. Based on this context, the concept of MSI is proposed as resource integration and facility sharing. Reference [3–6] explores the business model of MSI including substation, charging and replacement (energy storage) station, data center, 5G base stations, Beidou base station and other elements, as well as the operation mode for multi object business application scenarios such as

M. Tavana et al. (Eds.): IISA 2020, AISC 1304, pp. 211–222, 2021.
https://doi.org/10.1007/978-3-030-63784-2_27

power grid, government, users, enterprises, etc., as well as analyse the construction and operation mode of MSI, and points out the necessity and inevitable trend of this construction mode.

In July 2019, the first MSI of State Grid Corporation of China was completed and put into operation in Chongqing [7], and then several other pilot projects in Anhui, Jiangxi and Shanghai are successively implemented [8–10]. With the construction and operation mode of MSI gradually stepping into the stage of promotion and implementation, in order to comprehensively evaluate the construction effect of MSI and deeply analyze the direct and indirect benefits of data center, energy storage station, charging station, photovoltaic power station and substation integration construction mode on resource saving and equipment sharing, comprehensive and objective evaluation index system and evaluation method are needed to provide a decision for MSI construction decision basis in order to maximize the comprehensive benefits of MSI construction and operation.

The present research on comprehensive evaluation of MSI construction effect mainly focuses on the safety, economy and sociality assessment of single station such as substation, charging station, energy storage station and data center [11–15]. However, the comprehensive energy system benefit evaluation covering multi-energy interconnection and coordination of various links mostly focuses on engineering construction, low carbon energy saving and service quality [16, 17]. It is difficult to draw a comprehensive evaluation conclusion for multi-scenario and multi-element integration projects.

For the evaluation of the safety and benefits for it, this paper decomposes and reclassifies each integration link of MSI, and forms the index evaluation and comprehensive benefit evaluation model, put forward the evaluation system and method of MSI construction aiming at the operation safety, equipment energy efficiency, quality of power supply, resource sharing, power sharing, grid operation environment and regulation ability, comprehensive energy utilization efficiency and cost-benefit, social results, value-added benefits of multi-dimensional so as to provide basis for planning, design and construction of MSI, and maximize the comprehensive benefits of MSI construction and operation.

In this paper, a typical classification of different typical scenarios and the different needs of various evaluation subjects is first made; then the basic principles of evaluation system construction is put forward to ensure a concise, efficient and comprehensive evaluation work; afterwards it establishes the evaluation index system from three dimensions of operation evaluation, integration evaluation and economic and social evaluation; finally, the evaluation method of MSI construction effect in different scenarios is proposed in view of an improved optimal index weight obtained by row correction based on the cooperative game theory.

2 Analysis of Typical Integration Scenarios

The core of MSI is resource sharing and collaborative working. The traditional power node function of substation and other grid resources will be gradually expanded to the integration of energy flow, data flow and business flow. In the analysis of typical scenarios of MSI, three mainly fusion scenarios are considered including energy hub, data hub and business hub [6].

2.1 The Integration Scenario of Energy Hub

The MSI in this scenario aims to strengthen the role of the power grid as an energy hub as well as reducing the construction cost by sharing the station and power supply, also improving the power quality and saving the operation cost by improving the power complementarity among multiple elements. This paper mainly aims at the one or two elements of MSI in the grid resources including substation, energy storage station, charging station, DPV, etc.

The typical integration scenarios include "energy storage station plus distributed photovoltaic power station (DPV)", "DPV plus charging and replacement station".

2.2 The Integration Scenario of Data Hub

The MSI in this scenario aims to strengthen the data hub role of the power grid, transform the power grid side resources such as substations from the traditional power flow node to the hub node, and realize the sharing and mutual construction of energy and data through the integration construction with the data center.

The typical integration scenarios include "substation plus data center".

2.3 The Integration Scenario of Business Hub

The MSI in this scenario aims to strengthen the role of power grid as a business hub. On the one hand, it can solve the problems of difficult land acquisition and high cost of new 5G base stations and Beidou base station, on the other hand, it provides convenience for the grid side to cover 5G network, and creates conditions for users to enjoy Beidou positioning and 5G services, so as to achieve Win-win of multi-party cooperation. It mainly aims at MSI of power grid resources and one or two elements of non-grid resources such as 5G base stations and Beidou base station.

The typical integration scenarios include "substation plus 5G base stations plus Beidou base station", "data centers plus charging and replacement station plus DPV plus 5G base stations plus Beidou base station" etc.

3 Analysis of Typical Integration Scenarios

In the construction of MSI comprehensive evaluation system, the selected evaluation index should ensure the consistency, testability, comparability, mutual independence and integrity of the evaluation purpose. The basis setting indicators is as follows:

1) The influence of MSI on the power supply capacity of various power users, generators and power grid should be considered to reflect the supporting role of different facilities integration construction on power supply reliability.
2) Considering the promotion effect of MSI on resource conservation and environmental protection, the evaluation indexes of saving rate of land resources, facility

integration degree, energy conservation and emission reduction and human resource saving should be set.
3) The effect of MSI on the absorption of distributed generation (renewable energy generation) should be considered.
4) The promotion effect of MSI construction on the overall engineering economy of the project should be considered.

To sum up, the index construction is mainly include three dimensions: operation evaluation, integration evaluation and economic and social evaluation.

3.1 Operation Evaluation

The evaluation of MSI operation mainly includes operation safety, data security, equipment energy efficiency, power supply quality, power grid operation environment and power grid regulation ability. The main evaluation components include power supply reliability, operation reliability, safety isolation, classification authorization, identity authentication, intelligent defense [18], equipment energy efficiency, comprehensive voltage qualification rate, voltage ripple coefficient, instantaneous voltage drop and short-time interruption occurrence rate, reactive and active power regulation ability of power grid.

3.2 Integration Evaluation

The integration evaluation of MSI mainly includes energy conservation and emission reduction, resource sharing, comprehensive energy utilization, data interconnection and so on. The main evaluation components include annual fossil energy saving, annual carbon dioxide emission reduction, annual sulfur dioxide emission reduction, land resource saving rate, human resource saving rate, facility integration, power sharing, comprehensive energy efficiency, renewable energy utilization rate, safety interaction, linkage analysis ability, etc.

3.3 Economic and Social Evaluation

The economic and social evaluation of MSI is mainly carried out from the aspects of cost-benefit social results, value-added benefits. The main evaluation components include investment payback period, integration cost benefit, operation and maintenance cost, cost of energy, primary and secondary equipment investment reduction rate, industrial growth, employment promotion, market value added, user growth [17].

3.4 Construction of Evaluation System

The comprehensive benefit evaluation of MSI is a complex multi-attribute problem. Considering different application scenarios of MSI, the main evaluation components in typical scenarios are split, and the evaluation grade is classified according to the above comprehensive benefit evaluation indexes (Table 1).

Table 1. The comprehensive benefit evaluation system of MSI

First-level evaluation	Second -level evaluation	Main components	Energy integration	Data integration	Business integration
Operation evaluation	Operation safety	Power supply reliability	**	**	**
		Operational reliability	**	**	**
	Data safety	Safety isolation	–	**	**
		Classified authorization	–	**	**
		identity authentication	–	**	**
		Intelligent defense	–	**	**
	Energy efficiency of equipment		*	*	*
	Power supply quality	Comprehensive voltage qualification rate	**	**	**
		Voltage ripple factor	**	**	**
		instantaneous voltage drop and short-time interruption occurrence rate	**	**	**
	Power grid regulation capability	reactive power regulation ability of power grid	**	**	*
		active power regulation ability of power grid	**	**	*
	Power grid operation environment		*	**	*
Integration evaluation	energy conservation and emission reduction	annual fossil energy saving	*	*	*
		annual carbon dioxide emission reduction	*	*	*
		annual sulfur dioxide emission reduction	*	*	*
	resource sharing	land resource saving rate	**	**	**
		human resource saving rate	**	**	**
		facility integration	**	**	**
		power sharing	**	**	**
	comprehensive energy utilization	comprehensive energy efficiency	**	*	*
		renewable energy utilization rate	**	*	*
	data interconnection	safety interaction	–	**	**
		linkage analysis ability	–	**	**

(*continued*)

Table 1. (*continued*)

First-level evaluation	Second -level evaluation	Main components	Energy integration	Data integration	Business integration
Economic and Social evaluation	cost-benefit	investment payback period	*	*	*
		integration cost benefit	*	*	*
		operation and maintenance cost	*	*	*
		cost of energy	*	*	*
		primary equipment investment reduction rate	*	*	*
		secondary equipment investment reduction rate	*	*	*
	social results	industrial growth	*	*	**
		employment promotion	*	*	**
	value-added benefits	market value added	–	*	**
		user growth	–	*	**

Notes:①- :It means that it is not included in the evaluation for the time being; ②*: Indicates that the evaluation rating is a general evaluation index; ③: Indicates that the evaluation rating is the key evaluation index.

4 The Evaluation Method of Comprehensive Benefits of MSI

1) Construction of multi-attribute primitive matrix

Assuming the scheme set of multiple attribute decision making problem is $E = \{E_1, E_2, \cdots E_m\}$,attribute set is $F = \{F_1, F_2, \cdots F_m\}$. The attribute values of the *i*-th scheme from 1 ~ m with respect to the *j*-th index from 1 ~ n $x_{ij}(i = 1, 2, \cdots, m; j = 1, 2, \cdots, n)$ constitute the original multi-attribute matrix.

$$X = \begin{bmatrix} x_{11} & x_{12} & \cdots & x_{1n} \\ x_{21} & x_{22} & \cdots & x_{2n} \\ \vdots & \vdots & \ddots & \vdots \\ x_{m1} & x_{m2} & \cdots & x_{mn} \end{bmatrix} \quad (1)$$

2) Construction of standardized decision matrix

Since each index has different order of dimension and magnitude, the original matrix is dimensionless treated by vector normalization method, then the normalized selection matrix or grid Y is obtained.

$$Y = \begin{bmatrix} y_{11} & y_{12} & \cdots & y_{1n} \\ y_{21} & y_{22} & \cdots & y_{2n} \\ \vdots & \vdots & \ddots & \vdots \\ y_{m1} & y_{m2} & \cdots & y_{mn} \end{bmatrix} \tag{2}$$

Where y_{ij} represents as follows:

$$y_{ij} = x_{ij} \Big/ \sqrt{\sum_{i=1}^{m} x_{ij}^2}$$

3) Definition of index weight

MSI evaluation is a multi-attribute complex problem, which needs to consider many factors. When determining the weight, AHP is applied to make the index weight ranking more reasonable.

a. Establish hierarchical index structure

The subordinate relationship of the evaluation index system is comprised of the first level index, the second level index as well as the main components.

b. Establish judgment matrix according to experts' opinions

AHP is that experts transform the relative importance of elements into numerical values according to their own experience and express them in the form of judgment matrix.

Assuming that the upper level element A_k is related to the next level element $B_1B_2,\ldots\ldots B_n$, the importance of A_k and $B_1B_2,\ldots\ldots B_n$ should be judged to determine the weight of $B_1B_2,\ldots\ldots B_n$. Experts generally use the level of 1–9 as well as its reciprocal to express the relative importance of B_i and B_j under A_k, that is, if the importance of B_i to B_j is b, then the importance of B_j to B_i is 1/b.

According to the importance degree expressed by 1–9 scale and its reciprocal given by experts, the judgment matrix B is established as follows.

$$B = \begin{bmatrix} 1 & b_{12} & \cdots & b_{1n} \\ b_{21} & 1 & \cdots & b_{2n} \\ \vdots & \vdots & \ddots & \vdots \\ b_{nl} & b_{n2} & \cdots & 1 \end{bmatrix} \tag{3}$$

Where b_{ij} and b_{ji} are reciprocal.

c. Then the maximum eigenvalue λ_{max} of B is obtained, and the eigenvector v corresponding to λ_{max} is obtained.

$$Bv = \lambda_{\max} v \tag{4}$$

d. Normalize the feature vector and count the weight of each index

$$\omega_i = \frac{v_i}{\sum_{k=1}^{m} v_i}, \quad i = 1, 2, \cdots, n \tag{5}$$

e. Make consistency test of the judgment matrix

The calculation formula of consistency parameter CI is as follows.

$$CI = \frac{\lambda_{\max} - n}{n - 1} \tag{6}$$

Where n is the order of judgment matrix.
The formula for calculating the rate of the random conformance CR is as follows.

$$CR = \frac{CI}{RI} \tag{7}$$

Where RI is the index of the average random conformance.
The complete consistency of the first-order and second-order judgment matrix is always satisfied, but the complete consistency of order 3–9 needs to be tested, and the value of RI can be obtained according to Table 2.

Table 2. The value of the index RI based on average random conformance

Matrix	3	4	5	6	7	8	9
RI	0.58	0.90	1.12	1.24	1.32	1.41	1.45

The consistency of calculation results will meets the requirements When Cr ≤ 0.1; while it is considered opposite when Cr ≥ 0.1, then the judgment matrix needs to be adjusted accordingly.

4) Definition of the ideal point and calculation of closeness degree

After the positive and negative ideal solution are obtained by the method of successive approximation met-hod, the maximum or minimum value can be regarded as the best solution. If the number of evaluation targets changes, the calculation can be carried out same as before, and the evaluation results may be inconsistent with those before. In order to settle it, this paper propose the following calculation steps.

a. Calculate weighted normalized judgment matrix

The weighted normalized judgment matrix is formed by combining index weight vector W and normalized evaluation matrix Y showing as follows.

$$z_{ij} = w_j \cdot y_{ij} \tag{8}$$

Where i ranges from 1 to m while j 1 to n.

b. Calculate the absolute ideal solution

The positive ideal solution vector Z_j^+ as well as the negative ideal solution vector Z_j^- can be obtained by matrix z_{ij} referred to the following formula.

$$Z_j^+ = \begin{cases} 1, j \in T_1 \\ 0, j \in T_2 \end{cases} \tag{9}$$

$$Z_j^- = \begin{cases} 0, j \in T_1 \\ 1, j \in T_2 \end{cases} \tag{10}$$

Where T_1 is a very large index; T_2 is a very small index; 0 represents the minimum standard and 1 represents the highest standard.

c. Find the closeness degree

The closeness between each evaluation object and positive as well as negative ideal solution is showing as follows.

$$d_i^+ = \sqrt{\sum_{j=1}^{n} (z_{ij} - z_j^+)^2}, i = 1, 2, \cdots, m \tag{11}$$

$$d_i^- = \sqrt{\sum_{j=1}^{n} (z_{ij} - z_j^-)^2}, i = 1, 2, \cdots, m \tag{12}$$

5) Solver of the relative closeness degree and determination of the evaluation results

$$C_i = d_i^- / (d_i^- + d_i^+)', i = 1, 2, \cdots, m \tag{13}$$

Where the value of C_i represents the nearness degree with the ideal scheme. The smaller the value is, the greater the difference between the ideal scheme and the bigger the value is, the more consistent it is with the ideal scheme. The evaluation objects are sorted according to the C_i value, and the required results will be obtained.

5 Evaluation Case of MSI

In this paper, three integration station of State Grid Anhui and Chongqing are selected as examples to evaluate the comprehensive benefits. The basic information of integration station A、B and C is that station A is a data center, 5G communication base station and Beidou ground-based augmentation station; station B is a data center, 5G communication base station, DPV, energy storage and charging station; and station C is a data center.

According to the relevant data of integration station, the weight of each evaluation index is calculated. This paper will take the energy conservation and emission reduction index of station B as an example to illustrate the calculation process of weight.

The evaluation process is shown in Fig. 1.

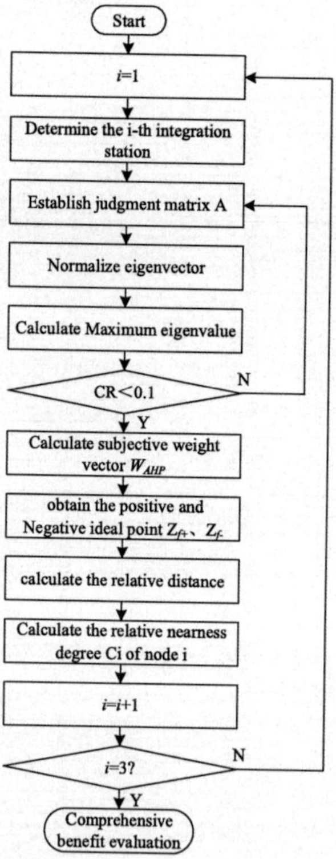

Fig. 1. Component mapping of economic and social evaluation

Energy conservation and emission reduction belongs to second-level evaluation index, including annual fossil energy saving, annual carbon dioxide emission reduction and annual sulfur dioxide emission reduction. The experts quantify the three main components and convert them into energy conservation and emission reduction judgment matrix B22 as follows.

$$B_{22} = \begin{bmatrix} 1 & 0.2 & 0.33 \\ 5 & 1 & 0.33 \\ 3 & 3 & 1 \end{bmatrix} \quad (14)$$

The maximum eigenvalue: $\lambda_{max} = 3.2871$, corresponding to the eigenvector $\omega = [0.1663\ 0.4214\ 0.8575]$, as well as $CR = 0.037 < 0.1$ passed the consistency test through calculation. The weights of annual fossil energy saving, annual carbon dioxide emission reduction and annual sulfur dioxide emission reduction are 67.8%, 18.4% and 13.8% respectively.

After getting the weight of each index of integration station A, B and C, the evaluation results are acquired as shown in Table 3.

Table 3. Comprehensive benefit evaluation results of three integration station

Integration station	Evaluation results	Ranking
A	1.436	2
B	1.497	1
C	0.654	3

It can be seen from Table 3 that the comprehensive benefit of integration station B is the best, which is in line with the reality.

6 Conclusion

This paper puts forward MSI index system and comprehensive benefit evaluation model to solve its operation evaluation and construction problem. The evaluation index system is established which is measurable, comparable and independent; In addition, an evaluation scheme based on ideal point method is proposed, the evaluation process is designed, and the index weight calculation method is given. Finally, the effectiveness of the proposed method is testified by an actual case.

Acknowledgments. This work is supported by State Grid Corporation Science and Technology Project (Research on Key Technologies and Business Models of Multi-Station Fusion, 5200-201941477A-0-0-00).

References

1. Li, X.L.: Accelerating the construction of smart grid is the inevitable requirement for energy transformation. China Energy News, 05. 11. (004) (2020)
2. Sun, Q., Ji, X.X., Wang, Z.Q., et al.: Research on theory and key technology system of smart grid development under new round of energy revolution. Ind. Control Comput. **32**(12), 91–92 +147 (2019)
3. Wang, B.Y., Zhang, Y., Liu, M.B., et al.: Research on the multi-station integration operation mode. Electric Power Inf. Commun. Technol. **17**(07), 41–45 (2019)
4. Yang, P., Gu, Y., Zhou, X.: Research on application scenario and construction operation mode of multi-station integration service. Data Commun. **03,** 1–3+6 (2020)
5. Meng, C., Liu, W.L., Yang, Q., et al.: Research on the operation optimization of edge data center under the background of "Multi-station Integration" [J/OL]. Adv. Eng. Sci. **52**(4), 49–55 (2020)
6. Hunag, Y., Zhang, C.Y., Liu, T.M.: "Three stations in one" mode of ubiquitous electric internet of things. Internet Things Technol. **10**(01), 44–47 (2020)
7. The first multi station integrated substation of State Grid of China put into operation. Rural Electrification, 8 (2019)
8. Anhui's first power multi station integration project will have operation conditions by the end of 2019. Transformer 56(12), 67, (2019)
9. Jiangxi's first "multi station integration" pilot project has been completed in Nanchang. Rural Electrification **02,** 14 (2020)
10. Shanghai Power Grid "multi station integration" demonstration project starts to provide effective resources and means for the construction of ubiquitous power Internet of things. Transformer, 56(05), 47 (2019)
11. Meng, L., Zhao, J.N., Chu, G.: Risk assessment of substation operation based on big data of equipment monitoring. Inf. Technol. **44**(06), 154–157+167 (2020)
12. Yang, B.Z.: research on evaluation of intelligent substation based on cyber-physics system architecture. huazhong university of science and technology (2019)
13. Zou, W.J.: Application and evaluation of electric vehicle charging station based on battery energy storage. Energy Conservation & Environ. Protection **02**. 106–107 (2019)
14. Chen, G.K., Qu, D.P., Hu, X.J., et al.: Cost benefit evaluation of electric vehicle charging/swapping station based on life cycle analysis. Electric Power Construction **37**(01), 30–37 (2016)
15. Mi, W.Q.: Introduction to data center security construction and evaluation indicators. Comput. Knowl. Technol. **04,** 109–111 (2016)
16. Wang, Q., Lu, Y.C., Liu, Y.J., et al.: Investment evaluation method of integrated energy system. Smart Power **48**(06), 20–27 (2020)
17. Zhang, LB., Zhi, R.N., Ding, T., et al.: Evaluation of the development of energy internet in china based on entropy weight matter-element model. J. Nanjing Univ. Aeronautics Astronautics (Social Sciences) **22**(01), 53–61 (2020)
18. Li, Y.L., Han, B.S.: Discussion on interstation safety of intelligent energy station in electric power industry. Electric Eng. **03**, 151–152 (2020)

Research on the Teaching Mode of Software Engineering in Colleges and Universities Based on Big Data Analysis

Huanqin Wu(✉) and Rucheng Ma

Guangdong University of Science and Technology, Dongguan 523083, China
2436481339@qq.com

Abstract. Under the development of modern society, with the popularization of the Internet and the rapid development of high-tech in the computer field, the Internet has been applied to all aspects of social production. The advancement of information technology in the era of big data has made significant contributions in data analysis, human-computer interaction, and e-commerce,the rapid development of information technology has a profound impact on the development of the software engineering profession. At the same time, the teaching of software engineering in colleges and universities should start from the current development of the industry, formulate more reasonable teaching models, change teaching methods, and send more comprehensive software professionals to the society. This article discusses the related issues of the teaching mode of software engineering in colleges and universities under the background of the development of big data, hoping to have some enlightenment for college teaching reform.

Keywords: Big data · University teaching · Software engineering major · Information age

1 Introduction

Faced with the increasingly complex Internet network structure and volume of data information, how the computer software industry faces the emerging new challenges, and strengthening the computing and data processing capabilities of the big data application center has become a decisive factor affecting the development of the contemporary computer industry. We know that software engineering is based on the practical application of computer science under modern information technology. The development and changes of modern information technology will inevitably lead to changes in the formulation of software engineering industry standards, technical implementation, and product requirements. The software engineering major is committed to cultivating talents who can meet the development needs of the industry, lead the future of the industry, and promote the software industry to keep up with the development of society. Therefore, it is of practical significance to pay attention to the influence of big data technology development on software engineering teaching [1]. The software engineer career planning classification chart is shown in Fig. 1.

M. Tavana et al. (Eds.): IISA 2020, AISC 1304, pp. 223–228, 2021.
https://doi.org/10.1007/978-3-030-63784-2_28

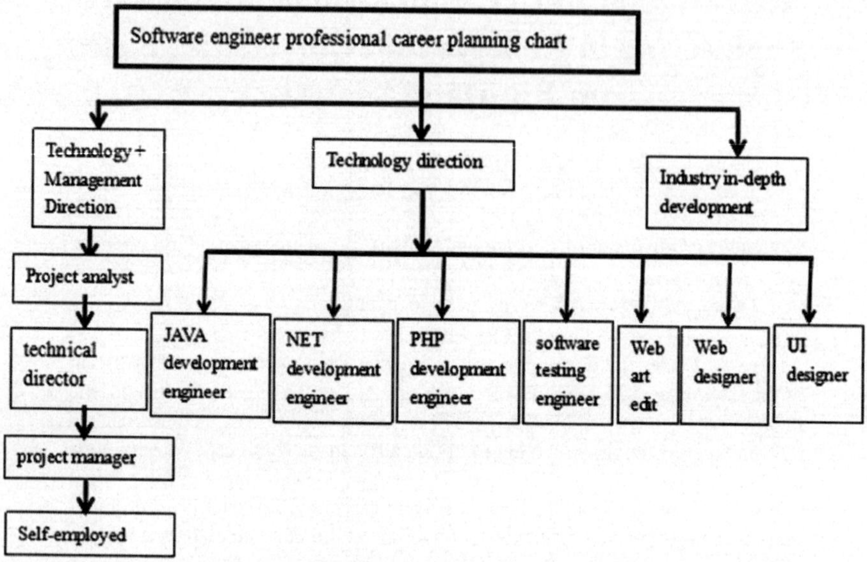

Fig. 1. Professional career plan of software engineer

2 The Impact of the Development of Big Data Technology on the Software Industry

2.1 Impact on Intellectual Property

Intellectual property issues in the information technology network environment mainly include copyright and trademark rights. For now, I think China's legal construction in this area has been put on a perfect agenda, but there is still a lot of room for improvement. Take the trademark rights incident that happened not long ago. The trademark registration incident of an UP owner named Jing Hanqing on the bilibili platform has attracted public attention, this UP owner was told his B station id name and his real name was registered as a trademark by a leather bag company. Therefore, "Jing Hanqing" cannot be reused. This kind of ridiculous incident quickly aroused the attention of official accounts such as the Central Committee of the Communist Youth League. At the same time, it also aroused nationwide discussion and attention on trademark rights. In the end, Jing Hanqing succeeded in defending rights. The purse company's delusional plan to profit by exploiting legal loopholes fell through.

This incident actually reflects two facts. The first is the fact that my country's legal construction related to trademark rights under the network environment is not perfect, and the nationals have not fully understood the laws related to trademark rights. From the fermentation process of the follow-up incident, I learned that it was not only Jing Hanqing who was squashed on the B station alone. Hundreds of thousands of UP's trademarks have been squashed. Among them, there have been people who have made a fortune and a purse. The reason why the company can use this unskilled method to profit. The first is that the use of my country's judicial system to establish laws related to online trademark rights is not yet complete, people's awareness of rights is not strong,Take advantage of China's widespread mentality of "breaking wealth and eliminating disasters" to execute blackmail. Second, it also reflects that my country's network supervision and laws are constantly improving. The reason why this incident has quickly gained official attention and aroused national discussion is that in recent years, my country's cybercrime cases have increased, and intellectual property-related cases have appeared more and more frequently, citizens have become more concerned about this issue. Judging from the response to this case, the development of information technology has made it more difficult to protect intellectual property rights, giving criminals an opportunity to take advantage of it, but it also reminds the judicial department to promote network supervision to pay attention to intellectual property protection.

2.2 Impact on Computer Crime

With the rapid development of computer technology today, network technology is no longer an unreachable professional term for the public as it used to be. More and more people are beginning to spend time and energy on learning related knowledge. At the same time, the Internet itself can also provide a lot of resources for learning computer technology. Under such an environment, the public has become more and more familiar with computer technology, and they have a deeper and deeper understanding of computer technology. However, while mastering computer technology, some computer professionals do not strengthen their own legal and moral awareness. They can steal personal information and spread it at will through their computer technology, or attack others' websites, which will affect the normal operation of relevant companies and enterprises [2].

Some hackers maliciously spread computer attacks on the Internet, using viruses, Trojan horses, worms and other means to spread through the computer, causing the user's computer program to crash, and the damage caused is difficult to estimate. Among them, the famous "panda burning incense", "slammer" and other viruses have brought great troubles to the economic and social development and the normal life of people. Studying the history of the development of network viruses, we can see that around 2000, network viruses developed rapidly. During this period of time, dozens of network viruses with major destructive capabilities and influence appeared. At the same time, this period of time was also a network. A period of rapid technological development. Therefore, the improvement of hacking technology and the development of the computer industry are synchronized. With the rapid popularization and continuous development of computer technology today, the level of hacker attack technology is also improving. The development of information technology has reduced the cost of cybercrime to a certain extent, increased the number of computer crimes, and increased the difficulty of combating computer crimes.

3 Teaching Reform Countermeasures of Software Engineering Specialty in Universities Based on Big Data Analysis

3.1 Innovate Teaching Content and Cultivate Students' Professional Ethics

Now, in the process of continuous development of the software industry, distributed network systems connect electronic products through the Internet. There are often more than one personal computer devices. In addition to personal computers and smart phones, the use of tablets is becoming more and more popular. The increase in the number and types of smart digital products provides more intrusion methods for computer attacks. The complexity of network connection also provides a variety of intrusion ports for attack programs, which increases the difficulty of maintaining computer network security. Such a development trend is inevitable. In the era of big data, the development of information technology links work from all walks of life to the computer network, and personal work cannot do without the support of electronic products and network, which inevitably leads to the increase in the number of personal computers. The complex network environment has increased the difficulty of building a network security defense system, software engineering practitioners are required to set up lifelong learning goals and improve their knowledge reserves according to the development of the times. The software engineering major needs to teach students the necessary professional knowledge, and more importantly, cultivate the concept of lifelong learning for students, broaden and extend their careers in the future.

Practitioners in the software engineering industry not only need to constantly update their knowledge base according to the development of information technology in the society, but they also need to regulate their own behavior, strengthen their own legal awareness and professional ethics, and they do not use their own professional knowledge to do things that violate ethics and laws. When teaching software engineering major in colleges and universities, we should not only pay attention to the improvement of students' professional ability, but also integrate professional ethics into daily teaching, so that students can take legal courses such as intellectual property law and understand the necessary industry regulations.

3.2 Introduce New Teaching Methods and Make Full Use of Online Teaching

With the current development of the times, in addition to physical courses, colleges and universities often have online courses in the process of offering courses. Mooc as the largest autonomous learning platform for universities in my country has a large number of learning resources and user groups. Teachers can upload learning videos and electronic materials on this platform, students can learn independently through videos and materials,and gain in-depth understanding of what they have learned through online examinations and student interaction. Considering that the objects of current university teaching have relatively strong autonomous learning ability and the fragmentation time utilization ratio is relatively high, Therefore, the online teaching of software engineering is an efficient and fast teaching method [3]. The current integration of online teaching and offline teaching is shown in Fig. 2.

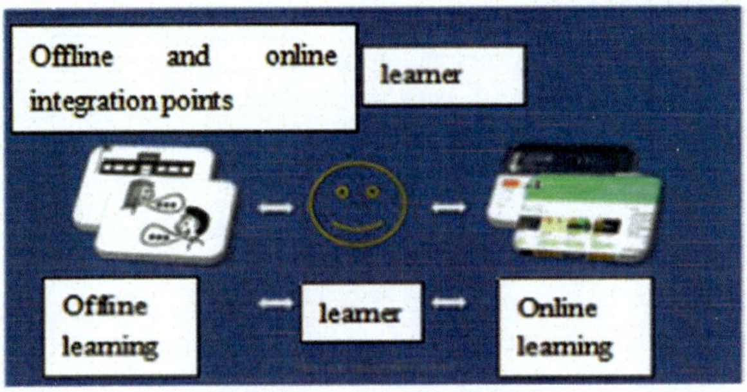

Fig. 2. Integration of online learning and offline learning

The society is developing, and higher and higher requirements are put forward for the application of talents, so colleges and universities must make certain changes according to the requirements of the talent market when training students. Obviously, the current teaching methods are no longer able to satisfy students' learning, At the same time, with the rapid development of computer network and other technologies, it is very realistic and necessary to establish a shared resource database for teaching in colleges and universities, because the establishment of a shared resource library allows students to enhance their theoretical knowledge and at the same time improve their ability to learn independently. At present, in addition to MOOCS, online teaching platforms of Chinese universities also have rain classes and some independent public accounts of universities, which can provide a lot of high-quality resources for students' learning. In the process of software engineering teaching reform, the introduction of online classrooms can provide a more flexible teaching model for teaching, and at the same time it can also increase the enthusiasm of software majors for classroom participation.

4 Strengthen School-Enterprise Cooperation and Enhance Students' Practical Ability

In the development of the new era, the social environment is constantly changing, and the industry's demand for talent training and the definition standards have changed. While constantly improving the software majors in universities in my country, it is necessary to introduce new educational concepts and think about ways to improve the talent training model under the new social environment. The practice of school-enterprise cooperation curriculum as a new education model appears in people's vision under this background. This new education method aims to cultivate new types of talents with practical ability, professional skills, and more in line with the professional development needs of modern society.

The curriculum practice of school-enterprise cooperation should be completed by the cooperation and exchange between schools and enterprises, in the design process, the professional needs should be the center, the students' practical ability should be the focus, and the professional, open and practical requirements should be reflected. According to the real industry norms and corporate environment, the setting of student group tasks during the learning process is placed in the context of completed corporate projects, with the goal of completing marketing tasks that meet industry and market needs. Students experience the real working environment of the enterprise in classroom teaching and combine enterprise management with classroom management, in this process, students play the roles of students and employees at the same time, guiding students to apply theoretical knowledge to industry work, achieve the unification of teaching and practice so that students can integrate industry norms, corporate culture, professional ethics and other factors into their ideological education while mastering professional skills. In the process of teaching, teachers change from traditional knowledge teaching to behavior guidance, and integrate teaching content into the process of practical operation by using the combination of work and learning, classroom production, and project task guidance [4].

5 Conclusion

In the era of big data, the development of computer software technology is changing with each passing day. The development of information technology promotes the progress of human society, but it also brings certain industry problems. These issues will affect the industry's demand for software engineering professionals. University software engineering educators should pay attention to industry changes, actively formulate teaching reform plans, respond to industry changes with more flexible methods and better preparations, and train to meet industry changes. Comprehensive talents needed for development.

References

1. Liu, M.: Research and practice on teaching reform of database system course. Wirel. Internet Technol. **16**(07), 88–89+104 (2019)
2. Li, J.: Research on the teaching mode of software engineering in colleges and universities based on big data analysis. Econ. Outlook Bohai Sea (03), 176 (2019)
3. Yang, W., Tong, L.: The exploration and practice of multi-dimensional teaching mode in the teaching of software engineering in colleges and universities. Sci. Educ. J. (Mid-Term) (12), 73–75 (2016)
4. Han, T.: Bilingual teaching model of software engineering for software service outsourcing in colleges and universities. Comput. Educ. (13), 67–69 (2010)

Application of Information Technology in New Energy Vehicles

Qing-an Li(✉)

Guangdong University of Science and Technology, Dongguan 523083, China
410518535@qq.com

Abstract. Today, more and more automobiles have entered thousands of households, marking that our country has realized the transformation from the original “bicycle kingdom” to the “automobile power”. The number of private cars in my country is currently growing rapidly. The blowout of car ownership has also brought about environmental pollution problems. For example, car exhaust emissions will cause serious environmental pollution. At this time, it is imperative to develop new energy vehicles. Today, when information technology is highly popular, the application of information technology in new energy vehicles has brought about the intelligence of new energy vehicles, effectively improve the efficiency and quality of new energy vehicle manufacturing.

Keywords: Automobile industry · Information technology · Intelligence · New energy vehicles · Application

1 Introduction

Since the founding of New China, the automobile industry has always been an important industrial industry in my country. In recent years, with the rapid development of our country’s economy [1], the increasing number of private cars has ushered in new opportunities for the auto industry. The rapid development of the automobile industry has also brought environmental pollution problems. With the popularization of the concept of ecological and environmental protection, traditional gasoline, diesel and other automobile energy sources urgently need to be replaced with new energy sources that have little environmental pollution. Therefore, we need to develop new energy vehicles. The combination of information technology and automobile industry has realized the intelligence of new energy vehicles and improved the income of production. This is where we will discuss.

2 The Significance of Information Technology in the Application of New Energy Vehicles

New energy vehicles are an important breakthrough in domestic new energy technology. They use electric energy, solar energy, natural gas and other energy in the power of automobiles, effectively changing the disadvantages of traditional energy such as large energy consumption and environmental pollution. There is already a

M. Tavana et al. (Eds.): IISA 2020, AISC 1304, pp. 229–234, 2021.
https://doi.org/10.1007/978-3-030-63784-2_29

certain market in our country. It can be said that its development prospects are broad. In particular, my country has become one of the world's major sales markets for the automotive industry, and there will inevitably be more demand for new energy vehicles in the future.

The 21st century is an information century. Realizing the informatization of new energy vehicles is an attempt to explore the development path of China's new energy vehicles and break the problem of the development path of my country's new energy vehicles [2]. It can effectively realize the intelligent production of new energy vehicles, increase output and production quality, and promote the effective popularization of new energy vehicles.

Nowadays [3], the informatization of new energy vehicles is mainly embodied in the informatization of automobile driving and information consumption control. It is a major opportunity for the development of transportation in the new era [4]. Since the power system and control system of the car are very large and complex, it will also bring a certain failure rate. Through the effective use of information technology, the failure rate can be effectively reduced and the safety of vehicle operation can be improved.

3 Key Points of Application of Information Technology in New Energy Vehicles

At present, the scale of use of new energy vehicles in my country continues to increase [5]. New energy vehicles convert other energy sources into electrical energy to drive the car to operate. Including its air conditioning, heat dissipation and transmission, it needs to be driven by electricity. To this end, it is necessary to use microcomputers to connect with systems such as environment perception and multi-level driving assistance to achieve intelligent and centralized control. Especially in highways, places with poor road conditions, or places prone to car accidents, information such as road conditions and routes can be automatically analyzed. When the information technology can be further used properly, further realize the control of all aspects of the new energy vehicles, even can achieve automatic driving.

4 Information Technology Application Measures in New Energy Vehicles

To apply information technology to new energy vehicles, we need to take measures in the following areas:

4.1 Realize the Intelligence of the Whole Vehicle

The purpose of adopting new energy is to avoid pollution caused by non-renewable energy consumption and exhaust emissions. The intelligentization of the vehicle is an important way to achieve this goal. Now we mostly adopt regenerative braking energy recovery technology and use energy recovery methods to achieve effective control of regeneration and composite paper motion. Now, the type of fuzzy control distributed drive has been achieved in the automotive industry. Specifically, it adopts regenerative braking to control the torque of the front and rear wheels, hydraulic pressure, and braking of the car, so as to improve the energy recovery capability of the vehicle. However, regenerative braking also has a weakness-the current generated can easily damage the battery. In this regard, some people suggest adopting a "layered control method". This can minimize the damage caused by the current to the battery [6].

In addition, a "single-pedal driving mode" is still being explored, and trials have been conducted on Chevrolet, Toyota, Jaguar and other brands. Its characteristic is that no matter what the vehicle speed, regenerative braking can be achieved when the door is loosened, so it is very convenient to realize regenerative braking.

4.2 Realize Intelligent Driving and Power Control

The current pilot of unmanned driving technology has become a manifestation of the intelligence of driving and power control. It uses high-precision positioning systems, machine learning algorithms, and big data technology for automatic control in specific applications [7]. In this link, the "highlight" lies in the intelligent path planning decision, so particle swarm and fuzzy logic algorithms are used for planning. In order to improve the safety of unmanned driving, we must pay attention to adopting local obstacle avoidance algorithms appropriately. It also requires the car to have an advantage in environmental awareness and be able to quickly avoid obstacles based on the specific conditions in the driving environment. In order to effectively reflect this goal, we must also pay attention to the combination of unmanned driving and distributed control to effectively control driving, braking, suspension and other systems to ensure operational safety. In addition, a model related to the loss analysis of the drive system can be established according to the situation, and the fixed torque distribution ratio can be arranged based on the specific loss, so that the coordination of vehicle dynamics control can be improved.

4.3 Realize the Intelligence of Networking

This is also an important aspect of realizing the informatization of new energy vehicles. Especially with the popularization of platforms such as big data and cloud computing, the collection and analysis of automobile data have become more efficient and convenient, which has further promoted the realization of unmanned driving. Judging from the current research progress at home and abroad, the American NACTO Association has launched an open digital platform for traffic planning and design and new technology development [8]. The British TfL agency provides support for dynamic and static traffic data analysis by studying a large amount of traffic data. There is also the Japanese ETC2.0 system, which can provide assistance to drivers and collect vehicle driving information to improve the overall safety of road traffic. At present, China is also actively building a new energy vehicle networking big data platform, adopting a three-level network architecture of state, local and enterprise. With the help of the big data platform, the driving information and road traffic information of new energy vehicles can be collected comprehensively, and the statistics, mining and analysis of them can provide strong support for the regulation of new energy vehicles. In this case, the market development speed of new energy vehicles can also be accelerated, so that people can truly feel the convenience of using new energy vehicles, thereby increasing purchase demand.

4.4 Establish a Monitoring System for Collision Avoidance and Collisions

This is an important measure to improve driving safety. Scanning for obstacles that may appear at any time through radar, sonar and laser beams, to monitor collision avoidance and collisions. For example, when you are about to face a collision, the system will alert the driver as soon as possible to make an emergency brake. It can also be combined with a GPS receiver to get in touch with the nearest rescue agency as soon as possible to provide accurate positioning, and its signal coverage is also relatively wide. In this way, even if a car accident occurs, the wounded can be treated in the shortest possible time, reducing casualties. At the same time, it can also provide uninterrupted service 24 h a day. See the Table 1 below for the advantages of collision avoidance and crash monitoring systems.

Table 1. Advantages of collision avoidance and crash monitoring system

Category	When facing a crash	Treatment after a crash	Casualty rate
Anti-collision and crash monitoring system	First time alert	You can immediately contact the nearest rescue agency	Lower
Normal system	No alarm	Difficult to seek help in the first place	Higher

It can be seen from Table 1 that the anti-collision and crash monitoring system is of great significance for improving the safety of the vehicle. Currently, new technologies in this area are constantly being updated, and an emergency system called TELEATD has been developed abroad. It can automatically open the door and automatically turn on the hazard lights when a traffic accident occurs, making it more convenient to use.

4.5 Developed an Electronic Anti-theft System

In order to prevent economic losses caused by car theft, an electronic anti-theft control system has also been developed. It can prevent the thief from starting the engine with usual mechanical or electrical methods, and the anti-theft code can be randomly generated. Due to the use of special communication means, the information transmitted each time is different, so it is more difficult to crack the anti-theft code with advanced electronic scanning methods. Now take the Santana 2000GSi car (Time Superman) to illustrate the car's electronic anti-theft system, which is an anti-theft control system that starts to work after the ignition switch is turned on. Figure 1 shows the basic composition of the system, which is mainly composed of pulse transponders, reading coils, anti-theft system ECU(Electronic Control Unit), engine ECU with variable code function and anti-theft alarm lights. Figure 1 Structure diagram of automobile electronic anti-theft system.

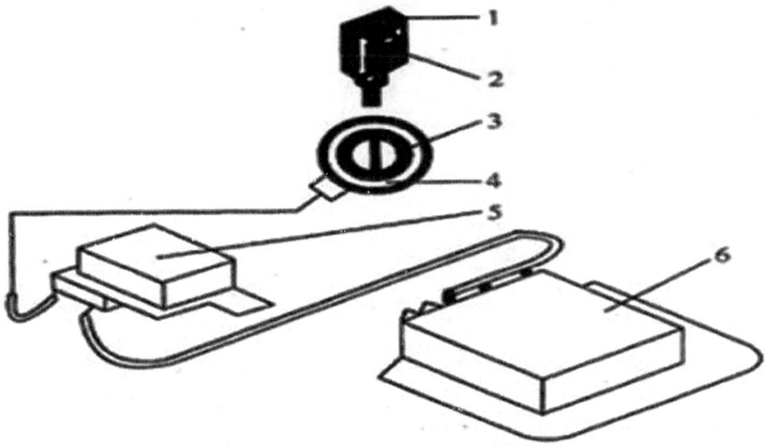

Fig. 1. Car electronic anti-theft system 1-car key; 2-pulse transponder; 3-ignition switch; 4-reading coil; 5-anti-theft ECU; 6-engine ECU

5 Conclusion

New energy vehicles are the future development trend of the automobile industry, which is conducive to avoiding environmental pollution caused by exhaust emissions and conforms to the principle of sustainable development. Information technology in new energy vehicles is currently being piloted and promoted. It can effectively achieve the intelligence of new energy vehicles and improve the safety of new energy vehicles. Therefore, its development prospects are broad, and it is worthy of our further exploration.

References

1. Fan, B.: The quality and service of new energy vehicles going to the countryside should keep up with [J]. China Energy News **7**(27), 90–92 (2020)
2. Rongliang, H.: Five suggestions for "new energy vehicles to the countryside." Mod. Logistics News **7**(20), 12 (2020)
3. Shengji, Y.: Considerations and responses to the security of the supply chain of the automobile industry chain. Autom. Aspects **7**(15), 89–90 (2020)
4. Wu, C., Yang, X., Wu, X.: Integrating infrastructure to expand new space for transportation development. China Commun. News **7**(16), 112–113 (2020)

5. Wei, P., Jia, W., Yuke, Li.: Analysis of the current situation of the recycling and utilization of electric vehicles in my country and relevant suggestions. Autom. Aspects **7**(15), 100–101 (2020)
6. Bergek, A., Berggren, C.: The impact of environmental policy instruments on innovation: a review of energy and automotive industry studies. Ecol. Econ. **7**, 112–123 (2014)
7. Lieven, T., Rietmann, N.: Forecasting the trajectory of electric vehicle sales and the consequences for worldwide CO2emissions. J. Cleaner Prod. **261**(7), 112–143 (2020)
8. Hossain Lipu, M.S., Hannan, M.A., Hussain, A., Hoque, M.M., Ker, P.J., Saad, M.H.M., Ayob, A.: A review of state of health and remaining useful life estimation methods for lithium-ion battery in electric vehicles: challenges and recommendations. J. Cleaner Prod. **9**, 115–133(2018)

Application

Study on the Attention Degree of Consumer Endowments to Children's Clothing

Xia Liu[1], Shihong Duan[2], Yanhong Hu[3(✉)], Fei Pei[1], Bisong Liu[1], Qianwen Chen[1], and Feng Qiao[1]

[1] China National Institute of Standardization, Beijing 100191, China
[2] University of Science and Technology Beijing, Beijing 100083, China
[3] Beijing Products Quality Supervision and Inspection Institute, Beijing 101300, China
326768505@qq.com

Abstract. The fit goodness and correlation coefficients are used to analysis their relationship. The results show that consumers attach special importance to the materials and appropriate age groups of children's clothing, reaching 88.13% and 79.68%, respectively; the purchase channels are mainly through the specialty stores, large stores/malls, and online purchases, respectively reaching 84.02%, 83.56%, and 74.66%; while the consideration on safety indicators focus on the irritating odors, harmful dyes, and the easy-loosen small parts, etc., reaching 88.58%, 67.12, and 59.82%, respectively. The significant relationship is significantly associated between the purchase channels, age, the occupation and the safety indicators of children's clothing.

Keywords: Attention degree · Children's clothing · Consumers · Endowments

1 Foreword

Children's clothing holds a significant position in China's textiles and clothing exports. With the continuous expansion of the children's clothing export from China, the increasingly strict safety standards for children's clothing have been made by European and American countries, taking stringent supervision on children's clothing as the "children's products". It shows that the children's clothing has been subject to the recall notifications or returned sometimes due to the main reasons of the danger of suffocation (irrational design of the straps and small parts, the length of the drawstrings of children's shirts and hats exceeding the standards), the chemical hazards (detection of dimethyl fumarate, formaldehyde, and illegal use of azo dyes, etc.), and the injury hazards (the length of the drawstrings at the waist or lower hem of children's clothing out of limits), etc. With the official implementation of the national standard for children's clothing—"Safety Technical Code for Infants and Children Textile Products", our country has been paying more and more attention to the development of children's clothing market. The characteristics of the consumers endowment may have a certain influence on the selected products. The different consumers may have different concerns about the quality and safety of the children's clothing. Many scholars have researched about consumer behaviors from different perspectives to reduce or avoid the

M. Tavana et al. (Eds.): IISA 2020, AISC 1304, pp. 237–248, 2021.
https://doi.org/10.1007/978-3-030-63784-2_30

related hazards. The concerns of consumers about children's clothing while purchasing and the differences in purchasing channels have been analyzed in this paper. The subjective evaluation of consumers would provide the basic research of risk assessment on children's clothing.

2 Survey Conditions

In this survey, 438 questionnaires were sent out and 438 were recovered by using the "questionnaire star" software. The effective rate of the questionnaire reached 100%. The main survey contents are the information of the respondents, respondents concerns, and the channels to purchase children's clothing, etc.

2.1 Personal Information

Table 1. Respondents age

Options	Subtotals	Proportions
18	0	0%
18-25	113	25.8%
25-35	236	53.88%
35-50	77	17.58%
50-65	11	2.51%
65 above	1	0.23%
Valid total persons		438

The main age of the respondents is between 18–50 years old, among which the proportion of 25–35 years old is the largest, accounting for 53.88%, then 18–25 years old, accounting for 25.8%, in addition, the proportion of 35–50 years old is 17.58%; t 50–65 years old people is 2.51%, accounting for 2.51%; the sex ratio of men and women is basically the same. The results are shown in Table 1 and 2.

Table 2. Sex ratio of male and female for the respondents

Option	Subtotal	Proportion
Male	215	49.09%
Female	223	50.91%
Valid total persons	438	

The proportion of bachelor's degree reached 72.37%, followed by junior college, accounting for 15.98%, master's degree and above accounted for 7.99%, high school etc. accounted for 3.42%, junior high school ect. accounted for 0.23%. See Table 3 for the specific proportion distribution.

Table 3. Education level distribution

Options	Subtotals	Proportions
Junior middle schools	1	0.23%
High school etc.	15	3.42%
University	70	15.98%
Undergraduate	317	etc 72.37%
Master degree	35	7.99%
Valid total persons	438	

Most of the respondents are company employees, accounting for 54.34%, professionals (including teachers, doctors, lawyers, etc.) accounting for 14.16%, public institutions, civil servants and government employees accounting for 10.27%, and students accounting for 10.05%. The proportion distribution of respondents shows that their cultural level is relatively high. Table 4 shows the detailed distribution proportion.

Table 4. Occupation distribution

Options	Subtotals	Proportions
Professionals	62	14.16%
Personnel in service industry	20	4.57%
Freelancers (artists, photographers and guides, etc.)	10	2.28%
Workers (construction workers, factory workers and urban sanitation workers, etc.)	16	3.65%
Company employees	238	54.34%
Civil servants, public institution etc.	45	10.27%
Students	44	10.05%
Housewives	1	0.23%
Retired	1	0.23%
Others:	1	0.23%
Valid total persons	438	

2.2 Respondents Attention to Product Information

This survey is mainly to understand the important product information that consumers pay attention to when purchasing children's wear products. After widely understanding the concerns of consumers when purchasing children's wear, the relevant product information is about brand, material, applicable age, color, factory name, address, origin domestic, origin foreign, qualified certification or relevant national standards are marked, priced and others, which is shown in Table 5 for the distribution details. The survey results show that the respondents pay the most attention to children's clothing information, which is material, accounting for 88.13%, applicable age 79.68%, price 71.69%, color 56.62%, factory name, factory site and certification mark 54.79%, brand 26.71%, origin domestic and origin foreign 21.46% respectively, 11.64%, the other 1.14%; the results show that consumers pay special attention to the material, age and price function of children's clothing.

Table 5. Respondents attention to the product information

Option	Subtotals	Proportion
Brand	117	26.71%
Appropriate age group	349	79.68%
Texture of material	386	88.13%
Marks of factory name, address and conformity certification	240	54.79%
Color	248	56.62%
Origin-domestic	94	21.46%
Origin-abroad	51	11.64%
Price	314	71.69%
Others	5	1.14%
Valid total persons	438	

2.3 Purchase Channels of Respondents

Through investigation, it is found that the main channels for purchasing children's wear at present include large stores, exclusive stores, online shopping, large chain supermarkets, small commodity wholesale markets, small supermarkets, roadside stalls, etc., which is shown in Table 6 for specific distribution. It can be seen from the table that 84.02% of the respondents buy and sell children's wear in exclusive stores, 83.56% in large stores, 74.66% in online stores and 42.92% in large supermarket chains.

Table 6. Purchase channels

Options	Subtotals	Proportions
Specialty stores	368	84.02%
Big stores and malls	366	83.56%
Small markets	47	10.73%
Online purchase	327	74.66%
Large supermarket	188	42.92%
Small supermarkets	24	5.48%
Street vendors	10	2.28%
Others	2	0.46%
Valid total persons	438	

The safety indicators the consumers consider when buying children's clothing mainly include easy-loosen small parts (such as the clothes buttons, zipper sliders, etc.), color fading, irritating odors, straps, formaldehyde content, sharp edges and tips, etc. See Table 7 for specific indicators. Consumers' considerations on various safety indicators focus on the irritating odors (88.58%), harmful dyes (67.12%), easy-loosen small parts (65.75%), sharp edges and tips (59.82%), etc.

Table 7. Proportion of the respondents considering safety indicators

Option	Subtotals	Proportion
Easy-loosen small parts (such as the clothes buttons, zipper sliders, etc.)	288	65.75%
Color fading	254	57.99%
Irritating odors	388	88.58%
Straps	129	29.45%
Formaldehyde	258	58.9%
PH-value	94	21.46%
Color fastness	102	23.29%
Plasticizers	99	22.6%
Heavy metals (lead, cadmium)	211	48.17%
Harmful dyes	294	67.12%
Sharp edges and tips	262	59.82%
Undesirable odor	231	52.74%
Others	5	1.14%
Valid total persons	438	

3 Relevance Analysis

3.1 Personal Information with Their Purchase Channels

Before performing relevance analysis, the data are first visualized to visually show relationship between education level, the occupation, age and their purchasing channels, as shown in Fig. 1, 2 and 3.

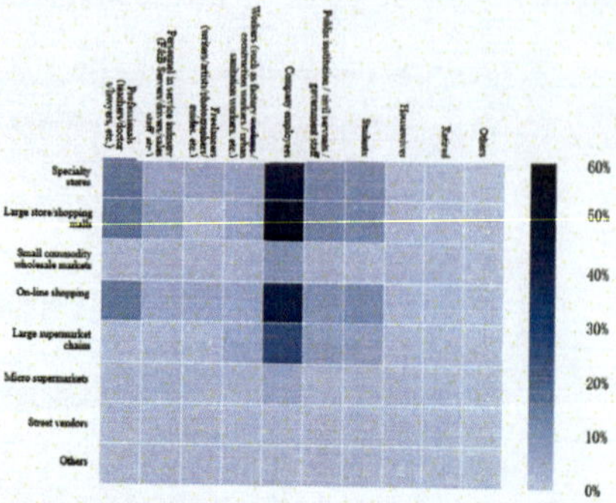

Fig. 1. Relevance of the occupations and purchasing channels

It shows that based on the relevance analysis on the education level, occupation and the respondents' age. From the results it can be seen that the most respondents choose to go shopping in large stores/malls and specialty stores, followed by online shopping and large chain supermarkets. With the increase in education level, more people would purchase children's clothing in big stores or malls" and "online shopping". However, the trend that the purchasing channels are changing with the education level, occupation and age is not obvious on the whole, (there's little variation in color range in each row). Therefore, it can be preliminarily judged the relevance for the education level, age, occupation and their purchase channels isn't obvious. Here the fit test goodness is used to test whether the mentioned relevance relation is significant as follows.

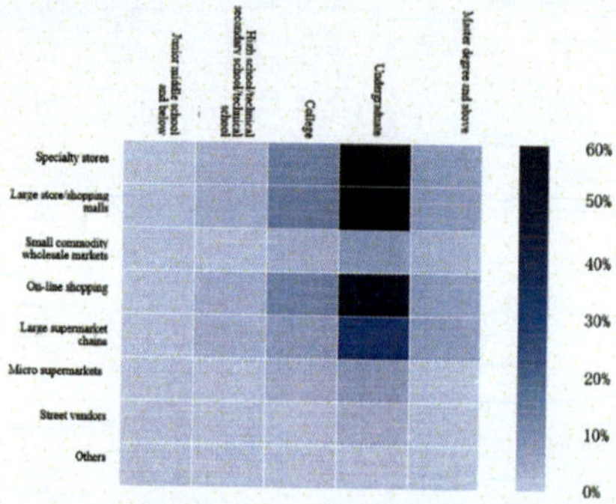

Fig. 2. Relevance between the education level and purchasing channels

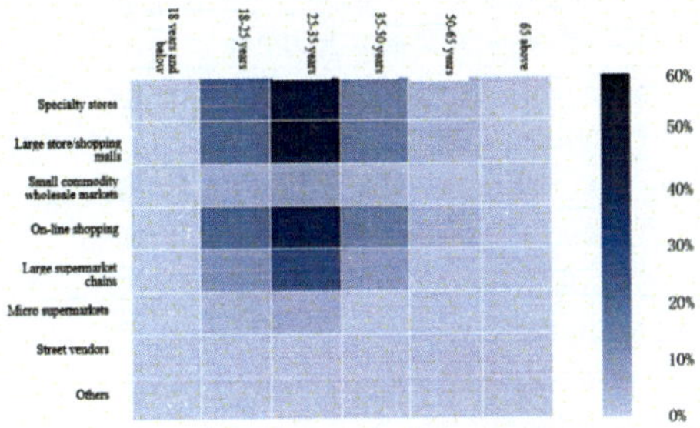

Fig. 3. Relevance between the ages of the respondents and their purchasing channels

Goodness of fit test is a statistical significance test using statistics. It is related to the overall distribution, by calculating expected frequency between the classified variables, and then by comparing the separated observation frequency, to determine the significant difference for the observed frequency and the expected frequency. Finally, it achieved the purpose of effective analysis of the classified variables.

$$\chi^2 = \sum \frac{(f_o - f_e)^2}{f_e}$$

$$f_e = \frac{RT * CT}{n}$$

In the above formula, f_e represents the expected frequency, f_o is the observed frequency, RT represents the total row where the unit is located, CT is represents the total column where the unit is located. N represents the sample capacity of the observed value.

Based on the above analysis and calculation, the goodness analysis is carried out for thc cducation level, age, occupation and purchase channel of the respondents, and the calculation results are shown in Table 8. The occupation has significant relationship for purchase channels.

Table 8. Results goodness

	Occupations	Education levels	Ages
χ^2	268.147	12.93	13.529
p-value	0	0.99321	0.95658

3.2 Personal Information and Their Attention

Before performing relevance analysis, the data are visualized to visually including the relationship of the education level, age etc. to the safety indicators of children's clothing, which are shown in Fig. 4, 5 and 6.

It can be found that based on the relevance analysis on the education level, occupation, and age, a lot of respondents pay more attention to the "Easy-loosen small parts (such as the clothes buttons, zipper sliders, etc.)", "Color fading", "Formaldehyde", "Irritating odors" and "Sharp edges and tips". However, the trend that the different POI (Point of Interest) are changing with the education level, occupation, and age is not obvious on the whole, (there's little variation in color range in each row). It is judged the relevance between the education level, occupation and the safety indicators of children's clothing is not obvious. Here the goodness is used to judge the above-mentioned relevance relation significant as follows.

Through the above analysis process, the goodness is carried out between the education level, age, occupation and children's clothing safety indicators of the respondents, and Table 9 shows the calculation results. There is a significant relationship of occupation and age and children's clothing safety indicators.

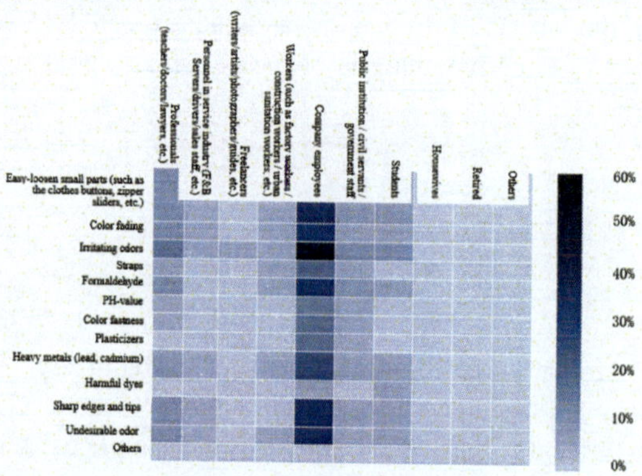

Fig. 4. Correlation between occupations and the attention to safety indicators of children's clothing

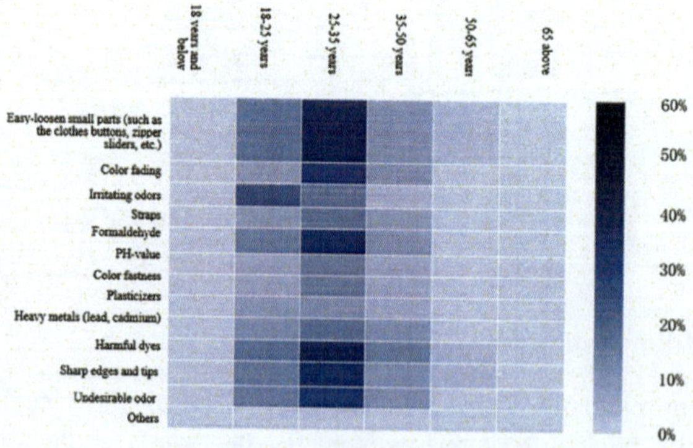

Fig. 5. Correlation between age and the attention to safety indicators of children's clothing

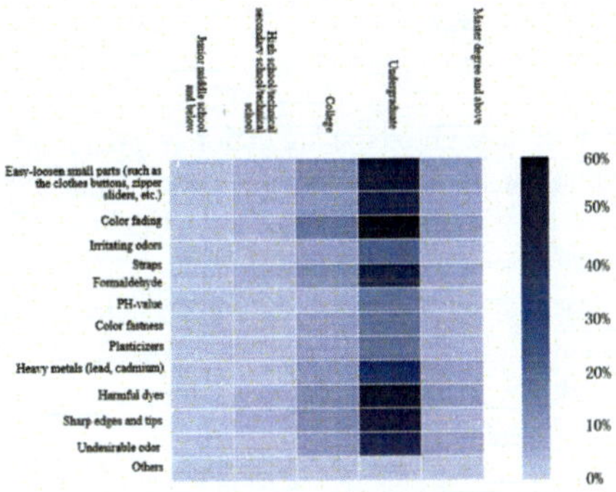

Fig. 6. Correlation between education level and the attention

Table 9. Fit test goodness

	Occupations	Education levels	Ages
χ^2	156.37	41.82	140.04
p-value	0	0.78	0

4 Conclusion

Through the questionnaire, the consumer's information and their attention to children's clothing and their purchase ways are researched. Researched results improved that consumers paid main attention to material or age group when they bought children's wear, reaching 88.13% and 79.68% respectively. In terms of purchase channels, consumers mainly focused on specialty stores, large stores and net shopping Main, 84.02%, 83.56% and 74.66% respectively; the safety indexes are mainly focused on irritating peculiar smell, harmful dye, loose small parts, etc., reaching 88.58%, 67.12 and 59.82% respectively. It analyzes the relation among the education level, age, occupation, purchase channel and relevant information of the respondents, and analyzes and tests the survey results. There is a significant relationship between the job and purchase ways of the consumers. There is a significant relationship between the job and age of the consumers and safety information of children's clothing.

Acknowledgements. This paper has been supported by the national project "Research on key technical standards for quality and safety control of consumer goods" (2016YFF02022600), and "Research on common technology for integrative services by internet plus" (2017YFF0209604), Central basic scientific research project "Research on Consumer Goods Safety Hazard Identification and risk Assessment based on scenario Simulation" (552018Y-5928), and "Research on the construction of consumer goods safety injury scenarios and risk assessment methods of intelligent service robots based on VR experiment" (552020Y-7462).

References

1. Ji, L., Niu, D., Wang, P.: Research on the prediction of photovoltaic generated power based on similar day aggregation and Bayesian neural network. Chin. J. Manage. Sci. **23**(3) (2015)
2. Liu, J., Zhang, M., Zuo, H., Xie, J.: Analysis of airline safety sensitivity based on improved Bayesian neural network. J. Safe. Sci. Technol. **10**(11) (2014)
3. Shen, P., Ling, L.: Research on the consumption behavior of peasant workers from the perspective of intergenerational endowment. Rural Econ. **2**, 35–40 (2014)
4. Flaherty, E.: Safety first: the consumer product safety improvement act of 2008. Loyola Consum. Law Rev. **21**, 372–384 (2008)
5. Liu, X., Luo, H.: Research on the methods of safety hazard identification for the consumer products based on product life cycle. In: The International Conference on Computer and Information Science, Safety Engineering (CAISSE), pp. 156–161 (2012)
6. Xie, G., Jiang, Y., Shi, C.: The influence of E - WOM dispersion on purchase intention: the moderating role of endowment effect. Chin. J. Manage. **16**(3), 425–438, 455 (2019)
7. Dhar, R., Wertenbroch, K.: Consumer choice between hedonism and utilitarian goods. J. Market. Res. **37**(1), 60–71 (2000)
8. Jinkins, D.: Conspicuous consumption in the United States and China. J. Econ. Behav. Organ. **127**, 115–132 (2016)
9. Heffetz, O.: A test of conspicuous consumption: visibility and income elasticities. Rev. Econ. Stat. **93**(4), 1101–1117 (2010)
10. O'cass, A., McEwen, H.: Exploring consumer status and conspicuous consumption. J. Consum. Behav. **4**(1), 25–39 (2004)
11. Zunxiong, L.: Efficacy Simulation of class data goodness-of-fit test. Stat. Decis. **34**(24), 86–87 (2018)

Application-Oriented Talent Cultivation to Meet the Needs of Cross-Border E-Commerce Positions—Take the Customer Service Position as an Example

Yi Jing(✉)

Wuhan Qingchuan University, Wuhan 430204, China
12506971@qq.com

Abstract. The "Internet + " era has ushered in the rapid development of cross-border e-commerce industry, and there is a great shortage of qualified personnel in this field. For the division of different positions of CBEC (shorten for cross-border e-commerce), exploring the application-oriented talent training path to meet the job requirements is the common needs of students, schools, and enterprises. Taking the frontline customer service position of CBEC operation as an example, the status quo of talents presents obviously insufficient stock and the level of talent structure is not complete. Therefore, the business content demand, capacity demand and skill demand of the position are analyzed, and school-enterprise cooperation order-based talent cultivation, establishment of off-campus training base, cooperation with well-known enterprises and other talent training paths are proposed.

Keywords: Cross-border E-commerce · CBEC customer service · Job requirements · Talent cultivation

1 The Talent Gap of CBEC and Job Division

With the development of the "Internet + " era, global trade has continued to advance, and CBEC has flourished. In recent years, the Council and relevant ministries and commissions have issued a number of policies to encourage the development of CBEC. In 2018, the scale of China's CBEC transactions reached 9.1 trillion yuan. It is estimated that this number can reach 11.8 trillion yuan in 2019, 12.7 trillion yuan in 2020. The average annual compound growth rate in the period of 2019–2023 is approximately 18.92%. By 2023, it will reach 23.6 trillion yuan [1].

As the cross-border market also matures, the development of CBEC platforms, the types of products sold by CBEC companies, and CBEC sales market have all shown a more diversified trend. The talent demand of CBEC companies is also increasing. At present, China's CBEC talent gap has exceeded 16 Mio., 86% of CBEC companies have a talent gap, and 82.4% of CBEC talents recruited by companies cannot meet their position needs, and talent-and-business needs cannot match [2]. This problem needs to be solved urgently.

M. Tavana et al. (Eds.): IISA 2020, AISC 1304, pp. 249–255, 2021.
https://doi.org/10.1007/978-3-030-63784-2_31

It is very important for application-oriented colleges to continuously innovate and improve the cultivation of CBEC talents and to provide application-oriented talents who can meet the needs of CBEC companies and positions. Under the new "Internet + foreign trade" model, CBEC job division is becoming more and more specific. An example of the division approach is shown in Fig. 1.

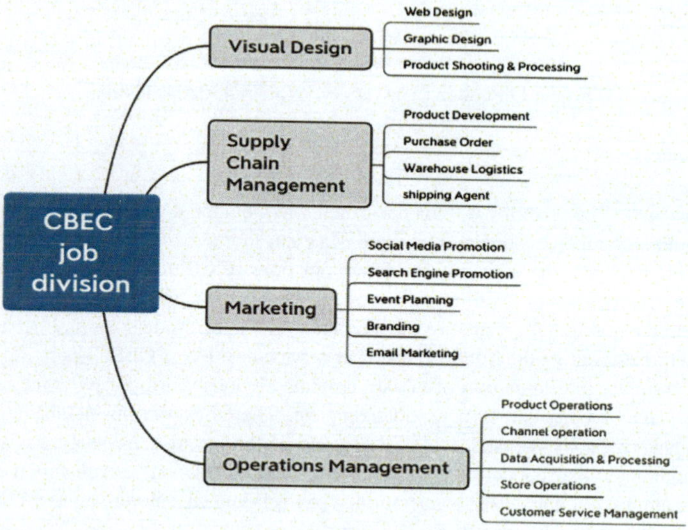

Fig. 1. The division of CBEC positions

2 Talent Status of CBEC Customer Service Position

Among all these job positions in the CBEC industry, its customer service is a front-line job which needs to communicate directly with customers. It occupies a basic and important position in the overall operation of CBEC, and its role cannot be underestimated. However, due to the rapid development of CBEC, the cultivation of talents in this field is obviously lagging behind, and there are some problems come into being.

2.1 Insufficient Talent Reserves

In the "Internet + " era, because CBEC has only developed in recent years, there is a huge shortage of professionals in this field. In 2019, the CBEC major has just be allowed to open in a very small number of 3-year vocational colleges mainly based on technology applications. All undergraduate colleges still do not have this major, but they are mostly established in majors such as business English, international trade, and e-commerce. In view of the CBEC talent cultivating direction, the lack of professional settings directly leads to a serious shortage of CBEC practitioners [3]. The position of CBEC customer service is usually served as a lead-in to this industry, so the proportion of talent gap in CBEC customer service position also appears to be particularly large.

2.2 Incomplete Levels of CBEC Customer Service Talent Structure

At present, most of the CBEC customer service staff are fresh graduates from colleges and universities. The customer service work is mainly in the initial stage, and there is an apparent lack of senior and middle-level participants with operational management experience. Most CBEC companies need intermediate-level composite talents, who are thought to be with more practical application value.

3 Demands for CBEC Customer Service Position

As the development of CBEC shows a trend of small orders, fragmentation, and rapid increase in order numbers, customer service staff engaged in CBEC are facing various complex situations. They must comprehensively pay attention to the influence of different languages, regions, climates, values, ways of thinking, behaviors, customs, cultures, consumption habits, even national policies, industry environment and other factors in CBEC field. Only by providing professional customer service can we adapt to the changing development of this field and the individual needs of customers. These require CBEC customer service staff to meet their business content requirements, ability requirements and skill requirements of this position.

3.1 Business Content Requirements

In actual work, the customer service staff of cross-border e-commerce is no longer limited to providing services to customers, and their work scope has been extended to cover all aspects of product marketing, cost control, team management and other aspects. CBEC customer service staff are working in an unpredictable competitive environment, who are dealing with their business in both B2B and B2C customer groups. They need to participate in the entire operation process of CBEC enterprises. At the same time, they also need to pay attention to the psychology and needs of customers, providing corresponding services proactively.

Table 1 has listed the customer service contents and purposes which need to be provided by CBEC customer service staff according to different order processes.

3.2 Ability Requirements

According to the current market demand, the CBEC industry has put forward higher and more professional requirements for the practical ability and comprehensive quality of customer service staff:

3.2.1 Professional Requirements

Cooperative communication and organizational coordination are the most important professional qualities to CBEC enterprises. As one of the hot positions for enterprises who urgently need talents, customer service position requires employees to communicate with customers online and establish good cooperative relationships. This requires employees to have knowledge and skills in fields of e-commerce, foreign

Table 1. Details of customer service while dealing with orders [4]

Processes	Customer service contents	Purposes
Funds not credited	Send emails to explain the possible delays between platform verification and customer payments, and give reasons	Prevent customers' mis-understanding caused by long funding verification
Preparation before shipment	Confirm the delivery address, contact information, personal tax number (CPF) or company tax number; remind and confirm the customs tax-collectible goods, etc.	Reduce lost shipments and being detained in the process of customs clearance
Cargo in transit	Proactively notify the status of packages; regularly send the update status of packages	Avoid disputes caused by buyers' not seeing the customer service information
Shipping may time out	Proactively extend the receiving time for buyers	Avoid disputes as buyers approach the final receipt confirmation date
Parcel delivery	Ask the buyer to confirm the receipt, and recommend the buyer to give a 5-star feedback	Shorten the payment cycle

languages, international logistics, cross-border marketing, international trading and computer technology.

3.2.2 Learning and Innovation Ability

It is necessary to have a strong learning ability and be able to flexibly apply the learned knowledge to meet job needs, by combining with platform sales characteristics and market development trends.

3.2.3 Honest and Responsible Working Attitude

In the context of the rapid development of CBEC, it is very important for practitioners to be able to bear hardships and work with seriousness and responsibility in order to meet the requirements of customer service.

Only with solid international business capabilities, language communication skills and practical innovation ability can customer service staff better sell products to foreign customers and open up a wider overseas market.

3.3 Skill Requirements

Traditional trade is often in large amounts, emphasizing the standardization of products rather than its individuality. The core purpose of CBEC customer service is to meet the needs of customers. With the increasingly fierce global competition, the current competition between sellers and companies is not only about price and quality of their products, but also about their service quality. In order to adapt to the development of the CBEC market, the core skills that CBEC customer service staff need are included as follows.

3.3.1 Rich Foreign Trade Professional Skills and Platform Operation Skills

As a CBEC customer service practitioner, they must be familiar with foreign trade rules and be aware of the entire procedures of CBEC transactions, such as payment, logistics, customs duties, tax refunds, etc.

3.3.2 Good Foreign Language Communication Skills

The most commonly used language in CBEC customer service is English. The foreign language requirement for this job position is CET4 in general, and staff need to master some necessary business English knowledge to deal with daily tasks. CBEC platforms are aimed at global buyers, and the language used is not only English. In the case of not being good at using other languages, it is suggested to customer service staff to use translation tools and translation software to translate non-English languages into English to make themselves understood.

3.3.3 Practical Skills in Network Technology and Communication Software

Only by being good at using computers and communication software for product description, image processing, store decoration, SEO optimization, etc. can we communicate more effectively with customers. In addition, with good communication the image of the store can be improved, and the buyers can feel that they have enjoyed high-quality service.

4 Talent Cultivation for CBEC Customer Service Position

Based on the above analysis, it can be concluded that the suggestions for the cultivation of applied talents based on the needs of CBEC customer service position are as follows.

4.1 Encourage Practical Skills Training for Customer Service Position in CBEC Entrepreneurship

In the past, DHgate, wish, AliExpress and other platforms that did not charge for registering personal accounts have already started charging. Therefore, it is also difficult for CBEC practical courses to introduce the operation of real platforms into class, and simulative teaching software is always difficult to reach the authenticity of platform operations, which is obviously inconsistent with the high practicality required for CBEC customer service position.

Nowadays, it is a good teaching model to encourage teachers to embed university students' innovative projects in teaching practice. In this way, you can use real CBEC platform accounts to help students initially build the foundation of the platform operation while posting commodities, especially for exercising the various business abilities required by the customer service position. It can be expected that the operation experience of real platforms will be very helpful for students to work in CBEC customer service position, or even to start a business in the CBEC field in the future.

At the same time, the practical teaching of entrepreneurship is also very much in line with the advocacy of "mass entrepreneurship, mass innovation", which means it is also worthy of promotion from the policy level.

4.2 The Cooperation Between School and Enterprise to Cultivate Talents

School-enterprise cooperation can be reflected in the aspects of talent cultivation, curriculum development, and joint development of practical training materials. For example, in terms of talent cultivation for CBEC customer service position, application-oriented colleges can cooperate with enterprises by signing agreements to establish CBEC training bases, and select a group of students to form a characteristic experimental class. On the basis of maintaining the original training plan, this class is added with some practical courses of CBEC customer service position taught by tutors from companies. At the same time, the graduates of this class can directly work in those cooperative companies without being required to do the internship.

In this way, students can learn the skills required by their future companies at school, which can fit with the needs of the companies to a large extent, save the cost of talent training for the companies, shorten the time for students to integrate into professional roles after graduation, and their job confidence and professional loyalty can be improved as well as their self-esteem. This is undoubtedly a trilateral win situation for students, enterprises and universities.

4.3 Establishing Off-Campus Training Base

The deep integration of school-enterprise cooperation has gradually promoted the transformation and development of CBEC education in application-oriented colleges and universities.

The off-campus practice, training, internship, and post-internship required for the training of CBEC talents can all be carried out in the training bases that have been established or will be established. For example, colleges and universities can use internship, practice, and credit replacement policy to promote students to go into those off-campus training bases for training, visiting, and internships during summer vacation.

Students can learn theoretical knowledge and platform skills on CBEC platforms and networks, thereby improving their practical skills, social skills and managing skills in the CBEC field. The talent cultivation of CBEC customer service position is quite suitable for this form.

4.4 Cooperation with Well-Known Companies

When application-oriented colleges and universities try to cultivate CBEC talents, they can complement the deficiencies in the curriculum system, teaching staff, and practical environment by cooperating with well-known leading companies or third-party operators of such enterprises in this industry. For example, the "One Hundred Cities, One

Thousand Schools, One Million Talents" campaign launched by the Ali Group in 2015 aims to unite with universities to train qualified CBEC talents. The mode of operation of this cooperation is to join the Ali credit system and plan a special base site at school. Students accumulate practical skills in various positions to obtain credits and corresponding salary. The number of Ali's third-party operating companies has also increased sharply in recent years, not only to reserve talents for CBEC companies led by Ali, but also to help colleges and universities cultivate application-oriented talents that meet the needs of job positions in CBEC.

Acknowledgments. This article is one of the results of the Hubei Province Education Discipline Planning 2018 key project: "Research on the Cultivation of Cross-Border E-Commerce Industry-Education Integration Talents in Application-Oriented Colleges under the New Model of Internet + Foreign Trade" (Project No.: 2018GA056, Project Leader: Yi Jing).

References

1. 2019 Global Cross-Border E-Commerce Market and Development Trend Research Report. iiMedia Research (2019).
2. Cross-Border E-Commerce Merchant Research Report. E-Commerce Research Center (2019).
3. Fang, H.: Research on the talent training model of cross-border E-commerce. Comput. Prod. Distrib. J. **4**, 282–283 (2019)
4. Jing, Y., et al.: Cross-border E-commerce customer service. Posts & Telecom Press Co., Ltd., Beijing (2019)

Check for updates

Research and Design of Stream Computing Framework Based on Storm

Huanbin Wang(✉) and Yangjun Gao

College of Equipment Management and Unmanned Aerial Vehicle Engineering of Airforce Engineering University, Xi'an 710051, Shanxi, China
whbkgd@163.com

Abstract. Firstly this paper introduces the basic concept of stream computing, studies the relevant characteristics of stream data and analyzes the differences between the stream processing system and the traditional data processing system in detail. Secondly, the composition, operation mechanism and characteristics of the typical stream computing system storm are analyzed in detail and the structure of storm cluster is studied. Finally, the workflow of storm cluster architecture is studied.

Keywords: Storm · Stream computing · Frame structure · Workflow

1 Introduction

In recent years, a new kind of dense data has emerged in the fields of web application, sensor monitoring and network monitoring. The data continues to arrive in the form of massive stream. Stream computing is aimed at the real-time calculation of stream data. Stream computing technology obtains massive data from different data sources in real time and obtains valuable information through real-time analysis and processing [1]. Stream computing can be used in many scenarios, such as web services, machine translation, advertising, natural language processing, climate simulation and prediction. In Baidu, Taobao and other large-scale websites, a large number of stream data will be generated every day, including the user's search content, user's browsing records etc. Through the real-time data analysis of stream calculation, we can understand the flow change of each moment and even analyze the real-time browsing track of users, so as to make real-time personalized content recommendation.

2 Stream Data Characteristics

The processing of convective data is suitable for situations with high real-time requirements and continuous data arrival. Generally, the stream data has the following characteristics [2]:

(1) The potential size of stream data is endless and the data can arrive quickly and continuously.
(2) The format of stream data is very complex, usually there are many data sources.

M. Tavana et al. (Eds.): IISA 2020, AISC 1304, pp. 256–262, 2021.
https://doi.org/10.1007/978-3-030-63784-2_32

(3) There is a large amount of data but it doesn't pay much attention to storage. Once the stream data is processed, the useless data will be discarded and the useful data will be archived and stored.
(4) The order of the general stream data is reversed or incomplete and the system cannot control the order of the newly arrived data elements to be processed.

3 The Processing Flow of Stream Calculation

The traditional data process needs to collect data and store it in the data management system such as relational database and then users interact with the data management system through query operation. The traditional data process implies two premises [3]: first, the stored data is the old static data and the stored static data is a snapshot of a certain time in the past which may not have timeliness when querying. Second, users need to actively issue queries to obtain results.

The processing flow of stream calculation generally includes three stages: data real-time collection, data real-time calculation and real-time query service. The basic architecture of data acquisition system for stream computing generally consists of the following three parts [4]:

(1) Agent: collect data actively and push the data to collector.
(2) Collector: receive data from multiple agents and realize orderly, reliable and high-performance forwarding.
(3) Store: store the data forwarded by collector (Fig. 1).

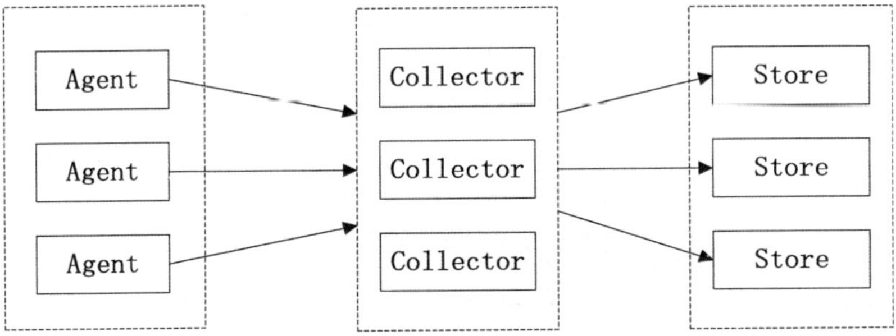

Fig. 1. Basic structure of data acquisition system

In the data real-time calculation stage, the collected data is analyzed and calculated in real time and the calculation results are fed back in real time [5]. The data processed by the stream processing system can be stored according to the situation, so as to analyze and calculate later. In the scenarios with high timeliness requirements, the processed data can be discarded directly if it has no value of preservation.

4 Difference Between Stream Processing System and Traditional Data Processing System

Due to the different background of requirements between the traditional data processing system and the stream processing system, the stream processing system is quite different from the traditional data processing system, including the following aspects:

(1) Stream processing system deals with real-time data, while traditional data processing system deals with pre stored static data [6].
(2) Users get real-time results through the stream processing system, while the results from a certain time in the past are obtained through the traditional data processing system.
(3) The real-time query service can push the real-time results to the users actively.

Real time query service means that users can query the results calculated by the stream calculation framework in real time and display or store them [7]. In the traditional data processing process, users need to actively send out queries to get the desired results. In the stream processing system, the real-time query service can constantly update the results and push the results required by users in real time. Although the traditional data processing system can update the results and push the results constantly by timing query, the results obtained in this way are still the results obtained according to the data at a certain time in the past, which is essentially different from the results provided by the real-time query service.

5 Storm Real-Time Computing System Composition

Storm is a distributed, scalable and fault-tolerant open-source real-time computing system which designs an abstraction of convection and computing [8]. Compared with the traditional off-line computing and batch data processing system Hadoop, storm is specially designed for real-time computing and stream data processing with a wide range of application scenarios, which is known as the real-time computing version of Hadoop. Storm mainly includes streams, outlets, bolts, topology and stream groupings.

(1) Streams: Storm describes the stream data as an infinite tuple sequence, which will be created and processed in parallel in a distributed way [9]. Each tuple is a pile of values, each value has a name and each value can be any type. Tuple should have been a key value map. Because the field names of tuple passed between components have been defined in advance, tuple only needs to fill in each value in order, so it forms the corresponding value list.
(2) Spout: Storm thinks that every stream has a source and abstracts this source as "spin". Usually, spin will read data from external data sources (queues, databases, etc.) and then package it into tuple form and send it to the stream. Spout is an active role. There is a next tuple function inside the interface and the storm framework will call it constantly.

(3) Bolt: Storm abstracts the state transition process of streams as bolt. Bolt can process tuple or send the processed tuple to other bolts as new streams [10]. Bolt can perform filtering operations, function operations, join operations and database operations, etc. Bolt is a passive role. In its interface, there is an execute method which will be called after receiving the message. Users can execute their own processing logic in this method.
(4) Topology: Storm abstracts the network composed of outlets and bolts into topology which can be submitted to storm cluster for execution. Topology can be regarded as a flow transformation diagram in which the node is a pivot or bolt and the edge indicates which stream the bolt subscribes to. When a tuple is sent by a spool or a bolt, it will send the tuple to each bolt that has subscribed to the stream for processing. Each processing component in topology contains processing logic and the connection between components indicates the direction of data flow [11]. Every component in topology runs in parallel. In topology, you can specify the parallelism of each component. Storm will allocate multiple threads in the cluster to calculate at the same time (Fig. 2).

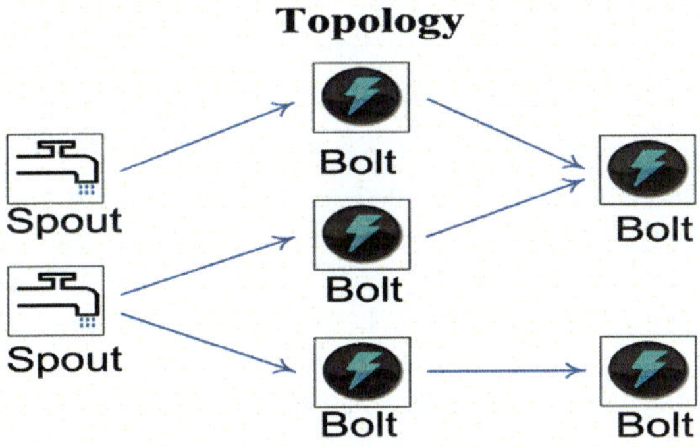

Fig. 2. Composition of Topology

(5) Stream Groupings: Stream groupings in storm are used to tell topology how to transfer tuple between two components (such as between the spin and the bolt or between different bolts). Each spin and bolt can have multiple distributed tasks. The time and method for a task to send tuple is determined by stream groupings [12]. The working mode of stream groupings is shown in Fig. 3:

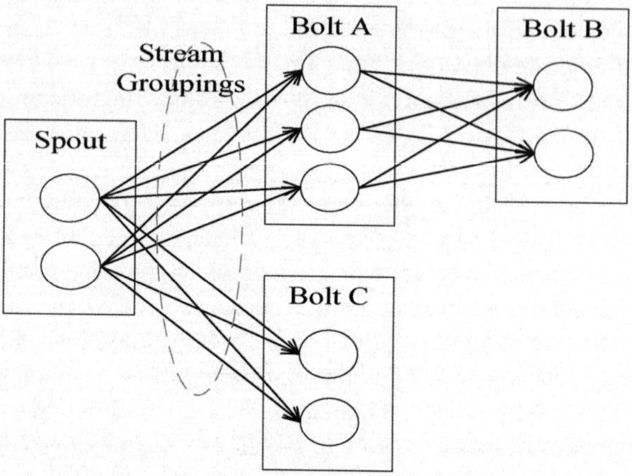

Fig. 3. Working mode of stream groupings

6 Storm Operation Mechanism

Storm has two modes of operation: local mode and distributed mode.

(1) Local operation mode
In storm's local mode, the system starts a process to serve storm and all threads of the simulated spin and bolt run in the same process. Because there is only one process in the local mode, it can't really simulate the storm distributed running environment so storm's local mode is usually used for development and testing. Topology in storm starter runs in local mode which is characterized by: all information in topology can be transparently and accurately captured [13].
(2) Distributed operation mode
The storm distributed model consists of machine clusters. When the user submits the calculation task to the control node in the form of topology, the corresponding code of topology will also be submitted. Nimbus process runs on the control node which is responsible for distributing the code submitted by the user to the work node and assigning the running process to the topology submitted by the user.

7 Storm Cluster Architecture and Working Mode

Storm cluster adopts the node working mode of "master-worker" [14].

1) The master node runs a background program named nimbus (similar to "job-tracker" in Hadoop) which is responsible for distributing code, assigning tasks to workers and monitoring faults within the cluster.
2) The worker node runs a background program named supervisor which is responsible for listening to the work assigned to its machine. According to the tasks

assigned by nimbus, it decides to start or stop the worker process. A worker node runs several worker processes at the same time.

Storm uses zookeeper as a distributed coordination component, responsible for all coordination between nimbus and multiple supervisors. With the help of zookeeper, if nimbus process or supervisor process is terminated unexpectedly, it can also read, restore the previous state and continue to work when restarting, making storm extremely stable (Fig. 4).

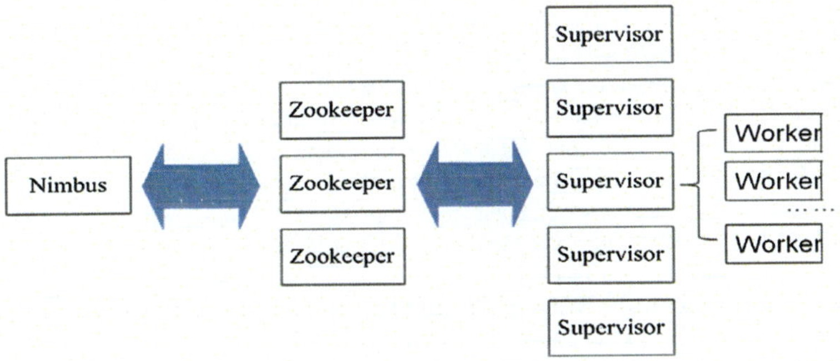

Fig. 4. Storm cluster architecture

(1) Worker: each worker process belongs to a specific topology. Each supervisor node can have multiple workers. Each worker runs one or more executor threads for each component (spin or bolt) in topology to provide task running services.
(2) Executor: executor is a thread generated in the worker process which will execute one or more tasks of the same component.
(3) Task: the actual data processing is completed by task. In the life cycle of topology, the number of tasks of each component will not change [15], while the number of executors is not necessarily. The number of executors is less than or equal to the number of tasks. By default the two are equal.

All topology tasks must be submitted on the storm client node. After submission, the nimbus node assigns them to other supervisor nodes for processing. The nimbus node first divides the submitted topology into multiple tasks and assigns them to the corresponding supervisor [16] and submits the task and supervisor related information to the zookeeper cluster. The supervisor will go to the zookeeper cluster to claim its own tasks and notify the relevant worker processes to handle the tasks.

8 Concluding Remarks

Storm real-time computing system has become an effective technology for real-time computing and analysis of big data. At present, it is widely used and its research value is increasingly prominent which has attracted the attention of many research institutions and enterprises. Learning and mastering the technical basis of storm real-time computing system is very important. In this context, this paper analyzes the storm stream computing framework and workflow. The research content will be helpful to the relevant researchers of storm computing.

References

1. Jin, Y., Wu, H.: Research on big data processing architecture based on storm and Hadoop. Mod. Comput. (Prof. Edn.) **04**, 9–12 (2015)
2. Wan, J., Liu, M., Hu, X.: Dual-JT: toward the high availability of JobTracker in Hadoop. In: 2012 IEEE 4th International Conference Cloud Computing Technology and Science (CloudCom), December 2012, pp. 263–268 (2012)
3. Yichen, L., Xuzhi, L., Zhen, Y.: Application of real-time stream calculation in aerospace ground data processing system. Microelectron. Comput. **9**, 23–26 (2014)
4. Sun, C.: Design and Implementation of Data Analysis System Based on Storm. Beijing University of Posts and telecommunications, pp. 33–38 (2014)
5. Iqbal, M.H., Soomro, T.R.: Big data analysis: apache storm perspective. Int. J. Comput. Trends Technol. **19**(1), 9–14 (2015)
6. Dawei, S., Guangyan, Z., Weimin, Z.: Big data stream computing: key technologies and system examples. J. Softw. **25**(4), 839–862 (2014)
7. Yuan, L., Xuanhua, S., Hai, J.: Runtime-aware adaptive scheduling in stream processing. Concurr. Comput. Pract. Exp. **34**(5), 562–574 (2015)
8. Huang, F.: Design and Implementation of Interactive Platform of Microblog Based on Storm. Sun Yat sen University, pp. 6–11 (2013)
9. Long, S.: Research and Implementation of Real-Time Big Data Analysis System Based on Storm. Shanghai Jiaotong University, pp. 17–26 (2015)
10. Mohamed, I.M.W., Abou, M.: Anomaly detection using Hadoop andMapReduce technique in cloud with sensor data. Int. J. Comput. Appl. **125**(1), 22–26 (2015)
11. Qu, G.: Design and Implementation of Real-Time Log Analysis System Based on Storm. Nanjing University, pp. 6–19 (2016)
12. Dawei, S., Guangyan, Z., Songlin, Y., et al.: Re-Stream: real-time and energy-efficient resource scheduling in big data stream computing environments. Inf. Sci. **319**, 92–112 (2017)
13. Qian, J., Xu, H., Dong, Y., et al.: Data flow window connection optimization algorithm based on minimum spanning tree. Comput. Res. Develop. **44**(6), 1000–1007 (2016)
14. Ning, G.Q., Wang, H., Shu, L.C., et al.: Towards load shedding and scheduling schemes for data streams that maintain quality and timing requirements of query results. Soft Comput. **20** (5), 1–16 (2015)
15. Li, C., Haihong, E., Meina, S.: Research and application of storm based real-time computing framework. Software **35**(10), 16–20 (2017)
16. Zhengdun, D., Jian, H., Futong, L.: An iterative topology scheme based on storm programming model. J. Chengdu Inst. Inf. Eng. **29**(1), 47–51 (2016)

Research on the Comprehensive Performance Evaluation of Large-Scale Event_Cases of China Chengdu International Intangible Cultural Heritage Festival

Xing Zeng[1,2](✉)

[1] Sichuan Vocational College of Cultural Industries, Chengdu 610213, China
gkittyzx@163.com
[2] School of Economics, Sichuan University, Chengdu 610065, China

Abstract. According to the stakeholder theory, this research evaluated the comprehensive performance of China Chengdu International Intangible Cultural Heritage Festival. Based on the survey, we indicate that the Inheritance Festival has received high satisfaction from the perspective of the organizing committee, exhibitors, professional visitors and related companies. Increasing the professional services and enhancing the professionalism of "Inheritance Festival are very important. We should promote on-line "Inheritance Festival". Inheritance Festival brand can be promoted through professional websites. And we have to invest abundant resources to strengthen infrastructure construction.

Keywords: Comprehensive performance · Large-scale event · Cultural festival

1 Introduction

Large-scale events are activities held in short duration to strengthen the recognition of tourist destinations, increase visitors' attractiveness, and increase economic outcome. Success depends on its uniqueness, status, and the significance which create public interest and attract people's attention.

The China Chengdu International Intangible Cultural Heritage Festival (referred to as the "Inheritance Festival") originated in 2007 and is co-sponsored by the Ministry of Culture of the People's Republic of China, the Sichuan Provincial People's Government, the National Committee of China's UNESCO, and UNESCO. The Chengdu municipal government, the China intangible cultural heritage protection center, and the Sichuan Provincial Department of Culture jointly organized the event. The Chengdu Municipal Bureau of Culture, Qingyang District government, and the International Intangible Cultural Heritage Expo Park implemented the event.

M. Tavana et al. (Eds.): IISA 2020, AISC 1304, pp. 263–270, 2021.
https://doi.org/10.1007/978-3-030-63784-2_33

2 Literature Review and Theoretical Basis

2.1 Large-Scale Event

The research on the impact and evaluation of large-scale activities has always been focus on tourism impact in Western countries. Getz believes that the evaluation of activities includes three basic types: ex-ante evaluation, process evaluation and ex-post evaluation. Existing research on large-scale event is mostly based on ex-post evaluation, and few research focus on the process of large-scale event. The impact of large-scale activities includes economic impact, environmental impact and social-cultural impact, such as the attitude of local residents, promotion of cultural heritage, protection of tradition, loss of comfort, public behavior etc. The positive influence includes gaining international prestige, improving the image, and increasing social cohesion. Negative impacts include loss of comfort, environmental damage, noise interference, and traffic jams. Large-scale activities have both tangible and intangible impacts. Intangible impacts include changes of social life and image of a place. Olav R. Spilling (1998) analyzed the long-term impact of the 1994 Winter Olympics in Norway on economic growth. John R. Madden (1999) made a detailed analysis of the impact of the 2000 Sydney Olympic Games on Australia's macroeconomics, industrial development, employment structure, and financial situation. In 2001, the Utah State Government organized a large number of researchers to study the impact of the 2002 Salt Lake City Winter Olympics on the state's economic impact, image impact, and subsequent development. It also investigated the destination of large-scale festivals for the first time through a questionnaire. The influence of image recognition and image cognition was analyzed.

2.2 Comprehensive Performance Evaluation of Festival Activities

Chen Zeyan outlined three types of evaluation subjects. Chen Jing proposed a city exhibition evaluation system. Liu Songping summarized the 16 most important items in the American exhibition quality evaluation index in 2004. The most important evaluation criteria for American exhibitions are "the proportion of participating decision-makers among exhibiting companies", "the proportion of exhibiting companies in target markets", "the number of exhibiting companies", while the Chinese exhibitions pay more attention to the "number of exhibiting companies", "Exhibition area", "Number of professional visitors", "On-site transaction volume and intentional transaction volume. In comparison, the US exhibition evaluation system is relatively more comprehensive, but it is not exactly suitable for China's exhibition evaluation. This is due to the development of China's exhibition industry and the current demand for China's exhibition evaluation market. Yang Fangping constructed the evaluation index system for brand exhibitions in 2010. Li Zhiling pointed out the reference and limitation of UFI certification model to China's exhibition evaluation in 2007.

3 Methodology

3.1 The Basis for Constructing the Evaluation System

Stakeholders are those individuals who would affect the achievement of the organizer's goals or can be affected by the process of the organizer's achievement of the goals. The definition not only considers individuals and groups that affect corporate goals as stakeholders, but also considers individuals and groups by the realization of corporate goals as stakeholders. Stakeholders for large-scale events should include the following categories (Table 1):

Table 1. Large-sale event stakeholder classification

Classification	Exhibition stakeholder
Principal stakeholders	Exhibition companies, exhibitors and professional visitors
Secondary stakeholders	Organizer
Supportive stakeholders	Exhibition-related enterprises (such as booth design, transportation and construction companies), exhibition industry associations
Related stakeholders	Government, hotel, catering, transportation and tourism agency

According to stakeholder theory and analytic hierarchy process theory, the "Inheritance Festival" stakeholder has principal stakeholders, including exhibition companies, exhibitors, and professional audiences; Secondary stakeholders, such as organizer; supportive Stakeholders, that is, exhibition-related enterprises, such as booth design, transportation and construction companies, and exhibition industry associations, etc.; Related stakeholders, that is, governments, hotels, catering, transportation, and tourism companies.

3.2 Index Design of the Evaluation System

The evaluation system is mainly composed of two levels of indicators including the exhibitor index, professional audience index. "exhibitor perspective" index consists of exhibitor profit, exhibitor quality, number of exhibitors, transaction, exhibitors activities. Fourteen secondary factors are the number of Exhibitor profit, booth construction costs, booth design costs, transportation cost, number of professional visitors, number of all exhibitors and performers, agreement, exhibitor goals, number of enterprise catalogue issued, news media absorption and application, pre-registration procedure, venue surroundings and health environment, venue security, intellectual Property Protection. The professional audience perspective index consists of visitors' activity, purchase plan, and exhibitor quality. Ten secondary factors are average duration spent in each booth, total Purchase plan, number of exhibitors, organization management, exhibitor size and reputation, service consulting, activity itinerary, venue environment, visiting intention, professional of the exhibition. In order to ensure the objectivity of the evaluation system, there are only three qualitative indicators including and the rest are quantitative indicators based on objective facts. Both qualitative and quantitative indicators should be analyzed through data based on literature research, questionnaire surveys and interviews (Tables 2 and 3).

Table 2. Factor analysis of exhibitors

Variable	Index
Exhibitor Profit	Transaction profit (Questionnaire Statistics)
	Booth construction costs
	Booth design costs (questionnaire statistics)
	Transportation Cost (Questionnaire Statistics)
Quality of exhibitors	Number of professional visitors (data)
Number of exhibitors	Number of all exhibitors and performers
Conclude a transaction	Agreement (Questionnaire Statistics)
Exhibitors' Activities	Exhibitor Goals (Questionnaire Statistics)
	Number of enterprise catalogue issued
	News media absorption and application
Service and management of the exhibition	Pre-registration procedure
	Venue surroundings and Health Environment
	Venue security (questionnaire statistics)
	Intellectual Property Protection

Table 3. Factor analysis of visitors

Variable	Index
Visitors activity	Average duration spent in each booth
Purchase Plan	Total Purchase Plan (Questionnaire Statistics)
Exhibitor Quality	Number of Exhibitors (Questionnaire Statistics)
	Organization Management (Questionnaire Statistics)
	Exhibitor size and Reputation (questionnaire statistics)
	Service Consulting (Questionnaire Statistics)
	Activity itinerary (Questionnaire Statistics)
	Venue environment (questionnaire statistics)
	Visiting intention (questionnaire statistics)
	Professional of the exhibition (questionnaire statistics)

3.3 Questionnaire Design and Formal Survey

As one of the main leading roles on the large-scale event, exhibitor refers to those who want to exchange market information, conclude a deal, promote the image and find cooperators; professional visitor refers to the group whose Business scope involves professionals design and development of the goods or services displayed at the exhibition. On the Inheritance Festival, the questionnaire survey includes both professional visitors and general audiences. During the "Inheritance Festival", a total of 140 face-to-face exhibitor questionnaire surveys were conducted. 800 questionnaire surveys were conducted for professional visitors.

3.4 Framework of the Evaluation System

See Fig. 1.

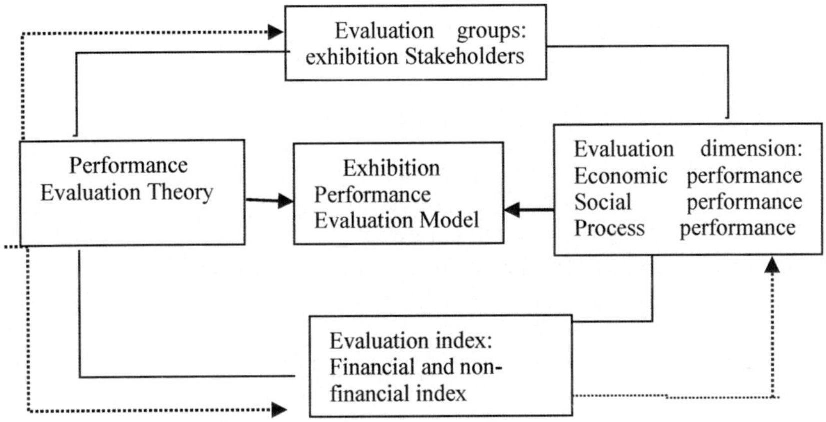

Fig. 1. Framework of the evaluation system

4 Data Analysis and Findings

4.1 Character of Exhibitor

The data shows that 42.86% of the exhibitors are individual entrepreneur, 35.00% are private enterprises, 3.57% are administrative institutions, 7.86% are foreign-funded enterprises and 0.71% are state-owned enterprises (Table 4).

Table 4. Character of exhibitor

Character of exhibitor	Proportion
Administrative institutions	3.57%
State-owned enterprises	0.71%
Private enterprise	35.00%
Joint-stock enterprises	6.43%
individual entrepreneur	42.86%
Foreign-funded enterprises	7.86%
Other	3.57%

4.2 Continuity of Participation

The survey shows that 74.70% of the exhibitors have participated in the previous Inheritance Festival, only 15.30% of the exhibitors have never participated in the event before, which means that the reputation of Inheritance Festival has increased year by year and it attracts more exhibitors and professional visitors (Table 5).

Table 5. Continuity of participation

"Have you participated in the previous "Inheritance Festival"?	Proportion
Never participated	15.30%
Have participated	84.70%

4.3 The Aims of Participating in the Inheritance Festival

39.73% of the exhibitors believe that the aim of products sales is achieved, 34.25% regard that the aim of increasing visibility is achieved, 11.64% of exhibitors show that the purpose of collecting information is reached, and 14.38% of the exhibitor regard that the aim of looking for agents and other purposes has achieved (Table 6).

Table 6. The aims of participating

Aims of participation	Proportion
Products sales	39.73%
Finding an agent	8.56%
Information collection	11.64%
Increasing reputation	34.25%
Other	5.82%

4.4 The Impression on the Image of "Inheritance Festival"

62.86% of exhibitors and visitors said that the Inheritance Festival was impressive. 22.14% of exhibitors and visitors said that the Inheritance Festival was fair. 10.71% of exhibitors and visitors said that the Inheritance Festival was ordinary. 1.43% of exhibitors and visitors said that the Inheritance Festival was poor. 2.86% of exhibitors and visitors have no comments (Table 7).

Table 7. The impression on the "Inheritance Festival"

Impression	Proportion
Good	26.43%
Well	36.43%
Fair	22.14%
Ordinary	10.71%
Poor	1.43%
Not related	2.86%

4.5 Evaluation of Relevant Aspects of "Inheritance Festival"

According to the survey, the exhibitors and visitors regards that the organization and reception, service consultation and intellectual property protection have excellent performance. However, event theme, promotion, and venue management need to be consistently improved (Table 8).

Table 8. Relevant aspects of "Inheritance Festival"

Relevant aspects	Good	Well	Fair	Ordinary	Poor
Organization and reception	48.75%	27.86%	18.57%	4.29%	0.71%
Service consultation	24.29%	49.29%	25.00%	1.43%	0.00%
Event theme	29.29%	34.29%	32.14%	3.57%	0.71%
Promotion	40.71%	17.14%	29.29%	10.00%	2.86%
Venue management	43.57%	19.29%	27.14%	9.29%	0.71%
Intellectual property protection	47.14%	24.29%	21.43%	3.57%	3.57%

4.6 Willing to Participate in Chengdu "Inheritance Festival" Successively

According to the survey, 48.57% of the exhibitors and visitors are willing to participate in successively. 17.86% of the exhibitors and visitors indicates that they will consider the participation. 14.29% of the exhibitors and visitors says that they haven't make decisions. Only 2.86% of exhibitors and visitors says that they are unwilling to attend next time (Table 9).

Table 9. Willing to participate in

Successive participation	Proportion
Willing	48.57%
General	17.86%
Fair	4.29%
Not sure	14.29%
Unwilling	2.86%

5 Conclusions and Implications

According to the survey, we can indicate that the Inheritance Festival has received high satisfaction from the perspective of the organizers, exhibitors, professional visitors and related companies.

Increasing the professional services and enhancing the professionalism of "Inheritance Festival are very important. On the one hand, it provides a place for exhibitors and professional visitors to communicate, display and conclude a deal. Suitable business atmosphere, organizational management and professional services on the event will gain satisfactions. On the other hand, suitable management should be strengthened, the participation of exhibitors and visitors should be increased, the professionalism and internationalization of "Inheritance Festival should be promoted, and more foreign exhibitors should be invited. Leading companies should be invited to ensure the professionalism of the event. Well-known brands should be promoted in the event to ensure and enhance the reputation of the event.

The organizer should continue to increase the promotion of the "Inheritance Festival" and cooperate with the tourism department to make appropriate tourism itinerary and custom travel plans in advance. The organizer should introduce short-term tourism project into the "Inheritance Festival" in Chengdu. It may be considered to establish an "Inheritance Festival brand image center. To make sure that audience can have better understanding of the "Inheritance Festival brand image, the organizer should provide favorable experience and suitable atmosphere. We also promote on-line "Inheritance Festival". Inheritance Festival brand can be promoted on professional websites. We can also create an "Inheritance Festival" on-line guest system, so that the audience can enjoy professional services anywhere, receive relevant information anytime, purchase products on line and achieve a real-time interactive experience.

Investing abundant resources will strengthen infrastructure construction. Taking catering service as example, superb catering service providers would provide customers with diverse and affordable catering services. We can also request for relieving congestion during the event. The exhibitors and visitors could obtain service conveniently by increasing the bus frequency, and increasing on-line taxi next to the venues. Exhibition management has a positive influence on the promotion of city image. According to the survey, the exhibitors and visitors are not satisfied with the organization and management of the exhibition, including service consultation, event theme arrangement, publicity, venue management and intellectual property protection. Some exhibitors and visitors indicate that visual identities were not convenient and distinct, which caused confusion between the exhibitors and visitors.

Acknowledgments. This research was partially supported by Sichuan Cultural Industry Development Research Center (WHCY2019A03); Sichuan Vocational College of Cultural Industries (18A001).

References

1. Getz, D.: Event Management and Event Tourism, p. 21. Cognizant Communication Corporation, New York (2005)
2. Fu, Y., Zheng, X.: Industrial merging of tourism industry and MICE industry: analysis of industrial value chain, approaches and countermeasures. J. Northwest A&F Univ. (Soc. Sci. Ed.) (2), 151 (2014)
3. Rittichainuwat, B., Mair, J.: Visitor attendance motivations at consumer travel exhibitions. Tour. Manag. **33**, 1236–1244 (2012)
4. Zhang, Y.: Understanding travel motivations of Chinese tourists visiting Cairns Australia. J. Hosp. Tour. Manag. **21**, 44–53 (2014)
5. Yuksel, A., Yuksel, F., Bilim, Y.: Destination attachment: effects on customer satisfaction and cognitive, affective and conative loyalty. Tour. Manag. **31**, 274–284 (2010)

Review on the Application of Machine Vision Algorithms in Fruit Grading Systems

Lin Kaiyan, Liu Chang, Si Huiping(✉), Wu Junhui, and Chen Jie

Modern Agricultural Science and Engineering Institute, Tongji University, Shanghai, China
sihuiping@tongji.edu.cn

Abstract. China is the largest fruit producer in the world, so fruit grading is one of important tasks in agricultural production. Because manual testing is time-consuming and laborious, it is reasonable to be replaced by machine vision technology. This review included three parts. The first part introduced the architecture of the fruit grading system, including image grasp and processing algorithms. And the system transferred images to computer for processing after captured by acquisition sensor. Then computer performed image processing operations, including preprocessing, segmentation, feature extraction and selection. The second part summarized the machine vision classification algorithms for fruit grading. Classification algorithms are classified as supervised machine learning, unsupervised algorithms, and deep neural network. The supervised machine learning algorithms are Naive Bayes, K-nearest Neighbor, Support Vector Machine and so on. The unsupervised machine learning algorithms included K-means Clustering and Principal Component Analysis. And the deep learning part mainly introduced Artificial Neural Network and Convolutional Neural Network. This part reviewed the characteristics and application potential of the above classification algorithms, and comprehensively compared the merits and demerits of assorted classification algorithms applied in fruit classification. The third one discussed the difficulties and challenges of fruit grading tasks based on machine vision technology and put forward some prospects. Finally, this paper summarized and analyzed the machine vision algorithms in fruit grading systems and concluded that the research of machine learning and deep learning combined with machine vision systems is the main development trend in the field of fruit grading and will be a hot research spot in future.

Keywords: Machine vision · Fruit grading · Machine learning · Deep learning

1 Introduction

China is the world's largest fruit producer and processing country, and has the reputation of "the world's fruit kingdom".Besides the food and vegetable industry, the fruit industry has developed to be the third largest agricultural industry in China. It has obvious advantages and competitiveness in the agricultural product processing industry [1]. In China, the exports of fruits account for a small proportion of total production. In order to improve the competitiveness of fruits in the international market, strengthening

M. Tavana et al. (Eds.): IISA 2020, AISC 1304, pp. 271–280, 2021.
https://doi.org/10.1007/978-3-030-63784-2_34

post-harvest handling is a key step. The practical experience of developed countries showed that fruit grading is not only one of the key factors for fruit commercialization, but is an important way to increase the output value.

Fruit is usually graded manually in post-harvest handling, having significant subjectivity and error problems. Along with the development of speedy and high-accuracy machine vision technology, robotization of the fruit grading process is supposed to decrease manpower and material resources and improve sorting efficiency and accuracy. Machine vision technology is gaining more and more attention in the fruit industry, particularly in quality inspection and defect classification. Through the years, a large number of studies have been carried out in the domain of fruit grading based on machine vision, including citrus, apple, oil palm, strawberry, mango, lemon, etc. [2–9]. Research algorithms involved image processing, spectral analysis, feature extraction, machine learning, deep learning, etc. [10–13]. This paper introduced the applications of machine learning in fruit grading. Firstly, the machine vision system composition and image processing algorithms were reviewed. Then its advantages and disadvantages were analyzed. Finally, this paper explored the future development trends and directions in this field, aiming to provide reference for the fruit grading research.

2 Fruit Grading System

Machine vision systems, which are composed of image acquisition sensors and image processing algorithms, have many applications in the agricultural and fruit industries, especially in the quality assessment and grading of fruits [14]. The fruit grading system can collect images, automatically calculate many features, for instance, texture, form, color, size and defects, then determine the quality and grade of the fruits.

2.1 System Construction

A fruit grading system is composed of camera, computer, light source, transmitting device, etc., which transfers the acquired target image to a computer for processing and making a grading decision with the extracted features. The image acquisition techniques used in in fruit grading systems are cameras, ultrasound, magnetic resonance imaging (MRI), and computerized tomography (CT) [15–17]. Megha et al. [18] designed a typical tomato classification system based on machine vision (Fig. 1), including conveyor belts, light sources, micro-controllers, DC gear motors, motor drives, UART bridges, computers and cameras. The motor controlled the movement of the conveyor belt to park the tomato directly under the camera for collecting images, which were then processed, analyzed and graded. Once the image was processed completely by the computer, the control signal was triggered to control the sorting mechanism via the RS232 serial output to move the tomato to the corresponding collection bins.

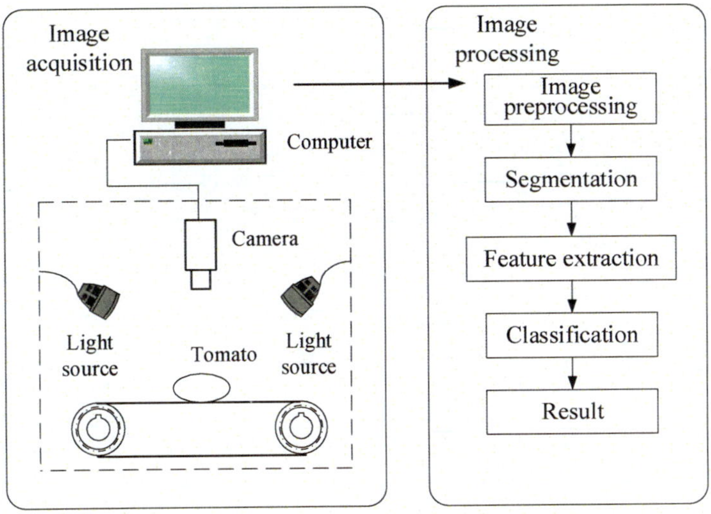

Fig.1. Tomato grading system

The process of acquiring images by sensors was affected by factors such as the environment, detection characteristics, lighting conditions, and target occlusion. To capture a better image, several kind of sensors were applied, such as black and white cameras, color cameras, spectral cameras and infrared thermal cameras and so on. Their merits and demerits are shown in Table 1. In the field of fruit image acquisition, the precision of hyperspectral imaging technology is higher than multispectral imaging. Bulanon et al. [19] used PCA(principal component analysis) to convert multispectral images and identified wavelengths of light that could improve fruit detection in the canopy background, enabling multi-spectral imaging for citrus detection under natural light conditions. Safren et al. [20] used hyperspectral cameras to capture apple images. First, the principal component analysis method was used to drop the dimension of the collected high-dimensional data, and then the ECHO(extraction and classification of homo generous objects) were used for apple image segmentation, with the classification accuracy of 88.1%.However, the feasibility of hyperspectral imaging is restricted due to time-consuming in data grasp and complex processes in image analysis.

Table 1. Fruit image acquisition sensors

Sensor Type	Merits	Demerits	References
Black and white camera	Provide grayscale information, less data, faster	Lack of color information	Edan et al. [21], Pla et al. [22]
RGB camera	Provide color information	Less accurate than black and white camera	Silwal et al. [23], Yao et al. [24]
Spectral camera	Provide spectral information and spatial information of objects	Longer time and high cost	Wang et al. [25], Peng et al. [26]
Infrared thermal camera	Perceived temperature, affected by background light	Affected by direct sunlight	Sahu et al. [27]

2.2 Image Processing

After capturing image, the computer preprocess the original image to improve the image quality for segmenting the target object. Then the corresponding features are collected, based on which the classification algorithms such as machine learning are applied to classify objects. The specific fruit grading machine vision system workflow is displayed in Fig. 2, in which the image procedure includes preprocessing, segmentation and feature extraction.

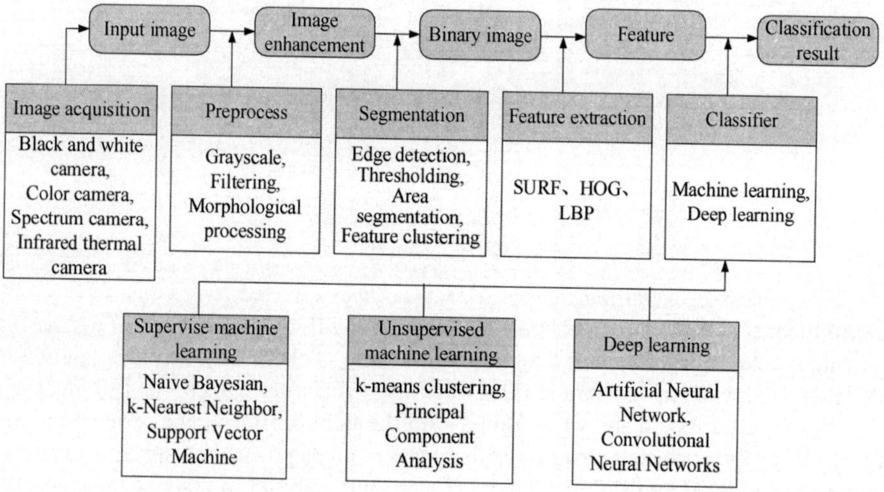

Fig.2. Working procedure of machine vision in fruit grading system

Image Preprocessing. If the image acquired by camera contains noise, it is not conducive to segmentation and processing. Image preprocessing is usually required to enhance effective digital information. The operations in image preprocessing are grayscale, filtering, and morphological processing and so on, which commonly uses a minor neighborhood of one pixel to produce a new luminance value in the output image. In addition, noise can be reduced by a simple filter, such as the averaging filter or the median filter.In the cherry tomatoes grading system developed in [28], the image subtraction method was firstly applied for background removal, followed by smoothing with a 15*15 Gaussian filter. Then the morphological opening operation was performed by the disk structural component with a radius of 9 to obtain the preprocessed data set.

Image Segmentation. It is very important to accurately segment the target area from captured image. The incorrect segmentation will affect the performance of the classifier. Among fruit image segmentation techniques, thresholding and clustering are widely used. The thresholding approaches are based on the pixel uniformity of object and easy to implement. Digital images can be divided into regions of interest and background based on image gray levels. In 2016, Lv [29] applied the Otsu dynamic threshold method to segment the apple image. It was found that the method is better than the fixed threshold segmentation. The recognition rates for fruits in case of unobstructed, overlapping, and obscured by branches and leaves reached 100%, 100%, and 86% respectively. Otsu method has the advantages of strong adaptability and no need to know the image structure in advance. However, with the number of clusters increases, more optimal threshold calculations are needed, which can be solved by using clustering methods. Clustering methods form clusters based on similar features of pixels to achieve image segmentation result. In [30], Ghabousian et al. used the fuzzy clustering to segment apple color images, eliminating shadow and brightness faults in the image region, and improving the accuracy of the classification result.

Feature Extraction and Selection. To provide accurate data for fruit quality assessment and grading, it is need to extract image features to describe the fruit after segmentation. In fruit grading procedure, features of color, texture, and shape are often used.Lu et al. [31] used R, G, and B color component values as features in RGB space for citrus detection. The final recognition accuracy of citrus was as high as 92.4%, which proved that it was efficient for color feature in detection. Kurtulmus et al. [32] successfully detected and classified citrus using Gabor texture features, which can effectively identify fruit texture areas in natural canopy.

The fruits features of color, size, form, and texture are often extracted to form feature sets for segmentation. The classification algorithms commonly used in machine vision include supervised/unsupervised machine learning, and deep learning methods, which play a very important role as classifiers.

2.3 Supervised Machine Learning Algorithms

Naive Bayesian. Naive Bayesian (NB) is a conditional independence hypothesis probability model based on predictive variables and features. Asadollahi et al. [33] used NB algorithm and seven other classifiers to classify tomato using the features of color, form and texture. The test results displayed that the classification accuracy of the NB classifier was 80.1%. The NB model can be trained with a variety of characteristics of fruit. Stegmayer et al. [34] constructed the model based on combination of feature selection methods and classifiers. Color, shape and texture features were combined to train NB for the classification of citrus, which contracted by three different types of diseases, obtaining the accuracy of 78.72%. On one hand, the classification accuracy of the NB algorithm is high with simple structure and less computing load. On the other hand, its disadvantage is that it is hard to learn the interaction between two features due to the assumption on the condition independence. Consequently, NB is suitable for a data set with less training samples.

K-Nearest Neighbour. The K-Nearest Neighbour (KNN) is a statistical classifier, in which, data are assigned to the nearest K class of adjacent samples by measuring the similarity with distance. In [35] a fruit grading system with KNN as a classifier was designed using the features of color, shape and size, which improved the accuracy of recognition. Rivera et al. [36] developed a mango peel mechanical damage assessment system based on hyperspectral imagery. They used KNN classify the input wavelength characteristics and decided a mango to be a complete or damaged class. The experimental results showed that KNN provided better classification performance than other classifiers and solved the problem of difficult detecting mechanical damage in the early stage of mango harvest. The KNN classifier is a lazy learning algorithm without training and is suitable for multiple classification of fruit-specific features. The downside is the heavy computing load, and the distance between each sample and the object to be classified needs to be calculated to get its neighbors. It also needs to be improved for real-time fruit grading systems with large numbers of samples.

Support Vector Machine. Support Vector Machine (SVM) is a statistical classifier based on supervised learning algorithm. It constructs a decision threshold to maximize the distance between sample points. In 2017, Moallem et al. [37] used a golden apple grading algorithm based on computer vision. The apples were divided into three categories using five texture features and three geometric features by SVM. Comparing with Multilayer Perceptron and KNN, the experiment results showed that the recognition rates of SVM are 92.5%, which was higher than MLP and KNN. And it also avoided the problems of over-fitting effectively. Mizushima et al. [38] studied an apple grading algorithm based on SVM and Otsu threshold algorithm. The image segmentation error was less than 2%. This algorithm automatically regulated the segmentation hyperplane calculated by linear SVM, which reduced the training time. Numerous studies have shown that SVM has excellent generalization capabilities and is superior to most machine learning algorithms in solving regression and classification problems. The main disadvantage of SVM is that it is difficult to train large-scale data.

2.4 Unsupervised Machine Learning Algorithms

K-Means Clustering. The principle of K-means clustering is to find the group marked by the variable K in the data, perform iterative work based on the provided features, and allocation each data point to K groups. As result the data points are clustered based on feature similarity. Xu [39] designed an automatic grading system for classifying strawberry of complex shape. The K-Means clustering algorithm was applied to cluster the shape characteristics of 144 strawberries, with 4 classes of cone, long cone, square and round, having accuracy rate of 90%. Jaffar [40] proposed a computer-aided photogrammetry method, in which K-Means clustering algorithm was used to classify maturity of palm fruit with color. The results of K-Means clustering algorithm are frequently influenced by the value of the initial center and the number of clusters, and often falling into local optimum.

Principal Component Analysis. The essence of Principal Component Analysis (PCA) algorithm is a multivariate statistical method, which is commonly used for dimensionality reduction. It calculates the main components and establishes as few new variables as possible. Combining with Raman spectroscopy, Feng et al. [41] used PCA and Hierarchical Cluster Analysis (HCA) to establish a fast and reliable classifier for citrus fruits with significant classification result. Vijayarekha et al. [42] proposed a multivariate image analysis technique based on multichannel PCA method. They used PCA to decompose the highly correlated pixel data existing in multi-spectral images and used it for defect segmentation of apple. The disadvantage of PCA is that the actual meaning of the obtained principal components is ambiguous, and is less clear than the original component meaning.

2.5 Neural Network and Deep Learning

Artificial Neural Network. The Artificial Neural Network (ANN) algorithm can simulate the structure of human brain neurons for distributed parallel information processing. In 2017, Jawale et al. [43] proposed a fruit bruise detection and grading system, which used ANN to accurately distinguish the degree of apple bruises and got satisfactory result. Awate et al. [44] used ANN to detect and classify grapes, apples, and pomegranates. They diagnosed diseases on fruits, which increased the yield of healthy fruits. ANN is robust to fuzzy data and t it is suitable for solving complex classification problems. However, when facing huge amount data, it may result to parameter expansion and massive calculation.

Convolutional Neural Networks. With the arrival of the wave of deep learning and Convolutional Neural Networks (CNN) algorithm, the characteristics of local connection and weight sharing of CNN can effectively reduce the complexity of the network and reduce the training parameters. Sa et al. [45] proposed a method for fruit detection based on CNN. They used a deep neural network to construct a faster RCNN to perform the detection of seven kinds of fruits. Mureşan [46] applied deep learning to classify fruits in 60 fruit categories, with the accuracy of 96.3% on the test set, and subsequently released the data set "Fruit-360'.Studies have shown that deep learning networks have better application potential in fruit grading tasks because of the pre-training weights and the openness of the deep learning framework. The disadvantage is that large data sets are required for long training time. Although data enhancement techniques can generate new training samples, more data sets are needed for higher complexity and accuracy requirements.

Table 2. Comparison of merits and demerits of classification algorithms

Classification algorithm	Merits	Demerits
NB	Works well for small sample data and can handle multi-classification problems	Unable to learn the interaction between variables
KNN	It is robust and requires fewer parameters to adjust	Sensitive to noise or extraneous data
SVM	Suitable for processing high dimensional data	The choice of kernel function is very difficult
K-means clustering	Fast convergence, scalability and efficiency for large-scale data	It is easy to get a local optimal solution, sensitive to noise and abnormal points
PCA	Can eliminate the correlation between raw data metrics	The actual meaning of the principal component is blurred
ANN	Ability to approximate complex nonlinear relationships	Facing large data sets is likely to cause parameter expansion problems
CNN	Weight sharing, no need to manually select features	Need a lot of data sets as input

3 Summary and Prospect

In this article, the application of machine vision algorithm in fruit grading system was reviewed in detail. Machine learning was divided into supervised learning and unsupervised learning. In fruit grading applications, the most commonly used supervised learning algorithms were NB, KNN and SVM. And unsupervised learning algorithms included K-means clustering and PCA. Last several years, with the rapid development of artificial neural network, deep neural networks represented by CNN and their variants have received great attention. Different classification algorithms show good results in different grade tasks of fruits, but they also have their own shortcomings. The comparison of merits and demerits of the above algorithms is shown in Table 2.

We introduced the general structure and image processing algorithms of fruit grading system, then compared the application of machine learning classification algorithms in fruit grading. Machine learning and deep learning combined with machine vision system research is the development trend in the domain of fruit grading. Machine learning and deep learning techniques are particularly important for agricultural fruit grading systems, especially for the fruit classification of color, size, shape, texture, and defects. Finally, we propose the following perspectives in the field of fruit grading systems: 1) It is important to choose the suitable sensors for data acquisition. It is necessary to strengthen the hardware research and consider the environmental impact factors for image acquisition. 2) It is necessary to design a fruit grading system capable of simultaneously processing different fruits. 3) Due to lacking of large data sets, public fruit data sets should be established for researchers using in training and verification. 4) Different complex deep learning models should be combined with classifiers for model optimization.

Acknowledgment. This work was supported by a Project named "Research on Evaluation Index and Standard System of Green Technology" (Project No. 2017YFC0212901) .

References

1. Sui, W., Jia, H., Jing, P., et al.: Research on utilization status and classification management of solid waste in fruit processing in China.Trans.CSAE**34**(Z), 172–180 (2018). (in Chinese with English abstract)
2. Li, L., Peng, Y., Li, Y.: Design and experiment on grading system for online non-destructive detection of internal and external quality of apple.Trans.CSAE**34**(9), 267–275 (2018). (in Chinese with English abstract)
3. Guo, Z., Chen, Q., Zhang, B., et al.: Design and experiment of handheld near-infrared spectrometer for determination of fruit and vegetable quality.Trans.CSAE**33**(8), 245–250 (2017). (in Chinese with English abstract)
4. Liu, Y., Wu, M., Sun, X., et al.: Simultaneous detection of surface deficiency and soluble solids content for Amygdalus persica by online visible near infrared transmittance spectroscopy. Trans. CSAE **32**(6), 289–295 (2016)
5. Naik, S., Patel, B.: Usage of image processing and machine learning techniques in agriculture - fruit sorting. CSI Commun. **37**(7), 25–27 (2013)

6. Unay, D., Gosselin, B., Kleynen, O., et al.: Automatic grading of Bi-colored apples by multispectral machine vision. Comput. Electron. Agric. **75**, 204–212 (2011)
7. Ohli, Y.A.: Computer vision based fruit grading system: design and implementation. J. King Saud Univ. Comput. Inf. Sci. **23**, 29–36 (2011)
8. Nandi, C.S., Tudu, B., Koley, C.: Machine vision based techniques for automatic mango fruit sorting and grading based on maturity level and size. Sens. Technol. Curr. Status Fut. Trends **II**(8), 27–46 (2013)
9. Blasco, J., Aleixos, N.: Computer vision detection of peel defects in citrus by means of a region oriented segmentation algorithm. J. Food Eng. **81**, 535–543 (2007)
10. Zhao, J., Peng, Y., Sagar, D., et al.: On-line detection of apple surface defect based on image processing method.Trans. Chin. Soc. Agric. Mach. **44**(Suppl), 260–263 (2013). (in Chinese with English abstract)
11. Huang, C., Fei,J.: Online apple grading based on decision fusion of image features.Trans. CSAE**33**(1), 285–291 (2017). (in Chinese with English abstract)
12. Gu, Y., Shi, G., Liu, X., et al.: Optimization spectral clustering algorithm of apple image segmentation with noise based on space feature.Trans.CSAE**32**(16), 159–167 (2016). (in Chinese with English abstract)
13. Kumar, A., Kumar, R.: Time-frequency analysis and SVM in automatic detection of defect from vibration signal of centrifugal pump. Measurement **108**, 119–133 (2017)
14. Naik, S., Patel, B.: A machine vision based fruit classification and grading: a review. Int. J. Comput. Appl. **170**(9), 22–34 (2017)
15 Palacios, T.P., Caballero, D., et al.: Applying data mining and computer vision techniques to MRI to estimate quality traits in Iberian hams. J. Food Eng. **131**, 82–88 (2014)
16. Li, S., Jiang, H., Wang, Z., et al.: An effective computer aided diagnosis model for pancreas cancer on PET/CT images. Comput. Methods Prog. Biomed. **165**, 205–214 (2018)
17. He, J., Leckey, C.A.C., Leser, P.E., et al.: Multi-mode reverse time migration damage imaging using ultrasonic guided waves. Ultrasonics **94**, 319–331 (2019)
18 Lakshmana, M.P.A.: Computer vision based fruit grading system for quality evaluation of tomato in agriculture industry. Procedia Comput. Sci. **79**, 426–433 (2016)
19. Bulanon, D.M., Burks, T.F., Alchanatis, V.: A multispectral imaging analysis for enhancing citrus fruit detection. Environ. Control. Biol. **48**(2), 81–91 (2010)
20. Safren, O., Alchanatis, V., et al.: Detection of green apples in hyperspectral images of apple-tree foliage using machine vision. Transactions of the ASABE **50**(6), 2303–2313 (2007)
21. Edan, Y., Rogozin, D., Flash, T., et al.: Robotic melon harvesting. IEEE Trans. Rob. Autom. **16**(6), 831–835 (2000)
22. Pla, F., Juste, F.: Feature extraction of spherical objects in image analysis:an application to robotic citrus harvesting. Comput. Electron. Agric. **8**(1), 57–72 (1993)
23. Silwal, A., Gongal, A.: Apple identification in field environment with over-the-row machine vision system. Agric. Eng. Int. CIGR J. **6**(4), 66–75 (2014)
24. Yao, Q., Zhang, C., Wang, Z., et al.: Design and experiment of agricultural diseases and pest image collection and diagnosis system with distributed and mobile device.Trans.CSAE**33** (z1), 184–191 (2017). (in Chinese with English abstract)
25. Wang, Z., Chi, Q., Guo, W., et al.: internal quality detection of apples during late developmental period based on near-infrared spectral technology. Trans. Chin. Soc. Agric. Mach. **49**(5), 348–354 (2018)
26. Peng, Y., Zhao, F., Li, L., et al.: Discrimination of heat-damaged tomato seeds based on near infrared spectroscopy and PCA-SVM method.Trans.CSAE**34**(5), 159–165 (2018).(in Chinese with English abstract)
27. Sahu, D., Potdar, R.M.: Defect identification and maturity detection of mango fruits using image analysis. Am. J. Artif. Intell. **1**(1), 5–14 (2017)

28. Innocent, N., Cedric, O., et al.: Tomato volume and mass estimation using computer vision and machine learning. J. Food Eng. **263**, 288–298 (2019)
29. Lv, J., Zhao, D.A., Ji, W., et al.: Recognition of apple fruit in natural environment. Optik **127** (3), 1354–1362 (2016)
30. Armin, G., Mousa, S.: Segmentation of apple color images utilizing fuzzy clustering algorithms. Adv. Dig. Multimedia **1**(1), 59–63 (2012)
31. Qiang, L., Jianrong, C., Bin, L., et al.: Identification of fruit and branch in natural scenes for citrus harvesting robot using machine vision and support vector machine. Int. J. Agric. Biol. Eng. **7**(2), 115–121 (2014)
32. Ferhat, K., Won, S.L., Ali, V.: Green citrus detection using 'eigenfruit', color and circular Gabor texture features under natural outdoor conditions. Comput. Electron. Agric. **78**(2), 140–149 (2011)
33. Hassan, A., Morteza,S.K., Mir,M.T.: Classification and evaluation of tomato images using several classifier. In: 2009 International Association of Computer Science and Information Technology, vol. 47, pp. 471–474 (2009)
34. Stegmayer, G., Milone, D.H., Garran, S., et al.: Automatic recognition of quarantine citrus diseases. Exp. Syst. Appl. **40**(9), 3512–3517 (2013)
35. Woo,C.S., Seyed, H.M.: A new method for fruits recognition system. In: International Conference on Electrical Engineering and Informatics, vol. 1, pp. 130–134 (2009)
36. Rivera, N.V., Gómez-, J., Chanona, J., et al.: Early detection of mechanical damage in mango using NIR hyperspectral images. Biosyst. Eng. **122**, 91–98 (2014)
37. Moallem, P., Serajoddin, A., Pourghassem, H.: Computer vision-based apple grading for golden delicious apples based on surface features. Inf. Process. Agric. **4**(1), 33–40 (2017)
38. Mizushima, A., Lu, R.: An image segmentation method for apple sorting and grading using support vector machine and Otsu's method. Comput. Electron. Agric. **94**, 29–37 (2013)
39. Xu, L., Zhao, Y.: Automated strawberry grading system based on image processing. Comput. Electron. Agric. **71**(1), 32–39 (2010)
40. Jaffar, A., Jaffar, R., et al.: Photogrammetric grading of oil palm fresh fruit bunches. Int. J. Mech. Mechatron. Eng. **9**(10), 18–24 (2009)
41. Feng, X., Zhang, Q., Zhu, Z.: Rapid classification of citrus fruits based on raman spectroscopy and pattern recognition techniques. Food Sci. Technol. Res. **19**(6), 1077–1084 (2013)
42. Vijayarekha, K.: Multivariate image analysis for defect identification of apple fruit images. In: 2008 34th Annual Conference of IEEE Industrial Electronics, pp. 1499–1503 (2008)
43. Jawale, D., Deshmukh, M.: Real time automatic bruise detection in (Apple) fruits using thermal camera. In: 2017 International Conference on Communication and Signal Processing (ICCSP), pp. 1080–1085 (2017)
44. Awate, A., Deshmankar, D., Amrutkar, G., et al.: Fruit disease detection using color, texture analysis and ANN. In: 2015 International Conference on Green Computing and Internet of Things(ICGCIoT), pp. 970–975 (2015)
45. Sa, I., Ge, Z., Dayoub, F., et al.: DeepFruits: A fruit detection system using deep neural networks. Sensors **16**(8), 1222–1228 (2016)
46 Mureşan, H., Oltean, M.: Fruit recognition from images using deep learning. Acta Universitatis Sapientiae Informatica **10**(1), 26–42 (2018)

Empirical Investigation on the Construction of E-Government Performance Evaluation Model Based on BSC

Fang Du[1], Yongming Cao[2(✉)], and Fei Du[3]

[1] School of International Studies, Sichuan University, Chengdu, China
[2] School of Public Administration, Sichuan University, Chengdu, China
308969898@qq.com
[3] Department of Law and Political Science, North China Electric Power University, Hebei, China

Abstract. This paper uses the balanced scorecard theoretical model to construct an e-government performance evaluation model, including a total of 4 dimensions and 20 indicator systems of finance, public satisfaction, internal business processes, and learning and growth. At the same time, the analytic hierarchy process method is used to assess weights and prioritize index system. This research provides theoretical basis and practical operation for the evaluation of e-government performance.

Keywords: Balanced score card · E-government performance · AHP method

1 Introduction

E-government is a systematic project that high-investment, high-risk, In order to make e-government fully play its role, performance evaluation tools are introduced to comprehensively evaluate e-government. Generally speaking, e-government performance is generally constructed evaluation the index system by using a combination of qualitative and quantitative method to assign weights and prioritize the index system, and finally obtain the weight of e-government performance as the basis for evaluation. Therefore, whether the selection of evaluation indicators is reasonable. This paper intends to introduce the Balanced Scorecard model, relying on related documents of e-government, and constructing an e-government performance index system according to a certain process to a certain process to scientifically and accurately evaluate e-government performance.

2 The Theoretical Framework of Balanced Scorecard

Balanced Scorecard (referred to as BSC) is based on a single financial indicator in the early 1990s to assess the problems of enterprises and start an evaluation study of the comprehensive performance of balanced financial and non-financial indicators. The Balanced Score Card originated from the research result of "Measuring the

M. Tavana et al. (Eds.): IISA 2020, AISC 1304, pp. 281–290, 2021.
https://doi.org/10.1007/978-3-030-63784-2_35

performance of future organizations", which was led and completed by The Norton Institute in 1990. The development process of the balanced score card particularly emphasizes describing the causal relationship behind the dimension, and achieves the final financial goal. The theoretical framework is shown in the Fig. 1 below.

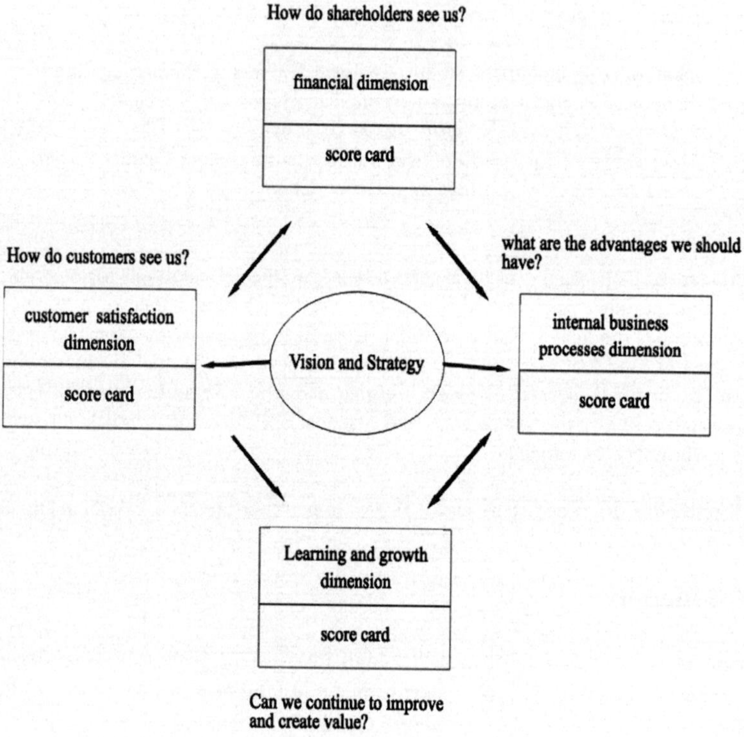

Fig. 1. The theoretical framework

According to the figure above, combined with e-government performance in this paper, its main contents include: first is public satisfaction. The main purpose of carrying out e-government is to construct a service-oriented government and improve the public service. The second is the financial dimension, that is the cost-benefit situation. E-government performance must measure the effectiveness to avoid the bad situation where the marginal cost is far greater than the marginal benefit. The third is the internal business processes dimension, which is mainly reflected in the smooth channels of e-government construction and the adaptability or scalability of the e-government management platform. The fourth is learning and growth dimension, which refers to business processes or the ability of government staff. It is worth explaining that framework has four characteristics, including specificity, measurability, explanatory and dynamic.

3 Construction of E-Government Performance Evaluation Model: Balanced Scorecard

3.1 Index System

This paper uses the theoretical framework of balanced score card, which divide the e-government performance into four dimensions, which including "financial dimension, public satisfaction dimension, internal business processes dimension and learning and growth dimension". It constructs evaluation indicators for each dimension, as shown below.

Firstly, the Financial Dimension. Cost is the core indicators to measure the financial dimension of the balanced score card, and the government's purpose is to serve the people. Therefore, the financial dimension of e-government mainly focuses on the public benefits brought by e-government, and how to provide efficient and convenient public products and services, including fund investment and infrastructure considerations, the indicators for this dimension are shown in the Table 1 below.

Table 1. Financial dimension indicator

Dimension	Indicator
Financial Dimension (B1)	Departmental financial disclosure (B11)
	Digitization of government information resources (B12)
	Departmental network coverage (B13)
	Computer ownership rate per capita (B14)
	Equipment localization level (B15)

Secondly, the Dimension of Public Satisfaction. The strategic goal on public satisfaction is to fully promote the customer satisfaction and trust. The main target of e-government services is not only the public, but also includes enterprises, the government itself and social groups. E-government should mainly focus on the service quality and business level brought by e-government. The indicators of this dimension are shown in the following Table 2.

Table 2. Public satisfaction dimension indicator

Public satisfaction dimension (B2)	Staff business level (B21)
	Staff service attitude (B22)
	Complaint and consultation response (B23)
	Public participation channels (B24)
	Public participation effect (B25)

Thirdly, Internal Business Processes Dimension. The goal is to improve management capabilities and techniques, still improve the efficiency of operating agencies. It refers to how the government improve the efficiency of internal processes and the operating efficiency of government. The details are shown in the Table 3 below.

Table 3. Internal business processes dimension indicator

Dimension	Indicator
Internal business processes dimension (B3)	Information disclosure (B31)
	Information disclosure effect (B32)
	Department statistics (B33)
	Institutional setup and functions (B34)
	Web design (B35)

Finally, Learning and Growth Dimension. It is the motive force which continuously promotes the above three dimensions to achieve excellent results. In the process of e-government development, the government should not only encourage and guide government staff, but also strictly manage employees, The details are shown in the Table 4 below.

Table 4. Learning and growth indicator

Dimension	Indicator
Learning and growth dimension (B4)	Online ability (B41)
	Business collaboration capability (B42)
	Training structure and facilities (B43)
	Annual training per capita (B44)
	Departmental e-government planning (B45)

3.2 Generation of E-Government Performance Evaluation Model

As shown in the above Fig. 2, The model form 4 dimensions and 20 indicators. (1) The Financial dimension including "Departmental financial disclosure, Digitization of government information resources, Departmental network coverage, Computer ownership rate per capital, Equipment localization level". (2) Public satisfaction dimension including "Staff business level, Staff service attitude, Complaint and consultation response, Public participation channels, Public participation effect". (3) Internal business processes dimension including "Information disclosure, Information disclosure effect, Department statistics, Institutional setup and functions, Web design". (4) Learning and growth dimension including "Online ability, Business collaboration capability, Training structure and facilities, Annual training per capital, Departmental e-government planning.

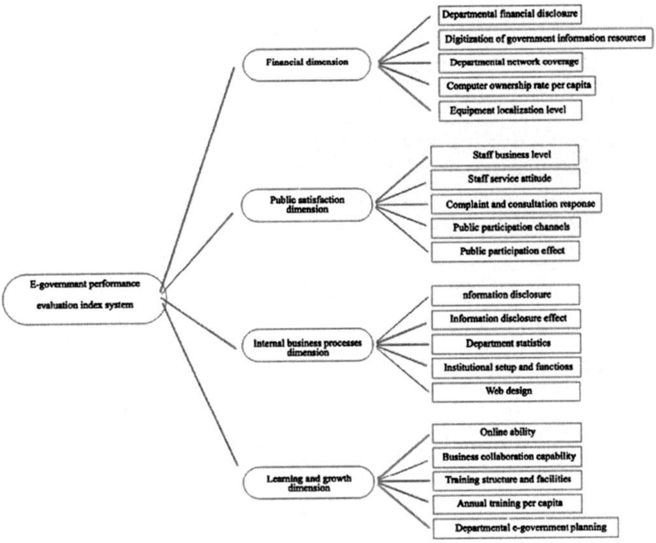

Fig. 2. Model

4 Weight Assignment and Prioritization of Evaluation Indicators Based on AHP

After the model is determined, it needs to be calculated and prioritized in order to more intuitively judge the weight and priority of each dimension and its index in the model. In this paper, the AHP method is used to calculate the qualitative and quantitative description, which proposed by the famous American operations researcher T.L. Satty in the 1970s. Specifically, the relevant elements are divided into the target layer, criterion layer and index layer, the Comprehensive hierarchy is introduced to objectively quantify the subjective judgment.

On this basis constructing a judgment matrix, and the hierarchical single ranking and consistency test are carried out. The weight assignment and prioritization of the indicators in each dimension finally provide a quantitative theoretical basis. Specifically, the hierarchical analysis structure is as follows (Fig. 3):

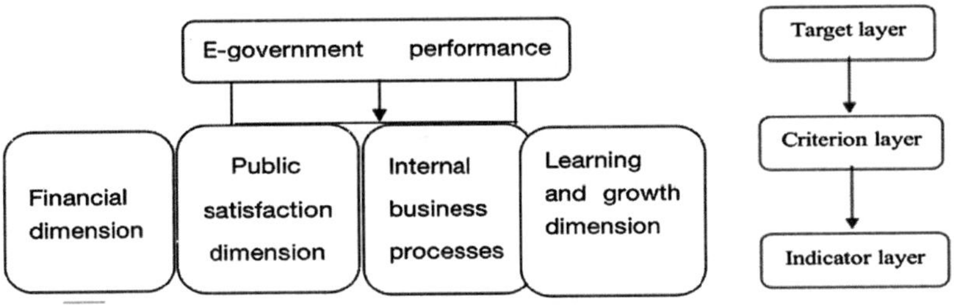

Fig. 3. Hierarchical analysis structure figure

E-government performance is committed to improving government efficiency, and changing the working methods, providing higher level of public services and ultimately achieve public satisfaction.

4.1 Judgment Matrix

After structured the analytic hierarchy model, each element in every layer is compared to construct a judgment matrix. When construct a judgment matrix, it needs to quantify the judgment based on a certain scale value, T. Star's 1–9 ratio scale method was introduced to convert qualitative ideas into quantified values, and the importance of the factors comparing each other through specific values are assigned and analyzed to judge, and the two factors can be judged (Table 5).

Table 5. Judgment matrix scale and meaning

Scale	Means
1	i element and j element are equally important
2	i element is slightly more important than the j element
3	i element is more important than the j element
4	i element is significantly more important than j element
5	i element is extremely more important than the j element
6	i element is slightly less important than the j element
7	i element is obviously less important than the j element
8	i element is significantly less important than j element
9	i element is extremely less important than the j element

(Note: These figures are based on people's intuition and judgment for qualitative analysis).

In fact, for more complex decision-making problems, the judgment matrix is formed after multiple experts fill out the consultation form. For e-government performance, the criterion layer A, B1, B2, B3, B4, relative to the target layer A, the judgment matrix is shown below (Table 6).

Table 6. Judgment matrix

A	B1	B2	B3	B4
B1	b11	b12	b13	b14
B2	b21	b22	b23	b24
B3	b31	b32	b33	b34
B4	b41	b42	b43	b44

After constructing the above judgment matrix, you can perform single ranking calculation on the judgment matrix.

4.2 Weight Assignment and Consistency Check

Theoretically, the single-level ranking can be reduced to the problem of calculating the largest eigen root of the judgment matrix and its eigenvector. Therefore, iterative method is generally used to obtain the approximate maximum eigenvalue and corresponding eigenvector. This research uses the common method to calculate the largest eigenvalue of the judgment matrix and the square root of the eigenvectors. Taking the judgment matrix A as an example, describe B1 (financial dimension), B2 (public satisfaction dimension), B3 (internal business processes dimension)) and B4 (learning and growth dimension) weight assignment and prioritization process.

(1) Calculate the judgment matrix

$$\overline{\omega} = \sqrt[n]{\prod\nolimits_{j=1}^{n} aij} \quad i = 1, 2, \ldots, n$$

$$\overline{\omega} = Y[(\overline{\omega})]_1, [(\overline{\omega})_2, \ldots, [\overline{\omega})_n^T$$

$$\sum_{j=1}^{n} \overline{\omega}_j \quad i = 1, 2, \ldots, n \quad \overline{\omega} = \mathbf{Y}\omega_1, \omega_2, \ldots, \omega_n \mathbf{Y}^T.$$

(2) Calculate the Maximum Eigenvalue λmax

$\lambda_{max} = \sum_{i=1}^{n} \frac{(B\overline{\omega})_i}{n\overline{\omega}_i}$, $(B\overline{\omega})_i$ is the i element of vector B.

(3) To judge whether all factors have satisfactory consistency, it is also necessary to introduce the C.R. value.

$$C.R. = C.I./R.I. \quad (C.R. < 0.1)$$

Among them, R.I. is the average random consistency index. For the judgment matrix of order 1–9, the R.I. value is shown in Table 7. The larger the C.I. value, the greater the degree of deviation of the judgment matrix from complete consistency, and the smaller the C.I. value (close to 0), the better the consistency of the judgment matrix. When C.R. < 0.1, it can be determined the satisfactory consistency and can be used as the basis for the next analysis. If there is no satisfactory consistency, it is necessary to reassign and calculate the judgment matrix until satisfactory consistency is reached.

Table 7. R.I. value of average random consistency index

Order	2	3	4	5	6	7	8	9
R. I	0	0.58	0.90	1.12	1.24	1.32	1.41	1.45

Table 8. Weight of A Layer.

E-government performance evaluation (A)	B1 (Financial dimension)	B2 (Public satisfaction dimension)	B3 (Internal business processes)	B4 (Learning and growth dimension)
B1	1	1/6	1/4	1/3
B2	6	1	2	2
B3	4	1/2	1	1/3
B4	3	1/2	3	1
Income weight	0.070	0.450	0.184	0.296
Single sort	4	1	3	2

(Note: λmax = 4.205; C.I. = 0.068; R.I. = 0.9; C.R. = 0.076).

5 Hierarchical Single Sorting and Total Hierarchical Sorting

According to the above calculation method, keep 3 decimal places, get the weight assignment and ranking of target layer A.

As shown in the Table 8 above, when the value of C.R. is 0.076, it has satisfactory consistency. Firstly, public satisfaction dimension acquire the highest weighting of 0.450, ranking first; the second is the growth and learning dimension, the weight is 0.296, the second ranking; the internal business processes dimension ranks the third, and the weight is 0.184; the last ranking is the growth and learning dimension, with a weight of 0.070. Public satisfaction is the most important, which fully reflects the main goal of providing public services, followed by the growth and learning dimension, the thirdly is internal business processes dimension, and finally is the financial dimension.

6 Weight Assignment and Prioritization of E-Government Performance Evaluation Model

According to the above Table 9, we can intuitively observe the weight assignment of each dimension and each index. We can scientifically evaluate e-government performance based on prioritization of indicators and dimensions.

Table 9. Weights of whole model

Aim	Dimension	Indicator	Weights
E-government performance evaluation index system (A)	Financial dimension (B1) 0.070	Departmental financial disclosure (B11)	0.075
		Digitization of government information resources (B12)	0.220
		Departmental network coverage (B13)	0.352
		Computer ownership rate per capital (B14)	0.133
		Equipment localization level (B15)	0.220
	Public satisfaction dimension (B2) 0.450	Staff business level (B21)	0.098
		Staff service attitude (B22)	0.096
		Complaint and consultation response (B23)	0.213
		Public participation channels (B24)	0.375
		Public participation effect B25	0.218
	Internal business processes dimension (B3) 0.184	Information disclosure (A31)	0.111
		Information disclosure effect (A32)	0.121
		Department statistics (A33)	0.199
		Institutional setup and functions (A34)	0.346
		Web design (A35)	0.223
	Learning and growth dimension (B4) 0.296	Online ability (A41)	0.329
		Business collaboration capability (A42)	0.286
		Training structure and facilities (A43)	0.189
		Annual training per capita (A44)	0.122
		Departmental e-government planning (A45)	0.074

7 Conclusion and Application

This paper uses the balanced scorecard model to construct the evaluation model, form an evaluation model with 4 dimensions and 20 indicators. At the same time, uses the analytic hierarchy process method to assign weights and prioritize the index system to scientifically and intuitively evaluate the e-government performance. Using this evaluation model, you can make a comprehensive and objective understanding of e-government, guide its overall development, strive to create good performance, and achieve the purpose of promoting excellent e-government. It is worth noting that the evaluation model is only an evaluation tool. The transcendental and universality contained in it need to be continuously practiced, summarized and revised in practice before the actual value of the evaluation model can be finally exerted. It is also a process of continuous circulation of theory and practice. The indicators in the evaluation model are only the "leaders" of the goal, and there is room for further improvement and improvement in the construction of their theoretical framework.

Acknowledgments. This work was supported by the "National Social Science Fund", China,

1. Grant No. 17AZD018. "The Principle of Chinese Frontier Science".

This research was supported by the "National Social Science Fund", China.

2. Grant No. SC19C022. "Theoretical and Empirical Research on Competency of the First Secretary of Poverty Alleviation in Sichuan Tibetan Areas from the Perspective of Rural Revitalization".

References

1. Xibin, L., et al.: Electronic Government Affairs and Government Management Innovation. Science Press, Beijing (2004)
2. Jieping, W.: Electronic Government Affairs Theory and Performance . Hunan University Press, Hunan (2007)
3. Liang, M., Liu, L.: How does e-government performance influence public use–an empirical study based on countries and cities around the world. J. Shanghai Admin. Coll. **20**(04), 83–95 (2019)
4. Lisha, W.: Research on the construction of government performance evaluation index system ——based on the district and county level government under the E-government environment. J. Liaoning Univ. (Philos. Soc. Scie. Edn) **46**(01), 88–97 (2018)
5. Shaohui, L., Junyi, Z.: Research hotspots and evolutionary path analysis in the field of government performance management—based on the measurement analysis of CNKI (2002–2015). Mod. Intell. **37**(02), 141–150 (2017)
6. Yun, Z.: Analysis of the problems and crux of the development of e-government in my country——taking the survey data of the Ministry of E-government in 2014 as an example. China Admin. **08**, 13–18 (2015)
7. Haifeng, D.: On the perfection of my country's e-government performance evaluation method ——taking Hangzhou's practice as a sample. Sci. Technol. Manage. Res. **35**(02), 181–185 (2015)
8. Hao, Q., Hongbo, L.: The latest progress and enlightenment of foreign e-government performance evaluation——taking Accenture and the United Nations as examples. E-government **02**, 80–86 (2013)

Analysis the Promotion of Computer to the Development of Garment

Zhong Nana(✉)

Guangdong University of Science and Technology, Dongguan 523083, China
7878634@qq.com

Abstract. With the progress of society and the development of computer technology, more and more industries began to use computer technology to carry out a new round of "industrial revolution" and "industrial innovation". Advances in computer technology have brought tremendous changes and impacts to more and more industries, and the apparel industry is no exception. Under the social background of the continuous advancement of computer technology, the continuous popularization of computers in the society and the theme of the times, the clothing industry has also quietly entered a high-tech period. Corresponding computer technical support and corresponding software cooperation have appeared in clothing design, clothing fabric production and design, clothing style design, clothing interior fabric structure design and other aspects, which have changed the development of the computer industry and promoted the development and progress of the clothing industry. Computers have provided a lot of applications and design methods that have not been fully utilized in the past to the development of the apparel industry, so that the development of the apparel industry has reached a new level. This article mainly introduces the related huge changes brought by the computer to the apparel industry, the development and application of apparel PAD and various types of software, such as; PHOTOSHOP, CorelDRAW and other applications in the apparel industry and the combination of apparel professional software. Under the background of the new era, with the rapid development of computer technology, the progress and development of the clothing industry are inseparable from the support of computer technology, so the research content of this article is very meaningful.

Keywords: Computer · Clothing design · Promotion function · Drawing software

1 Introduction

Apparel companies originated in a very early period. As early as a few centuries ago, the "garment industry" as one of the four major aspects of the "clothing, food, housing and transportation" of people and people's lives has already appeared, and it has been constantly in the process of development. Improve your own development path and way of development, develop in progress, and progress in development. Today's apparel industry is a labor-intensive industry. The current apparel industry employs more people, and people's daily life and apparel industry are inseparable. With the

M. Tavana et al. (Eds.): IISA 2020, AISC 1304, pp. 291–295, 2021.
https://doi.org/10.1007/978-3-030-63784-2_36

progress of society, people's requirements for the functionality of clothing have been greatly reduced, which is far less important than its fashion value and beauty value. Such consumption and product demands place higher requirements on the apparel industry. People are demanding continuous upgrades in the design, style, styling, and manufacturing process of clothing. Practitioners in the clothing industry must also come up with a higher technical level of design to meet this demand, especially in methods, means, citation and support of computer technology in all aspects of innovation. The computer provides the impetus for the garment industry to promote the development of the garment industry.

2 The Clothing Industry has Undergone Tremendous Changes with the Help of Computers

With the progress of society and the development of society, computer technology has also been constantly developing and progressing. More and more industries are beginning to rely on the development of computer technology. At the same time, the development of the clothing industry has gradually entered the computer era. The combination of the computer and the apparel industry has brought many changes to the development of the apparel industry. It plays a very important role in the management of the industry, not only in the design of clothing fabrics, but also in the design of clothing styles, fabrics, structures, and the necessary plate making and off-duty design. The combination of the two has greatly promoted the improvement of the work level of apparel practitioners. To a certain extent, they have reduced their labor intensity and brought about a revolution in the apparel industry under the background of the new computer era.

2.1 Computers have Changed the Creative Thinking of Clothing Designers

In the process of combining the computer with the apparel industry, it not only realized the conversion from "pen and paper" to "keyboard and mouse", but also opened up a brand new time and space for practitioners in the apparel industry. Computers bring designers into an environment that they might never be exposed to, let them understand a richer world through computers and networks, such a new and expanded world will provide these designers with many unprecedented new experiences, and these experiences will become the infinite inspiration for these designers in the process of clothing design. Practitioners in the clothing industry can obtain endless design and innovative thinking from the endless network environment. Such an environment will open the thinking of the practitioners in the clothing industry and enrich their imagination. Thus, with the help of endless inspiration and infinite thinking, it can better meet the needs of consumers and provide consumers with more abundant creative sources, so that the development of the clothing industry can more and more meet the needs of the times. The progress and development of the apparel industry practitioners is the soul of the apparel industry, whether a designer can design good, beautiful and fashionable clothing designs that meet people's functional needs and also their aesthetic needs depends on whether the designer has good inspiration. The computer's network environment helps designers and other practitioners expand their horizons through their own unique characteristics and functions, and provides conditions for their thinking to develop, changing the creative thinking of clothing designers.

2.2 Computers Reduce Manual Labor in the Clothing Industry

There is a lot of manual labor in the apparel industry, which makes the apparel industry have to use a lot of labor to achieve the normal operation of the industry. In a lot of labor, clothing plate making is a relatively representative part. Clothing plate-making plays an important role in the entire clothing manufacturing process. Clothing plate-making is a strong physical labor that consumes a lot of physical energy in all aspects of clothing production, but plate-making is not only a manual labor, but also a technical project. The technical content of clothing platemaking directly affects the development of the enterprise. In recent years, with the rapid development of computer technology, computer technology has continuously entered the clothing industry. For the content of clothing platemaking, the clothing industry has found a better solution with the help of computer technology. The development of CAD software has greatly reduced the difficulty of this work, liberated more and more strong workers from this work, and solved the technical difficulties of this work, thus successfully reducing the difficulty of this work and helping the garment industry to develop at a better and higher level. Advances in computer technology have enabled software such as CAD to help the apparel industry, thereby promoting the development and progress of the apparel industry.

3 Full Integration of Computer Drawing Software and Clothing Professional Software

In recent years, due to the development of the computer industry, more and more computer software has entered the design part of the clothing industry. In the design process of clothing, drawing is an indispensable part, and it plays an important role in the design of clothing styles, and to a certain extent plays a key role in the design of clothing. Among professional software in the apparel industry, such as CAD software, it has certain functions of clothing style design. However, compared with professional drawing software, it is not so handy to use. Therefore, more and more computer drawing software has gradually entered the design of the clothing industry and has been widely used.

3.1 Application of PHOTOSHOP Software in Clothing Design

PHOTOSHOP software plays an important role in the apparel industry. Its powerful drawing function provides great convenience for apparel designers, and is widely used in apparel design. This software can help designers complete a lot of configuration that cannot be done in the manuscript when designing the style of clothes and so on. In addition to the design of the style, this software also plays an irreplaceable role in the designer's color selection of clothing. In the traditional design, the designer usually scans the designer's manuscript through the computer to design. In this traditional way, compared to this software, the colors that the designer's manuscript can depict are relatively monotonous. By using this software, designers can choose more colors and collocations. This way combines the manuscript and the drawing software of the computer to play a common role. PHOTOSHOP software design diagram shown in Fig. 1, it can show the change of different colors.

Fig. 1. PHOTOSHOP software design clothes

3.2 Application of CorelDRAW Software in Clothing Design

The characteristic of this computer drawing software is that it conforms to many manual drawing thinking methods and drawing methods, and it can be closer to the manuscript of clothing design.It has intuitive, clear features, and this software can also be adjusted effectively in proportion to achieve different design effects. The CorelDRAW software design diagram is shown in Fig. 2.

Fig.2. CorelDRAW software design diagram

This picture clearly expresses the designer's idea, and it is very close to the characteristics of the manuscript painting. Therefore, this software can complete the drawing effects of different clothing, it can complete many drawing tasks in the clothing industry, help the design of clothing to be better completed, and thus promote the rapid development of the clothing industry.

4 Conclusion

As computers play an increasingly important role in various industries, computer technology has also been largely applied in the development of the apparel industry. Various new drawing softwares appearing with the computer make up for many past defects.The application of computers not only helps designers expand their thinking and get more inspiration, but also breaks through many traditional time-consuming and labor-intensive traditional models in the clothing industry, and injects new technical blood into the clothing industry. It has brought tremendous impact to the practitioners in the apparel industry. Under the impact of such technology, the apparel industry has constantly realized its own transformation. Whether at the technical level or at the thinking level, computer technology has served the purpose of helping the apparel industry to develop and innovate in the new era, realizing the transformation of the apparel industry.I believe that under the continuous development and innovation efforts, the clothing industry will go further and further.

References

1. Li, H., Guan, J.: Research on the trend of customized transformation of the clothing industry combined with computer network technology. Text. Ind. Technol. **49**(02), 73–74 (2020)
2. Liu, D., Zhu, J.: The effect of computer network on the development of clothing design-Comment on "“Computer Aided Clothing Design.”" Print. Dyeing Auxiliaries **35**(04), 71 (2018)
3. Tang, B.: Thoughts on the computer design teaching of costume painting. China Inf. Technol. Educ. **17**, 121–122 (2015)
4. Kai, W.: The promotion effect of computer network on the development of fashion design. China Collective Econ. **09**, 181–183 (2012)
5. Moretti, I.C., Junior, A.B., Colmenero, J.C.: Using the analytic hierarchy process for selecting prototypes in the development process of fashion garment products. Acta Scientiarum Technol. **39**(3), 367–374 (2017)
6. Abro, Z.A., Zhang, Y.F., Hong, C.Y., Lakho, R.A., Chen, N.L.: Development of a smart garment for monitoring body postures based on FBG and flex sensing technologies. Sens. Actuators A. Phys. **272**, 153–160 (2018)

Check for updates

Design of Positioning Orientation Movement APP Based on BDS

Xi Zhang, Wenquan Zhang(✉), Ying Li, Pengfei Liu, and Guangyue Li

Wuhan Mechanical College, Army Engineering University, Wuhan 430075, China
wenc99336@163.com

Abstract. With the construction and development of BeiDou Navigation Satellite System (BDS) in China, the positioning accuracy of BDS has been greatly improved and its application has become more and more extensive. However, the application of BDS in mass and professional directional activities is not precise enough. This paper presents a design scheme of positioning orientation movement APP based on BDS, can realize some functions such as positioning orientation, navigation, vital signs monitoring, activity release and management. It can also be used to instruct athletes in various skills, raise the quality of the training and effectively prevent the happening of the casualties and other emergencies.

Keywords: BDS · Orientation movement · APP

1 Introduction

The BeiDou Navigation Satellite System (BDS) is a satellite navigation system independently developed by China. When the BDS III global navigation System is fully completed in 2020, it will further improve its service performance and user experience. At present, the BDS is widely used in national defense, military, transportation, electric energy, agriculture, forestry, animal husbandry and fishery, etc., but it is not widely used in orientation movement. Orientation movement is a fashionable sport integrating fitness, intelligence, entertainment, social interaction and other values. It is widely carried out at home and abroad. The traditional orientation movement mainly uses the map, the compass and other equipment to select the road, by foot or by means of vehicle, looking for the target to reach several checkpoints plotted on the map in the shortest time possible [1]. In recent years, with the popularization of mobile phone, orientation movement APP has emerged based on mobile phone positioning [2], which has the following shortcomings: First, users generally cannot see satellite information, positioning data and other contents, leading to the majority of users believe that they use GPS for simple positioning and navigation, while ignoring the application value of the BDS; Second, the lack of knowledge is not conducive to the user to learn geography, mapping, logic, mathematics and other relevant knowledge and the ability to use these knowledge; Third, it has the shortcoming of single content and incomplete functions. Therefore, in order to give full play to the application value of BDS, let more

M. Tavana et al. (Eds.): IISA 2020, AISC 1304, pp. 296–301, 2021.
https://doi.org/10.1007/978-3-030-63784-2_37

people know the BDS benefit, use BDS, we design a positioning orientation APP, which can realize some main functions such as positioning based on BDS, orientation, navigation, monitoring, etc. It can be used to participate in various competitions, popular science, online fun sports, entertainment and other activities.

2 Overall Design

Based on the System consisting of BDS, mobile phone, PC, server and 4G/5G network, the Android SDK development platform and Java language is adopted to develop the Android Version of positioning orientation movement APP (see Fig. 1).

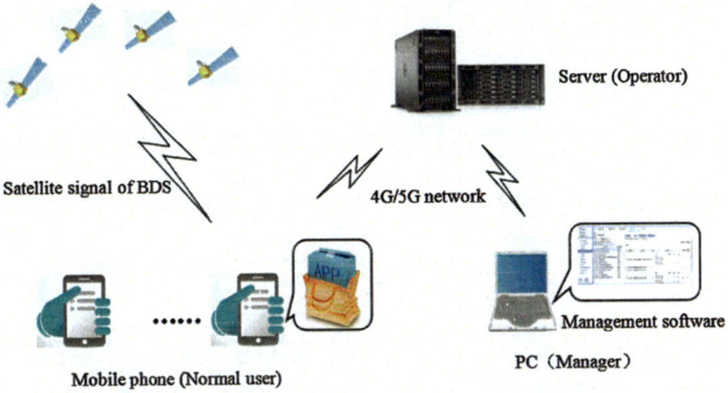

Fig. 1. Overall schematic of the system

3 Functional Design

The software architecture of APP includes five parts: positioning function module, navigation function module, monitoring function module, activity function module and user management module (see Fig. 2).

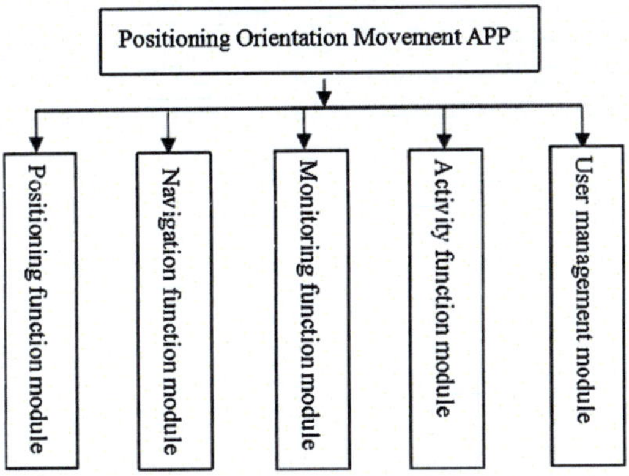

Fig. 2. Software function block diagram

3.1 Positioning

Positioning is the most basic function of APP, which first obtains the navigation data output by the BeiDou chip built in the phone, and then parses GSV, GSA, RMC, GGA and other statements to obtain PVT information and display according to the standard protocol.

Date: Display the current real-time positioning date information;

GMT + 8: It is the current positioning time. Because it is from the BeiDou satellite precision timing, so it can be used as the standard time;

Positioning mode: Including the three types of positioned, in positioned, and not positioned;

Coordinate transformation: Earth coordinates, Gaussian coordinates and space rectangular coordinates are displayed in three ways.

Speed: Display the current user's speed;

Number of satellites: The number of satellites searched or the number of satellites involved in positioning, and the number and distribution of BeiDou satellites received can be visually viewed;

Weather: Click to display the real-time weather data of the current positioning point.

3.2 Navigation

The navigation application of APP is mainly used for personal location service and various orientation movements organized by the society, such as marathon, city orientation, orienteering, etc.

(1) Map mode

In the map mode, the navigation interface obtains the map of the user's current position through 4G/5G network and displays the user's current position, as well as the user's motion trajectory. Under the motion trajectory, the number of steps and distance traveled by the user are displayed, as well as the user's motion time, average speed and heat consumption. According to the requirements of the general orientation movement [3], the user's current position is designed as the starting point, the point that must be passed according to the regulations of the race organization is the mandatory point, the place that does not pass is the avoidance point, and the destination specified by the race is the destination. The APP will automatically form a course, and the user can move according to this course. In orientation movement, the navigation function can reduce the registration of judges at each point, reduce manpower and increase the transparency and fairness of the race.

(2) No-map mode

When the mobile phone has no 4G/5G network signal, or map is not precise enough, especially in the case of outdoor environment, user can choose no-map mode for navigation. In no-map mode, user manually enters mandatory point (expressed as a Δ), avoidance point (expressed as a $\times$), and coordinates of target (expressed as a ☆). Then, the user can intuitively see the positional relationship between himself and each point, and travel in the direction of the reduction of the coordinate difference (ΔB, ΔL or ΔX, ΔY) with the target point (see Fig. 3).

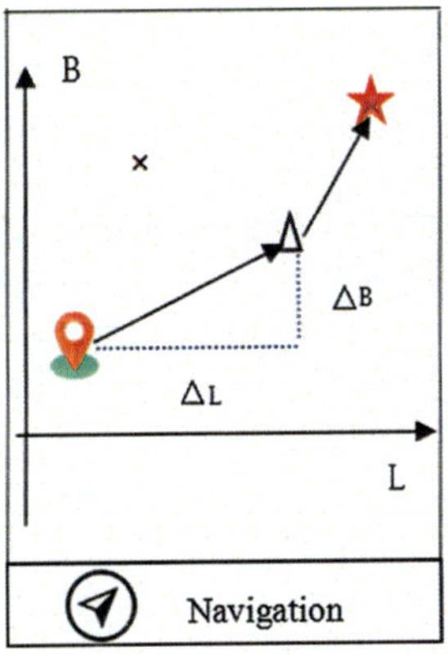

Fig. 3. Navigation schematic diagram of uncharted mode

The no-map mode is helpful to investigate the user's understanding of coordinates, calculation ability of relative position in coordinate system, orientation recognition ability, navigation and path planning ability, and can meet the knowledge requirements of related activities.

3.3 Monitoring

With the marathon gradually approaching to the public life, people's enthusiasm for running and exercise has been greatly enhanced. However, due to people's lack of understanding of their own physical conditions, accidents may easily occur [4, 5]. Therefore, the function of heart rate monitoring is designed on this APP, mainly to provide the standard value and the highest value, so as to provide users with reference and early warning. For this function, a smart bracelet with sensors is used, which transmits information to the mobile phone via Bluetooth and then sent to the APP. The specific functions are as follows:

(1) Monitoring

According to resting heart rate, aerobic heart rate, lactate threshold and maximum heart rate, seven heart rate intervals are divided. Life is less than or equal to 129; Warm-up is between 129 and 143; Fat is between 143 and 157; Cardiopulmonary function is between 157 and 171; Strength is between 171 and 185; Limit is between 185 and 199; Warning is greater than 199.

(2) Alarm

Different heart rate intensity intervals are optional. If the measured heart rate exceeds the set interval, an alarm will be given at any time to prompt the user to adjust the exercise intensity.

(3) Salvage

In terms of military training and adventure activities, the APP can be used as a tool for distress and rescue. Team members can share their location in real time. When there is danger, the injured could enter the APP and press "SOS' to ask for help, and inform other users of the team to determine their location and rescue immediately.

3.4 Activity

Orientation movement is a kind of outdoor leisure activity with strong interest, intelligence and ornamental value, mainly in the form of teamwork. Through the game can not only achieve the purpose of the exercise, improve the physical quality, but also can improve participants' skill of orienteering in recognition map, accurate positioning, judge direction, scientific select route in different environment. In recent years, with the improvement of people's level of economic and cultural life in our country and the development of undertakings of physical culture and sports, orientation movement gradually become a new kind of entertainment exercise way of life. This function is mainly used for the release and management of related activities organized by the event organizers, institutions and departments. Users can browse the latest orienteering activities on the activity interface, click to view information related to the activity, sign up and so on, open the APP when the selected activity starts, and begin the race.

In addition to orientation movement, users can also participate in various competitions, popular science, fun sports, entertainment and other activities online, as well as the release of some dynamic information.

3.5 User Center

(1) Manager mode

The organizer of orienteering applies to the platform to establish an activity, enters the manager mode after confirmation, and sends the link of the related activity to the platform. The platform displays the activity on the activity interface. The participants can enter the activity ID to register to become subordinate users. The organizer can see the situation diagram of the subordinate users on the user center interface, enter the ID of the subordinate users to monitor their movement track, heart rate, etc., and receive the monitoring alarm information of the subordinate users.

(2) Ordinary user mode

Ordinary users can choose WeChat, QQ and mobile phone number to log in. They can also use the ID given by the manager to log in to participate in activities and share the sports situation. At the same time, they receive instructions sent by the manager, including voice mode, picture mode, text mode.

4 Conclusions

The implementation of the BDS has driven the development and industrial upgrading of Satellite navigation in China, and the "BeiDou + " will promote the integration of the BDS with emerging technologies such as the mobile Internet, big data, cloud computing and Internet of vehicles. Under this big background, combining with the current situation and demand of Directional activities, we design a positioning orientation movement APP Based on BDS, to be suitable for the needs of the user and the content of the social activities. Users can carry out online or offline orientation movement after downloading and installing the APP on their mobile phones, which further enriches the forms and means of orientation movement and has good promotion and application value.

References

1. Mai, Q.: Design of directional motion Platform based on Internet of Things. Electron. World **18**, 162–185 (2017)
2. Zhao, X., Zhao, L.: Experimental study on WeChat platform in directional movement teaching in universities. Sci. Technol. Inf. **35**, 110–111 (2019)
3. Li, C.: Research on GPS orientation in Colleges and Universities. J. Chengdu Inst. Phys. Educ. **8**, 86–89 (2013)
4. Hou, Z., Lv, C., et al.: Application research of BeiDou Positioning Communication Technology in directional Motion. Mod. Electron. Technol. **21**, 38–40 (2014)
5. Xie, H., Wang, J.: Application of GPS technology in directional motion. Zhejiang Sports Sci. **21**, 61–64 (2009)

A Comparative Study of Yes/No Question Intonation in Chinese and English Speakers

Qing Guo(✉), Hongyan Wang, and Jeroen van de Weijer

Shenzhen University, Shenzhen, Guangdong, China
goo4867464@163.com

Abstract. Chinese speakers produced twelve yes/no questions in English as well as in Chinese, and English speakers produced the same in English. Compared to the native English speakers, the Chinese speakers tended to use a flatter intonation in their English and showed smaller fluctuations of pitch. This can be explained by first language interference: the flatter intonation was influenced by the final character "ma" at the end of Chinese yes/no questions and the smaller pitch fluctuations in general are due to the fact that Chinese intonation is restricted by lexical tone.

Keywords: Intonation · Yes/No questions · First language interference · Intonation patterns

1 Introduction

Languages can be categorized as either intonation languages or tone languages [1]. Pitch is distinctive at the lexical level in a tone language like Chinese, where it distinguishes lexical meaning. In intonation languages like English, lexical words have no specifications for pitch. Instead, pitch is used to produce intonation during the articulation of an utterance [2]. This study analyzes the intonation patterns in yes/no questions in Chinese and English and analyzes interference between L1 Chinese and L2 English.

1.1 Research Background

Intonation is the way pitch rises and falls during speech. Different ways of modeling intonation have been proposed such as the British analyses of intonation [e.g. 3] and the autosegmental-metrical theory, [e.g. 4] but all recognize the importance of tonal targets like High and Low. The common tone for English statements and wh-questions is a fall (HL) while the common tone for yes/no questions is a rise (LH) [5].

There are two types of yes/no questions (also known as polar questions), non-conducive questions ("open" or "real" questions), which have a final rising tone, and conducive questions ("leading" or "checking" questions) with a final falling tone [e.g. 6] Non-conducive questions are asked in situations where the speaker does not know the answer, and the question is asked to obtain this information. Conducive questions, on the other hand, are asked in situations where the speaker already knows the answer and expects the hearer to confirm these expectations.

M. Tavana et al. (Eds.): IISA 2020, AISC 1304, pp. 302–309, 2021.
https://doi.org/10.1007/978-3-030-63784-2_38

Mandarin Chinese is classified as a lexical tone language. There are four distinct tones, numbered Tone 1-Tone 4 and neutral tone [7]. Tone 1 is a high level tone, Tone 2 is a rising tone, Tone 3 is a dipping-rising tone and Tone 4 is a high falling tone, while the realization of "neutral tone" depends on the context. The effect of these lexical tones on intonation has long been a matter of research and here we investigate their interaction with intonation in yes/no questions.

Transfer refers to the interference of their mother language (L1) when speaking a second language (L2). L1 interference may either promote (positive transfer) or hinder (negative transfer) second language acquisition [8]. This study investigates the effect of L1 transfer in the realm of intonation, specifically for yes/no questions.

1.2 Research Questions

In this study, Chinese speakers' intonation patterns of yes/no questions will be analyzed. By comparing the Chinese speakers' intonation patterns of English yes/no questions with native English speakers' intonation, this study attempts to analyze how L1 interferes in Chinese speakers' intonation of yes/no questions in English. Concretely, the present study explores the following three questions: i) How do the intonation patterns of English yes/no questions produced by Chinese speakers differ from those produced by native English speakers?; ii) How do the intonation patterns of English yes/no questions produced by Chinese speakers differ from the intonation patterns of Chinese yes/no questions, produced by the same Chinese speakers? iii) How should L1 interference between Chinese and English be characterized with respect to intonation in yes/no questions?

2 Experiment

2.1 Materials

The materials used in this study were twelve English yes/no questions and twelve Chinese yes/no questions. All questions were extracted from online videos, movies, or TV shows where they were produced by native English speakers. The English yes/no questions were translated into Chinese and read by the Chinese speakers. All of the questions were from casual conversation and the contexts were provided to the speakers, for example "Should we invite Andrew to that party?" (English), and its translation: "我们应该邀请安德鲁来参加派对吗?" (Chinese).

2.2 Subjects

The English and Chinese yes/no questions above were recorded by five Chinese speakers (Due to the Covid-19 pandemic, it was very difficult to collect data of native English speakers, hence the low number of participants. This makes the results tentative and in need of future research.) (three males and two females) and one English speaker (male). They were asked to record the sentences at a natural pace and the contexts for all questions were provided. The Chinese speakers were all university students and the native English speaker was a teacher there. None of the Chinese speakers was bilingual in English, but all spoke fluent English.

2.3 Recording Procedure

Due to the Covid-19 quarantine situation, all recordings were made by the speakers themselves, with private headphones and/or phones. They were provided with the materials and asked to record in a quiet location. The English speaker recorded the twelve yes/no questions in English and the Chinese speakers recorded both yes/no questions in English and the same questions in Chinese. All recordings were submitted online and stored for analysis.

To analyze the intonation in the yes/no questions, the application Praat [9] was used, especially for pitch (measured in Hertz, Hz) and intensity (measured in Decibels, dB). Together, these reflect intonation patterns and word/sentence stress reasonably well.

For each recording, maximum and minimum pitch and intensity were computed. Averages of these were calculated in order to compare yes/no question intonation in English and Chinese. Finally, pitch contours were drawn for all utterances to aid visual inspection.

3 Results and Discussion

First, let us compare the intonation in English yes/no questions produced by native English speakers and by Chinese speakers.

After obtaining the statistics of maximum and minimum pitch of all sentences, mean pitch was calculated in order to calculate the average standard deviation. The means of these values can show if there was a difference between the two major vocal attributes (pitch and intensity) in Chinese and English speakers.

For humans, the average frequency range in speech varies from 80 to 260 Hz. The frequency of an adult female usually has a wider range than an adult male. Not only gender, but also age will affect the frequency of human speech. As for intensity, the decibel range for a normal human speaking voice is around 70 dB. When a person is talking in an elevated voice, the decibel range is around 76 dB [10].

Figure 1 below indicates that the male Chinese speakers have almost the same mean pitch (Hz) and average intensity (dB) as the male native English speaker when producing English yes/no questions.

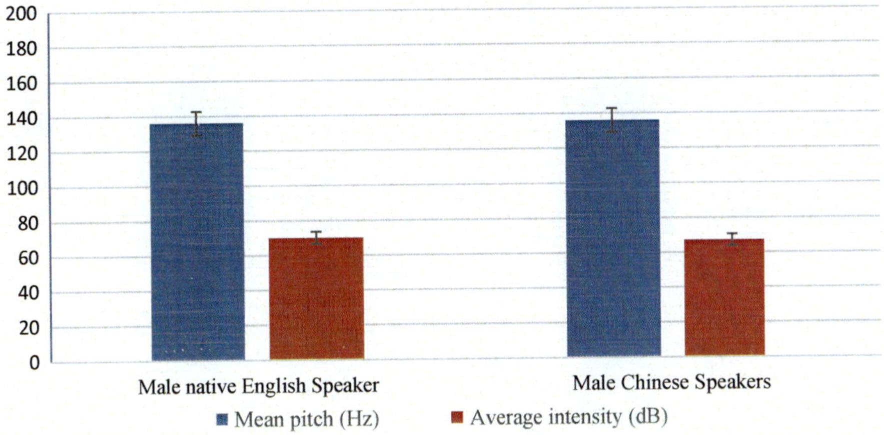

Fig. 1. Mean results of English yes/no questions uttered by male native English speaker and male Chinese speakers

The female Chinese speakers' data were not included because no female native English speakers' data was available.

Figure 1 shows that there is hardly any difference between the two major vocal attributes, pitch (F0) and intensity, in Chinese and native English speakers. A more detailed comparison will be made below.

The standard deviation is a measure of how widely distributed a range of numbers are. In this experiment, these numbers were the average pitch values of English yes/no questions. These values were calculated in Praat at steps of 10ms. This made it possible to calculate the average standard deviation in both English and Chinese yes/no questions.

In the experiment, the average standard deviation of pitch in English yes/no questions produced by English speaker was 31 while the one for Chinese speakers was 25. The female Chinese speakers were included, because there was not much difference in gender where average standard deviation was concerned. As a result, the fluctuation of pitch has a bigger range in native English speakers. In other words, Chinese speakers have a relatively "flat" intonation. This is also visually illustrated in Fig. 2 and Fig. 3 below.

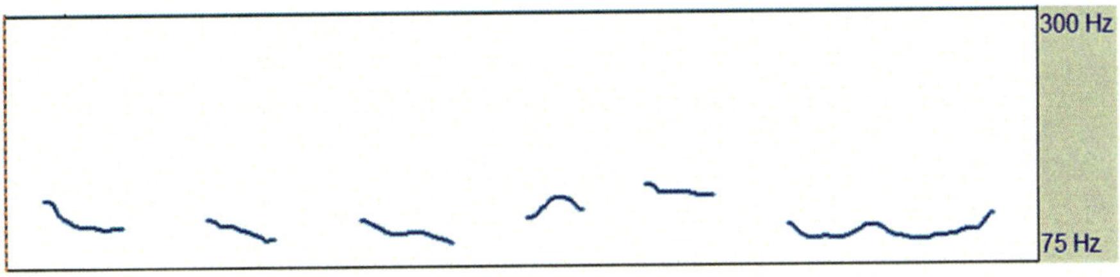

Fig. 2. The pitch contour of Q3 (Eng.) by a Chinese speaker.

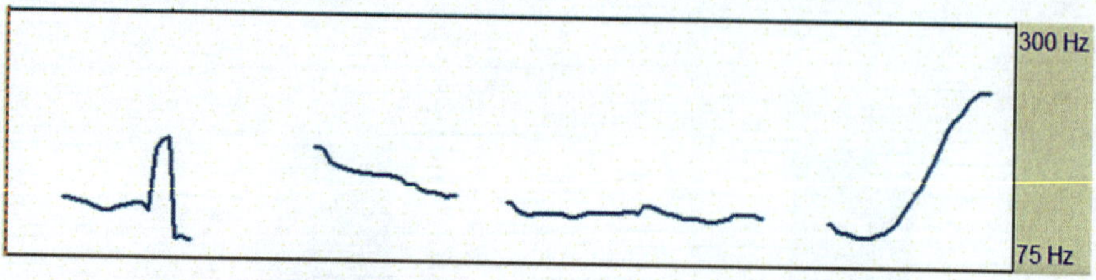

Fig. 3. The pitch contour of Q3 (Eng.) by a native English speaker.

Although the male Chinese speakers shared the same mean pitch with the male native English speaker when uttering English yes/no questions, there were a number of distinctions between their pitch patterns. In both materials, there were basically three kinds of intonation: rising, "flat", and falling tone.

Table 1 (which includes the patterns for the recorded voices in order to increase the sample size of native speakers) shows that pitch patterns of the native English speakers were either generally rising or falling, while the Chinese speakers also show a "flat" intonation, which was also suggested by the results before. We will return to the reason for the flat intonation pattern in Chinese speakers' English later.

Table 1. The pitch patterns of native and Chinese speakers.

Pitch patterns	Native speakers	Chinese speakers
Rising	75%	75%
Flat	0	6.67%
Falling	25%	18.33%

Further analysis showed a number of interesting points. The English speakers had the same tone in all twelve English yes/no questions (Q2, Q8, Q12 were falling while all others were rising), while the Chinese speakers had the same intonation only in Q7, Q9, Q10 and Q11. Given the context, Q2, Q8 and Q12 were conducive questions (see above; these are typically asked in situations where the speaker already knows the answer and expects the hearer to confirm their expectations), and they had a falling intonation. One female participant did use falling intonation in Q2, Q8, Q12, like the native English speakers. Most Chinese speakers, however, did not produce this (exceptional) pitch pattern that is used in yes/no questions of this type, probably since they had always been taught by textbook that yes/no questions have a rising intonation.

The rising intonation of the Chinese speakers was slightly different from that of the native speaker. This is shown below, where Fig. 2 and Fig. 3 show the pitch contour of Q3 ("Does that sound OK to you?") produced by a native English speaker and a Chinese speaker, respectively. These intonation contours are quite different: although they both have a rising intonation generally, the native English speaker's rise was much more pronounced than the Chinese speaker's one. About half of the Chinese speakers' rising intonation patterns resembled Fig. 2. We will return below to the reason for this difference.

Chinese speakers also put some of the sentence stresses in different locations compared to the native English speakers. Sentence stress is best shown by an intensity contour. The intensity contours of both native English speaker and Chinese speaker have almost exactly the same duration, but in the one of Chinese speaker, the duration of "you" is much longer than "OK", while in the other one by English speaker) the duration of "OK" is longer than "you". In other words, the Chinese speaker put a sentence stress on "you" while the native English speaker put it on "OK". According to the context, the sentence stress should be on "OK".

Next, let us compare the intonation contours of English and Chinese yes/no questions produced by Chinese speakers. Figure 4 shows that male and female Chinese speakers both had nearly the same mean pitch (Hz) and average intensity (dB) when producing English and Chinese yes/no questions. This shows that there were only marginal differences in English and Chinese yes/no questions when produced by Chinese speakers. The mean pitch differed by less than 10 Hz and the average intensity differed by less than 6 dB. Such differences are in general too small to notice [11].

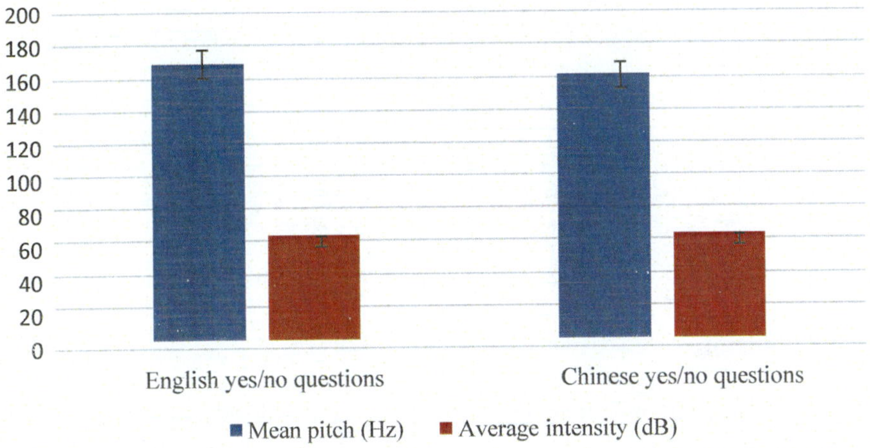

Fig. 4. Mean results of English and Chinese yes/no questions uttered by Chinese speakers.

Although differences are small, further analysis showed some interesting details. The average standard deviation in pitch in English yes/no questions produced by Chinese speakers was 25.33 and in Chinese yes/no questions also produced by Chinese speakers was 26, which indicates that there were no obvious distinctions between Chinese speakers' pitch fluctuations when producing English and Chinese yes/no questions. It is generally true in tone languages that intonation is flatter, so as not to interfere with lexical tones, which are crucial for lexical meaning. [e.g. 12].

Although Chinese speakers showed similar mean pitch and average intensity in their English and Chinese yes/no questions, there were many distinctions in pitch patterns between the English and the Chinese yes/no questions. In both kinds of utterances, there were also three kinds of intonation contour: rising, flat, and falling (see also above). Table 2 shows Chinese speakers tended to use a rising intonation in their English sentences, while they tended to use a flat intonation in their Chinese yes/no questions.

Table 2. The pitch patterns of English and Chinese yes/no questions by Chinese speakers.

Pitch patterns of YNQs	English	Chinese
Rising	75%	16.67%
Flat	6.67%	68.33%
Falling	18.33%	15%

Why did Chinese speakers produce such a flat intonation in Chinese yes/no questions? We can explain this by the fact that the final character in Chinese yes/no questions is almost invariably 吗 ma. This character has a neutral tone, which promotes a flattening out of the intonation at the end of the sentence. It suppresses the high rise that would normally be found at the end of English yes/no questions.

It turned out that the pitch patterns in the Chinese yes/no questions were determined by their final character, i.e. lexical tone trumps sentence intonation in Chinese. From Q1 to Q11, the last character was 吗 ma, but in Q12 the final character was 戏 xì, with a falling tone. Thus, Q1–Q11 received flat intonation made in 73% of all Chinese speakers' productions, while Q12 was realized with a falling intonation in 100% of the cases. We tentatively assume that the "flat" intonation is generally associated with Chinese yes/no questions unless there are reasons to diverge from this such as a final character with a falling or rising tone.

4 Conclusion

The aim of this study was to analyze L1 interference in Chinese English learners by making two comparisons. The first involved the differences in the intonation patterns of yes/no questions between Chinese speakers of English and native English speakers. The second involved the differences between the English and Chinese intonation patterns of yes/no questions produced by Chinese speakers.

Measured by average standard deviation, the Chinese speakers' pitch fluctuations were lower overall than those of the English speakers. Second, there were only marginal differences in pitch variation in English and Chinese yes/no questions produced by Chinese speakers. In terms of pitch pattern, the Chinese speakers tended to use a flat intonation while English speakers used more rises and falls. Chinese speakers used an even more flat intonation when they produced yes/no questions in Chinese.

In Chinese, when the last character had a neutral tone, the pitch pattern was commonly "flat". In this experiment, all Chinese speakers used a falling tone in Q12 (in Chinese) which ended with a character with Tone 4. Since most Chinese yes/no questions end with a toneless character, the speakers would transfer this flat intonation from Chinese to English when they produced the English yes/no questions. The comparison showed that Chinese speakers did not have as much pitch variation compared to the native English speakers. Larger pitch variations might affect lexical tone so that sentence pitch variation is generally limited in Chinese. Thus, L1 interference plays a role in intonational patterns in yes/no questions.

The current study has a number of limitations such as the extremely small sample size, especially the fact that only a single native English speaker participated. Besides, variables like age, gender and region, which might affect the participants' intonation, were not strictly controlled.

In order to gain representativeness, further studies need to involve a larger number and wider range of participants, and intonation might be a little bit different between British and American English, so this factor also needs to be controlled for.

References

1. Samarin, W.J., Pike, K.L.: Tone languages: a technique for determining the number and type of pitch contrasts in a language, with studies in tonemic substitution and fusion. Language **21**(4), 187 (1951)
2. Wells, J.: Sounds Interesting: Observations on English and General Phonetics. Cambridge University Press, Cambridge (2014)
3. Cruttenden, A.: Intonation, 2nd edn. Cambridge University Press, Cambridge (1997)
4. Ladd, D.R.: Intonational Phonology, 2nd edn. Cambridge University Press, Cambridge (2008)
5. Jones, D.: An Outline of English Phonetics. Forgotten Books, London (1922)
6. Brown, G., Currie, K., Kenworthy, J.: Questions of Intonation. Croom Helm, London (1980)
7. Duanmu, S.: The Phonology of Standard Chinese, 2nd edn. Oxford University Press, Oxford (2007)
8. Ellis, R.: Second Language Acquisition. Oxford University Press, Oxford (1997)
9. Boersma, P., Weenink, D.: Praat: Doing phonetics by computer. https://www.fon.hum.uva.nl/praat/. Accessed 20 June 2020
10. Baken, R.J.: Clinical Measurement of Speech and Voice. Taylor & Francis, London (1987)
11. Valenzuela, M.G.: A comparative analysis of intonation between Spanish and English speakers in tag questions, Wh-questions, inverted questions, and repetition questions. Revista Brasileira de Lingüística Aplicada **13**(4), 1061–1083 (2013)
12. Gussenhoven, C.: The Phonology of Tone and Intonation. Cambridge University Press, Cambridge (2004)

Design of Virtual Training and Maintenance System for Communication Equipment

Deyi Sang(✉), Xuepeng Jiang, Fuqiang Zong, and Xiao Zhang

Naval Aviation University, Beijing, China
sangdeyi@163.com

Abstract. In order to improve the training and maintenance level of military communication equipment, in view of the practical needs of communication equipment maintenance and support training, combined with the development of information technology, a communication equipment training and maintenance system based on virtual reality technology is built. Focusing on the functional analysis and module design of the system, starting from the top-level architecture of the system, the functions, system architecture, system processes, and data realization of the virtual training and maintenance system are elaborated. The functions of communication equipment maintenance support teaching, training, and assessment can be realized.

Keywords: Virtual reality · Communication equipment · Maintenance training · Simulation system

1 Introduction

The intact rate of communication equipment is an important factor affecting the generation of military combat effectiveness, and plays an important role in the construction of national defense and military informatization [1]. With the continuous increase of our military's informatization level, the complexity of communication equipment and the difficulty of maintenance continue to increase, the cost of communication equipment support has always been high, the contradiction between the lack of talents and a single training model is increasingly prominent, and there is an urgent need for low development costs and efficiency High, close to the actual training equipment and means of the army, innovate the training mode of communication equipment maintenance personnel, and promote the continuous improvement of the management, training and use of communication equipment [2]. With the development of simulation technology, the role of virtual reality technology that integrates realistic visual, auditory, and tactile sensations in the military field has gradually become more prominent. The communication equipment training and maintenance system based on virtual reality technology is cost-effective and time-sensitive [3]. The improvement of the operation level of maintenance personnel at the grassroots level and the construction of the maintenance support personnel system have a great role in promoting.

M. Tavana et al. (Eds.): IISA 2020, AISC 1304, pp. 310–314, 2021.
https://doi.org/10.1007/978-3-030-63784-2_39

2 Analysis of the Status of Communication Equipment Training and Maintenance

1) Insufficient teaching and training conditions

Traditional communication equipment maintenance teaching and training usually rely on training bases, military colleges and teaching institutions, and use a combination of classroom teaching and practical teaching [4]. The practical training as the core teaching content is highly dependent on communication equipment, but the configuration is practical. The cost of installing equipment is high, and high-tech communication equipment of inestimable value cannot be used as teaching aids directly for training and maintenance.

2) The cost of installation teaching is high

There are two major difficulties in the training of communication equipment training and maintenance: one is that there are certain risks in the disassembly and assembly of overweight and oversized components; the second is that precision communication equipment is expensive, and it is easy to cause huge losses if you are not careful. In the past, for similar key high-risk subjects, training methods such as theoretical explanation, movement demonstration, model work, and practical training were generally adopted.

3) The training evaluation model is outdated

The assessment of communication equipment training and maintenance is usually to set obstacles manually [5]. The examiner evaluates the maintenance skills of the reference personnel by observing whether the reference personnel's operation steps are correct, whether the maintenance procedures are standardized, and asking relevant technical knowledge [6]. Communication equipment failures are strange, but there are only a few that can be set manually, and these kinds of failures are usually familiar to the reference personnel, and it is difficult to check their true level.

3 Design of Virtual Training and Maintenance System for Communication Equipment

3.1 System Function Mode Design

1) Teaching mode

In order to achieve the same effect as the actual communication equipment training and maintenance, a variety of teaching methods should be used to facilitate the instructor's operation, such as system automatic explanation and demonstration, student self-learning, system-guided learning, etc. The learning content includes communication equipment principle, maintenance theory, maintenance charter, operation demonstration, etc. During the learning process, the instructor can monitor the learning progress, flexibly adjust the learning content, and the participants can get help at any time.

2) Training mode

The training mode is divided into two parts: learning and operation. The system should provide learning on the structure, principle, decomposition, and combination of target communication equipment in a virtual three-dimensional simulation maintenance environment, and carry out activities related to actual training and maintenance, including Fault detection and analysis, maintenance tool selection, disassembly and replacement, installation and commissioning, and high-level simulation throughout the process, just like operating on actual communication equipment.

3) Assessment mode

The assessment model is divided into two parts: theoretical knowledge assessment and maintenance skill assessment. The theoretical assessment should be able to randomly generate test papers for reference personnel of different positions and grades. Maintenance skill assessment involves the design of examination cases, the evaluation of operation steps and the formulation of scoring standards. The difficulty lies in the assessment of the final score. The system should comprehensively judge the final score of the reference personnel from the three dimensions of the completeness of the operation steps, the accuracy of the operation process and the length of the operation time. At the same time, the system must also be able to generate test cases with different tactical backgrounds, so that the trainees can carry out communication equipment maintenance theory and skill assessment in a realistic combat environment.

3.2 System Module Design

The communication equipment virtual training and maintenance system includes 7 modules (as shown in Fig. 1), and each module can independently complete certain learning and training assessment functions.

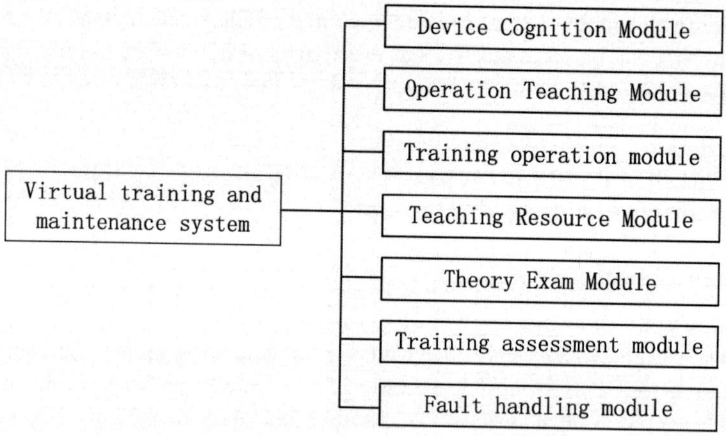

Fig. 1. Module design of virtual training and maintenance system

1) Device recognition module

This module can demonstrate the three-dimensional view and perspective view of the whole or individual parts of the corresponding communication equipment from various angles according to the learning content selected by the trainees, as well as pictures, text, audio and other auxiliary instructions and videos.

2) Disassembly and assembly of teaching modules

This module is mainly used for disassembly and assembly training, including disassembly and assembly demonstration and disassembly and assembly operations. The disassembly and assembly demonstration is used to show the disassembly and assembly sequence of various system components in the maintenance process. Trainees can use this module after learning the assembly and disassembly demonstration. Perform self-tests on disassembly and assembly operations.

3) Practice training module

This module is mainly used for the training of the disassembly and installation of components in various institutions of communication equipment.

4) Teaching resource module

This module stores the data resources of various types of communication equipment that trainees need to learn. It can be selected and learned by instructors and trainees at any time. In addition, related data resources can be imported and exported at any time to facilitate the reception of new communication equipment issued by superiors. The training and maintenance data package can realize the functions of resource browsing, retrieval and management.

5) Theoretical assessment module

The module can randomly generate theoretical test questions according to the teaching progress. This test question is different from the traditional paper test. It can be combined with the actual 3D model in a virtual simulation environment. This can not only improve the reference enthusiasm of the participants, but also Make the assessment truly combined with practice.

6) Practical training assessment module

This module can automatically generate practical test questions with a tactical background according to the reference personnel level, and at the same time record the entire maintenance review process. After the test, it can give relevant targeted guidance.

7) Fault setting module

The module has teaching and training functions for troubleshooting, and can systematically demonstrate the diagnosis and elimination of the faults of various system components of communication equipment.

3.3 System Software Architecture Design

The system consists of application layer, analysis layer, data layer and technical support layer (as shown in Fig. 2). The application layer is the way and interface of user interaction with the system. It is used to receive user instructions and feed back the processing results to the user in a graphical way or in the form of dialogue interaction. The analysis layer is the core of system application implementation and is used for training and maintenance. All activities related to operation; the data layer is mainly used for accessing, generating, maintaining and managing various modeling data during the application of the system; the technical support layer is the software and hardware supporting the operation based on the virtual maintenance system and its technology.

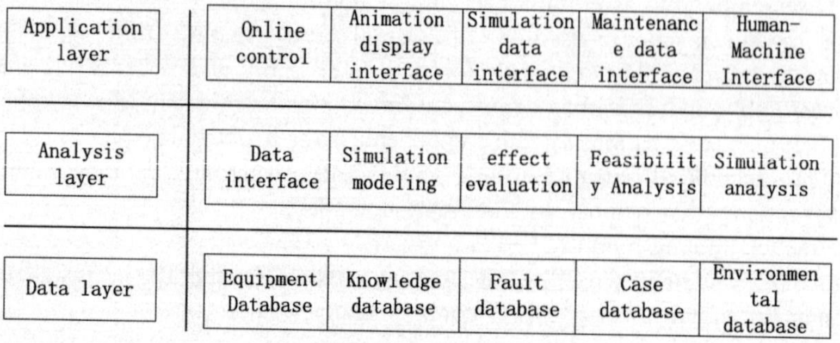

Fig.2. System software architecture

4 Conclusion

This article focuses on the functional analysis and module design of the system, starting from the top-level architecture of the system, to elaborate on the functional design, system architecture, system process and data realization of the virtual training and maintenance system, and to promote the development and promotion of the virtual training and maintenance system for military communications equipment Of great significance.

References

1. Zhang, Z., Liu, H., Zhang, R., et al.: Research on combat stress training of military college undergraduates based on virtual reality technology. J. High. Educ. Res. (12), 117–120 (2018)
2. Liu, D., Liu, X., Zhang, Y., et al.: The potential, progress and challenges of the application of virtual reality technology. Open Educ. Res. (8), 25–31 (2016)
3. He, J., Lai, Y.: Design and implementation of weapon equipment virtual training and maintenance system. Sci. Technol. Herald (12), 71–74 (2010)
4. Guo, Y., Li, Y., Tian, Q.: Design of maintenance test system for a certain optical terminal machine. Shandong Ind. Technol. (9), 111 (2014)
5. Huang, Z., Xiang, H., Tang, J., et al.: Design and implementation of equipment maintenance simulation training system. Comput. Eng. Des. (12), 4677–4680 (2012)
6. Liu, C., Zhang, Y., Zhang, J.: Application research of virtual reality technology in military equipment training. China Educ. Technol. Equipment (14), 44–45 (2017)

The Status and Strategy of Rebuilding the Library Space Under the "Internet+" Environment

Dong Na(✉)

Guangdong University of Science and Technology, Dongguan 523083, China
105520675@qq.com

Abstract. With the rapid development of modern social information technology and the interaction of economic and social information more frequently, all industries are in the background of the era of big data. In this case, The development level of computer software technology has become an important factor that affects and even determines the speed of social development. Facing the increasingly complex Internet network structure and traffic data information, how computer software technology is facing new and emerging challenges, strengthening the computing power and data processing capacity of the big data application center has become a decisive factor affecting the development of the contemporary computer industry. Based on the analysis of the problems in the field of computer software development in the era of big data, this paper gives a brief introduction to the software technology, hoping to help readers understand the relevant technology in the computer software industry.

Keywords: Internet+ · Library space · Rebuilding · Space transformation

1 Introduction

With the continuous development and progress of human society, space has evolved from the initial basic living space to a perfect social field with multiple functions.The pursuit of knowledge is the development demand derived from human beings' basic needs for food and clothing. The transformation of a society's pursuit of development demand represents the development level of this region. The library platform is the earliest space with information acquisition function, and it has made outstanding contributions to the development and progress of human society.However, in the current practical environment with the continuous development of "Internet+" technology, drawbacks such as insufficient information retrieval and transmission speed of traditional library space are gradually revealed. Therefore, it is necessary to integrate with "Internet+" technology to meet the needs of modern people for knowledge acquisition.

M. Tavana et al. (Eds.): IISA 2020, AISC 1304, pp. 315–320, 2021.
https://doi.org/10.1007/978-3-030-63784-2_40

2 The Status Quo of Library Space Construction Under the Background of "Internet+"

2.1 The Rise of "Internet+" Leads the Upgrade of Library Space Technology

Traditional library institutions meet the storage and recording requirements of information knowledge through the collection of paper books, and at the same time provide people with the conditions to obtain and disseminate information. The library is an indispensable infrastructure for a society, but its shortcomings as an information dissemination medium are also very significant. First of all, the collection of paper books needs a lot of space, and the cost of book purchase and collection is extremely high. The high cost of building a large library is invested in the replacement of the construction level of libraries in the underdeveloped areas, which causes certain conflicts to the flow of knowledge. In fact, it costs more for people to acquire knowledge inside the library, and they have to bear the time cost of transferring back and forth to the library, and the book retrieval is relatively high, which will definitely lead to the increase and decrease of the information transmission rate in the library space.In the development environment of "Internet+", the cost of information retrieval of modern libraries has been greatly reduced by mobile, Internet of things, big data and cloud computing technologies. It also provides the possibility of sharing resources between different library spaces, thereby promoting the modern technology level of library spaces to be greatly improved.

2.2 The Rise of "Internet+" Leads the Transformation of Library Space Structure

"Internet+" is a prefix-type load structure model, the "Internet + XX" model is not just a simple introduction of Internet technology into a certain field, but the integration of Internet technology with the development needs of different fields,this is a technology that sublimates the traditional structural model and leads the overall transformation of the social life model. In the process of rebuilding the library space, "Internet+" can drive the transformation of its space structure with its unique technical compatibility. For example, breaking the original book classification space structure of the library, so that information acquisition can be concentrated on the computer platform, Based on the Internet platform, the information resources of a library can also be shared with other libraries around the world, which plays an important role in the sharing of ancient books and solitary books that are not easy to move.In addition to basic book sharing, the integration of "Internet+" technology and the rebuilding of library space can also drive the development of library-derived spaces such as maker platforms, audiovisual reading experiences, and so on.

2.3 The Rise of "Internet+" Leads the Change of Library Space Concept

"Internet+" technology has a sufficient accumulation foundation in the field of knowledge dissemination. For example, the "curtain course" platform that has emerged in recent years has fully proved that "Internet+" technology can meet the flow of knowledge in society. "Internet+" technology provides feasible conditions for the practice of innovative thinking, all kinds of innovative and creative thinking in the process of rebuilding the library space can be carried out on the "Internet+" platform, so as to promote the level of library space thinking innovation. The process of reprinting knowledge on books can be seen as a representation of information, and people can quickly obtain specific information through reading, but the limitations of this information dissemination model are also very obvious. People can only obtain information if they have personal contact with the information carrier. The limitation of information dissemination for a long time makes the application of library space inefficient, especially after the book borrowing is completed, subsequent readers who need to read can no longer obtain the information in time, the "Internet+" technology directly makes this limitation change, and provides a feasible idea for the comprehensive innovation and development of the library space.

3 Reflections on the Reconstruction of Library Space Under the Background of "Internet+"

In terms of its specific application depth and application breadth, there are still certain deficiencies. In the process of rebuilding the library space, it is necessary to reflect on these issues in order to truly promote the "Internet+" and the construction of library space fusion.

From the perspective of "Internet+" and the practical application of library space rebuilding, there are mainly the following aspects that need to be reflected:

First of all, from the perspective of human care for the application of "Internet+" technology, the current use of "Internet+" technology in the process of rebuilding the library space in my country is still a simple application of its technology. For example, set up an electronic reading area (as shown in Fig. 1) and self-service search machine (as shown in Fig. 2) in the library, after these devices are put into application, there is no special person guidance or corresponding teaching content, this has caused certain obstacles for the elderly and other social groups who do not fully master the Internet technology to feel the "Internet+" library space. Therefore, it should be reconsidered whether the "Internet+" technology applied in the rebuilding of the library space really considers the humane needs of users.

Secondly, from the perspective of the service upgrade requirements for the rebuilding of library space, the application of "Internet+" technology just meets the requirements for the service upgrade of library space, however, due to the extensive application of Internet technology, a certain equipment foundation needs to be built, and many libraries cannot fully integrate their space construction with "Internet+" technology due to their weak corresponding construction foundation. For example, the construction of an online library requires the support of a memory-rich Internet server, but the cost of server rental and maintenance are relatively high, many libraries have not prepared corresponding server facilities in advance, as a result, the subsequent "Internet+" technology cannot be connected to the library space rebuilding process.

Fig. 1. The representative of Internet technology application of traditional library-electronic browsing room.

Fig. 2. Self-service borrowing machine of modern library

4 Analysis of the Thinking of Library Space Rebuilding Under the Background of "Internet+"

4.1 The Change of Concept of Library Space Construction

The fundamental requirement for the rebuilding of modern library space is 'innovation", how to rebuild the space module and structure of its crude oil, the development of its information dissemination function is the most important demand for the modern library space upgrade. Therefore, rebuilding the library space under the background of "Internet+" first requires the transformation of the previous space construction concept. First of all, from the perspective of the idea update, the construction of library space under the background of "Internet+" needs to be based on the needs of users at all levels and regions, maintain their rights to obtain information. Secondly, from the perspective of technological innovation, "Internet+" and library space are actually two different fields. However, the integration of "Internet+" technology is extremely strong. Therefore, in the process of rebuilding the library space, it is necessary to abandon the concept that the previous different areas cannot be integrated, and boldly introduce "Internet+" technology in the construction of the Internet space.

4.2 System Upgrade of Library Space Construction

Although "Internet+" is a brand-new social development model proposed by my country in recent years, the history of libraries applying interconnected information technology has been quite long. Many large libraries in my country have Internet access platforms. At present, most libraries in my country have also introduced digital technology for book retrieval. But first of all, the development of the library informatization level in my country has been very limited, basically focus on the above two areas, secondly, China's library Internet development of the market characteristics are too strong, "China HowNet" and other Internet book platforms pay too much for information, it deviates from the original commonweal characteristic of library. Based on this, in the process of "Internet+" promoting the rebuilding of library space, it is necessary to continue to deepen the depth of reform, so as to provide a breakthrough for the reform of the library space system and promote the optimization and upgrade of its overall structural system.

4.3 Innovation of Library Space Construction Strategy

In the current social background of rapid development of information technology, Internet technology has been integrated into my country's strategy of strengthening the country, so the rebuilding of library space based on "Internet+" technology should also be demonstrated at the strategic level of library development. Therefore, the library should first develop its own service system by relying on the network strategy, and promote the transformation and upgrading of its traditional service mode by using the technology and thinking of "Internet+". Secondly, the library also needs to apply "Internet+" technology to upgrade its own resource space, develop online reading, training and communication as the basic spatial model of the library, in order to provide the public with a more diversified library experience and promote the comprehensive strategic transformation of library development.

5 Conclusion

In summary, library space, as one of the basic social public organizations, carries the important role of information collection and dissemination. However, in the current real-world environment where the public's knowledge acquisition needs have greatly increased, it is necessary for library institutions to rely on "Internet + " technology to extend their own spatial structure, spatial construction strategy and spatial function, break the barriers of information knowledge dissemination, and satisfy citizens' rights to acquire basic information. I believe that through the joint efforts of all aspects of our society, the library space will be able to evolve towards the "Internet+" development direction of "Great Wisdom" in the future, thereby forming a two-way driven, collaborative and innovative "Internet+ library space" development model.

Acknowledgments. This paper is funded by Project of: Guangdong Library Association 2019 Guangdong Provincial Library Scientific Research Project–One of the results of the Research on Library Space Design and Space Service (GDTK1914).

References

1. Zhang, X.: Practice and exploration of space reengineering of discipline branches of colleges and universities driven by "Internet+" thinking——taking the branch of Peking University library school of foreign languages as an example. J. Univ. Libr. **38**(1), 51–56 (2020)
2. Liu, Z., Mao, G., Zhang, H., et al.: On library knowledge service and space construction——theoretical framework based on knowledge construction. Libr. Theory Pract. **95**(1), 12–17 (2017)
3. Liao, X.: The rise of the library virtual space in the new library construction wave-the second part of the urban library space variation. Library **108**(2), 103–104 (2011)
4. Pan, Y., Liu, L., Sun, M., et al.: Research on the space function layout of university libraries in the digital environment: taking the space of new buildings in Beijing, Jiangsu and Shanghai as the investigation objects.J. Univ. Libr. **35**(4), 46–53, 78 (2017)
5. Xusen, Z.: Design of library document personalized push system based on maximum mutual information coefficient. Mod. Electron. Technol. **09**, 146–149 (2020)
6. Wang, X., Wang, T.: Practice and thinking of public library space reengineering from the perspective of public cultural service —— taking the new library of Liaoning Province as an example. Library (04), 40–43+64 (2017)
7. Wu,H.Y.: Construction of library service innovation system under Internet Thinking. Libr. Sci. Res. (01), 65–68 (2017)
8. Meng, J.: Research on innovation and strategy of library space reconstruction in the context of "Internet+." Mod. Intell. **03**, 92–96 (2016)

Research of Blockchain-Based "Edge-Cloud" Collaborative Model, System Architecture and Overall Layout for Railway Application

Ma Jianjun[1], Wang Wanqi[2], Shen Haiyan[2(✉)], and Duan Jiaying[2]

[1] China Academy of Railway Sciences Group Co., Ltd., Beijing, China
[2] China Academy of Railway Sciences Group Co., Ltd., Institute of Electronic Computing Technology, Beijing, China
dzsshy@163.com

Abstract. With the advent of an explosive age of the data in perception of everything, interconnection of everything, and intelligence of everything as well as the high requirement for millisecond low latency business, Cloud Computing is facing with two main bottlenecks, namely bandwidth shortage and network latency. As the extension of and supplementary to Cloud Computing, Edge Computing offers quickly responding Intelligent Edge services. It is a new type of flexible computing model that is four-in-one including Cloud, Edge, Terminal, and Intelligence. Blockchain-based "Edge-Cloud" Collaborative Ability-based Model builds a flexible open eco system with built-in security attributes and security ability, and offers users with "credible, reliable, controllable, verifiable, auditable and traceable" services that cover the whole chain of "Terminal Protection – Device Verification- Cloud Identification" for the whole life cycle of data. "Edge-Cloud" Collaborative Ability-based Model satisfies the critical requirements for business agility connectivity, real-time business processing, data optimization, application intelligence, as well as security and privacy protection. As an indispensable key element and a mainstream model, it empowers the development of intelligent railway.

Keywords: Edge computing · Blockchain · Cloud computing · Edge-cloud collaboration · System architecture · Overall layout

1 Introduction

With the advent of an age of interconnection of everything, data generated by network edge devices experience an explosive growth. It is expected that 50 billion devices will be connected to the Internet by 2020, with over 10 billion [1] devices in China. The connected devices will generate 600ZB data, while only 10% are critical data and 90% are temporary data. Network edge analytics will be needed for 50% of data [2]; Sverdik research puts forward that China's data centers consume more energy than Hungary and Greece combined [3]. Cloud Computing offers a highly efficient computing platform for Big Data, but it has to face with two main bottlenecks [4], namely bandwidth shortage and network latency, for example, the businesses with high demand for real-time computing power that are extremely sensitive to network latency such as

M. Tavana et al. (Eds.): IISA 2020, AISC 1304, pp. 321–332, 2021.
https://doi.org/10.1007/978-3-030-63784-2_41

driverless cars and AR/VR can deploy small-sized or portable data centers at the 5G network edge and make use of localized processing of Edge Computing to satisfy URLLC and mMTC's demand for ultra-low latency [26]; Edge Computing emerges as the times require. It is a distributed open platform that integrates network, computing, storage, and application at the network edge side close to the things or the data source. It offers intelligent edge services [5, 6]. As a new type of flexible computing model that is four-in-one including Cloud, Edge, Terminal, and Intelligence and the expansion [7], it is the extension of and supplementary to Cloud Computing [8]. A survey shows that 78% of users are worrying about third party's unauthorized use of personal data [9]. With the General Data Protection Regulation (GDPR) in force, data security and privacy are particularly important considerations [10]. Blockchain-based "Edge-Cloud" Collaborative Ability-based Model builds a flexible open eco system with built-in security attributes and security ability, and addresses the computing power issue for "the last kilometer" of interconnection of everything. It offers users with "credible, reliable, controllable, verifiable, auditable and traceable" services [11, 12] that cover the whole chain of "Terminal Protection – Device Verification- Cloud Identification" for the whole life cycle of data. The three technologies will form a mutually coordinated and complementary development trend to satisfy the critical requirements for business agility connectivity, real-time business processing, data optimization, application intelligence, as well as security and privacy protection. It is an indispensable key element in the innovations for industrial applications [13].

The railway industry deeply implements the strategy of "A transportation power pioneered by railway development", and builds the first modern large-scale data center of China's railway, namely Railway Main (Cloud) Data Center, in Wuqing District of Tianjin City. Based on the "Railway Integrated Platform", the Data Center is a carrier for the unified management, flexible extension, distribution on demand and system-wide Big Data intelligent analytics for basic information resources such as computing, storage, and network of system-wide integrated application system. Faced with the issues such as vast amounts of data at the railway edge end and overloaded Cloud center, blockchain-based "Edge-Cloud" collaborative solution is designed after multi-dimensional research to amplify marginal utility and multiplier effect brought by integrated innovation, better maintain "Railway Main (Cloud) Data Center", and provide technical support to railway on security, collaborative command, accurate management and control and energy saving, and empower the development of intelligent railway [25].

2 Research of Edge Computing Value Model and Influencing Factors

2.1 Analytics of Edge Computing Value Model

In Fig. 1, the value model of edge computing is reflected through "core" (4 key abilities) and "periphery" (5 CROSS advantages) in an integrated manner. "Core" is new technology and new momentum, while "periphery" is new pattern, new experience

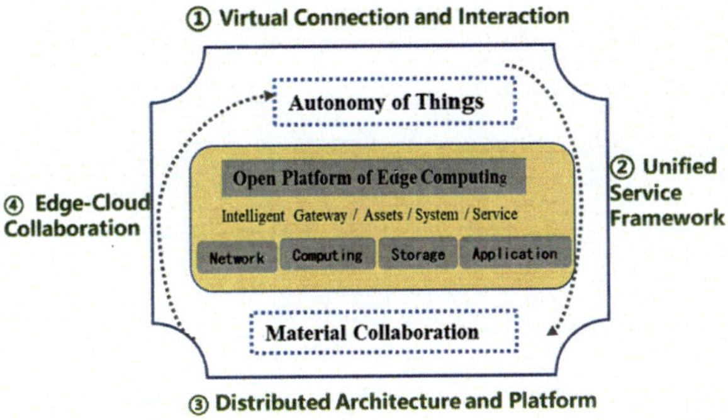

Fig. 1. CROSS value model of edge computing

and driving force. "Core" drives the continuous optimization and development of "periphery".

"Core" (4 key abilities). (1) Connect the physical world and the digital world by virtualization and reality. Promote the coexistence and interaction of the virtual world and the real world that are based on digital twin technology. (2) Unified development and deployment of the service architecture ability of operation. Support the rapid development/integration/verification/release/business arrangement of solutions as well as automatic deployment and centralized operation. (3) The abilities of model-driven intelligent distributed architecture and edge computing development platform. It realizes Independence and Cooperation of Things through the intelligent ability driven by knowledge model based on network, computing, storage, and application of the "platform". (4) "Cloud-Edge" network/business/application/intelligent collaborative ability. The edge side supports multiple network interfaces, protocols and topology, real-time business processing and deterministic latency, data processing and analytics, distributed intelligent as well as security and privacy protection [13].

"Periphery" (5 CROSS advantages). The real-time and reliable interconnection, flexible network extension, as well as operation and low- cost maintenance are guaranteed through Connection (compatible with massive and heterogeneous connections); business scenarios that have high demand for real-time performance such as control and operation are satisfied through Real-time (real-time business) to improve accurate management and control as well as immersive experience of terminal customers. Vast amounts of diversified heterogeneous data are aggregated, optimized, and normalized through Optimization (data optimization) to provide efficient services to Edge Application. The application of Smart (application intelligence) to Edge Intelligence of device predictive maintenance improves handling efficiency, cut down on costs and advance the transformation of new service models. The breadth and the difficulty for access control at the network edge side and threat protection are solved through Security (security and privacy protection) to guarantee the security of devices, network, data and applications as well as the completeness and confidentiality of critical data [13].

2.2 The Relationship and Influencing Factors Analytics of "Value Efficiency- Response Time"

Mckinsey Global Institute points out: value efficiency of edge computing only plays 20% of its role [14]. As indicated in Fig. 2, response time is a critical factor when value efficiency is determined. Other internal and external factors collaborate with each other.

$$S = \sum\nolimits_{i=0}^{n} \{Di, XiMi, YiNi\} \tag{1}$$

As in formula 1, S is the total value efficiency, Di (direct efficiency); (internal) factors = (diversity, connectivity, real-time, constraint, integrity, economy, reliability, and extendibility, etc.); (external) factor = (fluctuation, vibration, dust, explosion, and electromagnetic, etc.); Xi, Yi (the ith index effectiveness); Mi, Ni (the ith index weight), XiMi (internal value efficiency). YiNi (external value efficiency). The general response time limit range for daily maintenance [14]; device failure (millisecond – 1h), performance monitoring (1h- 1d), supply chain (1d- 1w), and predictive maintenance (1w- 1m). The second-level disposal value efficiency for failure is 100%. Disposal value efficiency falls to 80% within 10s, and falls to 50% within 20s. It falls faster successively. Only second-level real-time disposal of edge computing can guarantee operation security and enhance user experience.

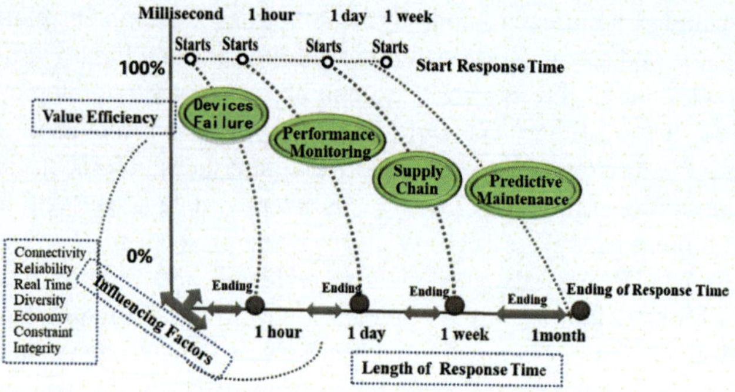

Fig. 2. The relationship and influencing factors analytics of "value efficiency- response time"

3 The Research of the Connotation, Ability-Based Model and Operating Principle of "Blockchain- Edge- Cloud" Collaboration

3.1 The Connotation of "Bionic Functions" -Based "Blockchain-Edge-Cloud" Collaboration

Blockchain, Cloud Computing and Edge Computing complement each other. Edge Computing and Cloud Computing have their merits. Cloud Computing is good at controlling the overall situation and focuses on non- real time and analytics of long-cycle Big Data. It can bring its superiority into full play in long-cycle maintenance and decision-making support; while Edge Computing focuses on local and regional situations as well as analytics of real-time and short-cycle data. It can realize better real-time disposal of local businesses [15].

To understand from bionic functions [5], "blockchain" is the lymphatic system that consists of lymphoid tissue, lymph, and lymphatics. It is the defense system of human body. "Cloud- Edge- Terminal" is the "nervous system" that consists of "brain- nerve center- neuron- nerve endings", that is, edge device perceives information and deliver it to edge computing center (neuron) to control and feedback the disposal results timely through edge network (nerve endings). Meanwhile, it delivers the results to "brain" (nerve center) for memorization and learning, and plays the role of the advanced intelligence for predictive reasoning (cognition + perception + logic (model, algorithm) + learning + talent + storage) when the body is in good health condition for intelligent analytics and decision making. It is thus clear that healthy and safe "lymphatic system + nervous system" play the role of the best functions of human body by collaboration (Fig. 3).

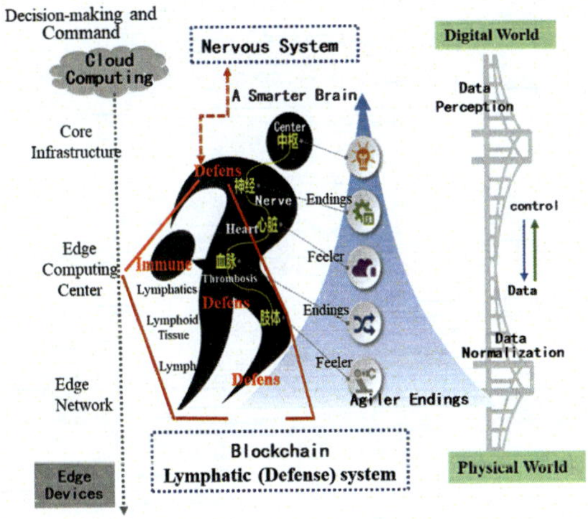

Fig. 3. The Connotation of "Bionic Function"-based "Blockchain-Edge-Cloud" collaboration

3.2 Blockchain- Based "Edge-Cloud" Collaborative Ability-Based Model and Its Operating Principle

The research designs blockchain-based "Edge- Cloud" collaborative ability-based model, data stream and operating principle [15]. It is a flexible model with complete abilities, and its collaborative abilities vary after cutting based on different application scenarios. Even the same type of collaborative ability varies after combination with different scenarios. Through complete collaboration of three levels (IaaS, PaaS, and SaaS), Edge IaaS and Cloud IaaS realize collaboration for network resources, virtual resources, and safe resources; Edge PaaS and Cloud PaaS realize data collaboration, intelligent collaboration, application management collaboration and business management collaboration; Edge SaaS and Cloud SaaS realize services collaboration [16].

Five typical characteristics: (1) Multi-source heterogeneous connection: it has wire connection and wireless connection, real-time connection and non-real time connection as well as a variety of abilities such as heterogeneous connection protocols; (2) Information normalization: it has normalization ability for multi-source heterogeneous information such as persistent information, intermittent information, time effective information, non-time-effective information, structural information, and non-structural information; (3) Distribution on demand. It has differentiation analytics and distribution on demand abilities for the needs such as computing, storage, and application on different locations and scenarios, and under different disposal requirements, as well as; (4) Flexible scheduling. It has abilities in flexible scheduling, arrangement and management for the resources, applications, and businesses of "Edge-Cloud". (5) Defense and trust. It offers blockchain-based "Edge-Cloud" security defensive system that includes data security, identity authentication, privacy protection and access control (Fig. 4).

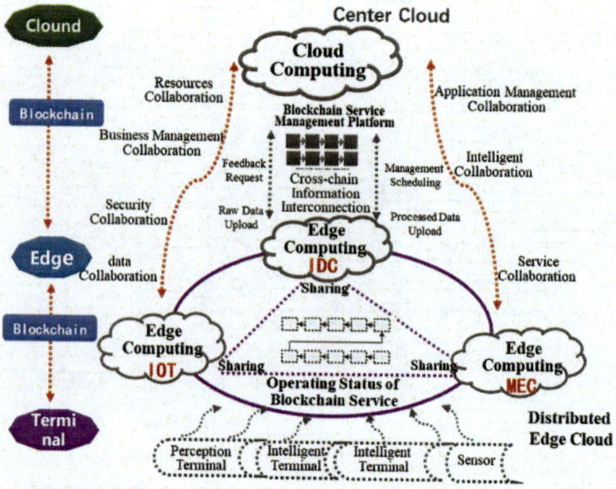

Fig. 4. Blockchain- based "Edge-Cloud" collaborative ability model

Seven major collaborative abilities and their operating principles. (1) Basic resources collaboration. Edge nodes have abilities of local and Cloud computing, storage, network, as well as virtual resources scheduling and management. (2) Data collaboration. After data are collected at the edge nodes, ① The raw data are directly uploaded to Cloud; ② The data are preprocessed and analyzed according to the rules or data models, and the results and the masked data are uploaded to Cloud; ③ Send "request" for service to Cloud; ④ "Edge- Cloud" establishes a "two-way" data transmission channel and forms a complete, orderly and controllable data flow path to realize an efficient and cost effective management of the whole life cycle of data, and explores values of data. (3) Intelligent Collaboration: after Cloud AI + centralized model training, it is sent to edge nodes for implementation to realize distributed intelligence; (4) Application Management Collaboration: Cloud offers application development, testing environment and application life cycle management. Edge nodes offer application deployment and operation environment; (5) Business Management collaboration: based on customers’ needs, Cloud realizes business arrangements such as application/digital twin/network; edge nodes offer modular and micro service applications; (6) Service collaboration: Cloud offers SaaS service distribution strategies and service abilities. Based on Cloud strategies, edge nodes realize part of ECSaaS service, and SaaS service on customers’ demand. (7) Security collaboration: blockchain-based decentralization, tamper proof, counterfeiting proof, traceability, collective maintenance, and consensus trust mechanism build a management platform for blockchain services in “Cloud” and a blockchain-based security defensive environment at “Edge Terminal” for scheduling, monitoring, consensus, and intelligence contract, to provide Blockchain-as-a-Service (BaaS) and realize cross-chain information interconnection and Cloud security defense.

4 The Research of Blockchain-Based “Cloud- Edge-Terminal” System Architecture for Railway Application

The traditional architecture and ordinary encryption defensive strategies cannot cope with the challenges brought by intelligence development. Blockchain-based “Edge-Cloud” collaborative system architecture can shoot the troubles such as reliable transmission and information recording in the unreliable network world. In terms of data sharing, credible data sharing between enormous amounts of non- credible nodes is realizable by making use of the transparent and lengthy blockchain to shoot the traditional troubles such as data loss, tampering and high cost. In terms of information security, the safe and credible new idea based on blockchain and decentralized network can shoot the troubles of the traditional network architecture such as single point of failure and vulnerability to attacks. Hence, around the development goal of master data center for railway application, the research designs an overall architecture for railway “Cloud- Blockchain- Edge- Terminal”, including collection layer, edge layer, transmission layer, consensus layer, platform layer, service layer, application layer, demonstration layer as well as supporting guarantee (standard specifications, security guarantee services, management services, operation and maintenance services, whole life cycle services of data) (Fig. 5).

By unified planning and design as well as normalized organization for implementation, the collaborative functions of all layers are given full play and the development of intelligent railway is empowered. The followings are the specific design of each layer:

Layer of Perception: It is the feeler of intelligent high-speed railway to perceive the physical world (a variety of sensor nodes). Based on the “total, complete, and real-time” collection principles and decentralized blockchain technology, it realizes the enormous amounts of access of the IoT at any scale and of any type such as environment and devices, obtains the information such as passengers’ transportation requirements, train’s operation status, operating status of equipment, and external environment conditions, satisfies the needs of data owners, demanding parties, equipment manufacturers, and constructing parties, and form a closed loop of the IoT.

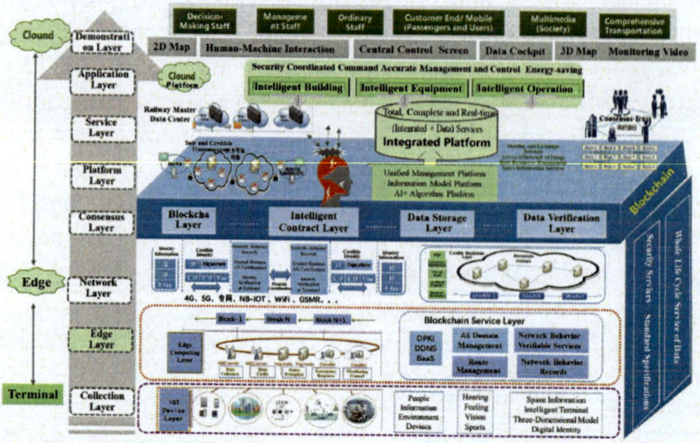

Fig. 5. Blockchain-based "Cloud- Edge- Terminal" system architecture for railway application

Edge Layer: An Intelligent Contract Layer that operates at the edge nodes. By taking advantage of the consistence between blockchain network and edge computing architecture, according to their computing power, edge computing devices (nodes) clean and store data, check data consistence, and process invalid and missing values. They also quickly process edge requests and enhance time effectiveness of "perception-computing- response" according to the predefined rules of intelligent contracts [17, 18].

Transmission Layer: It is a belt that establishes ties between all the essential factors of Intelligent Railway, mainly including IPV4/IPV6 technology, WiFi, 4G/5G, NB-loT, Ethernet, serial bus, parallel bus, and other means of communications. It takes advantage of blockchain cryptography and consensus mechanism to guarantee data transmission and access security to shoot the problems of the traditional network architecture such as single point of failure and vulnerability to attacks.

Consensus Layer: it mainly includes: data verification layer, data storage layer, blockchain layer, and intelligent contract layer. It quickly and sharply detects and controls illegal changes through monitoring chain; it makes use of blockchain structure to verify and store data; it makes use of distributed nodes and consensus algorithm to generate and update data; it makes use of intelligent contract and consensus mechanism to prevent data from being tampered and the completeness of data, and the users can safely and securely consume the data on the blockchain and deliver values.

Brain layer for intelligent high-speed railway: it mainly includes platform layer, service layer, application layer and demonstration layer. Around three major areas namely "intelligent building, intelligent equipment and intelligent operation", carried by the "integrated platform" of "railway main (Cloud) center", blockchain-based "Edge-Cloud" collaboration, model and operating principle make changes and bring reforms to the traditional architecture to adapt to the new technologies, new patterns and new ecology, and give full play to new dynamic, new experience and new platform. (1) Platform Layer and Service Layer: it includes information resources such as strategic decisions, transportation and production, operation and development, resources management, building management, comprehensive collaboration, and social data of intelligent high-speed railway. (2) Demonstration Layer (Intelligent Decision-Making Layer) and Application Layer. Users send requests for services to the platform. The platform verifies user identity and sends the verification result to intelligent contract of edge layer to check the credit accumulate points of users, and then judges if the request is reasonable, and feedbacks to the application layer. An all-round visual dynamic display and human-machine interaction based on AI + "Central control large screen", such as two-dimensional (three- dimensional) situation map, data cockpit, multimodal data representation, and monitoring video matrix, enhance existing applications and act as a catalyst to new innovative applications to provide accurate services for users including decision- making, management, basic levels, customer end/mobile end, multimedia (society), and comprehensive transportation.

5 The Overall Layout of Application Scenarios and Typical Case Study of "Edge- Cloud" Collaboration for Railway Application

5.1 Analytics of the Overall Layout of Application Scenarios of "Edge- Cloud" Collaboration for Railway Application

Based on the characteristics of Edge Computing and Cloud Computing, considering the actual needs of railway-related businesses, the main businesses of intelligent high-speed railway are analyzed and sorted in terms of three dimensions namely disposal efficiency, computing intensity and technology maturity, and three major layouts of application scenarios are designed on X axis and Y axis (see the Figure for details). "Response speed" develops from slow to fast from left to right on the X axis; "computing intensity" develops from low to high from bottom to top on the Y axis; "technology maturity" develops from embryo, growth, steady improvement, to maturity on Z axis [19] (Fig. 6).

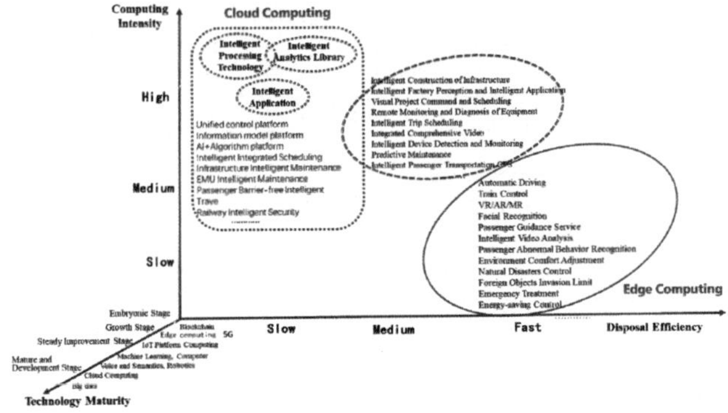

Fig. 6. The overall layout of application scenarios of three-dimensional "Edge- Cloud" collaboration for railway application

(1) Layout of "Cloud (Railway Master Data Center)": it is mainly designed for overall scheduling, in-depth learning, high computing intensity, long data cycle and other business scenarios, for example, simulation, intelligent analytics and predictive reasoning of intelligent processing technology, intelligent analytics base, railway integrated platform (management and control, model, and algorithm) as well as intelligent comprehensive scheduling, and integrated infrastructure operation and maintenance, EMU PHM and active operation and maintenance, barrier free intelligent travel, railway intelligent security, and comprehensive transportation. Presently, "5G+ Edge Computing" is "growing up". "Cloud Computing" is "maturing and developing". "Blockchain" is "sprouting". The technology portfolio at different stages match different application scenarios, and maximize values and benefits. (2) Layout of "Cloud-Edge Integration": it is mainly designed for "Edge-Cloud" interconnections, such as intelligent driving scheduling, C3 + ATO, intelligent ticketing analytics and forecasting, intelligent station brain, intelligent testing monitoring, predictive repair, and so on. (3) Layout of "edge terminal (regional center or railway administration/railway section)": it is mainly designed for local management and control, lightweight learning, short data cycle, quick disposal and other scenarios, for example, dynamic train tracking, facial recognition, passenger guidance service, abnormal behavior recognition, VR/AR/MR and other testing, monitoring, as well as management and control.

5.2 Case Study of "Railway Comprehensive Videos" Based on "Edge-Cloud" Collaboration

A total of over 3,000 cameras for comprehensive videos are deployed along Beijing-Zhangjiakou Highspeed Railway and its stations. Video data at TB level are generated every day. The videos at the edge nodes (regional centers) stream the data at real time. Firstly, through real-time response by edge computing, varied video analytics- based typical applications of local AI are realized such as expression recognition, behavior-based detection, tracking, hot spot management, and body attribute recognition. Secondly, through videos sharing and open services provided by (edge terminal) regional centers, the site videos are checked at real time for light weight learning and intelligent analytics to improve comprehensive utility efficiency of video data [20]; thirdly, based on the value density of video data and the capacity and needs of "Master Data Center", the alarm information of high values, and video stream data of typical scenarios are selected to access "Railway Data Service Platform"; permanent storage of alarm information; 50TB dynamic storage of the information of typical scenarios such as tunnel entrances, important bridge entrances; the "platform" combines historical data to make in-depth exploration, AI model training, predictive reasoning and iterative optimization thus to realize the demand upgrading of railway security from "visible" to "understandable" and empower the transformation of security model from "passive monitoring" and "after-the-event investigation" to "active recognition and early warning" (Fig. 7).

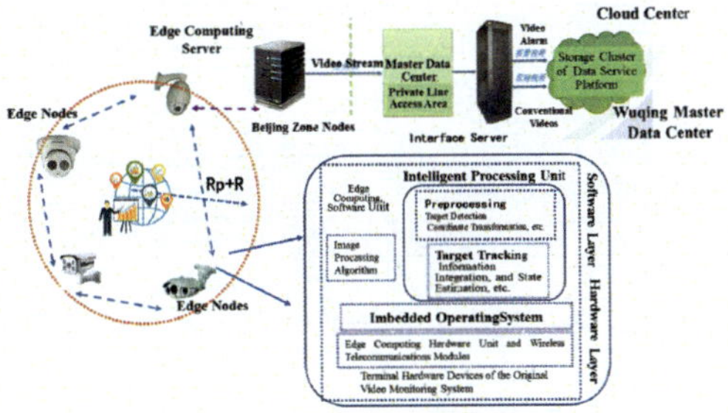

Fig. 7. Video Analysis and Disposal Topological Map Based on "Edge-Cloud" Collaboration

6 "Edge- Cloud" Collaborative Development Strategy for Railway Application

(1) Reinforce top-level design and planning, determine vision and aim, system architecture, deployment plan, implementation strategy and route for "Edge- Cloud" collaborative development for railway application; (2) accelerate the research of Edge Computing critical technologies by software definitions of machine, container, visualization, distributed computing, and light weight algorithm, etc. [21]; (3) break through various challenges one by one in the upgrading from Edge Computing to Edge Intelligence, the development from fragmentation to large scale, industry chain collaboration, and match between supply and demand; (4) based on the eco organization of "production, learning, research and utility", the researches of "5G + AICDE" integrated innovation, Edge Computing + X innovative applications, Edge Computing + Ecology co-construction and railway digital economy values [22].

The thesis analyzes digital values of "Edge Computing", "Value-Efficiency" relationship and influencing factors at multi levels and from multi dimensions; from the perspective of "Bionic Functions (nervous system + lymphatic system), the collaborative security defense and intelligent system, ability-based model as well as operating principle of blockchain-based "Edge-Cloud" are dissected; the system architecture and overall layout of railway application scenarios of blockchain-based "Edge-Cloud" collaboration for railway application are constructed; and analytics is made based on the case of "intelligent videos". At last, "Edge-Cloud" collaborative development strategies are put forward, which aim at exploring new ideas, new patterns, new architectures, new platforms, new dynamics, new experience and new formats that are suitable for the development of intelligent railway, as well as the best breakthrough points and solutions, and building a sound, open, flexible, collaborative, and autonomous edge computing eco system to empower the development of intelligent railway.

Acknowledgments. This work is supported by Scientific and Technological Research and Development Program (J2019X006) of China State Railway Group Co., Ltd.

This work is supported by Scientific and Technological Research and Development Program (J2018X008) of China State Railway Group Co., Ltd.

References

1. Evans, D.: The internet of things how the next evolution of the internet is changing every thing. CISCO White papers (2011)
2. Cisco Visual Networking. Cisco global cloud index: Forecast and methodology 2015–2020, CISCO White paper (2015)
3. Finnegan, M.B.: 787s to create half a terabyte of data per flight[EB/OL]. Accessed 03 Dec 2016
4. Armbrust, M.F., et al.: Above the clouds: a berkeley view of cloud computing. Eecs Dept. Univ. Calif. Berkeley **53**(4), 50–58 (2009)
5. Zhao, Z., Liu, F., et al.: Edge computing platform, application and challenges. Comput. Res. Dev. (2), 327–337 (2018)
6. Shi, W., Sun, H., Cao, J., et al.: Edge computing-an emerging computing model for the Internet of everything era. J. Comput. Res. Dev. **54**(5), 907–924 (2017) (in Chinese)
7. Shi, W., Caojie, Z.Q., et al.: Edge computing: vision and challenges. IEEE Internet Things J. **3**(5), 637–646 (2016)
8. Xie, C., Huang, T., et al.: Theory and Practices of Edge Computing. The People's Postal Service Press, Beijing (2019)
9. Groopman, J., Etlinger, S.: Consumer Perceptions of Privacy in the Internet of Things: What Brands Can Learn from a Concerned Citizenry. Technical report, Altimeter Group (2015)
10. Voigt, P., Von dem Bussche, A.: The EU General data Protection Regulation (GDPR). Springer, Berlin (2017)
11. Zhang, Y., Lu, Y., et al.: Blockchain and federal learning: integration and complementation. Commun. China Comput. Feder. (2), 17–22 (2020)
12. Xu, K., Xu, S., et al.: Infrastructure of blockchain-based decentralized credible internet. Commun. China Comput. Feder. (2), 29–34 (2020)
13. Edge Computing Consortium and Alliance of Industrial Internet. Edge Computing Reference Architecture 3.0 (2018)
14. Edge Computing Task Group. Introduction to Edge Computing in IoT. An Industrial Internet Consortium White Paper , IIC:WHT: IN24:V1.0: PB: 20180618
15. Zhang, J.: Edge Computing Methods and Engineering Practices. Publishing House of Electronics Industry, Beijing (2019)
16. Edge Computing Consortium and Engineering Alliance. Edge Computing and Cloud Computing White Paper (2018) [EB/OL]. Accessed 17 Dec 2018. https://download.csdn.net/download/jameswhite2417/10855185
17. J. Nanjing Univ. Inf. Sci. Technol. (Nat. Sci. Ed.) **11**(5), 596–600 (2019)
18. Shanghai Vonechain Information Technology Co., Ltd., IoT Edge Computing Solutions Based on Blockchain Technology. Automation Panorama: Edge Computing 2018 Special Issue, pp. 56–58 (2018)
19. Hype Cycle for Clound Computing (2017) (G00313602)
20. Zeng, D., Luo, J., et al.: The design of edge computing model based intelligent video monitoring system. Intell. Comput. Appl. (6), 254–261, 265 (2019)
21. Fang, J., Lei, K.: A summary of blockchain technology for edge artificial intelligence computing. J. Appl. Sci. **38**(1) (2020)
22. Wang, T.J.: The research and application of China's Intelligent high-speed railway system architecture. J. China Rail. Soc. (11), 17–22 (2019)

Study on the Special Vocalization Characteristics of Mongolian Long Tone

Menghuan Wang, Shuzhen Ma, and Axu Hu(✉)

Key Laboratory of China's Ethnic Languages and Information Technology of Ministry of Education, Northwest Minzu Univereity, Lanzhou 730030, Gansu, China
1415714973@qq.com

Abstract. This article analyzes two types of special vocalization phenomena in Mongolian long tone through acoustic and voice analysis, aiming to explore the relationship between the singing methods, acoustic characteristics and physiological mechanism of trill and falsetto in Mongolian long tone. This article selects the two most representative folk songs in Mongolian long tone, "Rich and Wide Alax" and "The Petite Chestnut Horse". Through analysis of these two folk songs, it is found that the three singing formants in the trill section are concentrated at 1000 Hz, 3000 Hz and 5000 Hz. There are four singing formants in the falsetto segment, which are concentrated near 1000 Hz, 3000 Hz, 4000 Hz and 5000 Hz.

Keywords: Mongolian long tone · Acoustic · Physiological · Sing formants

1 Introduction

Characters with few words and long tune, high and loud voice, and peaceful rhythm are typical characteristics of Mongolian long tone. It is Suitable for narrative and lyricism. Mongolian long tone are usually two sentences, describing the content of grasslands, horses, camels, cattle and sheep, blue sky, white clouds, rivers, lakes, etc. The Mongolian long tune tells the Mongolian people's perception of history, culture, humanities, customs, morals, philosophy, and art with their distinctive nomadic cultural characteristics and unique singing styles, so they are called "steppe music living fossils" [1]. There are a lot of vocal skills when singing long tone. These skills are not only found when the long tone are produced, singing was absorbed and developed by generations of long tone inheritors, and in the long-term practice, some of their singing skills were added, finally forming the characteristics of the current long tone folk songs. Therefore, the acoustic research on the beautiful timbre of Mongolian long-tune folk songs is the basic work of the intangible cultural heritage protection technology.

M. Tavana et al. (Eds.): IISA 2020, AISC 1304, pp. 333–339, 2021.
https://doi.org/10.1007/978-3-030-63784-2_42

2 Research on the Characteristics of Trill

2.1 Study on the Acoustic Characteristics of Trill

The trill in the Mongolian long tone is not deliberately made for the purpose of trill, but it is made automatically and can be used freely when the singer needs it. Of course, this requires the singer to master the correct vocalization method, and the trill will come out naturally. In order to have a scientific and objective understanding of the trill from the acoustic and vocal level, this part selects the two long tone of "*The Rich and Wide Alxa*" and "*The Petite Chestnut Horse*" for analysis.

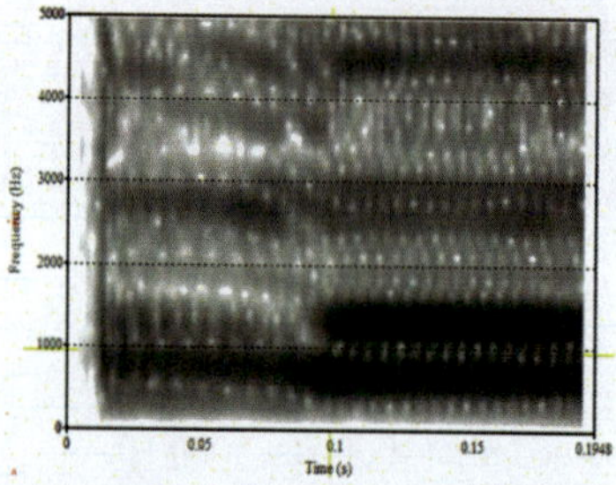

Fig. 1. Spectrogram of the trill section in "*Rich and Wide Alax*"

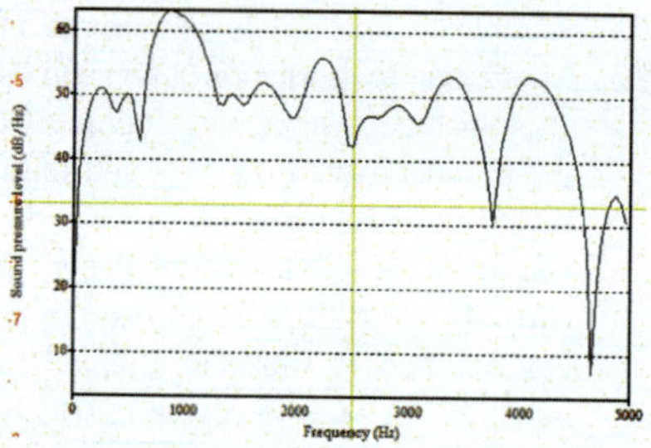

Fig. 2. Envelope spectrum of trill section in "*Rich and Wide Alax*"

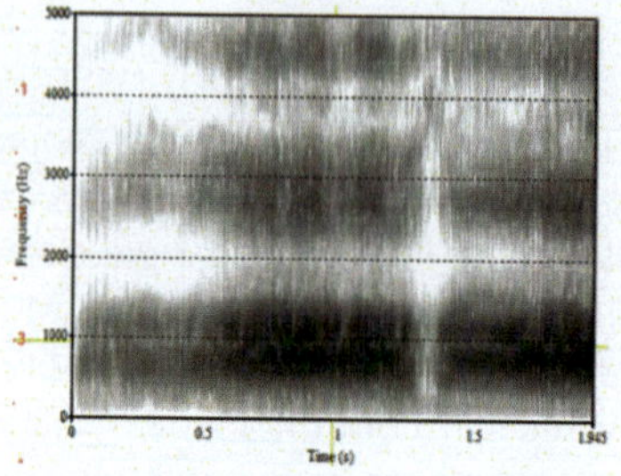

Fig. 3. Spectrogram of the trill section in "The Petite Chestnut Horse"

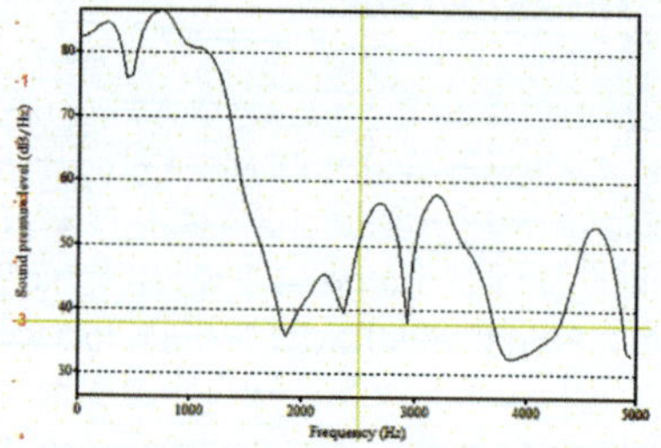

Fig. 4. Envelope spectrum of the trill section in "The Petite Chestnut Horse"

Figure 1 is a tremolo section at 16.221 s–6.4 s in "*Rich and Wide Alax*", and Fig. 2 is a spectral envelope at the beginning of the first syllable of "*Rich and Wide Alax*". Observation of the envelope spectrum of the first syllable shows that the energy fluctuates very uniformly. It is precisely because of the uniform fluctuation of the trill in the singing voice that the intensity of the formant of the voice is strengthened. Figure 3 is a trill section at 6.255 s–8.179 s of "*The Petite Chestnut Horse*". Figure 4 is a spectral envelope at the syllable of "*The Petite Chestnut Horse*" at 0.4 s. Observing the envelope spectrum of the syllable at 0.4 s, we can see that the energy drops sharply, indicating that the energy in the syllable of the tremolo segment is not always uniformly fluctuating. It is precisely because of this trill that Mongolian long tone are more pleasant. Observing trill segments spectrogram of the two long tone shows that the two have the same pattern of formant distribution, that is, two cloudy black bars are formed near 1000 Hz, and a light black bar is formed near 3000 Hz and 5000 Hz, respectively. Here it is reminiscent of singing formants. Singing formants are often accompanied by thrill. The appearance of thrill helps strengthen the singing formants.

2.2 Voice Characteristics of Tremolo

After analysing the thrill in the two long notes of "*Rich and Wide Alax*" and"*The Petite Chestnut Horse*" from the acoustic level, in order to understand more thoroughly, we analyze it from the voice side. Figure 5 and Fig. 6 are the relationship map between the fundamental frequency, the open quotient and the speed quotient of the thrill section of "*Rich and Wide Alax*" and "*The Petite Chestnut Horse*", respectively.

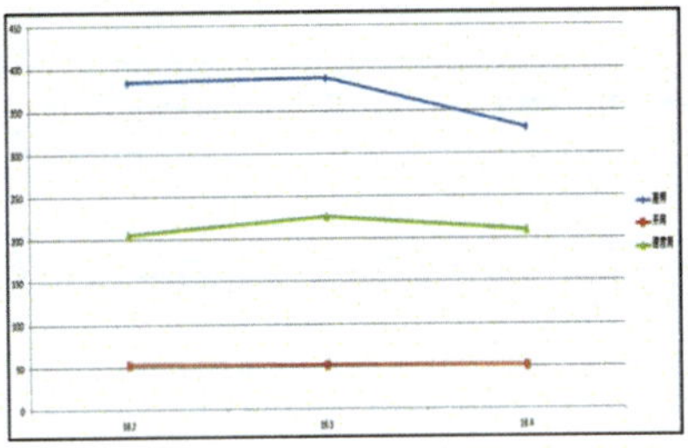

Fig. 5. The relationship map of the trill section of "*Rich and Wide Alax*"

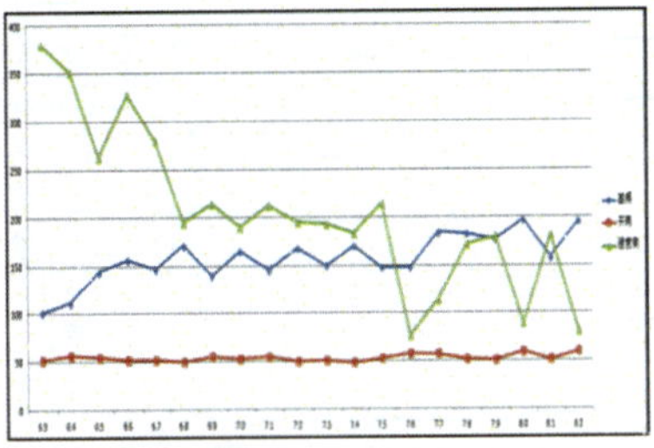

Fig. 6. The relationship map of the trill section of "*The Petite Chestnut Horse*"

Observing Fig. 5, we can see that there is basically no change in the open quotient during the entire trill process, and it basically shows a straight line. The open quotient value has been maintained at 50%. This is a very interesting and very special phenomenon. During a trill segment, the open phase of the vocal cords is equal to the close phase. The fundamental frequency and the speed quotient show a consistent change trend, and the two are positively correlated. Both are rising in the first half and falling in the second half. The fluctuation range of the speed quotient is 200% to 250%, and the speed quotient represents the high frequency energy distribution of the voice. The value of the speed quotient reaches the maximum at 16.3 s, indicating that the high-frequency energy of the voice rises sharply at this moment. The fundamental frequency fluctuates in the range of 300 Hz to 400 Hz. The normal fundamental frequency range for men when speaking is 110 Hz to 147 Hz. The fundamental frequency of the thrill section is much higher than this fundamental frequency range. However, the fundamental frequency and open quotient did not show any relevant characteristics. Similar to Fig. 5, the open quotient in Fig. 6 has basically no change throughout the trill process, showing a straight line, and the open quotient value has been maintained at 50%. From this, it can be determined that during the trill segment singing process the open phase of the vocal cords is equal to the close phase. The fundamental frequency and speed quotient of this long tone are completely different from the previous one, that is, the two are negatively correlated. Looking at the overall trend, the characteristic is that when the speed quotient decreases greatly, the open quotient shows an upward trend. The fluctuation range of the speed quotient is 50% to 400%, and the value reaches the maximum at 6.3 s, indicating that the high-frequency energy of the human voice at the moment rises sharply. The fundamental frequency fluctuates in the range of 100 Hz–200 Hz. The normal fundamental frequency range for men when speaking is 110 Hz–147 Hz. The fundamental frequency value of the thrill section is only slightly higher than this range compared to the previous. Moreover, the fundamental frequency and open quotient did not show any relevant characteristics.

3 Research on the Characteristics of Falsetto Voice

3.1 Study on the Acoustic Characteristics of Falsetto Voice

In Mongolian long tone singing, pure falsetto skills are almost uncommon. In most of the sung songs, true and falsetto voices are mixed. After mixing, the vocal range is wider and the sound is bright and smooth. True and falsetto voice make long tone full of charm in tone color, dynamic changes and so on. The falsetto voice in the two tracks of "*Rich and Wide Alax*" and "*The Petite Chestnut Horse*" were selected for analysis.

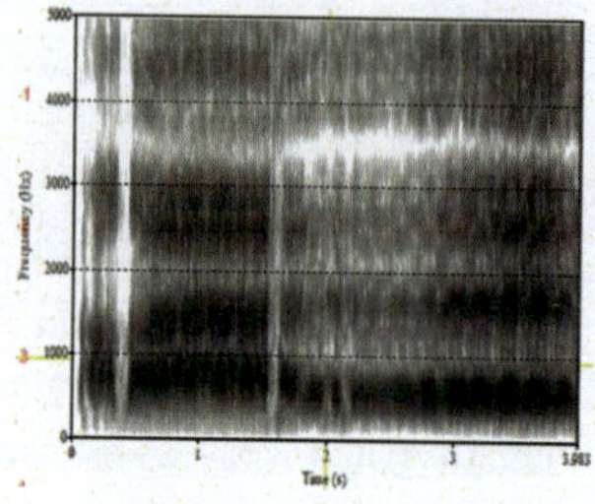

Fig. 7. Spectrogram of the falsetto section in "*Rich and Wide Alax*"

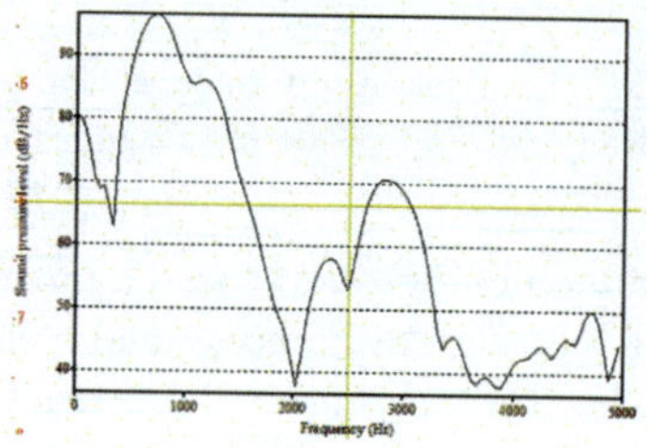

Fig. 8. Envelope spectrum of falsettol section in "*Rich and Wide Alax*"

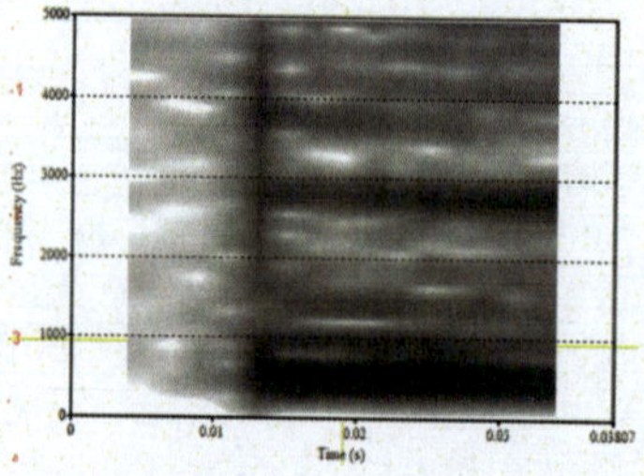

Fig. 9. Spectrogram of the falsetto section in "*The Petite Chestnut Horse*"

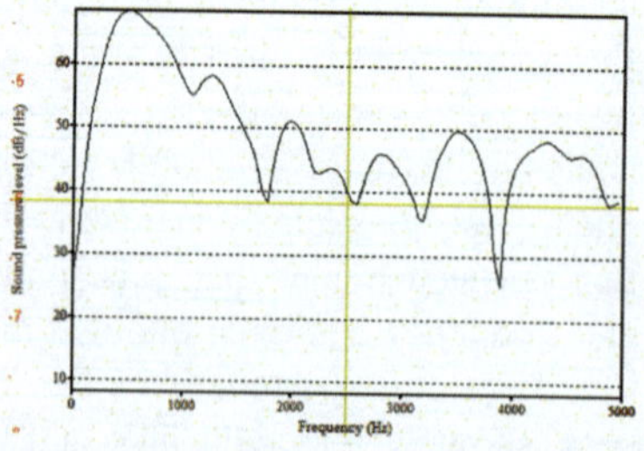

Fig. 10. Envelope spectrum of falsettol section in "*The Petite Chestnut Horse*"

Figure 7 is a falsetto voice at 13.933 s–17.904 s of "Rich and Wide Alax", and Fig. 8 is a spectral envelope at 0.1 s at the beginning of the falsetto voice of "*Rich and Wide Alax*". It can be seen from the spectrogram and the spectral envelope that the intensity of the falsetto voice is not high. This is the characteristic of falsetto voice. The pitch is increased but the intensity is not large. When singing long tone, the singer takes appropriate measures to make the vocal cords shrink as much as possible and emit a higher pitch. When the pitch rises to a certain limit, if the vocal cord contraction is changed so that the breath can only vibrate at the edge of the vocal cord, the pitch will continue to rise. At this time, the singing voice is high and sharp, which is generally called "falsetto voice". The sound obtained when the two vocal cords vibrate completely, even if they sing very high, will not be higher than the falsetto voice. This kind of sound is called "truth voice".

Figure 9 is a falsetto voice at 12.061 s–12.083 s of "The Petite Chestnut Horse", and Fig. 10 is a spectral envelope of the syllable at 0.02 s of the falsetto voice section of "*The Petite Chestnut Horse*". Observe the spectrogram and envelope spectrum can be seen that the pitch is higher than other segments but the intensity is not large. The "falsetto voice" squeezed in the throat by squeezing the neck sounded sharp, the pitch was indeed high, but the intensity was not high. Observation of the two long tones in the falsetto voice section of the spectrogram, found that the distribution of the formant peaks of the two are roughly the same, the only difference is that the "*The Petite Chestnut Horse*" falsetto voice section does not form a cloudy black bars near 2000 Hz.

3.2 Study on the Voice Characteristics of Falsetto Voice

After analysing the **falsetto voice** in the two long notes of "Rich and Wide Alax" and "The Petite Chestnut Horse" from the acoustic level, in order to understand more thoroughly, we analyze it from the voice side. Figure 11 and Fig. 12 are the relationship map between the fundamental frequency, the open quotient and the speed quotient of the **falsetto voice** section of "Rich and Wide Alax" at 13.933 s–17.904 s and "The Petite Chestnut Horse" in the period of 12.061 s–12.083 s, respectively.

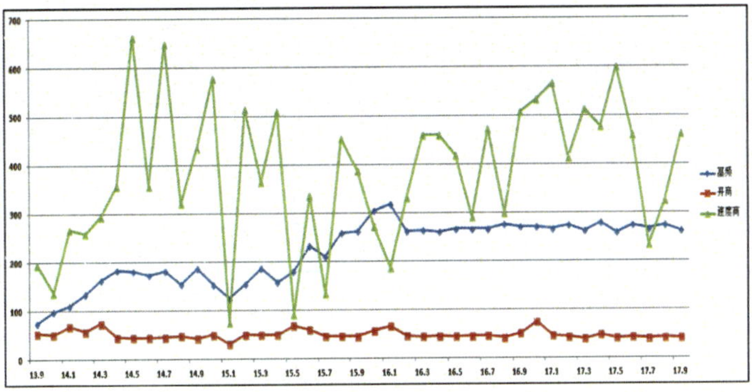

Fig. 11. The relationship map between the vocal parameters of the **falsetto** section of "Rich and Wide Alax"

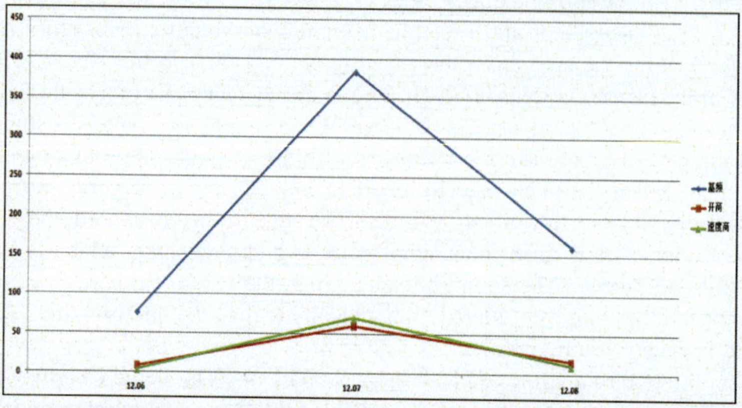

Fig. 12. The relationship map between the vocal parameters of the **falsetto** section of "*The Petite Chestnut Horse*"

Observing Fig. 12, we can see that in the whole process of **falsetto voice**, the open quotient basically shows a straight line, and the value of the open quotient has been kept at 50%. That is, when this **falsetto** segment is sung, the open phase of the vocal cords is equal to the close phase. The fundamental frequency and the speed quotient generally show opposite trends, and the two are negatively correlated, the fluctuation range of the speed quotient is 50% to 700%. The speed quotient represents the high-frequency energy distribution of the voice, and the value of the speed quotient reaches the maximum at 14.5 s, Indicating that the high-frequency energy has risen sharply at this moment. The fundamental frequency fluctuation range is 50 Hz–300 Hz, and this fluctuation range is much larger than male speech. However, the fundamental frequency and open quotient did not show any relevant characteristics.

Observing Fig. 12, we can see that in the entire process of falsetto, the voice characteristics in "The Petite Chestnut Horse" show the characteristics of positive correlation between the fundamental frequency, open quotient and speed quotient. The speed quotient fluctuates in the range of 0% to 50%. The value of the speed quotient reaches the maximum at 12.07 s, indicating that the high-frequency energy rises sharply at the moment, minimize the velocity quotient Corresponding to the time point of the value, you will find that the energy is very small at this time, and the velocity quotient and the sound intensity are positively correlated. Correlating the time point of the minimum value of the speed quotient with the spectrogram, it is found that the energy is very small at this time, and the speed quotient and intensity are positively correlated. The fluctuation range of the fundamental frequency is 50 Hz–400 Hz, and the fluctuation range is far higher than the fluctuation range of the fundamental frequency of speech. **Falsetto voice** is one of the singing skills. It can be seen that the minimum value of the speed quotient of the **falsetto** section is generally lower than the minimum value of the speed quotient of the thrill section, because the vocal cords leave faster than the contact. The **falsetto** voice segment is the highest note in the singing range, which can only be produced by special vocal training. This special vocalization method has a wider vocal range when performed by men. Natural **falsetto** voice, empty and pale. The trained **falsetto** voice has excellent expressive power in the singing process, this is what people call singing **falsetto**.

4 Conclusion

The following conclusions can be drawn from the acoustic level and vocal signal analysis of the long tone **falsetto** and trill segments: there are four singing formants in the **falsetto** segment, which are concentrated around 1000 Hz, 3000 Hz, 4000 Hz and 5000 Hz, respectively. The open quotient of the **falsetto** voice segment is about 50%, and there is no obvious change in the whole stage. At this time, the open phase of the vocal cord is equal to the close phase. The fundamental frequency is generally higher than speech, and there is no correlation between the fundamental frequency and the open quotient. And the relationship between the fundamental frequency and speed quotient of the two long tones shows diametrically opposite relationship and there is no correlation. There are three singing formants in the thrill section, which are concentrated near 1000 Hz, 3000 Hz and 5000 Hz; the open quotientof the thrill section is about 50%, and there is no obvious change in the entire phase. Other aspects are consistent with the false voice segment.

Acknowledgments. This work was fifinancially supported by Central University Fundamental Research Fund (No.31920200028) and Northwest Minzu Univereity 2020 Graduate Research Innovation Project (No.Yxm2020102).

References

1. Hu, H., Chen, J., Zheng, Y.: Mongolian Prosodic Feature Acoustic Parameter Database, 1st edn. Inner Mongolia University Press, Beijing (2001)
2. Wang, B., Yang, Y., Lv, S.: Standard Chinese sentences stressed syllable pitchstudy. Acta Acustica (3) (2002)
3. Ertai, Q.: Mongolian Grammar, 1st edn. Inner Mongolia people's publishing house, Beijing (1991)
4. Halazhen.: Ordos Wedding Big Songs, 1st edn. Inner Mongolia people's publishing house, Beijing (2011)
5. Yu, S.: Different types of singer resonance peak features singing tone and the mechanismof formation research. Vocal Music Study 4 (2010)

Experimental Study on Citation Tone of Dingxi Dialect in Gansu Province

Menghuan Wang, Shuzhen Ma, and Axu Hu(✉)

Key Laboratory of China's Ethnic Languages and Information Technology of Ministry of Education, Northwest Minzu Univereity, Lanzhou 730030, Gansu, China
1415714973@qq.com

Abstract. Based on the recorded corpus of field investigation, this paper uses experimental phonetics to conduct an experimental study on the citation tone of Dingxi dialect, and uses the acoustic analysis software Praat to extract the fundamental frequency parameters of the experimental materials, and analyzes frequency mode of the citation tone of Dingxi dialect. The experimental results show that Dingxi dialect has three tones of Pingsheng, Shangsheng, and Qusheng. Among them, the Pingsheng does not distinguish between yin and yang, and it is a rising tone with a tone pitch of 24; the Shangsheng is a falling tone with a tone pitch of 51; the Qushneg is a flat tone with a tone pitch of 33. Comparing the existing research results on Dingxi citation tone, it is concluded that the citation tone pattern obtained by the phonetic experiment is consistent with the conclusions obtained by traditional methods, but there are slight discrepancies in the description of tone values.

Keywords: Dingxi dialect · Citation tone · Fundamental frequency · Fifth degree

1 Introduction

Dingxi City is located in the middle of Gansu Province. It is connected to Lanzhou and Baiyin in the north, Pingliang and Tianshui in the east, Longnan in the south, Gannan and Linxia in the west. The dialects in Dingxi area belong to the northern dialects, and belong to the Longzhong of Zhongyuan Mandarin.

The research achievements on Dingxi dialect are mainly concentrated in Han Li's "Study Status of Dingxi Dialect", Sun Lixin's "Phonetic Features of Dingxi Dialect", Zhang Wenxuan's "The Phonetic Features of Dingxi Dialect of Three Tone Dialects" and Zhu Fulin's master thesis "Study on the Phonetics of Southern Dialect in Longzhong", although several papers generally agree that Dingxi dialect is a three-tone language, there is controversy in the description of tone value. For now, researchers mostly use the methods of field investigation and listening to phonetics to describe and summarize the tones of Dingxi dialect. Although this method can record the tone category well, due to the difference in personal listening ability and the difference in pronunciation, it will cause the problem that the tone value cannot be accurately recorded. This paper uses experimental phonetic methods to extract the objective

M. Tavana et al. (Eds.): IISA 2020, AISC 1304, pp. 340–345, 2021.
https://doi.org/10.1007/978-3-030-63784-2_43

physical parameters of each tone, which can more accurately describe the actual situation of the tone of Dingxi dialect, so as to solve the problem of the inconsistency of the phonetic value recording of the previous experts to fill the gap in the mouth and ears.

2 Experimental Methods

2.1 Pronunciation Partner

This study selected a 33-year-old male speak who has been living Anding District, Dingxi City. His family members all use Dingxi dialect. He is a native of Dingxi city. Above all, the pronunciation is normal and clear.

2.2 Recording Conditions

This experiment is recorded in a professional voice recording studio. The recording uses Adobe Audition 3.0 recording software. The sampling rate is set to 44100 s, monophonic, and the resolution is 16 bits. The recording equipment includes an eight-channel mixer, an external sound card and a high-sensitivity microphone.

2.3 Data Collection and Analysis

Data collection was performed with praat software to extract the fundamental frequency parameters of the selected experimental materials. The script program written in Matlab software normalizes the fundamental frequency parameters, and finally converts the normalized fundamental frequency into a five-degree value, using the T value formula proposed by Shi Feng: $\mathrm{T} = [(\lg x - \lg \min)/(\lg \max - \lg \min)] * 5$.

2.4 Experimental Materials

The experimental materials were selected from the "Chinese Language Resources Survey Manual-Chinese Dialects" published by the National Language and Characters Working Committee. A total of 66 tone words were selected to make an experimental pronunciation table (Table 1).

3 Experimental Results

3.1 Fundamental Frequency

Extract the fundamental ftequency of all selected samples, and normalize the processing to obtain the average value of fundamental frequency of the four tones. The data of fundamental frequency such as the maximum value, minimum value, and the average value of initial point, middle point an final point are plotted into a table for comparative analysis. The fundamental frequency parameter tables of Dingxi dialect are shown in Table 2, Table 3, Table 4 and Table 5.

Table 1. Pronunciation list

	Sample			
Tone	Ping	Shang	Qu	Ru
voiceless	[tuŋ^{55}][kai^{55}] [təŋ^{55}][fəŋ^{55}]	[tuŋ^{214}][ku^{214}] [kuei^{214}] [tɕiou^{214}]	[tuŋ^{51}][kuai^{51}] [pan^{51}][sɿ^{51}]	[ku^{214}][pai^{214}][ta^{55}] [tɕiɛ^{35}][tɕi^{35}]
Secondary voiceless	[tʰuŋ^{55}][kʰai^{55}] [tʰian^{55}] [tʂʰuən^{55}]	[tʰuŋ^{214}][kʰu^{214}] [tʰau^{214}] [tsʰau^{214}]	[tʰuŋ^{51}] [kʰuai^{51}] [tsʰuən^{51}] [tɕʰy^{51}]	[kʰu^{55}][pʰai^{55}] [tʰa^{214}] [tɕʰiɛ^{51}][kʰə^{51}]
Secondary voiced	[mən^{35}][luŋ^{35}] [niou^{35}][iou^{35}]	[mai^{214}][lau^{214}] [wu^{214}][iou^{214}]	[mai^{51}][lu^{51}] [yiŋ^{51}][luan^{51}]	[liou^{51}][məi^{51}] [iɛ^{51}][yɛ^{51}]
Voiced	[tʰuŋ^{35}][pʰi^{35}] [tʰaŋ^{35}][xuŋ^{35}]	[tuŋ^{51}][tsuei^{51}] [tɕin^{51}][xou^{51}]	[tuŋ^{51}][ti^{51}] [fan^{51}][ʂu^{51}]	[tu^{35}][pai^{35}] [xə^{35}][fa^{35}]

Table 2. The fundamental frequency table of Ping(Hz)

Tone				
Frequency	Ping			
	Voiceless	Secondary voiceless	Secondary voiced	Voiced
Initial point value	111.3	107.8	113.1	108.9
Middle point value	112.1	112.6	113.7	118.8
Final point value	131.6	128.3	136.3	133.9
Maximum value	137	132.2	142.4	143.6
Minimum value	105.9	101.9	108.1	103.6

According to the evolution of tones, Pingsheng can be divided into two categories: Yinping and Yangping. According to the difference between voiced and voiceless of the initial consonant, the voiceless are classified into Yinping, and the voiced are classified into Yangping. According to the data in the table, the maximum value of the fundamental frequency of Yinping is 137 Hz, the minimum value is 101.9 Hz, the vocal range is between 101.9 Hz–137 Hz, and the width of vocal range is 35.1 Hz; the maximum value of the fundamental frequency of Yangping is 143.6 Hz, the minimum value is 103.6Hz, the vocal range is between 103.6 Hz–143.6 Hz, and the width of vocal range is 40Hz. From the above data, it can be seen that voiceless, secondary voiceless of Pingsheng are rising tone, and the secondary voiced and voiced are also rising tone. Therefore, it can be explained that in the Dingxi dialect, the Pingsheng does not distinguish between yin and yang.

It can be seen from the table above that the maximum value of the fundamental frequency of the Shangsheng is 179.1 Hz, the minimum value is 75.8 Hz, the vocal range is between 75.8 Hz and 179.1 Hz, the width of the range is 103.3 Hz, and the span of the vocal range is relatively large. Among them, the fundamental frequency

Table 3. The fundamental frequency table of Shang(Hz)

Tone				
Frequency	Shang			
	Voiceless	Secondary voiceless	Secondary voiced	Voiced
Initial point value	165.1	162.4	140.5	129.5
Middle point value	137.2	127.4	135.0	127.9
Final point value	86.6	88.7	91.9	127.4
Maximum value	179.1	168.2	153.9	137.4
Minimum value	81.1	82.5	75.8	120.1

values of voiceless, secondary voiceless, and secondary voiced are generally consistent from the initial point to the final point, and all show a downward trend. The difference in change amplitude among them is also not large, so it can be seen that voiceless, secondary voiceless, and secondary voiced of Shangsheng in Dingxi dialect are trising tone. However, the change of the fundamental frequency of voiced of Shangsheng is not obvious, and the change between the maximum and minimum values of the fundamental frequency is also not obvious.

Table 4 shows that the maximum value of the fundamental frequency of Qusheng is 139.5 Hz, the minimum value is 111.8 Hz, the range is between 111.8 Hz and 139.5 Hz, and the width of the range is 27.7 Hz. In the Dingxi dialect, the change from the initial point value to the finial value of the fundamental frequency of voiceless, secondary voiceless, secondary voiced, and voiced is consistent. Therefore, Qushengin is a level tone in Dingxi dialect. Comparing the data of voiced of Shangsheng in Table 3, it can be seen that the Qusheng change pattern is consistent with the voiced of Shangsheng, both of which are levwl tone.

Table 4. The fundamental frequency table of Qu(Hz)

Tone				
Frequency	Qu			
	Voiceless	Secondary voiceless	Secondary voiced	Voiced
Initial point value	127.0	128.4	121.1	126.9
Middle point value	129.9	127.1	125.7	124.7
Final point value	129.7	126.4	126.9	125.8
Maximum value	139.5	137.1	130.2	130.0
Minimum value	111.8	113.1	118.3	117.6

Table 5. The fundamental frequency table of Ru(Hz)

Tone				
Frequency	Qu			
	Voiceless	Secondary voiceless	Secondary voiced	Voiced
Initial point value	107.4	104.0	110.9	104.5
Middle point value	109.2	111.9	108.7	112.9
Final point value	124.5	121.2	127.7	123.5
Maximum value	134.0	122.8	137.4	129.0
Minimum value	97.7	98.0	105.0	101.2

Analysis of the data in the above table shows that the maximum fundamental frequency of Rusheng is 137.4 Hz, the minimum is 97.7 Hz, the vocal range is between 97.7 Hz to 137.4 Hz, and the width of range is 40 Hz. The fundamental frequency changes of voiceless, secondary voiceless, secondary voiced, and voiced has the same trend, so it can be seen that the Rusheng in Dingxi dialect are not divided between yin and yang, Rusheng is a rising tone and all are classified as level tone.

Based on the above analysis, we can see that the change trend of the Pingsheng characters of voiceless, secondary voiceless, secondary voiced, and voiced are the same, all of which are in rising tone. Therefore, it is concluded that Pingsheng in Dingxi dialect has no distinction between Yin and Yang; The change of the fundamental frequency of voiceless, secondary voiceless, secondary voiced of Shangsheng is roughly the same, and they all show a downward trend. It can be seen that in the Dingxi dialect, the Shangsheng is falling tone; The fundamental frequency trend of the voiced of Shangsheng is consistent with the voiceless, secondary voiceless, secondary voiced, and voiced of Rushseng. Therefore, in Dingxi dialect, the voiced of Shangsheng and Qusheng are level tone. Finally, through the fundamental frequency data of Rusheng, it can be seen that the Rusheng in Dingxi dialect has no yin and yang. All sent into level tong, for rising tone.

3.2 Fifth Degree Value

The average data value of the fundamental frequency extracted from the experiment is brought into the T value calculation formula for calculation, and the final result is plotted in Excel. The T-value curve graphs of the three tones of Pingsheng, Shangsheng, Qusheng in Dingxi dialect are obtained.

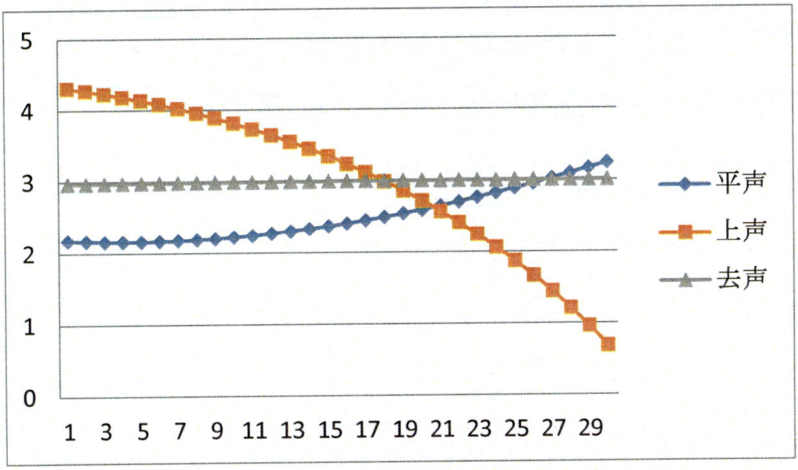

It can be seen from the figure above that the Pingsheng in Dingxi dialect is a rising tone with fifth degree value of 24, the Shangsheng is a falling tone with a fifth degree value of 51, the Qusheng is a level tone with the tone value of 33.

4 Conclusion

By analyzing the acoustic characteristics of the tones of Dingxi dialect, this paper compares the fundamental frequency parameters and the vocal range, concludes that Dingxi dialect has three tones: Pingsheng (24), Shangsheng (51), and Qu (33). This paper makes a preliminary study on the ciation tone with the method of experimental phonetic, so as to provide some reference for further study of Dingxi dialect.

Acknowledgments. This work was fifinancially supported by Central University Fundamental Research Fund (No. 31920200028) and Northwest Minzu Univereity 2020 Graduate Research Innovation Project (No. Yxm2020102).

References

1. Kong, J.: Basic Course of Experimental Phonetics. Peking University Press, Beijing (2013)
2. Wu, Z., Maocan, L.: Summary of Experimental Phonetics. Higher Education Press, Beijing (1989)
3. Zhu, X.: Shanghai Tone Experiments. Shanghai Education Press, Shanghai (2004)
4. Shi, F.: Tone analysis of Tianjin dialect double characters. Lang. Res. (01), 77–90 (1986)
5. Yang, Z.: Experimental Study on the Pronunciation of Liangzhou Mandarin. Yunnan Normal University, Kunming (2013)

The Study on the Role of Phonetic and Shape Information on Mongolian Students' Recognition of a Chinese Word

Axu Hu(✉), Menghuan Wang, and Gegen Tana

Key Laboratory of China's Ethnic Languages and Information Technology of Ministry of Education, Northwest Minzu Univereity, Lanzhou 730030, Gansu, China
1415714973@qq.com

Abstract. In this paper, we selected Mongolian college students as subjects to explore the activation of four kinds of competitive words include the pronunciation, glyph, meaning and irrelevant words and the role of voice and shape information in word recognition when Mongolian students recognize a Chinese character (second language), with the help of eye movement technology, under the different starting methods of psychological experiments. Mongolian college students are influenced by the characteristics of the characters inherent in the mother tongue when they are recognize the second-language words. The speech plays a mediating role in the Mongolian bilinguals' Chinese processing, which indicated when the two language systems mastered by the bilinguals are different, the second language processing is often influenced by the processing method of the mother tongue, which causes the phenomenon of language transfer. The speech will be used as the medium, the sound and meaning will be closely combined when the ideograph are processed by the subject whose native language is the alphabetic writing, so the result supports the theory of speech mediation.

Keywords: Chinese word · Phonetic information · Shape information · Recognition

1 Introduction

In recent years, people have paid more and more attention to the role of phonetic and shape information in word processing. Three influential theories have emerged, namely the straight through theory, the phonological mediation theory, and the dual–route theory. The straight through theory holds that semantic acquisition is the result of graphic processing first, and phonological activation comes after graphic. The theory of phonological mediation hold that semantic acquisition is mediated by the speech. The dual–route theory holds that the above two theories have their rationality. Many studies using English as material show that phonological plays a very important role in the process of acquiring semantics. But using Chinese characters as experimental material, it is concluded that semantic acquisition depends on the glyph information. Mongolian students are bilingual, they mastered two different writing systems, the mother tongue

M. Tavana et al. (Eds.): IISA 2020, AISC 1304, pp. 346–353, 2021.
https://doi.org/10.1007/978-3-030-63784-2_44

is Mongolian, which belongs to alphabetic writing, and the second language is Chinese that belongs to ideograph. In this paper, with the help of eye movement technology, under the different starting methods of psychological experiments, exploring the activation of four kinds of competitive words such as the pronunciation, glyph, meaning and irrelevant words and the role of voice and graphic information in word processing, when the Mongolian students recognize a Chinese character (second language). Focusing on Mongolian students recognize a Chinese word to discuss whether the accessibility semantics is obtained through the voice path or through glyph. So which theory is suitable for Mongolian college students to recognize a Chinese word?

2 Experimental Method

2.1 Participants

Subjects are thirty Mongolian college students (15 males and 15 females), Mongolian is their mother tongue. They have been learning Chinese systematically since the third grade. They participated in Chinese Putonghua proficiency test and all reached the second level. They had no color blindness and weak color, and there was no obstacle in listening and reading. The uncorrected visual acuity and corrected visual acuity were above 1.0.

2.2 Experimental Design and Equipment

A single-word recognition task were used in this experiment, two factors were considered. The first factor is the starting method, there are two types, which are listening and watching. The other factor is the type of the competitive word, which is similar to the target word in phonetic (homophone), similar in shape (orthographically similar) and similar in meaning (synonyms), and completely different from the target word. Real-time monitoring of the experiment was performed using SMI's Hi-Speed eye tracker. Use iView X software and Experiment Center software to record and extract various eye movement indicators during the experiment.

2.3 Experimental Materials

We select 30 words that from the "Xinhua Dictionary" published by the Commercial Press as the target words, and then select 120 competition words to match target words. Respectively, homophone (same pronunciation as the target word), orthographically similar (same shape as the target word), synonyms (same meaning as the target word) and irrelevant words (totally dissimilar to the target word). The experimental materials are divided into listening to a Chinese characters (15) and watching to a Chinese characters (15) as two start method. The four competitive words to match two starting methods are distributed in the 5 * 5 matrix lattice. It is made into a bmp format of 800×600 pixels by image processing software Photoshop, and then presented on the display screen.

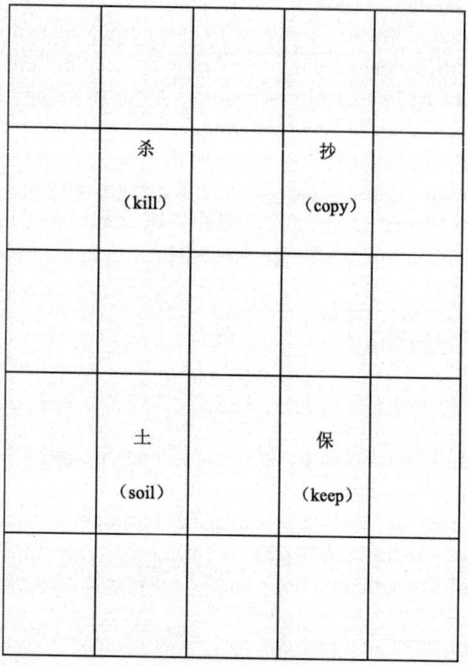

Fig. 1. Schematic diagram of the experimental material (if the target word heard is: 沙. The corresponding visual presentation words: homophone "杀", orthographically similar "抄", synonyms "土", irrelevant word "保". During the formal test, the border in the above picture does not appear.)

2.4 Experimental Procedure

This experiment was carried out in the eye movement laboratory and the sound insulation was very good. In this experiment, an eye movement indicator is controlled and recorded by a computer, and the stimulus is presented in the center of the display. After the subjects entered the laboratory, they sat about 100 cm away from the display and then to adjust their comfortable sitting posture. The subjects placed the chin on the bracket of the eye tracker, the forehead stuck to the upper edge of the eye tracker, and looked at the nine correction points on the display. When the correction effect is greater than 1.0, it was recalibrated. When the correction effect is less than 1.0, you can continue to do experiment. Divide the experiment into two groups, the first was started by listen a Chinese word. The guideline before the experiment is: after the experiment starts, you will see "+" on the screen. This is a Chinese word you will hear, and the competition words related to target stimuli will be presented on computer screen at the same time. You can quickly Scan these words and press the end key to enter the next set of experiments. This experiment has a total of fifteen sequences to be completed. The second group is started by watching a Chinese word. The guideline before the experiment is: start of experiment, you can see a "+" on screen, next a Chinese word will appears, and then the competition words related to the target stimuli will be presented on the computer screen at the same time, you can quickly scan these words, and then press the end key to enter the next set of experiments after scanning. There are 15 sequences to be completed in this experiment. Participants will practice before the experiment to understand the experimental task.

3 Conclusion and Analysis

3.1 Analysis of Experimental Parameters

Use iView X software and Experiment Center software to record and extract various eye movement indicators during the experiment.

(1) Division of areas of interest
Divide the four competing words into four equal areas of interest, they are located in the seventh, ninth, seventh and nineteenth grids in five*five grid (Fig. 1)
(2) Eye movement index selection and description
Select three eye movement index: First Fixation Duration, Total time in zone, and Fixation Count for analysis. The first indicator refers to the duration of the first fixation on the interest zone. The second is the sum of all fixation times in current interest area. The last is the number of all gaze points in interest area.
(3) Raw data processing.
Data processing standards: delete the original data whose gaze time is less than 80 ms and greater than 1000 ms. Perform the final statistical analysis on the deleted data.

3.2 Data Analysis of Eye Movement Indicators Under Auditory Activation

Table 1. Standard deviation and mean of eye movement indicators under auditory activation

Start method	First Fixation Duration (ms)		Total time in zone (ms)		Fixation Count (times)	
Listening	M	SD	M	SD	M	SD
Homophone	323.640	41.881	339.151	34.955	1.433	0.151
Orthographically similar	517.273	29.588	611.337	58.536	2.511	0.072
Synonyms	437.577	37.536	482.444	33.536	2.309	0.052
Irrelevant words	684.526	42.564	924.615	803.266	3.800	0.189

Use spss19.0 to statistics and Analyze the data of Mongolian college students in the first group of experiments (the starting method is to listen to Chinese word). We can see that in terms of the First Fixation Duration indicator, among the four competitive words, homophone (similar to the target word in phonetic), orthographically similar (similar in shape) and synonyms (similar in meaning) have all reached significant level.

(F = 7.421, p = 0.000 < 0.001; F = 5.129, p = 0.001; F = 4.235, P = 0.025 < 0.05.), but the irrelevant words did not reach the significance level. (F < 1, p = 0.658 > 0.05). Analysis of the indicator Total time in zone, as with First Fixation Duration, homophone (similar in phonetic), synonyms (similar in meaning) and orthographically similar (similar in shape) all reached a significance level, (F = 16.17, p = 0.000 < 0.001; F = 15.857, p = 0.000 < 0.001; F = 9.848, p = 0.000 < 0.001), but the unrelated word did not reach the significance level (F < 1, p = 0.916 > 0.005). Analyzing a indicator Fixation Count, we draw a conclusion that except for words that are not related to the target word, they did not reach the significance level (F < 1, p = 0.528 > 0.05), the other three competitive words (homophone, synonyms, graphic similar) all reach a significant level (F = 6.938, p = 0.000 < 0.001; F = 6.737, p = 0.000 < 0.001; F = 2.476, p = 0.035 < 0.05) (Tables 1 and 2).

Table 2. Standard deviation and mean of eye movement indicators under visual activation

Start method	First Fixation Duration (ms)		Total time in zone (ms)		Fixation Count (times)	
Watching	M	SD	M	SD	M	SD
Homophone	374.500	57.751	471.653	40.499	1.950	0.154
Orthographically similar	348.250	59.513	369.033	55.680	1.911	0.150
Synonyms	446.233	60.964	485.262	46.131	2.277	0.147
Irrelevant words	622.850	130.422	682.680	99.349	3.144	0.334

Use spss19.0 to statistics and analyze the data of Mongolian college students in the second group of experiments (the starting method is to watch a Chinese word). We can see that in terms of the First Fixation Duration indicator, homophone (similar in phonetic), orthographically similar (similar in shape) reached the significant level, (F = 4.719, p = 0.018 < 0.05; F = 6.266, P = 0.008), and synonyms (similar in meaning) reached a marginally significant level, (F = 3.187, p = 0.05). But irrelevant words was not reach a significant (F < 1, p = 0.615 > 0.05). Analyzing the indicator Total time in zone, we found that except for the words that are not related to the target word did not reach the significance level (F < 1, p = 0.686 > 0.005), the other three competitive words (homophone, synonyms, graphic similar) all reached Significant level (F1 = 11.758, p = 0.001; F = 10.308, p = 0.045 < 0.05; F2 = 13.727, p = 0.001). Analysis of the Fixation Count, we found that the effect of Fixation Count of homophone and orthographically similar was significant, F = 3.745, p = 0.033 < 0.05; F = 4.416, p = 0.025 < 0.05. And synonyms also reached a significant, F = 3.318, p = 0.045 < 0.05, but The Fixation Count of irrelevant words was not reach a significant level, F = 1.193, p = 0.319 > 0.05.

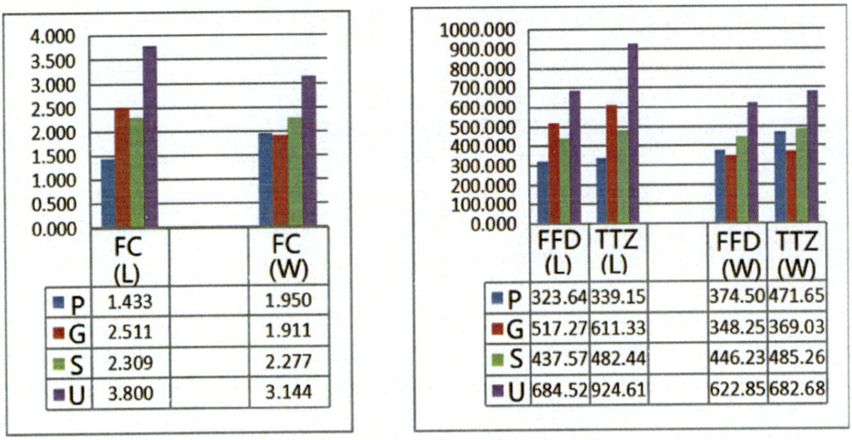

Fig. 2. Comparison of eye movement parameters under two start modes

(In the figure above, P stands for phonological, G for graphic, S for semantic, and U for related word. FFD, TTZ and FC are the abbreviations of First Fixation Duration, Total time in zone, Fixation Count; W means watching and L means listening).

The comparison of three indicators (First Fixation Duration, Total time in zone and Fixation Count) are shown in Fig. 2, when Mongolian college students recognize a Chinese character under different start-up modes. In Fig. 2, when Mongolian college students to listen a Chinese character, First Fixation Duration and Total time in zone of the homophone are the shortest, and Fixation Count is the least. Therefore, listening as the starting mode, the target word promote the activation of the homophone, which indicates the homophone is processed first, followed by the semantics, and then the glyph, the last activate is irrelevant word. When watching as the start-up mode, First Fixation Duration and Total time in zone of the orthographically similar are the shortest, and Fixation Count is the least. Therefore, watching as the starting mode, the target word plays a significant role in promoting the activation of the orthographically similar, which shows that the orthographically similar is processed first, followed by the homophone, and then acquiring meaning by the sound, and the last irrelevant word is activated when Mongolian students recognize the Chinese word in the way of watching.

4 Discussion

Under different starting methods (listening a Chinese word, watching a Chinese word), this article takes Mongolian college students as the research object to explore the role of speech information in the early and late processing, the activation order of the four competition words and the way of meaning acquisition when Mongolian college students' recognition of Chinese word. When listening a Chinese word is used as the starting method, the two indicators, First Fixation Duration and Total time in zone are all reached a significant level in homophone, synonym, and orthographically similar to the target word. Among the competition words, the first fixation duration of the homophone is the least, followed by synonym, then orthographically similar, and the first fixation duration of the unrelated words is the longest. Therefore, when Mongolian students recognize a Chinese character, the phonetic information plays an important role in word processing. The order of activation in word processing is phonological first, then semantic, followed by graphic, last is unrelated word. Therefore, the voice information directly activates the meaning of the word, and the activation of the word meaning does not necessarily have to pass through the glyph.

When watching a Chinese word is used as the starting method, the two indicators (First Fixation Duration, Total time in zone) of homophone and graphic similar are all reach a significant level, synonyms has a marginally significant level, but two indicators of unrelevant word are not significant. Analyze Fixation Count, the number of fixation points of graphic similar is the smallest, synonyms has the most number of fixation points, homophone lives in between. That is to say, When watching as start method, the activation order of the four words are: graphic similar is first, next by homophone, and finally the meaning of the word is obtained through voice information, and synonyms is finally activated. This indicates that the speech information has a key role in word processing. Firstly, speech representation system is activated by the glyph, and then meaning of the word is activated by speech information. The order of activation of the word processing is graphic first, next by phonological, then semantic activation, the last is irrelated word.

From above research results, the following conclusions can be drawn: Mongolian-Chinese bilinguals will be influenced by the characteristics of the native language when they recognize the second language characters. The speech plays a mediating role in the Mongolian bilinguals' Chinese processing, which is different from the cognitive style of Chinese monolinguals. This shows that when the two language systems mastered by bilinguals are different, the second language processing is often influenced by the processing method of the mother tongue, which causes the phenomenon of language transfer. When the subjects whose native language is alphabetic writing are used for the processing of ideograph, the voice is the intermediary, the sound and meaning are closely combined, so the result supports the theory of speech mediation.

Acknowledgments. This work was financially supported by Central University Fundamental Research Fund (No. 31920200028).

References

1. Schirmer, A., Tang, S., Penney, T.B., Gunter, T.C., Chen, H.: Brain responses to segmentally and tonally induced semantic violations in Cantonese. J. Cogn. Neurosci. **17**(1), 1–12 (2005)
2. Ye, Y., Connine, C.M.: Processing spoken Chinese: the role of tone information. Lang. Cogn. Process. **4**(5–6), 609–630 (1999). Special Issue: Processing East Asian Languages
3. Malins, J.G., Joanisse, M.F.: The roles of tonal and segmental information in Mandarin spoken word recognition: an eyetracking study. J. Mem. Lang. **62**, 407–420 (2010)

4. Huettig, F., Altmann, G.T.M.: Visual-shape competition during language-mediated attention is based on lexical input and not modulated by contextual appropriateness. Vis. Cogn. **15**(8), 985–1018 (2007)
5. Huettig, F., et al.: Using the visual world paradigm to study language processing: a review and critical evaluation. Acta Physiol. **137**, 151–171 (2011)
6. Huettig, F., McQueen, J.M.: The tug of war between phonological, semantic and shape information in language-mediated visual search. J. Mem. Lang. **57**, 460–482 (2007)
7. Xu, Y., Pollatsek, A.: The activation of phonology during silent Chinese word reading. J. Exp. Psychol. Learn. Mem. Cogn. **4**, 835–858 (1999)
8. Kroll, J.F., Dussias, P.E.: The comprehension of words and sentences in two languages. Blackwell, Cambridge (2004)
9. Kroll, J.F., De Groot, A.M.B.: Handbook of Bilingualism: Psycholinguistic Approaches, pp. 179–199. Foreign language teaching and research press and Oxford University Press, Oxford (2010)

Value Added and Socialized Science Information Service Based on Sci-Tech Novelty Retrieval

Meilong Ju(✉), Qing Zhang, Sumin Sun, Jian Xue, Lina Lou, and Chengguo Xin

Information Research Institute, Shandong Academy of Sciences, Qilu University of Technology, Jinan 250014, China
juml@sdas.org

Abstract. The Sci-Tech Novelty Retrieval is a kind of information consulting service, based on the manual and computer retrieval, using the method of comprehensive and contrast analysis, providing the objective facts for the novelty and advancement evaluation of research project approval, Sci-tech achievement appraisal, patent application, technology consulting and other innovation activities. The Sci-Tech Novelty Retrieval institutions have rich information resources, human resources, customer resources and brand resources to carry out innovative value-added information service. In this paper, we select some value-added information service modes which are suitable to carry out by the novelty search institutes, and make the contrastive analysis between traditional novelty search service and innovative value-added information service. For the customers' requirements, start the personal value-added services, such as the competitive intelligence, the patent novelty search, SCI submission, the perspective prediction and the thematic SDI, and expand the way of socializing the service of Sci-Tech Novelty Retrieval in universities, converting the passive Sci-Tech Novelty Retrieval service into the initiative one and then connecting with the international information consulting service.

Keywords: Sci-Tech Novelty Retrieval · Patent novelty search service · Competitive intelligence service · Market forecast of the project · Specific information service

1 Introduction

In recent years, with the continuous improvement of scientific and standardized degree of scientific research management and scientific research decision-making, Sci-Tech Novelty Retrieval has developed into an information service work with greater economic and social benefits. According to statistics, the repetition rate of scientific research projects in China is up to 40% [1], while the repetition rate with foreign countries accounts for about 30% in the other 60% [2], most of them are open technologies in foreign countries, which resulting in serious waste of human, material and financial resources [3]. Sci-Tech Novelty Retrieval can avoid the blindness of scientific research topics, reduce the low-level repetition of scientific research work, and improve

M. Tavana et al. (Eds.): IISA 2020, AISC 1304, pp. 354–360, 2021.
https://doi.org/10.1007/978-3-030-63784-2_45

the level of scientific research projects and scientific and technological achievements evaluation [4].

Sci-Tech Novelty Retrieval provides scientific and reliable basis for scientific research projects and achievements appraisal, provides rapid and accurate information service for economic development, and greatly promotes the construction and development of information resources. The Sci-Tech Novelty Retrieval department should make full use of its rich information resources and skilled retrieval means to help scientific researchers master the relevant information at home and abroad comprehensively and accurately, provide inspiration, reference and reference for scientific research work of scientific researchers, and establish a high-level information collection, sorting and analysis system for scientific research [5]. Sci-Tech Novelty Retrieval provides high-quality information services for enterprises, helps enterprises to grasp the development trend of the industry in time, masters new technology of the industry, etc., so that enterprises can adjust their business strategies and product R&D direction in time, so as to win in the fierce market competition.

The quality of Sci-Tech Novelty Retrieval and consultation work determines the admissibility of relevant contents of novelty search as the basis of literature evaluation, and also determines the scientific and objectivity of relevant scientific research management work [6]. How to improve the quality of novelty search work, innovate the service mode of novelty search work, and provide information analysis with certain reference value for scientific researchers and enterprises are the problems that the novelty search organization should consider.

Therefore, this paper attempts to make some discussions on the competitiveness of Sci-Tech Novelty Retrieval and the innovation of social services. This paper also discusses the shortcomings of the current Sci-Tech Novelty Retrieval work, and summarizes the Sci-Tech Novelty Retrieval institutions can carry out patent analysis, competitive intelligence analysis, project market forecast and other aspects. So that the China's Sci-Tech Novelty Retrieval work can have a comprehensive, coordinated and sustainable development.

2 Deficiency of Sci-Tech Novelty Retrieval Work

2.1 The Service Content of Sci-Tech Novelty Retrieval is Single

The traditional Sci-Tech Novelty Retrieval service mainly provides intelligence and information consulting services centering on the establishment of projects, the appraisal and reward of scientific and technological achievements, patent applications and other businesses. The service content is relatively simple, lacking in breadth and depth. The novelty judgment provided by the Sci-Tech Novelty Retrieval report is only a tiny link in the service of scientific activities. With the advent of the era of big data, people's demands for scientific and technological information processing are multi-faceted, in-depth and comprehensive. Facing massive data, people also put forward higher requirements for the level and efficiency of information processing. The novelty report should be transformed and developed from simple novelty judgment report to analytical and evaluative report. In-depth data mining should be carried out based on existing literature resources [7]. And according to the actual situation of the customer customized demand scheme, to provide customers with a series of higher level, more targeted information consulting services, such as patent analysis, competitive intelligence services, intellectual property evaluation, and so on, to play a more important role in the development of scientific and technological activities. This is not only conducive to the sustainable development of science and technology search, but also to meet the needs of the innovation for information.

2.2 Service Mode Relatively is Passive

After 30 years' development, Sci-Tech Novelty Retrieval work, faced with rich customer sources, has the ideology of "eating the old", almost all of which are waiting for customers' door-to-door entrusted service, in a passive state, to be carried out in cooperation with the project declaration, appraisal and other work of relevant departments. And with the development of scientific research management, some projects that were designated to need novelty retrieval in the past have been gradually withdrawn, resulting in the shrinking of some existing novelty retrieval markets. On the one hand, it is losing old customers; on the other hand, it is not attractive enough to attract new customers. In addition, in recent years, under the policy of vigorously encouraging scientific and technological innovation, not only the gold content of national projects is increasing, but also the independent innovation ability of small and medium-sized enterprises is increasing year by year. Therefore, based on the huge information consulting market contained in such rich scientific and technological activities, novelty search institutions should conform to the development of the times, dare to change their thinking and actively combine with customers, Relying on its own advantages of literature resources, actively push services, actively participate in multiple links of science and technology services, while improving the basic viability, we should seize the market as soon as possible, and complete the transformation and upgrading. To sum up, the traditional business of science and technology novelty search has been unable to meet the current needs of science and technology innovation services. Novelty search institutions must base themselves on their own reality, focus on the research of novelty search transformation and upgrading in the era of big data, and actively carry out comprehensive high-end information consulting services such as patent analysis, intellectual property evaluation, competitive intelligence services. In particular, as an important part of the intellectual property service industry, intellectual property appraisal can effectively reduce intellectual property risks for both government projects and enterprise projects, Actively avoid trade barriers, maximize project value, and provide intellectual property services and professional support for government, industry and enterprise decision-making. The novelty search organization can be a breakthrough as the direction of transformation and upgrading.

2.3 The Composite Quality of Novelty Searchers Needs to Be Improved

In the era of big data, interdisciplinary and interdisciplinary disciplines are rising rapidly, and new technologies and theories are emerging. It is increasingly difficult to grasp key information in massive information. Big data analysis and visualization technology not only deepens users' cognition and understanding of data, but also promotes information service to the level of knowledge cooperation and collaborative innovation. The professional quality and technical ability of novelty searchers put forward higher requirements. As a professional expert, data and intelligence analyst, the novelty searcher should not only have a certain professional background, but also have a strong ability of document interpretation, computer retrieval, foreign language translation, etc. In addition, we need to master the data management knowledge and skills of acquisition, mining, evaluation and use of scientific research big data. It is an urgent task for the novelty search organization to introduce the compound talents and improve the compound quality of the existing novelty search personnel.

3 Science Information Service Modes

3.1 Patent Novelty Search Service

With the development of knowledge economy, intellectual property plays an important role in the technological innovation of enterprises, and the service demand for patent information is increasing day by day. How to use the existing patent information, carry out patent strategy and avoid intellectual property disputes has become an urgent task for enterprises, especially small and medium-sized enterprises. It is reported that in 2019, the number of patent applications for invention in China is 1.401 million, which brings historic opportunities to the development of patent novelty search [8]. With rich information sources, senior experts and advanced information collection, analysis and processing technology, the novelty search organization can not only promote the sustainable development of the novelty search organization, but also provide better and all-round services and support for the enterprise's scientific and technological innovation to achieve better social benefits. Some novelty search institutions have carried out a series of novelty search value-added services, such as patent novelty search, patent infringement search and analysis, patent technology re innovation, and achieved good results. It not only promotes the novelty search work, but also meets the multi-faceted needs of enterprises in technological innovation.

3.2 Competitive Intelligence Service

As an important part of the core competitiveness of enterprises, the strength of information competitiveness determines the future life and death of enterprises to a large extent. Competitive intelligence is to infuse competitive consciousness on the basis of intelligence research, use legal and moral means, through long-term systematic tracking, collection, analysis and processing of all kinds of information that may have an impact on the development, decision-making and operation of enterprises, extract the key intelligence of the advantages, disadvantages and opportunities of the enterprise and its main rivals in the market competition, and provide it to the decision-making level of enterprises and Used by relevant departments. Competitive intelligence service is a higher-level service mode in the work of scientific and technological novelty search. It can not only promote the connection between scientific and technological information research institutions and enterprises, but also change the passive scientific and technological novelty search into the active information consultation, and broaden the field of information research, so that the research field gradually infiltrates into all levels and aspects of social and economic development. Using the advantages of scientific and technological novelty search institutions in information resources, information analysis and other aspects, to help enterprises with new strategic thinking, make forward-looking strategic decision-making, win competitive advantage and determine the future in the complex and changeable market economy environment.

3.3 Project Market Forecast

In today's highly information-based and economic globalization competition environment, the demand of enterprises for information is constantly increasing, from simple information service demand to high-quality, deep processing information consulting service demand. Technology novelty search is the best way for enterprises to search for information. Sci-Tech Novelty Retrieval institutions collect, integrate, analyze and compare information resources of regional industries or technologies with the help of rich resource advantages and strong professional consulting experts, So as to summarize the current situation of its development and the development direction of the existing problems, help the enterprises grasp the technical and market information at home and abroad at the fastest speed, provide the technical resources of the development industry, the development prospect of new products, and predict the development trend of technology, In order to formulate the enterprise's own development strategy, find the breakthrough point of technology and products, promote the research and development of new products and technologies, transform the past scientific and technological information research from indirect service to direct service, and transform the scientific and technological achievements into productivity as soon as possible, so as to gain competitive advantage and win the space and time for development.

3.4 Special Topic Setting Service

Under the situation of the development of information market, the information service of library develops to the information service of special topic and definite topic, and forms its own characteristic information service mode. If the scientific and technological novelty search is only limited to issuing a novelty search report to end the information service, then the information service is incomplete, not only does not give full play to the role of the collection resources, does not make full use of the information of the novelty search project, nor does the novelty search Department play a huge role in scientific research. The special information service is an interactive service. The information service personnel and the subject researchers should constantly exchange information, take their needs as the value scale of information development, timely understand the feedback information, track the current situation of the subject, clear the existing problems, the specific and expected results of the subject research, the schedule of the subject service arrangement, etc., Provide the latest information and trends at home and abroad at any time, and make special index and abstract. The novelty search institutions can also establish long-term cooperative relations with some scientific research institutions with strong strength and relatively concentrated research topics, Provide a series of regular and continuous scientific and technological information services from project selection, project approval to project tracking, project conclusion, achievement award and appraisal, We will develop scientific and technological novelty search into a “dynamic process”, combine scientific and technological novelty search with scientific research project management, and provide information services.

3.5 SCI Article Submission Service

SCI database has become an important index in scientific research performance evaluation in China. How to improve the hit rate of papers is very important, and novelty search is the most important link to improve the hit rate of SCI journals. Through novelty search, we can know the status of the research content of the thesis in the international academic research field, whether the research results are reported in the world, select topics that no one has studied, and open up new areas of exploratory research, so that there will be more opportunities to be selected. When using SCI database to search, we should determine the search points and search strategies according to the research contents of the papers to be submitted, then search them, and then determine 2–4 journals to be submitted according to the search results; The author’s country sources in some SCI journals are only limited to 1–2 countries. It’s better to choose the journals with Chinese authors; Only high-quality papers can occupy a place in international journals if we want to supplement and modify the latest progress materials.

4 Conclusion

Sci-Tech Novelty Retrieval is an information service work with strong vitality. It can not only improve the utilization rate of information resources, but also strengthen the social awareness of information and science and technology innovation. The novelty search institutions should make use of their own resource advantages to innovate the novelty search service mode, improve the height of novelty search service, actively provide high-quality services for scientific research, create conditions for scientific research and knowledge promotion in all walks of life, carry out personalized value-added services, promote the transformation of scientific and technological achievements, and promote the rapid development of scientific and technological undertakings.

References

1. Zhu, J.: Research on the acquiring users’ needs and providing precision information service by sci-tech novelty search training course. In: 2nd International Workshop on Education Reform and Social Sciences (ERSS 2019), pp. 34–38. Atlantis Press (2019)

2. Jia, T., Zhong, X.: Analysis on constructing fusion mechanism between novelty consultant's specialty and novelty search projects in comprehensive novelty search organization. J. Libr. Inf. Sci. (3), 9 (2017)
3. Lei, Z.: The mechanism, route, effects of the environment factors on scientific and technological novelty search service performance. Res. Libr. Sci. (3), 16 (2016)
4. Huili, W.: Research on coordinative innovation in sci-tech novelty retrieval. Inf. Res. (1), 15 (2016)
5. Yao, J., Le, X.: Semantic matching for sci-tech novelty retrieval. Data Anal. Knowl. Discov. **3**(6), 50–56 (2019)
6. Yan, C.: Review on studies of sci-tech novelty retrieval basing on co-word clustering analysis. Inf. Res. (2), 22 (2017)
7. Peixia, W., Hai, Y., Li, C., et al.: Using intelligent system to extract search terms for sci-tech novelty retrieval. Data Anal. Knowl. Discov. **32**(11), 82–93 (2016)
8. https://www.gov.cn/xinwen/2020-01/15/content_5469519.htm.

A Secure ECC-Based Authentication Scheme to Resist Replay Attacks for the IoT

Yuxiang Feng[1(✉)] and Wei Liu[2]

[1] School of Software Engineering, South China University of Technology, Guangzhou, China
yxfeng@scut.edu.cn
[2] Sunyeer Information & Technology Co., Ltd., Taoyuan, Taiwan

Abstract. Because of the resource-constrained characteristic of IoT devices, it is difficult to achieve user authentication in IoT by using classical Internet's authentication schemes. The reply attack is a classical attack on authentication protocols. The timestamp mechanism by maintaining clock synchronization is generally used in the exiting schemes for IoT. We propose a lightweight authentication scheme based on elliptic curve cryptography (ECC) in this paper. The proposed scheme introduces an improved challenge-response mechanism instead of the timestamp. Thus, storage and computation cost of generating and maintaining timestamp can be eliminated.

Keywords: Internet of Things (IoT) · Elliptic Curve Cryptography (ECC) · Authentication · Replay attack

1 Introduction

The future of Internet may be the Internet of Things (IoT), which enables everyone and everything to communicate and connect [1]. IoT produces a network between man's real life and digital world. Nowadays numerous sensors produce and manipulate sensitive data including health information, environment information and industrial information. The attacker will compromise the security of both devices and data in an IoT application without proper security scheme [2]. Thus, an authentication protocol is the key component of an IoT application [3]. Unfortunately, the exiting typical authentication protocols and communication schemes which are used in Internet environment are unsuitable for IoT because of the low computation and storage capacity of IoT devices [4].

In recent years, numerous lightweight authentication schemes which are designed for resource-constrained environment have been proposed. The timestamp mechanism is generally used in these schemes to resist the replay attack. Maintaining clock synchronization between sensors which causes more devices' energy consumption [5]. However, the energy consumption is the most important issue in IoT environment as the device's battery is always limited.

We proposed an ECC-based authentication scheme by using an improved challenge-response mechanism instead of the timestamp mechanism. The scheme does not need to keep clock synchronization between devices. Section 2 presents the relative

M. Tavana et al. (Eds.): IISA 2020, AISC 1304, pp. 361–370, 2021.
https://doi.org/10.1007/978-3-030-63784-2_46

work. The detail of the proposed scheme is illustrated in Sect. 3. We analyze the security and performance of the proposed scheme in Sect. 4 and 5. Finally, we conclude the paper in Sect. 6.

2 Related Work

Elliptic Curve Cryptography (ECC) is suitable for the IoT environment because of its low cost of computation [6]. Recently, several lightweight authentication schemes based on ECC for IoT have been proposed. Yao et al. [7] proposed a light weight ECC scheme based on attribute encryption. Moosavi et al. [8] focused on the requirements of the RFID systems in IoT environment, and proposed a lightweight authentication scheme using ECC. Kalra et al. [9] focused on the requirements of the embedded devices and cloud server's environment. The scheme is based on Hyper Text Transfer Protocol (HTTP) cookies to achieve mutual authentication for secure communication. Maurya et al. [10] proposed an ECC-based authentication scheme with a fuzzy extractor which ensures various security features. More recently, Shin et al. [11] proposed an ECC-based authentication scheme which is suitable for IoT environment which integrated the WSNs and 5G. In addition, using hash and XOR operations, Rahnama et al. [12] proposed a lightweight authentication scheme. Lopse et al. [13] proposed a symmetric cryptography-based scheme which supports device-to-device communication. Deebak et al. [14] proposed a biometric-based authentication scheme in e-healthcare applications. Dammak et al. [15] proposed a lightweight user authentication which using token technique to improve the robustness of authentication. The resistance of these schemes to replay attack depends on the timestamp mechanism.

Chen et al. [16] proposed an authentication scheme based on the self-certified public key mechanism and ECC which uses the nonce instead of the timestamp mechanism. Kumari et al. [17] proposed an improved scheme to solve the shortcomings of Kalra's scheme [9] without clock synchronization mechanism. Panda et al. [18] proposed an authentication scheme using the concept of password verifier. Alzahrani et al. [19] introduced the NFC into authentication protocol to improve the applicability of the protocol in mobile applications. These schemes have no energy consumption of maintaining clock synchronization. The proposed scheme introduces an improved challenge-response mechanism instead of the timestamp to resist the replay attack. The performance comparison between several schemes mentioned above and ours have been presented in Sect. 5.

3 The Proposed Scheme

3.1 System Model

The proposed scheme runs in a network model which contains three main components: the sensor node which has low storage and computation capacity, the gateway node (GWN) which is far more powerful than the sensor node, and the user node. Since the

GWNs are powerful nodes, some computation cost of the authentication scheme can be offloaded to the GWN. Figure 1 shows the detail of our system model.

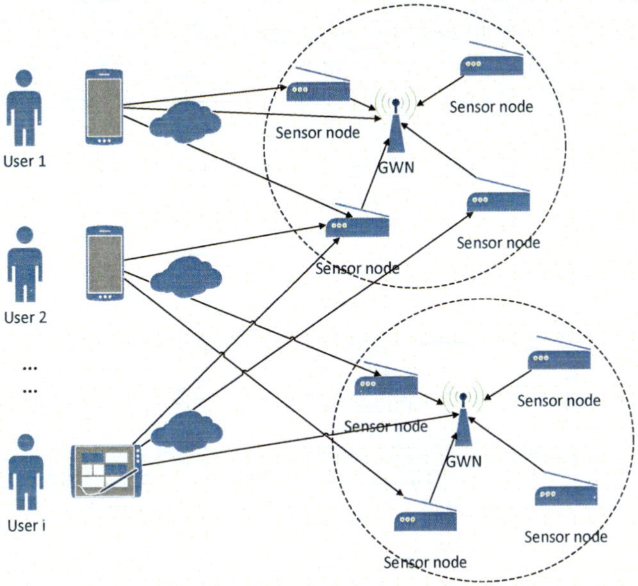

Fig. 1. The scheme's system model

3.2 Notations

The notations which are used in the scheme are presented in Table 1.

The proposed authentication scheme is developed as a two-phases scheme. The setup phase and authentication phase are described respectively.

Table 1. Notations used in the scheme

Notations	Descriptions
Ps	Parameters for the elliptic curve, $Ps = (q, GF(q), a, b, P, n)$
d, Q	The private key and the public key
U_i, S_j	The node of the user and sensor
ID_i, SID_j	The user and sensor's identity
$\rightarrow$	The unsecure communication channel
$\|\|$	The operation of message concatenation
ENC()	An encryption function of a symmetric-key encryption scheme, and DEC() is the decryption function
$signature_{GWN}$	The signature of GWN via ECC encryption scheme

3.3 Setup Phase

To resist the attacks on the ECDLP, the parameters of the elliptic curves should be carefully chosen. Before the system begins, the server shall select the elliptic curve parameters Ps for the application, and share those parameters to all the entities of the application, i.e. the sensor, the GWN, and the user.

In the setup phase, all entities need to generate their key pair with these parameters. Each sensor node S_j gets and stores its key pair and the GWN's public key in its memory. The GWNs are powerful nodes which have enough memory to store all sensor's public keys of its domain and its own key pair.

A user, U_i, need to register on the domain's GWN before he accesses the data of a sensor. The registration process is:

Step 1: U_i generates his own key pair $\{Q_i, d_i\}$.
Step 2: U_i exchanges the public key with GWN over a secure communication channel such as an off-line action.
Step 3: The GWN stores U_i's public key Q_i, and U_i stores the GWN's public key Q_{GWN}.

3.4 Authentication Phase

If the setup phase has been finished successfully, users can access the data from the sensors by initiating the authentication phase. In the authentication phase, the user and the sensor can negotiate a session key with the help of GWN. The detail of this phase is presented in Fig. 2.

The interactions in this phase are presented in detail as follows:

Step 1: First, the user U_i sends a message $\{ID_i \parallel SID_j\}$ to the GWN. It means user U_i informs the GWN of its intention to access the data of sensor node S_j.
Step 2: According to SID_j, GWN finds the public key Q_j in its memory. GWN returns to user U_i this public-key with the signature $signature_{GWN}(SID_j \parallel Q_j)$.
Step 3: In this step, U_i verify the signature from the GWN to confirm the public key Q_j of sensor node S_j. Then, U_i generates a nonce N_i, and sends $ciphertext_{Q_j}(N_i \parallel ID_i)$ to sensor node j.
Step 4: Sensor node j sends requirement to get U_i's public-key and a session key from GWN. S_j forwards U_i's nonce N_i to GWN so that the session key can be stamped with this nonce.
Step 5: The GWN sends U_i's public-key and the quadruple $\{ID_i \parallel SID_j \parallel N_i \parallel K_s\}$ back to S_j. This means that K_s is a session key generated by the GWN and tied to N_i ensure its freshness. This message is signed by GWN's private key and encrypted by S_j's public key, so that S_j can verify the source of this message and the attacker cannot use this message to cheat U_i.

Step 6: S_j verify Q_i and decrypt $\{ID_i \| SID_j \| N_i \| K_s\}$ in the message.S_j uses $\{ID_i \| SID_j \| N_i \| K_s \| signature_{GWN}(ID_i \| SID_j \| N_i \| K_s)\}$ from GWN, and generate a new nonce N_j, then encrypt them all and send to U_i.
Step 7: U_i performs the decryption and verification operations to retrieve the session key and the nonce, uses the session key to encrypt N_j, and send the result to S_j.

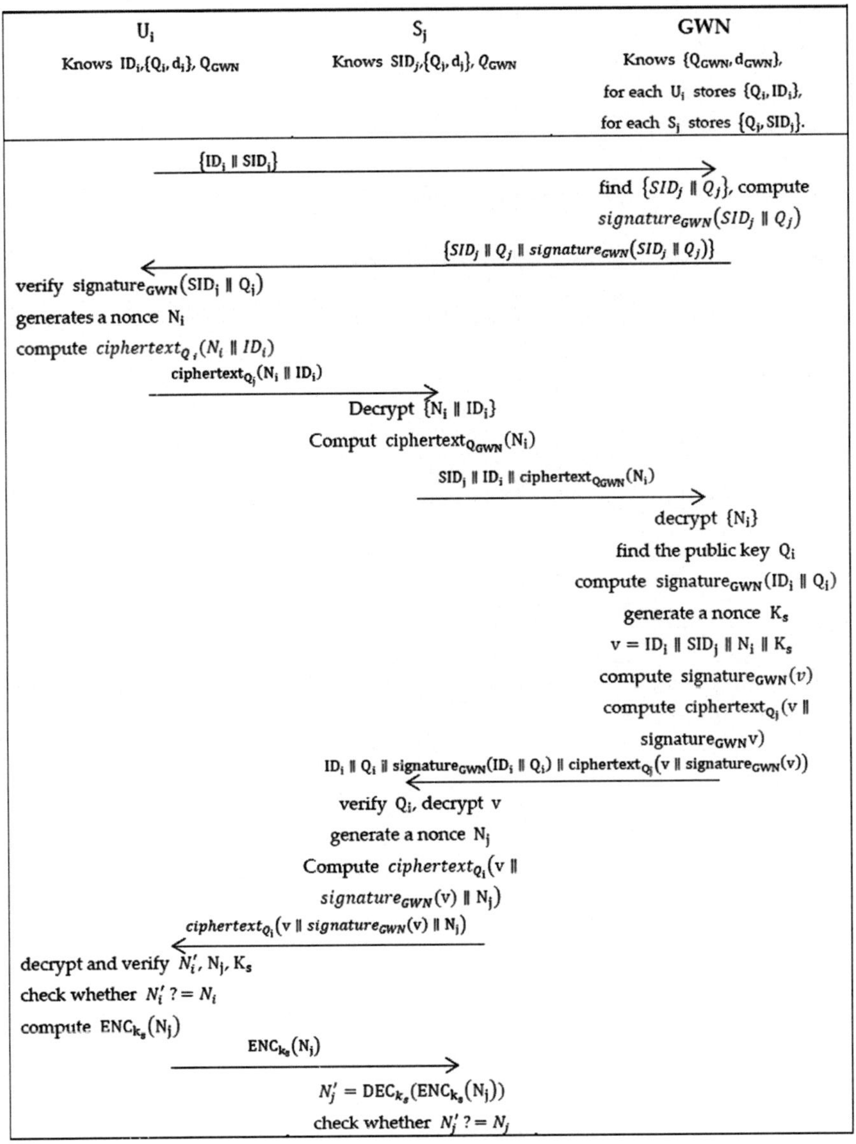

Fig. 2. The authentication phase of the proposed scheme

4 Security Analysis of the Proposed Scheme

We assume that the communication channel of the IoT application is unsecure. Thus, the attacker can intercept and forge all messages on these channels. The details and cryptographic functions of the proposed scheme are open. The user and sensor can only store its randomly selected nonce and secret key in secret. In addition, the attacker has enough computational capability and resources.

4.1 Resistance to Replay Attack

The attack on step 1, 2, 3 and 4 is useless because the message of step 6 is encrypted by the public key of ID_i so that only can be decrypted by ID_i.

In step 5, the attacker can replay the authentication message with a known K_s. This replay message's verification will not hold since a freshness identifier N_i is cryptographically integrated with K_s. Sensor node j can obtain assurance that K_s is fresh via checking the N_i. In step 6 and 7, the freshness identifiers N_i and N_j are cryptographically integrated with K_s too.

In order to pass the authentication, the attacker must change the nonce. For example, the attacker records a known K_s and the authentication message in step 5. In a new run, U_i generates a different nonce N_i'. To pass the authentication, the attacker has to forge a message: attacker $\rightarrow S_j : ID_i \| Q_i \| signature_{GWN}(ID_i \| Q_i) \| ciphertext_{Q_j}\left(ID_i \| SID_j \| N_i' \| K_s \| signature_{GWN}\left(ID_i \| SID_j \| N_i' \| K_s\right)\right)$. This authentication message's verification will not hold since the attacker doesn't have the private key of GWN, d_{GWN}, to compute a new legitimate signature of N_i'.

4.2 Resistance to Man-in-the-Middle Attack

In step 2 and 5, our proposed scheme exchanges the public key with the signature of GWN. The freshness identifiers N_i and N_j are cryptographically integrated with K_s. The attacker needs to forge N_i, N_j and $signature_{GWN}$ to perform man-in-the-middle attack. As mentioned before, it is impossible.

4.3 Provides Mutual Authentication

The public key of S_j is signed and sent to U_i by GWN in step 2, which is verified by U_i in step 3. U_i uses this public key to encrypt the nonce N_i and send to S_j in step 3. If N_i' sent back in step 6 equals to N_i, U_i can authenticates S_j's identity.

Moreover, the public key of U_i is signed and sent to S_j by GWN in step 5, which is verified by S_j in step 6. S_j uses this public key to encrypt the nonce N_j and send to U_i in step 6. If N_j' sent back in step 7 equals to N_j, S_j can authenticates U_i's identity.

Therefore, the mutual authentication can be established between U_i and S_j.

4.4 Provides Confidentiality

Each message necessary for authentication in the proposed scheme is protected by ECC encryption function or symmetric-key encryption function ENC() such as the AES. Therefore, the message necessary for authentication only can be accessed by the authenticated entities.

5 Performance Analysis of the Proposed Scheme

As the sensors in IoT are often resource-constrained, our scheme focuses on the efficiency of the highly constrained sensor side in the comparison.

5.1 Storage Cost

In our scheme, the memory requirement of sensor node always includes $\{SID_j, Q_j, d_j, DPs, Q_{GWN}\}$. In one run of the protocol, the sensor node needs to store $\{ID_i, Q_i, N_i, N_j, K_s\}$. The key length of ECC is 160 bits in our proposed scheme, and the length of ID value is 32 bits. The length of ECC's parameters Ps is 960 bits, the length of the nonce use in the run of the protocol is 128 bits and the key length of symmetric-key encryption function is 128 bits. The sensor's total storage cost in our scheme is 1472 bits, and 576 bits more in each run.

There is no storage cost of the clock synchronization in our scheme. The storage cost of each timestamp is typically about 152bits [20].

5.2 Computation Cost

In Table 2, the computational costs of several related schemes and ours are listed. Let T_M, T_A, T_F, T_X be the time of ECC point multiplication, ECC point addition, fuzzy extractor and symmetric encryption/decryption operation respectively. The one-way hash function, XOR operation and other simple functions take less and easier to implement than the operations mentioned above. Therefore, these functions' computation cost can be negligible.

According to [21], T_F approximately equals to T_M. Compared to T_M and T_F, T_A, T_x can be negligible. As the sensors are highly resource-constrained devices and the GWN has much more battery power than the sensors, the scheme should minimize the computation cost of the sensor nodes. As shown in Table 2, our scheme has more computation cost of GWN than other schemes, but less computation cost of sensor nodes than other schemes.

In addition, maintaining clock synchronization causes more computation and communication operations which lead to more computation cost of the devices. There is no computation cost of the clock synchronization in our scheme as no timestamps used in our scheme.

Table 2. Comparison of computation cost

Scheme	Sensor side computation cost	Other sides computation cost	Cost of clock synchronization
Kalra's scheme [9]	$4T_M$	$4T_M$	Yes
Maurya's scheme [10]	$3T_M + T_F + 2T_x$	$3T_M + 4T_X$	Yes
Shin's scheme [11]	$3T_M + T_F$	$3T_M$	Yes
Chen's scheme [14]	$5T_M + 3T_A$	$4T_M + 2T_A$	No
Kumari's scheme [15]	$4T_M$	$4T_M$	No
Panda's scheme [16]	$4T_M$	$4T_M$	No
Proposed scheme	$3T_M + 2T_A + T_x$	$7T_M + 2T_A + T_x$	No

6 Conclusions

IoT applications process the sensitive data for business or personal use. However, the devices in IoT are often characterized as resource-constrained. To avoid the cost of maintaining clock synchronization between the devices, we introduce the improved challenge-response mechanism instead of the timestamp mechanism. In comparison with the related schemes, our scheme has more computation cost of GWN than other schemes, but less computation cost of sensor nodes than other schemes. Moreover, security analysis is presented to verify the scheme's immunity against the typical threats. Hence, we argue that our scheme is applicable to the resource-constrained environment such as IoT.

In the future, the performance of our scheme can be further improved by using a more efficient signature algorithm. Additional, we would like to introduce the biometric-based scheme into our protocol so that the result of this paper can be applied in e-healthcare applications.

Acknowledgments. This work is partly supported by the Natural Science Foundation of Guangdong Province, China (2015A030310446).

References

1. Chopra, K., Gupta, K., Lambora, A.: Future internet: the Internet of Things-a literature review. In: 2019 International Conference on Machine Learning, Big Data, Cloud and Parallel Computing (COMITCon), Faridabad, India, pp. 135–139 (2019)

2. Lucia, O., Isong, B., Gasela, N., Abu-Mahfouz, A.M.: Device authentication schemes in IoT: a review. In: 2019 International Multidisciplinary Information Technology and Engineering Conference (IMITEC), Vanderbijlpark, South Africa, pp. 1–6 (2019).
3. Sicari, S., Rizzardi, A., Grieco, L.A., Coen-Porisini, A.: Security, privacy and trust in Internet of Things: the road ahead. Comput. Netw. **76**, 146–164 (2015)
4. El-hajj, M., Fadlallah, A., Chamoun, M., Serhrouchni, A.: A survey of Internet of Things (IoT) authentication schemes. Sensors **19**(5), 1141 (2019)
5. Wu, Y.C., Chaudhari, Q., Serpedin, E.: Clock synchronization of wireless sensor networks. IEEE Signal Process. Mag. **28**, 124–138 (2011)
6. Salas, M.: A secure framework for OTA smart device ecosystems using ECC encryption and biometrics. In: Advances in Security of Information and Communication Networks, pp. 204–218. Springer (2013)
7. Yao, X., Chen, Z., Tian, Y.: A lightweight attribute-based encryption scheme for the Internet of Things. Future Gener. Comput. Syst. **49**, 104–112 (2014)
8. Moosavi, S.R., Nigussie, E., Virtanen, S., Isoaha, J.: An elliptic curve-based mutual authentication scheme for RFID implant system. In: International Conference on Ambient Systems, Network and Technologies, vol. 32, pp. 198–206 (2014)
9. Kalra, S., Sood, S.: Secure authentication scheme for IoT and cloud servers. Pervasive Mob. Comput. **24**, 210–223 (2015)
10. Maurya, A.K., Sastry, V.N.: Fuzzy extractor and elliptic curve based efficient user authentication protocol for wireless sensor networks and Internet of Things. Information **8**(4), 136 (2017)
11. Shin, S., Kwon, T.: A privacy-preserving authentication, authorization, and key agreement scheme for wireless sensor networks in 5G-integrated Internet of Things. IEEE Access **8**, 67555–67571 (2020)
12. Rahnama, A., Beheshti-Atashgah, M., Eghlidos, T., et al.: A lightweight anonymous authentication protocol for IoT wireless sensor networks. In: 16th International ISC (Iranian Society of Cryptology) Conference on Information Security and Cryptology (ISCISC), pp. 39–44. IEEE (2019)
13. Lopes, A.P.G., Gondim, P.R.L.: Mutual authentication protocol for D2D communications in a cloud-based e-health system. Sensors **20**(7), 2072 (2020)
14. Deebak, B.D., Al-Turjman, F., Aloqaily, M., Alfandi, O.: An authentic-based privacy preservation protocol for smart e-healthcare systems in IoT. IEEE Access **7**, 135632–135649 (2019)
15. Dammak, M., Boudia, O.R.M., Messous, M.A., Senouci, S.M., Gransart, C.: Token-based lightweight authentication to secure IoT networks. In: 16th IEEE Annual Consumer Communications & Networking Conference (CCNC), Las Vegas, NV, USA, pp. 1–4 (2019)
16. Chen, H., Ge, L., Xie, L.: A user authentication scheme based on elliptic curves cryptography for wireless ad hoc networks. Sensors **15**, 17057–17075 (2015)
17. Kumari, S., Karuppiah, M., Das, A.K., et al.: A secure authentication scheme based on elliptic curve cryptography for IoT and cloud servers. J. Supercomput. **74**(12), 6428–6453 (2018)
18. Panda, P.K., Chattopadhyay, S.: A secure mutual authentication protocol for IoT environment. J. Reliable Intell. Environ. **6**, 79–94 (2020)
19. Alzahrani, B.A., Mahmood, K., Kumari, S.: Lightweight authentication protocol for NFC based anti-counterfeiting system in IoT infrastructure. IEEE Access **8**, 76357–76367 (2020)

20. Farash, M.S., Turkanovic, M., Kumari, S., Hölbl, M.: An efficient user authentication and key agreement scheme for heterogeneous wireless sensor network tailored for the Internet of Things environment. Ad Hoc Netw. **36**, 152–176 (2016)
21. Wazid, M., Das, A.K., Kumari, S., Li, X., Wu, F.: Design of an efficient and provably secure anonymity preserving three-factor user authentication and key agreement scheme for TMIS. Secur. Commun. Netw. **9**(13), 1983–2001 (2016)

Application Integration Using Context Model Based on CIM

Wang Liyan[1(✉)], Chen Lei[2], Du Jian[1], and Lin Haili[1]

[1] China Electric Power Research Institute, No. 15 Xiaoying East Road, Qinghe, Beijing 100192, China
lywang@epri.sgcc.com.cn
[2] State Grid Zhejiang Electric Power Co., Ltd., No. 8 Huanglong Road, Hangzhou 310007, Zhejiang, China

Abstract. Many utilities and control centers has adapted CIM standards as their data model in application integration. The CIM profile is a subset of CIM for specific domain. Object-oriented modeling is used to create CIM profile, and mapping process to XML Schema and XML messages is used in many scenario. This method has several weakness, mainly because object-oriented modeling is complicated, and XML is heavy-weight message format, not suitable for IOT and low-bandwidth environment. The context model is characterized as collection of attribute-value pairs, CIM model converting to context model and mapping to JSON is a much more efficient way than profile-XML mapping method. XML sample and context model mapping to JSON sample are illustrated too.

Keywords: CIM · Profile · Context model · XML · JSON

1 Introduction

CIM has been adopted as standard data model of electric devices and business procedures in utilities for years. CIM is defined by the Unified Modeling Language, providing a common way of data integration between applications in utilities [1]. Because of CIM is pretty big dataset, the CIM profile is introduced as subset of CIM, which can represent some aspect of business, such as operation, plan, and calculations of electric grid. CIM profile is usually composed of CIM classes, properties of these classes, relations between these classes.

CIM has been successful data model in electricity field for decades. At beginning, it was developed by IEC TC57 working group 13, which is mainly focused on Energy Management Systems for Transmission and Generation. Other extension based on CIM is made afterwards, such as diagram layout, measurement [2]. IEC TC57 working group 14 extended CIM to distribution field, built Distribution Information Exchange Model. CIM model and it extension are huge, and cover almost everything in electric grid, it has to be sliced to fit contextual scenario [3].

As showed in Fig. 1, CIM model is built by Unified Modeling Language, based on electric entities and business rules. CIM profiles, can be derived from CIM and its extension by CIMtool software. Those subset of CIM classes and relations is usually

M. Tavana et al. (Eds.): IISA 2020, AISC 1304, pp. 371–376, 2021.
https://doi.org/10.1007/978-3-030-63784-2_47

defined in XMI files, and can be mapped to XML Schema or XML RDF Schema, which describes the file format and syntax rules of messages between applications. In runtime environment, applications can create messages based on those schemas, and cooperate each other by exchanging those messages.

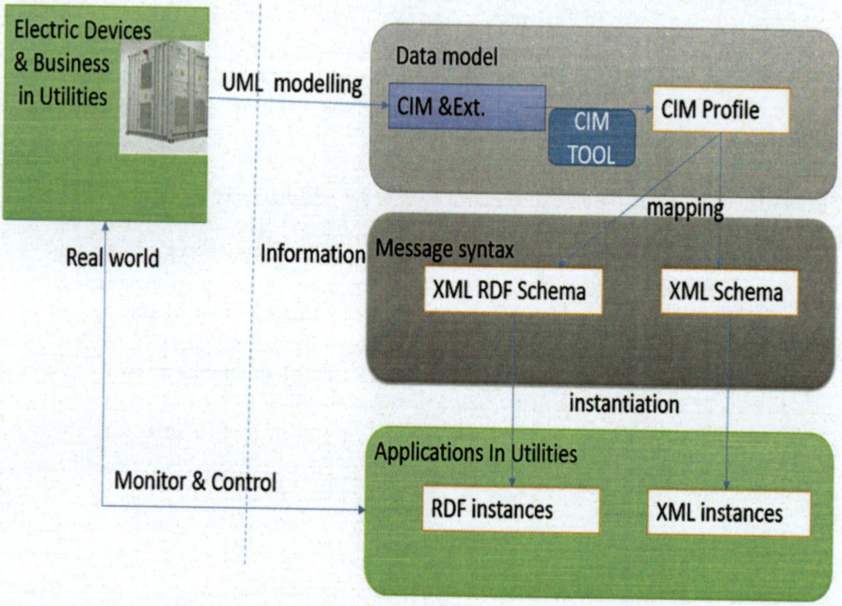

Fig. 1. Deriving a profile from an information model and mapping to XML

2 The Shortages of CIM Profile and XML

CIM profile is sliced CIM model, and slice CIM to profile is not easy. To build profile, the application scenario is studied, and use cases must be created and verified. Based on those discussion and study, the profile maker must decide which CIM classes are mandatory, which ones are optional, and which relationship of classes should be maintained. Though using profile is a tradition way to exchange messages between applications, this method has shortages listed below.

1. Profile is big and wasteful. Application scenario is complicated and constantly changing, profile has to contain more context to fit contextual variation. This makes profile big, and often left some classes is unused. In short, profile is not elastic.
2. Profile is slow. Due to complicated build process, profile is slow to build. And because of big size of XML format, building, transferring and parsing profile instances is slow. Profile is suitable for transferring a big bucket of data, and taking time to deal with it.

Profile is not suitable for low-bandwidth applications. As showed above, profile is big, and slow, it not capable of low bandwidth application. In recent years, the Internet of Things (IOT) has grown rapidly. In the IOT, the Intelligence Electronic Devices (IED) communicated with each other in a low-bandwidth situation. Since the IOT is the essential part of smart grid, profile is too big and too slow for the applications dedicated to exchanging messages between IED in utilities.

Fig. 2. (a) profile mapping to XML Schema (part); (b) XML instance (part)

As Fig. 2 showed, XML is heavy weight message, containing too much metadata information. A XML describing single power line may need hundreds kilobytes, building, transferring and parsing a XML are time-consuming. In the next section, the context modeling is introduced as more simple and quick solution of mapping and application of CIM.

3 The Context Model in Application of CIM

The context modeling permits the description and the structuring of the contextual information. Several models are available for different context domains [4]. There is a multitude of classifications that have been proposed for context models based on the data structure used for the description and exchange of context.

The context model is represented as a pair (attribute, value), which is also called Key-value models. The attribute represents the name of the contextual information. The value represents the current value of this information. This method has the advantage of being easily implemented.

CIM model can be represented by context model [6]. Since CIM is composed of classes, the context model of CIM is also composed by definition of classes and relationships between them. There are many ways to define a context model on CIM, the table below just one of them. Part of model listed in Table 1.

In the table above, the classes *IdentifiedObject*, *PowerSystemResource*, *Equipment* and their properties are listed. They are represented by the pairs of attribute, value, which are easy to follow. It is much simpler than the CIM profile, which has a net

Table 1. Partial context model definition of CIM

Entry	Attributes	Value type	Relationship	To
IdentifiedObject	Type	Structured lass		
Name	Type	Field	merberofclass	IdentifiedObject
	Datatype	String		
aliasName	Type	Field	merberofclass	IdentifiedObject
	Datatype	String		
mRID	Type	Field	merberofclass	IdentifiedObject
	Datatype	String		
Description	Type	Field	merberofclass	IdentifiedObject
	Datatype	String		
PowerSystemResource	Type	Structured class	Generalization	IdentifiedObject
Equipment	Type	Structured class	Generalization	PowerSystemResource
Aggregate	Type	Field	merberofclass	Equipment
	Datatype	Boolean		
inService	Type	Field	merberofclass	Equipment
	Datatype	Boolean		
…	…	…	…	…

structure (XML RDF mapping) or hierarchical structure (XML Schema mapping). On the contrast, context model can map to JSON format, which has a linear structure, makes message much smaller and efficient for the less metadata.

4 The Context Model Mapping to JSON

JSON is an open-standard file format, JSON is a text format for the serialization of structured data. It is derived from the object literals of JavaScript, as defined in the ECMAScript Programming Language Standard, Third Edition [ECMA-262] [7].

JSON is built on two structures:

1. A collection of name (attribute)/value pairs.
2. An ordered list of values.

The context model can easily mapped to JSON [8], since context model is also composed of attribute/value pairs. The effort of mapping mainly includes follow actions:

1. Datatype conversion, CIM primitive datatypes is not exactly same as JSON's, some basic type conversion is listed below (Table 2).

Table 2. CIM datatype mapping to JSON

CIM datatype	JSON datatype
Boolean	Boolean
Decimal	Number
Float	Number
Integer	Integer
Double	Number
String	String
Date	String

2. Select mandatory and optional items in CIM, and sort items by some algorithm, such as alphabetical order.


```
"measurement": {
    "parameter": [
        {
            "name": "Load",
            "desc": "配变台区的额定容量",
            "value": 600,
            "valueType": "int",
            "unitSymbol": "kW",
            "maxVal": 1000,
            "minVal": 100,
            "stepsize": 100
        }, {
            "name": "ImbA.strVal",
            "desc": "电流不平衡定值",
            "value": 20,
            "valueType": "int",
            "unitSymbol": "%",
            "maxVal": 80,
            "minVal": 5,
            "stepsize": 1
        }, {
```


Fig. 3. A sample of part of CIM data mapping to JSON file (part)

The Fig. 3 above is illustrated a part of CIM data mapping to JSON. JSON is an efficient way of messaging, lots of cost of building, transferring and parsing messages is saved. Because of JSON file is easy to modify, so is mapping to CIM items.

The context modelling helps to accelerate building a sliced CIM fit to application integration, and mapping to JSON makes messages small and efficient, which is a huge advantage in IOT low-bandwidth environment.

5 Conculsion

CIM is standard data model of electric devices and business procedures, and CIM profile is a subset of CIM. The profile can be mapped to XML/RDF or XML Schema syntax. The process of mapping may be long and complicated, because of case-study, demand analysis, object-oriented modelling and complexity of XML. The result of mapping, XML messages are heavy-weight ones, are not suitable for the IOT or low-bandwidth situation.

Acknowledgements. This Paper is sponsored by State Grid technology program 5400-201955454A-0-0-00.

References

1. Lambert, E., Quéric, A.: Use of CIM for EDF distribution automation. In: IEEE PES General Meeting, Providence, RI, pp. 1–4 (2010). https://doi.org/10.1109/PES.2010.5589894
2. Becker, D., Saxton, T.L.: CIM standard for model exchange between planning and operations. In: 2008 IEEE Power and Energy Society General Meeting - Conversion and Delivery of Electrical Energy in the 21st Century, Pittsburgh, PA, pp. 1–5 (2008). https://doi.org/10.1109/PES.2008.4596090
3. Würgler, E., McMoran, A.: CIM Diagram Layout Profile for graphics exchange. In: 2011 IEEE Power and Energy Society General Meeting, Detroit, MI, USA, pp. 1–5 (2011). https://doi.org/10.1109/PES.2011.6039444
4. Thorpe, S.: A case analysis of a Contextual Model of Trust for digital identities using UML 2.0 and context graph algorithms. In: 2010 Sixth International Conference on Information Assurance and Security, Atlanta, GA, pp. 15–20 (2010). https://doi.org/10.1109/ISIAS.2010.5604184
5. Ahmad, I., Fink, G.A.: Class-based contextual modeling for handwritten Arabic text recognition. In: 2016 15th International Conference on Frontiers in Handwriting Recognition (ICFHR), Shenzhen, pp. 554–559 (2016). https://doi.org/10.1109/ICFHR.2016.0107
6. McMorran, A.W., Lincoln, R.W., Taylor, G.A., Stewart, E.M.: Addressing misconceptions about the Common Information Model (CIM). In: 2011 IEEE Power and Energy Society General Meeting, Detroit, MI, USA, pp. 1–4 (2011). https://doi.org/10.1109/PES.2011.6039391
7. ECMAScript® 2019 Language Specification. https://www.ecma-international.org/publications/standards/Ecma-262.htm
8. Saidani, O., Rolland, C., Nurcan, S.: Towards a generic context model for BPM. In: 2015 48th Hawaii International Conference on System Sciences, Kauai, HI, pp. 4120–4129 (2015). https://doi.org/10.1109/HICSS.2015.494

Analysis of Laser-Induced Saturated Interferences on a Thermal Imager at Different Incident Angles

Li Shengpeng[1,2(✉)] and Xia Min[1]

[1] Huazhong University of Science and Technology, Wuhan 430074, People's Republic of China
763721668@qq.com
[2] Ordnance Sergeant College of Army Engineering University, Wuhan 430075, People's Republic of China

Abstract. This study focuses on saturated inference of low-energy laser on the thermal imager and quantitatively analyzes laser power densities and the corresponding saturated laser spots at different incident angles by taking into account the effects of interference laser, atmospheric turbulence, the lens and the photoelectric detector. The relative position of the laser spot on the photosensitive surface on the detector of the thermal imager from the target image is adopted as the evaluation index of laser-induced saturated interference effect. According to the present results, at an incident angle of 0, the target image is partly or completely covered by the interference laser spot, with obvious interference effect and low required laser power. As the incident angle increases, the interference effect is significantly affected and the interference effect is lower than the condition at an incident angle of zero. The interference can be achieved by controlling both size and position of the laser spot.

Keywords: Saturated interference · Laser spot · Off axis laser · Interference effect · Incident angle

1 Introduction

Laser-induced interference and failure mechanisms on a thermal imager are quite complex and generally can be divided into soft damage and hard damage, respectively [1]. Soft damage is a kind of temporary damage and the thermal imager can return to normal state. When the energy density of irradiated laser reaches a certain degree, the pixel value reaches the maximum [2], the components in the thermal imager are under saturated state so that the imager loses the capability of object detection for a short time [3]. By contrast, hard damage can be regarded as a kind of permanent damage beyond recovery; under large-power laser radiation, the components in the thermal imager are completely damaged so that the imager permanently loses the capability of target detection [4]. Since the detector is saturated under soft damage interference, soft damage can also be referred to as saturated interference. Soft damage interference generally aims at interfering the thermal imager's normal operating condition. When the thermal imager is at under saturated interference state, the pixels corresponding to

M. Tavana et al. (Eds.): IISA 2020, AISC 1304, pp. 377–383, 2021.
https://doi.org/10.1007/978-3-030-63784-2_48

the interference laser spot on the detector operate under saturated state, and the image of the interference laser spot is highlighted compared with the images in non-interference regions; at that time, the target image in the interference region is partly or completely submerged [5].

Currently, scholars have conducted a great deal of research on the mechanisms and effect of directional-laser saturated interference infrared imager [6] and mainly focused on close-range experimental calculation in laboratory and software simulation estimation. During these experiments, it was always assumed that the laser is dead against the thermal imager for radiation so as to evaluate the experimental effect. However, in actual laser interference, the direction of the radiated laser can hardly be in point-to-point alignment with the direction of the optical axis of the thermal imager, i.e., there exist certain angles between two directions. The existence of the interference angle significantly affects the laser interference, which has been poorly investigated so far. The author is mainly engaged in laser directional interference experiment and related teaching and research work. This paper theoretically analyzes the influence of incident angle change on laser interference effect, and gives the evaluation factor of saturation interference, which provides theoretical support for carrying out laser directional interference experiment in the field.

2 Analysis of Laser-Induced On-Axis Interference (a $\neq$ 0)

The experimental condition at an incident angle (a) of zero, corresponding to axial point-to-point interference, is easily realized at close range in the laboratory but can hardly be realized in outdoor experiment and requires accurate optical aiming system for auxiliary adjustment. Under fixed outdoor interference and non-mobile target (i.e., the thermal imager), the power density on the focal plane is related to the atmospheric transmission medium, the parameters of the interference laser source and parameters of the thermal imager [7].

2.1 Size of the Interference Laser Spot on the Detector Surface

Assuming that R denotes the distance between the interference point and the interference object, θ denotes the divergence angle of the interference source and f denotes the focal length of lens, the radiant power is focused on the focal plane of the thermal imager via the optical system. In actual interference, the interference point is generally far away from the interference object; otherwise, too close distance between these two objects can easily lead to interference failure. Based on the principles of optics, the interference point can be regarded as an infinitesimal point. Assuming that D_2 denotes the diameter of the laser spot after laser transmits through the atmosphere and arrives at the focal plane, and D_0 denotes the clear aperture of the optical lens of the interference object, the interference laser spot on the focal plane can be written as:

$$S_{laser} = \frac{1}{4}\pi f^2 \left(\theta + \frac{2.44\lambda}{D_0}\right)^2 \tag{1}$$

When the laser enters the thermal imager, the center of the interference laser spot is coincident with the detector center and the interference laser spot is located at the center of the view field [8].

2.2 Imaging Size of the Target

The imaging size of the target on the focal plane, denoted as S_{target}, can be calculated as:

$$S_{target} = \frac{\pi HWf^2}{R^2} \tag{2}$$

where H * W denotes the size of the target, R denotes the distance between the target and the thermal imager, d_0 denotes the size of the pixel on the detector surface, and IFOV denotes the spatial resolution.

Next, ρ is defend as the evaluation factor of laser interference effect and can be calculated as:

$$\rho = \frac{S_{laser}}{S_{target}} = \frac{(D_0\theta + 2.44\lambda)^2 R^2}{4D_0^2 HW} \tag{3}$$

When $\rho \geq 1$, the interference laser spot completely covers the target and cannot be observed, suggesting best interference effect; at that time, the interference object cannot normally operate. When $\rho < 1$, bright interference spot can be observed on the target image, which affects target reconnaissance and tracking; at that time, the interference object is under certain restrictions in terms of functions.

3 Analysis of Laser-Induced Off-Axis Interference ($a \neq 0$)

Off-axis interference experimental condition is quite harsh. The field close-range interference can be completed with the aid of photoelectric detection and tracking system. For example, the tracking and aiming system based on "cat's eye benefit" is greatly affected by the transmitting power, jamming distance and laser incidence angle [9]. During the field experiment, consistent axis can hardly be ensured at a far distance, i.e., there generally exists a certain angle. Even with fixed field interference and non-mobile target (the thermal imager), a certain angle error still exists.

3.1 Variation Rule of Laser Power Density on the Focal Plane

When the interference laser enters into the thermal imager at an angle of a, the laser interference energy into the lens decreases. At that time, the power density on the focal plane, denoted as p_f, can be written as:

$$p_f = \frac{3.352\tau_0\tau_1 P_0}{\pi R^2\theta^2}\left(\frac{D_0}{d_0}\right)^2 \cos\alpha \tag{4}$$

p_f is related to the parameters of atmospheric transmission, interference laser source and thermal imager as well as the laser radiation angel [10, 11]. When the distance target in far field is under interference, the laser energy required for the saturation of the laser imager is great, and the laser beam can be appropriately condensed.

3.2 Variation Rules of the Interference Laser Spot on the Focal Plane

When the interference laser enters into the thermal imager at an angle of a, the interference laser spot deviates from the pixel center of the detector, as the relation between deviation direction and position shown in Fig. 1.

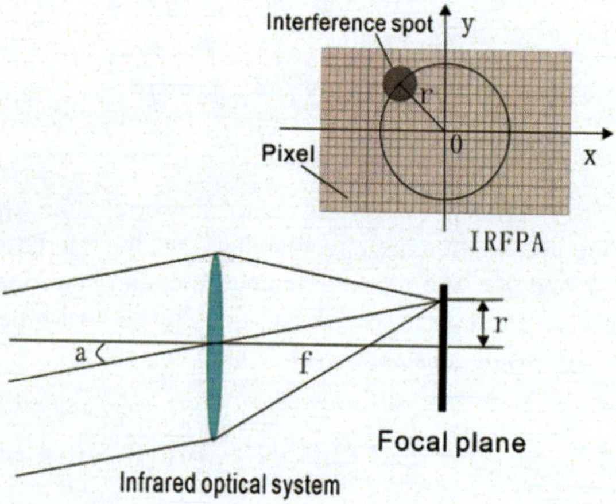

Fig. 1. Position of the interference laser spot.

The distance between the spot center and the detector center r can be written as:

$$r = f\tan\alpha \tag{5}$$

$$\alpha = \arctan\left(\sqrt{\frac{x^2+y^2}{f^2}}\right) \tag{6}$$

The focal distance of the thermal imager (f), the size of the infrared detector and the spacing between pixels are 57 mm, 384 * 288 and 25 um, respectively. At that time, -4.8 mm $\leq$ x $\leq$ 4.8 mm and -3.2 mm $\leq$ y $\leq$ 3.2 mm. Fig. 2 displays the effect of the incident angle of the laser on the interference laser spot.

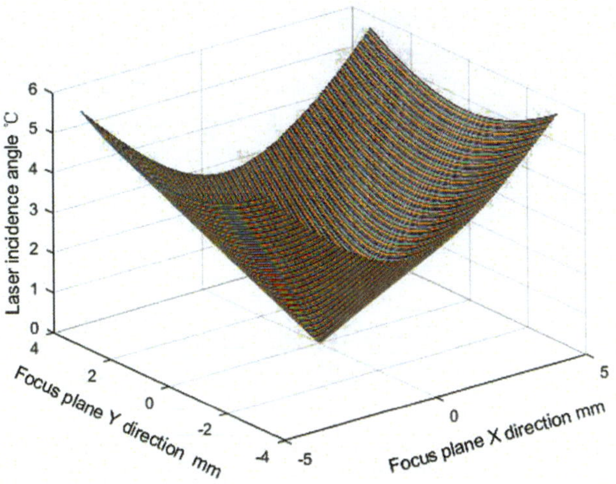

Fig. 2. Relation between the position of the interference spot and the laser's incident angle.

It can be seen from Fig. 2 that the saturation interference effect depends not only on the size and brightness of the laser spot, but also on the laser incident angle. When the incident angle is small, the spot is mainly located in the center of the field of view, which is basically consistent with the area where the target image is located. At this time, more target information is covered, and the interference effect is good; when the incident angle increases, the interference spot is gradually away from the target image, and the coverage is getting smaller and smaller, and the interference effect is not obvious [12–15].

Next, the effect of the incident angle on the interference under one-dimensional condition was considered. Let x = 0, the following expression can be derived:

$$y = f \tan \alpha \tag{7}$$

The position of laser spot is in direct proportion with the tangent value of the incident angle. As the incident angle increases, the value of y approximately shows linear motion. The laser interference effect varies slaserly with the position of the interference laser spot [16]. The distance between the spot and the target center can be used to reflect the occlusion degree of laser interference on the target, and to measure the jamming effect of laser [17–20]. Accordingly, η, defined as the evaluation constant of laser interference effect, can be written as:

$$\eta = \frac{r}{d_{\max}} \tag{8}$$

where d_{max} denotes the maximum distance between the single pixel at the edge of the target spot on the detector and the target center.

When $\eta \leq 1$, the interference bright spot exists on the target image and affects target reconnaissance and tracking, and the interference object function is restricted to certain degree. When $\eta > 1$, the interference bright spot is far away from the target image, which corresponds to the addition of a target image, and the target can be normally scouted; however, the newly-added interference spot can affect the target tracking function for image tracking system.

4 Conclusions

This study focuses the application background of the laser-induced saturated interference thermal imager and analyzes laser power density and the corresponding saturated laser spot under different incident angles. The relative position of the laser spot of the interference laser on the photo-surface of the detector of the thermal imager away from the target image is adopted as the evaluation index of laser saturated interference effect. Results show that, at an incident angle of zero the interference laser spot partly or completely cover the target image, the interference effect is significant, and the required interference laser power is low; as the incident angle increases, the position of the laser spot changes and the interference effect is greatly affected. The interference effect at a certain incident angle is lower than the value dead against the thermal imager. The interference at a certain incident angle can be realized by controlling both size and position of the laser spot. The present study can provide insight guidance for optimizing the interference strategy and effectively arranging the interference points under far-field laser interference so as to analyze and evaluate laser saturated interference effect.

References

1. Liu, L., Zhang, Y., Wu, Z., et al.: Research on technology development of airborne directional infrared countermeasure equipment. Electro-Optic Technol. Appl. **35**(3), 17–22 (2020)
2. Zhou, B., He, X., Liu, H., et al.: Research on Laser Irradiation Uncooled Microbolometer Based on Finite Element Analysis [EB/OL]. https://kns.cnki.net/kcms/detail/51.1125.TN.20191104.1059.002.html. Accessed 11 Apr 2019/07 Apr 2020
3. Han, M., Nie, J., Ye, Q., et al.: Damage proceeding and effects of damage on imaging capability of charge coupled device by 1.06μm continuous laser. Chin. J. Lasers **45**(9), 0901004 (2018)
4. Wang, S.: Study on laser-induced CCDdetector vulnerability and survivability and fussy synthetic evaluation on CCDjamming effects. A Ph.D.'s doctoral thesis from Changchun Institute of Optics, Fine Mechanics and Physics, Chinese Academy of Sciences, China (2002)
5. Zhang, Y., Zheng, R., Liu, J.: Analysis of pulsed laser disturbance and damage on satellite-borne detector. J. Electron. Inf. Technol. **28**(9), 1758–1760 (2006)
6. Zhou, J., Guo, J., Fu, Y.: Analysis of laser interference on remote target photodetectors. Appl. Optoelectronic Technol. **25**(4), 326–328 (2004)

7. Li, H., He, Y., Meng, Q.: Analysis of laser disturbance and damage to IR homing anti-warship missile.J. Aeronaut. Eng.Inst. **21**(3), 357–360 (2006)
8. Dang, H., Li, H., Zhao, E., Luo, W.: Research on laser jamming imaging guided missile. Electron. Opt. Control **23**(3), 41–44 (2016)
9. Chen, L., He, H., Wan, Y., et al.: The application of "cat-eye" effect in real-time evaluation of laser Directional jamming effect [EB/OL]. https://kns.cnki.net/kcms/detail/51.1125.TN.20191129.0933.002.html. Accessed 29 Nov 2019/04 July 2020
10. Sun, Y., Cheng, X, Wang, F.: Method of quality evaluation aimed at laser-disturbing image. Infrared Laser Eng. **36**(5), 659–662 (2007)
11. Wang, W., Jia, X., Han, Y., et al.: Infrared imaging modeling and simulation of DIRCM laser. Infrared Laser Eng. **45**(6), 0606005 (2016)
12. Qian, F., Sun, T., Guo, J., et al.: Evaluation algorithm of laser jamming effect based on distribution characteristics of feature points. Chin. J. Lasers **41**(5), 0509001 (2014)
13. Li, K., Shao, F., Jiang, G., et al.: Objective quality evaluation method of stereo image based on sparse representation. J. Optoelectron. Laser **41**(6), 0609004 (2014)
14. Gao, W.: Study on evaluation method of laser blinding jamming effect. Opt. Technol. **32**(3), 468–471 (2006)
15. Zhang, Y., Li, X., Hao, C.: Evaluation of laser jamming effect on satellite based on image feature correlation. Laser Infrared **47**(3), 352–356 (2017)
16. Wang, D., Zhang, H., Qin, X., et al.: Experiment study on jamming and damage thresholds of polycrystalline silicon detector irradiated by CW CO_2 laser. J. Appl. Opt. **36**(3), 475–479 (2015)
17. Tang, J., Luo, X.: Laser interference effect Evaluation method of based on laser spot and image features. J. Optoelectron. Laser **27**(11), 1220–1227 (2016)
18. Qian, F., Sun, T., Shi, N., et al.: Evaluation of laser jamming effect combined with spot and target characteristics. Opt. Precis. Eng. **22**(7), 1896–1902 (2014)
19. Wu, Y., Sun, X., Yan, F., et al.: Scale analysis of interference image based on edge intensity similarity. Acta Photonica Sinica **43**(1), 1–7 (2014)
20. Li, Z., Zhang, S., Zhou, J.: Study on the evaluation method of infrared anti-jamming effect based on image features. Laser Technol. **37**(3), 413–416 (2013)

Demand Analysis and Application Prospect of 3D Video Fusion Technology in Waterway Traffic Management

Zhaohui Wu[1], Haihua Wang[2], Changxing Ren[3], Jing Deng[1(✉)], and Xiaobo Wu[1]

[1] China Academy of Transportation Sciences, Beijing 100029, China
jingdeng1994@163.com
[2] Huzhou Port and Shipping Administration, Huzhou 313000, Zhejiang, China
[3] Zhejiang Port and Shipping Administration, Hangzhou 310011, Zhejiang, China

Abstract. Three-dimensional video fusion technology brings new opportunities for waterway traffic safety surveillance and smart management. This paper focuses on the application needs and innovation of three-dimensional video fusion technology in waterway traffic management. Firstly, a novel three-dimensional video fusion solution is proposed for waterway traffic safety monitoring on the basis of a summary of the mainstream technologies of video fusion at home and abroad. Secondly, the application requirements of three-dimensional video fusion technology are analyzed in detail by combing the characteristics of the water traffic safety management business. Finally, the smart management of waterway traffic based on 3D video fusion is expected to provide a reference for the deep integration and innovative development of technology and demand. In general, three-dimensional video fusion technology and system will help existing information infrastructure to play a greater role and help improve the governance of water transportation.

Keywords: Three-dimensional video fusion · Virtual-real fusion · Demand analysis · Intelligent management · Waterway traffic manage

1 Introduction

Waterway transportation plays an important role in the integrated transportation of China. The rapid development and popularization of new equipment and technologies, such as sensors, HD cameras, Internet of Things (IoT) and artificial intelligence, bring new opportunities and challenges to the safety and intelligent management of waterway traffic [1] based on surveillance video.

The intelligent management of waterway traffic based on video surveillance is always a challenging task with the characteristics of diverse scenes, complex areas, limited field conditions, great environmental impact, difficult to dynamic tracking, etc. On the one hand, different scenarios have different requirements for the monitoring area coverage, monitoring focus and acquisition accuracy. As the key monitoring areas of waterway traffic include various scenarios such as ports, important waterways, ship

M. Tavana et al. (Eds.): IISA 2020, AISC 1304, pp. 384–392, 2021.
https://doi.org/10.1007/978-3-030-63784-2_49

locks, anchorages, waterborne filling stations, and dangerous goods terminals, etc. On the other hand, different businesses have different requirements for video surveillance. As the waterway traffic management includes operation monitoring, daily inspection, safe operation, emergency disposal, asset management, tracing evidence, etc.

China has formed an inland river perception network basically with the rapid development of nearly 20 years, and the coverage of video surveillance has been greatly improved. It has realized the basic control of important ports, backbone channels, large tonnage and key ships' operation status, and effectively improved the management capacity of waterway safety. However, as the scale of video surveillance continues to increase, the traditional matrix video wall display mode exposes many outstanding problems, such as independent dispersion of surveillance scenes, large number of pictures, poor spatial relevance, prone to visual fatigue, etc. These problems greatly hindered the precise positioning ability and response efficiency of waterway monitoring.

Three-dimension video fusion technology [2] is an integrated monitoring system that uses virtual reality technology to achieve 3D panoramic video fusion. Through the integration and optimization of existing video resources, it integrates discrete surveillance videos with different field angles and 3D models of surveillance scenes, to form the spatial association between different video images in the scene. It helps to reduce the pressure on managers to understand real-time video, and improve the efficiency of daily waterway monitoring, forensics of violations and emergency response.

2 Related Works

Virtual-real fusion technology brings a new idea for the intelligent management and application of massive video. According to the different dimensions of the organization video using scene, the current methods related to the virtual-real fusion of multi-video streams and virtual scenes can be divided into four categories: video tag map, video image splicing, video superposition transition in 3D scene, video and 3D scenes fusion.

2.1 Video Tag Map Method

Video tag map method [3, 4] is respond to the demand of video surveillance for the effective organization of multi-site videos. Its fundamental method is place and watch the video content displayed in a 2D image state on the 2D basemap. At present, many institutions in the world have carried out research on video tag map method, such as American FX Palo Alto Laboratory, American Mitsubishi Electric Research Laboratory in the United States, Korean Electronics and Telecommunications Research Institute, and the French Atomic Energy Commission, etc.

2.2 Video Image Stitching Method

Video image stitching refers to a technique of spatially registering image sequences that have overlapping parts with each other to perform image transformation, resampling, and image fusion, so as to form a wide-angle or 360-degree panoramic image. In general, the result of splicing a small number of pictures is not conducive to dynamic interactive operation because its viewing angle is too narrow. The panoramic images covered by the 360-degree viewing angle can be displayed in cube, sphere and other ways, which can realize interactive operations and can easily provide virtual and real scenes for VR helmets and other devices.

Video image stitching method [5, 6] can realize the panoramic stitching fusion of multiple video streams, and can interact with the viewpoint movement and rotation near the shooting viewpoint. However, since the panoramic spliced image only uses a fixed virtual model of a spherical surface, the viewpoint is limited to the vicinity of the shooting viewpoint, and the interaction mode is limited.

2.3 Method for Superimposing Video Image into 3D Scene

The method for superimposing video image into 3D scene [7, 8] is a virtual-real fusion method based on 2D and 3D feature registration. This method allows users to view the superimposed results on the transfer path of the camera's viewpoint, which is superior to the indexing method of 2D plan views. At present, many commercial companies have used this method for the virtual fusion of video, such as Microsoft, Dutch Delft University of Technology, and The University of Leeds.

2.4 Video and 3D Scene Fusion Method

The principle of video and 3D scene fusion is to capture the video images of real object by camera, and register the video images to the virtual environment in real time with texture way. This method [9, 10] can enhance the virtual scene, representing the status of real objects and responding to interactions. This method is freer in the range of viewpoint selection, allowing users to observe the fusion result from any virtual viewpoint. Typical systems based on this method include the Video Flashing system of American Sanoff Corporation, the Ave system of the University of Southern California, the Augmented Earth of Georgia Institute of Technology, the immersive indoor monitoring system (HouseFly) of MIT.

In summary, the video tag map method is a 2D method, which is simple to implement but always needs to display video separately from the base image and other video, so as to truly reflect the perception effect of the fusion of video and 3D space. The video image stitching method is a 2D+ method, but the fixed virtual model of the spherical surface restricts the viewpoint to only the vicinity of the shooting viewpoint, so the freedom of interaction is limited. The method for superimposing video image into 3D scene is a pseudo 3D method, which realizes the superimposition and fusion display of multiple videos, but the viewing angle is limited to the camera path. The video and 3D scene fusion method allows users to observe the fusion result from any virtual point of view, which is better than the other three types of methods in terms of display effect and interactivity. The video and 3D scene fusion method can enhance virtual scenes, express the status of real objects and respond to interactions. It is an important direction for the future development of virtual-real fusion technology and has broad application prospects.

3 Three-Dimension Video Fusion Solution for Waterway Traffic Safety Surveillance

American Sandia National Laboratories has done a special study, and the results show that after only 22 min of staring at the scene video, the human eye will turn a blind eye to more than 95% of the activity information in the video. Therefore, the traditional video wall monitoring method makes it is difficult for the monitoring personnel to quickly recognize the monitoring scene. By fusing various surveillance camera videos in virtual reality scenes, 3D video fusion technology provides a dynamic evolution environment of virtual reality that integrates various videos, and constructs 3D surveillance scenes consistent with real scenes, so as to reduce the cognitive pressure of surveillance personnel.

In this work, a novel three-dimension video fusion solution is proposed for wateway traffic safety surveillance based on the virtual-real fusion technology, which belongs to the category of video and 3D scene fusion methods. This solution integrates a large number of real-world videos into virtual reality and constructs a virtual reality world that reflects the dynamic changes of the real world in real time, so as to quickly grasp the evolution status and historical changes of the real world. This solution can change the commonly used matrix video wall display mode, and achieve new virtual reality monitoring effects such as full coverage of large scene areas, automatic virtual patrols, remove building ceiling "God" viewpoint.

The comparison between the 3D video fusion monitoring system and the traditional matrix video monitoring system is shown in Fig. 1.

Fig. 1. Example of traditional matrix monitoring and 3D video fusion method

Figure 1(a) shows traditional matrix surveillance video images, 6-channel 2D surveillance video is very similar, so it is difficult to directly distinguish the location of each video in real space. Figure 1(b) shows the effect of 3D video fusion. The integrated monitoring of a wide range of scenes is realized by projecting 2D video into 3D space, and the position of each video in the real scene can be accurately recognized.

4 Demand Analysis of Waterway Traffic Management Based on 3D Video Fusion

In this section, the business characteristics of waterway traffic management are sorted out, and the demand for 3D video fusion technology in waterway traffic safety management is analyzed.

4.1 Features of Waterway Traffic Safety Management Business

Waterway traffic scenarios include ports, important waterways, ship locks, anchorages, waterborne filling stations, dangerous goods terminals, and so on. The main business involved in waterway traffic management includes facility management, daily inspection, emergency disposal, safe operation, etc.

Facility Management: The competent department conducts unified management on the important facilities and equipment of waterway transportation, including static account management and dynamic state information management of important facilities [11], to integrate all kinds of important equipment and facilities scattered in the management area, realize the effective association of dynamic and static information, and then accurately grasp the distribution and operation status of waterway transportation related facilities.

Daily Inspection: Carry out regular and random daily inspections on key ports, important waterways, ship locks, anchorages, waterborne filling stations, dangerous goods terminals and other key areas, to timely discover hidden safety hazards and irregular operations related to infrastructure, ships, meteorology, hydrology, etc. so as to promptly warn and remind waterborne vessels, operating units and management personnel.

Emergency Disposal: Emergency traffic incidents include waterway traffic accidents, storms, natural disasters, etc. In case of emergency traffic events, timely response and rapid handling are required. Combined with the preparation and drills of emergency plans, to improve the response capacity of emergency events and improve the efficiency of emergency disposal.

Safe Operation: According to the safety factors of waterway traffic operation and maintenance, through data monitoring, multi-source convergence, data processing, correlation analysis, early warning and reminder, emergency linkage and other processing, the safety of infrastructure, personnel, environment, ship operation is monitored to ensure the safe operation management of waterway traffic and improve the safety of waterway traffic operations.

4.2 Demand Analysis of Waterway Traffic Management for 3D Video Fusion

Combined with the technical advantages of 3D video fusion and the business characteristics of waterway traffic management, this section analyzes the requirements of waterway traffic management based on 3D video fusion, mainly including 3D panoramic virtual-real fusion display, 2D/3D bidirectional linkage, video data synchronization control, intelligent maritime automatic cruise, augmented reality multi-source information fusion display, intelligent maritime video data analysis.

Global Monitoring of Important Scenarios: Aiming at important waterway monitoring scenes such as ports, waterways, ship locks and anchorages, construct a virtual scene space environment. Through camera video fusion, it provides a top-view effect of a large-scale scene from the sky, helping users to grasp the overall situation changes of the target area from a macro perspective, grasp the overall situation changes of the monitoring area from a panoramic perspective, and realize continuous monitoring of a large area.

Automatic Cruise of Waterway Traffic: By setting the patrol control path in advance, the intelligent automatic patrol changes the manual single-click patrol mode to realize automatic video patrol, which greatly reduces the visual fatigue of the patrol and security control personnel, and improves the monitoring efficiency and camera utilization rate.

2D/3D Bidirectional Linkage of Emergency Disposal: Grasp the overall situation of the monitoring area in the 3D scene, and timely discover the emergency events to be handled. Based on 3D panoramic virtual-real fusion display, 2D map (GIS) global navigation display, real-time virtual-real fusion of local areas in the 3D map and other linkage technologies, 2D and 3D rapid linkage can be realized to discover and respond to emergency events.

Safe Operation Multi-source Information Fusion Monitoring: According to the requirements of waterway safety routine supervision business, data from multiple sources can be accessed as needed, such as vessel traffic information, weather, RFID, temperature, humidity and other sensor data, and these relatively scattered sensor data in space are fused in an organized manner and displayed in the 3D virtual-real fusion picture, to help users better understand the scene situation.

Forensics Traceback Controlled by the Same Clock: The partial playback mode of traditional single-channel video cannot visually and intuitively show the ins and outs of the entire event. Efficient traceability in the panoramic scene display mode ensures the overall playback of waterway safety historical events in the entire scene, describes the ins and outs of the event with the overall picture, and thus improves the ability to query historical events and the full scene recurrence of the event, saving a lot of manpower.

Intelligent Maritime Video Data Analysis: Combined with the video analysis server to realize the statistics of ships in key areas and the alarm function in the no-sail area. In the key navigation areas, by combining the surveillance video data of multiple deployment points, the number of ships can be counted in real time, and an alarm message is issued when the number of ships clearly exceeds the set threshold, to remind users of the necessary preventive deployment to effectively avoid possible navigational safety incidents.

5 Application Prospect of Waterway Traffic Management Based on 3D Video Fusion

5.1 Global Monitoring Technology for Large-Scale Water Areas, to Ensure "Fully Visible"

The waterway safety monitoring technology of 3D panoramic video fusion integrates scattered surveillance videos with 3D scene, restores the 3D scene with the video fragments, and provides the global observation perspective of a large area with "God's perspective", to realize dynamic display of large-scale monitoring scene in "one picture". Thus, it ensures that the monitoring areas is "visible and completely visible", and effectively reduce the pressure on managers to cognitive and understand the real-time video. An example of the top view global monitoring effect is shown in Fig. 2.

Fig. 2. Example of the top view global monitoring effect

5.2 Real-Time Fusion and Display Technology of Multi-source Sensor Information Based on Augmented Reality, to Ensure "Recognizable"

Real-time access to vessel traffic information, weather, RFID, temperature, humidity, and other sensor data, these relatively scattered sensor data in space are organized and fusion displayed in the 3D virtual-real fusion picture, and the multi-source sensor data are fusion displayed timely in the way of augmented reality (AR). Based on the business logic and management requirements, the video analysis algorithm is developed to timely alarm the abnormal information of the scene, and automatically schedule the alarm area picture to help users better understand the scene situation.

5.3 2D/3D Monitoring Association and Collaborative Supervision Technology, to Ensure "Manageable"

Realize the real-time linkage of 2D maps and 3D panoramas, and take into account the global dynamics and local details, to meet the needs of spatial scene monitoring at different scales. It supports the linkage of 3D scenes and dome cameras. It can automatically schedule the surrounding relevant dome cameras, and quickly locate and focus the picture to the location specified by the user, so as to realize multi-angle and all-round monitoring of the attention position. For important or temporary control points, the monitoring screen can be permanently placed on the interface, to assist users quickly understand and respond to situation changes of the control points.

6 Conclusion and Future Work

Smart management of waterway traffic will help to make the existing information infrastructure more effective by applying 3D video fusion technology, and improve the governance capacity of waterway traffic. Based on the review of the current status of 3D video fusion technology and applications at home and abroad, and combined with the characteristics of waterway traffic safety management business, this paper analyzes the application requirements of waterway traffic management based on 3D video fusion technology, and forecasts the application prospect of water traffic management based on 3D video fusion. The research results show that 3D video fusion technology is helpful to transform the traditional manual-based management mode without changing the field deployment, and realize "fully visible", "recognizable", "manageable" in the daily management of waterway traffic, which has a high application prospect in the waterway transportation industry.

Acknowledgements. This work was supported by the Key science and technology project in the transportation industry of Zhejiang (Grant No. 2019013), Key technology projects in the transportation industry in 2019 (Grant No. 2019-ZD2-005), and thc Construction Project of China Knowledge Centre for Engineering Sciences and Technology (CKCEST-2018-4-2, CKCEST-2019-1-9, CKCEST-2020-2-11).

References

1. Wu, Z.H., Wu, X.B., Wang, L.: Prospect of development trend of smart transportation under the background of building China into a country with strong transportation network. Transp. Res. **5**(4), 26–36 (2019)
2. Wu, Z., Wang, L., Fu, Z., Zhu, L., Dou, F., Xu, P.: VR+ BIM: perception and design optimization of highway. In: 2018 International Conference on Virtual Reality and Visualization (ICVRV), pp. 164–165. IEEE (2018)
3. Ivanov, Y., Wren, C., Sorokin, A., et al.: Visualizing the history of living spaces. IEEE Trans. Visual. Comput. Graphics **13**(6), 1153–1160 (2007)

4. Gay-Bellile, V., Lothe, P., Bourgeois, S., et al.: Augmented reality in large environments: application to aided navigation in urban context. In: IEEE International Symposium on Mixed and Augmented Reality, Seoul, South Korea, pp. 225–226. IEEE Computer Society Press (2010)
5. Zhong-Feng, L., Guang-Sheng, W., Sui-Ming, F.: Video image stitching and its application. China Cable Telev. (2003)
6. Kuranov, A.: Image and video stitching and viewing method and system (2008)
7. Veas, E., Grasset, R., Kruijff, E., et al.: Extended overview techniques for outdoor augmented reality. IEEE Trans. Visual. Comput. Graphics **18**(4), 565–572 (2012)
8. Milosavljević, A., Rančić, D., Dimitrijević, A., et al.: Integration of GIS and video surveillance. Int. J. Geogr. Inf. Sci. **30**(10), 2089–2107 (2016)
9. Kim, K., Oh, S., Lee, J., et al.: Augmenting aerial earth maps with dynamic information from videos. Virtual Reality **15**(2–3), 185–200 (2011)
10. Chen, S.C., Lee, C.Y., Lin, C.W., et al.: 2D and 3D visualization with dual-resolution for surveillance. In: Computer Vision and Pattern Recognition Workshops, Providence, RI, USA, pp. 23–30. IEEE Computer Society Press (2012)
11. Wu, Z.H., Li, Q., Zhu, L., et al.: Demand and innovative application of BIM in road facilities maintenance. Transp. Res. **6**(3), 116–128 (2020)

Development Strategies of Tianjin's Cultural and Creative Industry (CCI) Based on the Beijing-Tianjin-Hebei Coordinated Development

Nan Zhang(✉) and Xugao Qi

Tianjin University of Technology and Education, Tianjin 300222, China
Stephanie0504@163.com

Abstract. Under the guidance of the national strategy of Beijing-Tianjin-Hebei Coordinated Development in China, the development of Tianjin's cultural and creative industry (CCI) is facing unprecedented historical opportunities. This paper firstly combs the relevant literatures about CCI. Then, a SWOT analysis is made for Tianjin, which sorts out the strengths, weaknesses, opportunities, and threats of Tianjin's current industrial situation. The development strategies for Tianjin's CCI are finally proposed. These strategies include innovating management mechanism under the government's overall planning; focusing on distinctive local culture and positioning key industrial branches to form a linkage effect; supporting the creation of original works and forming a complete industrial chain of creativity, production and promotion; accepting high-end talents overflow from Beijing and building a talents cultivation mechanism; and building a new format of intelligent cultural and creative industry by in-depth integration of intelligent technology and cultural industry.

Keywords: Cultural and creative industry · Tianjin · Strategies · Beijing-Tianjin-Hebei Coordinated Development

1 Introduction

1.1 A General Introduction About the CCI and Tianjin

The cultural and creative industry (CCI) uses creation, creativity, and innovation as the basic means to create cultural content and creative achievements. It is an industrial cluster under intellectual property protection, meanwhile provides various cultural experiences and realizes consuming transactions [1]. The CCI is a product of the deep integration of culture, technology and economy. It has the characteristics of high added value, high integration, high influence, and high growth rate. The CCI is an important support for contemporary economy and society. Its development degree and scale are important indicators to measure the comprehensive competitiveness of a country or a city [2].

As an important economic center and an open coastal city in Bohai Rim region of China, Tianjin has profound cultural heritage and rich cultural resources, which are

M. Tavana et al. (Eds.): IISA 2020, AISC 1304, pp. 393–404, 2021.
https://doi.org/10.1007/978-3-030-63784-2_50

great potentials for the development of its CCI. During the "Thirteenth Five-Year Plan" period, Tianjin regarded CCI as an important part of its modern service industry. It actively supported and guided traditional superior cultural industry and promoted the rapid development of emerging cultural industry. At present, Tianjin's CCI is showing a healthy and rapid developmental situation.

Since 2014, Chinese government has promulgated a series of policies to push the CCI under Beijing-Tianjin-Hebei Coordinated Development (BTHCD). Under this special background, what opportunities and challenges will Tianjin face in the future? In line with the trend of BTHCD, what strategic measures should Tianjin adopt in order to make a breakthrough for its industrial development? This paper starts from a background analysis for the BTHCD, and then conducts a SWOT analysis for Tianjin based on a literature review, and finally puts forward strategies to promote its industrial development.

1.2 The Background of Beijing-Tianjin-Hebei Coordinated Development on CCI

The CCI is an important pillar industry for the economic and social development of Beijing-Tianjin-Hebei. It's also a strategic industry that needs priority support and development during the coordinated development of the three places. The theme of the Eighteenth National Congress of the Chinese Communist Party and the Third Plenary Session of the Eighteenth Central Committee, as well as the important speech delivered by President Xi when he was inspecting in Beijing, all advocated that the cultural exchange and cooperation among Beijing, Tianjin and Hebei should be developed to a higher level, a deeper degree, and a wider horizon. Therefore, the coordinated development of the cultural and creative industries in Beijing-Tianjin-Hebei has practical significance on promoting the transformation and upgrading of the economic industries, transforming the economic development mode, and enhancing the open cooperation of the industrial chains [3].

The "Strategic Framework Agreement for Cultural Area Collaborative Development of Beijing-Tianjin-Hebei" promulgated in August 2014 mentioned: "The region of Beijing-Tianjin-Hebei is geographically integrated, humanistic close and culturally connected. It has a deep historical origin and a suitable communication radius. It is one of the important regions in China with rich cultural resources, profound cultural heritages, distinctive cultural characteristics, and dynamic cultural developments. There are broad prospects for the cooperation of the three places." The agreement is a good interpretation for Beijing, Tianjin and Hebei in the cultural field. The cooperation among the three places has the natural advantages of "timeliness, residentialliness, and unitedness". The signing of the agreement will promote the complementary advantages of their cultural industries, and form a pattern for the integrated development of the regional culture and economy by using the respective advantage of the three places. In the following year, Chinese government promulgated the "Outline of Beijing-Tianjin-Hebei's Coordinated Development Plan", which continued to create favorable conditions for the coordinated development of the regional cultural and creative industries. At present, the coordinated development of the three places has achieved positive results. However, their coordinated development is still facing many problems.

Especially for Tianjin, how can it seize the opportunity and seek a greater development for its CCI? It requires the municipal government, the relevant departments, the related enterprises and all sectors of the society to brainstorm, to explore, and to create a future.

2 Review of Related Research

2.1 Research on the Concept of CCI

Nowadays there are different understandings for the concept of CCI. Some typical references are: In 1998, the British government firstly defined "creative industry" in the "Path Document of British Creative Industry" as "the social activities to create wealth and employment opportunities through personal intellectual creativity and the development and application for intellectual property [4]". In 2006, the Chinese government clarified the concept of CCI in the "Outline for National Cultural Development Plan" during the "Eleventh Five-Year Plan" period, and pointed out that the CCI is "an industrial cluster with intellectual property protection, which provides various cultural experiences and consuming transactions. It has cultural content and creative achievements by using creativeness and innovation as basic means". Chinese scholar Zhu Xiaoqing (2006) believed that the CCI was an emerging industry with creativity as the core, it could support and promote the development of various industries [5]. Feng Mei (2009) compared the core areas of the cultural industry and the creative industry, and pointed out that the CCI originated from the cultural industry, it was the highest form of the cultural industry's development [6].

Among the above definitions, the Chinese government's reference from the "Outline of the National Cultural Development Plan" is based on China's national conditions and is closely related to the research background of this paper, so it is adopted and applied by our research.

2.2 Research on China's CCI

Chinese scholars and foreign scholars made many researches on how to promote the development of the CCI in China. Keane (2004) believed that although China's economy had achieved rapid development, its CCI was still in a low developmental level. The key creative concepts was still far from enough, and innovation only stayed on the business model [7]. Liu Jinghua (2018) believed that there was still a gap of cultural and creative industries between China and other major developed countries. To enhance China's international competitiveness on its cultural and creative industries, it should optimize the financing channels, improve the talents training mechanism, perfect the industrial ecosystem and strengthen the intellectual property protection [8]. Lu Xiaoyan (2020) believed that the healthy and sustainable development of CCI was important to enhance a country's soft power. The development of CCI required cross-border thinking and the cultivation of creative talents; strengthening the development of cultural and creative products; and perfecting the legal supervision system [9]. Sun Liwen (2020) took 31 provinces and cities in China as research objects, and measured their cultural and creative industries in space and time through the niche model.

According to the measurement results and the problems from industrial development, expansion strategy, dislocation competition strategy and innovation factors driving strategy were proposed [10].

2.3 Research on CCI Clusters and Its Relationship with Urban Development

More and more foreign scholars began to focus on the research on CCI clusters and its relationship with urban development. Landry (2000) put forward the concept of "creative city". He believed that the concentration of CCI could not only promote the optimization of the urban human environment and functional layout, but more importantly, it could enhance the attractiveness of the city and enabled the city to renew in many aspects [11]. Caves (2002) believed that a large number of practitioners of CCI in a region had positively spawned a conventional industrial platform. The exchanges between enterprises in economic transactions had promoted the aggregation of production systems and geographical environments, which made the city more attractive to target talents. As the accumulation of talents further reduced various transaction costs, a series of international cities gathering cultural and creative industries were born [12].

In China, as the "Eleventh Five-Year Plan" of 2006 clearly proposed to vigorously develop CCI, the research on the clustering and development of CCI continued to increase in the following years. For example, Yu Wentao (2016) used Chinese provincial capitals and sub-provincial cities as samples, measured the production efficiency of the creative industry in China and constructed an empirical research model to explore the impact of creative industry clustering on the urban economy by DEA analysis tools [13]. Wang Yanan (2017) explored the relationship between the network structure of CCI clusters and the dissemination of innovative knowledge by social network analysis, and made an empirical analysis by taking the digital film industry cluster in Wuxi as an example [14]. Chen Hongxia (2018) analyzed the spatial agglomeration characteristics of China's cultural and creative industries based on the data of prefecture-level cities. She concluded that China's cultural and creative industries generally presented a point-like agglomeration state of "local concentration and overall dispersion", and a comparative analysis was made for the agglomeration characteristics of the eastern, central, and western areas [15].

However, it should be realized that, despite the rapid growth of research on CCI in China, the current research is still lacking in theoretical construction, and the interpretation of theories is relatively fragmented and has not yet formed a mature theoretical system. On the other hand, based on the situation of China, the development of its cultural and creative industries is obviously inseparable from the government's macro-control. Although the promulgation and implementation of government policies have played a great role in the development of the industry, the research which take government policies as an important background and explore its impact on the industry has not received enough attention from Chinese scholars.

The above research deficiencies have pointed out the directions for future research. First, in-depth theoretical research should be explored about the CCI's development and its cross-border integration with other industries by the combination of qualitative methods and quantitative methods; second, more attention should be paid to the impact of the macro policies of the Chinese government ("Regional Coordinated Development", "One Belt, One Road", etc.), and relevant strategies should be proposed to promote the industrial development.

Therefore, this paper takes the macro policy of "Beijing-Tianjin-Hebei's Coordinated Development" as a research background and conducts a SWOT analysis for Tianjin's CCI. In this way, it tries to lay a theoretical basis for the following strategic research. The development strategies for Tianjin's CCI are finally proposed.

3 A SWOT Analysis for Tianjin's CCI Under Beijing-Tianjin-Hebei Coordinated Development

The strategy of coordinated development in the cultural and creative fields of Beijing, Tianjin and Hebei provides historical opportunity for Tianjin's development, but it also brings challenges to the city. This paper will conduct a SWOT analysis for Tianjin's CCI, and sort out its strengths, weaknesses, opportunities, and threats. The final purpose of the paper is to put forward the development strategies for Tianjin.

3.1 Strengths

For the past few years, Tianjin's CCI has been developing rapidly, and its overall strength has increased significantly. The average annual added value of the cultural industry has maintained a growth rate higher than 20%. In 2016, the added value of Tianjin's cultural industry was more than 80 billion yuan, accounting for 4.49% of the city's GDP. At present, Tianjin has launched a total of 486 cultural and creative projects with a total investment of 181.3 billion yuan. The city has 22,640 cultural units, including 1086 cultural enterprises above designated size and 5 municipal state-owned cultural enterprise groups, which are initially showing leading roles among the industry [16].

At present, the CCI in Tianjin has formed a relatively sound system. The industry covers cultural creativity and design services, cultural information transmission services, radio and television services, crafts and arts, cultural arts and equipment production. It has basically formed a relatively comprehensive spatial layout covering the mountains, seas, cities and towns.

Tianjin has natural, historical, and cultural resources along the Haihe River, the South Canal, and the North Canal, as well as advanced manufacturing foundations along the coast. On this basis, Tianjin made full use of urban industrial heritages and style buildings and built a number of creative industrial parks or gathering areas. Currently, Tianjin has 35 cultural and creative industrial parks, including 9 national-level CCI clusters, including the National Animation Industry Comprehensive Demonstration Park and the National Binhai New Area Cultural Demonstration Zone, and 19 municipal-level cultural industry demonstration parks and 47 demonstration bases.

3.2 Weaknesses

The knowledge-intensive nature of CCI makes creative talents destined to become the core element of its development. The lack of high-quality talents, especially the lack of complex cultural and creative talents is the main obstacle for Tianjin. At present, compared with the most developed cities in the world and in China, Tianjin has a large gap in terms of the number of employees, as well as the level and the structure of its personnel. The employees in Tianjin's creative industry are mainly concentrated in cultural production. Not only are creative talents, professional marketing talents and management talents scarce, but compound talents who are proficient in various industrial links are even scarcer.

Judging from the overall situation of Tianjin's cultural and creative enterprises, the manufacturing capability is relatively strong while the originality is relatively weak. Therefore, there are few products with independent intellectual property rights, which causes the weak core competitiveness of the enterprises. As a result, Tianjin lacks influential industrial brands, and Tianjin's industrial products are far from well-known throughout China.

At the same time, there are some significant problems in the development of Tianjin's cultural and creative industrial parks. For example, it is difficult for various parks to form their own development characteristics, and there is a tendency to a homogenized competition. Some parks even have an alienation for its business forms. Although they are called cultural and creative industrial parks on the surface, there are actually lack of cultural and creative enterprises within the parks. At a result, the initial positioning and direction of the industrial parks are distorted to some extent.

3.3 Opportunities

In order to promote the integration of Beijing-Tianjin-Hebei's cultural and creative industries, the governments of the three places have reached consensus on coordinated development and division of labor cooperation. A series of policies to promote the development of Beijing-Tianjin-Hebei have been formed, which provide unprecedented opportunities for Tianjin's development.

On the one hand, Beijing as China's cultural center, technological innovation center, and foreign exchange center, is extremely rich in cultural resources. It is obviously the leader in the coordination of Beijing-Tianjin-Hebei. Beijing's cultural industry has the most concentrated human capital and significant comparative advantages in various fields such as cultural tourism, press and publication, culture and art, radio, film and television. Therefore, it occupies an undoubted core position in the coordinated development of the three places, which can realize the radiation effect on the surrounding areas and simultaneously, drive the development of the secondary cultural centers like Tianjin and Hebei.

On the other hand, Tianjin as a "port of national gate" and an "international city", will absorb more economic and cultural resources from Beijing under the background of coordinated development. The two-way flow of resources in the fields of finance, property and materials will be greatly enhanced. In other words, Beijing-Tianjin-Hebei's coordinated development will not only bring a large amount of investment to

Tianjin, but also greatly change its industrial pattern and bring broad space for its industrial upgrading and development. For example, as a specific measure of Beijing-Tianjin-Hebei's coordinated development, Beijing Service Center of Sino-Singapore Tianjin Eco-City began operations in June 2018. In addition to providing various policies and business services for Beijing's cultural enterprises located in the eco-city, the center also provided a platform for the cooperation and exchange between enterprises of Beijing and Tianjin by regularly holding various activities, which effectively promoted the growth and upgrading of Tianjin.

3.4 Threats

The development of CCI is inseparable from the support of the overall level of economy and the degree of service industry. At present, there are unbalanced and uncoordinated problems in Beijing-Tianjin-Hebei's economic development, which are detrimental to the coordinated development of the three places. Let's take only the figures of Beijing and Tianjin as an example: From the perspective of the overall level of economic development, the GDP of Beijing and Tianjin in 2017 were 2.8 trillion and 1.85 trillion respectively. At the same time, from the perspective of the industrial structure, the proportion of service industry in Beijing reached 80.6% in 2017, while that in Tianjin was only 61.5%, which means the industrial structures of the two places are quite different [17, 18]. Therefore, the unbalanced and uncoordinated economic development has made the coordinated development of Beijing and Tianjin trapped in an unfavorable external environment, which has hindered the coordinated development to a certain extent.

From the perspective of development level of CCI, Beijing has a number of key enterprises with national influence in various branches such as new media, animation, media, film and television. Their comparative advantages are outstanding in all aspects such as cultural intelligence, information technology and finance. On the other hand, although the annual average added value of Tianjin's CCI is relatively high, its innovation, technological content and human intelligence still need to be improved. Otherwise, the gap between Beijing and Tianjin will be greater, and the coordinated development between the two cities will be affected, which will be disadvantageous to the long-term development of Tianjin.

At present, there are still many problems in the BTHCD, which are manifested in the shallow cooperation among the three places but the deep cooperation is rare. The content and level of the coordinated development need to be further improved and optimized. One of the main reasons for these problems is the "separation of politics" phenomenon in the administrative management and taxation policy system of the three places. Each place only pursues its local benefits but ignores their mutual cooperation. Over time, not only will the current benefits of Tianjin be damaged, but it will also affect the industrial competitiveness of Beijing-Tianjin-Hebei.

3.5 The SWOT Matrix of Tianjin's CCI

The above SWOT analysis for Tianjin's CCI can be summarized as the following matrix (Table 1):

Table 1. SWOT analysis for Tianjin's cultural and creative industry (CCI)

Strengths—S S1 increasing power of overall industry S2 complete industrial system S3 rapid development of industrial parks	Weaknesses—W W1 be lack of leading creative talents W2 be lack of original ability W3 homogeneity of industrial parks
Opportunities—O O1 favorable policies for coordinated development O2 outflowing and leading function of Beijing O3 inflowing and taking on function of Tianjin	Threats—T T1 unbalanced development of the regional economy T2 greater gap between Tianjin and Beijing T3 restrictions from coordinated development

On the basis of clarifying Tianjin's advantages and disadvantages, it must be recognized that the BTHCD has brought favorable opportunities and meanwhile, severe challenges to Tianjin. The connotation of Tianjin's CCI must be upgraded, its development strategies must be innovated in order to adapt to the new historical positioning.

4 Development Strategies for Tianjin's CCI Based on the Coordinated Development of Beijing-Tianjin-Hebei

The trend of BTHCD requires a strategic top-level design for the CCI of Tianjin. Only in this way the historical opportunities could be seized and the leap-forward development of the city could be realized. The following content starts with three aspects: strategic positioning, strategic objectives and development focus. The development strategies for Tianjin's CCI are put forward by these three aspects.

4.1 Strategic Positioning

The development of CCI is an important breakthrough and growth point for promoting the development of Tianjin's modern service industry. In the next five years, Tianjin should take advantage of the BTHCD, vigorously develop its intelligent CCI and appropriately upgrade its traditional superior cultural industry. It can be expected that, with the expansion of the industrial scale and the improvement of the industrial level, the CCI will become one of the pillar industries in Tianjin's economic development.

4.2 Strategic Objectives

By the end of 2022, the added value of Tianjin's CCI exceeded 100 billion yuan, accounting for 8% of the city's GDP and 20% of the service industry. In the next five years, the annual growth rate of Tianjin's CCI should be no less than 25%, which is higher than the growth rate of the Chinese economy and service industry. Meanwhile, it will leap into the first echelon of Chinese cultural and creative industries.

4.3 Development Focus

Innovating Management Mechanism for CCI under the Government's Overall Planning. With the "Strategic Framework Agreement for Beijing-Tianjin-Hebei's Cultural Area Collaborative Development" as a guide and premise, Tianjin's municipal government should follow the idea of "government guiding, first planning, market operating, enterprises as main roles". The government needs to innovate and diversify cooperative management mechanism, and actively explores a variety of management models for CCI. For example, the government can cooperate with enterprises or sectors from the society to carry out group management, and jointly formulate and improve the overall development plan for Tianjin [19]. The plan needs to form a management system covering management and support content, industry standards and management procedures, etc. It also should include policy provisions providing supports for creative enterprises which are starting their businesses or in early growing stage. In addition, the government should advocate and promote the market operation for CCI, so that the market can play a leading role in land policy, investment financing, business development and network construction. In this way the sustainable development of Tianjin's CCI could be promoted.

Focusing on Tianjin's Distinctive Local Culture and Positioning Key Industrial Branches to Form a Linkage Effect. Taking advantage of the BTHCD, Tianjin should innovate development ideas while actively integrating into the cultural circle of Beijing. It should form a differentiated construction and dislocation development, which is connected but different from Beijing. Tianjin should innovate its industrial development based on the unique "Jin culture". Only by innovation can the transformation and upgrading of Tianjin's industry be finally achieved.

If Tianjin hopes to achieve innovative and sustainable development, firstly, it must clarify its characteristics and development directions. That is, to find out the uniqueness of Tianjin's CCI. For example, Tianjin still retains rich "concession culture" formed from the beginning of the semi-colonial and semi-feudal society. And Tianjin has rich local resources of the "Folk Culture". These are the distinguishing cultural characteristics of Tianjin which are different from other places. Secondly, from the perspective of the branches of Tianjin's CCI, animation and industrial design have obvious advantages over other branches. Therefore, how to use existing superior branches to drive the overall development of the industry? How to form a synergy effect between upstream and downstream industries such as creativity, production and marketing to revitalize the overall industry? This is an important strategic direction which should be seriously considered so that to find a breakthrough for Tianjin's industrial development.

Vigorously Supporting the Creation of Original Works and Forming a Complete Industrial Chain Including Creativity, Production and Promotion. The soul of the CCI is creativity, and its life is innovation. Only by enhancing the design and development capabilities of original works, and encouraging enterprises and individuals to create excellent creative works, can we develop originality, protect originality, and enhance the development connotation and overall level of Tianjin's industry. We must support original works with artistic creativity and high acceptance by the masses, so that Tianjin's CCI could show a vibrant development future. In addition, it is necessary to unite original talents, original enterprises and promotion platforms to form an endogenous operating mechanism for the self-development of the industry. In this way an upstream and downstream linkage industrial chain integrating creativity, production and marketing could be built. It's also important to form a creative brand with Tianjin's characteristics and increase its international influence.

Widely Accepting Beijing's High-End Talents Overflow and Building a Culture and Creative Talents Cultivation Mechanism. Human resources are the most important and active factors in the development of CCI. It is also the driving force behind the sustainable development of the industry. Under the background of coordinated development and facing the overflow of high-end talents from Beijing, Tianjin should accept these talents in an open and inclusive manner, and implement policies and regulations about talents attracting, talents supporting and talents stimulating. If so, these outstanding creative talents will undoubtedly become a new force to promote the development of Tianjin.

In addition, Tianjin should also cultivate its local talents, and build a talent cultivation mechanism through various channels. A database of creative talents information should be established to form a local talent information reserve and promote the flow of talents information among Beijing, Tianjin and Hebei; a joint training mechanism for creative talents should be constructed to build a creative talents team; high-level professional talents and compound high-end talents in various fields should be vigorously cultivated by international exchange, writing creation and original road show; professional technological talents and management talents should be cultivated and a talents exchange mechanism should be established; training and development services for talents should be provided by the respective advantages of the three places.

Building a New Format of Intelligent Cultural and Creative Industry by In-Depth Integration of Intelligent Technology and Cultural Industry. In December 2017, Tianjin Municipal Government promulgated the "Special Action Plan for Tianjin's Intelligent Cultural and Creative Industry", which put forward the guiding ideology, basic principles and development goals for Tianjin's intelligent cultural and creative industry (ICCI). It also explained ten key tasks for Tianjin's ICCI in 2018–2025. The document pointed out a new development direction for Tianjin and opened up a new development situation. It pointed out that it was necessary to use intelligent technology to transform and upgrade current cultural industry and build a new format of ICCI so that to realize the deep integration of intelligent technology and CCI [20]. Therefore, Tianjin should take its advantages of the National Independent Innovation Demonstration Zone and Binhai New Area to gather domestic and foreign innovation resources, actively develop intelligent cultural information service platforms, and break through a number of key technologies. The final objective is to take first mover advantage in the national intelligent cultural creative industry, and lay the foundation for entering the middle and even high-end of the national and global industrial value chain.

5 Conclusion

This paper firstly reviews the background of the BTHCD, then conducts a SWOT analysis for Tianjin based on a literature review. It finally puts forward the development strategies for the CCI of Tianjin. The focus of the strategies includes five aspects: innovating management mechanism, forming industrial linkage effect, supporting original works, building talent cultivation mechanism and promoting the formation of ICCI.

It could be expected that there are following trends for China's cultural and creative industries: technological innovations such as artificial intelligence and block chain will promote the transformation and upgrading of the industry; "regional coordinated development", "One Belt, One Road" and other related government policies will provide more favorable conditions for the industry's development; cross-border integrated development between CCI and other traditional industries will be further deepened; some industrial branches with innovative potential will stand out and win more attention and development space. According to the above expectations, this paper puts forward the development strategies for Tianjin's CCI. We hope it could provide theoretical and practical references for the sustainable development of Tianjin in the future.

Acknowledgements. This work is supported by the 2018 National Educational Science Planning Project 'Research on the Evaluation and Improvement Path of the Integration of Production and Education in Higher Vocational Colleges in Beijing-Tianjin-Hebei Region (No. DFA180311)' and the Research Project of Tianjin University of Technology and Education 'Research on the Derivative System of Small and Micro Enterprises of Science and Technology from the Perspective of Innovative Country (No.SK14-12)'.

References

1. People's Network. https://culture.people.com.cn/GB/22226/71018/4814170.html. Accessed 13 Sep 2019
2. He, Y.: Research on the problems and countermeasures of the coordinated development of the Tianjin and Hebei's cultural and creative Industry. Theory Modernization (3), 26–30 (2016)
3. Li, H., Li, Y., Li, Q.: Research on the coordinated development of Beijing-Tianjin-Hebei's cultural and creative industry based on the industrial chain. Commer. Econ. Res. **34**(2), 189–190 (2017)
4. Wang, J.: Research on the development strategy of cultural and creative industry in Nanjing. Master thesis of Southeast University (2017)
5. Zhu, X.: Discussion on the characteristics and development conditions of cultural and creative industry. New Horiz. (3), 23–25 (2006)
6. Feng, M.: Research on the Development of Chinese Cultural and Creative Industry, 1st edn. Economic Science Press, Beijing (2009)
7. Keane, M., Witting, A.: Brave new world: understanding China's creative vision. Int. J. Cult. Policy **10**, 265–279 (2004)

8. Liu, J.: Restrictive factors and realization ways of improving the international competitiveness of China's cultural and creative industries. J. Party Sch. Fujian Provincial Comm. Communist Party China (1), 109–115 (2018)
9. Lu, X.: How to optimize and innovate in cultural and creative industries. People's Forum (3), 138–139 (2020)
10. Sun, L.: Research on the evaluation of the development of China's cultural and creative industries based on the niche theory. J. Beijing Traffic Univ. (Soc. Sci. Ed.) **19**(1), 64–76 (2020)
11. Landry, C.: The Creative City: A Toolkit for Urban Innovators, 2nd edn. Earthscan Publications Ltd, London (2008)
12. Caves, R.: Creative Industry: Contracts Between Art and Commerce, 2nd edn. Harvard University Press, Cambridge (2002)
13. Yu, W.: Research on the concentration of creative industry and their production efficiency—based on the empirical analysis of provincial capitals and sub-provincial cities. Economist (6), 51–57 (2016)
14. Yanan, W.: The network structure of cultural and creative industry clusters and the flow of innovative knowledge—an analysis based on the perspective of social networks. Sci. Technol. Manag. Res. **11**, 158–163 (2017)
15. Hongxia, C.: The spatial agglomeration characteristics of cultural and creative industries and the comparison of the regional differences—an empirical research based on prefecture-level cities. Urban Dev. Res. **25**(7), 25–33 (2018)
16. Beijing-Tianjin-Hebei collaborative development document. https://gyxxh.tj.gov.cn/zhencwj/65058.htm. Accessed 09 Aug 2019
17. Beijing Municipal Bureau of Statistics. https://www.bjstats.gov.cn/zxfb/201802/t20180225_393332.html. Accessed 27 Feb 2020
18. People's Daily Online. https://tj.people.com.cn/n2/2018/0311/c375366-31329265.html. Accessed 11 Mar 2019
19. Li, Q., Ma, Z.: Investigation and thinking on the development of Tianjin's cultural and creative industries. Art Des. (Theory Ed.) (1), 26–28 (2016)
20. Tianjin Bureau of Industry and Information Technology. https://gyxxh.tj.gov.cn/zhencwj/65058.htm. Accessed 08 Sep 2019

Application Research of Interaction Design in Human-Machine Interface of Automobile

Caizhong Zhang(✉)

School of Visual Communication Design,
Shandong University of Art and Design, Jinan, Shandong, China
zhangcaizhong@sdada.edu.cn

Abstract. With the continuous advancement of technology, the products in people's lives are becoming more and more intelligent, but the existing automotive HUD (head-up display) on the market simply moves the driver's sight upwards, and does not have much content and other aspects. Many improvements. In response to this situation, using a user-centered design concept, this research focuses on the interactive design of automotive HUD, and enhances the driver's driving experience through "panoramic" and mode conversion methods, thereby improving the safety of vehicle driving.

Keywords: Interactive design · HUD · Mode conversion · Panoramic · On-demand display

1 Introduction

In recent years, automobile HUD has gradually become popular, and automobile interaction methods have been continuously developed [1]. However, the existing automotive HUD products cannot better satisfy the driver's driving experience and driving safety. Many products simply display the information on the instrument panel and the center controller by simply moving up the display, simplifying the instrument panel, navigation and other information and projecting it in front of the driver's body, ensuring that the driver's line of sight remains straight [2]. The focus of this research is to provide drivers with better functions without affecting their driving safety, thereby bringing a more intelligent driving experience.

2 Development Status of Automotive HUD

2.1 Market Status Analysis

With the advancement of science and technology and the development of "Internet of Vehicles", automobiles are no longer a simple means of transportation, and there are more and more scientific and technological products in the automotive field [3]. HUD is a typical one.

At present, the number and types of HUD products on the market are relatively small, and their functions are almost similar. There are two main types of HUD: HUD

M. Tavana et al. (Eds.): IISA 2020, AISC 1304, pp. 405–412, 2021.
https://doi.org/10.1007/978-3-030-63784-2_51

in the automotive aftermarket (Fig. 1) and HUD pre-installed by car manufacturers (Fig. 2). The automotive aftermarket HUD refers to a single HUD product, which can be installed and removed freely, and has a high degree of freedom. This also limits the realization of some functions of this type of product, and it has high requirements for the adaptability of the car model. The advantage of pre-installed HUD by automobile manufacturers is that it has a strong exclusive design, a high degree of product fit, and can form its own distinctive product characteristics, and has a better linkage with other systems in the car. Nevertheless, automotive HUD are faced with the dilemma of excessive functions and cumbersome display content, which affect the normal driving of the driver, or abandon some functions to avoid the driver being affected [4].

Fig. 1. Automotive aftermarket HUD

Fig. 2. HUD pre-installed by car manufacturers

2.2 Analysis of Existing HUD Products

This research has investigated the automotive HUD products on the market, and conducted a comprehensive analysis of these products, and selected several representative ones. Automotive aftermarket HUD: Halo, Carrobot, Carloudy; pre-installed HUD by car manufacturers: BMW.

These products are analyzed in terms of installation methods, main functions and interaction methods, as shown in Table 1.

Table 1. HUD product analysis

Features	Brand			
	Halo	Carrobot	Carloudy	BMW
Installation method	Sun visor	Sucker	Non-slip mat	Inline
The main function	Navigation, SMS, phone, music, driving recorder	Navigation, SMS, phone, music	Navigation, gas station information, speed, road conditions, parking information, diet, car wash index, mall	Mileage, navigation, control information, mobile computer, self-light sensing system
Interactive mode	Gesture, voice	Gesture, voice	Button, voice	Gesture, voice, button

Through research and analysis, it is found that the functional differences of current HUD products are not obvious, and most of the HUD functions are basically the same [5]. On the premise of avoiding the impact of the driver's line of sight, the content displayed by the HUD has certain limitations. As a result, HUD can only filter the most important information to display, resulting in the universal consistency of HUD functions. In this regard, the focus of this research is on the one hand the addition and updating of HUD functions, and on the other hand, how to deal with the problem of distracting the driver's sight caused by the cumbersome display content under the premise of satisfying more function points.

3 HUD Design Positioning and Design Goals

3.1 Design Positioning

According to the conclusions obtained from the survey, this research conducted a design positioning analysis of automotive HUD from six aspects: product concept, technical analysis, interaction mode, product highlights, visual design, and functional architecture.

(1) Product Idea
The product is positioned as a pre-installed HUD by car manufacturers, hoping to make more breakthroughs in function points, increase panoramic display methods, and through mode conversion methods, to better solve the problem of driver's distraction caused by excessive display content.

(2) Technical Analysis
With the development of OLED display technology, it has played a key role in realizing the overall visual operation of automobile windshield [6]. The improvement of self-light sensing technology makes the display of HUD not be affected by external light. In addition, the development of RealSense technology and the use of VR technology, 3D gesture operations have brought better interactive operation methods, and the projection of virtual images also provides drivers with an intelligent experience [7].
(3) Interactive Mode
Physical buttons generally can only represent a single functional command, which causes too many physical buttons to occupy a lot of space and touch positions, which will affect the driver's driving experience. Behaviors such as touch-screen operations will cause drivers to distract their sights in order to ensure the accuracy of operations, thereby affecting driving safety. This project adopts voice operation and gesture operation in interactive details, and uses somatosensory technology to complete virtual key operation.
(4) Product Highlights
According to the problems obtained from the research and analysis, this project adopts a panoramic and mode transformation method to solve it, which is also the core highlight of this project. Panorama refers to using the entire front windshield of the car as the display range of the HUD and displaying reasonably in some areas. Mode conversion is to display important information according to the needs of the driver, which not only satisfies the realization of the function, but also does not affect the driver's driving experience and safety due to the complexity of the displayed information.
(5) Visual Design
Cars have different light intensities under different environments and different lights. The environment when the car is driving is more complicated. It may suddenly enter a dim tunnel during the day or suddenly encounter the impact of severe weather. This requires the display of HUD [8]. For different conversions according to the intensity of light, the visual comfort of the driver in different environments needs to be considered; at the same time, in order to avoid the driver's distracted vision and driving fatigue, it must be visually rigorous and comfortable [9].
(6) Functional Architecture
The functional architecture matches the mode conversion [10]. In different modes, the functions of the HUD will also change, and some functions will be hidden or processed in the background. When a car is driving on a highway or a section with few people and vehicles, it will enter the high-speed mode. The HUD mainly displays the speed of the car, the road condition and the reminder of high-risk roads. When the car is driving in a town or a place with many people and vehicles, it enters the town mode. The HUD mainly displays information such as road conditions, pedestrian detection, and high-risk road warnings.

3.2 SWOT Analysis

Strengths: Compared with other HUD on the market, this project not only provides richer functions, but also solves the problem of filtering display content. Under the premise of not affecting the driver's sight, the on-demand display is adopted to satisfy the function points while effectively avoiding the driver's sight dispersion and shifting.

Weaknesses: This research is aimed at automobile manufacturers pre-installed HUD, so the type of vehicle is restricted. The product can only serve a certain type of car. Compared with the removable and transferable installation of HUD in the automotive aftermarket, it has a high Singleness and pertinence.

Opportunities: The HUD in the automotive aftermarket now has a larger share than the HUD pre-installed by auto manufacturers. The development space for pre-installed HUD by auto manufacturers is relatively large, and the market is less saturated. This creates an impact on the development of pre-installed HUD by auto manufacturers. Condition.

Threats: On the one hand, the cost of pre-installing HUD by auto manufacturers is relatively high [10]. On the other hand, HUD products are directly related to the development of technology. With the emergence of new technologies, product functions and architecture will also change.

3.3 Design Goals

Panoramic HUD is a native on-board head-up display device that can display information in the entire range on the premise that the front windshield of the car is used as an imaging device. The information displayed by the HUD is fed back on the front windshield panel according to the driver's needs [11]. Without affecting the driver's line of sight, the driver can understand the displayed information with the outside light. According to different driving environments, HUD automatically converts driving modes. For example: in towns or places with a lot of people and vehicles, HUD mainly displays driving conditions on the road, detection of pedestrians on the road, and warnings to prevent emergencies; on highways or sections with fewer people and vehicles, HUD mainly displays cars Speed, road conditions and high-risk road reminders. However, when the vehicle is not in the driving state or in the parking mode, the driver can perform autonomous control and all functions are available. Once the vehicle enters the driving mode, some functions of the HUD will be locked, hidden or run in the background.

4 HUD Gesture Interaction Design

4.1 Gesture Interaction Research

Gesture interaction, as a typical natural interaction method, has made major breakthroughs and applications on some other devices. But in the car information system, gesture interaction has been limited. First of all, gesture interaction is not as easy as physical buttons in terms of effective recognition of the controller. Second, in the case of car driving, the feedback of gesture interaction is not as obvious as that of physical buttons and vibration feedback of mobile devices. The advantage of gesture interaction is that it can solve the problem that physical buttons can only perform a single function, which leads to an increase in the number of physical buttons. At the same time, under the existing development and use based on other somatosensory devices, gesture interaction can also be used for reference and applied in the automotive field.

4.2 Problems with Gesture Interaction

In the research and data survey for gesture interaction, it is found that there are some inevitable problems related to human-computer interaction in automobiles [12].

In the environment of car driving, the driver's gesture interaction technology operation has certain recognition problems. Mainly reflected in: When the driver uses his right hand to complete a certain instruction (such as switching to the next song), he needs to wave his right hand to the right. When this action is completed, the driver needs to retract his right hand to the steering wheel again. This action of reclaiming the right hand is likely to be recognized by the recognition system as an action command to swing to the left, resulting in an incorrect operation.

Regarding gesture interaction recognition, similar actions will cause certain errors in the recognition system. For example, it is stipulated that moving the right hand to the right is the operation action to switch the next song, and the right hand rotating and twisting to the right is the operation action to switch the next function. When the driver performs the operation of switching the next song (that is, the right hand moves to the right When moving), if the driver's operation range is not large or there is a certain fuzzy operation, the recognition of the recognition device may be biased or the operation behavior cannot be judged.

Regarding the feedback of gesture interactive operations, the dynamic effects of the interface and the degree of integration of gesture interactive operations, in the entire operation process, whether the driver's operation is effective should be given a certain degree of operational feedback. When the driver uses the right hand to move to the right through gesture interaction to give a command instruction, when the driver's hand is doing such a movement, the interface will give a certain dynamic response, and what kind of dynamic response can perfectly match the gesture Interactive actions, and therefore bring a better driving experience to the driver, this is also a problem we should consider.

4.3 Gesture Summary Analysis

Gesture interactive recognition technology has certain recognition problems in the operation of the driver in a car driving environment. This research conducted some simulation studies using LEAP MONTION gesture detection, and found that when the user is performing some movement operations (as shown in Fig. 3), when the action is over, there will be a subconscious finger bending and palm contraction action. This action It can be regarded as a threshold for the end of an instruction (as shown in Fig. 4). But when the sensing area is close to the steering wheel, the driver's action threshold will be easier to ignore, so the position of the sensor has a great influence on the judgment of finger action.

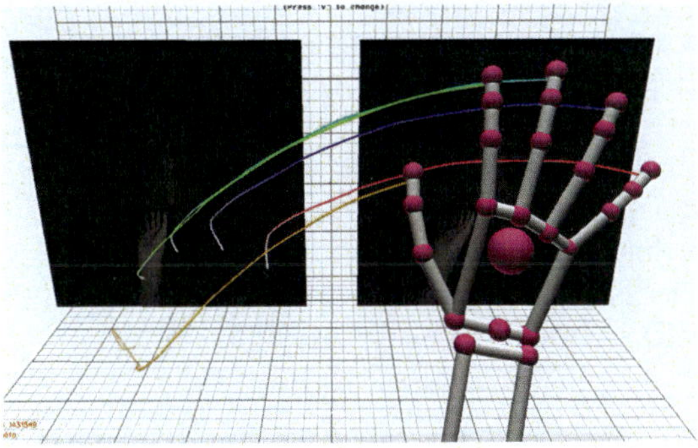

Fig. 3. Action flow

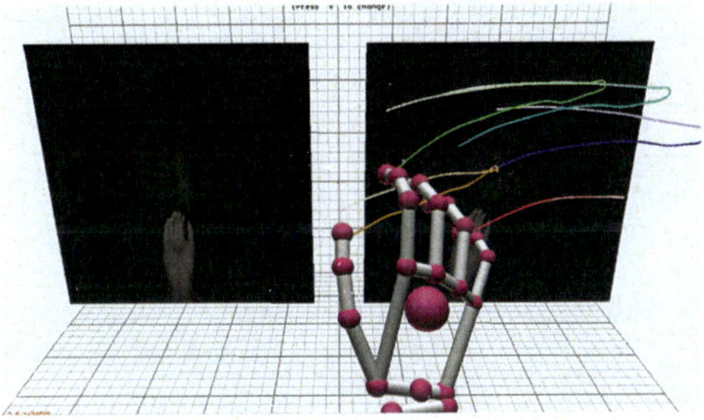

Fig. 4. End of action

In the operation of gesture interaction, similar actions will also cause certain recognition errors to the system [13]. There is a certain range of repetition between the two similar operations. If the driver's operation ends within this range, the recognition device may make a wrong judgment or fail to make a response. In response to this phenomenon, the driver's subconsciousness can be used to extend the operating range of oneself, which also produces the operating feedback that needs to be emphasized.

For example, when the user switches from music to navigation, that is, the right hand rotates to the right. The menu bar can also be designed in the form of rotation switching, so that the menu moves with the hand rotation action. When the driver's operating range reaches the correct recognition range, the menu bar completes the switch from music to the navigation interface.

5 In Conclusion

This project is to solve the problem of HUD display content, and at the same time conduct a more in-depth research on gesture design in automobile human-computer interaction. The paper first conducts market research, discussion and analysis on HUD, and based on this, conducts research on product positioning, function point analysis, and gesture interaction, and finally determines the design of the gesture scheme in HUD. After actual testing, its expression is more accurate. The effect is relatively satisfactory.

HUD is in the initial stage of development and requires long-term follow-up and more in-depth research. HUD should closely follow the development trend of technology, and in the future can be expanded to AR (augmented reality), VR (virtual reality), AI (artificial intelligence) and other emerging fields to realize the leapfrog development of HUD.

References

1. Park, H.S., et al.: In-vehicle AR-HUD system to provide driving-safety information. ETRI J. **35**(6), 1038–1047 (2013)
2. Wang, S., Charissis, V., Harrison, D.K.: Augmented reality prototype HUD for passenger infotainment in a vehicular environment (2017)
3. Gerla, M., et al.: Internet of vehicles: from intelligent grid to autonomous cars and vehicular clouds. In: Internet of Things. IEEE (2016)
4. Ohtsuka, S.: Head-up display (HUD) requirements posed by aspects of human visual system. In: 2019 IEEE International Conference on Consumer Electronics (ICCE). IEEE (2019)
5. Hun, L.S., Se-One, Y.: User interface for in-vehicle systems with on-wheel finger spreading gestures and head-up displays. J. Comput. Des. Eng.
6. Drezet, H., Colombel, S.: 62-1: invited paper: HMI concept for autonomous car. In: SID Symposium Digest of Technical Papers (2018)
7. Murata, Y., et al.: Proposal of an automobile driving interface using gesture operation for disabled people. In: ACHI, pp. 472–478 (2013)
8. Ryu, J.H., Choi, S.W., Lee, C.G.: The method of 3D information display for automobile HUD. J. Instit. Control **17**(1), 12–16 (2011)
9. Liu, H., et al.: Saliency difference based objective evaluation method for a superimposed screen of the HUD with various background. IFAC-PapersOnLine **52**(19), 323–328 (2019)
10. Liu, Y.C.: Effects of using head-up display in automobile context on attention demand and driving performance. Displays **24**(4–5), 157–165 (2003)
11. Maroto, M., et al.: Head-up displays (HUD) in driving (2018)
12. Sauras-Perez, P., Pisu, P.: A voice and pointing gesture interaction system for on-route update of autonomous vehicles' path. WCX SAE World Congress Experience (2019)
13. Kerdvibulvech, C.: A review of augmented reality-based human-computer interaction applications of gesture-based interaction. In: International Conference on Human-Computer Interaction. Springer, Cham (2019)

Architecture Design and Application Prospect of Predictive Maintenance Based on Multi-station Integration Edge Computing in Power Field

Fusheng Yuan[1], Qiang Li[2], Xian Sun[1], Zhuo Huang[1], Jing Chen[3], and Ying Zou[1(✉)]

[1] State Grid Information & Telecommunication Co., Ltd., Beijing Branch, Beijing 100031, China
zouying_neil@163.com
[2] State Grid Information & Telecommunication Co., Ltd., Beijing 102211, China
[3] Beijing Electric Power Economic Research Institute Co., Ltd., Beijing 100037, China

Abstract. Based on a brief description of predictive maintenance, current status of power equipment operation and maintenance, and edge computing technology, this paper makes a full prerequisite analysis from the aspects of basic environment, key elements of edge computing and key steps of predictive maintenance. For the architecture design of edge computing in predictive maintenance in power field, it is deeply integrated with the related architecture in Edge Computing Reference Infrastructure 3.0 to guide predictive maintenance in the field of edge computing, expand the application ideas, Under the background of vigorously constructing of electric Internet of Things (EIoT) and integration of transformer-substation & IDC-station for SGCC, guide the research and practical application of predictive maintenance of power field in the environment of edge computing.

Keywords: Edge computing · Power field · Predictive maintenance · Cloud computing and edge computing synergy · Architecture design

1 Introduction

By the end of 2018, With the development of China electric power, China's installed power capacity was 1899.67 million kW (including 1143.67 million kW of thermal power and 756 million kW of clean energy), 1.89 million km of transmission line circuit length, 4.03 billion kVA of substation equipment capacity, 136 million kW of trans-regional transmission, and 5012 kWh of per capita power consumption [1]. For such a huge power grid system, the traditional equipment (facility) operation and maintenance mode is mainly based on human, and the information acquisition method is traditional and single. It relies more on the on-site investigation and other manual methods to obtain the equipment and facilities related information. The perception of equipment status is still mainly based on power-cutoff repair and offline test, and advanced technical means

M. Tavana et al. (Eds.): IISA 2020, AISC 1304, pp. 413–425, 2021.
https://doi.org/10.1007/978-3-030-63784-2_52

and numbers such as online monitoring, live detection, robot and so on The utilization rate of data is low [2]. Moreover, all the equipment (facilities) involved are large in quantity and variety, and the operation and maintenance process is miscellaneous and the division of labor is detailed. It is necessary to introduce and carry out predictive maintenance.

Predictive maintenance is based on real-time data generated by the equipment collected by sensors, such as vibration, speed, power, noise, sound, strain, pressure, displacement, temperature, humidity, PH, salinity, image comparison, color difference ratio, etc., to analyze the operation status of the equipment with reference to these data, and predict the possible failure points and potential threats of the equipment, and then put forward the corresponding countermeasures Maintenance suggestion, according to the suggestion, the equipment with risk can be handled in advance to improve the reliability of the equipment [3, 4]. As early as the middle and later period of the last century, for power plants, chemical plants and other related manufacturing plants, state monitoring and diagnosis technologies such as vibration, strain, pressure, displacement and sound generation have been used to determine the possible causes of faults and to carry out predictive maintenance of equipment [5]. In this period, there are relatively few factors and basis to judge, and although there are computer tools, but more rely on human judgment of monitoring tool values. According to Cisco's white paper on predictive maintenance solutions (2017), the global annual loss caused by abnormal outage exceeds $600 billion. It is estimated that the annual expenditure of predictive maintenance will exceed $11 billion by 2022 [3]. At present, there are three kinds of maintenance methods for machinery and equipment and production facilities: the first is the reactive maintenance, the second is Preventive or Proactive Maintenance, and the third is the predictive maintenance. The nature and advantages and disadvantages of various methods are shown in Table 1.

In order to achieve real predictive maintenance, edge computing is indispensable. Content distribution network (CDN) is the rudiment of edge computing. It constructs data communication network by using Intelligent DNS, load balancing, content distribution and other technologies, effectively controls the data network traffic to end at edge nodes, and solves the problems of network bandwidth and users the stability and service quality of the network can be improved by the problems of traffic volume and sudden access [6]. By improving the cloud computing infrastructure, edge computing deepens and extends the original concept of CDN [7]. Edge computing is an open architecture that integrates the basic elements of distributed computing on the edge close to the computing needs, and provides computing power services nearby [8]. Edge cloud is to use the core of cloud and the ability of edge computing to build a distributed elastic edge cloud platform, and provide resource scheduling and computing power distribution services efficiently through the whole network collaboration of cloud edge end [9].

Table 1. Overview of maintenance methods of equipment and facilities

Name	Character	Advantages and disadvantages	
Reactive maintenance	Post maintenance: it is not maintained at ordinary times, and it will be dealt with when there is a problem	It is suitable for enterprises with little impact of downtime	Has caused losses and is also the most expensive maintenance
Preventive or proactive maintenance	Planned maintenance: no matter what the state of the equipment is, replace and upgrade regularly	The risk of major failure is avoided	There are plans and no targets, resulting in excessive maintenance and other investment waste, but also cannot avoid sudden failures
Predictive maintenance	Forecast maintenance: early warning, prevention in advance	Reduce unplanned downtime, improve equipment utilization and ensure continuous production. Reduce maintenance frequency, reduce maintenance cost, shorten maintenance time and improve operation and maintenance efficiency. Reduce the replacement of parts, fully utilize the old and extend the service life of the equipment	Equipment and facilities used to support predictive maintenance have a large initial investment, but this method is the most economical maintenance

2 Prerequisite Analysis

2.1 Key Conditions of Basic Environment

The SGCC (State Grid Corporation of China) vigorously promotes the multi-station integration data center station, which uses the surplus power supply, communication and site resources of the substation site to carry out the integration construction, operation and maintenance of the substation data center station, and finally realize the integration of the substation and the data center station in the aspects of building, power supply, communication, data and service [10]. The use of multi station integration of data center stations to form a distributed data center cluster is in line with the characteristics of distributed edge computing, which is the basic environmental key condition to support the edge computing of industrial Internet, power generation, power transmission and power distribution. At the same time, in the process of promoting construction and deployment of the power internet of things, the SGCC has formed the overall physical structure of power Internet of things (see Fig. 1). Based on the computing capability of "cloud edge collaboration", the architecture realizes deep interaction, flexible application, open sharing and extensive interconnection, and realizes the ability opening of terminal layer, network layer and platform layer, which is the physical basis to realize power domain predictive maintenance.

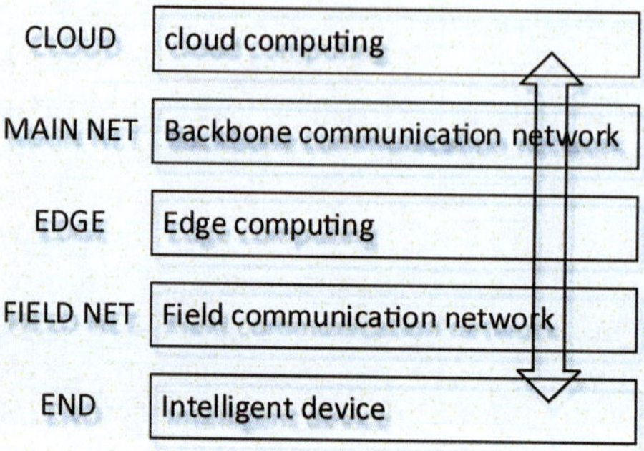

Fig. 1. General structure

2.2 Key Elements of Edge Computing

In the paper entitled "transformation of predictive maintenance enabling servitization, starting from the edge computing Internet of things" [11], Huawei puts forward that edge computing plays an important role in predictive maintenance in three aspects. First, it introduces edge computing architecture into the field of Internet of things, and integrates network, computing, storage, application and provides platform support for

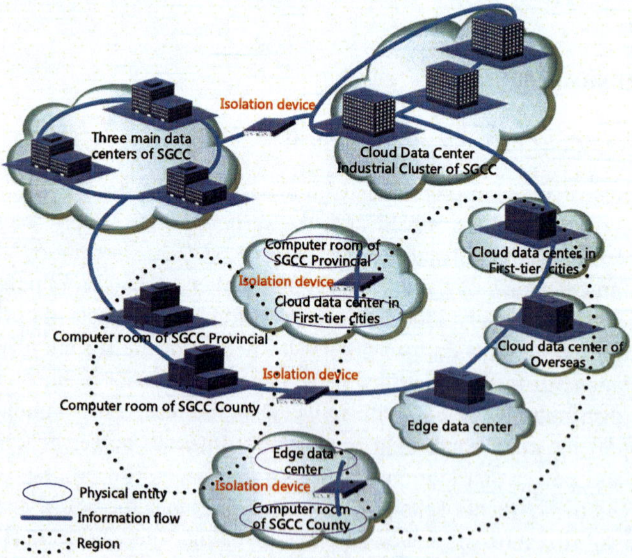

Fig. 2. Conceptual model of dual network convergence data center

corresponding domains at the network edge near the data source. The second is cloud-edge collaboration, edge computing is close to the source, and it is the acquisition unit required by the cloud, which can support the cloud big data analysis. The business arrangement output by cloud computing through can be distributed to the edge side, and the edge computing can continuously carry out business processing based on the optimized business arrangement. The third is the unified cloud management architecture that can support tens of millions of terminal access. Several important factors for edge computing to support predictive maintenance are abundant communication networks (including 5G power wireless VPN, power wireless PN, power optical cable, etc.), infrastructure supporting super computing power (cloud data center, edge data center), edge computing platform and cloud platform using virtualization technology. It is the inevitable choice of edge computing platform and cloud platform [12].

2.3 Predictive Maintenance Key Steps

In order to realize the predictive maintenance function of power equipment and production facilities [13], the first is to replace the sensing equipment (IOT terminal or components) with the required data acquisition function, build a communication channel meeting the data transmission requirements, and complete the collection, up transmission and distribution functions of key information and data. The second is to transfer the collected data to the edge computing equipment (or edge cloud) and cloud (the data that needs to be centralized and large-scale processing after edge processing), build a virtual monitoring model based on cloud edge collaboration in the edge and cloud, access and analyze the virtual storage data, and cooperate with the data experts and intelligent system, through the artificial way (data experts) Or intelligent (business choreography or data choreography) to direct and implement predictive maintenance. Third, on the premise of clear and easy to understand transmission data, ensure that data is pushed within the entire organizational structure involving predictive maintenance, so as to exert maximum influence on the whole implementation process.

3 Architecture Design

3.1 Overall Architecture Design

The three data centers of SGCC (the former as three disaster recovery centers) are respectively located in Beijing, Shanghai and Xi'an [14, 15]. As the carrier of disaster recovery in the same city of North China, East China and Northwest China, as well as the data center carriers of North China, East China and northwest China, the three data centers also realize mutual backup and dual activity. We can carry out the construction of cloud data center and edge data center in first-tier cities where the provincial, municipal and county companies of the SGCC have the conditions, and to cloud the data center of provincial, municipal and county companies, promote the seamless integration of the two data center networks of the SGCC information data center and the SGCC industrial data center under the security architecture, to realize cloud network integration, cloud edge collaboration and cloud interconnection, shown in Fig. 2.

The main body of the overall architecture of predictive maintenance in the power field based on edge computing should cover equipment side, edge side, cloud, display and application, as well as the corresponding standard system and security system. The access end of the equipment side includes sensor network, acquisition equipment, industrial control components and interface module group connected with the equipment local end. In addition to the computing resources, network resources and storage resources of the edge computing reference architecture, the edge side should also include the edge side device analysis module and the edge side device control module. At the same time, for the edge computing nodes with demand, it can be considered to sink the CDN to the edge side. The cloud includes cloud device analysis module, cloud device control module, basic information base, management information base, operation information base and maintenance information base corresponding to primary equipment and secondary equipment, as well as IaaS, PaaS and SaaS layers used to support cloud processing of the above equipment, show in Fig. 3.

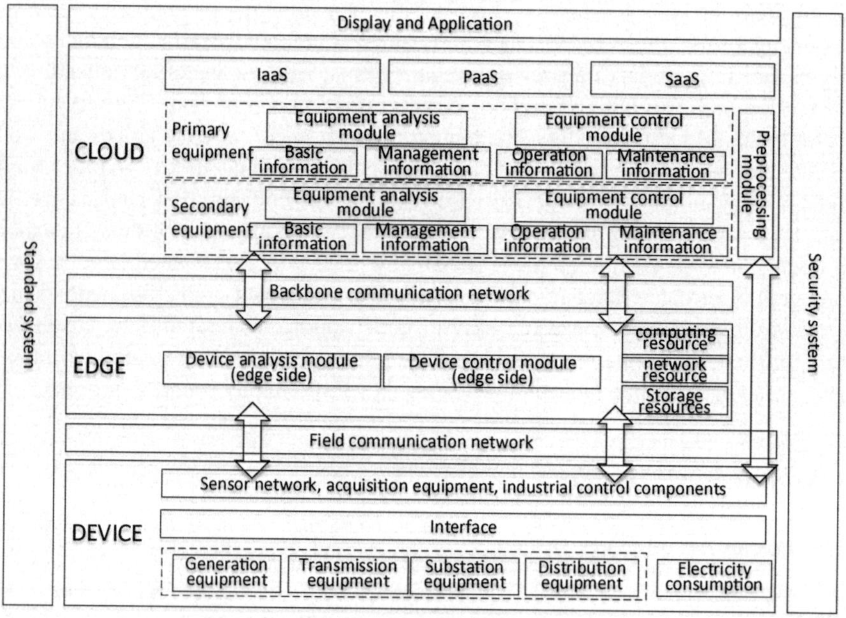

Fig. 3. Overall architecture

3.2 Edge Computing Architecture Design

Referring to the edge computing reference Infrastructure 3.0 [8], the functional implementation module structure of the edge computing architecture for predictive maintenance of power equipment includes cloud computing, edge computing, edge cloud computing, edge computing open platform, and massive (or large bandwidth) connected field communication networks. The data analysis module of the edge device quickly analyzes the data according to the established requirements, while the control data module of the edge device controls the intelligent device nearby according to the setting. The two modules are organically combined to achieve the connection, integration, control of the physical world and the digital world. Edge computing module (or pre-processing module, edge cloud computing module) connects intelligent primary device and intelligent secondary device through field communication network supporting mass connection, and achieves unification with cloud in terms of architecture, interface, management and other key capabilities, and becomes a part of cloud [9], which realizes the efficient collaboration between with edge computing with low delay, large access and long cycle, large computing power, show in Fig. 4.

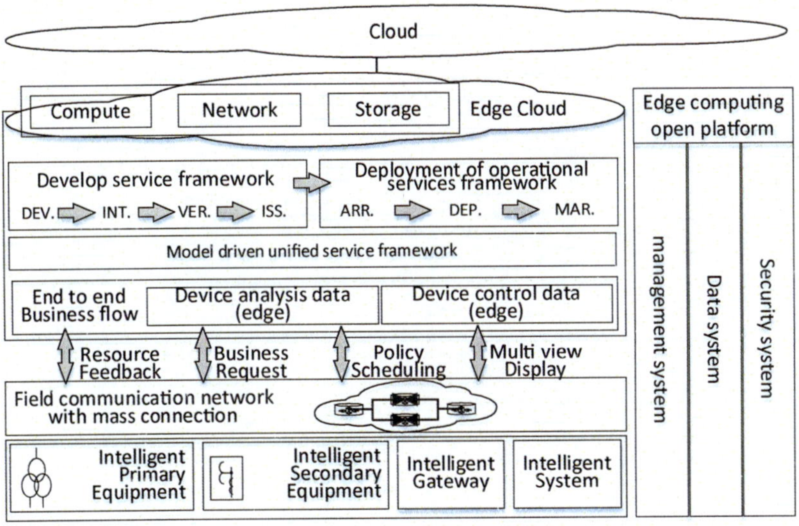

Fig. 4. Edge computing architecture based on predictive maintenance in power field

3.3 Communication Architecture Design

Edge computing communication link including field communication and backbone communication. Field communication is terminal access network communication using power wireless VPN and (or) power wireless PN, PLC network and other communication networks; backbone communication is a channel to realize large bandwidth and high-speed connection between edge data center and cloud data center. In the environment where the power wireless VPN can be used, the power wireless VPN [16] can be considered, but it should meet the requirements of power wireless VPN networking, operation and maintenance, and security protection. In the environment where power wireless PN can be used, it can be considered to use power wireless PN [17], but it should meet the requirements of power wireless PN networking, operation and maintenance, security protection. According to the bandwidth and delay requirements of the application itself, 5G [18, 19] features large bandwidth, low delay and large connection can be used to provide information transmission channel for edge computing, and realize the rapid transmission of data. For the application around the field of cloud platform construction, the pre-processing module with edge storage and edge computing power can be accessed through the field communication wireless network and (or) field communication wired network to perform the functions similar to the edge platform. The overall communication architecture is shown in Fig. 5.

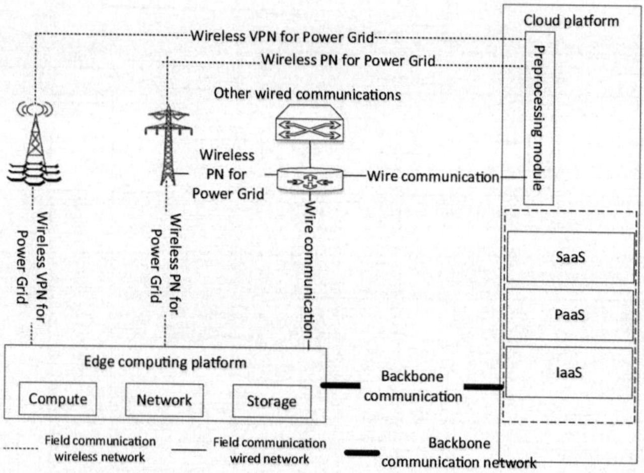

Fig. 5. Edge computing and cloud edge collaborative communication network architecture

3.4 Data Acquisition Architecture Design

Data acquisition is the first step in the process of predictive maintenance for equipment (system). The sensing equipment with required index acquisition function is equipped on the intelligent primary equipment, intelligent secondary equipment, intelligent gateway, intelligent system, etc., such as IOT terminals or components, etc., to build a sensor network, acquisition equipment and field communication network (see the communication architecture design for details) that meet the requirements of data acquisition and data transmission. Through the core component module of data acquisition to complete the collection, rough processing, upload and distribution of key information and data, show in Fig. 6.

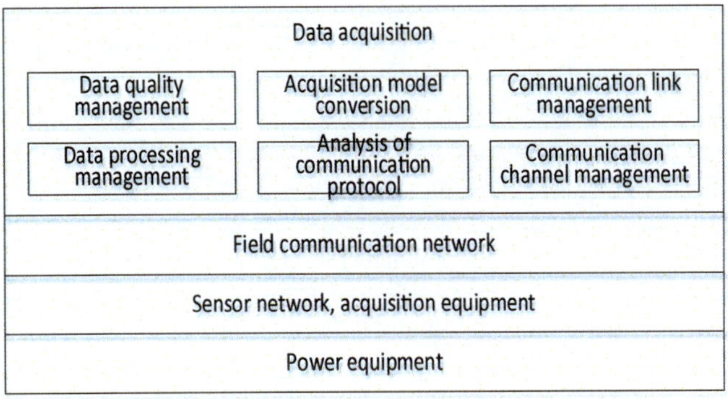

Fig. 6. Schematic diagram of data acquisition architecture

3.5 Data Processing and Execution Architecture Design

The intelligent means used in data processing generally include expert system, artificial neural network, decision tree, data mining, fuzzy theory, Petri theory, support vector machine, bionics theory, etc. [20]. In the data processing of predictive maintenance, deep learning technology can be realized in two ways, classification method and regression method. The classification method is to predict the following N steps in the

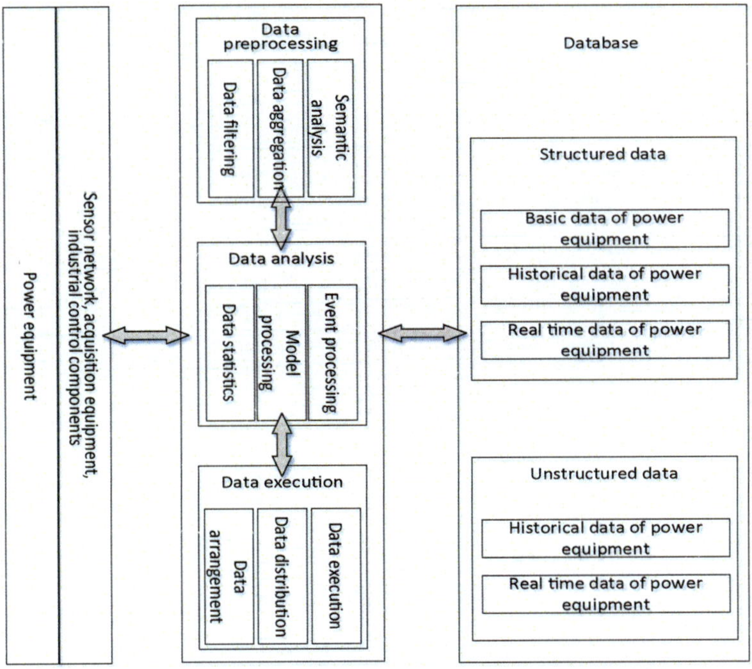

Fig. 7. Diagram of data processing and execution architecture

specified time series. The regression method is "predict how much time is left before the next problem" [21]. Data execution is to perform data cleaning and event data mining in real time through data analysis, data arrangement (business choreography), data distribution (policy control) and data execution (policy execution) [8]. Cloud edge cooperates with real-time execution of data cleaning and event data mining, and triggers pre-defined data modeling strategies according to the mining results, so as to discover the potential fault of equipment (system) in the first time In case of communication failure between the edge side and the cloud, the data can be stored and processed. After the communication is restored, the local uploaded data will be automatically transferred to the cloud to provide complete data for the cloud [11]. The diagram of data processing and execution architecture is shown in Fig. 7.

3.6 Security Architecture Design

Reference [8] has a relatively clear description on the design and application concept, design and application scenarios, design and application levels of edge computing security, which can maximize the security and reliability of edge computing architecture. The overall security of predictive maintenance based on edge computing involves equipment security, communication security, data security, application security and security situation system, security authentication system, security operation and maintenance system, security management system, etc. In terms of equipment security, it includes equipment software reinforcement, equipment security configuration, and sensor equipment security; in terms of communication security, corresponding to the communication security with power characteristics, it should include power wireless PN security [22], power wireless VPN security [16], sensor network

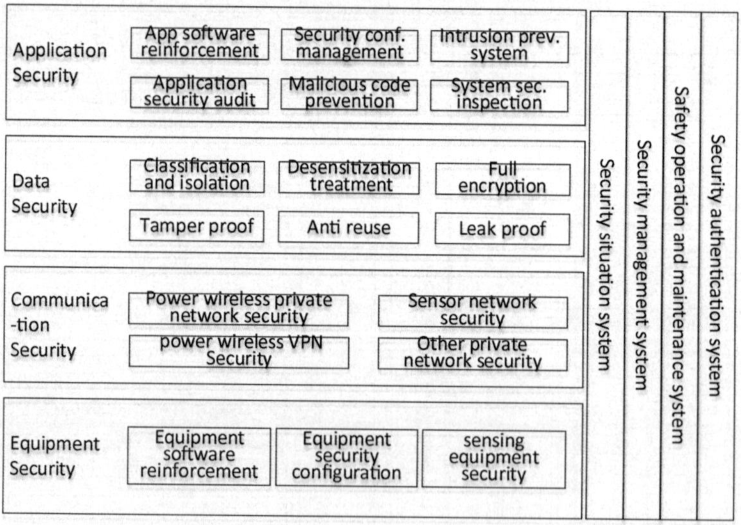

Fig. 8. Schematic diagram of security architecture

security and other private network security; data security includes the classification and isolation and desensitization processing, and the whole process encryption, anti-tampering, anti-reuse and anti-leakage related to data protection. In terms of application security, we can learn from the traditional application security, including application software reinforcement, security configuration management, intrusion prevention system, application security audit, malicious code prevention, system security detection, etc. Show in Fig. 8.

4 Application Prospect

In the whole power system, the main equipment (facilities) involved can be divided into primary equipment and secondary equipment, the primary equipment is all electrical equipment in the primary system composed of generator, transmission line, transformer, circuit breaker and other power transmission and power distribution equipment, the secondary equipment is the auxiliary equipment for the primary equipment in the power system to monitor, measure, control, protect and regulate [23]. The list of main equipment (facilities) of power system is shown in Table 2.

Table 2. List of main equipment of power system

Equipment category		Equipment name
Primary electrical equipment	Electric energy production equipment	Generator, motor, boiler, steam turbine, water turbine, gas turbine, fan, nuclear reactor, etc.
	Electric energy conversion equipment	Transformer, etc.
	Power transmission equipment	Bus bar, power cable, etc.
	On off circuit switch equipment	Circuit breaker, disconnector, contactor, fuse, etc.
	Gas insulated equipment	Insulator, wall bushing, etc.
	Limiting over current (voltage) equipment	Current limiting reactor, arrester, etc.
	Transformer equipment	Voltage transformer, current transformer, etc.
Power secondary equipment		All kinds of measuring meters, all kinds of relay protection and automatic device, automation system, DC power supply equipment, etc.

Use the characteristics of 5G high bandwidth, low delay and multi connection, and based on various mature communication modes established by the SGCC, and in combination with the multi-station integration multi-level distributed data center station of the SGCC, edge computing is carried out in photovoltaic power, wind power [24], thermal power, hydropower, nuclear power generation and other power fields, in the direction of Transmission line monitoring and control, power transformation and distribution, and the predictive evaluation and calculation of power supply and demand [25] and power demand response [26] in the field of power consumption are carried out.

The establishment from the physical world in the electric power field covers the research, design, construction, operation, maintenance, security and application to the digital world, so as to realize the physical and digital connection and physical and physical connection in the whole cycle and whole process of the electric power field. It is necessary to realize the transformation from "to be required when it is due", "to take compulsory courses and to do well in required courses" to "only those who need to be trained and those who have completed compulsory courses well".

Through the support of big data and AI, the linkage mechanism in the whole process of construction, operation and maintenance of power generation, power transmission and transformation is realized. At the same time, we should change our thinking and transform the preventive maintenance of power generation, power transmission, and power transformation & distribution equipment into the preventive maintenance of predictive maintenance system. The intelligent maintenance of the predictive maintenance system can make the maintenance of it truly intelligent and even unmanned. In the future, we only need to make maintenance plan for predictive maintenance system and carry out maintenance work.

5 Conclusion

As an important part of the new infrastructure, the comprehensive planning and large-scale construction of multi-station integration data center station provides edge computing with power, communication, location and computing power advantages. On the premise of giving priority to the cloud computing, edge computing, communication, data acquisition, data processing, data execution and end-to-end security capability involving predictive maintenance of edge computing, intelligent and efficient predictive maintenance of power equipment (systems and facilities) will become possible and even inevitable.

Acknowledgments. This work is supported by Science and technology project of the SGCC (Research on key technology and business model of multi-station integration, No. 5200-201941477A-0-0-00).

References

1. Mei, B.: 70 years' development achievements of power industry in New China. China Electr. Equip. Ind. (12), 11–19 (2019)
2. Hou, S., Ma, J.: "Da Yun Wu Yi" helps intelligent operation and maintenance. State Grid News, 19 December 2017
3. White Paper: CISCO Predictive Maintenance Solution. CISCO (2017)
4. Zhang, S., Su, S., Liu, C., et al.: Feature extraction method of sensing data for predictive maintenance of power generation equipment. J. Taiyuan Univ. Technol. **49**(1), 79–85 (2018)

5. He, Z., Wang, J.: The monitor of predictive maintenance of hydropower station (Edit according to the 1985's American Power Conference Collection). In: Mechanical & Electrical Technique of Hydropower Station, no. 01, pp. 59–62 (1989)
6. Zhang, C.G., Zou, Y., Wang, Y.R., et al.: Research on CDN network architecture design and safety protection for power grid. Int. J. Reason. Based Intell. Syst. **8**(1/2), 37–44 (2016)
7. Satyanarayanan, M.: The emergence of edge computing. Computer **50**(1), 30–39 (2017)
8. White Paper: Edge Computing Reference Infrastructure 3.0. Edge Computing Consortium (ECC) & Alliance of Industrial Internet (AII) jointly released (2017)
9. White paper on edge cloud computing technology and standardization. Jointly released by Alibaba cloud computing Co., Ltd., China Institute of electronic technology standardization, etc. (2018)
10. Zou, Y., Hou, R.: Multi station integration: exploring new formats of energy sharing. State Grid News, 21 May 2019
11. Huawei: The transformation of predictive maintenance enabling service starts from the edge computing Internet of things [EB/OL]. https://www.51cto.com/art/201709/551610.htm
12. Zhao, Z., Liu, F., Cai, Z., Xiao, N.: Edge computing: platforms, applications and challenges. J. Comput. Res. Dev. **55**(2), 327–337 (2018)
13. Predictive maintenance based on big data [EB/OL]. https://article.cechina.cn/15/0429/06/20150429064354.htm
14. Zhou, Y., Zhu, C., Huo, Y.: Construction planning and technical route of state grid cloud data center. Inf. Secur. Technol. (12), 39–44, 54 (2014)
15. Xu, Y., Zou, B., Huang, W., et al.: Research and implementation of bucket algorithm based disaster recovery data comparison tool. Electr. Power Inf. Commun. Technol. **12**(2), 20–23 (2014)
16. Zou, Y., Wang, Y., Wang, Y., et al.: Wireless VPN network structure and safety protection in the north and south call center of SGCC. Telecommun. Sci. **32**(2), 175–181 (2016)
17. Zhao, X., Bai, J., Ding, G., Xiang, L.: Optimization of electric power wireless communication technology. Telecommun. Sci. **35**(9), 158–164 (2019). V7
18. China Telecom 5G technology white paper. All rights reserved by China Telecom (2018)
19. China Mobile edge computing technology white paper. China Mobile edge computing open laboratory (2018). V9
20. Li, Z., Liu, M.: Review of intelligence fault diagnosis in power system. Electr. Eng. **8**, 21–24 (2010)
21. Perera, S., Alwis, R.: Machine Learning Techniques for Predictive Maintenance [EB/OL], 21 May 2017. https://www.infoq.com/articles/machine-learning-techniques-predictive-maintenance/
22. Zou, Y., Wang, S., Liu, Z., Shi, Q.: Research on P4PCDN automatic construction strategy based on network measurement. In: International Congress of Information and Communication Technology (ICICT 2017). ELSEVIER (2017). Procedia Computer Science 107 (2017):490–497.VC2
23. Pang, Q., Guo, W., Li, X.: Power Supply Technology, vol. 10, p. 98. Tsinghua University Press, Beijing (2015)
24. Wu, T., Liu, L., Wang, D.: Design of remote centralized intelligent monitoring system for wind farms. Electr. Power **51**(4), 161–167 (2018)
25. Li, B., Jia, B., Chen, S., et al.: Prospect of application of edge computing in the field of supply and demand. Electr. Power **51**(8), 1–9 (2018)
26. Li, B., Jia, B., Cao, W., et al.: Application prospect of edge computing in power demand response business. Power Syst. Technol. **42**(1), 79–87 (2018)

Application Research of a New Practical Transmission Device

Weiwen Ye(✉)

School of Mechanical and Electrical Engineering, Guangdong University of Science and Technology, Dongguan 523083, Guangdong, China
49510019@qq.com

Abstract. Workpiece automatic feeding equipment is a kind of substitute for manual transmission of workpiece to the detection position or transmission mechanism. At present, after the automatic workpiece feeder is loaded, manual monitoring and detection of the positive and negative direction of the workpiece are required. Manual, manual turnover of the workpiece, slow speed and low efficiency. The utility model provides a workpiece transmission device aiming at the problems of the existing technology, with novel structure, simple structure, high degree of automation, saving a lot of labor costs, reducing the damage of detection to artificial, solving the problems of high traditional labor cost, operation risk and other issues, in line with the national mechanization development strategy, has good promotion significance and value, high work efficiency and low cost. The difficulty of manual operation and the error rate of manual operation.

Keywords: High degree of automation · Fast speed · High efficiency · Saving labor cost · Reducing operation error rate

1 Introduction

The utility model relates to the technical field of workpiece conveying device, in particular to a workpiece transmission device, which includes a frame, on which is arranged a transmission mechanism for conveying the workpiece forward, a stop mechanism for stopping the workpiece on the transmission mechanism, a detection component for detecting the positive and negative directions of the workpiece, and a turnover mechanism for clamping the workpiece to overturn. The transmission synchronous belt is arranged around the rotating shaft rollers at the front and rear ends of the transmission support. The left and right sides of the transmission bracket can be disassembled and connected with side stops. The transmission synchronous belt is located between the two side stops, and the detection component is located above the transmission synchronous belt. The utility model has the advantages of novel and simple structure, saving a lot of labor cost, reducing the damage of detection to artificial, high working efficiency, reducing the difficulty of manual operation and the error rate of manual operation.

M. Tavana et al. (Eds.): IISA 2020, AISC 1304, pp. 426–431, 2021.
https://doi.org/10.1007/978-3-030-63784-2_53

2 In Order to Solve the Above Technical Problems, the Utility Model Adopts the Following Technical Solutions

The utility model provides a workpiece transmission device, which includes a frame. The frame is provided with a transmission mechanism for conveying the workpiece forward, a stop mechanism for stopping the workpiece on the transmission mechanism, a detection component for detecting the positive and negative directions of the workpiece, and a turnover mechanism for clamping the workpiece to overturn. The transmission mechanism, the stop mechanism and the overturning mechanism are all the same as the described one. The transmission mechanism includes a transmission bracket, a transmission synchronous belt and a rotating shaft roller which is rotationally connected with the front and rear ends of the transmission support. The transmission synchronous belt is arranged around the rotating shaft rollers at the front and rear ends of the transmission support. The left and right sides of the transmission bracket can be disassembled and connected with side stops. The transmission synchronous belt is located between the two side stops, and the detection component is located in the transmission Above the belt.

The transmission mechanism also includes a transmission stepping motor, and the output end of the transmission stepping motor is connected with one of the rotary shaft rollers.

Among them, the top of the frame is equipped with a gantry frame, the detection component includes a detection bracket, the detection bracket is fixedly connected with the gantry frame, and the bottom of the detection bracket is provided with two detection heads.

The stop mechanism includes a baffle plate, a stop bracket and a stop micro cylinder. The stop micro cylinder is installed on the top of the stop bracket. The piston rod of the stop micro cylinder is connected with the baffle plate. The baffle protrudes into the upper part of the transmission synchronous belt between the two side stops. The side stop is provided with a yielding groove of the yielding baffle, and the stop bracket is detachably connected with the frame.

Among them, the top of the stop bracket is provided with a fixing frame, the rear end of the fixing frame is provided with a rear baffle plate, both sides of the fixing frame are provided with side plates, the side of the two side plates close to each other is provided with a sponge cushion, and the blocking micro cylinder is arranged between the two sponge pads.

The turnover mechanism includes a turnover clamping manipulator and a moving component used to drive the turnover clamping manipulator to move left and right; the turnover clamping manipulator is located on one side of the transmission synchronous belt, the side stop is provided with a slot for giving way to the turnover clamping manipulator, the turnover clamping manipulator extends into the transmission synchronous belt to clamp the workpiece after passing through the slot; the turnover clamping manipulator is located at one side of the transmission synchronous belt The holding manipulator includes a rotating motor and a pneumatic finger, and the output shaft of the rotating motor is connected with the pneumatic finger.

The moving component includes a moving bracket, a moving servo motor, a moving driving screw rod and a moving slide block, the moving transmission screw rod is rotationally connected with the moving bracket, the moving servo motor is fixedly connected with the moving bracket, the output shaft of the moving servo motor is connected with the mobile transmission screw rod, and the moving slide block is sheathed on the periphery of the mobile transmission screw rod and connected with the mobile transmission The screw rod is driven and connected, and the rotating motor is arranged on the top of the moving sliding block.

The transmission bracket is provided with a slot, the bottom of the side stop is inserted in the slot, the bottom of the side stop is provided with a first magnetic iron block, and the bottom of the slot is provided with a second magnetic iron block matching with the magnetic suction.

3 The Utility Model Has the Following Beneficial Effects: Description of Drawings

Figure 1 is the structural diagram of a workpiece transmission device of the utility model.

Figure 2 is the structural diagram of the transmission mechanism, the stop mechanism and the pneumatic finger of the utility model.

Figure 3 is a structural decomposition diagram of the transmission mechanism of the utility model.

Specific i`entation mode.

In order to facilitate the understanding of those skilled in the art, the utility model is further described in combination with the embodiment and the attached drawings, and the contents mentioned in the implementation mode are not limited to the utility model. The utility model is described in detail below in combination with the attached drawings.

A workpiece transmission device, as shown in Fig. 1, Fig. 2 and Fig. 3, comprises a frame 1, which is provided with a transmission mechanism 2 for conveying the workpiece forward, a stop mechanism 3 for stopping the workpiece on the transmission mechanism 2, a detection component 4 for detecting the positive and negative directions of the workpiece, and a turnover mechanism 5 for clamping the workpiece to overturn. The transmission mechanism 2, the stop mechanism 3 and the turnover mechanism 5 are all arranged on the frame 1. The transmission mechanism 2 includes a transmission bracket 6, a transmission synchronous belt 7 and a rotating shaft roller 8 which is rotationally connected with the front and rear ends of the transmission support 6. The transmission synchronous belt 7 is arranged around the rotating shaft rollers 8 at the front and rear ends of the transmission support 6. The left and right sides of the transmission bracket 6 can be disassembled and connected with side stop 9. The transmission synchronous belt 7 is located between two side stops 9. The detection assembly 4 is located above the transmission timing belt 7. Specifically, when the utility model works, the external feeding device loads the workpiece to the transmission mechanism 2, and the transmission synchronous belt 7 of the transmission mechanism 2 drives the workpiece forward to the detection station of the detection assembly 4. At this time, the stop mechanism 3 extends into the transmission mechanism 2 to stop the transmission of the workpiece, and the detection component 4 detects the positive and negative directions of the workpiece, and the detection direction is the positive direction when the workpiece passes through the detection assembly 4 After passing, the stop mechanism 3 opens and exits the transmission mechanism 2 so that the workpiece can continue to transmit forward; when the detection assembly 4 detects that the motor end cover is in the opposite direction, the stop mechanism 3 extends above the transmission synchronous belt 7, and the turnover mechanism 5 extends into the transmission synchronous belt 7 to clamp the workpiece, and then the workpiece is sent to the transmission synchronous belt 7, and the workpiece is detected to be in the positive direction, The utility model has the advantages of novel structure, simple structure, high degree of automation, saving a large amount of labor cost, reducing the damage of detection to artificial, solving the problems of high cost of traditional labor, dangerous operation and other issues, in line with the national mechanization development strategy, and has good promotion significance and value, and work efficiency High rate, reduce the difficulty of manual operation and error rate of manual operation; the setting of side stop 9 prevents the workpiece from falling from both sides of the transmission synchronous belt 7, and works reliably.

The transmission mechanism 2 also includes a transmission stepping motor 10, and the output end of the transmission stepping motor 10 is connected with one of the rotary shaft rollers 8. Specifically, the transmission stepper motor 10 drives one of the rotary shaft rollers 8 to rotate to drive the transmission synchronous belt 7 to rotate and transfer the workpiece.

The top of the frame 1 is equipped with a gantry 11, the detection assembly 4 includes a detection bracket 12, the detection bracket 12 is fixedly connected with the gantry 11, and the bottom of the detection bracket 12 is provided with two detection heads 13. Specifically, the setting of the two detection heads 13 can detect the arrival of the workpiece at the detection position and the detection head 13 can be the sensor in the prior art or the CCD visual detector in the prior art.

The stop mechanism 3 includes a baffle plate 14, a stop bracket 15 and a stop micro cylinder 16. The stop micro cylinder 16 is installed on the top of the stop support 15. The piston rod of the stop micro cylinder 16 is connected with the baffle plate 14. The baffle 14 protrudes into the top of the transmission synchronous belt 7 between the two side stops 9. The side stop 9 is provided with the yielding position of the yielding baffle 14. The groove 17, the stop bracket 15 and the frame 1 are detachably connected. Specifically, when the stop mechanism 3 works, the stop micro cylinder 16 drives the baffle plate 14 to move forward from the yielding groove 17 to the top of the transmission synchronous belt 7 to stop the workpiece, and the stop bracket 15 can be detachable connected with the frame 1 through screws.

A workpiece transmission device described in the embodiment, the top of the stop bracket 15 is provided with a fixing frame 18, the rear end of the fixing frame 18 is provided with a rear baffle plate 19, both sides of the fixing frame 18 are provided with side plates 20, the side plates 20 close to each other are provided with sponge pads 21, and the stop micro cylinder 16 is arranged between the two sponge pads 21. Specifically, the side plate 20 of the fixed frame 18 and the rear baffle plate 19 are used to limit the installation of the micro cylinder 16, which has good stability. Through the setting of the sponge pad 21, the wear of the side plate 20 when blocking the micro cylinder 16 is reduced.

The turnover mechanism 5 includes a turnover clamping manipulator and a moving component for driving the turnover clamping manipulator to move left and right; the turnover clamping manipulator is located on one side of the transmission synchronous belt 7, and the side stop 9 is provided with a slot 22 for giving way to the turnover clamping manipulator, and the turnover clamping manipulator extends into the transmission after passing through the slot 22. The workpiece is clamped on the synchronous belt 7; the turning clamping manipulator includes a rotating motor 23 and a pneumatic finger 24, and the output shaft of the rotating motor 23 is connected with the pneumatic finger 24 through transmission. Specifically, when the workpiece needs to be turned over, the moving component drives the turnover clamping manipulator to extend from the slot 22 into the transmission synchronous belt 7, clamp the workpiece through the pneumatic finger 24, and then the moving component drives the turnover clamping manipulator to exit the transmission synchronous belt 7 from the slot 22, and then turns to the motor to drive the pneumatic finger 24 to turn over, thus driving the workpiece to turn over.

The moving component includes a moving bracket 25, a moving servo motor 26, a moving transmission screw rod 27 and a moving slide block 28. The moving transmission screw rod 27 is rotationally connected with the moving bracket 25, the moving servo motor 26 is fixedly connected with the moving bracket 25, the output shaft of the moving servo motor 26 is connected with the mobile transmission screw rod 27, and the moving slide block 28 is arranged The rotary motor 23 is arranged on the top of the moving sliding block 28. The utility model has the advantages of simple clamping structure and strong turning stability.

A workpiece transmission device described in the embodiment, wherein, the transmission bracket 6 is provided with a slot 29, the bottom of the side stop 9 is inserted in the slot 29, the bottom of the side stop 9 is provided with a first magnetic iron block 30, and the bottom of the slot 29 is provided with a second magnetic iron block 31 matched with the magnetic suction. Specifically, the first magnetic iron block 30 and the second magnetic iron block 31 are matched to ensure the stability, reliability and strong stability of the side stopper 9 after being inserted into the slot 29.

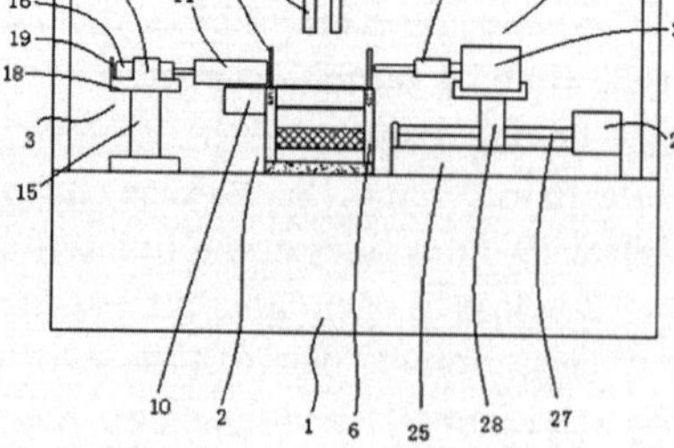

Fig. 1. Transmission device

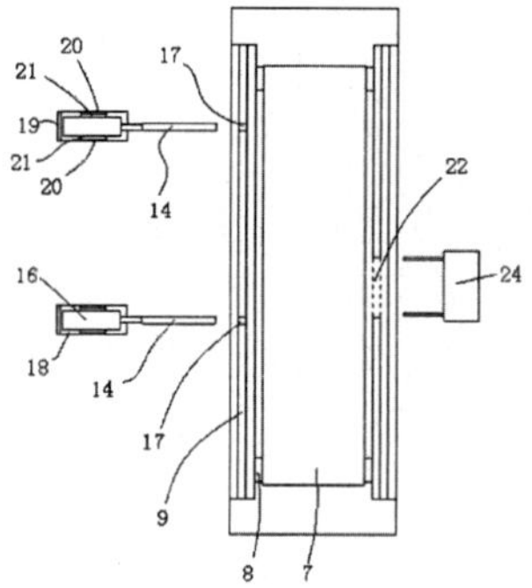

Fig. 2. Structure diagram

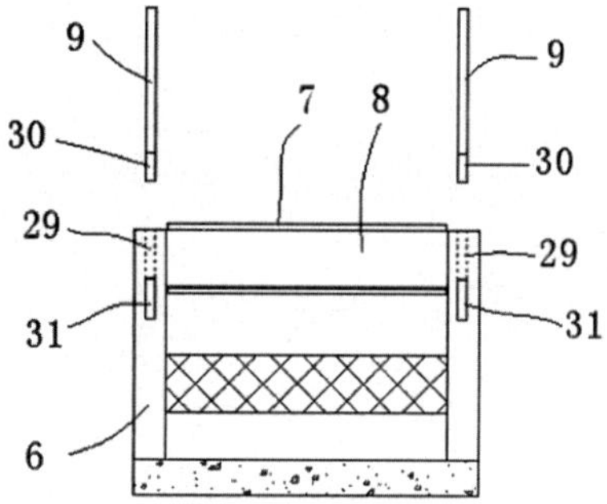

Fig. 3. Structure breakdown diagram

References

1. Gao, Y.: 3D Modeling and Manufacturing of Parts UG 3D Modeling. China Machine Press, Beijing (2010)
2. Li, J.: Fundamentals of Mechanical Design. China Machine Press, Beijing (2012)
3. Xu, X., Zhao, S.: Engineering Materials and Forming Technology. Metallurgical Industry Press, Beijing (2010)
4. Ye, W.: Accelerating agricultural mechanization in southern China with science and technology. South. Agric. Mach. **48**(07), 29–30 (2017)

Database Management Systems

Research on Information Management of Gas Engineering Project in Database

Jing Wan(✉)

Jiangxi Institute of Economic Administrators, Nanchang 330088, China
2960199289@qq.com

Abstract. For small and medium-sized enterprises adopting engineering project ledger to the gas project engineering management, this paper developed the gas engineering project information management system in database to fulfill the requirements for contract management, material management, schedule management, design alteration, engineering visas, payment management and budget and settlement management, which provide the reference to improve the information management level of gas engineering project for small and medium-sized enterprise.

Keywords: Database · Gas engineering project · Information management

1 Introduction

1.1 Gas Engineering Characteristics

(1) Large quantity. Taking one small and medium-sized enterprise as an example, it undertakes approximately 300 projects each year. (2) Less project loads, shorter construction period, lower cost but slender profits for construction units in single projects. Taking the case of industrial and commercial projects which cost RMB from several thousand to several hundred thousand and spend between ten days and a month, construction units should undertake projects as many as possible to make profits. (3) Difficulties in the on-site coordination relationship [1]. As one of the ancillary works of urban development and construction, gas pipeline installation should cooperate with municipal road construction and real estate construction, which requires communication and coordination with government departments, property companies and users. (4) Complicated construction formalities. Each small single project has complete construction management process and formalities, including technical and safety disclosure, the distribution, transportation and placing of materials and equipments, site surveying, groove digging, pipeline welding, pressure testing and connection, etc., which puts forward higher requirements for the real-time data collection and sorting in the case of limited time.

1.2 Current Situation of Project Information Management in Gas Enterprises

Compared to the well-known gas enterprises in China such as China Gas, CR Gas and ENN which have a mature project management system that can realize the whole

M. Tavana et al. (Eds.): IISA 2020, AISC 1304, pp. 435–441, 2021.
https://doi.org/10.1007/978-3-030-63784-2_54

process management from decision making to completion of acceptance of the project and collaborative management with relative departments of engineering management, material management, safety management and financial management, the statistics show that many state-owned enterprises or small and medium-sized private enterprises are lack of informatization tools. In addition to the SCADA system [2] for pipe network and the customer management system [3], other departments using the original Excel table for routine operational data statistics and separate accounts from each other which is difficult to realize data sharing. Therefore, in the face of the rapid development of customers, the paper developed the gas engineering project information management system based on the gas engineering characteristics and gas project management process to improve the efficiency and quality of engineering management for enterprises and realize data sharing and business control.

Access is one of develop software tools in database [4] which can directly modify, import, analyze and summary data from existing Excel report of each departments. Compared with the purchase of professional information system, it makes the cost greatly reduced and is simple in use and convenient which can quickly improve the work efficiency [5], and it is more suitable for many domestic state-owned gas enterprises and small and medium-sized gas enterprises for project information management.

2 Database Functional Design

2.1 Gas Project Management Process

Through the investigation of one small and medium-sized enterprise, its gas project management process is as follows: (1) Project establishing and project number: with the marketing department and the client arriving at the intention to sign the contract, the marketing department will collect detailed client information in time and organize the design department site reconnaissance and check on further relevant information. After the approval of site reconnaissance or assessment report and the conditions required to sign the contract, the marketing department will set up the project according to gas engineering project development sequence and number the project. (2) Project design and project budget: the design department is responsible for the engineering design according to the drawings provided by clients and relevant requirements. The design drawings are submitted to the engineering department to budget the project. (3) Construction assignments: the engineering department should complete the pipeline planning procedures before the engineering design and pipeline excavation procedures before the commencement of the site in principle based on the plan of the marketing department. After the completion of the budget, the engineering department should provide the design drawings, project budget to the construction unit and the supervision unit, sign the construction contract with the construction unit and timely type the contract information. (4) Commencement and completion: on the basis of the drawings and budget, the construction unit should organize site reconnaissance, review the pipe layout, type of gas appliance and its installation location, installation position of measuring instruments and gas pressure regulator, and other needed to validate after they receive the construction tasks. Sometimes it is necessary to gather the engineering department and the design department to make a technical clarification. The following procedures involve coordinating to determine the commencement date of the site, drafting the construction program and receiving materials and equipments from the materials list in drawings. With the client's requirements or the location of the pipe changed during the construction process, it is available to make design alteration or engineering changes, which the design alteration should be issued by the design department and sent to the construction unit, while the engineering changes should be made by the construction unit to submit the application to the supervision unit, filling in the reasons and contents of the change, and finally be signed for confirmation by the three parties of the design department, the engineering department and the supervision unit. In case of any work beyond the scope of the drawings due to site conditions, the construction unit may apply to the supervision unit for the spot visa, which should finally be signed for confirmation by both the engineering department and the supervision unit. After the completion of construction, the construction unit should finish the connection and ventilation of the pipeline and equipment under the supervision of the engineering department, the safety department and the pipe network operation department, submit the preliminary completion report of the project to the engineering department, return the excess materials to the material department after passing the acceptance, reorganize completion data, apply the transfer to the pipe network operation department, complete the preliminary examination documents of project settlement audit and submit them to the audit department for the completion of settlement, and transfer to the financial department for the finally procedure of fixed assets formation.

Based on the project management process above, the corresponding flowchart is drawn as shown in Fig. 1:

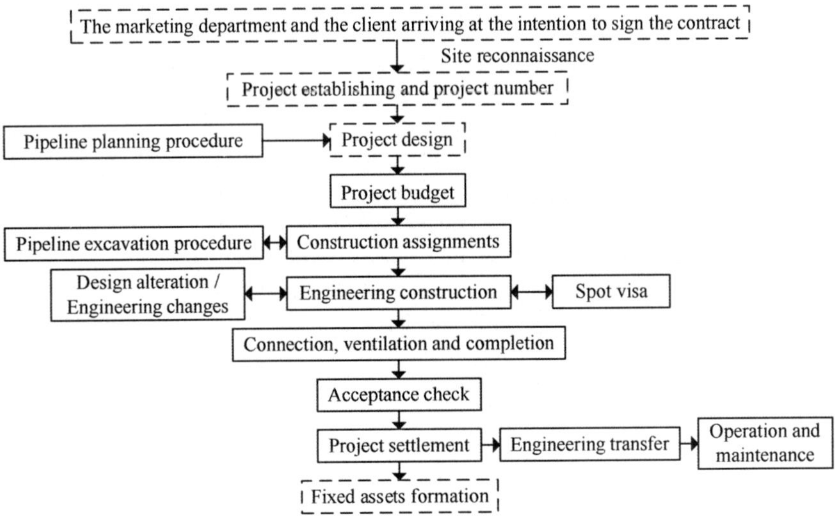

Fig. 1. Flow chart of gas engineering project management process

The work in the solid box of Fig. 1 is the key process of gas engineering project management, and this paper take it as the research and design content of the gas engineering project information management system in database. Combined with the flow chart, the project management process can be analyzed as follows: contract management, material management, schedule management, design alteration or engineering changes, engineering visas, payment management, budget and settlement management.

2.2 Data Table Design

Based on the seven managing requirements analyzed above, the paper designed seven data table, as shown below. (1) Contract information table: for querying, adding and modifying project number, commission contract number, project category, project name, the construction unit, planning/actual construction and completion date, site manager, designer, design and installation households/meters, gas engineering budget (total project cost, installation project cost, principal material and equipment cost), etc. (2) Materials information table: for querying, adding and modifying types and numbers of principal material and equipment in the project including PE pipe, seamless steel tube, welded steel tube, galvanized steel pipe, gas meter, flowmeter, regulator box/cabinet, etc. (3) Working schedule table: on the basis of monthly situation of the construction site, listing the amount of completed work of different projects, including project category (low and medium pressure gas pipe of municipal road engineering, garden pipe of inhabitant user, indoor installation of inhabitant user, industrial and commercial pipeline), stress level of work performed, pipe material, site manager, commencement date, completion date, budget cost, ventilation date, progress and difficulties, and summarizing statistics of month to date, current year cumulative, cumulative quantities, image progress (meters of different pipe diameters/installation households), etc. (4) Engineering change and visa table: the change information includes change type (change request/change), change time, reasons for change and content of change, the visa information includes visa type (initialed document/visa), date, content and visa fees, etc. (5) Connection and transfer information table: listing completion date, the construction unit, with or without underground facilities, receive time and return time of the engineering department and the pipe network operation department. (6) Payment table: listing details of charges during the construction period, including cost type (progress payment, project settlement accounts, design fee, supervision fee, excavation cost, quality deposit deduction and restitution), auditing date, payment date, payment amount, payee, agent, etc. (7) Budget and settlement table: listing gas engineering budget, project category, amount for review, preliminary audit account and date, audit account and date, auditor, design and installation households/meters, design flow of gas pipeline, riser installation method of inhabitant user (exterior wall/interior wall), municipal pipe installation method (directly-buried pipe, overhead pipe, undercrossing pipe), type of residential areas (newly built/occupied), type of buildings (multi-story, middle-height, high-rise, villa), installation mode of gas meter (outdoor/indoor), installation location (to front stove/to three-way connection), total change cost, total visa cost, valuation way, buried pipeline meters per household, cost per household, unit discharge cost, unit length cost, etc.

The above seven tables are associated with the "project number" (as shown in Fig. 2) to provide statistical basis for the later form query and report output.

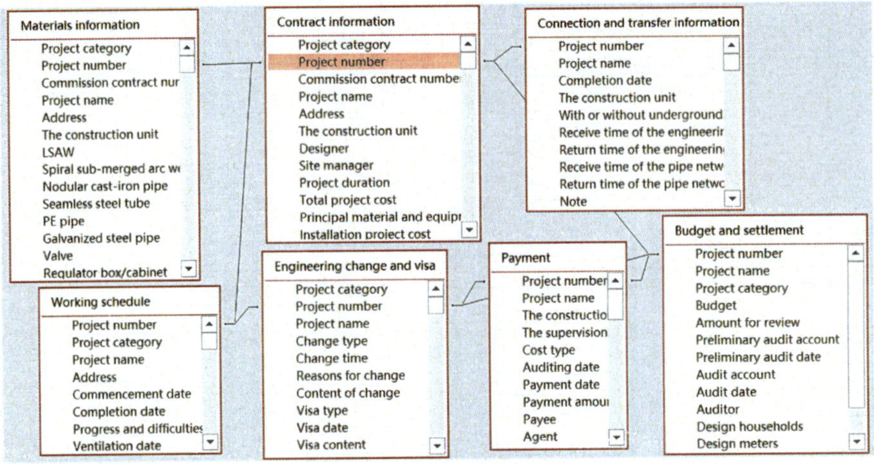

Fig. 2. Data relation

3 Form Design

3.1 Login Form Design

This form is designed for operating personnel of engineering management department of gas enterprises to authenticate, edit, modify and export data (as shown in Fig. 3). It is necessary to enter the correct user name and password in order to connect to the database, otherwise, error message will appear.

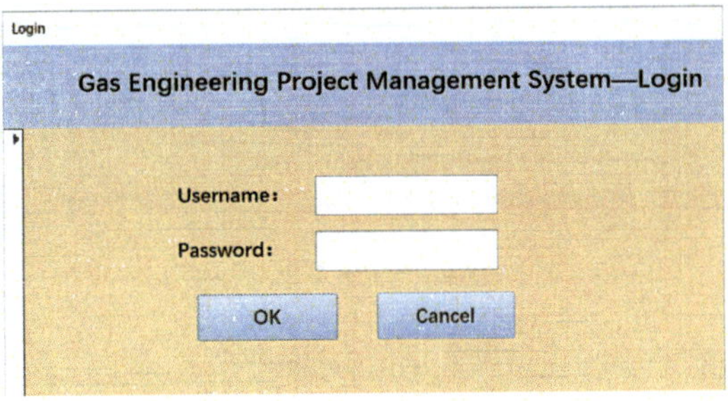

Fig. 3. Login form surface

3.2 Monthly Report Form Design

It mainly contains output of four types of reports: monthly construction commission report, monthly material planning report, monthly image progress report and monthly payment report, which is created according to the contract information table, materials information table, working schedule table and payment table by inputting the year and month (as shown in Figs. 4, 5, 6 and 7).

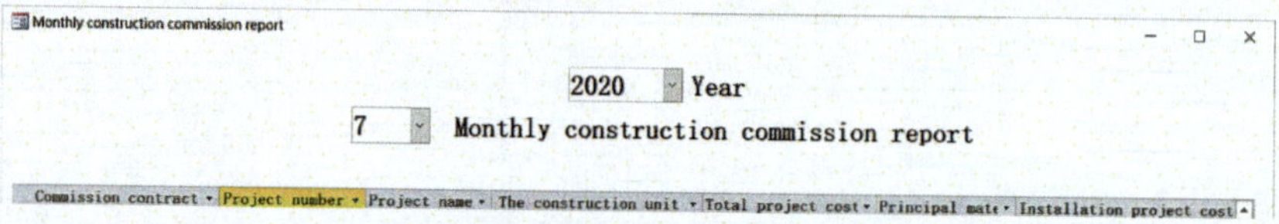

Fig. 4. "Monthly construction commission report" form surface

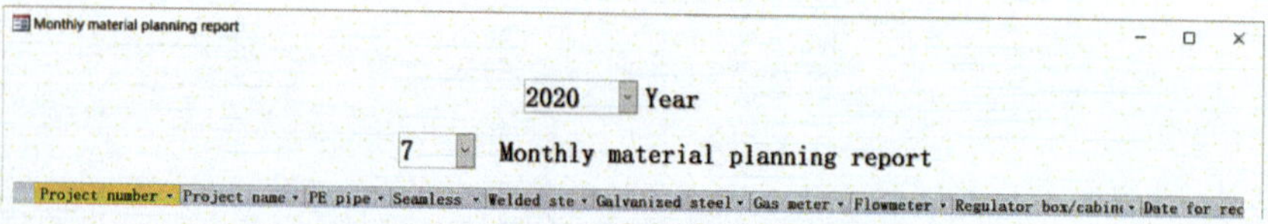

Fig. 5. "Monthly material planning report" form surface

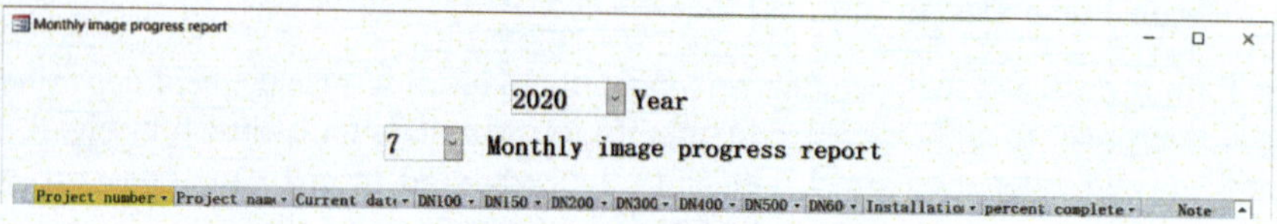

Fig. 6. "Monthly image progress report" form surface

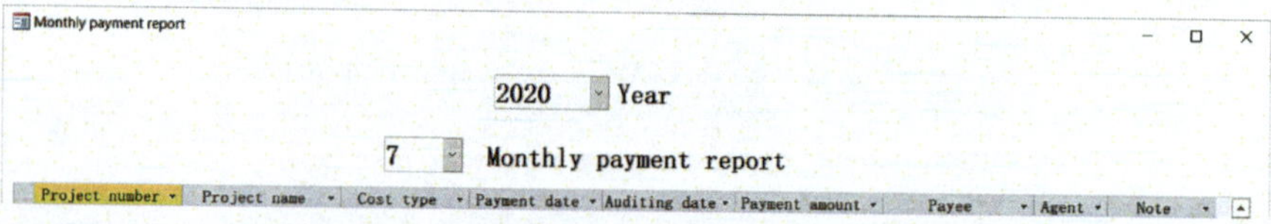

Fig. 7. "Monthly payment report" form surface

3.3 "Engineering Management" Form Design

This form integrated information of the above seven data tables into a "engineering management" form surface, which can query the engineering information of the project, including the contract information, material information, progress information, change and visa information, connection and transfer information, payment information, budget and settlement information through the "Project number" inputbox by inputting project number (as shown in Fig. 8).

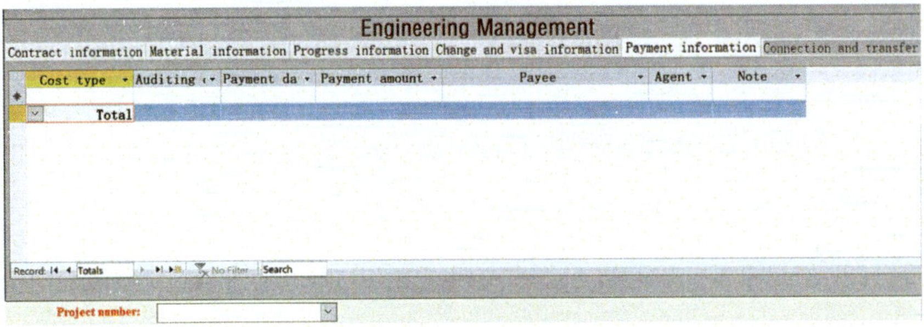

Fig. 8. "Engineering management" form surface

4 Conclusion

On the basis of one small and medium-sized gas enterprise as research object, this paper analyzed its project management process, adopted the access database technology to design data table and form, which can complete related business queries, engineering statistics and report outputs to promote efficiency for gas engineering project management of small and medium-sized gas enterprise directly within low cost. In the future the author will optimize and improve the running database from feedback of operating personnel of engineering management department and develop functional boundaries of engineering management database to achieve data collaboration with other gas operating departments.

References

1. Xu, L.: Analysis and design on management information system of city gas. J. Xingtai Polytech. Coll. **31**(1), 102–104 (2014)
2. Gaoruru: Design and implementation of gas management information system based on web. Shandong University, Weihai (2018)
3. Hui, X.: Design and implementation of intelligent gas management information system. Tianjin University, Tianjin (2018)
4. Alexander, M., Kusleika, D.: Access 2019 Bible. Tsinghua University Press, Beijing (2019)
5. Zhang, J., Sun, F.: Application of computer database technology in information management. China Comput. Commun. (20), 124–125 (2019)

A New Accountable Data Sharing Scheme

Jia Fan[1(✉)], Yunfei Cao[2], and Yili Luo[3]

[1] Sichuan Innovation Center of Industrial Cyber Security, Chengdu, China
june_jiafan@163.com
[2] Science and Technology on Communication Security Laboratory, Chengdu, China
[3] Southwest Jiaotong University, Chengdu, China

Abstract. At present, utilizing data sharing to promote the development of data and related industries has become more important. Most data sharing schemes use complex cryptography to ensure data security, but the scheme is inefficient, and almost without considering accountability. In this paper, we propose an efficient accountable data sharing scheme based on a trusted hardware device SGX (software guard extensions).

Keywords: Data sharing · Accountable · SGX

1 Introduction

Nowadays, people's demand for data sharing is more and more intense [1]. For example, in scientific research, it can provide basic data for relevant scientific research; in government affairs, it can enhance the government's credibility and improve the efficiency; in manufacturing, it can provide users with accurate products through data sharing and machine learning. However, there are many security issues, such as data confidentiality and accountability.

In order to solve the above problems, most existing data sharing systems apply a third party, and use cryptographic techniques such as encryption algorithms and access control techniques to ensure data security. In 2006, an encrypted data sharing scheme based on proxy re-encryption is proposed in [2], while the user cannot achieve manage the shared files directly. In 2010, in order to manage private health records in the cloud, Li et al. [3] applies attribute-based encryption for data sharing scheme. In 2015, based on attribute-based proxy re-encryption, an online-offline data sharing scheme [4] is proposed for mobile cloud environment application. However, user's offline computing overhead is very large in [4]. In 2016, a traceable attribute-based Signature is presented by Ahuja et al. [5] in data sharing environment. In 2017, authors in [6] presented a new sharing scheme in mobile cloud environment, which offers traceability. While in this scheme, the tracking algorithm is very complicated in computing.

In 2017, Severinse et al. [7] first presented a new data sharing protocol which is accountable. However, there are still two main drawbacks. First, in their scheme, users need to be online all the time to ensure the correctness of the log records. Second, this protocol only supports a single user to provide sharing data, which leads to impractical in real data sharing environment. In this paper, we present an improved method, solving the one user problem as well as the offline problems in Severinse's scheme.

M. Tavana et al. (Eds.): IISA 2020, AISC 1304, pp. 442–449, 2021.
https://doi.org/10.1007/978-3-030-63784-2_55

2 Severinsen's Scheme

In Severinsen's scheme, it is assumed that Software Guard Extensions (SGX) [8–10] is a fully trusted hardware, such that the key stored in SGX is secure, and the decrypt operation in it is always valid. The model of this scheme is shown in Fig. 1.

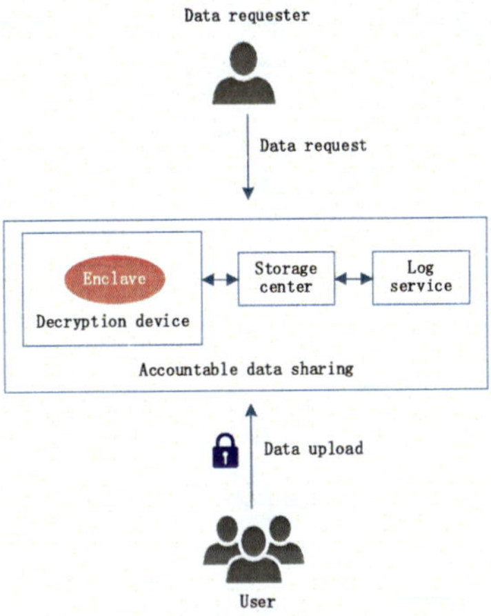

Fig. 1. Accountable data sharing scheme model

SGX (Enclave is contained in SGX) keeps the record of a nearest Merkle tree root. A decrypting requisition sends last root tree hash value H, current root hash H', a proof π of presence and a proof of extension of the log to the decryption device. Then, the decryption device performs a log check step. If the provided log is correct, then SGX would decrypt the ciphertext, and returns a plaintext.

3 Our New Accountable Data Sharing Scheme

3.1 The Process of Our Scheme

The process of our scheme is described in Fig. 2, and the details of the scheme is as follows:

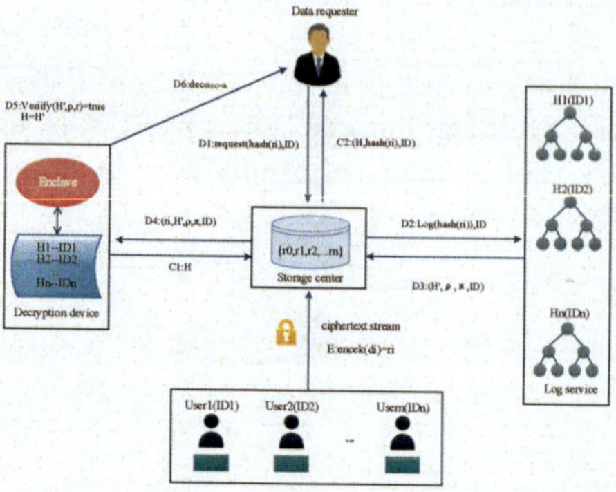

Fig. 2. Process of accountable data sharing scheme

(1) **Step E (Encryption Protocol)**
   1) User and enclave negotiate with the DH key exchange to produce a shared secret key *k*.
   2) User encrypts data *mi* in to ciphertext *ri*, using a symmetric encryption algorithm by shared secret key *k*.
   3) User uploads the ciphertext *ri* to the storage center.
   4) The storage center stores the hash index *hash(ri)* of the ciphertext.
   5) User stores all the hash value of his ciphertext records *R'= {Hash(r0),...,Hash (rn)}*.

(2) **Step D (Decryption Protocol):**
   1) D1: data request to the storage center.
      a) A data request concludes a hash identifier ID, as well as *hash(ri)*.
   2) D2: send to log service
      a) The storage center forwards this request to the log service,
   3) D3: log service generates proofs
      a) For a new request with *hash(ri)* and *H*, the log service adds *hash(ri)* to the existing Merkle tree, as the head of the new Merkle tree *H'*.
      b) The log service generates two proofs:
         i) The existence proof π: ensure that the new request is included in the new tree;
         ii) The extension proof ρ: ensure that the new tree is an extension of the old tree.
      c) The log service returns to the storage center with (H', π, ρ, ID).
   4) D4: Initial check for decryption device
      a) Storage center sends the ciphertext *ri* and (H', π, ρ, ID) to the decryption device.
      b) Decryption device needs all these data to check that the currently decrypted data is consistent with the version in the log.
   5) D5: Verification of proof and decryption
      a) If (π, ρ) *for (ri, H'ID)* is verified a valid proof, enclave updates H to H', stores H' in storage device, and decrypts the ciphertext with shared key k.

b) Otherwise, this protocol is stopped.
6) D6: Return decryption result
a) After the decryption step, the decryption device returns the decrypted result to the user.

(3) **Step C (Log Check):**
1) C1: Enclave runs sign operation
a) Enclave sign on *H'*, and sends *Sign(H')* as well as the *identity* to the storage center.
2) C2: Return signature to user
a) The storage center forwards *Sign(H') as well as the corresponding ciphertext Hash(ri), and it's* identity to the user.

3.2 User Offline

If user is offline, it is impossible to verify the decryption information (Sign(H′), Hash (ri)) received each time. Therefore, the client needs to have the storage capacity to store the decryption information sent by the decryption device. As shown in Fig. 4, if the user is offline for a long time, the client for this user stores multiple decryption information from decryption device. When user goes online, he can use the saved ciphertext record *R* or ciphertext hash record *R′* to identify the data and then knows which data has been decrypted. The Merkle root value *H′* in the current log service is calculated by hash(*ri*) and *H* to determine whether the log is correct.

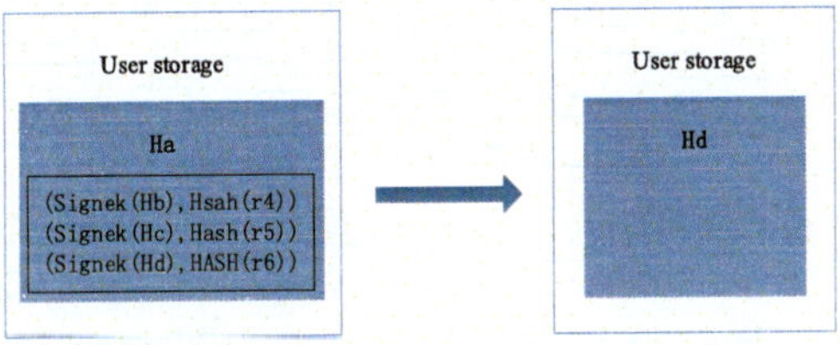

Fig. 3. Changes to client storage

Figure 3 shows the changes of client storage after the user changes from offline state to online state.

When the user is offline, he saves the log root value Ha, which was last calculated when the user was online, and the decryption information from the decryption device. Assume that users online at the moment,

The user runs the following steps:

(1) User verifies that whether Hb is issued by the decryption device.
(2) Calculates Hb by using its saved Ha and hash(r4).
(3) The user compares the calculated Hb′ with Hb. The Merkle root value update process in the log service is shown in the following Fig. 4.

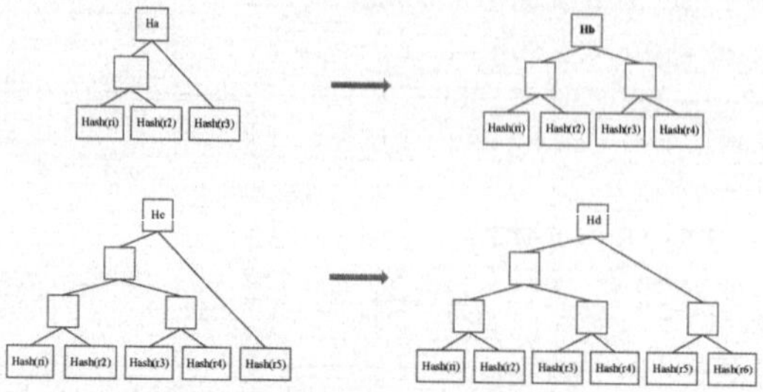

Fig. 4. The update process of Merkle root value

1) If Hb = Hb′, then
   a) The user updates his saved Ha to the currently calculated Hb.
   b) Delete the record of ($Sign_{sk}$(Hb), Hash(r4)).
2) User stops data sharing.
3) User ask the log service to take responsibility for the wrong log.

3.3 Multi-user

For multi-user case, if all users' decryption log information is stored in the same tree, it is cannot conduct log check and data decryption; Therefore, each user should have their own log tree. Currently, Intel only supports Enclave page cache for 128MBytec at most. To reduce the storage on enclave, we add an independent storage device to the decryption device to store the Merkle root value, and let enclave sign on *H*. Fig. 5 describes the process of data decryption in decryption device for multiuser case.

The detailed process is described as follows:

(1) When the decryption device received an decryption request on *(ri, H1′, π, ρ, ID)*, it uses the incoming *ID* value to find the corresponding *Sign(H1)* stored on the decryption device, and then it loads *Sign(H1)* into enclave.
(2) Enclave verifies *Sign(H1)*'s signature.
(3) If the signature verification passed, enclave checks the proof (π, ρ) provided by the log service. Otherwise, the protocol is stopped.
(4) If the proof verification is successful, then Enclave do decrypt operation, update the Merkle hash root *H1* to hash root *H1′*. Otherwise, the protocol is stopped.
(5) Enclave signs on the hash root value *H1′* and loads *H1′* to the storage device.

3.4 Scheme Analysis

In Table 1, we compare the features of Severinsen's scheme and our proposed scheme. It shows that our scheme solved the single user problem as well as the user offline problem in Severinsen's scheme.

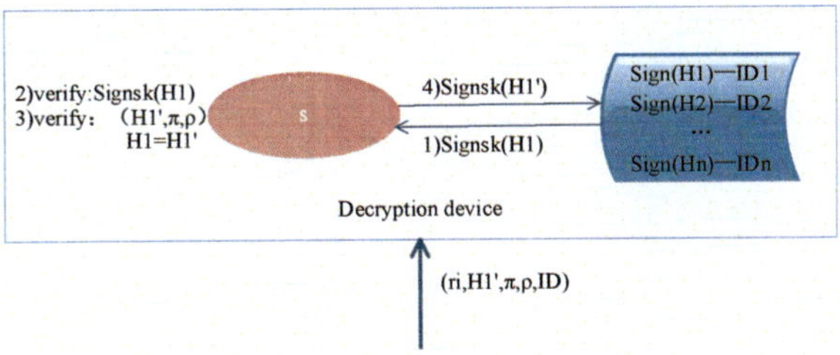

Fig. 5. The process of data decryption in decryption device

Table 1. Comparison of scheme features.

Scheme	Features			
	Number of users	Offline	Key management	Check log
Severinsen's scheme	Single user	No	Stored in enclave	Online
Our scheme	Multi-user	Yes	DH key exchange	Offline

4 Application on Medical Data Sharing

Medical data not only provides a lot of convenience for patients' treatment process, but also provides precious data for public medical research. However, at present, the sharing of medical data has become a difficult problem, mainly with the following problems:

1) Medical data is generally stored by medical institutions, but the storage equipment and management of each medical institution are different, resulting in scattered data storage and difficulty in sharing;
2) When users, doctors or medical institutions are sharing medical data, they need to spend a lot of time and resources on authentication and authorization audit;
3) Users cannot manage their medical data directly, and their personal privacy is greatly threatened.

In this paper, we designed a new medical data sharing platform. The application model of system is shown in Fig. 6. Our scheme aims to provide a common data sharing platform for medical data sharing, which enables users to manage their own data.

First, patients join in the medical data sharing system with real identity registration. After successful registration, the system will assign a unique identification ID to each user. The authentication platform will authenticate the user or institution that initiated the request, check whether the user has the authority to request data, and ensure that the patient's medical data is only viewed by the user with authority.

In our scheme, patients can obtain their use of medical data by checking the log records. Once patients discover that their medical data has been used in violation of

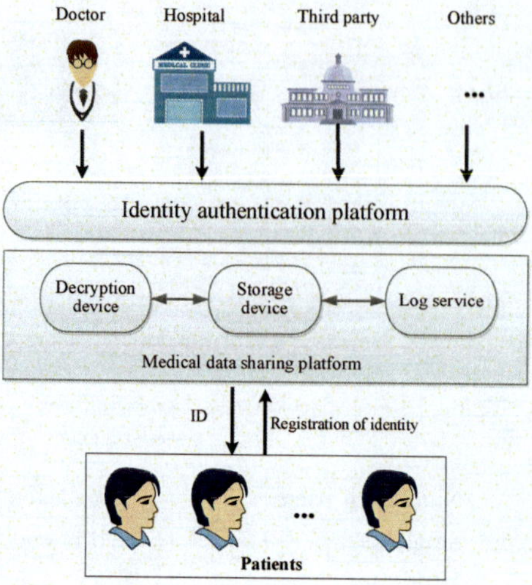

Fig. 6. Medical data sharing model

regulations, they can immediately stop sharing the data and hold themselves accountable for the violation through log recording.

Acknowledgement. This work was supported by the Key Research and Development Project of Sichuan Province of China under Grant 2020YFG0292.

References

1. Fienberg, S.E., Martin, M.E., Straf, M.L.: Sharing Research Data. National Academy Press, Washington, D.C. (1985)
2. Ateniese, G., Fu, K., Green, M., et al.: Improved proxy re-encryption schemes with applications to secure distributed storage. ACM Trans. Inf. Syst. Secur. **9**(1), 1–30 (2006)
3. Li, M., Yu, S., Ren, K., et al.: Securing personal health records in cloud computing: patient-centric and fine-grained data access control in multi-owner settings. In: International Conference on Security and Privacy in Communication Systems. Springer, Heidelberg (2010)
4. Shao, J., Lu, R., Lin, X.: Fine-grained data sharing in cloud computing for mobile devices. In: Computer Communications. IEEE (2015)
5. Ahuja, R., Mohanty, S.K., Sakurai, K.: A traceable signcryption scheme for secure sharing of data in cloud storage. In: IEEE International Conference on Computer & Information Technology. IEEE (2017)
6. Wang, Z.: Research on efficient traceable data sharing scheme in mobile cloud computing (2017)
7. Severinsen, K.M.: Secure programming with Intel SGX and novel applications (2017)
8. McKeen, F., Alexandrovich, I., Berenzon, A., et al.: Innovative instructions and software model for isolated execution. In: HASP@ ISCA, vol. 10, no. 1 (2013)

9. Anati, I., Gueron, S., Johnson, S., et al.: Innovative technology for CPU based attestation and sealing. In: Proceedings of the 2nd International Workshop on Hardware and Architectural Support for Security and Privacy, vol. 13. ACM, New York (2013)
10. Hoekstra, M., Lal, R., Pappachan, P., et al.: Using innovative instructions to create trustworthy software solutions. In: HASP@ ISCA, vol. 11 (2013)
11. Averill, J.D., Reneke, P., Peacock, R.D.: National Institute of Standards and Technology (NIST) (2004)
12. Merkle, R.C.: A certified digital signature. In: Advances in Cryptology (1989)
13. Diffie, W., Hellman, M.: New directions in cryptography. IEEE Trans. Inf. Theory **22**(6), 644–654 (1976)

The Automated Operation and Maintenance Solution for Cloud Data Centers Based on Multi-station Integration

Qiang Li[1], Fusheng Yuan[2], Jing Chen[3], Shengxi Shi[2], and Shilong Xu[2](✉)

[1] State Grid Information & Telecommunication Co., Ltd., Beijing 102211, China
[2] State Grid Information & Telecommunication Co., Ltd., Beijing Branch, Beijing 100031, China
xushilong@sgitg.sgcc.com.cn
[3] Beijing Electric Power Economic Research Institute Co., Ltd., Beijing 100037, China

Abstract. Multi-station integration, as a significant part of the power Internet of Things, can realize the in-depth integration of energy and information industries and the lean utilization of communication resources. This paper mainly proposes an automated operation and maintenance solution which is efficient, scalable and stable, according to characteristics and technical requirements of operation and maintenance for data center stations. Adopting hierarchical, modular design concept, the solution uses cloud computing and virtualization technology. This designed system can realize automatic management of resources and improve operation and maintenance efficiency.

Keywords: Multi-station integration · Data center station · Cloud computing · Automated operation and maintenance solution

1 Introduction

In the era of 5G, the energy revolution, marked by the large-scale development and utilization of new energy, continue to flourish [1]. As a network hub which connects energy production and energy consumption, the national grid plays an increasingly prominent role. In order to actively adapt to the new trend of energy revolution and technological development, Multi-station integration service, as one of the most important applications for the implementation of the ubiquitous power Internet of Things, has received wide attention from this field [2].

Multi-station integration refers to the integration of data center stations, charging stations, energy storage stations, 5G base stations, BeiDou base stations, photovoltaic stations, etc., on the basis of existing substations. By building a more powerful information processing and data storage platform for smart substations, the optimization of urban resource allocation and the improvement of data perception capabilities can be realized. As a newly developed station, data center station can be used to provide data storage, calculation and management services for power grid companies

M. Tavana et al. (Eds.): IISA 2020, AISC 1304, pp. 450–458, 2021.
https://doi.org/10.1007/978-3-030-63784-2_56

and the majority of social customers. In a word, building data center stations can help promote resources sharing and achieve mutual benefit [3].

With further research into Multi-station integration, the data center station system is faced with practical problems due to massive amounts of data and its complexity. How to realize the effective collection, storage, control, and access of heterogeneous data is the primary problem remaining to be solved. Currently, power grid companies have done some work in the collection and application of power data [4]. However, with regard to Multi-station integration services, there still exists problems like data models are inconsistent and massive heterogeneous data is difficult to achieve deep integration. Moreover, under the traditional architecture model, data centers have difficulty in data synchronization and fully resources utilization. The operation and maintenance methods cannot cope with the operation and maintenance requirements of data center stations for dynamic resource allocation and unified analysis of information. The operation and maintenance methods are unable to meet the needs of dynamic resource allocation and unified analysis of information.

To sum up, how to effectively integrate the resources of each station, maximize the potential of the data center station, and improve the efficiency of data sharing and processing has become an urgent problem for data center stations in the context of Multi-station integration. Hence, the main objective of this study is to propose an automated operation and maintenance solution which is efficient, scalable and stable. Therefore, our solution aims at realizing automatic management of resources and improving operation and maintenance efficiency.

Although the industry has proposed data center operation and maintenance solutions for different scenarios, and achieved some representative results [5–7], the current academic research on Multi-station integration is limited to a single station, such as the optimal sizing and locating of substation [8], capacity optimization design of energy storage stations [9]. However, there are few documents considering design data center stations' operation and maintenance. Therefore, we propose in this work an automated operation and maintenance solution which can realize automatic management of resources and improve operation and maintenance efficiency.

The rest of this paper is organized as follows. Firstly, we present related works. Secondly, we analyze the requirements of data centers. Thirdly, the proposed work is described. Finally, some potential problems worthy further investigation are given at the end.

2 Demand Analysis

The data center station has accumulated operation and monitoring data from the information systems of other stations. On the one hand, the business areas covered by the data system are diversified and complicated. On the other hand, technical architectures of various information systems are very different, and the development level is uneven too. Data between different systems are independent, which makes the problem of low efficiency of information exchange and sharing between systems extremely prominent.

(1) Inconsistent Business Data
As a facility for centralized processing, storage, transmission, exchange, and management of data information, data center stations will receive data in different formats and protocols from each station. For example, environmental monitoring stations collect environmental data and uploads relevant data to environmental monitoring analysts through the monitoring system of data center stations. BeiDou base stations receive the synchronization time signal of BeiDou satellite navigation system and communicate with the monitoring system in substations and data center stations, providing clock synchronization signals and positioning signals for the entire station and other equipment in the surrounding area. 5G signal stations use 5G signals for signal transmission between the monitoring system in other stations and terminal equipment. The construction of a standardized system helps to solve the problem caused by heterogeneous multi-source data.
(2) Difficulty in Operation and maintenance
Due to lack of overall planning for system construction in various fields in Multi-station integration and one single system corresponds to one business application, collected data cannot be effectively shared. In addition, as the amount of grid services and applications continues to increase, the traditional data center architecture cannot meet the requirements of centralized management of the operation and alarm information from various equipment systems as well as information systems.

As a result, high-value information contained in data has not been fully utilized [10]. Therefore, it is imperative to design an automatic operation and maintenance system based on cloud computing in the context of Multi-station integration.

3 Proposed Work

3.1 Technology Architecture

As a way of processing large-scale computing information, the cloud computing platform can realize the unified organization of various information and communication resources through technologies such as parallel programming, virtualization, high-speed networking, and rapid deployment.

In the context of Multi-station integration, the intelligent operation and maintenance system based on cloud computing can be divided into four levels, namely the basic resource layer, the virtualization layer, the basic platform layer, and the application management layer, as shown in the following Fig. 1.

1) Basic Resource Layer

The basic resource layer includes IT resources, such as servers, network equipment, storage equipment, and security equipment, as well as multi station resources such as power equipment, BeiDou equipment, and environmental monitoring equipment. These hardware facilities provide data collection, calculation, storage, and communication services for cloud based data centers.

The basic resource layer analyses data types, formats, and specifications from different information systems. In addition, it collects fused fault data, operating data, and environmental data according to the actual needs of Multi-station integration applications through data integration and cleaning.

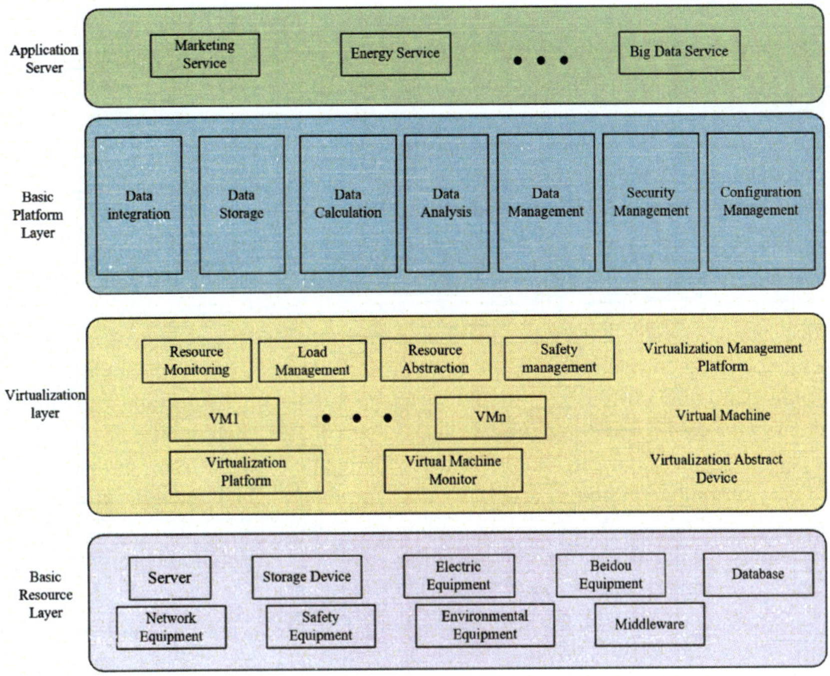

Fig. 1. The functional architecture of automatic operation and maintenance system for cloud data centers

2) Virtualization Layer

Compared with the hierarchical structure of traditional data centers, the use of virtualization technology can bring following advantages: a. Increase servers' resource utilization; b. Improve servers' security and isolate content from different origin; c. Realize storage virtualization [11].

In the proposed architecture, the virtualization layer consists of three parts, these are: virtualization abstraction devices, virtual machines, and the virtualization management platform. Among them, virtualization abstraction devices include a virtual machine monitor and a virtualization platform, which are used to realize the abstraction of network equipment, storage equipment, and monitoring equipment of each station, forming a logically unified computing resource pool. A virtual machine is the unit of a cloud data center. Application servers and database servers are deployed on virtual machines as an information system operating environment, aiming to provide a high-performance distributed computing environment for advanced applications such as large-scale storage, data mining, and decision-making assistance. Finally, the management of physical resources is realized through functions like resource monitoring, load management, resource abstraction, and security management provided by the virtualization management platform.

3) Basic Platform Layer

To form the operating environment of the data center, a server cluster needs to be constructed in units of virtual machines. The basic platform layer of cloud data centers, which are divided into a data processing platform and a centralized management platform, are supposed to have data collection, storage and computing capabilities.

The data processing platform includes four modules: data integration module, data storage module, data calculation module and data analysis module.

Specifically, the data integration module supports data access methods with different data collection frequencies. It is able to integrate multiple data types and heterogeneous data sources. The data storage module builds data storage systems such as distributed storage databases, distributed memory databases and distributed file systems for cloud computing. It supports storage of various structured data and massive real-time data, and provides a reliable distributed data processing and mining environment. The data computing module uses technologies such as stream computing, memory computing, and batch computing that integrate big data from multiple stations to realize the preprocessing of massive data. The data analysis module utilizes the R language, machine learning and data mining technology of big data to find valuable information from massive amount of information. It can comprehensively analyze relationships between various data.

The data processing platform includes three modules: data management module, security management module and configuration management module.

Specifically, the data management module realizes the monitoring and management of various data to ensure the normal processing of data. The security management module is used to ensure the legitimacy of access to the data collection system, data storage system and business application system. By this way, threats such as illegal reading, copying and modification can be avoided. The configuration management module is responsible for the overall configuration and management of each module and component. Also, it allocates computing and storage resources of the platform.

4) Application Layer

The intelligent operation and maintenance system for cloud data center provides multiple stations and users an interactive platform. It aims to create a new energy ecosystem featuring openness, sharing, and win-win cooperation. It has established an interactive channel for internal and external networks as well as horizontal and vertical data, aiming to support the efficient operation of businesses in various industries [12]. Specifically, there exists energy services, marketing services, and big data services.

3.2 Functional Architecture

Based on the technical architecture, the design of the system functional architecture should consider characteristics of the large number, large scale, and rich variety of services involved in Multi-station integration. Through modular design, it supports various applications and business functions required in enterprise-level cloud data center IT systems. The integrated and integrated management mode can effectively reduce operation and maintenance costs and improve operation and maintenance efficiency.

This paper proposes an intelligent operation and maintenance functional architecture for Multi-station integration. It allows different stations to collaborate efficiently and orderly on the same platform, ensuring stable operation of each station. The functional architecture consists several subsystems, such as unified operation and maintenance portal, operation and maintenance process management and so on, see Fig. 2.

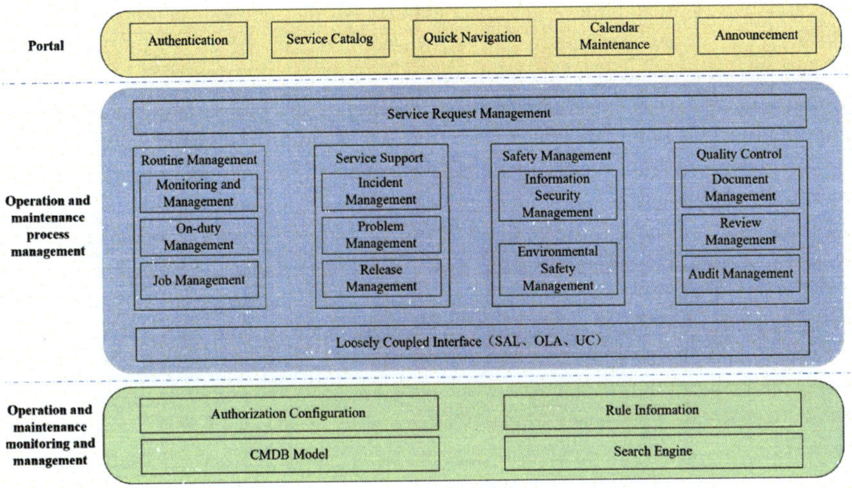

Fig. 2. The functional architecture of automatic operation and maintenance system for cloud data centers

1) Unified Operation and Maintenance Portal

The unified operation and maintenance portal, as a unified entrance to the intelligent operation and maintenance system for data center station, providing functions like identity authentication, service catalog, notification announcement, quick navigation, and calendar maintenance. What's more, this subsystem is based on the Echarts, calling APIs provided by servers to realize front-end data visualization.

2) Operation and Maintenance Process Management Subsystem

The operation and maintenance process management subsystem includes four core modules: routine management, service support, safety management, and quality management. Take the monitoring management function in routine management module as an example. It is based on real-time, full-coverage, and unified infrastructure monitoring. It can promptly detect and issue an alarm after troubleshooting, providing accurate and reliable information for supervision.

The operation and maintenance process management subsystem utilizes Customer Service Level Agreement (SAL), Operation Level Agreement (OLA), Supplier/Partner Contract Support Level Agreement (UC), etc., to connect the above five modules using loosely coupled interfaces [13]. In this way, an overall and orderly management order is established, while the flexible deployment of functional modules is maintained.

3) Operation and Maintenance Monitoring and Management Subsystem

The operation and maintenance monitoring and management subsystem monitors the service quality, operation status, and environmental information by building a unified operation and maintenance management system. It is easy to integrate the monitoring system, mail system, mobile APP, and data content of each station. Only after being authorized, the relevant components can be applied and tracked. The configuration management process mainly includes these steps: build and maintain data models, build a configuration management database (CMDB), maintain configuration item data, and audit configuration data. The construction of CMDB simplifies the data structure of the multi-source information system, so that the data content of different terminals, operating systems and multi-station equipment can get a unified management [14].

In addition, this solution introduces terminal equipment to realize real-time monitoring of various components in the system. Managers can view system operation information and alarm information through mobile APP anytime and anywhere, which saves labor costs and improves operation and maintenance efficiency.

Compared with current power data center systems, the technical architecture of this system integrates new services such as energy storage station, charging station and BeiDou base station Multi-station integration. It provides data center with location information and environmental information. Meanwhile, it uses 5G base stations to obtain more powerful information and communication capabilities, which significantly improves overall resource utilization of data center stations.

4 Open Research

4.1 Security Issues

Since the Multi-station integration system involves users' personal information, it is necessary to design an effective information security mechanism and make full use of identity authentication technology, key generation and update technology to ensure user privacy. What's more, data center stations store a large amount of information, leading to poor disaster preparedness and recovery capabilities. A dual-active data center can be taken into consideration to realize storage dual-active, server dual-active, network dual-active, and application dual-active [15].

4.2 Load Balancing

The technical architecture designed in this paper uses virtualization technology to integrate physical resources, make it convenient for the provision of flexible and scalable services. However, the high energy consumption caused by more and more servers in each station severely restricts cloud computing data centers' development. Therefore, it is necessary to find a balance between low energy consumption and high performance based on the concept of load balancing.

4.3 Cloud Edge Collaboration

Since the Multi-station integration system can obtain 5G services through 5G base stations, many emerging applications have higher requirements for high throughput and low latency. Therefore, people can consider introducing edge data centers closer to users to perform localized and real-time data processing. Therefore, the synergistic advantages of cloud computing and edge computing can be utilized to achieve unified resource scheduling and meet security and real-time requirements.

5 Conclusion

This paper analyzes problems faced by data center stations in operation and maintenance in the context of Multi-station integration. Next, it proposes an autonomous operation and maintenance solution for cloud data centers according to business needs. The proposed system can not only improve resource utilization and efficiency of operation and maintenance, but also solve the data fusion problem.

Acknowledgments. This work is supported by Science and technology project of State Grid Corporation of China (Research on key technology and business model of multi station integration, No. 5200-201941477A-0-0-00).

References

1. Zhou, W., Jiang, D.: Research on the management innovation of power grid enterprises based on the application of new technology of big cloud mobile intelligence.Enterp. Manag. (S2), 360–361 (2018)
2. Wang, B., Zhang, Y., Liu, M., et al.: Research on the multi-station integration operation mode. Electr. Power Inf. Commun. Technol. **17**(7), 41–45 (2019)
3. Duojie, C., Zhang, L., Wu, X., et al.: Analysis on the new mode of integrated development of transformer substation construction. Qinghai Electr. Power **38**(3), 24–26, 30 (2019)
4. Huang, W.: Building a power monitoring data center operation and maintenance system based on cloud computing. Plant Maintenance Eng. (15), 114–115 (2019)
5. Huang, S., Wang, H., Luo, Z.: Research on key technologies of cloud computing information and data center security in energy internet enterprise. Microcontrollers Embed. Syst. **20**(1), 27–29, 34 (2020)
6. Liu, D.: On architecture design of active-active smart campus data center based on cloud computing. J. Southwest China Normal Univ. (Nat. Sci. Ed.) **42**(5), 41–46 (2017)
7. Zheng, K., Xu, B., Xiao, Y., et al.: Interactive artificial intelligence platform solution for electric energy big data. South. Power Syst. Technol. **13**(8), 52–58 (2019)
8. Li, X., Wang, J., Ding, J., et al.: Optimal sizing and locating of multi-functional integrated substation based on voronoi diagram. Power Syst. Clean Energy **36**(2), 44–54 (2020)

9. Xu, W., Cheng, H., Bai, Z., et al.: Optimal design and operation of energy storage power station under multi-station fusion mode. Distrib. Utilization **36**(11), 84–91 (2019)
10. Len, X., Chen, G., Jiang, Y.: Data specification and processing in big-data analysis system for monitoring and operation of smart grid. Autom. Electr. Power Syst. **42**(19), 169–176 (2018)
11. Wang, P.: Discussion on the promotion value of network virtualization technology to traditional data center. China Comput. Commun. (17), 28–30 (2017)
12. Zhang, C., Liu, K., Cheng, F.: Research on global data center architecture facing grid online service channels. China Sci. Technol. Panorama Mag. (24), 117–119 (2018)
13. Lu, J., Shi, H.: Research on agile large-scale enterprise architecture based on chaos edge. Technol. Wind (13), 122–123 (2013)
14. Hang, F., Ou, W., Li, S., et al.: A Thought on the construction and data maintains of CMDB. China New Telecommun. **19**(14), 107–108 (2017)
15. Wang, H.: Research on "active-active data center plus three sites disaster recovery" to achieve all-round business continuity protection. Mod. Inf. Technol. **3**(02), 74–76 (2019)

Electronics Systems

Research on Component Level Test System of TCAS Circuit Board Fault Diagnosis System

Xiaomin Xie, Kun Hu(✉), Ying Hong, and Jianghuai Du

School of Mechatronic Engineering, Anhui Vocational and Technical College, Hefei 230011, China
357106305@qq.com

Abstract. TCAS circuit board automatic test system consists of two parts: hardware system and software system. Software is the kernel of the test system and coordinates and manages the hardware components. Hardware is the carrier of software implementation and supports the operation of the whole system. The test software mainly completes the generation of test codes, controls the operation of test hardware, completes the processing of test results, and provides an excellent interactive user interface for the whole test process. The hardware of the test system mainly completes the supply of excitation signals of the tested circuit board and the acquisition and conversion of response signals.

Keywords: TCAS · Fault diagnosis · Test software · Circuit board automatic test system

1 Introduction

TCAS (traffic warning collision avoidance system) provides an air separation guarantee independent of air traffic control, which plays an important role in safe flight and is an indispensable system for modern aircraft. Automatic Test Equipment (ATE) refers to all hardware and corresponding operating system software used to complete test tasks. Take the electronics industry as an example. In this field, automatic testing equipment is used to complete automatic testing of various integrated circuits, circuit boards, components and the whole equipment, so as to improve product quality and productivity.

The most direct purpose of adopting automatic testing equipment is to automate the testing process of products. The basic method to achieve this purpose is to integrate the resources (measuring instruments, excitation sources, changeover switches, power supplies, etc.) required for product testing into a unified system [1–5]. The testing process is controlled by the controller (computer) in the system through the execution of testing software. Its basic composition is shown in Fig. 1.

2 Scheme Design of TCAS Circuit Test System

In the system, the signal source provides various excitation signals (power supply, function generator output, D/A converter output, etc.) required for testing the object under test (UUT) and sends them to UUT. Measuring instruments (mainly digital

M. Tavana et al. (Eds.): IISA 2020, AISC 1304, pp. 461–468, 2021.
https://doi.org/10.1007/978-3-030-63784-2_57

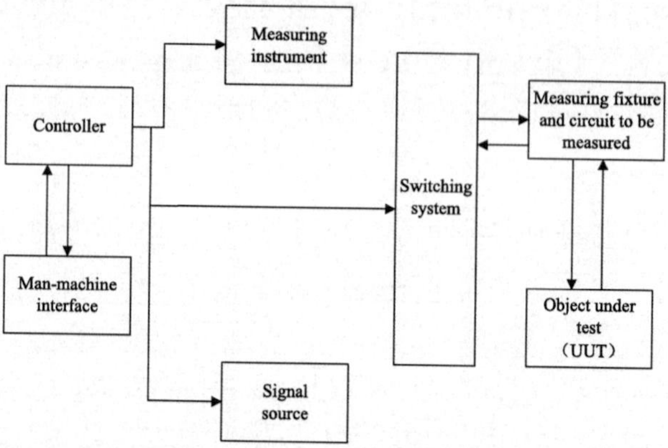

Fig. 1. Composition of automatic test equipment

multimeter, A/D converter, frequency/counter, oscilloscope, etc.) are used to measure the response of UUT measuring points after excitation is applied. The switching system is used to switch the signal to the required path according to the command of the controller. The controller is usually a general-purpose microcomputer or an embedded microcomputer, which is used to control the whole test process and process the measured data. The human-computer interface is a tool for operators to interact with ATE, mainly including CRT display, keyboard, printer, etc. Test fixture and adapter circuit are the interface between UUT and ATE, which ensure reliable mechanical and electrical connection and matching between UUT and ATE.

The test system designed in this paper belongs to a test system based on the idea of functional test system, which takes circuit board as the test object. The function automatic test system realizes the required detection by adding various excitation to some input terminals of the tested object (UUT, which can be various circuit boards, components, systems, etc.), and then measuring the response of the UUT. The response of the tested object is compared with the "standard" response of the "good" circuit board or component. If the two meet, the tested UUT works normally. If there is a difference between the two, the UUT has a fault.

The computer in ATE runs the test software, controls the excitation equipment, measuring instruments, power supply and switch components in ATE, adds the excitation signal to the place where it needs to be added, and measures the corresponding signal of the tested object at the appropriate point [6–8]. Then the test software analyzes the measurement result and determines the event that may be a fault, and then reminds the maintenance personnel to replace or replace one or several components. Since each UUT under test has different connection requirements and input/output ports, the connection of UUT to ATE usually requires corresponding interface devices, called interface adapters, which complete the correct and reliable connection of UUT to ATE and specify signal paths for each signal point in ATE to the corresponding I/O pins in UUT.

3 Hardware Design of Test System

TCAS processor mainly includes four boards: Video Memory, Memory Module I/II and Data Processor. Because the four boards have different functions, their test circuits are different.

(1) Test Design of Data Processing Board

The Data processor board is mainly composed of central processor and coprocessor, as well as peripheral module circuits of the processor, such as interrupt controller, bus controller, watchdog clock, etc. For the test of this board, the corresponding program and data memory are mainly accessed through its bus. The program memory stores the test programs of the processor and the peripheral modules of the processor, so that judgment and fault location can be carried out through the running results of the programs.

During the test, the computer sends a test control command to the controller, and the controller controls the corresponding memory to be connected with the interface board. At this time, the CPU on the Data processor board in the Data processor board slot on the interface board starts to run. The test block diagram of the data processing board is shown in Fig. 2.

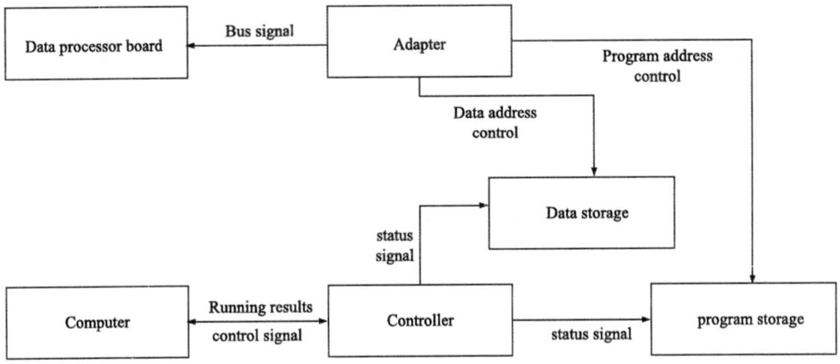

Fig. 2. The test block diagram of the data processing board

(2) Test Design of Data Memory Board

The data memory board includes Video Memory and Memory Module, among which there is FIFO (first-in, first-out memory) on the Video Memory board. For the testing of these boards, the data, address and control signals of CPU are simulated by special signal generator, and the RAM and FIFO are read and written. The test block diagram of the data memory board is shown in Fig. 3.

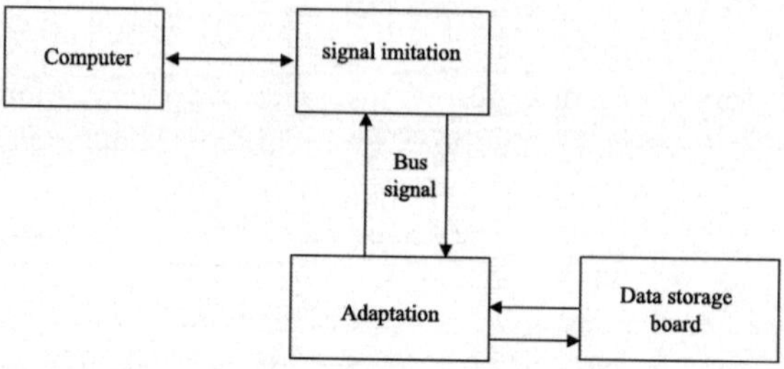

Fig. 3. The test block diagram of the data memory board

(3) Test Design of Program Memory Board

The program memory board test is basically the same as the data memory board. The main functions of the TCAS circuit board automatic test system hardware system are to complete the functions of sending test vectors to the circuit board to be tested, collecting response signals, interfacing with the circuit board to be tested, and switching each channel. The software in the test system controls the system through the hardware system. At the same time, the performance of the hardware system will directly affect the accuracy and reliability of the whole test system. Therefore, the hardware system is the basic environment of the test system, and the hardware system will directly affect the success or failure of the test system. The circuit board automatic test system adopts the idea of microcomputer instrument [9, 10]. The hardware part is actually a special microcomputer system, which consists of two parts: industrial control computer system and test hardware. Industrial personal computer is the carrier of the test hardware, the control center of the whole test system, and the junction of the tested object and the test system.

In order to enable the testing system to correctly and quickly complete the user's testing requirements, this testing system selects an industrial personal computer with Intel PentiumIII-1G CPU and 256M memory as the host of the testing system. The test hardware mainly includes A/D conversion interface, D/A conversion interface, air pump, needle bed clamp, data acquisition interface board, data acquisition card, power supply system and other parts.

4 Software Design of Test System

4.1 Software Platform Design

TCAS circuit board testing system software is an application software system running under Windows XP, which provides a friendly human-computer interaction interface for users by using graphical user interface. The software system of the automatic test system is used for parameter setting, resource allocation, generation of test vectors (excitation), collection and storage of response data, judgment, recording and display of test results, human-computer interaction and manual intervention of the test system. The software system is mainly composed of six main modules including user interface, system maintenance, task management, fault detection, fault diagnosis, data processing and analysis, as shown in Fig. 4.

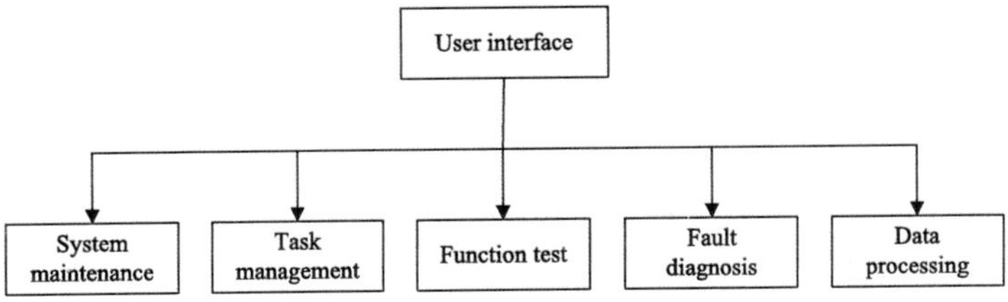

Fig. 4. Software module of circuit board test system

The main task of this system is to detect and locate the circuit board. The software system shall complete the following tasks:

(1) Provide the calling interface of all functions contained in the main window of the system for users to call various functions;
(2) Provide circuit board type selection function, and use open function for existing types;
(3) Provide a test task book editing window for users to fill in various test task books;
(4) Provide the task book saving function so that users do not have to repeat the task book editing process;
(5) Provide various types of test vectors or waveforms for system hardware;
(6) Realize different types of detection and diagnosis functions;
(7) Judging whether the tested circuit board has faults or fault positioning according to the test data.

A complete process of circuit board testing includes the following steps: system setup, channel configuration, task book filling, fault diagnosis and fault diagnosis results display. The software operation flow of the whole testing process is shown in Fig. 5.

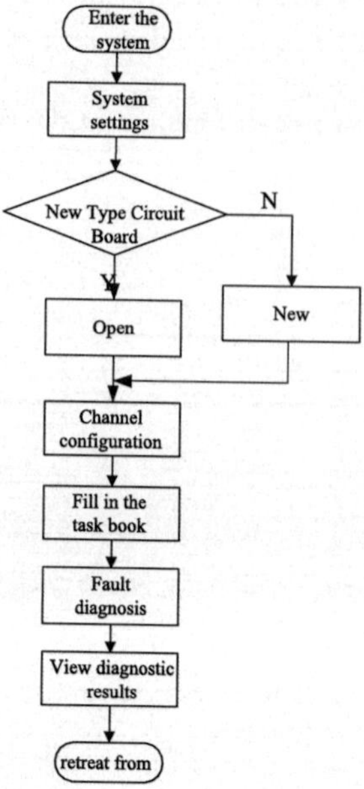

Fig. 5. Software flow chart of test system

4.2 Fault Diagnosis Program

When starting the fault diagnosis function, the fault diagnosis function tests the circuit board in detail, finds out the type of fault through the fault dictionary, and displays the fault location. The fault diagnosis process is shown in Fig. 6.

The establishment of the fault dictionary needs to simulate the possible faults in advance, collect the states of the detection points of the tested circuit board under the fault state, and store the states of the detection points into the fault dictionary as the judgment standard for the fault diagnosis. In actual fault diagnosis, the observation vector measured by the detection point is compared with the state in the fault dictionary to judge which fault type is closest to the observation vector, and the fault of the circuit board to be measured is considered to be that type of fault.

Because the simulation process of the fault is limited before the fault diagnosis, that is, the fault types in the fault dictionary cannot contain faults of this type, therefore, when a fault that has not been simulated is encountered, the test system will misjudge. The test system provides the function of user's manual judgment. The tester will choose whether the diagnosis result given by the system is correct or not. If the tester thinks it is wrong, the tester will fill in the fault type and the system will automatically add this state to the fault dictionary. In this way, as more and more fault boards are detected and diagnosed by the test system, its fault dictionary will be more perfect and the misjudgment rate will be lower and lower.

Because the diagnosis system is very complex when there are multiple faults, this test system only diagnoses single faults. In case of multiple faults, the fault diagnosis can be carried out after the fault diagnosed last time is eliminated; if the fault still

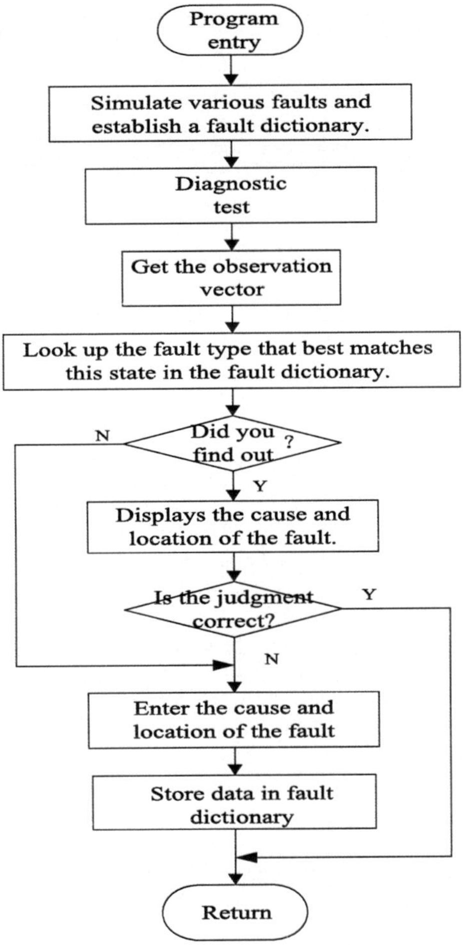

Fig. 6. Fault diagnosis flow chart

exists, the fault is eliminated again, and the process is repeated until the test system judges that the tested circuit board has no fault. Therefore, the fault diagnosis system of this test system is complete.

5 Concluding Remarks

TCAS automatic test system conforms to the development trend of test system, integrates the latest technologies and ideas, adopts the concept of integrated instrument and the architecture of automatic test, making the test system more powerful and convenient to use.

This paper introduces in detail the research and development process of TCAS circuit board automatic test system. This test system adopts common hardware interface, which is convenient to realize the universality of the test system. The function of the software system is complex, and the composition of the software system is designed, including various interfaces provided by the system to users and the working process of each core module. Hou will use the TCAS automatic testing device developed to test and diagnose the TCAS system and verify the function and performance of the testing device.

Fund Project. Natural Science Research Key Project of Anhui Province Higher School (KJ2019A0991);

Natural Science Research Key Project of Anhui Xinhua University (2018zr008).

References

1. Gao, L., Yang, Y., Liu, Y., et al.: Proof table based fault location method for process level channel in smart substations. Autom. Electr. Power Syst. **33**(4), 147–151 (2015)
2. Zhou, C., Wu, H., Hu, G., et al.: Non-intruding development of automatic test system based on IEC 61850 edition 2.0. Power Syst. Prot. Control **45**(14), 143–147 (2017)
3. Sharma, M.K., Vinesh, K.: Vague reliability analysis for fault diagnosis of cannula fault in power transformer. Appl. Math. Sci. **8**(18), 851–863 (2014)
4. Wang, L.: Design and implementation of transponder fault cable detection system. Beijing University of Posts and Telecommunications (2017)
5. Qiao, Z. -j, Lei, Y. -g, Li, N. -p: Applications of stochastic resonance to machinery fault detection: a review and tutorial. Mech. Syst. Signal Process. **122**, 502–536 (2019)
6. Li, J.-m., Li, M., Zhang, J.-f., Jiang, G.-j.: Frequency-shift multiscale noise tuning stochastic resonance method for fault diagnosis of generator bearing in wind turbine. Measurement **133,** 421–432 (2019)
7. Georgoulas, G., Climente, V., Antonino-Daviu, J.A., Tsoumas, I., Stylios, C., Antero, A., George, N.: The use of a multilabel classification framework for the detection of broken bars and mixed eccentricity faults based on the start-up transient. IEEE Trans. Industr. Inf. **13**(2), 625–634 (2017)
8. Zheng, Z., Pera, M.C., Hissel, D., Becherif, M., Agbli, K.S.: Li Y A double-fuzzy diagnostic methodology dedicated to on-line fault diagnosis of PEM fuel cell stacks. J. Power Sources **271**, 570–581 (2014)
9. Zheng, Z., Petrone, R., Pera, M.C., Hissel, D., Becherif, M.: Pianese C A review on non-model based diagnosis methodologies for PEM fuel cell stacks and systems. Int. J. Hydrogen Energy **38**, 8914–8926 (2013)
10. Aitouche, A., Olteanu, S.C., Bouamama, O.B.: A survey of diagnostic of fuel cell stacks systems. In: IFAC SAFEPROCESS, 8th IFAC Symposium on Fault Detection, Supervision and Safety of Technical Processes, Mexico City, Mexico, 29–31 August 2012 (2012)

Development of a HoloLens Mixed Reality Training System for Drop-Out Fuse Operation

Chibing Gong(✉), Deyun Ye, and Ruiling Xie

College of Electric Power Engineering, Guangdong Polytechnic of Water Resources and Electrical Engineering, Guangzhou 510925, China
spencergong@qq.com

Abstract. Drop-out fuse is the most commonly used type of protection switch for 10kV distribution lines and distribution transformers. The operations of opening and closing drop-out fuses are high-risk processes, and such live maintenance accidents can be fatal. This paper presents a framework of a training system for drop-out fuse operation. The developed system is based on HoloLens mixed reality and speech recognition technology to train students for drop-out fuse operation. The system has two modules: drop-out fuse structure and drop-out fuse operation. Three modes of operation are included in each module: learning, practice, and evaluation, which can be visited according to the students' level of skill. The proposed training system has low cost, occupies small space, can be used at any time, and guarantees the personal safety of students during the training operation.

Keywords: Mixed reality · HoloLens · Drop-out fuse · Training system

1 Introduction

Electricity is already ubiquitous in people's modern lives, such as industry, commerce, hospitals, schools, and houses. Drop-out fuses are often used as protection switches for 10kV distribution lines and distribution transformers. Opening and closing drop-out fuse are high-risk processes, and such live operations can be fatal. Therefore, training students' skills for operations is essential to ensure the safety of their lives in this dangerous work environment.

Virtual reality (VR), augmented reality (AR), and mixed reality (MR) have been employed and proved beneficial for training education in recent years. A VR training system was developed for live-line workers in a power distribution system [1]. VR was also used to maintain a high-voltage overhead power line in a training system, which greatly reduces the accident rate during live maintenance [2, 3]. A framework using VR and AR was established for experiential education of construction safety [4]. For the maintenance of high voltage systems, AR could increase safety and reduce the time and costs of interruption [5]. MR applications with HoloLens for medical training, such as Patiently and HoloSim, were discussed [6]. HoloLens was used to provide users with specific personalized evacuation routes in emergency response [7].

M. Tavana et al. (Eds.): IISA 2020, AISC 1304, pp. 469–476, 2021.
https://doi.org/10.1007/978-3-030-63784-2_58

The remainder of this paper is organized as follows: Sect. 2 describes drop-out fuse and its operation. A framework of the proposed training system is established in Sect. 3. Section 4 presents the development of the system, and Sect. 5 supplies a summary and conclusion.

2 Drop-Out Fuse and Its Operation

Drop-out fuse is a combination switch and fuse to protect transformers and to protect distribution lines from damage. Like a switch, it can be opened using a hot stick and as a fuse, it will be automatically opened when the fuse melts if the current is too large. An overcurrent led by a fault in the customer circuit or the transformer will result in the fuse to melt, separating the transformer from the line. It can also be opened and closed manually by a worker standing on the ground and using a long hot stick.

The drop fuse can be simply divided into three main parts: the body, the fuse tube, and the fuse.

The body includes an insulator with an open "C" shaped frame, which electrically isolates the conductive portion from the insulator. The interchangeable fuse is in the fuse tube that represents a switch. When the fuse blows, the fuse tube will open and disconnect the switch to ensure circuit isolation. The fuse can be replaced when a fuse tube is taken off.

Three drop-out fuses are located in the left, middle, and right positions over the transformer. When opening the fuses, the sequences of opening these fuses should be strictly followed according to the direction of the wind. If the wind direction is from left to right, the middle fuse should be opened first, then the right fuse, and finally the left one. The hook of the hot stick is carefully placed into the pull ring of the fuse tube, and then the stick is pulled out forcefully to complete the operation of opening the fuses. When closing the fuses, the left fuse should be closed first, then the right fuse, and finally the middle one. When the fuse is opened, the hook of the stick is located in the pull ring, and then the stick is pushed up the fuse tube to achieve the closing operation.

3 A Framework of the Training System

The training system aims to train students for drop-out fuse operation in a safe way. In order to achieve this goal, a framework has been developed with two modules: structure and operation of drop-out fuse. Three modes of operation are included in each module: learning, practice, and evaluation, which can be visited according to the students' level of skill, as illustrated in Fig. 1.

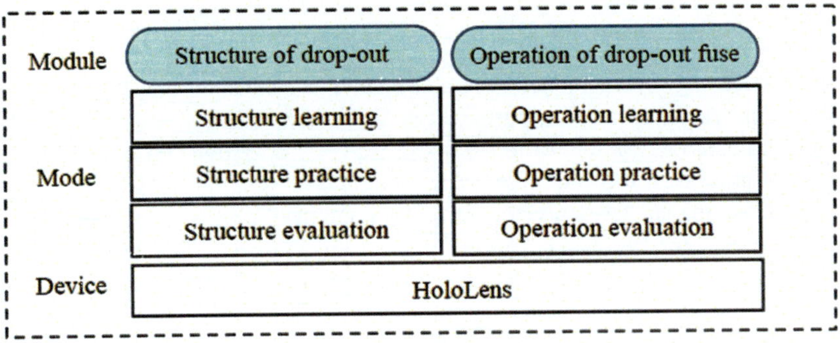

Fig. 1. A framework of the training system

3.1 Structure of Drop-Out Fuse

In the structure of drop-out fuse, there are three modes of operation, such as learning, practice, and evaluation.

In drop-out fuse structure learning, the structure of the fuse is explained for students to understand its principle. The entire fuse is broken down into ten parts that will be placed next to it while the location of each part in the fuse is highlighted. The fuse tube can be pulled apart by the pull ring to realize the opening operation. This opening operation separates the tube cap from the upper contact and disconnects the power, as seen in Fig. 2. The copper fuse in the tube can also be removed and replaced (Fig. 3).

Fig. 2. Opening operation of the fuse.

Fig. 3. Fuse tube.

Students learn by themselves and understand the structure of drop-out fuse through interactive learning in structure practice. Such learning content and learning progress are completely controlled by the students according to their situations.

The exam is designed to test students' level of understanding in structure evaluation. Each part of the fuse is randomly generated, and the learner is required to answer the name of the part by voice within a fixed time interval. The system automatically determines whether the answer is correct by voice recognition. The entire fuse is divided into a total of ten parts. When all are answered correctly, the score is 100.

3.2 Operation of Drop-Out Fuse

Operation learning, operation practice, and operation evaluation are designed in the module of operation of drop-out fuse.

During operation learning of the drop-out fuse, the operation consists of six basic sequential processes that are inspecting equipment, wearing personal protective equipment, setting up the fence, opening drop-out fuse, closing drop-out fuse, and cleaning up the site. Two people are required for this operation. One person monitors the process and the other performs the operation to ensure that is correct and safe. When the drop-out fuse is closed, the hook of the hot stick is moved carefully and inserted into the pull ring of the fuse, then the hot stick is pulled down strongly and quickly to open the fuse. In the case of opening the fuse, the hot stick is required to push upward to close the fuse, as shown in Fig. 4.

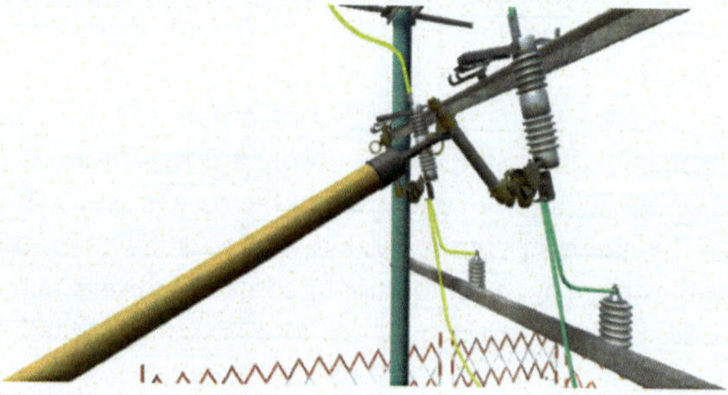

Fig. 4. Closing operation of the fuse by a hot stick.

In the operation practice mode, students can freely choose each training operation for practice, and understand various problems that need to be paid attention to in detail during the operation process. During the equipment inspection, it is required to carefully check whether the hard hat, safety glasses, work boots, and work gloves are all in good condition. Students can repeat the practice multiple times.

The system randomly presents one of the six basic processes in operation evaluation mode. The process is demonstrated and explained in a 3D environment for a student to speak the name of the process by voice. Finally, the system recognizes the sound to determine whether the answer is true and gives a score for valuation.

4 Development of the Training System

The development of the training system includes three main parts: instructional design for MR, the technical design of MR application, and MR implementation [8] (Fig. 5).

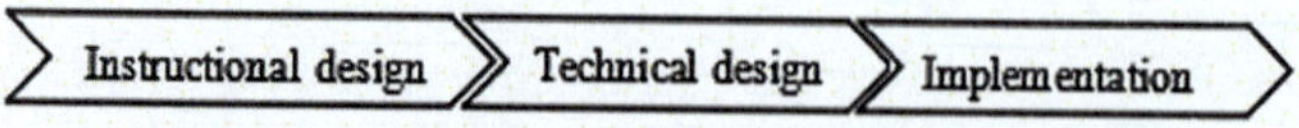

Fig. 5. The development process of the training system

4.1 Instructional Design for MR

During developing an MR application, the instructional design should be made based on specific teaching content. It should be noted that not all teaching contents are suitable for the MR application. The purpose of the MR application is to improve the teaching process and enhance the learning effect. If a good learning result is not achieved, it is not necessary to develop an MR application. Therefore, teaching content should be selected to suit for MR applications, teaching scenarios are designed, and resources are developed to support MR applications in teaching design.

The proposed training system simulates the operation of the drop-out fuse as the teaching content, and its purpose is to ensure the safety of students during the training process, which occupies less space, has low cost, and can be operated at any time to improve the training effect. The two teaching scenarios of drop-out fuse structure and operation are designed. The 3D digital model of drop-out fuse and a 10KV transformer with drop-out fuse scene need to be established.

4.2 Technical Design of MR Application

The goal of technical design is to carefully design details of various 3D models, different types of interactions, and to choose the appropriate hardware and related software for implementation.

The level of detail of the 3D model should be considered according to the instructional design. Some parts of the model will be built as a whole and will not be separated in the future, and certain parts must be disassembled. Otherwise, the 3D model will be modified in the subsequent implementation process. In the 3D model of the drop-out fuse, the structure will show the ten parts after disassembly, so the model will build detailed information of these ten parts.

There are three types of interactions to choose in technical design, which are passive, exploratory, and interactive [9].

In the passive type, there is little user interaction with the MR environment. The user can only watch 3D scenes with his eyes and listen to the sound with his ears, just like a movie, and cannot walk and not interact with 3D scenes. This kind of interaction is usually used for the explanation of knowledge. The passive interaction is adopted for the explanation of the drop-out fuse structure.

The exploration type can only allow users to move and look around and have a higher degree of a reaction than passive interaction, but it still cannot interact with other virtual objects. For example, a simple virtual museum belongs to this type of interaction. In the drop-out fuse operation learning, exploration interaction is used. The user can approach the transformer and watch around the device, but without other interactions during learning the entire process of opening and closing the drop-out fuse.

The interactive type allows users to visit, control and even alter the MR environment. In the MR training system, the virtual scene is converted by gaze, clicking a button by hand, etc., and the evaluation results are obtained by voice commands, which all use this interactive type.

Appropriate hardware should be chosen based on different application goals, such as equipment cost, immersion, and interactions. In this training system, Microsoft

HoloLens was selected as the hardware device since it has a big advantage over other commercial head-mounted displays from its hand gesture, and voice commands.

The HoloLens is based on an Intel 32-bit processor with a Holographic Processing Unit (HPU 1.0) that supports Universal Windows Platform (UWP) applications. It has 2 GB of RAM and 64 GB of flash memory. It also enables network connectivity via Wi-Fi 802.11 and Bluetooth 4.1 wireless technologies. It is equipped with 4 cameras to understand the environment, a depth camera, a video camera, and an inertial measurement unit to track the movements of the head. It has 4 microphones for sound and voice control [10].

Software for MR application includes two main parts: modeling and programming.

Autodesk 3DStudio Max is the most commonly used 3D modeling software. It is widely adopted in engineering, architecture and video game creation.

Unity is selected as the development engine for HoloLens to implement various interactions with MR applications. Unity is easy to learn and use, and has a free version. In Unity, a learning scene is built for drop-out fuse to achieve functions such as display models, parts disassembling and highlighting, and animations. The operation scene is also set up for drop-out fuse to realize the functions of operation learning and evaluation through gesture interaction and voice recognition. Then the Unity project is built to a Visual Studio solution, and finally, the solution is deployed to HoloLens through Visual Studio 2017.

4.3 MR Implementation

In the MR implementation, three tasks are completed, such as modeling the virtual world, programming, and testing.

According to the model requirements in the technical design, Autodesk 3DStudio Max software is used to build a single model of drop-out fuse and a virtualized scene of the entire 10KV transformer with drop-out fuse. Then these 3D models are imported into the scene of Unity project in the FBX format. During building these models, the number of faces should not be too much to ensure the performance of MR applications.

Fig. 6. 3D model of drop-out fuse

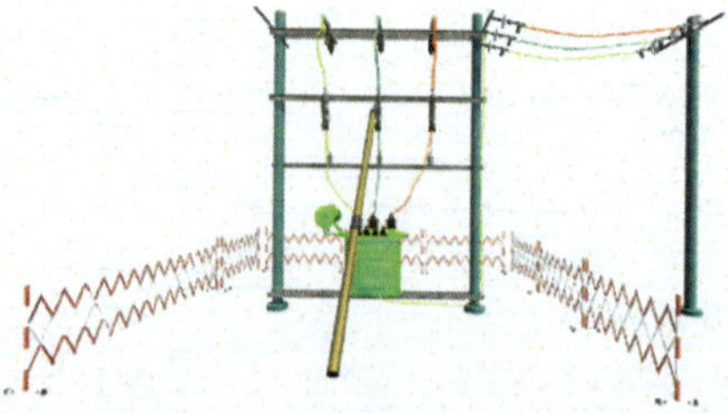

Fig. 7. Scene of the fuse with the transformer.

Figure 6 shows the 3D model of the drop-out fuse and the scene of the fuse with the transformer is depicted in Fig. 7.

In programming, C # language is used in the Unity 2017.4 development engine. During the explanation of the drop-out fuse structure, speech sounds, highlighting certain fuse parts and related animations are edited and controlled by Unity's Timeline (Fig. 8).

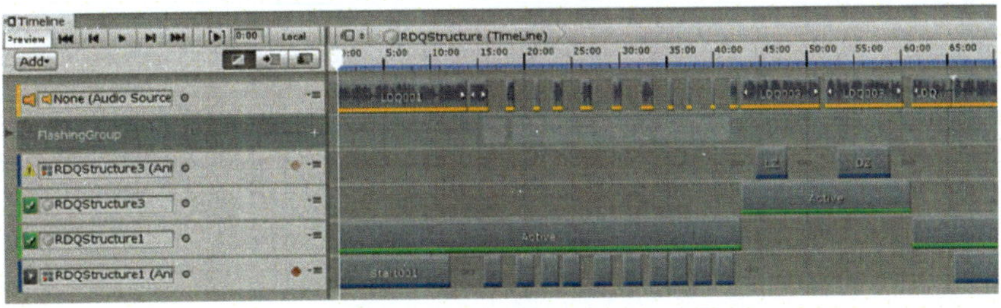

Fig. 8. Timeline in unity

Audio track, activation track, and animation tracks are added in the TimeLine. Audio track controls sequenced sounds that are set to play at a specific time. These sounds are the voice lecture for the structure of the drop-out fuse, and the 10 parts made up of the fuse are explained in detail. The activation track determines the objects that are dragged onto the track can be displayed in the scene at a specific time and not displayed at other times. The entire drop-out fuse is composed of these 10 objects. When these objects can be disassembled and visible, the part corresponding to a specific position in the drop-out fuse is also designed to be activated at the appropriate activation track time and will highlight and flash for two seconds so that students can recognize its position. All animations including the rotation animation of the entire fuser and the animation of each component being disassembled, are added to the animation track and played at the specified time.

When explaining a part of the drop-out fuse, this part is disassembled and displayed on the left side of the entire drop-out fuse. In order for students not only to know the part on the left but also the position of the part in the whole drop-out fuse, this part is highlighted with a blue outline glow effect.

Speech recognition is one of the hottest trends in computer technology. It is a very effective and intuitive way of communication and is supported in HoloLens. Some keywords in part names are predefined to improve the accuracy of speech recognition. In the structural test module of the drop-out fuse, a certain part appears randomly in front of the user, and the user speaks the name of the part which is judged whether the answer is correct.

The HoloToolkit 4.3 is imported into unity for implementing interactive functions in this training system with HoloLens. This toolset was developed by Microsoft and HoloLens developer community and it makes HoloLens development simple. The user wearing HoloLens can use an air tap gesture to click the button in the scene to get a response.

Finally, the HoloLens device was tested by some pilot people. The test results can check whether the expected objectives have been achieved, otherwise essential modifications can be required.

5 Conclusions

To reduce the accident rates of live-line operation for drop-out fuses, a training system for drop-out fuse operation was developed in this paper. This MR system was established with HoloLens and speech recognition technology for students training operations of opening and closing the fuse. It will be put into the practical training for students in the future and will be improved based on the feedback results.

References

1. Park, C., Jang, G., Chai, Y.: Development of a Virtual reality training system for live-line workers. Int. J. Hum. Comput. Interact. **20**, 285–303 (2006)
2. Ayala, A., Galván, B., et al.: Virtual reality training system for maintenance and operation of high-voltage overhead power lines. Virtual Reality **20**, 27–40 (2016)
3. Miguel, P., Arroyo, F., Ayala, A.: The use of a virtual reality training system to improve technical skill in the maintenance of live-line power distribution networks. Interact. Learn. Environ., 1–18 (2019)
4. Le, Q., Pedro, A., et al.: A framework for using mobile based virtual reality and augmented reality for experiential construction safety education. Int. J. Eng. Educ. **31**, 713–725 (2015)
5. Oliveira, R., Farinha, J., et al.: Augmented reality system for maintenance of high-voltage systems. In: Proceedings of 51st International Universities Power Engineering Conference (UPEC), Coimbra, pp.1–5 (2016)
6. Danilo, G., Ankur, J., et al.: Exploring mixed reality in specialized surgical environments. In: Proceedings the 2017 CHI Conference Extended Abstracts on Human Factors in Computing Systems, New York, NY, USA, pp. 2591–2598 (2017)
7. Sharma, S., Bodempudi, S., et al.: Emergency response using HoloLens for building evacuation. In: Proceedings of 21st International Conference on Human-Computer Interaction, Orlando, Florida, pp. 299–311 (2019)
8. Vergara, D., Rubio, M.P., Lorenzo, M.: On the design of virtual reality learning environments in engineering. Multimodal Technol. Interact. **11**, 1–2 (2017)
9. Aukstakalnis, S.D., Blatner, D.: Silicon Mirage-The art and science of Virtual Reality. Peachpit Press, Berkeley (1992)
10. Turini, G., Condino, S., et al.: A Microsoft HoloLens mixed reality surgical simulator for patient-specific hip arthroplasty training. In: Proceedings of Augmented Reality. Virtual Reality, and Computer Graphics, pp. 201–210. Springer, Cham (2018)

Comparative Study of Theoretical Analysis and Physical Analysis of Single Tube Amplifier Circuit

Hanhong Tan(✉), Zhoulin Chang, and Yanfei Teng

Guangdong University of Science and Technology, Dongguan 523083, China
12327399@qq.com

Abstract. Theoretical analysis of the single transistor amplifier is described in this paper. The influence of the variable resistor on the state of the circuit is analyzed. Its dynamic performance is calculated. Discrete components circuit is provided. The input and output waveforms are measured. The theoretical analysis is verified by experimental data. Finally, circuit schematic is drawn by using Protel DXP software. The welded single transistor amplifier is shown. The physical measured data verify the theoretical analysis. The theoretical analysis, experimental measurement data and the physical measurement data are compared.

Keywords: Single transistor amplifier · Variable resistor · Performance index · Protel

1 Introduction

The amplifying circuit is the earliest and most widely used electronic circuit for human applications, and is specifically divided into voltage amplification and current amplification (i.e., power amplification). For an amplifying circuit, in addition to obtaining an appropriate voltage amplification factor, it is required that the amplified waveform does not cause distortion. Studying the factors affecting the amplification factor and the condition that the output waveform is not distorted is two important aspects of understanding the operation of the amplifying circuit content.

2 Theoretical Analysis

The schematic diagram of the single-tube amplifier circuit is shown in Fig. 1:

M. Tavana et al. (Eds.): IISA 2020, AISC 1304, pp. 477–483, 2021.
https://doi.org/10.1007/978-3-030-63784-2_59

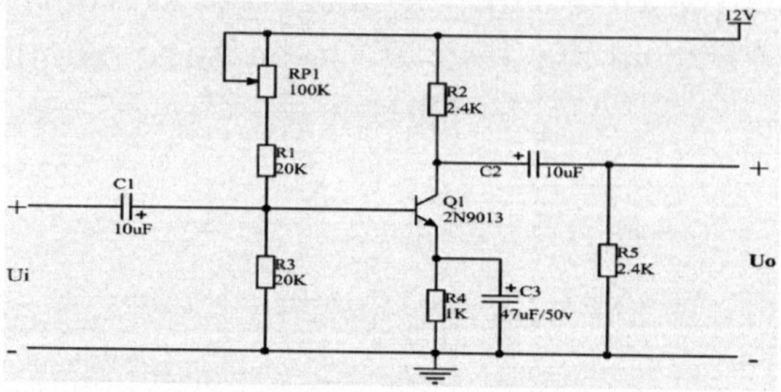

Fig. 1. Schematic diagram of single tube amplifier circuit

2.1 Static Working Point

Theoretical analysis of the circuit.

B pole potential calculation formula: $V_B = \frac{R3}{R1 + RP1 + R3} * V_{CC}$

If Q1 is in the amplified state, the current calculation formula

$$I_C \approx I_E = \frac{V_B - U_{BEQ}}{R4} = (V_B - 0.7)\,\text{mA},\ I_B = \frac{V_B - 0.7}{\beta}\ \text{mA}$$

Voltage calculation formula: $U_{BE} = 0.7$ V

$$U_{CE} = V_{CC} - I_C R2 - I_E R4 \approx V_{CC} - I_E(R2 + R4) \geq 1\,\text{V}\ U_C > U_B$$

When 1RP1 = 0, what is the working state of the analysis circuit? Analysis:

$$V_B = \frac{R3}{R1 + RP1 + R3} * V_{CC} = \frac{20}{20 + 20} * 12\,\text{V} = 6\,\text{V}$$

Assume that Q1 is in an amplified state, then

$$I_C \approx I_E = \frac{V_B - U_{BEQ}}{R4} = (V_B - 0.7)\,\text{mA} = (6 - 0.7)\,\text{mA} = 5.3\,\text{mA}$$

$$U_{CE} = V_{CC} - I_C R2 - I_E R4 \approx V_{CC} - I_E(R2 + R4) = 12 - 5.3 * (2.4 + 1) < 0$$

From the calculation results, the assumption is not true, the circuit is not in the amplified state; and the circuit current is large, it can be known that the circuit is in a saturated state.

Therefore, when RP1 = 0, the circuit is in a saturated operating state. When 2RP1 = 20 K, what is the working state of the analysis circuit? Analysis:

$$V_B = \frac{R3}{R1 + RP1 + R3} * V_{CC} = \frac{20}{20 + 20 + 20} * 12\,\text{V} = 4\,\text{V}$$

Assume that Q1 is in an amplified state. Then

$$I_C \approx I_E = \frac{V_B - U_{BEQ}}{R4} = (V_B - 0.7)\,\text{mA} = (4 - 0.7)\,\text{mA} = 3.3\,\text{mA}$$

$$U_{CE} = V_{CC} - I_C R2 - I_E R4 \approx V_{CC} - I_E(R2 + R4) = 12 - 3.3 * (2.4 + 1) = 0.78\,\text{V}$$

It can be seen from the calculation results that the circuit is in a critical saturation state.

When 3RP1 = 80 K, what is the working state of the analysis circuit? Analysis:

$$V_B = \frac{R3}{R1 + RP1 + R3} * V_{CC} = \frac{20}{20 + 80 + 20} * 12\,\text{V} = 2\,\text{V}$$

Assume that Q1 is in an amplified state. Then

$$I_C \approx I_E = \frac{V_B - U_{BEQ}}{R4} = (V_B - 0.7)\,\text{mA} = (2 - 0.7)\,\text{mA} = 1.3\,\text{mA}$$

$$\begin{aligned} U_{CE} &= V_{CC} - I_C R2 - I_E R4 \approx V_{CC} - I_E(R2 + R4) = 12 - 1.3 * (2.4 + 1) \\ &= 7.58\,\text{V} \geq 1\,\text{V} \end{aligned}$$

It can be seen from the calculation results that the assumption is established and the circuit is in an amplified state.

When 4RP1 = 100 K, what is the working state of the analysis circuit? Analysis:

$$V_B = \frac{R3}{R1 + RP1 + R3} * V_{CC} = \frac{20}{20 + 100 + 20} * 12\,\text{V} = 1.7\,\text{V}$$

Assuming Q1 is in an amplified state, then

$$I_C \approx I_E = \frac{V_B - U_{BEQ}}{R4} = (V_B - 0.7)\,\text{mA} = (1.7 - 0.7)\,\text{mA} = 1\,\text{mA}$$

$$U_{CE} = V_{CC} - I_C R2 - I_E R4 \approx V_{CC} - I_E(R2 + R4) = 12 - 1 * (2.4 + 1) = 8.6\,\text{V}$$

It can be seen from the calculation results that the assumption is established and the circuit is in an amplified state.

2.2 Dynamic Analysis

When the circuit is in an amplified state, The formula for calculating the voltage Amplification factor is $A_u = -\beta \frac{R2//R5}{r_{be}}$, among them $r_{be} = 300 + (1 + \beta) \frac{26\,\text{mV}}{I_E}$ 。

It can be seen from the above analysis that the larger the RP1 is, the smaller, the smaller, the larger, and the smaller. Therefore, it is concluded that the larger the adjustable resistor RP1 is, the smaller the voltage amplification factor of the circuit is when the circuit is in the amplified state.

3 Experimental Analysis

Use the laboratory components to build the circuit as shown in Fig. 2. If the input signal voltage peak-to-peak value is 30 mV, adjust the resistance to the middle position, view the oscilloscope input and output waveform, input voltage waveform and output voltage waveform as shown in Fig. 3.

Fig. 2. Single-tube amplifier circuit discrete component diagram

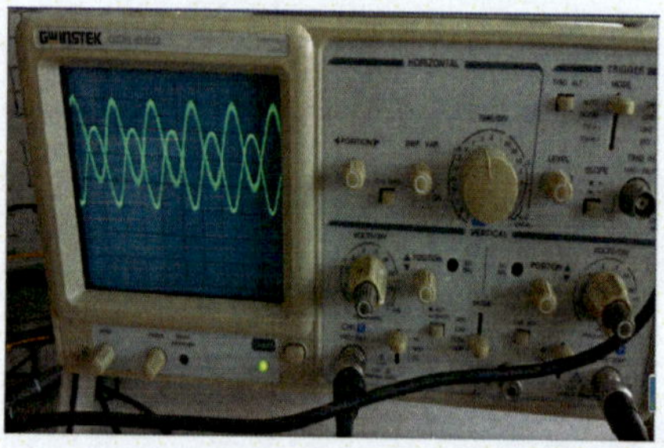

Fig. 3. Input and output voltage waveforms when the experimental circuit is in the amplified state.

As can be seen from Fig. 3, the input voltage peak-to-peak value is 3 * 5 mV = 15 mV. The output voltage peak-to-peak value is 3 * 1 V = 3 V, and the input voltage is opposite to the output voltage waveform, so the voltage amplification factor Au = −3 V/0.015 V = −200 is obtained. It can be seen from the experimental data that the circuit is in an amplified state at this time, and the experimental data verified the theoretical analysis.

(2) Without changing the input signal, turn the adjustable resistor clockwise to adjust the adjustable resistor to the rightmost position. The resistance of the adjustable resistor is the smallest. The input voltage waveform and output voltage waveform are shown in Fig. 4.

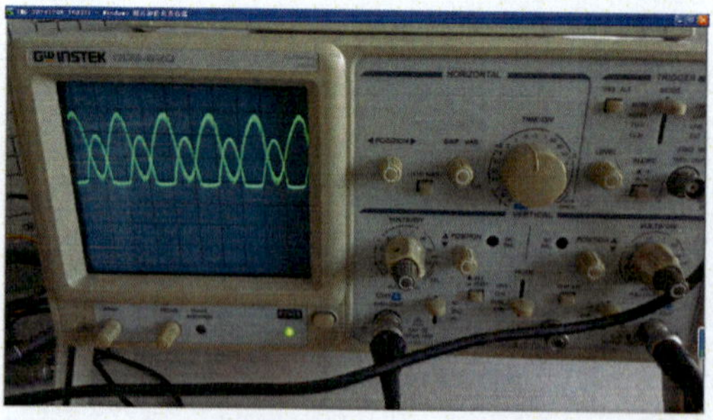

Fig. 4. Input and output voltage waveforms when the experimental circuit is saturated

As can be seen from Fig. 4, the output voltage is distorted and the bottom is flattened. It can be seen from the analysis that the distortion is saturation distortion, and the circuit is in a saturated working state. The experimental data is consistent with the theoretical analysis, and the experimental data validates the theoretical analysis.

4 Physical Analysis

Draw schematics using protel DXP software. The circuit schematic is shown in Fig. 5.

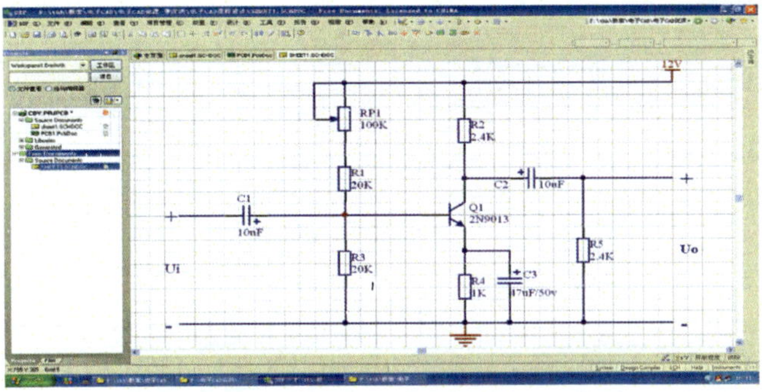

Fig. 5. Schematic diagram of a single tube amplifier circuit

Add a 12 V DC power supply to the circuit. A sinusoidal signal is input to the circuit Ui. Connect the Ui and Uo terminals to the oscilloscopes CH1 and CH2. Adjust potentiometer RP1, the adjustable resistor rotates clockwise from the leftmost end, and the circuit undergoes amplification and saturation. When the circuit is in the amplified state, the input and output voltage waveforms are shown in Figs. 6, 7 and 8. When the adjustable resistor is adjusted to the rightmost end, the adjustable resistor has the lowest resistance value and the circuit is in saturation state. The input and output voltage waveforms are shown in Fig. 9, 10 and 11.

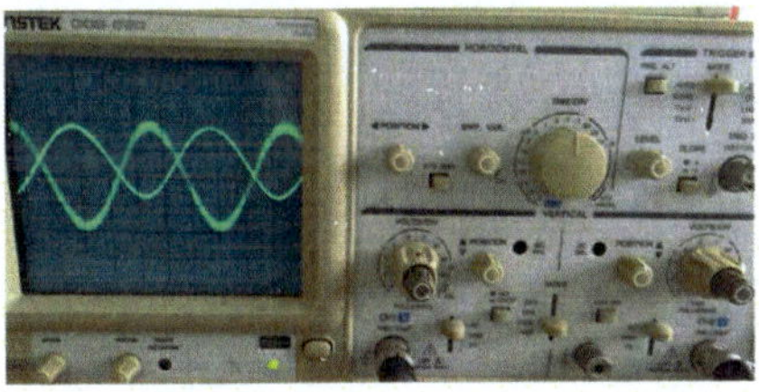

Fig. 6. Input and output voltage waveforms when the physical circuit is in the amplified state

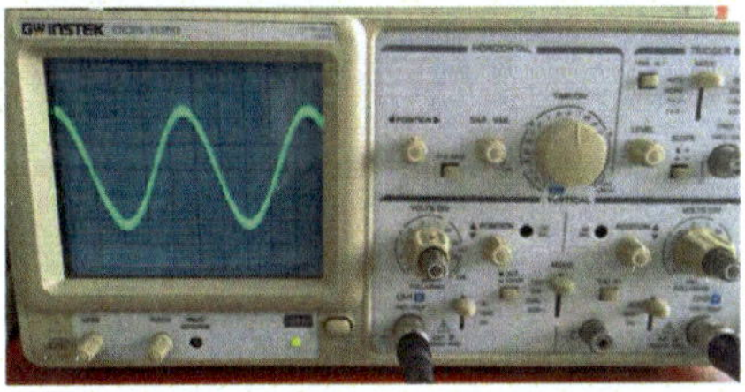

Fig. 7. Input voltage waveform when the physical circuit is in the amplified state

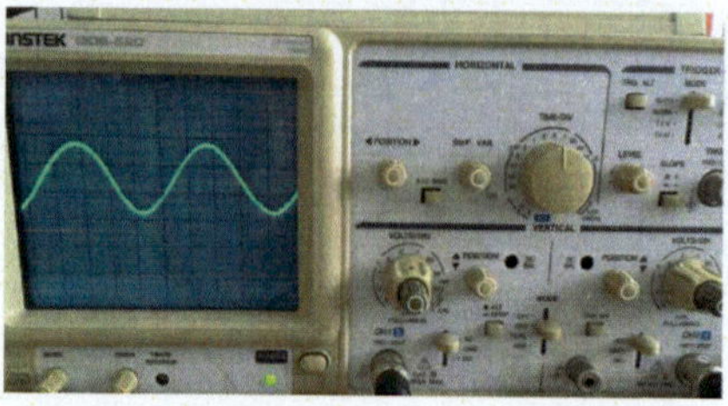

Fig. 8. Output voltage waveform when the physical circuit is in the amplified state

As can be seen from Fig. 6, the input voltage peak-to-peak value is 4 * 5 mV = 20 mV. The output voltage peak-to-peak value is 2.8 * 0.5 V = 1.4 V, and the input voltage is opposite to the output voltage waveform, so the voltage amplification factor Au = −1.4 V/0.02 V = −70 is obtained. It can be seen from the physical measurement data that the circuit is in an amplified state at this time, and the physical measurement data verifies the theoretical analysis.

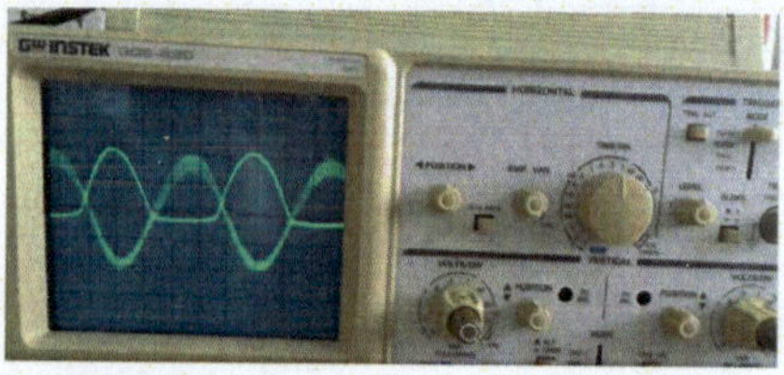

Fig. 9. Input and output voltage waveforms when the physical circuit is saturated

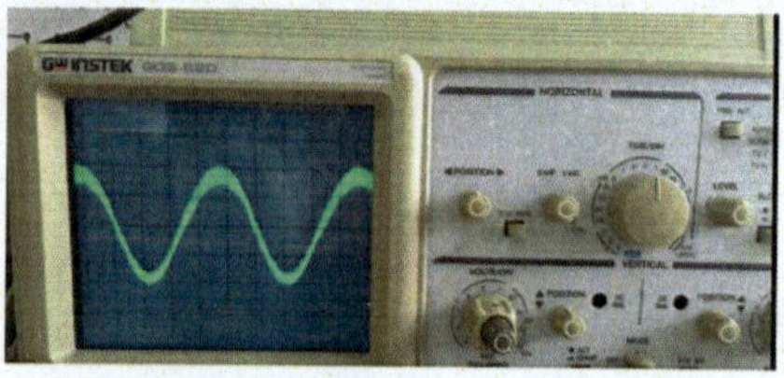

Fig. 10. Input voltage waveform when the physical circuit is saturated

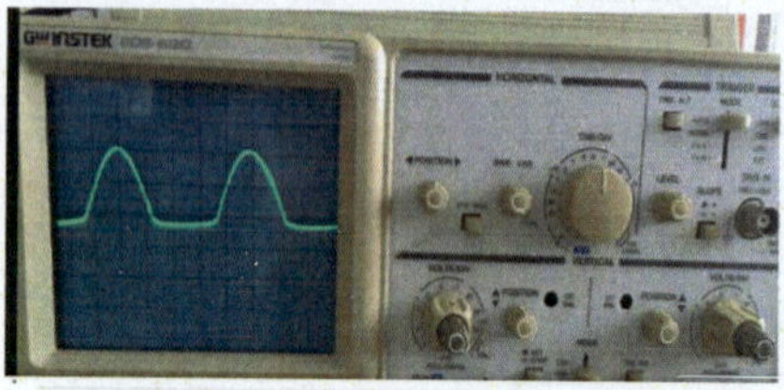

Fig. 11. Output voltage waveform when the physical circuit is saturated

As can be seen from Fig. 9, the output voltage is distorted and the bottom is flattened. It can be seen from the analysis that the distortion is saturation distortion, and the circuit is in a saturated working state. The physical measurement data is consistent with the theoretical analysis, and the physical measurement data verifies the theoretical analysis.

5 Conclusion

The argument of this paper is the electric vehicle braking energy recovery system. At the beginning of the article, the background and current status of the research and development of the electric vehicle braking energy recovery system are expounded. Then the principle of the electric vehicle braking energy recovery system is described, and then the comparative analysis is carried out. The control technology of the electric vehicle braking energy recovery system finally determines that the parallel braking energy recovery control technology is the best solution [5], through which the energy of the electric vehicle can be effectively recovered, and the cost of the technology is relatively low. The ability to endurance has also been greatly improved, in line with the requirements of this article, can be promoted and used.

References

1. Uehara, H., Konishi, D., Goya, K., et al.: Power scalable 30-W mid-infrared fluoride fiber amplifier. Opt. Lett. 4777–4780 (2019). https://doi.org/10.1364/ol.44
2. Strudwick, B.H., Koenis, M.A.J., Sanders, H.J., et al.: A tunable, fullerene-based molecular amplifier for vibrational circular dichroism. Chemistry (Weinheim an der Bergstrasse Germany) **02190**, 12560–12566 (2019). https://doi.org/10.1002/chem
3. Sun, S.-h., Niu, K.: The application of multisim in the analysis of single tube common shot amplifier. In: Education Teaching Forum, November 2018, no. 48 (2018)
4. Wang, Y., Shu, L.-z.: The application of proteus in electrical and electronic experiment. Teach. J. Taizhou Polytech. Coll. **17**(4) (2017)
5. Kong, M.-f.: Simplified analysis of amplifier circuit of composite tubes. J. EEE **40**(2) (2018)
6. Li, X., Shi, P.: Cooperative fault-tolerant tracking control of heterogeneous hybrid-order mechanical systems with actuator and amplifier faults. Nonlinear Dyn. **98**(1), 447–462 (2019). https://doi.org/10.1007/s11071-019-05203-2

Design of the Bus Support Capacitor in Servo Drive Controller Based on PMSM

Yao Yao[1,2](✉), Dawei Gu[1,2], Yebing Cui[1,2], Shuwei Song[1,2], and Fanquan Zeng[1,2]

[1] Shanghai Aerospace Control Technology Institute, Shanghai 201109, China
1392034655@qq.com
[2] Shanghai Engineering Research Center of Servo System, Shanghai 201109, China

Abstract. Bus support capacitor is an important part of the DC side of the servo drive controller, the design of capacitor has a great influence on the selection of the performance of the inverter. This paper focuses on the design method of the three-phase full-bridge inverter topology bus support capacitor based on permanent magnet synchronous motor. At the same time, the ripple current and stray inductance of the bus support capacitance are improved and designed to optimize the capacitance. By establishing a simulation model to verify, the results show that the designed capacitor can effectively suppress the DC side voltage ripple and ensure the stability of the bus voltage.

Keywords: Support capacitance · Servo drive controller · Stray inductance

1 Introduction

Inverter power supply based on PWM voltage has been widely used [1–3], For high-voltage high-current servo drive controllers, IGBTs are generally used as switching tubes, and the output voltage and current waveforms are controlled by controlling the turning on and off of IGBT. PMSM speed servo system double closed-loop control structure block diagram is shown in Fig. 1, The speed loop is connected in series with the current loop, the speed loop is the outer loop, and the current loop is the inner loop, By using $i_d^* = 0$ vector control strategy, The output of the speed loop controller is given as the cross-axis current i_q^*, The output of the current loop through the AC-DC axis is the voltage u_d and u_q of the AC-axis respectively, After the two voltages undergo coordinate transformation, space vector pulse width modulation technology is used to control the switching of the three-phase inverter power device, thereby controlling the amplitude and phase of the inverter output voltage. The current sensors collect the phase currents i_a, i_b and get the current feedback values i_d and i_q after coordinate transformation; The rotor position angle required for coordinate transformation is acquired by the position sensor; the speed feedback is calculated by the acquired position signal.

The power supply of the servo controller uses a lithium battery. When the servo system exists, when the IGBT tube is turned on and off instantly, it will cause bus voltage spikes and current spikes. To reduce the fluctuation of the drive output voltage,

M. Tavana et al. (Eds.): IISA 2020, AISC 1304, pp. 484–490, 2021.
https://doi.org/10.1007/978-3-030-63784-2_60

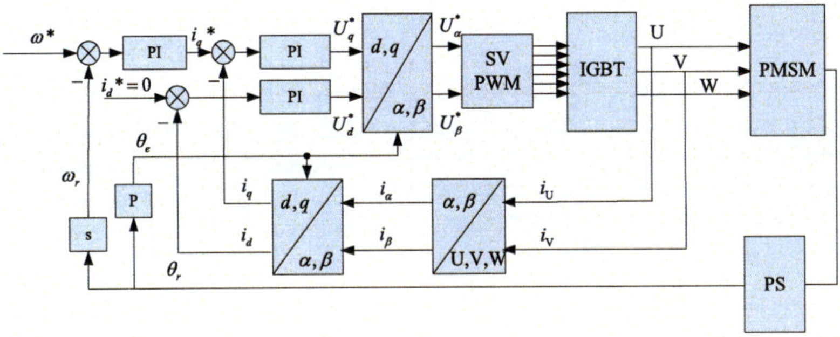

Fig. 1. Schematic diagram of PMSM speed servo system

a large-capacity filter capacitor must be added to the bus voltage input terminal as a temporary low-impedance power supply to reduce the power supply voltage fluctuation and reduce the current change rate of the input line.

The DSP calculates the speed command, FPGA as the interface sampling data and generates PWM waveform and output according to the control quantity transmitted by the DSP to realize the sequence control of the IGBT switch, which can realize the motor response to the command (Fig. 2).

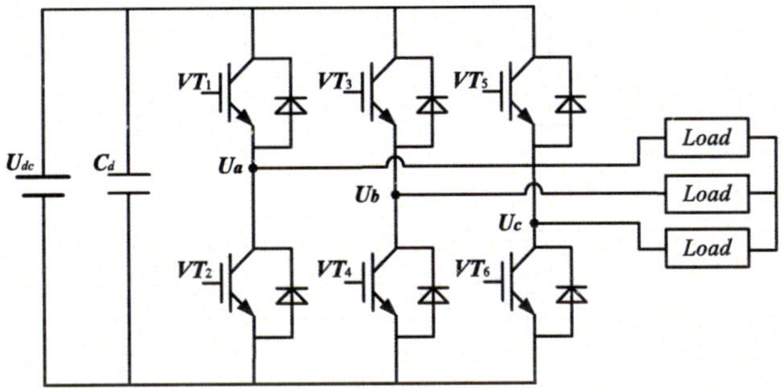

Fig. 2. Schematic diagram of three-phase full-bridge inverter topology

2 Design and Selection of Bus Support Capacitors

In the design process of organic film capacitors, parameters such as rated voltage, capacitance, ripple current, and stray inductance should be paid attention to. The size of stray inductance should be paid special attention. It is easy to cause a large voltage spike, which has an effect on the power device, such as overvoltage and increased output harmonics. Therefore, it is necessary to optimize the stray inductance of the busbar.

2.1 Rated Voltage

The organic film capacitor needs to have an overvoltage capacity of about 1.5 times. Considering the derating design of class I, the rated voltage of the design bus support capacitor is twice the output voltage of the lithium battery. For the servo drive controller, the bus input is used. The voltage is 300 V, and the rated voltage of the film capacitor is set to 600 V. At this voltage, the servo drive controller can meet the overvoltage requirements of the capacitor for a long time.

2.2 Capacitance Design

In practical applications, the bus support capacitor uses the carrier frequency of the switching device to charge and discharge. In a cycle, when the switching device is turned on, the lithium battery and the capacitor provide energy for the inverter bridge; when the switching device is turned off, The charging capacity of the bus support capacitor is charged by the lithium battery. Assuming that the maximum power transmitted by the DC bus is P_O, and the bus support capacitor is an ideal component, the energy required by the bus in one switching cycle is:

$$Win = \frac{P_0}{\eta} * \frac{1}{f_s} \tag{1}$$

Among them: For the efficiency of the inverter, η is generally 0.9, for the switching frequency, f_s is generally 10 kHz.

Due to the existence of the bus support capacitor, the amplitude of the design voltage ripple ΔU does not exceed 60 V, and the energy released by the capacitor in one switching cycle is:

$$Q = \frac{1}{2} C \left[(U + \Delta U)^2 - (U - \Delta U)^2 \right] \tag{2}$$

It is generally considered that within a PWM wave period, the bus support capacitor provides half of the energy required by the bus:

$$Q = \frac{1}{2} Win \tag{3}$$

The above relationship can be obtained by:

$$C = \frac{P_0}{4\eta f_s U \Delta U} \tag{4}$$

Substituting the value in the actual working condition, the size of the available capacitor is: considering the derating of class I, and taking into account the volume increase caused by the increase in the value of the bus support capacitor, the capacitor size is 800 uF. Figure 3 shows the three-dimensional model of the designed organic film capacitor.

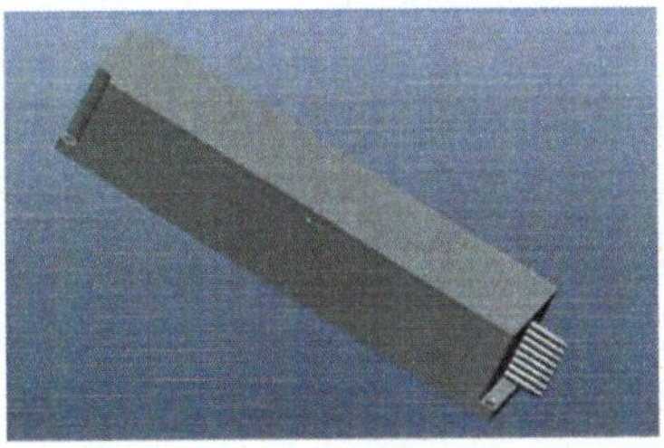

Fig. 3. Schematic diagram of three-dimensional design of bus support capacitance

2.3 Design of Ripple Current

According to literature [4–7], ripple current refers to the AC current component flowing through the DC bus capacitor. The maximum ripple current capacity that the capacitor can withstand depends on its working ambient temperature, its own inherent parameters and other factors. When the ripple current is too large, due to the existence of the equivalent series inductance and equivalent series resistance of the capacitor itself, a large heat loss will be generated, greatly reducing the life of the capacitor. Since a permanent magnet synchronous motor is used, its power factor is set to $cos\varphi = 1$; The effective value of the rated output current is I_N, the modulation ratio is M, and the phase angle of the output voltage and current of the inverter module is φ. Assuming that the DC output of the power battery is an ideal smooth curve, considering the limit conditions, all the AC components required by the inverter module are provided by the DC bus capacitors. Then there is the formula:

$$I_C = I_N\sqrt{2M\left[\frac{\sqrt{3}}{4\pi} + \left(\frac{\sqrt{3}}{4\pi} - \frac{9}{16}M\right)\right]} \quad (4)$$

At this time, when the modulation ratio M is 0.613, there is a maximum value $I_C = 0.65I_N$. This result can guide the construction of the simulation model.

2.4 Design of Stray Inductance

According to the set V_{DCnom} and Tvj, T_1 is double pulsed and T_2 is turned off, as shown in the left figure. During the passage of T_1, the current flowing through T_1 begins to increase, but the I_{Cnom}value is not reached. At this time, D_2 is in a freewheel state, $V_F = 0$. Due to the existence of the stray inductance L_δ of the converter circuit, a gap U_s will appear in the V_{CE} (as shown on the right), according to the formula:

$$Us = \mathrm{L}_\delta \cdot \frac{d_i}{d_t} \quad (5)$$

The emission control system uses GPS to realize synchronization. GPS uses CW25-TIM chip. The chip synchronization accuracy can reach 20 ns without cumulative error. At the same time, because the constant temperature crystal oscillator has the advantages of high short-time accuracy and low drift, the clock output of the constant temperature crystal oscillator can be corrected in real time by GPS second pulse in the way of cooperating with the constant temperature crystal oscillator. The principle of second pulse synchronization is shown in Fig. 4.

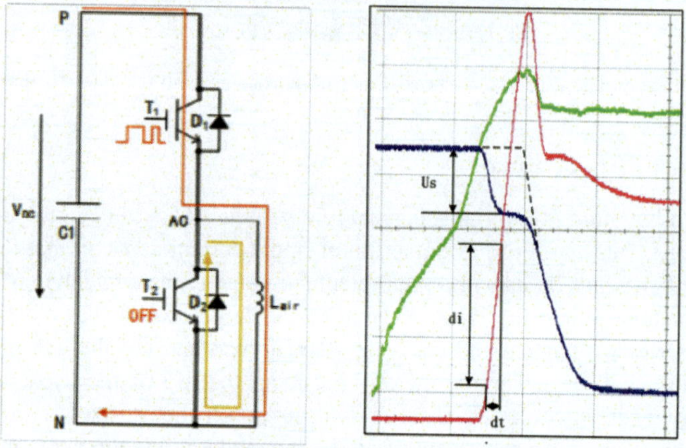

Fig. 4. Schematic diagram of the test rationale

Reading the data displayed on the oscilloscope through the cursor and calculating through the above formula can obtain $L_{\delta} = 59$ nH. The stray electricity of organic thin film capacitors can be affected by the busbar. By optimizing the busbar structure, the size of the stray inductance can be further reduced, and the current rise rate can be reduced to weaken the damage to the IGBT (Fig. 5).

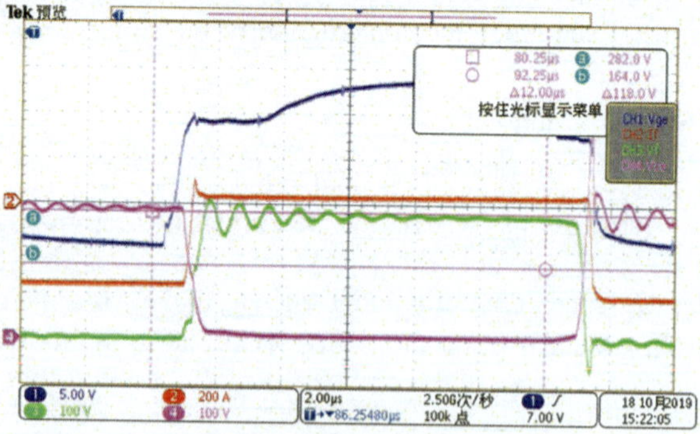

Fig. 5. Stray inductance test results

The capacitance design of the bus support capacitor should be designed taking into account the above 4 points, which can effectively suppress the bus voltage mutation, and the size of the structure should be considered in practical applications to avoid the design of the bus support capacitance is too large.

3 Test Verification

In order to verify the feasibility of the designed bus support capacitor, the Simulink simulation model was established in MATLAB, the lithium battery voltage was set to 300 V, the bus support capacitor size was set to 800 uF, and the rated voltage was set to 600 V, using a three-phase full-bridge inverter circuit Construct the speed servo double closed-loop control system, the simulation model is shown in the following Fig. 6:

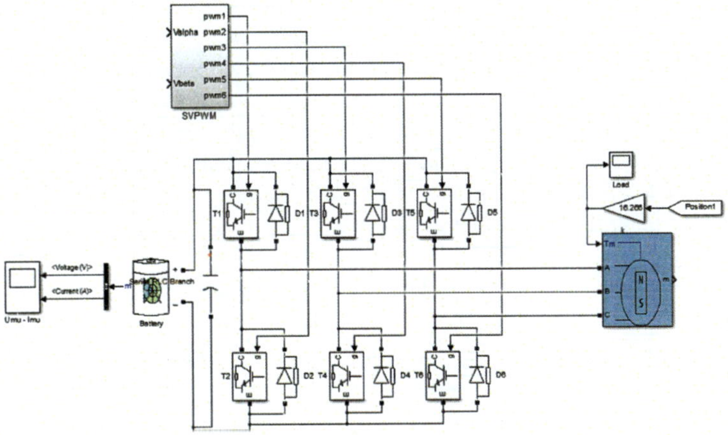

Fig. 6. Schematic diagram of simulation model

Running the simulation model, we can find that by using the bus support capacitor, the bus voltage can be better stabilized at 320 V, the ripple voltage is not more than 10 V, the fluctuation of the bus voltage is suppressed, and the capacity of the organic thin film capacitor cannot be too large, otherwise it will The volume is large, which is not conducive to structural design and installation (Fig. 7).

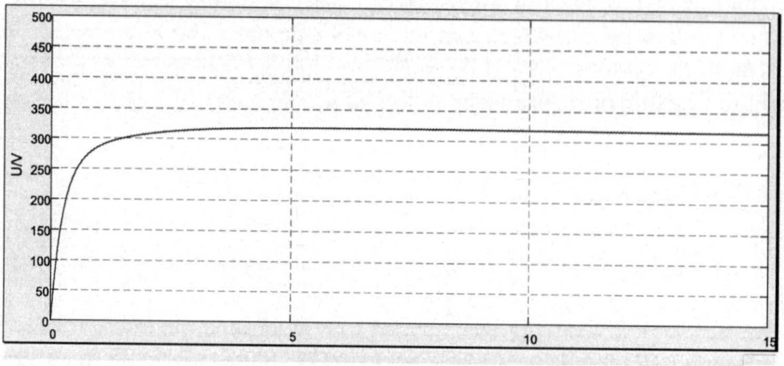

Fig. 7. Voltage waveform across the bus support capacitor

4 Conclusion

This paper analyzes the design method of the three-phase full-bridge inverter topology bus support capacitor based on permanent magnet synchronous motor, and gives the specific calculation method of the bus support capacitor and other design considerations, especially the need to consider the capacitor ripple current and Stray inductance guides the design of capacitors in practical applications. Through theoretical simulation of the designed capacitors, the results show that the bus voltage can be stabilized at 320 V, which can effectively suppress the DC side voltage ripple.

Acknowledgement. Fund: Shanghai Engineering Research Center of Servo System (No. 15DZ2250400).

References

1. Wen, H., Xiao, W., Wen, X., et al.: Analysis and evaluation of DC-link capacitors for high-power-density electric vehicle drive systems. IEEE Trans. Veh. Technol. **61**(7), 2950–2964 (2012)
2. Zhou, K., Wang, D.: Relationship between space-vector modulation and three phase carrier based PWM: a comprehensive analysis. IEEE Trans. Ind. Electron. **49**(l), 186–196 (2002)
3. Malinowski, M., Kazmierkowski, M.P., Hjansen, S., et al.: Virtual -flux -based direct power control of three-phase PWM rectifiers. IEEE Trans. Ind. Appl. **37**(4), 1019–1027 (2001)
4. Chen, Y.-j., Guangqiang, L.V., Geng, Y., et al.: Research on the volume optimization of support capacitor in EV driver. Electron. Des. Eng. **26**(12), 139–143 (2018)
5. Wang, J., Ji, B., Wang, H., et al.: An inherent zero voltage and zero current switching full bridge converter with no additional auxiliary circuits. J. Power Electron. **15**(3), 610–620 (2015)
6. Wang, G.L., Chen, X., Zeng, F.Q., et al.: Permanent magnetic servo motor space vector control system and MATLAB simulation. Flight Control Detect. **1**(3), 059–062 (2018)
7. Liu, H., Zhang, X., Duan, X., et al.: Research on space motion decoupling of axisymmetric thrust vector control servo mechanism. Flight Control Detect. **2**(4), 083–088 (2018)

Effect of Distributed Generation Grid-Connection on Line Loss in Low-Voltage Courts

Weiru Wang[1(✉)], Xincong Shi[2], Jie Hao[1], Mengzan Li[1], Xinyuan Liu[1], Yifan Zhang[1], and Jun Pi[1]

[1] State Grid Shanxi Electric Power Research Institute, Taiyuan, China
hdwangweiru@163.com
[2] State Grid Shanxi Electric Power Company, Taiyuan, China

Abstract. With the introduction of a series of photovoltaic subsidy policies by the state, rooftop photovoltaic (PV) of enterprises and residents has developed rapidly. The traditional passive single distribution network has become active bidirectional network, which brings new challenges to line loss management of distribution network. This paper analyses the change of power flow direction in distribution network containing distributed photovoltaic and the influence of different operation modes of photovoltaic on line loss. The traditional formula for calculating line loss rate in low-voltage courts is revised. Through the analysis of practical cases, it is concluded that in active distribution network containing distributed photovoltaic, the optimal utilization mode of photovoltaic power generation is spontaneous self-use, followed by the absorption within the court and the line loss rate in the court shows a downward trend with the increase of the absorption ratio.

Keywords: Distribution network line loss · Distributed generation · Photovoltaic power generation · On-grid energy

1 Introduction

Line loss is an important economic and technical index of power grid assessment. It not only reflects the rationality of a power grid structure and operation mode, but also reflects the planning and design, production technology and operation management level of power grid [1, 2]. Line loss of courts is an integral part of line loss management, especially in low-voltage courts of distribution network, the line loss is relatively large. To do well in line loss management is a key indicator to reflect the company's operating level [3], and it is particularly important to accurately reflect the true level of the comprehensive line loss in the courts.

The development of distributed generation of renewable energy has been greatly promoted by the enhancement of environmental awareness and the inherent requirements of sustainable development as well as the introduction of a series of policies to encourage distributed generation of renewable energy. The proportion of distributed generation in distribution network has gradually increased, and it has become an

M. Tavana et al. (Eds.): IISA 2020, AISC 1304, pp. 491–498, 2021.
https://doi.org/10.1007/978-3-030-63784-2_61

important power source in distribution network [4, 5]. In the process of line loss management, the traditional formulas for calculating the line loss of the courts with distributed photovoltaic power generation users will be unreasonable, obviously small, or even negative line loss. In addition, a large amount of photovoltaic grid-connected power brings new problems to the line loss management of courts.

This paper analyses the change of power flow direction caused by the addition of photovoltaic grid-connected power in the low-voltage courts with distributed photovoltaic power generation users, which causes the composition of power supply and sales in courts to change. The traditional formula for calculating line loss rate in the courts is amended, and the change of line loss in low-voltage courts with distributed photovoltaic power supply is analyzed through the practical case.

2 Influence of Distributed Photovoltaic Access on Line Loss in Low-Voltage Courts

2.1 Influence of Different Photovoltaic Operation Modes on Power Flow in Courts

If there are the distributed PV customers in the courts besides the general customers, the power flow direction in the courts may be changed by the addition of PV power supply, which can be divided into three cases:

1. PV power generation is entirely for self-use, with no electricity left over, as shown in Fig. 1. In this case, the direction of power flow in the court has not changed, and the composition of power supply and power sales in the court has not changed. But the current flowing through the distribution transformer decreases and the loss of the transformer decreases. The connection of distributed generation is helpful to reduce the line loss rate in the court.

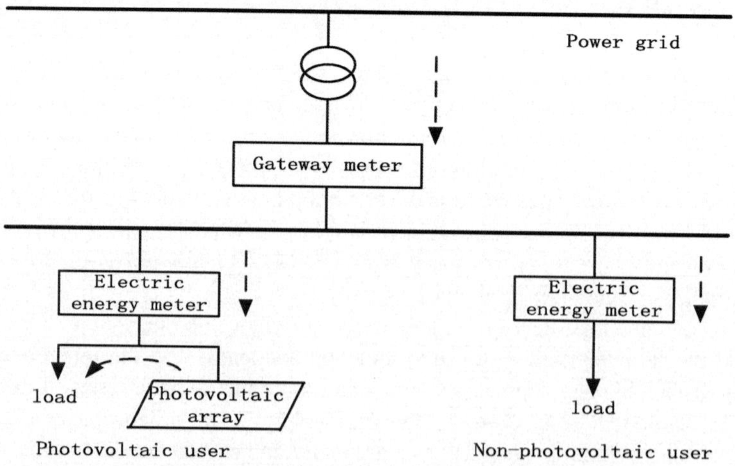

Fig. 1. Power flow direction when all distributed generation is self-use

2. The PV power generation has surplus electricity on the grid and the surplus electricity is only absorbed by other users in the court. In this case, the power flow direction in the court is not only from the power grid to users, but also from PV users to other nearby users in the court, as shown in Fig. 2. The power supply structure in this court has changed, which is equivalent to adding other power supply points in the court and shortening the power supply radius of some users. At this time, the power loss and line loss rate are reduced, and the more PV users, the greater the impact.

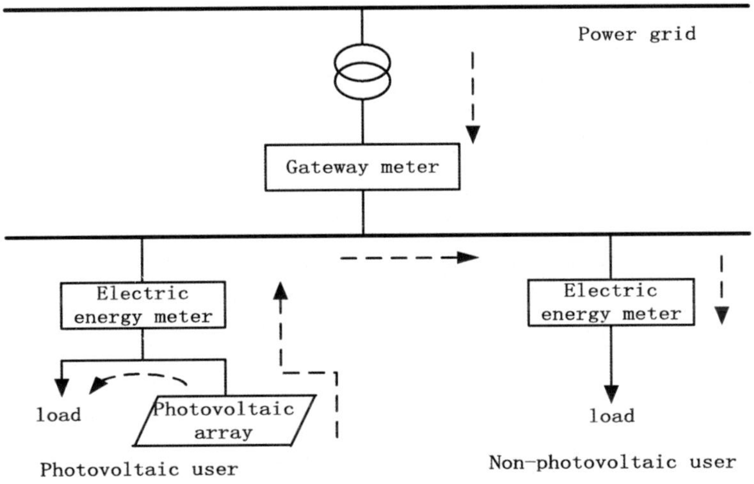

Fig. 2. Power flow direction when distributed generation is absorbed in the court

3. At a certain time, The PV power generation has surplus electricity on the grid and the surplus electricity cannot be completely consumed in the court. Some of the electricity on the grid is absorbed by other court, which is transmitted in reverse through the master meter of the court. That is, the reverse active power appears in the master meter of the court, as shown in Fig. 3. At this time, part of the PV users' power flow on the grid flows to other users in the court to meet the load demand of users, while the remaining part flows to the public grid after the reverse voltage boost through master table and the distribution transformer of the court, and is consumed between the courts. When the power of PV generation does not meet all the load demand in the court, the power grid will still supply power to the user. When the power of PV on the grid is less than a certain critical value, the loss of power and the line loss rate in the court will still decrease, but the decreasing range is smaller and smaller. When the power of PV on the grid is greater than the critical value, the loss of power and line loss rate will increase compared with that without PV users. In this paper, the selection of critical value is not discussed further, and only an example is given to illustrate such cases.

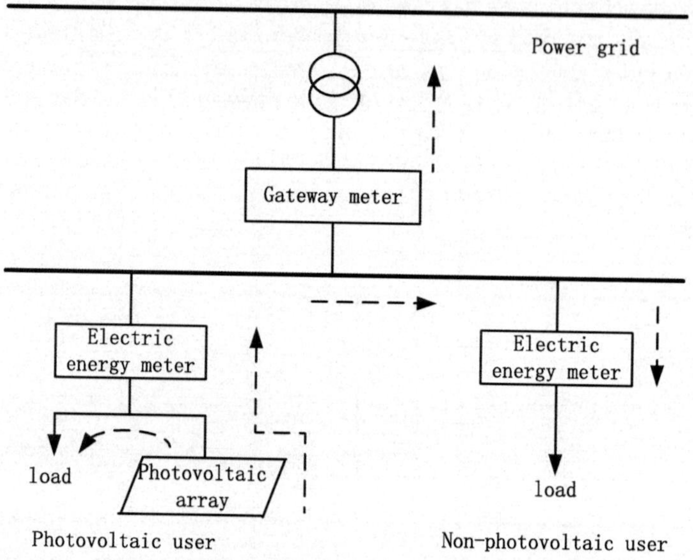

Fig. 3. Power flow direction when distributed generation cannot be absorbed in the court

According to statistics, the loss of distribution transformer accounts for 50%–60% of the total loss of the entire distribution network. If the generation capacity of distributed PV cannot be absorbed in the court, the power on the grid exceeding the critical value is sent out for consumption through the distribution transformer boost, which will greatly increase the loss of distribution transformer. In addition, when step-down transformer is used as booster for a long time due to the large amount of on-grid power of distributed power, the loss will increase and the heat will also increase, which will affect its service life.

2.2 Correction of Calculation Formula of Line Loss Rate in Court After Photovoltaic Access

In the traditional passive distribution network, the line loss calculation formula of the court is as follows:

$$\lambda = \left(1 - \frac{W_s}{W_g}\right) \times 100\% \tag{1}$$

Where:

λ– line loss rate of the court
Wg – total meter power in the court
Ws – sum of all user meters in the court

In the court with distributed PV power users, the power flow direction in the court may be changed due to the joining of PV power supply. In most cases, the line loss rate calculated by the traditional formula can not truly reflect the actual line loss rate in the court. There will be a small or even unreasonable negative loss. Therefore, when calculating the comprehensive line loss rate of the court containing PV users, it is necessary to consider the PV users' on-grid power and the reverse active power of the master table, which are also part of the power supply of the court. The modified formula for calculating the comprehensive line loss rate in the court is as follows:

$$\lambda = \left(1 - \frac{W_s}{W_g + W_f - \Delta W_f}\right) \times 100\% \tag{2}$$

Where:

λ – line loss rate of the court
Wg – total meter positive active power in the court
Ws – sum of all users positive active power in the court
Wf – sum of all PV users power on the grid in the court
ΔWf – the reverse active power of total meter in the court.

3 Practical Case Analysis

3.1 The Surplus of PV Power Generation on the Grid and the Surplus Power is Only Absorbed in the Court

Taking the No. 3 Wowo Hui court, Shanxi Province as an example, the distribution transformer capacity is 200 kVA. There are 75 residential households and one of these is PV household. This PV household was connected to the power grid on April 9, 2017, with a capacity of 5 kW. The relevant data of monthly line loss rates for this court from February 2017 to June 2017 are shown in Table 1.

Table 1. Relevant data of monthly line loss rates for the 3# Wowo Hui court

Date	Types	Positive active power (kWh)	Reverseactive power (kWh)	Loss of power (kWh)	Monthlyline loss rate (%)
February 2017	Total meter of the court	10566	0	413	3.91
	PV user	0	0		
	Non-PV user	10153	0		
March 2017	Total meter of the court	7265	0	358	4.93
	PV user	0	0		
	Non-PV user	6907	0		
April 2017	Total meter of the court	6207	0	261	3.88
	PV user	42	524		
	Non-PV user	6428	0		
May 2017	Total meter of the court	6877	0	259	3.34
	PV user	39	886		
	Non-PV user	7465	0		
June 2017	Total meter of the court	7880	0	262	3.05
	PV user	40	713		
	Non-PV user	8291	0		

As can be seen from Table 1, the average monthly line loss rate of this court was 4.04% in February and march of 2017 when there were PV users. When the PV user was reported to be installed in April 2017, the surplus of PV power generation was absorbed in the court except the user's own consumption, with an average monthly line loss rate of 3.27%. The loss of power and line loss rate are lower than that of uninstalled PV users.

3.2 The Surplus of PV Power Generation on the Grid and the Surplus Power is Transferred to Other Courts for Consumption

Taking the Chaoyang ND court, Shanxi Province as an example, the distribution transformer capacity is 200 kVA. There are 164 residential households and 7 PV household. And the total grid-connected capacity is 85 kW. Among them, 5 households connected to the grid in March 2017 with the capacity of 15 kW, 1 household connected to the grid in June 2017 with the capacity of 10 kW, and 1 household connected to the grid in December 2017 with the capacity of 50 kW. The data related of the monthly line loss rate of this court from January 2017 to January 2018 are shown in Table 2.

Table 2. Data related of monthly line loss rates for the Chaoyang ND court

Date	Types	Positive active power (kWh)	Reverse active power (kWh)	Loss of power (kWh)	Monthly line loss rate (%)
January 2017	Total meter of the court	1825	0	155	8.48
	PV user	0	0		
	Non-PV user	1670	0		
May 2017	Total meter of the court	4554	478	47	0.67
	PV user	209	2410		
	Non-PV user	6230	0		
July 2017	Total meter of the court	6199	538	178	2.12
	PV user	168	2195		
	Non-PV user	7511	0		
January 2018	Total meter of the court	2173	5669	493	5.59
	PV user	1018	6642		
	Non-PV user	1636	0		

As can be seen from Table 2, in January 2017 when there were no PV users in this court, the loss of power was 154 kWh and the monthly line loss rate was 8.48%. When 5 (15 kW) PV users were successively installed in March 2018, the remaining power generated by the PV users was connected to the grid and the extra power could not be fully consumed in the court. The power on the grid was less than the critical value, and it was transferred to other courts in reverse through the master table of the court to be consumed, and the power loss and line loss rate were reduced in this case. From June 2017 to December 2017, two PV households (60 kW) were installed again. When the PV power on-grid was greater than the critical value, and was transferred to other courts in reverse through the master table of the court to be consumed, the power loss and line loss rate increased.

4 Analysis of Other Factors Affecting the Line Loss Rate of Low-Voltage Court with PV

In practice, the access of distributed PV generation will also cause the following problems:

1. The problem of line limitation is prominent, and the line loss in the court is increased.

Without photovoltaic access, the original line is enough to support the entire court operation. However, after the PV generation connected, the current increases, the maximum current-carrying capacity of the line cannot meet the requirements, and the loss of the low-voltage line increases.

2. Improper matching of PV capacity with transformer capacity leads to the burning of the equipment.

According to working experience, if the photovoltaic capacity is not higher than 50% of the transformer capacity, the whole court will operate economically. If it is higher than this ratio, it may lead to wire burning, meter burning, transformer burning, and the loss of low voltage line and transformer increase.

3. The three-phase balance problem is not considered when photovoltaic is connected, which leads to the increase of line loss in the court.

Limited by geographical location and other factors, PV users concentrate on a certain phase, and the three-phase current difference is large, resulting in increased line loss.

5 Conclusion

Through the analysis, it can be concluded that in the low-voltage court with distributed power supply, the power flow direction will change due to the PV users' power flow on-grid. The optimal utilization mode of distributed power generation is spontaneous self-use, followed by consumption within the court, and the line loss rate of the court shows a downward trend with the increase of consumption ratio. However, when the distributed power through the distribution transformer after the reverse boost transmission to the superior distribution network, if the power on-grid is less than the critical value, the loss of power and line loss rate are lower than those courts without distributed PV. If the power on-grid is greater than the critical value, the distribution network loss increases, and the loss of power and line loss rate are rising, which is the most uneconomical way of consumption.

Therefore, in the process of distribution network planning and distributed power grid-connected site selection, the generation absorption of distributed power and the acceptance capacity of low-voltage power grid should be fully considered.

References

1. Ding, X., Luo, Y., Liu, W., et al.: Some suggestions on improving the calculation method of power loss in distribution network. Autom. Electr. Power Syst. **25**(13), 57–60 (2001)
2. Jiang, X., Sun, X.: Common problems and countermeasures in theoretical calculation of line loss in power supply enterprises. Distrib. Utilization **31**(7), 52–54 (2014)
3. Wan, S.: Analysis on the main ways of energy saving in urban distribution network. Distrib. Utilization **25**(2), 5–7, 19 (2008)
4. Hong, Y., Pen, K.: Optimal VAR planning considering intermittent wind power using markov model and quantum evolutionary algorithm. IEEE Trans. Power Delivery **25**(4), 2987–2996 (2010)
5. Meng, X., Piao, Z., Wang, Y., et al.: Loss reduction analysis of distribution network considering the influence of distributed power supply. Trans. CSAE **29**(No. 1 Supplement 1), 128–131 (2013)

Design and Realization of a GNSS Receiver Test Equipment

Wenquan Zhang(✉), Ying Li, Zhe Li, and Yuhai Li

Wuhan Mechanical College, Army Engineering University, Wuhan 430075, China
wenc99336@163.com

Abstract. Due to the high cost of the traditional GNSS receiver testing by an analog signal source of satellite navigation signals, a test equipment is designed based on the rapid signal detection toward the sky and the online automatic test method of interface signals, so as to realize the tests of overall positioning accuracy test and fault diagnosis of GNSS receivers. It has the features of rapid testing, ease for use and good practicability, thus reducing the test cost of GNSS receiver testing.

Keywords: GNSS receiver · Test equipment · Positioning accuracy test · Fault diagnosis

1 Introduction

In recent years, the satellite navigation industry begins to pay more attention to the research on performance analysis and evaluation of GNSS receivers. It is important to find appropriate methods for testing the performance of GNSS receivers [1]. At present, the mainstream method for the traditional performance test of GNSS receivers is using a GNSS satellite navigation signal simulator to build a test platform. Its advantage is that GNSS signals in different environments can be simulated indoors, thereby creating conditions for test and validation of the receiver performance in different scenarios and environments and greatly reducing the performance test duration. However, this method requires high professional competence of the testing personnel and needs high cost, and these disadvantages are not conducive to promotion and application among general users. General users are concerned most about whether the navigation positioning accuracy of GNSS receivers can meet the requirements, the availability of the equipment can be guaranteed and they can be supervised for rapid locating of fault position, detection and repair in the event of failure. Therefore, the positioning accuracy test method and fault diagnosis of GNSS receivers are the important directions for the design of GNSS receiver test equipment.

M. Tavana et al. (Eds.): IISA 2020, AISC 1304, pp. 499–505, 2021.
https://doi.org/10.1007/978-3-030-63784-2_62

2 Sky-Oriented Positioning Accuracy Test Method

Positioning accuracy refers to the degree of proximity between the location obtained by positioning solution of GNSS receiver and the real location [2], which is an important index to measure the performance and quality of GNSS receiver, and directly reflects the gap between the positioning data of GNSS receiver and the real data [3]. Aiming at the problem of high test cost of traditional satellite signal analog signal source, a rapid test method for sky (also known as real star [4]) is adopted, which can effectively reduce the performance test cost of GNSS receiver. Specific test method is: testing equipment acquisition to the location information of the measured GNSS receiver output compared with standard point of known location information, according to the circular location error probability (CEP) data processing method, if the horizontal positioning accuracy of 10 m or less (95%), vertical positioning accuracy of 10 m or less (95%), and the measuring receiver position precision meet the requirements of indicators, on the other hand, the very poor or not suitable to continue to use.

The calculation of horizontal positioning error ΔR is as follows [5], and the unit is m:

$$\Delta R = \sqrt{\frac{\sum_{i=1}^{n} (x_i - x_0)^2 + \sum_{j=1}^{n} (y_j - y_0)^2}{n - 1}} \tag{1}$$

Where, (x_0, y_0) is the true value of known standard point coordinates, (x_i, y_j) is the bit output of the tested GNSS receiver, and n is the total number of samples. These n positioning errors are sorted in the order from small to large. The $[n \cdot 95\%]$ results are taken as the positioning accuracy of the measurement.

Measured GNSS receiver in the measurement process, the requirements in the open, no tall around obstructions, receiver antenna and the standard point to, under the unified coordinate system data processing, excluding HDOP > 4/PDOP > 6 data values. The n value is at least 10. If the measured receiver outputs one frame of data per second, set the n value to 60, and the detection device can complete the test within 1 min.

3 Online Automatic Test Method of Components Based on Interface Signals

Different types of GNSS receivers do not have the same receiving and processing capabilities of navigation signals, and their structures and working mechanisms are also different. If the traditional offline method of "input excitation source, output signal voltage/current/impedance test" is adopted, the inspection and maintenance process is complicated and the test efficiency is low. In order to realize the rapid judgment and repair of faults of various types of receivers, this project realizes the online automatic test of GNSS receiver components based on the principle of functional integrity judgment, by building a fault tree model after sorting out the test and repair needs of different types of GNSS receivers and through applying the interface signal test.

4 Fault Diagnosis

In order to enable the detection equipment to quickly and accurately determine the fault location of the tested GNSS receiver, the FTA (fault tree analysis method) is used in this paper for fault diagnosis and positioning of the tested receiver. FTA uses fault tree as a tool to analyze various causes and ways of equipment system failure, and proposes an effective method for equipment system reliability research [6]. By collecting and classifying a large amount of fault information for a long time, the common fault phenomena of GNSS receiver can be classified into two categories: unable to start up and unable to locate. Based on the failure analysis of the tested GNSS receiver, the above two fault phenomena can be summarized into one result according to their impact on the system, that is, the receiver function is abnormal, which can be used as the top event of the fault tree. According to the construction of GNSS receiver, the fault tree can be established (see Fig. 1).

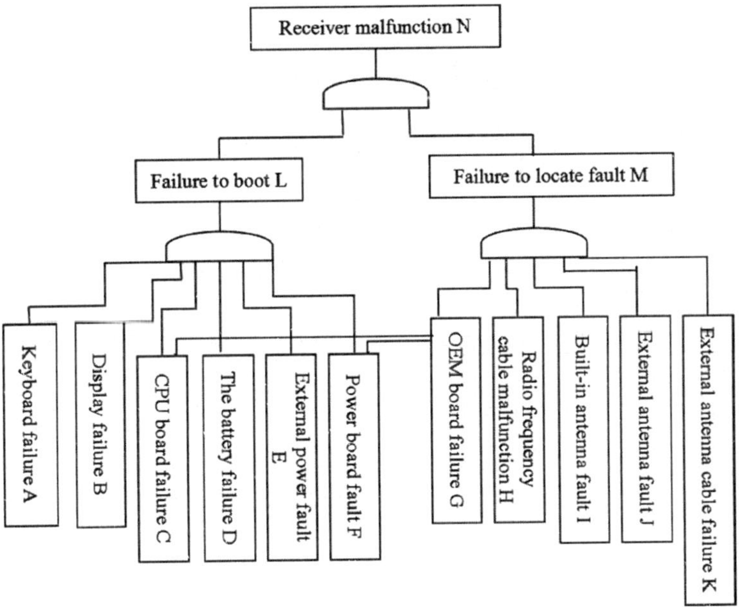

Fig. 1. Failure tree of GNSS receiver

5 Design of Test Equipment

5.1 Overall Design

This GNSS receiver test equipment primarily consists of a handheld terminal, a test adapter and a wireless router. It can not only check the positioning accuracy of the GNSS receiver, but also can test its components, such as antenna, cable, OEM board, CPU board and display screen. The test and repair guidance software is provided to manage the testing and repairing process. According to the teaching needs, it can provide the theoretical and skill assessment function (see Fig. 2).

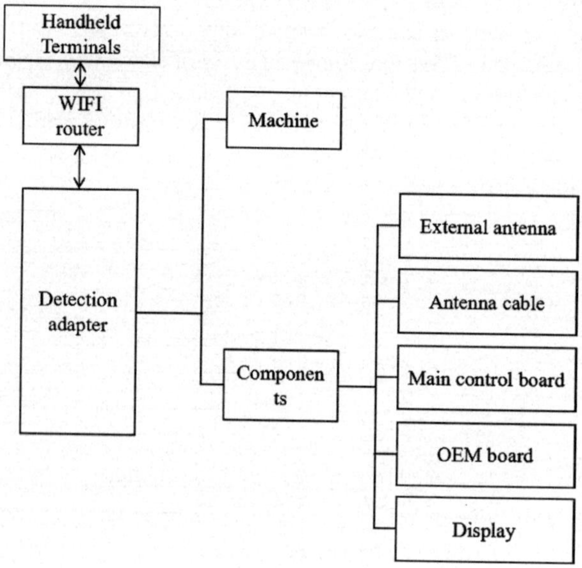

Fig. 2. Overall plan

The system workflow is detailed as follows:

(1) Wi-Fi networking
The handheld terminal, wireless router and test adapter are sequentially started. The handheld terminal test software searches for the test adapter with a fixed IP address, thereby establishing a wireless network connection;

(2) Complete test
After the handheld terminal test software sends the complete test command to the test adapter, the test adapter receives the positioning data of the tested receiver through the RS-232 serial port and forwards it to the handheld terminal. The measurement data are extracted after protocol analysis, and then the data analysis and calculation are performed to obtain the error result, thereby completing the positioning accuracy check of the receiver under test.

(3) Component test
After the handheld terminal test software sends the board test command to the test adapter, the test adapter separately tests the components of the receiver under test according to the fault diagnosis process and uploads the test results to the handheld terminal. The handheld terminal test software gives the fault location and repair guidance according to the test results, thus completing the functional test and repair of the components of the receiver under test.

5.2 Hardware Design and Realization

A GNSS receiver test equipment is composed of a handheld terminal, a test adapter and a wireless router. The handheld terminal adopts a three-proof tablet design, with built-in test and repair software, so it is easy and quick to use. The wireless router is used to establish the Wi-Fi network connection between the handheld terminal and the test adapter. It can support multiple terminal devices to be networked at the same time, and securely transmit data after encryption. The test adapter, which plays the role of connection and conversion between the handheld terminal and the receiver under test, must have the demand and test function of interfaces required for complete and component tests.

The test adapter consists of test board, battery, built-in positioning board, switch, indicator light, battery display screen, interface and casing (see Fig. 3).

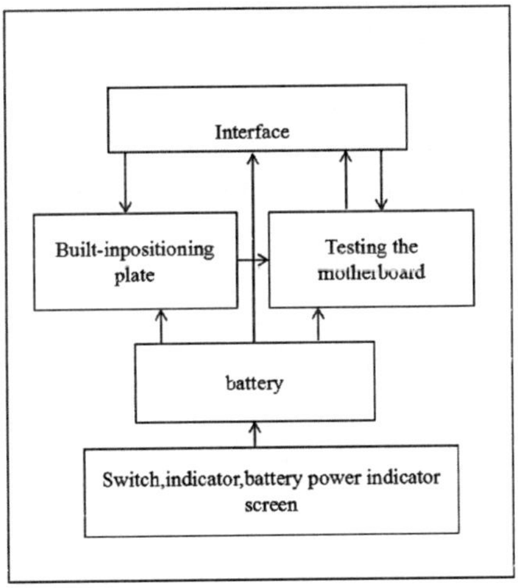

Fig. 3. Structural block diagram of test adapter

The test board as a core component of the test adapter is mainly used for completing wireless communication with the handheld terminal, data communication with the built-in positioning board, data communication with the receiver under test, and voltage and frequency test of components of the receiver under test. It is mainly made up of processor circuit, interface circuit, WIFI circuit, serial port communication circuit, relay switch circuit and power circuit.

The built-in positioning board, which provides the standard point coordinates when there is no accuracy check of known points and is connected with an external antenna for positioning and antenna testing, requires high positioning accuracy, small size and low power consumption. The built-in positioning board can simultaneously support two frequency points, namely BD2 B1 and GPS L1, for positioning. The positioning accuracy is less than 2m, the first positioning time is 1s, and three UART serial ports are used for data communication. The data format is standard NEMA0183.

The built-in battery, a 12V 7000mAh lithium battery for power supply to components of the test adapter, has the functions of overvoltage protection, overcurrent protection and high-temperature automatic protection. It has the features of small size and high safety.

5.3 Software System Design

A software system is developed on the Microsoft Visual Studio 2013 platform by using the fully object-oriented programming language C#. It integrates various design patterns such as factory method pattern, proxy pattern, state pattern and visitor pattern. The classic multi-thread processing series are adopted in the communication module to ensure the timely and efficient data communication. The classic inheritance method is utilized in various sub-modules to reduce the complexity of codes. In terms of database design, the lightweight database SQLite is used, and the data communication of the equipment is achieved through WIFI. After the software system establishes a WIFI network connection for communication, go to the selection interface of the GNSS receiver. Each receiver has the functions of performance test, fault diagnosis, repair guidance, repair management and assessment (see Fig. 4).

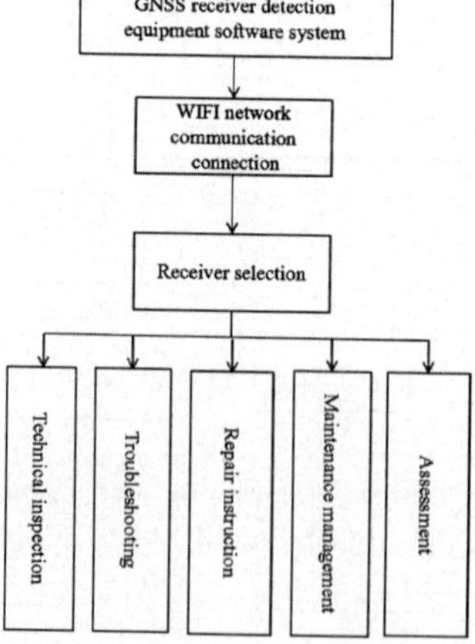

Fig. 4. Structural block diagram of the test equipment software system

The software system, which is built in the handheld terminal, is mainly used to establish a WIFI network connection for communication between the handheld terminal and the test adapter, respond to the user operation and control command, control the test adapter to send an excitation signal, collect the test data, handle abnormal situations, intelligently diagnose the receiver failure location and obtain the test results. Also, it enables the technical inspection, repair guidance, repair management and assessment and helps testing personnel and user to do the testing.

6 Conclusions

The GNSS receiver test equipment, which builds a test and repair platform by the architecture of "handheld terminal + WIFI wireless network + test adapter", can improve the portability of operation, support the simultaneous testing of multiple receivers and effectively enhance the flexibility of testing. The sky-oriented fast test method is applied to reduce the performance test cost of GNSS receivers. The online automatic test method for components based on interface signals can effectively solve the problems of complex tests and high requirement for the technical competence of repair personnel by the traditional key signal testing inside the circuit board or the offline testing method of the circuit board interface, thus greatly improving the repair efficiency of GNSS receivers.This GNSS receiver test equipment can have the advantages of small size, low cost, portability, simplicity and ease of software operation and practicability for field conditions. Moreover, the system can provide detailed technical information to operators and repair technicians and be used as a teaching and training platform for users and repair personnel of GNSS receivers.

References

1. Jin, X.: Research on Beidou Receiver Testing Method. Shanghai Jiao Tong University, Shanghai (2016)
2. Zhang, Q.J., Li, M., Wang, Na., et al.: Research on the test method of the big dipper ii civil equipment. Modern Telecommun. Technol. **2014**(7), 18–22 (2014)
3. Shan, G.: Design and Implementation of Satellite Navigation Terminal Detection Software. Huazhong University of Science and Technology, Wuhan (2013)
4. Wang, Bo., Li, F., Jiao, H., et al.: Measurement method and analysis of main indicators of beidou receiver. J. Navig. Positioning **3**, 19–23 (2015)
5. General specification for beidou/global navigation satellite system (GNSS) navigation equipment (2015)
6. Wang, F., Liu, G.: Fault tree analysis of inertial/GPS integrated navigation system. Aviation Maintenance Eng. **4**, 38–40 (2004)

Resonance Suppression of Position Servo System Based on Improved Notch Filter

Shuheng Chen[1,2(✉)], Wei Feng[1,2], Fanquan Zeng[1,2], Yaoyao Wang[1,2], and Jiangxianfeng Tian[1,2]

[1] Shanghai Aerospace Control Technology Institute, Shanghai 201109, China
a429357115@163.com

[2] Shanghai Engineering Research Center of Servo System, Shanghai 201109, China

Abstract. In order to solve the resonance problem of the position servo system under a flexible load, this paper proposes an improved mathematical model of the notch filter, and uses bilinear transformation to complete the design of the digital filter. After simulation analysis, the frequency characteristics of the filter are consistent with the theoretical design, and the parameter adjustment is precise and controllable. An experimental verification was carried out on the flexible nozzle load of the solid rocket. The experimental results show that: adding an improved notch filter to the control loop of the position servo system can effectively suppress the position resonance problem.

Keywords: Resonance suppression · Servo system · Notch filter

1 Introduction

Because the servo system transmission device inevitably has a certain elasticity, it cannot be completely eliminated. When transmitting torque, there will be different degrees of elastic deformation on the drive system, base, each axis and load, and all have certain natural frequencies. With the improvement of the performance of the servo system and the expansion of the bandwidth, it will inevitably overlap with its natural frequency. At this time, the phenomenon of mechanical resonance will occur [1]. This will affect the life of the parts, at the same time have a negative impact on the control performance and accuracy of the system, and even make the system lose stability. Therefore, the study of mechanical resonance suppression in this paper is of great significance.

At present, at home and abroad, the suppression methods for resonance are mainly divided into two ways: 1) Passive mode refers to the suppression of resonance through compensation or filtering after resonance is generated. Mainly include adding low-pass filters [2],notch filters [3], etc.; 2) Active mode refers to the use of state feedback method to suppress resonance from the source [4], mainly including pole configuration control strategy based on state feedback [5], etc.

In order to solve the problem of low-frequency resonance of the position servo system under a flexible load [6], this paper adopts the method of adding a notch filter between the position regulator and the speed loop to achieve the resonance suppression

M. Tavana et al. (Eds.): IISA 2020, AISC 1304, pp. 506–512, 2021.
https://doi.org/10.1007/978-3-030-63784-2_63

of the specific resonance frequency point. After experimental verification, this method is simple and effective.

2 Mathematical Model of Notch Filter

2.1 Analog Notch Filter

Using the early traditional notch filter, although the resonance peak can also be greatly reduced, the notch amplitude and width are controlled by the same parameter, which is difficult to adjust, the effect is not good.

$$G(s) = \frac{s^2 + \omega_n^2}{s^2 + 2\xi\omega_n s + \omega_n^2} \tag{1}$$

When the notch damping ratio ξ becomes smaller and smaller, the attenuation at the notch frequency is getting larger and larger, the change is more and more sharp, and the influence on the surrounding signal of the notch frequency is getting smaller. In other words, when the notch damping ratio ξ changes, the notch depth and width of the notch filter will change. One parameter affects both characteristics, which is obviously not conducive to the design of the notch filter. This is also the biggest flaw of the traditional notch filter.

In this paper, an improved notch filter is proposed. The center frequency, width and height parameters of this notch filter are determined by three parameters independently. The individual adjustment makes the notch more flexible and the effect is better. The transfer function is as follows.

$$G(s) = \frac{s^2 + k_1\omega_n s + \omega_n^2}{s^2 + k_2\omega_n s + \omega_n^2} \tag{2}$$

$$\begin{cases} k_1 = d \times (\frac{B}{\omega_n} \times \sqrt{\frac{1-a^2}{a^2-d^2}}) \\ k_2 = \frac{B}{\omega_n} \times \sqrt{\frac{1-a^2}{a^2-d^2}} \\ \omega_n = 2\pi f_0 \\ B = 2\pi f_B \\ d = 10^{D/20} \\ a = 10^{-3/20} \end{cases} \tag{3}$$

In the formula:

ω_n—notch center frequency, unit is rad/s. The center frequency should be set to the resonant frequency of the system. The accuracy of the notch center frequency is of great significance to resonance suppression. If the center frequency is not accurate enough, it will easily cause notch shift and lead to failure of resonance suppression.

B—notch width, the unit is rad/s. Characterize the width when the notch center frequency is the center and the amplitude of the frequency characteristic drops by 3 dB.

D—Notch depth, the unit is dB. Characterizes the depth of the amplitude drop at the center frequency of the notch.

In this way, each feature is adjusted by its corresponding parameter, which facilitates the design and maximizes the effect of the notch filter.

2.2 Notch Filter Discretization

If the notch filter needs to be implemented in software, it is necessary to complete the construction of the digital filter, that is, to convert the analog filter into a digital filter through some transformation after equivalent discrete sampling. This article uses the bilinear transformation method to complete the conversion from analog filtering to digital filtering. Including S plane to S1 plane, S1 plane to Z plane two changes.

$S \rightarrow S_1$ mapping relations:

$$s = \frac{2}{T}\frac{1 - e^{-s_1 T}}{1 + e^{-s_1 T}} \tag{4}$$

Making $z = e^{s_1 T}$, the $S \rightarrow Z$ mapping relationship available:

$$s = \frac{2}{T}\frac{1 - z^{-1}}{1 + z^{-1}} = \frac{2}{T}\frac{z - 1}{z + 1} \tag{5}$$

The above formula is the discrete mapping equation of bilinear transformation. After the transformation, the left half of the S plane is completely mapped into the unit circle of the Z plane.Substituting the above mapping relationship into the transfer function of the analog filter gives the following formula:

$$H(z) = \frac{y(n)}{x(n)} = \frac{b_0 + b_1 z^{-1} + b_1 z^{-2}}{1 + a_1 z^{-1} + a_1 z^{-2}} \tag{6}$$

The above formula is the discrete transfer function of the analog notch filter. The inverse Z transformation is derived to derive the corresponding differential differential equation:

$$y(n) = b_0 x(n) + b_1 x(n-1) + b_2 x(n-2) - a_1 y(n-1) - a_2 y(n-2) \tag{7}$$

The corresponding parameters are as follows:

$$\begin{cases} b_0 = \dfrac{4/T^2 + 2k_1\omega_n/T + \omega_n^2}{4/T^2 + 2k_2\omega_n/T + \omega_n^2} \\ b_1 = \dfrac{-8/T^2 + 2\omega_n^2}{4/T^2 + 2k_2\omega_n/T + \omega_n^2} \\ b_2 = \dfrac{4/T^2 - 2k_1\omega_n/T + \omega_n^2}{4/T^2 - 2k_2\omega_n/T + \omega_n^2} \\ a_1 = \dfrac{-8/T^2 + 2\omega_n^2}{4/T^2 + 2k_2\omega_n/T + \omega_n^2} \\ a_2 = \dfrac{4/T^2 - 2k_2\omega_n/T + \omega_n^2}{4/T^2 + 2k_2\omega_n/T + \omega_n^2} \end{cases} \tag{8}$$

3 Resonance Characteristics Analysis

This article uses a solid engine flexible nozzle for test verification,using a three-loop PI control algorithm, Id = 0 control strategy, the test environment is shown in the following Fig. 1:

Fig. 1. Test environment

The command signal for this experiment is sinusoidal signals with different amplitudes of 0.6°. The experimental results are shown below (Fig. 2).

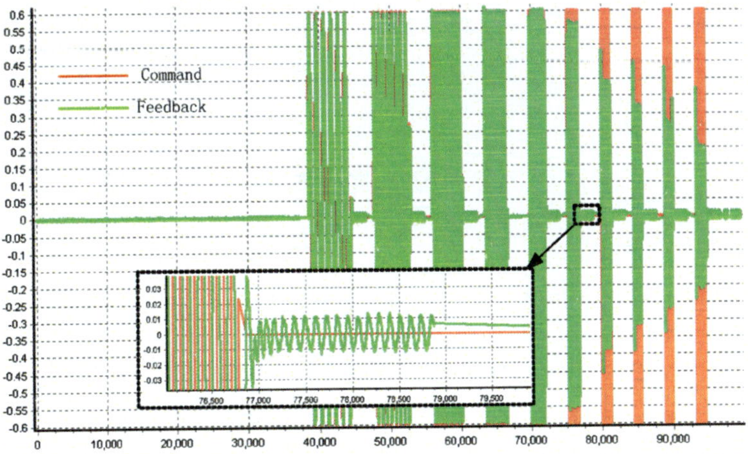

Fig. 2. Position command signal and feedback

It can be seen from the above figure that the position feedback signal has a position resonance signal with an amplitude of ±0.01° and a frequency of 8.7 Hz during the time period of 76900 ms–78800 ms, and the duration is more than 1.9 s.

4 Design and Simulation of Improved Notch Filter

The center frequency of the improved notch filter is 8.7 Hz (ω_n = 54.66 rad/s), the width B = 50.26 rad/s (f_B = 8 Hz), and the depth D = −4.2 dB. Using MATLAB software to analyze the frequency characteristics of the notch filter, the results are as follows (Fig. 3):

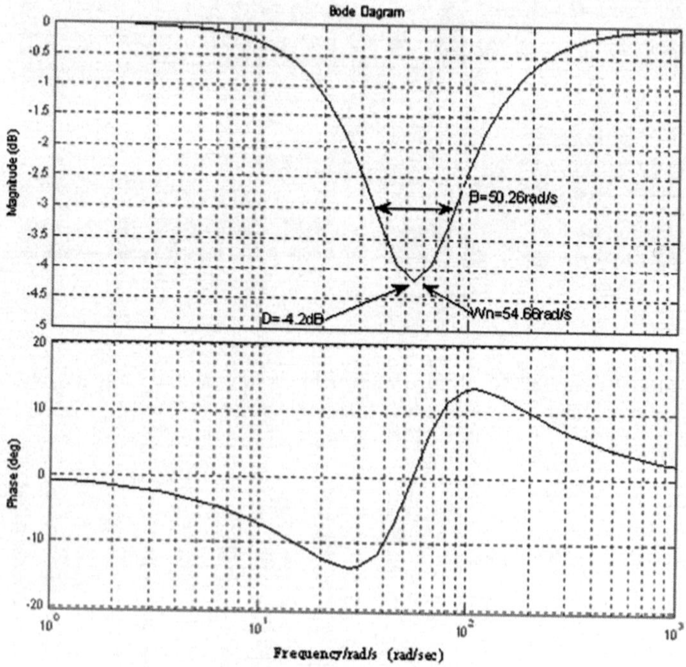

Fig. 3. Simulation result of improved notch filter

It can be seen that the frequency characteristics of the improved notch filter are highly consistent with the expected effects of the three parameter settings.

5 Experimental Verification

5.1 Design of Servo Control Algorithm Based on Notch Filter

Filter the specific frequency range by adding an improved notch filter between the position controller and the speed loop [7], as shown in the Fig. 4:

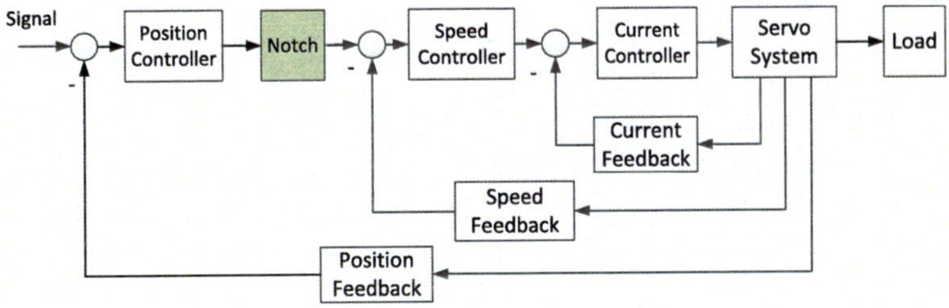

Fig. 4. Design of servo control algorithm based on notch filter

In MATLAB, SYSD = C2D (SYSC, Ts, METHOD) converts the continuous-time LTI model SYSC to a discrete-time model SYSD with sample time Ts. Among them, Ts = 0.00025 s, which is a three-loop calculation cycle in the servo system. The mathematical expression after discretization is:

$$y(n) = b_0x(n) + b_1x(n-1) + b_2x(n-2) - a_1y(n-1) - a_2y(n-2) \quad (9)$$

Calculated: $b_0 = 0.9952$, $b_1 = -1.9746$, $b_2 = 0.9796$, $a_1 = -1.9746$, $a_2 = 0.9746$.

5.2 Experimental Results

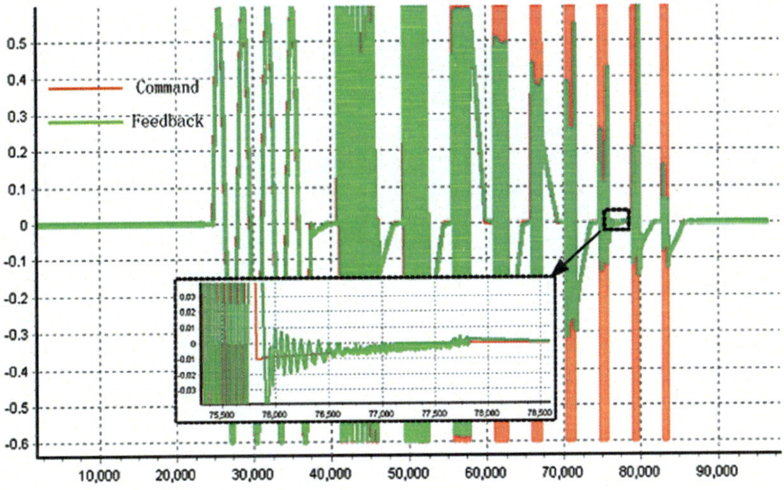

Fig. 5. Position command signal and feedback after using notch

Sending the sinusoidal angle signal of the same amplitude again, it can be seen that the position feedback resonance after the design of the notch filter rapidly attenuates from ±0.01° to almost zero within 1s. It can be seen that the notch filter has obvious effect on solving the position servo resonance problem (Fig. 5).

6 Conclusion

In this paper, for solving the resonance problem of flexible load servo system, an improved notch filter between the position controller and the speed loop is added, a mathematical model of the improved notch filter is proposed, and the design of digital filters is completed by using bilinear transformation. After simulation analysis, the filter frequency characteristics are consistent with the theoretical design, parameter adjustments are precise and controllable. Experiments were carried out on the flexible nozzle load of the solid rocket. The experimental results show that: adding an improved notch filter to the control loop of the position servo system can effectively suppress the position resonance problem.

References

1. Herreman, W., Duncan, A.J. Mechanical resonance. Phys. Educ.**18**(5), 205–206(2) (1983)
2. Szabat, K., Orlowskakowalska, T.: Vibration suppression in a two-mass drive systemusing PI speed controller and additional feedbacks—comparative study. IEEETrans. Ind. Electron. **54**(2), 1193–1206 (2007)
3. Ellis, G., Lorenz, R.D.: Resonant load control methods for industrial servo drives. In: Industry Applications Conference, 2000. Conference Record of the IEEE, vol. 3, pp. 1438–1445 (2000)
4. Zhang, G.: Speed control of two-inertia system by PI/PID control. IEEE Trans.Ind. Electron. **47**(3), 603–609 (2000)
5. Futami, S., Kyura, N., Hara, S.: Vibration absorption control of industrial robots byacceleration feedback. IEEE Trans. Ind. Electron.**IE-30**(3), 299–305 (1983)
6. Liao, H., Jin, Y., Peng, J., et al.: Analysis of self-oscillation in roll channel using describing function method. Flight Control Detect. **1**(2), 028–033 (2018)
7. Wang, G.L., Chen,X., Zeng,F.Q., et al.: Permanent magnetic servo motor's space vector control system and MATLAB Simulation. Flight Control Detect.**1**(3),059–062 (2018)

Design and Implementation of Digital DC Servo System Hardware

Rui Zhang(✉), Zhixin Cheng, Baomei Xu, and Xuebing Liao

Ordnance NCO Academy, Army Engineering University of PLA, Wuhan 430075, China
904972963@qq.com

Abstract. Aiming at the shortcomings of the complicated circuit and high failure rate of the analog servo control system, a digital DC PWM speed regulation system based on TMS320F28335 is designed. The design and implementation of hardware circuit are introduced. The circuit structure is simple and the control precision of the system is improved.

Keywords: Digital · DC motor · Servo system · Circuit design

1 Introduction

The DC servo system has the advantages of good starting performance and speed regulation performance, and large overload capacity. It widely uses in tank control systems such as tanks, armored vehicles, self-propelled artillery and naval guns [1]. However, the DC servo system currently used in the equipment is generally in the form of a motor amplifier-DC motor. The motor amplifier supplies power to the DC motor that needs to be adjusted. By adjusting the size and direction of the output voltage of the motor amplifier Control the speed and direction of the DC motor rotation to control the movement of the artillery turret and barrel. This structure requires more equipment, large size, high maintenance costs, low efficiency, slow motor response, and high noise and vibration. With the development of control technology, computer technology, power electronics technology and micro-motor technology, the control of DC servo systems is gradually shifting to digital control, and gradually realize automatic control.

2 Principle of Speed Regulation of DC Servo System

The DC servo system generally selects the separately excited DC motor, the mechanical characteristic equation of the motor such as [2]

$$n = \frac{U_d}{C_e\phi} - \frac{R}{C_e\phi C_t\phi}T = \frac{1}{C_e\phi}(U_d - I_d R) \quad (1)$$

In the formula,

U_d— Armature voltage (V).

M. Tavana et al. (Eds.): IISA 2020, AISC 1304, pp. 513–519, 2021.
https://doi.org/10.1007/978-3-030-63784-2_64

I_d—armature current (A).
n—motor speed (r/min).
φ—the magnetic flux of the motor (Wb).
R—Armature loop resistance (Ω).
T—Motor torque (N·m).
C_e—electric potential constant of the motor.
C_t—Motor torque constant.

It can be seen from the formula that the motor speed can be changed by changing the armature voltage, armature circuit resistance and excitation flux φ. The external resistance in the armature circuit can be changed to change the total resistance of the armature circuit to achieve the purpose of speed regulation. However, this type of speed regulation is a stepped speed regulation, the rate of change of the speed is large, and the low speed performance is not good, The efficiency is low; when the armature voltage is constant, the speed can be adjusted by adjusting the excitation flux, but the higher the speed, the smaller the motor output torque and the worse the commutation ability; change the armature voltage adjustment At speed, the magnetic flux and equivalent resistance of the motor are unchanged. No matter what speed, the motor can output rated torque, and the speed regulation range is relatively wide [3]. By comparing the three speed adjustment methods, in order to ensure that the motor has good mechanical characteristics, the method of changing the armature voltage is used to adjust the speed.

3 System Scheme Design

In order to improve the control accuracy, the servo system generally adopts closed-loop control. The current loop-speed loop speed regulation system is a widely used speed regulation system. The dynamic and static performance, speed regulation range and speed control accuracy of the servo system have been greatly improved, but the position control accuracy is still not accurate enough. In order to improve the accuracy of the position control, on the basis of the original current loop-speed loop, the position control loop is added to form a three-closed loop control servo system. The block diagram of the servo system is shown in Fig. 1.

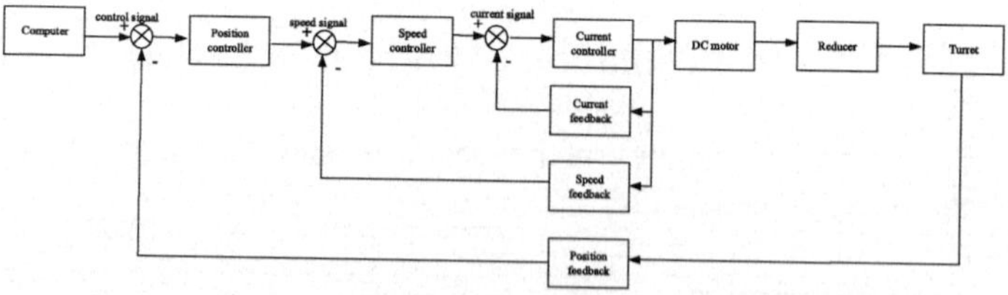

Fig. 1. Block diagram of servo system

The position feedback unit uses the photoelectric code wheel to measure the position of the turret or barrel in real time. The speed feedback unit uses the voltage signal output from the speed measuring motor coaxial with the DC motor to obtain the speed information of the motor. The current feedback unit uses the Hall element to measure the main circuit The current is converted into voltage and sent to the control system for calculation.

3.1 System Hardware Design

The hardware design of the system mainly includes the design of the main circuit, the design of the digital control unit, the design of the power drive unit and the design of other auxiliary circuits. The main circuit adopts H-bridge reversible PWM converter; the digital control unit mainly includes DSP and communication, A/D acquisition and auxiliary power module circuit, etc.; the power drive unit is mainly composed of the main power circuit, drive circuit and high-speed optocoupler isolation circuit; The auxiliary circuit mainly includes a current feedback circuit, a speed feedback circuit and a position feedback circuit. The overall hardware block diagram of the system is shown in Fig. 2.

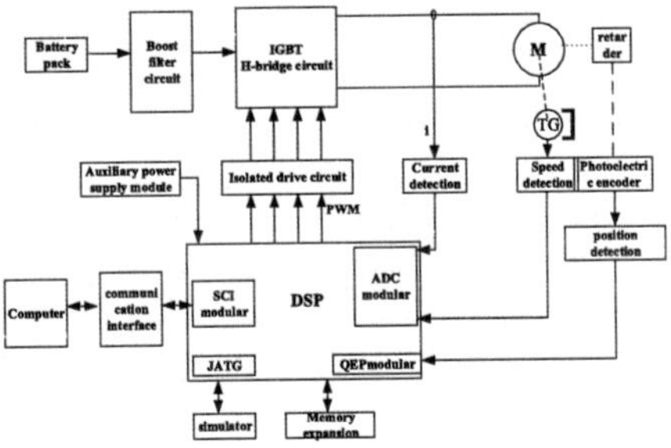

Fig. 2. Overall hardware block diagram of the system

3.2 Digital Control Unit Design

The digital control unit mainly includes the main circuit with TMS320F28335 as the main control chip and other peripheral circuits. TI's TMS320F28335 is the core of the entire controller. This chip is a 32-bit floating-point chip used by TI for high performance, high cost performance, and multifunction. It has high precision, low cost, low power consumption, and high performance. Higher peripheral integration, a large amount of data and program storage, A/D conversion is more accurate and fast [4].

Peripheral circuits mainly include auxiliary power circuits, SCI communication interfaces, memory expansion circuits, and power supply monitoring circuits. The auxiliary power circuit provides two voltages, 3.3V and 1.9V, required by the DSP chip to work. It uses the power chip TPS73HD301 specially designed by TI for the DSP control system. The input voltage of this chip is 5V, and the output has fixed 3.3V and 1.9V; SCI The communication interface circuit mainly completes the communication between the host computer and the DSP, uses the MAX232CSE chip of MAXIM Company to convert to the RS232 standard serial port for communication, and uses a differential method to send and receive data; the memory expansion circuit mainly focuses on the feedback current, voltage and communication The data in the process is saved, and the processing requirements of data and commands cannot be met during debugging. Therefore, the data storage space needs to be externally expanded. The high-speed SRAM chip IS61LV25616 is selected as the external expansion RAM. The chip is manufactured using high-performance CMOS technology. It has 256K of storage space; in order to prevent the program from looping, a power supply monitoring circuit is designed. The circuit uses the MAX706 chip to design. The MAX706 is a group of CMOS monitoring circuits that can monitor the power supply voltage and the working state of the DSP.

3.3 Main Circuit Design

The main circuit includes a battery pack, a boost filter circuit, a soft start circuit, a filter circuit, and a power conversion circuit. The equipment is generally powered by a battery pack or a built-in generator. The power supply voltage is generally 24V and cannot directly drive the motor. Therefore, a booster circuit is required. The booster circuit uses an isolated power conversion module. After the output ends of the modules are connected in series get the voltage U required for the motor to work; the instantaneous current of the motor is large. In order to prevent the IGBT module from being burned out, a soft start circuit is designed. When the motor is just started, R1 is connected to the circuit, which increases the starting resistance and the starting current. Becomes smaller, when the charging voltage of the capacitor C2 reaches the working voltage of the relay K1, the contact K1–1 pulls in, shorts R1, and the DC motor works normally; the power conversion circuit uses an H-bridge power conversion circuit, and the power switching device selects IGBT. By controlling the on and off of the switch tube, the control voltage Ud of the armature is obtained [5]. The average voltage of Ud such as

$$u_d = \frac{T_{on}}{T}U - \frac{T - T_{on}}{T}U = \left(\frac{2T_{on}}{T} - 1\right)U \tag{2}$$

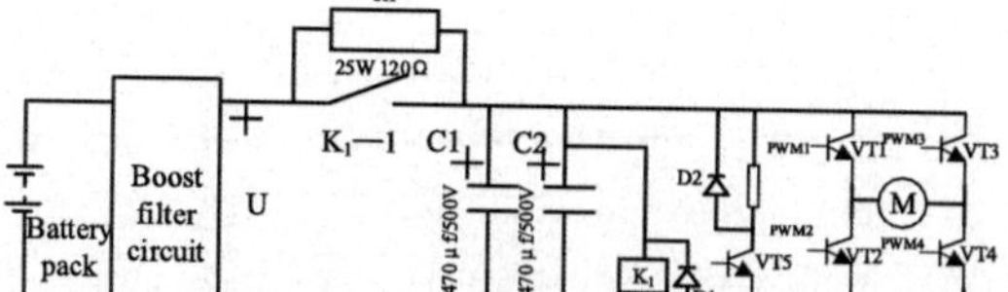

Fig. 3. Main circuit design schematic

In the formula, ton represents the on-time of the switch, and T represents the PWM period. The PWM speed can be adjusted by adjusting the ton time. When $t_{on} > T/2$, $U_d > 0$, the motor rotates forward, when $t_{on} < T/2$, $U_d < 0$, the motor reverses, when $t_{on} = T/2$, $U_d = 0$, the motor does not rotate, but there is still current in the armature winding, which improves the dynamic performance of the motor (Fig. 3).

3.4 Power Drive Circuit Design

Because the PWM amplitude of the DSP output is 3.3V, and the driving voltage of the IGBT is about 12V, a driving circuit needs to be added, and at the same time, the driving circuit also plays the role of isolation protection. Figure 4 shows a half-bridge drive circuit. Two such circuits can be combined into a complete H-bridge drive circuit. The driver chip selected in this paper is IR2110S driver chip of IR company, which has the advantages of small size of optocoupler isolation and high speed of electromagnetic isolation.

Pin 12 and pin 14 of the IR2110S chip are logic high-end and low-end inputs connected to the DSP output pins; pins 1 and 8 are low-end outputs and high-end outputs, connected to the main circuit's PWM2 and PWM1; pin 13 is the protection signal input When this pin is connected to high level, the output signals of the chip are

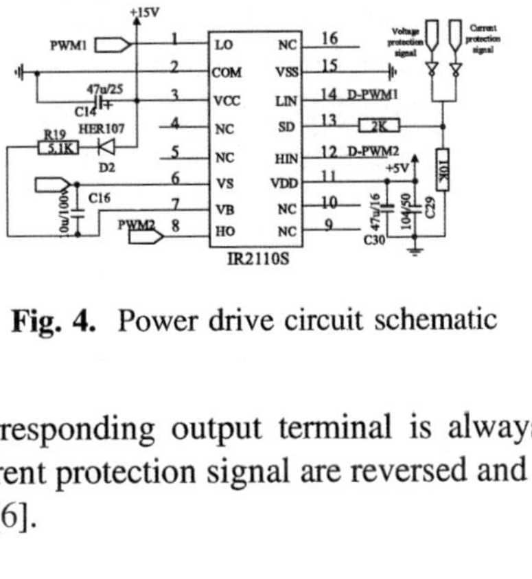

Fig. 4. Power drive circuit schematic

all blocked, and the corresponding output terminal is always low, and the voltage protection signal and current protection signal are reversed and connected to this pin to protect the main circuit [6].

3.5 Feedback Circuit Design

In order to improve the control accuracy of the system, three feedback circuits of current feedback, speed feedback and position feedback are designed. The current feedback directly uses the Hall element to take the current of the motor armature, which is converted into a voltage of 0–3.3V after being processed by the op amp It is collected by the A/D conversion module of the DSP; the speed feedback takes the voltage generated by the speed measuring motor coaxial with the motor armature, and after processing, it is sent to the A/D conversion module of the DSP, and the position feedback uses an incremental photoelectric code disk, Code wheel outputs 90° phase difference A, B two-way pulse signal and Z signal output one pulse per revolution, using the phase difference of A, B two-way pulse signal can distinguish the motor forward and reverse, the pulse output per second The number can be used to calculate the speed, and the Z signal is used for reference point positioning. The photoelectric code disk is directly installed on the non-load end of the motor rotor, so it will be subject to strong electromagnetic interference. In order to improve the position measurement accuracy, the pulses sent by the optical code disc need to be isolated by the optocoupler before entering the DSP for processing. In order to ensure the rapid response, high-speed optocoupler are used. The signal is sent to the DSP after the Schmitt trigger shaping process. The A and B pulse signals are connected to QEP1 and QEP2 of DSP respectively, and processed by the orthogonal coding unit of DSP. The quadrature coding unit detects and counts each rising and falling edge of the A and B signals, that is, the original signal is processed by 4 times the frequency, which enhances the counting accuracy. Pulse Z is connected to CAP3, which is connected to the capture unit of DSP [7].

3.6 Protection Circuit Design

In order to ensure the stable operation of the DC servo system and prevent damage to the IGBT and burnout of the motor windings when a fault occurs, the system has designed overvoltage and overcurrent protection circuits for the main circuit. The overvoltage protection circuit is shown in Fig. 5. During normal operation, the comparator output SV is high level. When VG is higher than the given value of Vref, the comparator output SV jumps from high level to low level. DSP collects SV After the low level, the PWM output is blocked to protect the main circuit. After the fault is eliminated, the SV turns to a high level, and the system can work normally. The

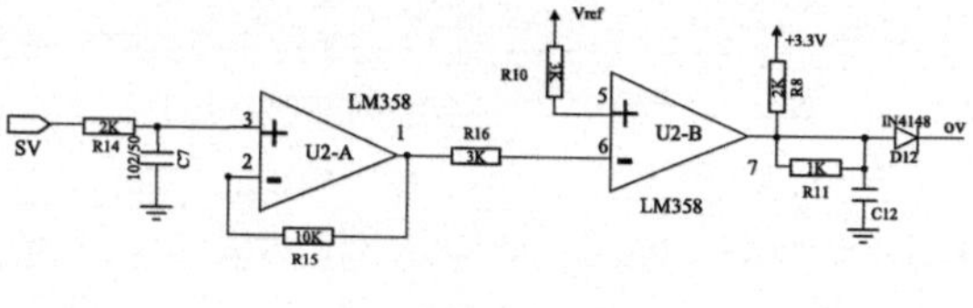

Fig. 5. Schematic of overvoltage protection circuit

function of the RCD circuit is delay protection to prevent the system from malfunction due to interference glitches. The delay time should not be too long to ensure that it can act in time under fault conditions and return quickly when recovering. The system can work normally.

Acknowledgments. This paper designs the hardware circuit of DSP-based digital DC PWM speed control system, which mainly includes the main circuit, the control system circuit, the drive isolation circuit, and the auxiliary circuit, which can realize the current feedback, speed feedback and position feedback of the servo system. The digital servo system not only simplifies the circuit structure but also saves the hardware cost, has a high steady-state accuracy, and the system is easy to control.

References

1. Zhang, Y.: Artillery control system and principle. Beijing Institute of Technology Press (2009)
2. Chen, B.: Electric drag DC control system. Machinery Industry Press (1993)
3. Wang, Y.: Research on ARM-based Digital DC Motor Controller. Zhejiang University (2007)
4. Rubaai, A., Ofoli, A.R., Cobbinah, D.: DSP—based real—time implementation of an adaptive fuzzy controller for the tracking control of servo·motor drives. In: 40th IAS Annual Meeting IEEE Industry Applications Conference (2005)
5. Chen, C., Fan, P.: Design and research of digital DC speed regulation system. Micro-Motors **45**(4), 52–55, 66 (2012)
6. Cheng, Z., Huang, L., Zhang, R., et al.: Development of multi-channel 400Hz high-precision AC power supply based on SPWM. Electr. Technol. **2016**(8), 65–68 (2016)
7. Xiaoming, W., Ling, W.: DSP Control of Electric Motor. Beijing University of Aeronautics and Astronautics Press, Beijing (2004)

A Small-Scale Current Sensor Scheme of Single-Loop Double-Winding Fluxgate

Xin Zhang, Aiming Zhao(✉), Yawei Shi, Ronghui Hu, and Shuaishuai Zhao

Shanghai DianJi University, Shanghai, China
zhaoam@sdju.edu.cn

Abstract. This subject designs a single-loop fluxgate current sensor with permalloy as the magnetic core for weak current measurement. The weak DC signal is converted into a voltage signal through the fluxgate and the voltage waveform data is processed and converted by Labview software. Using this design scheme to measure the depolarization current of the main insulation of high-voltage motors, the results show that this design scheme has higher accuracy and higher application value.

Keywords: Fluxgate · Single-loop · Current sensor · High accuracy · Labview

1 Introduction

For all kinds of weak measurands, they are generally converted into micro currents or micro voltages by corresponding sensors, and then their amplitudes are amplified by amplifiers to reflect the size of the measurements [1, 2]. The current weak current measurement scheme mainly uses the capacitance integration method and the transimpedance method [3]. The capacitance integration method has a slow response time, and the integration capacitor charge/discharge as a measurement cycle will lose current information [4–6]. The transimpedance method is to connect a resistor between the input and output of the operational amplifier. Long-term operation will result in poor stability and poor I-V linearity [7, 8].

It can be seen from the above that the weak current measurement scheme is unstable under long-term continuous measurement. In order to solve the above problems, this paper proposes a single-loop flux gate current sensor scheme for depolarization current measurement.

2 Principle Analysis

This article uses a single core dual winding scheme to design a simple and sensitive flux gate current sensor. The probe structure shown in Fig. 1. The periodically changing excitation signal excites the magnetic core and generates an excitation magnetic field [9, 10]. According to Faraday's law of electromagnetic induction,

M. Tavana et al. (Eds.): IISA 2020, AISC 1304, pp. 520–528, 2021.
https://doi.org/10.1007/978-3-030-63784-2_65

changes in the magnetic field cause the coil to generate an induced electromotive force. When the measurement current I_P is applied to the primary winding, the magnetic core generates a magnetic field H_P. Since the permeability of the magnetic core will change with the saturation degree of the magnetic core, which will affect the magnetic induction intensity, the induced electric potential will change accordingly, and eventually the secondary current I_S will change. Therefore, the change of the primary winding current signal can be obtained indirectly through the change of the voltage across the sampling resistor R_S.

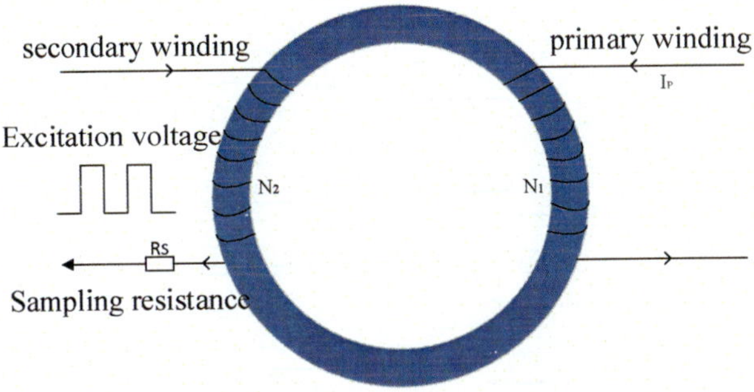

Fig. 1. Flux gate circuit with single core and double windings

When the ideal square wave signal is used as the excitation signal, and the sampling resistance R_s is much larger than the internal resistance of the coil, the loop equation of the excitation circuit is:

$$U(t) = I(t) * R_S + \mu(i) * N_2^2 * \frac{S}{l} * \frac{dI(t)}{dt} \tag{1}$$

Where l is the average magnetic path length of the toroidal core, N_1 is the number of turns of the primary winding N_2 is the number of turns of the secondary winding, $I(t)$ is the instantaneous current in the secondary loop, $\mu(i)$ is the dynamic permeability of the core, and S is the effective cutoff of the core area.

Figure 2(a) is the static magnetization curve of the magnetic core, and the magnetization curve is piecewise fitted, as shown in Fig. 2(b).

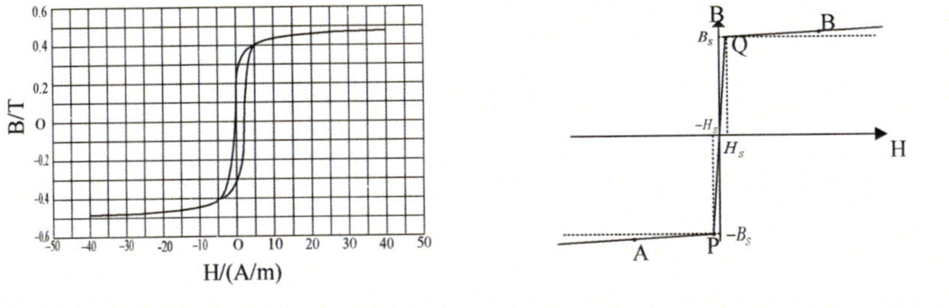

(a) Static magnetization curve of magnetic core (b) Magnetization curve after linear fitting

Fig. 2. Magnetization curve

When the primary winding does not apply the current signal to be measured, when $t = 0$, the square wave starts to be applied. At this time, the square wave is on the rising edge, and the magnetic core is in the negative saturation state. Under the action of a square wave voltage, the magnetic field of the magnetic core changes periodically along *A-P-Q-B-Q-P-A*, then the voltage equation of the segmented excitation loop can be listed as shown in Eq. 2.

$$\begin{cases} U_m = I(\mathrm{t}) * R_S + \mu_1 * N_2^2 * \frac{S}{1}, 0<t<t_1 \\ U_m = I(\mathrm{t}) * R_S + \mu_2 * N_2^2 * \frac{S}{1}, t_1<t<t_2 \\ U_m = I(\mathrm{t}) * R_S + \mu_3 * N_2^2 * \frac{S}{1}, t_2<t<\frac{T}{2} \end{cases} \Rightarrow \begin{cases} I = I_\infty\left(1 - 2e^{-\frac{t}{\tau_1}}\right), 0<t<t_1 \\ I = I_\infty + (-I_s - I_\infty)e^{-\frac{t-t_1}{\tau_2}}, t_1<t<t_2 \\ I = I_\infty + (I_s - I_\infty)e^{-\frac{t-t_2}{\tau_3}}, t_2<t<\frac{T}{2} \end{cases} \tag{2}$$

The current waveform of the excitation loop can be drawn by Eq. (2) as shown in Fig. 3, where I_S is the current in the secondary winding loop when the core is initially saturated.

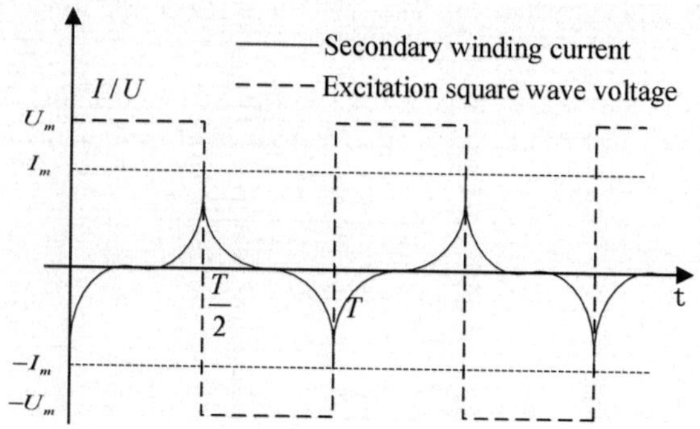

Fig. 3. Excitation loop current curve when $I_P = 0$

When the current to be measured is connected to the primary coil, the magnetic field strength in the magnetic core is:

$$H = \frac{N_1 I_P + N_2 I(t)}{l} = \frac{N_2[N_1 I(t) + \frac{I_P}{N_2}]}{l} \tag{3}$$

Let

$$N_1 I(t) + \frac{I_P}{N_2} = I'(t) \tag{4}$$

Then

$$I(t) = \frac{1}{N_1}[I^{'}(t) - \frac{I_P}{N_2}] \tag{5}$$

In the formula, $I''(t)$ is equivalent to the current curve of the excitation loop when the primary current is 0. Therefore, when the primary current $I_P \neq 0$, the secondary winding loop current curve $I(t) = I^{'}(t) - \frac{I_P N_1}{N_2}$. When the primary current change $\varDelta I_P$, the secondary winding current curve translates along the y axis by $\frac{I_P N_1}{N_2}$, and the voltage waveform of R_S translates along the y axis by $\frac{I_P N_1 R_S}{N_2}$. That is, when the primary current I_P is not 0, the voltage waveform of R_S shift ΔU_S at both ends of the sampling resistor is proportional to the current value $\varDelta I_P$ of the primary winding. As long as the magnitude of the voltage waveform change ΔU_S is measured and demodulated, an accurate measurement can be obtained indirectly current change value $\varDelta I_P$. The primary winding current value $\varDelta I_P$ and the voltage change ΔU_S across the sampling resistor can be expressed by a linear function.

Next use U to denote $\varDelta U_S$ and I to denote $\varDelta I_P$.

$$I = kU + b,\ k = \frac{N_2}{N_1 R_s} \tag{6}$$

Short-circuit the two ends of the primary winding, at this time $I_P = 0$, the voltage change across the sampling resistor is recorded as U_O. The value of b can be obtained:

$$b = -\frac{N_2}{N_1 R_s} U_0 \tag{7}$$

In summary, the amount of voltage change U_S across the sampling resistor and the current to be measured I_P of the primary winding can be expressed by the following formula:

$$I = \frac{N_2}{N_1 R_s} U - \frac{N_2}{N_1 R_s} U_0 \tag{8}$$

3 Design Scheme

Single-loop fluxgate current sensor is mainly composed of fluxgate probe, excitation signal and signal acquisition and processing. The secondary winding of the fluxgate probe is excited by the excitation signal, and the AD acquisition collects the voltage across the sampling resistor and transmits it to the PC for data processing and display. The overall structure is shown in Fig. 4.

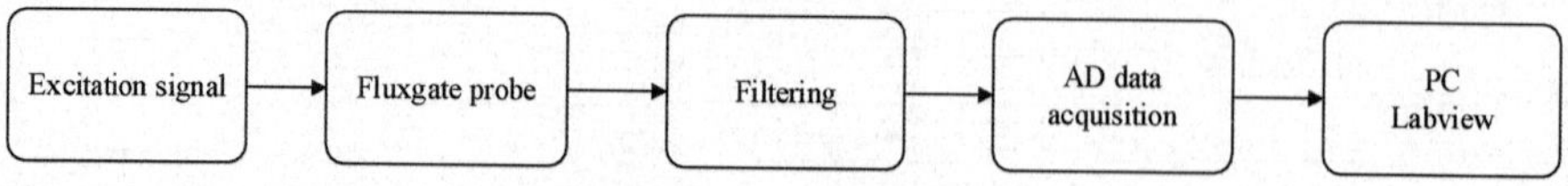

Fig. 4. The overall structure

3.1 Probe Design

In this paper, the magnetic core material of the fluxgate probe is selected from Permalloy *1J85*, which is representative of soft magnetic materials. Permalloy *1J85* has a coercive force between 875nT and 3000nT, and its magnetic permeability is two orders of magnitude higher than that of hard magnetic materials, and it is easily magnetized by a weak magnetic field to generate a large induced magnetic field.Part parameters of *1J85* are shown Table 1.

Table 1. Permalloy *1J85* some parameters

μ/(N/A)	μ_{max}/(N/A)	Hc/(nT)	B_S/(T)
(3–5.5) × 10^4	(1.1–2.6) × 10^5	875–3000	0.7

Increasing the number of turns of the fluxgate coil can improve the sensitivity of the sensor, but too much will increas the noise as well as the effective signal. Considering all aspects comprehensively, balancing various comprehensive performances, and repeatedly verifying through experiments, when the number of coil turns is too high, it will reduce the resolution of the probe, and finally determine the number of primary winding turns N_1 is 110 turns, secondary winding The number of turns N_2 is 1000 turns, and the sampling resistance is 1k ohms.The physical diagram of the fluxgate probe is shown in Fig. 5. The value vaveform shift value U_0 measured by shorting the primary winding is 31.0128mV.Substitute N_1, N_2 and R_S into Formula 8:

$$I = 0.0125U - 0.38736 \tag{9}$$

3.2 Excitation Signal

After the fluxgate probe parameters are determined, it is necessary to find the optimal excitation frequency of the fluxgate probe. Use high-precision RIGOL function signal generator and high-precision oscilloscope *TDS2048B* to find the best excitation frequency and amplitude. Figure 6 shows the voltage waveforms across the sampling resistor at excitation frequencies of 90 Hz, 110 Hz, and 130 Hz, respectively. (a) indicates that the core is oversaturated. The waveform of (b) is consistent with the theoretical waveform and meets the conditions. In (c), due to the high excitation frequency and large impedance, the waveform change is not obvious. After repeated verification, the excitation signal of the secondary winding selects a frequency of 110 Hz, and the voltage waveform at both ends of the sampling resistor meets the demand.

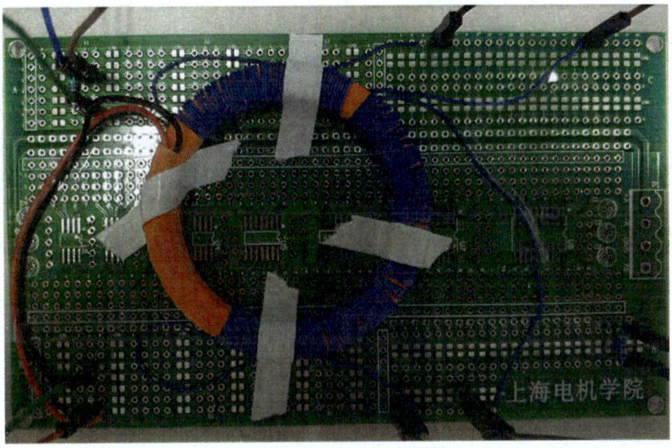

Fig. 5. Fluxgate probe

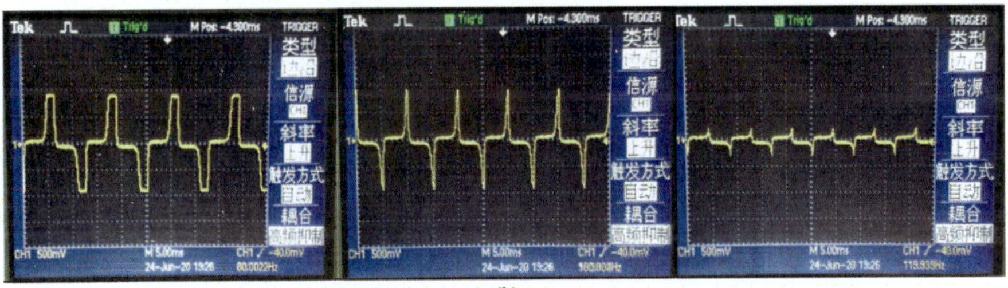

(a) (b) (c)

Fig. 6. Searching for the best excitation frequency

3.3 Voltage Acquisition and Data Processing

The voltage signal across the sampling resistor is sampled using the 16-bit AD data acquisition device *USB-DAQ-7606i*. The device supports 8 simultaneous bipolar analog input measurements, with a single channel sampling rate of up to 100Ksps. It can communicate with an industrial control computer via USB to meet experimental needs.

First of all, the voltage waveforms at both ends of the sampling resistor are collected through the AD data acquisition device, and the real-time data transmission is completed through the serial port and the industrial control computer, and the serial communication is completed through the standard I/O API VISA interface of Labview software. The positive peak value, negative peak value and amplitude difference are calculated by Labview software, and the variation of voltage waveform U is converted into the primary current value I by Eq. 8. The upper computer interface is written and drawn by the front panel of Labview. The configuration interface, voltage waveform, variation of voltage waveform and current value after conversion can be switched through the upper computer tab.

4 Experiment and Analysis

The square wave excitation signal has a voltage amplitude of 2.5V and a frequency of 110Hz. Use the fluxgate current sensor to measuring the depolarization current of the main insulation of a motor, and connect the Keithley high-precision electrometer *6517A* to measure the original current signal. The measurement data of the *6517A* electrometer is compared with the measurement data of the fluxgate current sensor as a standard current reference to measure the accuracy of the fluxgate current sensor.

In the experiment, the fluxgate current sensor and *6517A* electrometer were used to measure the depolarization current. According to the measurement data, Matlab software is used to draw a comparison curve of the measurement results, as shown in Fig. 7. It can be seen that the fluxgate current sensor has a good linearity in a certain range and has a higher accuracy which is enough to meet the depolarization current measurement work. The measurement data is shown in Table 2.

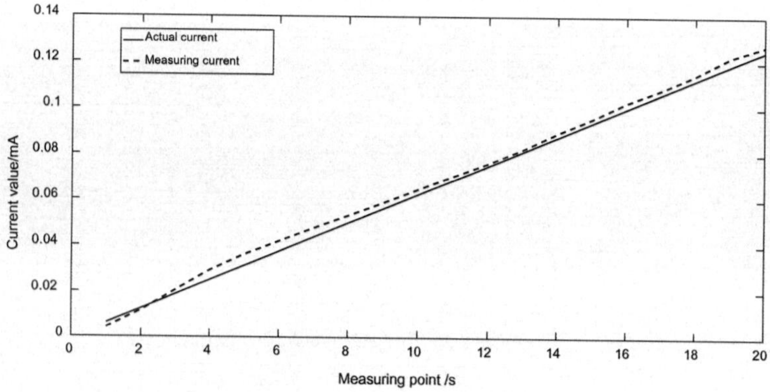

Fig.7. Measurement result comparison curve

Table 2. Measurement data

6517A electrometer/(mA)	Voltage waveform shift *U*/(mV)	Calculated current value/(mA)
0.00628	31.32720	0.00423
0.01245	31.92320	0.01168
0.01873	32.62640	0.02047
0.02495	33.32160	0.02916
0.03125	33.87920	0.03613
0.03745	34.34560	0.04196
0.04373	34.80080	0.04765
0.04995	35.24480	0.05320
0.05625	35.69920	0.05888
0.06253	36.16560	0.06471
0.06875	36.63200	0.07054
0.07495	37.08640	0.07622

(continued)

Table 2. (*continued*)

6517A electrometer/(mA)	Voltage waveform shift *U*/(mV)	Calculated current value/(mA)
0.08123	37.57520	0.08233
0.08745	38.14080	0.08940
0.09375	38.64000	0.09564
0.09995	39.20640	0.10272
0.10623	39.66160	0.10841
0.11250	40.17200	0.11479
0.11878	40.79280	0.12255
0.12498	41.18160	0.12741

5 Summary

This paper designs a single-loop fluxgate current sensor, and optimizes the circuit parameters through experimental optimization, and designs the first generation of experimental products. The device has a simple structure, flexible software design, powerful functions, low cost, and convenient debugging. Experiments show that the measurement accuracy, temperature characteristics and stability of the scheme have reached the expected design requirements, which can meet the measurement work of small range of weak current.

This solution does not consider the problems of geomagnetic infection and interference noise. In the follow-up study, the geomagnetic infection and noise suppression will be studied.

References

1. Yang, X., Wen, J., Chen, M., Gao, Z., Xi, L., Li, Y.: Analysis and design of a self-oscillating bidirectionally saturated fluxgate current sensor. Measurement **157**,107687 (2020).
2. Grim, V., Ripka, P., Bauer, J.: DC current sensor using switching-mode excited in-situ current transformer. J. Magnetism Magnetic Mater. **500** (2020)
3. Schoinas, S., El Guamra, A.-M., Moreillon, F., Passeraub, P.: Fabrication and Characterization of a Flexible Fluxgate Sensor with Pad-Printed Solenoid Coils. Sensors **20**(8), 2275 (2020)
4. Ando, B., Baglio, S., Crispino, R., Graziani, S., Marletta, V., Mazzaglia, A., Sinatra, V., Mascali, D., Torrisi, G.: A fluxgate-based approach for ion beam current measurement in ECRIS beamline: design and preliminary investigations. IEEE Trans. Instrum. Meas. **68**(5), 1477–1484 (2019)
5. Cao, J., Zhao, J., Cheng, S.: Research on the simplified direct-current fluxgate sensor and its demodulation. Meas. Sci. Technol**30**(7), 075101 (2019)
6. Liu, Y., Lin, Y., Lan, Q., Wang, D.F., Itoh, T., Maeda, R.: A high accuracy fluxgate DC current sensor applicable to two-wire electric appliances. Microsyst. Technol. Micro-Nanosyst. sInform. Storage Process. Syst. **25**(3), 877–885 (2019)

7. Weiss, R., Itzke, A., Reitenspiess, J., Hoffmann, I., Weigel, R.: A novel closed loop current sensor based on a circular array of magnetic field sensors. IEEE Sens. J. **19**(7), 2517–2524 (2019)
8. Tan, X., Chen, S., Yan, X., Fan, Y., Min, H., Wang, J.: A highly sensitive wide-range weak current detection circuit for implantable glucose monitoring, IEICE Electron. Express **13**(8), 20150616 (2016)
9. Pan, D., Li, J., Jin, C., Liu, T., Lin, S., Li, L.: A new calibration method for triaxial fluxgate magnetometer based on magnetic shielding room. IEEE Trans. Industr. Electron. **67**(5), 4183–4192 (2020)
10. Schoinas, S., Guamra, A.-M.E., Moreillon, F., Passeraub, P.: Fabrication and characterization of a flexible fluxgate sensor with pad-printed solenoid coils. Sensors (Switzerland) **20**(8), 2275 (2020)

ADRC-Based Wind Turbine Pitch Control Strategy

Jiabao He(✉) and Jianguo Li

Shanghai DianJi University, Shanghai, China
894265348@qq.com

Abstract. In order to make the output power of the wind turbine more stable, a pitch control strategy based on auto disturbance rejection controller is proposed. When the wind speed received by the wind turbine is higher than the rated wind speed of the unit, By adjusting the pitch angle, the aerodynamic torque of the fan is changed, thereby stabilizing the output power of the fan. The response speed of the auto disturbance rejection structure is very fast, the robustness is also strong, and it has strong resistance to external interference. The variable pitch ADRC controller is designed through theoretical analysis, and then compared with the traditional PID control for simulation and experimental analysis. The results show that under random wind conditions, the designed variable pitch auto disturbance rejection controller can better achieve the smooth power output of the wind turbine, which can effectively suppress the speed fluctuation under extreme operating conditions and avoid the overspeed shutdown of the fan. It improves the anti-interference ability and robustness of the system and guarantees the safety of the wind turbine. It is more suitable for the pitch control of large wind turbines than traditional PID control.

Keywords: Variable pitch · Auto disturbance rejection · PID · Stability · Wind turbine

1 Introduction

Wind energy is a kind of clean and renewable energy, which has become one of the key energy resources developed globally. With the rapid development of high magnetic energy permanent magnet technology, direct drive generators have been widely used. Direct-drive permanent magnet wind turbines have gradually become a hotspot for the development of the whole machine due to their advantages such as no speed-increasing gearbox, simple structure, high efficiency and low cost [1]. When the wind turbine is subjected to different wind speed characteristics, its output power also continuously shakes. When the wind turbines are connected to the grid, the jitter of the wind turbines's output power poses a major challenge to the integration, operation and control of the power system, especially when a wind farm injects into the grid with a lot of power, which may cause frequency deviation in the grid. Therefore, it is increasingly expected to accurately change the blade pitch angle to effectively control the wind turbine's output power [2]. The wind turbine is a nonlinear system with large inertia and strong coupling. The fluctuation of wind speed will cause the wind turbine to change the power and cannot be stabilized at the rated power.

M. Tavana et al. (Eds.): IISA 2020, AISC 1304, pp. 529–536, 2021.
https://doi.org/10.1007/978-3-030-63784-2_66

At present, there are still many deficiencies in the control algorithm of wind turbines. If conventional PI control is used, when disturbances such as wind speed changes occur, the performance of the controller will deteriorate or become unstable. The genetic algorithm PID controller used in literature [3] has many parameters, and there is no reasonable parameter adjustment scheme. It can only be determined by simulation and has not been widely promoted. Literature [4] designed a non-linear controller based on RBF neural network sliding mode variable structure. Although it can track instructions quickly, the algorithm is complicated. Literature [5] proposed a variable-gain-based wind turbine pitch control, which is not ideal for resisting the interference of external factors, but can better track the pitch angle through variable gain. Aiming at the problem of unsatisfactory control effect of traditional PID control when the disturbance is too large, control methods such as double fuzzy adaptive PID, neural network, chaos optimization PID, and H∞ have been introduced one after another, but they all have their own limitations [6–9].

In summary, in view of the above analysis, based on the theory of auto disturbance rejection control and the characteristics of fan operation, this paper designed a variable pitch auto disturbance rejection controller, which can better achieve the smooth power output of wind turbines Suppress the speed fluctuation under extreme working conditions, avoid the overspeed shutdown of the fan, improve the anti-interference ability and robustness of the system, and ensure the safety of the wind turbine [10].

2 Model Building of Wind Turbine

2.1 Wind Wheel Modeling

When the wind turbine works under different wind speeds, it is easy to find the power it gets in the wind according to the aerodynamic characteristics of the wind turbine:

$$P_M = \frac{1}{2} C_p(\lambda, \beta) \rho A V^3 \tag{1}$$

$$\lambda = \frac{r\omega}{V} \tag{2}$$

$$C_p(\lambda, \beta) = 0.22 \times \left(\frac{116}{\lambda_1} - 0.4\beta - 5 \right) \times e^{\frac{-12.5}{\lambda_1}} \tag{3}$$

$$\frac{1}{\lambda_1} = \frac{1}{\lambda + 0.08\beta} - \frac{0.035}{\beta^3 + 1} \tag{4}$$

Where: ω , r is the speed and radius of the blade, A is the area swept by the blade, β is the pitch angle to be adjusted, λ and V are the blade tip speed ratio and the wind speed received by the wind wheel, ρ is the air density of the system, C_p is the wind energy utilization rate of the wind turbine.

2.2 Modeling of the Transmission System

In order to simplify the model analysis, direct-drive wind turbines have rigid connections without a gearbox. The equation of the permanent magnet wind turbine system is:

$$J\frac{d\Omega}{dt} = T_m - T_e \tag{5}$$

In the test: J is the moment of inertia, kg·m^2; the wind turbine's speed is Ω; T_m is the wind wheel's torque; and the electromagnetic rotation of the generator is T_{e}.

2.3 Pitch Angle Actuator

The pitch actuators of large wind turbines are generally divided into two types: hydraulic pitch and electric pitch. The pitch actuator in this article is a section of inertia link as follows:

$$\theta = \frac{1}{1+\tau s}\theta_r \tag{6}$$

Where θ is the Current angle, θ_r is Fixed pitch angle, and τ is the time constant.

3 Structure and Composition of ADRC

The components of the auto disturbance rejection controller are a differential tracker (TD), a nonlinear feedback (NLSEF) and an extended state observer (ESO). Among them: The main function of TD is to arrange the transition process in real time and extract the given information in the system; ESO will expand "total disturbance" affecting the object into a new state variable, real-time monitoring and compensation; NLSEF will error, error differential and error integration of three signals to The control law formed by a combination of forms. Its structural block diagram is shown in Fig. 1.

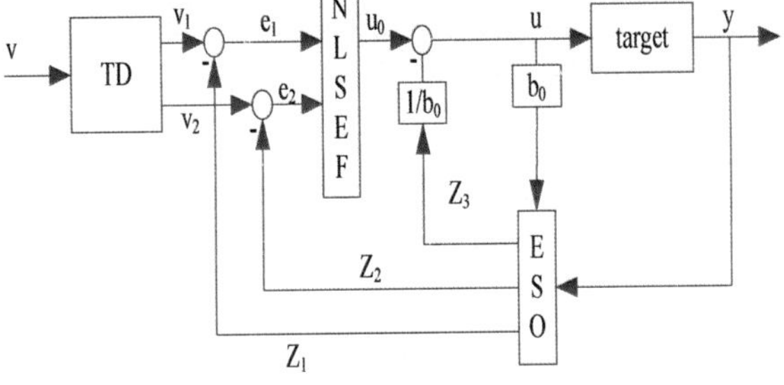

Fig. 1. Block diagram of an auto disturbance rejection controller

3.1 The Fastest Tracking Differentiator TD

Its discrete expression is:

$$\begin{cases} fh = fh(x_1(m) - v(t), x_2(m), r, h_0 \\ x_1(m+1) = x_1(m) + hx_2(m) \\ x_2(m+1) = x_2(m) + hfh \end{cases} \tag{7}$$

Among the parameters: V(t) is the target value h, h0 is the integration step, generally h can be equal to h0, but in order to reduce overshoot and reduce vibration, they are separated, generally h0 is larger than h, such as 20 times larger. When h0 is large, it can significantly reduce the oscillation, so it is also called the filter factor. Reducing h can suppress noise amplification. r is the factor of speed, The approach speed increases as it increases, but it is best to depend on the actual ability of the controlled object to bear. In the expression: the first expression of fhan function is:

$$u = fhan(x_1, x_2, r, h) \tag{8}$$

$$\begin{cases} d = rh \\ d_0 = hd \\ y = x_1 + hx_2 \\ a_0 = \sqrt{d^2 + 8r|y|} \\ a = \begin{cases} x_2 + \frac{(a_0 - d)}{2} \mathrm{sign}(y), |y| > d_0 \\ x_2 + \frac{y}{h}, \quad |y| \le d_0 \end{cases} \\ fhan = -\begin{cases} rsign(a), |a| > d \\ r\frac{a}{d}, \quad |a| \le d \end{cases} \end{cases} \tag{9}$$

The function of the fastest tracking differentiator is to play a buffer role, such as inputting a step signal, which will make the signal have a certain slope, so that the actuator will not enter the maximum horsepower as soon as it is started, resulting in the first and subsequent approaches Overshoot occurred when setting the value.

3.2 Extended State Observer

$$\begin{cases} e = z_1 - y, \mathrm{Ge} = \mathrm{fal}(e, 0.5, \delta), \mathrm{Ge}_1 = \mathrm{fal}(e, 0.25, \delta) \\ z_1 = z_1 + h(z_2 - \beta_{01} e) \\ z_2 = z_2 + h(z_3 - \beta_{02}\mathrm{Ge} + b_0 u) \\ z_3 = z_3 + h(-\beta_{03}\mathrm{Ge}_1) \end{cases} \tag{10}$$

The expression of the function fal is:

$$\mathrm{fal}(e, \alpha, \delta) = \frac{e}{\delta^{1-\alpha}} s + |e|^{\alpha} sign(e)(1 - s) \tag{11}$$

3.3 State Error Feedback Law

Three nonlinear feedback laws are recommended in the book:

$$\begin{cases} (1)u = \beta_0 \mathrm{fal}(e_0, \alpha_0, \delta) + \beta_1 \mathrm{fal}(e_1, \alpha_1, \delta) + \beta_2 \mathrm{fal}(e_2, \alpha_2, \delta) \\ \alpha_0 < 0 < \alpha_1 < 1 < \alpha_2 \; or \; 0 < \alpha_0 < \alpha_1 < 1 < \alpha_2 \\ (2)u = \beta_0 e_0 - \mathrm{fhan}(e_1, e_2, r_1, h_1) \\ (3)u = -\mathrm{fhan}(e_1, ce_2, r_2, h_2) \\ u = \frac{u_0 - z_3(t)}{b0} \end{cases} \tag{12}$$

4 Design of Variable Pitch ADRC

To design the ADRC wind turbine pitch, a balance point A is now randomly selected from the constant power operating state of the wind power generation system, and their corresponding parameters are respectively T_{r0}, ω_0, θ_0, and A is used as a reference point to perform Taylor series expansion of the wind turbine torque [11]:

$$T_r - T_{r0} = \frac{\partial T_r}{\partial \omega} \Delta\omega + \frac{\partial T_r}{\partial \theta} \Delta\theta + \frac{\partial T_r}{\partial v} \Delta v + h \tag{13}$$

Among them: h is to expand the high-order items. Make $\alpha = \frac{\partial T_r}{\partial \omega}$ $\beta = \frac{\partial T_r}{\partial \theta}$ $\gamma = \frac{\partial T_r}{\partial v}$.

From formula (5) and formula (13):

$$J\Delta\dot{\omega} = \alpha\Delta\omega + \beta\Delta\theta + \gamma\Delta v + h \tag{14}$$

Taking into account the dynamic characteristics of the pitch actuator, yes:

$$\theta - \theta_0 = \frac{1}{\tau s + 1} \theta_r - \theta_0 = \frac{1}{\tau s + 1} \Delta\theta_r \tag{15}$$

Therefore, Eq. (14) can be transformed into:

$$\Delta\ddot{\omega} = \frac{\tau\alpha - J}{J\tau} \Delta\dot{\omega} + \frac{\alpha}{J\tau} \Delta\omega + \frac{\beta}{J\tau} \Delta\theta + \frac{\tau s + 1}{\tau} (\gamma\Delta v + h) \tag{16}$$

In the formula: $\Delta\theta$——Pitch angle offset given value.

The coefficient b corresponding to $\Delta\theta$ in Eq. (16) will continuously change with the working condition of the wind turbine pitch system, and it is difficult to obtain its definite value. In view of this, the estimated value b0 is taken, and the error is attributed to the system disturbance. b0 fetch:

$$b_0 = \frac{1}{J\tau} \frac{\partial T_{aero}}{\partial \beta} |A \tag{17}$$

The disturbance component d of the system is expressed as:

$$d = \frac{\tau s + 1}{J\tau}(\gamma\Delta v_w + h) + (b - bo)\Delta\beta_r \tag{18}$$

Equation (18) becomes:

$$\Delta\ddot{\omega} = \frac{\tau\alpha - J}{J\tau}\Delta\dot{\omega} + \frac{\alpha}{J\tau}\Delta\omega + b_o\Delta\theta + d \tag{19}$$

Therefore, the ADRC based on this design can perform better wind turbine pitch control. Use Δw as the measurement input to create the corresponding expanded state observer. The purpose of ADRC control is to improve the pitch control effect and suppress the fluctuation of wind speed, so the given input Δwref and Δw can be ignored. The state error is calculated through the feedback of the rotation speed, so as to obtain the control quantity of the nonlinear control law. The block diagram of ADRC variable pitch is as follows (Fig. 2).

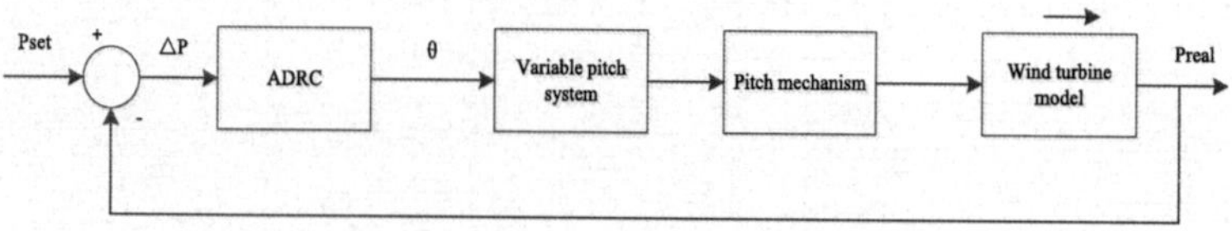

Fig. 2. Variable pitch structure diagram

5 Simulation Experiment and Result Analysis

A model of 2 MW permanent magnet synchronous generator is used in the paper, and used Matlab/simulink to build a wind turbine model based on PID control and ADRC control. The simulation is performed under random wind speed to analyze the simulation results.

Due to the large number of ADRC parameters, most of the parameter tuning methods are manually based on experience, and the Table 1 showes the results.

Table 1. Characteristic parameters of auto disturbance rejection controller

r	h	b_0	β_{01}	β_{02}	β_{03}	δ	β_1	β_2	a
5	0.001	1	90	1500	2500	0.5	−12	−7	0.5

The traditional PID and the ADRC controller designed's simulation results in this paper are shown in the Figs. 3 and 4.

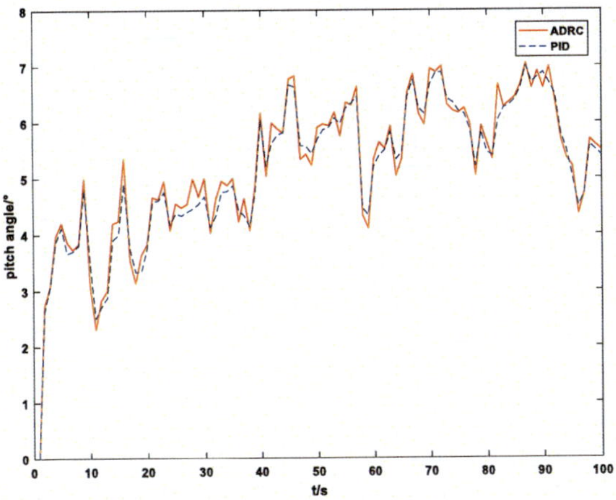

Fig. 3. Pitch angle response of PID pitch and auto disturbance rejection pitch

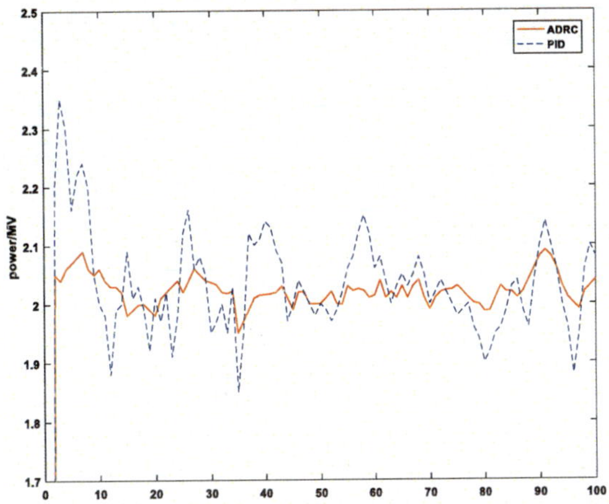

Fig. 4. Power response of PID pitch and auto disturbance rejection pitch

Figure 3 is the pitch angle curve of ADRC variable pitch control. The output power curves of the two control strategies shown in Fig. 4. Obviously, whether it is a pitch angle curve or a power control curve, ADRC is superior to traditional PID controllers in terms of control speed, accuracy, accuracy and anti-interference ability.

6 Conclusion

The biggest use of wind turbine pitch control is to stabilize the power of the wind turbine. Traditional PID or PI controllers are insufficient for nonlinear wind turbines in terms of anti-interference, robustness, and time lag, and it is difficult to obtain satisfactory control results. The article proposes an auto-interference-based The analysis of the control strategy through simulation results shows that the ADRC control proposed in this paper can guarantee the stability of the unit power more effectively than the traditional PID control. Through this control strategy, the wind turbine can be operated more safely and reliably, and the service life of the wind turbine is greatly extended.

References

1. Cao, M., Yu, M., Li, J., Wang, C., Xu, Y.: Auto disturbance rejection control of permanent magnet synchronous wind turbines. Electr. Eng. **08**, 1–7 (2017)
2. Han, J.: Auto disturbance rejection control technology. Front. Sci., 24–31 (2007)
3. Zhang, Z., Wang, P., An, B., Deng, X.: Pitch control of wind turbines based on genetic algorithm PID. Power Electron. **51**(07), 37–39 +85 (2017)
4. Jie, L., Dezhou, M.: Research on the control of wind turbine pitch system based on neural sliding mode control. J. Inner Mongolia Univ. Sci. Technol. **33**(01), 63–65 (2014)
5. Xunan, Q.: Simulation Research on Adaptive Controller of Direct Drive Wind Power Generation System. North China University of Water Resources and Hydropower, Zhengzhou (2018)
6. Geng, H., Yang, G.: Linear and nonlinear schemes applied to pitch control of wind turbines. Sci. World J. **2014**, 406382 (2014)
7. Meng, T., Bowen, Z., Lawu, Z., Hongzhi, Y., Yan, L.: Research on sliding mode variable structure independent pitch control based on RBF neural network. Power Syst. Protection Control **47**(04), 107–114 (2019)
8. Feifei, L.: Research on pitch system of wind turbine generator based on double fuzzy PID control strategy. Fuzzy Systems and Mathematics **33**(03), 29–34 (2019)
9. Keqilao, M., Jianguang, Ma., Dajiang, J.: Research on PID wind turbine pitch system optimization based on chaos. Renewable Energy **32**(03), 287–290 (2014)
10. Dazhou, Z., Jie, L.: Design and analysis of variable pitch auto disturbance rejection controller for direct drive wind Turbines. Renewable Energy **37**(12), 1875–1881 (2019)
11. Huangtian, T., Yuan, X., Lingli, S., Hao, L.: Application of teaching and learning optimization algorithm in wind turbine pitch control of auto disturbance rejection. Appl. Motor Control **46**(02), 120–125 (2019)

Research on Terminal Overvoltage Protection of Direct Drive Permanent Magnet Wind Turbine

Linzhao Hao[1](✉), Xia Liu[1], Chuan Jiang[2], Qinghua Zheng[1], and Wen Jing Li[1]

[1] Guangdong University of Science and Technology, Dongguan 523083, China
897464969@qq.com
[2] Jinhua Power Supply Company of State Grid Zhejiang Electric Power Co., Ltd., Hangzhou 321013, China

Abstract. With the rapid development of society and economy, energy crisis and environmental pollution become more and more obvious, wind power as a clean and renewable energy has been rapid development. Offshore wind power generation systems mainly use direct-drive permanent magnet wind power generation systems, which have a simple structure and high reliability, which have attracted the attention of the country and wind power companies. This paper proposes a new type of low-power RLC filter circuit for the PWM high-frequency pulse wave on the transmission cable between the converter and the generator, and designs the parameters of its resistance R, capacitance C and inductance L. A new type of low-power RLC filter circuit is proposed for the PWM high-frequency pulse wave on the transmission cable between the converter and the generator, and the parameters of its resistance R, capacitance C and inductance L are designed. Finally, based on the uniform transmission line Transmission theory, built on Matlab/simulink platform without filter circuit and RC, RLC, new RLC filter circuit simulation model. The simulation results show that the new R-L-C filter circuit effectively suppresses the generator terminal overvoltage caused by long cable reflection, and its power loss is also the lowest.

Keywords: PWM converter · PMSG · RC filter circuit · R-L-C filter circuit

1 Introduction

In recent years, wind power has attracted more and more attention from the country and wind power companies, and offshore wind power has also been extensively developed. My country's offshore wind power is mainly concentrated in coastal areas, such as Jiangsu and Zhejiang, and the East China Sea. It mainly uses a direct-drive permanent magnet wind power generation system, and the generator and the converter cannot be installed at the same location. They need to be connected by a long cable of tens of meters or hundreds of meters. When the cable between the converter and the generator is long, the PWM modulated rectangular pulse wave of the converter will be reflected at the generator end, and the reflected wave will be refracted to the converter end. If it is

M. Tavana et al. (Eds.): IISA 2020, AISC 1304, pp. 537–544, 2021.
https://doi.org/10.1007/978-3-030-63784-2_67

not suppressed in time, voltage oscillation will occur, and its peak voltage will reach twice the DC voltage of the converter. The IGBT voltage change rate dv/dt inside the converter is very large, which is likely to cause overvoltage at the converter and generator terminals. If not treated in time, it will cause winding heating, insulation aging, winding breakdown, and converter Problems such as IGBT breakdown.

There are three main methods to solve this problem: 1. Increase the insulation performance of the generator winding; 2. Design a filter circuit between the converter and the generator; 3. Use an active circuit to match the impedance of the cable line to achieve Suppress generator terminal overvoltage. The main factors for generating overvoltage at the generator end are the following two factors: 1. The breaking time of the IGBT; 2. The long cable length between the converter and the generator. In order to solve this problem, a new type of RLC filter circuit is designed between the converter and the generator to suppress the overvoltage at the generator end. The comparison simulation results show that the new R-L-C filter circuit effectively suppresses the overvoltage at the generator end and ensures This ensures the safe and reliable operation of generators and converters, and also ensures the stability of the system.

2 Analysis on the Mechanism of Overvoltage on Generator Side

The PWM pulse wave from the converter is propagated through the long cable between the converter and the generator. When the transmission cable is long, the impedance mismatch of the long power cable will occur, that is, the characteristic impedance of the generator is much greater than the impedance of the cable. According to the transmission theory of PWM high-frequency pulse waves, when the incident wave is transmitted to an uneven place, reflection and transmission will occur, and voltage superimposition will occur at the converter and generator terminals, resulting in overvoltage. The schematic diagram of voltage reflection at the impedance mismatch of transmission cable is shown in Fig. 1.

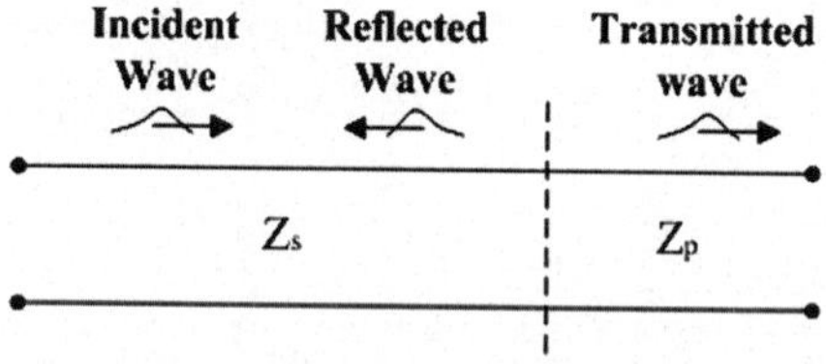

Fig. 1. Voltage reflection at uneven transmission line point

The peak voltage at the end of the generator mainly depends on the reflection coefficient δ, which is the ratio of the reflected wave voltage to the incident wave voltage on the transmission line. The reflection coefficient δ can be divided into the reflection coefficient δ_M on the generator side and the reflection coefficient δ_P on the voltage input side of the converter. The calculation of its reflection coefficient is shown in formula (1).

$$\begin{cases} \delta_{M} = \frac{Z_{M} - Z_{B}}{Z_{M} + Z_{B}} \\ \delta_{P} = \frac{Z_{n} - Z_{B}}{Z_{n} + Z_{B}} \end{cases} \quad (1)$$

Among them, Z_M is the generator characteristic impedance, Z_B is the characteristic impedance of the transmission cable and Z_n is the internal characteristic impedance of the converter.

$$Z_{B} = \sqrt{\frac{Z_{l}}{Y_{l}}} = \sqrt{\frac{R_{0} + jwL_{0}}{G_{0} + jwC_{0}}} \quad (2)$$

Among them, R_0 is the resistance of the transmission line, L_0 is the inductance of the transmission line,G_0 is the conductance of the transmission line, and C_0 is the capacitance of the transmission line. When the electric energy is transmitted without distortion and short-distance transmission, R_0, G_0 can be ignored. At this time, the voltage at the generator terminal is:

$$U_{M} = U_{line}(1 + \delta_{M} + \delta_{P}) \quad (3)$$

Among them, U_M is the generator terminal voltage, and U_{line} is the voltage of the transmission line.

3 Filter Circuit Design

(1) Topology analysis of filter circuit

Currently, there are mainly two filter circuits (RC and RLC) in use. The first-order RC filter at the generator end is shown in Fig. 2.

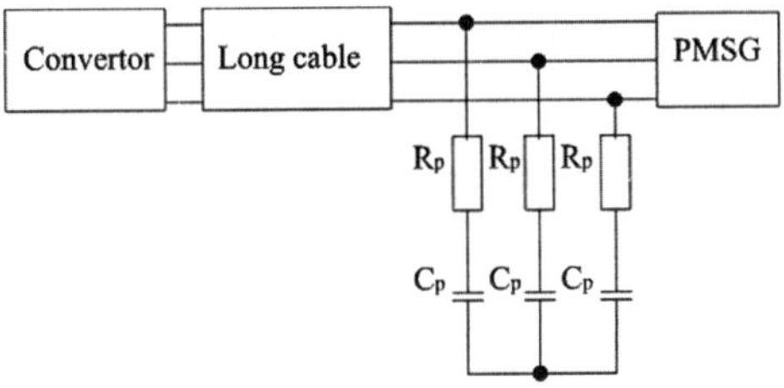

Fig. 2. First order RC filter circuit at generator end

In order to make the PWM pulse voltage wave output by the converter be absorbed by the RC filter circuit, the resistor R_P and the capacitor C_P must be selected with appropriate values to ensure that the R_P value matches the characteristic impedance of the cable. In actual engineering, the larger the value of the capacitor C_P, the better the overvoltage suppression effect at the converter and generator terminals.

The RLC filter circuit is installed at the converter end, and its structure is shown in Fig. 3. The RLC filter circuit of the generator-side converter can reduce the voltage change rate, increase the rise time of the PWM pulse and filter the PWM signal so that the output voltage waveform is similar to a sine wave. Therefore, the inductance L_P and the capacitance C_P in the RLC filter circuit are larger.

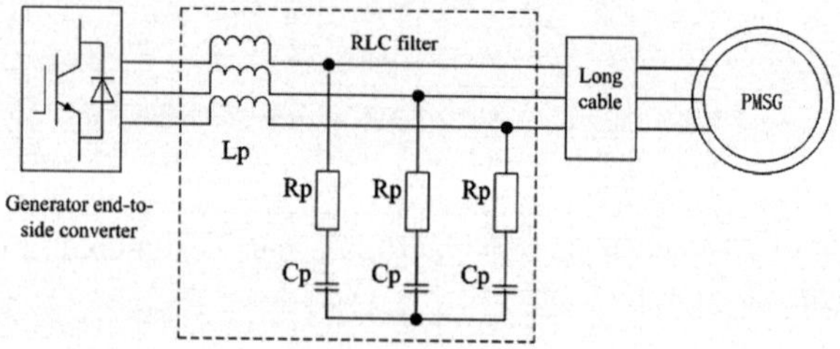

Fig. 3. Structure diagram of RLC filter circuit of generator-side converter

For these two filter circuits, the generator power is 10hp, the phase voltage is 230 V, the cable length is 500 m, the PWM rise time is 250 ns, and the filter parameters are selected as shown in Table 1:

Table 1. Parameter settings for two kinds of filters

Filter circuit type	R(Ω)	L(μH)	C(μF)
Generator side RC filter circuit	75	–	0.05
RLC filter circuit on converter side	65	20	0.01

Based on the analysis of the existing over-voltage filter circuit RC and RLC, this paper proposes a new low-power R-L-C filter circuit, in which R_P and L_P are in parallel, and then connected with the converter in series, while the capacitor C_P in the R-L-C filter circuit is connected in parallel with the generator. The filter circuit can effectively suppress the over-voltage of converter and generator, so as to protect the converter and generator, and ensure the safe and reliable operation of wind power generation system. The structure diagram of the new single-phase R-L-C filter circuit is shown in Fig. 4.

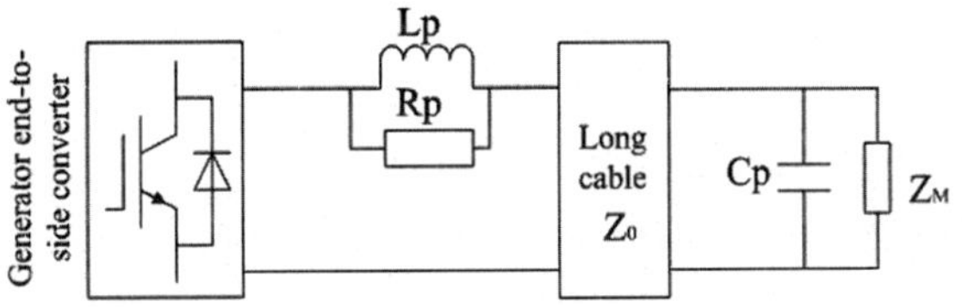

Fig. 4. Single-phase new R-L-C filter circuit structure diagram

(2) New R-L-C filter circuit parameter design
1) Design of resistance R_P parameter of new filter circuit

From Fig. 4, the input impedance Zin of the RL network can be obtained as:

$$Z_{in} = \frac{sL_P R_P}{sL_P + R_P} \tag{4}$$

The equivalent impedance of the new R-L-C filter is shown in (5) and (6):

$$Z_M(s)|_{s=jw} = \begin{cases} Z_B\angle 0^0, & w\rangle w_c \\ \infty, & w\langle w_c \end{cases} \tag{5}$$

$$Z_{in}(s)|_{s=jw} = \begin{cases} Z_B\angle 0^0, & w\rangle w_c \\ 0, & w\langle w_c \end{cases} \tag{6}$$

Among them, Z_M is the equivalent impedance at the generator end and Z_B is the impedance of the transmission cable.

According to formulas (4), (5) and (6), the filter circuit is close to the ideal state when $R_P = Z_B$.

2) Design of Capacitor C_p Parameter of New Filter Circuit

When the system is operating normally, ignoring the power transmission delay time of the transmission cable, the relationship between the voltage U_M at the generator end and the voltage U_{in} at the converter end is shown in Eq. (7).

$$U_M(s) = \frac{Z_M}{sZ_B Z_M C_P + Z_B + Z_M} U_{in}(s) \tag{7}$$

By Laplace transformation of Eq. (7), and substituting (8) into Eq. (7), we can get formula (9).

$$U_{in}(s) = U_{dc}/s \tag{8}$$

$$U_M(t) = \frac{U_{dc} Z_M}{Z_B + Z_M}\left(1 - e^{-\frac{Z_B + Z_M}{Z_B Z_M C_P}t}\right) \tag{9}$$

$$U_M(t_r) = 0.9U_{dc} \tag{10}$$

Substituting Eq. (10) into Eq. (9) can get:

$$C_P = \frac{(Z_B + Z_M)t_r}{Z_B Z_M \ln(0.1 - 0.9(Z_B/Z_M))} \tag{11}$$

3) Design of the Lp parameter of the inductance of the new filter circuit

In conclusion, when $R_P = 70\ \Omega$, $C_P = 4NF$, $Z_M = 3\ K\Omega$, $L_P = 0.23$ mH can be obtained by calculation.

Under the same voltage overshoot, the power consumption of the R-L-C filter circuit is much smaller than that of the RC and RLC filter circuits. The loss P1 of the resistance Rp at high frequency is shown in formula (12):

$$P_1 = 2f_s \int_0^{\frac{1}{2f_s}} \frac{U_{RL}^2}{R_P} dt \tag{12}$$

Among them, fs is the switching frequency of the converter, and URL is the RL network terminal voltage. P2 is the loss of resistance Rp at low frequencies, but because the low-frequency inductance of the inductance is very small, P2 can be ignored, so the loss of the new R-L-C filter circuit is equal to P1.

The parameters of the new R-L-C filter circuit are shown in Table 2:

Table 2. Parameter setting of RL-C filter

Filter circuit parameters	Parameter value
R_p	70 Ω
C_p	4 nF
L_p	3 mH

4 Simulation Analysis

Through the MATLAB/Simulink platform, the following are built: (a) no filter circuit simulation model; (b) RC filter circuit simulation model; (c) RLC filter circuit simulation model; (d) new R-L-C filter circuit simulation model. The simulation parameters are selected as follows: $U_{dc} = 600$ V, $Z_M = 2\ K\Omega$, impedance of converter $Z_n = 0.5\ \Omega$, switching frequency of converter is 10 kHz, length of transmission line is 500 m, distribution parameters of transmission line are $R_0 = 0.02\ \Omega$/km, $L_0 = 0.51\ \mu$H/m, $C_0 = 86$ pf/m. The simulation results are shown in Fig. 5.

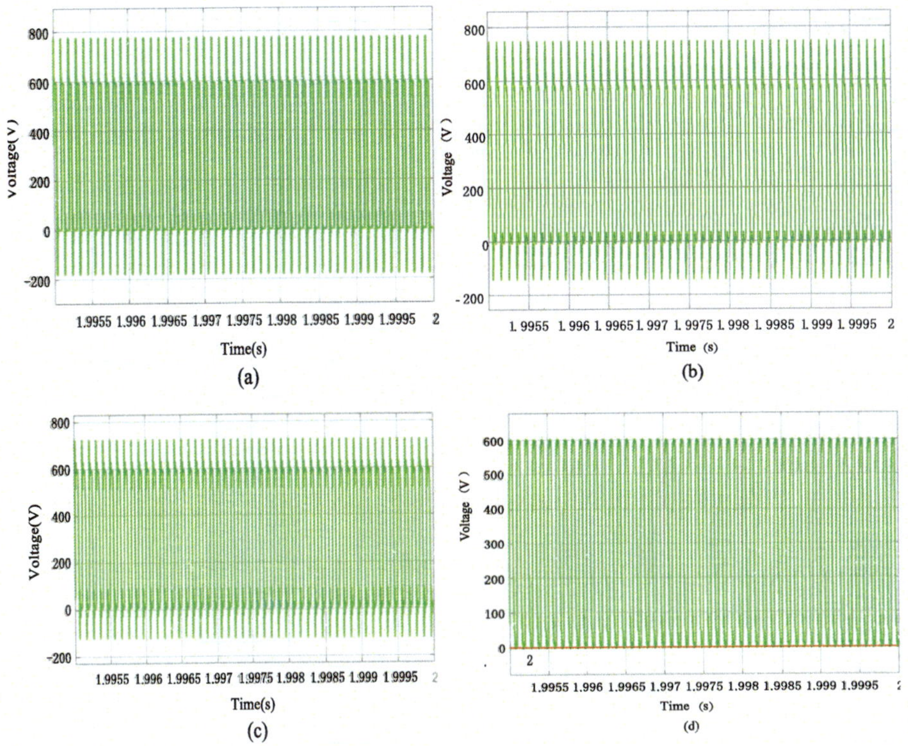

Fig. 5. Generator terminal voltage of four simulation models

Under the same other conditions, the four simulation models (no filter circuit, RC filter circuit, RLC filter circuit, new R-L-C three filter circuits) generator terminal voltage simulation results are shown in Fig. 5. In the first three cases, the voltage value of the generator terminal exceeds 700 V, and the new R-L-C filter circuit proposed in this article can completely suppress the overvoltage caused by the long cable, ensuring the safe and reliable operation of the system.

5 Conclusion

In this paper, the problem of over-voltage at generator terminal caused by reflection of high-frequency pulse wave caused by long cable in offshore direct drive permanent magnet wind power generation system is deeply studied. The generation mechanism of long cable overvoltage between generator and converter is analyzed. The existing RC filter circuit at generator side and RLC filter circuit at converter side are studied. Finally, a new type of R-L-C filter circuit is proposed, and its parameters and structure are analyzed in depth. Through simulation and comparative analysis, the new R-L-C filter circuit effectively suppresses the over-voltage, thus ensuring the safe and reliable operation of the system.

References

1. Li, G., Liang, J., Ugalde-Loo, C.E., et al.: Protection for submodule overvoltage caused by converter valve-side single-phase-to-ground faults in FB-MMC based bipolar HVDC systems. IEEE Trans. Power Delivery 1–6 (2020)
2. Liu, W., Li, G., Liang, J., et al.: Protection of single-phase fault at the transformer valve side of FB-MMC-based bipolar HVdc systems. IEEE Trans. Industr. Electron. **67**(10), 8416–8427 (2020)
3. Hao, L., Shan, H.E., Wang, W., et al.: Research on IGBT overvoltage protection circuit of direct drive wind turbine converter based on CR-CD circuit. Power Syst. Prot. Control **47**(09), 150–157 (2019)
4. Qian, J., Hui, W.W., Gao, Y., Zhang, C., Sun, Q.: Experimental design and research of RC filter circuit. Univ. Phys. Exp. **30**(05), 58–62 (2017)
5. Yuen, K., Chung, H.: Use of synchronous modulation to recover energy gained from matching long cable in inverter-fed motor drives. IEEE Trans. Power Electron. **29**(2), 883–893 (2014)
6. Liu, Y., et al.: Overvoltage mitigation of submersible motors with long cables of different lengths. In: International Conference on Electrical Machines and Systems, pp. 638–644. IEEE (2014)
7. Chunfei, F.: Research on the over-voltage of the motor port caused by long leads transmission. World Inverters **9**, 104–107 (2013)

Energy Systems

Analysis of Vehicle Energy Storage Brake Energy Recovery System

Zhiqiang Xu(✉)

Guangdong University of Science and Technology, Dongguan 523083, China
450521404@qq.com

Abstract. With the rapid development of China's automobile manufacturing industry, people's travel needs in life and work have been greatly satisfied. But what followed was a series of ecological problems. In order to improve China's ecological environment, vehicle electric energy storage braking energy recovery technology has become one of the key research objects in the automotive field. At present, many automobile companies have established a vehicle electric energy storage braking energy recovery system, which is specially used to strengthen the development and utilization of braking energy, and to some extent alleviate the development trend of energy loss. This paper mainly analyzes the vehicle energy storage braking energy recovery system.

Keywords: Vehicle · Electric energy storage · Braking energy recovery system

1 Introduction

With the transformation and upgrading of the automobile industry and the upgrading of automobile products, the travel of people in China has been inseparable from the common transportation tool of automobiles. From a macro perspective, cars have changed the way people live while changing the way people travel. Due to the increasing frequency of car purchase and use, air pollution in China has become increasingly serious. In order to purify the air environment and realize the strategic goal of developing green ecology in China, it is necessary to innovate the automobile braking energy recovery technology, realize the positive recycling of energy resources, and provide a more stable foundation for the development of the automobile industry.

2 Current Status of Vehicle Brake Energy Recovery Technology

The research on vehicle braking energy recovery technology originated from abroad, and three types of recovery systems have been established: hydraulic, flywheel and battery. Some studies have shown that loading a recycling system on a car can save about 30 percent of the power source [1]. Putting regenerative braking energy into use can save resources to a large extent and reduce the negative impact on the ecological environment. Because China's research on the ecological environment in the automotive field is rare, there is no concentrated advantage to develop vehicle braking

M. Tavana et al. (Eds.): IISA 2020, AISC 1304, pp. 547–551, 2021.
https://doi.org/10.1007/978-3-030-63784-2_68

energy recovery technology [2], so there is a big gap with the world's advanced level. Due to the imperfect construction of the domestic transportation infrastructure, many areas of travel can only choose cars. Although new modes of transportation such as subways, light rails and high-speed rails have already been put into use, buses still carry most people's travel. Because the bus needs to stop at the station or traffic lights, the driver often performs the brake operation, which greatly improves the energy consumption rate of the car and wastes a lot of unnecessary resources [3]. In order to improve and solve the above problems, the kinetic energy of the car can be stored during the normal running of the bus, and then put into use when it is necessary to restart. Since the research and development of new energy vehicles is still in its infancy, this power storage method is one of the core methods that can effectively cope with the above problems.

3 The Type of Electric Energy Storage Braking Energy Recovery System

Putting the electric energy storage braking energy recovery system into use can not only reduce the fuel consumption of the car, improve the driving performance of the car, but also improve the safety and environmental protection of the vehicle, and to a certain extent, protect the health of the traveler. In China, due to the late start of the development of this technology, many of the supporting technologies are not perfect. What can be done now is to convert the braking energy into other forms for storage, and then accelerate or restart. Release and use it during the process. There are three main forms of energy that are converted into other energy sources for storage: one is hydraulic energy storage; the other is flywheel energy storage; and the third is electrical energy storage [4].

The main problem of the energy storage of the flywheel is that the energy storage device is large in size, and the internal structure is very complicated. The whole is very cumbersome. Although energy conversion can be realized, there are still many hidden dangers from the perspective of portability and safety. The main problem of hydraulic energy storage is that the hydraulic system requires a very high degree of sealing, and it will cause serious friction during driving, which may cause damage to the system; the electrical energy storage method is superior to the above two in terms of overall performance. In this way, the operation of the electric energy storage recovery system is not complicated, it is easy to control it, and the internal structure is not complicated, and there is no high security risk [5].

4 The Basic Structure of the Electric Energy Storage Braking Energy Recovery System

The electric energy storage braking energy recovery system is mainly composed of three sections: one is an energy conversion module; the other is an energy recovery module; and the third is an electronic control module. Under the premise of ensuring the normal operation of the transmission of the original vehicle, the introduction of the braking energy recovery system in the form of electric energy storage can reduce the power energy loss of the vehicle during repeated starts and climbing uphill, and improve the performance of the vehicle. Driving safety provides a certain guarantee.

4.1 Energy Conversion Device

The car produces a lot of kinetic energy before braking, but it cannot be fully utilized. The energy conversion device converts the remaining part of the kinetic energy into an easily preserved form of energy, providing a power energy supplement for subsequent operations. Taking the introduction of the motor as an example, during the braking process of the automobile, the kinetic energy driving motor will start to work, and part of the kinetic energy is stored in the form of electric energy. Such an approach not only can share the energy loss with the brakes, reduce the wear of the brake pads during braking, but also reuse the waste energy, further improving the economic applicability of the vehicle.

4.2 Power Storage Device

After the kinetic energy of the vehicle passes through the energy conversion device, the converted electrical energy can be stored in the power storage device. Since the braking time of the automobile is short and the power of the battery is not large, it is difficult to accumulate the electric power within the required time, and the excessively intense electric storage process may damage the normal performance of the battery. In order to solve this problem, the currently widely used method is to use a super motor as a transfer station, but since the super motor itself cannot store a large amount of electric energy, it needs to be used in combination. Specifically, the motor, that is, the energy conversion device and a super capacitor are first spliced, the energy converted from the motor is stored in the super capacitor, and then the splicing of the super capacitor and the battery is used to realize the energy transfer. Fill the battery with electricity [6]. At this point, the entire power conversion and storage work has been completed.

4.3 Electronic Control Device

According to different functions, the electronic control device consists of three modules: one is the actuator; the second is the power control unit; the third is the sensor. The sensor is mainly used to monitor the driving situation of the car. The power control unit is mainly used for analysis and output of power signals, and performs storage and release operations of power storage by logical judgment. The actuator is an instrument that performs actual operation according to the signal output from the power control unit. The main function is to recycle the energy.

Fourth, Work flow of electric energy storage braking energy recovery system

(1) At the start, the sensor detects the throttle signal and the speed change signal, at which point the battery releases electrical energy to help the vehicle get off. While the vehicle engine is running, the energy regeneration system also generates energy to boost the vehicle's starting speed.

(2) When accelerating, each unit works the same as it did at the start.
(3) At constant speed, energy is neither released nor collected.
(4) As the vehicle taxis, the energy regeneration system begins to work, converting kinetic energy into electrical energy for storage.

When braking, if the required strength for braking is low, use the energy regeneration system to release part of the energy. If the required strength for braking is high, the original friction braking system must be activated to work with the energy regeneration system. The braking process mainly includes the following contents: First, the driver gently steps on the brake pedal, and then the sensor receives the signal and then passes the signal to the electronic control system for calculation processing, after which the energy regeneration system is activated, and the motor will activate the kinetic energy. Converted to electrical energy storage.

5 Conclusion

In summary, since people's demand for automobiles has changed, it is necessary to improve the performance of all aspects of the car to meet the actual needs of our people to pursue a better life. At present, automobiles have provided very convenient services for the travel of our people, but from the perspective of safety and health, there are still many problems in China's automobile manufacturing. With the development and application of vehicle energy storage braking energy recovery technology, the energy power of China's automobiles will be further utilized. Due to social history and limitations in science and technology, there is still a big gap between China's vehicle energy storage braking energy recovery and related technologies and the world's advanced level. Therefore, it is necessary to actively introduce external superior resources while increasing research and development efforts. International cooperation, improve the ability of technology research and development, improve the performance of vehicle electric energy storage braking energy recovery system, and thus improve resource utilization, and provide inexhaustible power for the construction of ecological civilization in China.

Acknowledgments. Dongguan Social Science and Technology Development Project of Dongguan Science and Technology Bureau in 2019: The Research and Realized on vehicle energy storage brake energy feedback system (2019507154530).

References

1. Huan, S.: Analysis of vehicle electric energy storage braking energy recovery system. Hebei Agri. Mach. **3**, 49 (2017)
2. Energy Weekly News. Technology - Oxidation Technology; New Findings from Z. B. Xu and Co-Researchers in the Area of Oxidation Technology Described (Design of Monitoring System for Brake Energy Recovery Storage Device) (2018)
3. Ruofei, W., Lixin, G., Ming, Z., et al.: Research on brake energy recovery control strategy of pure electric vehicles. Beijing Auto **5**, 32–36 (2015)

4. Wei, Z.: Research on hydraulic regeneration braking energy recovery system of new electric vehicle based on CPS. J. Jinhua Vocat. Tech. Coll. **6**, 60–64 (2015)
5. Wen, C.: Research and Implementation of Braking Energy Recovery System for Permanent Magnet Synchronous Motor Based on Super Capacitor. Anhui University (2017)
6. Wei Tongzhen, W., Lixin, H.L., et al.: Study on comprehensive recovery and utilization of braking Energy of AC/DC inverter drive system based on supercapacitor energy storage. Proc. CSEE **34**(24), 4076–4083 (2014)

Discussion on Human Body Energy Collection and Power Generation

Libo Yang(✉)

Guangdong University of Science and Technology, Dongguan 523083, China
654337610@qq.com

Abstract. It can collect the tiny energy generated by human activities, and convert it into electric energy through electrostatic induction and frictional electrification. Human activities are enormous every day, and a large amount can be obtained through cumulative effects. The electrical energy, and because it exists in the form of fabric, human beings will not be overly uncomfortable, can be used indoors and outdoors, and can be designed to be carried as they like, very economical and environmentally friendly. And safety can not only help solve the current non-renewable resource crisis, but also help people save electricity and economic costs, while reducing the use of fossil fuels, reducing pollution to air, water and soil, and reducing the exploitation of fossil energy. Thereby protecting the local land structure. Friction nanogenerators have the advantage of protecting non-renewable resources, economic value, protecting the ecological environment and protecting the soil structure. Fiber-based friction nanogenerators have been studied mainly for the main double helix structure and the coaxial core sheath structure.

Keywords: Human body · Energy harvesting · Power generation friction nanogenerator

1 Introduction

Science and technology continue to advance, so the demand for digital products is increasing, and the number of human beings is constantly rising. These require the use of electricity [1]. Although many wind power, water conservancy, and solar energy have been found. Ways to generate electricity and reduce the loss of natural resources, but the power generated by these forces is still difficult to fully meet the human needs for normal life, work and learning, so it is necessary to find a more efficient and energy-efficient way to obtain electricity [2]. Human beings will generate a lot of energy in the process of moving. If these energy can be collected and converted into electric energy, it will be able to obtain a large amount of electric energy, thus greatly reducing the consumption of natural resources and helping to protect the earth. The environment protects non-renewable resources and reduces pollution to the air. At present, the focus of human energy collection and power generation is mainly on the friction nano-generator [3]. This paper analyzes the principle of its work, and uses the fiber-based friction method generator as the research object [4].

M. Tavana et al. (Eds.): IISA 2020, AISC 1304, pp. 552–556, 2021.
https://doi.org/10.1007/978-3-030-63784-2_69

2 The Principle of Friction Nano-generator (TENG)

The friction nano-generator can be worn on human beings. The fabric with fiber mechanism becomes the main form of its existence, and has the same working principle. Both electrostatic induction and triboelectric interaction are required to convert low-frequency mechanical energy into electric energy. Store it up. Current friction nanogenerators are mainly divided into independent layers, single electrodes, horizontal sliding and separated. As shown in Fig. 1 below.

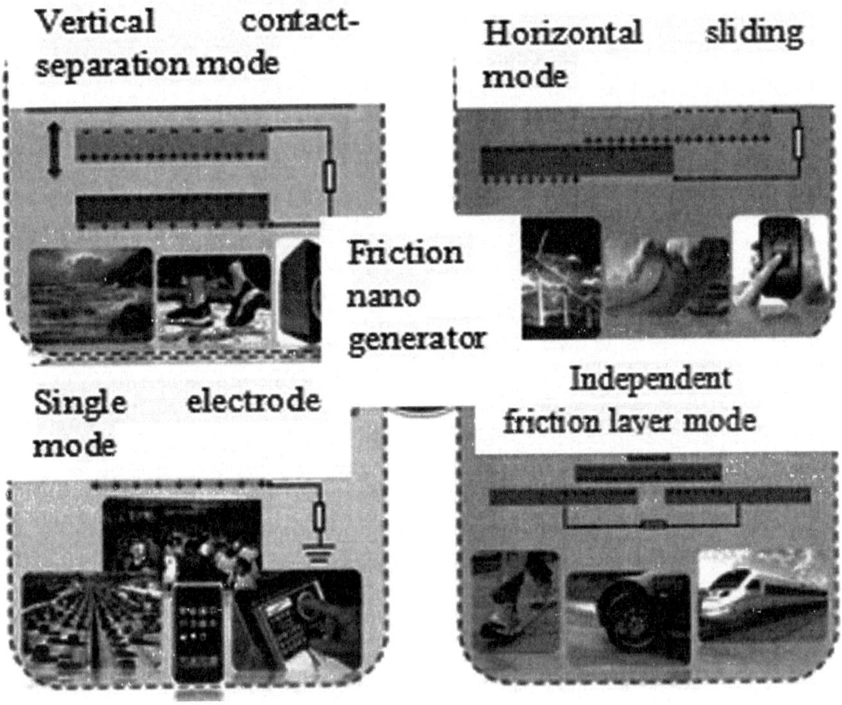

Fig. 1. Classification of friction nanogenerators

3 The Advantages of Friction Nanogenerator

Although it has been found that a certain amount of electricity can be obtained by means of solar energy, tidal energy, etc., the power thus obtained is still negligible. People in our country need to pay a certain amount of money for the electricity they use, and power plants need to use a lot of money when they manufacture electricity. At the same time, when burning coal mines and other materials that can be used for power generation, toxic substances such as sulfur dioxide and carbon monoxide are also generated [5], and various kinds of dust and the like are also introduced into the air, causing problems such as haze and contamination of the air. Moreover, when mining coal mines, it is necessary to damage mountainous areas and other places on a large scale, resulting in the destruction of the local environment. The friction nano-generator can reduce the use of coal mines and the like by converting the energy generated by the human body [6], thereby protecting the non-renewable resources, having economic value, protecting the ecological environment, and protecting the soil structure.

3.1 Protect Non-renewable Resources

Frictional nanogenerators can protect non-renewable resources because they can be converted into electricity by transforming the energy of the human body, which can be applied to the production, life, and work and study of human activities, reducing the power plant. The use of electricity, so the power plant's coal mine usage will also decline, and coal mines as a non-renewable resource, power plants are a major force in the use of coal mines, so if the power plant can reduce the use of coal mines, the amount of coal mining It will reduce, which can help protect non-renewable resources and avoid the problem of depletion of coal mines. Once the non-renewable resources are exhausted, it may trigger international wars and lead to competition for resources.

3.2 Have Economic Value

Frictional nano-generators also have extremely high economic value, because the energy generated by the human body does not need to spend extra money, turning it into electricity, it becomes free electricity, and fully utilizes the energy created by human beings, but currently China The electricity that the people get through the power plant needs to pay for electricity. Therefore, if people can obtain free electricity by friction nano-generators, they can reduce the amount of electricity used by the power plants, thus helping the people to save on electricity bills. Furthermore, in the process of purchasing and transporting coal mines, power plants need to use a large amount of cost. If the power supply to the users can be reduced, the power plant can reduce the procurement of coal mines and help the country save economic costs. At the same time, due to the reduced demand for coal mines, the demand for mining will be reduced, and the capital, manpower and material resources spent on coal mining projects will be reduced. These reduced human and material resources can be applied to other Among the areas that promote the development of our economy.

3.3 Protecting the Ecological Environment

In the process of burning, coal mines will produce a large amount of toxic gases such as sulfur dioxide and carbon monoxide. When they enter the air, they will damage the human respiratory health and cause acid rain and other problems. At the same time, exhaust gas is also generated in the process of transporting coal mines. These substances enter the air and cause problems such as haze. By rubbing nano-generators, this problem can be improved and the combustion and transportation of coal products can be reduced, so the environment can also be protected. As a safe and environmentally-friendly form of power generation, it is of great help to the strategic development goals of China's ecological sustainable development.

3.4 Protect the Soil Structure

Frictional nanogenerators also have the advantage of protecting the soil structure, because in the process of mining coal mines, it is necessary to mine forest land, etc., the land structure of these mining has been seriously damaged, and the land structure that was once destroyed, currently Still not getting a good recovery. Frictional nanogenerators can help to avoid further deterioration of the land structure in our country due to the serious damage caused by mining. The people use this kind of free electricity that does not need to be mined, and this part of the electricity is reduced to use and mine. An important alternative, so the reduction of coal mining can help protect our soil structure.

4 Fiber-Based Friction Nanogenerator

The fiber-based TENG that has been studied so far mainly has two structures, a double helix structure and a coaxial core-sheath structure.

4.1 Double Helix

The secondary structure is the earliest mechanism of the fiber base, and the two single wires are surrounded by a spiral. The current upgrade method for this is the infiltration-drying technique, which uniformly coats the carbon nanotube ink on the surface of the cotton thread to obtain a cotton thread having electrical conductivity, and then applies PTFE on the surface thereof to form a film. The friction dielectric layer needs to be annealed in order to improve adhesion, and the obtained two wires are screwed into a spiral structure, which can realize uninterrupted power generation, and combines flexible yarn capacitors to realize self-charging and self-storage energy. Ability. It can usually be placed under the arm to collect the mechanical energy generated during the swinging of the arm as electrical energy.

4.2 Coaxial Core Sheath Structure

It has only one single wire, and the core sheath structure is obtained by coating on the outside of the core yarn. It has high stability and can be well combined with weaving. The main material of the core wire is silicone rubber, PDMS and PU and other main, with high elasticity, so it can be in the human wrist, chest and neck.

4.3 Composite Friction Nanogenerator

In order to improve the power generation efficiency of coaxial fiber-based TENG, many scholars have combined various forms of nano-generators. Li et al. used the dual principle of piezoelectric and friction to design a fiber-based piezoelectric-friction composite nano-generator, which can collect more forms of mechanical energy and improve power generation efficiency.

5 Conclusion

China has a large population. In order to meet the people's demand for electricity, we need to continuously burn coal and other production power. With the advancement of technology, more and more equipment is needed to use electricity, under the influence of automation and life. People's demand for electricity is also growing, so it is necessary to continuously increase the supply of electricity. Fiber-based friction nano-generators have been studied mainly for the main double helix structure and the coaxial core sheath structure.

Acknowledgments. Dongguan Social Science and Technology Development Project of Dongguan Science and Technology Bureau in 2019: Research on core technology of human energy collection (2019507154529).

References

1. Chen, Y., Zhang, Z., Bai, Z., Guo, J.: Study on collecting human body kinetic energy by textile-based friction nano-generator. J. Text. Sci. Eng. **36**(02), 113–120 (2019)
2. Armbruster, M., Rist, M., Seifert, S.: Metabolite profiles evaluated, according to sex, do not predict resting energy expenditure and lean body mass in healthy non-obese subjects. Eur. J. Nutr. **9**, 2207–2217 (2019)
3. Energy: Findings from Princeton University Provide New Insights into Energy (Human body exergy consumption models' evaluation and their sensitivities towards different environmental conditions). Energy Weekly News (2019)
4. Huang, Z.: Energy harvesting technology and application of human footsteps. Electron. Technol. Softw. Eng. **02**, 217–218 (2019)
5. Wei, S., Hu, W.: A Review of research on human body energy harvesting technology based on piezoelectric vibration. Mach. Electron. **36**(10), 67–72 (2018)
6. Zhang, Z.: Research on LoRa wireless positioning system based on human energy collection and self-powered. East China Jiaotong University (2018)
7. Cheng, X.: Porous polypropylene piezoelectric electret device for human energy collection and self-powered sensing. Huazhong University of Science and Technology (2016)

DIY Parallel Corpora for Petroleum Production Engineering and Its Academic Application

Pengpeng Gao[1,2](✉)

[1] School of Foreign Languages, Xi'an Shiyou University, Xi'an 710065, China
ycsygp@126.com
[2] State Key Laboratory of Continental Dynamics (Northwest University), Xi'an 710069, China

Abstract. Corpus linguistics emerged as one of the new linguistic study areas and means in genre analysis, critical discourse analysis, and social activities. This paper has presented a method for DIY corpora construction and gives the reader detailed and repeatable steps that will be possible to repeat on a larger scale. The author also provides a channel to get the keyword or terminology in the text by AntConc. The data in the DIY corpora need to be further analyzed both qualitatively and quantitatively. DIY corpora are so significant to translation study and educational use and present great potential in an academic application.

Keywords: DIY parallel corpora · Petroleum production engineering · Academic application

1 Introduction

Corpus linguistics has been widely used in language research and teaching, and its research in genre analysis, critical discourse analysis, and sociolinguistics has received increasing attention. Then, the basis for carrying out these studies is to have a deep understanding of the keywords and terms of specific industries. Parallel corpus belongs to a bilingual corpus, which refers to a corpus that performs a full-text search on the source language and target language and compares and displays them. It can achieve parallel alignment at the level of text, paragraph, sentence, chunk, or phrase, and has essential guidance for translation teaching and learning effect. The two main approaches to corpus work are discussed as the 'corpus-based' and the 'corpus-driven' approach, and the theoretical positions underlying them explored in detail (Tognini-Bonelli 2002). Corpus linguistics is not a separate branch of linguistics, but rather "descriptive linguistics aided by new technology" and the "Corpus Analysis" consists of an overview of basic techniques in corpus annotation and analysis, including lemmatization, word-class tagging, parsing, and the use of frequency lists and concordances (Nelson 2000; Somers and Kennedy 2000). The corpora could be used to train a particular machine translation system to automatically translate the subtitles between German and Chinese (Xiao and Wang 2009). Using bilingual parallel corpus

M. Tavana et al. (Eds.): IISA 2020, AISC 1304, pp. 557–562, 2021.
https://doi.org/10.1007/978-3-030-63784-2_70

and retrieval tools in translation teaching is of great value (Wang 2004). A method of extracting sentence-level alignment by using an extended link-based bilingual lexicon method was discussed (Mohammadi and Ghasemaghaee 2010). Aligning sentences in bilingual parallel corpora based on punctuation was presented (Chuang and Yeh 2005). The basic knowledge of corpus linguistics and the essential operation of the corpus is described in detail (Liang et al. 2010).

This article focuses on petroleum production engineering operations, based on an authoritative database named SPE eLibrary (The Society of Petroleum Engineers), combined with industry expert proofreading, proposes a parallel corpus construction method for specific fields, which has important guiding significance for translation teaching and academic English research.

2 Corpora Resource Acquisition and Corpus Designs

Corpus resources are a significant and fundamental issue in the construction of corpora, and generally require consideration of several factors. Firstly, educational value or influence; secondly, corpus must be of high quality; thirdly, reliable representation. In other words, it is not that the more considerable the amount of data, the better the corpus is the most suitable corpus resource for the project. This article selects the SPE database as the source of the corpus because the corpus of this database is peer-reviewed and highly authoritative, good academic, as shown in Fig. 1.

Fig. 1. Corpus resource acquisition database of SPE eLibrary

The corpus resources download for the SPE database named 10.2118@1943XX-MS/PA with PDF format, which is needed to use the corresponding software to convert to plain text format (.txt or.doc). Use ANSI or UTF-8 encoding to save text, and then conduct sampling and pilot analysis, including text size analysis, text removal of noise, and related corpus cleaning, to prepare for large amounts of text collation. At the same time, gather the wisdom of industry experts to limit the retrieval time and technical fields of the corpus, formulate work plans, workflows, storage paths, etc. In this way, a pure source language text is created, which lays a good foundation for the subsequent development of a bilingual parallel corpus.

3 DIY Corpora Construction of a Methodology

Oil production engineering is a general term for various engineering and technical measures taken on oil reservoirs through production wells and injection wells by exploitation goals during oilfield development. What its studies focus on the theory, engineering design method, and implementation technology of various engineering and technical measures that can be used in oil reservoirs economically and effectively to improve oil well output and crude oil recovery. And also, it pays attention to the combination of theory and practice, emphasizes on training students' engineering awareness and engineering practice ability. And related technology is the key to innovation in the oil and gas industry. It is challenging to design and complete the technical documents for drilling and completion. The data package is complicated in the process of learning and translation. It is urgent to establish a DIY personalized corpus to assist in teaching and learning. This article gives the method and procedure of DIY corpora establishment shows in Fig. 2.

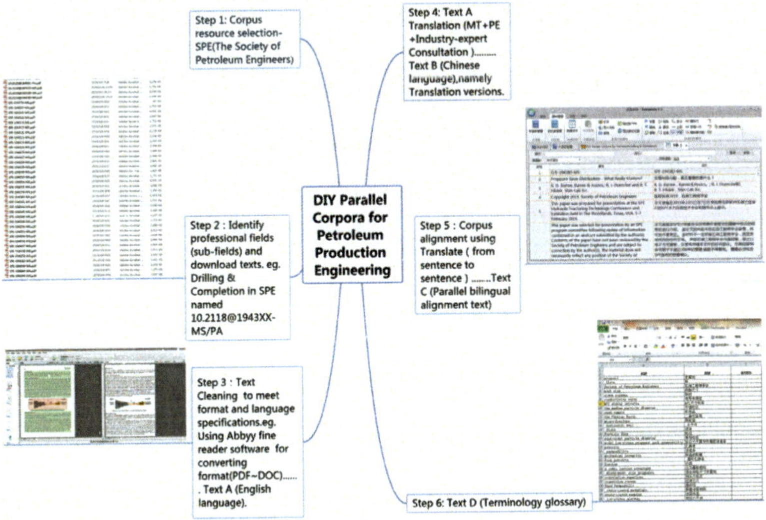

Fig. 2. DIY corpora construction of methodology and steps

The entire parallel corpus creation process is roughly divided into six steps as follows: (1) To determine the source of the corpus and select the authoritative database of SPE elibrary. For getting suitable materials to perform text retrieval based on the determined subject headings and core technology or tool features. (2) To download relevant documents from the SPE database according to the professional field and subdivision professional direction, mainly authoritative journal articles, conference abstracts, and related reports. (3) Text cleaning. (4) To translate the source language text to a target language (English to Chinese here) using the work pattern of Machine Translation + Post-editing + Industry-expert Consultation, with the considerations based on large-scale data processing. (5) Corpus alignment using the Translate platform for parallel bilingual text generation basically in Fig. 3. (6) Combining with subject knowledge, utilizing the retrieval software such as Antconc, Paraconc, and WordSmith Tools to extract terms, form a terminology glossary for translation study and academic English teaching &learning.

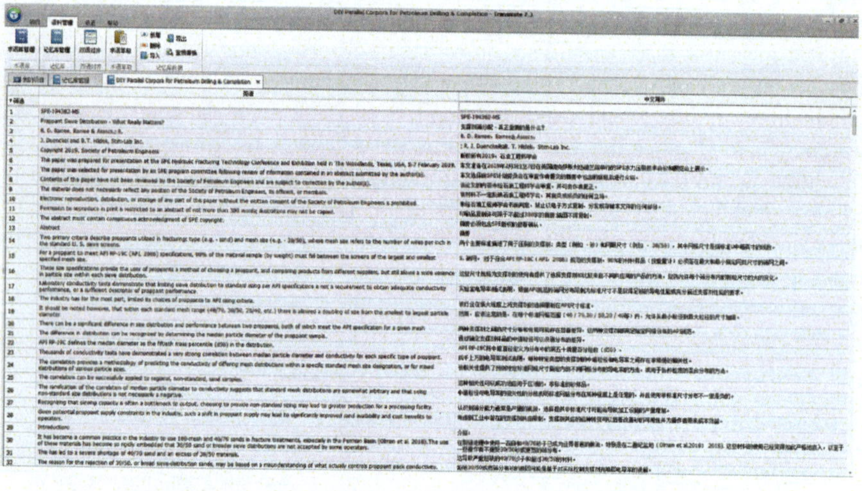

Fig. 3. Alignment of parallel bilingual text based on Translate Platform

4 Academic Application of DIY Parallel Corpora

4.1 Parallel Corpora and Translation

A corpus based on actual cases from translation practice will play an important role in automatic translation. Translators can use DIY specialized corpora, including memory and terminology, to use CAT platforms like SDL Trados Studio for machine translation and manual proofreading, which can greatly improve the efficiency and accuracy of the translation. At the same time, as the translation work progresses, the corpora further expand, which is more and more beneficial to the later translation. When processing data from an oil production project for translation, the initial translation can be performed quickly with the help of parallel corpora, especially when the text of source language is huge. Besides, the parallel corpora is a real translation practice, and it has important guiding significance to the translation.

4.2 Parallel Corpora and Academic Application

Parallel corpora can provide the teachers or students with a further list of domain-specific keywords and terms which are conducive to master and understand the core-knowledge of a subject like Petroleum Production Engineering and build the portfolio of English academic usage, also beneficial to knowledge internalization. When a corpus of a certain size is built, we can use AntConc software to perform word retrieval, vocabulary generation, and topic function analysis on the corpus. Using word frequency sorting, screening notional words such as in a specific field, a deeper understanding of the professional knowledge and culture expressed in the text. The notional words were selected by frequency such as fracture, stress, hydraulic, tensile, pressure, wells, fluid, injection, perforation, design, permeability, proppant, etc. demonstrated in Fig. 4 and also we can get the Word list of text in Fig. 5.

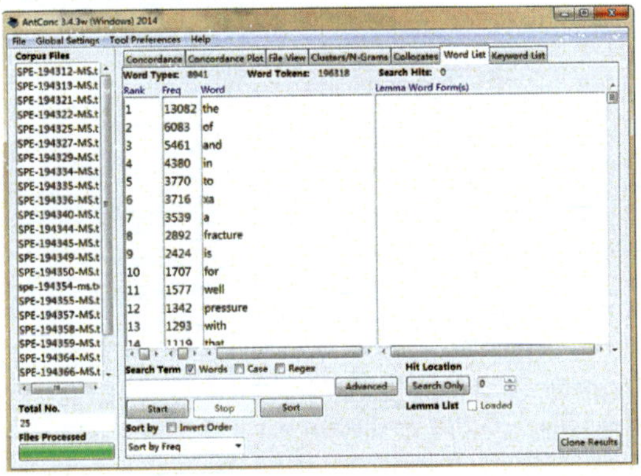

Fig. 4. Keyword frequency analysis in DIY Corpora by AntConc

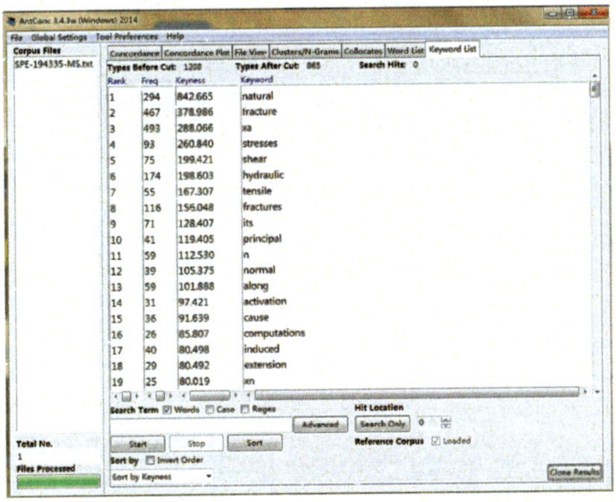

Fig. 5. Keyword List analysis in DIY Corpora by AntConc

5 Conclusion

Corpus linguistics plays an important role in n language research and teaching, and research in genre analysis, critical discourse analysis, and sociolinguistics has received increasing attention. This paper has presented a method for DIY corpora construction. It gives the reader detailed and repeatable steps to build DIY corpora, including corpus resource acquisition, corpus data cleaning, bilingual parallel text alignment, and data exporting and storage. The author also gives a channel to get the keyword or terminology in a text, which is so beneficial for teachers or students to master the technical vocabulary knowledge according to the analysis results by AntConc. The data in the DIY corpora need to be further analyzed both qualitatively and quantitatively. DIY corpora provide the terminology and usages of their specialist subjects, which is so significant to translation study and educational use and presents great potential in an academic application.

Acknowledgments. This research was supported by a China Foreign Language Assessment Fund Project (ZGWYCPJJ2018137B), Shaanxi Province Education Science "13th five-year plan" in the 2018 annual plan (SGH18H166).

References

Tognini-Bonelli, E.: Corpus linguistics at work. Computat. Linguist. **28**(4), 583 (2002)
Nelson, G.: An introduction to corpus linguistics. J. Engl. Linguist. **28**(2), 193–196 (2000)
Somers, H.L., Kennedy, G.: An introduction to corpus linguistics. Mach. Transl. **15**(3), 259–261 (2000)
Xiao, H., Wang, X.: Constructing parallel corpus from movie subtitles. In: International Conference on Computer Processing of Oriental Languages Language Technology for the Knowledge-Based Economy. Springer (2009)
Wang, K.F.: The use of parallel corpora in translator training. Comput.-Assist. Foreign Lang. Educ. 000(006), 27–32 (2004)
Mohammadi, M., Ghasemaghaee, N.: Building bilingual parallel corpora based on Wikipedia. In: Second International Conference on Computer Engineering & Applications. IEEE (2010)
Chuang, T.C., Yeh, K.C.: Aligning parallel bilingual corpora statistically with punctuation criteria. Chin. J. Comput. Linguist. **10**(1), 95–122 (2005)
Liang, M.C., Li, W.Z., Xu, J.Z.: Using Corpora: A Practical Coursebook. Foreign Language Teaching and Research Press (2010)

Two-Speed Pure Electric Vehicle AMT Transmission

Liu Wenguang[1(✉)], Bi Shanshan[1], and Su Zhaorui[2]

[1] School of Automotive and Transportation Engineering, Jiangsu University, Zhenjiang 212013, Jiangsu, China
liuzhangwang2000@163.com

[2] School of Mechanical Engineering, Jiangsu University of Science and Technology, Zhenjiang 212000, Jiangsu, China

Abstract. In order to improve the power of pure electric vehicles and reduce the power consumption of 100 km, a pure electric vehicle is matched with a two-speed AMT without clutch and synchronizer. By Analyzing the dynamic characteristics of the shifting structure, the shift dynamics model, the shift motor model and drive motor model are established. On the basis of shortening shift time and reducing the shift shock, a shift motor controller with permanent magnet synchronous motor (SPWM) modulation combined with three closed loop PID control is designed. Considering the frequent changes of vehicle state and the nonlinear and strong coupling of the drive motor, a fuzzy adaptive PI controller for the drive motor is designed. The simulation and test results show that the controller designed in this paper can effectively reduce the shift time and improve the smoothness of shifting.

Keywords: Two-speed automatic transmission · SPWM modulation · Three closed loop PID control · Fuzzy adaptive PI control

1 Introduction

Pure electric vehicle has the characteristics of low noise and emission, which is an important development direction of automobile industry at present. At present, the commonly used fixed transmission ratio transmission system cannot meet the requirements of vehicle economy and dynamic, so the multi-speed pure electric vehicles is the focus of domestic and foreign experts.

In order to reduce the impact of gear shifting, literature [1] takes the AMT of clutch and synchronizer as the research object. By designing the feedback controller of clutch and motor, the power interruption in the process of gear shifting is avoided. However, retaining the clutch and synchronizer makes the structure of the AMT complex, increasing the control difficulty, and the effect of automatic shift is poor. In order to improve shifting efficiency and reduce vehicle weight, the application prospect of clutchless AMT in pure electric vehicles was analyzed in literature [2]. Literature [3] proposed a synchronizer constraint control method, which can effectively shorten synchronizer synchronization time. In literature [4], a coordination controller is established to optimize the logic of the driving motor and the gear shifting motor and

M. Tavana et al. (Eds.): IISA 2020, AISC 1304, pp. 563–573, 2021.
https://doi.org/10.1007/978-3-030-63784-2_71

improve the gear shifting effect. Due to the large inertia moment of the driving motor and gear in the process of vehicle operation, the synchronizer has a long synchronization time and serious wear after the clutch is cancelled. In order to solve the above problems, literature [5] proposed a gearshift control method for pure electric bus without clutches and synchronizers. In literature [6–9], a variety of precise fork shifting controllers were designed, which avoided the occurrence of meshing but reduced the gear shifting speed.

In summary, the current research does not analyze the dynamic characteristics of the vehicle shifting process in detail, and designs the corresponding motor controller for the nonlinear and strong coupling characteristics of the PMSM. Therefore, on the basis of designing the structure and working mode of the transmission, a dynamic model describing the shifting process is established, and a fuzzy PI motor controller is designed to realize the strong coupling and nonlinearity between the PMSM and the motor, so as to ensure that the vehicle can quickly and smoothly complete the gear shifting operation.

2 Structure and Working Mode of Electric Gearbox

Figure 1 is a schematic diagram of the two-speed AMT structure.

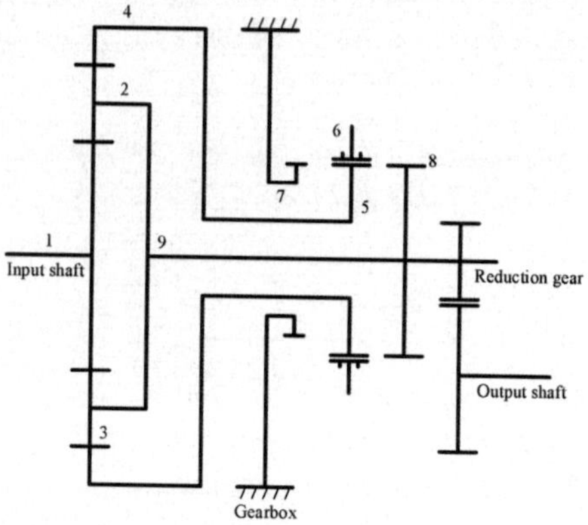

Fig. 1. Schematic diagram of single-row planetary gear two-speed AMT structure

In the Figure, 1 is the sun gear, 2 and 3 are planetary wheels, 4 is the ring gear, 5 is the fixed ring gear, 6 is the joint sleeve and 9 is the planet carrier. Besides, 7 and 8 are the first and second gear combined with the ring gear. This type of speed reducer switches between the first and second gears through a single planetary row, and use the synchronizer to reduce the shift shock. The working mode is divided into one gear speed reduction twisting and second gear direct transmission. In the deceleration and twist mode, the coupling sleeve 6 and the transmission housing 7 are consolidated by the coupling sleeve 6. In the direct drive mode, the carrier 6 and the planetary ring gear 4 are consolidated by the coupling sleeve 6.

3 Shift Dynamics Analysis

By analying the above two-speed gearbox structure, a schematic diagram of the dynamic model of the entire vehicle transmission system can be obtained, as shown in the Fig. 2.

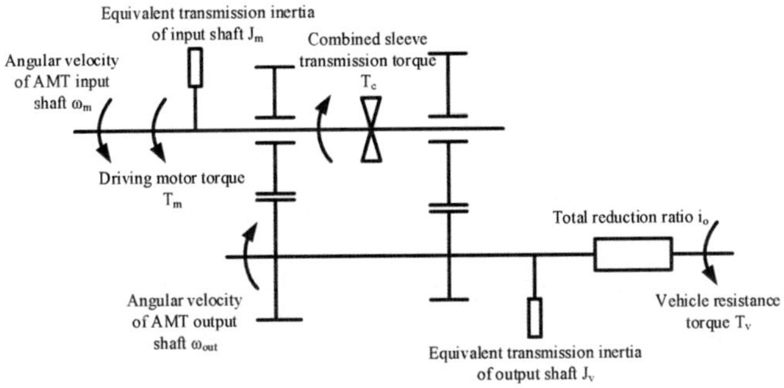

Fig. 2. Dynamics model of the vehicle transmission system

3.1 Picking Dynamics Analysis

Analyzing the force of the meshing teeth during the picking process, and the resistance is from the friction of the combined ring gear and the joint sleeve, which can expressed as:

$$F_f = \mu F_v = \frac{\mu T_c}{R} \tag{1}$$

In the formula, μ is the tooth surface friction coefficient; F_v is the contact surface positive pressure; T_c is the moment at the joint sleeve; R is the radius of the circle of the joint sleeve.

The calculation method of the torque transmitted at the joint sleeve and the resistance torque of the whole vehicle is as shown in Eq. (2):

$$\begin{cases} T_c = (T_m - J_m\dot{\omega}_m - c_{in}\omega_m)i_g \\ T_v = (T_c - J_v\dot{\omega}_{out} - c_{out}\omega_{out})i_o \end{cases} \tag{2}$$

In the formula, T_m is the output torque of the motor; J_m is the equivalent moment of inertia of the gearbox input shaft; ω_m and ω_{out} are the angular velocity of the input shaft and output shaft; T_v is the equivalent resistance torque of the vehicle; J_v is the equivalent moment of inertia at the output of the gearbox; c_{in}/c_{out} are the rotational damping coefficient of the input and output shafts respectively; i_o/i_g are the main reducer speed ratio and the current gear speed ratio.

Therefore, when AMT is in non-neutral, it has the following relationship:

$$\omega_m = i_g \omega_{out} \tag{3}$$

Since c_{in} and c_{out} are small, the above two values are ignored for simplifying the model, so Eqs. (1)–(3) are available.

$$F_f = \frac{\mu(i_g i_o J_v T_m + i_g^2 J_m T_v)}{R(i_o i_g^2 J_m + i_o J_v)} \tag{4}$$

3.2 Dynamic Analysis of the Gear

The force of the meshing teeth is analyzed during the gear stage. When the joint sleeve and the target gear gear ring are in contact, the fork resistance F_q can be expressed as:

$$F_q = F_N \sin\beta + F_s \cos\beta \tag{5}$$

$$\begin{cases} F_s = \mu_2 F_N \\ F_c + F_s \sin\beta = F_N \cos\beta \end{cases} \tag{6}$$

$$Fc = \frac{T_c}{R} \tag{7}$$

In the formula, μ_2 is the friction factor of the tooth end contact surface; β is the chamfer angle of the tooth end.

It can be seen from Eqs. (5)–(7) that the smaller the fork resistance F_q, the smoother the gearbox is, which means that T_c is as small as possible. According to Eq. (2), drive motor output torque T_m, the smaller the difference in speed between the sleeve and the engagement ring gear, the smaller the T_c. Therefore, when shifting gear, the driving motor should stop the output torque and control the difference between the clutch sleeve and the meshing ring gear.

3.3 Dynamic Analysis of Shifting Process

The two-parameter shift strategy allows the pure electric vehicle to achieve better power and economy than the single-parameter shift strategy. Therefore, this paper selects the vehicle speed and the current throttle opening as parameters, and designs a two-parameter shift strategy.

Analyzing the force situation during the driving process of the vehicle, and the driving equation of the pure electric vehicle can be obtained as follows:

$$F_t = \sum F \tag{8}$$

In the formula, F_t is the driving force of pure electric vehicles; F is the resistance received during driving, and there are rolling resistance, air resistance, ramp resistance, and acceleration resistance.

The pure electric vehicle two-speed variable speed drive can be divided into three operating phases: picking, motor synchronization and gear shift. The relationship between the torque and the speed of the electric shifting mechanism during the picking process is as follows:

$$T_{out} = T_f = T_m - T_{lost} \tag{9}$$

$$\omega_l = i_o i_g \omega_m \tag{10}$$

In the formula, T_{out}, T_f, T_m and T_{lost} are wheel drive torque, ground resistance torque, motor output torque and torque loss, respectively; ω_m and ω_l are the motor output speed and wheel speed respectively; i_g and i_O are the transmission ratio and main deceleration ratio respectively.

During the motor synchronization phase, the motor needs to adjust the speed to reach the meshing requirement. The speed that the motor needs to adjust is as follows:

$$\Delta\omega_{input} = \left(i_n - i_g\right) i_o \omega_m \tag{11}$$

In the formula, i_n is the gear ratio after shifting. In order to avoid gear collision caused by high speed difference during shifting, the gear meshing speed difference needs to be adjusted to (10–50) r/min during motor speed regulation.

The shifting impact of a pure electric vehicle j can be expressed as:

$$j = \frac{1}{\delta mr} \frac{d\left(T_m i_g i_o \eta_T - T_v\right)}{dt} \tag{12}$$

When the gear speed difference reaches a reasonable range, the vehicle will complete the forward operation and adjust the motor speed according to thc actual working conditions to ensure that the vehicle has the same dynamic performance before and after the shift. It can be known from Eq. (12) that the shifting degree of the pure electric vehicle is proportional to the first derivative of the motor output torque, and the more the motor output torque changes, the greater the impact.

The two-speed non-clutch AMT used in this article has a gear ratio of 14.699, which is the first speed and 4.716 is the second gear ratio. In order to be able to complete the shift in a short time, the speed from the first gear to the second gear PMSM needs to be increased rapidly and vice versa.

4 Design of Permanent Magnet Synchronous Motor Controller

4.1 Mathematical Model of Permanent Magnet Synchronous Motor

In order to analyze the PMSM system, the motor model is first established and the following assumptions are made: 1) The effects of magnetic leakage, eddy current, hysteresis and core saturation are not considered. 2) Rotor undamped winding. 3) The effects of alveolar, commutation and armature reaction were not considered. 4) The stator phase windings are the same and the three phases are symmetrical [10, 11].

Vector control is the best control method for PMSM,since the electromagnetic torque of the motor depends on the current space vector i_s of the stator, the phase and magnitude of i_s can be determined by controlling i_q and i_d, which also determines the electromagnetic torque of the motor. This paper uses the vector control method of i_d= 0, so the mathematical model of PMSM can be simplified to the following formulas:

$$\begin{cases} u_d = -P_m \omega_m L_q i_q \\ u_q = R_s i_s + L_q \dfrac{di_q}{dt} + P_m \omega_m \varphi_r \\ T_e = \dfrac{3}{2} P_m \varphi_r i_q \end{cases} \tag{13}$$

In the formula, L_q is the q-axis inductor; ω_m is the angular velocity; R_s is the internal resistance of the stator; i_q is the q-axis current component; u_q and u_d are the voltage components in the q-axis and d-axis directions, respectively; P_m is the number of motor pole pairs; T_e is the electromagnetic torque; φ_r is magnetic flux.

4.2 Fuzzy PI Control System Design

Although the traditional PID control does not rely on the establishment of accurate mathematical model, when the external conditions change, the controller parameters cannot be adaptively adjusted, resulting in deterioration of the control effect. Based on the fuzzy theory, the practical experience is transformed into mathematical expression [12]. A fuzzy PI permanent magnet synchronous motor control system with adaptive adjustment of PI parameters is proposed. The structure diagram of fuzzy control system is shown in Fig. 3.

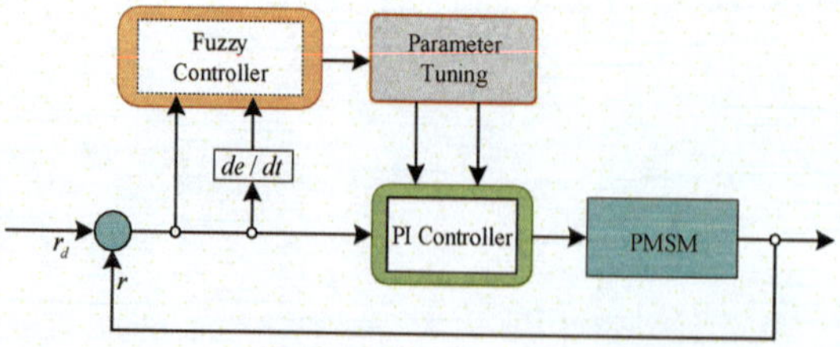

Fig. 3. Schematic diagram of fuzzy PID control system

The system input is the desired speed of the motor and the actual motor speed *r*. The speed difference signal *e* and the speed difference change rate *ec* are taken as the fuzzy controller input, and the controller output signals are the PID coefficient adjustment amounts *Δkp*, *Δki* and *Δkd*.

The fuzzy theory domain has a high number of levels, and the fuzzy control rules are relatively detailed, but the programming difficulty is high and the memory is occupied. This paper divides the rotational speed difference input (e), the rotational speed difference change rate input (ec), and the PID adjustment amount *Δkp* and *Δki* into seven fuzzy sets. They are PB, PM, PS, ZE, NS, NM and NB, respectively, where PB represents positive, PM represents positive, PS represents positive, ZE represents zero, NS represents negative, NM represents negative, and NB represents negative. In order to avoid runaway, the field of setting the speed difference and the speed difference change rate is [− 4 4], and the range of the PI parameter adjustment amount is [− 0.2, 0.2], [− 0.3, 0.3].

If the shape of the function is sharp, the system resolution is high and the control is sensitive, but the system is less robust. If the shape of the function is flat, the system has strong robustness, but the system sensitivity is low [13–15]. The membership functions include triangle, gaussian and trapezoid functions. In this paper, a combination of trigonometric and trapezoidal membership functions is used as input membership functions. When the input variable is large, the trapezoidal membership function is used. When the input variable is small, the triangle membership function is used to improve the sensitivity of the system. Figures 4 and 5 show the input variable e and the input variable ec membership function. Figures 6 and 7 show the membership functions of the output variables *Δkp* and *Δki*.

Fig. 4. Input variable e membership function

Fig. 5. Input variable ec membership function

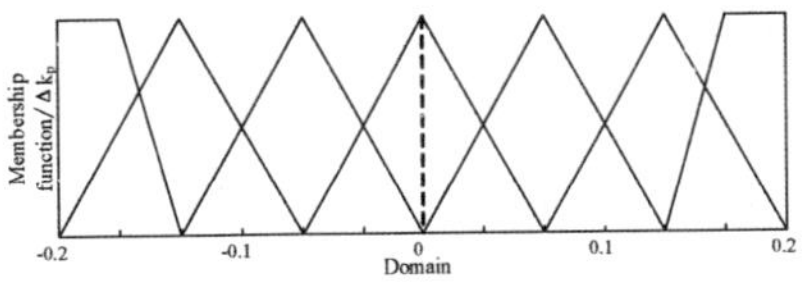

Fig. 6. Output variable *Δkp* membership function

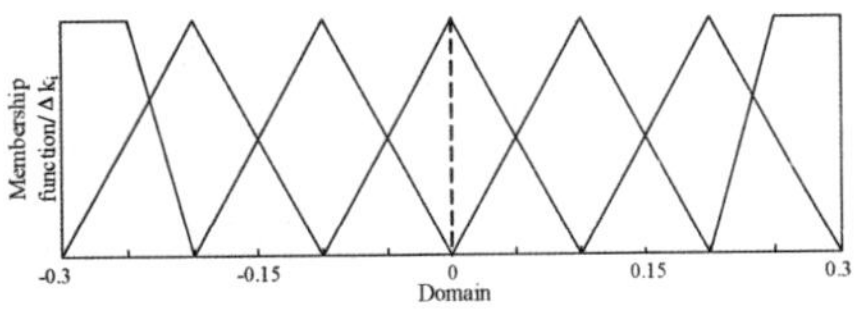

Fig. 7. Output variable *Δki* membership function

Tables 1 and 2 are the fuzzy control rule tables of the output variables *Δkp* and *Δki*, respectively.

Table 1. *Δkp* Fuzzy control rules

E	Ec						
	NB	NM	NS	ZE	PS	PM	PB
NB	PB	PB	PM	PM	PS	ZE	ZE
NM	PB	PB	PM	PS	PS	ZE	NS
NS	PM	PM	PM	PS	ZE	NS	NS
ZE	PM	PM	PS	ZE	NS	NM	NM
PS	PS	PS	ZE	NS	NS	NM	NM
PM	PS	ZE	NS	NM	NM	NM	NB
PB	ZE	ZE	NS	NM	NM	NB	NB

Table 2. *Δki* Fuzzy control rules

E	Ec						
	NB	NM	NS	ZE	PS	PM	PB
NB	NB	NB	NM	NM	NS	ZE	ZE
NM	NB	NB	NM	NS	NS	ZE	ZE
NS	NB	NM	NS	NS	ZE	PS	PS
ZE	NM	NM	NS	ZE	PS	PM	PM
PS	NM	NS	ZE	PS	PS	PM	PB
PM	ZE	ZE	PS	PS	PM	PB	PB
PB	ZE	ZE	PS	PM	PM	PB	PB

The fuzzy PI controller is built in MATLAB/Simulink. According to the fuzzy PI controller model built in Simulink, the fuzzy control planes of *Δkp* and *Δki* are obtained as shown in Fig. 8.

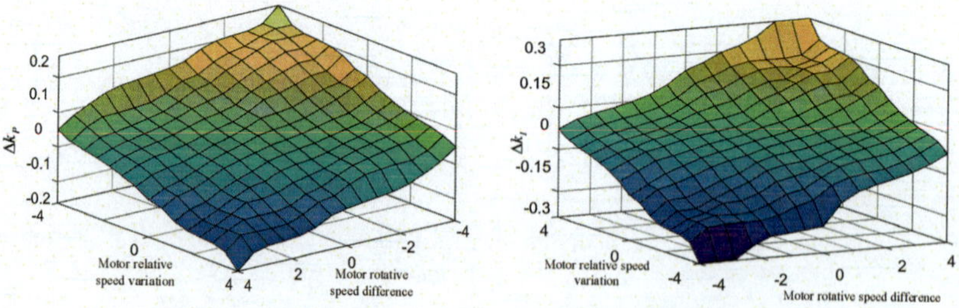

(a) Δ*kP* fuzzy rules surface (b) Δ*ki* fuzzy rules surface

Fig. 8. Fuzzy rules surface of *Δkp* and *Δki*

5 Simulation Analysis

The PMSM model and the fuzzy PI control system model are built based on MATLAB/Simulink, and the effectiveness of the control system is simulated and analyzed. Two simulation conditions were set: condition 1: the motor's target speed is 1000 r/min, and the target torque is 10 N/m; condition 2: the motor's target speed is 800 r/min, and the target torque is 7 N/m. The simulation results are shown in the following figure.

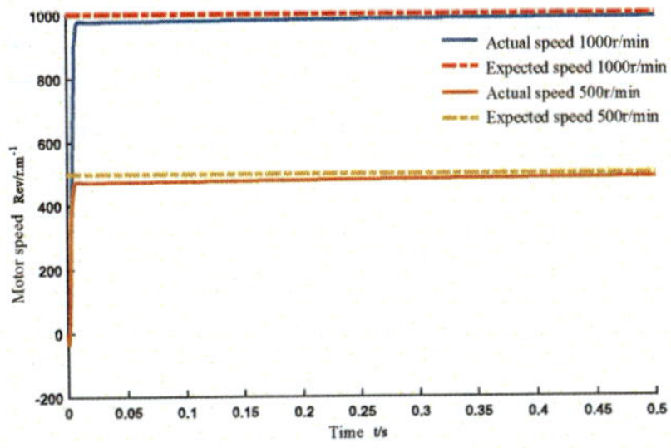

Fig. 9. Actual speed time domain response diagram

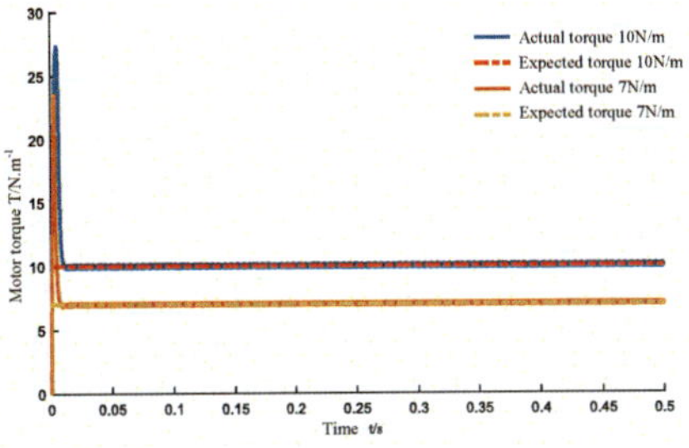

Fig. 10. Actual torque time domain response diagram

Figure 9 shows the controller designed in this paper can reach the target speed quickly, and the overshoot can be ignored. Figure 10 shows the actual time domain response of the motor torque. At the initial moment, the motor speed changes greatly, but the controller can control the motor to reach the required torque quickly and keep the torque output stable.

In actual driving conditions, the motor needs to switch multiple speeds in a short time. Set the speed switching condition, compare the control effect of traditional PID control and fuzzy PID control. The results are shown in Fig. 11.

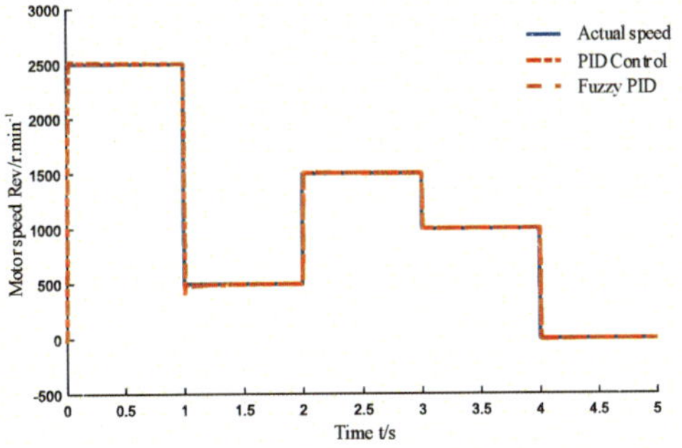

Fig. 11. Permanent magnet synchronous motor speed diagram

Figure 11 shows the traditional PID control system not only needs a longer time to reach the desired speed, but also has a large overshoot at the abrupt point of speed. The fuzzy PID control ensures that the PMSM can track the desired speed accurately while achieving the desired speed as fast as possible, thus ensuring the transmission's fast and stable shifting process.

6 Two-Speed AMT Bench Test

Based on dSPACE platform and two - speed AMT bench test device, this paper verifies the effectiveness of the drive motor controller. The test bench consists of a drive system, a test system, a drive system, a loading system, an industrial computer, and a dSPACE-based measurement and control system. In addition to the mechanical connection, the experimental bench uses CAN communication between the systems.

During the test, the measured data include the AMT input speed and torque, the AMT output speed and torque, and the position signal of the displacement actuator. This paper simulates the lifting and lowering process of the real vehicle at a certain speed, simulates the inertia of the vehicle with flywheel, and loads the simulated air resistance and road rolling resistance on the dynamometer. The test results are shown in the Figs. 12 and 13.

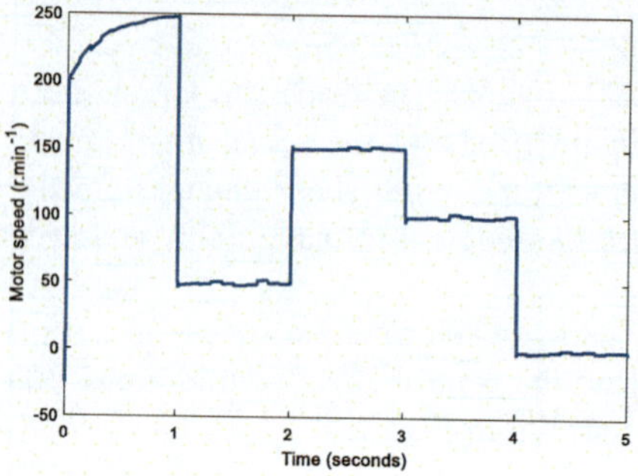

Fig. 12. Test motor speed time domain response diagram

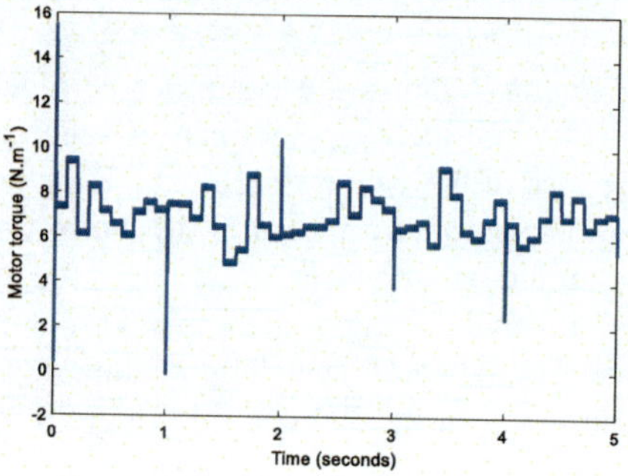

Fig. 13. Test motor torque time domain response diagram

It can be seen from Fig. 12 and 13 that the motor controller is designed in this paper to ensure that the vehicle is ready and smoothly completes the shifting operation.

7 Conclusion

In order to solve the problem that the traditional PID parameters of PMSM with non-clutch AMT cannot be self-tuned, a fuzzy adaptive PI motor speed controller is proposed, which has good adaptability to the nonlinear and strong coupling of PMSM. The simulation results show that the fuzzy adaptive PI controller is superior to the traditional PID controller in dynamic response, system overdrive and adjustment time, and can guarantee the accuracy, stability and real-time requirements of the two-speed AMT shift of pure electric vehicles.

References

1. Da-tong, Q., Bao-hua, Z., Ming-hui, H.: Parameters design of powertrain system of electric vehicle with two-speed gearbox. J. Chongqing Univ. **34**(1), 1–6 (2011)
2. Dong-peng, Y., Jie, W., Jun-zhi, Z.: Coordinated control strategies of HEV powertrain with AMT. J. Transp. Eng. **1**, 43–49 (2010)
3. Yong-dong, C.: Research on shifting law of electronically controlled mechanical automatic transmission. Wuhan University of Technology (2007)
4. Shuang, Z., Ya-bin, C.: Research on AMT shift schedule based on improving shift quality. Mach. Des. Manuf. **7**, 138–140 (2008)
5. Yang, W., Jun-qiang, X., Hui-yan, C.: A study on the mechanism and countermeasures for shift impact in AMT. Automot. Eng. **31**(3), 253–257 (2009)
6. Da-fang, W., Gang, L., Yi, J.: Gear-shifting control of clutchless automated mechanical transmission without synchronizer in short-distance pure electric vehicle. Chin. J. Highways **30**(2), 144–152 (2017)
7. Xian-Bing, L.: Research on shifting control strategy of pure electric vehicle equipped with two-speed AMT. Chongqing University (2017)
8. Hui, J.: Design and development on integrated motor and transmission drive system for light electric vehicle. Wuhan University of Technology (2011)
9. Hong, F., Guang-yu, T., Hong-xu, C.: A study on the torsional vibration control of motor transmission integrated drive system. Automot. Eng. **32**(7), 596–600 (2010)
10. Kalt, S., Erhard, J., Lienkamp, M.: Electric machine design tool for permanent magnet synchronous machines and induction machines. Machines **8**(1), 15 (2020)
11. DaWei, G., Yao, Y., Zhang, D.M., et al.: Matlab/Simulink based modeling and simulation of fuzzy PI control for PMSM. Procedia Comput. Sci. **166**, 195–199 (2020)
12. Lamamra, K., Batat, F., Mokhtari, F.: A new technique with improved control quality of nonlinear systems using an optimized fuzzy logic controller. Expert Syst. Appl. **145**, 113148 (2020)
13. Le, Z., Wei, W., Song, S., et al.: Optimisation of sparse array configuration using ambiguity function in automotive radar application. J. Eng. **2019**(19), 6152–6154 (2019)
14. Huitzil, I., Bobillo, F., Gómez-Romero, J., et al.: Fudge: fuzzy ontology building with consensuated fuzzy datatypes. Fuzzy Sets and Systems (2020)
15. Chen, X.-W., Zhou, Y., Zhang, J.-G., et al.: A synergetic strategy of automobile intelligent cruise system based on fuzzy control adopting hierarchical structure. Int. J. Adv. Rob. Syst. **16**(5), 1729881419877758 (2019)

Study on Urban Energy Internet and Its Influence Factor Analysis Model

Yifan Zhang[1(✉)], Jie Hao[1], Weiru Wang[1], Yunfeng Zhao[2], and Danyang Chen[1]

[1] State Grid Shanxi Electric Power Research Institute, Taiyuan, China
zyf_tju@126.com
[2] State Grid Shanxi Electric Power Company, Taiyuan, China

Abstract. Nowadays, energy problem has become increasingly prominent. How to systematically coordinate the relationship between various energy sources and to use energy efficiently are important issues for scholars in various countries. There is no doubt that Energy Internet is development trend of current grid. Also it is an important part of future power grid. The development of Energy Internet is summarized and the concept of Urban Energy Internet is proposed. This paper points out that Urban Energy Internet is the basic unit of national ubiquitous smart grid. A set of influence factors of Urban Energy Internet is put forward. An analysis model of the influence factors of Urban Energy Internet is constructed based on the Interpretative Structural Model and the Analytic Hierarchy Process method. The influence factors of Urban Energy Internet in an urban area are analyzed to provide a reference for the optimal allocation, the construction planning and operation control mode of Urban Energy Internet.

Keywords: Urban Energy Internet · Interpretative Structural Modeling Method · Influence factor · Analytic Hierarchy Process

1 Introduction

For a long time, social and economic development has relied too much on fossil energy. It leads to the problems of resource crisis and environmental pollution getting worse. To cope with energy challenges, it is necessary to coordinate the characteristics of economic development, the allocation of resource and the impact of environmental impact, and to focus on implementing clean energy replacement and electric energy replacement in order to accelerate the transformation of energy structure. Due to all above reason, Urban Energy Internet comes into being. In this paper, we summarize the development status of Energy Internet, put forward the concept of Urban Energy Internet, and point out that Urban Energy Internet is a basic unit of the national ubiquitous smart grid. In this paper, the typical features of Urban Energy Internet are obtained by comparing Urban Energy Internet and traditional distribution system. According to the composition and features of Urban Energy Internet, a set of influence factors is proposed. Based on the Interpretative Structural Model and the Analytic Hierarchy Process, an influence factors analysis model of Urban Energy Internet is constructed.

M. Tavana et al. (Eds.): IISA 2020, AISC 1304, pp. 574–583, 2021.
https://doi.org/10.1007/978-3-030-63784-2_72

2 Urban Energy Internet

2.1 Research Status at Home and Abroad

In 2008, members of the National Science Foundation studied a power grid structure based on clean energy power generation and distributed energy storage equipment, called Energy Internet. They pointed out a new concept—Energy Router, and carried out prototype implementation. The research team at the University of California proposed "information-centric energy network". In 2011, Japan showed the "Mark 1" digital grid router (DGR). The DGR can manage the power in a certain range of areas and dispatch regional power. The digital grid is built on the basis of the Internet, and the current synchronous grid is subdivided into asynchronous, autonomous and interconnected grids of different ranges. The energy distribution in the grid is completed by power router. On May 29, 2012, the European Union held a meeting in Brussels. Antonio Tajani, Vice-President of the Council of the European Union, clearly stated at the meeting that "the third industrial revolution will be revolved around the Energy Internet".

Energy Internet technology has caused widespread concern. Research on the Energy Internet started late and is now still in the theoretical research stage in China. In 2016, the National Development and Reform Commission (NDRC) issued the "Guiding Opinions on Promoting the Development of 'Internet+' Intelligent Energy", proposing a new mode of energy industry development. In the new mode, the Internet and energy production, transmission, storage, consumption and energy markets are deeply integrated. It is called "Internet+" smart energy. In general, domestic scholars' research on Energy Internet mainly focuses on three areas: conceptual models and development challenges, communication design and application (energy routers), technical frameworks and key technologies.

2.2 The Concept of Energy Internet

The concept of Energy Internet is first proposed in 2008 by Jeremy Rifkin. He believed the industrial model based on the large-scale utilization of fossil fuels laid in the second industrial revolution is coming to an end and pointed out that the energy utilization system "Energy Internet", which combines energy technology(ET) and information technology(IT), is about to emerge. It will finally achieve this result that fossil energy centralized utilization changes into clean energy distributed utilization.

The domestic scholars have given a preliminary definition of Energy Internet—It is a complex energy system using distributed clean energy as a primary energy source. And it use advanced energy storage technology, power electronics technology, smart energy management technology, intelligent fault management technology, communication technology and system planning analysis technology to realize the tight coupling of power system and other energy systems. Energy Internet has the characteristics of clean and efficient, dynamic random, wide-area sharing, multi-source interconnection, and interactive intelligence etc.

2.3 Urban Energy Internet

At present, China's urban areas account for over 50% of the national population, 75% of the country's resource consumption, 80% of China's GDP, and 70% of carbon emissions. The energy revolution is precisely the solution to the low-carbon economy and efficient use of energy. From the current development situation, the growth rate of power demand is higher than the growth rate of energy demand, and the dominant position of power in the energy structure is increasingly important. Energy Internet is the physical basis for electricity market transactions and an indispensable public service platform for future production and life. One of the four principles of power flow distribution is the principle of local priority, that is, local power generation should be prioritized for development and utilization. Urban Energy Internet is the node and important support of the Energy Internet in urban areas.

The National Energy Administration announced the list of the first batch of 56 "Internet +" smart energy (Energy Internet) demonstration projects, covering the city's Energy Internet construction, enterprise big data platform development, distributed micro-grid construction, and many other aspects. Tianjin, Jiaxing, Chengdu, Shanghai and other places have carried out the construction of energy interconnection demonstration projects. The main contents include: integrating multiple energy sources and optimizing utilization, building a smart-interactive power system coordinated by "power supply-grid-load-storage", building an urban energy big data sharing platform, efficient and comprehensive utilization of multiple energy sources of cold, heat and electricity, etc. (Table 1).

Table 1. Comparison of Urban Energy Internet and Traditional Urban Distribution system

Characteristics	Traditional urban distribution system	Urban Energy Internet
Green & friendly	Using fossil energy as a primary energy source; Renewable energy access has a large impact on the distribution network	The distributed electrical energy is mainly from renewable clean energy generation; Able to consume renewable energy on a large scale and friendly
Dynamic & random	Single business model in distribution and sales market; The operation mode is relatively fixed	Demand side load can respond actively; Multi-energy prices and strategies change in real time
Flexible & sharing	Distributed renewable energy is difficult to access and control	Breaking traditional energy barriers with Internet technology, Large-scale, high-proportional access to distributed power supplies, Support plug and play
Diversified & interconnected	Distribution network operates alone, lacking energy complementary; Limited by technology, low communication efficiency	Multi-energy systems are open and interconnected; Big data flow is tightly coupled to energy flow
Interactive & intelligent	Single user type with low participation; Lack of personalized electricity service	Focus on users thinking and improve users experience; Increase two-way interaction with users to provide customized services

Urban Energy Internet proposed in this paper refers to a regional energy interconnection network based on urban distribution system, covering urban areas, accepting distributed renewable energy widely and flexibly, supporting advanced energy storage, control, communication and management system, which can realize the close integration and flexible calls of urban power system and other energy.

3 The Interpretation Structure Model of Urban Energy Internet Influence Factors

Urban Energy Internet covers energy, communication, information and other aspects, and is a complex system affected by many factors. In order to study the relationship between each two factors of Urban Energy Internet and the degree of influence of various factors on Urban Energy Internet, we decide to use the method of Interpreting Structural Modeling (ISM) Method. The ISM Method was proposed by an American professor named J. Warfield in 1973 aiming at analyzing and solving problems related to structural modeling of complex systems. At present, a systematic and standardized operation method has been formed, which is widely used in various fields such as economy and engineering. The core of this method is that researchers decompose a complex system into several subsystems, and adopt knowledge technology and computer technology to construct a multi-level structural model.

3.1 The Influence Factor Set of Urban Energy Internet

Determining the research object and its Influence factor set is the first step of Interpretative Structural Modeling Method. All the influence factors of system is analyzed to determine the factor set. Due to the access of distributed clean energy and the access of equipment containing a lot of power electronics, Urban Energy Internet is more complicated than the traditional urban distribution system, and the factors influencing Urban Energy Internet are more dispersed and diverse.

The core of this paper is Urban Energy Internet, which is recorded as the factor F0. Through the study of the Energy Internet and related literature, combined with Urban Energy Internet influence factor knowledge model, Urban Energy Internet influence factor set is determined as shown in Table 2.

The energy factors include F_1–F_6; the environmental factors include F_7–F_{10}; the traffic factors include F_{11}–F_{13}; the user factors include F_{14}–F_{16}; the industrial factors include F_{17}–F_{20}; and the building factors include F_{21}–F_{23}.

Table 2. The influence factor set of Urban Energy Internet

	Influence factor		Influence factor		Influence factor
F_1	Types and distribution of clean energy	F_{11}	New energy vehicles population	F_{21}	Distribution of clean architecture
F_2	Total electricity demand	F_{12}	Total number of electric car charging station and gas station	F_{22}	Energy saving of clean architecture
F_3	Total consumption of fossil fuel	F_{13}	Population of buses per 10000 people and highway mileage	F_{23}	Total number of intelligent home
F_4	Wind and photovoltaic power station	F_{14}	User's load characteristics	F_{24}	Economic benefit of Urban Energy Internet
F_5	Total energy utilization efficiency	F_{15}	Total population and population structure	F_{25}	Power supply reliability of Urban Energy Internet
F_6	Clean energy penetration	F_{16}	User's energy consumption habits	F_{26}	Power quality of Urban Energy Internet
F_7	location conditions of city	F_{17}	Total GDP of city	F_{27}	Structure and planning of Urban Energy Internet
F_8	Weather condition of city	F_{18}	Annual output value of each industry	F_{28}	Load of Urban Energy Internet
F_9	Climate type of city	F_{19}	Distribution and planning of various industries	F_{29}	Equipment configuration of Urban Energy Internet
F_{10}	Air pollution index of city	F_{20}	Electricity demand of various industries		

3.2 Expert Experience Model

After the influence factor set is determined, we should find the directly influence factors for each factor in the set. So that a directed relationship between the factors is built and an expert experience model is formed.

3.3 The Connection Matrix

The Connection Matrix $C = (c_{ij})_{m \times n}$ shows the direct influence relationship. If $c_{ij} = 1$, it means f_i and f_j are directly connected; otherwise, $c_{ij} = 0$.

3.4 The Reachability Matrix

Based on the Connection Matrix and the relationship between all the 29 factors, the Reachability Matrix R is obtained according to the flowing formula.If

$$(C+I) \neq (C+I)^2 \neq \ldots \neq (C+I)^k = (C+I)^{k+1} = L(k<n-1) \tag{1}$$

Then

$$R = (C+I)^{k+1} \tag{2}$$

3.5 Stratifying Each Factor

$$P(f_i) = \{f_i | r_{ij} = 1\} Q(f_j) = \{f_j | r_{ji} = 1\} \tag{3}$$

$P(f_i)$ is Reachable Set representing all factors can be reached from influence factor f_i. $Q(f_i)$ is Advance Set representing all the factors that can reach influence factor f_i. Based on P (f_i) and Q (f_i), the set L_1 can be obtained.

$$L_1 = \{f_i | P(f_i) \cap Q(f_i) = P(f_i)\} \tag{4}$$

L_1 is the set of first-level factors. These factors can be reached from other Influence factors, but there is no factor can be reached from these factors.

The factors of L_2 can be obtained by the same steps. In order to stratify each factor, all above procedures should be repeated.

According to the method, Urban Energy Internet influence factor is divided into the following 9 levels as shown in Fig. 1.

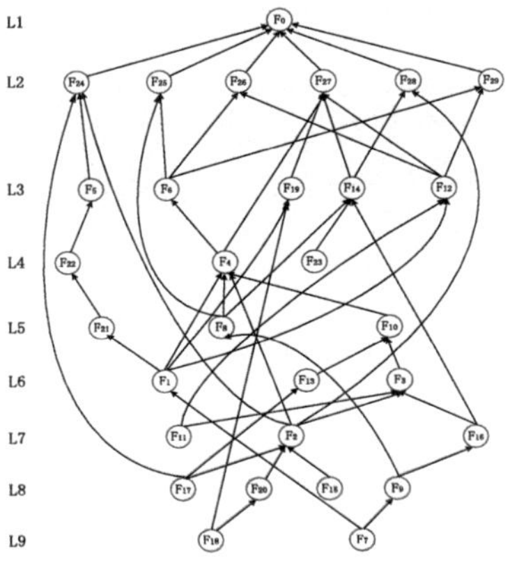

Fig. 1. The Hierarchical Structure of influence factors of Urban Energy Internet

4 Weight Determination of the Influence Factors of Urban Energy Internet

On the basis of constructing the Interpretation Structure Model of Urban Energy Internet influence factors, the Analytic Hierarchy Process method and related algorithms are applied to determine the weight of each factor.

The Analytic Hierarchy Process (AHP) is a weighted decision analysis method proposed by T.L.Saaty. It includes target level, criterion level and scheme level. We improve traditional AHP to form a new method. The method can adapt to the structural model of influence factors of Urban Energy Internet by extending the three levels of traditional AHP.

Assuming there are n elements in the ISM, if

$$R_{pq} = 1(p \leq n, q \leq n) \tag{5}$$

It means F_p directly affects F_q. Then, these r $(0 \leq r \leq n)$ factors directly affecting F_q can be determined. If $r = 0$, there's no factor affect F_q directly. If $r = 1$, it means $w_{pq} = w_q$. If $r > 1$, w_{pq} of the factor F_p can be get by following steps. Build the Judgment Matrix $U_{m \times m}$ and get the eigenvalue λ_{max} and Normalized Eigenvector W. Then the weight of each factor can be obtained by (6).

$$W = \frac{\sqrt[m]{\prod_{j=1}^{m} u_{ij}}}{\sum_{i=1}^{m} \sqrt[m]{\prod_{j=1}^{m} u_{ij}}} \tag{6}$$

u_{ij} is the element in Judgment Matrix U.

$$AW = \lambda_{max} W \tag{7}$$

$$C.I = \frac{\lambda_{max} - m}{m - 1} \tag{8}$$

$$W_{ik} = \sum W_{kq} \tag{9}$$

W_{kq} is the weight of factor F_k in L_i influencing on F_q. The weights of the factors in L_i and the comprehensive weight of each factor such as F_k can be get as follows.

$$W_k = \sum W_{ik} \tag{10}$$

Table 3. Calculation result 2 of influence factors of Urban Energy Internet

	Influence factor	Comprehensive weight	Normalized score(%)
F_1	Types and distribution of clean energy	0.2313194	6.64%
F_2	Total electricity demand	0.1215451	3.49%
F_3	Total consumption of fossil fuel	0.0283854	0.82%
F_4	Wind and photovoltaic power station	0.3027778	8.69%
F_5	Total energy utilization efficiency	0.05	1.44%
F_6	Clean energy penetration	0.2527778	7.26%
F_7	location conditions of city	0.553316	15.89%
F_8	Weather condition of city	0.2774306	7.97%
F_9	Climate type of city	0.3219965	9.25%
F_{10}	Air pollution index of city	0.3784722	10.87%
F_{11}	New energy vehicles population	0.1434549	4.12%
F_{12}	Total number of electric car charging station and gas station	0.1722222	4.95%
F_{13}	Population of buses per 10000 people and highway mileage	0.0094618	0.27%
F_{14}	User's load characteristics	0.1166667	3.35%
F_{15}	Total population and population structure	0.04051503	1.16%
F_{16}	User's energy consumption habits	0.0454566	1.31%
F_{17}	Total GDP of city	0.1499768	4.31%
F_{18}	Annual output value of each industry	0.0571817	1.64%
F_{19}	Distribution and planning of various industries	0.05	1.44%
F_{20}	Electricity demand of various industries	0.04051503	1.16%
F_{21}	Distribution of clean architecture	0.05	1.44%
F_{22}	Energy saving of clean architecture	0.05	1.44%
F_{23}	Total number of intelligent home	0.0388889	1.12%

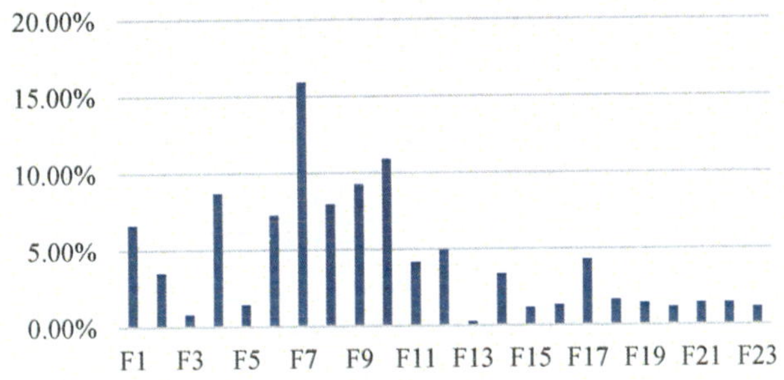

Fig. 2. A weight graph of the influence factors of some Urban Energy Internet

5 Examples

Using the algorithm and model proposed in this article to analyze a commercial-oriented Urban Energy Internet in China, the following results can be obtained (Table 3 and Fig. 2):

From the above results, we can get the following information:

The complementarity and compatibility of Urban Energy Internet is higher than traditional distribution system. Due to the high proportion of clean energy sources such as photovoltaics and wind power, meteorological factors such as light and wind will inevitably have an important impact on Urban Energy Internet. So no matter what type Urban Energy Internet it is, we need to pay close attention to the impact of bad weather such as lightning strike, strong wind and freezing rain on distribution systems.

Clean energy has a huge impact on Urban Energy Internet. The fundamental of green-friendly and multi-interconnectivity of Urban Energy Internet lies in the acceptance of clean energy. The resource endowment and geographical distribution of urban area have a direct impact on the centralized development and distributed utilization of clean energy. In addition, distributed generation and micro-grid in Urban Energy Internet play a significant part in the coordination and complementation of energy. The use of clean energy reduces the load pressure and the access to Urban Energy Internet of distributed power generation makes the system power flow no longer unidirectional. The bidirectional power flow puts forward new requirements for Urban Energy Internet planning and design, operation control, equipment configuration, and load forecasting.

6 Summary

Energy revolution will reshape the energy ecology. Energy Internet will be an important product of energy revolution and certainly be the main format of urban energy. This paper summarized the development status of Energy Internet and proposed the concept of Urban Energy Internet. It also analyzed the typical characteristics of both Urban Energy Internet and traditional distribution system. According to the characteristics, the collection of influence factors is summarized and proposed. Secondly, an explanatory structure model is proposed which is suitable for Urban Energy Internet influence factor analysis. At last, we can obtain the weight of each factor by extended AHP method. The result intuitively shows the influence of various factors on the Urban Energy Internet. Using the models and algorithms proposed in this paper can get practical conclusions, especially in the case that we lack the data during the development process of Urban Energy Internet. We applied the model and algorithm proposed in the study to analysis the influence factor weights of an Urban Energy Internet, and verified the effectiveness of the method. The method proposed in this paper is a new way for the analysis of Urban Energy Internet influence factors. It provides a reference for research on construction planning, optimized configuration, dispatching automation, operation management and control strategy of Urban Energy Internet.

References

1. CPC Central Committee, State council. Planning of new urbanization in China (2014–2020)
2. Javadian, S.A.M., Haghifam, M.R., Firoozabad, M.F., Bathaee, S.M.T.: Analysis of protection system's risk in distribution networks with DG. Int. J. Electr. Power Energy Syst. **44**(1), 688–695 (2013)
3. Zhou, J., Ding, J., Wang, Q., Song, Y., Li, Y.: Impact of PV integration on safety and stability of Yushu Grid and control strategy. Electr. Power Autom. Equip. **34**(6), 25–29 (2014)
4. Li, H., Bai, X.: Impacts of electric vehicles charging on distribution grid. Autom. Electr. Power Syst. **35**(17), 38–43 (2011)
5. Soroudi, A., Ehsan, M., Caire, R., Hadjsaid, N.: Possibilistic evaluation of distributed generations impacts on distribution networks. IEEE Trans. Power Syst. **26**(4), 2293–2301 (2011)
6. Warfield, J.N.: Participative methodology for public system planning. Comput. Electr. Eng. **1**(1), 23–40 (1973)
7. Zhou, X., Guo, C., Dong, S., Chen, W.: Expansion planning of urban multi-energy electricity-gas-heating distribution network incorporating electrical reconfiguration. Autom. Electr. Power Syst. **2019**(043(007)), 23–33 (2019)

Check for updates

Discussion on Safety Performance of Pressure Resistant Fuel Tank

He Yongao(✉) and Yin Wei

China Automotive Technology and Research Center, Tianjin, China
heyongao@catarc.ac.cn

Abstract. Pressure resistant fuel tank is a new type of fuel tank suitable for hybrid electric vehicle. In this paper, the safety performance requirements of pressure resistant fuel tank are studied through a variety of types of tests. The comparison test of different materials and different types of fuel tank shows that, in addition to the higher work pressure, better performance is needed in shape variable, temperature and pressure resistance, ventilation performance, etc. At the same time, it is found that the structure of pressure resistant fuel tank is different from that of traditional fuel tank.

Keyword: Pressure resistant fuel tank · Hybrid vehicle · Pressure alternative · Pressure relief performance · Ventilation performance · Separating explosion-proof

1 Introduction

With the increase of car ownership year by year, the world is facing the problem of non renewable energy exhaustion and environmental degradation. Energy and environment are the issues that must be considered to achieve sustainable development. At present, the governments of all countries are vigorously promoting the development of new energy vehicles in order to reduce the comprehensive fuel consumption and pollutant emissions of vehicles. China also lists the new energy vehicle industry as one of the seven strategic emerging industries in the 12th Five Year Plan. At present, new energy vehicles mainly include hybrid electric vehicles, pure electric vehicles and fuel cell vehicles, among which hybrid electric vehicles have become the priority development project of automobile manufacturers due to its high technology maturity and low dependence on supporting facilities. Because of lower comprehensive fuel consumption and more flexible driving mode, hybrid vehicles are widely used (Fig. 1).

The power system of hybrid electric vehicle is composed of internal combustion engine and battery. When the battery is used, the oil pump stops working, the gasoline in the fuel tank is constantly volatilizing, which makes the vapor pressure in the fuel tank much higher than that in the traditional automobile fuel tank. The internal pressure of traditional vehicle fuel is generally 6−10 kPa, while that of hybrid vehicle fuel tank is 35−45 kPa, therefore, the fuel tank of the hybrid vehicle must be resistant to high pressure.

M. Tavana et al. (Eds.): IISA 2020, AISC 1304, pp. 584–595, 2021.
https://doi.org/10.1007/978-3-030-63784-2_73

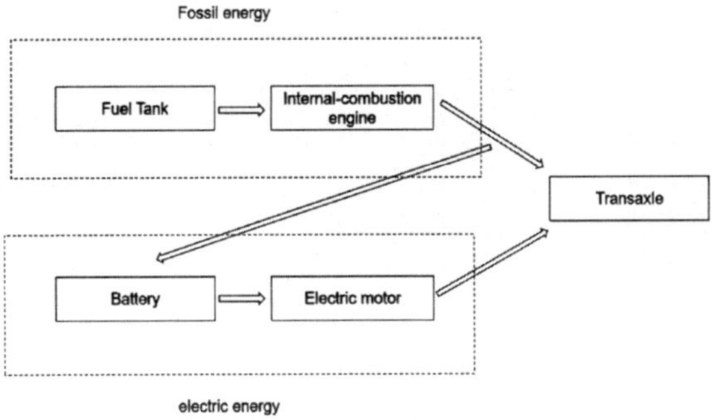

Fig. 1. Power system of automobile

According to the actual working conditions, the fuel tank mainly bears the following two kinds of pressures: firstly, the vehicle uses the battery to provide power, the fuel tank stops pumping oil, the internal pressure of the fuel tank is higher than atmospheric pressure, the fuel tank expands and deforms. Secondly, the vehicle uses the engine to provide power, the pump volume is higher than that of the gasoline, causing the atmospheric pressure higher than of the internal pressure, at this time, the fuel tank is compressed and deformed.

2 Material and Structure

Traditional fuel tank is mainly made of plastic, which is composed of HDPE + EVOH + LLDPE. It has the advantages of light weight, large freedom of shape, short die cycle and strong corrosion resistance, but also has the disadvantages of high permeability. The metal fuel tanks in traditional automobile fuel tanks are mainly divided into two types, i.e. commercial vehicle fuel tanks made of galvanized sheet or aluminum magnesium alloy, and fuel tanks made of spc3sz-s-30/30 or sctz270d 30n with main metal coating.

In order to solve the problem of large deformation of fuel tank, the following ways are generally adopted:

1) Using metal bracket or plastic support rod inside the plastic fuel tank to increase the strength, in the limit state, the support rod fails firstly to ensure the integrity of fuel tank shape.
2) Metal frame is added outside the plastic fuel tank for supporting if the first method is unavailable.
3) Change the fuel tank material, using 304 L, 316 L, 409 L, 439 l, 441 l and other carbon steel or stainless steel materials. At present, the mainstream hybrid vehicles adopt this technology form (Figs. 2, 3 and 4).

Fig. 2. Metal high pressure resistant fuel tank

Fig. 3. Plastic high pressure resistant fuel tank

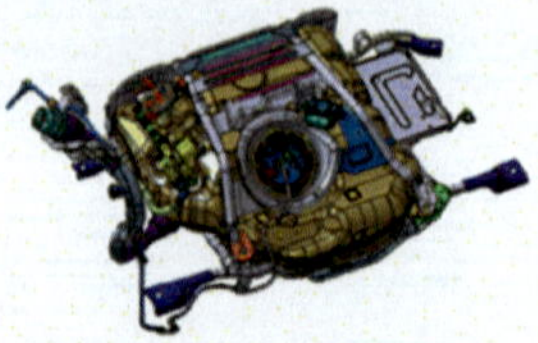

Fig. 4. Plastic high pressure resistant oil tank with metal frame

3 Study on Pressure Alternating Durability Test

We choose a plastic pressure resistant fuel tank with internal support, hereinafter referred to as A, and a carbon steel high pressure resistant tank with a wall thickness of 1.8 mm, hereinafter referred to as B, and a stainless steel high pressure resistant tank with a wall thickness of 1.0 mm, hereinafter referred to as C. Fix the fuel tanks of A, B and C on the tooling with reference to the actual vehicle status, seal each exhaust port, add water containing the rated solvent of dye into the fuel tank, add a pressure of 15 ± 1 kPa into the fuel tank. Check if there is leakage on fuel tank through water inspection. If there is no leakage, continue the test.

Mark the obvious variables and key parts of the fuel tank. The center of the fuel pump is 1 (corresponding to fuel tank A, i.e. A1), the center of the FLVV valve is 2 (corresponding to fuel tank A, i.e. A2), the left point of the fuel tank is 3 (corresponding to fuel tank A, i.e. A3), the middle point of the fuel tank is 4 (corresponding to fuel tank A, i.e. A4), and the right point of the fuel tank is 5 (corresponding to fuel tank A, i.e. A5).Point 6 is the point on the back of the tank, corresponding to point 1 on the top of the tank(corresponding to fuel tank A, i.e. A6), Point 7 is the point on the back of the tank, corresponding to point 2 on the top of the tank(corresponding to fuel tank A, i.e. A7), Point 8 is the point on the back of the tank, and it's at the bandage, corresponding to point 3 on the top of the tank(corresponding to fuel tank A, i.e. A8), Point 9 is the point on the back of the tank, corresponding to point 4 on the top of the tank(corresponding to fuel tank A, i.e. A9, if the fuel tank is saddle shaped, the thinnest part of the fuel tank is preferred), Point 10 is the point on the back of the tank, and it's at the heat shield, corresponding to point 5 on the top of the tank(corresponding to fuel tank A, i.e. A10). For the three fuel tanks, make sure that the key positions are comparable as much as possible.

The pressure alternative durability test is carried out on the fuel tank, which can simulate the internal pressure change of the fuel tank during driving (Fig. 5).

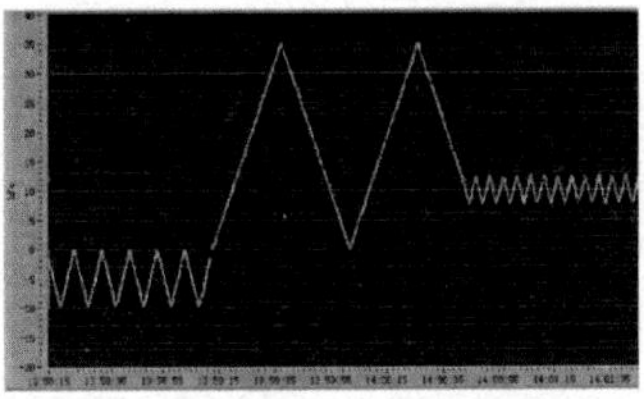

Fig. 5. Pressure waveform diagram

Phase 1:

a. 0 kPa to -10 kPa to 0 kPa, 50 times. Rate of pressure change: 2 kPa/s.
b. Pressure maintaining for 5 min.
c. 0 kPa to 35 kPa to 0 kPa, 20 times. Rate of pressure change: 1 kPa/s.
d. Pressure maintaining for 5 min.
e. 7.5 kPa to 12.5 kPa to 7.5 kPa, 350 times. Rate of pressure change: 2 kPa/s.
f. Pressure maintaining for 5 min.
g. Repeat cycle a to cycle f 100 times.

Phase 2: −15 kPa to 35 kPa to −15 kPa, 20 times. Rate of pressure change: 0.3 kPa/s. Sealing test shall be carried out after the test. During the test, if any dye liquid is found to flow out, proving that the fuel tank has leaked, the test shall be stopped immediately.

Take the deformation of each point before the test as the reference point (represented by A-0), and measure the change between phase 1 (represented by A-1) and phase 2 (represented by A-2) (“ + ” means The surface of the tank has expanded, “–” means The tank surface has shrunk). The test data are as follows (Table 1 and Fig. 6).

Table 1. Data of deformation

	Tank A (mm)			Tank B (mm)			Tank C (mm)		
	A-0	A-1	A-2	B-0	B-1	B-2	C-0	C-1	C-2
1	0.00	2.35	4.62	0.00	0.93	1.79	0.00	0.93	1.67
2	0.00	1.97	4.01	0.00	1.27	2.27	0.00	1.27	1.70
3	0.00	1.94	4.55	0.00	0.94	2.37	0.00	0.94	2.11
4	0.00	2.25	4.60	0.00	0.99	2.10	0.00	0.99	2.05
5	0.00	2.48	4.11	0.00	1.05	1.58	0.00	1.05	1.65
6	0.00	2.08	4.90	0.00	1.16	2.60	0.00	1.16	2.44
7	0.00	2.07	4.60	0.00	0.95	1.86	0.00	0.95	1.55
8	0.00	2.09	4.50	0.00	1.03	2.70	0.00	1.03	2.43
9	0.00	2.21	4.86	0.00	1.01	2.60	0.00	1.01	2.27
10	0.00	2.00	4.03	0.00	1.29	2.41	0.00	1.29	2.57

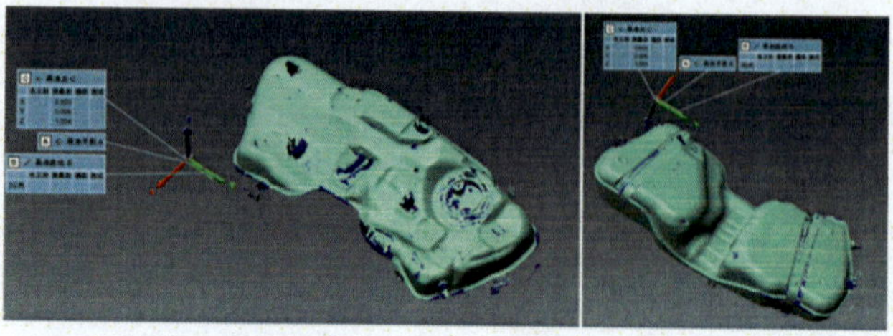

Fig. 6. 3D scan of oil tank

From the test results, it can be concluded that all three fuel tanks have withstood the pressure pulse test for a long time. Compared with the deformation of key points, it can be found that stainless steel products are better than carbon steel products than plastic products.

4 Deformation Test of Constant Temperature and Pressure

4.1 Select three new fuel tank D, E, F consistent with the above test, mark the same point, and carry out the sealing test.

4.2 PHASE 1: add 50% of the rated volume of water to the fuel tank, measure the deformation of each point as the reference point (expressed in D-0).

PHASE 2: keep the internal pressure of the fuel tank at 6 kpa, set the fuel tank at 60 °C for 6 h, and then measure the deformation of each point (indicated by D-1).

PHASE 3: keep the internal pressure of the fuel tank at 10 kPa, set the fuel tank at 60 °C. Measure the deformation of each point (represented by D-2).

PHASE 4: keep the internal pressure of the fuel tank at – 4 kPa, set the fuel tank at 60 °C.Measure the position of each point (indicated by d-3). Test data are as follows (" + " means The surface of the tank has expanded, "–" means The tank surface has shrunk) (Table 2).

4.3 It can be concluded from the test results that the stainless steel product is superior to the carbon steel product, which is better than the plastic fuel tank. The Deformation at fuel pump, fuel filler and general profile of plastic fuel tank exceeds 5 mm, and the deformation of heat shield or strap are even exceeds 10 mm. After the support bar is added, the tank shows higher pressure resistance than the original one, but there are still some gaps with metal high pressure resistant tank.

Table 2. Data of Deformation.

	TANK D(mm)				TANK E(mm)				TANK F(mm)			
	D-0	D-1	D-2	D-3	E-0	E-1	E-2	E-3	F-0	F-1	F-2	F-3
1	0.00	2.71	2.97	-1.56	0.00	1.8	1.96	-0.94	0.00	1.11	1.47	-0.23
2	0.00	2.83	3.49	-0.32	0.00	1.54	1.82	-0.51	0.00	1.25	1.42	-0.44
3	0.00	1.71	3.11	-0.95	0.00	1.38	1.47	-0.87	0.00	1.74	1.77	-0.16
4	0.00	1.82	2.79	-1.44	0.00	1.38	1.95	-1.01	0.00	1.87	2.07	-0.20
5	0.00	1.75	2.77	-1.63	0.00	1.00	1.74	-0.59	0.00	1.19	1.36	-0.26
6	0.00	2.77	3.78	-1.56	0.00	1.82	1.99	-0.74	0.00	1.31	1.33	-1.08
7	0.00	3.08	4.87	-1.25	0.00	1.46	1.92	-0.91	0.00	1.30	1.61	-0.94
8	0.00	3.26	4.12	-1.76	0.00	1.66	1.93	-1.13	0.00	1.88	2.03	-0.57
9	0.00	2.19	3.04	-0.02	0.00	1.81	2.02	-1.34	0.00	1.45	1.61	-0.74
10	0.00	2.07	3.50	-1.24	0.00	1.44	1.95	-0.60	0.00	1.80	1.98	-0.12

5 High Temperature Durability and Temperature Cycling Test

When hybrid electric vehicle is driven by electricity, the fuel tank is sealed, the high temperature durability is a new demand which is different from the traditional fuel tank. It may cause fuel tank rupture, fuel leakage, fuel injection, splashing and other serious consequences if the fuel tank can not bear high temperature and pressure for a long time. The purpose of temperature cycle test is to simulate the effect of expansion and contraction on fuel tank and support element.

As the temperature change has little effect on the metal fuel tank, a plastic fuel tank is selected for the test. Fix the fuel tank on the test fixture to simulate the real vehicle state, keep the internal pressure of the fuel tank at 45 kPa, set the fuel tank at 60 °C for 14 days. After the test, sealing test of oil tank are carried out, the result shows that there were no leakage of the fuel tank, but the fuel tank had obvious deformation. When the fuel tank is opened, it can be found that the deformation occurs at the connection of fuel tank, there is no obvious deformation at the support of fuel tank.

Select another plastic fuel tank and fix it on the test fixture to simulate the real vehicle state. Add 50% rated volume of antifreeze, test the fuel tank as the following steps: 80 °C for 15 h, room temperature for 1 h, −40 °C for 7 h, room temperature for 1 h. After 10 days of high and low temperature cycle test, sealing test of oil tank are carried out, the result shows that there were no leakage of the fuel tank, but the fuel tank had obvious deformation. When the fuel tank is opened, it can be found that the deformation occurs at the connection of fuel tank, there is no obvious deformation at the support of fuel tank (Fig. 7).

Fig. 7. Typical support structure

We found that the support frame plays an important role in the plastic high-pressure fuel tank. Dumbbell shape is widely used in the support. The increase of cross-sectional area at both ends can increase the binding force between the support and the fuel tank, and the smaller cross-sectional area in the middle can increase the deformation of the support. At the same time, the connection strength between the metal rod and PE layer can be improved by adding stiffeners on the external surface of the fuel tank shell, and the structural form of the fuel tank can be optimized by fillet connection to avoid stress concentration at the structural transition.

6 Pressure Relief Performance Test

This item is an obvious difference between pressure resistant fuel tank and traditional fuel tank. When filling the vehicle with fuel, if there is pressure inside the fuel tank, there is a risk of high temperature and high pressure gasoline vapor splashing or fuel filler cap flying out. Install the fuel tank that simulated real vehicle state, keep the internal pressure of the fuel tank at 45 kPa, open the pressure valve through the electronic control system, exhaust the fuel system, and start to release the pressure. Record the pressure curve during pressure relief. It can be concluded that the time when the internal pressure of fuel tank drops to 5 kpa should not exceed 10 s (Fig. 8).

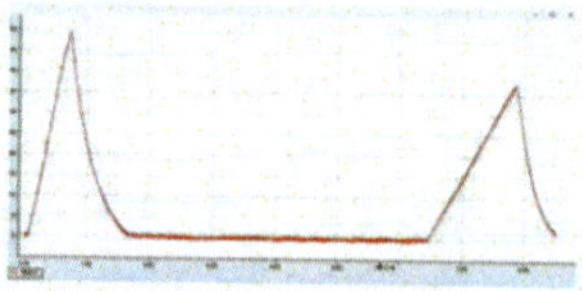

Fig. 8. Curve of pressure relief performance

7 Ventilation Performance Test

This experiment is to consider the impact of fuel sloshing on the ventilation performance of fuel tank during actual driving. The ventilation performance is divided into static ventilation performance and dynamic ventilation performance. The fuel tank is tested under the front and rear inclinations of 16.7° and left and right inclinations of 8.5° and horizontal conditions. It is required that the internal pressure of the fuel tank shall not exceed 10 kPa, otherwise the fuel system will generate overpressure, resulting in leakage. Also, there should be no liquid fuel spilled to the carbon tank, otherwise the carbon tank will be polluted, resulting in excessive hydrocarbon pollution.

7.1 Static ventilation performance test: fix the fuel tank on the test fixture to simulate the real vehicle state, add 105% of the rated volume of water into fuel tank. Open the FTIV through the electronic control module to make the fuel tank an unsealed state. Continuously inject air of 5 L/min into the fuel tank, connect the air pipe to the beaker filled with water, observe whether there is continuous bubble overflow; slowly rotate and reverse test bench, The tilt angles of front, rear, left and right are 16.7°, 16.7°, 8.5° and 8.5°. Record the internal pressure of the fuel tank. During the test, there shall be continuous bubbles in the vent pipe, and the internal pressure shall not exceed 10 kPa.

7.2 Dynamic ventilation performance test; fix the fuel tank on the test fixture to simulate the real vehicle state, add the rated volume of water into fuel tank. Seal the oil tank and conduct the test as follows:

a. Tilt forward from 0° to 35° in 1 s, keep for 2 S, and return to horizontal position in 1 s, keep for 2 S.
b. Tilt backward from 0° to 35° in 1s, keep for 2 S, return to horizontal position in 1s, keep for 2 S.
c. Incline from 0° to 10° in 1 s, keep for 2 S, return to horizontal position in 1 s, keep for 2 S.
d. Incline from 0° to right to 10° within 1 s, maintain for 2 S, return to horizontal position within 1 s, maintain for 2 s.
e. Repeat the above cycle 75 times. During the test, the internal pressure of the fuel tank shall be less than 35 kPa.

7.3 Six axis test

The six-axis test bench can meet the test bench with six free degrees, it can achieve higher frequency and more obvious deformation. The test bench inputs the road spectrum of Belgian roads, angle steel roads, washboard roads, twisted roads and other road conditions collected by the actual vehicle into the system to simulate the ventilation performance of the fuel tank. Due to the difference of installation positions and installation forms of different vehicle models, the feedback to different road conditions. Therefore, the actual use of the fuel tank can be simulated to the maximum extent by placing sensors on the fuel tank (Fig. 9).

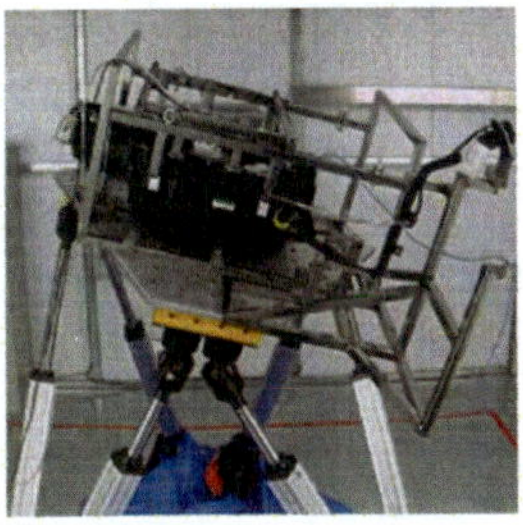

Fig. 9. Six axis test of oil tank

8 Other Performance Requirements

The common standards for fuel tank are GB 18296-2019 *safety property requirements and test methods for automobile fuel tank and its installation* [1] and ECE R34 *uniform provisions concerning the approval of vehicles with regard to the prevention of fire risks* [2]. Compared with the requirements of the national standard and the European standard, we carried out low-temperature impact performance and fire resistance performance tests on the plastic pressure resistant fuel tank according to the difference test.

The requirements of 5.3 in GB 18296-2019 are adopted in the low-temperature impact performance test, but the impact point needs to select other positions such as the connection of the support to inspect whether the impact will cause damage. The fire resistance test adopts the requirements of ECE R34 Annex 5 *testing of fuel tanks made of a plastics material 5. Resistance to fire* (Figs. 10 and 11).

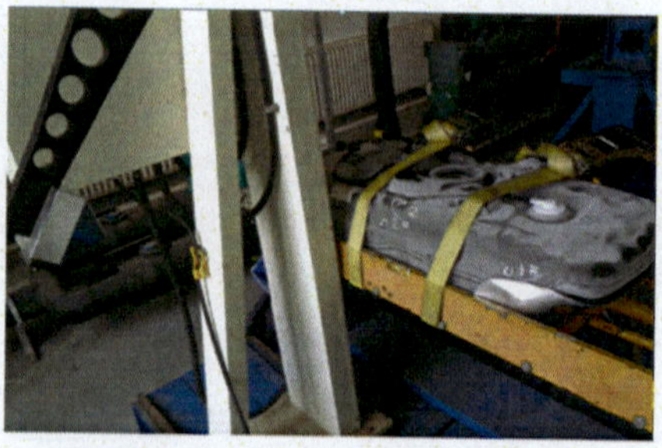

Fig. 10. Low temperature impact test

Fig. 11. Fire resistance test

According to the test, the performance of the pressure resistant fuel tank is not different from that of the traditional fuel tank in terms of low temperature impact resistance and fire resistance, so there is no special requirement.

9 Study on Explosion Proof Performance of Metal Pressure Resistant Fuel Tank

In the past, metal fuel tanks were mostly used in commercial vehicles, military vehicles and other large vehicles. Now, more and more pressure resistant fuel tanks of hybrid vehicles are made of stainless steel or carbon steel and other metal materials. Metal fuel tanks are prone to deflagration which will cause extremely serious consequences. Therefore, relevant standards clearly stipulate that the relevant fuel tank products must adopt barrier explosion-proof technology.

According to 6.7 of JT/T 1178.1-2018 *Safety specification for commercial vehicle for cargos transportation - Part 1: Goods vehicle* and 6.7 of JT/T 1178.2-2019 *Safety specification for commercial vehicle for cargos transportation - Part 2: Towing vehicle and trailer*, the fuel tank of gasoline trucks and oil traction vehicles shall adopt barrier explosion-proof technology, which shall conform to JT/T 1046-2016 *Explosion suppression safety technical requirements for road transportation vehicle fuel tank and liquid fuel transportation tank*. In the national military standard of the people's Republic of China GJB 8455–2015 *General specification for explosion suppression materials used in fuel tanks* and JT/T 1046-2016 *Explosion suppression safety technical requirements for road transportation vehicle fuel tank and liquid fuel transportation tank* [3], the performance requirements for barrier and explosion-proof materials are specified.

It is specified in the standard that the original structure and function of the fuel tank shall not be changed after the barrier explosion-proof material is added, the volume reduction rate of the fuel tank shall not be more than 6%, and the compatibility test indexes of the fuel tank before and after the barrier explosion-proof material is added shall not be changed (the compatibility test includes chromaticity, acidity, solvent cleaning gum, solid particle pollutant, etc.). At the same time, the volume resistivity, compression strength, combustion performance, vibration collapse deformation debris quality of the barrier explosion-proof material are clearly defined.

After the fuel tank is filled with explosion-proof materials, the fuel tank shall also meet the following explosion-proof performance:

a. When the fuel tank is subject to the combustion and explosion pressurization test, the combustion and explosion pressurization value shall not be greater than 0.14 MPa;
b. During the static explosion test of the fuel tank, the liquid fuel in the fuel tank should not explode twice;
c. When the fuel tank is roasted, the liquid fuel in the fuel tank should not explode twice;
d. When the fuel tank conducts penetration test of armor piercing warhead, the duration reduction rate of the high temperature zone of oil and gas explosion in the fuel tank shall not be less than 80% (Figs. 12, 13, 14 and 15).

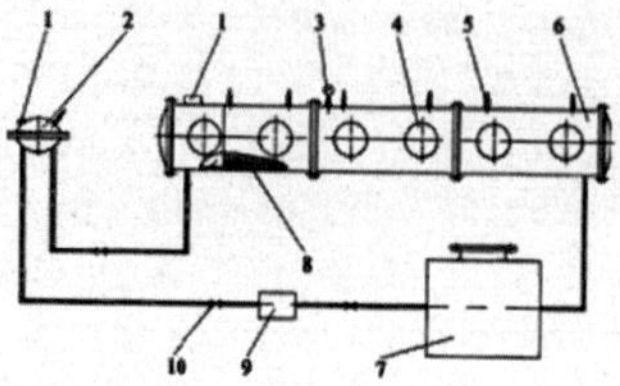

1-Ignition device 2-Explosion vessel 3-Pressure gauge 4-Observation window 5-Pressure sensor 6-Test vessel

7-Batching container 8-sample 9-Circulating pump 10-valve

Fig. 12. Deflagration test device

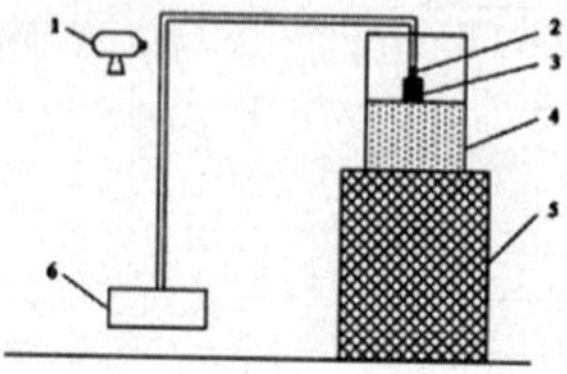

1-High speed camera 2-detonator 3-gunpowder

4-Test drum 5-Bracket 6-Electric detonator

Fig. 13. Static explosion test device

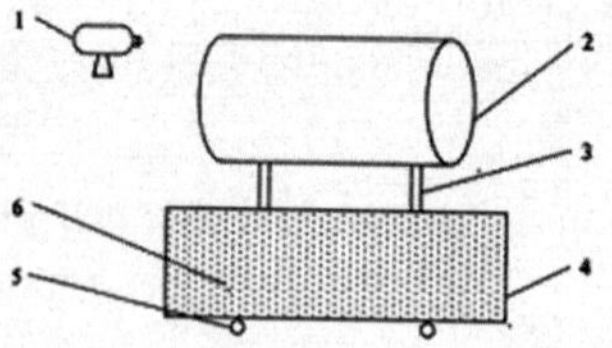

1- High speed camera 2-tank 3-Metal 4-grating

5-Oil tank 6-pulley 7-Aviation kerosene

Fig. 14. Baking and burning test device

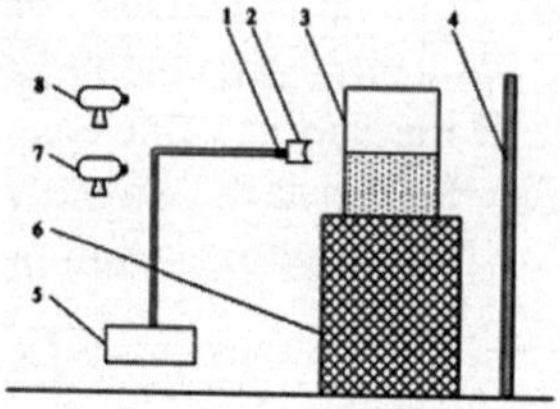

1-detonator 2-Armor breaking warhead 3-Test drum 4-baffle 5-Electric detonator 6-Bracket 7-infrared camera

8-High speed camera

Fig. 15. Penetration test device of armor piercing warhead

10 Conclusion

Compared with the traditional fuel tank, the pressure resistant fuel tank used in hybrid electric vehicle has changed obviously in structure and safety performance. Through the difference test, it can be concluded that the pressure resistant fuel tank needs to bear more working pressure and produce less deformation at the same time. The requirements of temperature resistance, pressure resistance and ventilation are much higher than the traditional fuel tank. In order to meet the requirements, the fuel tank needs to be improved in material or structure, the manufacturing process and testing standards of the pressure resistant fuel tank should be improved as well.

References

1. GB 18296-2019 safety property requirements and test methods for automobile fuel tank and its installation
2. ECER34 uniform provisions concerning the approval of vehicles with regard to the prevention of fire risks
3. JT/T 1046–2016 Explosion suppression safety technical requirements for road transportation vehicle fuel tank and liquid fuel transportation tank

Analysis on Vibration Characteristics of Spring Passive Valve High-Pressure Long-Distance Slime Paste Pipeline Transportation

Lyu Fuyan, Xiaohui Hou, Li Sun, Li Chunzhi(✉), and Jia Xuankai

College of Mechanical and Electronic Engineering,
Shandong University of Science and Technology, Qingdao 266590, China
skdlcz@163.com

Abstract. Coal slime is high concentration and viscosity, which has viscoelastic characteristic when transporting in pipe of high pressure and long distance. The paste is compressed by the double-cylinder plunger pump in long pipe. When the pump switched, the pressure disappears instantly and paste flows back, forming a reverse impact on the pumping equipment, which causes severe vibration of the pumping system. In order to reduce the reversed flow and vibration, a spring passive check valve is installed at the pump outlet, but its effect is not clear. This paper analyzes the effect of the hydraulic check valve on the vibration characteristics based on the test data of the long distance pipeline in project, and carries out the actual vibration measurement and analysis. The vibration characteristics of the spring passive check valve dense paste transportation pipeline are obtained based on the viscoelastic paste axial vibration model of the pipeline, which provides a reference for the selection of coal slime paste transportation pipeline parameters and the method of reducing the shock and vibration of the pipeline system.

Keywords: Viscoelastic · Pipe transporting · The spring passive valve · Shock vibration

1 Introduction

High concentration viscous solid waste refers to the solid-liquid two-phase waste or by-product with high solid content, large viscosity and fine particles produced in the process of industrial production and municipal sewage treatment, which generally does not have fluidity under normal temperature and pressure, and flows in the form of plunger when transported in high-pressure pipeline, also known as dense paste [1, 2]. For example, coal slime in coal industry, dewatered sludge in the water supply and drainage industry, oil residue, oil sludge and other general industrial solid wastes in the petrochemical industry, involving more than 20 industries [3]. Dense paste pipeline transportation is a rapidly developing technology in the pipeline transportation industry. Pipeline transportation technology is moving towards “high pressure, large concentration, long distance, large diameter”. When the slime is transported in the high-pressure pipeline, it will be a paste-like plunger flow. It belongs to the three-phase state of gas, solid and liquid, and has obvious compressibility under high pressure.

M. Tavana et al. (Eds.): IISA 2020, AISC 1304, pp. 596–602, 2021.
https://doi.org/10.1007/978-3-030-63784-2_74

When the reversal of the S-tube valve of the double-cylinder piston pump causes pulsation of the pumping pressure, the slime paste in the pipeline will elastically recover and deform, and reverse flow will occur, referred to as backflow, and "reverse impact" will occur at the same time, causing the pump body actuator and its S-bend directional valve to withstand a strong impact force, and the pumping equipment is subjected to strong shocks to produce severe axial and lateral vibrations, driving all pipelines to vibrate (Fig. 1).

Fig. 1. The export state diagram of dense paste pump flow

During the transportation of slime paste pipelines, the pipeline vibrates violently, which has attracted the attention of many researchers. For slime paste, Daqing Zhang [3, 4], Chen Xin [5], Xueyi Zhao [6], Xuankai Jia [10], Lyu Fuyan [11], Wujisng She [12] et al. studied the shock and vibration of pipe from different aspects. The research on the vibration characteristics of high-concentration viscous paste pipelines has been widely concerned by scholars, while the research on the vibration of high-concentration viscous paste pipelines has only tested and analyzed the vibration phenomenon during impact feeding. Analysis of the vibration characteristics of the paste in the pipeline, but the research of the check valve on the pipeline vibration caused by its reverse impact is still in the blank stage, so the impact of the check valve on the pipeline vibration is urgently needed to be studied.

2 Analysis of the Flow State of Slime Paste and Its Pipeline

Coal slime is the state of non-segregation, non-stratification and non-dehydration after removing most of the water from the coal washing liquid in the coal washing process through solid-liquid separation equipment, with high solid content (70 ± 3%) and fine particles (d50 = 50 μm), Viscosity is as high as tens of thousands of pas. There is a very thin layer of low-concentration liquid between the outside of the plunger and the inner wall of the pipe, which is called the "boundary layer". This layer has a high water content, and it can be observed with naked eyes that it has obvious water precipitation out of the surface layer with higher brightness formed at the surface [8, 9], and a slip phenomenon occurs. Since the wall boundary layer provides the full velocity of the paste flow, it is called full slip [10]. When the applied external force is greater than the yield stress of the paste, the flow core area moves forward as a whole and belongs to the structural flow paste.

The slime with a mass concentration of 71.69% was measured by the elastic modulus experiment. When the pressure is 20 MPa, volume elastic modulus value up to 3.08 GPa.

3 Test Platform Operating Conditions

The slime transportation pipeline system includes: large-diameter double-cylinder piston pump, scraper conveyor, storage bin, positive pressure feeder, spring passive check valve, multi-function feeder and pipeline accessories. Carrying out tests, carrying out actual measurement and operation of the pipeline.

The test points include 3 pressure points, 1 angular displacement point, 3 vibration points. Test point layout as shown in Fig. 2.

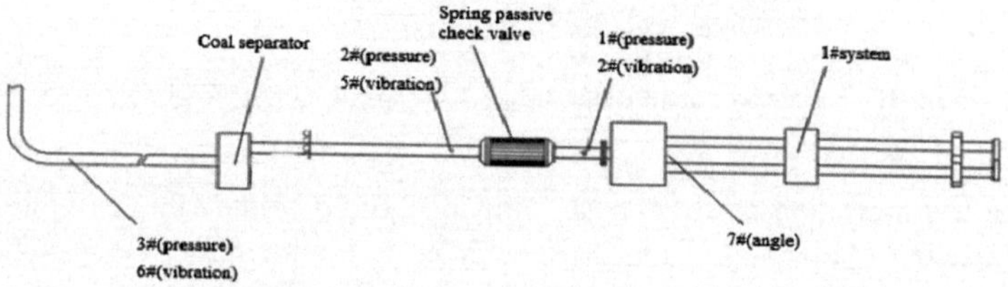

1#、2#、3#: Pressure Sensor; 4#、5#、6#: Vibration sensor(4# is divided into axial and radial); 7#: Angle sensor

Fig. 2. The position of sensors

1# is used to measure the pump outlet pressure. 1# and 2# measured the pressure loss before and after the check valve and the pressure change curve before and after the change. The measuring point 3# is used for comparing the phase relationship with the measuring points of 1# and 2#. Vibration sensors are placed at the corresponding pressure measurement points to measure the relationship between pressure changes and vibration, Place vibration sensors at the corresponding pressure measuring points to measure the relationship between pressure changes and vibration. 7# is a magnetic angle sensor, measure the commutation signal of the S swing valve cylinder.

4 Pump Commutation Pressure and Vibration Analysis

As shown in Fig. 3, the pressure sensor measured when the plunger pump pumped 55%, The transfer pump piston cylinder replacement direction, S tube valve swing cylinder swing pressure before and after the outlet pressure pulsation as shown in Fig. 3. The greatest pressure is 3.94 MPa, the minimum pressure is 2.295 MPa. Instantaneous pressure difference is 1.649 MPa.

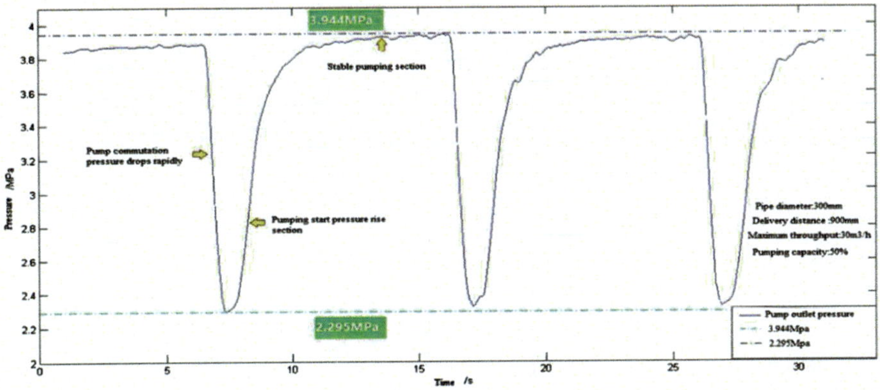

Fig. 3. Diagram of pressure pulsation at the outlet when the pumping capacity of the plunger

5 Pumping Volume vs. Pipeline Vibration Analysis

Taking 30% of pumping volume as an example, Fig. 4 is the time-domain waveform diagram of pressure and vibration along the pipeline corresponding to two commutation cycles, and the time-domain waveform of vibration at each measuring point. Vibration, 5# axial vibration, 6# axial vibration and 4# radial vibration.

Statistic the pressure change of each pressure measuring point and the effective value and peak-to-peak value of each vibration signal before and after the S-tube valve changes. As shown in Table 1, Table 2 and Table 3.

It can be seen from Table 1 to Table 3 that the amplitude of the pressure change of each pressure measuring point before and after the S-tube valve is changed to increase with the increase of the pumping volume occurs when the pumping volume is less than 60%. When the pumping volume is greater than 60%, the pressure amplitude change of each pressure measuring point before and after the S-tube valve is changed is kept in the smaller range of 5.2 MPa–5.6 MPa, 3.4 MPa–3.5 MPa and 3.2 MPa–3.3 MPa. As can be seen from Fig. 5, except for the 6# vibration measurement point, when the pumping volume is less than 60%, the effective value and peak-to-peak value of the vibration measurement point increase as the pumping volume increases when pumping. When the amount is greater than 60%, the effective value of each measuring point remains within a small change interval.

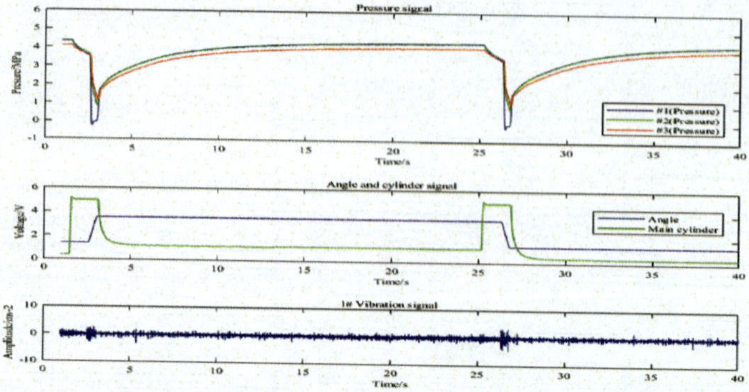

(a) Time-domain waveform of pressure and vibration at pumping capacity 30%

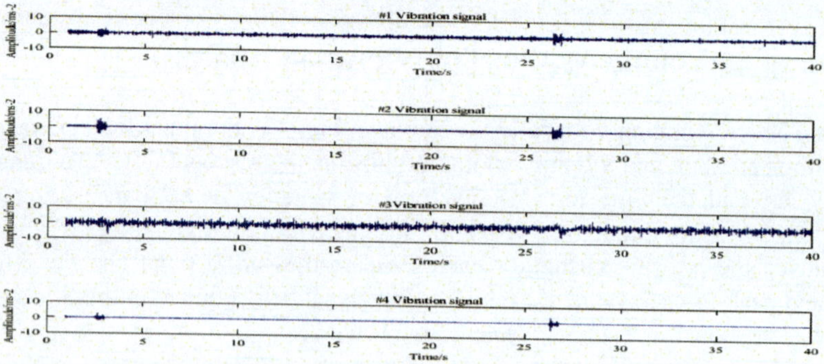

(b) Time-domain waveform of vibration at pumping capacity 30%

Fig. 4. Vibration pressure and vibration time-domain waveform at 30% pumping volume

Table 1. Pressure change at pump actuator move. Unit: MPa

Pumping capacity	1#		2#		3#	
	Forward	Reverse	Forward	Reverse	Forward	Reverse
30%	3.4	0	3.4	0.6	3.4	1.0
45%	4.2	0	4.2	1.2	4.1	1.4
60%	5.2	0	5.2	1.8	5.1	2.1
75%	5.5	0	5.5	2.0	5.4	2.3
90%	5.6	0	5.6	2.2	5.6	2.5

Table 2. RMS of vibration signal. Unit: ms^{-2}

Pumping capacity	4#	5#	6#	4#
	Axial	Axial	Axial	Radial
30%	0.3502	0.2550	0.6976	0.1038
45%	0.4934	0.3940	0.6100	0.2739
60%	0.6995	0.6136	0.6501	0.5224
75%	0.7516	0.7513	0.6919	0.6179
90%	0.7261	0.7610	0.7636	0.7680

Table 3. Vpp of vibration signal. Unit: ms^{-2}

Pumping capacity	4#	5#	6#	4#
	Axial	Axial	Axial	Radial
30%	8.0683	10.3532	11.9014	8.4674
45%	12.9524	16.5116	10.9062	16.0725
60%	36.5365	23.6224	10.6294	28.9839
75%	37.9733	26.2841	8.8257	27.0568
90%	13.6427	22.1659	10.8096	28.5224

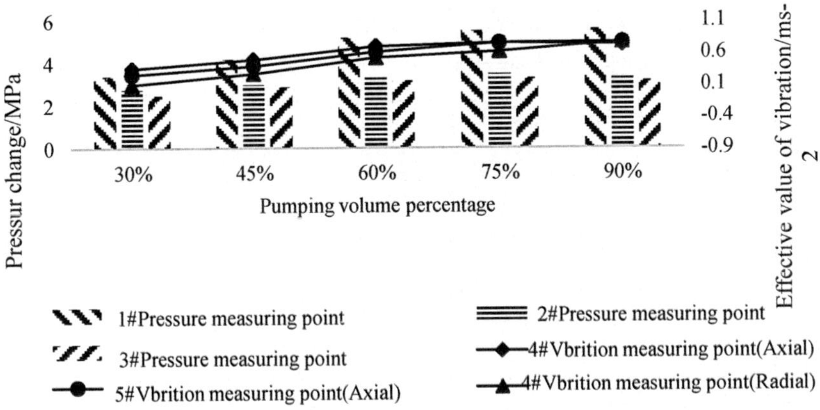

Fig. 5. Relationship between effective value of vibration signal and pressure change

6 Conclusions

Through the work completed, this article has the following research conclusions: In the pipeline of spring passive check valve, when the pumping volume is less than 60%, the effective value of vibration increases with the increase of the pumping volume. When the pumping amount increases from 30% to 60%, the effective value of the pump outlet pipeline vibration Increased from 0.3502 ms^{-2} to 0.6995 ms^{-2}, when the pumping volume is greater than 60%, the effective value of vibration remains unchanged with the increase of the pumping volume, and remains in the interval of 0.735 ms^{-2}–0.755 ms^{-2}. The axial vibration of the pump outlet pipe is greater than its radial vibration in the case of a spring passive check valve pipe.

Acknowledgments. This work is supported by the Natural Science Foundation of Shandong Province (ZR2018BEE014), and China National Science Foundation (61803374; 51874308), and Project of Shandong Province Higher Educational Young Innovative Talent Introduction and Cultivation Team [Performance enhancement of deep coal mining equipment].

References

1. Zhou, Z., Zhao, Z., Yang, X., et al.: Research on the basic characteristics of papermaking waste sludge-one of the researches on energy utilization technology of gasification treatment of papermaking waste sludge. Papermak. Sci. Technol. **20**(6), 14–17 (2001)
2. Jing, Y., Yang, Q., Jing, Y.: Basic properties and engineering characteristics of red mud. Shanxi Archit. **27**(3), 80–81, 108 (2001)
3. Zhang, D.: Vibration Test and Fatigue Life Analysis of Cement Concrete Pump Truck, pp. 73–75. Chang'an University, Xi'an (2003)
4. Zhang, D., Lv, P., He, Q., et al.: Experimental study on the dynamic strength of concrete pump truck. Vibr. Shock **24**(3), 111–113 (2005)
5. Xin, C.: Experiment and Research on Impact Test of High-Concentration Viscous Materials Pipeline. China University of Mining and Technology, Beijing (2007)
6. Zhao, X.: Experimental study on pipeline transportation characteristics and pipeline vibration state of viscous materials. [Publisher Unknown], Beijing (2009)
7. Stefan, J., Lars, H., Tor-Arne, H., et al.: Flow conditions of fresh mortar and concrete in different pipes. Cem. Concr. Res. **39**(11), 997–1006 (2009)
8. Chen, L., Duan, Y., Zhao, C., Yang, L.: Rheological behavior and wall slip of concentrated coal water slurry in pipe flows. Chem. Eng. Process. **48**(7), 1241–1248 (2009)
9. Choi, M., Roussel, N., Kim, Y., Kim, J.: Lubrication layer properties during concrete pumping. Cem. Concr. Res. **45**(5), 69–78 (2013)
10. Jia, X.: Shock and Vibration Characteristics of Thick Paste Long-Distance Pipeline System. China University of Mining and Technology, Beijing (2014)
11. Lyu, F.: Experimental study on pressure shock and vibration of long-distance slime transportation pipeline. Ind. Mine Autom. **41**(5), 91–95 (2015)
12. Yu, W., Wang, H., Chen, E., Ye, C., Zeng, B.: Simulation study on the three-dimensional dynamic flow field stability of the check valve. Rocket Propul. **41**(01), 82–89 (2015)

Intelligent Systems

Research on Intelligent Fault Diagnosis of Board Circuit Based on Expert Case Reasoning

Xiaomin Xie(✉), Kun Hu, Ying Hong, Boli Yu, and Jianghuai Du

School of Mechatronic Engineering, Anhui Vocational and Technical College, Hefei 230011, China
357106305@qq.com

Abstract. Aiming at the matching conflict problem in the inference engine of traditional expert system, a method of applying expert case reasoning to board level circuit fault diagnosis is proposed. This method focuses on the index strategy and matching principle of expert case reasoning, and gives the method of case representation, organization and storage, and maintenance measures of case library. According to the research, the combination of hierarchical multi-level indexing mechanism and k-nearest neighbor matching principle and the use of expert case-based reasoning for fault diagnosis of board-level circuits can solve the bottleneck problems such as matching conflicts. Taking the failure of power supply circuit in airborne electronic equipment as an example, it is concluded that expert case reasoning has the advantages of fast diagnosis speed and high accuracy.

Keywords: Expert case reasoning · Index strategy · K-nearest neighbor matching principle · Fault diagnosis of board-level circuits

1 Introduction

Expert case-based reasoning is an artificial intelligence technology based on expert knowledge for diagnostic reasoning [1–5]. Case is used to represent expert knowledge and a reasoning method for solving and relearning problems. It is pointed out that when people solve new problems, they often use similar situations accumulated in the past to deal with them, and make appropriate modifications and improvements on the basis of this method to solve new problems. Expert case-based reasoning diagnosis is a method to solve similar case problems by using knowledge representation or experience information of past cases. This method has great advantages in knowledge acquisition, solution efficiency, knowledge accumulation and solution quality. Expert case-based reasoning is developing towards a mode of creating intelligent systems. This mode mainly uses the existing experience and knowledge of the system for diagnosis and reasoning, and is more comfortable in diagnosis than traditional expert systems [6–8]. When solving new problems, the traditional rule-based reasoning system uses the rules in the rule base to match. The matching is successful and a solution is obtained. Otherwise, there is no solution. In view of the shortcomings of the traditional reasoning

M. Tavana et al. (Eds.): IISA 2020, AISC 1304, pp. 605–613, 2021.
https://doi.org/10.1007/978-3-030-63784-2_75

system, the ECBR system searches the existing case base according to the given conditions to obtain a matching case. There are two possibilities for search results: one is to find cases that exactly match the input features, so that the problem can be solved quickly and accurately; The second is that we can't find a completely matched case, but we can find a case or set of cases with great similarity to its characteristics, then the ECBR system can correct or learn according to the unmatched parts, and then we can get a new solution to solve the problem. The workflow block diagram is shown in Fig. 1.

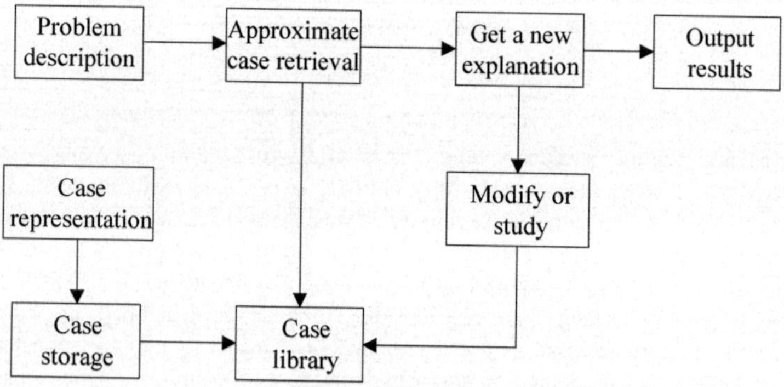

Fig. 1. Case reasoning work flow chart

2 Establishment and Addition of Case Library

In board level circuit fault diagnosis, a case solution will be given for feature extraction and matching of the given fault phenomena in the case library. Maintenance personnel usually need to fill in some test parameters and failure information such as damage of main components before and after diagnosis when troubleshooting board-level circuits. These fault data information constitute the fault cases of maintenance personnel or experts during diagnosis.

2.1 Representation and Organizational Storage of Cases

(1) Representation of cases
In the board level circuit fault diagnosis case base, the case content is mainly divided into three parts: problem description, solution and feedback information. The problem description includes two parts: the external performance and the characteristic attributes of the case, mainly referring to the phenomena and basic characteristics when a fault occurs. Among them, the influence of different characteristic elements on case problems is also different. The case problem representation description based on feature attributes is shown in Fig. 2.

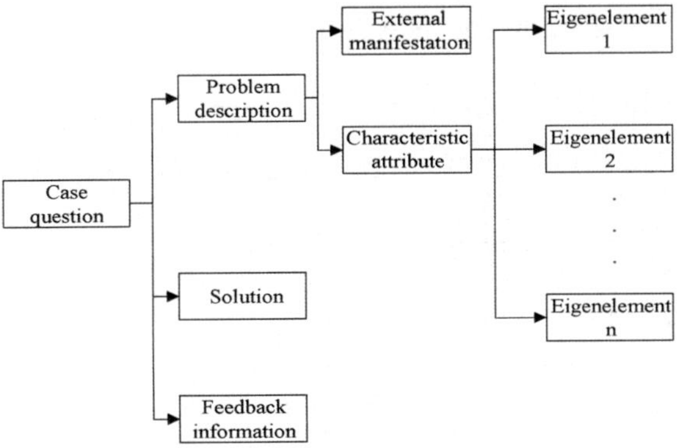

Fig. 2. Cases based on feature attributes that describe the figure

(2) Organizational Storage of Cases
In the board level circuit fault diagnosis system, cases are organized and stored according to the strategy of fault classification to achieve the purpose of building a fault case library [9]. Among them, the data table technology is used to design the fault case category table, fault case table, fault case feature table and fault key feature table. The relationship between these four data tables is shown in Fig. 3.

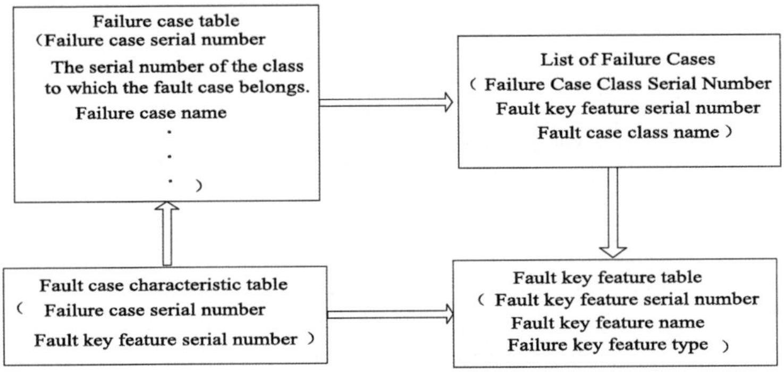

Fig. 3. The relationship between the four data table figure

2.2 Addition of Case Library

For the board level circuit fault diagnosis case library, to add a new case, a board level circuit fault problem needs to be formally described as a fault case, and the description to be completed includes: board type, case serial number, fault type (e.g. power circuit abnormality), fault key feature attributes, repair personnel and time, repair personnel and time, and solutions, etc.

3 Expert Case Reasoning Mechanism

3.1 Workflow Based on Expert Case Reasoning

Working process based on expert case reasoning: when a board-level circuit fails, the characteristic information of the problem can be extracted from the current failure phenomenon, and the extracted information is provided to the diagnosis system in the form of case organization, thus forming a target case (case to be diagnosed). Target cases are indexed and matched in the case base through the case retrieval mechanism. If a case completely matching the target case can be found in the case library, the solution of the target case is directly output from the case library; If the matching is not complete, the correlation matching and similarity calculation can be carried out according to the extracted characteristic parameters, and the preliminary diagnosis scheme can be determined through comparison [10, 11]. The solution and strategy provided by this scheme do not exactly match the solution of the current problem. The reason is that although the preliminary diagnosis scheme is very similar to the target case, there are still some differences. Therefore, according to the needs of solving the current target case, the specific situation should be appropriately revised and improved to confirm the final solution. The workflow of expert case-based reasoning is shown in Fig. 4.

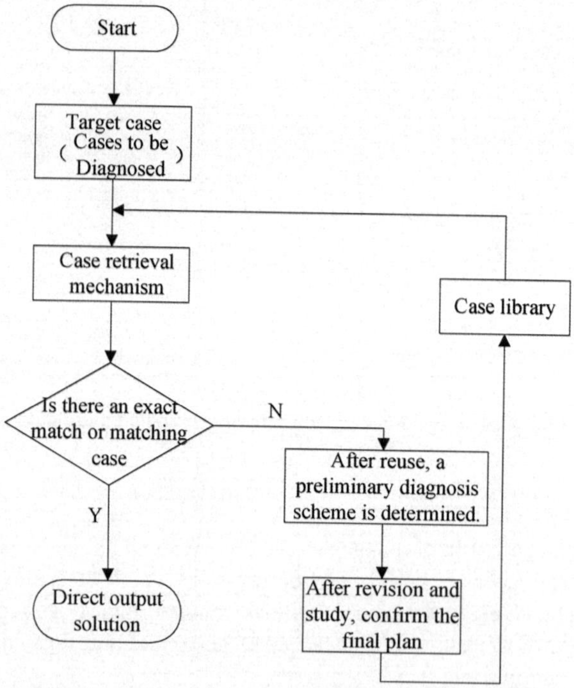

Fig. 4. The workflow of expert case-based reasoning

3.2 Case Retrieval

Case retrieval is to search and match one or a group of cases from the current case library that are most similar to the target case in terms of characteristic attributes, thus obtaining the solution of the target case. Among them, the efficiency of case retrieval depends on case indexing strategy and case matching process. This article focuses on the analysis of case index and case matching.

(1) Index of cases
According to the representation and organization structure of cases in the case base, case index usually adopts the method of hierarchical multi-level index. Case-based reasoning system induces, summarizes and extracts feature attributes of cases from a large number of cases, and organizes some cases into a hierarchical representation structure according to these feature attributes. This indexing method can organize cases hierarchically and hierarchically in tree form, and can summarize the corresponding characteristic attributes from most cases. According to the difference of case contents and attributes, hierarchical multi-level index divides the case library into tree-like or mesh-like structures and makes them be expressed in a structured way. The hierarchical multi-level index structure diagram is shown in Fig. 5.

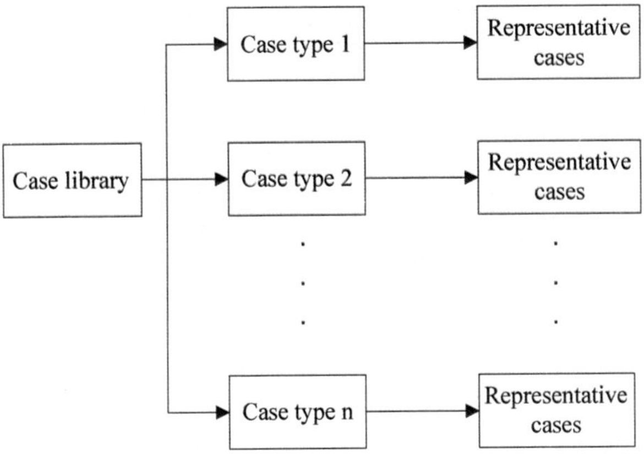

Fig. 5. The hierarchical multi-level index structure diagram

(2) Case matching
The case matching process is mainly based on the similarity between the target case (case to be diagnosed) and a series of cases in the case library. At present, the k- nearest neighbor model, which is a widely used similarity matching comparison model, is used. This matching model mainly classifies new cases by calculating the similarity analysis of the cases. Its premise is to assume the cases as points in European space and ignore cases with great differences.

K-nearest neighbor model

The nearest neighbor retrieval method is to search the case method closest to the current target case from the case library. Each characteristic attribute of the case is assigned or a corresponding weight is calculated. In case retrieval, the weighted sum of the matching degree between the input target case and each feature attribute of the cases in the current case library is obtained according to the weight value to select the best matching case. Nearest Neighbor Strategy: First, the case retrieved from the case retrieval is called a similar case, and then a case with the smallest difference in similarity matching degree is retrieved from the similar case, then the case is the best matching case. The k-nearest neighbor model is calculated according to the standard Euclidean distance method, i.e. the cases in the case base are assumed to correspond to points in the N-dimensional space, and the characteristic attributes of the cases requiring matching are consistent.

Let any case X be expressed as the following feature vector:

$$(b_1(x), b_2(x), \ldots, b_n(x))$$

Where, the $b_r(x)$ attribute value of case x is represented. The distance between the two cases is X_i,X_j

$$d(X_i, X_j) = \sqrt{\sum_{r=1}^{n} \left[b_r(X_i) - b_r(X_j)\right]^2} \tag{1.1}$$

The selection of K in k- Nearest Neighbor Weighted Thought is to assign a larger weight to the nearest neighboring cases. Among them, Euclidean distance method is mainly used to find out some cases closest to case X, and then k- nearest neighbor formula is used to calculate the similarity between these similar cases and target case X. In case-based reasoning expert system, the weights of feature attributes in a case are given by experts in related fields. The similarity of k- nearest neighbor model is defined as follows:

$$S_{ij} = \frac{\sum_{r=1}^{k} m_r P'_{ij}}{\sum_{r=1}^{k} m_r} \tag{1.2}$$

Among them, S_{ij} represents the similarity matching degree between case I and case j, m_r represents the weight value of the r-th feature information in the whole case library, m_r weight value is generally selected according to the weighted sum in the case library, P'_{ij} represents the similarity of the r-th feature attribute of case I and case j, and k represents the number of all feature attributes of the case.

4 Application of Case Reasoning System Based on Experts

In the board-level circuit case-based reasoning system, its core work is to search for similar fault cases, complete the reasoning and diagnosis of input fault phenomenon problems, find fault cases similar to fault problems, and use the searched fault case diagnosis scheme as the guidance scheme for new fault problems (target cases). Similar fault case retrieval in expert case reasoning system mainly includes fault problem description, case index and case matching.

In the aviation instrument and meter part, if the power indicator light does not illuminate or flashes intermittently, the fault signal source can be located in the power circuit module part. After entering the fault diagnosis interface, select the “power circuit” control button to enter and enter the more detailed fault phenomena. At this time, multimeters and related electronic instruments are used to measure the relevant characteristic parameter values in the power supply circuit, such as voltage value of 25 V, current of 2.5 mA, frequency of 38 Hz and output power of indicator lamp of 5.8 W. According to this fault phenomenon and fault data information, the diagnosis steps using case-based reasoning are as follows:

(1) Fault problem description
Input the fault key characteristic attributes of airborne electronic equipment;
Key feature name: circuit module, location of occurrence, abnormal phenomenon
Key characteristic values: power supply circuit, indicator lamp, not on or intermittent on/off
(2) According to the case index strategy, the characteristic attributes and similar candidate case sets are obtained according to the relevant fault phenomenon descriptions given, as shown in Table 1.

Table 1. Feature attributes and similar candidate case sets

Case number	Voltage (V)	Current (mA)	Frequency (Hz)	Output power (W)
xxm-a	24.8	2.5	52	6.9
xxm-b	26.5	3.1	45	7.4
xxm-c	27.7	3.5	49	8.2
xxm-d	25.8	2.9	52	7.6

The key characteristic weights of this type of fault are:
Voltage: 0.22 Current: 0.27 Frequency: 0.45 Output Power: 0.15
The characteristic attribute value of the target case fault is:
Voltage: 25.0 V Current: 2.5 mA Frequency: 48 Hz Output Power: 6.5 W

(3) Case matching process
According to the requirements, the k-nearest neighbor formula is used to calculate the similarity between the case to be diagnosed (power circuit failure problem of airborne electronic equipment) and the candidate case set, as shown in Table 2.

Table 2. Case similarity

Case number	Similarity
xxm-a	12.256
xxm-b	4.649
xxm-c	6.879
xxm-d	8.878

According to the case matching results, the case similarity in the candidate case set is shown in Table 2. According to the previous matching principle, the closer it is to the target case, the smaller the similarity is. Therefore, the case number xxm-b is the most similar to the target case in this article. After experts in the field of diagnosis (senior maintenance experts) reuse and partially revise the diagnosis results (solutions) of case xxm-b, the diagnosis solution obtained is the final solution of the target case in this system. Among them, the detailed fault information of case xxm-b, such as the cause of the fault, the type of the fault and the diagnosis scheme, is as follows:

The type of the failed board: ACP1167015
Case number: xxm-b
Fault type: power supply circuit
The characteristic weights are respectively:
voltage: 0.20 current: 0.25 frequency: 0.43 output power: 0.12
Characteristic attribute value:
voltage: 26.0 V current: 3.0 mA frequency: 43 Hz output power: 7.2 W
Possible cause of failure: unstable voltage or frequency

Solution:

1. Check whether the voltage of the power supply circuit reaches the actual required value range with a multimeter.
2. Check again whether the power frequency changes normally within the specified range.

According to the detailed fault information of the most similar case xxm-b, and through reuse and partial correction study, the final solution is obtained: crystal oscillator fault is the main cause of inverter power control board fault in power controller. Measures taken: Only the crystal oscillator needs to be replaced.

5 Conclusions

The expert case reasoning thought proposed in this paper fully takes into account the experience of fault cases accumulated by electronic equipment maintenance personnel in the actual maintenance process of board-level circuits, and represents, organizes and stores these cases and establishes a perfect case library system. In the process of diagnostic reasoning, the first step is to extract fault information knowledge from board level circuits, such as the magnitude and period of electrical signals and other parameter values. According to the extracted fault information, searching and matching with the experience cases in the case base; If there is a matching case, the expert case reasoning diagnosis is successful, and the fault information and its solution are output. Using expert case-based reasoning can overcome the bottleneck problems such as matching conflicts in traditional system reasoning. Taking the failure of power supply circuit in airborne electronic equipment as an example, it is concluded that expert case reasoning has the advantages of fast diagnosis speed and high accuracy.

Fund Projects. Natural Science Research Key Project of Anhui Province Higher School (KJ2019A0991);

Natural Science Research Key Project of Anhui Xinhua University (2018zr008).

References

1. Wu, D.: An electronic commerce recommendation algorithm joining case-based reasoning and collaborative filtering. In: Proceedings of 2015 International Industrial Informatics and Computer Engineering Conference (IIICEC 2015), pp. 1222–1225 (2015)
2. Melville, P.: Recommender systems. In: Sammut, C., Webb, G.I. (eds.) Encyclopedia of Machine Learning, pp. 829–838. Springer, Boston (2010)
3. Li, L.: Research on the intelligent decision support system of group events based on case reasoning. Economics and Management College, Xi'an Science and Technology University, Xi'an (2012)
4. Martin, A., Emmenegger, S., Hinkelmann, K., Thönssen, B.: A viewpoint-based case-based reasoning approach utilising an enterprise architecture ontology for experience management. Enterp. Inf. Syst. **11**(4), 551–575 (2017)
5. Thönssen, B., Witschel, H.-F., Hinkelmann, K., Martin, A.: Experience knowledge mechanisms and representation. Learn PAd—Model-Based Social Learning for Public Administrations—EU FP7-ICT-2013-11/619583, Pisa, Italy (2016)
6. Agarwal, S., Sureka, A.: Using KNN and SVM Based One-Class Classifier for Detecting Online Radicalization on Twitter. In: Natarajan, R., Barua, G., Patra, M.R. (eds.) ICDCIT 2015. LNCS, vol. 8956, pp. 431–442. Springer, Cham (2015)
7. Wei, X., Stillwell, D.: How smart does your profile image look? Estimating intelligence from social network profile images. In: Proceedings of the Tenth ACM International Conference on Web Search and Data Mining, pp. 33–40. ACM (2017)
8. Ahmed, S., Danti, A.: Effective sentimental analysis and opinion mining of web reviews using rule based classifiers. In: Behera, H.S., Mohapatra, D.P. (eds.) Computational Intelligence in Data Mining, pp. 171–197. Springer, India (2016)
9. Gao, H., Zhang, Y., Zhou, X., Li, D.: Intelligent methods for the process parameter determination of plastic injection molding. Front. Mech. Eng. **13**(1), 85–95 (2018)
10. Jiang, Z., Jiang, Y., Wang, Y., Zhang, H., Cao, H., Tian, G.: A hybrid approach of rough set and case-based reasoning to remanufacturing process planning. J. Intell. Manuf. **30**(1), 19–32 (2016)
11. Li, Z., Liu, L., Barenji, A.V., Wang, W.: Cloud-based manufacturing blockchain: secure knowledge sharing for injection mould redesign. Procedia CIRP **72**, 961–966 (2018)

Analysis of Key Technical Problems in Internet of Vehicles and Autopilot

Yongqiu Liu(✉)

Guangdong University of Science and Technology, Dongguan 523083, China
banbianqiu@126.com

Abstract. At present, China's key technologies for car networking and autonomous driving are still not developed. Although many years of efforts have made some progress, if you want to safely put car networking and autonomous vehicles into the market, and a wide range There are still many problems in the promotion of the land. Among them, the wireless technology advancement roadmap problem, the V2X investment problem, and the 5G technical issue. If a car wants to realize the Internet of Vehicles and autopilot technology, it needs to pay attention to the three steps of perception, decision-making and execution. 5G technology can be the finishing touch in these three steps, and it is very important to improve the efficiency of the three steps. Technology. This paper uses perception and decision making as the main aspect of analysis. In addition to considering the 5G technology issues, it is also necessary to fully consider the issue of information security and privacy, as well as hardware configuration issues.

Keywords: Vehicle networking · Automatic driving · Key technology · Problem analysis

1 Introduction

China has a high degree of attention to networked automatic driving, and requires its frequency band of 5905–5925 MHz, with a dedicated frequency resource of 20 MHz bandwidth, which is very important for intelligent networked direct communication technology (C-V2X). Ford is expected to sell its C-V2X technology in China for the first time in 2021. Geely also announced that it is expected to launch models supporting C-V2X and 5G technology in the same year. In the 5G era, the relationship between autonomous driving technology and 5G is very close, not only the attention of the Ministry of Transport of China, but also the attention of the Ministry of Industry and Information Technology, and in order to further realize the combination of autonomous driving and car networking, the intelligent traffic and digital construction of the network has also become an extremely important task. This paper analyzes the problems existing in the Internet of Vehicles and Autopilot [1].

M. Tavana et al. (Eds.): IISA 2020, AISC 1304, pp. 614–618, 2021.
https://doi.org/10.1007/978-3-030-63784-2_76

2 Car Networking and Autonomous Driving

Autopilot is a technique in which a vehicle can be safely driven under the control of a computer [2]. This technology can help liberate people's hands. If the technology matures, the safety factor of the car will be greatly improved, the number of traffic accidents will be greatly reduced, and people can carry out other activities in the car, saving people's time. At the end of 2108, China and the United States had already opened test photos for unmanned vehicles, but current driverless cars can only support simple bicycle intelligence, and the computer's processing unit and vehicle sensors can realize the most surrounding environment of the vehicle. Measurements and evaluations are carried out, but intelligence is driven on roads with simple traffic conditions [3]. If used in complex traffic conditions, it is difficult to make correct judgments, which may lead to traffic accidents. Vehicle networking technology can build pedestrians, signal lights, other vehicles and other information into a network, so that vehicles can better analyze the road conditions, can more fully understand the surrounding road conditions, and need to have certain intelligent equipment on the road to cooperate [4]. Therefore, it is necessary to build a smart road and help the driverless car become a smart car. The car networking technology is shown in Fig. 1.

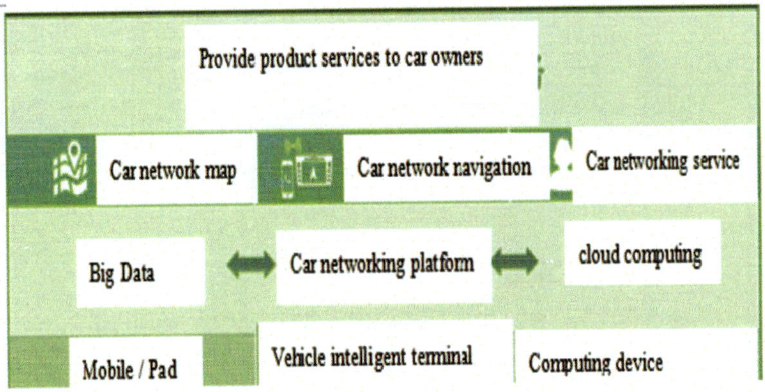

Fig. 1. Vehicle networking technology

3 The Limitations of Car Networking and Autopilot Key Technologies

At present, China's key technologies for car networking and autonomous driving are still not developed. Although many years of efforts have made some progress, if you want to safely put the Internet of Vehicles and autonomous vehicles into the market, and carry out on a large scale. There are still many problems in the promotion. Among them, there are wireless technology promotion roadmap problems, V2X investment issues, and 5G technical issues [5].

3.1 Wireless Technology Evolution Roadmap Problem

When the auto-driving and car-connected vehicles Ganggang began research, there was no new technical means such as 5G. Therefore, the original wireless technology evolution did not take into account the 5G technology in the original infinite technology evolution roadmap. The roadmap remains on a 4G basis and needs to be updated and redesigned.

3.2 Investment Issues of V2X Technology

In the process of researching the Internet of Vehicles and autonomous driving technology, it is necessary to spend a lot of energy and funds, and it is necessary to pay labor costs for the personnel involved in the research. Therefore, it is necessary to continuously invest in V2X technology, but currently China is using this technology. The investment is still not in place, resulting in fewer talents willing to invest in this technology research.

3.3 G Technical Issues

With the development of technology, 5G technology is about to bring revolutionary progress to human society, so many things need to be related to 5G, especially the products of cutting-edge technology, and car networking and autonomous vehicles are cutting-edge research directions. And it has not yet been put into the market, so it is necessary to integrate 5G technology in consideration of its future market development prospects [6]. However, all previous studies have not considered 5G technology, so it is necessary to solve this problem as soon as possible, otherwise it will be produced. Vehicles will become "outdated" vehicles, and their gains in the market will be severely affected.

4 5G in the Car Network Automatic Driving Application

If a car wants to realize the Internet of Vehicles and autopilot technology, it needs to pay attention to the three steps of perception, decision-making and execution. 5G technology can be the finishing touch in these three steps, and it is very important to improve the efficiency of the three steps. Technology [7]. This paper uses perception and decision making as the main aspect of analysis.

4.1 Perceptual Step

At present, the sensing steps of the automatic driving technology are mainly based on satellite positioning methods [8], but this method is subject to weather and nearby signal interference, which makes the stability difficult to maintain, so the safety performance of the vehicle will be greatly reduced, and it is possible It will threaten the safety of the people inside the vehicle, and it may also threaten the safety of the vehicles and pedestrians on the road they are driving. If you want to realize the vehicle networking technology of the vehicle, avoid it is only a single smart vehicle, you need to ensure that it can match the high-precision map direction, and can have network-assisted differential positioning, which can enhance the object it perceives. Accuracy, these higher requirements, it is very difficult to implement on the basis of 4G, but under 5G technology, it will be much easier to implement, because 5G technology has the advantages of large capacity, high speed and safe and reliable performance. Therefore, the vehicle traveling on the road can be perceived more quickly, and the delay can be greatly improved.

4.2 Decision Steps

The current 4G technology does not perfectly implement the vehicle network technology, because there is a delay in the communication between the vehicle and the vehicle, which causes a delay in the decision making process. At the time of execution, it is possible to make problems that are not suitable for the environment at the time, resulting in a security incident, and the current 4G technology can not achieve remote operation of people and vehicles, but also due to delay problems, resulting in sudden When the situation arises, the computer can't make the right decision like humans, and when the remote manipulator receives a dangerous signal, the danger may have occurred due to the delay. The remote person can't do it. Anything to avoid problems. However, after combining 5G technology, the problem of delay can be obviously solved, and the synchronous communication between vehicles can be realized. The decision made is also adapted to the environment at that time, and there is no delay in human remote operation. Realize the safe manipulation of the vehicle by the human distal end.

5 Car Networking and Automatic Driving Need to Be Further Solved

In addition to considering the 5G technology issues, it is also necessary to fully consider the issue of information security and privacy, as well as hardware configuration issues.

5.1 Information Security and Privacy Issues

Since the Internet of Vehicles and the self-driving vehicles need to be networked, the driving path of the user who uses the vehicle is exposed to the Internet, and all the things associated with the Internet have the possibility of being exposed, and other information of the owner. It will also be exposed to the Internet, which will cause the privacy of the owner and the security of their information to be threatened. If the person who is badly ill has access to the information and use it illegally, it may lead to serious consequences, so it also needs information security. Paying close attention to privacy issues requires not only key, but also a more secure protection system.

5.2 Hardware Configuration Problem

At present, the research on autonomous vehicles in China is still limited to 4G. Many hardware devices are researched and designed on the basis of 4G. Therefore, it is difficult to completely match them under 5G technology, and it cannot be completely The advantages of 5G technology are exerted. Therefore, it is necessary to continuously research new hardware configurations and further improve the safety performance of vehicles. In the selection of materials, etc., economic and safety issues should also be fully considered, and 5G cannot be satisfied. All hardware devices have been upgraded and upgraded, and they are well prepared for the coming of the 5G era.

6 Conclusion

Autopilot technology is a technology that is being paid attention to all over the world. China attaches great importance to this technology, because if autopilot technology can realize the combination of safe vehicle networking technology and automatic driving, it can not only liberate people's hands and help people drive. Doing other things on the road to help people save time, can also improve the safety factor of driving, reduce the number of traffic accidents, and also reduce the problem of road congestion through the Internet of Vehicles technology, especially during the peak hours of commuting, if To solve this problem, the time spent by our people on the road can be greatly reduced. People can devote more time and energy to life and work and improve the quality of life. Therefore, car networking and autonomous driving technology are important to human beings. Meaning, the current problems need to be resolved as soon as possible.

Acknowledgments. Key Research Platforms and Scientific Research Projects of Guangdong Province in 2018 (Special Innovation Projects): Research on Key Technologies of Information Transmission in Vehicle Networking Application (2018KTSCX262).

References

1. Szabolcsi, R.: Optimal PID controller based autopilot design and system modelling for small unmanned aerial vehicle. Rev. Air Force Acad. **09**, 43–58 (2018)
2. Pengfei, F.: Key technology of vehicle network automatic driving based on 5G communication. Electron. Technol. Softw. Eng. **16**, 29–30 (2019)
3. Vehicle and road coordination! The first domestic self-driving 5G vehicle networking demonstration island was built. Auto Parts **07**, 27 (2019)
4. Wei, L.: A "traffic evolution" will be: 5G brings new kinetic energy to the internet of vehicles and autopilot. Big Data Era **04**, 6–15 (2019)
5. Yunlong, B., Kaixin, Y., Xiaowei, C., Haibo, D., Jinyu, G.: 5G + V2X car network automatic driving. Comput. Knowl. Technol. **15**(08), 129–132 (2019)
6. Ying, W., Jinwang, W.: Key technical issues of car networking and autonomous driving. Electron. Prod. World **24**(05), 20–22 (2017)
7. Zhiyong, Z.: Research on LoRa Wireless Positioning System Based on Human Energy Collection and Self-powered. East China Jiaotong University (2018)
8. Xiaofeng, C.: Porous Polypropylene Piezoelectric Electret Device for Human Energy Collection and Self-powered Sensing. Huazhong University of Science and Technology (2016)

Research and Development Housing Rental System with Recommendation System Based on SpringBoot

Yaozhang Li[1], Sheng Gao[1(✉)], Weisheng Wu[1], Peifeng Xie[1], and Hao Xia[2]

[1] Faculty of Mathematics and Computer Science, Guangdong Ocean University, Zhanjiang, China
minix@139.com

[2] Faculty of Electronics and Information Technology, Guangdong Ocean University, Zhanjiang, China

Abstract. With the rise of the floating population in the past few years, the housing demand of the society is also increasing. In order to improve the combination of Internet and traditional group properties and explore the feasibility of recommendation system in the housing rental industry, the system adopts spring Boot and other frameworks for research and development. The system realizes the functions of house information editing, house information auditing, house information publishing and so on. In the process of using the system, the system will collect the user's browsing records, construct user feature vectors, recommend house listings and potential roommates for the tenant through the recommendation algorithm, and improve the rental efficiency.

Keywords: Housing rental industry · Recommendation system · System development · SpringBoot

1 Introduction

1.1 Background

With the development of the Internet industry, traditional industries combine with the Internet industry to improve and upgrade, which has an impact and brings convenience on social life. As for the housing rental industry, the improvement is the provision of online sales functions for the rental industry. Users can contact the landlord orletting agency through the Internet; they can also research of the house more information through the Internet to improve information transparency.

With the gradual increase of data, if you search houses directly, it will lead to a lot of houses are difficult to be found which are suitable for user. Existing housing rental software such as: 58 city, Airbnb does not have a good solution to this problem. With the development of economy in recent years, the increasing number of young floating population has led to the increasing demand for rental housing. In view of the above problems, the housing rental industry has not been effectively resolved during the combination of the Internet industry, leading to the loss of potential users and difficulty

M. Tavana et al. (Eds.): IISA 2020, AISC 1304, pp. 619–627, 2021.
https://doi.org/10.1007/978-3-030-63784-2_77

in expanding the market. Therefore, in the combination of housing rental industry and Internet industry, there is much room for growth. A recommendation system can improve the conversion rate of potential users and improve the efficiency of housing rental.

1.2 Literature Review

The recommendation system is a potential solution to the problem of information overload in the Internet [1]. when we get information through Internet, recommendation system play a import role [2]. The recommendation system has been widely used in many industries. In the e-commerce industry, recommendation system recommend product information and suggestions for users, simulation of virtual salesperson to help users complete the purchase process [3, 4].

The recommendation system consists of three modules: information collection module, information analysis module and recommendation algorithm module [5, 6]. The information collection module is responsible for collecting user behavior records, user preferences and item information; the information analysis module is responsible for translating the information collected by information collection module into data, which will be used in the recommendation algorithm module; the recommendation algorithm module is responsible for calculations between user data and item data to provide users with recommended items. The current types of recommendation algorithms include: content-based recommendation [7], collaborative filtering recommendation [8], hybrid recommendation [9], utility-based recommendation [10], knowledge-based recommendation [11] and so on.

In the real estate development, floating population is not the main development target of real estate developers; therefore in the relationship of housing supply, the living space of the floating population can not be satisfied [12]. At this time, the government needs to take a lead in real estate development to satisfy the living space demands of the floating population. Through the recommendation system, the government can collect the rental demand of different users to assist the government to provide a data for real estate development. Combined with the Intelligent City to provide housing suggestions for the floating population, so as to achieve a virtuous cycle of mutual promotion from demand to supply [13].

2 Development Technology and System Architecture

This chapter mainly introduces the development technology, application framework and system architecture of this system. This system uses an object-oriented development method and uses Client-Server for improving system operating efficiency and reducing coupling. The following will introduce the development technology and system architecture in detail.

2.1 Development Technology

This system uses front-end and back-end separated Client-Server for development. This article mainly introduces back-end development technology. Spring Boot is the main development technology of the system. Its main function is to reduce the complexity of the project and integrate other Java frameworks. In order to achieve front-end and back-end separation, SpringMVC framework divides the whole system into three lawyers: Model lawyer, View lawyer, Controller lawyer. The system uses the SpringMVC framework to code the main logic program for back-end interaction with the front-end. In order to improve the efficiency of the recommendation algorithm, this system use non-relational database—redis. Compared to the more widely used relational database, the data types in redis are more suitable for characteristic value data operations and storing. And use Spring Data Redis framework to interact with redis.

2.2 System Architecture

The system uses the object-oriented development method. In order to reduce the coupling degree of the system, the C/S development mode is used to realize the separation of front-end and back-ends. The system is divided into view layer, model into and control layer by SpringMVC framework. Such a structure can reduce the coupling degree of the system while retaining the strong expansibility of the system (Fig. 1).

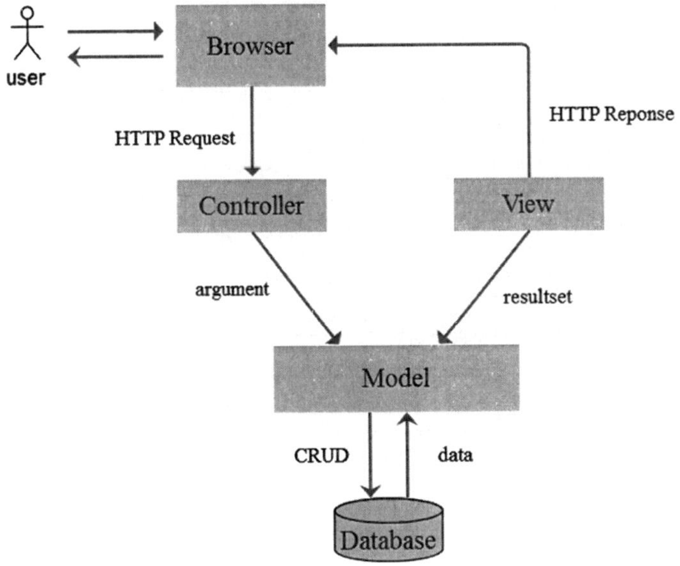

Fig. 1. Architecture of system

As can be seen from the above figure, during the use of the system, the user sends a request to the Controller layer through the View layer. After the request is processed by the Controller layer, the parameters are calculated in the Model layer and the result is returned to the View layer. After the Model layer calculations, the database stores data to achieve the Model layer upgrade. The data in the Model layer is finally passed to the front-end code and the result is finally returned to the user.

2.3 Algorithm Analysis

In order to realize the function of recommending houses, we need to characterize the house data and construct a feature matrix of the house data. By design, the housing data consists of the following attributions (Table 1).

Table 1. Housing attributions

Attribution	Description	Participate in operation
House ID	A unique identifier of the house in the database	Yes
Name	Name of the house	No
Address	The address of the house	No
Type	The type of the house	Yes
Rent	The rent of the house	Yes
Area	The area of the house	Yes
Url	The location of the house image in the server	No
WIFI	The house have WiFi or not	Yes
Washing machine	The house have washing machine or not	Yes
Kitchen	The house have kitchen or not	Yes
Decorate	The house can be decorated by yourself or not	Yes
First time rent	The house is rented for the first time or not	Yes
Ready to move in	The house can move in immediately or not	Yes

As you can see from the table, most attributions can be represented by 0 or 1 in the database storage. Attributions like Rent Area differ from the interval [0, 1], which need to be normalized. Delete other attributions like House ID and Url which do not participate in the operation.

$$\mathrm{x}_i = [x_{i1}, x_{i2}, x_{i3} \cdots x_{ij}] \tag{1}$$

$$X = \begin{bmatrix} x_1 \\ x_2 \\ x_3 \\ \cdots \\ x_i \end{bmatrix} = \begin{bmatrix} x_{11}, x_{12}, x_{13} \cdots x_{1j} \\ x_{21}, x_{22}, x_{23} \cdots x_{2j} \\ x_{31}, x_{32}, x_{33} \cdots x_{3j} \\ \cdots \\ x_{i1}, x_{i2}, x_{i3} \cdots x_{ij} \end{bmatrix} \tag{2}$$

In the above formula, i represents the number of features of the house, and j represents the length of the feature vector-attributions of the housing. The feature matrix X is composed of this vectors.

Because Euclidean distance is not obvious in the high-dimensional data calculation, the cosine distance is used here instead of Euclidean distance. Since the magnitude of each value in the feature matrix is in the interval [0, 1], the cosine distance is in [0, 1]. Therefore, the cosine distance between the two vectors is the largest, that the two vectors are the most similar. The similarity formula of the two vectors is as follows.

$$sim(x_i, y) = \cos(x_i, y) = \frac{x_i \cdot y}{||x_i|| \bullet ||y||} \tag{3}$$

In this system, the recommendation algorithm has the problem of cold start, because the new user has no user behavior records, there is no data for recommendation. To solve this problem, we will assume a feature vector for this user, take the average of all users as the preference of this user. In the subsequent process, users will modify their feature vectors according to their behavior records, thereby achieving the purpose of personalized recommendation.

3 System Development

3.1 Requirements Analysis

According to the above analysis, the users who use the system can be divided into two types: the publisher (hereinafter referred to as the landlord) and the general user (hereinafter referred to as the tenant). The landlord edits the listing information through the system. In order to ensure the normal operation of the system and the credibility of the information, the information published by the landlord will be identified by the system administrator, and the administrator decides whether the information can be published. After the information published, the tenants can obtain the listing information directly through the system or through the system recommendation. The user case diagram of each role in the system is as follows (Fig. 2).

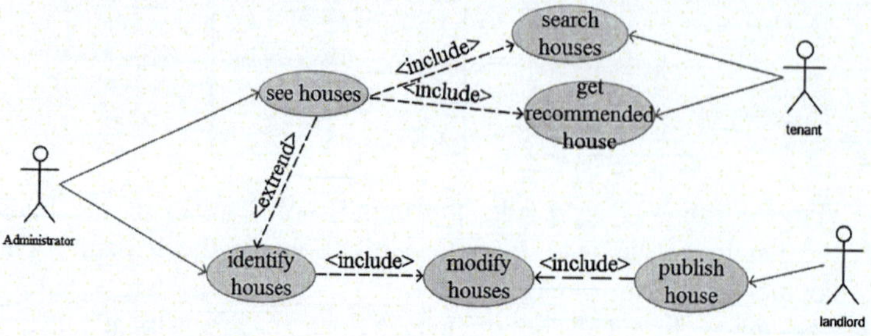

Fig. 2. User case diagram of system

In the above process, the system needs to collect user behavior records. Modify the user's preference characteristics according to the user's behavior record to achieve the user's personalized recommendation.

3.2 Recommendation Algorithm

According to the above algorithm analysis, two data structures are needed to represent the stored data of the housing and the feature vector of the housing. In the Java language, Class is generally used as the data structure of the data set, and the class used to store the data is called the Data Class, and the class used for the feature vector is called the Feature Class. The following uses the E-R diagram to show the relationship between the two classes (Fig. 3).

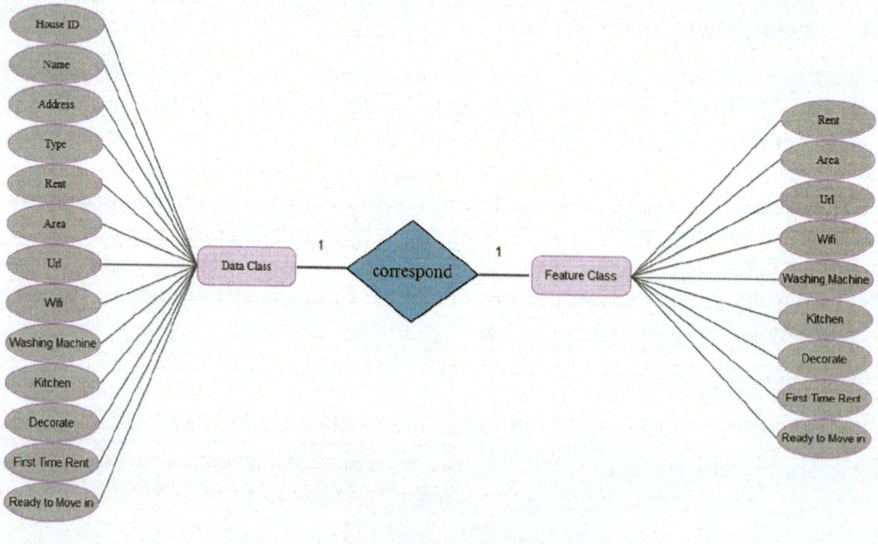

Fig. 3. E-R diagram between two classes

The house data needs to be deleted and normalized to participate directly in the calculation. Using the mapping function in the Java language, the utility class BeanUtil is used to map the attributions of the Data Class object to the Feature Class object. Then the data will be normalized.

The housing ID and the mapped feature class objects are stored in the form of <key, value> using the map data type. Cosine distance calculations are performed on the user's preference characteristics and the values in the map one by one, sorted by size, then the listing with the largest cosine distance will be recommended. In order to achieve the cosine operation of the two feature class objects and improve the readability of the code, we use the mapping function in Java to obtain the same attribution value of the two Feature Class objects for operation. The process is shown as follow (Fig. 4).

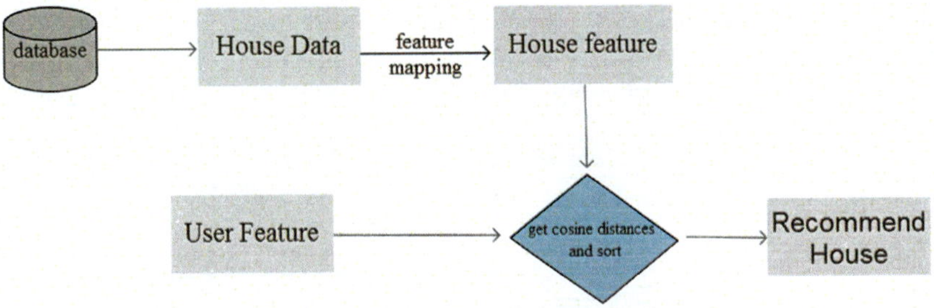

Fig. 4. Recommend housing process

When a user browses a house, system can obtain the information of the users who browsed the house in the recent period, and obtains their preference characteristics by reading the browsing history of this housing. After cosine calculation, system will recommend potential tenants according to the similarity (Fig. 5).

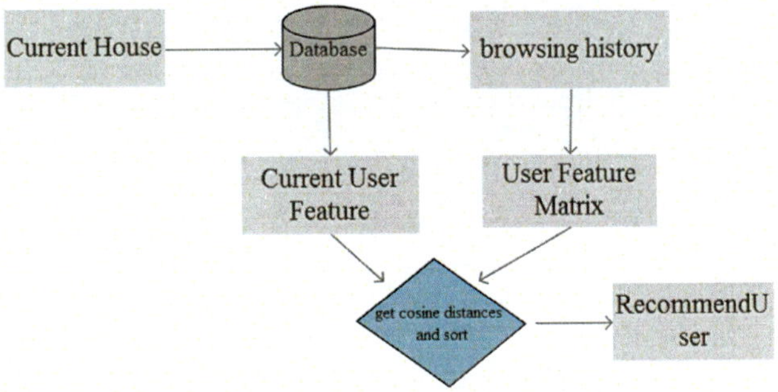

Fig. 5. Architecture of system

4 Evaluation and Future Work

4.1 Evaluation

In the above analysis and research, the system realized the functions of publishing, reviewing, and searching for housing information. A complete rental process is provided in the rental supply. The system's recommendation system is a large amount of housing information and user information, can be used as a supplement to the search system. While the recommendation system provides users with personalized services, the system discovers the user's points of interest for users, thereby guiding users to discover their information needs and improve the efficiency of the system. In addition, the function of recommending renters to users can also shorten the distance between users and improve system adhesion.

However, in the preparatory stage of the project, due to insufficient data collection and fewer user records, the system recommended housing may not be accurate enough. During the project development phase, distributed computing was not implemented. When the system data reached a high level, the server will be overloaded, affecting the use of the system.

4.2 Future Work

In future work, in order to further improve the accuracy of the recommendation system, we will use a semi-supervised classification algorithm to classify users according to their preferences to achieve more accurate recommendations. At the same time, in order to improve system performance and prepare for large-data operations, the distributed cluster operation and maintenance of the system will be realized.

In terms of exploring the user's housing and personal living economic conditions, we conduct data mining and big data analysis through the user data collected by the system, and research the relationship between the user's housing situation and personal living habits and economic conditions. Do data analysis and data foundation for urban construction during when the migrant population is gradually growing.

Acknowledgments. This research is supported and funded by the following projects. Guangdong Ocean University Innovative Entrepreneurship Training Project (CXXL2019247) and Guangdong Ocean University Innovative Team for College Students (CXTD2019003).

References

1. Perugini, S.: Recommender Systems Research (2005)
2. Cosley, D., Lam, S.K., Albert, I., et al.: Is seeing believing? How recommender system interfaces affect users' opinions. In: Proceeding of the SIGCHI Conference on Human Factors in Computing Systems, vol. 5, pp. 585–592 (2003)
3. Wei, K., Huang, J., Fu, S.: A survey of E-commerce recommender systems. In: International Conference on Service Systems and Service Management. IEEE (2007)
4. Chen, B.S.: An electronic commerce recommender system based on product character. Adv. Mater. Res. **267**, 909–912 (2011)

5. Jian-Guo, L., Tao, Z., Qiang, G., et al.: Overview of the evaluated algorithms for the personal recommendation systems. Complex Syst. Complex. Sci. **6**, 1–10 (2009)
6. Jianguo, L., Tao, Z., Binghong, W.: Research progress of personalized recommendation system. Adv. Nat. Sci. **019**(001), 1–15 (2009)
7. Lops, P., Gemmis, M.D., Semeraro. G.: Content-based recommender systems: state of the art and trends. In: Recommender Systems Handbook. Springer US (2011)
8. Wei, S., Ye, N., Zhang, S., et al.: Item-based collaborative filtering recommendation algorithm combining item category with interestingness measure. In: International Conference on Computer Science & Service System. IEEE (2012)
9. Shih, Y.Y., Liu, D.R.: Hybrid recommendation approaches: collaborative filtering via valuable content information. In: Hawaii International Conference on System Sciences. IEEE Computer Society (2005)
10. Liang, S., Liu, Y., Jian, L., et al.: A utility-based recommendation approach for academic literatures. In: IEEE/WIC/ACM International Conference on Web Intelligence and Intelligent Agent Technology. ACM (2011)
11. Felfernig, A., Gula, B., Leitner, G., et al.: Persuasion in knowledge-based recommendation. In: Persuasive Technology, Third International Conference, Persuasive, Oulu, Finland. Springer-Verlag, June 2008
12. Hongping, L., Jianfeng, L., Gangqiao, Y.: Study on the status quo of housing in small towns based on different economic development levels in China. In: International Conference on Management Science and Engineering. IEEE (2009)
13. Intelligent city. Natl. Munic. Rev. **32**(2), 68–82 (2016)

Design of Intelligent Recommendation System of Smart Library Under Big Data Environment and Its Application Research in Applied University

Weining Huang(✉)

Guangdong University of Science and Technology, Dongguan 523083, China
13100327@qq.com

Abstract. With the development of society and the progress of the times, the era of big data has gradually become a new stage of development. "Big data" is a collection of data that is difficult to obtain, manage and analyze in a short time under the calculation of conventional software. The so-called "big data environment" is a large-scale data integration that can be developed by the support of the current network technology and computer technology. The recommendation system of the smart library is to be established and founded under such an environment. The recommendation system of smart library is designed to provide more beneficial help for the education system of application-oriented universities on the basis of perceiving students' preference in reading and learning. In the context of big data, the intelligent library recommendation system designed in this paper is to collect data on students' learning and reading needs with the help of big data. Integrate, process and analyze on the basis of collection, and finally send the results presented after this process to students to form a specialized smart library recommendation system suitable for applied universities. The establishment of this system can better meet the needs of students of applied universities, recommend students to adapt to their unique learning resources, and help the data-driven development of applied universities to be carried out better.

Keywords: Big data · Smart library · Intelligent recommendation system design · Applied university

1 Introduction

The idea of the smart library comes from the concept of "smart earth" put forward by IBM President and CEO Samuel J. Palmisano in 2008. The so-called smart library is to introduce big data technology, cloud computing technology and Internet of things technology into the construction of the library. The advantage of this is that it can transform and upgrade traditional libraries with storage as the main function. To build an intelligent analysis and management resource. A new type of library that can effectively share knowledge on a larger level and accurately perceive the needs of readers. The intelligent recommendation system design of the smart library is mainly

M. Tavana et al. (Eds.): IISA 2020, AISC 1304, pp. 628–634, 2021.
https://doi.org/10.1007/978-3-030-63784-2_78

supported by big data,thereby forming a system that can provide readers with personalized reading and learning resources. The completion of this process must first record and predict the readers' information such as resource requirements, and provide the basis for the data for the next stage. Because of the powerful information collection, data analysis and processing functions of big data,it can help the library to collect huge amounts of information,it can also perform relevant data analysis and process data efficiently. Provide readers with a higher level of intelligent recommendation services. This paper conducts relevant research and analysis on the design of smart library intelligent recommendation system under big data environment and its application in applied universities, Pay more attention to the practicality of the intelligent recommendation system of smart libraries, and explore how to help students of applied universities to make better use of this resource of the library.

2 Goal Setting of Intelligent Recommendation System of Smart Library in Big Data Environment

Under the background of big data, the setting of the goal of the intelligent recommendation service of the applied university smart library should consider two principle issues. The first is that the intelligent recommendation system of the smart library in the application university should mainly serve the students of the application university. Therefore, the first consideration for the goal setting of the intelligent recommendation system of wisdom library should be the satisfaction of students' needs for collection resources, the ability to accurately identify students' personalized needs, and help them generate the most personalized resources. The second is that the target setting of the intelligent recommendation system of the smart library should fully consider the utilization rate of resources in the library, which should be one of the main goals of the intelligent recommendation system.

(1) Precise positioning and actively meet the students' needs for library resources
The library as a resource treasure house of every applied university, contains the richest knowledge of the entire school and a variety of academic information. It should be the first place for students to acquire knowledge. But in recent years, with the development and progress of the society, science and technology have been constantly improved, students have access to resources and information greatly increased, the status of the library is declining. For students, although the library has rich knowledge resources and information resources, However, the time and energy required for the retrieval and search of these resources make them more willing to turn to help, which is not as rich as the resources of the library, but the search method and the process of obtaining information are relatively simple methods. For example, Baidu Encyclopedia, Zhihu, Tianya and other websites are increasingly showing the style that covers the library.
Therefore, in the libraries of applied universities, we should pay more attention to the integration of information resources and data in the library. Further complete the relevant analysis and processing of data, develop new services and systems, and enhance the service value of applied universities. The resources of books,

newspapers and periodicals in the libraries of applied universities are complex. In order to solve this problem, we can transform and upgrade the directory system so that this directory system completes the tasks of the corresponding instructions, and then according to the students' the query records these preference data and pushes them for other resources they may need. This kind of intelligent recommendation system has greatly improved the previous search mode based on consultation, changed the catalogue system of application-oriented university libraries, made the library resources play the role of helping application-oriented university students to the greatest extent, and made the contents of the collection closely linked with the needs of application-oriented university students. For this reason, big data is involved in the construction of intelligent recommendation system of smart library, which can help the library resources to integrate, store and analyze to the maximum extent, so that this framework can provide the most suitable content for readers or college students.

(2) Strengthen the connection between library resources and education of applied universities

The advantage of big data is not simply to use some programs to solve some theoretical problems and methods. The most important thing is to use these data to better apply the technology of big data to the practice we want to carry out. Help us to complete our practical tasks more efficiently. The many advantages of big data in the intelligent library of the applied university can make the library of the applied university better use these technologies to promote the combination of the library itself and the education of the applied university. From the perspective of improving the resource utilization efficiency of libraries of applied universities, the intelligent recommendation system of smart libraries of applied universities mainly needs to solve the problem of the relevance between the library resources and the curriculum settings of applied universities.

In the wisdom of an applied university library intelligent recommendation system, we should be on campus through the use of big data processing technology of data processing and analysis, fully considering the course content, the students improve library resources with the matching degree of these content, in order to help students to find more useful resource. In this way, it can not only effectively improve the matching degree of the resources of the library in the applied university with the curriculum of the applied university, it can also continuously increase students' interest in collection resources,realize the combination of the curriculum of the applied university library and the application university and the self-study of the students. Help students of applied universities to better their academic performance, get a better learning experience, and promote the improvement of teaching effects of applied universities.

3 The Key Problems of Intelligent Library Intelligent Recommendation System Design and Its Application Research in Applied University Under Big Data Environment

(1) Openness and standardization of big data
During the operation of the intelligent recommendation system of the smart library, the system usually mines according to the reader's retrieval preferences, Dig into the need for reading some resources and make records to help the intelligent recommendation system of the smart library in the next step of the process, to achieve the push of relevant useful information. This requires big data to have certain standards and open rows, and both standards and open rows must be satisfied. Without any of the properties, we can not successfully complete the intelligent recommendation of the smart library system setup and construction. In the process of related configuration, the application-oriented university should contact each data with the business of each system to strengthen the unity and standardization of the school system. At the same time, it is also necessary to ensure that the recommendation of resources in the library cannot be limited to a certain retrieved record, remove some miscellaneous interference, improve the value of received data, and achieve better intelligent recommendation, so as to greatly enhance the value-added role of libraries in application-oriented universities through big data, and promote them to give play to their own value.

(2) Continuously track students' reading behavior
In the process of building the intelligent recommendation system of the smart library, we should continue to track and perceive the behaviors related to the learning and reading completed by students in the library with the help of the resources in the library. In the process of collecting big data in this respect, we can use some terminal devices, such as monitoring equipment, smart machines that borrow and return books autonomously, reading sensors, inductive bookshelves and resource positioning systems. The use of these devices can maximize the library's tracking and recording of students' learning and reading behavior. Collect the reader's use of the library and related academic characteristics. Whether it is the collection and acquisition of static data or the capture of dynamic data, we must pay attention to the standardization of these data and the construction of contextualization in the process of system setup to ensure that these data can be given corresponding value, Let these data be able to complete the construction of an intelligent recommendation system to meet the needs of students' personalized reading and learning.

4 Construction of Intelligent Recommendation System Model of Smart Library Under Big Data Environment

The following Fig. 1 is a schematic diagram of the main modules of the smart library, which helps us to explore the construction of the memory model.

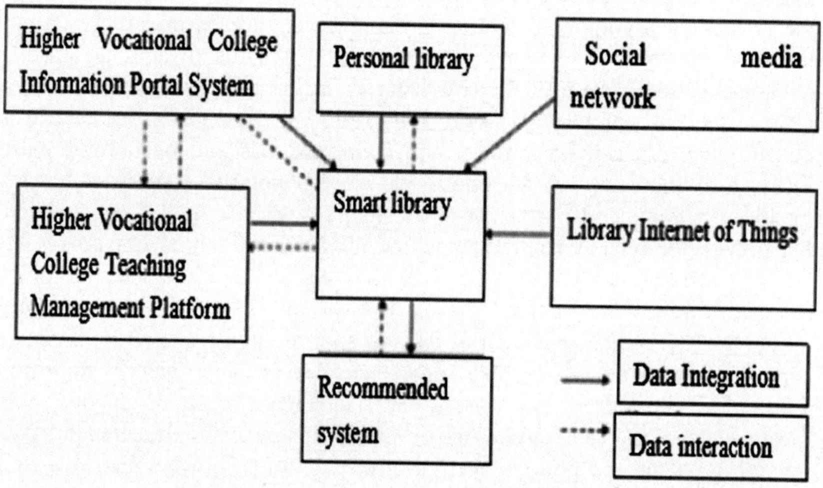

Fig. 1. Schematic diagram of smart library intelligent recommendation system module

(1) Related data collection
The most basic part of the operation of the intelligent recommendation system of the smart library is the collection of related data. First, we need to clarify what data is required for the intelligent recommendation system of the smart library: Selected courses and their online courses on the student management cycle on the teaching management platform, exchange of information in the Open Platform Forum, students' preference for course selection and related textbooks and supplementary teaching information in the teaching affairs information management system, student-related information collected from social media networks (binding student social media accounts on smart library related platforms) The log file of the OPAC system server of the library of the applied college, Contains the reader's attribute characteristics and borrowing information, as well as the reader's book query, recommendation and other log information, the location and library usage information collected from the library iot sensor, readers can link to the third party e-book database (such as Super Star ebook) through "my library". The e-book search, reading and downloading log information are all the information we need. The collection of these data is very important for the operation of intelligent recommendation system of smart library.

(2) Intelligent recommendation system process
For datasets from various sources, we will load and integrate into HCatalog through big data processing system, collect, process and analyze, as well as realize and visualize the final content. That is to provide personalized content recommendations based on readers' interests. Every time the user visits the educational information system platform, the Hadoop system will analyze the following: 1) Attribute information of books used by users (inquiries, loans, downloads, etc.); 2) The data of course selection for this school year in the teaching management platform; 3) Reader's book borrowing, recommendation history and personal attributes of readers in OPAC system; 4) Operation log information for multiple data. After processing the existing data, the system will provide readers with the following suggestions based on their interests and needs: First, when relevant books have been collected in the library of higher vocational colleges, the system will reserve the book for the book within the borrowing period specified by the reader according to the book ID; the second is to send the book information that readers are interested in to the library's editorial department and book recommendation system. As shown Fig. 2:

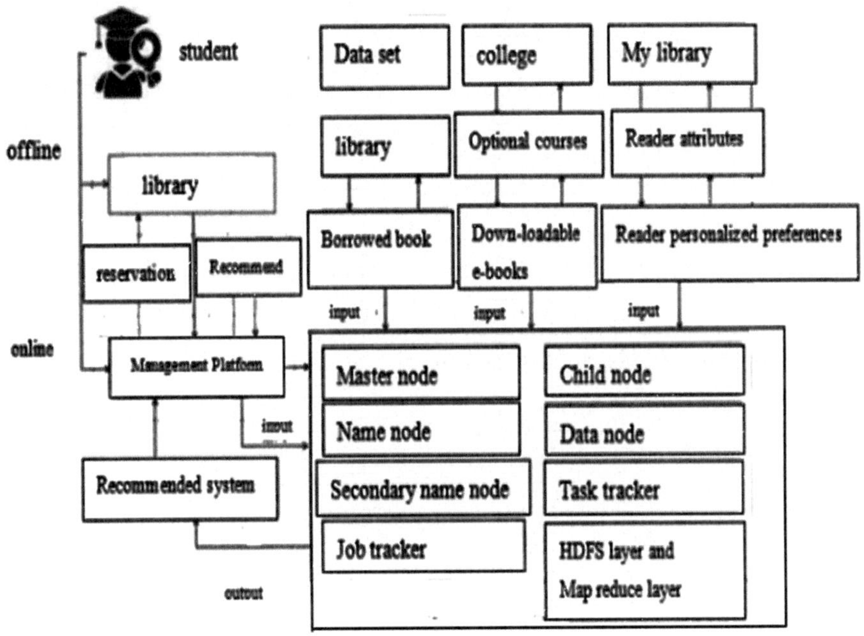

Fig. 2. Flow chart of intelligent recommendation system of smart library

5 Conclusion

In the context of big data, it is necessary to design the intelligent recommendation system of smart library and study its application in applied universities. This research can not only help to improve the utilization of library resources, but also help students of applied universities to obtain more useful resources in the library, help students improve the efficiency of learning, cultivate interest in learning, and thereby improve applied universities. teaching effect.

References

1. Tang, Y., Liu, X., Li, J.: Research on space reengineering of smart library and innovation of digital humanities service. Library (05), 74–80 (2020). http://kns.cnki.net/kcms/detail/43.1031.G2.20200527.1018.024.html
2. Fu, A., Wang, D., Hu, J., Zhang, J., Yin, S.: Fudan University teachers and students Chinese electronic journal resource access behavior data set. Libr. J. 1–6 (2017). http://kns.cnki.net/kcms/detail/31.1108.G2.20200526.1135.004.html
3. Kang, L., Lui, H., Ren, B.: Research on library wisdom management from technical ethics to system ethics. Digit. Libr. Forum **05**, 66–72 (2020)
4. Li, N.: Probe into the intelligent service mode of university library under the "Internet+" environment. Talent **15**, 249 (2020)

A top-N Recommendation Approach Based on Reliable Users

Dongyan Jia(✉), Shengnan Gao, Jiayin Feng, Jinling Song, and Gang Wang

School of Mathematics and Information Science and Technology, Hebei Normal University of Science and Technology, Qinhuangdao 066004, China
187398475@qq.com

Abstract. In order to improve the recommendation accuracy and robustness of recommendation algorithm, a top-N recommendation algorithm based on reliable users is proposed in this paper. We first present a method to choose reliable users for target user using the degree of trust between users. Then based on the rating information of reliable users, we can choose candidate recommendation items using some approaches. Finally, we can provide top-N recommendation results for target user using the strategy of average aggregation. Using MovieLens dataset, we carried out a comparative experiment. The results show that the proposed algorithm improves the recommendation recall. More importantly, it has better robustness.

Keywords: User profile injection attacks · Top-N recommendation · Robustness · Reliable users · Cold start user

1 Introduction

Collaborative filtering technology [1, 2] is one of the most widely used recommendation technologies, which can solve the problem of information overload very well. The top-N recommendation is the most important task of recommender system [3–6]. However, there are profile injection attacks or shilling attacks [7, 8] in the recommender systems. With the existence of attack profile, the recommendation quality of existing top-N recommendation algorithms will be harmed. Therefore, how to provide reliable top-N recommendation for target user has become a key issue to be solved.

Deshpande [9] is one of the first research scholars to put forward the problem of the top-N recommendation. However, this method is unsatisfactory for the cold start users. In order to improve the rating sparsity issue, researchers in [10] introduced real-time contextual data into matrix factorization. A top-n recommender system is proposed by Alqadah et al. [11]. This system is designed by using biclustering neighborhood. However, this method has not taken the influence of shilling attacks to the recommendation result into account. To improve the recommendation recall, Yang et al. [12] proposed to combine the method of matrix factorization and k-nearest neighbors with trust information between users. This method has solved the sparsity problem of explicit trust information, but the recommendation recall is vulnerable to attack profiles. Combining the trust relationship between users with the top-N recommendation

M. Tavana et al. (Eds.): IISA 2020, AISC 1304, pp. 635–644, 2021.
https://doi.org/10.1007/978-3-030-63784-2_79

process, Jamali et al. [13] proposed two recommendation methods. one is the random walking method and the other is trust-based collaborative filtering recommendation. However, the recommendation quality of the two methods are both unsatisfactory when there are attack profile in the system. Using the association rule mining technology, a top-N recommendation method for environment-aware recommender system is proposed by Cremonesi et al. [14]. To improve the probability of the target users to discover new items, a top-N recommendation method based on clustering technology is proposed in [15]. However, the diversity of the recommendation is at the cost of decreasing the recommendation accuracy. In addition, many algorithms are proposed based on implicit feedback (e.g., view, click) [16–18]. These methods can achieve better performance for top-N recommendation, but there is certain vulnerability in the face of the attack profiles. In recent years, some scholars have applied deep learning to top-N recommendation filed, and made some progress [19, 20].

To improve the problem above, a top-N recommendation method based on reliable users is proposed (UTR). The highlights are as follows:

- According to the degree of trust between users, we propose a reliable users selecting method to screen reliable users for target user. The ratings information of reliable users is used as the basis of recommendation.
- Based on the ratings of reliable users, three selecting strategies for candidate recommendation items are proposed, including item popularity, diversity recommendation users and user preferences.
- We design a top-N recommendation method and carry out the experiments using the MovieLens dataset. The results show that our algorithm can achieve better recommendation performance.

2 Reliable Users Selecting

The rating database of collaborative filtering recommender systems includes m users and n items, which is denoted by $U=\{u_1, u_2, \ldots, u_m\}$ and $I=\{i_1, i_2, \ldots, i_n\}$. Each user can rate some items, so the rating matrix is described as a $m \times n$ matrix R, where $R_{i,j}$ is the rating of user u_i on item i_j. therefor, the rating set of user u_i and item i_j can be described as $R(u_i)=\{R_{i,1}, R_{i,2}, \ldots, R_{i,n}\}$ and $R(i_j)=\{R_{1,j}, R_{2,j}, \ldots, R_{m,j}\}$, respectively.

Definition 1 (Reliable users). Let $T_{trust}(T_{trust}\in[0, 1])$ be the trust threshold between users, for the target user $u_a \in U$ and user $u_b \in U$, if the degree of trust of u_a to u_b $trust_{a,b} \geq T_{trust}$, then the user u_b is called as the reliable user of user u_a. So the reliable users of user u_a can be described as $RU(u_a)=\{u_b | u_b \in U, u_b$ is the reliable user of $u_a\}$.

Where the $trust_{a,b}$ is computed by using the multidimensional trust model proposed in paper [21].

According to the idea of definition 1, we give the assumption as follows: for the target user $u_a \in U$, if user $u_b \notin RU(u_a)$, then the ratings of user u_b cannot be used for the recommendation of user u_a. In other words, we mainly use the ratings information of reliable users to give the top-N recommendation results for the target user.

3 The Proposed Method

3.1 Candidate Recommendation Items' Selection

The existing strategies for candidate recommendation items' selection are relatively single and vulnerable to attack users, which resulting in poor recommendation quality. To tackle this problem, we select candidate recommendation items from different angles in this paper, such as item popularity, diversity recommendation users and user preferences.

3.1.1 The Selection Method Based on Item Popularity

In real life, people usually like to pursue popular and fashionable things. So the basic idea of the selection method based on item popularity is to use some popular items as candidate recommendation items.

Definition 2 (item popularity). For user $u_a \in U$ and item $i_j \in I$, let $n(i_j)$ be the number of reliable users have rated the item i_j, and $|RU(u_a)|$ be the number of reliable users, then the item popularity of item i_j is computed as follows:

$$DIP_j = \frac{n(i_j)}{|RU(u_a)|} \tag{1}$$

Definition 3 (popular items). Assuming that the item popularity of item i_1, i_2, i_3, ..., i_n is represented as DIP_1, DIP_2, DIP_3, ..., DIP_n respectively, we select the top M items with the highest item popularity as popular items { PI_1, PI_2, ..., PI_M }.

Let $\hat{r}_{pa,j}$ be the predicted rating for user u_a on popular item PI_j, which is defined as follows:

$$\hat{r}_{pa,j} = \begin{cases} 0, & R_{a,j} \neq \phi \\ \dfrac{\sum_{v=1}^{n(PI_j)} R_{v,j}}{n(PI_j)} \times \sin(\frac{\pi}{2} \times DIP_j), & \text{otherwise} \end{cases} \tag{2}$$

where $R_{v,j}$ and $R_{a,j}$ are the explicit rating for u_v and u_a on PI_j respectively; $n(PI_j)$ is the number of reliable users who have rated PI_j.

3.1.2 The Selection Method Based on Diversity Recommendation Users

When encountering problems, we usually listen to the opinions of some experienced users. Therefore, the selection method based on diversity recommendation users takes the items recommended by diversity recommendation users as candidate recommendation items.

Definition 4 (diversity of rating item). For $u_b \in RU(u_a)$, let $I(u_b) = \{i_j | R_{b,j} \neq \phi, i_j \in I\}$ be the item set rated by user u_b, then the diversity of rating item for user u_b is defiend as follows:

$$D_b = \frac{2}{|I(u_b)|(|I(u_b)| - 1)} \sum_{i_j \in I(u_b)} \sum_{i_t \in I(u_b)} nsim(i_j, i_t) \quad (3)$$

$$nsim(i_j, i_t) = \frac{1 + sim(i_j, i_t)}{2} \quad (4)$$

where $|I(u_b)|$ is the number of items that rated by user u_b; $sim(i_j, i_t)$ is the rating similarity between i_j and i_t calculated by using Pearson correlation coefficient.

Definition 5 (diversity recommendation user)**.** For $\forall u_b \in RU(u_a)$, the lower the $D_b$is, the higher the diversity of user u_b has, then we call the user u_b as diversity recommendation user.

Definition 6 (preference item set)**.** For user $u_b \in RU(u_a)$ and item $i_j \in I$, let R_{max} be the highest rating in the rating database, and $R_{b,j}$ be the rating for user u_b on item i_j, if $R_{b,j}= R_{max}$, then we call the item i_j as the preference item of user u_b. therefore, the preference item set of user u_b can be defined as $FI_b=\{i_j| i_j \in I$ and $R_{b,j}= R_{max}\}$.

In this paper, we use the top 5 users with the highest diversity of rating item as diversity recommendation users $\{u_1, u_2, u_3, u_4, u_5\}$ and the item set $I_e = FI_1 \cup FI_2 \cup FI_3 \cup FI_4 \cup FI_5$ as the candidate recommendation items. For the u_a and the candidate recommendation item $i_j \in I_e$, let D_i be the diversity of rating item for diversity recommendation user u_i, and $R_{i,j}$ be the rating for u_i on i_j, then the corresponding predicted rating is defiend as follows:

$$\hat{r}_{ei,j} = \begin{cases} 0, & R_{a,j} \neq \phi \\ R_{i,j}^{1-D_i}, & \text{otherwise} \end{cases} \quad (5)$$

If the target item i_j is corated by many diversity recommendation users, the predicted rating for user u_a on item i_j is defiend as follows:

$$\hat{r}_{ea,j} = \frac{\sum_{u_i \in DU(i_j)} trust_{a,i} \times \hat{r}_{ei,j}}{\sum_{u_i \in DU(i_j)} trust_{a,i}} \quad (6)$$

where $DU(i_j)$ is the diversity recommendation user set who have rated the item i_j; $trust_{a,i}$ is the degree of trust of user u_a to user u_i which is calculated by using the multidimensional trust model proposed in paper [21].

3.1.3 The Selection Method Based on User Preferences

Usually, if u likes i, it is likely that u will like items which is similar to i. So for the target user u_a, we can firstly get the preference item set FI_a according to the ratings of user u_a. Then based on the ratings of reliable users, the item similarity between any items is computed using Pearson correlation coefficient.

We set the similarity threshold T_{sim}, and select the items with similarity greater than T_{sim} as candidate recommendation items. So we can get the set of candidate recommendation item $CS(u_a)$ for u_a. For $i_j \in CS(u_a)$, using the ratings of reliable users, the predicted rating for u_a on i_j can be computed by the formula as follows:

$$\hat{r}_{sa,j} = \begin{cases} 0, & R_{a,j} \neq \phi \\ \overline{R}_a + \frac{\sum\limits_{u_c \in N(u_a)} (R_{c,j} - \overline{R}_c) \times S(u_a, u_c)}{\sum\limits_{u_c \in N(u_a)} |S(u_a, u_c)|}, & \text{otherwise} \end{cases} \quad (7)$$

where $\overline{R}_a$ and $\overline{R}_c$ are the average rating of u_a and its neighbor u_c respectively; $S(u_a, u_c)$ is the rating similarity between u_a and u_c; $N(u_a)$ is the trusted neighbors set of user u_a.

3.2 TOP-n Recommendation Algorithm

For the target user $u_a \in U$ and item $i_j \in I$, let the item set CP, CE and CS be the candidate recommendation items getted by using the selection method based on item popularity, diversity recommendation users and user preferences respectively, $\hat{r}_{pa,j}$,$\hat{r}_{ea,j}$ and $\hat{r}_{sa,j}$ be the predicted rating calculated by using the selection method above-mentioned, then the final predicted rating $\hat{r}_{a,j}$ for u_a on i_j is defined as follows:

(1) if the item i_j does not belong to any set of CP, CE and CS, then $\hat{r}_{a,j} = 0$;

(2) if the item i_j belongs to only one set of CP, CE and CS, then $\hat{r}_{a,j}$ is computed as follows:

$$\hat{r}_{a,j} = \begin{cases} \hat{r}_{pa,j}, & i_j \in CP, i_j \notin CE, i_j \notin CS \\ \hat{r}_{ea,j}, & i_j \in CE, i_j \notin CP, i_j \notin CS \\ \hat{r}_{sa,j}, & i_j \in CS, i_j \notin CP, i_j \notin CE \end{cases} \quad (8)$$

(3) if the item i_j belongs to any two sets of CP, CE and CS, then $\hat{r}_{a,j}$ is computed as follows:

$$\hat{r}_{a,j} = \begin{cases} \frac{\hat{r}_{pa,j} + \hat{r}_{ea,j}}{2}, & i_j \in CP, i_j \in CE, i_j \notin CS \\ \frac{\hat{r}_{ea,j} + \hat{r}_{sa,j}}{2}, & i_j \in CE, i_j \in CS, i_j \notin CP \\ \frac{\hat{r}_{pa,j} + \hat{r}_{sa,j}}{2}, & i_j \in CP, i_j \in CS, i_j \notin CE \end{cases} \quad (9)$$

(4) if the three sets CP, CE and CS all contain the item i_j, then $\hat{r}_{a,j}$ is calculated as follows:

$$\hat{r}_{a,j} = \frac{\hat{r}_{pa,j} + \hat{r}_{ea,j} + \hat{r}_{sa,j}}{3} \quad (10)$$

Based on the above ideas, a top-N recommendation algorithm based on reliable users UTR is proposed, which is represented as follows:

Algorithm 2 UTR

Input: target user u_a, rating matrix R;

Output: the top-N recommendation results $top\_N(u_a)$ for user u_a。

Begin

(1) complete the initialization of variables and the selection of the reliable users for target user;

(2) get the popular item set CP and calculate the predicted rating for every item in CP ;

(3) get the candidate recommendation item set CE and calculate the predicted rating for every item in CE;

(4) get the candidate recommendation item set CS and calculate the predicted rating for every item in CS;

(5) calculate the predicted rating for u_a on every candidate recommendation item and get the final top-N recommendation results.

4 Experimental Evaluations

4.1 Experimental Dataset

In the experiment parts, the MovieLens dataset is used, which include 100,000 ratings of 943 users on 1682 movies. Every rating is integer between 1 and 5, where 1 indicates dislike and 5 indicates most liked. We divide the dataset into training set and test set.

4.2 Evaluation Metrics

In order to evaluate the recommendation precision of algorithm, recall is used, which is calculated by counting the number of hit. For user u_a and item i_j to be recommended for the user u_a in the test set, if i_j is included in the top-N recommendation results given to the user u_a, it is called a hit. The larger the recall is, the higher the recommendation precision is. Let N_{total} be number of recommended items, and N_{hit} be the number of hits, then the recall is calculated as follows [9]:

$$recall = \frac{N_{hit}}{N_{total}} \tag{11}$$

In addition, to evaluate the robustness of algorithm, we use top-N recommendation shift(TRS) metric, which is calculated by counting the number of attack item to be recommended. The larger the TRS is, the worse the robustness of the recommendation method is. TRS reflects the probability of attack item to be recommended, which is calculated as follows:

$$TRS = \frac{N_{hit1} - N_{hit2}}{m} \tag{12}$$

where N_{hit2} and N_{hit1} are the number of hits for attack item to be recommended before and after the system is attacked, respectively. m is the number of users in the recommender system.

4.3 Experimental Results and Analysis

In the experiment, we compare the performance of UTR with the top-N recommendation method based on k-neighbors (CF-User). In the training set, there is 10 cold start users. In addition, we determine the value of parameters T_{trust} and T_{sim} by experiment, which T_{sim}=0.6,T_{trust}=0.4.

4.3.1 Recommendation Recall Analysis

In order to evaluate the recommendation recall of UTR and CF-User, using training set, we give the top-N recommendation results for every user in the test set. With the value of N increasing gradually, the results of experimental data comparison are shown in Fig. 1.

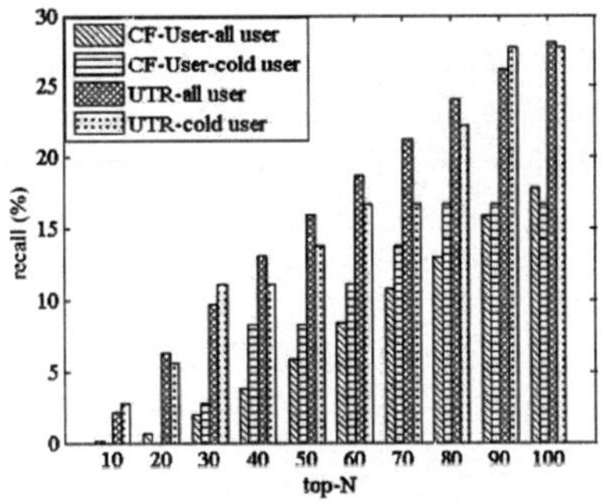

Fig. 1. Comparison of recall

As shown in Fig. 1, no matter which training set is used, with the increasing of the number of recommendation items, the recommendation recall of UTR and CF-User for all users and cold start users are also improving. It can be seen that the more items recommended to the target user, the higher the hit is. However, when the number of the recommendation items N is fixed, the recommendation recall of UTR for all users and cold start users are obviously higher than CF-User. Let N be 80, take the comparison result of the recall as example, recall of UTR improves 45.64% and 24.98% for all users and cold start users, respectively. This indicates recommendation performance of UTR is better than CF-User, the reason is that UTR uses multiple candidate recommendation items selecting methods to improve the diversity of candidate recommendation items and the recommendation recall.

4.3.2 Recommendation Robustness Analysis

To evaluate the robustness of UTR and CF-User, we inject average attack profiles and bandwagon attack profiles into the training set. Where the filler size is 5%, the attack size is 1%, 2%, 5% and 10% respectively. As the attack size increases gradually, the comparison results of *TRS* are shown in Fig. 2.

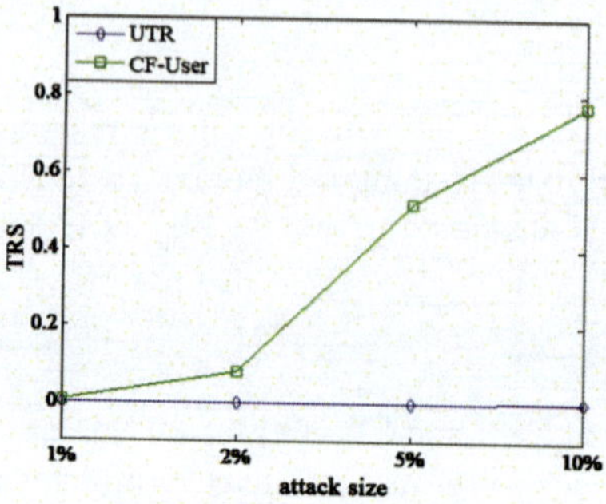

a) average attack

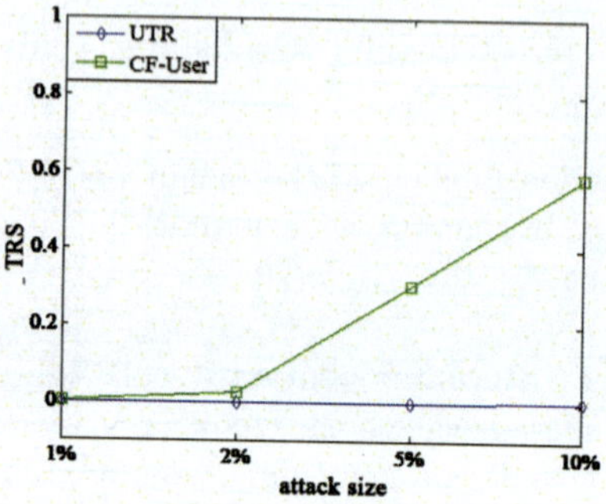

b) bandwagon attack

Fig. 2. Comparison of TRS with different type of attack using training set T_1

As shown in Fig. 2, no matter what type of attack exists in the system, the TRS of CF-User algorithm gradually increases with the increase of the attack size. For example, when the attack size is 10%, the maximum TRS of CF-User can reach 0.7923. It can be seen that CF-User has very weak robustness. However, as the attack size increases, the TRS of our algorithm UTR is close to 0. This indicates our UTR algorithm has better robustness and stability. This is because UTR filters out most of attack profiles and only use ratings of reliable users in the recommendation process, which improves the robustness and quality of recommendation.

5 Conclusions

The existing top-N recommendation approaches are vulnerable to attack profiles. How to improve the performance of recommendation becomes a key problem to be solved. In this paper we proposed a top-N recommendation algorithm based on reliable users. First, a method of reliable users' selection is proposed by using the trust relationship between users. Second, in order to select the candidate recommendation item set, we propose many selection methods from three aspects, such as item popularity, diversity recommendation users and user preferences. Finally, we use mean fusion strategy to get the predicted rating of every candidate recommendation item. The top-N items with highest predicted rating are the recommendation result for target user. As we all know that, there will be new users and new items in the system, and the incremental recommendation method will be our next work.

Ackowledgments. This research was supported by Qinhuangdao's Scientific and Technological Support Project of Key Research and Development Plan (No. 201703A018), and Ocean Special Project of Hebei Normal University of Science and Technolog (2018HY020).

Conflict of Interest. We all declare that we have no conflict of interest in this paper.

References

1. Zhang, F., Nicholas, J.Y., Lian, D., Xie, X., Ma, W.Y.: Collaborative knowledge base embedding for recommender systems. In: Proceedings of the 22nd ACM SIGKDD International Conference on Knowledge Discovery and Data Mining, pp. 353–362 (ACM)
2. Yi, H., Zhang, F.: Robust recommendation method based on suspicious users measurement and multidimensional trust. J. Intell. Inf. Syst. **46**, 349–367 (2016)
3. Cheng, H.-T., et al.: Wide & deep learning for recommender systems. In: Proceedings of the 1st Workshop on Deep Learning for Recommender Systems, Boston, MA, USA, pp. 7–10 (2016)
4. Covington, P., Adams, J., Sargin, E.: Deep neural networks for youtube recommendations. In: Proceedings the 10th ACM Conference on Recommender Systems, Boston, MA, USA, pp. 191–198 (2016)
5. Gomez-Uribe, C.A., Hunt, N.: The netflix recommender system: Algorithms, business value, and innovation. ACM Trans. Manag. Inf. Syst. (TMIS) **6**(4), 1–19 (2016)
6. Okura, S., Tagami, Y., Ono, S., Tajima, A.: Embedding-based news recommendation for millions of users. In: Proceedings the 23rd ACM SIGKDD International Conference on Knowledge Discovery and Data Mining, Halifax, NS, Canada, pp. 1933–1942 ((2017))
7. Santhiya, C., Indira, K.: Identification of profile-injection attacks in recommendation system. In: International Conference on Intelligent Data Communication Technologies and Internet of Things (ICICI), pp. 1442–1448 (2018)
8. Gunes, I., Kaleli, C., Bilge, A., Polat, H.: Shilling attacks against recommender systems: a comprehensive survey. Artif. Intell. Rev. **42**(4), 767–799 (2014)
9. Deshpande, M., Karypis, G.: Item-based top-n recommendation algorithms. ACM Trans. Inf. Syst. **22**(1), 143–177 (2004)
10. Yu, S., Nicholas, J.Y., Xing, X., Kieran, M., Rui, Z.: Collaborative intent prediction with real-time contextual data. ACM Trans. Inf. Syst. **35**(4), 30:1–30:33 ((2017))
11. Alqadah, F.K., Reddy, C., Hu, J., Hatim, F.A.: Biclustering neighborhood-based collaborative filtering method for top-n recommender systems. Knowl. Inf. Syst. **44**, 475–491 (2015)
12. Yang, X.W., Steck, H., Guo, Y., et al.: On Top-k Recommendation using Social Networks. In: Proceedings of the sixth ACM Conference on Recommender Systems, pp. 67–74. ACM, New York (2012)
13. Jamali, M., Ester, M.: Using a trust network to improve top-n recommendation. In: Proceedings of the 3rd ACM Conference on Recommender Systems, pp. 181–188. ACM, New York (2009)
14. Cremonesi, P., Garza, P.: Top-n recommendations on unpopular items with contextual knowledge. In: Proceedings of 3rd Workshop on Context-aware Recommender Systems. ACM, New York (2011)
15. Aytekin, T., Karakaya, M.: Clustering-based diversity improvement in Top-N recommendation. J. Intell. Inf. Syst. **40**(3), 1–18 (2013)

16. Polato, M., Aiolli, F.: Boolean kernels for collaborative filtering in top-N item recommendation. Neurocomputing **286**, 214–225 (2018)
17. Bayer, I., He, X., Kanagal, B., Rendle, S.: A generic coordinate descent framework for learning from implicit feedback. In: Proceedings of the 26th International Conference on World Wide Web. International World Wide Web Conferences Steering Committee, pp. 1341–1350 (2017)
18. He, R., McAuley, J.: VBPR: visual bayesian personalized ranking from implicit feedback. In: Proceedings of the Thirtieth AAAI Conference on Artificial Intelligence, pp. 144–150 (2016)
19. Feng, X., He, X., Wang, X., et al.: Deep item-based collaborative filtering for top-N recommendation. ACM Trans. Inf. Syst. **37**(3), 1–25 (2019)
20. Shuai, Z., Lina, Y., Sun, A.: Deep learning based recommender system: a survey and new perspectives. ACM Comput. Surv. **52**(1), 1–38 (2019)
21. Jia, D., Zhang, F.: A robust collaborative recommendation algorithm incorporating trustworthy neighborhood model. J. Comput. **9**(10), 2328–2334 (2014)

A Safety Distance Automatic Control Algorithm for Intelligent Driver Assistance System

Guiru Liu(✉) and Lulin Wang

School of Computer and Information, Anhui Polytechnic University, Wuhu 241000, China
liuguiru@ahpu.edu.cn

Abstract. In this paper a safety distance automatic control model (SDACM) and algorithm are introduced to solve the problem that the traditional vehicle safe distance control models have poor safety distance control. The algorithm uses road recognition and learning-aided module to enhance the self-adaptability for different brake system features, road conditions and the drivers. Based on the vehicle speed and braking performance, the proper period closed-loop safe distance control model are introduced to improve estimation and control accuracy of safety distance. The algorithms have been applied on Chery intelligent autonomous vehicle, through simulation and test under complex urban road condition, the minimum safe distance is maintained within 1.0–2.0 m, control accuracy is ±0.35 m. The results show that the control model and algorithm achieve much better control accuracy in complex urban conditions and effectively improve driving safety, comfort and road traffic efficiency.

Keywords: Safe Distance Model · Minimum Safe Distance Control Model · Collision Avoidance Algorithm

1 Introduction

With the rapid development of automotive active safety technology and the increasing demand of vehicle safety, many experts and scholars have deep researched on vehicle collision prevention [1, 2]. Some safety distance models have been proposed, including fixed vehicle distance method, headway method and driver estimation model method [3].

The safety distance model dynamically estimates the alarm and the braking intervention distance in real time according to the operating status of the subject and target vehicle and the related factors to avoid collision [5]. The reliability and accuracy of the distance estimation are the basis of early warning and brake intervention. It directly affects the driving safety, driving comfort and road driving efficiency [6].

Fixed safety distance model (FSDM) uses the preseted distance as judgment basis and gives the alarm when the measured safety distance is less than the preseted value. Assuming D is the safety distance, the model can be expressed as D = S, S is selected from a preset safety distance list based on the speed of the subject vehicle, but the

M. Tavana et al. (Eds.): IISA 2020, AISC 1304, pp. 645–652, 2021.
https://doi.org/10.1007/978-3-030-63784-2_80

algorithm lacks flexibility and does not consider the factors such as vehicle driving environment and driver characteristics [3].

Free slip time model (FSTM) calculates the safety distance based on the preset random slip time. Assuming D_s is the safety distance, the model is expressed as:

$$D_s = V_b \times T_s + L \tag{1}$$

where v_b is the speed of the subject vehicle; T_s is the free slip time of the subject vehicle: L is the appropriate parking distance. The algorithm is simple and only suitable for urban traffic environment with constant car-following, which is not suitable for stationary target vehicles and has poor adaptability. When the value of L is set too large, the vehicles in the adjacent lane will overtake the subject vehicle in most scenes which reduce the driving efficiency. If the value of L is set too small, rear-end collisions are easy to occur. In addition, the free slip time parameter can only be preset as an empirical value. For different reaction time for each driver, the estimation value of the safety distance deviates greatly [4].

The driver forecast safety distance model (DFSDM) estimates the safety distance based on the subjective judgments of the relative speed and slipping time T_g [3]. Assuming D_{gs} is safe distance; the model can be expressed as:

$$D_{gs} = V_r \times T_g + \frac{a_f \times T_g^2}{2} + m \tag{2}$$

where V_r is the relate speed of the subject and target vehicle; a_f is the deceleration of the target vehicle; m is the minimum tolerable distance for driver between the subject and target vehicle after the time T_g. The model emphasizes the subjective feelings of the driver. However, the parameter m and T_g is difficult to adapt the various conditions. The estimation accuracy of the safety distance model is difficult to guarantee.

Distance keep safe distance model (DKSDM) estimates early warning safety distance and brake intervention distance based on the preset free slip time, the subject vehicle speed, the deceleration, the relative speed [3, 4]. Assuming D_w and D_z are the early warning and brake intervention distance under stationary target vehicle conditions. D_w and D_z are expressed as:

$$D_w = v_b \times T_s + \frac{v_b^2}{2a_b} + d \tag{3}$$

$$D_z = \frac{v_b^2}{2a_b} + d \tag{4}$$

where v_b is the speed of the subject vehicle; a_b is the brake deceleration estimation value of the subject vehicle; d is the minimum safety distance when the relative speed has been eliminated.

Assuming D_f and D_b are the early warning and brake intervention distance under driving target vehicle conditions. D_f and D_b are expressed as:

$$D_f = v_b \times T_s + \frac{v_b^2}{2a_b} - \frac{v_f^2}{2a_f} + d \tag{5}$$

$$D_b = \frac{v_b^2}{2a_b} - \frac{v_f^2}{2a_f} + d \tag{6}$$

where v_f is the speed of the target vehicle. The DKSDM model considers the driver response time, the brake deceleration of the subject and target vehicle. Based on the pre-test data, relevant standards and empirical statistics, the relevant parameters can be set. But, the parameters cannot be modified under actual running environment. Especially in extreme conditions, the safety distance may be too small to occur the collision or too large to reduce the driving efficiency of the road [3].

However, the traditional models have some common problems. The parameters will be determined based on the priori data. At the same time, the differences between the driver, the driving road surface and the braking performance of the vehicle are not considered comprehensively. The adaptability of the models is poor, the alarm and braking distance are larger or smaller and cannot take into account the safety and road traffic efficiency [4].

In order to solve the problems of the traditional safety distance model, this paper presents a safety distance automatic control model (SDACM) and algorithm which will take into account braking performance of different subject and target vehicle, differences among drivers and pavement adhesion coefficient. For different driving conditions, different safety distance estimation model will be proposed to effectively control the minimum safety distance. Based on the dynamic closed-loop control of the safety distance, the model constantly adjusts the braking deceleration of the vehicle to keep the minimum safety distance a certain range when the relative speed has been eliminated. It not only can ensure driving safety, but also can ensure the driving comfort and road traffic efficiency.

2 Description of the Proposed Safe Distance Control Model

Assuming the subject vehicle only decelerates to avoid the collision when the front target vehicle is stationary. The early warning safety distance D_w and brake intervention distance D_z estimation model under this condition is expressed as:

$$D_w = v_b \times j \times T_d + v_b \times k \times T_z + \frac{v_b \times v_b}{2 \times l \times a_b} + d \tag{7}$$

$$D_z = v_b \times k \times T_z + \frac{v_b \times v_b}{2 \times l \times a_b} + d \tag{8}$$

where T_d and j are the driver response time and coefficient; T_z and k are the braking coordination time and coefficient; l is estimation value of the subject vehicle brake deceleration. l is the parameter of the pavement adhesion coefficient. a_b can be adjusted by l to adapt the current pavement [10, 11]. Equations (7) and (8) are the estimation models of initial warning and brake intervention distance. When the subject vehicle comes into the brake intervention state, if the vehicle maintains a constant deceleration during braking process, the value of d may be too large or too small and deviate from the 1–2 m safety distance. The minimum safe vehicle distance closed loop control model will be designed to ensure that the minimum safety distance is less deviation.

Assuming D_h is the safety distance estimation value; D is the relative distance which is cycle measurement by radar. Minimum safety distance automatic control model under this condition is expressed as:

$$D_h = \frac{v_b \times v_b}{2 \times l \times a_b} + d \tag{9}$$

Ideally, D should be close to D_h and maintain within a certain tolerance scope. If D is too large, d may be too large. The brake deceleration may be smaller to avoid driver and passenger uncomfortable caused by the urgent deceleration. If D is too small, d may be too small. It easy increases the risk of collision. In this model, the relative distance is measured by the millimeter-wave radar in real time, and the brake deceleration of the subject vehicle is adjusted by l to achieve the stable and precise control of the minimum safety distance.

Assuming the subject vehicle only decelerates to avoid the collision when the front target vehicle is driving condition and the final state of the subject and target vehicle is stop, the D_w and D_z estimation model under this condition is expressed as:

$$D_w = (v_f - v_b) \times j \times T_d + (v_f - v_b) \times k \times T_z + \frac{v_b^2}{2 \times l \times a_b} - \frac{v_f^2}{2 \times m \times a_f} + d \tag{10}$$

$$D_z = (v_f - v_b) \times k \times T_z + \frac{v_b^2}{2 \times l \times a_b} - \frac{v_f^2}{2 \times m \times a_f} + d \tag{11}$$

where m is the coefficient of a_f. The braking deceleration a_f of the model is adjusted by m to match the braking performance of the target vehicle; The driver's response time and braking system coordination time are evaluated by self-study module. T_d and T_z are adjusted by j and k according to the output of the self-study module.

Based on the pavement adhesion coefficient which is obtained by the output of the pavement identification module, a_b and a_f are adjusted by l and m to match the current driving pavement. Through the adaptive adjustment, the safety distance estimation is more reasonable, accurate and suitable for practical application scenarios, which not only can maintain a high driving efficiency, driving comfort and driving experience while avoid collision situations [12].

Equation (10), (11) are only suitable to estimate the initially early warning and braking intervention distance. If the deceleration of the subject vehicle does not adjust in real time, the minimum safety distance will deviate from the ideal distance. It may be too large or too small and cannot maintain a high collision avoidance probability and driving efficiency [13]. In order to maintain the minimum safety distance in the ideal range, a closed-loop adaptive minimum safety distance control model is introduced for the driving conditions after the subject vehicle braking intervention. After the vehicle starts to brake and decelerate, d is affected mainly by the braking distance of subject vehicle and front target vehicle. The brake distance of the subject vehicle is adjusted by changing the braking deceleration a_b to keep d in the ideal range.

When the target vehicle is in driving condition, the minimum safety vehicle distance control model under this condition is expressed as:

$$D_h = (v_b \times t - \frac{l \times a_b \times t^2}{2}) - (v_f \times t - \frac{m \times a_f \times t^2}{2}) + d \tag{12}$$

Where t is radar measurement cycle. In the model, a_f is basically the same except the adjustment according to the output of the pavement identification module. v_b is adjusted by change l within radar's distance measurement period to make the measurement value D is close to D_h. This estimation method does not need to consider the final speed of the front target vehicle, which is more practicable in the actual scene and keeps the value d within a reasonable range. a_b and a_f are updated according to the speed of subjective vehicle. The closed-loop prediction model adjusts the relative safety distance by change l and m to keep the value d within a reasonable range. Equation (12) is the minimum safety distance control model after the vehicle braking intervention. The closed-loop control process is very similar to the driver's manual collision avoidance process and can kept smaller safety distance, ensure driving comfort and road traffic efficiency [14].

The relevant parameters of Eq. (7–12) as follows: The value of T_d is in the range of 0.4 s–1.5 s, typical value 1s; The value of T_z is in the range of 200–400 ms, typical value 300 ms; The value of a_b is in the range of 3.0–9.8 $\mathrm{m \cdot s^{-2}}$, typical value 5.5 $\mathrm{m \cdot s^{-2}}$ for car, typical value 3.6 $\mathrm{m \cdot s^{-2}}$ for truck; the preseted value of coefficient j, k, l, m is 1, the variation range is 0.5–1.5; the value range is 1–2 m.

3 Performance Study

For different driver and brake system, T_d and T_z vary greatly and cause too early or too late brake intervention. It not only increases the risk of collision, but also not effectively improves the road traffic efficiency [16, 17]. After braking intervention, the vehicle begins to decelerate. The minimum safety distance is the key of each estimation model. It directly affects the risk of rear-end collision and road traffic efficiency. Because of the different influencing factors, the minimum safety distance is quite different. The test results under stationary target vehicle conditions are shown in Table 1.

As Table 1 shows that the minimum safety distances get closer and closer when the relative speed is getting smaller and smaller. So, it is not risky to keep the minimum safety distance within 1–2 m. Second, the detection period of radar is 50 ms, the running time of the control strategy is 50 ms, the total time is 100 ms. If the vehicle with 10 km/h speed is closing, the free slip distance is 0.3 m during this period, radar detection error is 0.5 m, it is safe to keep the minimum safety distances in 1–2 m. If the distance is too small, especially less than 1m, the risk of collision becomes greater. If the distance is more than 2.5 m, the distance will be too large. It not only reduces the traffic efficiency of the road, but also increases the risk of collision because of the overtaking [18, 19].

Table 1. The minimum safety distance under the stationary condition of the target vehicle

Safety distance model	The minimum safety distance					Subjective evaluation
	100	80	60	40	20	
FSDM	4.87	4.81	4.73	4.24	3.89	Common
FSTM	0.65	0.38	−0.15	0.34	0.85	Poor
DFSDM	0.84	−0.15	−0.24	−0.36	0.43	Poorer
DKSDM	3.13	2.84	2.52	2.33	2.24	Better
SDACM	1.83	1.74	1.73	1.56	1.15	Best

As can be seen from Table 1. The minimum safety distance of the FSDM model is larger to 4m. It ensures safety, but has low driving efficiency. The minimum safety distance of the FSTM model is too small because of the difference of the driver response time. Especially at high speed conditions, it easily leads to collision [20, 21]. Based on the subjective prediction of people, the minimum safety distance of the DFSDM model is too small. Considering the driver response time, the brake deceleration and a priori data, DKSDM model has a more reasonable safety distance. But, it may also be too large or too small under ice and snow conditions with smaller adhesion coefficient [22, 23]. The proposed SDACM control model not only considers the driver response time, but also takes into account the brake coordination time, the driver response time, the brake deceleration and the road conditions. Its relevant parameters are calibrated by pavement identification module and self-learning module to adapt the driver and driving conditions. At the same time, the closed-loop control method is introduced. The safety distance is controlled by periodically adjusting the braking deceleration based on the pavement adhesion coefficient and distance deviation. The minimum safety distance will be kept in 1–2 m when the subject vehicle speed is equal to 0 km/h. The proposed SDACM control model not only avoids the collision accident, but also improves the road traffic efficiency.

4 Conclusions

The paper analyzed the influencing factors of safety warning distance and brake intervention distance and studied the existing safety early-warning models and algorithms. Combined with the driver's driving experience and the kinematics analysis of braking process, this paper presents a safety distance automatic control model and algorithm. Through theoretical analysis, simulation and actual scene test, the minimum safety distance can be kept in the range of 1.0–2.0 m when the target vehicle is stationary or running conditions. It not only ensures driving safety, but also improves the road traffic efficiency. It is close to the driver's subjective feelings and actual brake intervention process. Through the test and verification, the algorithm is adaptive and effective.

Acknowledgements. This work was funded partly by National Key R&D Plan (2017YFB010 2600), Anhui Science and Technology Major Project Plan (16030901032), Anhui Provincial Natural Science Foundation (TSKJ2015B12), Key Laboratory of Computer Application Technology, Computer and Information Science, Anhui Polytechnic University (JSJKF201514).

References

1. Wang, C., Xu, Y.X., Fu, R., Guo, Y.S., Yuan, W.: Potential dangerous target identification algorithm for lane change warning system. J. Changan Univ. (Nat. Sci. Ed.) **35**, 98–105 (2015)
2. Niu, R.X., Xia, J.T., Wang, X.H., Mei, T.: Tentacle algorithm of obstacle avoidance and autonomous driving for intelligent vehicle. J. Traffic Transp. Eng. **10**, 53–58 (2010)
3. Liu, G., Hou, D.Z., Lin, K.Q., Yang, D.G., Lian, X.M.: Warning algorithm of vehicle collision avoidance system. J. Tsinghua Univ (Sci. Tech.) **44**, 697–700 (2004)
4. You, F., Zhang, R.H., Wang, H.W., Wen, H.Y., Xu, J.M.: Warning model for safety analysis of overtaking behavior based on longitudinal safety spacing. J. South China Univ. Technol. (Nat. Sci. Ed.) **41**, 87–92 (2013)
5. Qiu, X.P., Yu, D., Sun, R.X., Yang, D.: Cellular automata model based on safety distance. J. Transp. Syst. Eng. Inf. Technol. **15**, 54–60 (2015)
6. Dang, H.S., Han, C.Z., Duan, Z.S.: Determination of distance for vehicle collision avoidance warning and braking. J. Chang'an Univ. (Nat. Sci. Ed.) **11**, 89–91 (2002)
7. Xu, L.H., Luo, Q., Wu, J.W., Huang, Y.G.: Study of car following model based on minimum safety distance. J. Highw. Transp. Res. Dev. **6**, 72–78 (2010)
8. Pei, X.F., Liu, Z.D., Ma, G.C., Ye, Y.: Safety distance model and obstacle detection algorithms for a collision warning and collision avoidance system. J Autom. Saf. Engrgy **3**, 26–33 (2012)
9. Liu, Y.T., Zhang, J., Shi, Z.K.: Improvement and verification of safety distance model when vehicles start. J. Northwest. Polytechnical Univ. **32**, 725–729 (2014)
10. Zhang, B., Zhang, J.Z., Liu, Z.D.: Road identification method for ABS based on vehicle body deceleration estimation. China J. Highw. Transp. **24**, 109–113 (2011)
11. Pei, X.F., Liu, Z.D., Qi, Z.Q.: Road identification method for vehicle longitudinal control system. China J. Highw. Transp. **27**, 177–182 (2014)
12. Pei, X.F., Qi, Z.Q., Wang, B.F., Liu, Z.D.: Vehicle frontal collision warning/avoidance strategy. J. Jilin Univ. (Eng. Technol. Ed.) **44**, 599–604 (2014)
13. Li, L., Zhu, X.C., Dong, X.F., Ma, Z.X.: A research on the collision avoidance strategy for autonomous emergency braking system. Autom. Eng. **35**, 168–174 (2015)
14. Li, L., He, J.P., Liu, W.G., Zhu, X.C.: Threat assessment algorithm based on characteristics of driver emergency braking behavior. J. Tongji Univ. (Nat. Sci.) **42**, 109–114 (2014)
15. Guo, Y.S., Wang, C., Fu, R., Yuan, W., Yu, P.C.: Driver's reaction time under city road conditions. China J. Highw. Transp. **26**, 135–142 (2013)

16. Lv, J.E., Zhu, L.H., Zhang, R.S., Wei, Y.F.: Effects of Driver's reaction time on safe driving. J. Transp. Syst. Eng. **14**, 80–86 (2014)
17. Wang, P.W., Yu, G.Z., Wang, Y.P., Wang, D.: Cooperative active collision avoidance algorithm based on sliding mode control. J. Beijing Univ. Aeronaut. A **40**, 268–273 (2014)
18. Hou, D.Z., Liu, G., Gao, F., Li, K.Q., Lian, X.M.: A new safety distance model for vehicle collision avoidance. Autom. Eng. **27**, 186–199 (2005)
19. Tang, Y.S., Jiang, Z.W., Bai, Y., Fang, Y.: Model of vehicle safety distance for collision avoidance and simulation study. J. Liaoning Univ. Technol. (Nat. Sci. Ed.) **28**, 324–326 (2008)
20. Gao, Z.H., Wu, T., Zhao, H.: Model of driver's braking moment in virtual car following collision avoidance scenes. J. Jilin Univ. (Eng. Technol. Ed.) **44**, 1233–1239 (2014)
21. James, R.W., Gabriel, A., Stewart, W., et al.: Extending time to collision for probabilistic reasoning in general traffic scenarios. Transp. Res. Part C **51**, 66–82 (2015)
22. Vicente, M., Joshue, P., Jorge, G., et al.: A fuzzy aid rear-end collision warning/avoidance system. Exp. Syst. Appl. **39**, 9097–9107 (2012)
23. Emeli, A., Andras, V., Mario, D.F.: The effects of a driver assistance system for safe speed and safe distance-a real-life field study. Transp. Res. Part C **19**, 145–155 (2011)

Emotional Cues Recognition in Natural Speech by Chinese Speakers

Yifei Wang(✉)

Inner Mongolia University for Nationalities,
Tongliao 028000, Inner Mongolia, China
tongliaowyf@163.com

Abstract. We carried out an experiment about the word *ah's* position in Chinese sentences in order to find out the effect on emotion expression in speech. We predict the positions of particles in Chinese and other languages imply some subtle information of speakers' in natural speech. 240 Chinese sentences based on 6 MSP sentences with ah are presented to 40 Chinese listeners. The study aim to find out the information in speech itself instead of pitch and melodies.

Keywords: Natural speech · Emotion · Recognition · Cues · Particles · Chinese sentences

1 Introduction

1.1 Research Background

Emotions exist in speech, which is one of the fundamental ways of conveying information in between people. As a special form of sound, speech contains acoustic properties which carry important effective features. Human ears can capture these subtle details [1]. These special sensing mechanisms of human beings are difficult to be learned by machines. Emotions in speech are transmitted from one communicator to another during an interaction by intonation, pitch and melody. Researchers explore all the features contained in these factors [2]. Even though we know that words are not sole component of speech there are researchers try to find the detailed information implied in the expressed sentences. In research we explore the smallest meaningful unit in sentence.

1.2 Research Questions

Since emotions play a crucial role in the whole human experience and they have a huge effect on the human decision making process, the study of human speech in the emotion indication and recognition is important in human social relation as well as in machine emotion recognition [3]. We ask what are the detailed indication in oral expressions and the way that expressions convey different emotion cues. This research aimed to explore the natural speech of sentences with particles, such as *ah (啊)* , *heh (呵)* , *eh (呃) oh (哦)* Chinese, at different places. We assume that the word with these

M. Tavana et al. (Eds.): IISA 2020, AISC 1304, pp. 653–659, 2021.
https://doi.org/10.1007/978-3-030-63784-2_81

particles before or after, must mean something different from the words without these particles. In this paper, we present the experiment we have done with the particle *ah*, in our pilot study. we ask if there is emotion indication related to the place of the particle, and if so, what kind of emotions the particle indicate when it is at the different positions. We hope this research will carry some information back to the study of language itself.

2 Experiment

2.1 Speakers

We take the Chinese version of Harvard Sentences, Mandarin Speech Perception (MSP) Test – Sentences [4]. Speakers are adult female speakers. In this experiment, we select the recordings from first 5 sentences in the first group.

2.2 Materials

We take the Chinese version of Harvard Sentences, Mandarin Speech Perception (MSP) Test - Sentences. As in Harvard Sentences, all the English vowels and consonants are included in the 10 natural meaningful sentences, in the MSP sentences there are 10 neat sentences, each with 7 Chinese characters bearing the 4 Chinese tones in every sentence, and within the 10 sentences neutral tone is included. "节假日不用门票 (jie jia ri bu yong men piao)" (Table 1).

Table 1. The first 3 original sentence samples and their versions of inserted *ah1–ah4* into different positions.

version	Varieties of sentence 1	Varieties of sentence 2
Ah 0	Jintian de yangguang zhenhao	Jiejiari buyong menpiao
Ah 1	*Ah* Jintian de yangguang zhenhao	*Ah* Jiejiari buyong menpiao
Ah 2	Jintian de *Ah* yangguang zhenhao	Jiejiari *Ah* buyong menpiao
Ah 3	Jintian de yangguang *Ah* zhenhao	Jiejiari buyong *Ah* menpiao
Ah 4	Jintian de yangguang zhenhao *Ah*	Jiejiari buyong menpiao *Ah*

We select 5 sentences from one group of MSP list and modify the recording by inserting the ah we synthetized in neutral tone into 4 different places. For example, the sentence "*Jintian de yangguang zhenhao*" is generated into 4 varieties as "*ah Jintian de yangguang zhenhao*", "*Jintian de ah yangguang zhenhao*", "*Jintian de yangguang ah zhenhao*", and "*Jintian de yangguang zhenhao ah*". In order to make this experiment effective doable, we select each version from the 5 sentences as the final stumili. We regroup the modified ssentences with ah at different positions into the way we presented in Table 2. The sentence at the diagonal line is in one group. The first sentence with *ah0* (original sentence) and the first sentence with *ah1* and the second sentence with *ah2* ......to the fifth sentence with *ah5*, will be presented to listeners in part one. So in this part, none sentences are played more than twice. There are five parts in the whole experiment. In the following part, we make the stumili the same way from the diagonal line. In part two, the sentences will be S1ah0, S2ah1, S3ah2, S4ah3, S5ah4, and S0ah5, as shown in the dark shade in Table 2. Then we produce an experiment with six sentences in 5 parts. In each part the six sentences are randomized. The original S0 in every part is selected from the MSP test material, in this way, all of the new generated sentences are played only once. The whole sound files in this experiment takes 40 min.

Table 2. The organization of original five sentences with 4 *ah*varieties.

S	S0	S1	S2	S3	S4	S5
Variation number	**Ah 0**	Ah 0	Ah 0	Ah 0	Ah 0	Ah 0
	Ah 1	**Ah 1**	Ah 1	Ah 1	Ah 1	Ah 1
	Ah 2	Ah 2	**Ah 2**	Ah 2	Ah 2	Ah 2
	Ah 3	Ah 3	Ah 3	**Ah 3**	Ah 3	Ah 3
	Ah 4	Ah 4	Ah 4	Ah 4	**Ah 4**	Ah 4
	Ah 5	Ah 5	Ah 5	Ah 5	Ah 5	**Ah 5**

2.3 Listeners

Listeners are selected from the research volunteers in universities. They are all young adults with no hearing problems. They are given a brief instruction for this experiment. After a short try out of the earphones and fill in the information on the answer sheet, the participants start to judge their perception response. In the answer sheet, we leave the listeners six possibilities for their judgment of the emotions in the sentences [5].

As we know, *Ah*, is always a monomorphemic and can be free standing in Chinese sentence structure. It has been assumed that it can be inserted more or less randomly into any sentence and that it does not bear grammatical relationships to other phenomena in the language [6]. It is isolated from other parts of the sentences and it can be inserted almost anywhere in a sentence without affecting the structure of the sentences. But why we Chinese do this insertion when we are in the natural speech? Is it just a kind of habits? Or it really mean some implied emotion and even we ourselves have not noticed that. In this experiment, we use the MSP sentence, with normal tones, and neutral speech. We will find if any of the insertion of ah can imply some degrees of emotion indication. This experiment is carried out in quiet rooms individually with 40 paid volunteer participants.

3 Results and Findings

Listeners given the instruction of this experiment are sitting in the quiet room. Colors for degrees of emotion are presented in the instruction. Six basic state of emotions colored in red, orange, yellow, light green and green indicate the state of emotion of surprised, happy, normal, disappointed, hesitate and sad, as well as the colored blue in the situation of listener have not any perception of the speaker' emotion. All the colors are arranged chromatically in order to let listeners get familiar with the selection choices while they are listening [7] (Table 3).

Table 3. Emotions states and expected emotion density provided in the instruction for the listeners.

Emotions	Matching space	Emotion Density	Color Degrees
Surprised		Very strong	
Happy		Strong	
Normal		Mild	
Disappointed		Minus strong	
Hesitate		Strong	
Sad		Very strong	

Six groups of sentences with one group of original sentences included are presented to lsiteners. In these six groups of sentences, ah0–5 are the target for recognition. Listeners are required to choose one possible emotions of the speakers included normal state emotion. 240 feedback of every state of emotion in the six types of *ah* positions in total.

Figure 1 presented the result of 40 listeners recognize the emotion state according to the sentences with ah at the different positions. From the six groups of recognition, differences of emotion states are guessed by listeners. Normal emotion state is recognized by listeners in *ah0* (in blue)state, which is the sentence without *ah* inserted. This result with 78% recognition inferred us that sentences without ah at any position in sentence expression are regarded as the normal sentences, or the normal emotion state.

Ah1 is the *ah* at the beginning of the sentences, in the sentences except *ah0*. *Ah1* is mostly regarded as the emotion state of surprised, happy and disappointed, these strong or very strong states except sad. Interestingly, this result shows that when people are sad they normally keep it to the end. As we can see, the green bar in Fig. 1, it is highly recognized in *ah 5*. It is the *ah* at the end of the sentences.

As we can tell from Fig. 1, other emotion states are recognized as well from ah0 to ah 5, hesitate state appears the situation when ah is inserted in the middle of the sentences and in the end of the sentences.

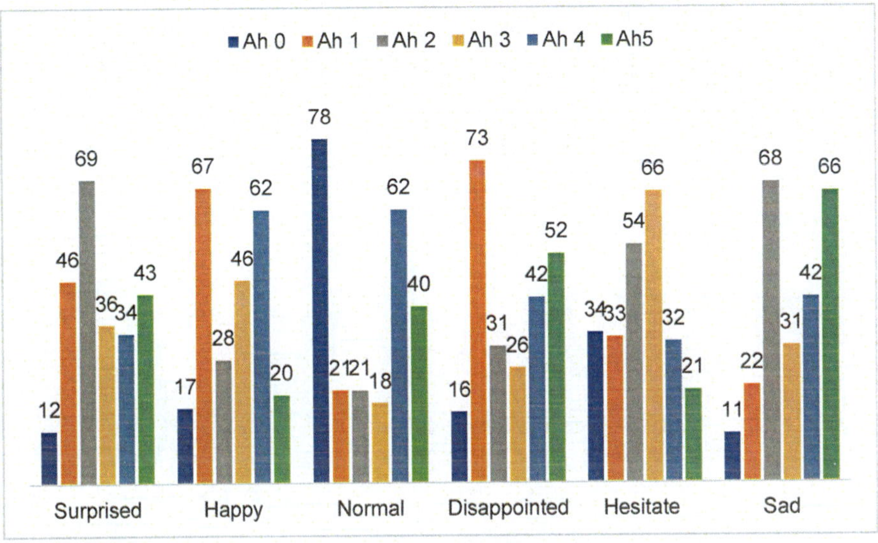

Fig. 1 Emotion perception of the sentences with *ah0–ah5*.

This individual *ah's* results indicate the possibilities of the perception of ah in natural speech. The reason for the positions of ah will be analyzed by the word neighbors, which is about in what kinds of words when ah is followed or next to. Generally, we can tell that the morphological features can be a kind of cues to identify and to recognize the state of emotions in natural speech. These features can be traced to small words, such as particles [8] (Fig. 2).

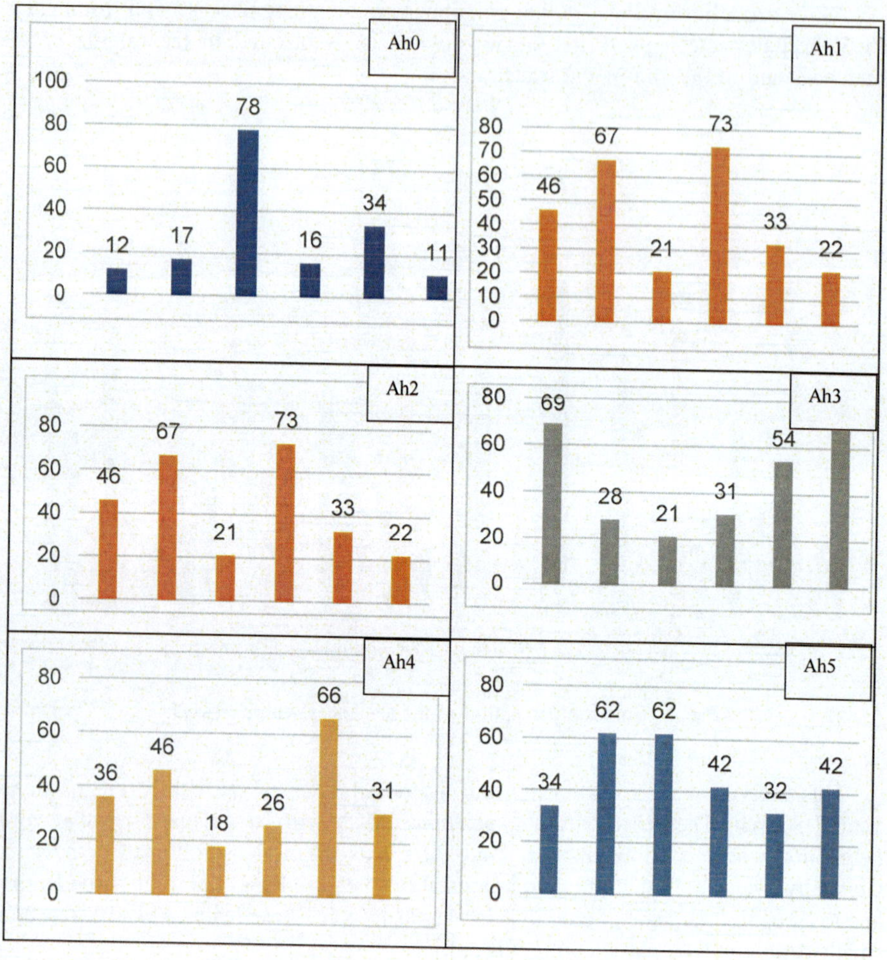

Fig. 2. Individual sentences state of *ah0–ah5*.

References

1. Aijmer, K.: *Oh* and *ah* in English conversation. Corpus Linguistics and Beyond. University of Gothenburg, pp. 61–86 (1987)
2. Bhaykar, M. Yadav, J. Rao, S.K.: Speaker dependent, speaker-independent and cross-language emotion recognition from speech using GMM and HMM. In: 2013 National Conference on Communications (NCC), pp. 1–5 (2013)
3. Corive, R., Douglas-Cowie, E., Tsapatsoulis, N., Votsis, G., Kollias, S., Fellenz, W., Taylor, J.G.: Emotion recognition in human-computer interaction. IEEE Signal Process. Mag. **18**(1), 32–80 (2001). https://doi.org/10.1109/79.911197https://doi.org/10.1109/79.911197
4. Fu, Q.J., Zhu, M., Wang, X.: Development and validation of the Mandarinspeech perception test. J. Acoust. Soc. Am. **129**(6), EL267–EL273 (2011)

5. Bao, H., Xu, M., Zheng, T.F.: Emotion attribute projection for speaker recognition on emotional speech.In: INTERSPEECH 2007, Antwerp, Belgium, pp. 758–761 (2007)
6. James, D.: The use of *oh, ah, say,* and *well* in relation to a number of grammatical phenomena. Paper Linguist. **11**(3–4), 517–535 (1978)
7. Kim, J., Lee, S., Narayanan, S.: A detailed study of word- position effects on emotion expression in speech. In: INTERSPEECH-2009, pp. 1987–1990 (2009)
8. Ververidis, D., Kotropoulos, C.: Emotional speech recognition: resources, features, and methods. Speech Commun. **48**(9), 1162–1181 (2006). https://doi.org/10.1016/j.specom.2006.04.003https://doi.org/10.1016/j.specom.2006.04.003

Research on Fault Diagnosis Expert System of On-board Radio of a Certain Armored Vehicle Based on CLIPS

Changhong Gong(✉), Xiao Ming, and Lingxiang Xia

Ordnance Non-commissioned Officer School, Army Engineering University, Zhengzhou, China
qicheng77611@163.com

Abstract. This article briefly introduces the concept and grammatical composition of the expert system language CLIPS. For a certain type of armored vehicle car radio failure occurrence mode, a failure analysis model is established using CLIPS, a knowledge base and inference engine are written, and CLIPS embedded programming is implemented in combination with VC++. This example is used to illustrate the application of CLIPS in armored vehicle fault analysis.

Keywords: CLIPS · Expert system · Knowledge base · Inference engine · VC++

1 Introduction

Armored vehicles integrate high-tech technologies such as machinery, electronics, firepower, hydraulics, optics, information, and navigation, and put forward higher requirements for the maintenance of armored vehicles. In order to identify faults quickly and accurately, it is necessary to rely on the fault diagnosis expert system for decision-making. The rapid development of expert systems in recent years has also provided an effective means for analyzing and studying armored vehicle faults. Expert system is the predecessor of today's artificial intelligence. It is an intelligent program that can use a lot of knowledge and inference engine to solve complex problems that experts can solve [1]. The domain knowledge is stored in the computer in a certain form of knowledge representation, and the inference engine uses the knowledge to infer and simulate experts to solve problems. The knowledge base and inference engine are core components of the expert system. In addition to these two modules, it generally includes man-machine interface, global database, interpreter, etc. [2].

This paper mainly uses knowledge-based reasoning network and expert system fault diagnosis method for a certain type of armored vehicle car radio, and uses CLIPS in armored vehicle fault analysis by combining Visual C++ and CLIPS technology.

M. Tavana et al. (Eds.): IISA 2020, AISC 1304, pp. 660–666, 2021.
https://doi.org/10.1007/978-3-030-63784-2_82

2 CLIPS-A Language for Creating Expert Systems

2.1 CLIPS Overview

CLIPS is an acronym for "C Language Integrated Production System". It is an expert system language designed by NASA/Johnson Space Center in C language. It supports object-oriented, process-oriented, and rule-based programming. Since its launch in 1986, CLIPS has been widely used in computers, electronics, aerospace, chemical engineering, geology and other fields, with complete functions, simple syntax, good portability, and low cost. It can be used in Windows, Linux, embedded Develop applications on different platforms. Especially CLIPS has been published as open source software, researchers can modify the source code according to their own needs, which provides convenience for users' secondary development. In addition, the CLIPS system has good program compatibility, and programs developed with CLIPS are well compatible with languages such as C, C++, and JAVA. The system developed with CLIPS can be embedded as a subroutine in other systems. Programs written in other high-level languages can be called by CLIPS-based systems in the form of external functions.

Rule-based CLIPS is mainly composed of the following parts [3]:

1. Fact list: contains the data required for inference, and a fact library is formed by custom facts;
2. Knowledge base: contains all rules, and multiple rules are formed by custom rules;
3. Inference engine: inference based on a certain inference mechanism and overall control of operation.

2.2 Introduction to CLIPS Basic Syntax

CLIPS contains three main components: custom template structure, custom fact structure and custom rule structure.

2.2.1 Custom Template Structure

CLIPS uses a knowledge representation method that combines framework and semantic network. Before the fact is created, CLIPS must be informed of a list of legal slots for the given relationship name. The custom template structure (deftemplate) is used to describe the fact that they share the same relationship name and contain common information. The general format is:

```
(deftemplate < relation_name>
      (slot -1)
      (slot-2)
      ...
      (slot-N))
```

The slot represents a single-field slot. The multislot represents a multi-field slot, mainly using ISA to describe the relationship between the frames.

2.2.2 Custom Fact Structure

The facts (that is, the initial knowledge) that are known to be correct before the program runs can be defined using a custom fact (deffacts) structure. The general format is as follows:

```
(deffacts < deffacts name > [ < optional comment >]
      <facts > * )
```

2.2.3 Composition of Rules

Production rule system, as IF-THEN rules, is used in CLIPS. It is based on the large number of causal relationships between various knowledge blocks in the memory pattern of the human brain. The characteristic is modularity, that is, the rules are independent of each other. The IF part is called the condition part, the antecedent, or the left side of the production rule. It states the conditions that must be met to apply this rule. The THEN part is called the operation part, the result part, the latter, or the right side of the production rule. If the condition part of the rule is triggered during the execution process, the action part is executed. CLIPS rule format is as follows:

```
(defrule rule_name "optional_comment"
  (pattern_1);           The left part of the rule, IF part.
  ...
  (pattern_N);
  =>
  (actions_1);           The right part of the rule, THEN part.
  ...
  (actions_N))
```

If all the patterns of a rule match facts, the rule is activated and put on the agenda. The agenda is a collection of activations which are those rules which match pattern entities.

3 Application of CLIPS for Armored Vehicle Failure Analysis

3.1 Model Design

When analyzing the faults of armored vehicles, experts generally associate all possible factors according to the fault phenomenon, infer the main cause of the fault and give corresponding treatment methods. We can simulate expert thinking characteristics and classify each fault phenomenon as a module according to the hierarchical classification method, and the various causes that cause this fault phenomenon are its submodules, corresponding to each submodule output a processing method.

In this paper, a certain type of armored vehicle on-board radio station is used for analysis and program construction. The common fault phenomena of this radio station are mainly manifested in 6 types. The program is divided into 6 modules, and the causes of the fault phenomenon are written into sub-modules. The model of a certain type of armored vehicle vehicle radio fault analysis is shown in Fig. 1.

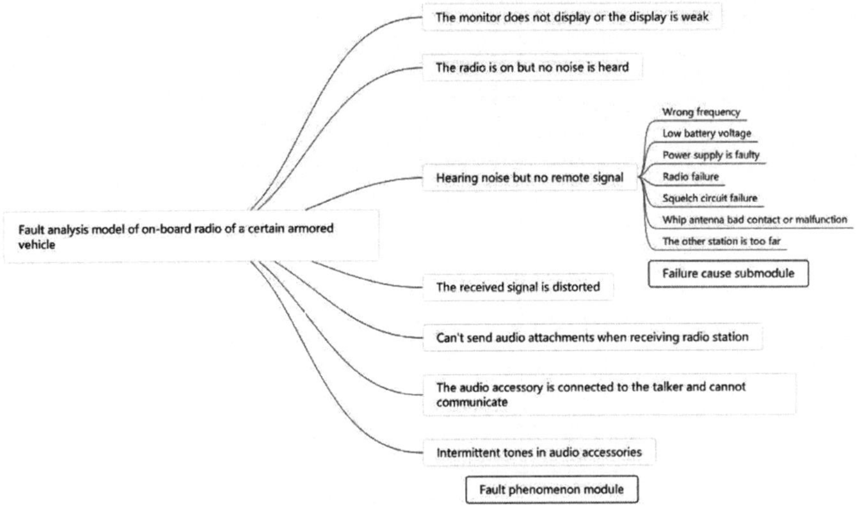

Fig. 1. Diagram of the failure analysis model of an on-board radio of a certain armored vehicle

3.2 Reasoning Process

The system model is designed by analyzing the fault model of the radio station, which is shown in Fig. 2. At the entrance of the program, an inquiry method is set up and the user makes corresponding responses according to the facts to enter different modules.

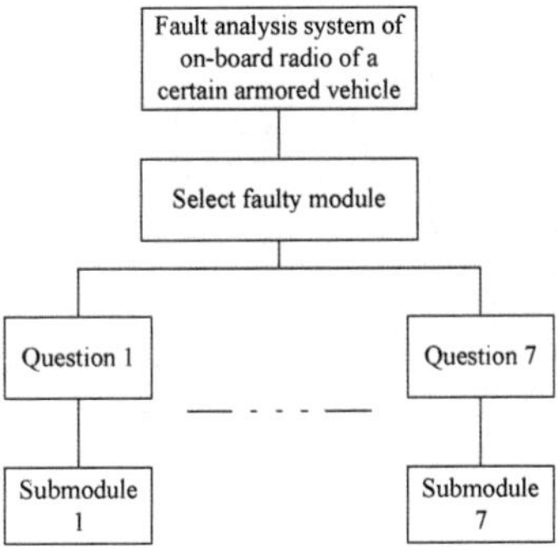

Fig. 2. Flow chart of system failure

First, define a selection function (defrule selection_state) at the entrance of the system. The user selects the corresponding sub-module through the fault phenomenon.

When programming a program, some functions are common. Once the function is compiled, it can also be called in other rules. The selection function is to ask the user to choose and generate according to the user's choice new facts for reasoning. The general syntax is as follows:

```
(deffunction selection (?ques $?allowed_-values)
           (printout t ?ques)
        (bind ?ans (read))
        (if (lexemep ?ans)
        then(bind ?ans (lowcase ?ans)))
  (while (not (member ?ans ?allowed_values)) do
        (printout t ?ques)
        (bind ?ans (read))
        (if (lexemep ?ans)
        then (bind ?ans (lowcase ?ans))))
?ans)
```

?ques is a single variable that represents a question, and $?allowed_values is a wildcard multi-field variable, that means the answer to the question has multiple choices. In this custom function, a loop structure function is used, which is to repeatedly output questions and ask the user to answer when the user's input does not match the answer given until the user inputs are legal.

With the above entry program and custom function, you can start to build a knowledge representation of the station's fault. Since the composition of each module and its submodules is the same, the CLIPS procedure is the same for each module. We use one of the modules to explain the reasoning process of the system. Taking question 3 "Hearing noise but no remote signal" as an example, the inference network is shown in Fig. 3.

The following describes the writing of CLIPS rules in detail through the description of one of the rules.

```
(Defrule frequency-block
      (no remote single)
      =>
      (if (yes-or-no-p
             "Check if the frequency is correct (yes|no)?")
      then(if(yes-or-no-p
                   "Check if the operating frequency is on the preset
channel (yes|no)?")
             then(print "re-enter the specified frequency")
             else(print "Check communication with nearby transmitters")
      else(assert(wrong frequency)))
```

Rule writing is based on knowledge with the characteristics of a decision tree, and the current node is used as the root node for decision: if the user selects "Yes", the relevant node is turned to, otherwise it is turned to another node.

On the right side of the rule, there are often multiple actions that can be executed, how to control the execution order of actions? CLIPS mainly uses data-driven principles and limited control rules implemented by RHS actions. The activated rules to be executed are saved in the agenda. If there are multiple "actions" on the right, a rule is arranged in order; multiple rules are executed in priority order. Priority rules refer to

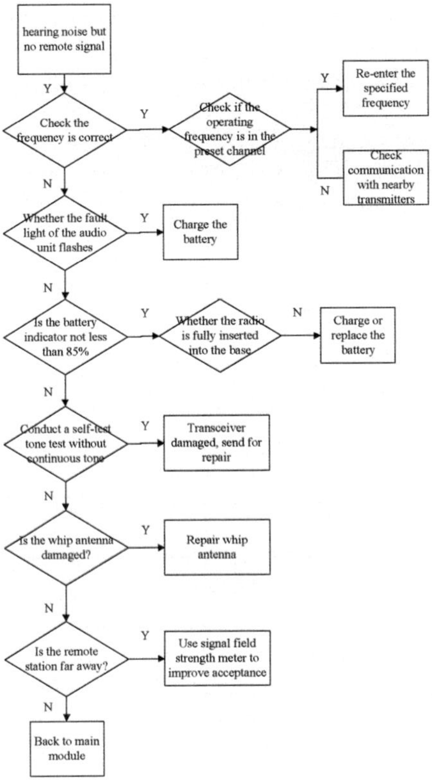

Fig. 3. Problem 3 inference network diagram

multiple rules that are activated at the same time. The priority (salience) size is arranged, and the rules with the same priority size are sorted according to conflict resolution.

3.3 Mixed Programming of CLIPS and VC++

CLIPS is not good at data collection, numerical calculation and only provides a text environment to interact with users. CLIPS lacks the ability to develop graphical user interfaces. VC++ can make up for the shortcomings of CLIPS. In this way, combining VC++ and CLIPS programming can realize a powerful and friendly expert system [4].

There are two main methods for nesting the use of CLIPS and VC++ mixed programs: direct embedded and DLL. Here mainly introduces the mixed programming of DLL way.

We can download the software package released with the CLIPS development environment on the Internet, which contains a clips.dll file, which is a dynamic link library provided by the CLIPS developer. This dynamic link library encapsulates CLIPS core commands such as Load, Reset, Run, etc. Most of the core functions of CLIPS can be realized by calling the functions in this dynamic link library. The developers of CLIPS have implemented a class: CCLIPSWrap in a file named Clipsmfc which can be found on websites. This class encapsulates all the functions of clips.dll, making it easier and more convenient to use, and conforms to the habits of the C++ language.

When we call the dynamic link library, we need to perform various initializations on CLIPS. The functions used are described as follows. The following briefly introduces the usage of this class.

```
ClipsInitialize(......) //Initialize CLIPS environment
ClipsClear(... ...) //Clear CLIPS workspace
ClipsLoad(... ...) //Load system knowledge base
ClipsReset(... ...) //Start CLIPS and wait for execution
ClipsRun(... ...) //Perform CLIPS
```

4 Conclusion

The use of expert systems for armored vehicle fault diagnosis is a very important application. This article combines the current common thinking on fault diagnosis and establishes a knowledge base with a hierarchical classification method. At the same time, it uses a very effective method to solve such problems-decision tree to establish Reasoning mechanism. However, with the upgrading of armored equipment, the expert system is facing a lot of unknown knowledge and information. How to use the existing expert system to expand knowledge and expand functions is a problem that needs to be considered in the future.

References

1. Zixing, C., Guang, X.: Artificial Intelligence and Applications, pp. 131–134. Tsinghua University Press, Beijing (2000)
2. Chuanding, X., Jiaming, Y., Qingsheng, R.: Principles and Applications of Artificial Intelligence, pp. 291–338. Donghua University Press, Shanghai (2005)
3. Giarratano, J., Riley, G.: Expert Systems-Principles and Programming, pp. 235–236. Thomson Learning, London (2001)
4. Guangdong, Z.: Design of fault diagnosis system for airborne equipment based on CLIPS and VC++ mixed programming. Comput. Program. Skills Maint. **21**, 31–33 (2015)

Research on Course Recommendation System Based on Artificial Intelligence

Fuqiang Zong(✉), Deyi San, and Weicheng Cui

Naval Aviation University, Yantai, China
641940546@qq.com

Abstract. In order to improve the cultivation of talents in colleges and universities, based on artificial intelligence technology, a military school student curriculum recommendation system based on syllabus and talent training programs is built, which closely follows the current needs of military personnel and focuses on the development planning of military students. By collecting the personalized data of the students to carry out the classification and layering structure and labeling, and use these data to construct the personalized data set of the students, so as to provide basic data for the construction of the curriculum recommendation system. The system will scientifically recommend courses based on student characteristics, syllabus and talent training programs, use artificial intelligence algorithms to obtain the final recommendation results, and present them to the user terminal according to rules. Provide help to improve the development of college students and the efficiency of talent training.

Keywords: Talent training · Course recommendation · Artificial intelligence

1 Introduction

Under the background of national "double first-class" and "high-level" policy guidance, and the increasingly complicated international situation, college teaching systems and teaching concepts are facing unprecedented impact and changes, and colleges and universities urgently need to improve the direction and science of talent training. The curriculum recommendation system based on the syllabus and talent training program will play an increasingly important role. At the same time, the advent of the era of artificial intelligence has greatly promoted the progressive development of higher education in the direction of data and intelligence. Under this background, relying on artificial intelligence technology and the campus cloud desktop system will enable students to choose courses autonomously and become a new trend in the development of colleges and universities. This article will provide help through data preprocessing, syllabus, talent training program, and constructing a recommendation model to achieve individualized course selection for students, improve the development of college students and the efficiency of talent training.

M. Tavana et al. (Eds.): IISA 2020, AISC 1304, pp. 667–671, 2021.
https://doi.org/10.1007/978-3-030-63784-2_83

2 Data Preprocessing

The personalized data sets of college students can be summarized as static data and dynamic data [1]. Static data includes basic information, course grades, questionnaires, majors, development directions, etc.; dynamic data includes access data, learning data, behavioral data, book borrowing, etc. These data are used to embed "invisible" probes on key pages of the system to use the SDK Can be collected in the same way. The pre-processed data contains a lot of error data, which needs to be filled, replaced, merged, split, loaded, and anomaly processed by ETL to form structured and labeled data.

Structured refers to transforming the display form of behavior data from unstructured data to structured data, and performs classification and statistics; labeling refers to labeling behavior data according to business scenarios, surrounding devices and deeply integrating with business scenarios [2].

We label behavior data through three data dimensions: time, frequency, and result. The time dimension of behavior data mainly focuses on the time period and duration of student learning behavior. The frequency of behavior data mainly focuses on the number and trend of certain specific behaviors. Among them, the frequency has a large positive correlation with the interest of students. The results mainly focus on whether to complete collection, recommendation, course selection, etc., used to judge the results of students clicking to browse. The result data can be divided into favorites, canceled collections, recommended, selected courses, canceled courses and unselected courses, etc. [3].

3 Personalized Data Sets

The personalized data set is a model established by analyzing the characteristics of students' interests, behaviors, and their own attributes [4]. Through the investigation of students and the analysis of student behavior, students are divided into different groups, and then some typical characteristics are abstracted from the groups, and recorded with structured information to summarize the overall characteristics of students.

The personalized data set of students is the core foundation of course selection recommendation. According to the historical behavior of the students, many scene labels are aggregated, and the order of the scene labels is adjusted according to the current student's personalized data set model. If the student selects the "sports and leisure" tab, the system will reorder the recommended courses according to the dimensions of gender, grade, interest, and sensitivity in the student's personalized data set.

Student personalized data sets have their own characteristics and limitations, and are time-sensitive. Therefore, it is necessary to continuously update and modify itself based on the basic data of student personalized data sets, and to be good at figuring out new ones from known data. The tags make the personalized data sets of students more and more vivid and three-dimensional, and play the value of their reference guidelines.

The main functions of the student personalized data set include: (1) accurate recommendation, analysis of potential learning objects of the course, and notification of specific students using the message mechanism; (2) data statistics, such as the most popular Top100 courses; (3) easy to use Data mining builds an intelligent recommendation system, for example, using association rules to speculate that students who like golf courses may also like business administration courses; (4) effect evaluation, improve algorithms through recommendation effect evaluation, and improve recommendation effects; (5) course analysis, through students Personalized data sets and feedback of recommendation results, analyze the richness and rationality of the curriculum, and then provide strategic support for school teaching.

4 Collaborative Filtering Recommendation Algorithm

In the field of personalized recommendation, Collaborative Filtering, although it is one of the most widely used technologies, but due to its serious project sparsity and cold start problems, a large number of scholars have conducted continuous research on this. Lika and others have proposed an application Known classification algorithms create user groups and combine semantic similarity techniques to identify users with similar behaviors to alleviate cold start problems. Ji et al. proposed a factor matrix decomposition model that combines user and project content information to alleviate cold start problems. Recommendation algorithm. Forsati et al. proposed a matrix factorization algorithm that dynamically fine-tunes regularization parameters. It has been widely concerned and proved the effectiveness of the matrix factorization algorithm.

Matrix Factorization is to decompose the user item matrix into user hidden feature vector matrix and item hidden feature vector matrix using the idea of dimensionality reduction, and then calculate the missing items of the prediction scoring matrix through the dot product of the two hidden feature vector matrices. The typical matrix Decomposition algorithms include: Probabilistic Matrix Factorization, Maximum Margin Matrix Factorization, Nonnegative Matrix Factorization, Regularization Singular Value Decomposition, Bayesian Probabilistic Matrix Factorization, SVD++, etc. However, the problem of cold start of the project caused by the sparsity of the user's project rating matrix is still urgently needed to be solved In order to further improve the effectiveness of the recommendation algorithm, in recent years, many scholars have proposed to use different types of information sources to solve the problem of cold start of the project. Yang et al. proposed a matrix decomposition model of social trust network, using additional trust Data to solve this problem. Gurini et al. proposed a social network recommendation algorithm that incorporates sentiment analysis. During the matrix factorization process, the user's emotional information extracted from the content generated by the social platform is used to recommend the target user. Followed preferred users.

Unlike recommendation algorithms such as SMS that measure the similarity of students, this paper uses Coupled Object Similarity to capture student attribute information during matrix decomposition to improve the recommendation effect. Through target constraints, the process of matrix decomposition is constrained by student attribute information to learn the hidden between students Feature relationships to make the recommendation results more interpretable.

5 Build a Recommendation Model

Constructing a personalized course selection recommendation model includes the following three core links:

The first step is to establish a recommendation model for students' course selection. Its dimension is based on historical behavior data, hobbies and disciplines. The historical behavior data is based on students' historical browsing, click, course selection, comment, sharing, collection, attention and other touch points. Recommend online correlation, online similarity, offline correlation, offline similar behavior; hobbies are based on the classification of personalized data sets of students and multi-platform interoperability of the platform; subject majors are based on the students' subject majors and curriculum relevance to achieve recommendations [5].

The second step is the recommendation model matching algorithm. The implementation strategy of the algorithm includes recommendation based on student information, recommendation based on course characteristics, recommendation based on historical knowledge, and custom supplement strategy. Use algorithms to get a list of courses that match student behavior;

The third step is a ranking algorithm for the model recommendation results. Based on the student interaction logs, the model is trained with feature weights, and the ranking algorithm is used to automatically match personalized recommendations.

A simple implementation of the course recommendation model. First, the algorithm offline trains a course set similar to a course. When a student initiates a request online, a real-time click recommendation strategy is invoked. The strategy first obtains the list of courses the student has clicked in real time, and then takes each Similar courses of the course, so as to obtain a candidate set, and then through online ranking to predict the click rate, and take the TopN course to get a recommended result set [6].

Acknowledgments. The key to the course selection system in universities is to intelligently recommend courses that meet the actual needs of each student. If you want to realize the "thousands of people" course selection interface, the background needs to establish complex behavior data preprocessing, data storage, data modeling and portrait construction processes. Simply preprocessing a certain dimension of data can only achieve personalized recommendations If you want to improve the overall effect of personalized recommendation, you must cover the student's comprehensive behavior trajectory, even its offline behavior.

By building an intelligent recommendation model for students' course selection, using data mining, machine learning and other technologies, the "one thousand faces" is changed to "one thousand faces", which can improve the satisfaction of students in selecting courses, improve the reasonableness of course settings, and enhance the experience of selecting courses, Shorten the path for students to choose courses, stimulate students' enthusiasm for learning, and give full play to the functional value of university courses.

References

1. Zhao, G., Chen, X., Li, S., Wu, A.: Preliminary analysis of self-portraits of college students based on data analysis. Digit. Technol. Appl. **08**, 233–234 (2017)
2. https://blog.csdn.net/sfm06sqvw55dft1/article/details/78739738[N]

3. Minli, S.: Research and design of student course selection system. Digit. Technol. Appl. **1**, 176 (2016)
4. Shiwei, L., Kun, Z., Yanqiu, Y., et al.: Analysis of the application of big data mining technology in online education platform. Inf. Commun. **9**, 138–139 (2016)
5. Ting, Y.: Research and analysis of the course selection management system of Baoshan University (2016)
6. Liang, Z.: Research on the key technology of college course selection system. Microcomput. Appl. **32**(6), 36–38 (2016)

Design of ZigBee-Based Control System of Urban Intelligent Street Light

Fan Gao and Hua Jin(✉)

Yanbian University, Yanji 133002, Jilin, China
hjin@ybu.edu.cn

Abstract. The traditional urban street light system has limitations in humanization and energy saving. This paper puts forward a design scheme of ZigBee-based control system of urban intelligent street light. Residential street lights and urban lighting lights adopt time-sharing control mode, humanized control mode and holiday mode respectively, and the time of turning lights on and off is determined according to natural light intensity and sunset time. The brightness of the street light in the residential is adjusted according to whether there are pedestrians or not, which brings more convenience to people who go out. Through the experiment, this scheme has achieved the expected effect and has a remarkable effect in energy saving. It will play an active role in accelerating the construction of intelligent cities and improve the quality of life.

Keywords: Street light control · Edge computing · Light · ZigBee · Intelligent city · Energy saving

1 Introduction

The seriousness of energy shortage has affected the economic development of countries all over the world, and the concept of energy conservation and environmental protection has become the consensus of all countries in the world [1]. At present, lighting electricity accounts for about 13%–14% of the country's total power generation, of which road lighting electricity accounts for about 20%–30% of the total lighting power consumption of a country, and the energy utilization rate is less than 65%, and energy waste is serious [2]. According to statistics, among the cities, only Shanghai, Beijing, Shenzhen, Guangzhou, Chongqing, Tianjin and other key cities have 2.07 million street lights, and the state attaches great importance to the energy saving of road lighting [3].

Nowadays street lights in most cities are switched on and off according to season and sunrise and sunset time [4], however, this way of timing is easily affected by the weather [5], for example, in cloudy days, turning the lights off too early and turning the lights on too late may bring inconvenience to people; while in sunny days, turning the lights off too late or turning the lights on too early may cause energy waste. Urban road lighting is an important part of social public facilities, and intelligent lighting is an inevitable trend in the development of the Internet of things [6]. The system uses ZigBee wireless communication mode, Internet of things technology and energy-saving control strategy to achieve centralized control and effective management of lighting equipment, which can greatly save power resources and bring more

M. Tavana et al. (Eds.): IISA 2020, AISC 1304, pp. 672–679, 2021.
https://doi.org/10.1007/978-3-030-63784-2_84

convenience to people's travel, not only improve the level of public lighting management, but also save energy and protect the environment.

2 Overall Design of the System

2.1 Overall Idea of the System

According to the size of the city, the street light coverage is divided into several areas, and the illumination values of different areas are collected and transmitted to the main controller through wireless communication. The controller calculates whether to turn the street light on or off according to the illumination value of each area and the sunrise and sunset time of the day to switch the street light on and off. The brightness of the light shall be adjusted according to the traffic flow during the turning on of the light. The urban rope lights adopts the weekday/holiday mode; The edge computing method is adopted for street light in the community, if someone walks under the street light, the brightness will be turned up, otherwise the brightness will be kept low. Different control strategies are implemented for street lights, street lights in the community and urban rope lights to ensure that they can bring convenience to the passengers and save power at the same time. Intelligent street light system is one of the best entrances and service ports for smart cities. Develop urban intelligent street light systems, address the need for intelligent street light management, providing more livelihood guarantees for the construction of smart cities [7].

2.2 System Architecture

The system is composed of terminal module, communication module and control module. The overall system architecture is shown in Fig. 1. The terminal module includes light acquisition and infrared detection of human body, and sends the collected light data to the controller in time through the network. A light detection point is set in each area to acquire the light simulation value. The pyroelectric infrared sensor is installed on the street light of the residential to detect whether there are pedestrians around the street lights when the street light is turned on, providing humanized lighting for pedestrians by adjusting the brightness. The communication module is composed of Zigbee router cluster, and each router is in charge of data forwarding among Zigbee terminal nodes, coordinators and other routers. The lighting control module makes decisions according to the collected data and control strategy, and transmits the decision results to the terminal module through the wireless communication module, and controls the street light according to the instructions.

In this system, light information and human body infrared signal are mainly collected, the amount of data collected is not large, and the requirement of data transmission rate is not high; ZigBee meets the networking between the spacing distances of street lights and the cost is lower than NB-IOT. In conclusion, ZigBee wireless communication mode is adopted in this design.

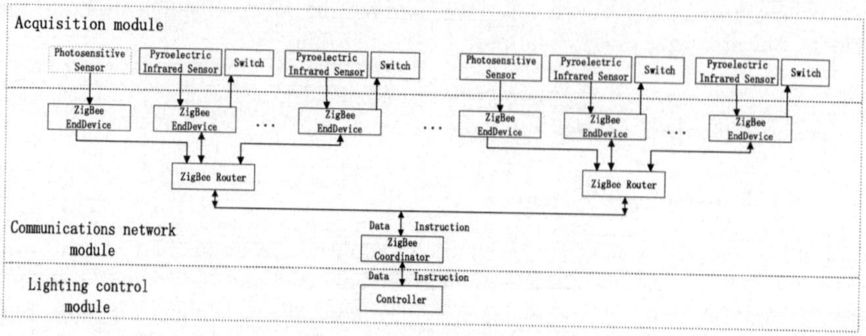

Fig. 1. Structure diagram of system

3 Hardware Design of the System

3.1 Communication Module

ZigBee is a short-distance, low-power, low-rate, low-cost wireless access technology [8, 9]. The module works in the free 2.4 GHz frequency band, and the digital I/O interface is all led out. The mesh topology has a more flexible way of information routing, which completes the mesh propagation of messages through the routing table [10], suitable for the construction of large-scale wireless sensor networks.

Therefore, this system adopts network topology. As the core of the whole ZigBee network, coordinator establishes network, transmits network beacon and manages all nodes of the whole network, stores information of each node and provides routing information between related nodes in the network [10]. The data collected by each terminal node is transmitted to the coordinator through other nodes and routers, and then transmitted to the controller module for decision-making; the switch command of street light is transmitted to each terminal node through the coordinator and router.

3.2 Sensor Module

The sensors used in this system are photosensitive resistance sensor and pyroelectric Infrared Sensor.

The pyroelectric Infrared Sensor is an automatic control module based on infrared technology, which can detect the specific wavelength of about 10 um infrared emitted by human body. When there are pedestrians passing by the street light, the pyroelectric Infrared Sensor will generate high level by detecting the specific infrared wavelength emitted by human body, while it will keep low level when there is no one.

Photoresistor is a special resistor made of semiconductor materials such as cadmium sulfide or cadmium selenide, and its working principle is based on internal photoelectric effect; As the light intensity increases, the resistance value decreases rapidly, and the resistance value can be as small as 1 KΩ or less. When there is no light, the photoresistor is in a high resistance state, and the resistance can generally reach 1.5 MΩ.

3.3 Street Light Terminal Module

The street light terminal module is composed of pyroelectric Infrared Sensor, LED street light and CC2530 controller, and the illumination acquisition module is composed of photoresistor sensor and CC2530 board. The connection between each sensor pin and the CC2530 development board interface is as follows. The pyroelectric infrared sensor's OUT pin is connected to the P0.1 pin of CC2530. The photoresistor sensor's DATA pin is connected to the P0.7 pin of CC2530. The VCC of the LED street light is connected to the P0.7 pin, the timer 1 of CC2530 is used for PWM control to adjust the brightness of street light.

4 Software Design of the System

4.1 The Work Flow of the System

The flow of data and instructions in this system is shown in Fig. 2, and the dotted line in the figure represents ZigBee communication. The photoresistor sensor acquires natural light, converts it into an analog light value through the AD conversion of the Zstack of the ZigBee terminal node, and then sends it to the coordinator by unicast. The coordinator is connected with the controller through the serial interface, and the controller sends the light on and off instructions to the coordinator through the exit, and then the coordinator finally reaches all the terminal street light nodes in the area through the forwarding of the router cluster, the terminal node opens its P0.7 pin, which is used as the output pin to power the LED street light to turn on the connected street light.

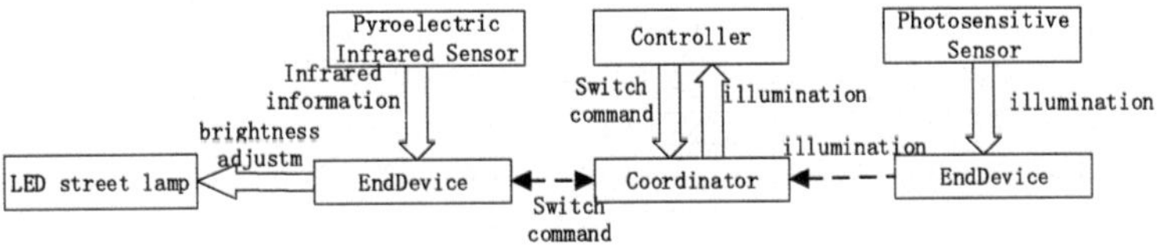

Fig. 2. Data and instruction flow chart

The main flow chart of the system is shown in Fig. 3. Among them, (a) represents the workflow of the controller. After receiving the illumination value, the controller passes the street light control strategy and the street light switch state, comprehensively judges and issues the command to open (flag = 1) or close (flag = 0) the street light. After experiments, when the simulated light value is less than 30, the street light needs to be turned on. Therefore, in this system, the threshold value of turning on the light is set as 30, and the threshold value of turning off the light is set as 31. In Fig. 3, (b) is the working flow chart of the street light terminal module in the community. It realizes the operation of turning on and off the light according to the instructions from the master controller, and adjusts the brightness of the light according to whether there are pedestrians.

When the pedestrian passes by the street light in the community, the pyroelectric infrared sensor generates high level by detecting the specific infrared wavelength of the human body. In the case of no one, the state of low level will be maintained. Then, combined with PWM (pulse bandwidth modulation), the brightness will be turned up when there is a pedestrian and turned down when there is no pedestrian.

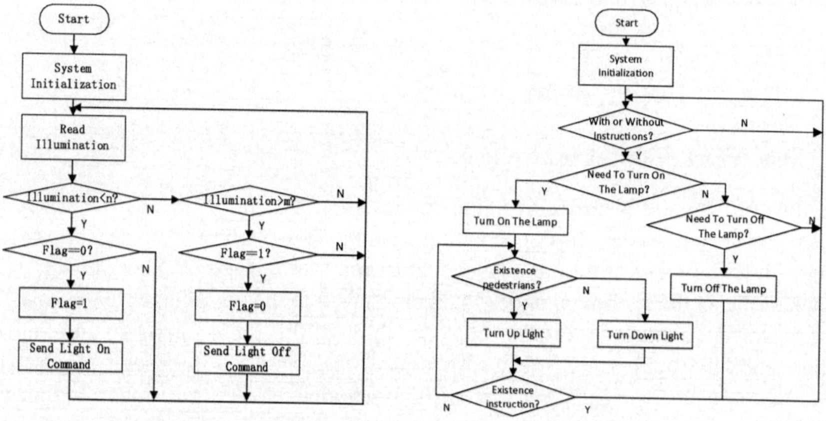

(a) (b)

Fig. 3. System work program flow diagram

4.2 Street Light Control Strategy

According to the practical function of the system, the urban street light can be divided into three types: street light, residential street light and urban lighting light. Because the demand for different functions of street light is different, different control strategies are designed.

The street lights on both sides of the urban road are mainly used for vehicles and pedestrians, and the traffic flow changes obviously in different time periods. For example, 1) the rush hour from turning on the lights to 22:00 is the rush hour with large traffic flow and large number of pedestrians; 2) the traffic flow from 22:00 to 1:00 of the next day is less; 3) the traffic volume is particularly low from 1:00 to 5:30 the next day; 4) from 5:30 the next day to turn off the lights It's rush hour. The strategy of each time period is shown in Table 1, and the critical time can be adjusted according to the specific situation.

The street light in the community adopts the humanized mode based on the above time-sharing control mode and adopts edge cloud. Edge cloud is a computing technology to overcome the edge of limited computing resources and ensure scalability and flexibility [11]. When there are pedestrians passing the street light, the light will be gradually turned on; when no pedestrian is passing, the street light will be gradually adjusted to the initial brightness. In order to prevent more pedestrians, the terminal repeatedly dimming to shorten the life of the light tube, this design should maintain this brightness for a period of time after pedestrians pass by, if there are pedestrians again during this period, overload the time.

Table 1. Time sharing control strategy table.

Classification	Time slot			
	After light on - 22:00	22:00–1:00	1:00–5:30	5:30 - turn off the lights
Road light	100% brightness	70% brightness	70% brightness interval light	100% brightness
Street light	100% brightness	70% brightness and 100% brightness for pedestrians	70% brightness interval lights, pedestrians adjust to 100% brightness	100% brightness

The urban lighting adopts the holiday mode to make a schedule in advance, which includes different light on time and off time on weekdays, weekends and festivals, as shown in Table 2. Considering that it is still bright just after sunset, so the preset light on time should be set later than the sunset time. The computing method of sunset time T1 is as Formula 1 [4], and n1–n3 is the delay time. The light off time is set at a fixed time T2, but weekends and festivals are later than usual, n4–n5 is the delay time. The delay time (n1–n5) can be set according to the specific situation.

Table 2. City lighting schedule.

Time	Date		
	Weekdays	Weekend	Festival and holiday
Light on time	T1+n1	T1+n2	T1+n3
Light off time	T2	T2+n4	T2+n5

$$Tsunset = \frac{(180 + 15Z - Lo) - \arccos(\tan(\frac{10547}{450} \times \cos(\frac{2\pi(T+9)}{365}) \times \frac{\pi}{180}) \times \tan(\frac{La \times \pi}{180})) \times \frac{\pi}{180}}{15} \tag{1}$$

In the formula, Z represents the time zone, Lo represents longitude, La represents latitude, and T indicates the day of the year.

5 Test and Analysis

5.1 Test Scheme Design

The system should be sensitive to different meteorological environments, so the test is mainly divided into sunny and cloudy days. For the test site, it should choose the open higher place, this test chooses the roof of the teaching building as the lighting test point. After several tests, the critical value of illumination is set at 30. Some of significant test results are shown in Table 3.

5.2 Test Result Analysis

Three conclusions are drawn from the test results:

The energy saving effect can be obtained. During the experiment, the street lights are turned on regularly (16:00), but during the test period, the ambient lighting conditions are good. The lighting time of this system is generally later than the actual lighting time, with an average of 15.5 min later. At the same time, the End Device uses the infrared sensor to test whether there are pedestrians, and if there are pedestrians, increase the brightness, otherwise maintain a low brightness, which will further save power resources.

The response to different weather conditions is reasonable. On cloudy days, lights are turn on earlier than on sunny days, and if it is overcast, turn on lights earlier than on cloudy days, bringing convenience to people's travel.

It can save human resources. There is no need to frequently modify the turning-on and off times due to changes in weather.

Table 3. Comparison of this system with the current system.

Date	Opening time		
	Current system	In this system	Disparity
11.24 (Cloudy)	16:00	16:16	+14
11.26 (Overcast)	16:00	16:14	+12
11.27 (Sunshine)	16:00	16:19	+17
11.30 (Sunshine)	16:00	16:18	+16

6 Conclusions

Based on CC2530 board, combined with ZigBee communication technology, infrared detection technology and light detection technology, this system puts forward a control scheme of ZigBee-based control system of urban intelligent street light. The road light adopts time-sharing control mode, the residential street light adopts humanized mode, and the urban brightening light adopts holiday mode. The road light decides whether to turn on or off the street light according to the ambient brightness, the residential street light adjusts the street light brightness according to whether there are pedestrians, and the urban brightening light is determined with reference to the sunset time. After testing, the system can effectively turn the lights on and off according to the illumination value, and the energy-saving effect is obvious.

The shortcoming of this system is that it can't give an accurate warning of street light failure, and it needs to be further developed and improved. Because of the complex environment and the different temperature in winter and summer, it has a great influence on the photoresistor, and it is still necessary to further study the operation of the photoresistor in different environments.

Acknowledgments. This research project was supported by the Education Department of Jilin Province ("Thirteen Five" Scientific Planning Project, Grant No. JJKH20180898KJ), Jilin Education Science Planning Project ("Thirteen Five" Plan, Grant No. GH170043).

References

1. Zhao, Y., Chen, R., Wan, Y., Bao, S.: Intelligent dimming system design of LED street light for smart city. Tech. Innov. Appl. **28**, 37–38 (2019)
2. Qiao, Z., Liu, Y., Liao, Y., Wang, X.: Research on urban road lighting energy-saving potential. Power Electron. **50**(12), 38–41 (2016)
3. Tang, J., Xu, F.: Analysis of the scene based solution of intelligent street lamp under NB-IoT. Guangxi Commun. Tech. **04**, 22–25 (2018)
4. Yu, J., Zong, W., Wang, T., Zhang, G., Cheng, H.: Research and implementation of smart urban lighting control system. Comput. Eng. Des. **39**(03), 836–841 (2018)
5. Gong, Y., Lu, Z., Qiao, Y., Wang, Q.: An overview of photovoltaic energy system output forecasting technology. Power Syst. Autom. **40**(04), 140–151 (2016)
6. Chen, X., Wang, Z.: Design of intelligent street lighting control system based on ZigBee. Mod. Electron. Tech. **12**, 72–75 (2019)
7. Chen, H.: Smart city IOT solutions based on intelligent street lighting and heterogeneous network communication gateway. Electr. Tech. Intell. Build. **04**, 61–64 (2017)
8. Liu, Q., Zhaoyou, Z., Xiangui, L.: Design of ship intelligent illumination system based on ZigBee. Mod. Electron. Tech.**42**(22), 137–139+144 (2019)
9. Sun, J.: Design and implementation of wireless sensor network based on ZigBee. Mod. Electron. Tech. **39**(15), 18–20+24 (2016)
10. Niu, P., Luo, D., Liu, L., Guo, Y., Li, S.: Self-adaption road lighting system based on ZigBee networking technology. Mod. Electron. Tech. **42**(04), 121–124 (2019)
11. Kim, H.S., Lee, H.C.: Development of edge cloud platform for IoT based smart factory implementation. J. Korea Soc. Comput. Inf. **24**(5), 49–58 (2019)

Intelligent University Identity Identification System Based on FaceNet and FSRNet

Zewen Zheng, Yinghuai Yu(✉), and Chengkun Song

Guangdong Ocean University (S), Zhanjiang, Guangdong, China
yuyinghuai@126.com

Abstract. University identification is one of the important ways to ensure the security management of today's open campus. It can effectively detect and identify the suspicious person's identity and ensure the security of the university campus. However, traditional university identity recognition systems have slow detection speeds, relatively low recognition rates, and inability to process low-resolution images. In view of the above problems, this paper proposes a university identity recognition system based on FaceNet and FSRNet. Combining virtual sample generation technology, face rectification algorithm, and improved MTCNN face detection. It effectively improves performance indicators such as detection speed and recognition rate, and provides solutions for low-resolution face images. The experimental results show that compared with the traditional university identity recognition system, the implementation method proposed in this paper improves the recognition rate of unconstrained natural environment by 1–2%, reduces the system face detection time, and effectively resolves low-resolution Facial face image for information restoration.

Keywords: Virtual samples · Face rectification · Face detection · Super-resolution reconstruction · Identity recognition

1 Introduction

Today, most college campuses are open, and there are often mobile personnel entering and leaving the campus, which makes the environment of the university campus increasingly unsafe. In the daily safety management work, the campus security department's investigation of the identity of suspicious personnel is time-consuming and costly. Coupled with the low security awareness of college students, the campus security situation is increasingly complex and severe [1, 2].

For the application of face recognition technology in the field of identity recognition, Literature [3] proposed a three-dimensional face identification technology based on key points and local features, with strong algorithm recognition ability. However, due to the non-rigid deformation of the face caused by the change in expression of the algorithm, the recognition effect is poor. Literature [4] proposed a dormitory identity recognition system based on face recognition, but its face collection module simply requires students to collect samples of different poses of the head, resulting in major flaws in the design. Literature [5] proposed an improved campus face recognition security system based on the Alex Net algorithm. The deep learning algorithm was

M. Tavana et al. (Eds.): IISA 2020, AISC 1304, pp. 680–688, 2021.
https://doi.org/10.1007/978-3-030-63784-2_85

used to improve and optimize the three major processes of face recognition. However, no further processing such as face rectification is performed on the detected faces and the recognition effect needs further improvement.

In a word, the current application system based on face recognition technology in the field of identity recognition have different degrees of defects in terms of algorithm recognition speed, face image processing, system integrity and so on. Therefore, we proposes a college identity recognition system based on FaceNet [6] and FSRnet [7], which uses virtual sample generation technologies to effectively solve the single-sample problem. Second, we use the optimized MTCNN cascade neural network to detect faces to ensure the speed of the algorithm. For the multi-pose problem, this paper uses face rectification algorithm based on general analysis and affine transformation to correct and align the face image. Finally, we also use FaceNet and FSRNet for feature extraction and super-resolution reconstruction of human faces, which improves system recognition rate and ensures the practical application value of the system.

2 Overall System Design

The overall design of the system is mainly composed of a training module and a recognition module. As shown in Fig. 1.

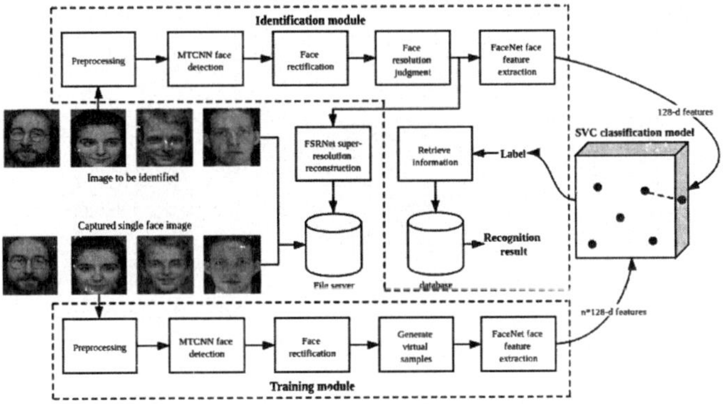

Fig. 1. Overall design block diagram of the system

2.1 Virtual Sample

The training images used by the system are mostly positive face single-sample images, and the change of face attitude caused by the difference in shooting angle in practical application will be characterized by poor robustness.

Literature [9] uses attitude and rotation algorithms to simulate the shooting angle of the image and achieve good results. However, the algorithm itself will cause cracks in the image. In view of this deficiency, we use a bilinear interpolation algorithm to fill the broken image area. The details are as follows: We set image P size to $m \times n$, then use the following algorithm to calculate each pixel point $(x,y)\{x = 1, 2, \ldots, m;\ y = 1, 2, \ldots, n;\}$ to get the transformed virtual image P' coordinate(x',y'):

$$\begin{cases} x' = x \\ y' = y + (1 - f(y,u))(\lceil \frac{n}{2} \rceil - y) + (\lceil \frac{n}{2} \rceil - 1)u \end{cases} \quad (1)$$

Where $\lceil . \rceil$ represents rounding up. $f(y,u)$ is the rotation factor of image P scaling, and f as shown in the following formula:

$$f(y,u) = -\frac{2u}{\Pi}(y-1) + 1 + u \quad (2)$$

Where u is the rotation value of the image shooting angle (left rotation is positive and right rotation is negative). When the image is fractured due to stretching, set the fracture point coordinate be (X,Y), gray value be g and the width of fracture area be $dist$. The following formula can be obtained from the bilinear interpolation algorithm:

$$g(X,Y) = \frac{g(X - dist, Y) \times d + g(X + dist, Y) \times (dist - d)}{dist} \quad (3)$$

Since the fracture interval width is greater than or equal to 1, use d represents the current fracture point position $\{d = 1, 2, \ldots, dist\}$. The specific effect is shown in Fig. 2.

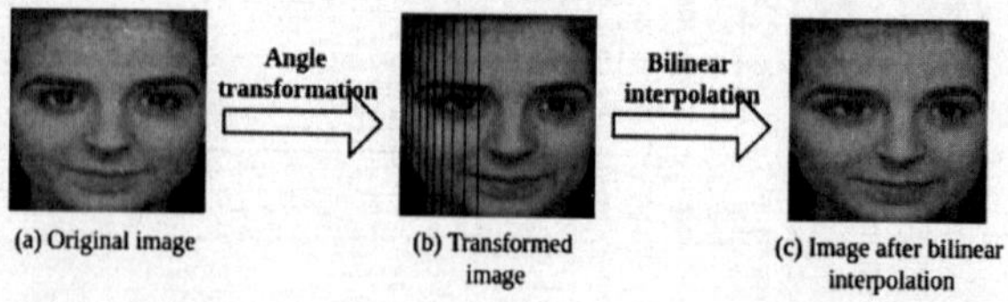

Fig. 2. Effect of improved Angle transformation algorithm

Besides, through the combination of formulas and digital image technologies. System will create 12 virtual samples (excluding the original image). As shown in Fig. 3.

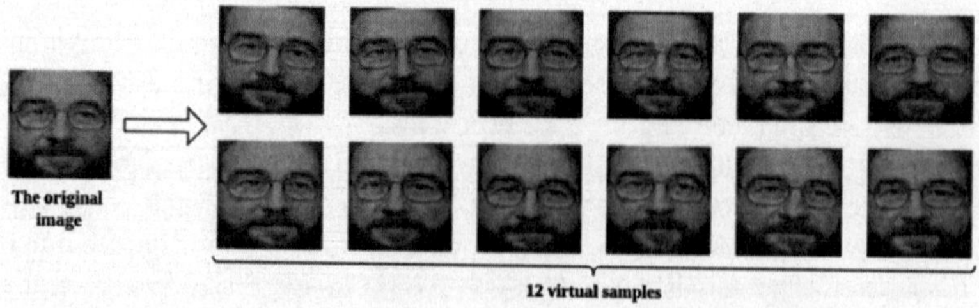

Fig. 3. Generated 12 virtual samples from the original image

2.2 Face Rectification

In the applications of identity recognition, image samples collected by the system often appear on the side face, side head, etc. At this point, the face shape needs to be normalized, i.e. face correction. The face rectification algorithm used in this paper is mainly used for general analysis and affine transformation between key points and standard points ([43, 61], [117, 61], [80, 94], [50, 125], [110, 125]) of face.

According to the known conditions, let the coordinate matrix of the feature point and standard point be respectively set as p, q, then $p \in R^{5\times2}$, it is written in mathematical form:

$$argmin_{s,\theta,t} \sum_{i=1}^{5} \|sRp_i^T + T - q_i^T\|^2 \tag{4}$$

Equation 4, formulas in the analysis of the objective function, which is suitable for rotation matrix $R = \begin{bmatrix} cos\theta & -sin\theta \\ sin\theta & cos\theta \end{bmatrix}$. Written in matrix form:

$$argmin_{s,R,T}\|sRp^T + T - q^T\|_F \tag{5}$$

$$R^T R = I$$

$\|\cdot\|_F$ means *Frobenius* norm, eliminating the influence of translation T and scaling coefficient s in E so that the formula is converted into the following mathematical form:

$$R = argmin_{\Omega}\|\Omega A - B\|_F \, subject\ to\ \Omega^T\Omega = I \tag{6}$$

In Eq. 6 A, B for processing the data obtained after point set, using the singular value decomposition to solve Eq. 6 can get the following results:

$$\begin{gathered} M = BA^T \\ svd(M) = U\sum V^T \\ R = UV^T \end{gathered} \tag{7}$$

Introduce homogeneous coordinates to R and extend a dimension based on the original coordinates to get affine transformation matrix M. By using M for affine transformation, the original image can be corrected to a positive face image. As shown in Fig. 4 in the figure below:

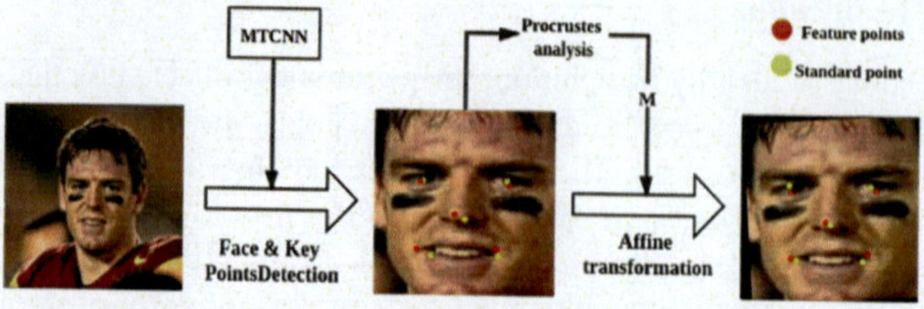

Fig. 4. Structure of face rectification algorithm

2.3 Improved MTCNN Face Detection

MTCNN uses cascaded deep convolutional neural network to achieve, which has good robustness to the illumination, face posture and angle existing in the natural environment. In the specific implementation to the MTCNN face detection algorithm, there are two key parameters that determine the detection duration: *minsize*, *factor*. Where *minsize* represents the smallest detectable image and usually standing for 20 in practice in order to ensure a higher detection rate, but it will reduce the system response speed. Therefore, we propose an algorithm to optimize for *minsize* that makes *minsize* dynamic change with the input image size. The formula is as follows:

$$minisize = 96 \times 0.709^{n-11} n \leq 11 \tag{8}$$

Where n is the maximum value of the image size that can be reached by the power of 2. And *96* represents the initial value of *minsize*. 0.709 represents the decreasing coefficient and is equal to $sqrt(2)/2$. The *minsize* generated by our method reduces the detection time of the system, which is especially obvious in large-size images.

2.4 Face Recognition and Super-Resolution Reconstruction of Face Image

In this paper, we use FaceNet as a face feature extraction network. FaceNet was originally composed of the inception deep neural network structure and the triple Let Loss function. However, Szegedy [11] et al. proposed an inception-resnet model in 2016, which combines Resnet and Inception network structure and improves the performance of the model. Therefore, this paper selects the inception-resnet-v1 depth model as the system feature extraction network structure.

In terms of depth model training, Wen [12] proposed a loss function of model training by combining softmax loss and central loss. By using this method system can get better effect of facial feature extraction, loss function is defined as follows:

$$L = L_s + \lambda L_c = -\sum_{i=1}^{m} log \frac{e^{W_{y_i}^T x_i + b_{y_i}}}{\sum_{j=1}^{n} e^{W_j^T x_i + b_j}} + \frac{\lambda}{2} \sum_{i=1}^{m} ||x_i - c_{y_i}||_2^2 \tag{9}$$

Where L_s and L_c indicate Softmax loss and central loss respectively. m represent the number of batch training samples and n represent the number of categories on Softmax.

$x_i \in R^d$ is the i-dimensional feature of the full-connection layer dimension as the characteristic vector of d. $c_{y_i} \in \mathbb{R}^d$ represents the clustering center belonging to y_i class in L_c. And factor λ is used to balance the L_s and L_c.

Because the images recognized by the identification system are captured in non-binding situations, the image resolution may be too low due to factors such as distance or lens jitter. For low-resolution face images that may be collected by the system. We adopt the face super-resolution network FSRNet based on facial prior knowledge and deep learning proposed in reference [7]. The loss function is:

$$\mathcal{L}_F(\Theta) = \frac{1}{2N}\sum_{i=1}^{N}\{||\tilde{\mathrm{y}}^{(i)} - y_c^{(i)}||^2 + ||\tilde{\mathrm{y}}^{(i)} - y^{(i)}||^2 + ||\tilde{\mathrm{p}}^{(i)} - p^{(i)}||^2\} \tag{10}$$

In Eq. 10, N is the number of training images, $\tilde{\mathrm{y}}^{(i)}$ and $\tilde{\mathrm{p}}^{(i)}$ respectively represent standard of image label and prior information of the ground-truth. $p^{(i)}, y^{(i)}$ represent the high-resolution image and estimated prior information of the *i-th* image.

3 Experimental Analysis

In order to verify the experimental effect, we use FDDB, LFW and CELEBA data sets for separate experiments. Some data sets are shown in Fig. 5.

Fig. 5. Sample of the experimental data set FDDB(top), LFW(middle), and CelebA (bottom)

In order to test the improvement of recognition rate of face rectification in the unconstrained natural scene. We used LFW for the experiments and selected 9158 images of 1678 individuals as the experimental data set, among which 1678 were SVC single training samples and the remaining 7480 were the test set.

Experimental results are shown in Table 1. It indicates that the face rectification algorithm used in this paper has improved the recognition rate compared to the unprocessed sample. Especially in the 15° to 30° deflection Angle, the face recognition rate has increased by nearly 2%.

Table 1. Recognition results of face rectification in various face positions

Human face gesture	Test sample size	Recognition accuracy	
		No face rectification	Face rectification
Pitch (0°–15°, down)	1721	86.81%	87.74%
Pitch (15°–30°, down)	1007	86.10%	88.08%
Pitch (0°–15°, upward)	3160	86.27%	87.06%
Pitch (15°–30°, upward)	2247	85.94%	87.09%
Yaw (0°–15°, To the left)	2257	87.42%	88.66%
Yaw (15°–30°, To the left)	851	87.31%	89.07%
Yaw (0°–15°, To the right)	2556	87.13%	88.46%
Yaw (15°–30°, To the right)	3044	86.86%	87.78%
Roll (0°–10°, To the right)	2327	85.60%	86.68%
Roll (0°–10°, To the left)	3755	87.22%	87.58%

For the experimental verification of face detection speed of improved MTCNN, this paper adopts the FDDB evaluation data set to test and processes it to obtain multi-scale image data set. The experimental results are shown in the Fig. 6.

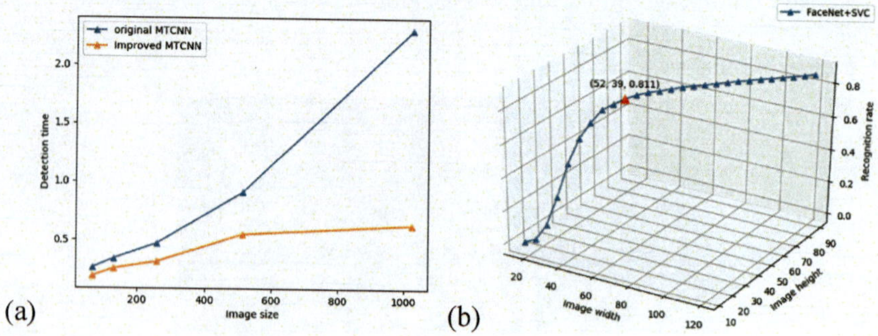

Fig. 6. **a** Experimental results of improved MTCNN. **b** Results of FaceNet recognition on different scale data image

The Fig. 6a shows the speed performance of the improved MTCNN is significantly improved with the increase of image size compared to the original MTCNN. Besides, through down sampling of the LFW and calculation of the corresponding scale recognition rate, we selected the image size corresponding to the recognition rate of 81% as the resolution threshold (Fig. 6b). And we use it to performed super-resolution reconstruction experiment on the CelebA. As shown in Fig. 7.

Fig. 7. Reconstruction of low-resolution face image by FSRNet

The experimental result shows that compared with the LR image, the face reconstructed by FSRNet has enhanced facial features and is very close to the original image, which also illustrates the effectiveness of FSRNet in rebuilding LR faces.

4 Conclusion

In this paper, we propose a university identification system based on FaceNet and FSRNet, which combines virtual image generation technology, improved MTCNN face detection and face rectification algorithm to improve the traditional identity recognition. Besides, we also provide a new idea for low-resolution face processing. The experiment shows that this system in unconstrained nature scene showed higher application value and robustness, which provides a practical implementation idea and effectively ensures the security of the campus environment.

Acknowledgments. The research is supported by Project of Innovative and Entrepreneurship for College Students in Guangdong Province (S201910566089), Team of Innovation of Guangdong Ocean University Students (CXTD2019004), and Project of Enhancing School with Innovation of Guangdong Ocean University (GDOU2016050222).

References

1. Huang, Y.: Application of face recognition technology in campus security management. Technol. Mark. 26(8), 213–213, 215 (2019)
2. Gu, J., Shi, M.: Application of face recognition technology in campus security system. Jiangxi Build. Mater. **21**, 294–295 (2015)
3. Guo, M.L., Da, F.P., Deng, X., Gai, Z.Y.: Based on the key points and local characteristics of 3D face recognition. J. Zhejiang Univ. (Eng. Sci.) **51**(03), 584–589 (2017)
4. Su, J.Y., Liu, J.F., Yang, M.H.: Intelligent access control system based on face recognition. Softw. Guide **18**(04), 32–35 (2019)

5. Zhang, Y., Dai, M.L., Lei, G.P.: An improved face recognition system based on deep learning algorithm – a case study of intelligent campus security system. China CIO News. **05**, 136–137+139 (2018)
6. Schroff, F., Kalenichenko, D., Philbin, J.: FaceNet: a unified embedding for face recognition and clustering. In: Proceedings of the IEEE Conference on Computer Vision and Pattern Recognition, pp. 815–823 (2015)
7. Chen, Y., Tai, Y., Liu, X., Shen, C., Yang, J.: Fsrnet: endto-end learning face super-resolution with facial priors. In: CVPR, pp. 2492–2501 (2018
8. Zhang, Y.Y., Zhang, H.L., Chen, L.L.: Multi-pose face image generation based on second-generation id card. Electr. Sci. Technol. **28**(06), 64–67+71 (2015)
9. Xu, X.Y.: Single sample face recognition method based on virtual image. Comput. Eng. **38** (01), 143–145 (2012)
10. Huang, F.F.: Face Recognition Research Based on LBP. Chongqing university (2009)
11. Szegedy, C., Ioffe, S., Vanhoucke, V., et al.: Inception-v4, inception-resnet and the impact of residual connections on learning [EB]. arXiv preprint arXiv:1602.07261, 2016.
12. Wen, Y., Zhang, K., Li, Z., et al.: A discriminative feature learning approach for deep face recognition. In: European Conference on Computer Vision, vol. 2016, pp. 499–515. Springer (2016)

A Novel Group Key Management Protocol Based on Secure Key Calculation Code

Lijun Zhang(✉) and Cheng Tan

Science and Technology on Communication Security Laboratory, Chengdu, China
41049250@qq.com

Abstract. This paper presents a novel group key management protocol based on secure key calculation code, which achieves the security requirements of forward security, backward security and collusion resistance. Compared with LKH (logical key hierarchy) and OFT (one-way function tree) protocol, the performance of our protocol is more efficient in crucial rekeying process when group members change dynamically. In fact, the required communication of rekeying is almost the same as that in the most efficient protocol CKC (code for key calculation), but we repair the security vulnerability found in CKC. In addition, we propose an encryption manner named "one message one key" by use of key evolution. It enables that every group member employs a distinct key when encrypting a new message, which improves the security level furthermore.

Keywords: Group communication · Cryptographic protocol · Key evolution

1 Introduction

With the development of broadband wireless networks, group communication is widely used in TV broadcast, video conference, group chat of instant messaging, online education and military communication scenarios. Because of public access to wireless signal, group communication data needs to be encrypted using a group session key. Only authenticated group members can decrypt data ciphertext correctly. In actual group communication, group members will probably change dynamically which means members will join or leave. In order to ensure security of communication all the time, data encryption should provide forward security (i.e., group member could not construct the message decryption key after leaving), backward security (i.e., a new member could not get access to the decryption key before he joins the group) and the key independence (i.e., these updated keys must not be predictable). A greater security requirement is the ability to resist collusion attack. People outside the group including those who have withdrawn from the group cannot reconstruct the current decryption key even if they cooperate with each other by exploiting their own key information. To achieve these security requirements, group key management protocol is responsible for generating and distributing group session key when a group is established, using this session key when the group members are communicating, and updating the key when group members dynamically change or the key validity period expires [1, 2, 4].

M. Tavana et al. (Eds.): IISA 2020, AISC 1304, pp. 689–697, 2021.
https://doi.org/10.1007/978-3-030-63784-2_86

Group key management protocols can be divided into two categories: network independent and network dependent [8, 9]. The network independent and centralized key management protocols are especially suitable for group key management in most Internet communication scenarios in which a key center (or a key server) takes the responsibility of key generation, distribution and update while many clients form a group to communicate with each other. Here we only focus on this kind of protocols.

1.1 Related Work

Besides meeting the security requirements, A group key management protocol also needs to consider the resources consumption related to key generation and distribution. These resources include the amount of key calculation, key storage and communication of key materials. Compared with key calculation and storage, people are more concerned about the communication efficiency during key generation and update process, especially under the circumstance of a large number of group members or precious bandwidth resource. Therefore, researchers mainly study how to reduce the communication overhead. The crucial part is to design a key update protocol with higher communication efficiency. Here are such three representatives:

LKH Protocol. Wong et al. [6] proposed a group key management protocol based on logical key hierarchy. Every group member stores all logical keys on the binary key. When the key is updated, the key server uses logical keys at these intermediate layers to encrypt the key materials and then multicasts. The group members receive these key materials and decrypt them using the corresponding logical keys, and finally calculate the updated group session key. Compared with the early pairwise communication, it reduces key update traffic from O(n) to O(logN), where n is the total number of group members.

OFT Protocol. Balenson et al. [3, 5] proposed a one-way function tree key management protocol. The logical keys are not completely generated by key server as in LKH, while they are calculated from the logical keys at their left and right child nodes. The key update process can save half communication traffic of that in LKH.

CKC Protocol. Hajyvahabzadeh [7] exploited the idea of key calculation in OFT and introduced key calculation code. All logical keys at intermediate node can be directly calculated from the group session key Gk and node code. Once group members received the updated group session key, they can derive all the necessary logical keys. Hence CKC became the most efficient protocol currently in the process of rekeying. In paper [9], the author declaimed the protocol is secure in the sense of forward and backward security. However, in 2016, Sun et al. [10] pointed out that the CKC protocol is subject to brute force attack and collusion attack due to the vulnerability of short node code. The group session key can be recovered by guessing the node code and the leaven group members can collude to obtain subsequent group session keys. In practical application, group key managements in wireless sensor networks are studied intensively [11–16].

2 CKC Protocol and Attack

2.1 CKC Protocol Description

The protocol is mainly divided into three parts: key generation, rekeying of members joining and leaving respectively. The characteristic is that logical keys at the intermediate nodes are calculated by node code and group session key Gk. Every node of logical key is coded, i.e., the code of a child node is concatenated by a random string of predetermined length on the basis of its parent node code. The group session key Gk is root node. As shown in Fig. 1, the code of key k14 is 0 and code of k12 as its child node is 03 while the leaf node k1 is 037.

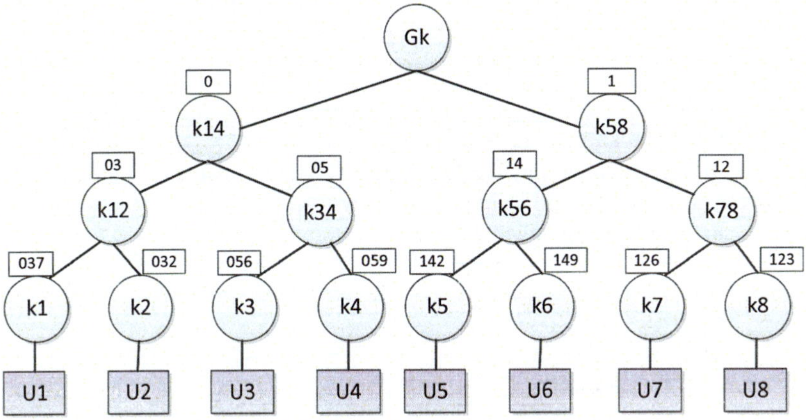

Fig. 1. The node codes in CKC protocol

(1) Rekeying of new group member joining. Assuming newcomer is U8. Firstly, group members from U1 to U7 will receive the message informing new member joining sent by the key server. Then every member of them employs the same one-way function f to update group session key $Gk' = f(Gk)$ and then update logical keys $K'_{middle\_code} = f\,(Gk' \oplus \text{middle\_code})$ at the intermediate node of binary key tree, where $\oplus$ means exclusive-or operation, and middle_code means the code of corresponding logical key node. For example, the updated key $k12' = f(Gk' \oplus 03)$ and $k14' = f(Gk' \oplus 0)$.

(2) Rekeying of a group member leaving. Assuming that member U1 leaved, the key server firstly generates a new random group session key Gk′, and then sends Gk' to the rest of group members. Since all the logical keys at the path from U1 to root node will be affected, i.e., all these keys need to be updated. These new logical keys are encrypted by the top unaffected keys and then sent to the remaining members. For example, Gk′ is encrypted by k2 when the key server sent Gk′ to U2. Gk′ is encrypted by k34 and sent to member U3 and U4. Similarly, Gk′ is encrypted by k58 and sent to member U5, U6, U7, U8. After receiving the Gk′,

every remaining group member also uses $K'_{middle\_code} = f(Gk' \oplus middle\_code)$ to update all intermediate logical keys.

2.2 Attack of CKC

Sun et al. proposed two attacks against the CKC protocol, indicating that this protocol does not have forward security and the ability to resist collusion attack.

Guessing Code Attack. If child node code is only appended a digit number at the code of its parent node, then the group member could calculate the possible logical key by guessing the code of other logical nodes, because one appended digit number only induces 10 possible key values. When members change, the key server will use a certain logical key to encrypt a new group session key. As an attacker, a group member can use these possible logical keys to decrypt this new group session key, he can determine the correct logical key after some simple comparisons.

Collusion Attack If a leaven member A tells member B the logical key code he knows. Then B calculates the corresponding logical keys of these known codes. Later when B leaved the group, he can use these known logical keys to decrypt the ciphertext of new group session key, and decrypt the group message. In this way, group member A and B realize a collusion attack.

3 Our Scheme

3.1 Secure Key Calculation Code (SKCC)

Firstly, we also encode all the logical key nodes of binary key tree. Here we use hexadecimal string to encode the key node. The code of child node is appended a random hexadecimal string of fixed length L on the basis of its parent node code. In order to prevent guessing code attack as in CKC, the value of L is at least 16, i.e. each layer code in binary key tree will increase at least 128 bits. As shown in Fig. 2, every capital letter represents a random hexadecimal string of length L. For example, A_0 and B_0 are both such random strings of length L. In practice, the unicast keys (such as k1, k2, etc.) between a group member and the key server need not be encoded, thereby it will reduce the code length of logical key node.

3.2 Group Construction

The key server selected some clients to form a communication group, and generates a random group session key Gk. This key server uses the unicast key ki and a shared encryption algorithm E to encrypt the session key Gk and the code of bottom node in binary key tree, then sends this ciphertext to each group member. For example, the ciphertext sent to U1 is $E(k1, Gk\|A_0A_1)$, where the symbol $\|$ means the concatenation of strings.

Every group member decrypts the ciphertext to obtain the group session key Gk and the underlying bottom code, and parses in turn all codes of logical key nodes from

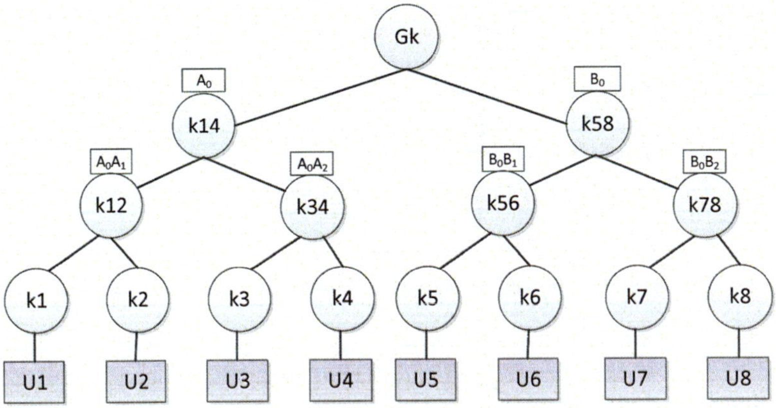

Fig. 2. Secure key calculation code

the leaf code to the root node according to the length L. For example, member U1 parses out that the code of parent node K12 is A_0A_1, while code of k14 is A_0.

Every group member calculates the corresponding logical keys according to the parsed node codes and stores these keys. For example, member U1 computes $k12 = H(Gk||A_0A_1)$ and $k14 = H(Gk||A_0)$, where H is a cryptographic hash function.

3.3 Rekeying of New Group Member Joining

When a new member joins, the key server will generate a new group session key $Gk' = H(Gk)$, by using a hash function H, and locate this new member in a suitable position of the key tree. Assuming this new member is U8, the key server will send Gk′ and code B_0B_2 corresponding to the location to U8, where this unicast encryption key is k8.

Once U8 received the ciphertext $E(k8, Gk'||B_0B_2)$, he will decrypt and parse out Gk′ and code B_0B_2, then calculate the key $k78 = H(Gk'||B_0B_2)$ and $k58 = H(Gk||B_0)$.

When the original group members receive the message sent by the key server that a new member U8 has joined, they will update group session key $Gk' = H(Gk)$ by local computation. Then they update all logical keys by $k_{middle\_node} = H(Gk'||middle\_node)$, where middle_node is the corresponding code of logical key node.

3.4 Rekeying of a Group Member Leaving

When a group member leaved, the key server will randomly generate a new group session key Gk′. More importantly, it is necessary to update all the logical node codes known by this leaving member and replace the corresponding logical keys. These new node codes are encrypted by the top logical key that is not affected by the leaving member. It means the new group session key and updated node codes will be encrypted with the top key and sent to the corresponding group members. Here the plaintext is Gk′ and code of the parent node of top key.

Taking member U1 leaving as an example, the codes and key values of k14 and k12 need to be updated. The unaffected top keys are k2, k34 and k58, i.e., the binary key tree is divided into three branches. The new codes generated by the key server for k14 and k12 are C_0 and C_0C_1 respectively. k2 is used to encrypt string $Gk'||C_0C_1$ and the ciphertext $E(k2, Gk'||C_0C_1)$ is sent to member U2. k34 is used to encrypt string $Gk'||C_0$ and the ciphertext $E(k34, Gk'||C_0)$ is sent to members U3 and U4. k58 is used to encrypt Gk' and the ciphertext $E(k58, Gk')$ is sent to members U5, U6, U7 and U8.

The member U2 decrypts the ciphertext to get Gk' and C_0C_1, then updates the code of k12 and k14 to C_0C_1 and C_0 respectively. The updated key values of $k12 = H(Gk'||C_0C_1)$ and $K14 = H(Gk'||C_0)$.

The members U3 and U4 decrypt their ciphertext to get Gk' and C_0, then update the code of k34 and k14 to C_0A_2 respectively. The updated key values of $k34 = H(Gk'||C_0A_2)$ and $K14 = H(Gk'||C_0)$.

The members U5, U6, U7 and U8 decrypt the ciphertext to get Gk'. Then U5 and U6 calculate the updated key value $k56 = H(Gk'||B_0B_1)$ and $k58 = H(Gk'||B_0)$. U7 and U8 calculate the updated key value $k78 = H(Gk'||B_0B_2)$ and $k58 = H(Gk'||B_0)$.

3.5 Analysis of Security and Performance

Backward Security. The newly joining member does not have the previous group session key and logical node code. He cannot obtain the corresponding logical key information from group communication, so our scheme maintains backward security as in CKC protocol.

Forward Security. When a member has leaven from group, the key server will update the codes and key values of all logical key nodes known to that leaving member. Since the appended string of a child node is of length at least 16 random bytes, it will not work for the leaving member to obtain the other unrelated logical keys by guessing the node codes. Because this is equivalent to guessing a 128-bit key value. Under the condition that the validity of group session key is set reasonably, our scheme can resist guessing code attack. The leaving member cannot obtain subsequent new group session key, which achieves forward security.

Collusion Resistance. Even if the leaving member tells other group members the node codes they know, these leaked key node codes will become invalid since all the known key node codes of the leaving members will be replaced immediately. The collusion of members will also fail to obtain the updated logical key node codes and key values of the binary key tree.

Execution Performance. Execution efficiency is mainly measured by the storage of all necessary keys and communication in rekeying process, of which the communication account is more important. Table 1 summarizes security comparison between our scheme SKCC and several other known centralized group key management protocols. The symbol "Y" means the corresponding requirement is satisfied while "N" is not. Table 2 lists efficiency comparison of these protocols. It can be seen that our scheme has repaired security weakness of CKC scheme. In terms of protocol execution efficiency, our scheme and the CKC scheme have almost the same highest communication efficiency when members dynamically change. In addition, the SKCC scheme also has a great advantage in group construction. It only needs to send to every group member group session key Gk and bottom node code of the binary key tree, eliminating the transmission of intermediate layer logical keys as in the LKH and OFT schemes. This method greatly reduces the amount of key materials communication during group construction.

Table 1. Security comparison of centralized group key management protocols

Protocol	Forward security	Backward security	Collusion resistance
LKH	Y	Y	Y
OFT	Y	Y	Y
CKC	N	Y	N
SKCC	Y	Y	Y

Table 2. Performance comparison of centralized group key management protocols

Protocol	Member joining		Member leaving	Key storage	
	Unicast	Multicast	Multicast	key server	Member
LKH	(d + 1) K	(d−1) K	2dK	(2n−1) K	(d + 1) K
OFT	(d + 1) K	(d + 1) K	(d + 1) K	(2n−1) K	(d + 1) K
CKC	$K + C_0$	0	(d−1)K	$(n + 1)K + nC_0$	$2K + C_0$
SKCC	$K + C_1$	0	$(d-1)K + C_1$	$(n + 1)K + nC_1$	$2K + C_1$

In Table 2, parameter n represents the total number of group members. d represents the depth of the key tree d = logn. K represents the length of the group session key and logical key. C_0 means the code length of the bottom node of key tree in the CKC scheme while C_1 is code length of the bottom node of key tree in our SKCC scheme.

4 The Design of "One Message One Key"

In the current group key management protocols, every group member uses the same group session key Gk to encrypt message. Within the specified period of key validity, this group session key keeps the same. Obviously, if the group members can use different group session keys for encryption during the group session, the security of communication message will be further improved. Based on idea of key evolution, we design the following key variation method.

When every group member employs the group session key Gk to encrypt one message, the current encryption key k is derived from the hash function F and Gk, where F is specified in advance. That is, k = F(Gk||ID||N), where ID is the identity of message sender, and N represents the N-th message sent by this sender. The message sender uses this derived key k to encrypt this message M and sends E(k, M)||ID||N to other group members. After receiving E(k, M) ||ID||N, the group members use ID||N and Gk to calculate the same decryption key k of this message and decrypt ciphertext. It can be seen that the encryption key of every message sent by every group member is different. The message sender only needs to increase N value when sending one new message, in this way, we achieve the security ability of "one message one key".

5 Conclusion

In this paper, we proposed a novel group key management protocol based on secure key calculation code, which is more efficient and secure than previous protocols. It has a great advantage of communication efficiency when group key is generated and updated in the process of group construction and members changing dynamically. Our protocol is especially suitable for an environment with constrained communication resources. We consider the distinct encryption key generation for every group message as our future work, which further improves the security of group communication.

Acknowledgments. We are very grateful to the reviewers for their valuable comments on our earlier version, which help us a lot to improve the quality of this paper.

References

1. Harney, H., Muckenhirn, C.: Group Key Management Protocol (GKMP) Specification, RFC 2093, Internet Engineering Task Force (1997)
2. Harney, H., Muckenhirn, C.: Group Keg Management Protocol (GKMP) Architecture, RFC 2094, Internet Engineering Task Force (1997)
3. McGrew, D.A., Sherma, A.T.: Key Establishment in Large Dynamic Groups Using One-Way Function Trees. Technical report TR-0755, World Academy of Science Engineering and Technology (1998)
4. Wallner, D., Harder, E., Agee, R.: Key management for multicast: issues and architectures, RFC 2627, Internet Engineering Task Force (1999)
5. Balenson, D., McGrew, D., Sherman, A.: Key Management for Large Dynamic Groups: One-Way Function Trees and Amortized Initialization. Draft-balenson-groupkeymgmtoft-00.txt, February 1999. Internet-Draft
6. Wong, C.K., Gouda, M., Lam, S.S.: Secure group communications using key graphs. IEEE/ACM Trans. Networking **8**(1), 16–30 (2000)
7. Hajyvahabzadeh, M., Eidkhani, E., Mortazavi, S.A., Pour, A.N.: A new group key management protocol using code for key calculation: CKC. In: IEEE International Conference on Information Science and Applications, pp. 1–6 (2010)
8. Mapoka, T.: Tshepo. Group key management protocols for secure mobile multicast communication: a comprehensive survey. Int. J. Comput. Appl. **84**, 28–38 (2013)
9. Seetha, R., Saravanan, R.: A survey on group key management schemes. Cybern. Inf. Technol. **15**(3), 3–25 (2015)
10. Sun, Y., Qian, Y., Song, J., Miao, Y., Chen, M.: NCKC: Non-code-aided key calculation for group key management. In: 2016 International Wireless Communications and Mobile Computing Conference (IWCMC), pp. 333–338 (2016)

11. Chen, Y.R., Tzeng, W.G.: Group key management with efficient rekey mechanism: a Semi-Stateful approach for out-of-Synchronized members. Comput. Commun. **98**, 31–42 (2016)
12. Hsu, C.-F., Harn, L., Mu, Y., Zhang, M., Zhu, X.: Computation-efficient key establishment in wireless group communications. Wireless Netw. **23**(1), 289–297 (2017)
13. Harn, L., Hsu, C.-F., Li, B.: Centralized group key establishment protocol without a mutually trusted third party. Mob. Networks Appl. **23**(5), 1132–1140 (2018)
14. Zhang, Q., Gan, Y., Liu, L., Wang, X., Luo, X., Li, Y.: An authenticated asymmetric group key agreement based on attribute encryption. J. Network Comput. Appl. **123**, 1–10 (2018)
15. Albakri, A., Harn, L.: Non-Interactive group key pre-distribution scheme (GKPS) for end-to-end routing in wireless sensor networks. IEEE Access **7**, 31615–31623 (2019)
16. Jiao, R., Ouyang, H., Lin, Y., et al.: A computation-efficient group key distribution protocol based on a new secret sharing scheme. Information **10**(5), 175 (2019)

Personalized Custom Clothing for Intelligent Interaction Design

Yanxue Wang and Zhengdong Liu(✉)

Yan-Xue Wang, No. A2, East Yinghua Street, Chaoyang District, Beijing 100029, People's Republic of China
jsjlzd@bift.edu.cn

Abstract. With the development of the digital age, the traditional clothing industry has gradually developed into a digital industry. Major Internet companies and clothing companies have launched clothing digital personalized customization platforms. However, due to the short development time and the platform is still in the exploratory stage, the user experience is more or less not very satisfactory during the application process. Thus, aiming at the problems in the current clothing digital platform that lack interaction, consumer participation is low, and it is impossible to express some specific questions about their needs professionally and clearly. Combined with the construction of the current clothing electronic sales platform, analyzed and described the invisible needs of users and the customization features of the product, and optimized the interaction design of the existing customization platform. Thereby designed a smart interactive platform for personalized clothing customization. And optimize the number of intelligent interactions on the platform. Provide a reference for design personalized custom clothing platform.

Keywords: Digitization · Personalized custom clothing · Intelligent interaction · Network customization

1 Introduction

As today's digital age, all industries are closely related to digital technology. The core of digital technology is developed based on computer technology, Internet technology, multimedia technology, etc. As one of the pillar industries of China's national economy, Clothing will inevitably evolve from traditional industrial technology to digital technology. At present, there are two main modes to realize the digitalization of clothing: One is the rapid coordination of the value chain, which is "Internet + manufacturing", Mainly include MTM intelligent manufacturing system, eMTM electronic tailor-made system, and p-MTM personalized clothing customization body shape analysis and intelligent revision system; The other is an Internet + clothing digital personalized customized operation mode for niche users and designers, It's dominated by consumers, by placing orders online, the platform delivers customer orders to various designers, designers customize according to customer requirements. In this way, the production cost is reduced, inventory is reduced, and efficiency is greatly improved.

M. Tavana et al. (Eds.): IISA 2020, AISC 1304, pp. 698–709, 2021.
https://doi.org/10.1007/978-3-030-63784-2_87

Traditional clothing customization generally uses manual customization, and the designer measures the size of the user. Then according to the data, the steps of making a board-cutting material-making a sample-modifying- achieving. Such customized methods often require a lot of manpower and material resources, the cost of time is also very high, the clothing produced is usually more formatted. The digital customization is based on the traditional clothing customization and garment production, and establishes a digital model through the Internet, computers, cloud computing, etc. to change the original customization approach. Users generally place orders through online customized platforms or mobile customized APP, and communicate with designers to explain their design needs, and conduct big data analysis through Internet technology. User participation is high, communication costs are low, and production efficiency is high. Most of the domestic clothing customization platforms focus on the O2O model to tap the fundamental motivation for consumer customization, it's in the exploratory research stage. Li Lu et al. [1] found that customer satisfaction depends on the quality of a site recommendation, the relationship between them was positively correlated; Shi Meihong et al. [2] focused on the suit customization virtual display system, and proved that the virtual display of the website can improve the participation and satisfaction of user customization, thereby improving the core competitiveness of suit value; Shen Lei et al. [3] started from the O2O customization model to explore the influencing factors that affect user satisfaction. However, there were relatively mature cases in foreign countries as early as the 90 s of last century. American Levi's company launched a jeans customization program, Nike launched NIKE ID service, etc. With the development of the Internet, the customization model will be applied to various fields, which is particularly important for the apparel industry. Fiore et al. [4] based on improving user pleasure, proving that 3D display, 3D fitting and other designs can increase the interaction between web pages and users, thereby increasing users' sense of pleasure and increasing consumer desire; Lee et al. [5] found that the ease of use of large-scale customized websites can increase consumers' desire to buy by using the technological acceptance model.

But for traditional online personalized customization [6], what most platforms use is to let users determine the style they want according to product parameters. This choice requires the user to have a certain professional foundation and aesthetic, but when most users make choices, they often do not have a very concrete concept of the clothes they want. Therefore, users are often not very satisfied with the actual clothing produced [7]. So this article starts from user experience, by analyzing the characteristics of the clothing customization platform; researching the key technologies of online customization; and designing the interactive model of the customization platform. To improve customer satisfaction, conversion transaction platform to provide a reference basis [8].

2 Determination of Parameters of Digital Clothing Personalized Customization Platform

Clothing personalization mainly involves the following aspects: (1) Virtual display of clothing: including dynamic and static, the display of static clothing also includes text introduction of fabrics, details, designer design concepts, styles, etc. (2) Clothing customization module: Not only for users, designers can also continuously expand their own clothing customization database and constantly update their own styles; (3) Virtual experience and purchase module: The most important thing is to establish a virtual human model based on the user's body type. It is necessary to calculate other body size according to some basic body data of the user and make a size recommendation. The factory orders and manufactures according to the requirements of customization and size, and provides after-sales service, so as to refine the customization needs of users and meet the personalized customization needs of customers as much as possible. The model is shown in Fig. 1 [9, 10].

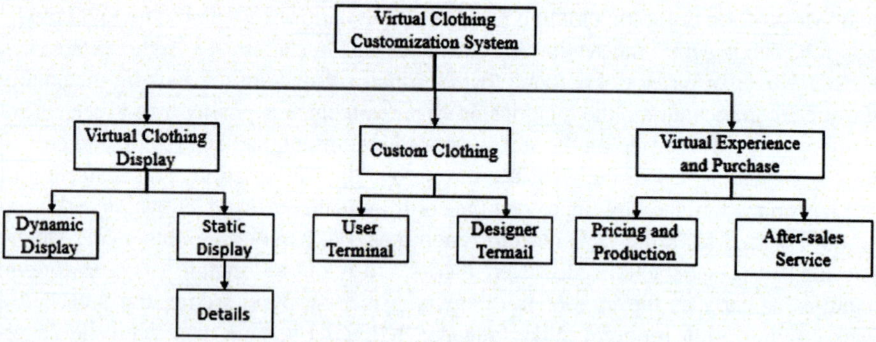

Fig. 1. Virtual clothing customization system

2.1 Production and Display of Virtual Clothing

The most important thing in the production of virtual clothing is the degree of restoration with real clothing. This platform is based on CLO3D software to produce and display virtual clothing, need to pay attention to in the production process: One is to determine the arrangement point in Fig. 2(a), to sew the garment through the two-dimensional version, and to set the basic parameters of the fabric, such as latitude and longitude, drape, elasticity, etc., Fig. 2(b).

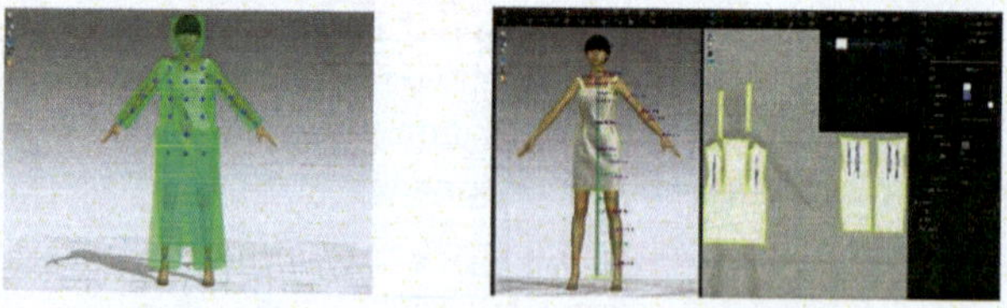

(a) Arrangement point distribution (b) Plate stitching and fabric parameter setting

Fig. 2. Virtual clothing production

The second is the adjustment of the model. After the production of the virtual clothing, you need to test the wearing comfort through the try-on image, transparent image, grid mode, etc. to check the pressure change, fit degree and stretching state of the clothing on the human body. To restore the clothing to the actual wearing condition as much as possible, as shown in Fig. 3:

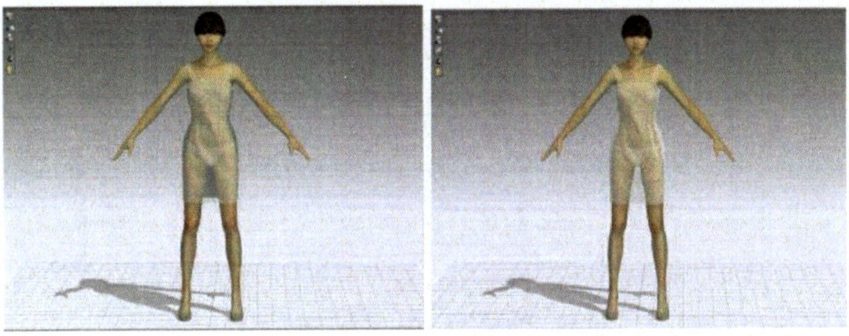

(a)Transparent (b) Texture

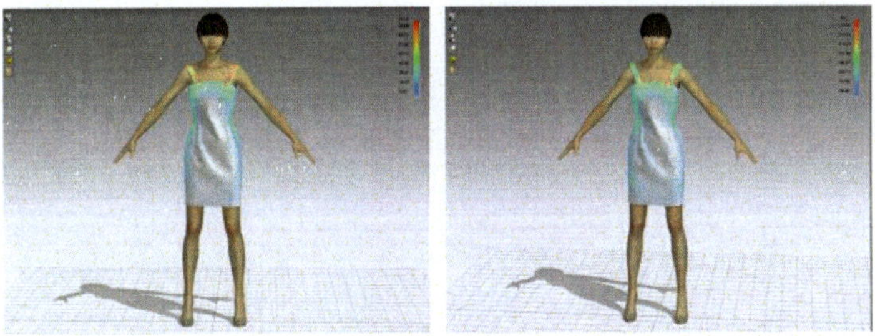

(c) Pressure unit (kpa) (d) Stress unit (%)

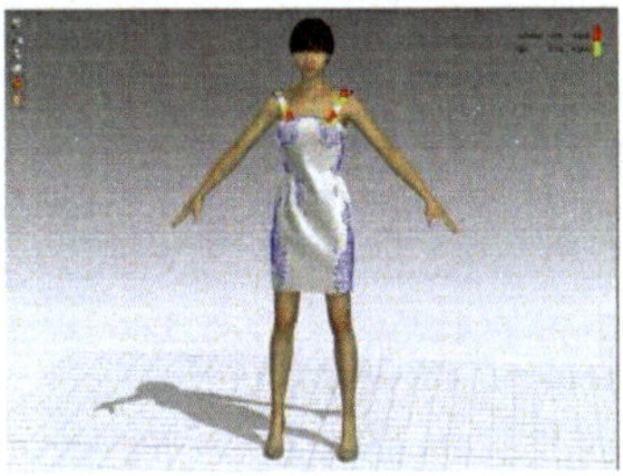

(e)Try on

Fig. 3. Simulated try-on state

The third is the simulation of virtual dynamic display. In the process of dynamic display of virtual models, it is necessary to establish different simulation state models such as aerodynamics, linear and non-linear lifting forces, wind power, etc., to simulate the wearing state of clothing under natural conditions And corresponding to the virtual display scene, the overall dynamic display is integrated, see Fig. 4.

Fig. 4. Virtual clothing dynamic simulation capture

2.2 Description of the Parameters of the Clothing Customization Platform

When personalizing clothing, it is necessary to understand, analyze, and classify users' clothing needs. It is necessary to determine the priority of the clothing customization parameters first, so as to understand which parameters of the clothing are highly valued and which are secondary when the user customizes. When customizing the plan for different users [11], the plan needs to be refined according to different needs, and the determined plan should be classified, coded and saved. If it is found that the plan has problems, it needs to be re-analyzed. Therefore, it is often a repetitive cross-cutting process when acquiring clothing customization needs from users.

For some clothing parameters, due to some professional restrictions, customers often cannot express clearly what they want. At this time, we need to use some means to expose the user's invisible needs, such as conversion, analogy, metaphor and other methods. For example, when making a cheongsam or shirt, it is difficult for users to describe the collar in professional terms. At this time, we can use the form of transformation or analogy to transform the form of the collar into what the user can. The way it is understood makes the user's invisible needs transform into explicit needs in the mode of intelligent interactive customization [12, 13].

Decompose the user's demand for clothing parameters. Assuming that the final customized target is 'A', then the part corresponding to target 'A' is 'B', and each part corresponds to its own parameter, so each part can be regarded as a separate target, and further subdivided. By constantly refining the characteristics of clothing [14], to meet the needs of users to the greatest extent. The specific customized structure diagram is shown in Fig. 5.

The specific correspondence is as follows:

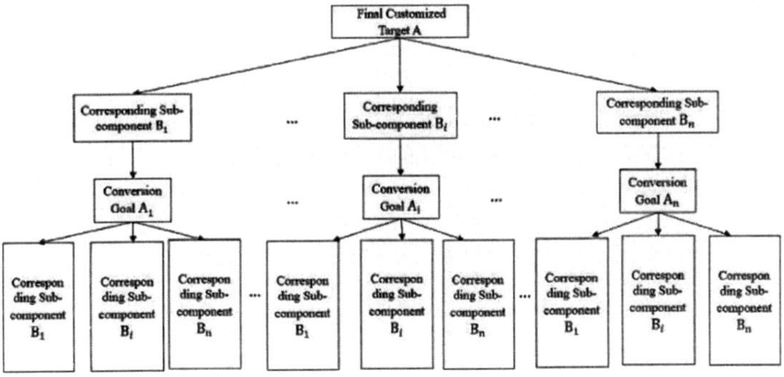

Fig. 5. Custom structure diagram

User-customized clothing:

$$A = [B_{11}\ B_{12}\ B_{13} \ldots B_{1n}] \tag{1}$$

Corresponding subcomponent:

$$B11 = [b_{11}\ b_{21}\ b_{31} \ldots b_{m1}]^T \tag{2}$$

Combine it:

$$A = \begin{bmatrix} b_{11} & \cdots & b_{1n} \\ \vdots & \ddots & \vdots \\ b_{m1} & \cdots & b_{mn} \end{bmatrix} \tag{3}$$

Equation (3), the target 'A' has primary sub-component, each primary subcomponents can be divided into secondary subcomponents. At this time, the primary sub-component can be regarded as a new target 'A', 'b' is corresponding characteristic.

Take a shirt as an example. The final customized target shirt is AT, The style is BT11, the color is BT12, the pat- tern is BT13, and the collar type is BT14. The customization process is as follows:

The final customized target shirt:

$$AT = [B_{T11}\ B_{T12}\ B_{T13}]$$

Corresponding sub-component:

$$BT11 = [\text{Tight}\ \ \text{Slim}\ \ \text{Loose}\ 0]^T$$

$$BT12 = [\text{Warm}\ \ \text{Neutral}\ \ \text{Cold}\ 0]^T$$

$$BT13 = [\text{Plain Spot Stripe Printing}]^T$$

$$BT14 = [\text{Standard Open - angle } 0\ 0]^T$$

Combine it:

$$A = \begin{bmatrix} \text{Tight} & \text{Warm} & \text{Plain} & \text{Standard} \\ \text{Slim} & \text{Neutral} & \text{Spot} & \text{Open = angle} \\ \text{Loose} & \text{Cold} & \text{Srtripe} & 0 \\ 0 & 0 & \text{Printing} & 0 \end{bmatrix}$$

Clothing customization intelligent interactive platform needs to consider all possibilities of user customization. Therefore, we to calculate the maximum number of questions and answers to back up the interactive data- base. In this interaction design, because the user interacts with the parts and characteristics of the clothing when selecting the clothing, the interaction object is relatively abstract and there is no clear function expression [15]. Based on this, the system selects the golden section iterative method to optimize the number of intelligent interactions [16].

The golden section iterative method determines the number of iterations by continuously narrowing the selected interval and searching for the lowest point. For clothing [17, 18], the characteristic attributes corresponding to each part are independent. Assuming that all attributes of a part are M, the first golden section of this attribute is di- vided into the first 0.382 M attributes and the last 0.618 M attributes. Select the specified characteristic attribute from it, remove the unselected feature attributes, and then goldenly divide the selected feature attributes until $M < 2$ stops. The number of iterations is N, then the minimum value N_{min} is $\log_{0.382}\frac{2}{M}$, and the maximum value N_{max} is $\log_{0.618}\frac{2}{M}$. Therefore, the interval of the number of exchange questions and answers for a part is $\left[\log_{0.382}\frac{2}{M}, \log_{0.618}\frac{2}{M}\right]$. M is convergent. Therefore, no matter how much the user requires to customize a piece of clothing, this method can get the only customized style.

2.3 Establishment of Virtual Model and Fitness Evaluation

When evaluating the fit of clothing, it is necessary to combine the needs of users with the objective fit of the clothing itself. First, the user must establish a virtual model. Since the specific values of all the main control parts of the user cannot be obtained, the other values are calculated by combining the training of the BP neural network. Neural network can be seen as a fitting process. That is through some samples, continuously changing the weight and error between the input layer, hidden layer, and output layer to get a model. A process in which the output layer and the input layer can be mapped to each other. Therefore, we take some basic data input by the user as input layer data and record it as $X = (x_1, x_2,\ldots, x_n)$, Initial weight is $\omega(0)$, $\omega i(t)$ is the'i' input weight in the 't' iteration. The expected output value of other parts is expected to be $Y = (y_1,y_2,\ldots, y_n)$, the actual output is $Z = (z_1,z_2,\ldots,z_n)$, the error limit is μ, error function is

$E = \frac{1}{2}\sum_{j=1}^{n}(z_i - y_i)^2$. Therefore the error between the actual output of the node and the target output is:

$$\|Z - Y\| = \sqrt{\sum_{j=1}^{n}(z_i - y_i)^2} \tag{4}$$

If $\|Z - Y\| > \mu$, adjust the weight:

$$\omega_i(t+1) = \omega_i(t) + \sum_{j=1}^{n}\frac{\partial E}{\partial j}\frac{\partial z_j}{\partial \omega_i} = \omega_i(t) + \sum_{j=1}^{n}(y_j - z_j)r_{ij} \tag{5}$$

Bring the readjusted weights back into the calculation. If each sample satisfied $\|Z - Y\| > \mu$, completed the training. Construct the trained model into 3D human body model system, store the relevant data in the database, and obtain the prediction results, as shown in Fig. 6.

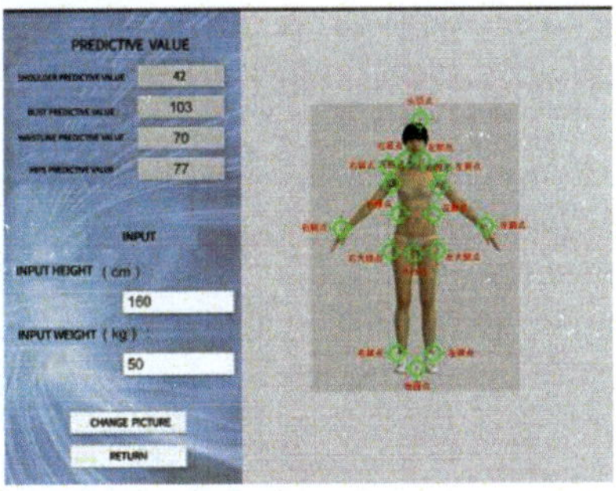

Fig. 6. 3D human body data prediction model

Obtain the human body model created by the system after calculating the data of each part, collect the user's wearing preferences and personalized requirements, and blur it. Then according to the size of the clothing in the pattern library and the amount of clothing selected by the customer [19], the size of the clothing that the customer wears after personalized customization is recommended, using CLO3D to simulate and predict the fit. As shown in Fig. 7.

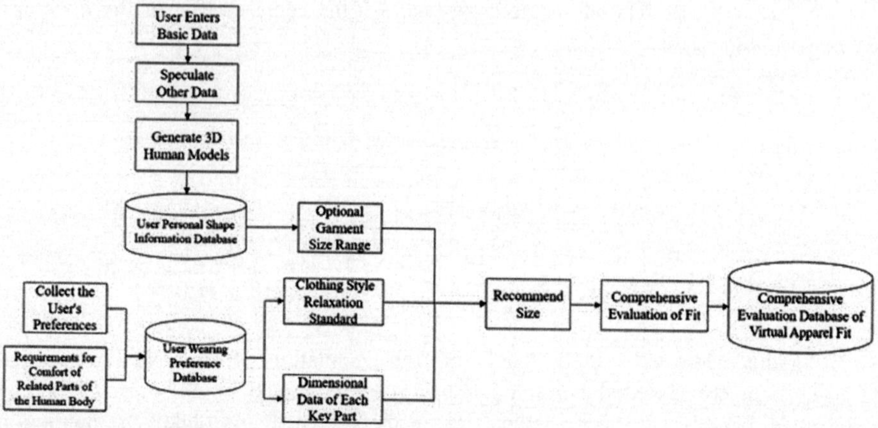

Fig. 7. User fit evaluation flow chart

3 Design of Intelligent Interactive Platform for Clothing Customization

The clothing personalized customization platform needs to integrate the user's personalized clothing customization and fit, help users to combine styles that meet user requirements under massive resources. Therefore, the platform needs a lot of reasoning, calculation and data processing [20]. This article starts from improving the interactivity and guidance of digital clothing customization, combined with the Agent interaction model to make the platform more intelligent to operate.

3.1 Work Process

When the user enters the clothing personality customization platform to register, the platform will provide the user with an information questionnaire, including the input of basic information and the user's preference survey. If user does not choose to fill it at the beginning, the system will collect and recommend the user's preferences based on the user's future search information on the platform. The workflow of the platform is shown in Fig. 8. After the user customizes the determined style, the platform will recommend the number according to the user's body type. The background will publish the virtual model based on the user's body after wearing the selected clothing, and evaluate the fit. User can further communicate and modify with the designer according to the effect of the display and related data, until they are satisfied, and confirm the order.

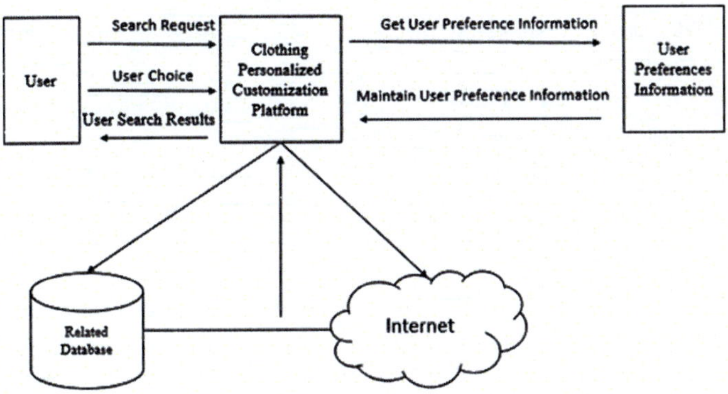

Fig. 8. System workflow

3.2 Implementation Model of Personalized Customization Platform Based on Agent Clothing

Combine the Agent to build the interactive platform, and modeling using Agent technology. Agent has the characteristics of autonomy, initiative and predictability, and can constantly perceive changes in itself and the external environment. Multi-agent system is composed of multiple independent agents, each agent in the system is independent, and each agent can interact with other agents. The clothing personalized customization platform uses a horizontal layered agent model, as shown in Fig. 9, Agents are divided into two categories, one is used to provide information, It is responsible for collecting and managing data, and the other is responsible for executing the commands of the clothing personalization platform. The specific division of labor of each agent is as follows [21, 22]:

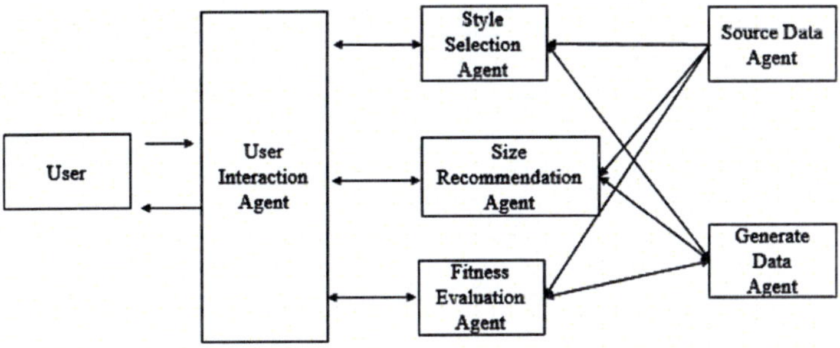

Fig. 9. System structure

User Interaction Agent: Responsible for handling the user's personalized choices and interacting with various functions in the platform, such as body shape acquisition, style preference, clothing comfort, etc.

Source Data Agent: Responsible for collecting and sorting out the various databases generated in the system;

Generate Data Agent: Responsible for processing and collecting evaluation results generated by various parts of the system;

Style Selection Agent: Responsible for establishing the mapping between personalized choices and styles based on the user's self-selected preferences or search results on the home page, and recommend suitable clothing styles for users through user interaction;

Size Recommendation Agent: Based on the body size of several key parts input by the user, calculate the customer's body characteristics, and according to the user's wearing preferences collecting, recommend the appropriate model for the user;

Fitness Evaluation Agent: According to the results of size recommendation, style selection and the judgment of the fit degree of clothing parts, evaluate the fit of the selected clothing.

4 Conclusion

With the continuous advancement of technology, in the garment manufacturing industry, the application of digital technology will penetrate into all aspects of the garment field. Clothing digitization will be a major development trend in the future apparel manufacturing industry. This platform combines the question and answer interaction mode with the clothing customization platform, and describes the clothing parameters through the acquisition of clothing customization requirements and the mining of invisible requirements. Based on this, an intelligent interactive platform for clothing customization for designers, users, and factories has been established, and using an iterative method has been optimized for the number of interactions. This article has certain application value to the user experience of the construction of digital clothing platform. In the future, combined with various technical links such as intelligent ordering, intelligent inventory, and intelligent marketing, a comprehensive online and offline, virtual and actual clothing sales platform can be created to upgrade traditional clothing customization.

References

1. Li, L., Xie, H., Wang, S.: The influence of RA on customer satisfaction in clothing network customization Sound. Silk (10), 32–36 (2014)
2. Shi, M., He, X., Zhu, X., et al.: Personalized suit customization and virtual display system Design and realization. Wool Spinning Technology (10), 36–42 (2015)
3. Lei, S., Na, L.: Research on influencing factors of customer satisfaction under O2O mode. Bus. Res. **5**, 148–153 (2016)
4. Fiore, A.M., Jim, H.J., Kim, J.: For fun and profit: Hedonic value from image interactivity and responses toward an online store. Psychol. Mark. **22**(8), 669–694 (2005)
5. Lee, H.H., Chang, E.Y.: Consumer attitudes toward online mass customization: An application of extended technology acceptance model. J. Comput. Mediated Commun. **16**(2), 171–200 (2011)

6. Gao, C., Hu, S.: Research on the virtual display of digital clothing and the network customization business system. Shanghai Text. Sci. Technol. **42**(4), 56–60 (2014)
7. Zhou, H., Xu, Y., Chen, Y.: Modular design method of clothing styles. J. Text. **36**(8), 104–109 (2015)
8. Qian, S.: Research on Intelligent Clothing Style Design System. Donghua University, Shanghai (2004)
9. Wu, Y., Hou, L., Lai, R.: A module dynamic programming method for product family design. Comput. Integr. Manuf. Syst. **19**(7), 1456–1462 (2013)
10. Wang, Y., Chen, S., Tan, X.: Modern clothing marketing based on digital technology. Sci. Technol. Econ. Guide **18**, 260–263 (2017)
11. Xu, L.: Basic characteristics analysis of clothing digital technology. J. Text. **26**(5), 140–142 (2005)
12. Yang, Y., Sun, Z., Kong, Y., Zhang, Y.: Research on the "tracking" experience marketing model of digital customized clothing. Wool Text. Technol. **47**(4), 66–70 (2019)
13. Xu, C.: Digital Clothing Customization System for Special Figure. Donghua University, Shanghai (2012)
14. Yan, J., Zhen, J., Xie, Z., Zhou, C.: Research on online personalized product customization model based on consumer innovation. China Sci. Technol. Forum **10**, 109–114 (2016)
15. Shi, W., Liu, Y., Gong, H.: Application of golden section method in unconstrained multiple optimization problems. Northeast Normal Univ. J. Nat. Sci. Ed. **35**(2), 11–14 (2003)
16. Technology - Digital Technology; Researchers at National Yunlin University of Science and Technology Publish New Data on Digital Technology (Dimensions of Customer Value for the Development of Digital Customization in the Clothing Industry). J. Eng. (2020)
17. Jost, P.J., Theresa, S.: Company-customer interaction in mass customization, p. 220 (2020)
18. Yan, W., Chiou, S.C.: Dimensions of Customer Value for the Development of Digital Customization in the Clothing Industry **12**(11) (2020)
19. Dou, R., Huang, R., Nan, G., et al.: Less diversity but higher satisfaction: an intelligent product configuration method for type-decreased mass customization **142** (2020)
20. Qi, Y., Mao, Z., Zhang, M., et al.: Manufacturing practices and servitization: The role of mass customization and product innovation capabilities **228** (2020)
21. Salman, K.A., Branko, K.: Variability and validity: flexibility of a dimensional customization system, p. 109 (2020)
22. Qiu, J.X., Xu, Y.Q., Zhang, M.: Conversion of information flow in digital manufacturing for clothing industry **1037**, 1535–1539 (2010)

Application of Automatic Text-Classification Algorithm Based on Feature Extraction for Intelligent System of Transportation Knowledge Service

Li Zhang[1(✉)], Han Zhang[1], Peihong Yang[2], and Zhaoqiang Cai[2]

[1] China Academy of Transportation Sciences, Beijing 100029, China
lelezhang@126.com
[2] Department of Transport of Qinghai Province, Highway Monitoring and Response Center of Qinghai Province, Xining 810003, Qinghai, China

Abstract. Through analyzing the data scale and characteristics of transportation science and technology information, an automatic text-classification algorithm based on text feature vector space model is presented. In the process, the keywords are determined by introducing the prior probability of each feature word and the TF-IDF value, and the CLC (Chinese Library Classification) classification number is determined by comparing the co-occurrence of keywords with the professional vocabulary data. In practice, the efficiency and accuracy of information classification indexing have improved perceptibly in application of intelligent system of transportation knowledge service.

Keywords: Text-classification algorithm · Feature vector extraction · Transportation knowledge service

1 Introduction

With the rapid development of transportation technology, the amount of scientific and technological information in the transportation field has exploded. To support the innovation of transportation technology, the transportation industry has integrated scientific and technological information resources since the beginning of this century, and established a series of public information service systems such as Transportation Science & Technology Information Resource Sharing Platform and Transportation Knowledge Service System to provide one-stop information search and consulting services. In-depth processing of scientific and technological information and subdivision of professional fields are not only the basis for providing accurate information discovery services, but also the premise for providing information analysis and consulting services. At present, the scientific and technological information classification and indexing in the transportation field mainly uses manual indexing, the indexing efficiency is low, and the classification standards are greatly affected by people. General automatic classification tools are often unable to meet the requirements of subdivision of disciplines or professional fields in the field of transportation. Therefore, according to the scale and characteristics of transportation science and technology information,

M. Tavana et al. (Eds.): IISA 2020, AISC 1304, pp. 710–716, 2021.
https://doi.org/10.1007/978-3-030-63784-2_88

combined with the classification precision and accuracy requirements of transportation knowledge service system, this paper studies and designs a classification method that combines vector space model and professional dictionary, and develops an automatic classification indexing tool to achieve automatic classification indexing or manual auxiliary indexing of transportation science and technology information, thereby improving the efficiency and accuracy of information classification indexing.

2 Related Work

Technology projects, achievements, documents and other information is essentially text automatic classification, that is, under a given classification system, it is automatically determined to the corresponding predefined category according to the text content [1]. Since the 1950s and 60s of the 20th century, the foreign library and information field has begun to explore automatic classification technology, and the research in this field in China began in the 1980s [2]. From the current research results in this field, there are two common classification methods. One is the rule-based method, which mainly includes dictionary-based classification methods and expert system-based classification methods [3]. This kind of method needs to build a topic classification vocabulary. Matching the topic words will make the classification results more accurate, but it needs to constantly adjust the classification inference rules according to the development of the discipline and the change of the vocabulary in the discipline field. The other type is machine learning methods based on statistics, that is, using classified documents to build a knowledge base, extracting feature data for learning, identifying the characteristics of various categories of documents, and building a classification model on this basis to classify the text to be classified, mainly includes nearest neighbor classification algorithm (KNN), support vector machine (SVM), center vector method, naive Bayes, decision tree method, neural network, etc. The core of this kind of method is to train the document classification model based on the feature vector space. Many studies have shown that the nearest neighbor classification algorithm (KNN) and support vector machine (SVM) are better algorithms in the automatic text classification algorithm [4]. The network method has a good advantage in the training of large-scale massive data [5]. In addition, in order to improve the performance of text classification, scholars at home and abroad also introduced semantic analysis technology, and proposed methods such as latent semantic analysis, ontology semantic mapping, and knowledge base semantic expansion [6], which are used in conjunction with machine learning methods to achieve better classification results.

3 Algorithm Model

3.1 Data Analysis

The scale of transportation science and technology projects, achievements, and documents waiting for classification information is about hundreds of thousands. The thesaurus has been compiled and published in the field of transportation, and has been revised and upgraded three times successively. In the field of transportation, a discipline classification standard has also been formed in the long-term scientific and technological work, as shown in Table 1.

Table 1. Discipline classification table of transportation

Code	Discipline name	Code	Discipline name
JT	Transportation	JT0299	Other disciplines of transport
JT01	Transportation infrastructure engineering	JT03	Automobile utilization engineering
JT0101	Road engineering	JT0301	Automobile utilization engineering
JT0102	Bridge engineering	JT0303	Ship engineering
JT0103	Tunnel engineering	JT0304	Marine engineering
JT0104	Traffic engineering	JT0399	Other disciplines of vehicle operation engineering
JT0105	Port and waterway engineering	JT04	Transportation planning and management
JT0199	Other disciplines of transportation infrastructure engineering	JT05	Transportation economy
JT02	Transport	JT06	Transportation safety
JT0201	Road transportation	JT07	Green transportation
JT0202	Waterway transportation	JT08	Intelligent transportation
JT0203	Comprehensive transportation and logistics	JT99	Other disciplines of transportation
JT0204	Urban public transport		

3.2 Building Vocabulary

Build three types of vocabulary: self-built vocabulary, professional vocabulary and mapping vocabulary. By extracting titles, abstracts, keywords, and CLC classification numbers from 136 journal papers in the field of road and waterway transportation, a self-built word list is established. The self-built vocabulary needs to count the parameters of keyword characters, frequency, main frequency, standard fitting degree, classification aggregation degree, time change trend and other parameters. Standardizes the parameters using Sigmoid function, Determine the parameters according to the data distribution so that the standard value of the median is around 0.5. Assigns weights to each standardized parameter and calculates the fitting degree of the original keywords. Uses gradient descent method to calculate its optimal weight distribution scheme and uses it as a priori probability for each word in subsequent calculations. The professional vocabulary is established based on the Chinese Thesaurus for Highway and Waterway Transportation [7], that is, a mapping table of theme-words and discipline categories is established. The mapping vocabulary is to establish a correspondence table between the discipline classification and the CLC classification number, as well as a mapping table of theme-words and CLC classification numbers generated according to the theme-words and discipline categories mapping table.

3.3 Identify Keywords

Treats each piece of science and technology information as a text document, segments it, and excludes the invalid words and clusters the words combined with the self-built vocabulary to represent the document as a vector space, that is

$$D = D(T_1, W_1; T_2, W_2; \ldots\ldots; T_n, W_n) \tag{1}$$

W_i is the weight of the i-th feature word T_i obtained after processing in document D, where T_i is the word in the self-built vocabulary. Calculates the TF-IDF value [8] of each feature word, and introduce the prior probability P(T_i) of each word in the self-built vocabulary, and bring it into the Bayesian formula. Since keyword extraction can be regarded as a constant value for the same piece of information, the prior probability of each word and the corresponding TF-IDF worth product can be directly calculated to obtain the weight W_i of each word. Sorts W_i and extracts the first k words as the keywords of the document. The value of k is before 3 to 8, that is $3 \leq \text{k} \leq 8$.

$$W_i = P(T_i) \times (\frac{tf(T_i, D)}{Sum(D)} \times \log(\frac{N}{n(T_i) + 1})) \tag{2}$$

Where P(T_i) is the prior probability of the feature word T_i in the self-built vocabulary, $tf(T_{i,} D)$ is the number of times the feature word T_i appears in the document D, $Sum(D)$ is the total number of feature words in the document D, N is the total number of documents, and $n(T_i)$ is the number of documents containing the feature word T_i.

3.4 Classification Algorithm

Clusters the self-built vocabulary and combines the clustering results of word segmentation to form the final clustered vocabulary. For document D, calculates the keyword co-occurrence and constructs a keyword co-occurrence matrix M. Where $M[K_i][K_j]$ represents the number of times that the keyword K_i and the keyword K_j appear together in the same paragraph in the document. The keyword co-occurrence matrix is compared with the standard data in the clustering vocabulary to extract m CLC classification numbers with the highest frequency. If m is less than the specified number of classification numbers, the words of the professional vocabulary are added as standard points to the cluster vocabulary to form a combined feature vector of each keyword. The similarity calculation formula is used to calculate the closest standard point and extracts the CLC classification number of the standard point data. Finally, the discipline classification number of the document is formed according to the mapping table of the discipline classification and the CLC classification number.

4 Development and Application

Based on the relational databases SQLSERVER 2008 and Mysql 5.7, Python 3 is used to develop an automatic classification tool for transportation science and technology information. The main functions are shown in Fig. 1.

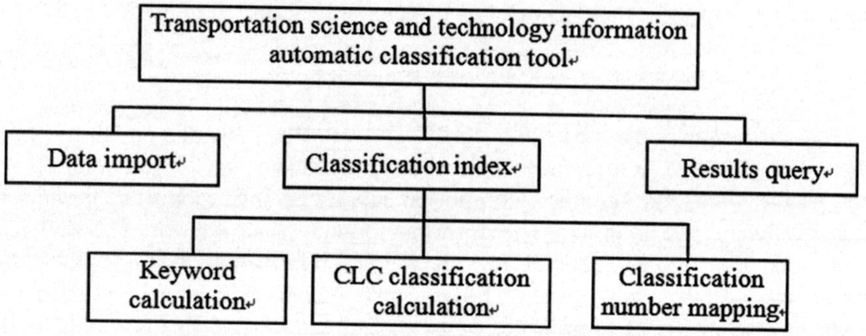

Fig. 1. Main functions

In the development of automatic classification tools, the IK Analyzer plug-in professional vocabulary is used to segment the document, and the theme-words in the professional vocabulary are used as high-priority available information in the segmentation process. Uses word2vec to generate the word vector space model of the document, clusters the self-built vocabulary and word segmentation results, and calculates the similarity between feature vectors.

Uses the transportation science and technology information automatic classification tool to classify five types of information, such as scientific and technological achievements, journal papers, conference papers, scientific and technological project research reports, and scientific and technological news in the field of highway and waterway transportation. The characteristics of various types of information are shown in Table 2. The results show that the average accuracy of information classification is more than 80%, which can basically meet the needs of information classification. Among them, the accuracy of short text classification is slightly lower, the main reason is that the keyword extraction is more difficult. The classification results of typical disciplines such as transportation infrastructure engineering, transportation, vehicle operation engineering, transportation planning and management, transportation economics, etc. are ideal. The interdisciplinary fields such as transportation safety, green transportation, and intelligent transportation involve many fields, and feature words and semantic relations are relatively complex. The accuracy is slightly lower due to the completeness of self-built vocabulary and professional vocabulary.

Table 2. Characteristics of transportation science and technology information

Science and technology information category	Data structure	Data scale (pieces)	Single message length
Scientific and technological achievements	Achievement name; keywords; brief technical description	55800	About 800 Chinese characters
Journal papers	Title; keywords; abstract; full text	208000	5000–10000 Chinese characters
Conference papers	Title; name of maternal literature; full text	105000	2000–8000 Chinese characters
Scientific and technological project research reports	Project name; keywords; abstract; full text	1030	10000–100000 Chinese characters
Scientific and technological news	Title; full text	10000	About 1000 Chinese characters

5 Conclusions

This article aims at the problems of low efficiency and large influence by people in manual classification indexing of transportation science and technology information, such as projects, achievements, documents, etc. Based on the Thesaurus and discipline classification standard already in the transportation field, with the goal of improving the precision and accuracy of discipline subdivision in the field of transportation, an automatic classification method combining vector space model and professional dictionary is designed, and transportation science and a transportation science and technology information automatic classification indexing tool is developed. This tool can basically meet the needs of automatic classification indexing and manual auxiliary indexing of transportation science and technology information. In addition, the accuracy and perfection of the self-built vocabulary is very important for the accuracy of the automatic indexing results, and the length of the information text has a certain influence on the classification results. In the future works, the intelligent system of transportation knowledge service is to improve keyword extraction, classification algorithms, and optimize the self-built vocabulary by combining journal papers and thesaurus. Thereby, the completeness and accuracy of the self-built vocabulary and short text information classification may get continuous improvement.

Acknowledgements. This work is supported by the Construction Project of China Knowledge Centre for Engineering Sciences and Technology (CKCEST-2018-4-2, CKCEST-2019-1-9, CKCEST-2020-2-11).

References

1. Wang, H., et al.: Research on automatic classification for chinese bibliography based on machine learning. J. Libr. Sci. China **36**(11), 28–39 (2010)
2. Cheng, Y., Shi, J.: Research on the automatic classification: present situation and prospects. J. China Soc. Sci. Tech. Inf. **18**(1), 20–26 (1999)
3. Ming, X., Ying, S.: Development of research on automatic classification. New Technol. Libr. Inf. Serv. **5**, 25–28 (2000)
4. Liu, J.: Design and Implementation of the Technical Text Categorization System, 5th edn. Henan University of Technology, Zhengzhou (2013)
5. Limin, G.: Study of automatic classification of literature based on convolution neural network. Libr. Inf. **6**, 96–103 (2017)
6. Ba, Z., et al.: Research on automatic classification of digital document based on semantic extension. J. Mod. Inf. **35**(6), 70–74 (2013)
7. Institute of Scientific and Technical Information of China. Chinese Thesaurus: Engineering Technology, vol. 12 Transportation), 9th edn. Scientific and Technical Documentation Press, Beijing (2014)
8. Salton, G., Buckley, C.: Term-weighting approaches in automatic text retrieval. Inf. Process. Manag. **24**(5), 513–523 (1988)

Research on Intelligent Decision Support System Framework for Deep-Sea Emergency Response

Kun Lang[1(✉)] and Mingming Zhang[2]

[1] School of Maritime Economics and Management, Dalian Maritime University, Dalian, Liaoning, China
kun.lang@dlmu.edu.cn
[2] Navigation College, Dalian Maritime University, Dalian, Liaoning, China

Abstract. In order to make effective emergency decisions timely, this paper presents an intelligent decision support system framework for deep-sea emergency response. There are two kinds of intelligent methods integrated in the presented system. One is the emergency situation analysis method based on Bayesian network, which is used to evaluate the emergency situation quantitatively by calculating the occurrence probability of emergency situation risk at each level. The other is the emergency assistant decision-making method based on case-based reasoning, which is applied to make assistant decisions. The presented system can provide scientific basis for decision making and perform well in practice.

Keywords: Deep-sea emergency response · Decision support system · Emergency situation · Case-Based Reasoning

1 Introduction

With the development of international trade and business cooperation, there are more and more demands for maritime transport. As the navigation environment becomes increasingly complex, the traffic safety risk is growing. The known accidents, such as Sewol, oil tanker Sanchi and MH370, have caused serious losses.

In addition, there are some more complex characteristics in the deep-sea environment than the shallow water, such as complex and changeable emergency situation, special environment, and complex implementation conditions. Once an accident occurs, it will result in enormous losses. Therefore, it is urgent to research on the issue of deep-sea emergency response.

It is well known that the response speed is a key factor in the process of emergency response. It is necessary to make accurate judgements and effective emergency decisions rapidly. In order to reduce the inevitable risk in the process of emergency response, advanced intelligent technologies can be used to improve the risk analysis capability and provide effective and reliable emergency strategies from the overall perspective. As a result, how to analyze the emergency situation and make the assistant decision intelligently becomes the research focus.

M. Tavana et al. (Eds.): IISA 2020, AISC 1304, pp. 717–724, 2021.
https://doi.org/10.1007/978-3-030-63784-2_89

In recent decades, there are some research achievements in the related field. Liao, Z.L. et al. developed an environmental emergency response and preparedness system based on case-based reasoning (CBR) technology, and verified the effectiveness through the example of oil spill accidents [1]. Wang, J.Y. et al. proposed an emergency decision support system based on CBR to solve the emergency decision problem in crisis management [2]. Zhang, Y.J. et al. presented an emergency assistant decision-making prototype system based on CBR [3]. Zhang, J.H. designed an emergency preparedness information system based on CBR and RBR [4]. Wang, Q.Q. et al. proposed a kind of knowledge reasoning method based on the theory of category and the theory of typed category to acquire new knowledge to assist decision-makers in making decisions [5]. Liu, J.D. et al. designed a maritime oil-spilling emergency aid decision support system based on CBR [6]. Huang, L. et al. introduced a platform structure for maritime oil spill response ability visualization to provide decision support in the oil spill emergency [7]. Ai, B. et al. presented an intelligent decision method to make the emergency response plan in the maritime search and rescue, which contains a resource scheduling model based on GSAA and a regional task allocation method based on space-time features [8]. Yu, W.H. et al. designed a kind of intelligent decision support system for the maritime search and rescue to help RCC make decisions timely [9]. Wu, B. et al. designed an emergency simulation system to improve the maritime emergency response ability in the maritime search and rescue [10]. Si, D.S. et al. developed an improved BN model based on kernel density estimation and weighted averaging strategy to analyze accident causation factors of container ship collisions, and achieved the accident causation chain [11]. Chen, C.H. et al. presented an information mining system based on genetic algorithm. Intelligent computational techniques were used to detect anomalies to improve the capacity of maritime situational awareness [12]. Snidaro, L. et al. fused different information based on the theory of Markov logic networks to realize the situation assessment, and proposed a mechanism to assess the level of complex events [13]. Wang, Z.L. et al. used tools of big data processing and real-time data processing to evaluate the security situation of cloud computing platform, and converted the situation composed of multi-source heterogeneous data into a security situation indicator with a protection level. Experiments results illustrated the effectiveness of this method [14]. Goerlandt, F. et al. applied the Bayesian Network modeling technology to the study of probability risk quantification, and presented a risk analysis framework of the maritime transport system, which was applied to a case study of maritime oil spill [15]. Roed, W. et al. discussed the applicability of HCL framework for the offshore industry, and proposed a new probability assignment method based on specific features of the framework. This method can greatly simplify the assignment process, meanwhile it can retain the required flexibility to reflect the studied phenomena [16].

Based on the above research achievements, in this paper, an intelligent decision support system framework for deep-sea emergency response will be proposed. By analyzing the process of deep-sea emergency response, we will develop an emergency situation analysis method and an assistant decision-making method, which will be integrated in the decision support system.

2 System Design

The emergency decision support system is designed to record the emergency information, analyze the emergency situation risk and assistant the decision-making.

According to the general system requirements, there should be an interface supporting different input and output modes and four functional modules with good performance of information management, information transmission, data analysis and simulation calculation in the emergency decision support system. The functional modules are user information management, emergency information management, emergency situation analysis and emergency assistant decision-making, respectively.

System users can be divided into three kinds of roles: decision-maker, operator and visitor according to different responsibilities. The decision-maker is set as the system administrator with the highest authority. When a marine emergency incident takes place and the system receives the emergency information, decision-makers can inquire the latest data of the emergency in real time, obtain emergency situation analysis results, and make emergency response decisions based on historical similarity cases and related operation instructions. Decision-makers can also adjust the decision scheme in real time according to the feedback of the implementation of the decision. System operators can record the emergency information and inquire relevant information and results without making decisions. Visitors can just inquire limited emergency information.

There are two key functions in the intelligent decision support system for deep-sea emergency response. They are emergency situation analysis and assistant decision-making.

After the relevant information is obtained, the emergency situation will be analyzed. A quantitative model of situation analysis is used to identify and evaluate the emergency situation based on marine environmental data, emergency target data and rescue force data of the emergency scene. There are three risk levels. As a result, the emergency risk probability at each level will be acquired to provide decision basis for decision-makers.

According to results of the emergency situation analysis, conditions and constraints of deep-sea emergency operation can be determined, including natural environment conditions (such as wind, wave and flow), navigable environment condition (such as the depth of the water, rocks, obstructions), and so on. According to limitations of the rescue force and the current emergency situation, assistant decisions can be generated to provide decision support for decision-makers.

The structure of the proposed intelligent decision support system for deep-sea emergency response is shown in Fig. 1.

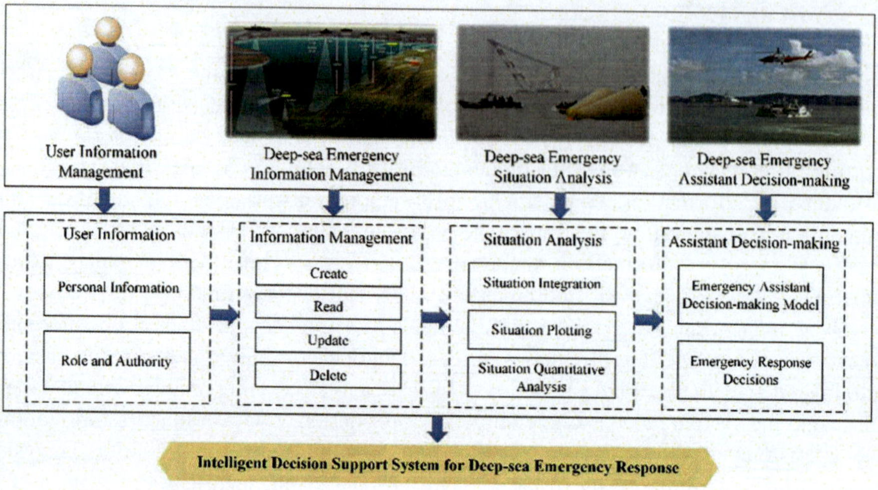

Fig. 1. Structure of the decision support system for deep-sea emergency response.

3 Core Technologies

3.1 Emergency Situation Analysis

At present, there are some common methods of emergency situation analysis and assessment, including neural network, Bayesian network and other machine learning methods. As the deep-sea emergency environment is complex and changeable, a variety of factors should be considered. In order to improve the accuracy of results, this paper designs a kind of emergency situation evaluation method suitable for the characteristics of the deep-sea emergency to identify and evaluate the emergency situation of the emergency scene.

In this paper, the emergency situation intelligent analysis is based on the real-time dynamic analysis of emergency data. Firstly, extract the main situation influencing factors. Then, build a quantitative description model of emergency situations. Lastly, apply the proposed intelligent method to identify and evaluate the emergency situation of the emergency scene.

Based on the analysis of previous successful emergency cases, this paper analyzes and extracts the main influencing factors of emergency environmental situation, emergency rescue force situation and emergency target situation, respectively. The relevant situation can be quantitatively described according to the extracted influencing factors. In this research, the complex and changeable emergency situation of the deep-sea can be plotted and displayed apparently based on the high-precision electronic chart display system. It is convenient for decision-makers and relevant organizations to identify the current emergency situation visually.

In this proposed method, the quantitative description model is based on the theory of Bayesian network and the fuzzy comprehensive evaluation method.

Firstly, construct a Bayesian network (BN) model. The extracted emergency influencing factors are set as the nodes of the Bayesian network. Since there is no historical data available for training and learning in this research, the structure of Bayesian network can only be determined according to the domain background knowledge of this field.

Secondly, allocate parameters of the BN model. This research applies an algorithm of node conditional probability table based on weighted distance to allocate parameters automatically. The weighted distance between the parent node and the child node is used to determine the conditional probability distribution of the intermediate node. The calculation formula is as follows:

$$P_j = \frac{e^{-RZ_j}}{\sum_{j=a}^{f} e^{-RZ_j}} \quad P_j \in [0, 1] \tag{1}$$

where R represents the distribution index, and P_j and Z_j represent the probability and the weighted distance, respectively, when the state value of the child node is equal to j.

Thirdly, reason and analyze quantitatively. The variable elimination method is used for accurate reasoning calculation of the model. Hence, quantitative results of marine environment state, emergency rescue force state and emergency target state can be obtained.

Finally, analyze the overall emergency situation. The global situation probability risk level can be obtained based on the above reasoning results. In this paper, the risk of deep-sea emergency situation is divided into three levels, namely, low risk, medium risk and high risk. It is obvious that the boundary between any two neighboring levels among these three levels is fuzzy. Therefore, the fuzzy comprehensive evaluation method is introduced to analyze the overall emergency situation.

As a result, the proposed system can provide quantitative situation analysis results, and the occurrence probability of emergency situation risk at each level can be given.

3.2 Emergency Assistant Decision-Making

Nowadays, there are more and more artificial intelligence technologies applicated to solve the problems in the field of management and decision successfully. In this research, an intelligent emergency assistance decision-making method for deep-sea emergency response is proposed based on the theory of CBR.

CBR is a significant intelligent knowledge-based method to solve problems. The process to solve problems based on CBR is usually divided into 4 steps, namely, case retrieve, case reuse, case revise and case retain, which is called 4R model. In some other researches, CBR is defined as 5R model, in which the step of case represent is added as the first step. The workflow is shown in Fig. 2.

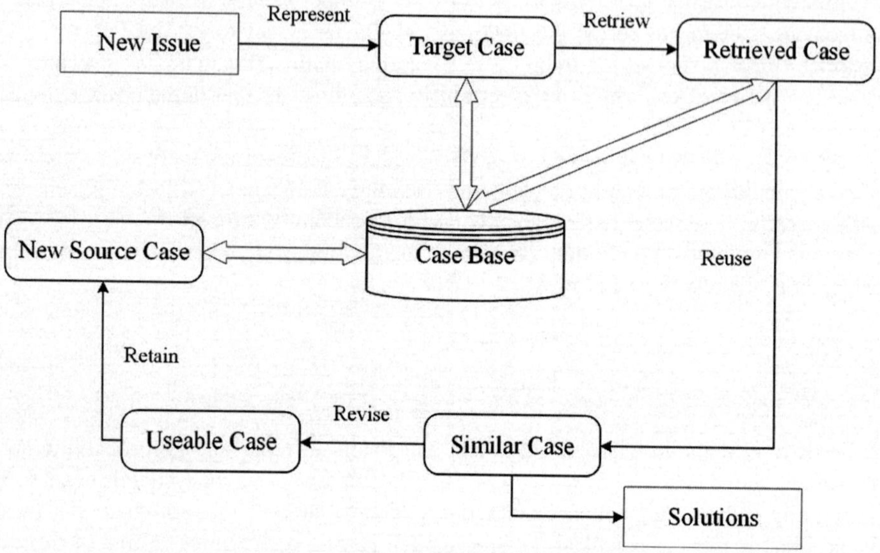

Fig. 2. Case-based reasoning model.

When a new issue comes up, it is represented as a target case. Then previous similar cases are retrieved in the case base. After that, solutions of similar cases are reused and revised according to new characteristics of the target case. Then the target case is judged whether it is useable. If it is a useable case, it will be retained as a new source case in the case base. The method of CBR imitates human's learning through past experience to realize the learning of new knowledge.

In this research, the emergency assistance decision-making method based on CBR is designed as a 5R model.

Firstly, represent the emergency case. Through analysis of characteristics of deep-sea emergency cases, the knowledge representation method based on the theory of framework and object oriented is chosen to represent the deep-sea emergency.

Secondly, retrieve similar cases. In this paper, the nearest neighbor algorithm based on the structural similarity is utilized to retrieve previous similar cases in the deep-sea emergency case base.

The traditional nearest neighbor algorithm can be expressed as Formula (2):

$$\mathrm{Sim}(X, Y) = \sum_{i=1}^{n} Sim(X_i, Y_i) * \omega_i \tag{2}$$

where $Sim(X_i, Y_i)$ stands for the similarity of the i-th characteristic attribute between the target case X and the existing case Y in the case base, and ω_i is the weight of the i-th characteristic attribute.

In this research, as the deep-sea emergency case has a sudden characteristic, it is hard to obtain all the information at once. Therefore, the structural similarity should be considered when calculating the global similarity between deep-sea emergency cases. As a result, the global similarity should be calculated by Formula (3) instead.

$$\mathrm{Sim}(X, Y) = S_{str} * \sum_{i=1}^{n} Sim(X_i, Y_i) * \omega_i \tag{3}$$

where S_{str} represents the structural similarity between the target case X and the existing case Y in the case base.

Thirdly, reuse similar cases. If similar cases have been retrieved through the second step above, and the similarity between the target case and historical cases is big enough, solutions of historical cases can be applied directly to the target case. However, this situation is less common. Therefore, the next step is usually taken.

Then, revise solutions of similar cases. According to characteristics of the target case, solutions should be revised accordingly. Generally speaking, the case revision relies on the expertise and experience of experts.

Finally, retain the target case as a new source case. After the case reuse and case revise, the target case can be retained as a new source case in the emergency case base if it is useable. Otherwise, it will be abandoned.

As a result, the proposed system can provide a corresponding reference solution for the deep-sea emergency and assist decision-makers to make the emergency response decision.

4 Conclusion

This research presents an intelligent decision support system framework for deep-sea emergency response. In order to make accurate judgements and effective emergency decisions rapidly, the intelligent emergency situation analysis method and emergency assistant decision-making method are integrated in the proposed system. The system is designed considering various characteristics of the deep-sea emergency, so that it can provide scientific basis for decision making in the process of deep-sea emergency response and perform well in practice.

Acknowledgments. The study was supported by "National Key R&D Program of China" (No. 2018YFC0309600\03) and "the Fundamental Research Funds for the Central Universities" (No. 3132019313).

References

1. Liao, Z.L., Mao, X.W., Liu, Y.H., et al.: CBR respond and preparedness system development for environmental emergency. Civ. Eng. Environ. Syst. **28**(4), 301–323 (2011)
2. Wang, J.Y., Wang, J.T.: A study of emergency decision support system based on case-based reasoning. J. Manag. Sci. **16**(6), 46–51 (2003)
3. Zhang, Y.J., Zhong, Q.Y., Ye, X., et al.: Research on method of emergency aid decision-making based on CBR. Appl. Res. Comput. **26**(4), 1412–1415 (2009)

4. Zhang, J.H., Liu, Z.Y.: Case-based reasoning and rule-based reasoning for emergency preparedness information system. J. Tongji Univ. **30**(7), 890–894 (2002)
5. Wang, Q.Q., Rong, L.L., Yu, K.: A knowledge reasoning method for emergency decision-making knowledge discovery. Oper. Res. Manag. Sci. **19**(1), 21–29 (2010)
6. Liu, J.D., Zhu, F.X., Zhang, Y.J., et al.: A marine oil-spilling emergency auxiliary decision-making system based on case-based reasoning. In: 9th International Conference on Natural Computation, pp. 1753–1757 (2013)
7. Huang, L., Peng, X., Zhang, F., et al.: A cloud-based platform for marine oil spill emergency response capability visualization. In: 5th International Conference on Transportation Information and Safety, pp. 810–815 (2019)
8. Ai, B., Li, B.S., Gao, S., et al.: An intelligent decision algorithm for the generation of maritime search and rescue emergency response plans. IEEE Access **7**, 155835–155850 (2019)
9. Yu, W.H., Xue, J.K.: Intelligent decision support system of maritime search and rescue based on JADE. Comput. Appl. Softw. **28**(8), 237–239 (2011)
10. Wu, B., Yan, X.P., Wang, Y., et al.: Maritime emergency simulation system (MESS) - a virtual decision support platform for emergency response of maritime accidents. In: 4th International Conference on Simulation and Modeling Methodologies, Technologies and Applications (2014)
11. Si, D.S., Zhang, Y.J., Lang, K.: Causation analysis of container ship collision accidents based on improved BN. China Saf. Sci. J. **29**(10), 27–31 (2019)
12. Chen, C.H., Khoo, L.P., Chong, Y.T., et al.: Knowledge discovery using genetic algorithm for maritime situational awareness. Expert Syst. Appl. **41**(6), 2742–2753 (2014)
13. Snidaro, L., Visentini, I., Bryan, K.: Fusing uncertain knowledge and evidence for maritime situational awareness via Markov logic networks. Inf. Fusion **21**, 159–172 (2015)
14. Wang, Z.L., Xue, R., Chen, C., et al.: A real-time security situation assessment system designed for cloud platform. In: International Conference on Artificial Intelligence and Computer Science, pp. 300–308 (2016)
15. Goerlandt, F., Montewka, J.: A framework for risk analysis of maritime transportation systems: a case study for oil spill from tankers in a ship-ship collision. Saf. Sci. **76**, 42–66 (2015)
16. Roed, W., Mosleh, A., Vinnem, J.E.: On the use of the hybrid causal logic method in offshore risk analysis. Reliab. Eng. Syst. Saf. **94**(2), 445–455 (2009)

An Intelligent TEV Sensor for Partial Discharge Detection of Cable Terminals

Chen Shen[1], Xiaochun Bai[1], Yan Jing[1], Jiangang Ma[1], Ming Ren[2], Tianxin Zhuang[1], and Changjie Xia[2](✉)

[1] Electric Power Research Institute of State Grid Shaanxi Electric Power Company, Beijing, China
[2] Xi'an Jiaotong University, Xi'an, China
xcjl9960521@stu.xjtu.edu.cn

Abstract. Power cable, as an important power equipment widely used in urban power transmission and distribution, its insulation state influences the reliability of urban power supply. Partial discharge (PD) detection is an effective approach to realize early fault detection and diagnosis. Transient earth voltage (TEV) is an electrification detection technology with the advantages of high sensitivity and strong anti-interference ability, which has emerged as a powerful PD detection method of cable terminals or joints. However, different with other power equipment, due to the wide distribution and underground installation of attachment monitoring points, it is difficult to apply the current TEV to achieve the real-time monitoring. In this paper, a distributed TEV intelligent sensing network for PD detection of cable terminals is proposed. On the basis of analyzing the time-frequency characteristics of the internal discharge signals at the cable terminations and joints, the intelligent TEV sensing unit based on wireless network is designed.

Keywords: Partial Discharge (PD) · Transient Earth Voltage (TEV) · Wireless sensing network · Fault diagnosis

1 Introduction

With the continuous upgrading of urban distribution network, the power cable has gradually become the inevitable trend of future development because of its advantages such as occupying less urban space and being less affected by severe weather damage. As the main artery of the power distribution, the insulation state of power cable is directly related to the reliable operation and power supply quality of urban transmission and distribution system [1, 2]. The cable terminal, as a special insulation structure for body insulation recovery layer by layer, is a weak link and a high fault incidence in the operation of the cable system. According to statistics, in the absence of external damage and other factors, more than 70% of power cable operation faults are caused by cable accessory faults [3]. Partial discharge (PD) is considered as an important cause of insulation aging of electrical equipment and a common characteristic parameter for insulation state assessment and fault diagnosis. Therefore, realizing the real-time PD

M. Tavana et al. (Eds.): IISA 2020, AISC 1304, pp. 725–732, 2021.
https://doi.org/10.1007/978-3-030-63784-2_90

detection and online monitoring for cable terminal is of great significance for reducing insulation fault and maintaining the safe operation [4].

TEV is defined as the transient voltage pulses to the earth generated by the leakage electromagnetic pulse at the discontinuity of the cable terminal metal shielding layer during PD. With advantages of high sensitivity, strong anti-electromagnetic interference ability, wide frequency band and easy installation, TEV is a preferred method for detecting the local discharge of cable accessories [5, 6]. With the advanced of the power grids, the conventional manual PD detection method can no longer meet the demand of low labor cost and real-time online monitoring. Therefore, although TEV technology has certain ability to prevent and evaluate the insulation failure of cable terminals, its application effect is not ideal. The main causes of these problems are the lack of reasonable application mode of TEV detection technique rather than the detection technology itself. Therefore, in order to better apply TEV detection technology in cable insulation fault diagnosis, it is necessary to further innovate the application mode of TEV detection method [7, 8]. In this paper, an intelligent TEV sensor for PD detection of cable terminals is designed.

2 Coupling Module of TEV Sensor

The charges excited by PD first accumulate and rapidly form the current that diffuses over the metal surface to the earth point. For the actual structure of the cable terminals, there are usually unsealed parts, e.g. the edge of the copper belt and the connection of the stress cone, where the high frequency PD signals will leak from and generate transient voltage pulses to the earth, and these transient voltage pulses are defined as TEV signals.

The coupling module of the TEV sensors can be shown as Fig. 1. In this figure, C_1 represents the equivalent capacitance between the metal shielding layer (such as grounding copper belt) and the sensor (as shown in Fig. 2), C_2 is the filter capacitance of the module, C_3 is the signal coupling capacitance, R is the sampling resistance, u_i and u_o respectively represent TEV signals transmitted to the surface of the metal shielding layer and output of the coupling module.

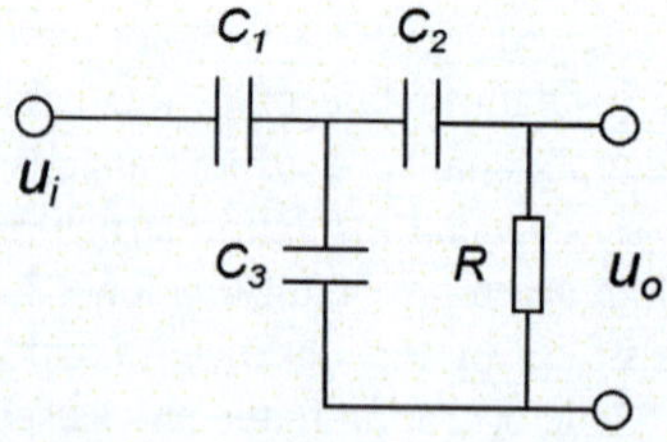

Fig. 1. Circuit schematic of the TEV sensing unit

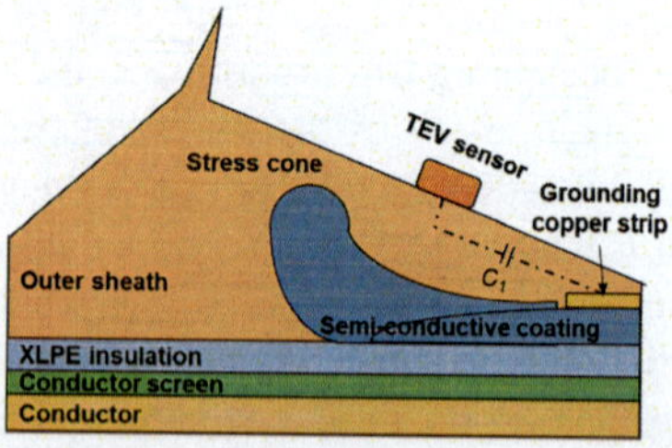

Fig. 2. Cable terminal structure and wireless TEV sensor placement diagram

Through circuit analysis, the transfer function of the module can be given by the formula (1):

$$T(s) = \frac{u_o(s)}{u_i(s)} = \frac{1}{1 + \frac{C_2}{C_1} + \frac{(C_1 + C_2 + C_3)}{C_1 C_3 R \cdot s}} \tag{1}$$

The sensitivity of the coupling module can be increased by reducing the denominator of the transfer function. When C_1 is far greater than C_2 and meets $C_3 = kC_1$, the above equation can be simplified as:

$$T(s) = \frac{u_o(s)}{u_i(s)} = \frac{1}{1 + \frac{1+\mathrm{k}}{\mathrm{k}C_1 R \cdot s}} \tag{2}$$

To meet high sensitivity requirements, it needs to make $\frac{(1+k)}{wkC_1 R} << 1$, where ω is the center frequency of the signal, $\omega = 2\pi f$, and sampling resistor R should meet formula (3):

$$R >> (1 + \frac{1}{\mathrm{k}}) \frac{1}{wC_1} \tag{3}$$

TEV coupling module adopted in the study is shown in Fig. 3. The capacitance C_1 between the bottom copper plate and the terminal grounding shield is calculated by formula (4):

$$C_1 = \frac{\varepsilon_0 \varepsilon_r \pi r^2}{d} = \frac{3.3 \cdot 8.85 \cdot 10^{-12} \cdot \pi \cdot 0.015^2}{3 \cdot 10^{-3}} \approx 6.9\,\mathrm{pF} \tag{4}$$

Where C_2 is the filter capacitor, 1 pF, C_3 is the coupling capacitor, 100 pF, therefore the value of k can be gotten by formula (5):

$$k = \frac{C_3}{C_1} = \frac{100\,\mathrm{pF}}{6.9\,\mathrm{pF}} \approx 14.5 \tag{5}$$

The center frequency of TEV signal is under 100 MHz ($f_{\max} = 10^8$ Hz), therefore R should meet formula (6):

$$R > > (1 + \frac{1}{\mathrm{k}}) \frac{1}{wC_1} = (1 + \frac{1}{14.5}) \cdot \frac{1}{2 \cdot 3.14 \cdot 10^8 \cdot 85 \cdot 10^{-12}} = 20\,\Omega \tag{6}$$

In order to test the performance of the model of the TEV coupling module, the discharge platform (as shown in Fig. 4) in the laboratory is used for measurement test.

Fig. 3. Schematic of the TEV sensing unit

In Fig. 4, the power supply for the stability of the power frequency high voltage source, through the water resistance measured terminal Ω (500 k) on the cable core. In the test, a resistance - capacity voltage divider (with a partial voltage ratio of 2000:1) was used to monitor the test voltage. The TEV coupling module is attached to the surface of the silicone rubber coat. The coupling capacitance is used to coupling the pulse current of the measurement loop, and A (sensitivity 5 V/A, bandwidth 250 MHz) Roche coil is used to synchronously detect the discharge pulse. In the test, TEV, pulse current and voltage signals were collected by LECORY 64MXS-B digital oscilloscope (bandwidth of 600 MHz, sampling rate of 10 GS/s).

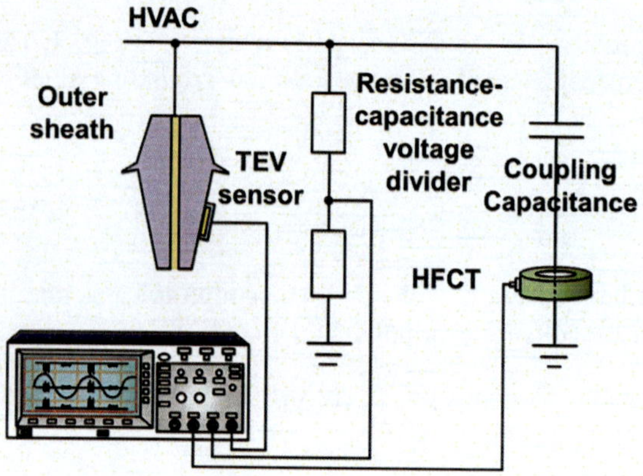

Fig. 4. Testing circuit

Figure 5 shows the TEV signal measured when the test voltage is 11.5 kV, and it can be observed that the TEV signal has the characteristic of oscillation attenuation. The spectrum distribution of discharge pulse can be obtained through spectrum analysis (as shown in Fig. 6). As can be seen from Fig. 6, TEV signal duration is about 0.6 s, and the main frequency distribution is between 1 MHz and 30 MHz. The frequency response of TEV coupling module was measured by a spectrum analyzer (GWIN-STEK, GSP-930, measuring range 9 KHz–3 GHz and frequency resolution 1 Hz). The frequency band of the coupling module is nearly 450 MHz, covering all frequency components of TEV signal and meeting the measurement.

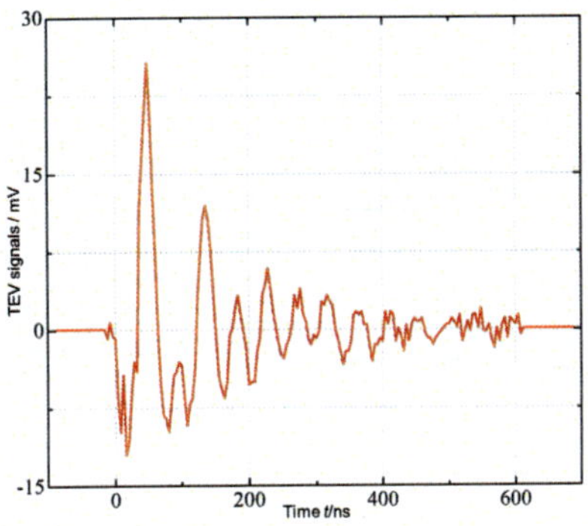

Fig. 5. PD pulse waveform measured by the TEV coupling module

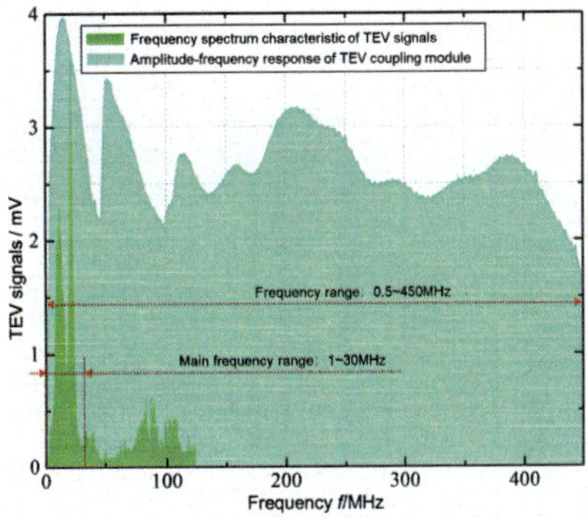

Fig. 6. Frequency response of the measured TEV signal and TEV coupling module

3 Signal Processing Circuit

As can be seen from the Sect. 2, the main frequency of TEV signals can reach tens of MHz, so the traditional TEV local emission detection system has a high requirement on the sampling rate. On the other hand, low cost, low energy consumption and compact structure, which first requires to reduce the sampling rate of the system, are also needed to be met for the distributed wireless TEV sensing network. It has been pointed out that TEV signal envelope containing partial discharge fault characteristics and phase information is sufficient for condition evaluation of power equipment.

The TEV signal conditioning process is shown in Fig. 7, which is based on the band-pass filter, low-noise amplifier and detector. The original high-frequency signals induced by TEV coupling module are filtered and amplified in turn in the circuit, and then converted into low-frequency double exponential waves by special detection circuit. The detection circuit is mainly composed of diode (D), capacitor (C) and resistor (R). Detection principle is when the signal amplitude is greater than the diode is to the voltage, the diode conduction (equivalent resistance r') and capacitance C charging circuit, the time constant $\tau_1 = r'C$ is far less than the discharge circuit of $\tau_2 = rC$ time constant, so the high frequency oscillation of voltage can be quickly stored in the capacitance C and by r slow release, to reduce the effect of signal frequency.

The adjusted TEV signal can be converted into digital quantity through A/D module and collected by the data acquisition system. The TEV signal and original signal after conditioning are shown in Fig. 8. In the figure, the duration of the original pulse signal is about 0.6 s, and the modulated low-frequency signal lasts about 50 s, which can be amplified by a low-noise amplifier and collected by a collection unit with a lower sampling rate. In this paper, the sampling rate of TEV perception unit is 1 MS/s, and the resolution of A/D module is 12 bits.

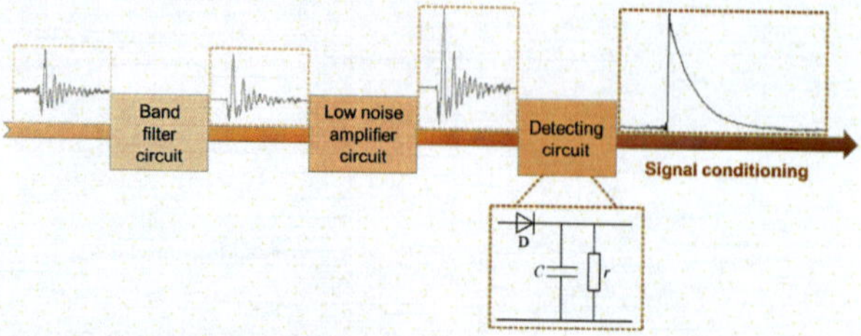

Fig. 7. TEV signal preprocessing procedure

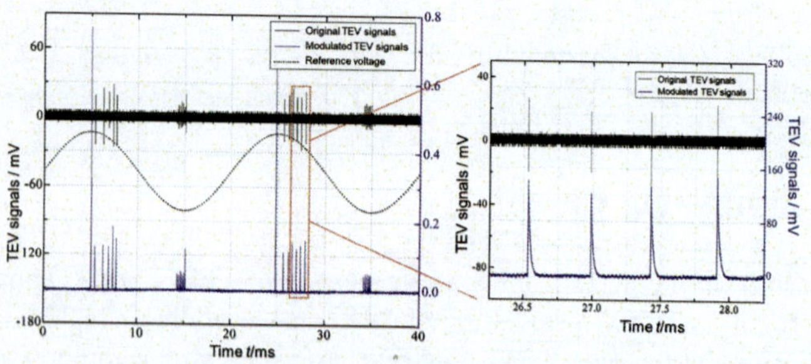

Fig. 8. Signal modulation of TEV signals.

4 Wireless TEV Sensor Performance Calibration

This paper designed a TEV local emission calibration system based on the principle of TEV signal generation and local emission detection. The quasi-system consists of TEV signal source, matching resistance, wireless TEV sensor, signal transmission cable and terminal computer (see Fig. 9).

A pulse generator is used as TEV signal source to output periodic Gaussian pulses with a rising time of 11.7 ns (or 30 MHz) to simulate TEV signals generated by electromagnetic waves on the metal surface. Metal plate by 50 Ω matching resistance grounding. The TEV sensor measures the signals on the metal surface and sends the results to the terminal computer via Wi-Fi. The amplitude range of pulse output by the pulse generator is 20 mV to 80 mV. The input-output (sensitivity) characteristics of the sensors are shown in Fig. 10. It can be seen that there is a high linear fitting between the amplitude of the original signal and the amplitude of the output signal of the sensor, and the fitting expression is $A_{\text{TEV}} = 24.04 + 5.08 \cdot A_{\text{ori}}$, A_{TEV} and A_{ori} Respectively output signal and original signal amplitude of TEV sensor (unit: mV).

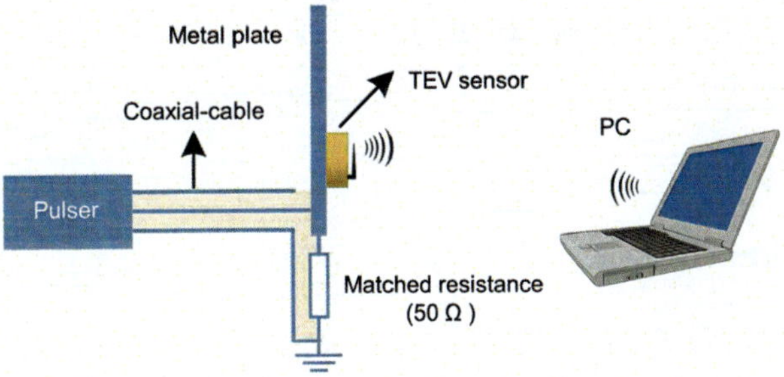

Fig. 9. The application method of TEV system

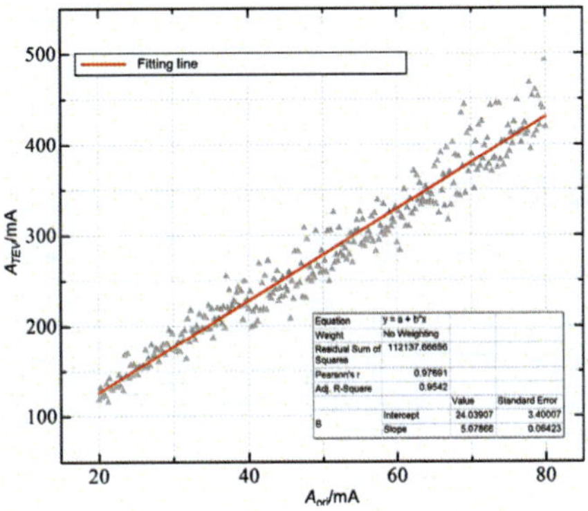

Fig. 10. Amplitude calibration curve for wireless TEV sensor

The traditional pulse current method measures the discharge intensity by measuring the perceived discharge (unit: pC), while TEV detection usually uses dBmV to represent the discharge intensity. The relation between TEV signal (mV) and output result (dBmV) is: $\mathrm{dBmV} = 20\log_{10}(\mathrm{mV})$. According to the input-input relation of the wireless sensor, it can be obtained as follows:

$$D = 20\log_{10}\left(\frac{A_{\mathrm{TEV}}\text{-}24.04}{5.08}\right) \tag{7}$$

Where D is the dBmV value output by the wireless sensor.

According to the above calibration results, the technical parameters of the TEV sensors are shown in the Table 1, where the relative error is estimated by the relative error between the measured value of the wireless sensor (ATEV) and the standard value (Aori). The sensor parameters designed in the study meet the requirements of field detection applications.

Table 1. Characteristic parameters of the TEV sensors

Parameter	Minimum detected electric quantity	Bandwidth	Relative error
Value	5 pC	1–30 MHz	$\leq 5\%$

5 Conclusion

In this paper, a partial discharge on-line monitoring system of cable terminal based on distributed TEV wireless sensor network and its application strategy are proposed, which can provide an effective application method of TEV sensors. The design of wireless sensor TEV coupling partial discharge signals can be effectively, has the good response, high SNR, low power consumption, the characteristics of high linearity and strong anti-jamming capability.

References

1. Sheng, B., et al.: Partial discharge pulse propagation in power cable and partial discharge monitoring system. IEEE Trans. Dielectr. Electr. Insul. **21**(3), 948–956 (2014). https://doi.org/10.1109/TDEI.2014.6832236
2. Takahashi, T., Okamoto, T.: Effective partial discharge measurement method for XLPE cables based on propagation characteristics of high frequency signal. In: 2012 IEEE International Conference on Condition Monitoring and Diagnosis, Bali, pp. 221–224 (2012). https://doi.org/10.1109/cmd.2012.6416415
3. Luo, J., Qiu, Y., Yang, L.: Statistical analysis of operational failure of power cables of 10 kV and above. High Volt. Eng. **29**(6), 14–16 (2003)
4. Sarathi, R., Umamaheswari, R.: Understanding the partial discharge activity generated due to particle movement in a composite insulation under AC voltages. Int. J. Electr. Power Energy Syst. **48**, 1–9 (2013)
5. Luo, G., Zhang, D.: Recognition of partial discharge using wavelet entropy and neural network for TEV measurement. In: 2012 IEEE International Conference on Power System Technology (POWERCON), Auckland, pp. 1–6 (2012). https://doi.org/10.1109/powercon.2012.6401331
6. Cai, Y., Guan, Y., Liu, W., He, J.: Study of transient enclosure voltage coupling to secondary cables in a gas-insulated substation. IEEE Trans. Power Delivery **33**(2), 761–768 (2018). https://doi.org/10.1109/TPWRD.2017.2688406
7. Zhuang, T., Ren, M., Gao, X., Dong, M., Huang, W., Zhang, C.: Insulation condition monitoring in distribution power grid via IoT-based sensing network. IEEE Trans. Power Delivery **34**(4), 1706–1714 (2019). https://doi.org/10.1109/TPWRD.2019.2918289
8. Li, X., Zhou, K., Wan, L., et al.: Development of cable terminal partial discharge status monitoring device based on TEV method. Power Syst. Prot. Control **12**, 98–103 (2013)

The Remote Voice Detector Design by a Laser Monitor

Tongliang Fan(✉) and Yandong Sun

Department of Maritime Command, China Coast Guard Academy, Ningbo 315801, China
libufan432@163.com

Abstract. Laser detecting sound technology has been widely concerned and studied with non-contact and traceless features. Based on the principle of laser eavesdropping, a phototransistor laser detector is designed. First, the phototransistor converts the laser signal into an electrical signal. Then, the electric signal passes through the filter circuit, the signal amplifies the circuit, the Audio Frequency amplifies the circuit, the output restores the sound. The function of the designed laser detector is tested.

Keywords: Eavesdropping · Voice detection · Laser monitor

1 Introduction

With the rapid development of modern science and technology, laser has been widely used in people's daily life and military field, and the laser eavesdropping technology has been gradually developed and perfected. This eavesdropping method has a long range and is not easily interfered with. The most important feature is that no equipment is needed around the eavesdropping target [1, 2]. The basic principle of laser eavesdropping is to shoot a laser beam at an object which is susceptible to vibration caused by sound pressure, and then to receive the vibration signal in the direction of the reflection of the light beam, and to solve the signal to achieve sound reduction [3]. This type of eavesdropping has a long range and is not easily interfered with. One of the most important features is that no equipment is needed around the eavesdropping target.

As a new monitoring technology, laser monitoring field has been paid more and more attention by scholars at home and abroad. It is based on the infrared light incident on the object produced by the sound pressure vibration, and the real-time audio monitoring is realized by extracting the audio information from the Echo. Compared with other monitoring methods, laser monitoring has the advantages of simple operation, strong secrecy and simple structure. This paper is based on the basic principle of laser voice monitoring. The visible red laser transmitter with power of 100 mW and wavelength of is selected as the transmitting device. The reflective laser beam is aligned to the phototransistor using the type phototransistor as the light sensor. The reflected light will deviate from its original position due to the vibration of the glass, resulting in the change of the light spot area. Using the photoelectric effect of the

M. Tavana et al. (Eds.): IISA 2020, AISC 1304, pp. 733–741, 2021.
https://doi.org/10.1007/978-3-030-63784-2_91

phototransistor, the slight change of the laser beam is converted into different voltage signals, and then the converted electric signals are passed through the conversion circuit, the audio signal amplifying circuit and the filter circuit, finally, the sound signal is restored.

2 The Principle of Laser Detector

The basic principle of laser monitoring is to use a well-hidden infrared light as a transmission medium to illuminate the target object to generate vibration and modulate and demodulate the signal so as to output audio signal. The device has the advantages of long transmission distance, good stealth and simple structure [5].

The analysis of the beam reflection principle of laser detective is shown in Fig. 1. The reflective surface mainly refers to the surface of the object that is easy to launch, such as the window glass of the target room, the smooth desktop and the file cabinet, etc. And the reflection surface can vibrate according to the vibration of sound waves. When setting the transmitting and receiving position, be sure to follow the light reflection rule. Generally speaking, the transmitter of laser detector includes the laser transmitter and optical aiming device. The receiver mainly includes the photoelectric conversion system, signal processing and sound output device.

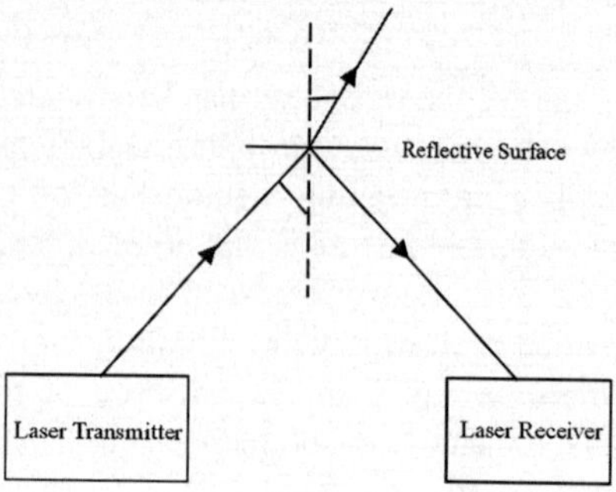

Fig. 1. Principle of reflection

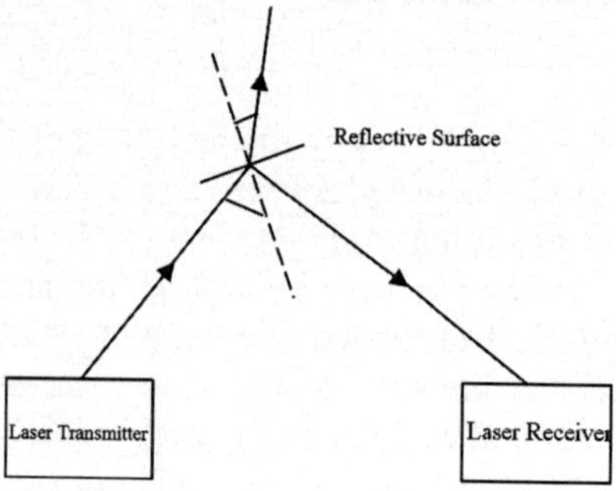

Fig. 2. Principle of reflection after deformation

The laser detector launcher fires a laser beam at the reflective object in the target room. When the target is far away, the system can use telescopic targeting system to aim at the target, and the reflected laser can be focused on the phototransistor. Sound pressure is created when sound travels through an air medium to a solid surface. Different particles will produce sound pressure of different strength. The vibration of the solid material surface is directly related to the local sound pressure, so the reflective surface will produce weak deformation, as shown in Fig. 2. The incident direction of laser emission remains unchanged, and the incident Angle will change.

$W = sq$ represents the total energy flow, where q is the cross-sectional area of the laser, and s represents the energy density. When incident light passes through a transparent glass, it will reflect and refract, and the energy of the light will be relatively split in two. We assume that the cross-sectional area of the reflected beam is the same as that of the incident beam. Both are q_1; the energy density of the incident light is s_1, and its energy flow is $W_1 = s_1 q_1$. The energy density of reflected light is s_2, and its energy flow is $W_1' = s_2 q_1$; The cross-sectional area of the transmitted beam is $q_2 = q_1 \cos r / \cos i$, and its energy flow is

$$W_2 = s_2 q_2 = s_2 (q_1 \cos r / \cos i) = W_1' (\cos r / \cos i) \tag{1}$$

According to the law of conservation of energy,

$$W_1 = W_1' + W_2 \tag{2}$$

That is,

$$W_1 = W_1' + W_1' (\cos r / \cos i) = W_1' (1 + \cos r / \cos i) \tag{3}$$

And we derive that:

$$W_1 = W_1' (1 + \cos r / \cos i) \tag{4}$$

The glass refractive index is n, and the solution can be:

$$W_1' = \frac{W_1}{1 + \sqrt{1/\cos^2 i - \tan^2 i / n^2}} \tag{5}$$

Then

$$\frac{W_1'}{W_1} = \frac{1}{1 + \sqrt{1/\cos^2 i - \tan^2 i / n^2}} = \frac{1}{1 + \frac{1}{\cos^2 i}(1 - \frac{1}{n^2}) + \frac{1}{n^2}} \tag{6}$$

3 Design Scheme

The hardware design of laser detector mainly includes two parts: transmitter and receiver. The scheme design of the two devices will be discussed and analyzed in the following.

3.1 Scheme Design of Transmitter

The transmitter of the laser detector is composed of a laser emitter and an optical aiming system. The schematic diagram of the transmitter is shown in Fig. 3. The laser transmitter emits infrared laser for the following reasons: (1) Infrared laser is not visible, and it is not easy to expose the position of laser detector transmitter. (2) The propagation velocity of infrared laser in the air is hardly affected by environmental factors, and the scattering phenomenon does not need to be considered. (3) Infrared laser transmitter is simple, practical, portable and maneuverable. (4) The wavelength of infrared laser cannot be too long and will be affected by temperature. (5) The infrared laser does not need to be equipped with other cooling devices, so it is relatively sensitive when working.

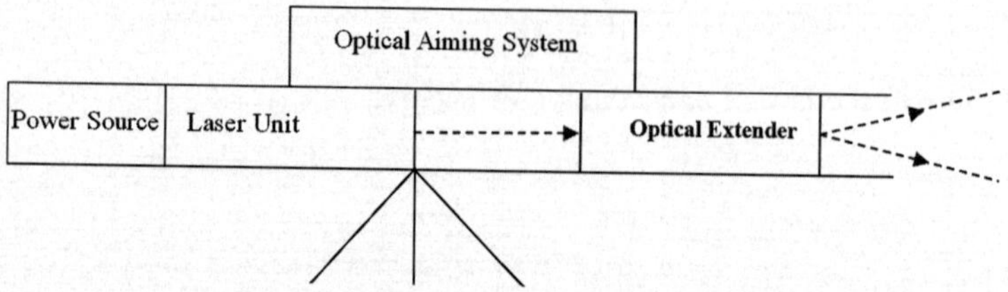

Fig. 3. Schematic diagram of laser eavesdropping transmitter

When the target of eavesdropping is far away, the optical aiming system is used to aim at the target object in the distance, so that the infrared laser can be directly shone to the center of the target object as far as possible, so as to achieve the best reception effect.

3.2 Scheme Design of Receiver

The main components of laser detector receiver are photoelectric detector and LM386 audio power amplifier. In the design of detecting light, we add a monochrome filter to filter the interference caused by the surrounding environment, so as to make the laser beam that reflected on the phototransistor more pure and improve the signal-to-noise ratio of the circuit of the whole system. Due to the monochromatic nature of infrared laser, only infrared light can pass through monochromatic filter, no matter it is incident light or reflected light.

The optical principle of the receiver is shown in Fig. 4, and the block diagram of the receiver's electrical signal processing is shown in Fig. 5.

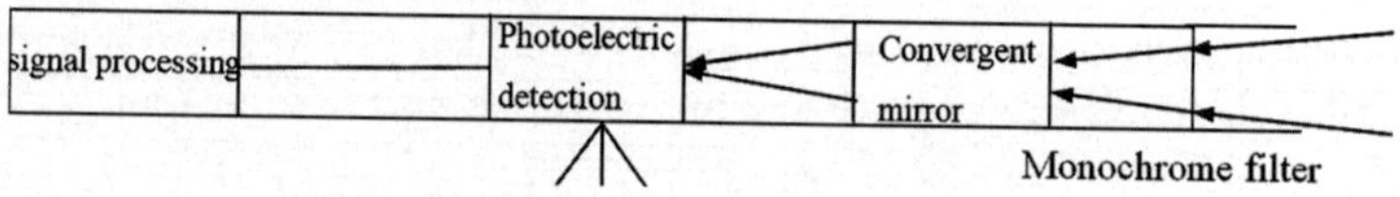

Fig. 4. Optical schematic diagram of receiver

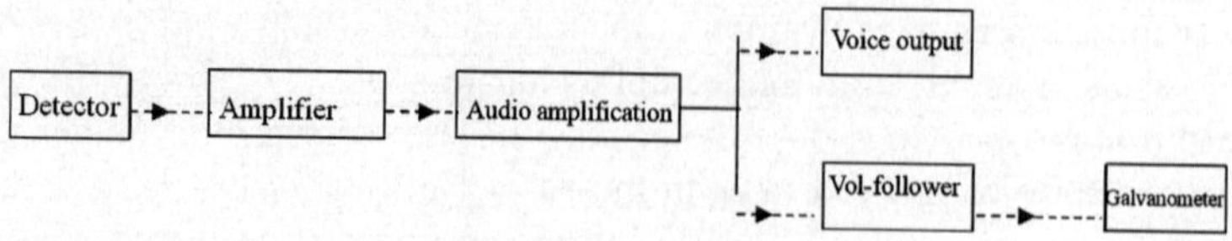

Fig. 5. Block diagram of receiver electrical signal processing

4 Hardware Design

According to the receiving principle of laser detector, we mainly use NPN phototransistor and operational amplifier LM386. The circuit diagram is shown in Fig. 6. External 9 V DC power supply is connected to supply power for the whole receiver. Switch are used to control power supply. The resistance R2 mainly plays the role of grounding, so that the current flowing through the phototransistor is small enough to play the role of protecting elements. Q1 phototransistor, has the obvious photoelectric effect, which will reflect the laser light signal into the electrical signal. Variable potentiometer can be used to control the volume of the eavesdropping sound by adjusting the size of the resistance. Usually we set the resistance is very low, so that the operational amplifier LM386 to maximize the multiple, so as to have a significant eavesdropping effect. Audio power amplifier LM386 is to amplify electrical signals. Between 1 pin and 8 pin in LM386, 10uF capacitance can make the amplifier gain up to a maximum of 200 times. The resistor R1 and the capacitance C1, with a capacity of 0.01 uf, form a filter circuit. The capacitance C1 blocks the DC current, and the total impedance changes with the frequency of the AC signal.

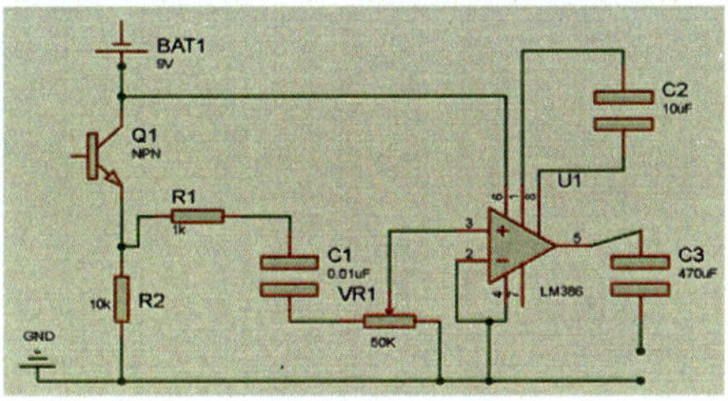

Fig. 6. Hardware circuit design of receiver

We can calculate a transition frequency of $f_0 = 1/2\pi(R1 + VR1')C1$, about 312.5 Hz to 15 kHz. The whole amplifier circuit consists of three parts: input stage, intermediate stage and output stage. The input stage is composed of a capacitance of 0.01 uF, a 1 kΩ resistance and a variable potentiometer 50 kΩ, which can block the DC signal of the power supply, stabilize the amplified AC signal, and adjust the volume. The intermediate stage consists of LM386 and a capacitance of 10 uF which connected to a pin 1 and a pin 8. *C*2 determines the inherent gain of the amplifier circuit. The output stage consists of a complementary symmetrical power amplifier circuit which has a capacitor 470 uF and an audio output unit. The capacitor *C*4 is connected to the output end of the audio power amplifier. As an audio output coupling capacitor, the output of the audio is coupled to the AC signal, which is mainly used for filtering and audio output. An external speaker is connected to the audio jack for output. Breadboard welding is used to design and manufacture the hardware circuit of laser detector. The list of components used to make detector is shown in Table 1.

Table 1. Components list

Laser detector parts list
IC_1: LM386 1-W audio amplifier integrated circuit
Resistors: $R_1 = 1\,\text{k}\Omega(K), R_2 = 10\,\text{k}\Omega, VR_1 = 50\,\text{k}\Omega$
Capacitor: $C_1 = 0.01\,\mu\text{F}, C_2 = 10\,\mu\text{F}, C_3 = 470\,\mu\text{F}$
Inductor: $Q_1 = NPN$ phototransistor or cadmium sulfide component
Power supply: 3–9 V battery or battery pack

The hardware circuit design material is shown in Fig. 7, and the specific physical components are shown in Table 2.

Fig. 7. Hardware circuit

Table 2. Physical components

1 9 V battery	2 audio port	3 C3 (470 uF)	4 Switch
5 Phototransistor	6 variable resistance	7 *LM*386	8 C2(10 uF)
9 resistance(1 kΩ)	10 C1(0.01uF)	11 resistance(10 kΩ)	

5 Testing and Analysis

The main point of using laser detector is to debug the equipment. The purpose of debugging is to find the errors in the circuit and solve them in time. Then, "Can the detective equipment be made to realize the eavesdropping function and the requirements for the detective environment? The actual effective distance of the detective?" This has become an urgent problem for us to solve. Therefore, the debugging of the equipment is a delicate task. So we can tune the system, based on the sound of the light beam projecting the element. When no light beam hits the component, we can hear a sound that sounds like a crack. When the light beam hits the whole photosensitive component, you can hear a loud sound. This method is suitable for the debugging of invisible laser and can be used to determine whether there is reflected light on the receiving device. After debugging, we began to test the laser detector. The test was conducted indoors under a bright environment. We first put a mobile phone that could play music in the room, simulating our eavesdropping target. Because the distance of system test is relatively close, we do not need to use the auxiliary telescopic equipment. First of all, adjust the position of the laser transmitter. Any slight angle change will affect the whole debugging result. Therefore, all parts of the laser transmitter must be stable and immobile. The semiconductor laser emits a 670 nm red laser directly into the center of the phone's interface and aims the reflected light at the phototransistor. After adjusting the position, turn on the switch of the receiver and the speaker. If the phone does not play music, there will be no vibration. Then we can eavesdrop on the reflected light and hear a faint hiss. When turning on the phone to play music, we can eavesdrop normally through the speaker. In order to achieve the best test effect, we changed the variable and repeated the experiment many times.

Because in the actual eavesdropping process, the environment is complex and changeable, the eavesdropping effect will be interfered by a variety of environmental factors, and the laser beam emitted by the laser will be interfered by many kinds of obstacles. Therefore, it is necessary to test the effect of our equipment. First, we tested the eavesdropping effect of laser detector in the case of obstacle interference. We did it indoors. With the same conditions, we tested it separately by changing the types of obstacles set in front of the laser receiver. In the first group, we didn't put any obstacles as a comparison test; in the second group, we tested the photoelectric transistor in front of the receiving device by shielding it with a black cloth. In the third group, we tested the device by adding a clear piece of colorless glass in front of the phototransistor. In the fourth group, we added a piece of colored glass in front of the photoelectric transistor of the receiving device for testing. Respectively compare the distances under the interference of different obstacles. The test results are shown in Table 3.

Table 3. Comparison of obstacle tests

Obstacle group	No obstacle	Black cloth	Clear glass	Coloured glass
First	20.5 m	2.4 m	15.6 m	18.5 m
Second	19.6 m	1.3 m	13.2 m	16.5 m
Third	18.0 m	2.0 m	14.8 m	15.8 m

Secondly, we conducted environmental tests. We set the following environmental conditions for several environmental variables: (1) under the condition of the same other conditions, indoor and outdoor tests were conducted; (2) other conditions being the same, we conducted the test in the bright environment and the dark environment indoors; (3) other conditions being the same, we conducted the test outdoors with wind and without wind. The maximum distance they could detect was compared, and the test results are shown in Table 4.

Tests have shown that it is increasingly difficult to detect distant targets. As the distance between the transmitter and the target increases, it is increasingly difficult to successfully capture the reflected beam from the target window.

Table 4. Comparison of environmental factor tests

Environment	Group 1		Group 2		Group 3	
	Inside	Outside	Bright	Dark	Windy	No wind
First	20.5 m	18.7 m	20.5 m	24.6 m	10.4 m	18.7 m
Second	19.6 m	17.5 m	19.6 m	30 m	8.8 m	17.5 m
Third	18.0 m	15.6 m	18.0 m	25.8 m	11.0 m	15.6 m

Because a little bit of ambient vibration can affect the direction of the reflected light path. We had asked for the light beam to be aimed at the phototransistor. But in the actual test, I found that it was better to shine the edge of the light spot onto the surface of the phototransistor. In this way, when the detection distance is relatively far, the reflected light spot is more divergent, and it is no longer a highly concentrated bright spot. However, we only need to capture the edge of the reflected light spot, which greatly improves the eavesdropping effect. In the test results of changing environmental variables, we can conclude that laser detector is also vulnerable to interference from external factors, especially in the following cases:

(1) When there is wind outside, the glass will vibrate because of the wind blowing. So the change of reflected laser light is not only caused by the vibration of the target sound, but also by the vibration of the wind blowing, so the sound finally restored will be accompanied by noise.
(2) The other light sources for indoor interception effect can also impact. In the process of testing, the effect can be found in dark environment effect is better than in bright environment. If there are the stove or bright lights in the target room, it will produce a huge waves with extremely low frequency. This results in a loud hum of an alternating current in the speaker output.
(3) When there are other vibrations occurring in the room where the target is located, such as the sound of people moving, or the noise caused by cars outside, the changes in the position and energy of the beam will lead to significant voltage changes.

6 Conclusion

Compared with the traditional detective technology, the advantages of laser detective technology are that it does not need to approach the target, is not easy to be detected, and is easy to operate. Based on the laser eavesdropping technology detector, in order to achieve good eavesdropping effect, and better applied to the actual work, we still have a lot of work to be done in the next step. First of all, to ensure that the monitoring of the covert, visible light certainly can't be used, you can use infrared laser, and replace the corresponding photoelectric detector. Secondly, the requirements of the light path are stricter, the reflected light must be received quickly and accurately. What more, noise and environmental factors must be eliminated, in order to achieve the real-time monitoring purposes.

Acknowledgements. This work was supported by the online open course of China Coast Guard Academy No KF2020002. It was also supported by teaching reform subject of China Coast Guard Academy No YBKT202042.

References

1. Mims, F.M.: Beware of laser eavesdropping. Am. J. Phys. **55**(10), 871–872 (1987)
2. Solomon, J., Prigo, R.: Eavesdropping with a laser. Am. J. Phys. **55**(4), 381 (1987)
3. Fan, Z.: The improvement and realization of laser eavesdropping. Laser Infrared **38**(2), 145–148 (2008)
4. Ning, Q., Ma, X.: Experimental research on laser eavesdropping. Phys. Exp. **29**(12), 38–41 (2009)
5. Liu, Y., Mu, Y., Li, Y., Li, X.: Laser eavesdropping system based on four quadrant detector. Piezoelectric Acousto-optic **36**(4), 675–680 (2014)
6. Lin, Y., Zhang, G., Li, Z.: Design and optimization of a Cat's eye retro reflector. Acta Opt. Sin. **22**(10), 1245–1250 (2002)

GDA-Based Tutor Module of an Intelligent Tutoring System for the Personalization of Pedagogic Strategies

Adán Gómez(✉), Laura Márquez, Heider Zapa, and María Florez

Departamento de Informática Educativa, Universidad de Córdoba, Córdoba, Colombia
aagomez@correo.unicordoba.edu.co

Abstract. Goal-driven autonomy (GDA) is a goal reasoning method which allows autonomous systems to introspectively monitor the results of their decisions and generates new goals as needed. The objective of this paper is to present a model of a GDA-Based Tutor Module of an ITS (Intelligent Tutoring System) for the Personalization of Pedagogic Strategies ITS using the theoretical assumptions of the Metagogic metamodel. In this research, the GDA controller allows the tutor module to determine when new goals should be selected and decides which goals should be pursued at each learning lesson. An overview regarding the conceptual model of GDA is provided and how each component was implemented in the Metagogic-Based ITS tutor module. This enables the ITS tutor module to incorporate additional techniques for responding to unforeseen situations. Moreover, an illustrative example about an ITS that facilitates healthcare protocols teachings for the early diagnosis of gestational and congenital syphilis is described.

Keywords: Goal-driven autonomy (GDA) · Tutor module · Intelligent tutoring systems · Metagogic · Planning

1 Introduction

An Intelligent Tutoring System (ITS) is an educational software, that aims to give individualized teaching or feedback to learners in an interactive environment [1]. According to Gómez et al. [2], an ITS infers the pedagogical strategies that can be executed according to the learner's performance trace and his needs. In this way, the ITS dynamically selects and adapts these pedagogical strategies to the student's learning styles. The amount and complexity of mechanisms that structure the personalizing and adaptation of pedagogical strategies make of this process a difficult and time-consuming task [3].

The instructional planning is a central issue when traditional and adaptive educational software are developed, especially an ITS [4]. [5] states that the pedagogical strategies allow the system to adapt and to improve its tutorial strategies taking into account the needs and interests of the learner. [6] affirm that in an ITS the pedagogical model, which is part of the tutor module, is in charge of establishing the learning objectives and aiming the most appropriate pedagogical strategies to guide and

M. Tavana et al. (Eds.): IISA 2020, AISC 1304, pp. 742–750, 2021.
https://doi.org/10.1007/978-3-030-63784-2_92

facilitate the instruction and learning of a determined student [7]. Pedagogic strategies refer to abstract teaching methods, for facilitating the instruction and learning of students. A main characteristic of the process of personalizing and adaptation of pedagogical strategies in this type of systems is its complexity. Thus, some ITS have been developed for addressing personalized adaptation of pedagogical strategies from many approaches [8–11]. METAGOGIC is a MOF-based metamodel for pedagogic strategy modeling in ITS [12]. This metamodel conceptualizes the terms most used in the design of an ITS pedagogical module. Instructional planning, assessment of instruction and advice on learning activities are the tasks involves in the modelling of instructional strategies which are described in this Metamodel using concepts and relationships [2].

Planning, in an intelligent system, enables to find the actions needed to achieve a goal [13]. Planning plays an important role in ITS. According to [14], when an autonomous system plans, it requires to trigger expectations and detect possible discrepancies for the formulation of goals and actions constituting a reasoning cycle for the system. Goal Driven Autonomy (GDA) is a goal reasoning method that enables autonomous agents to direct the focus of their planning tasks in order to be more self-sufficient. GDA focuses on the explanation of discrepancies to formulate new goals [15].

However, the current works do not incorporate components of reasoning, or techniques of planning goal driven autonomy in order to monitor and control in an ITS the personalizing and adaptation of pedagogical strategies. Also, Boulehouache et al. [16] and Chi et al. [17] affirm that ITS uses fixed pedagogical strategies which limit adaptability or employ hard-coded pedagogical strategies that seek to implement existing or pedagogical model. Therefore, most ITS employ just one set of pedagogical strategies when making a decision due to the problem of adding new pedagogical strategies. Boulehouache et al. [16] build a Self-Switching Multi-Strategic Pedagogical Agent which integrates ITS components, the Component Based Agent paradigm, and Autonomic Computing principles for triggering the appropriate pedagogical strategy. Nevertheless, this model does not employ goal-based autonomous reasoning mechanisms to detect discrepancies and to generate new goals in the pedagogical strategies personalization and selection process.

The objective of this paper is to present a model of a GDA-Based Tutor Module of an ITS for the Personalization of Pedagogic Strategies ITS using the theoretical assumptions of the Metagogic metamodel [12, 18]. The phases of development of this model described in this paper are: i) analysis of the GDA Controller Stages; ii) analysis of the packages that structure Metagogic; iii) Design of the model; iv) Building of the prototype according to the described module. This paper is structured as follows: In the second chapter, we explain the Metamodel of Personalized adaptation of Pedagogical Strategies in intelligent tutoring systems (Metagogic). Then, in the third chapter an overview of Goal Driven Autonomy is presented. The next chapter shows the Model of a GDA-based Tutor Module for the Personalized Adaptation of Pedagogic Strategies in ITS. The fifth chapter describes the Illustrative Example of the model. Finally, the conclusions of this study are presented.

2 State of Art About Planning Techniques in ITS

In the literature, many studies state that an ITS is composed by three main modules: tutor module, student module and expert module and an interface [4]. The tutor module is also defined by other researchers as "Instructional Planner" [19]. According to [20], the integration of planning techniques in pedagogical modules of an ITS could facilitate the reasoning and definition of plans to achieve a specific goal. In the last ten years have been developed planning techniques for ITS from different approaches [21–28]. Nevertheless, these studies do not use mechanisms of goal reasoning, or techniques of planning goal driven autonomy that allow to ITS to manage its processes of personalization and adaptation of pedagogical strategies. Goal Driven Autonomy is planning method that involves integrated planning, execution, and goal reasoning. This method has been widely implemented in metacognitive architectures [29], formalism of the notion of metacognitive expectations, and plans, in strategy games [16] on intelligent systems [13] to establish hierarchical plans and hierarchical expectations to enable high-level planning. However, GDA mechanisms never have been used in ITS to develop planning processes.

3 Goal Driven Autonomy (GDA)

An autonomous agent is a computational system that has a set of goals and drives with complete autonomy in an unstructured, dynamic environment [30]. A goal-driven autonomous system must generate plans to achieve their goals and execute them in dynamic and complex environments [15, 18, 29]. Thus, automatic planning is one of the most important techniques for finding and generating a sequence of actions to take, in order to change the current state into a one that contains the agent's goal, predict changes to the continuous and discrete state across the execution. In that way, building more reliable and autonomous plans [31]. Hierarchical Task Network Planning (HTN) is it's one of the most used planning paradigms in autonomous agents [32]. The GDA Controller is structure into four-steps: i) detects some kind of anomaly, ii) attempts to explain why the anomaly occurred, iii) uses the explanation to formulate a new goal, and iv) decides which goal(s) to achieve [31]. The system $\sum = (S, A, E, \gamma)$ where S represents the states, A represents actions, E represents exogenous events and a state transition function $\gamma = S_x(A \cup E) \rightarrow 2^S$, which describes how execution or occurrence transforms the environment from one state to another.

According to [13, 18, 31] the GDA model receives as input $P = \left(M_{\sum}, s_c, g_c\right)$ which is a planning problem, where $M_{\sum}$ is a model of $\sum$, s_c is the current state, and $g_c \in G$ which is the current goal that will be achieved by a set of states $S_g \subseteq S$. It gives this problem to the Planner that generates a sequence of actions $A_t = [a_t, \ldots a_{t+n}]$ and a sequence of *expectations* $X_t = [x_t, \ldots x_{t+n}]$, where each $x_i \in X_t$ is a set of constraints that are predicted to hold in states $[s_{t+1}, \ldots, s_{t+n+1}]$ when executing A_t in s_t using $M_{\sum}$. The Controller sends a_t to $\sum$ for execution and retrieves resulting state s_{t+1}, as it performs four knowledge-intensive tasks: *Discrepancy Detection, Explanation Generation, Goal Formulation* and *Goal Management.*

4 GDA-based Tutor Module for the Personalized Adaptation of Pedagogic Strategies in ITS

The Tutor module is the principal module of an ITS [33]. The selection process of the most adequate pedagogical strategies facilitates the knowledge acquisition of a specific learner, is in charge of the pedagogical model included into the tutor module [34]. The process of personalizing and adaptation of pedagogical strategies in an ITS, consists in the capacity that an ITS has to adapt instructional plans in a personalized way according to some characteristics of the learner. This personalized adaptation process is made according to learning theories, teaching strategies and the pedagogical knowledge rules contained in the pedagogical model [12]. In this model, an instructional plan is a set of individual learning resources related to the learning lesson [12], generated by a GDA controller to accomplish students' learning objectives. In addition, this model integrates the theoretical assumptions proposed by Flórez et al. [7] about a tutor module based on Pedagogical Content Knowledge.

In this research, the GDA controller allows the tutor module to determine when new goals should be selected and decides which goals should be pursued at each learning lesson. For this, the tutor module must know the student profile, the courses in which the student is enrolled and associated with the resources of these courses.

The functionality of the GDA controller in the tutor module of an ITS is described in the Fig. 1. The ITS captures the information which constitutes the **Student Profile** $S_p = \langle S_i, S_s^l, S_d, S_s^p \rangle$ which is a 4-tuple where S_i is the student identifier, S_s^l is the Student Learning Style, S_d is the Pedagogical Dimension assigned to the Student by the system and S_s^p is the Pedagogical Strategy assigned to the Student by the system. Also, the information of the **Selected Course by the Student** $C_s^s = \langle C_i, C_l^c, C_l^s \rangle$ which is a 3-tuple where C_i is the Course Identifier, C_l^c is the Current Course Lesson and the C_l^s is the structure of the Current Course lesson. These S_p and C_s^s are stored in the *User Package* and the *Planner.* The Planner in this research incorporates news mechanisms, which include the processing of data that will be used by GDA Controller. Both S_p and C_s^s are used to build the Domain D as well as the Problem P (using PDDL Language) which receives the Planner through a file generated by the *User Package.*

An initial goal g_0 is given to the *Planner* by the *Metacore Package.* The *Planner* using D, P and g_0, generates a plan, $\pi = \langle a_i, a_{i+1} \ldots a_n \rangle$. Where a_i is an action that is stored in the *Metacore Package* to be later executed by the ITS Graphic User Interface. In this moment, the *Planner* generates expectations, x, which are given to GDA Controller *Discrepancy Detector.*

This *Discrepancy Detector* uses an ontology to compare ITS world facts and inferences from these facts to the expectations x. The ITS world current state, S, is a set of facts f, which are represented as triples $f = \langle S_p, C_s^s, R_s \rangle$ where R_s is a **Selected Resource** in a given time. This match process allows to detect if exists a discrepancy between the current state and the expected state. Thus, in this research, an expectation x

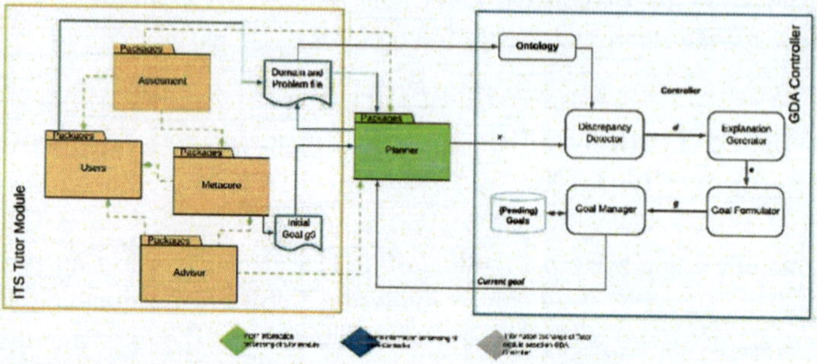

Fig. 1. GDA-based tutor module of an intelligent tutoring system for the personalization of pedagogic strategies.

consists in the activated resource and completed activity *(the completion of an activity consists of the achievement of an expected minimum score)*. A discrepancy d, is the contradiction of the previously described *(Deactivate resource and uncompleted activity)*.

The *Explanation Generator* gives an explanation for the detected discrepancy. This GDA component generates a hypothesis as explanation e that describes d, taking into account an ITS world current state, S and a discrepancy d.

The *Goal Formulator* uses this explanation to build a new goal g_n to pursue, with $g_n \in G_p$, where G_p represents the pending goals of the ITS GDA Controller (the ITS first pending goal is the initial goal g_0). This goal g_n constitutes from the response to an explanation e and a discrepancy d in the ITS world current state S.

Then the *Goal Manager* update G_p adding g_n which may also ensure other edits (e.g., to remove and to modify goals). The Goal Manager will select g_n and turn it into g_c to be given to the Planner.

5 Illustrative Example

Below, the *ITS Fichas y Protocolos en Salud* is presented as an illustrative example of the previously described model. *ITS Fichas y Protocolos en Salud* is an ITS that facilitates the teaching to nursing program students of Universidad de Córdoba-Colombia, about healthcare protocols for the early diagnosis of gestational and congenital syphilis based on the Colombian Ministry of Health. This ITS can be implemented in higher education institutions and others institutions providing health services. The Fig. 2 presents a screenshot about the student profile which is structured by the information captured in the first login of the user in the system. This information and the course information selected by learner is internally codify to PDDL language to build the domain as well as the problem that the planner needs to start reasoning process.

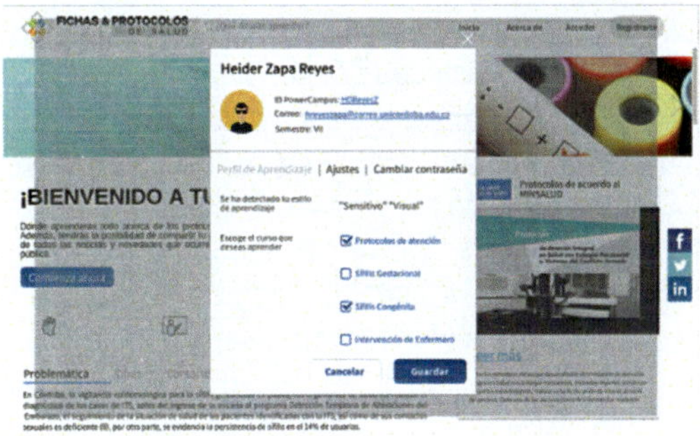

Fig. 2. Student profile in the ITS.

Figure 3 shows the result of the ITS reasoning process. The pedagogical strategies selection process has finished and a pedagogical resource is presented to the learner waiting for its response, which will determine the start of the GDA process.

Fig. 3. Result of the ITS reasoning process.

6 Conclusions and Future Work

In this paper, it was presented an approach for giving autonomy into a Tutor Module of an Intelligent Tutoring System for the Personalization of Pedagogic Strategies. This research allowed the integration of the GDA controller components to the packages of the Metagogic-Based ITS tutor module. The GDA controller enables the tutor module to determine when new goals should be selected, and to decide which goals should be pursued at each learning lesson. For this, the tutor module must know the student profile, and courses to which the student is enrolled and associated with the resources of these courses. An overview about the conceptual model of GDA is provided, and it was explained how each component was implemented in the Metagogic-Based ITS tutor module. This enables the ITS tutor module to incorporate additional techniques for responding to unforeseen situations. There is an illustrative example regarding an ITS that facilitates the healthcare protocols teachings for the early diagnosis of gestational and congenital syphilis. This ITS will allow solving the existing problems in learning the protocols of early detection of the infection, in the timely identification of the contacts, to improve the efficiency in the evaluation time of medical consultations and the ignorance of public policy by part of health professionals.

Currently, the ITS *Fichas y Protocolos en Salud* has a limited ontology of discrepancies, explanations, and goals. In a forthcoming research we will be looking to increase the size of this ontology for enabling the ITS to react to more types of facts. Also, it is necessary to give ITS meta-reasoning capabilities about the monitoring of plans and replace in execution time. In a next paper, a planner that allows the ITS to have meta-planning characteristics will be described.

References

1. Almurshidi, S.H., Abu-Naser, S.S.: Design and Development of Diabetes Intelligent Tutoring System (2016)
2. Gomez, A., Fernando, M., Piñeres, C.: Meta-Modeling Process of Pedagogical Strategies in Intelligent Tutoring Systems Personalization of pedagogical strategies in Intelligent Tutoring Systems View project. (2018). https://doi.org/10.1109/ICCI-CC.2018.8482046
3. Latham, A., Crockett, K., McLean, D.: An adaptation algorithm for an intelligent natural language tutoring system-computer interface intelligent tutoring systems interactive learning environments teaching/learning strategies. Comput. Educ. **71**, 97–110 (2014)
4. Rishi, O.P., Rekha, G., Madhavi, S.: Distributed case based reasoning for intelligent tutoring system: an agent based student modeling paradigm. World Acad. Sci. Eng. Technol. **5**, 273–276 (2007)
5. Gomez, A.A., Caro, M.F.: Meta-Modeling process of pedagogical strategies in intelligent tutoring systems. In: Proceedings of 2018 IEEE 17th International Conference on Cognitive Informatics and Cognitive Computing, ICCI*CC 2018 (2018). https://doi.org/10.1109/ICCI-CC.2018.8482046
6. Aguilar, R., Muñoz, V., González, E.J., Noda, M., Bruno, A., Moreno, L.: Fuzzy and multiagent instructional planner for an intelligent tutorial system. Appl. Soft Comput. J. **11**, 2142–2150 (2011). https://doi.org/10.1016/j.asoc.2010.07.013
7. Gómez, A.A., Flórez, E.P., Márquez, L.A.: Design of the tutor module for an intelligent tutoring system (ITS) based on science teachers' pedagogical content knowledge (PCK). In: Villalba-Condori, K.O., Aduríz-Bravo, A., Lavonen, J., Wong, L.-H., Wang, T.-H. (eds.) CISETC 2019. CCIS, vol. 1191, pp. 141–157. Springer, Cham (2020). https://doi.org/10.1007/978-3-030-45344-2_12
8. Viccari, R.M., Ovalle, D.A., Jimenez, J.A.: ALLEGRO: teaching/learning multi-agent environment using instructional planning and cases- based reasoning (CBR). CLEI Electron. J. **10** (2007). https://doi.org/10.19153/cleiej.10.1.4
9. Espinosa, M.L., Sanchez, N.M., Valdivia, Z.G., Perez, R.B.: Concept Maps Combined with Case-Based Reasoning in Order to Elaborate Intelligent Teaching/Learning Systems, April 28 (2008). https://doi.org/10.1109/isda.2007.33.
10. Prentzas, J., Hatzilygeroudis, I., Garofalakis, J.: A web-based intelligent tutoring system using hybrid rules as its representational basis. In: Lecture Notes in Computer Science (Including Subseries Lecture Notes in Artificial Intelligence and Lecture Notes in Bioinformatics), pp. 119–128. Springer Verlag (2002). https://doi.org/10.1007/3-540-47987-2_16

11. Milrad, M., Mitrovic, A., Nakabayashi, K., Wong, S.L., Yang, S.: A two-layer reasoning framework for a teaching strategies engine using SWRL. Asia-Pac. Soc. Comput. Educ. (2009)
12. Fernando, M., Piñeres, C., Jiménez-Builes, J.A., Caro, M.F., Jiménez, J.A.: MOF-based Metamodel For Pedagogical Strategy Modeling in Intelligent Tutoring Systems (2014). https://doi.org/10.1109/ColumbianCC.2014.6955365
13. Cheng, V., Li, C.-H.: Combining supervised and semi-supervised classifier for personalized spam filtering. In: Zhou, Z.-H., Li, H., Yang, Q. (eds.) PAKDD 2007. LNCS (LNAI), vol. 4426, pp. 449–456. Springer, Heidelberg (2007). https://doi.org/10.1007/978-3-540-71701-0_45
14. Molineaux, M., Kuter, U., Klenk, M.: DISCOVERHISTORY: Understanding the Past in Planning and Execution (2012)
15. Cox, M.T., Dannenhauer, Z.A.: Perceptual Goal Monitors for Cognitive Agents in Changing Environments (2017)
16. Boulehouache, S., Ouareth, S., Maamri, R.: A self-switching multi-strategic pedagogical agent. J. King Saud Univ. - Comput. Inf. Sci. (2019). https://doi.org/10.1016/j.jksuci.2019.06.009.
17. Chi, M., Vanlehn, K., Litman, D., Jordan, P.: Empirically evaluating the application of reinforcement learning to the induction of effective and adaptive pedagogical strategies. User Model. User-adapt. Interact. **21**, 137–180 (2011). https://doi.org/10.1007/s11257-010-9093-1
18. Muñoz-Avila, H., Jaidee, U., Aha, D.W., Carter, E.: Goal-driven autonomy with case-based reasoning. Lecture Notes in Computer Science (including Subseries Lecture Notes in Artificial Intelligents Lecture Notes in Bioinformatics) LNAI, vol. 6176, pp. 228–241 (2010). https://doi.org/10.1007/978-3-642-14274-1_18
19. Zhang, Z., Geng, X., Jiang, Y., Yang, Y.: An intelligent tutoring system(ITS) for tactical training based on ontology. In: Proceedings - 2009 International Conference on Information Engineering and Computer Science. ICIECS 2009 (2009). https://doi.org/10.1109/ICIECS.2009.5362927.
20. Elbih, H., Elkawkagy, M., Tawfic, H.: Dialog Based Planning in Intelligent Tutoring System Proposed Approach. **4**, 19–26 (2019)
21. Garrido, A., Morales, L., Serina, I.: On the use of case-based planning for e-learning personalization. Expert Syst. Appl. **60**, 1–15 (2016). https://doi.org/10.1016/j.eswa.2016.04.030
22. Rahati, A., Kabanza, F.: Automated planning of tutorial dialogues. In: IEEE 2010 International Conference on Autonomous and Intelligent Systems. AIS 2010 (2010). https://doi.org/10.1109/AIS.2010.5547015
23. Aguilar, R., Muñoz, V., González, E.J., Noda, M., Bruno, A., Moreno, L.: Fuzzy and multiagent instructional planner for an intelligent tutorial system. Appl. Soft Comput. **11**, 2142–2150 (2011)
24. Fournier-Viger, P., Nkambou, R., Nguifo, E.M., Mayers, A., Faghihi, U.: A multiparadigm intelligent tutoring system for robotic arm training. IEEE Trans. Learn. Technol. **6**, 364–377 (2013). https://doi.org/10.1109/TLT.2013.27
25. Marinov, M., Valova, I.: Application of planning techniques in knowledge-managed tutoring systems. In: 2016 15th International Conference on Information Technology Based Higher Education and Training, ITHET 2016 (2016). https://doi.org/10.1109/ITHET.2016.7760745

26. Vannaprathip, N., Haddawy, P., Schultheis, H., Suebnukarn, S., Limsuvan, P., Intaraudom, A., Aiemlaor, N., Teemuenvai, C.: A planning-based approach to generating tutorial dialog for teaching surgical decision making. In: Nkambou, R., Azevedo, R., Vassileva, J. (eds.) ITS 2018. LNCS, vol. 10858, pp. 386–391. Springer, Cham (2018). https://doi.org/10.1007/978-3-319-91464-0_44
27. Cloude, E.B., Taub, M., Azevedo, R.: Investigating the role of goal orientation: metacognitive and cognitive strategy use and learning with intelligent tutoring systems. Lecture Notes in Computer Science (including Subseries Lecture Notes in Artificial Intelligents Lecture Notes in Bioinformatics). LNCS, vol. 10858, pp. 44–53 (2018). https://doi.org/10.1007/978-3-319-91464-0_5
28. Gómez-Conesa, A.A., Carrillo, F.X.M.: Lumbalgia ocupacional. Fisioterapia **24**, 43–50 (2002). https://doi.org/10.1016/s0211-5638(01)73017-9
29. Cox, M.T., Alavi, Z., Dannenhauer, D., Eyorokon, V., Munoz-Avila, H., Perlis, D.: MIDCA: A metacognitive, integrated dual-cycle architecture for self-regulated autonomy. In: Thirtieth AAAI Conference on Artificial Intelligence (2016)
30. Jiming, L., Jianbing, W.: Multiagent Robotic Systems. CRC Press, Boca Raton (2018)
31. Dannenhauer, D.: Self monitoring goal driven autonomy agents, http://preserve.lehigh.edu/etd.2564 (2017)
32. Molineaux, M., Klenk, M., Aha, D.W.: Goal-Driven Autonomy in a Navy Strategy Simulation (2010)
33. Rongmei, Z., Lingling, L.: Research on internet intelligent tutoring system based on MAS and CBR. In: 2009 International Forum on Information Technology and Applications, pp. 681–684 (2009)
34. da Silva, C.B.: Pedagogical model based on semantic web rule language. In: 2012 12th International Conference on Computational Science and Its Applications, pp. 125–129 (2012)
35. Barros, H., Silva, A., Costa, E., Bittencourt, I.I., Holanda, O., Sales, L.: Steps, techniques, and technologies for the development of intelligent applications based on semantic Web Services: a case study in e-learning systems. Eng. Appl. Artif. Intell. **24**, 1355–1367 (2011). https://doi.org/10.1016/j.engappai.2011.05.007

Automatic Lane Recognition for Surveillance at Road Intersections

Fanlei Min, Guan Wang, Liantao Wang(✉), and Jing Liu

College of Internet of Things Engineering, Hohai University, Changzhou, China
ltwang@hhu.edu.cn

Abstract. Aiming at lane information recognition under the intersection monitoring environment, this paper proposes a method which can realize the detection and recognition of lane lines and direction arrows. We extract the lane background by a multi-frame averaging algorithm and implement our method on the basis of the background image. The detection of lane line parameters is realized by Hough Transform (HT), and then according to the lane line parameters, we localize the position of the direction arrows by projection segmentation on an aerial view image of the road. The recognition of the direction arrows is realized by image similarity matching based on Normalized Cross Correlation (NCC), and the aerial view of the road is obtained by the perspective trans-formation (PT) combined with Least-square method. The effectiveness of pro-posed method is verified in real intersection surveillance video.

Keywords: Lane recognition · Perspective transform · Intersection surveillance

1 Introduction

Traffic signs on pavement contain a lot of important information, and their intelligent identification is of great significance for the construction of Intelligent Transportation System (ITS). With the development of ITS, the recognition of pavement information has aroused wide interests of scholars, and the research interests mainly focus on the detection of lane lines and the identification of indicators such as direction arrows.

Among them, the detection of lane lines has been studied widely. In [1], a technique utilizing the context-aware information and Maximum Stable Extreme Region detector is used for lane marking detection. [2] and [3] both use the modified Hough Transform for lane line detection, the difference between the two is that [2] establishes two models of straight lines and curves, and fits the straight and curved lanes based on the feature points of the straight line and curves detected by the HT respectively. [3] uses the spatial clustering algorithm for clustering in the second stage of its algorithm, and then performs curve fitting. On the other hand, for the recognition of direction arrows, some are implemented based on image processing methods [4], and more are trained on large-scale data sets to obtain classifiers [5, 6]. Of course, in addition to the above-mentioned studies that focus extensively on lane lines and direction arrows, there are also scholars who work on the detection of traffic signs such as zebra crossings, hatched areas, stop line, dotted lines and words [7, 8].

M. Tavana et al. (Eds.): IISA 2020, AISC 1304, pp. 751–759, 2021.
https://doi.org/10.1007/978-3-030-63784-2_93

Most of the studies mentioned above involve the field of driver assistance, which are mainly used for safe driving. However, there are still relatively few researches on pavement information recognition in the context of traffic surveillance at road intersections. Compared to the field of assistant driving, lane recognition in a surveillance video environment captured by a static traffic camera has a relatively fixed background. But the more complicated external environment of the intersection, i.e., vehicle occlusion, illumination variation, shadow of trees, dust and wear, also brings great difficulties to identification. In addition, the identification of road information from a traffic monitoring perspective is of great significance to intelligent traffic flow detection, auxiliary decision-making, and the identification of vehicle violations.

Therefore, for this application scenario, we propose the detection and recognition method of lane information in road intersection monitoring environment. The content of this paper includes the detection of lane lines and the recognition of direction arrows, and the ultimate goal is to achieve the determination of the lane type. On uncovered roads, lane line detection and arrow recognition are relatively easy to achieve, but at busy city intersections, there are many passing vehicles, resulting in objects to be identified being blocked. Therefore, we first use the video sequence to extract the background image of the lane, and then perform lane information detection and recognition on the background image. In addition, the lane markings will be deformed in the monitoring perspective, we have used the least squares method to estimate the perspective transformation matrix and restore the lane markings for identification.

The rest of this paper is organized as follows. Section 2 introduces the steps of the proposed method and describe the algorithm used in each step in detail. Then, the experimental setup and results are presented in Sect. 3. Finally, we summarize the research contents in Sect. 4.

2 The Proposed Method

In this section, we discuss all the steps of the proposed approach illustrated in Fig. 1. First of all, a multi-frame averaging algorithm is used for creating the lane background model; Then, the Hough Transform is applied to the detection of lane lines; In the next part, we use the least squares method to estimate the perspective transformation matrix, and get a top view image of the lane that is more in line with the realistic proportions; Finally, based on the projection segmentation and similarity matching, the detection and recognition of the guidance arrows are realized.

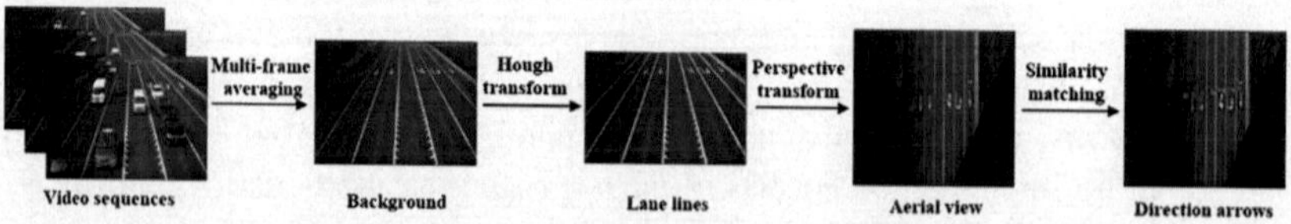

Fig. 1. Overview of the proposed lane marking detection approach.

2.1 Background Modeling

By establishing a background model of the lane area, it can effectively avoid the difficulty in identifying lane information caused by complex foreground. Multi-frame averaging algorithm is a simple and efficient background extraction technology under fixed background. It is characterized by high computation speed, strong back-ground simulation ability and good adaptability to background changes. Moreover, our research object is the static road signs, only one background image is needed. Therefore, the averaging algorithm is used to create the lane background model.

Assuming that f_i is a frame of size $n_1 \times n_2$, the background model extracted from N frames shown in (1) can be obtained according to the averaging method.

$$B_n(x,y) = \frac{1}{N}\sum_{i=n}^{n-N+1} f_i(x,y)(x = 0, 1, \ldots, n_1 - 1, y = 0, 1, \ldots, n_2 - 1) \tag{1}$$

2.2 Hough Transform Based Lane Line Detection

Hough transform is used to obtain the parameters of the lane model. Before the transformation, a 90% area of the background model near the end of the camera illustrated in Fig. 2a is selected as our Region of Interest (ROI) to exclude the influence of the far-end blurred area and repeated arrows on subsequent detection effects. And median filtering is used to remove image noise. Then, the established background model is binarized via Otsu algorithm, and the image edges are extracted by canny operator.

After the above pre-processing, HT is then performed to transform the edge points in the image space into the parameter space. The peak points in the parameter space are determined through the voting process. Finally, the peak points in the parameter space are converted to the image space to obtain the end point coordinates of the lane lines and the expression of the endpoint is as follows:

$$\begin{cases} y_1 = ax_1 + b, \\ y_2 = ax_2 + b. \end{cases} \tag{2}$$

where (x_1, y_1) and (x_2, y_2) represent the end coordinates of each lane line detected.

The starting and ending points of the line detected by HT are used for lane line fitting, and the lane line model expression of the intersection area is as follows:

$$x = ay + b \tag{3}$$

2.3 Perspective Transformation with Least-Squares Method

The different shooting angle and distance leads to the various distortion of traffic signs on road surface. Therefore, perspective transformation is utilized for image rectification, so as to eliminate its impact on object detection and recognition. Be-cause of that the camera is not calibrated and the perspective transformation matrix is unknown, we use the structured information of the lane lines to solve the transformation matrix. That is, the detected lane lines are restored to parallel lines.

In this section, we take the end coordinates of the lane line detected by HT as the original coordinates of the perspective transformation, and combine the least-square method to obtain the perspective transformation matrix.

The essence of perspective transformation is to project an image onto a new image plane, which is a mapping process from 2D (*x*, *y*) to 3D (*X*, *Y*, *Z*) and then to another 2D (X', Y') space. The general formula is:

$$[X \quad Y \quad Z] = T\begin{bmatrix} x \\ y \\ z \end{bmatrix} = \begin{bmatrix} t_{11} & t_{12} & t_{13} \\ t_{21} & t_{22} & t_{22} \\ t_{31} & t_{32} & t_{33} \end{bmatrix}\begin{bmatrix} x \\ y \\ z \end{bmatrix} \tag{4}$$

where (x, y) represents the coordinates of the original image, $\left(X' = \frac{X}{Z}, Y' = \frac{Y}{Z}\right)$ represents the coordinates in the target image after the transformation, and $T = \begin{bmatrix} T_1 & T_2 \\ T_3 & t_{33} \end{bmatrix}$ represents the transformation matrix, where T_1, T_2 and T_3 represent the linear transformation, translation transformation and perspective transformation of the image respectively, and $t_{33} = 1$ can realize the full-scale transformation of the image.

Therefore, we need at least 4 pairs of corresponding points before and after the transformation to compute T. In the lane environment, our goal is to restore non-parallel lane lines in the background model to parallel lane lines. Hence, for more robust and accurate results, we use more than 4 pairs of coordinates to calculate T, and construct the overdetermined system of equations shown in (5):

$$\begin{bmatrix} x_1 & y_1 & 1 & 0 & 0 & 0 & -x_1X_1' & -y_1X_1' \\ 0 & 0 & 0 & x_1 & y_1 & 1 & -x_1Y_1' & -y_1Y_1' \\ \vdots & \vdots & \vdots & \vdots & \vdots & \vdots & \vdots & \vdots \\ x_n & y_n & 1 & 0 & 0 & 0 & -x_nX_n' & -y_nX_n' \\ 0 & 0 & 0 & x_n & y_n & 1 & -x_nY_n' & -y_nY_n' \end{bmatrix}\begin{bmatrix} t_{11} \\ t_{12} \\ \vdots \\ t_{32} \\ t_{33} \end{bmatrix} = \begin{bmatrix} X_1' \\ Y_1' \\ \vdots \\ X_n' \\ Y_n' \end{bmatrix} \tag{5}$$

where $(x_i, y_i)(i = 1, 2, \ldots n)$ represent the points on the input image, $(X_i', Y_i')(i = 1, 2, \ldots n)$ represent the corresponding points on the output image.

The least square method is used to solve the overdetermined equations, and the solution is the global transformation coefficient of perspective transformation. The transformation process based on the perspective transformation matrix is applied to the global image. And the bilinear interpolation algorithm is used for interpolation.

An example of the complete process of PT is illustrated by Fig. 2.

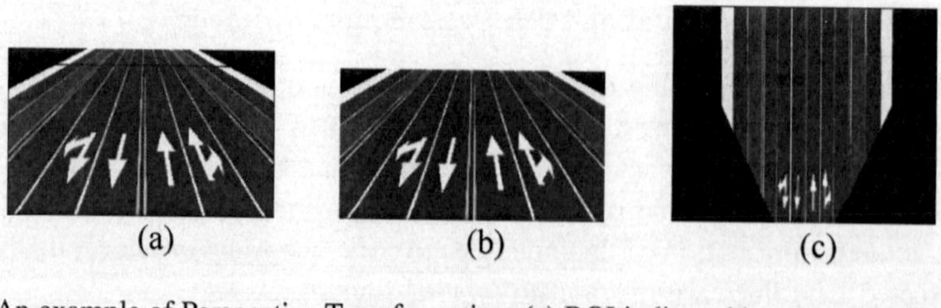

(a) (b) (c)

Fig. 2. An example of Perspective Transformation. (a) ROI indicated by a box. (b) The original coordinates of perspective transformation shown by green '*'. (c) The transformed results.

2.4 Arrow Marking Recognition

As stated above, lane line parameters are obtained by HT, and then the perspective distortion of the monitoring image is rectified. In this section we describe the detection and recognition of direction arrow on road surface. And the direction arrow recognition process can then be divided into 4 parts: image binarization, projection segmentation, candidate region filtering and similarity matching.

1) Image binarization: Due to the existence of buildings and trees, there are often shadows on roads, which results in uneven illumination and that it is difficult to reconstruct the real lane background in a short video sequence. We thus adopt an adaptive threshold algorithm [9] to binarize the bird's-eye view of road, so as to reduce the influence of shadows on the recognition of lane markings.
2) Projection segmentation: We apply the projection segmentation method, which is often used in character segmentation, to the positioning of direction arrows. Compared with the method of sliding window, the proposed method is able to improve the efficiency of detection and the accuracy of recognition. Positioning projection can be further divided into two steps: a) Vertical projection: According to the lane line parameters obtained by (3) and the PT matrix, the linear equation of each parallel lane line in the transformed image is determined, and the vertical projection is carried out between adjacent two lane lines to determine the starting and ending column coordinates of the area where arrow located; b) Horizontal projection: On the basis of vertical projection, the horizontal projection determines the starting and ending row coordinates of the candidate region then.

Table 1. Direction arrow length corresponding to the road type.

Road grade	Expressway		Arterial road, Secondary trunk road, Branch road	
Design road speed(km/h)	100	80,60	60,50,40	40,30,20
Arrow length(m)	9	6	4.5	3

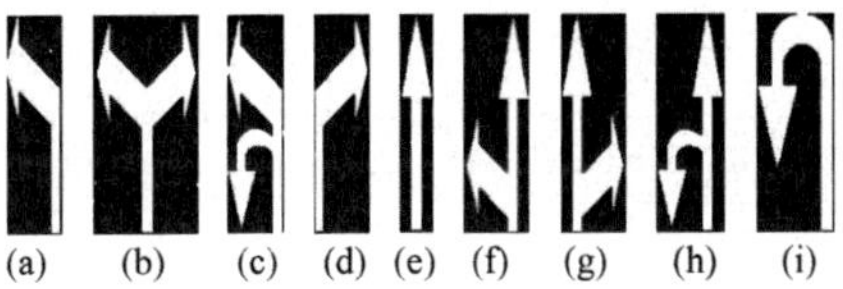

Fig. 3. Arrow templates (a) L. (b) L_R. (c) L_TR. (d) R. (e) S. (f) S_L. (g) S_R. (h) S_TR. (i) TR.

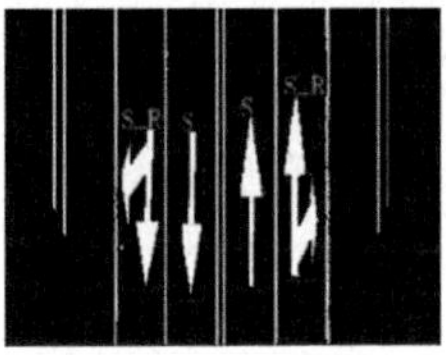

Fig. 4. Identification results

3) Candidate region filtering: Due to the influence of dirt and other traffic signs, there are candidates that are not direction arrows. However, the aspect ratio and area of the direction arrow vary depending on its type. According to Code for layout of urban road traffic signs and markings GB51038 - 2015, the length of the direction arrow can be divided into 3.0 m, 4.5 m, 6.0 m and 9.0 m in terms of the road grades and design speeds shown in Table 1. In addition, according to Code for design of urban road engineering, the width of a vehicle lane is generally 3.25–3.75 m. Therefore, we select the ratio of the length of the arrow to the width of the lane as the filter of the target arrow. Furthermore, considering that there may be wear or grease on the direction arrow, a extensive range of thresholds is selected for the selection of candidates.
4) Similarity matching: The discrimination of road direction arrows is realized by similarity matching. And it can be further divided into the following three steps: a) Standard template library building: We construct a standard template library with labeled arrow as shown in Fig. 3, including turn left (L), turn left or right (L_R), turn left or turn around (L_TR), turn right (R), go straight (S), go straight or left (S_L), go straight or right (S_R), go straight or turn around (S_TR) and turn around (TR); b) Image-size normalization: We scale the template arrow to the same size as the candidate target to facilitate the similarity calculation between the two; c) NCC coefficient calculation: The similarity between the template and the candidate is measured via NCC. The NCC algorithm is a matching algorithm based on gray values, which is simple to calculate. In addition, NCC is not sensitive to signal strength, so it tends to be more robust. Assume that the size of the candidate image is M × N, the calculation method of NCC coefficient value in the proposed method is shown in (6):

$$NCC(x,y) = \frac{\sum_{x=1}^{M}\sum_{y=1}^{N}(T(x,y)-\bar{T})(I(x,y)-\bar{I})}{\sqrt{\sum_{x=1}^{M}\sum_{y=1}^{N}(T(x,y)-\bar{T})^2}\cdot\sqrt{\sum_{x=1}^{M}\sum_{y=1}^{N}(I(x,y)-\bar{I})^2}} \quad (6)$$

where $I(x,y)$, $T(x,y)$ are the pixel values of the candidate target image and the template image respectively, $\bar{I}$ and $\bar{T}$ represent the average pixel values in candidate target image and template image, respectively, which are given by:

$$\bar{T} = \frac{1}{M\times N}\sum_{x=1}^{M}\sum_{y=1}^{N}T(x,y), \bar{I} = \frac{1}{M\times N}\sum_{x=1}^{M}\sum_{y=1}^{N}I(x,y)$$

The pixel values of candidate target image and the template image are calculated by integral graph, and the NCC values of the current candidate target and all the templates are obtained and compared. The category of the template with the maximum value and greater than the threshold is considered to be the category of the arrow.

For the convenience of observation, we indicate the detection results with a blue box, and mark the recognition results of the arrow type with green text. Figure 4 shows an example of the detection and classification results of multiple direction arrows.

3 Experiments

In order to verify the effectiveness of the proposed approach, we conduct background modeling, lane line detection, perspective transformation and arrow recognition experiments on the surveillance collected by traffic camera in intersections with different situations. And the experimental results of the four-part experiments corresponding to different road environments are illustrated in the 2nd–5th columns of Fig. 5.

Datasets. The video dataset used in the experiment is the realistic traffic video collected by traffic camera and was taken at different times from 8 am to 9 am. To validate the recognition performance in various road environments, we select four challenging video sequences in the video dataset as experimental objects. Figure 5(a) shows a representative frame of each video sequence (150 frames) used in experiments, and these four video sequences vary in: (1) rows 1–4 are in different illumination conditions; (2) rows 1, 2 and 3, 4 belong to two intersections in opposite directions, and have different shooting angles; (3) row 1 is in a relatively general road environment; (4) row 2 has a distinct process of illumination change on the road surface (light intensity increases with time); (4) row 3 has a large traffic flow; (5) the right side of the road is always in the shadow of shade tree as shown in the last row.

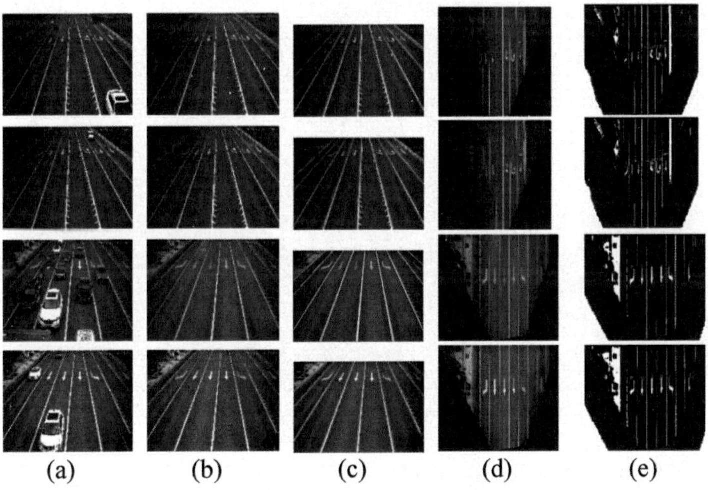

Fig. 5. Experiment Results. From top to bottom: General situation; Obvious illumination change process; Numerous vehicles; Constant shadow alongside the road. From left to right: (a) One representative frame corresponding to the above four various conditions; (b) Reconstructed backgrounds; (c) Lane lines detection results; (d) Perspective Transformation results; (e) Direction arrows recognition results.

Background Modeling Experiments. 150 frames are used for creating lane background. Figure 5(b) shows the background image extracted by multi-frame averaging method, we can observe that in different environments, the complete background can still be restored. However, it is undeniable that when the road surface is blocked by parked vehicles or in a shadow state for a long time, the method is difficult to separate the static noise.

Lane Lines Detection Experiments. As shown in Fig. 5(c), after image enhancement processing such as median filtering, the linear detection algorithm based on HT is able to detect lane lines of multi-lane highways completely.

Perspective Transformation Experiments. The image perspective transformation combined with HT and Least-squares obtains a practical top view of the road in above conditions illustrated in Fig. 5(d). Compared with Inverse Perspective Mapping [10] which require camera parameters and placement angle, the proposed method is much less dependent on external equipment, and still can get expected results.

Arrow Detection And Recognition Experiments. For the recognition of arrows, we use the adaptive threshold algorithm to binarize the aerial view of the road, so as to avoid the influence of uneven illumination caused by shadows. In addition, the candidates are acquired by projection segmentation between adjacent lane lines, and the candidate regions are selected by the ratio of the length of the standard arrow to the width of lane, thereby eliminating non-arrow signs on road surface. The experimental results shown in Fig. 5(e) demonstrate that by the above process, the proposed method can detect and classify different types of arrows in multi-lane environment correctly, even if the detection targets are worn or shaded (see the last two rows of Fig. 5), and can exclude the disturbance of non-arrow signs outside of the driveway.

4 Conclusions

In this paper, we propose a method for the recognition of lane lines and direction arrows under intersection monitoring environment. A robust perspective transformation method are applied before lane line detection and arrow recognition. The proposed method is proved to be useful in recognizing lane information in intersections. Experiment results demonstrate that the proposed method is capable of recognizing multi-lane lines and direction arrows under different light conditions, shooting angles, and in the presence of shadows and long-term occlusions. Also, it performs well in road environments with a lot of traffic. In future work, we also need to verify the performance of the proposed method in a wider range of data.

References

1. Chen, T., Lu, S.: Context-aware lane marking detection on urban roads. In: International Conference on Image Processing, pp. 2557–2561 (2015)
2. Deng, G., Wu, Y.: Double lane line edge detection method based on constraint conditions hough transform. In: International Symposium on Distributed Computing, pp. 107–110 (2018)
3. Niu, J., Lu, J., Xu, M., Lv, P., Zhao, X.: Robust lane detection using two-stage feature extraction with curve fitting. Pattern Recognit. **59**, 225–233 (2016)
4. Maier, G., Pangerl, S., Schindler, A.: Real-time detection and classification of arrow markings using curve-based prototype fitting. In: 2011 IEEE Intelligent Vehicles Symposium (IV), pp. 442–447 (2011)
5. Hoang, T.M., Nam, S.H., Park, K.R.: Enhanced detection and recognition of road markings based on adaptive region of interest and deep learning. IEEE Access **7**, 109817–109832 (2019)
6. He, U., Chen, H., Pan, I., Ni, A.: Using edit distance and junction feature to detect and recognize arrow road marking. In: 17th International IEEE Conference on Intelligent Transportation Systems (ITSC), pp. 2317–2323 (2014)
7. Poggenhans, F., Schreiber, M., Stiller, C.: A universal approach to detect and classify road surface markings. In: 2015 IEEE 18th International Conference on Intelligent Transportation Systems, pp. 1915–1921 (2015)
8. Chen, T., Chen, Z., Shi, Q., Huang, X.: Road marking detection and classification using machine learning algorithms. In: Intelligent Vehicles Symposium, pp. 617–621 (2015)
9. Bradley, D., Roth, G.: Adaptive thresholding using the integral image. J Graph. Tools **12**(2), 13–21 (2007)
10. Borkar, A., Hayes, M., Smith, M.T.: Robust lane detection and tracking with Ransac and Kalman filter. In: International Conference on Image Processing, pp. 3261–3264 (2009)

Network Systems

A Multi-factor Reputation Evaluation Model of Vehicular Network

Huang Yue[1(✉)], Qin Guihe[2], Liu Tong[1], Huang Wei[3], and Meng Chengxun[4]

[1] Center for Computer Fundamental Education, Jilin University, Changchun, China
huangyue@jlu.edu.cn
[2] College of Computer Science and Technology, Jilin University, Changchun, China
[3] Communication Science and Technology Co., Ltd. of Changchun FAW, Changchun, China
[4] Jilin Hangsheng Electronics Co., Ltd., Jilin, China

Abstract. A multi-factor reputation evaluation model of Internet of vehicles is proposed according to the malicious vehicle node and false information in the IOV (Internet of Vehicles). The model comprehensively considers the influence of traffics, information, environment and other factors on vehicle node reputation, and establishes a model of vehicle reputation evaluation. Facing with the characteristics of information of multiple applications, quantifies the influence of different factors and types of information on vehicle reputation.

Keywords: Vehicle · IOV · Reputation evaluation · Multi-factor

1 Introduction

In recent years, with the development of IOT, ITS, IOV has attracted more and more attention by researchers. With the rapid developing of technology on Internet of Vehicles, a large number of new applications have been transferred, such as the application of driving safety for drivers, the application of traffic efficiency and management, and the application of entertainment for drivers [1]. In order to prove the reliable operating of these applications, IOV with the help of comprehensive linking and information interaction between cloud platform, realizing the Vehicle-to-everything (V2X), etc., and it will become the important way of information service IOV applications in the future, especially for instant application service of traffic information, traffic status, safety alarm, such as short-range communication for vehicle-vehicle communication. And even vehicle-vehicle communication also easy to attack by the malicious false information of the vehicle, a malicious vehicle can directly through the way of Vehicle-to-Vehicle (V2V) release some false information, caused traffic disorders and even threaten the life and property safety and causing more serious traffic accidents. Particularly, more important is that the high dynamic nature of vehicle nodes in IOV and the openness of the internet make the attacks above are easy to come true and difficult to detect. Therefore, distinguishing malicious vehicle nodes and

M. Tavana et al. (Eds.): IISA 2020, AISC 1304, pp. 763–770, 2021.
https://doi.org/10.1007/978-3-030-63784-2_94

detecting wrong information to ensure the dependability and information reliability of vehicle nodes in the IOV has become an important aspect to ensure the security of the IOV [2].

In order to solve the problem of malicious vehicle nodes and false information, researchers proposed many security solutions, mainly including reputation mechanism, cryptography based authentication mechanism and Trusted Platform Module (TPM) embedded hardware and mobile terminal of trusted enhancement architecture [3–6].

This context proposes a multi-factor reputation evaluation mechanism for the IOV, which comprehensively considers the influence of various factors on the reputation of vehicles. Facing with the multi-application of the IOV, that is, there are various types of communication information between vehicles. We consider the different influences of different types of information on vehicle reputation, and updating vehicle reputation according to information types to better reflect the credibility of vehicle nodes. In consideration of the influence on various factors on the vehicle node reputation, it is said that combine the historical reputation of the vehicle node and the influence of RSU and time on communicating information transmission and factors of location on the vehicle reputation, so that it can describe the vehicle reputation comprehensively and objectively.

2 Framework of Vehicles Reputation on IOV

2.1 Composition of Evaluation Model

In recent years, there are many researchers have proposed many credibility evaluation methods and models for the related IOV fields based on different theoretical bases and perspectives, mainly focusing on the credibility calculation that based on different methods. According to the calculation ways of reputation, the reputation model can be divided into these types: reputation based on Gaussian probability distribution, reputation model based on β distribution, and reputation model based on subjective logic [7–9]. In our scheme, the system model is mainly composed of three parts: certification association (CA), road side unit (RSU) and communication vehicle.

(1) The authority CA is mainly responsible for awarding identity certificates for vehicles which send in the communication range, ensuring that each vehicle has only got one correct identity (ID), which is reliable and will not disclose the privacy of vehicle users.

(2) The RSU usually deployed in various sections of roads or crossroads, through the cable way to communicate between RSU, so that the RSU can quickly get the information such as road conditions, accidents or the vehicle communication, so the main purpose in this reputation evaluation model of RSU communication to the scope of information collection, for the credible evaluation reputation of vehicle nodes, and this model confirmed that the RSU is completely can be trusted.

(3) The communication vehicle is the direct user of the system, and the communication vehicle in this model can be divided into message sender and message receiver. The message sender is directly broadcasting the accident observed by themselves, road condition and other information or actively forwarding the information he desired to be true. Message receiver is calculated through the credibility of the data which collected in the reputation of each message sender according to final reputation decision which received by the message sender to send information, and in the current vehicle after the fact situation of the communication information of the actual situation to get feedback and take to vehicle nodes reputation updating.

This credibility evaluation for the operation of the process is if the message send to the information of the receiver to get vehicles, you will decide which receives the message sender to send information, and you have to collect various influence factors of reputation firstly, passing the credibility mechanism of calculation and get the communication reputation, then the corresponding communication of message sender reputation sort, selecting the higher vehicle communication reputation, and communicate with it.

2.2 Establishment of Evaluation Model

There are many kinds of information services in the IOV, different business needs correspond to different information types, and different information types need different processing mechanisms provided by the Internet of vehicles. In general, Internet of vehicles applications can be divided into active road safety applications, traffic efficiency and management applications, and online entertainment applications. Different application services correspond to different information types. In the application scenario of Internet of vehicles, the information of communication between vehicles can be divided into three categories: safe driving information, traffic management information and network entertainment information. Because the importance of each kind of information in the application of Internet of vehicles is different, the influence of different kinds of Internet of vehicles information that the vehicle actively sends right or wrong on its own reputation should also be different. Therefore, we set each reputation value V to include three parts: safe driving information reputation value (V_s), traffic management information reputation value (V_m) and network entertainment information reputation value (V_e), as shown in the following formula:

$$V = a_1 V_s + a_2 V_m + a_3 V_e \tag{1}$$

Among them, the parameters a_1, a_2, and a_1, a_2 are the weight of information reputation value of safe driving, traffic management and network entertainment respectively, that is $a_1 + a_2 + a_3 = 1$.

2.3 Composition of Reputation Value

While calculating the reputation value of these communication between two vehicles, we consider the influence of two major factors on the reputation value of the vehicle of the message sender after the current communication: node reputation value (V_J) and environment reputation value (V_E) are as follows:

$$V_i = \alpha V_J + \beta V_E \tag{2}$$

Where the parameter α and β are partly represent the weight of node reputation value and environment reputation value in this communication reputation value, and $\alpha + \beta = 1$.

(1) Node reputation value (V_J)

The node reputation value reflects the reputation of the message sender. When calculating the node reputation value of message sender S, message receiver A needs to consider the following: the historical reputation value of message sender (V_{H_S}) and the reputation value of RSU (V_{R_S}). The specific calculation processes are as follows:

$$V_J = b_1 V_{HS} + b_2 V_{RS} \tag{3}$$

For them, $b_1 + b_2 = 1$. Descriptions of each parameter are as follows:

Historical reputation value (V_{H_S}): the achievement of the historical reputation value of message sender S by message receiver A can be divided into two situations. One is that if A has historical communication with S, and the historical reputation value of S is picked up by A, A can directly use the historical reputation value of S stored by itself. If A and S have not communicated before, then A receives and uses its own historical reputation value sent by S. In this way, as A cannot determine whether S has modified its own stored historical reputation value, A should weaken the influence of the historical reputation value sent by S, that is, reduce the weight value of a_1. In any way, the historical reputation is affected by the time factor, that is, the historical reputation will be updated as time increases.

RSU stored reputation value (V_{R_S}): It is because that RSU is deployed in each road segment, it knows the interaction of information such as each accident event, and can receive messages from vehicles in the communication range and know the fact of messages. If there is no historical communication between A and S, A needs to receive its own reputation value sent by S. Then message receiver A can compare the historical reputation value (V_{H_S}) of message sender S with the reputation value (V_{R_S}) stored in RSU:

When $\left|V_{H_S}^t - V_{R_S}\right| > \gamma$ (t represent time) shows that the credibility of $V_{H_S}^t$ S is low, then the system needs to automatically adjust $V_{H_S}^t$ of the weight b_1, that is, to reduce the impact of historical reputation value on the total reputation value. The reputation value of vehicles stored in RSU makes the final reputation value more accurate and prevents malicious vehicles from changing with their own reputation value.

(2) Environmental reputation value (V_E)

The environmental reputation value reflects the real situation of the message sent by the message sender in this communication. Therefore, we consider that the environmental reputation value is affected by the timeliness of the message and the geographical location of the message sender, that is, the longer the message generation time is from the event occurrence time, the lower the environmental reputation value; the farther the geographic location of the message generation which from the current occurrence geographical local space, the lower the environmental reputation value will be.

We have divided the information in the Internet of vehicles into three categories: safe driving, traffic management and online entertainment. Each kind of information has timeliness. In the Internet of Vehicles, the best effective period T' of these three kinds of communication information is different. And we assume that when the distance between the geographic location of the event and the geographic location of the information generation is in the range of 0-s, we consider that the information is thoughtful in the distance influencing factors. The effective range of communication messages of collision vehicles is usually about 300 m, so length for 300 m is assumed, s = 300 m.

We define V_t and V_l to represent the time and distance in influencing factors, that is:

$$V_E = V_t + V_l \tag{4}$$

t_1 and t_2 partly represent the time when the accident and other information are generated and the time when the message is generated by the message sender s, (X_{Send}, Y_{Send}) represents the location coordinate where the message is generated by the message sender S, (X, Y) represents the geographic location coordinate where the accident and other information are generated. The specific description is as follows:

To make $\Delta t = t_2 - t_1$, that is.

$$\begin{cases} V_t = 1, & \Delta t < T' \\ V_t = \frac{T'}{\Delta t}, & \Delta t > T' \end{cases} \tag{5}$$

To make $\Delta s = \sqrt{|X_{Send} - X|^2 + |Y_{Send} - Y|^2}$, that is.

$$\begin{cases} V_l = 1, & \Delta s \le 300\,\text{m} \\ V_l = \frac{S}{2\sqrt{|X_{Send}-X|^2 + |Y_{Send}-Y|^2}}, & \Delta s > 300\,\text{m} \end{cases} \tag{6}$$

To sum up, we can calculate the node reputation value (V_J) of the sending vehicle S and the environmental reputation value of this communication V_E, and then get the reputation value of the sending vehicle S of this communication V_C. While the message which received multiple same messages by the vehicle A, it can select the vehicle node with higher reputation value as the service node; when the vehicle receives one message, it can determined whether it is received by judging to the reputation value of the vehicle sending the message reaches its own reliable threshold.

2.4 Calculation Parameters

Four factors that affect the reputation value of communication are described in the previous section. The importance of various types of information is also different from these four factors. Therefore, a multi factor judgment matrix is constructed, as shown in Fig. 1.

$$\begin{array}{c} \\ a_1 \\ \vdots \\ a_i \\ \vdots \\ a_n \end{array} \begin{array}{c} \begin{array}{ccccc} a_1 & \cdots & a_j & \cdots & a_n \end{array} \\ \begin{bmatrix} a_{11} & \cdots & a_{1j} & \cdots & a_{1n} \\ \vdots & & \vdots & & \vdots \\ a_{i1} & \cdots & a_{ij} & \cdots & a_{in} \\ \vdots & & \vdots & & \vdots \\ a_{n1} & \cdots & a_{nj} & \cdots & a_{nn} \end{bmatrix} \end{array}$$

Fig. 1. Multi-factor judgment matrix

Table 1. Judgment matrix of "reputation value"

	V_{HS}	V_{RS}	V_t	V_l
V_{HS}	1	1/3	3	3
V_{RS}	3	1	5	5
V_t	1/3	1/5	1	1
V_l	1/3	1/5	1	1

Several comparison judgment matrixes are constructed by using the 1–9 scale method as follows, as shown in Table 1, 2, 3, 4 and 5.

Table 2. Judgment matrix of "historical reputation value"

	V_s	V_m	V_e
V_s	1	3	4
V_m	1/3	1	2
V_e	1/4	1/2	1

Table 3. Judgment matrix of "RSU stored reputation value"

	V_s	V_m	V_e
V_s	1	2	4
V_m	1/2	1	2
V_e	1/4	1/2	1

Table 4. Judgment matrix of "time influencing factors"

	V_s	V_m	V_e
V_s	1	5	7
V_m	1/5	1	3
V_e	1/7	1/3	1

Table 5. Judgment matrix of "distance influencing factors"

	V_s	V_m	V_e
V_s	1	1/2	3
V_m	2	1	4
V_e	1/3	1/4	1

The sum-product method is used to solve the matrix as follows: First, use the formula $\overline{a}_{ij} = \frac{a_{ij}}{\sum_{i=1}^{m} a_{ij}}$, normalize the data from the same column. Then use the formula $w_i = \frac{\sum_{j=1}^{m} b_{ij}}{m}$, take the mean value of each row of the new matrix and get the weight w_i of each factor.

$$w_1 = \begin{bmatrix} 0.252 \\ 0.555 \\ 0.097 \\ 0.097 \end{bmatrix}, w_2 = \begin{bmatrix} 0.623 \\ 0.240 \\ 0.137 \end{bmatrix}, w_3 = \begin{bmatrix} 0.571 \\ 0.286 \\ 0.191 \end{bmatrix}, w_4 = \begin{bmatrix} 0.723 \\ 0.193 \\ 0.083 \end{bmatrix}, w_5 = \begin{bmatrix} 0.320 \\ 0.557 \\ 0.123 \end{bmatrix}$$

Verify the consistency of each matrix as follows: Calculate the maximum eigenvalue of each matrix $\lambda_{max} = \sum_{i=1}^{n} \frac{(AW)_i}{nW_i}$; Calculate consistency indicator $CI = \frac{\lambda_{max}-n}{n-1}$, when $CI = 0$, it shows that the matrix is consistent, the larger the CI, the more inconsistent the matrix is. Then look up the Table 6 by the order n of the matrix, the random consistency index RI can be obtained; Finally, calculate the consistency ratio $CR = CI/RI$, when $CR < 0.1$, it shows that the degree of matrix inconsistency is within the allowable range, the corresponding weight vector is available. For judgment matrix of "reputation value", CI = 0.0144. According to the random consistency index, n = 4, RI = 0.9, then CR = 0.016 < 0.1, pass the consistency test. Similarly, the rest of the matrices pass the consistency test.

Table 7 can be obtained by the above calculation:

Table 6. Random consistency indicator ***RI***

n	1	2	3	4	5	6	7	8
RI	0	0	0.58	0.90	1.12	1.24	1.32	1.41

According to the above table, various parameters of formula 1–3 can be obtained:

$$\begin{aligned} V &= 0.559V_s + 0.319V_m + 0.134V_e \\ V_i &= 0.807V_J + 0.194V_E \\ V_J &= 0.312V_{HS} + 0.688V_{RS} \end{aligned} \tag{7}$$

Table 7. Multi-factor weight table

	V_{HS}	V_{RS}	V_t	V_l	
	0.252	0.555	0.097	0.097	
V_S	0.623	0.571	0.723	0.320	0.559
V_m	0.240	0.286	0.193	0.557	0.319
V_e	0.139	0.191	0.083	0.123	0.134

3 Conclusion

The existence of malicious nodes in IOV will seriously affect the network performance. Once there are malicious nodes break and actively sharing false information, they may even cause incalculable consequences. This paper proposes a multi-vehicle networking reputation evaluation model, this model can accurately reflect the vehicle under different scenarios share the influence of different kinds of information about their reputation, and increase the decision accuracy of vehicle receiving true information, to ensure that the vehicle information receiving credibility, enhancing the credibility of the whole car networking environment. The next research we consider is applying the mechanism into more realistic area of IOV for simulation, and improving the operation efficiency of the scheme on the premise to ensure the operating efficiency of the reputation mechanism.

Funding. Funded by Science and technology research and planning project of Jilin Provincial Education Department, JJKH20190163KJ.

References

1. Schmittner, C., Ma, Z., Reyes, C., Dillinger, O., Puschner, P.: Using SAEJ3061 for automotive security requirement engineering. In: International Conference on Computer Safety, pp. 157–170 (2016)
2. Li, C.C.: Research on Secure Mechanism in Internet of Vehicles for Information Security Issues. Beijing Jiao tong University (2019)
3. Sumra, I.A., Hasbullah, H.B., Manan, J.-L.A.: Using TPM to ensure security, trust and privacy (STP) in VANET. In: 5th National Symposium on Information Technology: Towards New Smart World, NSITNSW, Art. no. 7176402 (2015)
4. Mahapatra, P., Naveena, A.: Enhancing identity based batch verification scheme for security and privacy in VANET. In: Proceedings-7th IEEE International Advanced Computing Conference, IACC 2017, Art. no. 7976822 , pp. 391–396 (2017)
5. Zhu, L., Zhang, Z., Xu, C.: Security and privacy preservation in VANET. In: Secure and Privacy-Preserving Data Communication in Internet of Things, pp. 53–76 (2017)
6. Marmol, F., Perez, G.: TRIP: a trust and reputation infrastructure-based proposal for vehicular ad hoc networks. J. Netw. Comput. Appl. **35**(3), 934–941 (2012)
7. Jaramillo, J.J., Srikant, R.: DARWIN: distributed and adaptive reputation mechanism for wireless ad-hoc networks. In: Mobicom 2007: Proceedings of the Thirteenth ACM International Conference on Mobile Computing and Networking, pp. 87–97 (2007)
8. Feng, J., Xiao, D., Yang, B.: Reputation system for wireless sensor networks based on β distribution. J. Comput. Appl. **1**(1), 111–117 (2007)
9. Laniepce, S., Achemlal, M., Bouabdallah, A., et al.: A cross-layer reputation system for routing non-cooperation effects mitigation within hybrid ad-hoc networks. In: International Conference on Wireless Communications & Mobile Computing, pp. 296–300 (2010)

Design of Network Information System Equipment Health Management Software based on Combat Readiness

Yuwen Liu, Hongtu Cai(✉), Pengfei Ma, Yonghui Xu, and Yaoze Han

Army Artillery and Air Defense College, Hefei 230031, China
sypycht@126.com

Abstract. Based on the analysis of the functional requirements of the network information system health management software, the paper designs the network information system equipment health management software based on the combat readiness from the information resource layer, technology application layer and user operation layer, and combines the health status assessment, fault intelligent diagnosis, residual life prediction and maintenance decision support to provide the key technologies needed for the realization of network information system equipment health management software.

Keywords: Combat readiness · Network information system · Health management · Maintenance support

1 Introduction

The future combat is based on network information system [1], Once the key equipment in the system fails, it will not only cause huge economic losses, but also have a significant impact on the performance of the combat effectiveness and even the combat results. Therefore, it is more and more urgent for the army to operate the network information system equipment safely, stably, in a long period and at full load. So it is necessary to build the network information system health management software to manage the technical status of the network information system equipment, prevent the occurrence of faults and accidents, and maximize the operational efficiency of the network information system equipment.

2 Functional Requirement of Health Management Software of Network Information System based on Combat Readiness

Equipment health management is not a simple management activity, but a kind of technical work around improving equipment health status, improving equipment combat readiness and prolonging equipment life cycle [2]. The software functions

M. Tavana et al. (Eds.): IISA 2020, AISC 1304, pp. 771–777, 2021.
https://doi.org/10.1007/978-3-030-63784-2_95

mainly include health status analysis, fault diagnosis and positioning, remaining life prediction and maintenance decision-making.

2.1 Health Status Analysis

Analyzing the health status of equipment is the basic condition of implementing equipment health management. Through technical inspection, on-line monitoring, functional detection and other technical means to obtain the state characteristic parameters of the equipment, and use information fusion, data mining, intelligent computing and other methods to process these characteristic parameters, determine the health level of the equipment, judge whether the current technical state can complete the combat task and achieve the expected goal. The final results of health status analysis and evaluation provide decision basis for optimizing equipment maintenance strategy and making equipment task plan.

2.2 Fault Diagnosis and Location

Diagnosis and location of equipment fault is a key work in equipment health management. The state characteristic signal is the real reflection of the dynamic characteristic of the equipment. The main means of fault diagnosis is to identify the fault symptoms according to the equipment health state standard data, the state characteristic signal and the fault discrimination criteria. Through fault diagnosis, we can find out the causes of equipment performance degradation, identify the fault mode, determine the location, size, nature and hazard degree of the fault, so as to provide support for the rapid replacement of equipment and the rapid recovery of combat capability.

2.3 Residual Life Prediction

Predicting the remaining service life of equipment is an important part of equipment health management. Through the real-time monitoring of the operation status of the equipment and the whole process collection of health data, and reasoning based on the current and past status of the equipment, analyze the future development trend of the equipment, predict the remaining service life, and judge whether it can achieve the predetermined operational purpose in the future. Through the prediction of remaining life in advance, it can provide support for taking measures in advance to avoid failure risk and prevent major accidents.

2.4 Maintenance Decision Support

It is the core part of health management and the ultimate goal of health management to make a scientific maintenance decision-making scheme, optimize the allocation of maintenance support resources and ensure that the equipment is in a good technical state. According to the nature, size, location, type, development speed and hazard degree of equipment failure, scientifically formulate maintenance plan, optimize the allocation of maintenance support resources, determine the best maintenance time, maintenance strategy and maintenance spare parts demand quantity, and provide help for safe and stable operation of equipment and successful completion of scheduled tasks.

3 Design of Network Information System Equipment Health Management Software Organization Structure based on Combat Readiness

3.1 Design Principles

In terms of design, unified technical standards must be adopted, including data format, information exchange protocol, information transmission protocol, information processing standard, man-machine interface specification, system safety standard, etc. According to the unified technical specifications for modular design, so that the system has component plug-in assembly capacity, can be composed and deployed as needed. At the same time, the software interface should be simple, intuitive, stable, easy to operate, and meet the needs of equipment maintenance.

3.2 Software Architecture

According to the previous analysis, the architecture of equipment health management software can be divided into three layers: information resource layer, technology application layer and user operation layer. Its structure is shown in Fig. 1.

The information resource layer is composed of text information, health standard database, system fault information database, intelligent diagnosis reasoning database and maintenance decision support database. It is mainly used for information data storage, retrieval, addition, deletion, access and exchange, providing data support for the functional application of the upper layer and realizing information sharing. The technology application layer includes core technology layer and general technology layer, general technology layer mainly provides general application functions for system operation, including equipment information, management and maintenance, auxiliary tools, help description and other function. The core technology layer is mainly composed of four professional components, including health status analysis and evaluation, remaining service life prediction, system intelligent fault diagnosis, condition maintenance decision support, etc. The user operation layer through human-machine interface and technology should realize computer interaction, provide functional application services for user operation.

The component of health status analysis and evaluation consists of status information collection module, feature parameter processing module, status monitoring and analysis module and evaluation modeling and simulation module. The evaluation modeling and simulation module includes various resources such as model, algorithm, data and environment needed for the development and operation of system health analysis and evaluation component.

The system intelligent fault diagnosis component is mainly composed of fault information acquisition module, diagnosis knowledge learning module, diagnosis rule

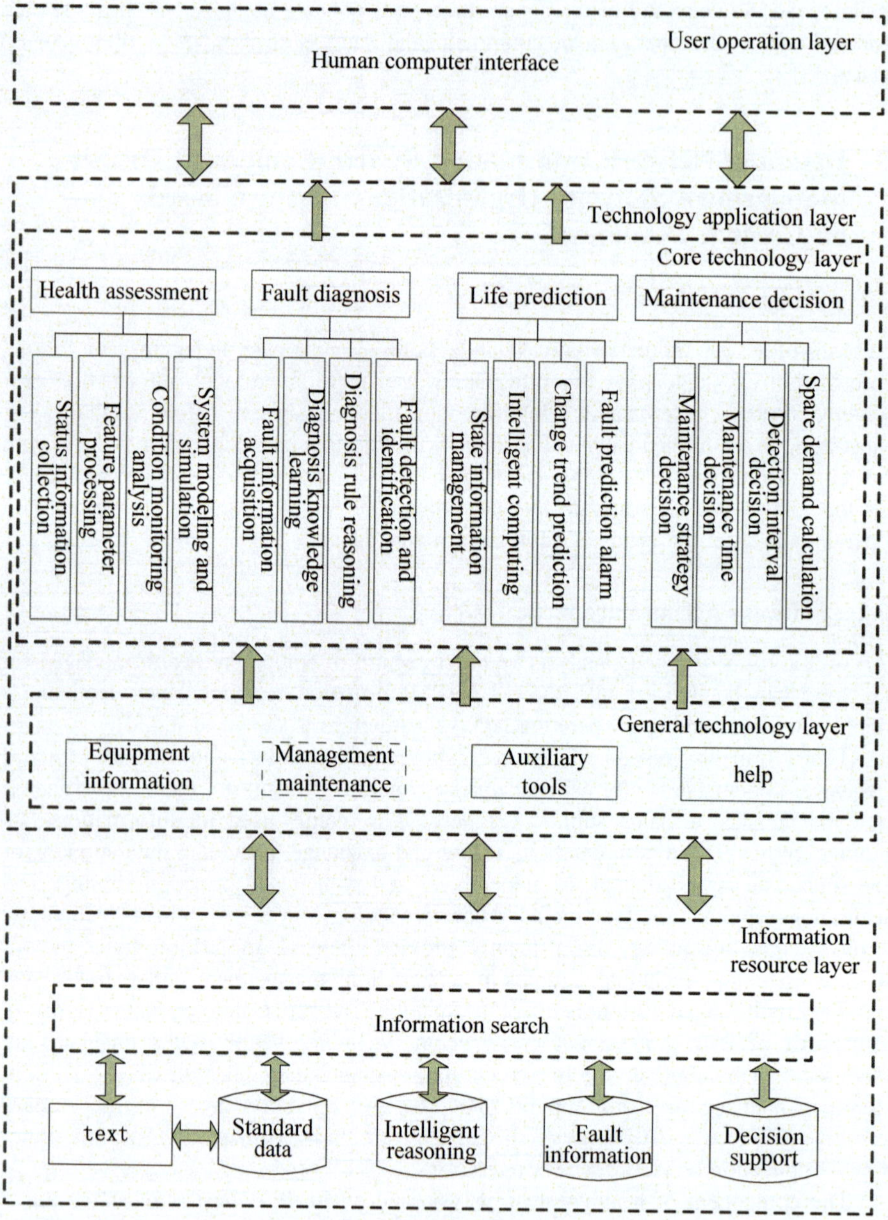

Fig. 1. Network information system equipment health management software architecture

reasoning module, fault detection and identification module and explanation reasoning mechanism module. The diagnosis knowledge learning module is used to store a large number of professional domain expert knowledge and machine learning algorithm.

The remaining service life prediction component is mainly composed of state information management module, change trend prediction module, artificial intelligence calculation module and fault prediction and alarm module. The state information management module includes not only the health state characteristic data of the system, but also the fault state characteristic parameters of the system.

The state maintenance decision support component is mainly composed of maintenance strategy making model, maintenance time decision model, inspection interval decision model and spare demand calculation model. The maintenance strategy is mainly based on the actual status of the system, diagnosis and prediction information to determine the specific maintenance behavior of the system.

4 The Key Technology for Realization of Network Information System Equipment Health Management Software

Equipment health management is a system engineering with multiple disciplines and technologies crossing each other. In the software design and implementation, the key technologies involved mainly include health status assessment technology, residual life prediction technology, intelligent fault diagnosis technology and condition based maintenance decision-making technology [3–5].

4.1 Health Assessment Technology

Health assessment is an important content to evaluate whether the system works normally and whether the combat technology performance is intact, and it is also the precondition to implement fault diagnosis prediction and condition based maintenance decision. Through scientific and effective health assessment, we can timely and accurately understand the current technical status of the equipment. For equipment health assessment, there are many kinds of assessment methods, including health threshold judgment method, case-based reasoning and reasoning algorithms of various models. Generally, the common methods can be divided into classical evaluation methods and machine learning algorithms. The classical evaluation methods mainly includes model method, analytic hierarchy process, fuzzy evaluation method and grey evaluation method. The machine learning algorithm mainly includes artificial neural network, Bayesian network, support vector machine, Markov model and so on.

4.2 Intelligent Fault Diagnosis Technology

The development of equipment fault diagnosis technology has generally gone through three stages. At present, it belongs to the stage of intelligent fault diagnosis. In the stage, knowledge processing is the core, and the mathematical model, system signal and fault knowledge are combined by using expert system, neural network and other technical means, so as to get the cause, type, size, nature, harm degree and specific parts of the fault. The commonly used intelligent fault diagnosis technologies include intelligent fault diagnosis technology based on genetic algorithm, intelligent fault diagnosis technology based on neural network, intelligent fault diagnosis technology based on support vector machine, intelligent fault diagnosis technology based on expert system, intelligent fault diagnosis technology based on Petri net and intelligent fault diagnosis technology based on rough set theory.

4.3 Residual Life Prediction Technology

Prediction refers to the prediction and estimation of the state that has not yet occurred or has not been determined. Residual life prediction is also called fault prediction, which is different from other maintenance support modes. It is a higher level maintenance support mode, and has a direct impact on the operational efficiency of equipment. Residual life prediction technology refers to the prediction of residual life or normal working time of equipment by using reasoning model or intelligent calculation on the basis of comprehensive analysis of equipment state detection parameters, usage, working environment, current conditions and prediction, which can be basically divided into three types: mechanism model based prediction technology, data mining based prediction technology and knowledge integration based prediction technology.

4.4 Condition Based Maintenance Decision Technology

Condition based maintenance decision technology refers to the technology to determine the specific decision content of equipment maintenance mode, maintenance force scheduling and maintenance spare parts demand based on the health status of equipment and existing support conditions. Condition based maintenance decision is an important function of equipment health management, which is the key to improve the safety, reliability, readiness and mission success of equipment. It mainly includes the selection of equipment maintenance strategy, optimization of maintenance task allocation and determination of maintenance spare parts support quantity. The commonly used decision-making technologies of condition based maintenance are mainly based on fuzzy multi-attribute decision-making method, based on proportional risk model, based on reliability real-time evaluation, and based on expert experience and data mining.

5 Conclusion

The network information system equipment health management software can evaluate the health status of the equipment timely according to the state characteristic parameters of the equipment, accurately predict the time and location of the fault, and scientifically give the maintenance decision recommendations. Through the design of network information system equipment health management software, it not only lays a solid foundation for reducing equipment maintenance cost, improving equipment combat readiness and safety reliability, but also provides technical support for eliminating faults, preventing deterioration and improving equipment task success.

References

1. Chun, C., Jichun, Z., Wenbo, W.: Army Command Information System. PLA Press, Beijing (2017)
2. Ju Jianbo, H., Shenglin, S.Z.: Research on equipment health management driven by big data. J. Ordnance Equipment Eng. **38**(6), 73–75 (2017)
3. Jinglin, W., Zeli, L., Guo, Z., Yong, H.: Research on PHM technology scheme of aircraft electromechanical system. Comput. Meas. Control **24**(5), 163–166 (2016)
4. Yan, W., Biao, C., Yingkun, C.: Health management of large spacecraft control system. Space Control Technol. Appl. **42**(5), 42–46 (2016)
5. Yuguang, X.F., Junyuan, W.G., Yuchao, W.: Research on missile equipment health management and its key technologies. J. Ordnance Equipment Eng. **38**(1), 7–11 (2017)

A Network Attack Recognition Method Based on Probability Target Graph

Ying Liu[1] and Yuefeng Zheng[2](✉)

[1] College of Humanities & Science of Northeast Normal University, Changchun, China
[2] College of Computer, Jilin Normal University, Siping, China
bdxyl627@163.com

Abstract. Aiming at the current network security problems, we propose a network attack recognition method based on probability target graph. Based on the target graph, this method replaces the state nodes with the target nodes and adds the observation nodes, and constructs a new structure named probability target graph (PTG). The probability distribution of observed actions is calculated by background knowledge, and then the probability value of each target node is calculated. The target with the highest probability value is the target to be attacked by the intruders. Due to the uncertainty of the network environment, attackers will also deliberately hide their behaviors, so by adding probability values to the target graph, some observable actions can be effectively identified. At the same time, we build the knowledge base based on the causal network, which provides help for the alarm correlation analysis, and can more effectively analyze the attack plan and predict the next actions.

Keywords: Planning recognition · Causal network · Intrusion detection · Alarm association

1 Introduction

With the rapid development and application of networking, mobile, Internet of things and big data technology, information security is becoming more and more important. While greatly promoting the development of productivity, people are increasingly dependent on the information network, which makes the country and society face increasingly severe information security threats. The current intrusion detection system often underreports due to the unsound detection mechanism or failure to detect unknown intrusions, which seriously affects the detection performance. Therefore, the system must be able to infer the attacker's true intention by analyzing the attack behavior. In order to further develop intrusion detection, it is necessary to introduce the planning recognition method in artificial intelligence.

In 2001, Geib and Goldman first applied the planning identification method in the field of intrusion detection [1]. They adopt the method of planning recognition based on plan execution which proposed previously, and it can deal with a wide range of planning recognition problems because it does not make too many restrictive assumptions. In 2007, Wang Lei proposed a planning recognition model based on

M. Tavana et al. (Eds.): IISA 2020, AISC 1304, pp. 778–785, 2021.
https://doi.org/10.1007/978-3-030-63784-2_96

behavior state diagram [2], which simulated and demonstrated the attack events of a specific intrusion detection problem and detected the intrusion behavior in a closed virtual state. Since then, many scholars had been keen on the research in this field, including misuse detection method based on conditional probability, anomaly detection method based on Bayesian inference, anomaly detection method based on Bayesian network, and anomaly detection method based on pattern prediction. In 2010, Peng Wu and Yao Shu-ping proposed an intrusion intention identification algorithm based on probabilistic reasoning, which greatly promoted the application of probabilistic programming identification method in intrusion detection [3]. The network attack planning and identification algorithm [4] proposed by Zhu-ge Jian-wei et al. It can effectively track the monitored system and identify the opponent planning. However, the method can't effectively identify the attacker's high-level targets and predict the attacker's next action. Yin Ji-li et al. also proposed an opponent planning and identification method based on complete target graph [5] on the basis of the target graph, which makes the actions directly related to the targets, can predict the opponent's next action according to the opponent's action, and can identify different targets with different completion degrees.

2 Background

Here are a few concepts related to this article:

Intrusion Detection is the detection of network intrusion behavior. It collects and analyzes network behaviors, security logs, audit data, other information available on the network [6].

Planning Recognition refers to the process of deducing the target or planning of an agent from the observed action or action effect of the agent [7].

Alarm Association is to analyze all kinds of alarms according to a certain strategy, and construct an attack scene by identifying and integrating alarms with different degrees of correlation, so as to obtain the attack strategy and intention [8].

Causal Network can be represented as a directed acyclic graph, which is actually an And/Or graph. In the graph, nodes are collections of variables, and side tables are causal relationships between variables. The root node of causal network represents the final target of attack planning [9].

3 The Algorithm of Probability Target Graph

The alarm association is used to process the alarm information firstly, which can reduce the alarm quantity, enhance the alarm semantics, identify the false alarm and its trigger source, and improve the alarm accuracy. In addition, the target graph is extended and the probability value is added according to the knowledge base, so as to find the most likely attack target of the intruder.

3.1 Related Definition

Event: An event is a description of the attacker's behavior. Security events triggered by attackers can be obtained by analyzing the alarm information provided by intrusion detection system.

Target: Target is an attacker's intent, such as damaging network systems, stealing confidential data, tampering with home page information, and so on. An attack event usually does not exist in isolation, the previous behavior is to prepare for the later attacks, such as the attacker to carry out network port scan is to obtain the open port and service program of the target host and other information, in order to further exploit the vulnerability of the target host service program to prepare for the attack.

In practice, an attack event may cause multiple alarm messages, which represent the redundant relationship between the alarm events of the same attack event. Usually, the alarm events representing the same attack behavior are fused to achieve the simplification effect.

3.2 Framework of the Algorithm

The design process of the overall framework of the algorithm is shown in Fig. 1, and the specific implementation steps are as follows:

Alarm Correlation Analysis. According to the alarm set and the established attack causal network, the target and strategy of the attacker are identified and the targets are predicted in the future. The causal network is actually an and/or graph, which is used in this article to represent the background knowledge, as shown in Fig. 2. In the graph, only one node at the top is the total target, and its lower nodes are the sub-targets. The realization of the upper level targets depends on the completion of the lower level sub-targets. Among them, there are two relations between each node in each layer: one is the relation of and, the other is the relation of or. Here we call it by and operations and or operations. When all the sub-targets are realized, the upper target can be realized, each sub-target performs and operation. When any one of the sub-targets achieves its upper target, each sub-target performs or operation. Based on the information obtained from observation, the probability distribution of each sub-target and the total target can be calculated according to the structure of causal network with prior probability. The calculation method is based on the literature [10].

AND operation: For the upper node, if its lower nodes are AND relationship, that is $P(g_1) \wedge P(g_2) \wedge, \ldots, \wedge P(g_n)$, the probability of this node is $P(G) = P(g_1) * P(g_2) * \ldots * P(g_n)$, $n \geq 0$.

OR operation: For the upper node, if its lower nodes are OR relationship, that is $P(g_1) \vee P(g_2) \vee, \ldots, \vee P(g_n)$, the probability of this node is $P(g) = Max\{P(g_1), P(g_2), \ldots, P(g_n)\}$, $n \geq 0$.

If an upper node and a lower node both exist OR operation and exist AND operation, AND operation is performed firstly, and then performed OR operation. Such as $P(G) = Max\{P(g_1), P(g_2), P(g_3) * P(g_4)\}$.

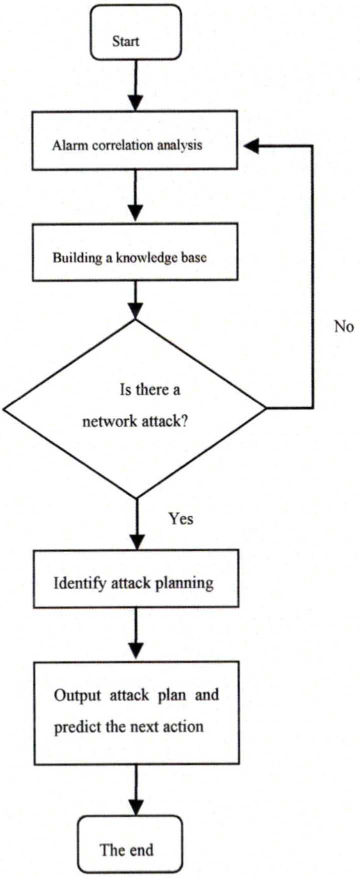

Fig. 1. Overall frame design flow chart

Recognition of Network Attacks. In this paper, the causal network is used as the knowledge base, through the probability value of each node can be calculated and the upper level target can be predicted. On the basis of literature [10], the probability target graph is constructed, and the probability value is added to the graph to provide a basis for finding the optimal plan (the plan most likely to be executed by the network attacker). The probability target graph is shown in Fig. 3. At the same time, this paper also proposes the algorithm flow of constructing the probability target graph and identifying the optimal planning, as shown in Fig. 4.

Construction of Probability Target. The probabilistic target planning graph similar to the reference [10], it defined as an tuple $< Q, O, A, Pa, Pg, G, P, E >$,Where, Q is the set of state nodes; O is the set of observation nodes, and the observation made by the time step i is denoted as O_i; A is the set of action nodes, $a_i \in A$, a_i is the action of the time step i; P_a is the set of predicted action nodes; P_g is the set of predicted target nodes;

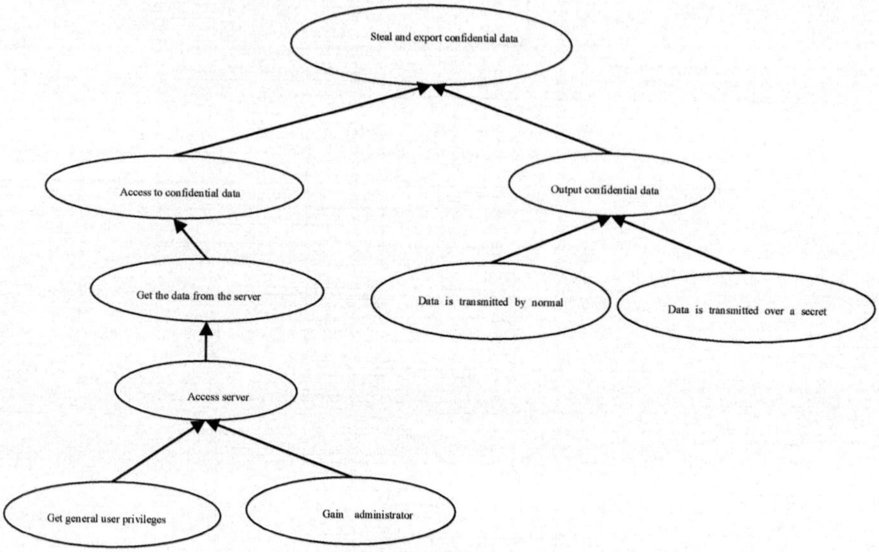

Fig. 2. Causal network diagram

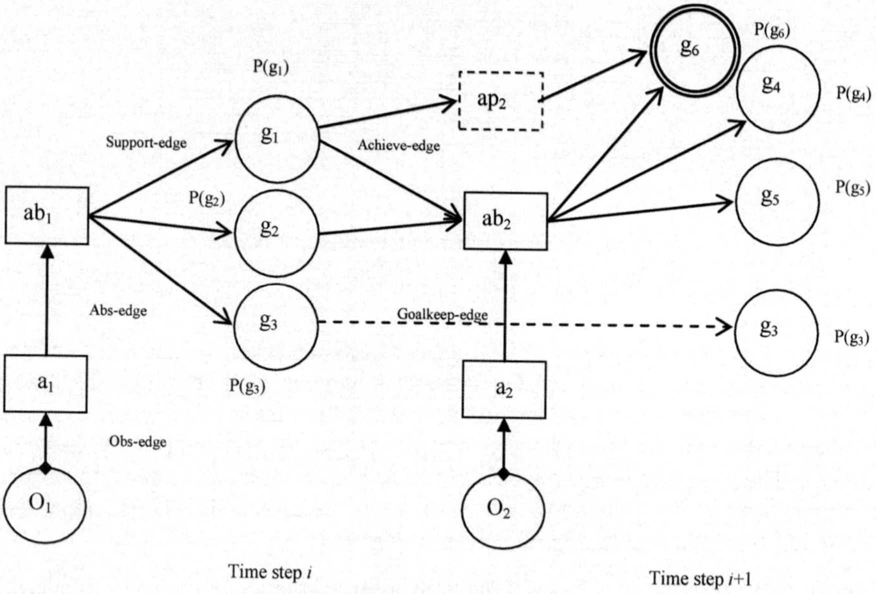

Fig. 3. Probability target graph

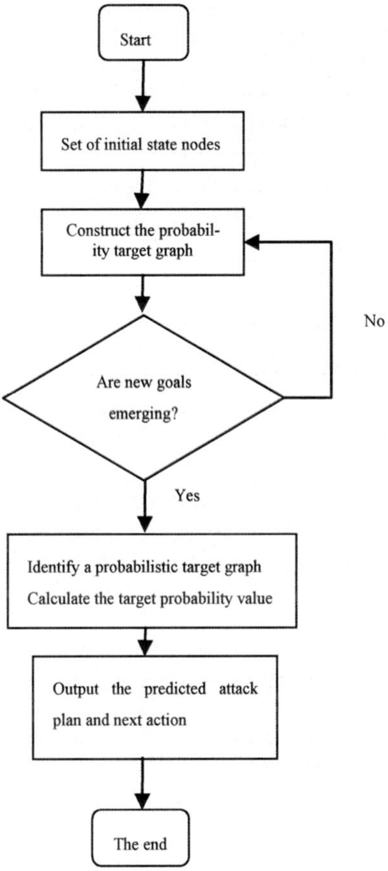

Fig. 4. The construction and identification flow chart of probability target graph

G is the set of target nodes. The target reached by the time step i is denoted as g_i, $g_i \in G$, and P represents the probability distribution of each node. E is a set of edges, including:

Support-edge: represents an abstract action in time step i supports the target of the time step i.

Achieve-edge: represents an abstract action in time step $i + 1$ achieves the target of the time step i.

Abstract-edge (Abs-edge): represents the concrete action abstract in the abstract action.

Observation-edge (Obs-edge): represents the observation event of the observation node for a specific action.

Target keep-edge: indicates that the target node has not changed from the time step i to the time step $i + 1$.

The probability target graph is divided into observation layer, concrete action layer, abstract layer and target layer. It starts at the action layer and ends at the target layer. The probability distribution of observed actions is calculated by background knowledge, and then the probability value of each target node is calculated. Finally the optimal plan can be found.

Causal link: suppose a_i and a_j are respectively the observation actions of time step *j* and time step ($i < j$). If there is causal link $b_i \rightarrow b_j$, if and only if:an effect of a_i satisfies a premise if a_j.

The construction algorithm of the probability target graph is described as follows: the input of the algorithm is a target planning graph containing only the set of initial state nodes, that is T0 = {Q0,∅,∅,∅,∅,∅,P0,∅}.The initial state can be estimated by empirical rules through the automatic sensing mechanism of network environment. Given the initial state, action *a* is obtained according to the observation, and the observation node is connected with the observed action node by the observation edge. Query the knowledge base, instantiate action *a*, and connect the prerequisite and consequence state edges. By comparing the consequence state node and the knowledge base, the next action and the target are predicted. The consequence state node and the prediction action node are connected by the prediction action edge, and the consequence state node and the prediction target node are connected by the prediction target edge. Continue the next expansion by observing the action. For the state nodes that have not changed at the time step *i* and the time step *i*+1, connect with a continuous edge.

Recognition of Probability Target Graph. When a new target appears in the probabilistic target graph, the identification process of attack planning is entered. The description of the algorithm are as follows: For each newly realized target node g_i in the time step *i*, find the state node connected to g_i with the target description edge, and then find the action to obtain the state node, add the action to the attack planning, and according to the causal link, find an attack plan after the loop.

In modern society, network security is crucial. More and more people are interested in the research of network security. This paper provides a better method for the effective identification of network security. It makes causal correlation at the target level to understand the attack behavior and predict the subsequent target. At the same time, we propose a network attack identification method based on probability target graph, which can effectively analyze the attack plan and predict the next action. In the future, we will combine this method with response planning to form a comprehensive network security protection system with recognition and response.

Acknowledgments. This research was funded by the "13th five-year" Science and Technology Project of Jilin Education Department: JJKH20190356KJ.

References

1. Geib, C.W., Goldman, R.P.: Plan recognition in intrusion detection systems. In: DARPA Information Survivability Conference and Exposition (DISCEX-2001). Anaheim, California, pp. 46–55 (2001)
2. Lei, W.: Research and Implementation of the Program Recognition Model Based on Behavior State Diagram. Northeast normal university, ChangChun (2007)

3. Wu, P., Shu-ping, Y.: Research on intrusion intention identification based on probabilistic reasoning. Comput. Sci. **37**(1), 79–82 (2010)
4. Jian-Wei, Z., Hui, H.X., Zhi-Yuan, Y., et al.: A network attack plan recognition algorithm based on the extended target graph. Chin. J. Comput. **29**(8), 1356–1366 (2006)
5. Jili, Y., Ying, L., Jinyan, W., et al.: A recognition approach for adversarial planning based on complete target graph. In: 2007 International Conference on Computational Intelligence and Security, pp 286–290 (2007)
6. Lazarevic, A., Kumar, V., Srivastava, J.: Intrusion detection: a survey. Managing Cyber Threats, 19–78 (2008)
7. Wenxiang, G., Ying, L.: Recognize and Respond to Adversary Planning, pp. 275–276. Science Press, Beijing (2016)
8. Xiao, F., Li, X.: Security alert correlation: a survey. Comput. Sci. **37**(5), 9–14 (2010)
9. Liu, Y., Wen-xiang, G.: An effective recognition method for network attack. OPTIK **124** (2013), 4823–4826 (2013)
10. Ying, L., Wenxiang, G.: A plan recognition algorithm based on the probabilistic goal graph. In: 2011 International Conference on Network Computing and Information Security, pp. 359–361 (2011)

A Systematic Review Study on Research Challenges, Opportunities, Threats and Limitations in Underwater Wireless Sensor Networks (UWSNs)

Syed Agha Hassnain Mohsan[1(✉)], Sardar Shan Ali Naqvi[2], Farhad Banoori[3], Muhammad Irbaz Siddique[4], Muhammad Muntazir Mehdi[5], Frederick Nii Ofie Bruce[6], and Alireza Mazinani[7]

[1] Department of Electrical Engineering, COMSATS University Islamabad (CUI), Islamabad, Pakistan
Hassnainagha@zju.edu.cn
[2] North China Electric Power University, Beijing, China
[3] South China University of Technology, Guangzho, China
[4] Beijing Jiaotong University, Beijing, China
[5] Northwestern Polytechnic University, Fremont, USA
[6] Beijing Institute of Technology, Beijing, China
[7] Beihang University, Beijing, China

Abstract. Wireless Sensor Networks have high stature in underwater observation, exploration and monitoring applications. Qualitative and effective research has been carried out in UWSNs from last few decades. Ambition behind this research systematic review study is to provide a comprehensive survey on latest research in underwater wireless sensor networks. Herein we provide a brief discussion on major research challenges and development in UWSNs. We provide a periodical review on UWSNs issues and potential challenges highlighted in previous studies. In this paper, we investigate research methods to gain attention of research fraternity towards future technologies and challenges of UWSNs on the basis of existing approaches. Thus it is our foremost requirement to provide state-of-art analysis of existing UWSNs.

Keywords: Underwater wireless sensor networks · Research questions · Security · Challenges

1 Introduction

An extensive research has been carried out on terrestrial sensor networks in different aspects and nowadays Underwater Wireless Sensor Network (UWSN) is generating considerable attention among recent researchers. UWSN is an emerging technology for premier research in underwater environment. It is a fusion of wireless technology having intelligent computing, smart sensing and communication abilities. UWSNs are used in numerous applications such as mineral exploration, ocean seismic observation,

M. Tavana et al. (Eds.): IISA 2020, AISC 1304, pp. 786–797, 2021.
https://doi.org/10.1007/978-3-030-63784-2_97

pollution monitoring [1], military applications, and surveillance [2]. UWSN has successfully proven its stature in target detection, tracking, underwater robotics, disaster prevention, underwater chemicals and marine population monitoring, oceanographic data collection. Though UWSN is providing cutting-edge solution but many challenges exist at the same time [2]. The researchers face challenges [3] regarding shadow zones, limited bandwidth, propagation delay, high bit error rates, connectivity losses, attenuation, limited energy, harsh geographical atmosphere and 3D topology [4]. The unpredictable condition of ocean environment generates these serious challenges and issues in designing and deployment of UWSNs. USWNs pose many optimization and conceptual challenges like location coverage and tracking, tsunami detection [5] and smart disaster management [6]. In this paper, we have investigated several challenges and presented our work summary on it. We have discussed several constraints and particularities of Underwater Wireless Sensor Networks (UWSNs). Based on our systematic study, we have investigated some potential threats and security challenges for UWSNs. We have briefly discussed the feasible countermeasures to tackle these security challenges in existing security mechanism [7].

To realize underwater potential applications, we can involve many tools and design principles from latest ongoing research. Current research on UWSNs is focusing on connectivity, communication, self-organization and low energy consumption. This existing research has been facing severe security threats in network [7, 8] as it did not take security problems into consideration. UWSNs are vulnerable to numerous malicious attacks and security threats [9]. We have briefly discussed these security problem of UWSNs in our systematic review study.

2 Relevant Work

In this section, we have summarized challenges in UWSNs in existing research literature. We provide an overview of UWSN technology, enhancing functions and issues highlighted in recent research publications. Authors in [7] presented a survey on current acoustic sensor network applications. They particularly focused on recent underwater deployments, applications and UASNs deployment for control and monitoring of underwater domains. Researchers have given a detailed survey on securing UASN in [10]. Authors discussed available technologies and technical challenges to establish UWSN solution for monitoring applications. They have summarized underwater positioning systems, communication problems and highlighted marine spatial planning tools. Authors in [11] covered a huge spectrum of UWSNs applications by giving a comprehensive survey to gather recent deployments in UWSNs. Existing challenges and issues in aquatic environment have been unveiled in a review paper [8]. They focused on UWSN architecture and localization technologies. Recently, authors in [12] investigated a fault-resilient localization scheme for UWSNs. A survey of UWSNs threats and security concerns is presented in [13].

Table 1. Research work

Reference #	Focused key challenges	Year
Ref. 14	Research challenges in UASN	2005
Ref. 15	UWSN for aquatic Applications	2005
Ref. 16	Challenges and applications of USN	2006
Ref. 17	Challenges of building UWSN	2006
Ref. 18	Prospects and problems in communication	2008
Ref. 19	Oceanographic monitoring	2010
Ref. 20	Monitoring in USN	2011
Ref. 21	Monitoring marine environment	2011
Ref. 22	Secure communication	2012
Ref. 1	Applications of WSN	2014
Ref. 23	UWSN security issues and novel trends	2015
Ref. 24	UWSN security framework	2015
Ref. 25	Eavesdropping attacks	2016
Ref. 26	Network layer attack	2016
Ref. 27	Routing attacks in USN	2017
Ref. 4	Threats and security issues	2018
Ref. 6	Challenges and security issues	2018
Ref. 9	Challenges and security issues	2019
Ref. 3	Review of recent issues and challenges	2019
Ref. 28	Undewater localization techniques and Challenges	2020
Ref. 29	Applications of WSN	2020
Ref. 30	Sensor coverage strategy	2020
Ref. 31	Systematic review on UOWC	2020

In this article, authors have presented recent security challenges and mechanism. Authors in highlighted potential applications for seismic monitoring and tsunami detection [5]. Kiran, J. Sasi, et al. gave a review on research challenges and applications of USNs [13]. Research community has investigated several research directions in time synchronization, data scheduling, and short range acoustic communications. Research are working on novel networking paradigms for large scale mobile UWSNs for aqueous environment exploration. These novel networking paradigms address research challenges in mobile UWSNs along with technological solutions [15]. Authors have presented the intrinsic properties and challenges of topology control in UWSN [7]. Table 1 shows previous research work in UWSN. Considering all above research discussion, our research study aims at providing comprehensive UWSN constraints, challenges, threats and current security mechanism.

3 Challenges and Opportunities

UWSNs are effected by several constraints as shown in Fig. 1. These constraints play a key role in performance of UWSN. Any problem in these constraints can lead to serious issues and challenges in network.

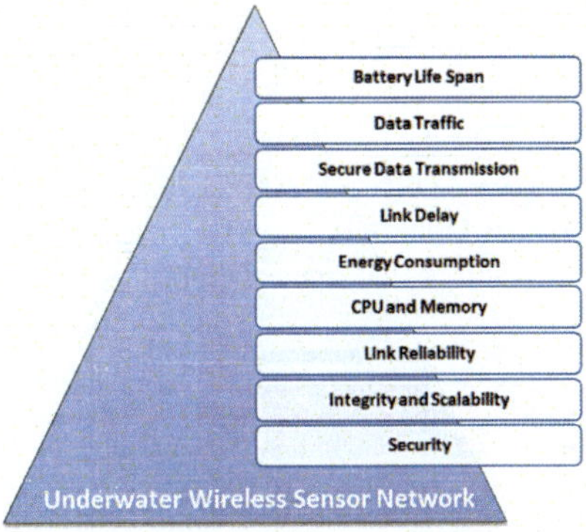

Fig. 1. UWSN constraints

In this promising era of UWSNs, we can explore the unfathomed world beneath ocean. Along with exciting applications and opportunities, there are several challenges which exist. We have outlines some specific particularities below.

Unpredictable Underwater Environment: Underwater environment is highly unpredictable and harsh so human activities are not feasible in UWSNs. Uneven water depths, extreme water pressure and vast areas are only suitable for unmanned exploration which makes it difficult to design and deploy UWSNs.

Intricate Network Design and Deployment: It is quite challenging to deploy a network beneath water surface which works wirelessly and reliably. Constrained communication can be achieved through tethered technology but it incurs high cost to deploy, maintain and recover devices to meet volatile aquatic conditions.

Unscalability: Present underwater monitoring and exploration demands either an extremely expensive device or small scale underwater network. Current technologies are not suitable to cover a large area. Thus a scalable UWSN technology is imperative to explore a large underwater area.

Dynamic Network Topology: Locating nodes beneath water surface becomes a crucial task due to continuous mobility by water currents. Traditional localization and positioning techniques are not sufficient. Empirical observation shows that underwater objects flow at a speed of 36 km/h which generates a highly dynamic topology. Thus, unpredictable conditions of ocean change the location of nodes and make dynamic topology which causes unreliable data transmission.

Physical Damage to Equipment: Underwater conditions can cause physical damage to network devices. Sensors used in underwater devices are vulnerable to underwater harsh condition such as salt accumulation and algae collection on camera lens can lower its efficacy.

Cost: Cost and energy requirement has high impact in any network. The cost required to replenish drained batteries in UWSN is high.

Unsecure Environment and Vulnerability: For some special applications such target tracking and security monitoring, USWN becomes insecure. Underwater sensor nodes share the open underwater acoustic channel to monitor hostile objects in sea. Consequently, these nodes are highly vulnerable to any malicious attack and potential threat to disrupt network services. Generally, marine organism can physically damage UWSNs nodes at harsh deep sea environment where it is difficult to protect these nodes from physical damages.

4 Security Threats

Underwater Wireless Sensor Networks are vulnerable to different malicious attacks. These attacks are divided into two categories as follows:

4.1 Passive Attacks

These attacks are made by malicious nodes in order to steal data from network without any disruption. It includes interfering, message replay, stealing information, eavesdropping, distortion and impersonation. The attacker can predict communication path, capture and exchange packets and can determine locations. As network is not affected by these passive attacks so usually it is difficult to detect. These attacks can be reduced by data encryption.

4.2 Active Attacks

These attacks are made to destroy, delete or alter the transmitted data in network. Active attacks can cause data interception of modification. It can harm with internal and external attacks. It is easy to detect and external attacks as it is because of nodes which do not belong to network. While internal attacks are more severe and their detection is difficult. These attacks can be defended by authentication and encryption. Active attacks can be classified as shown in Fig. 2.

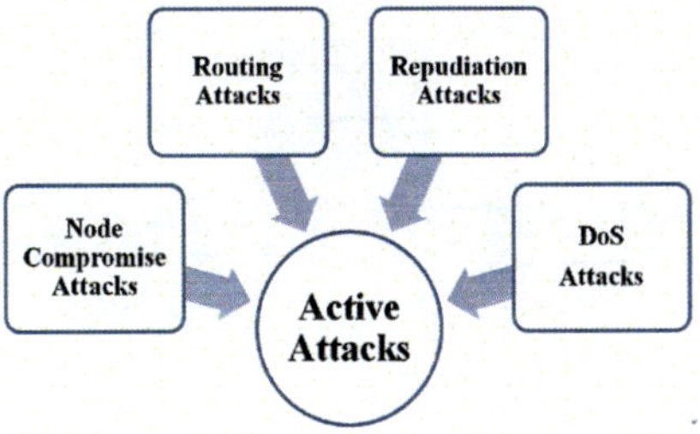

Fig. 2. Active attacks

5 Research Methodology

In this section, we discuss the subjects selection process of our findings and presented the framework and approaches we used to identify our results.

5.1 Approaches Selection

Our investigation starts with selecting the research articles providing a survey or review on challenges in Underwater Wireless Sensor Networks (UWSNs). In order to find initial list of target research articles, we used Google Scholar search. Google Scholar helped us to obtain relevant research articles, peer-reviewed publications, abstracts, preprints and research surveys and technical reports. Google Scholar gave us confidence to complete results based on articles collected from Google search, ResearchGate and academic publishers such as IEEE, Springer and ACM. It is worth noting that Google Scholar gives full-text search against our keywords and validates that obtained research articles are relevant to our performed queries. To achieve significant coverage of research work related to challenges of UWSNs we performed several queries on Google Scholar. First we search challenges in UWSNs, later we checked while combining research survey, study, new trends and security issues. We collected research articles and considered only those published in recent years in order to maintain good results and state-of-the-art approach. Basically, we selected the recent research articles related to security, challenges, new trends and issues in UWSNs. We made lists of relevant papers focusing same potential features and started reading abstracts and conclusions to achieve primary study related to this review.

Our research methodology includes research questions, sources and keywords. A basic overview of our research procedure is displayed in Fig. 3.

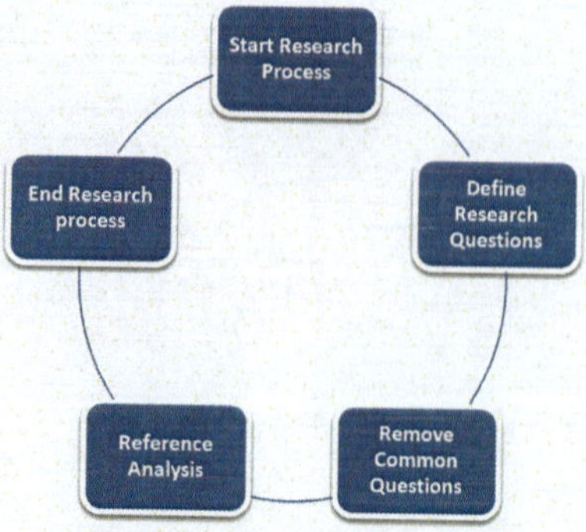

Fig. 3. Research overview

5.2 Research Questions

We have formulated a set of research questions (RQs) and motivation behind it. These questions will be helpful for researchers to identify the missing gap in this research field. RQs and motivation is highlighted in Table 2.

RQ1. Find out the theoretical properties of UWSNs in existing and future technologies and communication devices.
RQ2. Find out good simulations tools, best network architecture design, suitable routing protocols and proper communication model.
RQ3. Design new routing protocols and techniques to meet real world requirements.
RQ4. What are the potential threats and expected physical damages?
RQ5. Give timely response to security threats to secure sensitive data transmission.
RQ6. Test each solution against possible challenges in real world environment to achieve required performance of designed UWSN system.

Table 2. Research Keywords

Research questions	Motivation
RQ1	It will help to find out UWSNs potential application
RQ2	It will help to understand complete designing process of UWSNs
RQ3	It will help to design UWSN in order to meet real world requirements to
RQ4	It will help toidentify all risks
RQ5	It will help to find solution to cope with threats and security attacks
RQ6	It will help to shape the model to meet required performance criteria

5.3 Search Strategy

We have defined a good research strategy which can help researchers to retrieve relevant research literature. Our research strategy includes research method, search terms and data resources. Our search strategy starts with basic steps to identify search terms and data resources:

a) Preliminary search to retrieve previous relevant potential literature.
b) Check research papers published in good journals and leading conferences.
c) Consult with researchers to find relevant data from conferences and journals.
d) Trial search based on prior defined research questions.
e) Using own learning and experience related to UWSNs.

5.3.1 Search Method

Our search strategy includes two search methods: manual search and automatic search. For manual search, we selected papers published in specific venues as listed in Table 3.

Table 3. Research Keywords

Journal/Conference	Venue	Acronym
J1	Journal of king saud university – computer and information sciences	JKSUCI
J2	International journal of distributed sensor networks	IJDSN
J3	Journal of Sensors	J. Sens.
J4	Wireless communications and mobile computing	WCMC
J5	Journal of network and computer applications	J NETW COMPUT APPL
J6	International journal of computers and applications	IJCA
J7	EURASIP journal on wireless communications and networking	EURASIP JWCN
J8	International journal of computer science and mobile computing	IJCSMC
C1	Future technologies conference	FTC
C2	International conference on innovations in information technology	IIT
C3	International conference on identification, information and knowledge in the internet of things	IIKI
C4	Euromicro conference on software engineering and advanced applications	SEAA
C5	International conference on emerging trends in expert applications & security	ICETEAS

For automatic data search, we used electronic data resources provided in Table 4 to obtain relevant papers.

Table 4. Research keywords

Electronic database	Search items	Web address
ACM Digital Library	Keywords, Paper title	http://portal.acm.org
Google Scholar	Keywords, Paper title	https://scholar.google.com/
ScienceDirect	Keywords, Paper title, abstract	https://www.sciencedirect.com/
ResearchGate	Keywords, Paper title, Author	https://www.researchgate.net/
Sci-Hub	Keywords, Paper title	https://sci-hub.tw/
Elsevier	Keywords, Paper title, abstract	https://www.elsevier.com/en-au
IEEE eXplore	Keywords, Paper title, abstract	www.ieeexplore.ieee.org
ISI Web of Science	Keywords, Paper title, abstract	http://www.webofknowledge.com
SpringerLink	Keywords, Paper title	http://www.springerlink.com

5.3.2 Search Terms

We used search terms to match with keywords, paper titles and abstracts in electronic data sources. Keywords are given in Table 5.

Table 5. Research keywords

Keywords	Synonyms
WSNs	Wireless sensor network
USN	Underwater sensor networks
UWSN	Underwater wireless sensor network
UWSN Protocols	Underwater wireless sensor network protocols
UWSN Challenges	UWSN issues, threats, security attacks
AUV	Autonomous underwater vehicles
UWSN Survey	Review and study on UWSN

We defined inclusion and exclusion criteria to rectify this primary study to validate our research queries.

Inclusion Criteria:

- Any paper which declares its key findings related to challenges in UWSNs.
- Any paper which is published after 2005.
- Any paper which is published in English language only.

Exclusion Criteria:

- Articles extended by another article which we have selected already in our list. It gives us to find representative article towards our approach.
- Articles which do not contain complete information about our desired research work.
- Article which do not meet specific required details against our approach.

- Articles representing short paper, only abstract, an editorial, poster summary, panel summary or workshop summary. Such articles are missing with sufficient informative data.
- Articles such as white papers and technical reports as Computer Science community does not rectify such articles.

6 Research Validity and Limitations

We have found relevant research articles by using scientific keywords and references. We selected many papers from Google Scholar. In our exclusions, we selected relevant papers after reading title, abstract, conclusion and future aspects. A possibility of missing some recent papers or data still remains by using this keyword search, inclusion and exclusion methods.

Data Sources: Our major sources for data collection were Google Scholar and ResearchGate. Though these sources are beneficial to search relevant data as they automatically fetch data from different databases, technical sources and academic publishers against single data query, but still we take it as a threat to validity as these sources suffer from certain limitations such as lack of search facility, partial control on content body and vulnerability to spam. To tackle this, we will include more sources, scientific search engines and publishers while extending our current work. Enabling this approach in extending our work, we are confident to get broader coverage of potential issues in UWSNs and will provide more statistical analysis.

Data Collection Process: While our data collection and analyzing process, we assigned a single researcher to read article and collect information from title, abstract and conclusion. It is a potential threat to validity of our findings and we tried to mitigate this concern through group discussion. We are aware that a single researcher can inject certain amount of inconsistencies in selected data. In future extended work, we will perform detailed statistical analysis through assigning this task to more than one researcher.

Selection of Potential Challenges: We tried to identify potential challenges of UWSNs. This investigation comes from preliminary analysis of different articles. Though we used an effective way to get appropriate analysis, still there can be more challenging factors. To alleviate this threat, one research member was assigned this task to keep a good record of possible challenges and issues discussed in each selected article. In the end, this approach provided confidence while validating our approach.

Time Span: In inclusion criteria, we considered articles published after 2005. Although it gives us right approach to find latest articles and key topics but it can affect the completeness of our search results as we did not include papers published before 2010. In future, we expect to involve more research recourses and considering articles published at a wide range of years.

7 Discussion and Future Topics

We have highlighted some key research areas in our systematic study. However, research fraternity needs to focus on some areas: topology control, time synchronization, data processing, self-healing, self-calibration, energy consumption, power control, localization, data encryption, security, routing protocols, communication and debugging. UWSNs will offer potential features in real life applications and will put more impact in future. Our main objective was to identify key challenges for future research. It is concluded from our study that researchers are putting relevant efforts to handle challenges in design, deployments and analysis of UWSNs. Moreover, we have pointed a limitation of security concerns and researcher must give specific concern to potential threats in any UWSN. As a precondition, research community should focus on data encryption to avoid proliferation of security attacks. Future research should contemplate to involve Internet of Underwater Things (IoUT) to properly achieve real world applications.

References

1. Xu, G., Shen, W., Wang, X.: Applications of wireless sensor networks in marine environment monitoring: a survey. Sensors **14**(9), 16932–16954 (2014)
2. Indu, S.D.: Wireless sensor networks: issues and challenges. Int. J. Comput. Sci. Mob. Comput. (IJCSMC) **3**, 681–685 (2014)
3. Awan, K.M., et al.: Underwater wireless sensor networks: a review of recent issues and challenges. Wireless Commun. Mob. Comp. 2019 (2019)
4. Coutinho, R.W.L., et al.: Underwater wireless sensor networks: A new challenge for topology control–based systems. ACM Comput. Surveys (CSUR) **51**(1), 1–36 (2018)
5. Casey, K., Lim, A., Dozier, G.: A sensor network architecture for tsunami detection and response. Int. J. Distrib. Sens. Netw. **4**(1), 27–42 (2008)
6. Coutinho, R.W.L., et al.: Underwater sensor networks for smart disaster management. IEEE Consum. Electr. Mag. **9**(2), 107–114 (2020)
7. Yang, G., Dai, L., Wei, Z.: Challenges, threats, security issues and new trends of underwater wireless sensor networks. Sensors **18**(11), 3907 (2018)
8. Yang, G., Dai, L.E., Si, G.N., Wang, S.X., Wang, S.Q.: Challenges and security issues in underwater wireless sensor networks. In Proceedings of the International Conference on Identification, Information and Knowledge in the Internet of Things, 19–21 October 2018, Beijing, China. (in press)
9. Yang, G., et al.: Challenges and security issues in underwater wireless sensor networks. Procedia Comput. Sci. **147**, 210–216 (2019)
10. Jiang, S.M.: On securing underwater acoustic networks: a survey. IEEE Commun. Surv. Tutor. (2018)
11. Jiang, S.: Wireless Networking Principles: From Terrestrial to Underwater Acoustic. Springer, Heidelberg (2018)
12. Das, A.P., Thampi, S.M.: Fault-resilient localization for underwater sensor networks. Ad Hoc Netw. **55**, 132–142 (2017)
13. Kiran, J.S., et al.: Review on underwater sensor networks: applications, research challenges and time synchronization. Int. J. Eng. Res. Technol. (2015)

14. Akyildiz, I.F., Pompili, D., Melodia, T.: Underwater acoustic sensor networks: research challenges. Ad Hoc Netw. **3**(3), 257–279 (2005)
15. Cui, J.H., Kong, J., Gerla, M., et al.: Challenges: building scalable and distributed underwater wireless sensor networks (UWSNs) for aquatic applications. Channels **45**(4), 22–35 (2005)
16. Heidemann, J., Ye, W., Wills, J., Syed, A., Li, Y.: Research challenges and applications for underwater sensor networking. In: Proceedings of the IEEE Wireless Communications and Networking Conference (WCNC 2006), pp. 228–235, April 2006
17. Cui, J.-H., Kong, J., Gerla, M., Zhou, S.: The challenges of building mobile underwater wireless networks for aquatic applications. IEEE Netw. **20**(3), 12–18 (2006)
18. Lanbo, L., Shengli, Z., Jun-Hong, C.: Prospects and problems of wireless communication for underwater sensor networks. Wirel. Commun. Mob. Comput. **8**, 977–994 (2008)
19. Albaladejo, C., et al.: Wireless sensor networks for oceanographic monitoring: a systematic review. Sensors **10**(7), 6948–6968 (2010)
20. Le Sage, T., et al.: Development of a wireless sensor network for embedded monitoring of human motion in a harsh environment. 2011 IEEE 3rd International Conference on Communication Software and Networks. IEEE (2011)
21. Perez, C.A., Jimenez, M., Soto, F., Torres, R., Lopez, J.A., Iborra, A.: A system for monitoring marine environments based on wireless sensor networks. In: Proceedings of the IEEE OCEANS, pp. 1–6. IEEE, Santander, Spain, June 2011
22. Dini, G., Lo Duca, A.: A secure communication suite for underwater acoustic sensor networks. Sensors **12**, 15133–15158 (2012)
23. El-Rabaie, S., Nabil, D., Mahmoud, R., Alsharqawy, M.A.: Underwater wireless sensor networks (UWSN), architecture, routing protocols, simulation and modeling tools, localization, security issues and some novel trends. Netw. Commun. Eng. **7**, 335–354 (2015)
24. Ateniese, G., Capossele, A., Gjanci, P., Petrioli, C., Spaccini, D.: SecFUN: security framework for underwater acoustic sensor networks. In: Proceedings of the MTS/IEEE OCEANS, 18–21 May 2015, Genoa, Italy, pp. 1–9 (2015)
25. Wang, Q., Dai, H.N., Li, X., Wang, H., Xiao, H.: On modeling eavesdropping attacks in underwater acoustic sensor networks. Sensors **16**, 721 (2016)
26. Ioannou, C., Vassiliou, V.: The Impact of network layer attacks in wireless sensor networks. In: Proceedings of the 2016 IEEE International Workshop on Secure Internet of Things (SIoT), 26–30 September 2016, Heraklion, Greece, pp. 20–28 (2016)
27. Dargahi, T., Javadi, H.H.S., Shafiei, H.: Securing underwater sensor networks against routing attacks. Wirel. Pers. Commun. **96**, 2585–2602 (2017)
28. Su, X., et al.: A review of underwater localization techniques, algorithms, and challenges. J. Sens. **2020** (2020)
29. Kandris, D., et al.: Applications of wireless sensor networks: an up-to-date survey. Appl. Syst. Innov. **3**(1), 14 (2020)
30. Yao, L., Xiujuan D.: Sensor coverage strategy in underwater wireless sensor networks. Int. J. Comput. Commun. Control **15**(2) (2020)
31. Mohsan, S.A.H., Hasan, M.D.M., Mazinani, A., Sadiq, M.A., Akhtar, M.H., Islam, A., Rokia, L.S.: A Systematic review on practical considerations, recent advances and research challenges in underwater optical wireless communication. Int. J. Adv. Comput. Sci. Appl. (IJACSA) **11**(7) (2020)

Synchronization Behavior of a Class of Oscillator Networks with Weighting Exponent

Junqing Feng[1(✉)], Guohong Liang[1], and Lixin Yang[2]

[1] Air Force Enigeering University, Xi'an 710051, China
fjq1983@163.com
[2] College of Science, Shannxi University of Science and technology, Xi'an 710021, China

Abstract. Based on the second-order Kuramoto-like oscillatory model,A complex network model with weighting exponent is presented in this paper. Moreover,the impact of exponent on synchronizability of oscillatory network is investigated vianumerical simulations.It shows that the characteristics of coupling have great effects on the synchronizability.On the contrary,the odd weighting-frequency expon ent reduces the synchronizability of oscillatory network.

Keywords: Weighting exponent · Oscillatory network · Synchronizability

1 Introduction

In recent years, complex networks with oscillators as nodes have become a hot topic for scholars due to their wide application.Power network is composed of power plants, substations, large-scale user centers and high-voltage transmission lines [1–3] in different geographical locations. The scale of power system is expanding, and the network topology is becoming more and more complex. The theory of complex network provides a new theoretical tool for the study of network. Some important achievements have been achieved in the study of power network by using the theory of complex network, such as the analysis of the structural characteristics of the actual network, the modeling of the actual network, the synchronous stability of the power network [4–6] and so on. The initial network is regarded as a complex network with no right and no direction. Although there are many achievements in using the complex network theory to study the internal mechanism of network cascading failure, there are few achievements in using the network as a high-dimensional dynamic system to study the network synchronization ability [7].

In the study of dynamics, the most widely used model is Kuramoto oscillator model and its deformation form. The dynamics of the network nodes are described by the Kuramoto like phase oscillator model In reference [7], and the dynamic equations of the network nodes are obtained by the energy conservation theorem of the generator. In fact, the interaction between nodes will be affected by their own characteristics, among which the most responsive is the maximum power of nodes. Therefore, in this paper,

M. Tavana et al. (Eds.): IISA 2020, AISC 1304, pp. 798–802, 2021.
https://doi.org/10.1007/978-3-030-63784-2_98

the author proposes a new network model, that is, the coupling strength between nodes with the coupling index term is more practical. Based on this, the synchronization behavior of this model is further studied.

2 Oscillator Network Model

Let θ_i represents the output phase of the generator node, Then the phase of each node is $\theta_j(t) = \Omega t + \phi_j(t)$, $\phi_j(t)$ is the phase deviation.

Based on the energy conservation theorem, the energy conservation equation of nodes is $P_{source} = P_{diss.i} + P_{acc.i} + P_{trans.i}$, Where P_{source} describes the input power of node i, $P_{acc.i}$ is the cumulative frequency of machine rotation, $P_{diss.i}$ is the loss power of rotor, $P_{trans.i}$ is the transmission power between two nodes.

Due to the rotating friction of the machine, the loss power of the rotor is $P_{diss.i} = k(\theta_j)^2$, k is the coefficient of friction. The accumulated power of machine rotation can be expressed as $P_{acc.j} = \frac{1}{2} I \frac{d}{dt}(\theta j)^2$, I is the moment of inertia of the mechanical rotor. Transmission power between two nodes i and j is directly proportional to the sine value of the phase difference between two nodes and the transmission capacity of the transmission line. Assuming that the phase angle of the voltage is the same as that of the rotor, the transmission power between node i and node j is $P_{trans.ij} = P_{max.ij} \sin(\theta_i - \theta_j)$. So the energy conservation equation of the node is $P_{source.i} \cong I\ddot{\phi}_j\dot{\phi}_j + k(\dot{\phi}_j)^2 - \sum_{i=1}^{N} P_{max.ij} \sin(\phi_i - \phi_j)$.

Suppose that $\left|\dot{\phi}_j\right| \ll \Omega$ then the above formula is

$P_{source.i} \cong I\Omega\ddot{\phi}_j\dot{\phi}_j + k(\Omega)^2 + \left[I\ddot{\phi}_j + 2k\Omega\right]\dot{\phi}_j - \sum_{i=1}^{N} P_{max.ij} \sin(\phi_i - \phi_j)$. If the first derivative of the coefficient is constant, then $\ddot{\phi}_j \ll \frac{2k\Omega}{I}$, then

$I\Omega\ddot{\phi}_j \cong P_{source.i} - k\Omega^2 - 2k\Omega\dot{\phi}_j - \sum_{i=1}^{N} P_{max.ij} \sin(\phi_i - \phi_j)$..

In order to facilitate the research, it is assumed that the transmission capacity of transmission lines is equal, which is represented by the same parameter P_{max}. So the above formula can be written as $\ddot{\phi}_j \cong [\frac{P_{source.i} - k\Omega^2}{I\Omega}] - \frac{2k}{I}\dot{\phi}_j - \frac{P_{max}}{I\Omega}\sum_{i=1}^{N} \sin(\phi_i - \phi_j)$.

The above equation is transformed into $\ddot{\phi}_j \cong P_j - \alpha\dot{\phi}_j + K\sum_{i=1}^{N} a_{ij}(\phi_i - \phi_j)$, then

$P_j = \frac{P_{source.i} - k\Omega^2}{I\Omega}$, $\alpha = \frac{2k}{I}$, $K = \frac{P_{max}}{I\Omega}$. Element a_{ij} can describe the topological structure of the network. When node i and j are connected, $a_{ij} = 1$. Otherwise $a_{ij} = 0$. $P_j > 0$ is represent the generator node, $P_j < 0$ is represent the load node.

For the sake of simplicity, the above formula can be rewritten as

$$\left.\begin{array}{l}\dot{\phi}_i = \omega_i \\ \dot{\omega}_i = \boldsymbol{P_j} - \alpha\omega_i + \boldsymbol{K}\sum\limits_{i=1}^{N} \boldsymbol{a_{ij}}(\phi_i - \phi_j)\end{array}\right\}$$

If coupling index is introduced, the model of oscillator network with coupling index can be expressed as

$$\left.\begin{array}{l}\dot{\phi}_i = \omega_i \\ \dot{\omega}_i = \alpha\omega_i + \boldsymbol{P_i} + \boldsymbol{KP_i}\sum\limits_{i=1}^{N} \boldsymbol{a_{ij}}(\phi_i - \phi_j)\end{array}\right\}$$

In order to quantitatively explain the synchronization degree of the whole phase oscillator, the complex order parameter is introduced by $\boldsymbol{r(i)e^{i\psi(t)}} = \frac{1}{N}\sum\limits_{i=1}^{N} \boldsymbol{e^{i\phi_j(t)}}$,The amplitude $\boldsymbol{r} = |\boldsymbol{Z}|$ of the complex order parameter is called "order parameter", $\psi(\boldsymbol{t})$ is the average phase, $\boldsymbol{r}$ is the center of mass of all phase oscillators on the unit circle, starting from the center of the circle to each node, forming a series of vectors. The sum of these vectors is r, which can describe the degree of synchronization of the whole network. The higher the degree of synchronization, the greater the $\boldsymbol{r}$. Figure 1 is the schematic diagram of the order parameters.

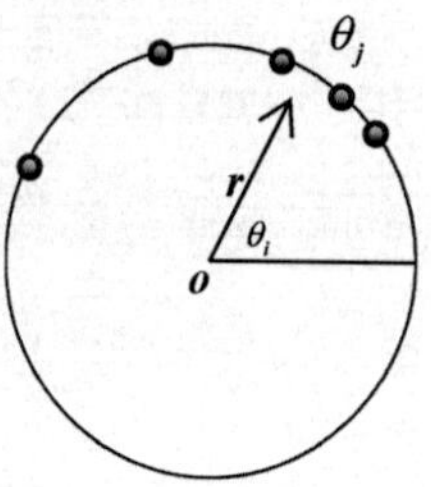

Fig. 1. Schematic diagram of sequence parameters

3 Numerical Simulation

With the help of numerical simulation, the influence of coupling index on networkynchronization ability is analyzed. Assume that the value of power $\boldsymbol{P_i}$ is the following linear expression $\boldsymbol{P_i} = -1 + \frac{2(i-1)}{N-1}, \boldsymbol{i} = 1, 2, \ldots N.$ $\boldsymbol{P_i} \in [-1, 1]$.

Assuming that the network has three nodes, we study the case that the weight coefficients are odd or even. Figure 2 shows the change of order parameter $\boldsymbol{r}$ with coupling strength K when the number of network nodes $\boldsymbol{n} = \mathbf{3}$ and the coupling index is odd.

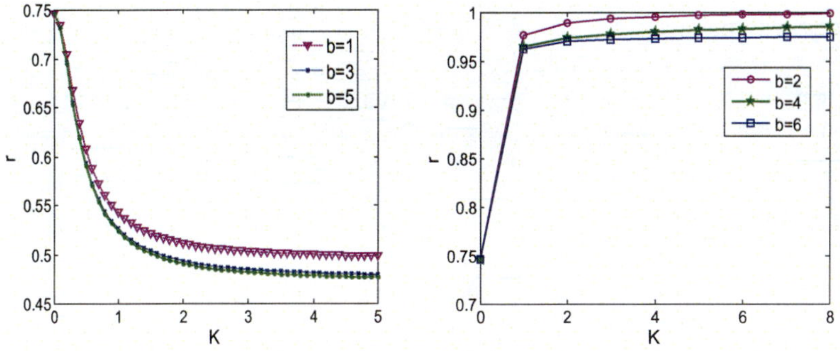

Fig. 2. The node numbers $n = 3$, order parameter r versus the coupling strength K

It can be seen from Fig. 2 that the synchronization pattern of oscillator network is completely different due to the different parity of coupling index. Therefore, we further expand the network scale, observe the evolution process of the order parameters, and discuss the influence of the weight coefficient on the network synchronization ability. Assuming that the number of nodes in the network is, the simulation results are shown in Fig. 3 below

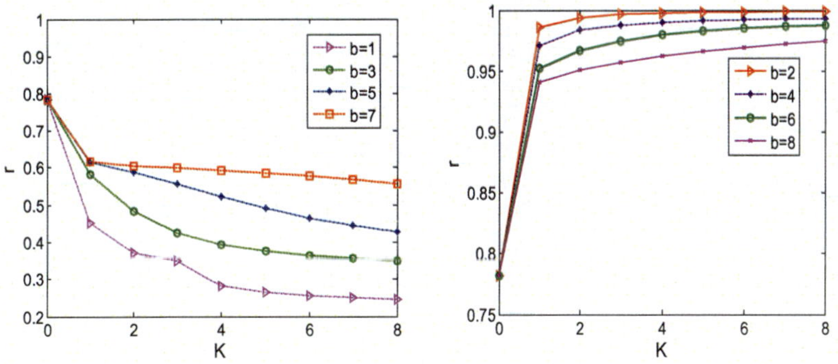

Fig. 3. The node numbers n =**20**, order parameter r versus the coupling strength K

It can be seen from Fig. 2 and Fig. 3 that when the coupling index is even, there is a critical point in the system. When the coupling strength is greater than the critical value, the order parameter will increase with the increase of coupling strength, and the system can achieve better synchronization state. Further by observing our model, we can find that when the coupling index is odd and the power is positive, the coupling strength between the oscillators is a positive value, then the nodes attract and couple each other; when the power is negative, the coupling strength between the oscillators is a negative number; therefore, the interaction between the nodes is repulsive coupling. When the coupling index is an odd number, there are both mutual attraction and mutual repulsion among the oscillators. Because of the mutual competition between the attractor and the repulsor in the process of system synchronization, on the unit circle, the two clusters of oscillators are in the relative position, that is to say, the overall order parameter is the joint action of the order parameters of the attractor and the repulsor, then the corresponding oscillator network The ability to synchronize is reduced.

4 Conclusion

Because the characteristics of the oscillator will affect the synchronization ability of the network, this paper proposes a kind of network model with weight coupling index. By means of numerical simulation, the synchronization behavior of networks with weighted coupling index is discussed. Continuously adjust the coupling index value, and simulate the evolution process of the order parameter with the coupling strength. We find that: when the frequency value form is certain, when the weight re coupling index value is odd, the synchronization ability of the network is weak; when the weight index is even, the order parameter of the network is gradually close to 1, and the network can achieve a better synchronization state, that is, the synchronization ability of the oscillator network is affected by the weight index The impact is relatively large.

References

1. Filatrella, G., Nielsen, A.H., Pedersen, N.F.: Phase model with feedback control for power grid. Eur. Phys. J. B **61**, 485 (2008)
2. Florian, D., Michael, C., Francesco, B.: Synchronization in complex oscillator networks and smart grids. PNAS **110**, 2005–2010 (2013)
3. Delellis, P., di Bernardo, M.: On adaptive bounded synchronization in power network models. IEEE Int. Symp. Cricuits Syst. **6**, 1640–1643 (2012)
4. Lozano, S., Buzna, L., Diaz-, A.: Role of network topology in the synchronization of power systems. Eur. Phys. J. B **85**, 231–239 (2012)
5. Carareto, R., Baptista, M.S., Grebogi, C.: Natural synchronization in power-grids with anti-correlated units. Commun. Nonlinear Sci. Numer. Simul. **18**, 1035–1046 (2013)
6. Witthaut, D., Rohden, M., Zhang, X.Z.: Critical links and nonlocal rerouting in complex supply networks. Phys. Rev. Lett. **116**, 138701 (2016)
7. Girvan, M., Newman, M.E.J.: Community structure in social and biological networks. Proc. Natl. Acad. Sci. USA **99**, 7821–7826 (2002)

Impact of Circular Field in Underwater Wireless Sensor Networks

Syed Agha Hassnain Mohsan[1(✉)], Muhammad Hammad Akhtar[2], Md. Israq Aziz[3], Md. Mehedi Hasan[4], Maryam Pervez[2], Asad Islam[4], and Farhad Banoori[3]

[1] Department of Electrical Engineering, COMSATS University Islamabad (CUI), Islamabad, Pakistan
Hassnainagha@zju.edu.cn
[2] National University of Science and Technology (NUST), Islamabad, Pakistan
[3] South China University of Technology, Guangzhou, China
[4] Beihang University (BUAA), Beijing, China
Hassnainagha@zju.edu.cn

Abstract. Underwater Wireless Sensor Networks (UWSNs) face challenges regarding high propagation delay, limited bandwidth, 3D topology and excessive energy consumptions. In this paper, we propose a routing scheme with circular field for an efficient collection of data packets by using two mobile sinks. Moreover, we have compared results of our scheme with previous implemented schemes which are used to measure the usage of mobile sink in the collection of data packets. We have compared the proposed scheme with current state-of-the-art routing protocols. The statistical significance of this work was analyzed in MATLAB. Marked observations to emerge from our results include an improvement in lifetime, increased throughput, increment in alive nodes and balanced energy consumption. In our view, these results strengthen the validity of our proposed circular field. We observe a significant increase in received packets because maximum nodes are alive till 1500 rounds which provides maximum communication and less chance of creating void holes.

Keywords: Underwater wireless sensor networks · Mobile sink · Routing scheme · Circular field

1 Introduction

A considerable research has been carried out on terrestrial sensor networks in different aspects but currently research community is attracted towards a new era of research in Underwater Wireless Sensor Networks (UWSNs). Human beings are unable to work at high pressure ocean environment for a long time. On the other hand, Terrestrial Wireless Sensor Networks (TWSNs) are feasible but it cannot be considerable for UWSNs because signals transmission is elusive due to higher attenuation of aquatic environment. Underwater Wireless Sensor Network (UWSN) consists of various components such as sensors and vehicles deployed in acoustic area for data collection and collaborative monitoring tasks. Generally UWSNs differ from TWSNs in terms of

M. Tavana et al. (Eds.): IISA 2020, AISC 1304, pp. 803–814, 2021.
https://doi.org/10.1007/978-3-030-63784-2_99

sensors cost, dense deployment of these sensors, communication method and maximum power required for this communication. USWNs are being used in several applications of underwater environment such as observation of chemical pollution, seismic activities, submarine movements, monitoring of marine creatures, distributed tactical surveillance, disaster prevention and military applications. However, UWSNs face challenges regarding narrow bandwidth, 3D topology, long propagation delay, media access control, attenuation, harsh geographical atmosphere, power constraints, losses of connectivity and high bit error rates. Though, research community has provided several techniques to cope with these challenges but still gap lies for research because of variable characteristics of ocean.

This technology has already achieved a great stature among researchers. With the progress in underwater technologies, a growing research inclination has been noticed by industry as well as academia in underwater communications [1–3]. Currently, acoustic signals are in great consideration for underwater applications as they are more feasible for underwater communications with low absorption rate [4]. Long transmission range can be achieved by acoustic communication. However, it faces time-varying signal distortion and transmission losses. In UWSNs, reliability and stability of network has paramount importance. The maintenance of network lifetime in UWSNs is big issue as its difficult to replace those batteries which provide power to sensor nodes. Recently, researchers are working in underwater wireless power transfer to tackle these challenges. In addition, TWSNs routing protocols cannot be implemented to UWSNs because of limited bandwidth, long propagation delay, energy constraints, high mobility and non-uniform 3D topology. Researchers have suggested several routing protocols for UWSNs. Some routing protocols can enhance the stability duration of network on the basis of throughput while others increase throughput on a compromise over packet delay [4]. In these protocols clustering is considered as the pragmatic method for load balancing in UWSNs because sensor nodes are categorized in different groups. Among these sensor nodes, cluster head is used as leading head [5]. Water parameter such as conductivity, permeability and permittivity [6, 7] are different from air medium. It is noticed when both sinks are located at adjacent positions then throughput is affected due to same sensor nodes [8].

Considering above discussion, we were inspired and implemented a routing protocol which uses two mobile sinks without covering same transmission area [9]. The overriding objective of this study is to seek an intelligent use of mobile sinks. We have implemented a previous routing scheme Towards Efficient Sink Mobility (TESM) [9] with considering the circular field while previously it was taken as rectangular field. Authors in [9] discussed that high MUR represents the effective use of mobile sinks and we have compared our obtained MUR results with SEEC and TESM. We have organized our paper as: in Sect. 2, we have made relevant research discussion. Section 3 will focus on motivation behind this present study. We have evaluated the impacts of circular network field in Sect. 4 while results discussion has been carried out in Sect. 5. In last sections, we have presented conclusion and future work.

2 Background and Significance

Since many years' researchers have been focusing to design several UWSNs protocols for qualitative and effective research analysis. Research community is working with the aim of paving a way for new era of underwater actuation and monitoring applications. UWSNs will help us to know our ocean with its envisioned landscape of applications. Researchers have suggested various routing scheme to achieve better results as authors in [10] highlighted capabilities and limitations on the basis of a survey on UWSN with data aggregation. In previous studies, routing scheme named as weighting depth and forwarding area division depth base routing protocol (WDFAD-DBR). Researchers achieved reliable results in sparse deployment by minimizing void holes issue. They proposed balanced energy consumption on the basis of division criteria of forwarding regions. A SEEC protocol is proposed in [8]. In [11] authors presented an energy-balanced algorithm based on energy level partition and depth threshold. In this protocol, mobile sinks can collect data at sparse region and they implied clustering on dense regions. Authors of [12] suggested an energy efficient scheme based on multi-level cluster selection criteria which aims to choose cluster head according to high residual energy and less radius from sink. In [13], network lifetime of UWSNs is maximized by the assistance of sink mobility with geographic routing protocol. A reliable energy efficient routing protocol based on clustering is simulated in [14] for UWSNs. Authors proposed that clustering technique is useful to reduce energy resources consumption for a sensor network. Authors of [17] have used a hybrid approach to maximize network lifetime and minimize the energy holes In order to collect data, these mobile sinks can connect to individual sensor node or full cluster. In research study discussed in [15], authors utilized two mobile sinks to gather data from sparse regions. In this research study, authors maximized the network lifetime in AEDG as AUV is sent to gather data from gateway nodes. The usage of multiple mobile sinks to enhance network lifetime is discussed in [16]. Authors validated that a better network lifetime can be achieved through assigning a predefined path to mobile sinks for movement. While research study in [16] focuses on relation between ocean currents and energy hole problems. This study introduces drift speed calculations. It mentions that we can calculate drift speed at any time if we know the current location of sensor nodes. Mobile sink changes its speed after sensing the ocean current accordingly. It is worth noting that throughput of this proposed scheme is also better. In another article [17], authors presented current research challenges of intrinsic properties, topology control and potentials in UWSNs through a topology control algorithm. In another research study discussed in [10], Hop ID is given to specific sensor nodes according to hope count from sink. In [18], authors proposed a mechanism for data collection involving large scale multihop sensor networks.

3 Motivation

As discussed in [9], authors introduced a novel routing scheme TESM for efficient sink utility in which two mobile sinks are involved to gather required data. Authors introduced a novel metric Mobile Sink Utility Ratio (MUR) which is useful to measure the utility of mobile sinks while data gathering. They have used a rectangular field in their proposed scheme while we focused on circular field in this current study. We have compared our results with routing scheme for efficient sink utility and other previously implemented routing schemes with regard to network lifetime, throughput, residual energy, stability and instability period.

4 Scheme with Circular Field

An efficient use of mobile sinks in UWSNs appears as maximum throughput with minimum energy consumption. In this study we have considered routing scheme TESM discussed in [9] with a different circular field. We achieved good results which will be discussed in next sections.

4.1 Network Model

We consider a network circular field with a logical division of two equal parts. Circular field has a diameter of 100 m. We opted this logical division for an efficient movement of mobile sinks. It is worth noting, we have used three sinks: one is static while other two are dynamic. Two mobile sinks are placed in each divided portion of circular field to gather data. The static sink is located above central position of circular network field. Network model can be seen in Fig. 1.

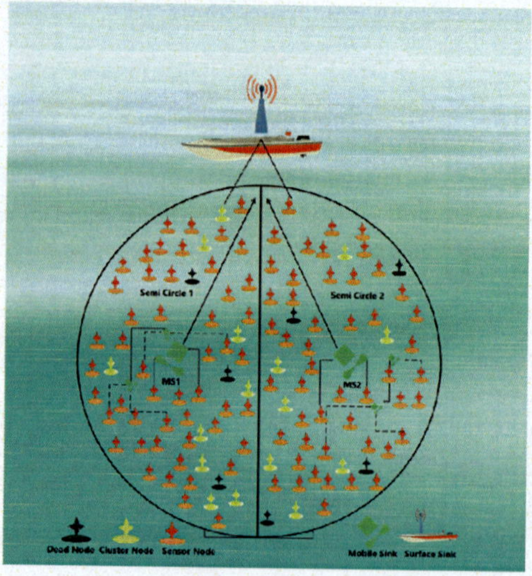

Fig. 1. Network model

4.2 Energy Consumption Model

We represent transmission distance as D and B represents the total size of packet. B_d and R_{disp} show bit duration and radio dissipation respectively. The total energy required for sending B bits packets at transmission distance D can be calculated by following equation in [9].

$$E_{trans}(B,D) = (B * R_{disp}) + (B * B_d) \quad (1)$$

While we can calculate the energy consumed during packet reception by using Eq. 2 [12].

$$E_{rcv}(B,D) = B * R_{disp} \quad (2)$$

We can also obtain the residual energy **Eris** for each sensor node from Eq. 3.

$$E_{ris} = E_{int} - (E_{rcv} + E_{trans}) \quad (3)$$

Here ***Eint*** shows the initial energy level of any sensor node.

4.3 Circular Network Field Configuration

Here, we follow the procedure provided in literature [9] to implement routing scheme with a circular network field. For this purposed work, we divided the circular field into two equal regions. We deployed two mobile sinks MS1 and MS2 at center point of each region at same distance from center of circle. We deployed nodes randomly in circular field. MS1 is present in semi-circle 1 while MS2 is present in semi-circle 2. We move both sinks at an angle of 45° directing upward towards base station deployed at terrestrial region. Corresponding sink receives data from nodes when sink appears within transmission range of these nodes. We have defined transmission range and some others parameters in next section. We need to find sparse and dense areas after network deployment. Cluster is formed where nodes are dense. Random selection of cluster head is made according to highest energy. It is worth noting, in every round each member is provided opportunity to become cluster head. Cluster head can communicate to mobile sink and base station. Remaining nodes without cluster can communicate each other and can send packets to mobile sink as well. The coordinates of mobile sinks can be found from below equations.

$$X = r_i \cos\theta_i \quad (4)$$

$$X = r_i \sin\theta_i \quad (5)$$

We divide circular network field into two semi-circle regions where radius of each region is r = 50 m. We represent the origin of circle as $O(x, y)$.

$$o(x, y) = (0, 0) \tag{6}$$

While considering radius r = 50m we can calculate the area of two semi-circles by following Eq. 7.

$$A_i = \frac{\pi r^2}{2} \tag{7}$$

i = 1,2 for two sub regions. As area calculation for both sub-regions is expressed in above equation, we can also calculate the total area of the whole circular network field. Equation 8 expresses the division criteria for each sector of circular network field.

$$S_n = \int_0^{\pi} \left(\frac{\theta}{360}\right) * \pi * r^2 \tag{8}$$

We represent both mobile sinks as MS1 and MS2. Mobile sinks move 45^0 in upward direction in a tilt way towards base station.

$$MS1 = \sum_{t=0}^{n} (\theta + 45^0/360) * \pi * r_i^2 \tag{9}$$

Similarly.

$$MS2 = \sum_{t=0}^{n} (\theta + 45^0/360) * \pi * r_i^2 \tag{10}$$

Here n depicts total sectors while t = 0 represents that MS1 and MS2 are at center position of first and second sub-region.

5 Results Discussion

To evaluate performance, we compare our results with previously implemented routing protocols SEEC, TESM and DBR. Simulation work was carried out by designing a complete UWSN environment through MATALB [9]. We have summarized simulation parameters in Table 1.

Table 1. Simulation parameters

Parameters	Values
Data rate	16 Kbps
Number of nodes	100
Packet size	50 Bytes
Center frequency	30 kHz
Initial energy of nodes	5 J
Receive power of packet	0.1 W
Transmission power of packet	2 W
Running rounds	3500
No. of sinks	2
Radius of circle	50 m
Transmission range	50 m

For evaluation performance of routing protocols, we consider primary metrics of network stability and instability period, packets received per round, network residual energy, network throughput, mobile sink utility ratio and packets received at sink.

5.1 Network Lifetime and Stability Period

By considering circular field, network lifetime is improved as it can be seen in Fig. 2. SEEC did not use two mobile sinks efficiently as their movement is in sparse regions only. Considering this transmission range, we do not suggest to use mobile sinks based on sparse area. We have restricted both MS1 and MS2 in each semi-circle. Both mobile sinks are moved upward in a tilt position towards base station. Due to this movement, a greater stability period performance is obtained as compared to SEEC, DBR and TESM. The restriction on movements of both mobile sinks in each semi-circle increases the stability period. we can clearly see that our network lifetime results of our proposed work are better comparatively with previous schemes. Figure 2 validates more nodes are alive in our work than SEEC, DBR and TESM.

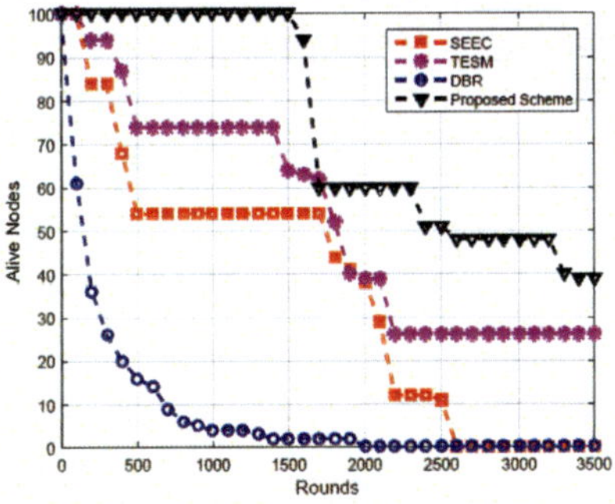

Fig. 2. Network lifetime

5.2 Network Residual Energy

In SEEC, clustering is achieved in top 4 dense regions while using only two mobile sinks in sparse regions. In DBR, energy consumption is more due to selection criteria because forward nodes are selected on lower depths only. In our results, both mobile sinks are restricted in each potion of circular field. If number of dense regions is exceeding highly than sparse regions, mobile sinks movement will increase energy consumption only. The distance between nodes is larger in sparse regions whereas it is smaller in case of dense regions. Residual energy of our proposed work compared with SEEC, TESM, DBR is plotted in Fig. 3.

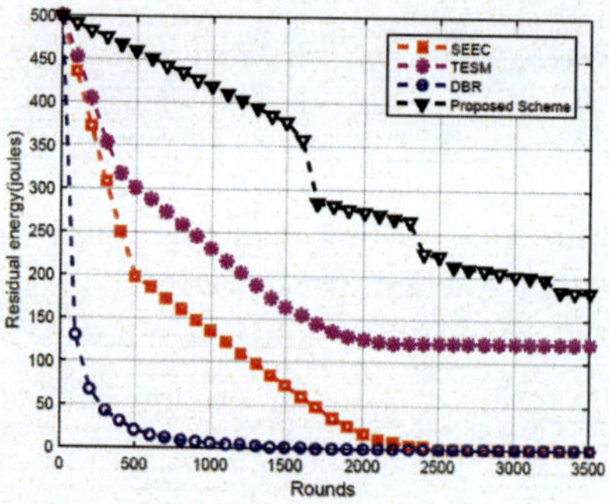

Fig. 3. Residual energy

5.3 Network Throughput

In Fig. 4, the received packets per round at any sink are more than SEEC, TESM and DBR.

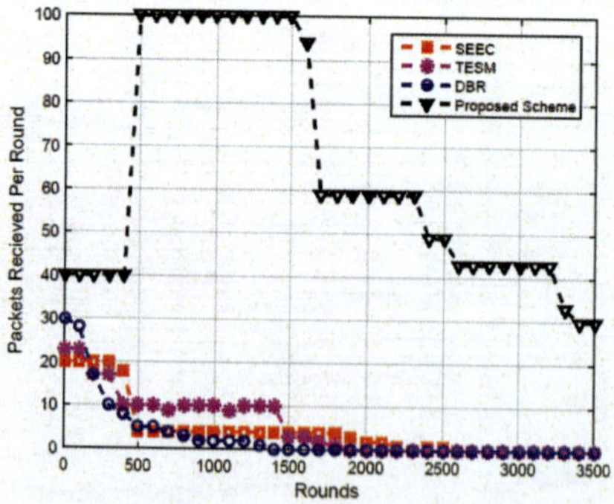

Fig. 4. Packets received per round

We can notice an increased throughput in our proposed work with circular field as sensor are omni-directional as well. In SEEC, mobile sinks movement is according to sparse region and in TESM it is based on selection criteria opted for forwarding nodes.

Figure 5 is showing the overall received packets at sink. Therefore, It gives the addition of received packets in both previous round and next round. Graph shows that we received more packets at sink than SEEC, TESM and DBR.

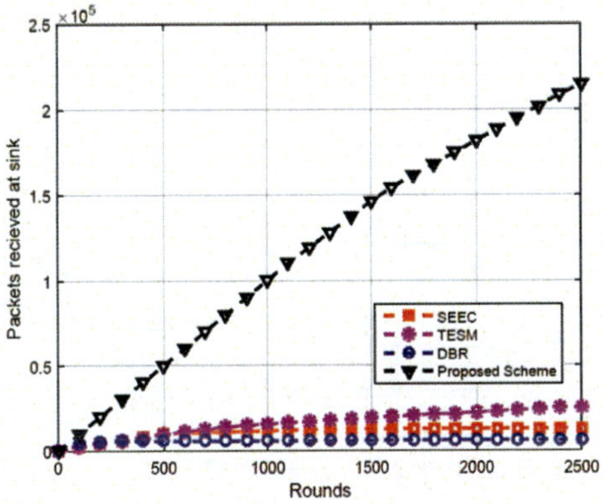

Fig. 5. Packets received at sink

5.4 Mobile Sink Utility Ratio

Figure 6 is showing the usage of mobile sink is high for data packets collection as compared to SEEC and lower than TESM. We did not compare our results with other routing schemes as only SEEC and TESM use mobile sinks among all schemes. Therefore, our proposed work is compared with SEEC and TESM to properly analyze the efficient utilization of mobile sinks. In our results, motion pattern of mobile sinks

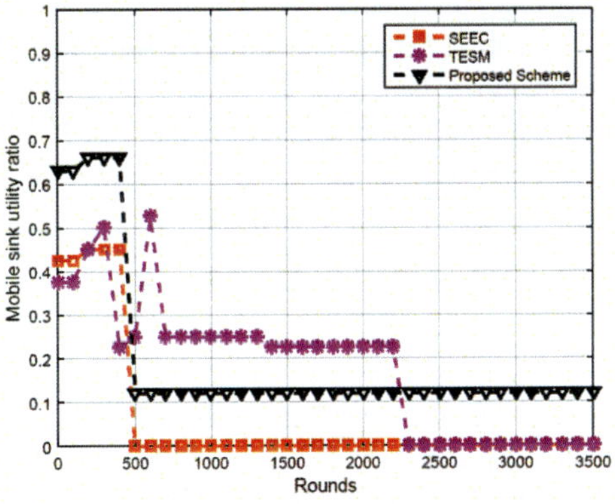

Fig. 6. Packets received at sink

give high MUR than SEEC but less than TESM because mobility pattern performance of TESM is better. In our case, high MUR results as high throughput while we notice the overall packets collected in SEEC by mobile sinks are lower which results as low MUR.

5.5 Dead Nodes

We have shown stability period of implemented routing scheme TESM [8] with a circular field. We achieved good performance in terms of stability period as compared to previous scheme of SEEC, DBR and TESM. In our work, network remains stabilize for a long time with a minimal consumption of energy. Our dead nodes are less than previous scheme as it is presented in Fig. 7.

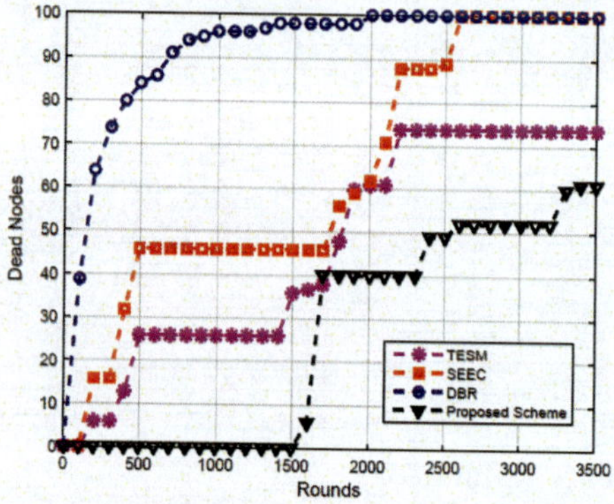

Fig. 7. Stability and instability period of network

6 Conclusion

Results presented in this work illustrate that optimum and efficient use of mobile sinks is crucially important in enhancing network lifetime and throughput. Our work has led us to conclude that a proper distribution of network field and shape is pivotal. In addition, the results of any routing scheme can be improved if we monitor the proper use of resources in the protocol. In this study, we compared our results while considering circular field with previous implemented schemes. To sum up our work, the proposed circular field is better in throughput, network lifetime and balanced energy consumption.

7 Future Work

In future, we are aiming to consider sparsity control in order to improve network lifetime. Though proposed work is energy efficient and we achieve throughput but it will cause delay. We are currently investigating to reduce this delay. We will evaluate the delay in each round in future research work. We will concentrate to minimize interference in dense area which will make reliable delivery of data packets in any network. Holding time phenomenon towards reducing packet drop ratio is deferred to future work.

References

1. Li, N., Cürüklü, B., Bastos, J., Sucasas, V., Fernandez, J.A.S., Rodriguez, J.: A probabilistic and highly efficient topology control algorithm for underwater cooperating AUV networks. Sensors **17**, 1022 (2017)
2. Li, N., Martínez, J.F., Meneses Chaus, J.M., Eckert, M.: A survey on underwater acoustic sensor network routing protocols. Sensors **16**, 414 (2016)
3. Mohsan, S.A.H., Hasan, M.M., Mazinani, A., Sadiq, M.A., Akhtar, M.H., Islam, A., Rokia, L.S.: A systematic review on practical considerations, recent advances and research challenges in underwater optical wireless communication. Int. J. Adv. Comput. Sci. Appl. (IJACSA) **11**(7) (2020)
4. Yan, H.; Shi, Z.J.; Cui, J.H.: DBR: depth-based routing for underwater sensor networks. In International Conference on Research in Networking, pp. 72–86. Springer, Heidelberg (2008)
5. Han, G., Jiang, J., Bao, N., Wan, L., Guizani, M.: Routing protocols for underwater wireless sensor networks. IEEE Commun. Mag. **53**, 72–78 (2015)
6. Hassnain, S.A., Mughal, M.J., Naqvi, Q.A.: Layered chiral spheres with zero backscattering. In: 2019 Photonics & Electromagnetics Research Symposium-Fall (PIERS-Fall). IEEE (2019)
7. Hassnain, S.A., Mughal, M.J., Naqvi, Q.A.: Analysis of effective medium parameters on polarizability of homogeneous chiral sphere. In: 2019 Photonics & Electromagnetics Research Symposium-Fall (PIERS-Fall). IEEE (2019)
8. Sher, A., Javaid, N., Azam, I., Ahmad, H., Abdul, W., Ghouzali, S., Niaz, I.A., Khan, F.A.: Monitoring square and circular fields with sensors using energy-efficient cluster-based routing for underwater wireless sensor networks. Int. J. Distrib. Sens. Netw. **13**, 1550147717717189 (2017)
9. Yahya, A., et al.: Towards efficient sink mobility in underwater wireless sensor networks. Energies **11**(6), 1471 (2018)
10. Goyal, N., Dave, M., Verma, A.K.: Data aggregation in underwater wireless sensor network: Recent approaches and issues. J. King Saud Univ.-Comput. Inf. Sci. **31**(3), 275–286 (2019)
11. Feng, P., et al.: Improved energy-balanced algorithm for underwater wireless sensor network based on depth threshold and energy level partition. EURASIP J. Wirel. Commun. Netw. **1**, 1–15 (2019)
12. Wan, Z., Liu, S., Ni, W., Xu, Z.: An energy-efficient multi-level adaptive clustering routing algorithm for underwater wireless sensor networks. Clust. Comput. **22**, 1–10 (2018)
13. Ahmed, F., Wadud, Z., Javaid, N., Alrajeh, N., Alabed, M.S., Qasim, U.: Mobile sinks assisted geographic and opportunistic routing based interference avoidance for underwater-wireless sensor network. Sensors **18**, 1062 (2018)
14. Huang, X., Sun, S., Yang, Q.: Data Uploading Strategy for Underwater Wireless Sensor Networks. Sensors **19**(23), 5265 (2019)
15. Coutinho, R.W.L., et al.: Underwater wireless sensor networks: a new challenge for topology control–based systems. ACM Comput. Surv. (CSUR) **51**(1), 1–36 (2018)

16. Chen, Y.S., Lin, Y.W.: Mobicast routing protocol for underwater sensor networks. IEEE Sens. J. **13**, 737–749 (2013)
17. Ma, M., Yang, Y.: SenCar: an energy-efficient data gathering mechanism for large-scale multihop sensor networks. IEEE Trans. Parallel Distrib. Syst. **18**(10), 1476–1488 (2007)
18. Wang, X., et al.: Has4: a heuristic adaptive sink sensor set selection for underwater auv-aid data gathering algorithm. Sensors **18**(12), 4110 (2018)

Investigation and Design of Multi-wavelength LED Based Optical Communication System

Syed Agha Hassnain Mohsan[1(✉)], Mirha Malik[2], Shanzay Khan[2,3], Asad Islam[4], Hammad Akhtar[2], Laraba Selsabil Rokia[5], and Muzammil Zubair[3,6]

[1] Department of Electrical Engineering, COMSATS University Islamabad (CUI), Islamabad, Pakistan
Hassnainagha@zju.edu.cn
[2] National University of Science and Technology (NUST), Islamabad, Pakistan
[3] South China University of Technology, Guangzhou, China
[4] Beihang University (BUAA), Beijing, China
[5] East China University of Science and Technology, Shanghai, China
[6] Northwestern Polytechnical University, Xi'an, China

Abstract. Diverse need proper communication techniques as sound waves cannot propagate through water. Research fraternity has designed several underwater communication systems to solve this communication problem. Researchers have opted several approaches for audio signal communication through water. Designed communication systems vary in complex circuitry, audio signal quality, system components and cost. In this present study, we seek to experimentally demonstrate the feasibility of light emitting diodes (LEDs) and solar cell receiver for audio signal transmission through water. Audio signals were transmitted through water by imposing over blue, green and red LEDs. We have examined light scattering and our findings validate LEDs with shorter wavelength show less scattering. This paper also outlines an approach for out-of-sight communication. In our experimental trials, the maximum propagation distance is 25 m by using 450 nm blue light emitting diode with 20-W optical power. This study also includes discussion of collimating lens effect on audio signal transmission.

Keywords: Solar cell · Light emitting diode · Audio signal · Underwater communication · Diver

1 Introduction

In this emerging world of technology, it has become imperative to develop new methods for data transmission. Visible light has gained attention from research community as a good source for communication. It is considered as a reliable technology for data transmission as compared to other technologies. Both LED and LD are good sources are visible light communication. LED and LD have been widely used for communication purpose from several years. LEDs are low cost and lightweight. LEDs feature low power consumption and simple design. Recently, light emitting diodes have been widely used for several underwater communication systems. In this study,

M. Tavana et al. (Eds.): IISA 2020, AISC 1304, pp. 815–825, 2021.
https://doi.org/10.1007/978-3-030-63784-2_100

we discuss the advantages of LEDs for diver-to-diver communication in underwater environment. There is a growing need to design communication system with good performance, higher data rate and large propagation distance to facilitate diver to diver communication as shown in Fig. 1. Generally scoba divers are trained to use hand gestures for communication. They also use underwater writing board to deliver important information. In this study, we consider LED based underwater communication system as it plays an essential role by offering an efficient and secure communication.

Fig. 1. Underwater divers [1]

It is a challenging task to propagate audio signal in underwater environment. Researchers are working to design new communication systems and to improve existing technologies for diver to diver communication. Several underwater communication systems have been developed. Such communication systems have different complexity, cost, components and voice quality and distance range. Acoustic signals are considerable for underwater data transmission but it looks inappropriate for diver to diver communication due to complexity. Some challenging factors appear in terms of weight and cost. Many researchers have used optical systems for underwater communication with data rate in Gbps [2] but these systems are not suitable due to modulation, demodulation techniques and complicated data processing methods. It is very essential to maintain diver to diver communication in case of emergency and deliver any information in urgency. Divers should keep a communication link with each other and surface for both safety and logistical support. Researchers are working to design underwater communication system which can bring a promising revolution to support diver to diver communication with best performance. However, these researcher must ensure that a communication system for this purpose must be reliable, simple, low weight and cost effective. Basic requirements for an efficient optical communication system are displayed in Fig. 2.

Fig. 2. Basic requirements

Underwater Optical wireless communication proves to be a good solution for high data rates, low energy consumption and cost effectiveness. Research communication has shown inclination to design optical communication based on LEDs due to high efficiency, low power consumption, long lifetime and secure communication for indoor visible light communication systems. Some characteristics of LEDs are shown in Table 1.

Table 1. LED characteristics

Characteristics	LED
Modulation bandwidth	<200 MHz
Optical spectral width	25–100 nm
Output beam divergence	Wide (0.50)
Cost	Low
Temperature dependency	Little
Lifetime	Long
Coherence	Incoherent
Reliability	High
Eye safety	Safe
Circuit design	Simple

2 Relevant Work

Jan and Peter in [8] used phosphor converted green LEDs for optical communication. Researcher in [10] used a promising technique to increase underwater transmission by considering a freeform lens. In [3] authors have used a good approach for optical communication based on LEDs and photo sensor technology. William C. et al. achieved Mbps data rate by introducing a software defined radio system by using laser and LED [4]. Usually underwater communication is different from air due to varying conductivity, permittivity and permeability [5, 6] parameters. A blue LED based transmitter and photodiode received was designed in [7] to transmit 1 Mbps data at 10 m distance. A high speed UOWC system was designed by using blue GaN-based micro LEDs [8]. Researchers in [9] designed optical communication system through GaN-based micro LEDs array in a series connection. Significant literature discussed has been made for underwater optical communication system in a Doctoral thesis by Giovanni Giuliano [10]. An underwater optical communication system based on RGB LEDs to achieve 2 Mbps data rate at 1 m distance [12]. We have comprised Table 2 which shows a comparison of research work implemented by research fraternity with different data rates and distances.

Table 2. Relevant studies

First Author	Distance (m)	Data rate (Mbps)	Source
Mangrio [12]	1	2	LED
Cossu [13]	2.5	58	LED
Cochenour [14]	3.6	5	LED
Simpson [15]	7.7	5	LED
Chancey [16]	12.0	10	LED

3 Underwater Communication

Current underwater optical system consider acoustic waves as EM waves cannot travel well. However, acoustic waves have limitation in aspects of limited bandwidth and slow speed. Acoustic communication is preferred for long distance but optical communication gives better performance for short distance. Light signals lose energy and change direction due to water particles. These challenging factors limit the range of communication system. In optical region, waves are highly faded due to scattering and absorption. The extinction coefficient c of water medium is the attenuated beam over the complete path length. Extinction coefficient is the transmission wavelength which minimizes the propagation loss and it varies for different types of water. It is obtained by adding both absorption and scattering coefficients a and b [10, 11].

$$c(\lambda) = a(\lambda) + b(\lambda) \tag{1}$$

Table 3 shows the color frequencies and respective wavelengths. Different color lights get absorbed at different water levels, where every water level has related spectrum characteristics.

Table 3. Light colors characteristics

Colour	Frequency	Wavelength
Blue	631–668 THz	450–475 nm
Green	526–606 THz	495–570 nm
Yellow	508–526 THz	570–590 nm
Orange	484–508 THz	590–620 nm
Red	400–484 THz	620–750 nm

Table 4 provides the scattering, absorption, and extinction coefficient for different types of types of waters. Thus, we conclude demonstrating a UOCS closer to shore is more challenging than wide open ocean.

Table 4. Optical properties of ocean water types

Water Type	a (m−1)	b (m−1)	c (m−1)	−10 dB distance (m)
Pure water	0.041	0.02	0.043	53.55
Clear sea water	0.114	0.037	0.151	15.25
Coastal water	0.179	0.220	0.399	5.77
Harbour water	0.366	1.829	2.159	1.05

4 Proposed System

Our designed system size is compatible with underwater divers requirement as it has small size, low cost and less weight. Figure 3 presents the schematic diagram of our proposed system.

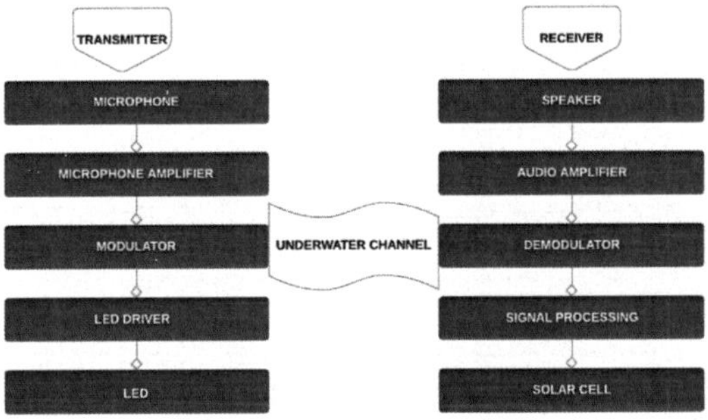

Fig. 3. System components

Our proposed audio transceiver system can efficiently transmit an audio signal via light and reproduces an input signal at receiver side. We used LEDs as our incident light sources, to transmit an audio signal, and compared performance based on the quality of output signal in terms of different wavelengths LEDs and distance between transmitter and receiver. We have used blue, green and red LEDs of 450 nm, 520 nm and 658 nm respectively with 10-W and 20-W optical power ratings. In this system we have used a solar panel as receiver. The experiments were performed in lab test tank with water salinity of 900 mg/L and common room temperature 25°. Our designed system provides transmission in a free underwater optical link. Our designed receiver component consists of solar panel, an audio amplifier and a speaker. We used common silicon solar panel of 55 * 55 mm displayed in Fig. 4 as receiver to collect large number of photons emitted from incident LED light. The lightweight solar panel is used to convert light emitting from LEDs into electrical signals. Amplification is performed on electric signals received from solar panel by using an amplifier based on LM386 IC and audio signals are obtained by the speaker. This designed system and components are lightweight and divers can carry easily.

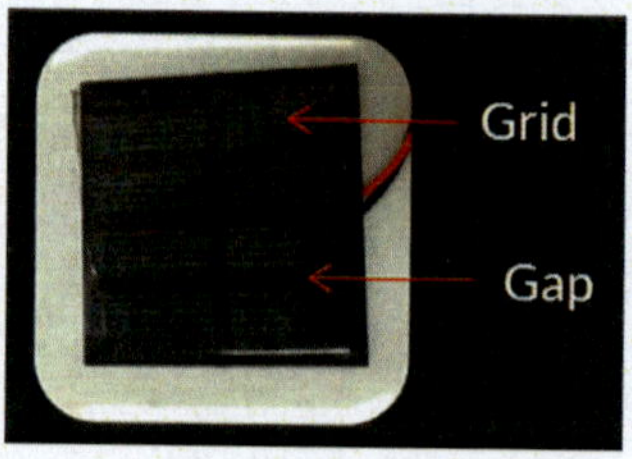

Fig. 4. Solar cell

5 Experimental Setup

In order to investigate the use of solar cell as an optical receiver, initially we performed a basic experiment to check frequency response. We consider a white LED of 15 mW as an optical transmitter in air medium awhile we used a solar cell on receiver side.

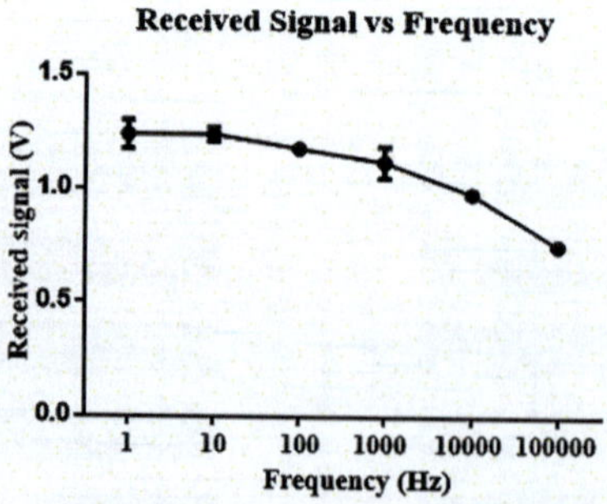

Fig. 5. Frequency response

Figure 5 is showing the frequency response of solar cell which validates that low frequency optical signals can be received by using silicon solar cell receiver.

6 LEDs Without Collimating Lens

First, we investigated our results without using any lens. Graph displayed in Fig. 6 gives the received voltages when LEDs of different wavelengths illuminate the solar cell.

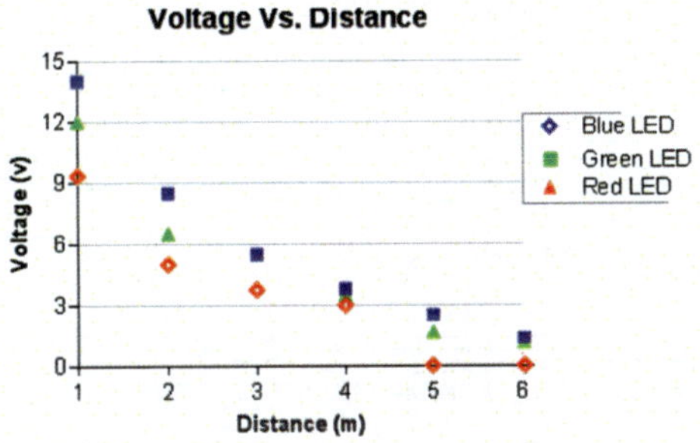

Fig. 6. Generated voltages (10-W)

Voltage variation is a factor of distance between transmitter and receiver. At these distances we receive good audio sound. In our experiments, we have used 450 nm blue LED, 520 nm green LED and 658 nm red LED. We performed experiments at optical powers 10-W and 20-W respectively. We were able to receive audio signal at 7 m, 6.2 m and 4.8 m for blue, green and red LED respectively with 10-W. Similarly, Fig. 7 gives voltage results across distance with 20-W LEDs. We calculated that audio signal can be received up to 7.6 m, 6.55 and 5.3 m for blue, green an d red LEDs respectively.

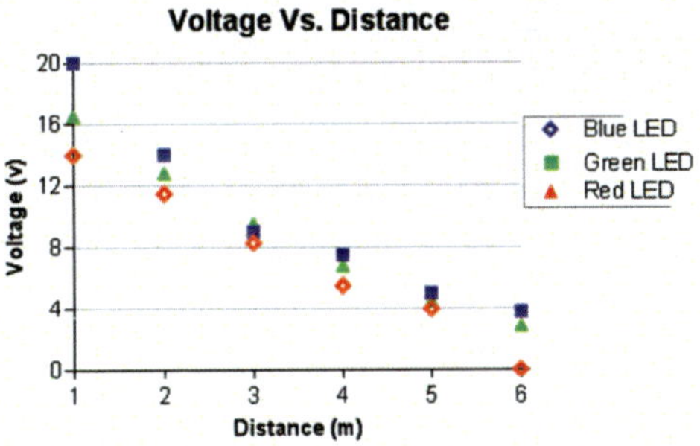

Fig. 7. Generated volts (20-W)

7 LEDs with Collimating Lens

In previous section, we obtained results with LEDs to transmit audio signal without collimating lens. It is useful if we want to transmit data at few centimeters. However, if we want to transmit data at larger distance then we can use collimating lens which decreases divergence angel and light can propagate to larger distance. We repeated same experiment by considering a collimating lens adjusted in front of LEDs in order to reduce divergence angel.

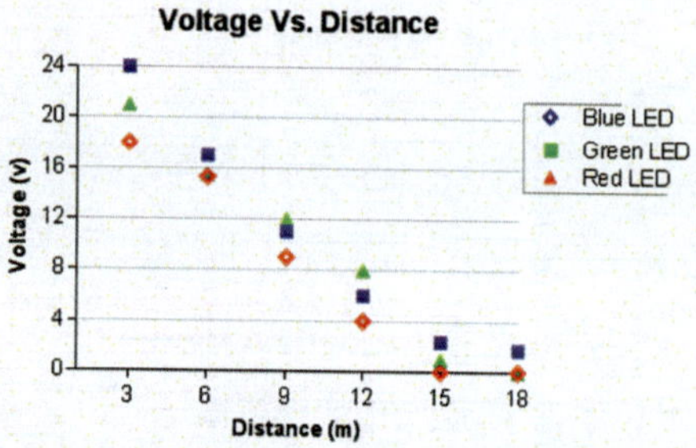

Fig. 8. Generated volts with collimating lens (10-W)

Light can propagate at larger distance with involving collimating lens. We were able to hear audio signal at 18.25 m, 15 m and 12.4 m for blue, green and red LEDs with optical power 10-W as shown in Fig. 8. We repeated same experiment with 20-W LEDs and received voltage results are shown in Fig. 9. We concluded that audio signal can travel at a larger distance of 25 m, 19.2 m and 16 m for blue, green and red LED respectively.

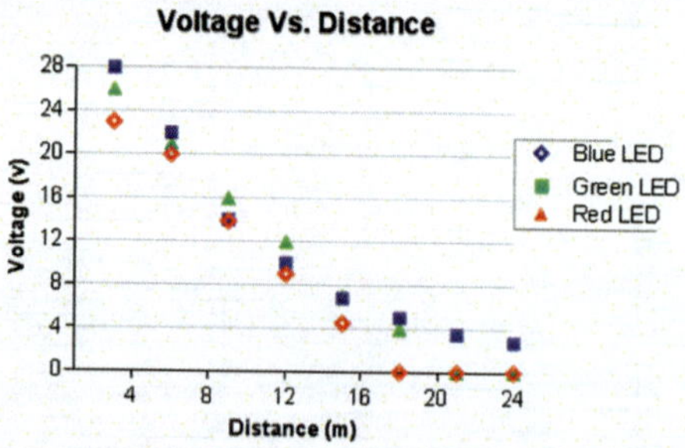

Fig. 9. Generated volts with collimating lens (20-W)

It is noticed that a high optical power is required for red and green LED as compared to blue LED for conversion of optical signal into electric signal. That is why we were unable to receive audio signal even after receiving some voltage for green LED. From all these experiments, we can conclude that red LED covers minimum distance. Scattered light due to molecules shows an inversely proportional behavior against wavelength. So blue light can be transmit over larger distances than green and red light emitting diodes. It is also worth noting that higher optical power is required for receiver to convert red and green light into electrical signals. We conclude that received audio signal captured by red and green LED light is weaker and it is a major limitation factor to hear voice. In order to confirm the high bandwidth capability of any optical systems, we used red LED of 658 nm. The frequency can be obtained by using Eq. 2, in which speed of light is c, frequency is f and wavelength is λ.

$$f = c/\lambda \tag{2}$$

8 Non-line-of-sight Communication

An essential feature to use this communication system is that information can be received even both transmitter and receiver are out of phase. When incident light propagates through water, light photons face both absorption and scattering. The scattered photons carry same data as non-scattered photons. Thus, we can transmit data even when LEDs and receiver solar panel are out of sight. We have repeated same experiment in this section with respect to lateral angel. Result graphs are shown in Fig. 10 and 11.

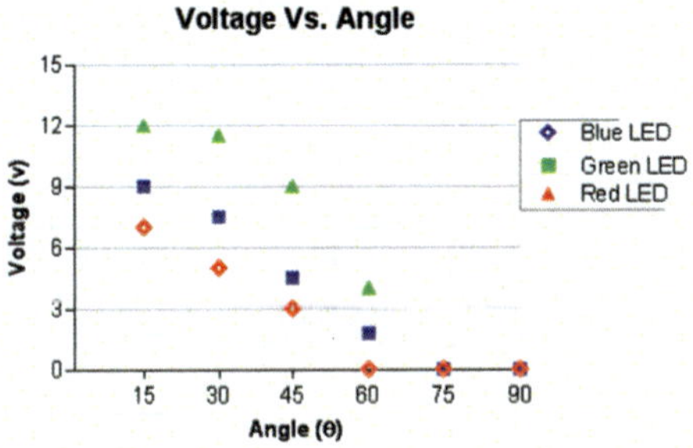

Fig. 10. Generated volts at different angles (10-W)

Our results validate that some photons are captured by solar panel even when an LDs are solar panel are directed away. We can hear clear sound at 45°. We receive audio signals beyond this angel but quality is not good at 60°. We performed this experiment by changing angular position of our receiver solar panel.

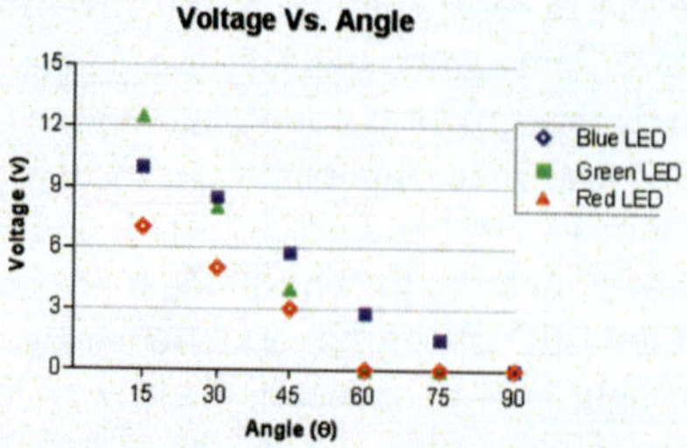

Fig. 11. Generated volts at different angles (20-W)

9 Conclusion

We conclude from our experiments that solar cell can receive low frequency signals. Light emitting diode can transmit audio signals from 30 Hz to 30 kHz. We investigated that distance variation has greater influence on received audio signal quality. Our proposed system is lightweight and cost effective. Moreover, divers can use this system to transmit data for underwater communication. Our findings include that receiver solar can capture scattered photons. Results verify that communication is possible even when receiver and transmitter are being out of sight up to 60°. Proposed system can be improved to capture more light photos with less power consumption. For the future work, we will consider laser diodes of different wavelengths. If such system is further improved then it can be used for underwater exploration task in AUVs.

References

1. Yagimli, M.: Design of laser based underwater communication system. J. Naval Sci. Eng. **7** (2), 1–0 (2011)
2. Lu, H.H., et al.: An 8 m/96 Gbps underwater wireless optical communication system. IEEE Photonics J. **8**(5), 1–7 (2016)
3. Silva, B.M.C.: Underwater optical communication: an approach based on LED (2015)
4. Cox, W.C., Simpson, J.A., Muth, J.F.: Underwater optical communication using software defined radio over LED and laser based links. In: 2011-MILCOM 2011 Military Communications Conference. IEEE (2011)
5. Hassnain, S.A., Mughal, M.J., Naqvi, Q.A.: Layered chiral spheres with zero backscattering. In: 2019 Photonics & Electromagnetics Research Symposium-Fall (PIERS-Fall). IEEE (2019)
6. Hassnain, S.A., Mughal, M.J., Naqvi, Q.A.: Analysis of effective medium parameters on polarizability of homogeneous chiral sphere. In: 2019 Photonics & Electromagnetics Research Symposium-Fall (PIERS-Fall). IEEE (2019)
7. Brundage, H.: Designing a wireless underwater optical communication system. Dissertion, Massachusetts Institute of Technology (2010)
8. Tian, P., et al.: High-speed underwater optical wireless communication using a blue GaN-based micro-LED. Opt. Exp. **25**(2), 1193–1201 (2017)
9. Arvanitakis, G.N., et al.: Gb/s underwater wireless optical communications using series-connected GaN micro-LED arrays. IEEE Photonics J. **12**(2), 1–10 (2019)

10. Giovanni, G.: Underwater optical communication systems. Dissertion University of Glasgow (2019)
11. Mohsan, S.A.H., Hasan, M.M., Mazinani, A., Sadiq, M.A., Akhtar, M.H., Islam, A., Rokia, L.S.: A systematic review on practical considerations, recent advances and research challenges in underwater optical wireless communication. Int. J. Adv. Comput. Sci. Appl. (IJACSA) **11**(7), 162 (2020)
12. Mangrio, H.B., et al.: Effects of modulation scheme on experimental setup of RGB LEDs based underwater optical communication. Wireless Pers. Commun. **106**(4), 1827–1839 (2019)
13. Cossu, G., et al.: Experimental demonstration of high speed underwater visible light communications. In: 2013 2nd International Workshop on Optical Wireless Communications (IWOW). IEEE (2013)
14. Cochenour, B., Mullen, L., Laux, A.: Phase coherent digital communications for wireless optical links in turbid underwater environments. In: OCEANS 2007. IEEE (2007)
15. Simpson, J.A., et al.: 5 Mbps optical wireless communication with error correction coding for underwater sensor nodes. In: OCEANS 2010 MTS/IEEE SEATTLE. IEEE (2010)
16. Chancey, M.A.: Short range underwater optical communication links (2005)

Research on High-Speed Data Acquisition System Based on PCIE

Xinxin Sun, Jun Yang(✉), Juan Li, Fenxian Tian, and Shengkai Wang

School of Information Science and Engineering, Yunnan University, Kunming 650091, Yunnan, China
1015462008@qq.com

Abstract. As a third-generation I/O technology, PCIE is widely used in data acquisition systems. The traditional I/O and PCI bus bandwidth is small, the transmission rate is slow, and the compatibility is insufficient. This paper uses the excellent transmission performance of the PCIE interface to realize high-speed communication between the peripheral device and the host, and utilizes the reconfigurability of the FPGA technology. Strong, scalable, high reliability, low cost, using hardware description language verilog HDL design to achieve ADC module, FPGA timing control and PCIE module, DMA mode for data transmission between FPGA and PC, and finally low The PCIE high-speed data acquisition system was validated on ALTERA's Cyclone IV GX family of chips. After testing, the data acquisition system captures an average reading speed of 190 MB/S, which meets the design requirements. It realizes PCIE high-speed data acquisition system with low-cost FPGA, which has certain practical value and practical significance.

Keywords: FPGA · PCIE · DMA · High speed data acquisition

1 Introduction

Data acquisition systems are widely used in radar, image, military, industrial and mechanical fields. As the application range of data acquisition systems continues to expand, the requirements for data acquisition systems will become higher and higher, which will lead to complex operations and frequent data transmission on some external devices. Between high-speed exchanges, a new bus technology PCI Express was introduced to replace PCI, PCI-X and AGP [1]. The high integration of FPGA, high speed and low power consumption, and short development cycle are the key points of this design based on PCIE high-speed data acquisition system [2]. In this design, the data transmission between FPGA and PC adopts the direct memory access technology (DMA) based on PCIE bus for transmission.

This paper presents the FPGA implementation plan, using the low-cost Altera company's Cyclone IV GX series of fully programmable system-on-chip as a design platform, the signal conditioning circuit first, then design the ADC module, FPGA timing control unit and DDR2 SDRAM cache Finally, the PCIE interface is used for transmission with the PC. Finally, the PCIE high-speed data acquisition system with practicality, stability and versatility has been completed. The PCIE high-speed data

M. Tavana et al. (Eds.): IISA 2020, AISC 1304, pp. 826–835, 2021.
https://doi.org/10.1007/978-3-030-63784-2_101

acquisition system of this design has faster instruction processing speed and less logic resource occupation than the data acquisition system designed by the traditional single chip, and the system is highly integrated and reconfigurable, with certain application value and practical significance.

2 PCI Express

The PCIE bus uses a serial connection and uses data packets for data transmission. This structure effectively removes some sideband signals present in the PCI bus [3]. Unlike the PCI bus connection, the PCIE bus uses an end-to-end connection. In the PCIE bus, in the process of receiving and sending data packets, it needs to pass through multiple layers, including the transaction layer, the data link layer, and the physical layer. The transaction layer is the upper layer. Its main function is to transmit and receive transaction layer packets (TLPs) [4]. The PCIE protocol and configuration packet information are designed at the transaction layer according to the PCIE interface. The data link layer is the middle layer, and its main responsibility is to provide a reliable mechanism for the exchange of TLPs. The physical layer interfaces the data link layer with signaling technology for link data exchange, and is responsible for framing, deframing, scrambling, decompression, 8b/10b encoding, and encoding of TLPs and DLLPs.

At present, many companies provide PCIE solutions, which are divided into FPGA-based and ASIC-based solutions according to their implementation. The effectiveness and compliance of these two solutions are beyond doubt. Because PCIE IP cores are automatically generated in the configuration of FPGA chips [5], FPGAs enable designers to create designs that perfectly match their needs and have greater flexibility. Therefore, this article adopts the FPGA chip of Cyclone IV GX introduced by FPGA manufacturer Altera, which integrates a customizable PCIE IP hard core module and uses × 1 channel PCIE interface.

3 System Design

The designed PCIE high-speed data acquisition system uses a modular design idea. The system first inputs the external analog signal, and then adjusts the analog signal to the appropriate receiving frequency of the AD7606 through the signal conditioning circuit,

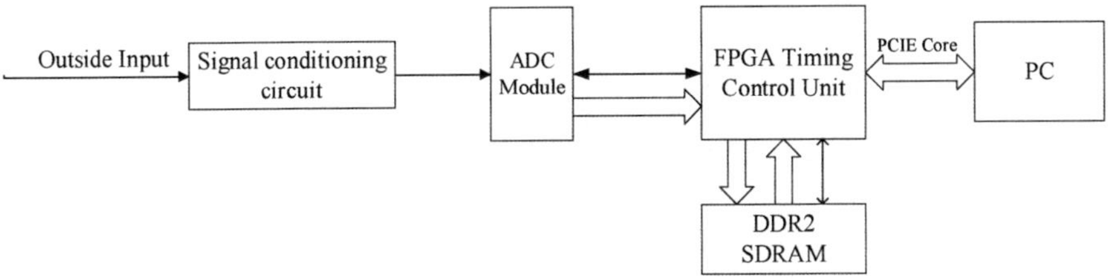

Fig. 1. PCIE high-speed data acquisition system design block diagram

the ADC module performs A/D analog-to-digital conversion, and FPGA timing The control unit is finally connected to the PC via DMA transmission via the PCIE bus. Finally completed the PCIE high-speed data acquisition system. FIFO buffer capacity cannot keep up with the data transmission of the PCIE bus, then the FPGA memory interface is hung with DDR2 SDRAM particles, as DMA buffer. The design block diagram of this system is shown in Fig. 1.

3.1 ADC Module

This paper designs an improved ADC high-precision, low-latency control firmware to solve the problems of low ADC sampling accuracy and high transmission delay in traditional PCIE-based high-speed data acquisition systems. The ADC module implemented in this design is implemented with the AD7606 chip. By studying the working principle of the AD7606 analog-to-digital conversion chip, the control logic of the AD7606 is designed, and programmed with the verilog hardware description language to finally realize the A/D analog-to-digital conversion function.

Firstly, the input signals of clk, rst_n, ad_data, ad_busy, first_data and output signals such as ad_convstab, ad_rd, ad_cs, ad_reset are designed. Then the verilog language design is performed, and finally the RTL diagram of the A/D module is obtained after compilation. The module generation diagram of the AD7606 is shown in Fig. 2.

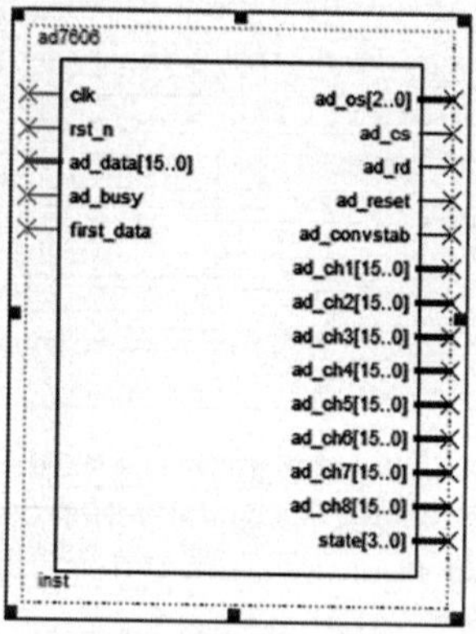

Fig. 2. AD7606 module generation diagram

3.2 FPGA Timing Control Unit

FPGA adopts Altera's Cyclone IV GX series chips, and implements ADC controller, FIFO controller, clock management module, and PCIE control module through FPGA. After the FPGA receives the signal collection, it sends the collected data to DDR2 SDRAM. After the host computer sends out the collection instruction, the data in DDR2 SDRAM will be transmitted to the PC through the PCIE bus through the FIFO in FPGA. In this design, DDR2 SDRAM is plugged into the FPGA as a cache, which provides great convenience for high-speed data acquisition. The block diagram of FPGA's timing control unit is shown in Fig. 3.

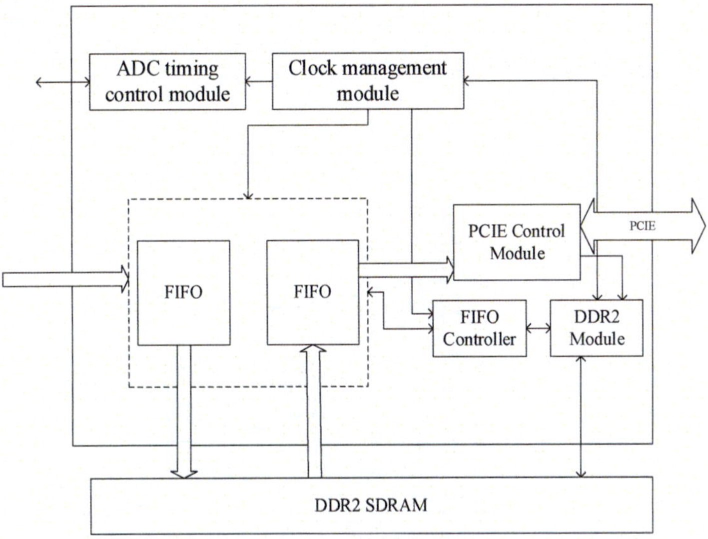

Fig. 3. FPGA timing control unit

3.3 PCIE Core Module

The PCIE core module designs a DMA controller based on the PCIE transaction layer, data link layer, and physical layer, which can perform batch transfer independently of the processor. A DMA controller is designed inside the FPGA, and the DMA mode is used to achieve high-speed transmission between the system and the PC. The main function of the DMA controller is to receive and send transaction-level data packets with various transaction types. The PCIE IP core module is transmitted through the 64-bit Avalon-ST interface. As shown in Fig. 4, the basic control signals of the PCIE IP core.

Fig. 4. Basic control signals of PCIE IP core

The PCIE IP core module is generated by MegaWizard in Quartus II. Figure 5 is a diagram of PCIE IP core RTL components. It can be seen from the PCIE IP core component diagram generated in Fig. 6 the generated signals can be divided into: RX data reception signal, TX data transmission signal, clock signal, configuration signal, and interrupt signal. Figure 6 is a PCIE module structure diagram. This article designs the PCIE core module based on the PCIE IP core. The signals of the RXPROC, RSSTINF, TXPROC, and FIFO modules cooperate with each other to realize the PCIE bus function and the PCIE core module. The RSSTINF module is that the PCIE IP core provides signals to be integrated in a certain order and input to the RXPROC module in the form of FIFO. FIFO is a buffer module. The following focuses on RXPROC and TXPROC.

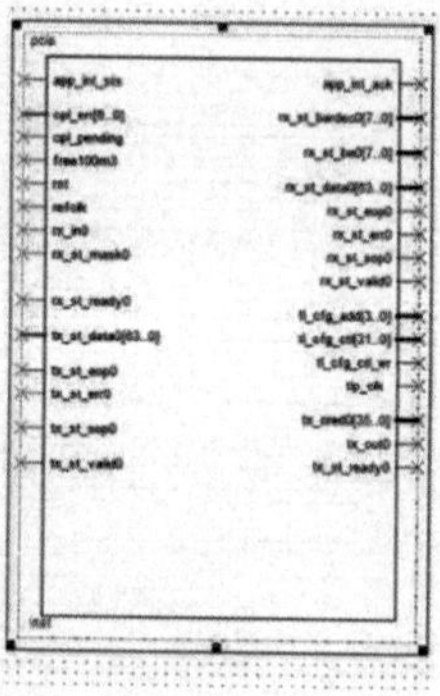

Fig. 5. PCIE IP core component diagram

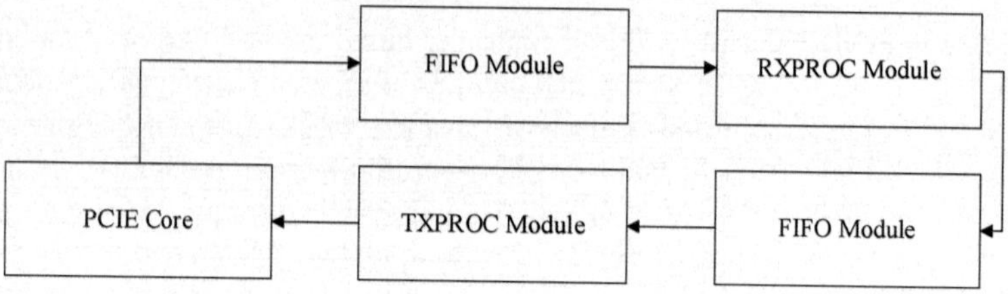

Fig. 6. PCIE module structure diagram

3.4 RXPROC Module

The RXPROC module is a processing module for RX data packets. It is responsible for the DMA read operation state machine and parsing the data packets transmitted by the IP core. The DMA read operation state machine is shown in Fig. 7. The RXPROC module mainly implements the processing of receiving TLP data packets. It splits the control signal collection rxqd transmitted by the RSSTINF module through the FIFO module to obtain rx_data, rx_be, rx_bardec, rx_valid, rx_err, rxsop, rxeop signal, and the packet header format. And packet length information. Then parse the relevant contents of the packet header format: tc, fmt, tysel, td, ep, attrib, lenth, lenth, reqid, tag, fbe. Memory read, memory write, complete packet format and packet length are achieved through the RXPROC processing module. Based on the above, the Verilog hardware description language is written and the RTL components that generate RXPROC are run as shown in Fig. 8.

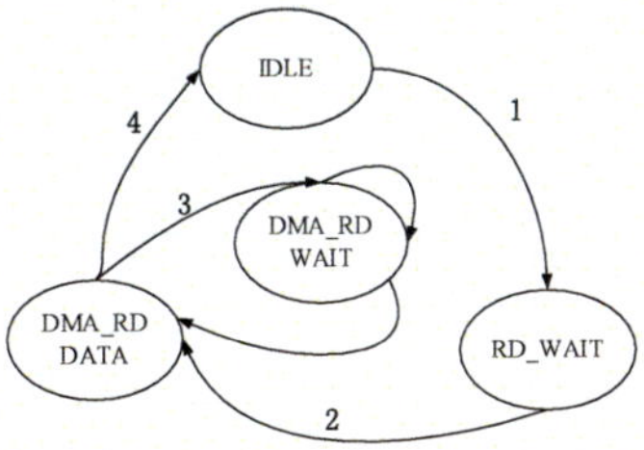

Fig. 7. DMA read operation state machine

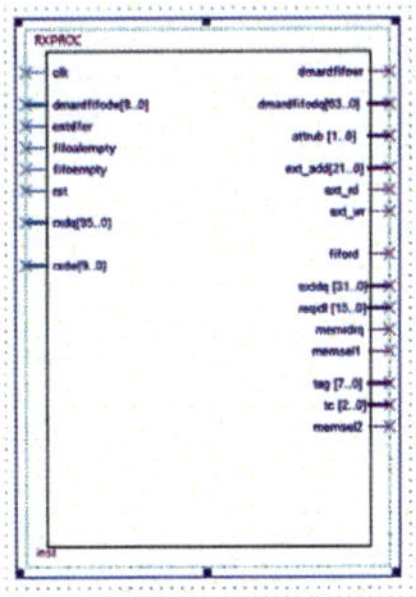

Fig. 8. RXPROC Module component diagram

3.5 TXPROC Module

The TXPROC module is a processing module for TX data packets. It is responsible for the realization of the DMA write operation state machine and the request state machine in the DMA read operation. The DMA write operation state machine is shown in Fig. 9. When responding to the bus operation command through the PCIE transaction layer TLP interface, the user also encapsulates the data into a corresponding format TLP packet, and then sends it through the PCIE IP core. TLP is called the transaction layer data message. The PCIE IP core module provides logic control signals. When a processor or other PCIE device accesses the PCIE device, the transmitted data message is first encapsulated into one or more TLPs through the transaction layer. In order to send out through the various layers of the PCIE bus, finally generate several TLP data packets of memory read, memory read completion, and memory write. PCIE data transmission is DMA logic control. The method of data transmission in PCIE data acquisition system is to use DMA transmission DMA register, interrupt, control signal line and other methods for transmission. Program according to the above DMA control logic and TX data transmission working principle According to the different types of TLP packets sent to design different processing methods, write Verilog hardware description language. After compiling, run the RTL component that generates the TXPROC module as shown in Fig. 10.

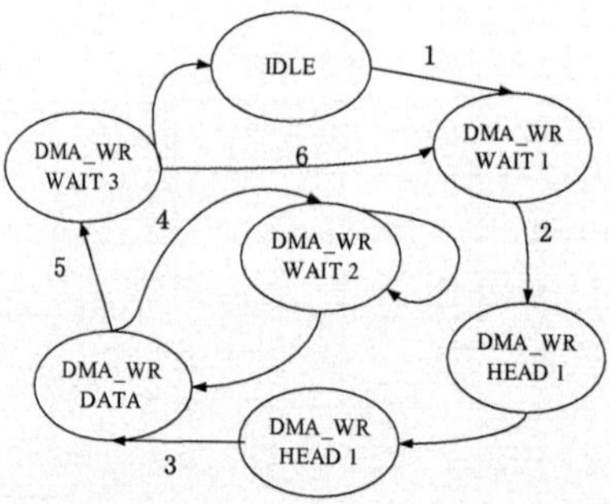

Fig. 9. DMA write operation state machine

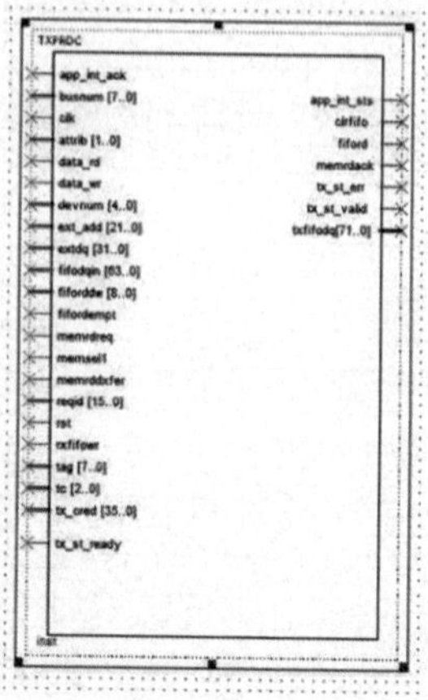

Fig. 10. TXPROC Module component diagram

4 Hardware Driver Design

After the PCIE data acquisition hardware design is completed, write a driver program to allow the operating system in the PC to recognize the PCIE data acquisition module and perform corresponding hardware operations on the data acquisition module. Write command control word, interrupt response and DMA to read collected data to FPGA respectively.

This paper designs the PCIE device driver to realize the following parts: Initialize the device module, device open module, read and write and control module, interrupt processing module and DMA read and write module. Use DriverStudio and DDK software developed by Microsoft to design the hardware driver. The DDK is used to develop the underlying driver of the system. The code is written in assembly. The DriverStudio combined with the DDK combines the access written by the driver into a class in an object-oriented manner, plus the DriverWizard designed by the driver code generation wizard. And DriverAgent, DriverWorks and other powerful tools, which simplifies the difficulty of driver development, reduces workload, and has good flexibility.

5 Test Results and Analysis

5.1 Simulation Test

PCIE high-speed data simulation diagram is shown in Fig. 11. PCIE has finished initializing, the tx_st_sop0 signal changes from high level to low level, that is, a TLP data packet is started. The tx_stream_read0 signal starts to write data to the data port, and the header of the TLP is the first two data. When rx_st_sop0 changes from low level to high level, the start of the TLP packet begins. The purple line of the rx_steam_ready0 signal is at a high level. When it is at a high level, latching is interesting. Both tx data transmission and rx data reception operations are running normally. After testing, the PCIE data acquisition simulation diagram shows that the correct reading and writing is completed. The result of data transmission on the ST port is consistent with expectations, and the data transmission of the PCIE bus realizes the requirement of high-speed acquisition.

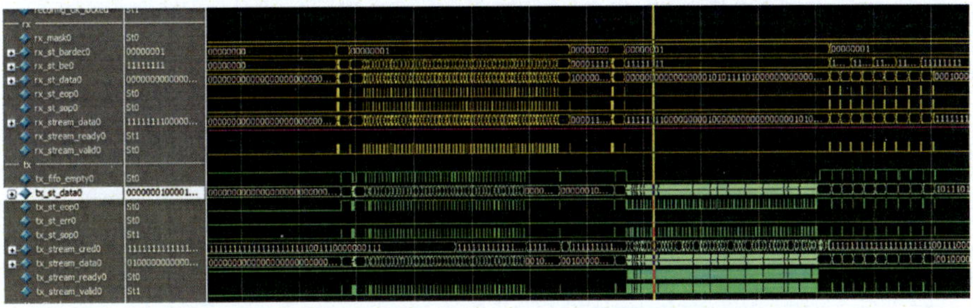

Fig. 11. PCIE high-speed data acquisition simulation diagram

Figure 12 is a PCIE speed test chart. The maximum speed of PCIE when it is in the x1 lane is 2.5 Gbps. It can be known from the simulation results that the predicted results are consistent with the simulation results.

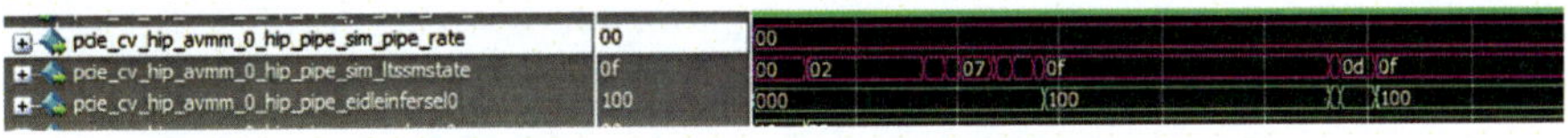

Fig. 12. PCIE speed simulation diagram

5.2 PC Test

As shown in Fig. 13, the DMA read data speed is 173 MB/s and the write data speed is 207 MB/s. The average speed is 190 MB/s. According to the function of the PCIE bus protocol, in the case of the $\times$ 1 channel PCIE bus interface, the transmission speed limit of the physical layer of the PCIE bus is 2.5 Gbps, and the PCIE bus uses 8b/10b encoding because its effective data rate is on the physical line. 80% of the rate, so the transmission speed must be lower than 2.5 Gbps. If the data packets are calculated for TLPs and DLLPs during the transmission process, the effective rate in practical applications will be lower. In addition, FPGA receiving data needs to pass through the receiving buffer module and receiving processing module, the data transmission speed will be slightly lower than the FPGA sending data speed. After testing, the actual data transmission rate is 190 MB/s, which is 1520 Mbps after conversion.

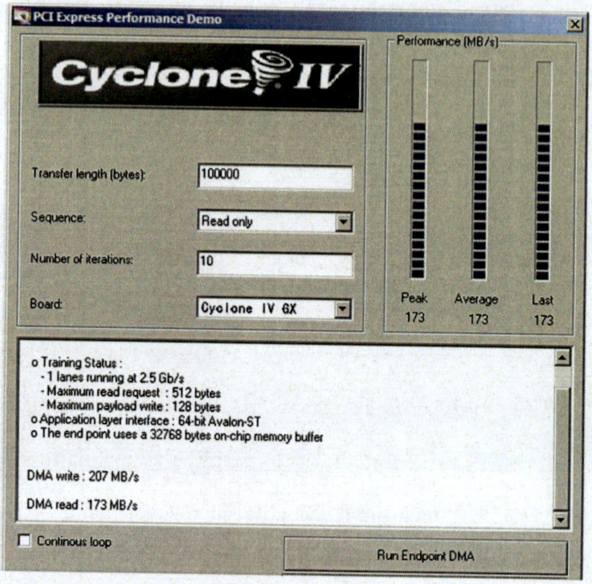

Fig. 13. PC Test

5.3 Performance Analysis

The simulation results of modelsim and the data of PCIE host computer test show that the PCIE bus used in this paper has a higher transmission rate than other bus interfaces. The specific comparison rate table is shown in Table 1.

Table 1. PCIE and other bus transmission rate table

Bus interface	Transmission rate
PCIE	1520 Mbps
RS485	1 Mbps
Gigabit Ethernet	1024 Mbps
SATA 3.0	600 Mbps
USB 2.0	480 Mbps

It can be seen from Table 1 that the rate of PCIE reaches 1520Mbps under low-cost FPGA, which is in line with the expected effect of this design, thus solving the problem of slow and unstable transmission rate of low-cost FPGA.

6 Conclusion

This paper designs a PCIE data acquisition system based on FPGA chips' high-speed parallel processing capability and low-cost Altera's Cyclone IV GX series chips. It first introduces the principle of the PCIE high-speed data acquisition system, and then designs each system module. Finally, After meeting the timing requirements, install the FPGA development board into a PC with a PCIE card slot. It can be seen from the simulation results of modelsim and the test results of the host computer that the system has achieved the expected design goals and the requirements of high-speed data collection. The PCIE high-speed data acquisition system in this paper has faster instruction processing speed and less logic resource occupation than the data acquisition system designed by traditional single chip microcomputer, and has high system integration, good stability, low cost and certain application value. And practical significance.

Acknowledgments. The author, Xinxin Sun, thanks the postgraduate project of Yunnan university school of information for supporting the publication of this paper "research on key technologies of intelligent home robot voice control system based on FFT dual-core processor".

References

1. Zhou, B., Qiu, Z.: Design of PCI express real-time measurement and control platforms for MEMS gyroscopes. QinghuaDaxueXuebao/J. Tsinghua Univ. **57**(12), 1310–1316 (2017)
2. Zhao, J., Zeng, X., Guo, Z.: Design and implementation of high speed PCIe cipher card supporting GM algorithms. Dianzi Yu Xinxi Xuebao/J. Electron. Inf. Tech. **41**(10), 2402–2408 (2019)
3. Zhao, Y., Liu, X., Yang, J.: A resource and timing optimized PCIe DMA architecture using FPGA internal data buffer. IEICE Electron. Exp. **16**(1), 20180858 (2019)
4. Li, A., Song, S.L., Chen, J., et al.: Evaluating modern GPU interconnect: PCIe, NVLink, NV-SLI, NVSwitch and GPUDirect. IEEE Trans. Parallel Distrib. Syst. 31(1), 94–110 (2020)
5. Ghodhbani, R., Horrigue, L., Saidani, T., et al.: Fast FPGA prototyping based real-time image and video processing with high-level synthesis. Int. J. Adv. Comput. Sci. Appl. **11**(2), 12 (2020)

Impact of Transmission Power Control Mechanism in Underwater Wireless Sensor Networks

Syed Agha Hassnain Mohsan[1(✉)], Asad Islam[2], Syed Basharat Hussain[3], Alireza Mazinani[2], E. L. Hacen Alioune[4], Md. Israq Aziz[5], and Naqeeb Ullah[6]

[1] Department of Electrical Engineering, COMSATS University Islamabad (CUI), Islamabad, Pakistan
Hassnainagha@zju.edu.cn
[2] Beihang University, Beijing, China
[3] North China Electric Power University, Beijing, China
[4] Beijing Institute of Technology (BIT), Bejing, China
[5] South China University of Technology, Guangzhou, China
[6] Engineering and Management Sciences (BUITEMS), Balochistan University of Information Technology, Quetta, Pakistan

Abstract. Underwater Wireless Sensor Networks (UWSNs) face challenges regarding high propagation delay, limited bandwidth, 3D topology and excessive energy consumptions. In this paper, we propose transmission power control mechanism for Underwater Wireless Sensor Networks (UWSNs). We experimentally investigate the impact of transmission power and propose a control mechanism to enhance the performance of underwater wireless sensor network. We consider a homogenous network in which each sensor node has four different power levels as a concentric circle. In this proposed mechanism, source nodes will adjust its transmission power according to the location of destination node. This paper aims to provide a mechanism which is incorporated in SEEC. This study also outlines the mathematical modeling for proposed idea. Moreover, we have compared results of our scheme with previous implemented schemes. The statistical significance of this work was analyzed in MATLAB. Marked observations to emerge from our results include an improvement in lifetime, increased throughput, higher residual energy, increment in alive nodes and balanced energy consumption. In our view, these results strengthen the validity of our proposed power control mechanism. We observe a significant increase in received packets because maximum nodes are alive till 1500 rounds which provides maximum communication and less chance of creating void holes.

Keywords: Underwater wireless sensor networks · Mobile sink · Transmission power · Concentric circles

M. Tavana et al. (Eds.): IISA 2020, AISC 1304, pp. 836–846, 2021.
https://doi.org/10.1007/978-3-030-63784-2_102

1 Introduction

We have witnessed the penetration of wireless sensor networks (WSNs) into every avenue of our lives, ranging from environmental monitoring, biological medicine, national defense, underwater vehicles to marine observations [1]. A considerable research has been carried out on terrestrial sensor networks in different aspects but currently research community is attracted towards a new era of research in Underwater Wireless Sensor Networks (UWSNs). Human beings are unable to work at high pressure ocean environment for a long time. On the other hand, Terrestrial Wireless Sensor Networks (TWSNs) are feasible but it cannot be considerable for UWSNs because signals transmission is elusive due to higher attenuation of aquatic environment.

Underwater Wireless Sensor Network (UWSN) consists of various components such as sensors and vehicles deployed in acoustic area for data collection and collaborative monitoring tasks. UWSN is made up of individual sensor nodes and many autonomous vehicles which perform data collection, store and forwarding operations. Generally UWSNs differ from TWSNs in multiple stringent design constraints such as sensors cost, dense deployment of these sensors, communication method and maximum power required for this communication. The major challenges in deployments of UWSN are limited battery sources, memory and computational power. As the life span of each sensor node is limited, so the sensor nodes which stop working will increase over a large deployment time. It is clear that the problem of limited battery sources is important and it is a big challenges for research fraternity to achieve long operation time without effecting performance of UWSN. USWNs are being used in several applications of underwater environment such as observation of chemical pollution, seismic activities, submarine movements, monitoring of marine creatures, distributed tactical surveillance, disaster prevention and military applications. However, UWSNs face challenges regarding narrow bandwidth, 3D topology, long propagation delay, media access control, attenuation, harsh geographical atmosphere, power constraints, losses of connectivity and high bit error rates. In this new paradigm, UWSN should fulfill, despite transmission power and energy limitations, the communication requirements of several advanced technologies and manufacturing methods. Though, research community has provided several techniques to cope with these challenges but still gap lies for research because of variable characteristics of ocean.

This technology has already achieved a great stature among researchers. A growing research inclination has been noticed by industry as well as academia in UWSNs [2, 3] with the progress in underwater technologies. Currently, acoustic signals are in great consideration for underwater applications as they are more feasible for underwater communications with low absorption rate [4]. Long transmission range can be achieved by acoustic communication. However, it faces time-varying signal distortion and transmission losses. In UWSNs, reliability and stability of network has paramount importance. A key challenging issue for UWSN lifetime is the maintenance as its difficult to replace those batteries providing power to sensor nodes. Recently, researchers are working in underwater wireless power transfer to tackle these challenges. In addition, TWSNs routing protocols cannot be implemented to UWSNs because of limited bandwidth, long propagation delay, energy constraints, high mobility and non-uniform 3D topology. Researchers have suggested several routing protocols for UWSNs. Some routing protocols can enhance the stability duration of network on the basis of throughput while others increase throughput on a compromise over packet delay [4]. In these protocols clustering is considered as the pragmatic method for load balancing in UWSNs because sensor nodes are categorized in different groups. Among these sensor nodes, cluster head is used as leading head [5]. Usually underwater communication is different from air due to varying conductivity, permittivity and permeability [6, 7] parameters. A SEEC protocol is proposed in [8] and we have implemented our idea on it. It is noticed when both sinks are located at adjacent positions then throughput is affected due to same sensor nodes.

Power consumption plays a vital role in efficient performance of UWSN. One major issue in power constraint of UWSN as it is unrealistic to recharge or change batteries during sensor's life span. It is not feasible to recharge these batteries frequently. Maximum power consumption occurs while data communication [9]. Power adaptation on the basis of environment conditions is a good approach towards efficient power consumption. Two major factors, wireless link quality and distance, affect the power consumption in any wireless sensor network. The link quality depends on climate conditions and physical barriers. Transmission power is controlled to obtain a good link providing successful data delivery. On the other hand, it is preferred to adopt minimum transmission power which can extend the WSN lifetime. Reliability insurance and power optimization appears as an important concern while dealing with WSN [10]. In this context, we have provided transmission power control mechanism. Reducing transmission power can have counterproductive effects to network performance. We are interested to understand better spectrum efficiency of UWSN through efficient control of transmitted power. It has the potential to minimize power consumption in UWSN. However, it still faces issues in real world deployments. Such mechanism must be robust against dynamic and complex wireless properties [11]. Through MATLAB simulation, we analyze the performance of UWSN is better through power control mechanism in terms of throughput, residual energy and network stability.

Authors in [12] have also presented adaptive transmission power control schemes to adjust energy in WSN. In [13], distributed algorithms are proposed to lower communication energy consumption in WSN through reducing transmission power of sensors. The appropriate selection of transmission range is discussed in [14] which greatly influences network connectivity and energy efficiency. Power consumption control problem is briefly discussed in [15] where authors have addressed dynamic power management.

Considering above discussion, we were inspired and implemented a power control mechanism in SEEC. The overriding objective of this study is to seek an intelligent use of transmitted power. We have implemented a previous routing scheme SEEC with incorporating a power control mechanism. We have compared our obtained results with SEEC, DBR, EEDBR and TESM. We have organized our paper as: in Sect. 2, we have made relevant research discussion. Section 3 will focus on motivation behind this present study. We have evaluated the impacts of power control mechanism in Sect. 4

while results discussion has been carried out in Sect. 5. In last sections, we have presented performance tradeoff, conclusion and future work.

2 Background and Significance

Since many years researchers have been focusing to design several UWSNs protocols for qualitative and effective research analysis. Research community is working with the aim of paving a way for new era f underwater actuation and monitoring applications. UWSNs will help us to know our ocean with its envisioned landscape of applications. Different topologies, power control mechanisms, power management algorithms and adaptive techniques have been implemented in Wireless Sensor Networks (WSNs). Authors in [16] have presented a self-learning power control system for WSN. Power control for interference mitigation and better spectral efficiency is discussed in [12]. A topology control technique for efficient energy consumption is proposed in [17] based on power management and power control. Power control for real life WSNs and channel allocation is briefly discussed in [18]. A dynamic power management technique to enhance energy efficiency is discussed in [19]. Researchers have suggested various routing scheme to achieve better results as authors in [20] highlighted capabilities and limitations on the basis of a survey on UWSN with data aggregation. In [21] authors presented an energy-balanced algorithm based on energy level partition and depth threshold. In this protocol, mobile sinks can collect data at sparse region and they implied clustering on dense regions. Authors of [22] suggested an energy efficient scheme based on multilevel cluster selection criteria which aims to choose cluster head according to high residual energy and less radius from sink. In [23], geographic routing protocol is used to maximize network lifetime of UWSNs through sink mobility. An energy efficient routing protocol based on clustering is simulated in [24] for UWSNs. Authors proposed that clustering technique is useful to reduce energy resources consumption for a sensor network. Authors of [25] have used a hybrid approach to maximize network lifetime and minimize the energy holes. Researchers have proposed a routing protocol AEDG which involves effective data collection through AUV [26]. A recent systematic review on UOWC has focused on research challenges of underwater wireless communication [27]. In this research study, authors maximized the network lifetime in AEDG as AUV is sent to gather data from gateway nodes. The usage of multiple mobile sinks to enhance network lifetime is discussed in [28]. Authors validated that a better network lifetime can be achieved through assigning a predefined path to mobile sinks for movement.

3 Network Design

An efficient use of power consumption in UWSNs appears as maximum throughput and enhanced lifetime. In this study we have considered routing scheme SEEC and incorporated with transmission power control mechanism. We achieved good results which will be discussed in next sections.

3.1 Network Model

we consider a homogenous network model in this work. In addition, location information of all sensor nodes is given to each sensor node in this network. In our proposed idea, each sensor node forms four concentric circles around it. These concentric circles represent different power levels. As source node maintains the location information of destination node so it will adjust the transmission power which will be in either of these concentric circles. We have discussed the concentric circles formation process in mathematical modeling section. Network model design of our proposed mechanism is shown in Fig. 1.

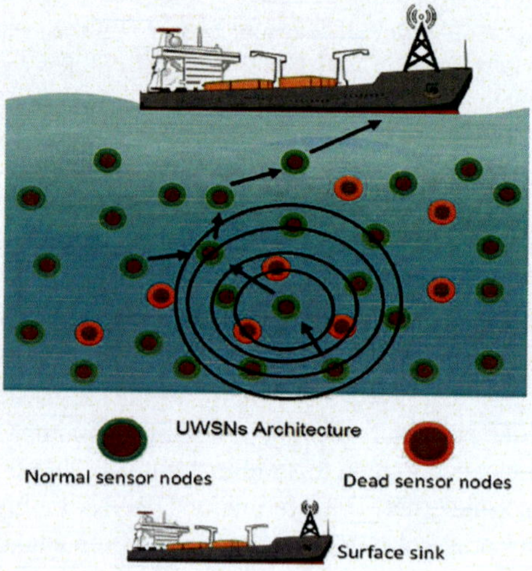

Fig. 1. Network model

3.2 Mathematical Modeling

In UWSN, approximately 0:68 dBm power is required for sending packets up to distance of one meter. We consider $\boldsymbol{R_{sn}}$ as the range of sensor node. The maximum transmission power for each sensor is calculated as:

$$\boldsymbol{Trans_{pwr}} = \boldsymbol{I} * \boldsymbol{R_{sn}} \tag{1}$$

Here we take $\boldsymbol{I}$ = 0:68, as we have divided the total transmission power into four different levels so we can calculate the radius of each concentric circle around the sensor node from following equation.

$$\boldsymbol{R_{dist}} = \boldsymbol{i} * (\boldsymbol{R_{sn}}/\mathbf{4}) \quad \mathbf{1} \leq \boldsymbol{i} \leq \mathbf{4} \tag{2}$$

We can calculate transmission power for calculated four radii as follow:

$$\boldsymbol{Trans_{pwri}} = \boldsymbol{I} * \boldsymbol{R_{dist}} \tag{3}$$

If we consider the transmission range of the sensor node $\boldsymbol{R_{sn}}$ ***as*** 50 m then Table 1 shows the calculated radii and power magnitude at each radius.

Table 1. Research keywords

R_{dist} (Meter)	Power (dBm)
12.5	8.5
25	17
37.5	25.5
50	34

We have incorporated the proposed transmission power control mechanism in SEEC and MATLAB simulation results will be discussed in next section.

4 Simulation Results

4.1 Network Lifetime

We have noticed an improvement in SEEC's network lifetime after incorporating our proposed idea of controlled transmission power mechanism. In SEEC, network was dead early as sensor nodes were sending packets with a predefined transmission power value.

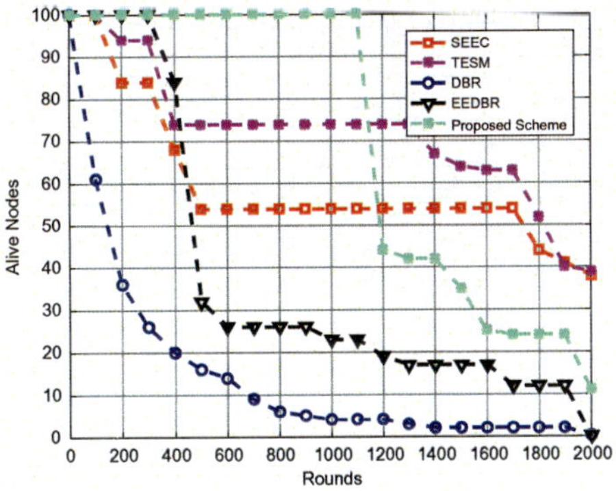

Fig. 2. Network lifetime

While in controlled transmission power mechanism, source node adapts the transmission power on the basis of Euclidean distance from the destination node.

Whenever any packet is transmitted, every consumption will vary every time. We can see a clear increase in network lifetime. In Fig. 2, nodes in SEEC were dead early while all nodes were alive till 1100 rounds after incorporating controlled mechanism.

4.2 Stability Period

We cannot obtain same number of sparse and dense region each time through randomly deploying sensor nodes. Network stability is compromised whenever dense regions are maximums and sensor nodes communicate with same transmission power within these dense regions. On the other hand, network stability period is improved if we adapt controlled transmission power mechanism as shown in Fig. 3. In controlled mechanism, besides controlled transmission power, stability period is better than SEEC, TESM [29], DBR and EEDBR through clustering in dense regions and mobile sinks movements in sparse regions.

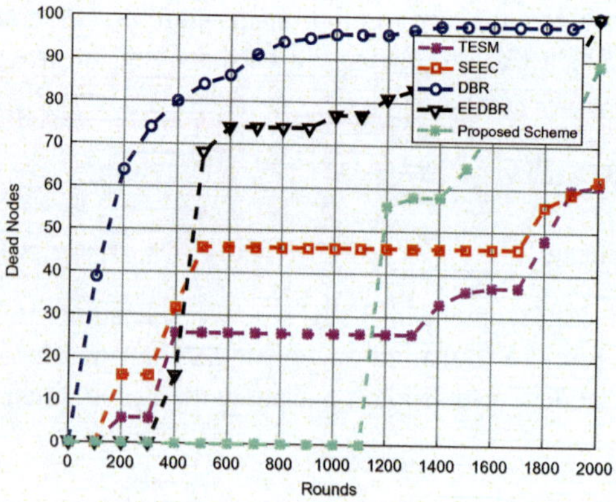

Fig. 3. Network stability period

4.3 Network Residual Energy

The transmission power factor was constant in energy consumption equation of SEEC. While sending a packet, same energy was being consumed without considering distance between two communicating nodes. In controlled transmission power mechanism, transmitted power adjustment is made on the basis of distance between source and destination node. Low transmission power in needed in case of less distance. Low requirement of transmission power results into low energy tax providing higher network residual energy. In our proposed work, energy consumption varies its return value after transmitting the packet which also gives higher network residual energy. Figure 4 shows the network residual energy for proposed scheme along with SEEC, DBR, EEDBR and TESM.

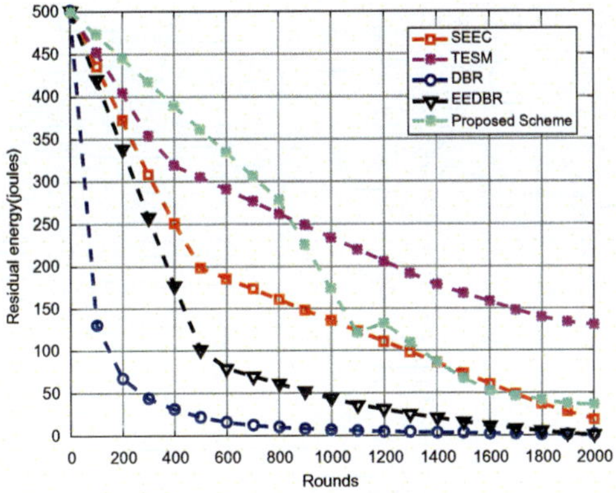

Fig. 4. Network residual energy

4.4 Network Throughput

In Fig. 5, the received packets per round at any sink are more than SEEC, DBR, EEDBR and TESM. We can notice an increased throughput in our proposed work with controlled transmission power mechanism. In SEEC, less packets are received as mobile sinks movement is according to sparse region and in TESM it is based on selection criteria opted for forwarding nodes.

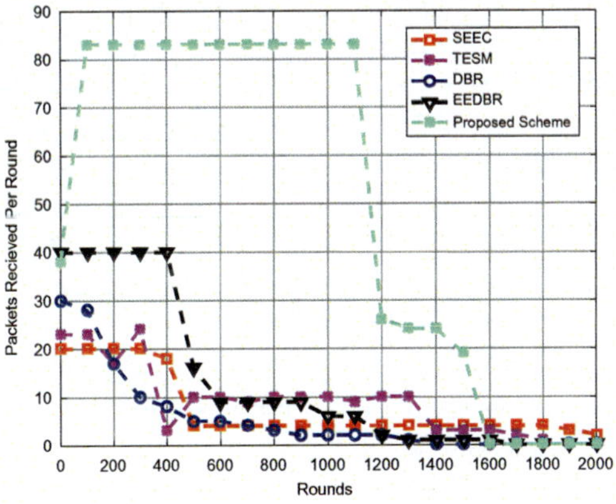

Fig. 5. Packets received per round

Figure 6 shows that the total received packets at sink are higher with controlled mechanism in SEEC. It is because of increased network lifetime by using transmission power control mechanism. EEDBR receives packets at the cost of high residual energy. In SEEC, total received packets are lower than our proposed scheme. It is because in SEEC, network is dead early and predefined transmission power also remains the same.

It can be seen in Fig. 5 that received packets per round are better in our proposed controlled mechanism than SEEC, TESM, DBR and EEDBR.

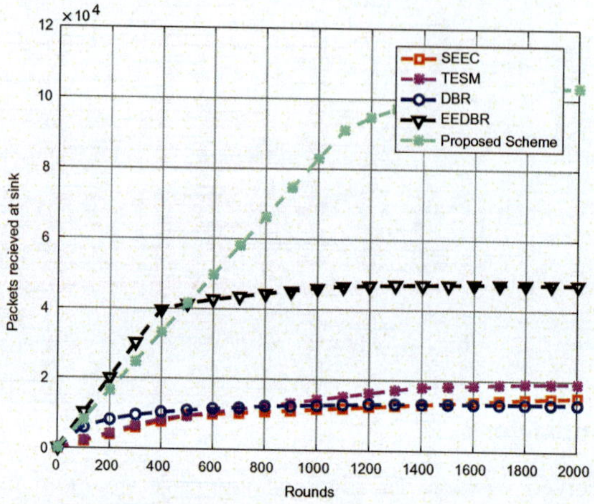

Fig. 6. Packets received at sink

5 Conclusion and Future Work

Results presented in this work demonstrate that optimum and efficient use of transmission power in mobile sinks is crucially important in enhancing network lifetime and throughput. Our work has led us to conclude that a proper control mechanism for power transmission is pivotal. In addition, the results of any routing scheme can be improved if we monitor the proper use of resources in the protocol. In this study, we compared results of power control mechanism incorporated in SEEC with previous implemented schemes TESM, DBR and EEDBR. To sum up our work, the proposed idea is better in throughput, network lifetime and residual energy. Taking complex underwater environment and economic concerns into consideration, we conclude that UWSN should be capable to adjust itself according to environment conditions. Thus, we can achieve this through proper protocol, power control and topology design which is self-adaptive according to the changing environment conditions. However these design characteristics should take energy efficiency into account which is crucially essential for normal operation of UWSN. In future, we are aiming to consider sparsity control in order to improve network lifetime. Holding time phenomenon towards reducing packet drop ratio is deferred to future work.

6 Future Work

In future, we are aiming to consider sparsity control in order to improve network lifetime. Though proposed work is energy efficient and we achieve throughput but it will cause delay. We are currently investigating to reduce this delay. We will evaluate the delay in each round in future research work. We will concentrate to minimize interference in dense area which will make reliable delivery of data packets in any network. Holding time phenomenon towards reducing packet drop ratio is deferred to future work.

References

1. Luan, Z.: Calculation and simulation of transmission reliability in wireless sensor network based on network coding. Int. J. Online Biomed. Eng. (iJOE) **13**(12), 150–161 (2017)
2. Li, N., Cürüklü, B., Bastos, J., Sucasas, V., Fernandez, J.A.S., Rodriguez, J.: A probabilistic and highly efficient topology control algorithm for underwater cooperating AUV networks. Sensors **17**, 1022 (2017)
3. Li, N., Martínez, J.F., Meneses Chaus, J.M., Eckert, M.: A survey on underwater acoustic sensor network routing protocols. Sensors **16**, 414 (2016)
4. Yan, H.; Shi, Z.J.; Cui, J.H. DBR: Depth-based routing for underwater sensor networks. In: International Conference on Research in Networking, pp. 72–86 (2008). Springer, Heidelberg
5. Han, G., Jiang, J., Bao, N., Wan, L., Guizani, M.: Routing protocols for underwater wireless sensor networks. IEEE Commun. Mag. **53**, 72–78 (2015)
6. Hassnain, S.A., Mughal, M.J., Naqvi, Q.A.: Layered chiral spheres with zero backscattering. In: 2019 Photonics & Electromagnetics Research Symposium-Fall (PIERS-Fall). IEEE (2019)
7. Hassnain, S.A., Mughal, M.J., Naqvi, Q.A.: Analysis of effective medium parameters on polarizability of homogeneous chiral sphere. In: 2019 Photonics & Electromagnetics Research Symposium-Fall (PIERS-Fall). IEEE (2019)
8. Sher, A., Javaid, N., Azam, I., Ahmad, H., Abdul, W., Ghouzali, S., Niaz, I.A., Khan, F.A.: Monitoring square and circular fields with sensors using energy-efficient cluster-based routing for underwater wireless sensor networks. Int. J. Distrib. Sens. Netw. **13** (2017). https://doi.org/https://doi.org/10.1177/1550147717717189
9. Khemapech, I., Miller, A., Duncan, I.: A survey of transmission power control in wireless sensor networks. In: Proceedings of PGNet, pp. 15–20 (2007)
10. Lavratti, F., et al.: A transmission power self-optimization technique for wireless sensor networks. ISRN Commun. Netw. **2012**, 1–12 (2012)
11. Fu, Y., et al.: Practical control of transmission power for wireless sensor networks. In: 2012 20th IEEE International Conference on Network Protocols (ICNP). IEEE (2012)
12. Chincoli, M., et al.: Power control in wireless sensor networks with variable interference. Mob. Inf. Syst. **2016**, 10 (2016)
13. Chen, X., Rowe, N.C.: Saving energy by adjusting transmission power in wireless sensor networks. In: 2011 IEEE Global Telecommunications Conference-GLOBECOM 2011. IEEE (2011)
14. Gao, M., Foh, C.H., Cai, J.: On the selection of transmission range in underwater acoustic sensor networks. Sensors **12**(4), 4715–4729 (2012)
15. Mokrenko, O., et al.: Dynamic power management in a wireless sensor network using predictive control. In: IECON 2014–40th Annual Conference of the IEEE Industrial Electronics Society. IEEE (2014)
16. Chincoli, M., Liotta, A.: Self-learning power control in wireless sensor networks. Sensors **18**(2), 375 (2018)

17. Nickray, M., Afzali-Kusha, A., Jäntti, R.: Simultaneous power control and power management algorithm with sector-shaped topology for wireless sensor networks. EURASIP J. Wireless Commun. Netw. **2015**(1), 118 (2015)
18. Barceló, M., et al.: Joint routing, channel allocation and power control for real-life wireless sensor networks. Trans. Emerg. Telecommun. Technol. **26**(5), 945–956 (2015)
19. Sinha, A., Chandrakasan, A.: Dynamic power management in wireless sensor networks. IEEE Des. Test Comput. **18**(2), 62–74 (2001)
20. Goyal, N., Dave, M., Verma, A.K.: Data aggregation in underwater wireless sensor network: recent approaches and issues. J. King Saud Univ. Comput. Inf. Sci. **31**(3), 275–286 (2019)
21. Feng, P., et al.: Improved energy-balanced algorithm for underwater wireless sensor network based on depth threshold and energy level partition. EURASIP J. Wireless Commun. Netw. **2019**(1), 1–15 (2019)
22. Wan, Z., Liu, S., Ni, W., Xu, Z.: An energy-efficient multi-level adaptive clustering routing algorithm for underwater wireless sensor networks. Clust. Comput. **22**, 1–10 (2018)
23. Ahmed, F., Wadud, Z., Javaid, N., Alrajeh, N., Alabed, M.S., Qasim, U.: Mobile sinks assisted geographic and opportunistic routing based interference avoidance for underwater-wireless sensor network. Sensors **18**, 1062 (2018)
24. Huang, X., Sun, S., Yang, Q.: Data uploading strategy for underwater wireless sensor networks. Sensors **19**(23), 5265 (2019)
25. Ma, M., Yang, Y.: SenCar: an energy-efficient data gathering mechanism for large-scale multihop sensor networks. IEEE Trans. Parallel Distrib. Syst. **18**(10), 1476–1488 (2007)
26. Khan, J.U., Cho, H.S.: A distributed data-gathering protocol using AUV in underwater sensor networks. Sensors **15**(8), 19331–19350 (2015)
27. Mohsan, S.A.H., Hasan, M.M., Mazinani, A., Sadiq, M.A., Akhtar, M.H., Islam, A., Rokia, L.S.: A systematic review on practical considerations, recent advances and research challenges in underwater optical wireless communication. Int. J. Adv. Comput. Sci. Appl. (IJACSA) **11**(7), 162 (2020)
28. Chen, Y.S., Lin, Y.W.: Mobicast routing protocol for underwater sensor networks. IEEE Sens. J. **13**, 737–749 (2013)
29. Yahya, A., et al.: Towards efficient sink mobility in underwater wireless sensor networks. Energies **11**(6), 1471 (2018)

Design and Implementation of Deep-Sea Emergency Response System

Wei Cong(✉)

China Transport Telecommunications & Information Center, Beijing, China
congwei@ctticsh.cn

Abstract. This paper proposes a design framework which aims at solving some key issues still existing in the field of deep-sea emergency response, such as lack of monitoring data on the operation site, ineffectiveness of locating and searching, low intelligence of situational analysis and decision-making process, insufficient technical cooperation guarantee during operation process. This solution integrates technologies of real-time monitoring of key data at the scene of deep-sea rescue and salvage operations, integration of multi-source information, collaborative search and positioning of underwater targets, real-time analysis of deep-sea emergency data, intelligent scenario parsing, intelligent auxiliary decision support for emergency response and phased array antenna beamforming and 5G for marine communication. Through the sea trial, it shows that the system has the functions of hierarchical response, autonomous judgment and auxiliary decision-making. It can be deployed on the deep-sea salvage operation ships and shore-based command center to improve the SAR and dispatch level of deep-sea emergency response.

Keywords: Deep-sea · Emergency response · Rescue and salvage · ROV · Auxiliary decision-making

1 Introduction

As the world economy becomes more closely connected, international trade, especially marine transport, is getting heavier, the navigation environment becomes more and more complex, and the pressure on maritime safety is increasing. The occurrence of major accidents such as "MV SEWOL", "MV SANCHI" and "Malaysia Airline MH370" [1] has raised new demands for deep-sea emergency response. This paper at aims at solving some key issues still existing in the field of deep-sea emergency response, such as lack of monitoring data on the operation site, ineffectiveness of locating and searching, low intelligence of situational analysis and decision-making process, insufficient technical cooperation guarantee during operation process [2–8], designed and implemented a deep-sea emergency response system that integrates data monitoring and acquisition, search and positioning, auxiliary decision-making, smart antennas and networking communications. The system comprehensively uses real-time monitoring and multi-source information integration technology during deep-sea rescue and salvage operations as well as technologies of underwater target collaborative search and positioning, deep-sea emergency data real-time analysis, situation intelligent

M. Tavana et al. (Eds.): IISA 2020, AISC 1304, pp. 847–856, 2021.
https://doi.org/10.1007/978-3-030-63784-2_103

analysis, intelligent assistant decision support for emergency response, phased array antenna-based beamforming, 5G marine communication, etc. [9–12]. The marine trial of deep-sea salvage ships, rescue helicopters and shore-based command centers show that the system is a comprehensive system that integrates dynamic data collection and fusion, hierarchical processing, that is, using marine meteorological observation equipment, Side-scan Sonar, and Ultra-short Baseline, ROV (with forward looking sonar, optical camera), audio and video and other types of equipment, centralized processing of the obtained information, to provide on-site visual command capability. Through real-time interaction and visualization of relevant data in the shore-based command center, the efficiency of SAR and dispatch level of deep-sea emergency response is effectively improved.

2 General Architecture

The deep-sea emergency response system involves multiple cooperative units such as water surface efforts, underwater efforts, air efforts situated on-scene and shore-based command center, etc., with characteristics of space spanning widely, nodes quantity increasing, and complex networking. Therefore, the system utilizes a traditional shore-based, on-site two-level command center mode. The shore-based command center is responsible for overall coordination and assignment of operational tasks; the on-site command center is responsible for the organization and coordination of each unit of the mother ship, summarizing and reporting, issuing specific operation commands according to the dynamics on site. The deep-sea emergency response system displays the static, dynamic data and real-time audio and video to the two-level command centers simultaneously and gives auxiliary decision-making, strengthening the shore-based command center's ability to grasp the dynamics of the emergency site. The effectiveness of command center to coordinate, dispatch emergency resources is key for emergency response. It is necessary to ensure smooth communication and information sharing between the command centers at all levels and between each rescue unit to ensure that all participating units being able to fully grasp the on-site situation. Note that all above considerations depend on the communication capabilities of the system.

The overall structure of the deep-sea emergency response system is shown in Fig. 1. The system is divided into five components according to functions: hydro-meteorological forecast subsystem, collaborative search subsystem, target positioning subsystem, auxiliary decision-making subsystem and communication network subsystem.

The deep-sea emergency response system is mainly integrated on the deep-sea salvage operation mother ship. How to arrange all kinds of equipment reasonably is the key to the design of deep-sea emergency response system. In this paper, based on the characteristics of each system and equipment, combined with the actual situation of the rescue operation mother ship, a comprehensive design of each component of the system is carried out.

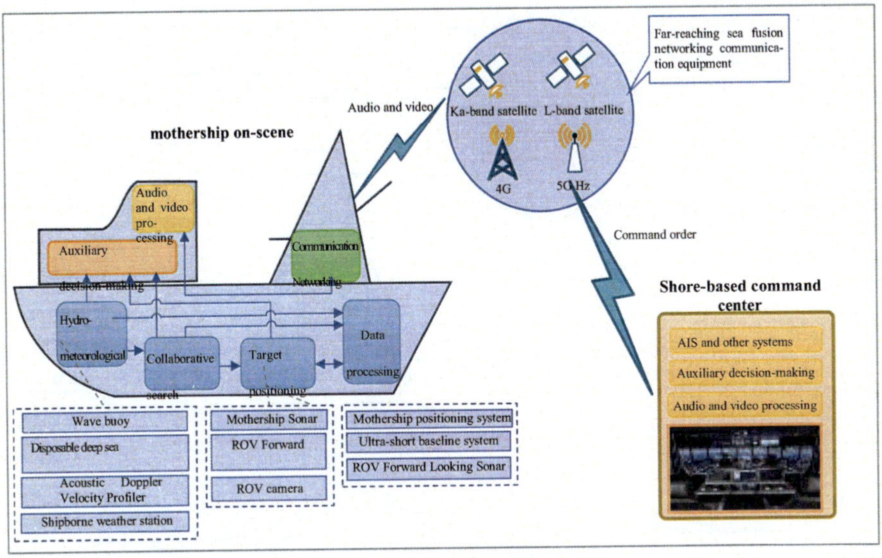

Fig. 1. Overall structure of the deep-sea emergency response system

3 Design of Hydro-Meteorological Forecast Subsystem

The hydro-meteorological forecast subsystem can accurately collect and forecast hydro-logical and meteorological information near the incident area before and during emergency action decision-making (Fig. 2).

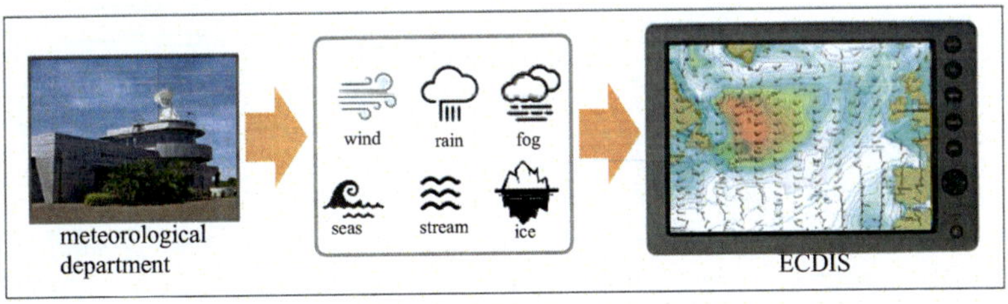

Fig. 2. Overlay hydro-meteorological forecast information on electronic chart

It assesses the degree of safety for marine operations under boisterous weather. It aims at planning navigation path and work scheme of the salvage ships, combining environmental factors, equipment parameters, ship loading status and other factors that greatly impact on navigation and operation safety and efficiency. By considering the different platforms, equipment and operation types requirements for hydro-meteorological conditions, this paper selects the corresponding impact indicators and integrate the their corresponding impact weights as well as the uncertainty of forecast data, utilizes the relevant theories of fuzzy mathematics and the risk-vulnerability model in risk analysis. Finally, a judgment set, seaworthiness, association of probability suitable for operating conditions is established, which can be applied in an operational marine environment safety assessment system with different platforms and equipment.

4 Collaborative Search Subsystem

The collaborative search subsystem is used to perform the two-stage tasks, one of rough exploration of the target area in the accident area, the other of fine exploration of the ROV area. The main functions include delineation of the suspected target area, ROV exploration path planning and tracing back the exploration process, as shown in Fig. 3.

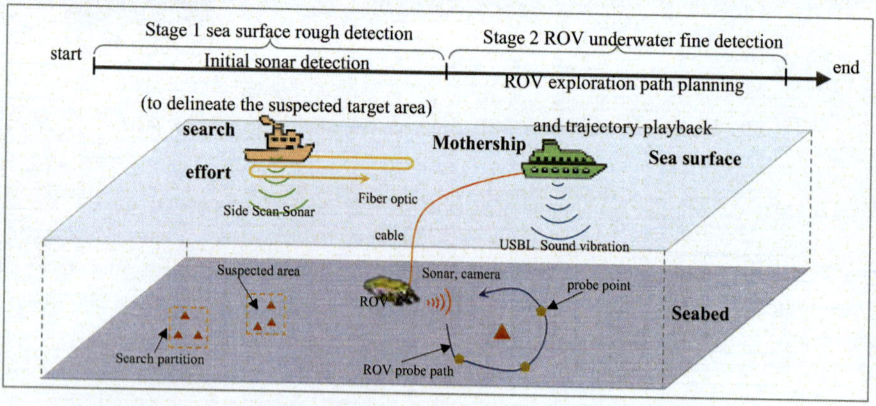

Fig. 3. Collaborative search subsystem

In the rough exploration stage, the goal is to maximize the probability of target discovery. First, the search path is planning according to the probability of target distribution in the accident area. Then, along the search path, the mother ship searches using side-scan sonar to obtain a preliminary topographic map within the search range. In order to suppress factors such as noise, which may impact the quality of side-scan sonar detection, this paper uses a deep convolutional neural network to eliminate noise and other factors to ensure the acquisition of high-quality topographic maps.

According to the above maps, and using the prior information of the wrecked target, the suspected target area can be delineated. Finally, as per the number of suspected targets and the structural features of the topographic map, it can determine the search partitioning scheme for suspected targets.

In the stage of fine exploration of the accidental sea area, in order to fully grasp the surrounding environment information of the target and provide environmental data support for the subsequent search and salvage work, the mother ship releases ROV to probe the surrounding environment of the suspect target, combining the seabed terrain and hydrographic information, to set multiple probe points. ROV determines the optimal probe path based on these probe points. In the process of performing the exploration task, ROV utilizes sonar and visible light equipment to collect the environmental data, determine the target of interest (such as the masts near the shipwreck, the engine, etc.) and mark the sonar image and visible light image on the probe trajectory, so the probing process can be traced back.

5 Target Positioning Subsystem

The target positioning subsystem consists of three modules: detection information acquisition and enhancement processing module, underwater target positioning module and data integration transmission and caching module. All modules work together to realize the positioning of underwater targets and surrounding information's acquisition, transmission, processing, integration and sending function, as shown in Fig. 4.

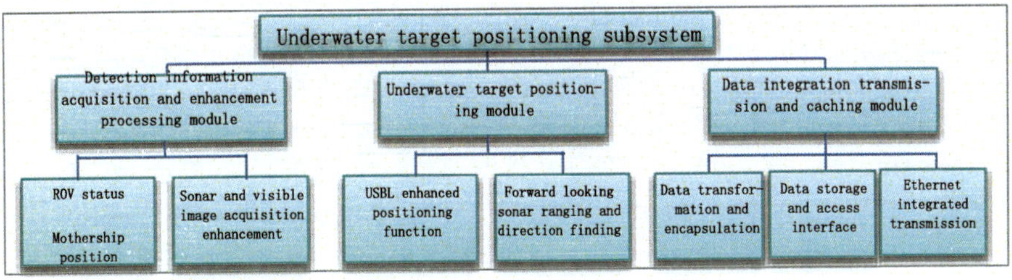

Fig. 4. The composition of the target positioning subsystem system

The collection of underwater targets and surrounding environment information is realized by the visible light sensor and multi-beam sonar mounted on the ROV; the transmission is realized by the communication chain of the ROV - ROV control room - mother ship - shore-based command center; the processing is enhanced by the sonar image enhancement algorithm and the visible light image enhanced algorithm; integration and transmission are realized by Ethernet bus data integrated transmission system. The positioning of the underwater target is realized by the USBL enhanced positioning system and the forward-looking sonar ranging and direction finding, as shown in Fig. 5.

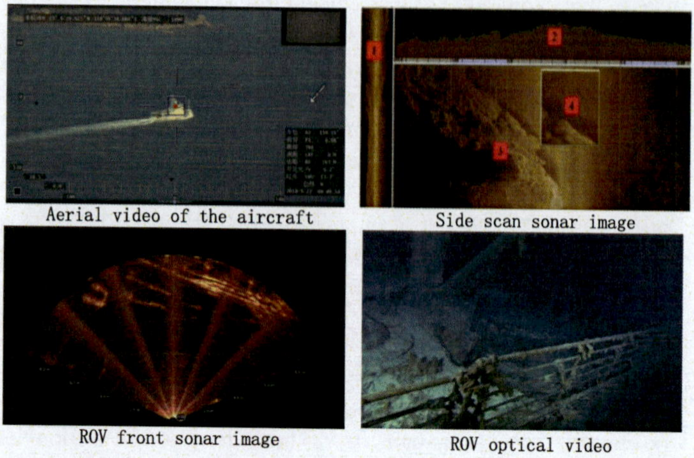

Fig. 5. Images of underwater targets and their surroundings

6 Auxiliary Decision-Making Subsystem

The auxiliary decision-making subsystem of deep-sea emergency response integrates data analysis, scenario assessment and auxiliary decision-making. Deep-sea emergency response data is complex and has a strong correlation with time and space. Therefore, multivariate data modeling for deep-sea complex scenes and visualizing and efficiently expressing them based on high-precision electronic charts are key issues.

To start with, the entire project's data is collected by the hydro-meteorological support subsystem, the collaborative search subsystem and the target positioning

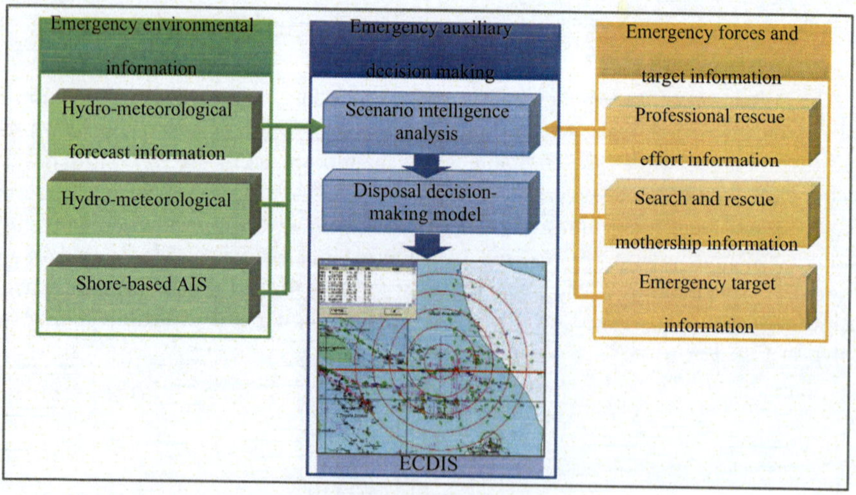

Fig. 6. Structure diagram of auxiliary decision subsystem

subsystem and stored in the shipborne database for sharing. Then, data from different sources is divided into three categories according to their business logics: emergency environmental data, emergency target data, emergency efforts data. These data are analyzed and integrated as per their temporal and spatial characteristics. Finally, utilize the electronic chart display system with high-precision charts covering emergency site to integrate and display them effectively, as shown in Fig. 6.

Emergency scenario intelligence analysis is based on real-time dynamic analysis of emergency data, extracting the main influencing factors, and then using intelligent methods to quantitatively analyze the situation on the emergency scene. The auxiliary decision-making subsystem analyzes and extracts the emergency environment, emergency force, emergency target and the main influencing factors of task execution, and quantifies the emergency situation as per the extracted influencing factors. Based on the high-precision electronic chart display system, the complex and changeable emergency situation of the deep sea is differentially plotted and displayed to facilitate emergency personnel to intuitively understand the current emergency scenario. The decision-making model of deep-sea emergency is constructed through the consideration of intelligent analysis results of the emergency scenario, the conditions and constraints of deep-sea emergency operations, the limitations of emergency efforts and eventually to assist to make emergency decision under multi-constraints. In terms of system architecture, the data from the emergency on scene can be rapidly integrated, analyzed and visualized, ensuring that the personnel thereof can grasp the situation of emergency site intuitively and quickly; at the same time, the emergency site data is transferred by the communication subsystem to the shore-based database. Then, the shore-based command center integrates the target information, emergency efforts information and emergency environmental information to make a comprehensive analysis of the emergency scenario and generate related emergency plans. In terms of system function, a hierarchical response modal covering data analysis, situation assessment and auxiliary decision-making is constructed, including functions of high-precision electronic chart display, visual representation of emergency data, emergency scenario analysis and auxiliary generation of emergency plan.

7 Communication Network Subsystem

The communication network subsystem can provide high-speed communication links for command centers, mother ships, helicopters, drones, individual soldiers and underwater robots, ensuring the timely delivery of all kinds of search and rescue instructions as well as the rapid uploading of all kinds of collected data. Through the deployment of deep and far sea fusion network communication equipment (referred to as Fusion Communication System, FCS) on mother ships, all information is gathered thereof in a wired and wireless manner. With the help of L and Ka-band satellite communications, the information exchange between the mother ship and the shore-based command center is carried out and deep-sea integrated network communication is achieved, as shown in Fig. 7.

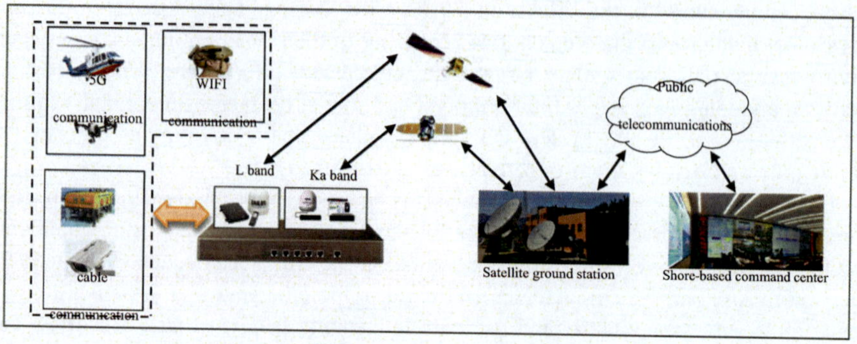

Fig. 7. General diagram of integrated network communication

8 Emergency Response Procedure

After receiving the accident (alarm) report, the shore-based center needs to further verify the accident information, including information on the accident ship and accident location. Subsequently, the auxiliary decision-making subsystem is activated to dispatch the professional rescue efforts for emergency response, analyze passing ships nearby, and look for additional available rescue efforts. Assign search and rescue ships and aircrafts carried the relevant equipment to rush to the accident location; when the search and rescue forces arrive at the emergency scene, utilize the communication network to establish an audio and video connection between the search and rescue mother ship's on-site and the shore-based command center, so as to promptly feedback the on-site emergency response dynamics; the on-site command plans the search area. The helicopter (including unmanned aerial vehicle) conducts aerial search. The mother ship's side scan sonar searches according to the search area and plots the path and the suspected target area simultaneously; the shore-based command center based on the suspected target area to determine the further fine search plan; start the ROV search and plot the path simultaneously. Once the ROV finds the target, it confirms and transmits the relevant video around the target from multiple angles back to the shore-based command center, search and positioning terminates and the emergency response ends. The procedure is shown in Fig. 8.

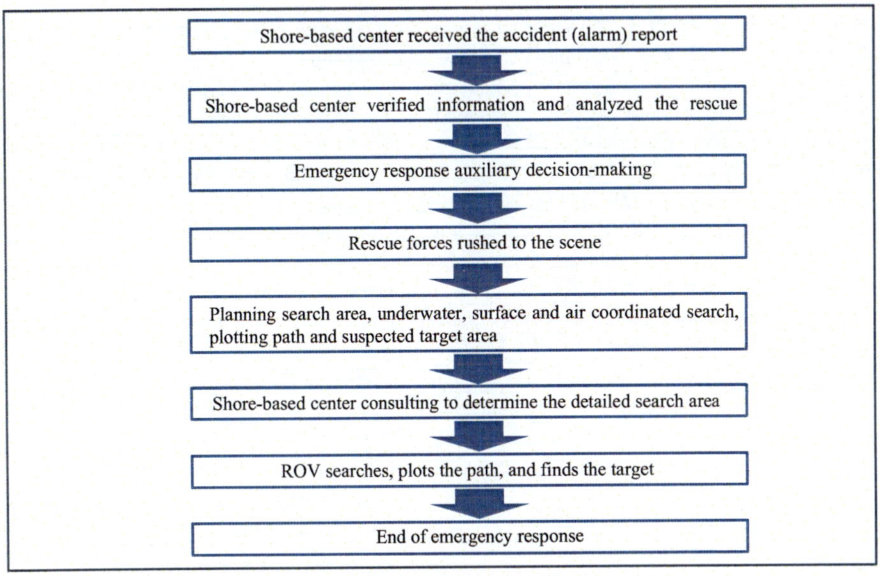

Fig. 8. Emergency response flow chart

Acknowledgments. The study was supported by "National Key R&D Program of China" (No. 2018YFC0309600\06).

References

1. Jacobs, T.: Search for flight MH 370 shows durability of AUVs. J. Pet. Technol. **68**(7), 26 (2016)
2. Seokhoon, Y., Chunming, Q.: Cooperative search and survey using Autonomous Underwater Vehicles (AUVs). IEEE Trans. Parallel Distrib. Syst. **23**(3), 364–379 (2012)
3. Vedachalam, N., Ramesh, R., Ramesh, S., et al.: Challenges in realizing robust systems for deep water submersible ROSUB6000. In: 2013 IEEE International Underwater Technology Symposium (UT), Tokyo, p. 10, March 2013
4. Grasmueck, M., Eberli, G.P., Viggiano, D.A., et al.: Autonomous Underwater Vehicle (AUV) mapping reveals coral mound distribution, morphology, and oceanography in deep water of the straits of Florida. Geophys. Res. Lett. **33** (2006)
5. Yoerger, D.R., Jakuba, M., Bradley, A.M., Bingham, B.: Techniques for deep sea near bottom survey using an autonomous underwater vehicle. Int. J. Robot. Res. **26**, 416–429 (2007)
6. Selvakumar, J.M., Ramesh, R., Subramanian, A.N., et al.: Technology tool for deep ocean exploration – Remotely Operated Vehicle. In: Twentieth International Offshore (Ocean) and Polar Engineering Conference, Beijing, China, June 2010
7. Ramesh, S., Vedachalam, N., Subramanian, A.N., et al.: Technology development in India for gas hydrates exploration and extraction feasibility studies. In: 2015 IEEE Underwater Technology (UT), pp. 1–6 (2015)

8. Fard, R.N., Tedeschi, E.: Power system design considerations for a seafloor mining vehicle. In: 2018 IEEE Energy Conversion Congress and Exposition (ECCE), pp. 1164–1171 (2018)
9. Juan, L., Jianxin, Z., Lijuan, Y., Chaojiong, H., Bingjian, Z.: Target search of multiple autonomous underwater vehicles in an unknown environment. J. Harbin Eng. Univ. **40**(12), 1951–1957 and 1972 (2019)
10. Paull, L., Saeedi, S., Seto, M., et al.: AUV navigation and localization: a review. IEEE J. Oceanic Eng. **39**(1), 131–149 (2014)
11. Fricke, G.M., Black, S.R., Hecker, J.P., et al.: Distinguishing adaptive search from random search in robots and T cells. In: Proceedings of the 2015 Annual Conference on Genetic and Evolutionary Computation, Madrid, Spain, pp. 105–112 (2015)
12. Song, J.H., Yun, Y., Luo, L.B., et al.: Application of artificial intelligence in the field of deep sea robots. Electron. World **6**, 185–186 (2019)

Research on Component Classification Strategy Based on Convolutional Neural Network

Mei Guo, Min Xiao(✉), and Wenfen Zhang

College of Software and Communication Engineering, Xiangnan University, Chenzhou 423000, Hunan, China
xnxyxm@163.com

Abstract. With the continuous improvement of technical system in recent years, the rise of software component technology drives the current development of the manufacturing industry, the construction of component library is also carried out for the application of current software construction technology, the application of software component technology and the construction and expansion of component library, both improve the efficiency of the construction and retrieval in the current software reuse process, but in recent years, the software component technology is in the process of application, due to the lack of comprehensive and comprehensive management strategies, there are some problems in the application of this technology. The application of convolutional neural network technology can better improve the application effect and application efficiency of software construction technology, therefore, this paper will conduct a detailed study on the construction of classification strategies based on convolutional neural networks.

Keywords: SoftWare reuse · Construct classification · Convolutional neural network

1 Introduction

The rise of software construction technology has promoted the rapid development of the current manufacturing industry, and also provided more new technical content for current software development. In the application process of software component technology, according to the software component technology needs to build component library, firmware library can better store detailed construction content, it can also accelerate the efficiency of component-based software development, reasonable application of construction technology through proper firmware retrieved from component library, one of the important technical contents involved in the construction of the component library is how to reasonably classify the content stored in the firmware library, only in this way can the efficiency of constructing retrieval be effectively improved, the most effective component classification strategy can better help software component technology to quickly apply for product production, reduce the influence of external factors, Improve the accuracy of faceted information.

M. Tavana et al. (Eds.): IISA 2020, AISC 1304, pp. 857–862, 2021.
https://doi.org/10.1007/978-3-030-63784-2_104

And study in this paper, the process of building classification strategy for research work, and put forward in order to utilize convolution neural network classification of fixed job applications, in the study process of the construction of the classification model based on neural network, the technology of component classification and comparison between the advantages and disadvantages of analyzed through model building and content, thus put forward using build classification strategy based on convolutional neural networks, it can better improve the efficiency of the firmware testing.

2 Comparison of Advantages and Disadvantages of Component Classification Technology

2.1 Comparison of Artificial Intelligence Methods

During the early development of artificial intelligence technology, it was actually only applied as a separate technical content. With the continuous development of artificial intelligence technology in the later period, artificial intelligence methods are also born within the artificial intelligence technology system. In the early stage, the artificial intelligence method was also effectively applied as a unit neural network, and did not form a diversified network structure under the current technical system, so many specific tasks could not be completed in time. In addition, due to the limitations of early hardware level and corresponding technical conditions, artificial intelligence technology and artificial intelligence methods can only deal with many simple linear problems and linear classification problems in the application process. For many non-linear problems, in fact, they cannot be reasonably applied, nor can they achieve good results. Therefore, in the early research process, the application of artificial intelligence method to software component technology and software component classification can only belong to the theoretical research content.

With the development of convolutional neural network technology and the continuous application of artificial intelligence methods and artificial intelligence technology in recent years, it can be applied to more fields on the basis of convolutional neural network technology, and it can also handle many nonlinear problems. And with the continuous improvement of convolutional neural network technology, the application of intelligent technology and artificial intelligence methods has been applied to many current image processing, speech recognition and other functions.

Therefore, in the application process of artificial intelligence methods, convolutional neural network technology is also used as the basis of application. Therefore, compared with artificial intelligence methods, the application process of convolutional neural network models and methods has better technical application effects. It can analyze the construction features layer by layer, so as to effectively realize the construction classification.

2.2 Hypertext Method Comparison

Hypertext method compared with the method of artificial intelligence, in the application process, in fact more simple hypertext method is the method of using hyperlinks, the existing relationship between the information and contact, through different settings and the content of the text, the combination of the nonlinear structure, and form a network of text.

The super method has its own unique advantages in the process. Firstly, it can better deal with the nonlinear information comprehensively. Secondly, a relatively hierarchical text network can be formed through the hyperlink method. But in the process of using it, the biggest problem that exists is how to determine the keywords in a text. Even in the process of information classification and construction of classification, a lot of cost and resource consumption are reduced. However, the splitting effect of the classification work carried out according to the super tender method includes the accuracy, scientificity, and rationality of the classification results themselves.It does not reach a certain standard, so sometimes it is necessary to use some artificial factors from the outside to carry out certain interventions to achieve the construction classification. The construction strategy of the convolutional neural network mentioned in the research process of this article can actually reduce the interference of external human factors, and it can also better reduce the participation of manual screening work, so it can greatly improve information classification and accuracy of construction classification [1].

2.3 Comparison of Information Science Methods

In the early application and development of software engineering technology, one of the methods with the highest frequency of classification and counting is the information science method.This scientific method has achieved very good application results in the early application process, because in the process of component classification, due to the flexibility of facet classification and the characteristics of multiple views and multiple descriptions, when searching the firmware information inside the component library, it often requires a comprehensive classification of faceted information, and the traditional information science classification method can be classified according to the information existing in the component itself during the application process, in the early greatly improved the efficiency of classification, so there is a very good effect. The information science classification method is shown in Fig. 1.

But information science method in use process, it is more dependent on the characteristics of the component itself, because the firmware library of classifying components, according to the characteristics of the building and the building of a controlled vocabulary, so in scientific methods of information in the application process is in view of the above several screening and selection of the contents, such ability can realize component retrieval. But it also requires the interference of external human factors. In the application process of the convolutional neural network construction classification method studied in this paper, it can avoid the influence of too many human factors, and it can greatly improve the accuracy of documents and construction retrieval [2].

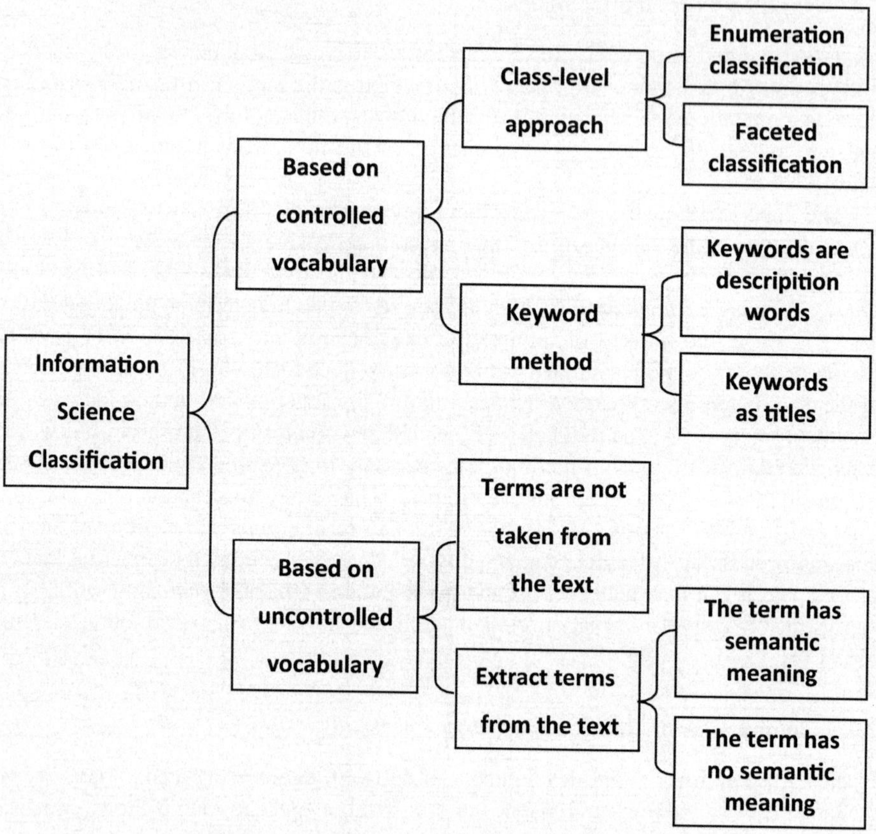

Fig. 1. Diagram of information science classification method

3 Research on Construction Classification Strategy Based on Convolutional Neural Network

3.1 Convolutional Neural Network

Convolutional neural network is actually composed of three levels, the first level is the convolution layer, the second level is the appearance of Qun Linjieshu and the third level is the pooling level. The combination of these three levels can form a neural network model. As convolutional neural network is a network application model composed of multiple layers, it has automatic learning algorithm at each layer in the application process. So in the final process of information output, it is actually the result of each level after processing the external information, so compared with other entry methods studied in this paper, the convolutional neural network method is in progress. It can process information visual signals layer by layer during application.

Therefore, the application frequency of convolutional neural networks in current image recognition is constantly increasing, and the application effect is also continuously improving. Moreover, convolutional neural networks have achieved very good application results in the field of natural language and element recognition in recent years. Therefore, the application of convolutional neural network to the current firmware classification can achieve a very good classification effect compared with other classification methods, and can also improve the current construction retrieval efficiency.

3.2 Construction Classification Algorithm Based on Convolutional Neural Network

The construction and classification algorithm of convolutional neural network is actually completed through three levels of internal information processing. First, the information and the mapped data will be convolved with the input data under the instruction of the filter, after the data is processed by the algorithm of the convolution layer, the data information will be subjected to local feature extraction, then the processed data information will directly enter the pooling layer of the convolutional neural network, in the pooling layer, the data information will be weighted, biased and downsampled. Finally, all the feature information is transmitted to the connection layer to be displayed to everyone through mapping. Figure 2 shows the structure diagram of the convolutional neural network model.

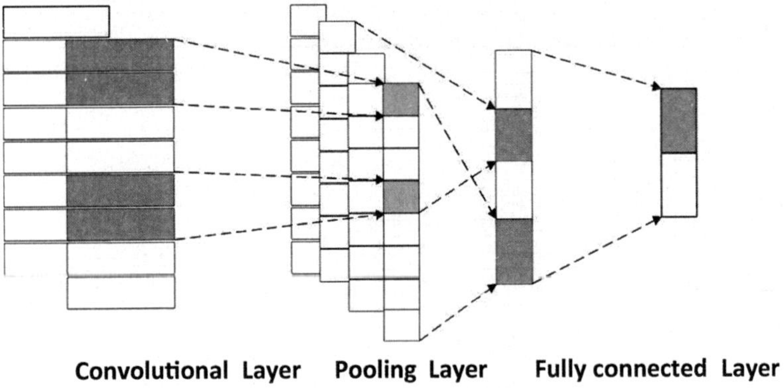

Fig. 2. Convolutional neural network model structure diagram

4 The Conclusion

This article conducts detailed research on the construction of classification strategies based on convolutional neural networks. During the research process, firstly, a comparison of multiple component classification methods was carried out. Secondly, a detailed introduction to the construction of classification based on convolutional neural networks. Through the research in this paper, it can provide some help for the application of current convolutional neural network construction classification strategy.

Acknowledgments. This paper is funded by Project of:

1. Scientific Research Fund of Hunan Provincial Education Department, Research on Intelligent Medical Logistics Distribution based on Particle swarm optimization, (No. 19C1708).

2. Chenzhou Municipal Science and Technology Project, Research on key technologies of intelligent parking in smart city, (No. Zdyf201913).

3. Chenzhou Municipal Science and Technology Project, Research on AR system based on Chenzhou Mining Fair, (No. Zdyf201911)

4. Hunan Provincial Department of Education Fund, Computer Application Technology Innovation and Entrepreneurship Education Center, (No. [2018] 380)

5. Hunan Provincial Department of Education Fund, Big Data and Processing Innovation Entrepreneurship Education Base, (No. [2019]333)

6. Scientific Research Fund of Hunan Provincial Education Department, Research on system architecture and key technologies of smart home for the elderly based on Internet of things, (No. 17A201)

References

1. Zhang, F.: Research and application of component-based software reuse technology. North University of China (2018)
2. Zhang, C., Yue, H., Wang, B., Liu, Y., Luo, S.: Deep learning pattern recognition method for PD ultrasonic signals based on similarity Matrix blind source separation and convolutional neural network. Power Grid Technol. (06), 1900–1907 (2019)
3. Wan, X., Song, H., Luo, L., et al.: Pattern recognition of partial discharge image based on one-dimensional convolutional neural network. In: International Conference on Condition Monitoring & Diagnosis (2018)
4. Sandler, M., Howard, A., Zhu, M., et al.: Inverted residuals and linear bottlenecks. In: The IEEE Conference on Computer Vision and Pattern Recognition (2018)
5. Yi, W., He, J., Zou, M.: Memory mechanism of dialogue system based on recurrent neural network. Comput. Eng. Des. (11), 3259–3264 (2019)
6. Zhang, T.: Hardware-based neural network acceleration. World Commun. (08), 77–79 (2018)

Research on Multi-label Clothing Image Classification Based on Convolutional Neural Network

Ying Hong(✉)

Basic Course Department, Beijing Institute of Fashion Technology, Beijing, China
jsjhy@bift.edu.cn

Abstract. Online shopping is a new and convenient experience for people with the development of the internet. Massive images are uploaded on the web. We want to recognize the clothing images and classify them according to different categories with computer. This is also a focus of research in clothing field. In this paper, we crawled the clothing images from the internet. The dataset contains six categories and it was split into training set and validation set. In our work, we built a convolutional neural network model and trained a multi-label classifier to predict both clothing type and color. We find from experiment that the overfitting will appear if we don't have enough data and it will cause the testing data accuracy lower than the training data accuracy. Data augmentation technique is a good way to eliminate the overfitting. It is demonstrated by experiment that convolutional neural network model has a good effect on clothing image classification and recognition.

Keywords: Image classification · Convolutional neural network · CNN · Deep learning

1 Introduction

More and more people share and record their life experiences on network with the development of the internet. Massive images emerge on the web. But it is difficult to accomplish searching or classifying images according to its content [1]. So, the research on how to use computer to recognize and classify images automatically is necessary. Nowadays, image recognition and classification are research hotpots. But how to improve the accuracy of image intelligent recognition and classification is the difficulty.

The purpose of image recognition and classification is to confirm and distinguish specific elements in picture. The main research content is image feature extraction and classifying image using classifier [2]. Traditional image classification usually designs feature extraction manually and confirms something existing in image using super vector machine (SVM) etc. The accuracy of traditional image classification is too low to meet actual needs [3]. Deep learning technology is neural network model which can

M. Tavana et al. (Eds.): IISA 2020, AISC 1304, pp. 863–870, 2021.
https://doi.org/10.1007/978-3-030-63784-2_105

learn and extract image features automatically. Comparing with traditional methods it improves accuracy and recall rate greatly of image recognition and classification.

Accompanied by e-commerce improvement, there are massive clothing images in the internet. The companies upload lots of clothing images on their website and people show their pictures on social media. Clothing image classification is a useful research point now. It is helpful to people to search their interesting clothing images quickly and precisely. In this paper, we researched clothing image classification by building a convolutional neural network. Then we analyzed and compared the results to adjust the network for getting better effect.

The scenes of the clothing images are complex. They always contain background information. It is difficult to classify these images by traditional algorithms such as SIFT [4], HOG [5], etc. In 1986, the back propagation (BP) neural network applied gradient descent algorithm and engaged the research of neural network [6]. Then in 1989 Yann Lecun proposed LeNet-5 [7] convolutional neural network. Many neural network models such as AlexNet [8], VGGNet [9], GoogleNet [10], ResNet [11] emerged along with large scale visual recognition challenge were held. ImageNet [12] is an open images dataset used in image classification and recognition. It promotes the research of deep learning.

2 Related Technologies and Theories

Image classification based on deep learning get the features through learning automatically from massive data. Representation of features and classifier are optimized. The accuracy of classification is improved greatly.

2.1 The Principle of Convolutional Neural Network

The convolutional neural network is a feedforward neural network [13]. It consists of input layer, convolutional layer [14], pooling layer [15], fully connected layer and output layer. Data is preprocessed at input layer, such as normalization. The classification results are got after several times convolution and pooling.

Convolutional layer consists of a lot of convolution units. The parameter of each unit is optimized from backpropagation algorithm. Some special size convolutional kernels are used to operate with the image which is inputted from the upper layer.

The function of pooling layer is to reduce the number of parameters and the amount of computation. It can also prevent overfitting in the network [16]. The common pooling methods are average pooling and max pooling. The latter has been shown to work better than the former in the practice [17]. Max pooling selects the max value in a little field. If the other values in this field has been changed, the result of pooling is constant. Pooling layer always is behind of convolutional layer. It receives and compresses the feature map from convolutional layer. Hence it can reduce computation and extract the main feature. In this paper, we use max pooling to select the max value as the value after pooling in this field.

Convolutional layer and pooling layer perform a linear transformation on incoming data. Activation function such as Sigmoid, ReLU is used in the architecture to add nonlinearity to the data [18]. In this paper, we use ReLU function as activation function. The equation of ReLU function is shown as below.

$$f(x) = \max(0, x) = \begin{cases} 0 & if\, x < 0 \\ x & if\, x \geq 0 \end{cases} \tag{1}$$

In Eq. (1), we can see if the input is less than 0, the output equals 0. If the input is greater than 0, the output equals to the input.

Fully connected layer is an essential component of the convolutional neural network. It classifies the dataset in according with the features extracted by the front layers.

2.2 Deep Learning Framework

There are several powerful frameworks can be used in deep learning such as TensorFlow [19], Theano, Keras, Torch and Caffe [20]. In our work, we choose Caffe as our deep learning framework to train the model. Caffe is developed by Berkeley AI Research (BAIR) and constantly improved by community contributors. It is simple and friendly to users. Model designing and optimization are defined by configuration without coding. The switch between CPU and GPU is convenient [21].

3 Experimental Design and Result Analysis

3.1 Dataset

The dataset we used in our work was crawled from the internet. Our dataset includes 3126 images which are divided into six categories. The data categories are black coat (532 images), red dress (530 images), blue shirt (522 images), black dress (513 images), blue coat (503 images), red shirt (526 images). We split the dataset for training and testing in the experiment and we used 80% of the images as the training data and 20% as the testing data.

3.2 The Network Architecture for Multi-label Classification

In our convolutional neural network architecture, the first convolutional layer has 32 filters with a kernel size of 3 × 3 and ReLU is used as activation function. Max pooling and batch normalization are applied in our network. Dropout with a probability of 25% is used to reduce overfitting. Then the first pooling layer has pool size 3 × 3. It is followed by two convolutional layers. Each of them has 64 filters with a kernel size of 3 × 3 and applies ReLU as activation function. Then the second pooling layer has pool size 2 × 2. It is followed by two convolutional layers. Each of them has 128 filters with a kernel size of 3 × 3 and ReLU is used as activation function. And the third pooling layer has pool size 2 × 2. The fully-connected (FC) layers has 1024 channels. The network architecture for multi-label classification is shown in Table 1.

Table 1. The network architecture for multi-label classification.

Layer (type)	Output shape
Input layer	96 × 96, 3
Conv1	3 × 3, 32
Pooling layer1	max-pooling, 3 × 3
Conv2	3 × 3, 64
Conv3	3 × 3, 64
Pooling layer2	max-pooling, 2 × 2
Conv4	3 × 3, 128
Conv5	3 × 3, 128
Pooling layer3	max-pooling, 2 × 2
Fc	1024

3.3 Results and Discussion

In our work, we used training data to train the network and verify the classification effect of our model on testing data. We ran it for 80 epochs. The loss and accuracy curves are shown in Fig. 1 and Fig. 2.

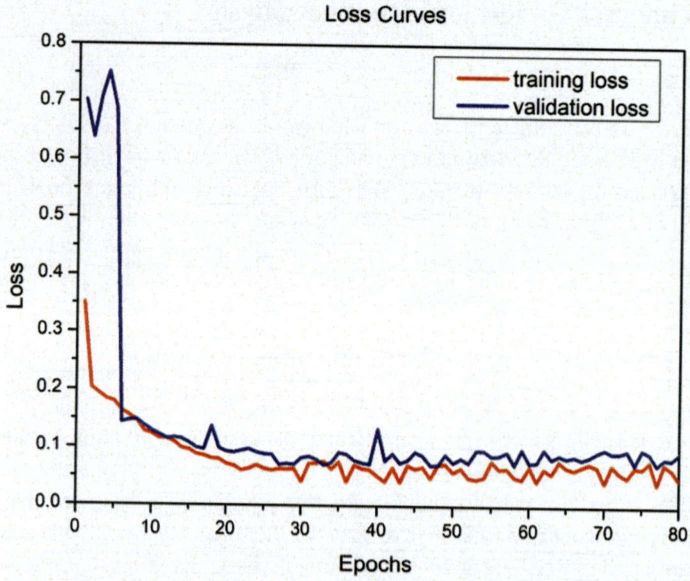

Fig. 1. The multi-label classification loss curves on training data and validation data

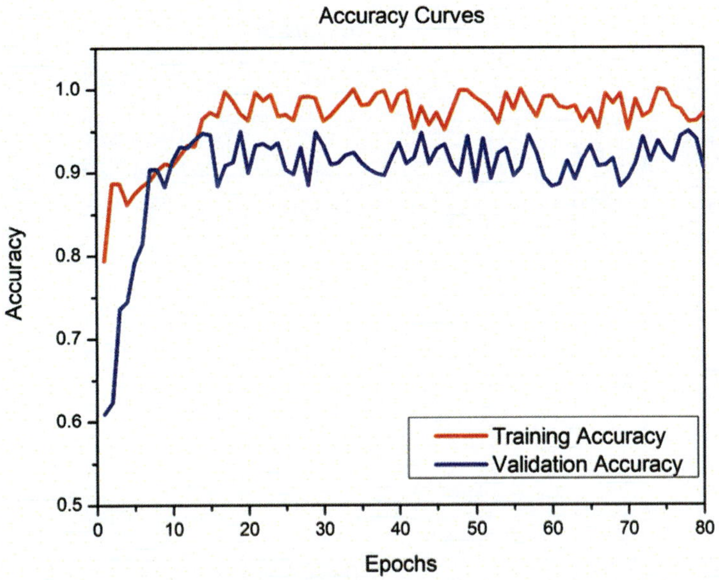

Fig. 2. The multi-label classification accuracy curves on training data and validation data

We can see from the above curves that there is a difference between the training and validation loss. It is obvious on Fig. 2 that the network can get better accuracy on training data. It shows that overfitting has appeared. The major reason of overfitting is that we don't have enough data to train our convolutional neural network. We don't have to look for new images. We can use data augmentation technique to create more images by making minor alterations to our existing images such as changing their size, rotation, zooming, shearing, flipping, changing the brightness level etc.

Then we build our convolutional neural network model. The procedure is the same as we mentioned above, but we use the data after applying data augmentation technique while training.

The Fig. 3 and Fig. 4 shown below are the multi-label classification loss and accuracy curves that we got after using data augmentation technique on our dataset. The accuracy on training set we achieved is 97.8% and on testing set is 96.3%. We can see from the Fig. 4 that the testing accuracy is improved comparing with it in the Fig. 2. It means that our network has a good effect on testing data classification.

We also can find that the accuracy on validation data is almost the same as the accuracy on training data. It proofs that our neural network model works very well and it can achieve an ideal result on new test images after applying data augmentation technique on our data. It is also demonstrated that data augmentation technique is an effective way to eliminate the overfitting for deep Learning. Because this technique can enlarge your dataset when you have limited data and can promote the network performance.

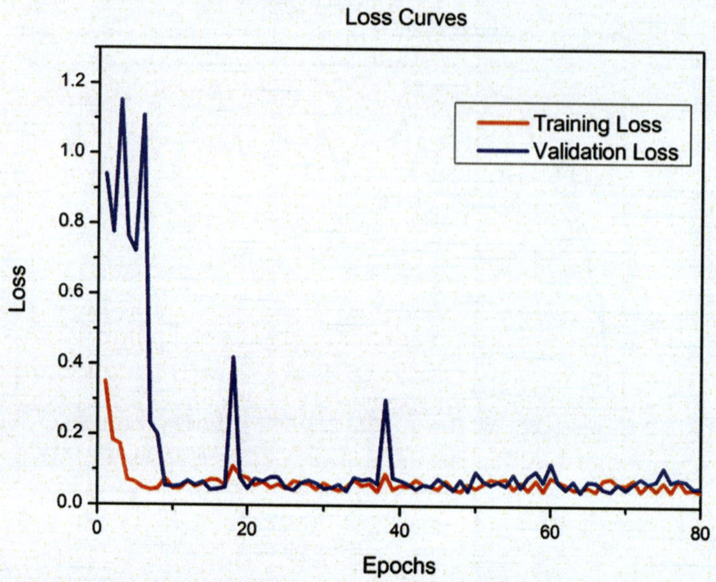

Fig. 3. The multi-label classification loss curves after applying data augmentation

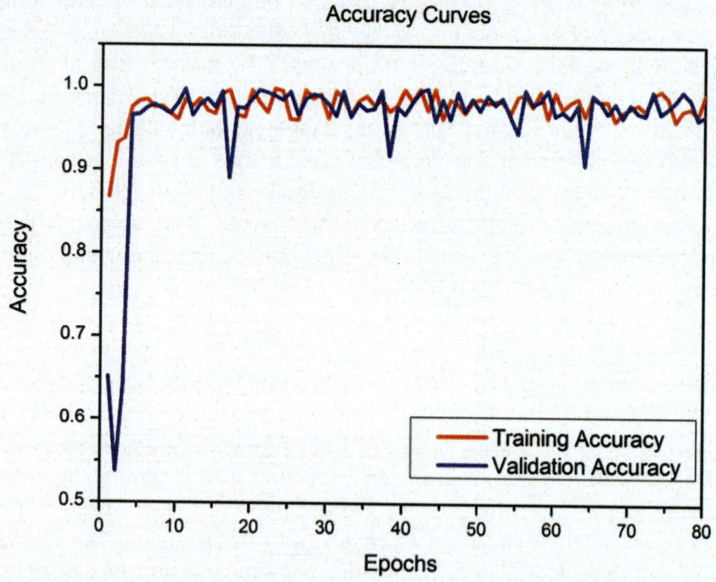

Fig. 4. The multi-label classification accuracy curves after applying data augmentation

4 Conclusion

In this work, we design a convolutional neural network architecture to complete multi-label classification. It was demonstrated from our experiment that convolutional neural network can achieve good results on multi-label clothing image classification. It is also proved by experiment that the dataset is an important factor for training in convolutional neural network architecture. So, how to get more dataset is the key question in multi-label clothing image classification. In addition, data augmentation technique is a good way to get enough data and eliminate the overfitting. The other question is it always take a long time to train our model. The good news is that more and more researchers open their models which has been trained on massive images. We can use these models to start our work.

Acknowledgments. This research was financially supported by Scientific Research Project of Beijing Municipal Education Commission (KM202010012008).

References

1. Sukhbaatar, S., Bruna, J., Paluri, M., et al.: Training convolutional networks with noisy labels. Comput. Sci. (20), 265-278 (2015)
2. Nair, V., Hinton, G.E.: Rectified linear units improve restricted Boltzmann machines. In: Proceedings of the 27th International Conference on Machine Learning, pp. 807–814 (2010)
3. Gong, Y., Jia, Y., Leung, T., Toshev, A., Ioffe, S.: Deep convolutional ranking for multilabel image annotation. In: Proceedings of International Conference on Learning Representations, pp. 1819–1837 (2014)
4. Lowe, D.G., Lowe, D.G.: Distinctive Image Features from Scale-Invariant Key-points. Int. J. Comput. Vis. **60**(2), 91–110 (2004)
5. Dalal, N., Triggs, B.: Histograms of oriented gradients for human detection. In: 2005 IEEE Computer Society Conference on IEEE Computer Vision and Pattern Recognition, CVPR 2005,vol. 1, pp. 886–893 (2005)
6. Hecht-Nielsen: Theory of the backpropagation neural network. In: International Joint Conference on Neural Networks, pp. 445–445. IEEE (2002)
7. Lecun, Y., Bottou, L., Bengio, Y., et al.: Gradient-based learning applied to document recognition. Proc. IEEE **86**(11), 2278–2324 (1998)
8. Krizhevsky, A., Sutskever, I., Hinton, G.E.: ImageNet classification with deep convolutional neural networks. In: International Conference on Neural Information Processing Systems, pp. 1097–1105.Curran Associates Inc. (2012)
9. Simonyan, K., Zisserman, A.: Very deep convolutional networks for large-scale image recognition.arXiv preprint arXiv:1409.1556,2014
10. Szegedy, C.,Liu, W., Jia, Y., et al.: Going deeper with convolutions. In: Cvpr (2015)
11. He, K., Zhang, X., Ren, S., et al.: Deep residual learning for image recognition. In: Proceedings of the IEEE Conference on Computer Vision and Pattern Recognition, pp. 770–778 (2016)
12. Deng, J., Dong, W., Socher, R., et al.: ImageNet:a large-scale hierarchical image database. In: 2009 IEEE Conference on Computer Vision and Pattern Recognition.CVPR 2009, pp. 248–255.1EEE (2009)

13. Wu, F., Wang, Z., Zhang, Z., et al.: Weakly semi-supervised deep learning for multi-label image annotation. IEEE Trans. Big Data **1**(3), 109–122 (2017)
14. Kavukcuoglu, K., Sermanet, P., Boureau, Y.L., et al.: Learning convolutional feature hierarchies for visual recognition. In: Advances in Neural Information Processing Systems, pp. 1090–1098 (2010)
15. Scherer, D., Müller, A., Behnke, S.: Evaluation of pooling operations in convolutional architectures for object recognition. In: International Conference on Artificial Neural Networks, pp. 92–101.Springer,Berlin (2010)
16. Wei, Y., Xia, W., Lin, M., Huang, J., Ni, B., Dong, J., Zhao, Y., Yan, S.: HCP: a flexible CNN framework for multi-label image classification. IEEE Trans. Pattern Anal. Mach. Intell. **38**(9), 1901–1907 (2016)
17. Oquab, M., Bottou, L., Laptev, I., Sivic, J.: Learning and transferring mid-level image representations using convolutional neural networks. In: Proceedings of IEEE International Conference on Computer Vision and Pattern Recognition, pp. 1717–1724 (2014)
18. Girshick, R., Donahue, J., Darrell, T., Malik, J.: Rich feature hierarchies for accurate object detection and semantic segmentation. In: Proceedings of IEEE International Conference on Computer Vision and Pattern Recognition, pp. 580–587 (2014)
19. Abadi, M., Barham, P., Chen, J., et al.: TensorFlow:a system for large-scale machine learning. In: OSDI 2016, pp. 265–283 (2016)
20. Jia, Y., Shelhamer, E., Donahue, J., et al.: Caffe: convolutional architecture for fast feature embedding. In: Proceedings of the 22nd ACM International Conference on Multimedia, pp. 675–678.ACM (2014)
21. He, K., Zhang, X., Ren, S., et al.: Spatial pyramid pooling in deep convolutional networks for visual recognition. IEEE Trans. Pattern Anal. Mach. Intell. **37**(9), 1904–1916 (2015)

Partition-Based Energy Efficient Routing Protocol for Wireless Sensor Networks

Jia Yanfei[1], Xing Liyun[1(✉)], Guo Shuaichao[2], and Zhao Liquan[2]

[1] College of Electrical and Information Engineering, Beihua University, Jilin 132013, China
516104653@qq.com

[2] College of Power Engineering, Northeast Electric Power University, Jilin 132012, China

Abstract. The design of routing protocol method of a wireless sensor network is very important for network performance. It wants to make the sensor network have a long lifetime, a partition-based cluster structure routing protocol is proposed. The proposed protocol uses a cluster structure instead of a chain structure. It selects the wireless sensor node that the current energy has the largest value to be used as the cluster head sensor node in each partition. The minimum spanning tree method is used to form the routing for connecting the clusters and sink sensor node. The simulation results presented in this paper prove that our improved protocol method exhibits better performance than existing algorithms in the aspect of prolonging network lifetime.

Keywords: Routing protocol · Wireless sensor node · Sensor network lifetime · Cluster head node

1 Introduction

Many sensors that monitor various data in the target area consist of the wireless sensor networks (WSNs) They are connected to each other within a self-organizing network to transmit detected data to the sink node for data analysis and processing [3–5]. WSNs are applied in many fields, such as intelligent housing systems, reconnaissance, environmental detection, and traffic monitoring. However, the amount of energy which can be accommodated by sensor nodes is limited, and it is also difficult to replace or charge batteries with more complex applications. Energy has become a key factor which dictates the limit of WSN performance. Therefore, researchers have been endeavoring to make wireless sensor nodes has a better energy efficiency in an effort to make the WSNs has a longer longevity [1, 2].

Routing protocol design has been a major part of WSN development. A good protocol design not only enables a balanced network load, but also make the energy consumption associated with transmitting data reduce [3, 4]. For a WSN, the routing protocols have two classes that are cluster structure routing protocol and a chain structure routing protocol. The LEACH protocol [5] is an established clustering routing protocol which has emerged in the last few years. It selects some nodes as cluster head (CH) nodes based on the probability threshold condition. Other ordinary sensor nodes

M. Tavana et al. (Eds.): IISA 2020, AISC 1304, pp. 871–876, 2021.
https://doi.org/10.1007/978-3-030-63784-2_106

form the cluster based on the signal strengths that are received from the cluster head nodes. The LEACH protocol method has a very simple structure, and many cluster structure routing methods that are based on the LEACH protocol have been proposed, that have LEACH-C, TL-LEACH, and P-LEACH [6, 7]. The HEED protocol [8] is another cluster routing protocols: it introduces the energy that the sensor node has in the current round and average minimum energy that sensor node has in the current round into the selection of the temporary cluster head and final cluster head, respectively. The DSBCA protocol [9] introduces the distance from the wireless sensor node to the base station and the density of the nodes into construction of the load balance routing protocol.

Among chain structure routing protocols, the PEGASIS protocol [10] can be regarded as the most established chain routing algorithm. It uses a greedy algorithm to generate a long chain so as to connect each sensor, whereby these nodes only communicate with their neighboring nodes. Based on the PEGASIS protocol, the HMPBC routing protocol [11] has been proposed: it divides the detection area into 20 partitions of the same size; the wireless sensor nodes communicate to the other wireless sensor nodes by a single chain routing in each partition; and, the chain head nodes transmit sensor data to base station using the minimum spanning tree algorithm.

2 Proposed Protocol

The HMPBC protocol uses single chain routing to transmit data in each partition. With the exception of the initial sensor node, every sensor node must receive data, aggregate data and forward data. This will consume much energy, particularly due to data aggregation. To solve this problem, we propose a Partition-Based Cluster Structure (PBCS) algorithm that uses the cluster routing structure instead of the chain routing structure in each partition. It uses the wireless sensor node that has the highest residual energy to be cluster head, and broadcasts the cluster head sensor node information to the other wireless sensor nodes in the same partition. The other normal wireless sensor nodes become one of the sensor node of the cluster according to the received information in order to form the cluster structure routing. In our proposed method, the cluster head is the only sensor node that consumes energy during data aggregation in a partition. Only one aggregation is required for transmitting data once, and so energy consumption in relation to aggregating data is reduced. For the data forwarding between different partitions, we introduce the residual energy of cluster head and the distance between different cluster heads into a minimum spanning tree method to form the routing of cluster heads. For example, if a current wireless cluster head sensor node is marked as wireless sensor node i, the distance between itself and other cluster heads will be computed. It selects cluster head j that has largest E_{rem}/D_{dis}(E_{rem} expresses the energy that sensor node j has in current round; and D_{dis} is the distance that sensor nodes i is far away from the wireless sensor node j) as the next hop node.

Although our proposed PBCS protocol incurs lower energy consumption during data aggregating, it allows for increased data transmitting distance. This increases the energy consumption involved with transmitting data. In the next section, we analyze the energy consumption within our proposed PBCS protocol with cluster structure and the HMPBC protocol with single chain structure. We consider two nodes in the same position, whereby the initial distance between two nodes equals d. For simplicity, we set $d = 0$, and the change of distance as Δd. The increased energy consumption for transmitting data can be computed using Eq. (1):

$$\Delta E_{TX}(l, \Delta d) = \begin{cases} l \times \varepsilon_{fs} \times \Delta d^2, & \Delta d < d_0 \\ l \times \varepsilon_{mp} \times \Delta d^4, & \Delta d \geq d_0 \end{cases} \quad (1)$$

The increased energy consumption for aggregating data can be computed as follows (2):

$$E_{ag}(l) = l \times E_{DA} \quad (2)$$

In our proposed method, Eq. (1) refers to increased energy and (2) refers to reduced energy. If they are equal, then:

$$\Delta d = \begin{cases} \sqrt{\frac{E_{DA}}{\varepsilon_{fs}}}, & \Delta d < d_0 \\ \sqrt[4]{\frac{E_{DA}}{\varepsilon_{mp}}}, & \Delta d \geq d_0 \end{cases} \quad (3)$$

In our quantitative analysis, we employ 400 sensor nodes in a 200 m × 200 m detection area. For the partition strategy, we divide the detection area into 20 partitions, each being the same size. The size of each partition is 40 m × 50 m, and ε_{fs}, ε_{mp} are 10 pJ/bit/m^2 and 0.0013 pJ/bit/m^2, respectively.E_{DA} is 5 nJ/bit, and l is 1 bit. Based on the above suppositions, we can obtain:

$$\Delta d = \begin{cases} \sqrt{\frac{E_{DA}}{\varepsilon_{fs}}} \approx 22.36, & \Delta d < d_0 \\ \sqrt[4]{\frac{E_{DA}}{\varepsilon_{mp}}} \approx 44.28, & \Delta d \geq d_0 \end{cases} \quad (4)$$

This means that if the increased energy for transmitting data is equal to the reduced energy for that without aggregating data, Δd is 22.36 m when $\Delta d < d_0$ and Δd is 44.28 m when $\Delta d \geq d_0$. According to Eq. (5), we can argue that the mean distance d_{toCH} that the normal wireless nodes are away from the wireless cluster head is 17.85.

$$d_{toCH}^2 = \int_{x=0}^{x=x_{\max}} \int_{y=0}^{y=y_{\max}} (x^2 + y^2)\rho(x, y)dxdy = \frac{M^2}{2\pi k} \quad (5)$$

Whereby x is the length of the partition, y is the width of the partition, the partition area is M; optimal number of cluster heads is expressed as k; and ρ represents the density of the nodes in the partition. Based on Eq. (5), the mean distance between the normal sensor nodes and cluster head sensor node is less than 22.36 m (as obtained using Eq. (4)). This means that the average value of Δd is not larger than 17.85 m. Therefore, we can conclude that the energy increase stemming from transmitting data is smaller than the energy reduction stemming from avoiding aggregating data. Based on the above analysis, we conclude that our proposed PBCS protocol method has lower energy consumption.

3 Simulations and Discussion

In our simulation, we apply the supposition described in Sect. 3. We employ 400 sensor nodes in the 200 m × 200 m detection area, and divide the detection area into 20 partitions, each of the same size and with the same number of wireless sensor nodes. The size of each partition is 40 m × 50 m. The based station is deployed in the center of the detection area. The other parameters are shown (in which *Eo* denotes the energy that each sensor node has before the first round in the network deployment phase; E_{elec} is used to express the energy that is consumed by the transmitter or receiver of the sensor node for 1 bit data; *Data* denotes the packet size; *EDA* denotes the energy consumption of the sensor nodes during data aggregation; *DA* represents the data aggregation rate; εfs and εmp respectively denote the amplifier energy that is consumed in the free space and multipath fading channel models; *do* denotes the threshold adopting the two channel models). The running time is 3000 rounds.

We test our introduced PBCS protocol, the PEGASIS protocol, the HMPBC protocol using MATLAB simulation software. Figure 1 shows the relationship with the number that the survival wireless sensor nodes is and rounds. The sensor node starts to die after about 951, 1744 and 2118 rounds for the PEGASIS, HMPBC and PBCS protocols, respectively. Among the WSNs, our proposed PBCS protocol has the best stability. The last survival sensor node dies at about 1611, 1892 and 221 rounds (PEGASIS, HMPBC and PBCS, respectively). This shows that wireless sensor network has the longest lifetime based on our proposed PBCS protocol. The relationships about the number of rounds and the energy that is residual for the different protocols are shown in Fig. 2. The sensor networks based on our proposed protocols still have more residual energy than networks based on PEGASIS and HMPBC protocol in the same rounds. Therefore, our proposed protocol network consumes less energy than networks based on PEGASIS and HMPBC protocols in the same rounds.

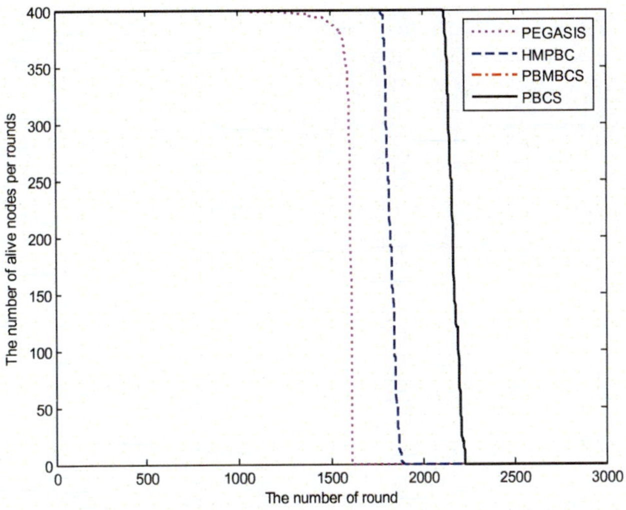

Fig. 1. The number of survival nodes for different rounds

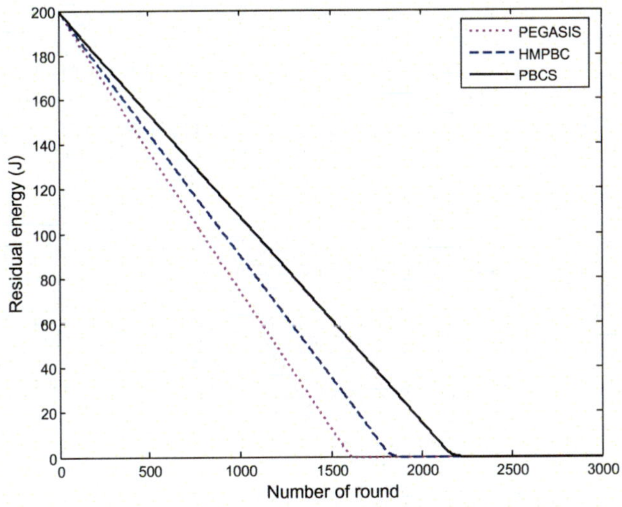

Fig. 2. The residual energy of WSNs

4 Conclusion

We have proposed the PBCS protocol. It replaces the chain structure of each partition with a cluster structure in the HMPBC protocol. The PBCS protocol make the additional aggregation energy that is consumed by the sensor nodes reduce; meanwhile, the PBCS algorithm introduces the remaining energy that the cluster head nodes have into the minimum spanning tree algorithm. The simulation results show that the PBCS algorithms is an improvement in aspect of network stability period, energy consumption and network lifetime, compared to the PEGASIS and HMPBC protocols. Development of a heterogeneous WSN routing protocol is also an important research direction, and we intend to carry out further work in this field.

Acknowledgments. The authors were supported financially by science and technology project of Jilin Provincial Department of Education (JJKH20180336KJ, JJKH20190642KJ); Research Fund for the Doctoral Program of Beihua University (2017-1424).

References

1. Liu, X.X.: Atypical hierarchical routing protocols for wireless sensor networks: a review. IEEE Sens. J. **15**(10), 5372–5383 (2015)
2. Rani, R., Kakkar, D.: Distance based enhanced threshold sensitive stable election routing protocol for heterogeneous wireless sensor network. Wirel. Pers. Commun. **85**, 2643–2656 (2015)
3. Shaikh, F.K., Zeadally, S.: Energy harvesting in wireless sensor networks: a comprehensive review. Renew. Sustain. Energy Rev. **55**, 1041–1054 (2016)
4. Jingjing, Y., Mengchu, Z.: Recent advances in energy-efficient routing protocols for wireless sensor networks: a review. IEEE Access **4**, 5673–5686 (2016)
5. Heinzelman, W.R., Chandrakasan, A.: Energy-efficient communication protocol for wireless microsensor networks. In: Hawaii International Conference on System Sciences, Hawaii, USA, pp. 3005–3014 (2000)
6. Loscri, V., Morabito, G., Marano, S.: A two-levels hierarchy for low-energy adaptive clustering hierarchy (TL-LEACH). In: VTC-2005-Fall. 2005 IEEE 62nd Vehicular Technology Conference, Dallas, TX, US, pp. 1809–1813 (2005)
7. Abdul, R., Musbah, A., Chaitrali, J.: P-LEACH: energy efficient routing protocol for wireless sensor networks. In: 2016 IEEE Long Island Systems, Applications and Technology Conference (LISAT), Farmingdale, NY, USA, pp. 1–5 (2016)
8. Younis, O., Fahmy, S.: HEED: a hybrid, energy-efficient, distributed clustering approach for Ad hoc sensor networks. IEEE Trans. Mob. Comput. **3**, 366–379 (2004)
9. Ying, L., Huan, Q., Weiqun, L.: Load-balanced clustering algorithm with distributed self-organization for wireless sensor networks. IEEE Sens. J. **13**, 1498–1506 (2013)
10. Lindesey, S., Raghavendra, C.S.: PEGASIS: power-efficient gathering in sensor information systems. In: Proceedings of the Aerospace IEEE Conference, Big Sky, MT, USA, vol 3, pp. 1125–1130 (2002)
11. Wang, C.M., Zhang, Y., Wang, X.W.: Hybrid multihop partition-based clustering routing protocol for WSNs. IEEE Sens. Lett. **2**, 1–4 (2018)

Design of Remote Control Signal Hijacking Device Based on Microcomputer System

Tongliang Fan(✉) and Yandong Sun

Department of Maritime Command, China Coast Guard Academy, Ningbo 315801, China
libufan432@163.com

Abstract. This design takes 51 single-chip microcomputer as the core of signal hijacking device, at the same time, a brief introduction of the design hardware and appropriate software are given. The whole system consists of key interface circuit, indicator module, infrared signal receiving module, crystal oscillator module and single-chip control module. And software system includes master control modules, signal transmitting modules, signal hijacking function. The design can not only learn and imitate the instructions of the original remote controller, but also prevent the original infrared remote controller from controlling the receiving device.

Keywords: Infrared remote control · Signal hijacking · Single-chip

1 Introduction

With the constant development of science, people's life is more and more inseparable from the participation of electronic devices. In the course of law enforcement, we have seen an increasing proportion of crimes committed using electronic devices. Among all kinds of electronic devices, infrared remote control has most widely used remote control means which is small size, low power consumption, powerful function, low cost. Based on the research of infrared remote controller, the infrared signal hijacker is made. TC9012F infrared remote controller is widely used because of its cheap price and strong practicability. The research and implementation of this paper is mainly based on the signal hijacking device of 51 single-chip microcomputer, SM0038 integrated infrared receiver is used as the receiving device of infrared remote control system, which is composed of key input module, infrared signal receiving module and minimum single-chip microcomputer system.

2 System Design Layout and the Basic Principle

The design is composed by the following module: the receiving module, the transmitting module and the hijacking module. The whole structure is illustrated in Fig. 1. The receiving module is the basis of the whole function, on which the function of signal transmitting or signal hijacking can be realized. The design can realize the learning of the hijacked signal, and then transmit the signal with the same effect as the

M. Tavana et al. (Eds.): IISA 2020, AISC 1304, pp. 877–884, 2021.
https://doi.org/10.1007/978-3-030-63784-2_107

hijacked signal to replace the hijacked signal control receiver. The hijack function can be used to cut off the connection between the hijacked signal and the receiving device in the process of being hijacked. The function of the receiving part of hijacked signal is realized by the integrated receiving head of SM0038. The signal transmitting part relies on the single chip microcomputer system (MCU) and the infrared transmitting unit to realize the expected function. Signal hijack part also needs MCU to control.

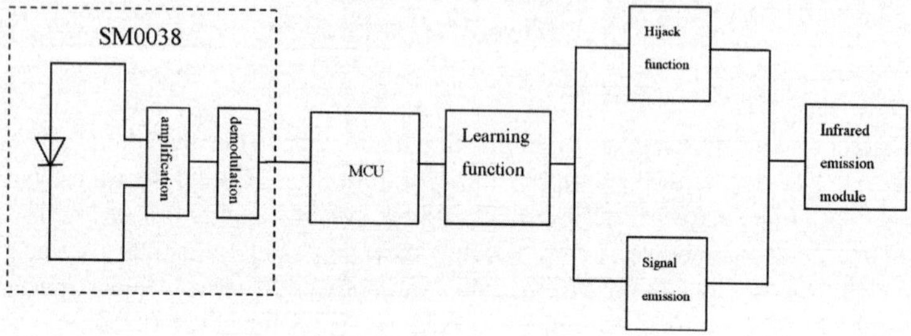

Fig 1. The system structure frame

2.1 Transmission Principle

This design is based on the hijack of the 9012 remote control, which adopts the infrared communication NEC protocol. The pulse signal sent by TC9012 is composed of the boot code, the user code, the data code and the inverse code of the data code. The remote coded pulse sent by the 9012 remote control can form the remote control signal after pulse amplitude modulation to the 40 kHz carrier, which is then amplified and filtered by the driving circuit and transmitted by the infrared transmitting tube. The boot code, as a flag to received data, consists of a high level of 4.5 ms and a low level of 4.5 ms. The user code and the data code judge the code value by the time interval between the pulses. The first 16 bits are the user code, and the last 16 bits are the 8-bit data code and the inverse of the 8-bit data code. When it sends "0" code, the carrier transmission lasts for 0.52 ms, and the non-transmitting carrier lasts for 0.52 ms. Therefore, the pulse time interval of "0" code is 1.04 ms. When sending "1", a carrier of 0.52 ms is transmitted, and the non-transmitting carrier is 1.68 ms, so the pulse time interval of "1" code is 2.25 ms. When the button of the remote control is pressed, the remote control will send out a string of binary code as shown in the Fig. 2, that is, a frame of data.

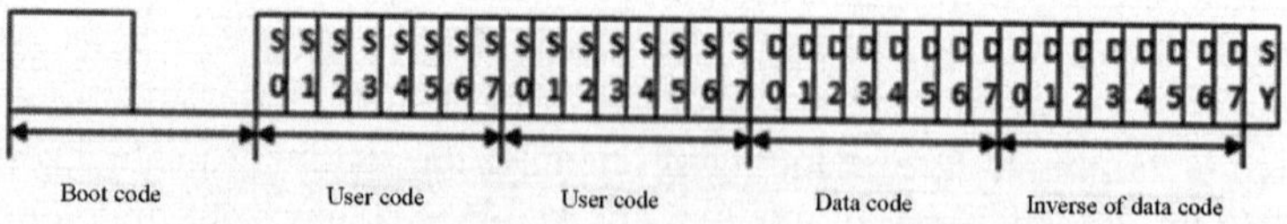

Fig. 2. Frame format of remote control signal coding

2.2 Reception Principle

The signal sent by 9012 remote control is a frame of data that is composed of binary codes of "0" and "1". When the infrared receiving module receives the boot code, it represents the beginning of the received data frame. The user code sent by TC9012F is four-digit hexadecimal data, and the user code sent by all buttons of the same remote control device is the same. The data code is two bits of hexadecimal data, which is the symbol that distinguishes each key from other keys on the same remote control. The last two bits of hexadecimal data represent the inverse of the data code, which is compared with the data code to determine whether the received data is correct. The sum of the data code and its inverse should be 0FFH. The core of the design is to use the AT89C51 minimum single-chip microcomputer system to decode the hijacked signals. After receiving the infrared remote control signal through the receiving module, the signal is transmitted from the output terminal to the external interrupt of the AT89C51 single-chip microcomputer (INTO, the input terminal P3.4 pin). The MCU only needs to analyze the data code of the infrared signal sent by the 9012 infrared remote control through its internal program, then it can carry out further learning or hijacking of the hijacked signal.

3 The System Hardware Design

With AT89C51 as the main control core of the signal hijack, the expected functions are completed by combining the key input module, the indicator lamp module and the infrared signal receiving module. The overall system structure is as follows:

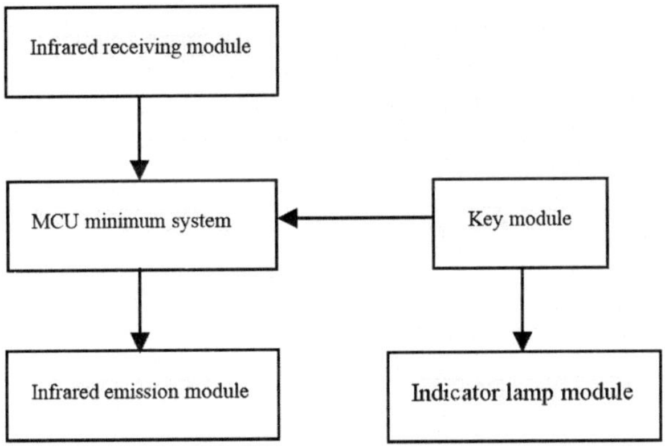

Fig. 3. System architecture

This design has three control keys. The circuit has three working states: "learning" state, "controlling" state and "hijacking" state. "Learning" is connected to the P3.2 / INTO (external interrupt 0) pin of the MCU; "Control" is connected to the P1.0 pin of the MCU; "Hijack" the P3.0 /RXD (serial input) pin of the MCU. Each key is externally connected with a 10 KΩ pull-up resistor. The I/O port connected by each button is at a high level when no switch is pressed. When the switch is pressed, the I/O port is pulled to a low level. This scheme of one I/O port corresponding to one switch has the advantages of simple and reliable circuit and simple design program, but occupies more resources of I/O port. In view of the design requires less keys, so this way is enough to meet the demand.

When the study button is pressed, the yellow light is on, indicating that the system is ready to learn. When the infrared receiving module detects the hijacked signal, the yellow light goes out and the blue light comes on, indicating that the system has finished learning. At this point, you can either choose to hijack the signal or transmit the learned signal. If the hijacking function is selected, press the hijacking key and the blue light flashes. At this time, the hijacked infrared signal transmitter loses control of the receiver. If the transmitting function is selected, press the transmitting button and the blue light will be on to indicate the transmitting signal. Press the launch button again to turn off the blue light, indicating that the initial state is restored. You can continue to press the button to send signals. Every time you press the button, a signal will be sent.

According to the coding method of 9012 infrared remote control, we choose to demodulation the hijacked signal with the SM0038 integrated receiver head. The SM0038 integrated receiver consists of infrared receiving tube and amplifying circuit, which can receive infrared signal modulated by pulse code. It has the advantages of small size, good sealing, high sensitivity and low price. The demodulation frequency of the SM0038 integrated receiving head is 38KHZ. When the infrared pulse signal of 38 KHZ is received, the output is low level.Otherwise the output is high level. SM0038 has three external pins: VCC, GND and pulse signal output pin OUT. The external pins are shown in the Fig. 3. The VCC is connected to the power supply +5 V and filtered by the capacitor to avoid the interference of the power supply; GND is connected to the ground line of the system (0V); the output of the pulse signal is connected to the interrupt input pin of the CPU, namely the P3.4 pin of 89C51.

4 System Software Design

Signal hijacker is an integrated infrared receiver of SM0038 to receive infrared signals, after decoding and amplification, signals output from the OUT port, then transmit to the single chip microcomputer P3.4 pin. The learning function is controlled by connecting the learning key in P3.2 port. After learning, the characteristics of the hijacked signal are stored in the MCU for the next instruction. After successful learning, we can choose to hijack the hijacked device or send the learned signal.

The main function of the main program is such as: initialize the single chip microcomputer, scan the key cycle, and perform different functions according to the button action. The flow chart of the main program is as follows (Fig. 4):

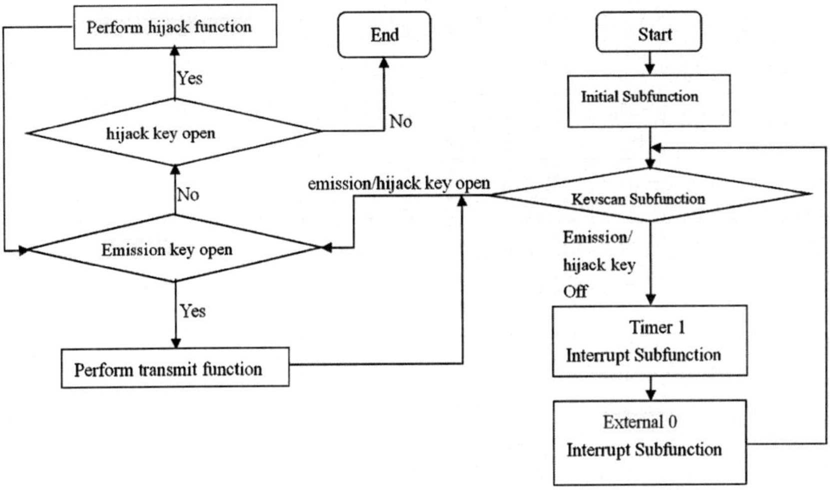

Fig. 4. The program flow diagram of signal hijacking

When performing the learning function, turn off all interrupts. At the same time, the operation control bit TR1 of timer 1 is set to 0, that is, the timer 1 is turned off. The starting bit is reset to 0. When the learning key is pressed, the external interrupt port P3.2 of the MCU changes from high level to low level. Turn on the learning status indicator and turn off the launch indicator. After the button is released, wait for the infrared receiver to receive the signal. Judge whether the remote control signal input port P3.4 is at a high level. If it is, it means it has received the signal and started to record the hijacked signal. Temp is defined as the number of the low level times when the starting bit is received, and the initial value is assigned to 0. The loop while function, *nop*(); Function represents the time of an empty cycle of a machine instruction. Since we choose the crystal oscillator with the frequency of 11.0592 MHZ, the time to run an empty instruction is 1μs. Looping once, temp increases by 1. Similarly, define Top as the number of times that a high level was received in the starting bit. By comparing received the time of the high level with being hijacked in the start bit high level code, determine whether the received signal for us is to hijack the signal from the remote control. If not, we will set array[0] to 0, the external interrupt 0 interrupts, open the global interrupt control, directly turn on emission light, go out learning status indicators, the learning process is over; If the learned signal is indeed intended to hijack, use the TAB array to store the high level pulse width data and the low level pulse width data.

By calling the key function sub function, we can realize the signal transmitting function. First, the key state is judged according to the pin voltage value. When the key is pressed, the P1.0 port is at low level. The delay function is used to delay 10 ms to eliminate keyboard jitter and reduce misjudgment and interference. Use the while function to determine the status of the key. After the key is released, open timer 1 interrupt. Since the TMOD is set to 0×22 in the initial sub function, the gate control bit is 0, and the timer start and stop are only controlled by TR0 or TR1 in the TCON register, we set the timer run control bit TR1 to be 1, that is, start timer 1. Temp is the number of times the low level in the starting bit is received, and the time to send the starting bit is taken out of loop. Turn off the remote output, turn on the interrupt of timer 1, and start timer 1. Judge whether the learned signal is a signal to be hijacked, and if not, terminate the transmission; if so, the data in the TAB array is sent out through the boot code of the hijacked signal using for loop. According to the encoding mode of the 9012 remote control, since the user code of each remote control is the same, only the boot code, data code and the inverse code of the data code can be transmitted to realize the control of the receiving device. Since every bit of code is composed of transmitting pulse and non-transmitting pulse, and only "0" code and "1" code are distinguished by duty cycle. So even address does not emit pulse, and odd address transmits modulation pulse. After each data is sent, turn off the infrared emission indicator, turn on the timer interrupt 1, and start the timer 1.

By calling the key function sub function, we can also achieve the signal hijacking function. First, the key state is judged according to the pin voltage value. When the hijack key is pressed, the P3.0 port is at low level. The delay function is used to delay 10ms to eliminate keyboard jitter and reduce misjudgment and interference. Use the while function to judge the status of the key. If the level of port P3.0 is kept at high level, the timer 1 will be interrupted. The principle of the hijacking function is to send a set of coded pulses containing only the guidance code, so as to realize the interference to the hijacked signal. At last the receiving device cannot recognize the received hijacked signal or the received hijacked signal is scrambled code. The rest of the statement is the same as the transmitting program. Only when the odd address, it does not send a modulation pulse.

5 Testing and Analysis

The design of the hardware circuit uses bread board welding. According to the pre-determined scheme, Weld resistance, capacitance, switch, diode, MCU and other components to the bread board fixed position. The pins between components are connected by wires, and a layer of insulating glue is added to prevent short circuit. The power VCC is connected together. Hardware circuit design and production of physical objects are shown in Fig. 5.

Fig. 5. Circuit diagram of signal hijacking

When performing the learning function, turn on the receiving device and the signal hijacker, Open the USB serial port assistant, and press a key on the remote control to transmit the signal. And the message number of the key will be displayed in the window of the USB serial port assistant. Press the study key and the yellow light of the study indicator LED1 will be on. Press a key on the remote control, and when the jack learning ends, the LED2 blue light will light up. Meanwhile, the signal number of the key will be displayed in the window of the USB serial port assistant.

When the hijacking function is performed, on the basis of the end of the learning function, press the hijacking key and the LED2 blue light flashes. At this time, you can judge: If the blue light flashes too fast, it means that what you have learned is not the signal to be hijacked; if the blue light flashes normally, the hijack function can continue. In the state of hijacking, pressing any button on the remote control can result in two situations: one is that the signal receiving device of the hijacked equipment cannot be recognized; the other is that the received signal is scrambled code, which cannot enable the receiving device to perform the expected function. Therefore, when you press the button, the window of USB serial assistant may not have a new signal code, or it appear scrambled code.

When performing the transmitting function, on the basis of the completion of the learning function, press the emission key, and the LED2 blue light will be on. At this time, the learned hijack signal has been sent out. The same number of the hijacked signal can be observed in the window of USB serial assistant.

6 Conclusion

The research and implementation of this paper is mainly based on the signal hijacking device of 51 single-chip microcomputer, SM0038 integrated infrared receiver is used as the receiving device of infrared remote control system. The design is mainly aimed at

TC9012F as the core of the 9012 infrared remote controller. It is composed of keystroke input module, indicator module, infrared signal receiving module, crystal oscillator module and single-chip control module. The hijack device adopts the principle of measurement of pulse width, can learn remote encoding, and launch the imitate signal, and when the hijack is learning and emulating the original remote command, it can prevents the original infrared remote control from controlling the receiver. The experimental results prove the feasibility of the scheme.

Acknowledgements. This work was supported by the online open course of China Coast Guard Academy No KF2020002. It was also supported by teaching reform subject of China Coast Guard Academy No YBKT202042.

References

1. Fan, T.-L., Ye, W.: Design on Security Alarming System of Vehicle terminal equipment. ITA, p. 03005 (2016)
2. Sun, Y.: Design of intelligent remote controlling calendar **19**(SUPPL.2), 298–304 (2011)
3. Hao, W.: Design and implementation of temperature and humidity gathering communication system based on ZigBee technology. J. Yancheng Inst. Technol. **26**(1), 64–68 (2013)
4. Wan, J., Zou, C., Zhou, K., Lu, R., Li, D.: IoT sensing framework with inter-cloud computing capability in vehicular networking. Electron. Commer. Res. **14**, 389–416 (2014)

Design and Implementation of Simple Radio Signal Transmitter

Yandong Sun and Tongliang Fan(✉)

Department of Maritime Command, China Coast Guard Academy, Ningbo 315801, China
libufan432@163.com

Abstract. To complete the transmission of information, a radio transmitter must be used. This design uses a common-emitter triode amplifier circuit as an amplifier. The capacitance three-point oscillation circuit is used as the modulation circuit of the radio transmitter.

Keywords: Radio signal transmitter · Communication · Circuit

1 Introduction

The wireless transmitter has the advantages of low cost, simple and convenient use, few constraints, and can be widely used in many military fields. For example, using amplitude modulation technology for FM broadcasting can avoid lightning or other electromagnetic interference, and using single side band technology for communication at isolated points such as ships and airplanes can save frequency band and make more effective use of transmitting power. This design is a device that converts the surrounding speech signal into electrical signal, amplifies it through the common emitter triode amplifier circuit, modulates it through the triode to match the tuning circuit, and then transmits the signal through the antenna. This design is suitable for analog and digital narrowband, sprcad spectrum radio links, Cordless telephone, handheld data terminals, Wireless Lan and other applications [1–4].

2 System Design Layout

Design and manufacture of radio transmitter, framework diagram, functional requirements and selection of main components of radio transmitter.

The radio transmitter is mainly composed of two circuits: One is the audio signal acquisition and amplification circuit, which mainly uses the electret microphone to collect nearby signals and amplify them to the next stage, including a common emitter triode amplifier circuit. When there is no AC signal input, the circuit enters a static working state supported by the DC power. When there is an AC signal input, the triode amplifies the base signal and transmits it to the next stage. The other is a capacitive three-point oscillation circuit, which is mainly tuned to the amplified base signal and radiated out the modulated signal through the antenna.

M. Tavana et al. (Eds.): IISA 2020, AISC 1304, pp. 885–892, 2021.
https://doi.org/10.1007/978-3-030-63784-2_108

Most of the components of the design are on the pre-audio amplifier circuit. The pre-audio amplifier circuit adopts the common emitter triode amplifier circuit. It is well known that the common emitter amplifier is the most common amplifier application. There are three main features: 1. Output signal and input signal are phase inversion; 2, the current and voltage gain are large; 3. It is often used as the middle stage of amplifying circuit.

Circuit requirements in radio transmitters, however, because of the received signal is very weak, need to promote the amplifier output. Ans amplifying circuit between the electret microphone input and tuned circuit, as the middle class, just share the same characteristics of the emitter amplifier circuit. So common emitter amplifier circuit is very appropriate. In the pre-audio amplifier circuit, the 2N3904 triode is used. In the common emitter triode amplifier circuit, the triode is NPN type and rated power is relatively small. So 2N3904 triode meets the design requirements, the zoom effect of $\beta = 100$–400 also meets the design needs. More satisfactory is that it is widely used. The problems encountered are easier to solve.

3 Practical Circuit

This chapter mainly introduces the two circuit combinations of radio transmitter. The first part is the pre-audio frequency amplifier circuit, which is mainly the common emitter triode amplifier circuit. The second part is the rear radio frequency circuit, which is mainly the capacitive three-point oscillation circuit. The two parts are inter-related and interact with each other to realize functions together.

3.1 Radio Signal Transmitter Circuit

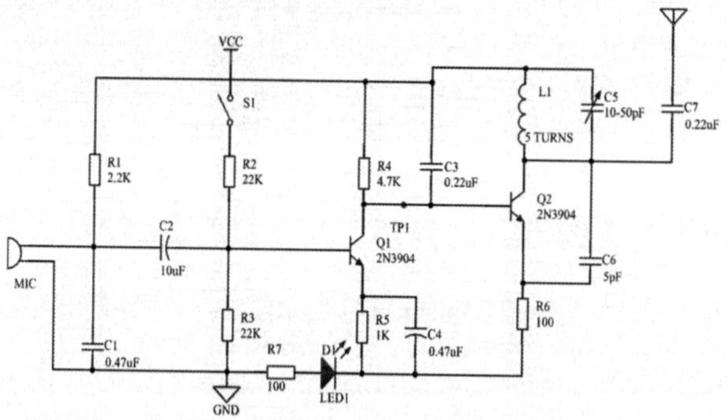

Fig. 1. Circuit of wireless transmitter.

Figure 1 shows the general circuit principle diagram of the radio transmitter. The whole circuit is composed of two parts: the audio frequency amplifier circuit controlled by transistor $Q1$ and the oscillation circuit controlled by transistor Q2. As you can see from the diagram, the audio preamplifier takes up most of the wiring and components in the circuit. Triode *Q*1 will send the amplified signal to the base electrode of *Q*2. When the power is connected, *C*3 and the adjustable capacitance *C*5 and *C*6 will be charged. *C*3 is the coupling capacitance, so that the AC signal will pass through and the DC signal will be insulated. For the rear radio frequency segment, *L*1 is the working inductance, *C*6 and variable capacitance *C*5 are the working capacitors. Coil *L*1 generates a magnetic field. As the base voltage increases, the transistor is switched on and the internal resistance is connected in parallel to both sides of the 5pF capacitor *C*6. When the voltage of transistor *Q*1 base reaches the working voltage, the signal will be abnormally disordered. So at this point, if near the working voltage, base voltage of triode continue to get bigger, 5 pf capacitor will reverse to stop the moving emitter voltage. As energy is run out of the capacitor that doesn't keep emitter voltage, base-emitter voltage is reduced, the transistor *Q*2 is cut off, the current does not flow into the coil any more, so magnetic field fades. This generates a voltage in the opposite direction, charging capacitor *C*5 in the opposite direction. This reverse voltage will also result in the voltage drop of emitter resistor R6 and the charging of 5pF capacitor *C*6, so that the transistor enters a stable cut-off state. When the 5pF capacitor *C*6 is charged, the emitter voltage of transistor drops to where one of the transistors begins to operate, and the current flows into the coil and resists the fading magnetic field. The voltage on the coil is reversed, resulting in a decrease in collector voltage, which is transmitted through the 5pF capacitor *C*6 to the emitter, causing the transistor to enter a deeper conduction and short-circuit the 5pF capacitor *C*6. The process goes on in cycles. Therefore, *Q*2 generates a stable AC oscillating signal here.

3.2 Pre-audio Amplifier Circuit

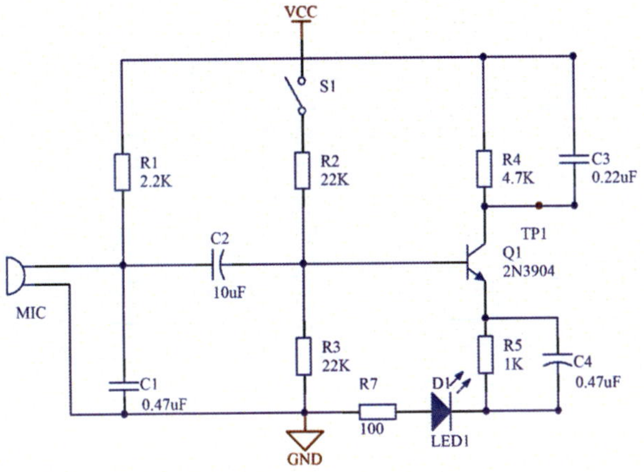

Fig. 2. Pre-audio amplifier circuit

The pre-amplifier circuit is shown in Fig. 2: C1 and C2 act as filters to separate the AC signal and DC signal. R2 and R3 are indispensable bias resistors for the Common Emitter Amplifier. The power voltage Vcc is then loaded onto the base electrode of triode Q1 to form the bias voltage and bias current. Q1 is triode, R5 is emitter resistor and negative feedback resistance. Assume that the input AC signal increases suddenly, which causes the transistor base current increases suddenly, the voltage drop of feedback resistor R5 also increase. Namely the launch voltage of triode emitter increases. This will cause the transistor forward biased voltage, between the base and emitter voltage, decreases. At last base current decreases. C4 is a shunt capacitor in parallel with R5 of the transistor emitter, playing a role of negative feedback. Its function is "Separate DC, Through AC", so that the circuit has AC negative feedback but no DC negative feedback, so as to improve the stability of the amplifier circuit and suppress noise interference.

3.3 Rear Radio Frequency Circuit

The rear radio frequency circuit is shown in Fig. 3: This is an oscillation circuit based on a capacitive three-point oscillation circuit. The radio frequency section mainly consists of transistor *Q*2, hand-wound coil *L*1 and adjustable capacitor *C*5. The coil and the capacitor form a tuning circuit that will oscillate at a frequency in the FM radio band. The frequency depends on the capacitance setting of the tunable capacitor. Since this oscillating circuit is controlled by transistor *Q*2, which in turn is controlled by transistor *Q*1, variations in the audio preamplifier cause modulation of the radio frequency portion. Coil *L*1 and capacitance are the necessary components of the oscillation circuit. Only when the phase and amplitude conditions are met, the oscillation circuit starts to vibrate. The modulated current is output through the collector and then transmitted through the antenna after coupling with the capacitor *C*7.

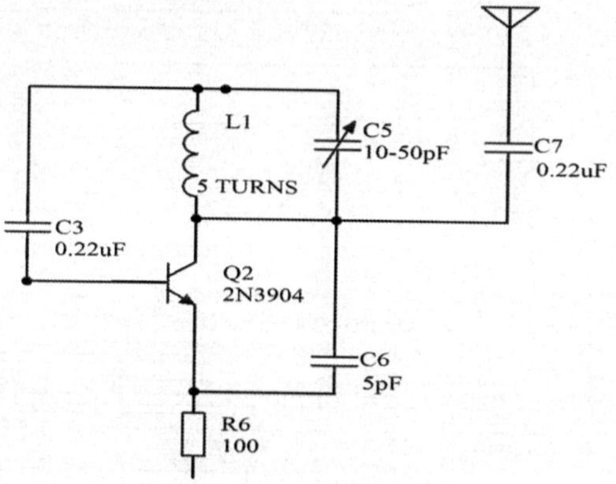

Fig. 3. Tuned circuit

There are two conditions for starting vibration of capacitance three-point oscillation circuit: one is the phase condition, the other is the amplitude condition. The phase condition is that the reactance properties of the triode base electrode and the components connected with the emitter electrode in the capacitor three-point oscillator circuit should meet the requirements. The equivalent circuit of the oscillator circuit is shown in Fig. 4. C junction capacitance is the polar capacitance between the base and the collector. The base electrode of $Q2$ is connected with the C junction capacitance and the inductor coil $L1$, and the $Q2$ emitter stage is connected with the C junction capacitance and $C6$, so the starting phase condition is satisfied. The second condition for starting vibration of an oscillating circuit is that the amplitude condition. It is required that the loop gain must be greater than 1, and the formula for the loop gain is $T(j\omega) = \dot{A}\dot{F} = T(\omega_{osc})^{ej\phi_T(\omega)}$, Then the amplitude condition for starting vibration of the oscillation circuit can be transformed into $T(\omega_{osc}) = |\dot{A}\dot{F}| = AF > 1$. In this formula, $A = \frac{V_{output}}{V_{input}}$ is the amplitude of the amplifier gain. F is the feedback coefficient of the oscillation circuit, and the formula is $F = \frac{V_{feedback}}{V_{input}}$. If F is too small, the oscillation circuit is not easy to vibrate. If F is too large, the input impedance of the triode will cause the frequency instability of the tunable circuit, and it will distort the oscillation waveform, even cause the circuit not to vibrate in severe cases. In order to satisfy the starting vibration of the oscillating circuit and stabilize the quality of the signal, F is usually 1/8–1/2.

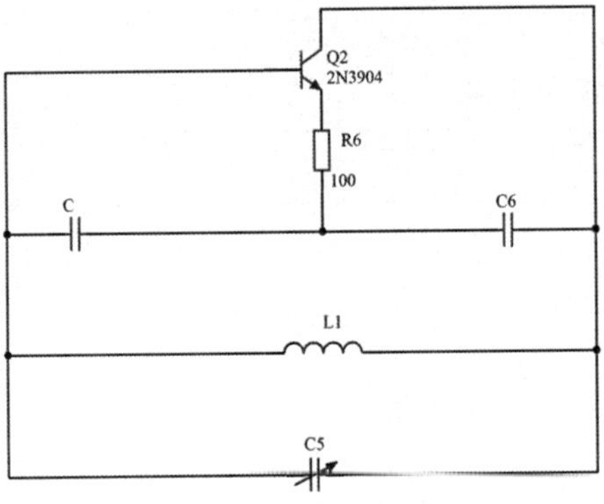

Fig. 4. Equivalent circuit of oscillation circuit

In the tunable circuit, coil $L1$ and tunable capacitor $C5$ constitute the tunable circuit. Coil and capacitor jointly decided the oscillation frequency of the oscillation circuit.

4 Testing and Analysis

The radio transmitter can pick up the voice signal in the room and send it by the antenna after amplification and modulation. This chapter mainly compares and explains the test results, obtains the data through the test, analyzes the data, examines the advantages and disadvantages, finds the problems, and consults the improved methods. The hardware circuit design material is shown in Fig. 5, and the specific physical components are shown in Table 1.

The hardware circuit design material is shown in Fig. 5, and the specific physical components are shown in Table 1.

On the basis of the functional test, three different conditions are set to test what conditions will have an impact on the transmission effective distance: The antenna length is different; there is no obstacle; there is no electromagnetic interference.

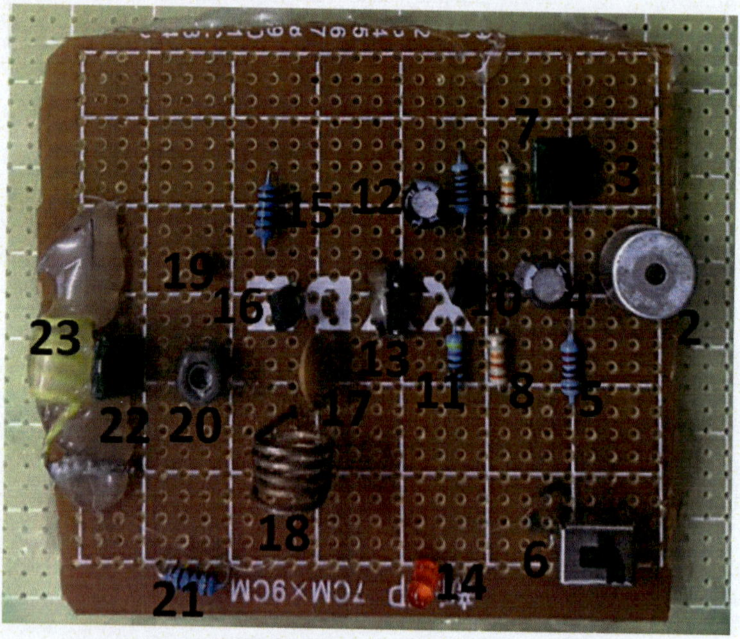

Fig. 5. The practicality picture of radio transmitter

In the actual design, in order to compare the influence of the antenna to the signal transmission, we wound a long, thick antenna and a thin, short antenna. The effective distances at which they can transmit audio signals are compared. The test data are shown in Table 2.

Table 1. Physical components

1 9V battery	2 Electret microphone	3 C1 (0.47uF)	4 C2 (10uF)
5 R1(2.2 kΩ)	6 Switch S1	7 R2(22 kΩ)	8 R3(22 kΩ)
9 R4(4.7 kΩ)	10 $Q_1 = NPN$ 2N3904	11 R5(1 kΩ)	12 C4 (0.47uF)
13 Test point (TP1)	14 LED D1	15 R6(100 Ω)	16 $Q_1 = NPN$ 2N3904
17 C3 (0.22pF)	18 L1	19 C6 (5pF)	20VC(0.22pF)
21 R7(100 Ω)	22 C7 (0.22uF)	23 Antenna	

Table 2. Performance comparison of different antennas

Antenna Type / Number of Measurements	A quarter wavelength antenna (unit: m)	A half - wavelength antenna(unit: m)
1	47.0	30.75
2	48.5	31.5
3	47.75	30.0
4	47.0	32.25

The data in Table 2 show that under the condition of wild, when antenna length is close to a quarter wavelength, its propagation distance is more farther, due to the characteristics of the antenna. It was proved that only antenna length is close to a quarter wavelength, transmission efficiency and effect is best, so the test result is accord with the theoretical value.

In the actual design, in order to compare the influence of obstacles on signal transmission, I transmitted under the conditions of no obstacles and in the room, meanwhile received outside the room. The effective distances at which they can transmit audio signals are compared. The test data are shown in Table 3.

Table 3. Comparison of different obstacles.

Obstacle Type / Number of Measurements	No obstacles (unit: m)	Room transmit, room reception (unit: m)
1	27.75	12.0
2	28.5	11.25
3	27.75	11.25
4	27.0	12.75

In the actual design, in order to compare the influence of electromagnetic interference on signal transmission, it is well known that there will be a large number of interference signals in thunderstorm, which have a great impact on the transmission of radio signals. The tests were conducted in the absence of signal interference and in the thunderstorm. The effective distances which they could transmit audio signals were compared. The test data are shown in Table 4.

The data in Table 4 show that: in thunderstorms, the transmission efficiency is very low and the transmission distance is also extremely shortened. Lightning produces a lot of radio noise, and the transmission environment is very different from normal. Clouds,

Table 4. Comparison under different disturbances.

Antenna Type / Number of Measurements	No signal interference (unit: m)	Thunderstorm (unit: m)
1	27.75	9.75
2	28.5	9
3	28.5	10.5
4	28.5	9

airflow and suspended particles in the air have a great influence on the transmission process. There are many obstacles in the transmission process, and lightning strike or fire may even occur, resulting in the damage of equipment.

5 Conclusions

Compared with the traditional radio signal transmitter, the advantage of this design is simple circuit structure, reliable performance and good signal quality, so it is suitable for near distance communication. In order to achieve better effect, we still have a lot of work to do. Firstly, the propagation distance can be further improved by adding power amplifier circuit. Secondly, using a noise-reducing microphone circuit, one microphone collects positive signals and the other microphone reversely connects to the circuit to generate negative signals. When two electrical signals are added to each other, the noise is eliminated. The signal quality may be improved from the input of pre-audio amplifier circuit.

Acknowledgements. This work was supported by the online open course of China Coast Guard Academy No KF2020002. It was also supported by teaching reform subject of China Coast Guard Academy No YBKT202042.

References

1. Fan, T.-L., Ye, W.: Design on Security Alarming System of Vehicle terminal equipment. ITA, p. 03005 (2016)
2. Shitanda, I., Takada, K., Sakai, Y., Tatsuma, T.: Amperometric biosensing systems based on motility and gravitaxis of flagellate algae for aquatic risk assessment. Anal. Chem. **77**, 6715–6718 (2005)
3. Rosmini, M.R., Signorini, M.L., Schneider, R., Bonazza, J.C.: Evaluation of two alternative techniques for counting mesophilic aerophilic aerobic bacteria in raw milk. Food Control **15**, 39–44 (2004)
4. Yang, L., Li, Y., Griffis, C.L., Johnson, M.G.: Interdigitated microelectrode (IME) impedance sensor for the detection of viable Salmonellatyphimurium. Biosens. Bioelectron. **19**, 1139–1147 (2004)

Multi-station Integration-Based Coordinated Control for Voltage Support in Active Distribution Network

Ma Weijing[1], Jin Qiang[1](✉), Li Qiang[2], Chen Jing[3], Yuan Fusheng[4], Xu Jian[5], and Chen Yufeng[5]

[1] State Grid Economic and Technological Research Institute Co., Ltd., Beijing 102209, China
jinq1980@126.com

[2] State Grid Information &, Telecommunication Co., Ltd., Beijing 102211, China

[3] Beijing Electric Power Economic Research Institute Co., Ltd., Beijing 100037, China

[4] State Grid Information &, Telecommunication Co., Ltd., Beijing branch, Beijing 100031, China

[5] Beijing Sifang Engineering of Relay Protection Co., Ltd., , Beijing 100080, China

Abstract. With the incereasing penetration of distributed generaitons (DGs), regulations of voltage along feeder lines in active distribution network (ADN) are becoming unprecedentedly challenging. Fortunately, technological advances in distributed energy resources (DER) provide promising and powerful ways to cope with these challenges. In this paper, we propose a coordinated control scheme for voltage support in ADN based on multi-station integration, which integrates various DER stations as an integrated station (IS). First, we study the voltage impact mechanism of IS. Then we put forward a local and node-centered topology information model, and propose a distributed adaptive voltage-reactive power control approach, and an expanded voltage-reactive power control approach for backup. Finally, the feasibility and effectiveness of the proposed scheme are verified by Matlab simulation.

Keywords: Multi-station integration · Coordianted control · Voltage support · Active distribution network · Node-centered topology information model

1 Introduction

With the booming of distributed generation (DG) technologies and the expectation for green lifestyle, distributed energy resources (DERs) have been playing an increasingly important role in energy supplies. To access and consume DERs more effectively and efficiently, it is inevitable trend for distribution network to transform from radial passive network to active distribution network (ADN) characterized by multi-source access, flexible control, active management and whatnot [1].

M. Tavana et al. (Eds.): IISA 2020, AISC 1304, pp. 893–903, 2021.
https://doi.org/10.1007/978-3-030-63784-2_109

The impact of large-scale and high-penetration access of DGs on distribution network is twofold. On one hand, it provides numerous control methods and increases control freedom for the active distribution network. The access of DGs helps to alleviate the low voltage problem in distribution networks and improve the voltage level of active distribution network. On the other hand, it also has a negative impact on voltage-reactive power control, power quality and stability [2]. When the connected power capacity increases above a certain threshold, it will cause overvoltage problems. It is pointed out in considerable amount of literature that the voltage at the access point and neighboring areas increases when DG injects power into the distribution system, and decreases when it stops working. Moreover, the intermittent operation of DGs is reflected in voltage volatility and deviation from the rated voltage of distribution network. This is the major challenge of voltage control in active distribution network [3].

In the existing research, according to the dependence degree on the communication system, voltage control methods for ADN can be divided into two main categories, local autonomous control and communication-based cooperative control. According to the way of information exchange between the participating parties, the latter category can be further divided into centralized control, distributed control and decentralized control. Among many control methods, centralized control and decentralized control have been extensively studied [4].

Although there have been many fruitful research results on voltage control with DGs, it becomes more complicated and computationally intensive to handle the growth in number, capacity and type of accessed DGs as well as their intermittence. Integrating various neighboring DER power stations as a whole, so-called integrated station (IS) or station of stations, acts as a promising solution concept to overcome these drawbacks. Different from many existing concepts, say microgrid or virtual power plant, integrated station is originally designed for the supporting function of supplementary DERs for one pre-existing facility to increase the energy and space efficiency, and economic profit consequently. For example, a PV station can be built on the rooftop of an existing distribution substation, a battery storage station can be built in a substation with large peak-valley difference in loads, and a small-scale data center can be built using the free space in substation's computer room. Recently, integrated station has been extensively studied for various applications in many fields, such as planning and operation of multi-energy system, integrated substation, distribution network and the like.

Motivated by the advantages of integrated station, this paper propose a voltage control scheme based on integrated station in active distribution network. Considering the uncertain factors of DGs and load fluctuations, on the basis of analyzing the influence of integrated station on voltage of feeder lines and building a local node-centered topology information model, this paper propose a distributed acaptive voltage-reactive control approach, and an expanded voltage-reactive power control approach for backup. Finally, Matlab-based simulation verifies the feasibility and effectiveness of the propose scheme.

2 Analysis of the Influence of Integrated Station on Voltage

The simplified circuit of a typical distribution network is shown in Fig. 1. Assume that each node may have load and power access, such as integrated station. For example in the node N, the P_{LN} represent the load active and the Q_{LN} represent reactive power of the node, and the P_{GN} and Q_{GN} represent the active and reactive power of the integrated station on the node N. P_{LN} and Q_N represent the active and reactive power injected into the node.

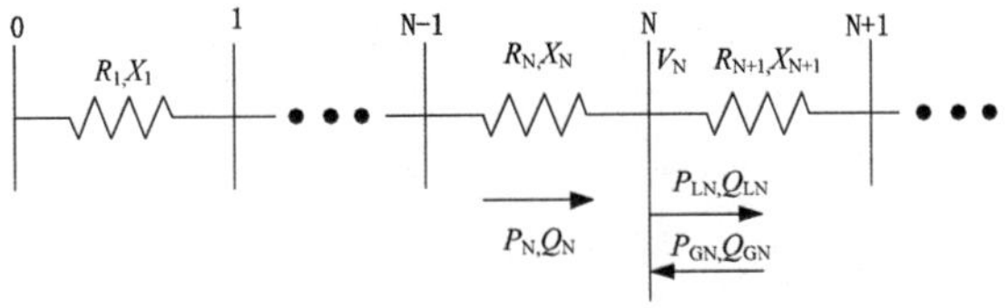

Fig. 1. Typical distribution line

According to the DistFlow power flow equation, the voltage relationship between node N and $N-1$ can be expressed as:

$$V_{N-1}^2 = V_N^2 + 2(P_N R_N + Q_N X_N) + \left(R_N^2 + X_N^2\right)\frac{P_N^2 + Q_N^2}{V_N^2} \tag{1}$$

Since the third term on the right side of the Eq. (1) is far smaller than the second term, ignoring the third term, Eq. (1) can be simplified as:

$$V_{N-1}^2 = V_N^2 + 2(P_N R_N + Q_N X_N) \tag{2}$$

Equation (2) reflects thc voltagc relationship between node N and node $N-1$, and the voltage relationship between node N and node 0 can be derived, as shown in the following equation:

$$V_N^2 = V_0^2 - 2\sum_{n=1}^{N}(P_n R_n + Q_n X_n) \tag{3}$$

2.1 Analysis of the Influence of Access Position on Voltage Regulation

Assuming that the power changes in the line are ΔP_G and ΔQ_G, positive means reducing power output, and negative means increasing power output [5].

Assuming that the changed power at the downstream nodes of the node N is $m\Delta P_G$ and $m\Delta Q_G$, and the changed power at a upstream node i is $(1-m)\Delta P_G$ and $(1-m)\Delta Q_G$, the voltage at the node N changes from V_{N0} to $V_N^2(m)$ expressed as

$$V_N^2(m) = V_{N0}^2 - 2m\Delta P_G \sum_{n=1}^{N} R_0 - 2(1-m)\Delta P_G \sum_{n=1}^{i} R_n - 2m\Delta Q_G \sum_{n=1}^{N} X_n - 2(1-m)\Delta Q_G \sum_{n=1}^{i} X_n \tag{4}$$

From Eq. (4), we can arrive at the following conclusion:

Conclusion 1: If the connected power source absorbs power, the line voltage will decrease; if output power, will increase the voltage.

Further simplifying formula (4) can be obtained:

$$V_N^2(m) = A - mB \tag{5}$$

where A and B are described in detail as

$$\begin{cases} A = V_{N0}^2 - 2\left(\Delta P_G \sum_{n=1}^{i} R_N + \Delta Q_G \sum_{n=1}^{i} X_N\right) \\ B = 2\left(\Delta P_G \sum_{n=i}^{N} R_N + \Delta Q_G \sum_{n=i}^{N} X_N\right) \end{cases} \tag{6}$$

It can be seen from Eq. (6) that when ΔP_G, ΔQ_G, R_N, X_N and i are consistent, A and B are also definite values. Combining Eq. (5), $V_N^2(m)$ and m have a linear relationship, and the larger m is, the more obvious the adjustment effect is.

In particular, when $m = 0$, it means that the changed power is at the upstream node, so that

$$V_N^2(0) = A = V_{N0}^2 - 2 \enspace \Delta P_G \sum_{n=1}^{i} R_n + \Delta Q_G \sum_{n=1}^{i} X_n\Bigg) \tag{7}$$

When m = 1, it means that the changed power is at the downstream node of node, so that

$$V_N^2(1) = A - B = V_{N0}^2 - 2 \enspace \Delta P_G \sum_{n=1}^{N} R_n + \Delta Q_G \sum_{n=1}^{i} X_n\Bigg) \tag{8}$$

Subtract formula (7) and formula (8) to get

$$V_N^2(1) - V_N^2(0) = -B = -2 \enspace \Delta P_G \sum_{n=1}^{N} R_n + \Delta Q_G \sum_{n=1}^{N} X_n\Bigg) \tag{9}$$

From Eqs. (7) (8) (9), the following conclusion can be obtained:

Conclusion 2: For the same power change, when it occurs upstream of the node, the sensitivity to voltage regulation is low and the impact on voltage is small, and when it occurs downstream, the sensitivity to voltage regulation is high and the impact on voltage is greater.

Combining conclusion 1 and 2, the following corollaries can be made:

Corollary 1: When distributed generation is connected at the upstream node along the line, more capacity can be absorbed.

Therefore, when considering the issue of increasing the consumption of distributed generation, the Absorbing capacity can be increased by adding distributed generation upstream.

Corollary 2: When distributed generation is connected downstream of the line, it can achieve the effect of regulating node voltage with a smaller capacity.

Therefore, when considering the use of distributed generation to solve voltage problems, the ability to adjust voltage reactive power control can be increased by arranging distributed generation at low-voltage points.

Corollary 3: Combining Corollary 1 and 2, when the voltage of each node of the distribution network is low in the initial state, it tries to solve it by connecting to renewable energy. Considering the influence of the volatility of distributed generation, it can be realized by dispersing in each node of the distribution network.

2.2 Sensitivity Analysis of Access Capacity to Voltage Regulation

In formula (4), considering that the adjustable power of ΔP_G and ΔQ_G are variable, the change rate of k times is changed according to the step length of ΔP_G and ΔQ_G, and the relationship function between voltage and the adjustable power multiple k is

$$V_N^2(k) = V_{N0}^2 - kC \tag{10}$$

where

$$C = 2\left(\Delta P_G \sum_{n=1}^{i} R_n + \Delta Q_G \sum_{n=1}^{i} X_n\right) - 2m\left(\Delta P_G \sum_{n=i}^{N} R_n + \Delta Q_G \sum_{n=i}^{N} X_n\right) \tag{11}$$

When m, R_N, X_N and i are fixed, then C is also a certain value.

Construct the sensitivity observation function:

$$Sen_N(k) = \frac{V_N^2(k+1)}{V_N^2(k)} = \frac{-kC - C + V_{N0}^2}{-kC + V_{N0}^2} \tag{12}$$

The following conclusions can be obtained from formula (12):

Conclusion 3: The greater the output power of the power supply is, the lower the sensitivity of the adjusted voltage is, that is $Sen_N(k) < Sen_N(k-1)$.

3 Distributed Voltage Control Method

3.1 Node Topology Information Model

The node topology information model includes two parts: a static information model and a dynamic information model. The static information model includes the topological connection relationship between nodes in the active distribution network, the line parameters between the current node and adjacent nodes, and the voltage setting value of the current node The dynamic information model includes the real-time status of the current node and adjacent nodes, the real-time voltage value of the current node, the real-time adjustable power of the current node, the real-time voltage value of the adjacent node, and the real-time adjustable power value.

In the node topology information model shown in Fig. 2, C_0 is the root node of the current node controller, $C_1 \sim C_n$ are all adjacent nodes. All adjacent nodes and the root node C_0 need to configure line parameter information (R_i, X_i). Communicating to obtain real-time voltage and real-time node schedulable capacity (V_i, S_i). Among them, the real-time data set (V_0, S_0) of the root node is calculated by the node controller or obtained by actual measurement.

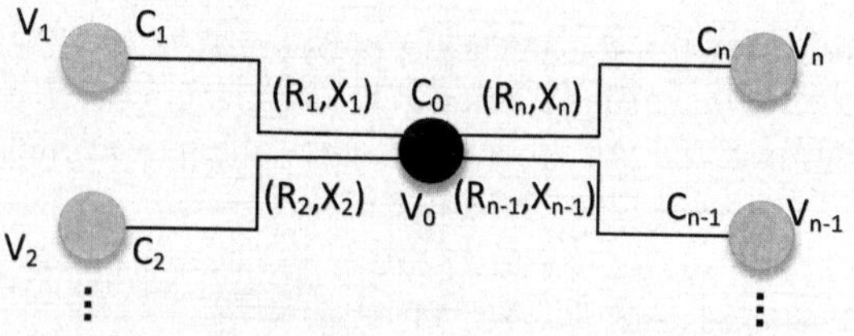

Fig. 2. Node topology information model

3.2 Adaptive Voltage Control Method

(1) Identification method for node control authority of voltage latching.

In order to avoid control disorder caused by multi-node control in distributed control, according to Corollary 3, a method for identifying node control authority of voltage latching is proposed, as follows:

When the following two conditions are met, it means that the voltage control authority is available, that is, the control authority is allowed.

① The voltage of the root node is above the upper bound (condition 1-1) and the voltage of the root node is above the voltage of all adjacent nodes (condition 1-2).
② The voltage of the root node is below the lower bound (condition 2-1) and the voltage of the root node is below the voltage of all adjacent nodes (condition 2-2).

When the condition 1-1 is satisfied, and the condition 1-2 is not satisfied, sending the adjustable capacity information of the root node to the voltage node whose voltage is above the root node, and control authority is set to limited.

When the condition 2-1 is satisfied, and the condition 2-2 is not satisfied, sending the adjustable capacity information of the root node to the voltage node whose voltage is below the root node, and control authority is set to limited.

In other states, the control authority is prohibited.

As shown in Fig. 3(a), if the root node voltage V_0 in the information model is above the upper bound V_{set-H} and above all adjacent node voltages, then the control authority of the root node is set to allowed.

In Fig. 3(b), the root node voltage in the information model is above the upper bound V_{set-H}, but the adjacent node voltages V_1 and V_4 are above voltages V_0, then the control authority of the root node is set as limited. And send the adjustable capacity of the root node C_0 to nodes C_1 and C_4.

In other states, the control authority of the root node is prohibited.

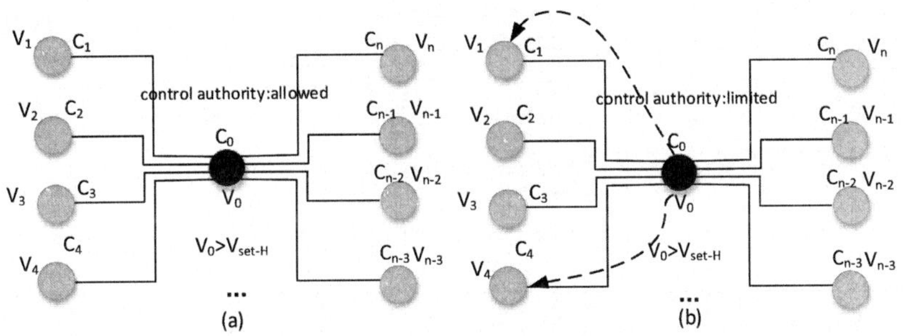

Fig. 3. Sketch of determining control authority when node voltage is above the upper bound

As shown in Fig. 4(a), if the root node voltage V_0 in the information model is below the lower bound V_{set-L} and below all adjacent node voltages, then the control authority of the root node is set to allowed.

In Fig. 4(b), the root node voltage in the information model is below the lower bound V_{set-L}, but the adjacent node voltages V_2 and V_4 are below voltages V_0, then the control authority of the root node is set as limited. And send the adjustable capacity of the root node C_0 to nodes C_2 and C_4.

In other states, the control authority of the root node is prohibited.

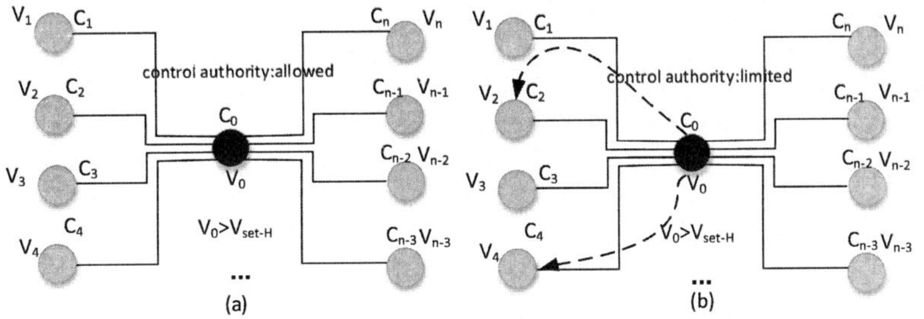

Fig. 4. Sketch of determining control authority when node voltage is below the lower bound

(2) Calculation method of node control power.

The calculation of node adjustable capacity needs to accumulate the adjustable capacity of all distributed power adjustable objects connected to the node, including the increasable active power (P_{dis}), the diminishable active power (P_{cha}), the increasable reactive power (Q_{dis}) and the diminishable reactive power (Q_{cha}). The expressions are shown in Eqs. (13) and (14).

$$P_{dis} = \sum P_{s,dis}(i), P_{cha} = \sum P_{s,cha}(i) \tag{13}$$

$$Q_{dis} = \sum Q_{s,dis}(i), Q_{cha} = \sum Q_{s,cha}(i) \tag{14}$$

According to conclusion 1, when the integrated station absorbs power, the voltage decreases; Output power, the voltage rises. However, in order to ensure the penetration of the active power of the renewable energy, when the voltage needs to be increased, the active power of the renewable energy is first released, and then the adjustable reactive power of the node is adjusted. When the voltage needs to be reduced, first adjust the adjustable reactive power of the node.

3.3 Expanded Voltage Distributed Control Method

Considering possible capacity constraints of single node adjustment,the extended voltage distribution control method is proposed [6], as follows.

1) When the voltage adjustment authority of this node is allowed and the voltage needs to be increased but the adjustment capacity is 0, an enlarged adjustment support request is sent to the upstream node with the highest adjustable capacity. The upstream node refers to the neighboring node with a higher voltage than this node. The upstream node must not adjust before receiving the support signal, because the voltage adjustment authority of the node is allowed, and the adjustment authority of adjacent nodes must be restricted or not allowed.
2) When the voltage adjustment authority of this node is restricted, and the adjustment support signal sent by the downstream node is received, the power adjustment is increased according to its own adjustment capacity. Downstream nodes refer to adjacent nodes whose voltage is lower than this node.
3) When the voltage regulation authority of this node is allowed and the voltage needs to be reduced, but the adjustable capacity is 0, then an expanded regulation support request is sent to the downstream node with the highest adjustable capacity. The same reason as (1), the downstream node must not make adjustments before receiving the support signal
4) When the voltage adjustment authority of this node is restricted, and the adjustment support signal sent by the upstream node is received, the power adjustment will be reduced according to its own adjustment capacity.

3.4 Distributed Voltage Control Process

As shown in Fig. 5, when the node-i's adjustment authority is allowed and the voltage needs to be increased, or the adjustment authority is restricted and the downstream node is requested for support, the active power of renewable energy is released first; When the output is limited, increase the reactive power; when the reactive power increases to the adjustable upper limit, increase the active power such as ESS; when the ESS active power is limited, send a support request to the node with the largest upstream adjustable capacity.

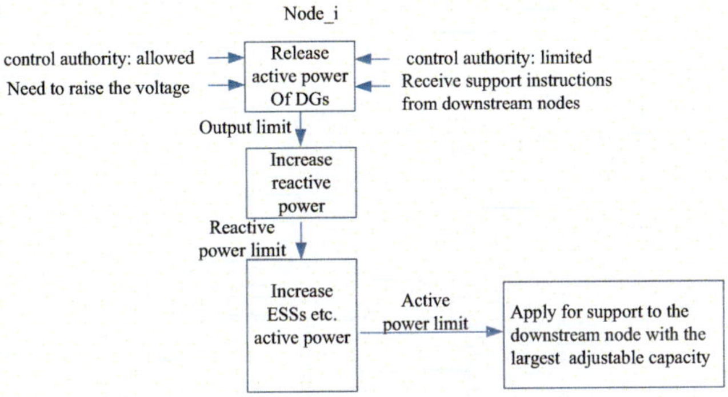

Fig. 5. Control steps to raise the voltage

As shown in Fig. 6, When the node Node-i's adjustment authority is allowed and the voltage needs to be reduced, or the adjustment authority is restricted and the upstream node's support request is requested, the reactive power will be reduced first; When the reactive power is reduced to the adjustable upper limit, reduce the active power such as ESS; When the active power of ESS is limited, the active output of renewable energy is limited; When the output of renewable energy is limited, a support request is sent to the downstream node with the largest adjustable capacity.

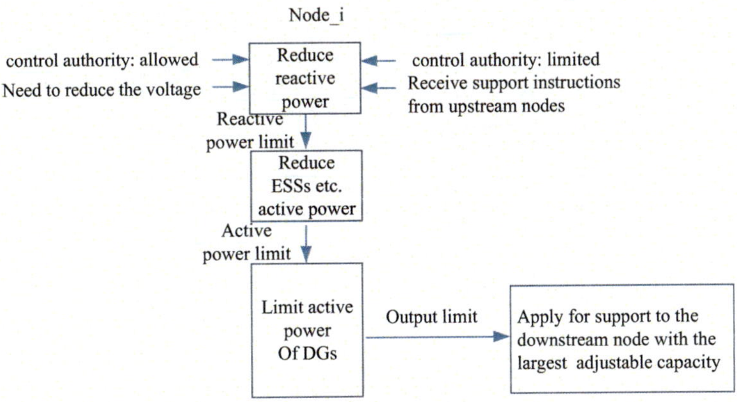

Fig. 6. Control steps to lower the voltage

4 Simulation Analysis

Using IEEE 13-node distribution system as shown in Fig. 7(a), arranging tunable capacitor banks at nodes 3, 4, and 5, and deploying integrated stations at nodes 5, 6, 8, 11, 12, and 13, a simulation model with integrated station access is formed as shown in Fig. 7(b).

The simulation model in Fig. 7(b) is built based on Matlab, using distributed voltage control method, aiming at keeping node voltage greater than 9.8 kV by adjusting the power of each controllable node. The maximum capacity of the integrated station and the tunable capacitor are set to 80 kW and 100 kVA respectively.

Figure 8 shows voltage of each node before and after adjustment. The circle marks the tunable capacitor node participating in the adjustment, and the adjusted reactive power is shown in the left floating subfigure. The rectangle marks the regional autonomous integrated station nodes participating in regulation, and the regulated active power is shown in the right floating subfigure.

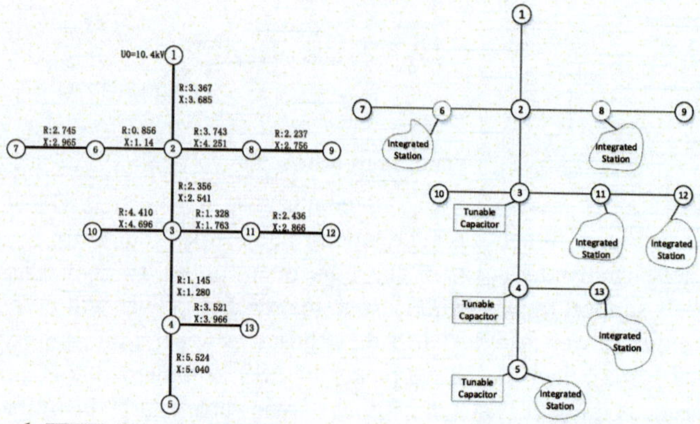

(a) Standard IEEE 13-node distribution system (b) modified model by integrated stations

Fig. 7. Standard and modified IEEE 13-node distribution system

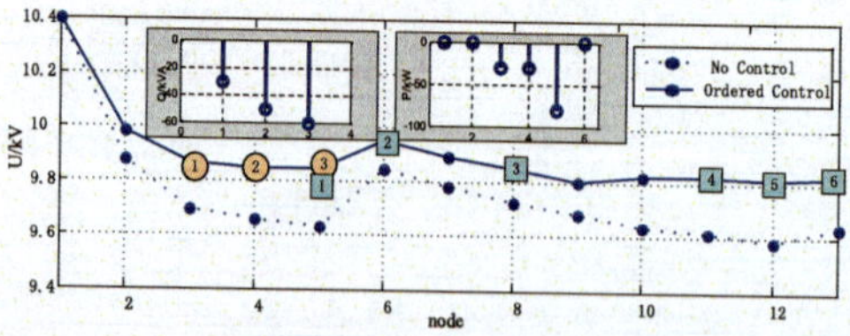

Fig. 8. Resulting curves after distributed voltage control

Simulation analysis: After the function of integrated station is put into use, the voltage of each node also be significantly increased, and meet the requirement of voltage greater than 9.8 kV.The reactive power of nodes 3, 4, and 5 show an increasing effect, respectively −30 kVA, −50 kVA and 60 kVA, so that the voltage of these three nodes can be raised to a basically consistent level. The nodes of the three integrated stations participating in the regulation are nodes 8, 11 and 12, and the corresponding power is −30kW, −30 kW and −80 kW respectively. Between nodes 11 and 12, the voltage of node 12 is low, it can be seen that the node is adjusted to the maximum power value first, and then a power reinforcement command is sent upstream, and node 11 outputs −30 kW of power; Node 8 is not the node with the smallest adjacent voltage, but the node with the smallest voltage (node 9) has no power adjustable ability, so node 8 as an upstream node outputs power of −30 kW. This verifies the feasibility and effectiveness of the method proposed in this paper.

5 Conclusion

This paper proposes a coordinated control scheme for voltage support in ADN based on integrated station. First, we study the voltage impact mechanism of IS. Then we put forward a local and node-centered topology information model, and propose a distributed adaptive voltage-reactive power control approach, and an expanded voltage-reactive power control approach for backup. Finally, the feasibility and effectiveness of the proposed method are verified based on simulation analysis.

Acknowledgments. This work is supported by Science and technology project of State Grid Corporation of China (Research on key technology and business model of multi station integration, No.5200-201941477A-0-0-00).

References

1. Ma, Z., Liu, Y., Shang, Y., Su, J.: CIGRE 2016 development trends of future power system and active distribution system. Proc. CSEE **37**(1), 27–35 (2017)
2. Luo, D., Tian, X., Liu, C., Li, R., Jiao, Y., Li, Y.: A survey of research on distributed renewable energy in distribution network. Power Syst. Clean Energy **33**(8), 101–108 (2017)
3. Chen, X., Zhang, Y., Huang, X.: Review of reactive power and voltage control method in the distribution network. Autom. Electr. Power Syst. **40**(1), 143–151 (2016)
4. Chai, Y., Guo, L., Wang, C., Liu, J., Lu, C., Jin, W., Pan, J.: Distributed voltage control in distribution networks with high penetration of PV. Power Syst. Technol. **42**(3), 738–746 (2018)
5. Chen, X.: Planning and voltage control of distribution systems with distributed generation integration. South China University of Technology,Guangzhou (2017)
6. Wang, Y., Wu, W., Zhang, B., Zhi jun, E., Yao, W.: Active and reactive power coordinated robust optimization for active distribution networks.Autom. Electr. Power Syst. **40**(9), 29–34 (2016)

Optimization Systems

Scheme to Optimize and Improve the Stepped Structured Sleeve Hydraulic Cylinder

Taiping Xie[1(✉)] and Zhiqiang Zhao[2]

[1] Hope College, Southwest Jiaotong University, Chengdu 610400, Sichuan, China
scncstp@126.com
[2] China National Heavy Duty Truck Group Co., Ltd., Jinan 620032, Shandong, China

Abstract. Based on the full analysis of the advantages, disadvantages and processing problems of the current stepped structure and through structure sleeve hydraulic cylinders, this paper proposes an improvement scheme for the stepped structure sleeve hydraulic cylinders. This scheme can better solve the problems existing in the structure and processing technology of the current stepped structure sleeve hydraulic cylinder.

Keywords: Stepped structured sleeve hydraulic cylinder · Structure and processing technology · Optimization and improvement

1 Introduction

Current sleeve-type hydraulic cylinders usually adopt a stepped structure and a through structure. There are advantages and disadvantages to each of these two types, but both have the problem that it is difficult to balance machining accuracy and structural strength. This makes it difficult to improve the quality of this type of hydraulic cylinder, which makes it difficult to meet the requirements for the quality of hydraulic cylinders with large total strokes for mechanical equipment such as tipping mechanisms of dump trailers, mining support columns, and other mechanical equipment. How to solve this problem? Based on the analysis of the advantages, disadvantages, and processing problems of the current ladder-structured and straight-through sleeve hydraulic cylinders, the author proposes an optimization and improvement scheme for the structure and processing technology of the ladder-structured sleeve hydraulic cylinders. Thus, this problem can be better solved.

2 Analysis of the Advantages, Disadvantages and Processing Problems of the Current Stepped and Through Structure Sleeve Hydraulic Cylinders

The inner sleeve of the sleeve hydraulic cylinder is a combination of the cylinder barrel and the guide sleeve to form a whole inner cylinder barrel (the step structure is a combination of the cylinder barrel and the guide sleeve in a welded form, and the

M. Tavana et al. (Eds.): IISA 2020, AISC 1304, pp. 907–915, 2021.
https://doi.org/10.1007/978-3-030-63784-2_110

straight structure is the cylinder barrel and the guide sleeve as a whole). However, because it is installed in the outer cylinder and the inner cavity of the inner sleeve is also equipped with an inner sleeve or a plunger rod, in order to take into account the installation of the inner cylinder barrel or plunger rod at the same time. Therefore, the inner and outer diameters of the inner cylinders at all levels will be strictly restricted.

The stepped structure and straight-through structure commonly used in the inner cylinder tube have advantages and disadvantages. As shown in Fig. 1, the inner cylinder tube of the stepped structure has higher strength, mature technology, and related seals have been standardized. However, during the processing of the inner and outer circles of the sleeves at all levels, due to the need to change the tooling (the cylinder tube and the guide sleeve are welded to form an inner cylinder tube blank, the two ends of the boring machine are clamped first) and bore holes; then, use a three-jaw chuck with an extended jaw to clamp the inner hole, or use an expansion mandrel to clamp the inner hole, and then use a three-jaw chuck to clamp the expansion mandrel to process the center frame of the guide sleeve; finally, the inner hole is clamped, and the groove of the guide sleeve is corrected and processed by using the guide frame center position. The positioning accuracy of the extended jaw clamping and the secondary clamping of the expansion mandrel are relatively good. The coaxiality of the central position of the guide sleeve is relatively large), the coaxiality of the inner hole of the inner cylinder tube and the groove of the guide sleeve is difficult to ensure, which is likely to cause the poor coaxiality of the adjacent two inner cylinder tubes. It may cause malfunctions such as stuck cylinders or chaotic cylinders. As shown in Fig. 2: The processing technology of the inner cylinder of the through structure is different from the step structure. After boring of the inner cylinder of the through structure, although the through hole can be clamped through the clamp, the central position of the guide sleeve end can be processed, thereby reducing the coaxiality deviation, but the tooling must be replaced during the machining process. The coaxiality deviation is not fixed. It uses non-standard L-shaped support rings and T-shaped support rings, which is difficult to purchase on site and difficult to replace during maintenance; repair station must prepare spare parts or ask suppliers to provide spare parts. In addition, during processing, the through structure needs to be turned out of a deeper groove for the installation of seals and limit retaining rings based on the bored inner hole. This directly reduces the strength of the sleeve. Hydraulic cylinder is used for mining support column. In order to ensure safety performance, general customers will specify the material and type of the seal. The groove of the seal cannot be changed by itself. Only high-strength steel can be used to ensure the strength of the groove of the seal. The ladder structure can use different steels for the cylinder barrel and the guide sleeve to weld different kinds of steel. The material of the guide sleeve can be high-strength steel to ensure the strength at the groove of the seal. The through structure can only use high-strength steel as a whole, which has a higher cost.

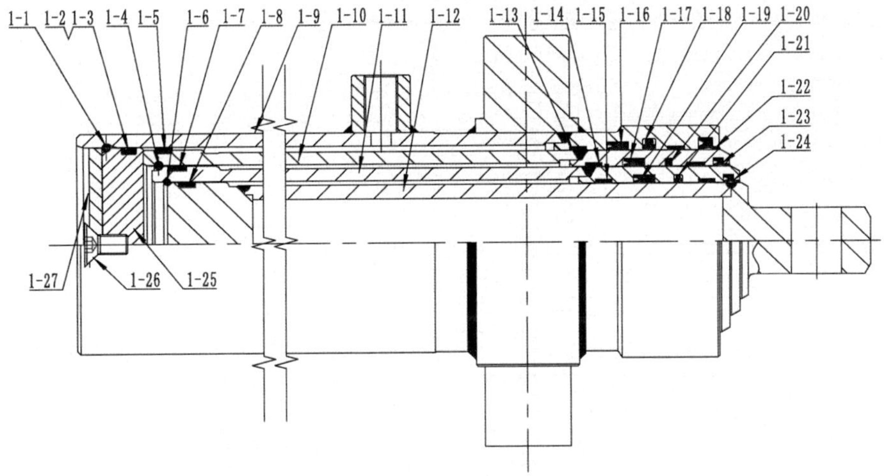

Fig. 1. Diagram of a stepped hydraulic cylinder

1–1 Wire retaining ring 1–2 Seal ring 1–3 O-ring 1–4 Wire retaining ring for holes 1–5 Support ring for holes 1–6 Wire retaining ring 1–7 Support ring for holes 1–8 Support ring for holes 1–9 Outer cylinder 1–10 First-level inner cylinder 1–11 Secondary inner cylinder 1–12 Plunger rod 1–13 Support ring for rod 1–14 Support ring for rod 1–15 Support ring for rod 1–16 Y-ring 1–17 Y-ring 1–18 Step seal 1–19 Y-ring 1–20 Step seal 1–21 Step seal 1–22 Anti-dust ring 1–23 Anti-dust ring 1–24 Anti-dust ring 1–25 End cap 1–26 Countersunk head screws 1–27 Blocking board.

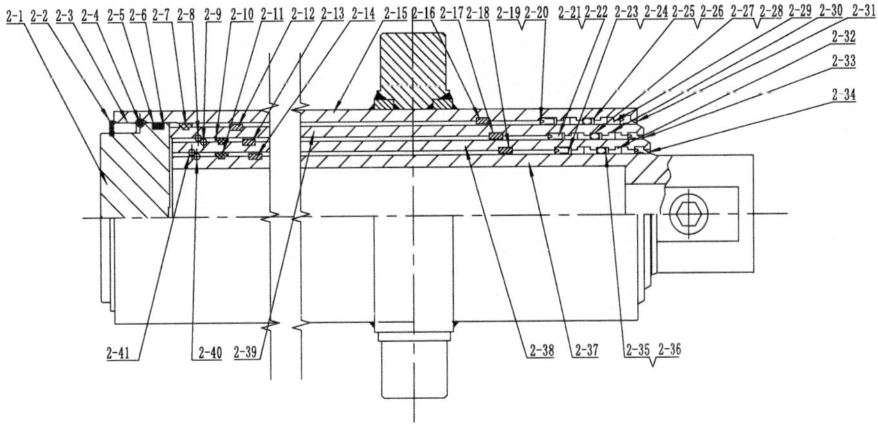

Fig. 2. Drawing of a through structure hydraulic cylinder

2–1 End cap 2–2 Spring retaining ring 2–3 Bushing 2–4 Wire retaining ring 2–5 Seal ring 2–6 O-ring 2–7 T-shaped support ring 2–8 Wire retaining ring 2–9 Wire retaining ring 2–10 T-shaped support ring 2–11 T-shaped support ring 2–12 Limit ring for shaft 2–13 Limit ring for shaft 2–14 Limit ring for shaft 2–15 Outer cylinder 2–16 Limit ring for holes 2–17 Limit ring for holes 2–18 Limit ring for holes 2–19 Y-ring 2–20 Seal ring 2–21 Y-ring 2–22 Seal ring 2–23 Y-ring 2–24 Seal ring 2–25 O-ring 2–26 Seal ring 2–27 O-ring 2–28 Seal ring 2–29 L support ring 2–30 L support ring 2–31 Anti-dust ring 2–32 L support ring 2–33 Anti-dust ring 2–34 Dust ring 2–35 O-ring 2–36 Sealed anti-extrusion ring 2–37 Plunger rod 2–38 Secondary inner cylinder 2–39 First-level inner cylinder 2–40 Wire retaining ring 2–41 Wire retaining ring.

As shown in Fig. 3: The L-shaped support ring must be equipped with a cutout to fit into the groove. There must be a gap in the cutout to fit into the groove. The cutout will leave an extrusion gap S1. As shown in Fig. 4: In order to prevent the hydraulic cylinder from issuing a card during use, there must be an assembly gap between the motion pairs, and the extrusion gaps S2 and S3 will be left. The above structure is only suitable for low-pressure light-load hydraulic systems; however, if it is used in medium and high pressure heavy-duty hydraulic systems, the O-ring will be embedded in the gap between the coupling parts of the motion pair due to elastic deformation after bearing the pressure load. As shown in Table 1 (new hydraulic design manual: Beijing Institute of Technology Press), the maximum clearance without retaining ring is usually more than 7 MPa for sports seals. If the O-ring is subjected to unilateral pressure, a sealing ring retaining ring should be added to the low-pressure cavity side. Although the use of the sealing ring retaining ring can prevent the O-ring from extruding, it will increase the frictional resistance of the motion pair and easily cause creep [1].

Table 1. 26.2-5

O-ring hardness Hs / Pressure (MPa)	≤0.35	0.35-7	7-10.5	10.5-14	14-21
70	0.40	0.25	0.15	0.1	0.04
90	0.70	0.60	0.50	0.4	0.25

The shaft Yx ring (a common Y-ring) is generally not used for the seal ring, but when the pressure is greater than 16MPa, a seal ring is also required to prevent the bottom from being cut . As shown in Fig. 3 and 4, when the L support ring is supported by the extrusion gaps S1, S2, and S3, the Y-ring and O-ring will be damaged due to the stress concentrated on the load P; it is necessary to add a sealing ring and a retaining ring to prevent damage. If a retaining ring is installed, the deepened groove needs to be widened to reduce the bearing capacity of the inner cylinder and the structure is more complicated. The larger the clearance reserved by the sealing ring retaining ring, the worse the protection effect on the sealing ring; the smaller the clearance reserved by the sealing ring retaining ring, the greater the frictional resistance.

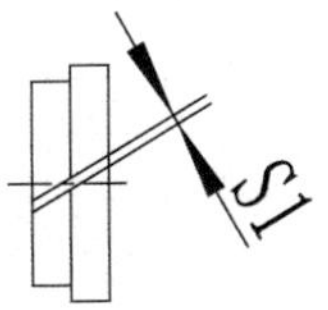

Fig. 3. Cutout of L support ring

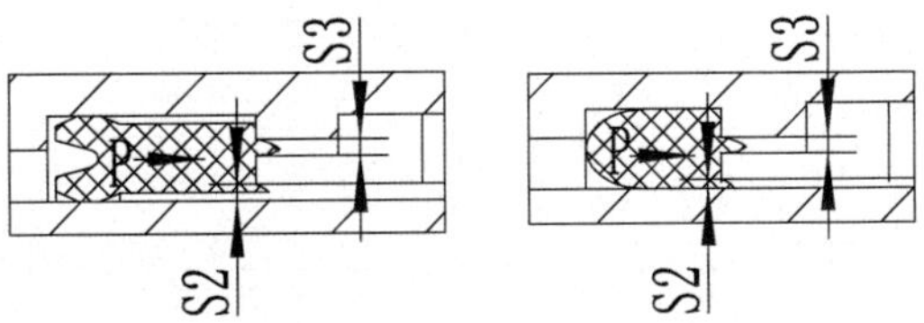

Fig. 4. Schematic diagram of the seal being squeezed into the extrusion gap under medium and high pressure

As shown in Fig. 1: The stepped structure usually uses a more pressure-resistant and wear-resistant step seal for the secondary seal. The seal is a combination of a modified polytetrafluoroethylene slip ring with a filler material and a rubber ring serving as an elastomer. The slip ring has good self-lubricating performance and low friction resistance. It has the smallest dynamic friction resistance in the existing various reciprocating motions and has smooth motion without creep. Its compact structure and small footprint. The sealing performance is good. According to the use conditions, the proportion of the slip ring filling material can be changed, or the type of filling material can be changed, and the required physical and chemical performance indicators have been obtained [1]. As shown in Fig. 2, the through structure cannot process the grooves required by the seal. Usually, O-rings are used. Their sealing performance and service life are not as good as the seal, and the friction resistance is greater than the seal.

3 Stepped Structured Sleeve Hydraulic Cylinder Optimization and Improvement Scheme and Processing Technology Design

Aiming at the coaxiality deviation caused by changing the tooling of the stepped structure during processing, the author of this article has made corresponding improvements to the existing processing technology, as shown in Fig. 5. The specific processing technology is as follows: Before the inner cylinder is bored, the boring machine stop (boring and clamping process end) is processed, the machining center is positioned in the middle of the work-piece, and the boring machine is improved. Use the MODEL T2225 boring machine spindle chuck 5–7 to drive the work-piece 5–8 to rotate, the boring bar 5–9 feed works, and then use the flat key 5–3 to prevent the boring bar from rotating. Use the standard sample to pass the center frame 5–13 and calibrate the tailstock chuck 5–12. During processing, the work-pieces 5–8 are positioned by the center frame after clamping, thereby eliminating the bending and offset of the work-piece during the top clamp and machining.

The turning tool assembly is installed on the boring machine guide 5–34, the tailstock 5–14 is equipped with the auxiliary tool bar 5–19, and the boring head 5–11 is installed, as shown in the left figure of Fig. 6. 18 Correction of the auxiliary tool holder 5–19 to ensure the straightness of the boring; the reversing valve 6–4 is set to the left, and the cylinder 5–21 pushes the auxiliary tool rod 5–19 to the left to dock with the boring tool rod 5–9. As shown in Fig. 9: The lead angle α is reserved at the shaft end of

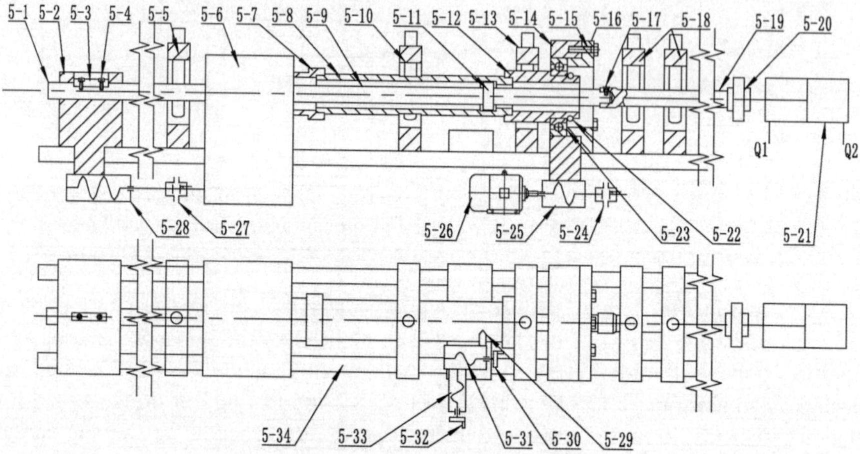

Fig. 5. Machine improvement chart

the boring bar 5–9. When the auxiliary bar 5–19 and the boring bar 5–9 are docked, guide the auxiliary bar 5–19 to center; as shown in the figure in Fig. 8, after the auxiliary cutter bar 5–19 is docked with the boring cutter bar 5–9, the directional valve 6–4 is centered, the cylinder 5–21 is self-locking, and then the screw 5 is tightened with a hexagon socket head -17 fastening. As shown in the right figure of Fi. 8, after the docking, the directional valve 6–4 is set to the right, the cylinder 5–21 is relieved of pressure, and the auxiliary cutter rod 5–19 follows the boring cutter rod 5–9.

5–1 Boring bar 5–2 Boring tool holder 5–3 Flat key 5–4 Hexagonal screws 5–5 Center frame 5–6 Spindle box 5–7 Spindle chuck 5–8 Workpiece 5–9 Boring bar 5–10 Center frame 5–11 Boring head 5–12 Tailstock chuck 5–13 Center frame 5–14 Tailstock 5–15 Coupling bolt 5–16 Bearing cover 5–17 Hexagon cylinder head set screw 5–18 Center frame 5–19 auxiliary tool bar 5–20 elastic sleeve pin coupling 5–21 cylinder 5–22 wire retainer 5–23 angular contact bearing 5–24 meshing safety clutch 5–25 screw 2–26 motor 5–27 Engaging safety clutch 5–28 lead screw 5–29 turning tool 5–30 axial handle 5–31 axial lead screw 5–32 radial handle 5–33 radial lead screw 5–34 lathe guide.

As shown in Fig. 5: The screw rod 28 drives the boring tool rod 5–9 to enter the tool, and is equipped with a limit block limit. The meshing safety clutch 5–27 ensures that the boring head 5–11 automatically slips and stops the transmission when the boring head 5–11 reaches the limit block. At this time, the boring head is just at the retraction slot to avoid damaging the limit step. At this time, the connection between the boring bar 5–9 and the auxiliary bar 5–19 has passed the tailstock, and the boring bar 5–9 stops feeding. After boring, without changing the tooling, use the axial handwheel 30 and radial handwheel 32 to feed directly. After the center of the car, remove the hexagon socket head set screws 5–17 and tighten the center frame 5- 18. Withdraw the boring bar 5–9, and separate the auxiliary boring bar 5–19 from the boring bar 5–9; the screw rod 5–25 drives the tailstock 5–14 to the right to the mounting limit block, and the workpiece is unloaded. The meshing safety clutch 5–24 ensures that the tail screw 5–25 will slip when the tailstock chuck 5–12 exits the center frame 5–13, and avoids that the tailstock chuck 5–12 exits the center frame 5–13. The elastic sleeve shaft pin coupling 5–20 can provide handicap through elastic deformation, and is used to connect the auxiliary cutter rod 5–19 with the cylinder 5–21. The elastic sleeve pin coupling can compensate the relative displacement of the two axes, can reduce the effect of the straightness deviation of the cylinder piston rod and the auxiliary tool rod from 5–19, and improve the positioning accuracy of the double center frame from 5–18; at the same time, it has shock absorption and cushioning performance, small shape and large bearing capacity [2]. Because the boring tool holder is much longer than the turning tool holder, during machining, uneven workpiece hardness or uneven cutting depth during rough machining will cause torque fluctuations caused by cutting forces and the vibration of the transmission system itself. 5–9 vibration. This causes the boring head to deviate from the axis, reduce the straightness of the inner hole, increase the ovality of the inner hole, and easily cause malfunctions such as card issuance, random cylinders, and wrong cylinders. The longer the boring bar, the more the torsional vibration characteristics cannot be ignored. Because the MODEL T2225 boring machine boring bar 5–9 only performs feed movement, the vibration damping principles such as the impact vibration damping principle and the shaved boring rod vibration damping principle are not applicable; therefore, this paper uses elastic couplings to twist the vehicle transmission Improvement of vibration characteristics [3]. An elastic coupling was selected as a shock absorber to improve the torsional vibration characteristics of boring processing. Installing an elastic coupling on the auxiliary tool bar can reduce the torsional vibration of the boring tool bar. The elastic coupling acts as a shock absorber and improves the torsional vibration characteristics of boring processing.

Then the inner sleeve turning lathe is corrected with a center frame to machine the grooves. Because the center frame and the inner hole are not changed in the same process, the deviation of coaxiality is within a very small range, and the deviation of the coaxiality between the bore of the cylinder barrel and the groove of the guide sleeve is also within a very small range; At the same time, this can save the process of replacing the central position of the guide sleeve end of the tooling and processing, reduce processing equipment, and improve production efficiency.

Although the through structure can be used for positioning and clamping when changing the tooling, processing the central end of the guide sleeve to ensure coaxiality, but as shown in Fig. 6: the retaining ring groove steps and the inner sleeve, The problem of the verticality of the rod body, or even if the parallelism of the two ends of the stop ring is increased by plane grinding, the fit of the stop ring cannot be guaranteed, which will cause stress concentration on one side. As shown in Fig. 6, the load F1 and F2 are located outside the groove. If the retaining ring groove is too shallow, it will easily come out when squeezed. If it is deepened, the strength at the groove will be weakened. The retaining ring groove is a load-bearing area, and it is.

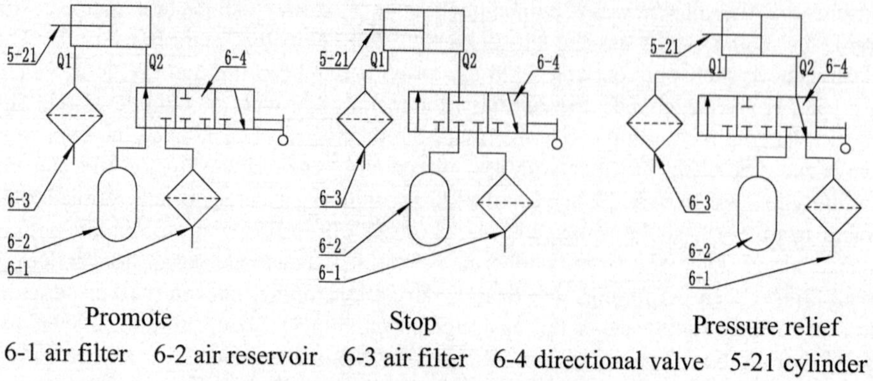

Fig. 6. Simplified diagram of the cylinder air circuit

6–1 air filter 6–2 air reservoir 6–3 air filter 6–4 directional valve 5–21 cylinder.

easy to form a dangerous section. The stepped structure can borrow the self-aligning principle of spherical washers. As shown in Fig. 8: the guide sleeve limit step is changed to a 45° inclined step, the cylinder limit step is changed to an R step, and the loads F3 and F4 are divided into equal radial and axial stress during bonding. This can be automatically corrected to avoid stress concentration; because the radial and axial stress directions are perpendicular and equal in size, the best force separation effect has been achieved. At this time, the limit step has the strongest anti-deformation ability, and there is no need to process grooves at the limit step, and the strength will be higher. The through structure adopts the limit retaining ring to limit the position. The limit retaining ring is assembled and fastened, and the pre-tightening amount is reserved. If R or inclined steps are processed in advance, it is extremely easy to be misplaced, because the retaining ring must be cut for assembly, and it is extremely difficult to process after assembly (Fig. 7).

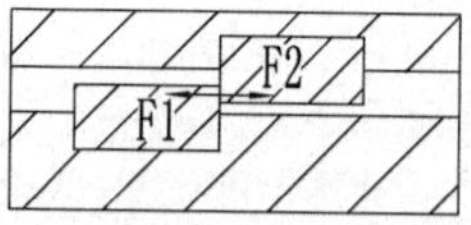

Fig. 7. Stress diagram of limit ring

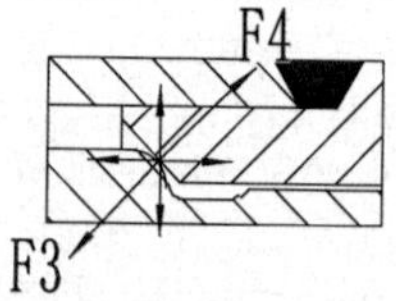

Fig. 8. Stress diagram of inclined step with R step

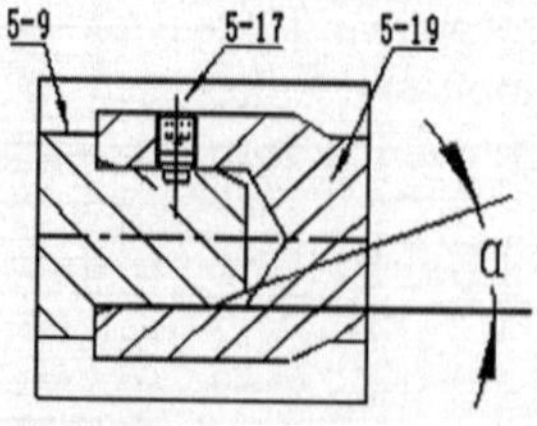

Fig. 9. Connection diagram of boring bar and auxiliary bar

4 Conclusion

In this paper, the structure optimization of Stepped Structured Sleeve Hydraulic Cylinder is adopted, and the principle of self-aligning of spherical washers is used for reference. This circumvents the influence of the vertical concentration of the traditional stepped rectangular limiting step and the groove of the guide sleeve, and the straightness of the rectangular limiting stop ring of the straight structure and the groove of the guiding sleeve. And through optimization and improvement of processing technology and processing equipment. This solves the problem that the coaxiality of the inner hole of the inner cylinder tube and the groove of the guide sleeve caused by the replacement of the fixture during the processing of the stepped structure is coaxial, and the structural strength and processing accuracy are taken into account. In addition, this can effectively improve the structural strength and machining accuracy of such hydraulic cylinders, thereby improving their bearing capacity and service life.

References

1. New Hydraulic Design Manual. Beijing Institute of Technology Press
2. Mechanical Design Manual, 4th edn, vol. 2. Chemical Industry Press
3. Liu, X.: The influence of elastic couplings on its torsional vibration characteristics in power transmission systems. Low Carbon World.Journal. Accessed 20 June 2018

Optimization Algorithm of Pigeon Flight Path Location Based on Least Square

Cong Wen, Zhenping Lan(✉), Yijie Zhang, Ping Li, and Yuru Wang

Information Science and Engineering, Dalian Polytechnic University, Dalian, China
lanzp@dlpu.edu.cn

Abstract. Beidou satellite navigation system is a satellite positioning and navigation system independently developed by China. It is an active exploration of science and technology and traditional sports events to apply it to the folk sports event carrier pigeons competition. The flight paths of carrier pigeons are changeable, so they need relative accurate positioning to get the flight paths of the carrier pigeons, which are used to maintain the fairness of the competition. In this paper, an optimal algorithm based on the least square method is proposed for locating the flight paths of carrier pigeons. It can optimize the flight paths of carrier pigeons and get the relatively accurate flight paths of carrier pigeons.

Keywords: Beidou satellite navigation system · Least square method · Positioning · Optimization algorithm · Flight Path

1 Introduction

With the continuous development of science and technology, the emergence of Beidou satellite navigation system fills in a blank for China's satellite navigation. The research focus of this project is to apply Beidou satellite navigation system to the traditional pigeon racing competition. Due to the high prize money, in order to ensure the objective fairness and justice of carrier pigeon competition and avoid fraud, it is necessary to compare the flight paths of some winning carrier pigeons with the existing aircraft route or the inherent operation path of high-speed rail transit and air route. In order to obtain the relative accurate flight paths of carrier pigeons, the observation stations can be set up in a certain area, the observation equation can be established, the target position can be estimated by solving the equation, the Least Square Estimation [1] of the target can be obtained by using the least square principle, the error value of the target in the theoretical route can be simulated by using the gradual change model, and the observation equation can be processed by dealing with the interference of environmental factors, information delay and other problems. Finally, the relative accurate flight path of carrier pigeon is obtained.

M. Tavana et al. (Eds.): IISA 2020, AISC 1304, pp. 916–924, 2021.
https://doi.org/10.1007/978-3-030-63784-2_111

2 The Problem of Pigeon Flight Path Location

The purpose of applying the Beidou satellite navigation system to the pigeon racing competition is to ensure the fairness of the competition, that is, to accurately locate the flight path of the pigeon in the process of racing, and to place the designed Beidou pigeon ring on the pigeon ankle. Pigeon ring takes BD/GPS [2, 3] dual-mode chip as the core component, and uses data flow for real-time data feedback to achieve accurate positioning.

Because the satellite positioning system has certain errors, plus the carrier pigeon flight path is changeable, and there is a certain distance between the satellite and the ground monitoring station in the flight process, under the interference of physical factors, the result of satellite positioning will be greatly different from the actual situation, and the ideal result can not be obtained, so it needs an algorithm for the optimization of carrier pigeon flight path to carry out the secondary optimization of the satellite measurement results, and a relatively accurate result is obtained. In this paper, the algorithm of pigeon flight path optimization based on the least square method is proposed.

3 The Basic Idea of Location Optimization Algorithm Based on Least Square

At present, the Least Square method is widely used in the positioning algorithm, a more ideal positioning algorithm, commonly used in curve fitting algorithm, is a good method to solve the pigeon flight path [4–6].

Method 1: For remote areas, it defines key areas to locate pigeon flight position on the pigeon racing route, and sets up fixed observation points in the algorithm to solve the problem. The observation stations can use the cellular network with wide coverage and mature technology, which can allow relatively large error values and reduce investment costs [4]. For example, airport, railway station and national highway are used to set up important stations, and the error value of such stations is allowed to be within 50 m. In these stations, intermittent monitoring can be carried out to check whether there are pigeons participating in the competition entering the key area. Once there are pigeons competing in the competition, intermittent positioning tracking can be used to locate the flight path of pigeons. If ranking is obtained in the competition, the flight path can be compared with the existing one.

Method 2: For urban areas, the whole process monitoring of carrier pigeons flight. Due to the rich observation sites in urban areas, the whole process monitoring of homing pigeon is tested and realized. However, the urban environment is complex and changeable, which greatly interferes with the homing pigeon's flight path positioning [5]. For example, during the homing pigeon's flight, the actual environment and observation sites have a great interference with the positioning, so that the positioning of a certain time point will have a big deviation, such as losing the information transmission at a point In method 2, the flight path of carrier pigeon needs to be optimized to reduce the impact of error as much as possible.

4 Algorithm Description

In this paper, a variety of ranging algorithms combined with equations are used to solve the positioning problem of carrier pigeons in the process of moving and stationary.

4.1 Location Algorithm Based on Ranging

The location algorithm based on ranging is suitable for method 1. This method is simple and effective by setting up fixed observation points to locate the target.

In positioning, it can be divided into two-dimensional coordinate and three-dimensional coordinate positioning. Here we mainly introduce two-dimensional coordinate ranging [6].

First, determine the coordinates of the target as m (x, y), the coordinates of the observation points in the monitoring area (x_1, y_1), (x_2, y_2)…… (x_n, y_n), n observation points can be set, then the distance between the target and the observation point is d_1, d_2,……d_n.

$$(x_n - x)^2 + (y_n - y)^2 = {d_n}^2 \tag{1}$$

From Formula 1. Subtract from the top to the bottom, and use the least square principle to obtain the least square estimation of the target position, namely,

$$X = (A^T A)^{-1} A^T b \tag{2}$$

$$A = -2 \begin{bmatrix} x_1 - x_n & y_1 - y_n \\ x_2 - x_n & y_2 - y_n \\ \vdots & \vdots \\ x_n - 1 - x_n & y_n - 1 - y_n \end{bmatrix} \tag{3}$$

$$b = \begin{bmatrix} d_1^2 - d_n^2 + x_n^2 + y_n^2 - x_1^2 - y_1^2 \\ d_2^2 - d_n^2 + x_n^2 + y_n^2 - x_2^2 - y_2^2 \\ \vdots \\ d_{n-1}^2 - d_n^2 + x_n^2 + y_n^2 - x_{n-1}^2 - y_{n-1}^2 \end{bmatrix} \tag{4}$$

$$X = \begin{bmatrix} x \\ y \end{bmatrix} \tag{5}$$

$$d = d_{real} + v_{noise} \tag{6}$$

Where d_{real} is the real distance and v_{noise} is the superimposed noise [2].

This algorithm can be applied to the initial monitoring of carrier pigeon's flight path in key areas. Through this method, fixed observation points can be set, and four or more observation points can be set, which can be applied to three-dimensional space position. This method can be applied to method 1.

4.2 Location Algorithm Based on RSSI (Rreceived Signal Strength Indicator)

RSSI's implementation is after the baseband receiving filter of the reverse channel. RSSI is the base station side indicator [7].

Using RSSI technology, gradual change model and simplified gradual change model are generally used in wireless signal transmission, namely,

$$\mathrm{P(d)} = \mathrm{P}(d_0) - 10\mathrm{nlg}\left(\frac{d}{d_0}\right) + \mathrm{XdBm} \tag{7}$$

Where P(d) is the received signal strength, P (d0) is the received signal power at d_0, d is the actual distance, d_0 is the reference distance, where n is the path loss value, X is a Gaussian random variable in dBm with an average value of 0 [2].

As shown in Fig. 2, three observation stations are set, the target location is at observation station 3, and the estimated location is 12.8683 m away from observation station 3. In method 1, the error between the actual position and the estimated position is within the allowable value, and the effect is good. The simulation error value is obtained.

RSSI algorithm model obtains the distance between the target and the observation point by setting up the observation point to observe the target for many times. Through this distance, it can determine the position of the target relatively accurately. If the target is in the observation point for a long time, it can mark the target, provide an important basis for the subsequent determination, and reduce the input cost, which is a very feasible method.

RSSI algorithm model is applicable to the division of key areas. In this model, various environmental values of observation points, such as noise, reference distance, obstacles and other factors, can be set, so as to obtain a relatively accurate RSSI value. Then, the specific position of the target can be calculated by the least square method, so as to determine whether the target has entered the key area, which is suitable for method 1 (Fig. 1).

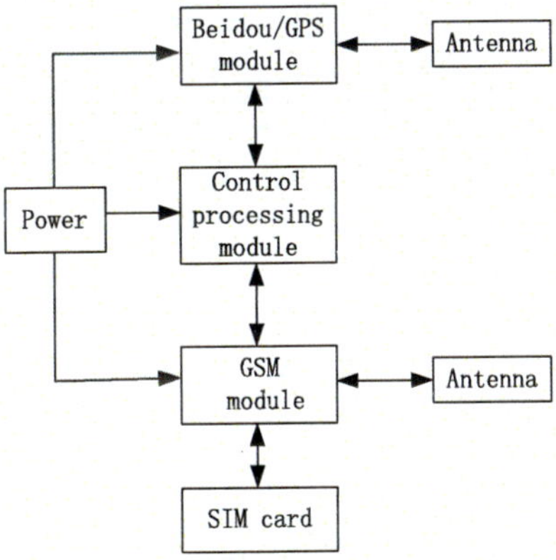

Fig. 1. Composition of Beidou pigeon ring

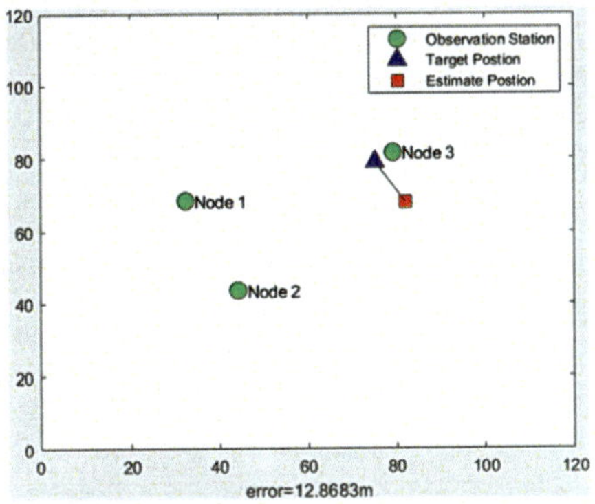

Fig. 2. RSSI simulation error value

4.3 Optimization Algorithm for Mobile Target

For positioning the moving target, as the double multiplication and maximum likelihood method can only estimate the position of the target, when obtaining the speed and acceleration of the target, the time needs to be processed to get the speed and acceleration. Here, for the flying speed and acceleration of the carrier pigeon, a range value can be made from the empirical value [8].

For the moving target, we can set an ideal flying speed value of a homing pigeon, and then take this value as a reference value, put it into the motion measurement equation to solve, use the least square method to dynamically locate the moving target, fit and predict the motion trend of the target, and get an optimization result.

Assuming that the target moves in a straight line at a constant speed, and set up an observation station (Xi, Yi) to build the target motion mode l [9, 10].

$$\mathrm{X(k)} = \Phi \mathrm{X(k-1)} + \Gamma \mu(\mathrm{k}) \tag{8}$$

$$\mathrm{Z}(k) = \sqrt{(x(k) - xi)^2 + (y(k) - yi)^2} + v(k) \tag{9}$$

$$\Phi = \begin{bmatrix} 1 & T & 0 & 0 \\ 0 & 1 & 0 & T \\ 0 & 0 & 1 & 0 \\ 0 & 0 & 0 & 1 \end{bmatrix}, \Gamma = \begin{bmatrix} \frac{T^2}{2} & 0 \\ T & 0 \\ 0 & \frac{T^2}{2} \\ 0 & T \end{bmatrix} \tag{10}$$

Here, the mean square deviation of μ (k) is Q = wdiag ([1, 1]), w ≪ 1, μ (k) element represents the interference factors such as noise interference, air resistance, etc. that the computer receives in the process of simulating the actual movement of the target.

In the simulation of the target non-uniform motion processing, according to the empirical value to take a relatively stable acceleration value of the target, the empirical value here refers to the acceleration value obtained after a large number of experiments on the target. The acceleration condition is added to the initial velocity of the target, and the time and period of acceleration calculation are set to simulate the trajectory of the target accurately.

Because the computer has given the target, and the observation station and the target are independent of each other. In order to measure the distance information between the observation station and the target, it is necessary to link the two. Therefore, the Z(k) function is set. Because there is noise interference in the measurement process, in order to explore the possible motion model of the target, the v (k) noise function is added to simulate the possible motion of the target model, where v (k) is the amount of noise in dB.

5 Simulation Verification

In the simulation process, the parameters of the observation station, the initialization position of the target, the related motion conditions, the noise parameters, the sampling time and frequency are set first [11–13]. Generally, the flying speed of a carrier pigeon is between 70–110 km/h, and the speed here is 80 km/h, i.e. 22 m/s.

The noise driving matrix and the state transition matrix are set up. The state equation is established by using the measurement equation and various parameters to simulate the movement of the target.

It simulates the real trajectory of the target, uses the least square method to estimate the observed target, gets the target position under the least square method, and then does the position deviation between the two methods [14, 15].

Through the simulation of the moving target, it can be seen that the interference of noise has a great influence on the result. In order to optimize the flight path of the target, methods must be used to optimize the noise, control the noise and minimize the error.In order to control the noise effectively, multiple observation stations can be used to collect the target information, and the observation equations of N observation stations can be set up based on the principle of least square positioning.Using the least square method for simulation, we can get Fig. 4 (Fig. 3).

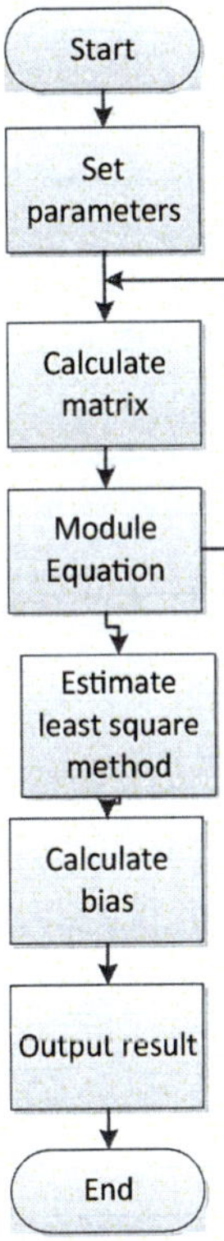

Fig. 3. Simulation flow chart

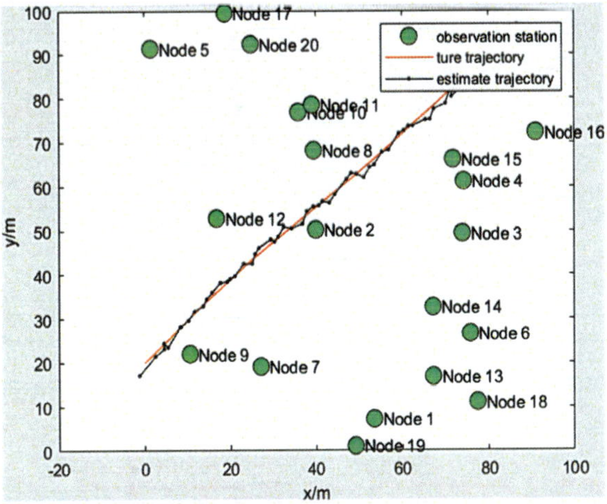

Fig. 4. Comparison of real trajectory and predicted traject

According to the figures, it can be seen that the more observation stations are, the smaller the error between the target track and the real track is. It is an effective and feasible method to increase the control noise of observation stations, and it is feasible to optimize the target track by using the least square principle.

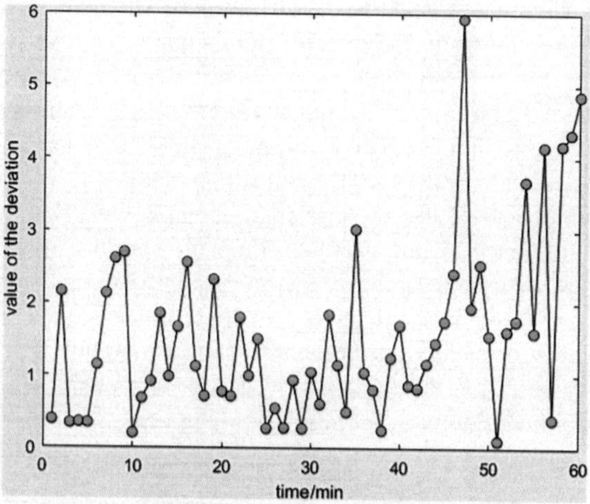

Fig. 5. Estimated time deviation

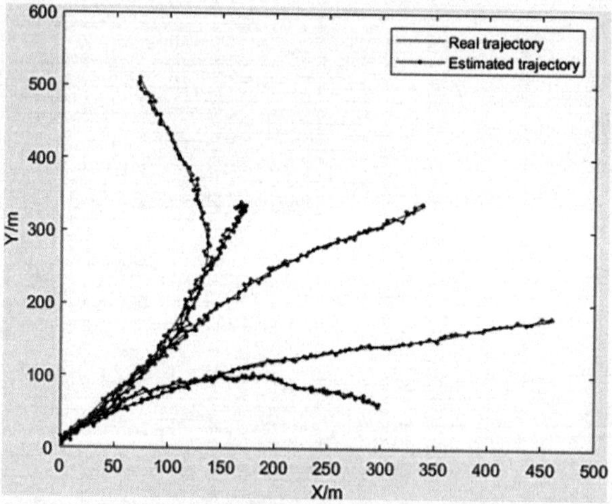

Fig. 6. Track comparison of multiple targets

6 Conclusion

In this paper, aiming at the location and tracking problem of pigeon racing track, different schemes are put forward for different environments such as remote and open areas and dense cities. The least square method is used to fit and predict the moving trend of the moving target, so as to prevent the wrong moving track of the target caused by the disconnection of the network, and the process noise of different targets is set through the noise matrix, and the noise square is analyzed In order to reduce the error of the trajectory, the noise interference is reduced. The curve is compared with the current traffic route to determine whether carrier pigeons cheat in the event, so as to ensure the fairness of the game. This method can also be applied to other mobile target location, not limited to this (Figs. 5 and 6).

Acknowledgments. This project is supported by the national innovation and entrepreneurship plan for college students 2019. In addition, We would like to thank the Institute of integrated measurement and control technology of Dalian Polytechnic University for its strong support.

References

1. Zeng, W., Fang, X., Liu, J.: Weighted global least squares algorithm with inequality constraints. J. Surv. Map. **10**, 1013–1018 (2014)
2. Chen, Z., Tang, L.: Study on GPS leveling fitting based on robust total least squares. Map. Geogr. Inf. **4**, 84–87 (2018)
3. Qin, N.: Study on the application of global least square surface fitting in GPS Height Fitting (2015)
4. Chen, G., Ren, Z., Sun, H.: Least square curve fitting and MATLAB realization. Autom. Mil. Eng. **03**, 107–108 (2005)
5. Huang, X., Wang, Y., Miao, P.: Principle and Application of Target Location and Tracking: MATLAB Simulation, 12th edn. Electronic Industry Press, Beijing (2018)
6. Gao, Q.: Research on Passive Target Location and Tracking Technology of Multiple UAVs. Nanjing University of technology, Nanjing (2017)
7. Xinxin, L., Kun, C.: Hybrid ranging weighted positioning algorithm based on RSSI and TDOA. Comput. Dig. Eng. **39**(4), 25–28 (2011)
8. Zhang, C.: Research and Design of Mobile Target Location and Tracking System Based on GPS/BDS. Suzhou University, Suzhou (2015)
9. Chambers, J.A.: Mixed modality recursive least squares and radial basis function-based scheme for transit time estimation in multiphase flow. Electron. Lett. **31**(3), 179–180 (1995)
10. Chee, C.S., Wang, Y.: Least squares estimation of a k-monotone density function. Comput. Stat. Data Anal. **74**, 209–216 (2014)
11. de Chaumaray, M.D.R.: Weighted least-squares estimation for the subcritical heston process. J. Appl. Prob. **55**(2), 543–558 (2018)
12. Meng, J.F., Fu, X.: A method of air moving target location and tracking and speed course estimation. J. China Acad. Electron. Sci. **01**, 62–66 (2018)
13. Zhang, C., Gu, J., Zhang, C., et al.: Design of mobile target location and tracking system based on GPS/ DS. Mod. Electron. Technol. **38**(17), 121–123 (2015)

14. Wai, K.L., Ling, S., Li, D.: Asymptotic theory on the least squares estimation of threshold moving-average models. Econ. Theory. **29**(3), 482–516 (2013)
15. Zhang, D., Liu, M.Q., Zhang, S.L., Fan, Z., Zhang, Q.F.: Mutual information based weighted fusion for target tracking in underwater wireless sensor networks. Inf. Electron. Eng. Front. Engl. Vers. **004**, 544–556 (2018)

Multi-colony Ant Algorithm Applied to the Yangtze Gorges Ship Lock Arrangement Optimization

Ruijie Liu(✉), Qiang Lin, Lijuan Wang, Lin Li, and Can Wang

School of Digital Technology, Dalian University of Science and Technology, Dalian, China
lrj95119@163.com

Abstract. The Yangtze River is the largest river in China, and the navigable resources are abundant, with the development of the economic belt along the Yangtze River, a large number of ships have been backlogged, which result in the Yangtze Gorges ship locks navigable capacity insufficient contradictions obviously and increasingly. In case of special weather or ship locks maintenance, ships will pile up. Based on the analysis of influencing factors of the ship lock navigable capacity, this paper draws the conclusion that the maximum value problem about the lockage tonnage every time is actually a problem of the ship lock arrangement optimization, establishes the Multi-Colony Ant Algorithm for the ship lock arrangement optimization, furtherly designs and implements the corresponding algorithm, and tests the algorithm through two examples. The experimental results show that the Multi-Colony Ant Algorithm is effective to the arrangement optimization of the Yangtze Gorges ship lock, it can improve the practical navigable capacity of the Yangtze Gorges ship lock.

Keywords: Navigable capacity · Multi-colony ant algorithm · Ship lock arrangement

1 Introduction

The Yangtze River is the largest river in China, with a total length of more than six thousand kilometers, and the navigable resources are abundant. With the rapid development of economy, the flow of vessels in the Yangtze River is increasing, meantime the great pressure has been brought to the Yangtze Gorges ship locks, the navigable capacity of the Yangtze Gorges ship locks is insufficient increasingly. The vessels waiting for navigable through the ship locks had been appeared generally. The average waiting time has from hours per ship of 2007 rapidly increased to 30 to 60 h per ship of 2012, the longest 5–7 days even. That does not include special cases such as the bad weather, the Yangtze River flow for excess and the ship lock overhaul. Although the ship lift completed in 2015 reduces the waiting time for the ship to pass through the dam (about 40 min) [1], there will still be a number of ships have been stranded in the Yangtze Gorges, causing potential safety hazards to the Yangtze Gorges Project, in the meantime, the water transport cost of the whole channel is also high and efficiency is low. Therefore, it is of very significance to study the relevant technical

M. Tavana et al. (Eds.): IISA 2020, AISC 1304, pp. 925–931, 2021.
https://doi.org/10.1007/978-3-030-63784-2_112

issues to improve the navigable capacity of the Yangtze Gorges ship locks, so as to give full play to the comprehensive benefits of the Yangtze Gorges Project and ensure the sound and rapid development of the economy and society along the river.

Domestic and foreign scholars have done a lot of useful research on the ship lock arrangement [1–9]. In this paper, a Multi-Colony Ant Algorithm is applied to the Yangtze Gorges ship lock arrangement optimization. The corresponding algorithm is designed and implemented, and a better optimization result is achieved.

2 Influencing Factors Analysis of Ship Lock Navigable Capacity

There are many factors that affect the navigable capacity, but we can analyze the factors by the designed ship lock formula. The calculation formula in the Code for Master Design of Shiplocks (JTJ 305–2001) is shown in Eqs. (1) and (2).

$$P_1 = nNG \tag{1}$$

P_I is the annual deadweight tonnage of lockage vessels.
n is the average daily number of passing lockage.
N is the annual navigable days.
G is the average tonnage of once passing lockage.

$$P_2 = (n - n_0)\frac{aNG}{b} \tag{2}$$

P_2 is the annual volume of freight transport passing through the ship lock.
n_O is the times of the non-cargo ship passing the ship lock in the day and night.
a is the coefficient of ship loading.
b is the unbalanced coefficient of monthly traffic volume.

n_O, a and b are determined according to the statistical data of the ship lock in a specific channel. But when the ship lock is actually in operation, obviously these parameters are random. Therefore, when a certain random situation occurs (i.e. the value of parameter N, n_O, a and b are given), the navigable capacity can be improved by increasing the average tonnage of once passing lockage and the average daily number of passing lockage. That is the maximum value problem about the average tonnage of once passing lockage, namely,

$$P_{actual} = \sum_{i=1}^{nN} \max(G_i) \tag{3}$$

Pactual is the actual annual deadweight tonnage of lockage vessels.
Gi is the *ith* time average tonnage of once passing lockage.

3 Determination of the Parameter *Gi* and the Ship Lock Arrangement Problem Posing

The determination of the *ith* passing tonnage *Gi* with the actual operation of the ship lock also needs to refer to the calculation of the average passing tonnage *G* when design the ship lock. At present, the standard ship type coefficient method or the fleet arrangement method [8] is usually used to calculate the average tonnage of once passing lockage.

The standard ship type coefficient method is to multiply the design maximum ship type by a certain coefficient to get the average tonnage of once passing lockage. This method is more random and difficult to determine the coefficient. Obviously, this method is not suitable for determining the parameter *Gi*, because the waiting vessels are known when the ship lock is actually in operation.

The fleet arrangement method is to determine the average tonnage of once passing lockage by combining different ship types and fleets. However, there are many problems when calculate the average tonnage of once passing lockage, such as the variety of the ship types and the fleet types in actual operation, and the arrival of various ships to the ship lock is random, so it is difficult to reasonably determine the combination of design the ship types and the proportion of design the ship types to pass the ship lock in the whole year.

The idea of the fleet arrangement method is very suitable for determining the parameter *Gi*, this is actually a problem of arranging as many passing ships (small areas) as possible in a given limited area (the ship lock chamber), so as to improve the navigable capacity, which is called the arrangement problem of the ship lock. The arrangement of the ship lock must meet the following constraints:

1) Waiting vessels cannot be overlapped each other.

2) Waiting vessels cannot beyond the boundary of the ship lock chamber.

3) Waiting vessels can't arbitrarily be rotated, vessel head direction is parallel to lock chamber side.

4 Multi-colony Ant Algorithm for the Ship Lock Arrangement

The Multi-Colony Ant Algorithm [11, 12] is extended from the basic ant colony algorithm [10], which is a stochastic optimization technique proposed by Leonor Melo et al. Compared with the basic ant colony algorithm, which has only one ant colony, the Multi-Colony Ant Algorithm is composed of multiple independent ant colonies, each of which is a complete ant colony in the basic ant colony algorithm, that is, a group of artificial ants through mutual cooperation to complete the processing of the designated problem. This kind of cooperation is realized by simulating that ants in the real world leave a certain amount of so-called pheromone material on their path. This substance can be perceived by other ants, and affect the next path of other ants. The higher the pheromone concentration on the path, the greater the probability of other ants passing the path again. The lower the pheromone concentration, the less the probability of other ants passing the path again. Therefore, the group behavior of a large number of ants shows a positive feedback phenomenon.

In addition to the basic ant colony algorithm, the Multi-Colony Ant Algorithm can cooperate with each other through individuals, and also can cooperate with each other through pheromones among colonies, so as to improve the efficiency of the algorithm and the probability of finding the optimal solution. There are many strategies for colony cooperation. Next, we further explain the Multi-Colony Ant Algorithm by improving actual navigable capacity of the Yangtze Gorges ship lock.

4.1 Description of This Algorithm

The Multi-Colony Ant Algorithm for the Yangtze Gorges ship lock arrangement is to divide the ant colony model into several independent colonies, in which a certain number of artificial ants in each colony randomly assign an initial state, where the initial state is that the artificial ants randomly select a waiting vessel from all the waiting vessels. Then, the artificial ant randomly selects the next waiting vessel, Until the ship lock chamber is full. Each step of this process needs the individual artificial ants in the colony to communicate with each other through pheromones.

In each step of the transition, the state transition rule used is shown in formula (4):

$$S = \arg\max\left\{[\tau_h(i)]^{\delta}\cdot[\eta(i)]^{\delta}\right\} \quad \text{if } q \le q_0 \tag{4}$$

If $q \le q_0$, which waiting vessel to be selected determined by the maximum value of the $\left\{[\tau_h(i)]^{\delta}\cdot[\eta(i)]^{\delta}\right\}$.

q is random value, it belongs to [0, 1].

q_0 is a initial parameter, $0 \le q_0 \le 1$. q_0 indicates what is the probability of the Multi-Colony Ant Algorithm doing a random search.

i is the number of the waiting vessel, $1 \le i \le n$, n is the quantity of waiting vessels.

h represents the *hth* colony, $1 \le h \le C$, C is the number of colonies.

k represents the *kth* ant, $1 \le k \le m$, m is the quantity of ants.

mhk is the set of the waiting vessels which already selected by the *kth* ant in the *hth* colony, all ants cannot select them again.

$\tau_h(i)$ is the pheromone trail associated with the waiting vessel i, the ant in the *hth* colony selected the waiting vessel i.

$\eta(i)$ is the heuristic information, it is the utilization rate of the waiting vessel i to the ship lock chamber.

δ and β are a parameter that adjust the relative importance between the pheromone trail and the heuristic information, $\delta \ge 0$, $\beta \ge 0$.

If $q > q_0$, which waiting vessel to be selected determined by the formula (5).

$$p_{hk}(i) = \begin{cases} \dfrac{[\tau_h(i)]^{\delta}\cdot[\eta(i)]^{\delta}}{\sum\limits_{j \notin m_{hk}} [\tau_h(j)]^{\delta}\cdot[\eta(j)]^{\delta}} & \text{if } i \notin m_{hk} \\ 0 & \text{otherwise} \end{cases} \tag{5}$$

$p_{hk}(i)$ is the probability that the *kth* ant in the *hth* colony selects the waiting vessel i, $0 \leq p_{hk}(i) \leq 1$.

j is the waiting vessel number, $1 \leq j \leq n$.

The locally updated according to the rule (6).

$$\tau_h(i) = (1-\rho) \cdot \tau_h(i) + \rho \cdot \tau_0 \quad (6)$$

Where ρ is a coefficient, which controls pheromone trails persistence and evaporation, $0 \leq \rho < 1$.

τ_0 is the initial pheromone associated with waiting vessel, this value is the chamber arrangement utilization rate as the heuristic information, $\tau_0 \geq 0$. It is also the initial value assigned to $\tau_h(i)$.

The globally updated according to the rule (7).

$$\tau_h(i) = (1-\alpha) \cdot \tau_h(i) + \alpha \cdot \Delta\tau_h(i) \quad (7)$$

α is an initial parameter, $0 \leq \alpha \leq 1$.

$\Delta\tau_h(i)$ is the chamber arrangement utilization rate associated with the iteration best ant in the *hth* colony.

The updating rules of pheromone trail concentration among colonies is shown in formula (8).

$$\tau_h(i) = \tau_h(i) + \lambda \cdot \Delta\tau_{best}(i) \quad (8)$$

Where, λ is an initial parameter, $0 \leq \lambda \leq 1$.

$\Delta\tau_{best}(i)$ is the chamber arrangement utilization rate associated with the best colony, its chamber arrangement utilization rate is highest in all colonies after a specified number of cycles (e.g. R times).

The Multi-Colony Ant Algorithm for optimizing the arrangement of ship locks is described as follows:

1) Parameter initialization.

Set $\alpha = \lambda = \rho = 0.1$, $R = 10$, $C = 4$, $\delta = 1.0$, $q_0 = 0.9$, $\beta = 2.0$, $m = 20$.

Calculate τ_0 and set $\tau_h(i) = \tau_0$ as the initial pheromone.

2) For all artificial ants ($C*m$ ants) select their initial waiting vessel according to a certain strategy (random selection strategy is used here).

Do.

For every artificial ant.

From the initial selection of waiting vessel.

Repeat.

For every artificial ant According to the transfer rules (4) and (5), the next waiting vessel is selected, and the corresponding pheromone concentration of every colony is locally updated according to formula (6).

Until the ship lock chamber cannot be arranged.

3) Find out the current optimal ship lock arrangement in every colony, and perform the global update according to formula (7) for all the waiting vessels corresponding to the optimal ship lock arrangement.

4) Repeat the above two steps (2) and (3) to judge whether the number of the repetitions reaches R. If it is, find out the current optimal colony in all colonies, the corresponding pheromone concentration of other colonies is updated according to formula (8).

5) Judge the convergence condition, if not, continue to repeat the above step (2), (3) and (4). Otherwise, the lockage arrangement is given and the program is finished.

4.2 Parameters

The parameters in the Multi-Colony Ant Algorithm for the Yangtze Gorges ship lock arrangement are chosen as follows:

$\alpha = \lambda = \rho = 0.1$, $R = 10$, $C = 4$, $\delta = 1.0$, $q_0 = 0.9$, $\beta = 2.0$, $m = 20$.

4.3 Computational Results

The Multi-Colony Ant Algorithm for the Yangtze Gorges ship lock arrangement has been designed and implemented with VC+ + in Win10. In the calculation process, waiting vessels are replaced by an envelope, the envelope is a rectangular region, which defines the minimum circumscribed rectangle of each geometry. The ship lock chamber of the Yangtze Gorges regards as the rectangular with 280 m length and 34 m width. The following, two examples are given to verify the effectiveness of the algorithm.

The first example uses the data which is the 2th instance's data in [6], the ship lock chamber utilization rate is 95.42%, the ship lock chamber utilization rate is 95.52% using the Multi-Colony Ant Algorithm, the ship lock chamber arrangement is showed in Fig. 1 (the number that showed in the figure is the waiting vessel number).

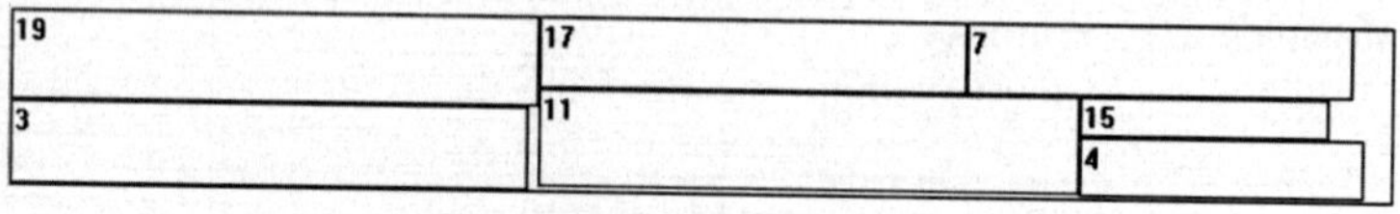

Fig. 1. Lock chamber arrangement of the first example.

The second example uses the instance is the instance 2 in Sect. 5.2 given in [1]. The ship lock chamber utilization rate with weight can reach 71%, the ship lock chamber utilization rate is 94.69% using the Multi-Colony Ant Algorithm without weight, the ship lock chamber arrangement is showed in Fig. 2.

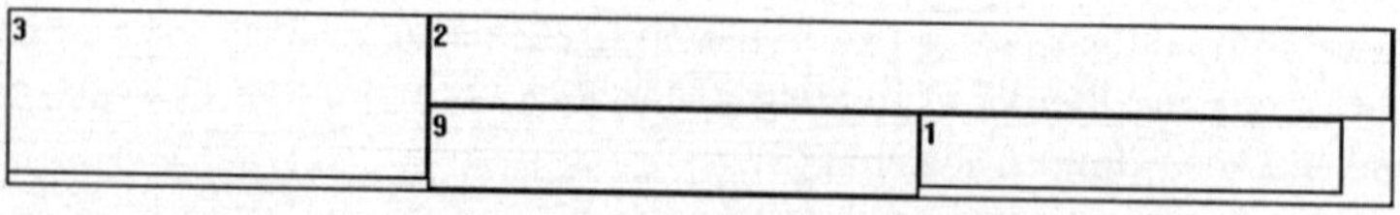

Fig. 2. Lock chamber arrangement of the second example.

5 Conclusions

The experimental results of the above examples show that the Multi-Colony Ant Algorithm is effective for the Yangtze Gorges ship lock arrangement optimization, this paper also provides the idea and the method for improving the practical navigable capacity of the Yangtze Gorges ship lock and solving the optimization problem.

Acknowledgment. Authors thank Leonor Melo, Aljanaby A and Zeng Feifan et al. for the related research on the Multi-Colony Ant Algorithm and the ship lock arrangement.

References

1. Zeng, F.F.: Packing Problem Research of Navigable Co-scheduling System for Ship-lift at Three Georges Project and Lock. Central South University of Forestry and Technology, Changsha (2016)
2. Deng, W.F.: Cascade hub Research of Navigable Co-scheduling System for Ship-lift at Three Georges Project and Lock. Central South University of Forestry and Technology, Changsha (2017)
3. Zhou, C.F.: Design of ship arrangement algorithm for an expanded non-rectangular lock chamber. J. Dalian Maritime Univ. **43**(2), 27–32 (2017)
4. Zhang, Y.F.: Research on traffic flow prediction and ship lock scheduling model of in-land river. Zhejiang Gongshang University, Hangzhou (2018)
5. Liu, Y.F.: The Optimization Research of ship lock operation-Verified by Qing-yuan Water Conservancy Lock. South China University of Technology, Guangzhou (2016)
6. Liu, R.F., Qin, F.S., Zhai, Y.T., Wang, L.F.: Ship lock arrangement in yangtze gorges based on simulated annealing algorithm. Comput. Modernization (11), 65-67 (2013)
7. Liu, W.F.: Application of greedy algorithm in the arranging of lock chamber. Com-put. Knowl. Technol. **7**(10), 2417–2418 (2011)
8. Liao, P.F., Zhang, W.S.: Analytical model for average lockage tonnage of waterway lock. J. Southeast Univ. (Nat. Sci. Ed.) **40**(1), 207–212 (2010)
9. Sun, B.F., Qi, H.S., Zhang, X.T., Cai, X.F.: Dimensionality reduction quickly arranging algorithm of lock chambers in co-scheduling of yangtze gorges dam and gezhouba dam system. Comput. Technol. Dev. **16**(12), 19–21 (2006)
10. Dorigo, M.F., Caro, G.D.S., Gambardella, L.M.T.: Ant algorithms for discrete optimization. Artif. Life **5**(2), 137–172 (1999)
11. Melo, F.L., Pereira, S.F., Costa, T.E.: MC-ANT: a multi-colony ant algorithm. Artif. Evol. LNCS **5975**, 25–36 (2010)
12. Aljanaby, A.F., Ku-Mahamud, K.R.S., Norwawi, N.M.T.: Interacted multiple ant colonies optimization approach to enhance the performance of ant colony optimization algorithms. Comput. Inf. Sci. **3**(1), 29–34 (2010)

Infrared and Visible Image Fusion Based on Multi-scale Decomposition and Gradient Optimization

Siyan Zhang[1,2](✉), Guangming Zhou[2], and Di Chen[3]

[1] College of Aerospace Science and Engineering, National University of Defense Technology, No. 109 Deya Road, Changsha 410073, People's Republic of China
final08888@163.com

[2] Beijing Institute of Astronautical Systems Engineering, Beijing 100076, China

[3] Beijing Institute of Precision Mechatronics and Controls, Beijing 100076, China

Abstract. The novel infrared and visible image fusion algorithm based on multi-scale transform (MST) decomposition and gradient optimization is proposed in this paper. First, infrared and visible images are divided into the basic layer and the detail layer at the same time. Second, the common "weighted-average" fusion rule is performed in base layer fusion. And last, gradient optimization is used for detail layer fusion. The results of qualitative and quantitative experiments show that the image generated by this algorithm retains more texture details and clear targets. Compared with other five commonly methods used MST decomposition, the image obtained by this algorithm has better visual effect and objective measurement value.

Keywords: Image fusion · Multi-scale decomposition · Gradient optimization

1 Introduction

For the past few years, the methods of image acquisition have become more and more diversified. Different sensors are usually used to gain images of the same scene, such as visible light and short-wave infrared, so as to obtain more accurate information. However, different types of sensors will form their own images, and these images will also have a lot of redundant information, which is not conducive to further image processing. Therefore, more and more scholars begin to study how to fuse the feature content of multi-source images into a single and rich image. These technologies are widespread use in medical imaging [1, 2], remote sensing [3], target recognition [4–6], detection [7, 8] and other fields. The fusion of infrared and visible image is a research direction [9, 10]. It combines the advantages of infrared sensor and visible imaging, and makes up for each other's deficiency. So as to obtain many information details and useful features in the scene [11–16]. Images can be fused from the pixel level, feature level and decision level. Pixel-level image fusion is directly carried out on the original data layer, with high accuracy and rich details. Therefore, it is also the basis for the

M. Tavana et al. (Eds.): IISA 2020, AISC 1304, pp. 932–940, 2021.
https://doi.org/10.1007/978-3-030-63784-2_113

other two levels. This paper also focuses on the study of pixel level image fusion. Traditional image fusion methods based on MST include pyramid and wavelet methods [17]. First, basic information and detail information are obtained through image decomposition. Secondly, the information of the basic and the detail layer are fused according to the corresponding fusion rules. Different fusion rules may result in different fusion effects. Therefore, the selection of rules are crucial. At present, the common infrared and visible light image fusion include fusion rules based on "average" [18] and based on the gradient transfer and total variation gradient transfer fusion (GTF) method [19], etc. We propose a new fusion algorithm based on MST in this study. The resulting image has the advantages of both visible and infrared light image. First, we use the symbiotic filter [20] to decompose the source image and obtain the basic and detail information. Second, the common "weighted-average" fusion rule is performed in base layer fusion. And last, gradient optimization is used for detail layer fusion.

2 The Proposed Method

2.1 Image Decomposition

A co-occurrence filter [21] is introduced, which has better smoothing effect and more edge texture information than gauss filter and bilateral filter. To do this, we use the co-occurrence filter in formulas 1 and 2. Here I_1 and I_2 are infrared and visible images:

$$B_1 = I_1 * G \tag{1}$$

$$B_2 = I_2 * G \tag{2}$$

Among them, Base layer B_1 and B_2 are decomposed from source image I_1 and I_2. G is the co-occurrence filter, which is used for smoothing source images to gain the base level information.

$$D_1 = I_1 - B_1 \tag{3}$$

$$D_2 = I_2 - B_2 \tag{4}$$

The corresponding detail layers D_1 and D_2 can be obtained from the source image using formulas 3 and 4.

2.2 The Base Layer Fusion Method

We use the weighted average strategy to accomplish the base layer fusion; firstly, we need to calculate the weight map of the infrared and visible images. The weight map is as follows:

$$S_p = \sum_{V=0}^{255} N_v(I_p - V) \tag{5}$$

where N_V is the total number of pixels in image I whose pixel values are equal to V. In this paper, S_I is normalized to [0 1]. Then we can get the base layer fusion of image:

$$B_F = S_I * B_I + S_V * B_V \tag{6}$$

2.3 The Detail Layer Fusion Method

In order to preserve more edges and details, we use gradient optimization method in the detail layer. Firstly, we calculate the gradient of both the infrared and visible image.

$$G_1 = G(I_1) \tag{7}$$

$$G_2 = G(I_2) \tag{8}$$

Where I_1 and I_2 are the infrared and visible images, respectively. G(.) is the gradient map of the input image.

Then, to make the fusion result reserve more detail information of the visible image and infrared image, we need minimize the following formula:

$$G_F^* = \min(\|G_F - G_1\|_2 + \lambda * \|G_F - G_2\|_2) \tag{9}$$

Where λ is the parameter which is used as the regularization. λ is set to 0.5 in order to balance the two terms of the formula. G_F is the gradient map of the fused detail layer. Then we can get the final detail image by its gradient map by solving the Poisson equation:

$$D_F^* = \min_{D_F} \|divD_F - \nabla G_F\|_2^2 \tag{10}$$

div means the divergence of the fused detail layer. By solving the above equation, we can get the final detail fusion image D_F. And the final fusion image is obtained:

$$I_F = B_F + D_F \tag{11}$$

2.4 Objective Performance Evaluation

The subjective evaluation method is a method that relies on human eyes to evaluate the quality of the fusion image subjectively. However, this method of quality evaluation mainly depends on the observer, which will result in strong subjectivity, one-sided and poor repeatability. Moreover, when the observation conditions change, the evaluation results may be different. Therefore, the subjective evaluation method needs to be combined with objective quantitative evaluation criteria to comprehensively evaluate the fusion images. Here, we use six metrics to fully evaluate the results of various fusion algorithms, they are EN [22], MI [23], SD [24], SF [25], SSIM [26] and VIF [27].

3 Experiment Result

We give some image pairs which is used for experiment, from the following images (Fig. 1).

Fig. 1. Image pairs

We use some other methods as the comparison with the proposed method, From the experiments, we can see that the proposed can not only keep the infrared object, but also retain most of the details of the visible image (Figs. 2, 3, 4 and 5).

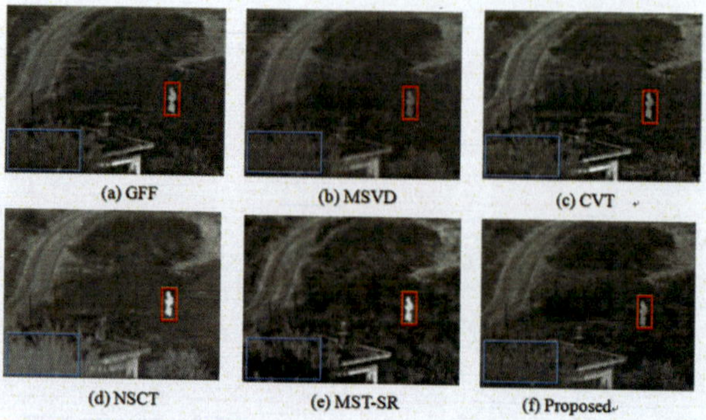

Fig. 2. 'Camp' experiment results

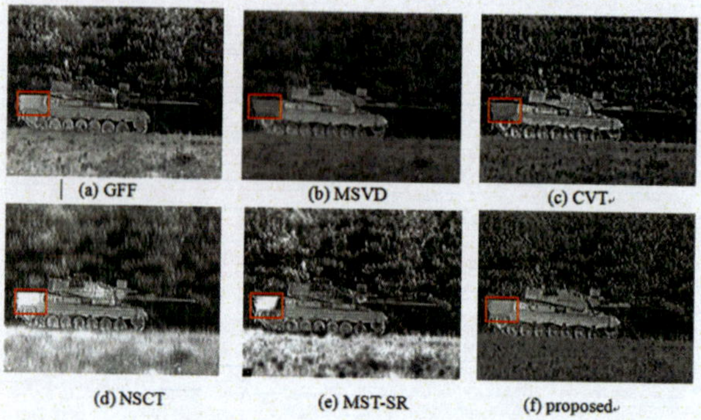

Fig. 3. 'Tank' experiment results

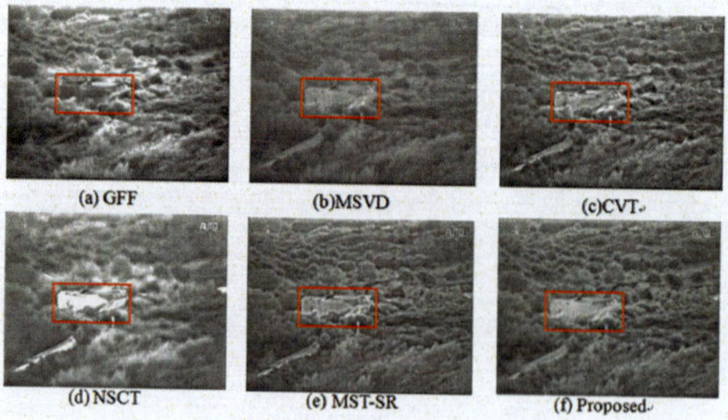

Fig. 4. 'Bunker' experiment results

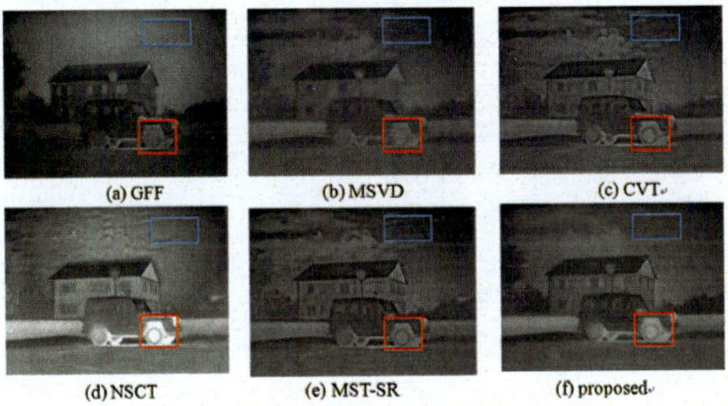

Fig. 5. 'Marne 04' experiment results

Here is the objective judgment of all the six methods:

Table 1. Objective assessment of 'Camp'.

	GFF	MVSD	CVT	NSCT	MST-SR	Proposed
EN	6.3633	6.2388	6.7105	6.7148	6.8377	**6.8409**
SF	8.6932	6.8678	12.2490	7.7465	9.9310	**11.0948**
AG	3.4943	2.6873	4.8500	3.4880	4.3251	4.3802
MI	1.0522	1.0942	1.0457	1.5033	1.4088	1.4317
MSSIM	0.6186	0.6522	0.6289	0.6093	0.5307	0.5827
$Q^{AB/F}$	0.4981	0.3262	0.4673	0.3768	0.4559	**0.5139**
VIF	0.2811	0.3147	0.5301	0.3030	0.3162	**0.5407**

Table 2. Objective assessment of 'Tank'.

	GFF	MVSD	CVT	NSCT	MST-SR	Proposed
EN	6.3633	6.2388	6.7105	6.7148	6.8377	**6.9409**
SF	8.6932	6.8678	12.2490	7.7465	9.9310	**11.0948**
AG	3.4943	2.6873	4.8500	3.4880	4.3251	**4.8802**
MI	1.0522	1.0942	1.0457	1.5033	1.4088	1.4317
MSSIM	0.6186	0.6522	0.6289	0.6093	0.5307	0.5827
$Q^{AB/F}$	0.4981	0.3262	0.4673	0.3768	0.4559	**0.5139**
VIF	0.2811	0.3147	0.5301	0.3030	0.3162	**0.4407**

Table 3. Objective assessment of 'Bunker'.

	GFF	MVSD	CVT	NSCT	MST-SR	Proposed
EN	6.3633	6.2388	6.7105	6.7148	6.8377	**6.9409**
SF	8.6932	6.8678	12.2490	7.7465	9.9310	**11.0948**
AG	3.4943	2.6873	4.8500	3.4880	4.3251	**4.9802**
MI	1.0522	1.0942	1.0457	1.5033	1.4088	**1.5317**
MSSIM	0.6186	0.6522	0.6289	0.6093	0.5307	**0.6827**
$Q^{AB/F}$	0.4981	0.3262	0.4673	0.3768	0.4559	0.4139
VIF	0.2811	0.3147	0.5301	0.3030	0.3162	**0.5407**

Table 4. Objective assessment of 'Marne 04'.

	GFF	MVSD	CVT	NSCT	MST-SR	Proposed
EN	6.3633	6.2388	6.7105	6.7148	6.8377	6.7409
SF	8.6932	6.8678	12.2490	7.7465	9.9310	**11.0948**
AG	3.4943	2.6873	4.8500	3.4880	4.3251	**4.8802**
MI	1.0522	1.0942	1.0457	1.5033	1.4088	1.4317
MSSIM	0.6186	0.6522	0.6289	0.6093	0.5307	**0.6827**
$Q^{AB/F}$	0.4981	0.3262	0.4673	0.3768	0.4559	0.4939
VIF	0.2811	0.3147	0.5301	0.3030	0.3162	**0.5407**

The bold font in the above Tables 1, 2, 3 and 4 means the maximum value among all the indicators.

4 Conclusion

The novel infrared and visible image fusion algorithm based on multi-scale transform (MST) decomposition and gradient optimization is proposed in this paper. The fusion method of "weighted average" is adopted in the base layer, and the gradient optimization method is used in the fusion of the detail layer. Compared with other five commonly used MST decomposition methods, the image obtained by this algorithm has better visual effect and objective measurement value, which has good practical value in the field of target recognition.

Acknowledgments. I would like to thank the teachers and students for their help in the paper preparation and experiment.

References

1. Bavirisetti, D.P., Bavirisetti, D.P.: Multi-sensor image fusion based on fourth order partial differential equations. In: Proceedings of the International Conference on Information Fusion, pp. 1–9 (2017)
2. Beaulieu, M., Foucher, S., Gagnon, L.: Multi-spectral image resolution refinement using stationary wavelet transform. In: Proceedings of the IEEE International Geoscience and Remote Sensing Symposium, pp. 4032–4034 (2003)
3. Burt, P.J., Adelson, E.H.: The Laplacian pyramid as a compact image code. IEEE Trans. Commun. **31**(4), 532–540 (1983)
4. Chipman, L.J., Orr, T.M., Graham, L.N.: Wavelets and image fusion. In: Proceedings of the IEEE International Conference on Image Processing, pp. 248–251 (1995)
5. Dogra, A., Goyal, B., Agrawal, S.: From multi-scale decomposition to non-multi-scale decomposition methods: a comprehensive survey of image fusion techniques and its applications. IEEE Access **5**, 16040–16067 (2017)
6. Du, J., Li, W., Lu, K., Xiao, B.: An overview of multi-modal medical image fusion. Neurocomputing **215**, 3–20 (2016)
7. Du, Q., Xu, H., Ma, Y., Huang, J., Fan, F.: Fusing infrared and visible images of different resolutions via total variation model. Sensors **18**(11), 3827 (2018)
8. Eskicioglu, A.M., Fisher, P.S.: Image quality measures and their performance. IEEE Trans. Commun. **43**(12), 2959–2965 (1995)
9. Fernandez-Beltran, R., Haut, J.M., Paoletti, M.E., Plaza, J., Plaza, A., Pla, F.: Remote sensing image fusion using hierarchical multimodal probabilistic latent semantic analysis. IEEE J. Sel. Top. Appl. Earth Obs. Remote Sens. **11**(12), 4982–4993 (2018)
10. Huang, Y., Bi, D., Wu, D.: Infrared and visible image fusion based on different constraints in the non-subsampled Shearlet transform domain. Sensors **18**(4), 1169 (2018)
11. Jin, X., Jiang, Q., Yao, S., Zhou, D., Nie, R., Hai, J., He, K.: A survey of infrared and visual image fusion methods. Infrared Phys. Technol. **85**, 478–501 (2017)
12. Kong, S.G., Heo, J., Boughorbel, F., Yue, Z., Abidi, B.R., Koschan, A., Yi, M., Abidi, M. A.: Multi-scale fusion of visible and thermal IR images for illumination-invariant face recognition. Int. J. Comput. Vis. **71**(2), 215–233 (2007)
13. Kong, W., Yang, L., Zhao, H.: Adaptive fusion method of visible light and infrared images based on non-subsampled Shearlet transform and fast non-negative matrix factorization. Infrared Phys. Technol. **67**, 161–172 (2014)
14. Kong, X., Liu, L., Qian, Y., Wang, Y.: Infrared and visible image fusion using structure-transferring fusion method. Infrared Phys. Technol. **98**, 161–173 (2019)
15. Lewis, J.J., O'Callaghan, R.J., Nikolov, S.G., Bull, D.R., Canagarajah, C.N.: Pixel- and region-based image fusion with complex wavelets. Inf. Fusion **8**(2), 119–130 (2007)
16. Li, H., Li, X., Yu, Z., Mao, C.: Multifocus image fusion by combining with mixed-order structure tensors and multi-scale neighborhood. Inf. Sci. **349**, 25–49 (2016)
17. Li, H., Manjunath, B., Mitra, S.K.: Multisensor image fusion using the wavelet transform. Graph. Models Image Process. **57**(3), 235–245 (1995)
18. Li, S., Kang, X., Fang, L., Hu, J., Yin, H.: Pixel-level image fusion: a survey of the state of the art. Inf. Fusion **33**, 100–112 (2017)
19. Li, S., Kang, X., Hu, J.: Image fusion with guided filtering. IEEE Trans. Image Process. **22**(7), 2864–2875 (2013)
20. Li, S., Yang, B., Hu, J.: Performance comparison of different multi-resolution transforms for image fusion. Inf. Fusion **12**(2), 74–84 (2011)

21. Zhang, P., Yuan, Y., Fei, C., Pu, T., Wang, S.: Infrared and visible image fusion using co-occurrence filter. Infrared Phys. Technol. **93**, 223–231 (2018)
22. Ma, J., Jiang, J., Liu, C., Li, Y.: Feature guided Gaussian mixture model with semi-supervised EM and local geometric constraint for retinal image registration. Inf. Sci. **417**, 128–142 (2017)
23. Ma, J., Liang, P., Yu, W., Chen, C., Guo, X., Wu, J., Jiang, J.: Infrared and visible image fusion via detail preserving adversarial learning. Inf. Fusion **54**, 85–98 (2020)
24. Ma, J., Ma, Y., Li, C.: Infrared and visible image fusion methods and applications: a survey. Inf. Fusion **45**, 153–178 (2019)
25. Ma, J., Yu, W., Liang, P., Li, C., Jiang, J.: FusionGAN: a generative adversarial network for infrared and visible image fusion. Inf. Fusion **48**, 11–26 (2019)
26. Ma, J., Zhao, J., Jiang, J., Zhou, H., Guo, X.: Locality preserving matching. Int. J. Comput. Vis. **127**(5), 512–531 (2019)
27. Min, Z., Wang, J., Meng, M.Q.H.: Joint rigid registration of multiple generalized point sets with hybrid mixture models. IEEE Trans. Autom. Sci. Eng. (2019). https://doi.org/10.1109/TASE.2906391

Novel Bayesian Network Incremental Learning Method Based on Particle Swarm Optimization Algorithm

Yunnan Ling[1], Neng Yang[1], Haitao Yu[1], and Yungang Zhu[2(✉)]

[1] College of Software, Jilin University, Changchun, China
[2] College of Computer Science and Technology, Jilin University, Changchun, China
zhuyungang@jlu.edu.cn

Abstract. Traditional bayesian network learning mainly uses batch learning method, that is, assuming that the training data can be acquired in one time, the whole learning process will end after learning these samples. But most real-world data is generated sequentially and incrementally over time. Therefore, it is necessary to improve batch learning into incremental learning to adapt to the reality. Many scholars have studied the incremental learning method [1–3] in the bayesian network field. In this paper, the discrete particle swarm optimization (PSO) algorithm is introduced into the structure learning of bayesian network, and a new incremental learning method based on PSO is proposed. Finally, the control variable method is used to compare the parameters of the algorithm, the number of iterations of a single learning and the number of particles, etc., to explore the influence of these factors on the incremental learning effect of bayesian network.

Keywords: Bayesian network · Batch learning · Structure learning · Incremental learning · Particle swarm optimization

1 Introduction

The construction of bayesian network is an important research content in the field of bayesian network. It has become a trend to use heuristic intelligent optimization algorithm to learn bayesian network, among which the structural optimization method based on genetic algorithm has made great progress. Representative research achievements in this field include: 1) learning bayesian network structure based on genetic algorithm, and setting repair operator to ensure the closure of bayesian network [4]; 2) structure learning was carried out under incomplete data sets based on genetic algorithm, and a new fitness function was proposed based on the maximum expectation [5]; 3) the genetic algorithm adaptation function based on bayesian network equivalence class representation is used, and the greedy operator is introduced to improve the convergence rate [6]; 4) a new genetic coding method of bayesian network structure based on genetic algorithm [7]; 5) an improved genetic algorithm called matrix genetic algorithm was used for bayesian structure learning [8]. But the problem with these studies is that they ignore the fact that the data is mostly obtained linearly over time.

M. Tavana et al. (Eds.): IISA 2020, AISC 1304, pp. 941–947, 2021.
https://doi.org/10.1007/978-3-030-63784-2_114

Traditional bayesian network structure learning is often a batch learning model, that is, it is assumed that learning samples can be acquired at one time, after learning these samples, the learning process is terminated, and no new knowledge will be learned. However, in practice, the training samples are not always obtained at one time, but are generated continuously and incrementally over time, and the information reflected by the samples may also change over time. Incremental learning is an online learning process that updates the existing learning results in time sequence. It does not discard the existing learning results, but constantly updates and refines the learned results with new samples, which is suitable for this kind of situation.

2 A BPSO-Based Incremental Learning of Bayesian Network

2.1 Bayesian Network Coding

Bayesian network is a directed acyclic graph. If the traditional adjacency matrix is used, the structure will have a ring. In order to avoid this problem, this paper adopts a new encoding method [13]. In the first part of this method, one-dimensional array is used to store the sequence of each node, the sequence is numbered from small to large, and only the node with small sequence number can be used as the parent of the node with large sequence number. For example, for a set of nodes $\{X_1, X_2, X_3, ..., X_n\}$, if X_i is the parent of X_j, then we have $1 \leq i < j \leq n$. This sequence is a kind of topological sequence of bayesian network, which can be set artificially according to the actual application. This method can avoid the appearance of the ring, save the step of detecting the ring, and improve the efficiency of the algorithm. The second part uses a two-dimensional array to store the parent number of each node. In order to reduce the search space, this paper adopts the method of limiting the number of parents of each node [9]. Let the maximum number of parent nodes of a node be 2.

In order to facilitate the particle velocity to act on the current position of the particle, we still use the method of position coding for the velocity. But only the second part of the position code, the parent part of the node, is left, because the speed can only change the parent of a node, not the number of the node itself. The specific definition of particle velocity and position operation has been given in literature [9].

2.2 BIC Scoring Function

The scoring function in bayesian network, as the fitness function in PSO, is used to calculate the score of bayesian network represented by each particle. This score can measure the quality of A bayesian network, and the higher the score is, the more the network matches the data set D. In this paper, *BIC* (Bayesian Information Criterion) scoring function is adopted as the fitness function, and the formula is as follows:

$$F_{BIC}(D) = \sum_{i=1}^{n} \sum_{k=1}^{r_i} \sum_{j=1}^{q_i} N_{ijk} \ln \frac{N_{ijk}}{N_{ij}} - \frac{1}{2} (\ln N) |\theta| \tag{1}$$

Where: the first term is the likelihood degree of the optimal parameter logarithm, which is used to measure the fitting degree between the bayesian network and data set D; the second term is about the penalty of the structural complexity, which is conducive to selecting a relatively simple network.

The formula can be decomposed into the summation form of the local fitness of each node:

$$F_{BIC}(D) = \sum_{i=1}^{n} F_{BIC\_i}(D) \tag{2}$$

Where:

$$F_{BIC\_i}(D) = \sum_{k=1}^{r_i} \sum_{j=1}^{q_i} N_{ijk} \ln \frac{N_{ijk}}{N_{ij}} - \frac{1}{2}(\ln N)|q_i(r_i - 1)| \tag{3}$$

That is, the local fitness of each node can be calculated separately and then summed to obtain the total score of the network. In addition, the nodes with the same number in different networks can be compared according to their local fitness, which facilitates some updating operations of location and speed.

2.3 Algorithm Framework

According to the above analysis, this paper improved particle swarm velocity updating formula, the formula of the improvement can be well combined with the characteristics of the old and new data, and combining with the design structure of incremental learning new strategies, this paper select the best match with the new data after many iterations (that is, under the new data has the highest score *BIC*) network as the final training results, the bayesian network is higher compatibility for new data.

The speed update formula of the classical particle swarm optimization algorithm is:

$$V_{k+1} = \omega V_k + c_1 r_1 (P_{best_k} - X_k) + c_2 r_2 (G_{best_k} - X_k) \tag{4}$$

Where: P_{best_k} means the historical optimal network structure of the current particle, and the optimality is reflected in the structure's highest *BIC* score, that is, the structure has the highest matching degree with the data set. Similarly, G_{best_k} represents the optimal structure of the historically optimal structure of all particles in the population.

Based on this feature, this formula is improved as follows:

$$V_{k+1} = \omega V_k + c_1 r_1 (O_{best} - X_k) + c_2 r_2 (N_{best} - X_k) \tag{5}$$

This formula makes the particles reflect the correlation and complementarity of the data information of the before and after time periods. This part is called as a subroutine in the incremental structure learning new strategy.

The specific steps of the algorithm are as follows:

1) Reads the first set of data and processes the data into a processable format. Arrange the nodes in a topological sequence.
2) Initializes the particle, including the initial position and initial velocity of the particle.
3) Carries out the first traditional structure learning and selects the global optimal bayesian network as O_{best}.
4) read a new batch of data sets, re-initialize a batch of particles and use the new data set for traditional structure learning, and select the global optimal bayesian network as N_{best}.
5) initializes a batch of particles and iteratively invokes an incremental learning subroutine (using an improved speed update formula) for incremental learning. The improved velocity update formula is used to update the particle velocity and obtain the new particle position. The new *BIC* score of particles was calculated by using the new data set, and the optimal network structure of individuals was updated.
6) select the bayesian network with the highest *BIC* score in the population as the final result of this incremental learning. Assign the result to O_{best}.
7) determines whether the termination condition has been met (that is, whether a new data set has been entered or the maximum number of iterations has been reached).If so, output the incremental learning end result. Otherwise, go back to step**4**.

When processing data, nodes have been prearranged according to the topological sequence. As a result, the network with illegal ring will not appear, eliminating the step of checking the ring division.

3 Experimental Results and Analysis

In this paper, the Aisa network is used to experiment on the model, and 3000 data are input in 6 times to simulate the process of incremental learning. The experiment is realized by the method of control variables. The following experimental results are obtained by taking the mean value after 5 experiments.

3.1 The Effect of Random Perturbation on Network Learning

In the experiment to explore the influence of random disturbance on network learning effect, we set the maximum parent node as 6, the number of particles as 10, and the number of iterations as 30. Comparing the cases with and without random disturbance, the experimental results are as follows: N(Normal) curve represents the experimental results without random disturbance, and R(Random) curve represents the experimental results with Random disturbance (Fig. 1).

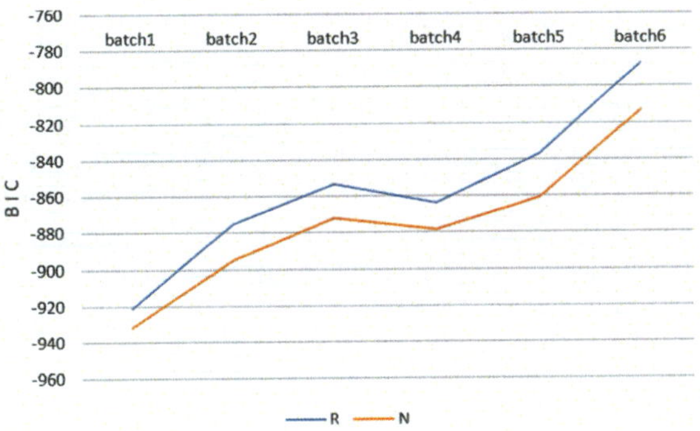

Fig. 1. The effect of random perturbation on network learning

Experimental results show that adding random perturbation to the velocity can improve the accuracy of bayesian structure learning. We analyze that because in the iterative process, the particle is prone to fall into local optimality, so that the advance speed is small or even no. If you add random perturbations, then even in the later iteration, the particles are very active and can expand the search range, so you can get better results.

3.2 The Influence of the Number of Particles on the Effect of Network Learning

In the experiment exploring the effect of the number of particles on network learning, we set the maximum parent node as 6 and the number of iterations as 30. Experiments were conducted with different particle Numbers and the results were as follows (Fig. 2):

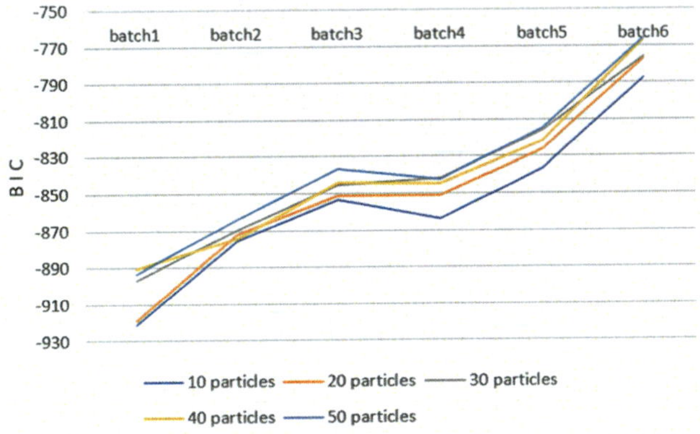

Fig. 2. The influence of the number of particles on the effect of network learning

The experimental results show that there is a positive correlation between the number of particles and the learning effect. If the number of particles is small, the searching ability of particles is weak, which leads to the convergence of particles. However, when the number of particles reached a certain number, the fitness value of increasing the number of particles again did not change significantly. For example, when the number of particles is 50, the fitness value of the trained bayesian network hardly improves when the number of particles is 40. On the contrary, the increase of the number of particles will result in a longer running time of the program. Therefore, considering the factors of learning efficiency and effect, setting the particle number to 40 will achieve the approximate optimal result.

4 Conclusions and Future Works

This paper proposes a new incremental learning method of bayesian network based on particle swarm optimization (PSO) algorithm. Based on the characteristics of PSO, this paper combines PSO with the learning of bayesian network structure to realize a new incremental learning method.

In the future, this topic should pay more attention to how to accurately construct a bayesian network in the absence of data, and even to construct a better bayesian network without a given node order.

Acknowledgments. This work is supported by National College Students' Innovation and Entrepreneurship Training Program (201910183045).

References

1. Yasin, A., Leray, P.: iMMPC: A Local Search Approach for Incremental Bayesian Network Structure Learning (2011)
2. Friedman, N., Goldszmidt, M.: Sequential update of Bayesian network structure (1997)
3. Tian, F., Huang, L., Yu, J.: Bayesian network incremental learning method with hidden variables. Chin. J. Electron. (11), 7–10 (2005)
4. Zhou, B., Tian, X.: Bayesian network structure learning algorithm based on genetic algorithm and reinforcement learning. Microcomput. Appl. (S1), 51–54 (2007)
5. Wang, F., Liu, D., Wang, S.: The research of Bayesian network structure incremental learning based on genetic algorithm. J. Comput. Res. Dev. (09), 1461–1466 (2005)
6. Chen, F., Wang, X., Rao, Y.: Learning Bayesian networks using genetic algorithm. J. Syst. Eng. Electron. **18**(1), 142–147 (2007)
7. Lee, J.H., Chung, W.Y., Kim, E.T., Kim, S.H.: Structure learning of Bayesian networks using dual genetic algorithm. IEICE Trans. Inf. Syst. **91-D**(1),32–43 (2008)
8. Lee, J.H., Chung, W.Y., Kim, E.T., Kim, S.H.: A new genetic approach for structure learning of Bayesian networks: matrix genetic algorithm. Int. J. Control Autom. Syst. **8**(2), 398–407 (2010)
9. Liu, X., Jia, H., Liu, D.: Bayesian network structure learning based on particle swarm optimization algorithm. J. Chin. Comput. Syst. (08), 1516–1519 (2008)
10. Daly, R., Shen, Q., Aitken, S.: Learning Bayesian networks: approaches and issues. Knowl. Eng. Rev. **26**(02), 99–157 (2011)

11. Friedman, N., Nachman, I., Pe'Er, D.: Learning Bayesian network structure from massive datasets: the "sparse candidate" algorithm
12. Chickering, D.M.: Learning Bayesian networks is NP-complete. Networks **112**(2), 121–130 (2000)
13. Zhu, Y.: Research on some problems in Bayesian network learning and its application in information fusion. Jilin University (2012)

Feed Formulation Cost Optimization Based on the Improved Genetic Algorithm

Yao Juan, Xu Wang, Zhang Cheng, and Tian Fang(✉)

Huazhong Agricultural University, Wuhan, Hubei, China
fangzhihai_2003@mail.hzau.edu.cn

Abstract. The design of feed formulation algorithm aims to help feeders to obtain more cost-saving new formulas satisfying livestock demands for nutritional ingredients through further scientific calculation based on the original formula. The comprehensive requirements for contents and nutrients of many fodders were transformed and digitalized, and a group of optimal solutions were obtained using the improved genetic algorithm, namely the most cost-saving formula satisfying the demand for optimal individual growth.

Keywords: Improved genetic algorithm · Feed formulation

1 Feed Formulation and Genetic Algorithm

Feed formulation is the basis for production management of the cultivation industry, where the cost of fodders accounts for 60%–70% of total cost. The present researches on feed formulation software mainly include linear programming, goal programming, genetic algorithm, etc. For instance, Liu Q et al. from Xi'an University of Technology improved encoding scheme, penalty function and search strategy based on artificial fish swarm algorithm of symbiotic system and proposed the optimization algorithm [1]. Xu D S et al. used genetic algorithm to improve real number encoding, crossover operator and mutation operator, thus being able to solve problems of many factors and high nonlinearity degree [2]. The extensive application of the improved genetic algorithm to delay time optimization of tray selection, dynamic programming of EV charging station, ammonia coverage optimization of SCR system and ensured scheduling of shipboard aircraft [3–6] indicated that genetic algorithm had outstanding performance in the optimal solution.

2 Genetic Algorithm

Just as its name implies, genetic algorithm derives from computer simulation study for biosystem. Selection, mating and mutation in the evolution theory of species can all be simulated into logic operation and arithmetic operation between concrete data in the program, the individual with the best phenotype will be finally screened out through several generations of reproduction process of "survival of the fittest in natural selection", namely the optimal formula (approximate). The genetic algorithm was

M. Tavana et al. (Eds.): IISA 2020, AISC 1304, pp. 948–956, 2021.
https://doi.org/10.1007/978-3-030-63784-2_115

proposed in the last century, so many paradigms and models have come into being. In the common traditional genetic algorithm routine, the approximate optimal solution is obtained through gradual implementation of various steps such as data coding, generation of initial population, fitness calculation, selection, crossover, mutation, iterative reproduction and final selection. The concrete meanings of the steps will be introduced first, followed by the algorithm improvement in this study.

2.1 Data Coding

This is a coding operation of original data, which is convenient for logic operation and arithmetic operation of data. So far, many different coding methods have been proposed and can be largely divided into three major types: binary coding method, floating-point coding method and symbolic coding method. When binary coding method is used to solve continuous variable optimization problem, large Hamming distance may exist in binary coding with adjacent integers, and there lacks fine tuning function for data accuracy. The floating-point coding method is proposed also to solve these defects of binary coding method. Demonstrative calculation indicates that floating points are featured by faster expression and more consistent between operations, and moreover, higher accuracy can be realized.

2.2 Generation of Initial Population

Genetic algorithm is specific to numerous individuals, which constitute a population. The initial population is set and taking as starting point for evolution until the evolutionary process is terminated by one evolutionary stopping criterion, and thus the last generation is obtained. The distribution scope of the space occupied by the optimal solution in the whole problem space can be mastered, and the initial population can be set within this distribution scope.

2.3 Fitness Calculation

Biologists use the term—fitness—to measure the adaptability of one species to its survival environment. This concept of fitness is used in genetic algorithm to measure the degree to which individuals in a population may reach or approach or help to find the optimal solution. For individuals with high fitness, the probability for them to be inherited to the next generation will be higher.

2.4 Selection

Selection operation refers to selecting individuals relatively adaptable to the environment from the population according to their fitness values. Generally speaking, selection will contribute to reproduction of a large number of individuals with high fitness values to the next generation, and the number of individuals with small fitness values reproduced to the next generation will be small, and they may even be eliminated. The traditional selection method is roulette model.

2.5 Crossover

In genetic algorithm, crossover operation is the main method used to generate new individuals and the key deciding the algorithm convergence performance. During crossover operation in the standard genetic algorithm, individuals to be crossed over are selected out first according to the preset crossover probability, a crossover pairing pool is then formed, completely random and equal-probability one-to-one pairing of the individuals in the pairing pool is conducted, crossover point of each pair of parent individuals is finally determined, and their gene segments are exchanged to generate new child individuals. Arithmetic crossover is applicable to data obtained through floating-point coding. For arithmetic crossover method used by the standard genetic algorithm, when the corresponding crossover probability is satisfied, the followings will hold.

$$\text{child1} = rnd * prarent1 + (1 - rnd) * parent2 \tag{1}$$

$$\text{child2} = rnd * prarent2 + (1 - rnd) * parent1 \tag{2}$$

Where rnd is true random number of [0, 1].

2.6 Mutation

In genetic algorithm, mutation operation is an auxiliary method of generating new individuals and also an essential operation step. Mutation operation can enhance local searching ability of genetic algorithm and further evolve individual coding structures obtained through the crossover operation so that it is possible to obtain an individual closer to the optimal solution. Moreover, population diversity can be maintained and premature phenomenon can be prevented.

2.7 Iterative Reproduction

The first-generation population will be obtained through the first six operations. Iterative reproduction operation will repeat fitness calculation, selection, crossover and mutation operations of this population, so as to realize population reproduction generation by generation.

2.8 Final Selection

Through enough but limited reproductions, the final superior population is obtained through iterations, and then the individual having the most outstanding phenotype is selected, thus obtaining the optimal solution (approximate).

3 Improved Genetic Algorithm

3.1 Data Encoding

For floating-point data in the formula, the concrete operations were implemented by the program in way of floating-point encoding.

3.2 Generation of Initial Population

As there were only one group of data, namely one individual, at the beginning, a population could not be formed, and then an initial population needed to be constructed. According to use level of each ingredient in the original formula, an initial population with a certain amount was constructed through the random generation method (that the optimal solution might have an interval, and actual livestock feed quality each time were both considered) based on this use level (fluctuation at this level not exceeding 40%). Here, each individual in the population should be phenotype adaptable to the environment, namely data combination needed to satisfy various nutritional requirements.

3.3 Fitness Calculation

As the design of feed formula aims to save the fodder cost as much as possible while nutritional requirements are satisfied, the reciprocal of total fodder cost is taken as the fitness, which conforms to such requirement: the higher the fitness, the better the individual phenotype.

3.4 Selection

When roulette model in standard genetic algorithm is used to perform selection operation of single fitness in a feed formula, many indexes in the fitness function remain to be calculated, especially during the multi-iterative process, this will impact the program running efficiency; More seriously, due to selective probability, that the optimal individual can enter the next round of competition cannot be guaranteed. To solve the abovementioned defects, stochastic tournament was used to substitute the roulette model in order to reduce the calculated quantity in computer processing. In the selection operation of the tournament, there was only comparison operation but not arithmetic operation between individual fitness values. To be specific, 4 individuals were randomly selected from the population each time to compare their fitness values, and the 2 individuals with the highest fitness values among them were inherited to the next population; This process was repeated until the next generation of population was generated. Moreover, in order to ensure reserving the temporary optimal solution in early-phase operation, elitist approach was also added. When a new generation was generated each time, the present optimal solution was copied to the new generation first as it was, other steps were unchanged, and thus the optimal solution generated at any time could join in the new iteration.

3.5 Crossover

As for the set of child generation method of crossover operation in the standard genetic algorithm, child individuals generated by crossover are always located on the ligature of parent individuals in the rectangular plane coordinate system. In this way, the search space of child individuals will be continuously shrunk as the iteration proceeds, thus leading to prematurity. Different random numbers were used in this study to perform crossover operation of each component of parent vector as follows:

$$\text{child2}[j] = prarent2[j] + rnd[j] * (parent1[j] - parent2[j]) \tag{3}$$

$$\text{child1}[j] = prarent1[j] + rnd[j] * (parent2[j] - parent1[j]) \tag{4}$$

Where parent1[j] and parent2[j] represent components of parent1 and parent2. Child1 [j] and child2[j] are components of child vectors child1 and child2. Rnd[j] is true random number within the interval [−2, 2]. The child individuals generated by this were not restricted to the ligature of parent individuals, but instead, they could be dispersed in a broader space. As the search space of child individuals was expanded, prematurity could be avoided and global optimization efficiency could be improved. In the end, survival of the fittest was simulated through simple child fitness calculation, namely, if child phenotypes failed to satisfy the given requirements, children could not survive and would be abandoned. When the quantity of children reached the preset population size, the next crossover operation would be carried out.

3.6 Mutation

The uniform mutation in the standard genetic algorithm is especially applicable to initial operation phase of genetic algorithm. With uniform mutation, search points can freely move in the whole search space, so as to enhance population diversity, and the algorithm can process more patterns. However, for critical local search in later operation phase, it is difficult for uniform mutation to reach a very good effect. Here, Gaussian mutation was used to improve uniform mutation in the standard genetic algorithm. According to characteristics of normal distribution, Gaussian mutation also lays the emphasis on searching one local region nearby the original individual, and it can realize local searching of the key search region very well. Finally, the survival of the fittest was still simulated through individual fitness calculation through simple mutation, namely if individual phenotypes could not satisfy the given requirements after mutation, it would be considered that they could not survive, and they would be abandoned and individuals before the mutation would be reserved.

3.7 Iterative Reproduction

Just as the standard genetic algorithm, repeated fitness calculation, selection, crossover and mutation were needed to realize the reproduction of populations generation after generation.

3.8 Final Selection

The optimal solution of the last generation was selected through iteration.

4 Parameter Configuration and Core Code of the Improved Genetic Algorithm

4.1 Parameter Configuration

De_Variable: 8
Po_Size: 500.
Ov_Probability: 0.85.
Va_Probability: 0.5.
Ev_Algebra: 300.

Cross operation key generation progeny formula:

$$\text{child1} = prarent1 + random\_number * (parent2 - parent1) \tag{5}$$

$$\text{child2} = prarent2 + random\_number * (parent1 - parent2) \tag{6}$$

4.2 Gauss Mutation Core Code

```
l2=[0 for x in range(De_Variable)]
    for i in range(Po_Size):
        for w in range(De_Variable):
            l2[w]=nextpopulation[i][w]
        for j in range(De_Variable):
            for f in range(2):
                if f==0 :
                    v0=np.random.randint(1,10000)
                else:
                    v1=np.random.randint(1,10000)
            v2=np.random.randint(0,9999)
            r1=float(v0)/10000
            r2=float(v1)/10000
            p=float(v2)/10000
            if p<Va_Probability :
                sign=np.random.uniform(0.0,1.0)
                sigma=0.01*(Range[j].getupper() -
Range[j].getlower())
                if sign!=0 :
```

```
                    l2[j] = (l2[j]+ sigma * math.sqrt(-2 *
math.log(r1) / 0.4323)*math.sin(2 * 3.1415926*r2))
                else:
                    l2[j] = (l2[j]- sigma * math.sqrt(-2 *
math.log(r1) / 0.4323)*math.sin(2 * 3.1415926*r2))
                if l2[j]>Range[j].getupper() :
                    ben=np.random.randint(0,9999)

l2[j]=(float(ben))/10000*(Range[j].getupper()-Range[j].ge
tlower())+Range[j].getlower()
                if l2[j]<Range[j].getlower() :
                    ben=np.random.randint(0,9999)

l2[j]=(float(ben))/10000*(Range[j].getupper()-Range[j].ge
tlower())+Range[j].getlower()
```

A new individual L2 [J] is obtained by mutation, and j represents the individual number in the population.

5 Simulation Experiment

The set parameters in tables of background MYSQL database were acquired by the program, evolution operation of the genetic algorithm was performed combining the initial formula input at the front end (here data input at analog front end in the given initial formula in the program was used), and the final optimal formula (approximate) was obtained through evolution of 300 generations. The initial formula was milk cow feed formula, which was formed by 8 raw materials: mixing concentrate, alfalfa hay, wet sugar beet pulp, wheat straw, carrot, full cottonseeds, rumen protected fat and corn silage. Nutritive indexes included five nutrients such as digestible protein, milk energy, calcium, etc. The use level of the initial formula was as follows (Table 1).

Table 1. The use level of the initial formula.

Additive amount (kg)	11.00	2.00	5.00	3.00	5.00	1.00	0.30	20.00

The total price was RMB 36.99.

The formula obtained through the improved genetic algorithm and the price are shown in the following Table 2:

Table 2. Five times results based on the improved genetic algorithm.

	1	2	3	4	5
x0	10.16401878	10.17004082	10.1670067	10.16104989	10.16967765
x1	1.20429517	1.20446826	1.20014818	1.21500101	1.20333129
x2	6.99988201	6.99984253	6.99922121	6.99511883	6.99921894
x3	4.19902550	4.19933405	4.19707144	4.19952710	4.19611799
x4	3.02057660	3.00196721	3.00001946	3.00363716	3.01491874
x5	0.96572311	0.96433302	0.97258059	0.96795867	0.96472331
x6	0.18000094	0.18007276	0.18001423	0.18003349	0.18006188
x7	18.03527276	18.00887049	18.01529701	18.0447869	18.00402065
Total price of optimal formula	33.112446	33.107235	33.112534	33.118832	33.108384

The above are corresponding to the results obtained through random debugging for five times, and it could be seen that the final cost was maintained at RMB 33.11 and floating range was 0.01.

6 Conclusion

The improved genetic algorithm was used for optimization from construction of initial population, selection of fitness function, crossover operator and mutation operator, and the optimal feed formula was obtained. Hence, the algorithm is of fast operation and easy promotion.

7 Problems and Prospects

In practical application, we need to consider the confidence level, that is to say, we need to consider how much is appropriate that we set the success probability to meet the nutrient needs [7]. In addition, the normal loss of feed will also affect the feeding effect, so the actual application needs to calculate the loss of feed components according to the time of feed placement.

References

1. Liu, Q., Li, Y., Qing, M.Y., Tomohiro, O.: Application of artificial fish swarm algorithm to optimization of feed formula based on symbiotic system. J Comput. Appl. **36**(12), 3303–3310 (2016)
2. Xu, D.S., Yang, F., Wang, B.H., Li, X.C.: Application of improved genetic algorithm to feed formula design. Chin. Agric. Sci. Bull. **28**(08), 24–28 (2012)

3. Liu, Y., Wang, N.J., Luo, X., Meng, X.L.: Ensured scheduling method of shipboard aircraft via the improved genetic algorithm. J. Natl. Univ. Def. Technol., 194–205 (2020)
4. Wang, G.Y., Qi, J.Z., Liu, S.Y., Shuai, S.J., Wang, Z.M.: Ammonia coverage optimization of SCR system based on multi-objective genetic algorithm. Automot. Eng. **42**(3), 279–285 (2020)
5. Li, J.H., Cao, W., Zhao, D.G., Jiang, Y., Zhou, Q.H.: Delay time optimization of try selection based on improved genetic algorithm. Comput. Integr. Manuf. Syst. (2), 340–355 (2020)
6. Zang, H.Y., Fu, Y.T., Chen, M., et al.: Dynamic programming of EV charging station based on improved adaptive genetic algorithm. Electr. Power Autom. Equip. (1), 163–170 (2020)
7. Sun, W., Zhang, Z.: Models of least cost ration formulation dealing with nutrient variability. J. Anim. Nutr. (3), 31–37 (1996)

Vehicle Disposition and Routing Optimization of Reverse Logistics Based on Improved Genetic Algorithm

Shengping Zhao[1(✉)], Jingrui Li[2], and Jing Wang[3(✉)]

[1] Yuxi Normal University, Yuxi 653100, China
papernew@163.com
[2] Honghe University, Mengzi 661100, China
[3] Nantong Institute of Technology, Nantong 226002, China

Abstract. The goals of reverse logistics are resource recovery, protect the living environment and dispose of waste reasonably, etc., which meet the current sustainable development requirements of the circular economy. This paper combines circular economy and reverse logistics to Vehicle disposition and routing optimization. This article aims at saving distribution costs. First, it analyzes the characteristics of reverse logistics, simplifies partly to establish a reverse logistics distribution model, and secondly, uses the saving algorithm to create the initial population. Finally, on the basis of genetic algorithm, combined with local search algorithm to improve, complete the configuration and path optimization of reverse recovery vehicle with distribution cost as the main constraint.

Keywords: Reverse logistic · Improved genetic algorithm · Saving algorithm

1 Introduction

With the increasing emphasis on the circular economy, the field of reverse logistics has become the focus of attention from all walks of life in society. The uncertainties of recycling products in the reverse process, the complexity of the recycling processing system, and the dispersion of recycling points have added many difficulties to the smooth development of reverse logistics. However, recycling activities also bring value to the environment, create considerable economic benefits and added value for companies [1].

The research on reverse logistics abroad is mostly on the integration of positive and reverse logistics, vehicles deliver goods to customers while bringing back some goods. Fisher divides the VRP problem into three stages: 1) Apply simple heuristic algorithm, greedy algorithm and other algorithms. 2) Heuristic algorithms based on mathematical programming or other algorithms are combined with TSP problems. 3) Use artificial intelligence and modern heuristic algorithm [2–4].

Domestic research on reverse logistics path optimization is still in its infancy. Huo et al. [5, 6] adopted C-W saving heuristic algorithm and nearest neighbor method to insert and sort each recycle bin. Wang et al. [7, 8], combined with the TSP problem and

M. Tavana et al. (Eds.): IISA 2020, AISC 1304, pp. 957–963, 2021.
https://doi.org/10.1007/978-3-030-63784-2_116

used the heuristic algorithm to analyze the shortest path, which proved the practicability and effectiveness of the optimal return load design on the shortest path. This paper analyzes and simplifies the path planning of reverse logistics, using genetic algorithm combined with local search algorithm for modeling and optimization.

2 Methodology

The network characteristics of reverse logistics are opposite to that of forward logistics, and the network process is convergent. The reverse logistics network structure is shown in Fig. 1. The complexity of the reverse logistics system is mainly reflected in the complexity of its cost factors, which consumes a lot of costs in the distribution link. Therefore, the efficient arrangement of vehicles in reverse logistics and the optimization of the distribution path have become the key points of reverse logistics network planning.

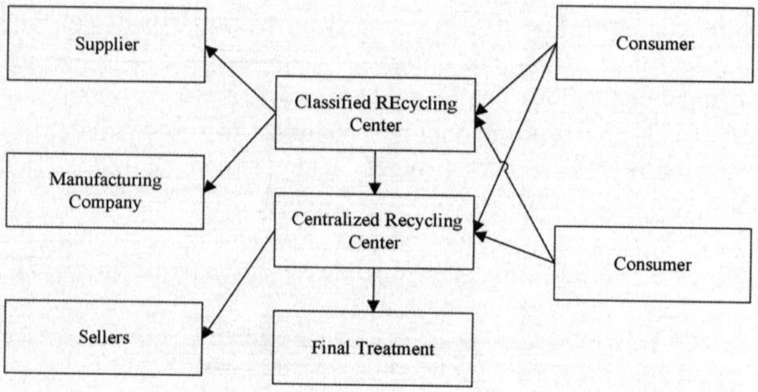

Fig. 1. Reverse logistics network structure

2.1 Model Analysis

This article combines the concept of circular economy to plan and analyze the problem of certain vehicles. The basic assumptions of the model are as follows [9].

1) The model contains a recycling center and multiple recycling points. Each car departs from the recycling center. After completing the recycling task through each customer point, it finally returns to the recycling center.
2) The relative position coordinates of the recycling center and each recycling customer point are known, and the goods to be recycled at each recycling customer point are cleared by the vehicle once.
3) The number of vehicles that can be deployed in the recycling center is known, and no clear limit on the recycling capacity of the recycling center.
4) The weight of the goods at each customer point is known.

5) All vehicle models are unified, the unit transportation cost is fixed. The maximum load is known, and the recovery volume of a single recovery point cannot exceed the maximum load constraint of a single vehicle.

The model parameters are shown in Table 1 below.

Table 1. Model parameters

Parameter	Definition
P	Location set: P = {i},i = 0 Recycling Center; i = 1,2…n, Consumer
Q	Consumer
V	Logistics vehicle
N	Number of logistics vehicles
D_{ij}	Distance
L	Maximum load
W_i	Total weight recovered at the i-th consumer
LW	Overweight program
X_{ijn}	1: The nth car drives from point i to point j for recycling
	0: The nth car does not drive from point i to point j for recycling
Y_{in}	1: The nth car recycles the goods at consumer i
	0: The nth car does not collect the goods at consumer i

2.2 Reverse Logistics Vehicle Configuration and Path Optimization Model

Combining the above parameters and variables, the vehicle configuration and path planning of reverse logistics are modeled with the goal of minimizing transportation costs. The unit operating cost of vehicles is the same, so the distribution distance is used to represent the operating cost. Increasing the total vehicle recovery exceeds the maximum load of the vehicle as a penalty function, $\alpha = 100$. F represents the objective function of the model.

$$F = \min\left(\sum\nolimits_{i\in P}\sum\nolimits_{j\in p}\sum\nolimits_{n\in V} D_{ij}X_{ijn} + aLW\right) \tag{1}$$

$$\sum_{n\in V} Y_{in} \leq 1; i \in Q \tag{2}$$

$$\sum_{i\in P} X_{ijn} = Y_{jn}; j \in P, n \in V \tag{3}$$

$$\sum_{j\in P} X_{ijn} = Y_{in}; i \in P, n \in V \tag{4}$$

$$\sum_{i\in Q} W_i Y_{in} \leq L_n^{\max}; n \in V \tag{5}$$

Where, n is the number of logistics vehicle, i is the number of consumer. Its constraints are: the vehicle can transport the goods at the collection point in one go; the same vehicle goes through the same collection point and each collection point serves at most once; the vehicle load cannot exceed the rated maximum load.

3 Genetic Algorithm

Genetic algorithm is a random global search optimization algorithm that imitates the natural selection of natural selection in natural world and the genetic mechanism. It simulates the evolutionary process of organisms: from the initial reproduction, crossover, mutation and selection, and finally realizes random optimization and selection, thereby generating new and more adaptable offspring. After repeated iterations, the optimal or satisfactory solution to the problem is found. Genetic algorithm has good parallel processing effect, the result is usually the global optimal solution.

3.1 Initial Distribution Plan

In order to improve the optimization speed and effect of the model, the saving algorithm (C-W Algorithm) is used to initialize the population. Assuming that each vehicle only serves one recycling point, n mutually independent recycling routes can be generated. By comparing the savings value of the combined individual routes, find the vehicle configuration and route planning scheme corresponding to the optimal solution, as the initial population. The Path merging based on C-W Algorithm is shown in Fig. 2.

$$S_{ij} = d_{0i} + d_{0j} - d_{ij} \tag{6}$$

Where, d_{0i}, d_{0j}, d_{ij} are independent recycling route.

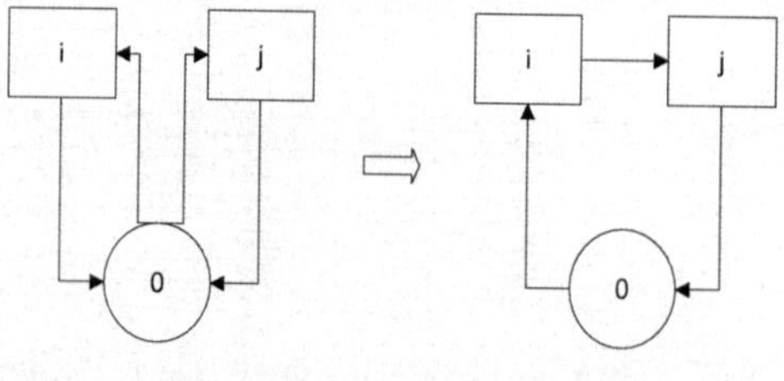

Fig. 2. Path merging based on C-W algorithm

3.2 Initial Population

Encode the scheme and generate chromosomes to get the initial distribution plan. Nteger coding is selected for the coding method of chromosome, in the chromosome, cusnum + i (i = 1,2,3…n) represents the recycle bin, cusnum represents the consumer, and cusnum is the number of consumers. When the number of vehicles used is less than the maximum adjustable vehicle, use cusnum +j at the end of the chromosome to complete. Taking 10 consumers as an example, the chromosome is shown in Fig. 3. Based on the above analysis, the length of the encoded chromosome N = Number of consumers + Number of recovered vehicles −1.

4	7	10	11	2	5	1	9	12	6	3	8
4	7	10	11	2	5	1	12	9	6	3	13

Fig. 3. Chromosome Structure Diagram

4 Results

In this paper, the solomom example is used for modeling and analysis. There is a recycling station, 100 customer points, The coordinates of the recycling station and recycling point are known, also the amount of goods to be recycled at each customer point. The recycling Center can be equipped with 20 vehicles, and the maximum load of each vehicle is 250. Chromosome length = 119, The population size = 100, Maximum iterations = 200, Pc = 0.9, Pm = 0.1, GGAP = 0.5. Selection and crossover choose roulette Whell Selection and Order Crossover (OX) separately.

The initial distribution plan is shown in Fig. 4a. The initial distribution plan obtained with the CW Algorithm greatly saves the optimization time. The optimal plan after optimization with the genetic algorithm is: number of vehicles used: 9, total distance traveled by the vehicle: 981.574, number of violations of the constraint path: 0, optimization time: 198.763912 s. As shown in Fig. 4b.

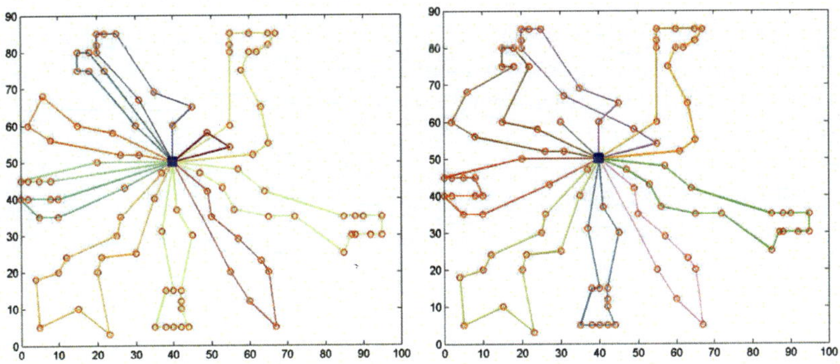

a) b)

Fig. 4. Distribution plan: a) Initial distribution plan; b) Distribution plan optimized by genetic algorithm

In Fig. 4b there is only one consumer point in the recycling route 8, the overall effect of the distribution plan is not good enough. Therefore, based on the genetic algorithm combined with the local search algorithm for optimization, in order to get a better distribution plan. The result is shown in Fig. 5. The optimization result: the number of vehicles used: 8, the total distance traveled by the vehicle: 920.3714, number of paths that violate constraints: 0, Optimization time: 467.278199 s.

Compare the two distribution plan shown as Fig. 4b and Fig. 5, the effectiveness of the route planning has improved, the order of consumer access is also more reasonable. Overall, the number of vehicles and the objective function have decreased, but the optimization time increased significantly.

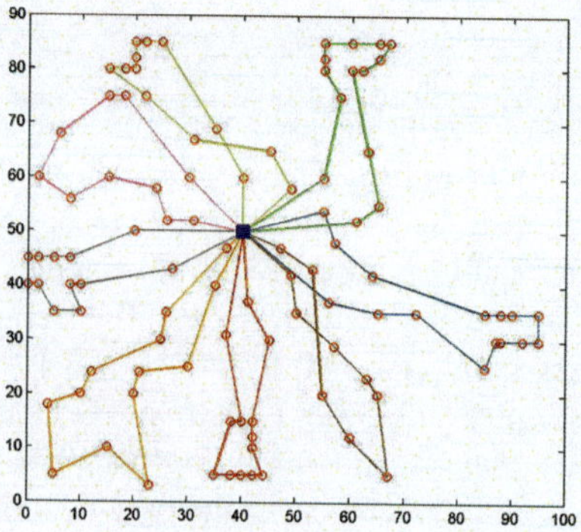

Fig. 5. Distribution plan optimized by improved genetic algorithm

5 Conclusion

This article combines the goals of sustainable development in a circular economy environment and the concept of green and efficient logistics in reverse logistics, with the goal of saving resources, reducing costs and improving the efficiency of reverse logistics. Establish a model for reverse logistics to solve the problems of optimization of recovery paths and vehicle configuration in reverse logistics.

In the process of modeling establishment, the combination of C-W Algorithm and genetic algorithm provides a more reasonable initial population for genetic algorithm, which greatly saves optimization time. Adding a local search algorithm to improve the genetic algorithm can provide a better delivery plan, but sacrifices optimization time. In general, both models can meet the requirements of reverse logistics vehicle configuration and recovery path optimization, and have certain theoretical and practical significance. Future research should also consider practical issues such as volume requirements, time requirements, and inconsistent logistics vehicle models.

Acknowledgments. Supported by Yunnan Fundamental Research Projects(grant NO. 2018FD094).

References

1. Dantzig, G.B., Ramser, J.H.: The truck dispatching problem. Manag. Sci. **6**(1), 80–91 (1959)
2. Dethloff, J.: Vehicle routing and reverse logistics: the vehicle routing problem with simultaneous delivery and pick-up. OR-Spektrum **23**(1), 79–96 (2001)
3. Liu, X.: Research on Reverse Logistics Transportation Strategies with Separable Backload. Dalian Maritime University (2009)
4. Bing, X., Bloemhof-Ruwaard, J.M., Vorst, J.G.A.J.: Sustainable reverse logistics network design for household plastic waste. Flex. Serv. Manuf. J. **26**(1–2), 119–142 (2014)
5. Jiazhen, H., Lei, Z.: Heuristic algorithm for solving traveling salesman problem with pickup and delivery. J. Tongji Univ. (Nat. Sci.) **34**(1), 134–138 (2006)
6. Erera, A.L., Morales, J.C., Savelsbergh, M.: The vehicle routing problem with stochastic demand and duration constraints. Transp. Sci. **44**(4), 474–492 (2010)
7. Da Qingli, W.F.: Transportation strategy of single vehicle in reverse logistics. J. SE Univ. (Nat. Sci. Ed.) **01**, 156–160 (2006)
8. Jayant, A., Gupta, P., Garg, S.K.: Reverse logistics network design for spent batteries: a simulation study. Int. J. Logistics Syst. Manag. **18**(3), 343–365 (2014)
9. Calvete, H.I., Galé, C., Oliveros, M.-J., Sánchez-Valverde, B.: A goal programming approach to vehicle routing problems with soft time windows. Eur. J. Oper. Res. **177**(3), 1720–1733 (2005)

Pattern Recognition Systems

Two-Stage Fusion of Local Binary Pattern and Discrete Cosine Transform for Infrared and Visible Face Recognition

Zhihua Xie(✉), Ling Shi, and Yi Li

Key Lab of Optic-Electronic and Communication, Jiangxi Science and Technology Normal University, Nanchang, China
xie_zhihua68@aliyun.com

Abstract. Combination the corresponding advantage of infrared and visible images, the fusion of visible-light and near-infrared features is becoming one of the key research directions of uncontrolled face recognition. In this paper, a new face fusion recognition method based on near-infrared and visible-light imaging is proposed to make full use of the complementarity in two modalities. Firstly, for near infrared imaging, the low frequency part is retained for discrete cosine transform (DCT) coefficients and the block histogram of local binary pattern (LBP) is jointly applied to represent the discriminative features. Then, LBP features from visible faces are gotten to make up the incapable of local representation in near-infrared imaging. Finally, the three features are sent to the novel two-state fusion module to leverage the performance of face recognition. The experimental results show that the combination of DCT and LBP can extract the complementary features from infrared and visible images, and the two-stage fusion strategy lifts the robustness of the multi-modality face recognition. Especially, the recognition rate of the proposed fusion recognition algorithm has been improved significantly, with regard to the other methods based on statistical feature fusion under the circumstance of small samples.

Keywords: Discrete cosinc transform · Local binary pattern · Fusion strategy · Face recognition · Two-stage fusion strategy

1 Introduction

Biometrics is one of the most paramount directions in computer vision application [1]. In biometrics, face recognition has become one of the most attractive recognition method for its strong concealment, non-aggression and more consistent with individual recognition habits. However, there are still shortcomings in the application of robust face recognition [2]. The main reason is that unconstrained conditions such as posture, illumination and facial expression have a great impact on face recognition. While the visible light image has abundant information, it is also sensitive to light. Compared with other unrestricted conditions, the variable illumination has the great influence on the recognition. In recent years, some research has made efforts to solve this problem. Firstly, some algorithms were used to improve the performance on illumination variation. Gross et al. proposed a proposal for estimating and compensating illumination

M. Tavana et al. (Eds.): IISA 2020, AISC 1304, pp. 967–975, 2021.
https://doi.org/10.1007/978-3-030-63784-2_117

areas in faces [3], which could partially alleviate the influence of illumination. Secondly, infrared imaging is applied for face recognition. The near-infrared image can suppress the ambient light, while isn't affected by the ambient temperature.

Li et al. developed a fast near-infrared face recognition system using LBP histogram features [4]. The near-infrared image has high robustness to posture and expression while eliminating the influence of illumination. Although with good performance in face recognition, they lack sufficient detail information compared to visible-light images. Thus, there is room for improvement. Raghavendra et al. proposed two methods for merging of visible and near-infrared images [5]: feature-level fusion and decision-level fusion. These methods fused sub-band coefficients via particle swarm optimization (PSO). Similarly, in [6], the genetic algorithm was used in fusion, while the literature [7] introduced the use of the algorithm in nuclear space for feature fusion as well as real numbers instead of binary to improve the accuracy of the algorithm. Literature [5–7] all applied complex algorithms to find an optimal fusion strategy. Zhongli Ma et al. presented a face recognition method to fuse infrared and visible faces via two-dimensional linear discriminant analysis [8]. Kai Guo et al. implemented a deep VGG network to extract infrared and visible features respectively, and then use a simple adaptive layered fusion method for face recognition [9], which requires the training samples and computational complexity higher.

Among the tasks of face-oriented identification, the information in infrared and visible imaging has the character of complementarity: on one hand, near-infrared can compensate for the lack of sensitivity of visible light; on the other hand, visible light can improve the local details of near-infrared facial representation. In this paper, we proposed a two-stage fusion strategy that effectively extracts the local high-frequency (texture) and complementary features in visible and infrared facial images. Experiment results show the novel fusion recognition algorithm can make full use of the complementarity features, and outperform the traditional fusion methods in recognition performance, especially in the case of single sample, the recognition rate can be significantly improved.

2 The Proposed Method

In recent years, the representation complementarity of infrared and visible imaging has been successfully utilized to improve the robustness of unrestricted face recognition [6–8]. The main reason is due to the characteristics of both of them: the latter is clear but sensitive to illumination while the former is relatively blurred but not sensitive to light. In order to effectively fuse different types of features, this paper proposes a two-stage fusion recognition strategy based on feature-level and decision-level. Our method proceeds as follows in Fig. 1. In addition, in fusion recognition, this paper directly uses near-infrared and visible-light images which aren't preprocessed with multi-model images such as alignment and configuration. Meanwhile, to adapt to the authentication of small samples (single sample), instead of statistical training based on 2DLDA, deep network and other extraction features, we use the pre-designed methods (ULBP and DCT) to extract feature. Both of them are based on different frequency information and are not sensitive to location, which are suitable for multi-modal images fusion recognition without strict alignment [7].

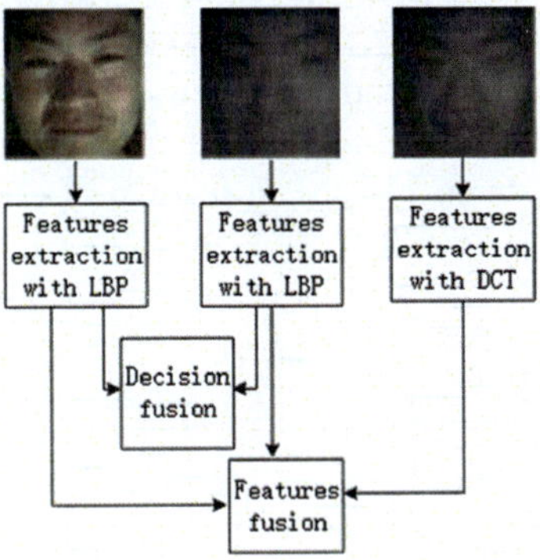

Fig. 1. The framework of the fusion recognition system

2.1 Uniform LBP

The traditional LBP compares the central pixel value with its neighborhood values in a square or circular field. As one of its improvements, ULBP is classified by the difference of the number of 0/1 hops in the encoding, which greatly diminishes the number of modes [10]. In this paper, we utilize ULBP [11] to extract visible local features. It's calculated by:

$$\mathrm{LBP}_{P,R}^{u2} = \begin{cases} \sum_{p=0}^{P-1} s(g_p - g_c)2^p, & \text{if } U(LBP_{P,R}) \\ p+1, & \text{otherwise} \end{cases} \tag{1}$$

then:

$$U(LBP_{P,R}) = |s(g_{p-1} - g_c) - s(g_0 - g_c)| + \sum_{p=0}^{P-1} |s(g_p - g_c) - s(g_{p-1} - g_c)| \tag{2}$$

$$s(x) = \begin{cases} 1, & x \geq 0 \\ 0, & x < 0 \end{cases} \tag{3}$$

where g_c is the center pixel value of the circle with radius R, g_p represents the value of the neighborhood pixels in the circle, P is the number of pixels around the center pixel in the circle, and the number of hops of 0/1 equals to (2) in the statistical code, (3) is given as a threshold function. By this way, we adopt the mode $\mathrm{LBP}_{8,1}^{u2}$.

As can be seen from Eq. (1), differential features are extracted by LBP. In terms of signal processing, the differential feature is considered as the high-frequency information of the signal, which is the detailed feature of the image [10].

2.2 Discrete Cosine Transform for Features Extraction

Discrete cosine transformation (DCT) has been widely used in speech and video compression due to its own characteristics [12]. Moreover, for the image with strong correlation, the generation of the DCT transform is a coefficient matrix of the same size where the larger coefficient is concentrated in the upper left corner of the matrix while the lower right corner of the matrix is smaller or zero, so that the Z-shaped Scanned data is sparse. It is feasible to use the sparsity of DCT to represent and reduce dimensionality of image features. Supposing an image *f(x, y)* with the size of M * N, the two-dimensional discrete cosine transform can be defined by [13]:

$$C(u,v) = \alpha(u)\alpha(v)\sum_{x=0}^{M-1}\sum_{y=0}^{N-1} f(x,y) \times \cos\left[\frac{\pi(2x+1)u}{2M}\right]\cos\left[\frac{\pi(2y+1)v}{2N}\right] \tag{4}$$

Where $\alpha(u)$ and $\alpha(v)$ are DCT coefficients.

The DCT coefficient has the characteristics of energy concentration, which is especially suitable for effectively extracting the low frequency or integral features contained in the image [13].

2.3 The Two-Stage Fusion Strategy

The recognition rate can be improved by the complementarity features between near-infrared imaging and visible-light imaging [14, 15]. To effectively integrate the contributions of different modalities to face recognition, we propose a two-stage fusion recognition system. Fig. 2 shows the detail process of the system.

Step 1: Decision fusion is firstly applied based on the ULBP features in both images: various ULBP features are sent to chi-square distance based nearest neighbor classifiers to obtain category labels (label 1 and label 2). If the label 1 and the label 2 are the same, the second stage fusion is not carried out and the label is the final identified result.

The nearest neighbor classifier (NNC) is chosen on the chi-square distance. It can calculate the features of each pair of test images (*Te*) and training images (*Tr*) with the same method, and then get the matching score matrix from both of these features using the chi-square distance. The calculation process is as follows.

$$\chi^2(i,j) = \frac{1}{2}\sum \frac{(Te_i - Tr_j)^2}{Tr_i + Tr_j} \tag{5}$$

Step 2: Turn to the feature fusion of the second stage if the two labels in the step1 are different. Respectively, the features of near-infrared and visible image (H_1, H_2) are extracted using ULBP, and then the low-frequency features (L) of the near-infrared image are extracted by DCT. Furthermore, in order to achieve effective feature fusion, all features to be merged. H_1, H_2 and L require to be normalized as H_1^N, H_2^N and L^N. In this paper, all features are normalized using norms. If the extracted features are stored as $F = [F_1, F_2, F_3, \ldots, F_n]^T$, the normalized features $N = [N_1, N_2, N_3, \ldots, N_n]^T$ can be expressed as:

$$N_i = \frac{F_i}{norm(F)} \tag{6}$$

$$norm(F) = F_1 + F_2 + \ldots + F_n \tag{7}$$

By normalizing, all features H_1, H_2 and L are combined into one output feature which is sent to the same classifier as the first stage to get the label 3.

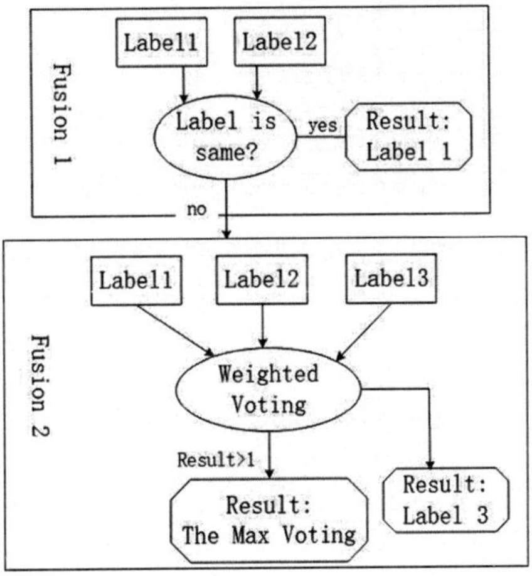

Fig. 2. The framework of two-stage fusion strategy

Finally, the implement of the simple decision-making layer fusion, that is, a minority is subject to a majority in voting, is based on Label 1, Label 2 as well as Label 3. If the number of votes is the same, label 3 is the final recognition result.

3 Experimental Results and Analysis

To evaluate the good performance of the fusion face recognition system, all experiments are implemented on the HITSZ LAB2 face database [9] which was established by Harbin Institute of Technology (HIT). The database, capturing a total of 2,000 face images with the resolution 200 × 200, was collected from 50 volunteers. Everyone contains 20 color visible face images and 20 near-infrared face images. Table 1 shows the specific distribution of database samples. These faces were acquired under the following illumination variations: (a) natural illumination (b) natural illumination + left illumination (c) natural light + right illumination (d) natural illumination + simultaneous illumination from the left side and right side lights.

With a partial illumination changing, Fig. 3 contains partitional face images in the database. In Fig. 3, the upper two lines are visible light images while the lower two lines are near infrared images. During this scenario, the lighting conditions are: among the first and the third line, the first 4 images are under the left and right side illumination while the last 4 images correspond to the left side illumination; likewise, among the second and the fourth line, the first 4 images are under the natural illumination whereas the last 4 images correspond to the right illumination.

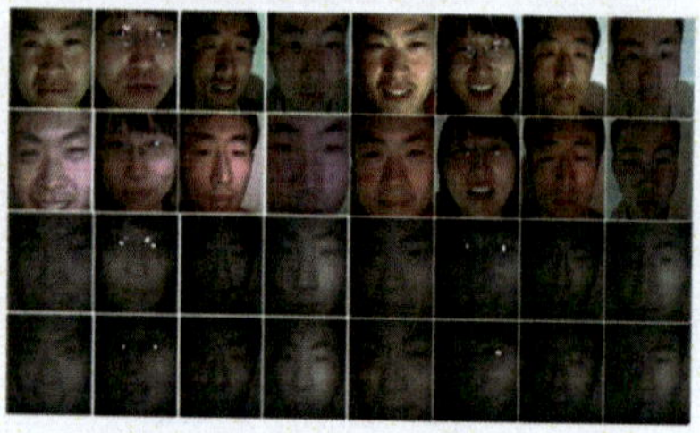

Fig. 3. Visible and near-infrared images in different illumination conditions in the HITSZ Lab2 database

In this paper, the original RGB face images are converted into the grayscale maps whose resolutions are reduced to the 100 × 100 based on the down sampling method. What's more, according to the fusion strategy shown in Fig. 2, we extracted ULBP features and DCT features from infrared face images as well as ULBP features from visible faces. Afterwards, these three kinds of features were respectively applied to two-stage decision-making and feature fusion classification. The experiments of this paper were conducted, where other parameters were set, the LBP block mode was 2 × 2 and the DCT low-frequency feature extraction scanned the first 100 coefficients as the shape of 'Z' retaining. In order to test the method proposed, the fusion recognition is performed in following two situations: a. The number of training samples is much larger than the number of testing samples, that is, the first 4 subjects under different illumination conditions are chosen as the training samples and the rest are used as the test samples. b. The number of training samples is much smaller than the number of test samples, that is, the first one under the same condition is captured as the training sample and the remaining are used as the testing samples.

Under the condition a, the recognition rates of several different algorithms are listed in Table 1. As shown in Table 1, in the case of sufficient training samples, it's evident that the proposed two-stage fusion recognition method is powerful and effective with the verification rate of 100%, which is higher than the traditional 2DLDA-based fusion method. Our approach is basically the same as the VGG fusion method which requires complex deep network training. Likewise, under the condition b, the recognition rates of several different algorithms are shown in Table 2. It can be seen from Table 2, for the condition of a single sample, the two-stage fusion recognition method proposed in this paper achieved a recognition rate of 96.13%, which is higher than the recognition rate of 93.13% in the first-stage fusion recognition. Meanwhile, there is no improvement for the recognition rate of the first-stage ULBP feature fusion, indicating that in the case of a single sample, the one-stage ULBP fusion does not involve the complementary features of infrared and visible-light images. Table 2 also manifests that the recognition rate of our proposed method outperforms not only the traditional recognition rate based on statistical feature fusion of 92.75% and the recognition rate based on VGG fusion of 91.50%. The main reason is that DCT combined with LBP fully extracts effective features of the near-infrared images of which the visible face features based on LBP can

Table 1. The result of different algorithms on the HITSZ LAB2 database (a)

Images types	Method	Recognition rate
Visible-light	2DLDA1	95.00%
Visible-light	2DLDA2	94.50%
Visible-light	2DLDA1 + 2DLDA2	96.25%
Visible-light	ULBP	99.50%
Visible-light	DCT	96.50%
Near-infrared	2DLDA1	95.00%
Near-infrared	2DLDA2	96.00%
Near-infrared	2DLDA1 + 2DLDA2	95.00%
Near-infrared	ULBP	97.50%
Near-infrared	DCT	95.00%
Fused	Ref [8]	98%
Fused	Ref [9]	99.5%
Fused	First-stage fusion	100%
Fused	**Our approach**	**100%**

Table 2. The result of different algorithms on the HITSZ LAB2 (b)

Images types	Method	Recognition rate
Visible-light	2DLDA1	85.75%
Visible-light	2DLDA2	84.62%
Visible-light	2DLDA1 + 2DLDA2	86.62%
Visible-light	ULBP	93.13%
Visible-light	DCT	84.50%
Near-infrared	2DLDA1	77.25%
Near-infrared	2DLDA2	86.00%
Near-infrared	2DLDA1 + 2DLDA2	84.13%
Near-infrared	ULBP	93.13%
Near-infrared	DCT	84.75%
Fused	Ref [8]	92.75%
Fused	Ref [9]	91.50%
Fused	First-stage fusion	93.13%
Fused	**Our approach**	**96.13%**

effectively make up for the lack in local feature extraction. Due to insufficient samples, other learning-based fusion methods degrade obviously in performance. In addition, in the case of a single sample, the advantages of the proposed method demonstrate that without statistical learning on a large number of samples, the fusion of DCT and ULBP features extracts robust fusion features in infrared faces and visible faces.

4 Conclusions

In this paper, a two-stage face fusion recognition system is developed based on DCT and ULBP. With respect to visible-light images, the illumination invariant texture features are extracted by LBP, which effectively compensates for the lack of local features for near-infrared faces. By jointing decision and feature fusion, the two-stage multi-mode fusion face recognition method can make full use of the complementary features from infrared and visible face images. The experimental results illustrate that the proposed face fusion recognition method achieves a significant improvement in both multi-samples and single-sample cases.

Acknowledgments. This paper is supported by the National Nature Science Foundation of China (No. 61861020), Science & Technology Project of Education Bureau of Jiangxi Province (No. GJJ190578).

References

1. Lu, G.F., Wang, Y.: Feature extraction using a fast null space based linear discriminant analysis algorithm. Inf. Sci. **193**, 72–80 (2012)
2. Xu, Y., Zhu, Q., Fan, Z.Z., et al.: Using the idea of the sparse representation to perform coarse-to-fine face recognition. Inf. Sci. **238**, 138–148 (2013)
3. Gross, R., Brajovic, V.: Audio-and Video-Based Biometric Person Authentication, pp. 10–18. Springer, Heidelberg (2003)
4. Li, S.Z., Chu, R.F., Liao, S.C., et al.: Illumination invariant face recognition using near-infrared images. IEEE Trans. Pattern Anal. Mach. Intell. **29**(4), 627–639 (2007)
5. Raghavendra, R., Dorizzi, B., Rao, A., et al.: Particle swarm optimization based fusion of near infrared and visible images for improved face verification. Pattern Recogn. **44**(2), 401–411 (2011)
6. Bebis, G., Pavlidis, I.: Infrared and visible image fusion for face recognition. In: Proceedings of SPIE, vol. 5404, pp. 585–596. The International Society for Optical Engineering (2004)
7. Desa, S.M., Hati, S.: IR and visible face recognition using fusion of kernel based features. In: International Conference on Pattern Recognition [S. 1.], pp. 1–4. IEEE (2008)
8. Ma, Z., Wen, J., Liu, Q., et al.: Near-infrared and visible light image fusion algorithm for face recognition. J. Mod. Opt. **62**(9), 745–753 (2015)
9. Guo, K., Wu, S., Xu, Y.: Face recognition using both visible light image and near-infrared image and a deep network. CAAI Trans. Intell. Technol. **2**(1), 39–47 (2017)
10. Meibin, Q., Zhonghe, T., Jianguo, J.: Face recognition based on adaptive threshold local feature fusion. J. Hefei Univ. Technol. **41**(4), 468–472 (2018)

11. Ojala, T., Pietikinen, M., Menp, T.: Multiresolution gray-scale and rotation invariant texture classification with local binary patterns. IEEE Trans. Pattern Anal. Mach. Intell. **24**(5), 971–987 (2002)
12. Lian, Z., Er, M.J., Liang, Y.: A novel efficient local illumination compensation method based on DCT in logarithm domain. Pattern Recogn. Lett. **33**(13), 1725–1733 (2012)
13. Zhihua, X.: Infrared face recognition combining DCT and features selection. J. Front. Comput. Sci. Technol. **8**(5), 614–621 (2012)
14. Ross, A., Jain, A.: Information fusion in biometrics. Pattern Recogn. Lett. **24**(13), 2115–2125 (2003)
15. Shunjie, Z., Xuezhi, Y., Zhangyu, D., et al.: Fusion algorithm of SAR and visible image for feature recognition. J. Hefei Univ. Technol. **41**(7), 900–907 (2018)

Illustration and Identification of Skin Lesions in Dermoscopic Dataset Using Pattern Recognition Techniques

Syeda Tooba Haider[1(✉)], Syed Basharat Hussain[1], Amreen Batool[2], Laraba Selsabil Rokia[3], Syed Agha Hassnain Mohsan[4], Muhammad Abubakar Sadiq[5], and Mazahir Saleem[6]

[1] North China Electric Power University, Beijing, China
1504472115@qq.com
[2] Tiangong University, Tianjin, China
[3] East China University of Science and Technology, Shanghai, China
[4] Ocean College, Zhejiang University, Hangzhou, China
[5] Beihang University (BUAA), Beijing, China
[6] National University of Science and Technology (NUST), Islamabad, Pakistan

Abstract. This paper proposes a generic method of description of skin lesion types and recognition of Melanocytic lesions in Dermoscopic images. Scientific judgments show that cancerous lesions demonstrate various characteristics such as streaks, texture or pigmented networks as compared to normal or benign lesions. Our aim is to confirm those judgments by carrying out proper segmentation, features extraction and classification. Segmentation phase of PH2 Public dataset was done by means of active contours. Features extraction stage included implementation of Local Binary Patterns and its extension to study textural features more deeply. Finally, K nearest neighbor and Support vector machine were used for classification. This choice of methods is motivated by scientific importance of differential configurationally structures in the melanomas. The experiment of the offered algorithm is done on both sets of images, manually segmented (segmentation done by physicians followed by Features extraction and classification done by us) and automatic segmented images (all the steps performed by us) using performance metrics of sensitivity, specificity and Accuracy. The highest result of the system was attained by manually segmented images, i.e. SE = 89.2, SP = 84, Accuracy = 89.2. Whereas, images with automated segmentation showed SE = 89.3, SP = 75.4, Accuracy = 88.2.

Keywords: Melanocytic lesions · Segmentation · Classification · Local binary pattern · Dermoscopy

1 Introduction

Cancer is one of the key reasons causing deaths around the globe [1]. On the basis of origin, skin lesions can be characterized into melanocytic and non-melanocytic. The occurrence of melanoma type skin lesions is seen increasing rapidly and it became a major reason for deaths of people around the globe. Usually, visual examination is

M. Tavana et al. (Eds.): IISA 2020, AISC 1304, pp. 976–983, 2021.
https://doi.org/10.1007/978-3-030-63784-2_118

done by Dermatologist to detect cancerous skin lesions but scientific diagnostic accuracy in this method is a bit inadequate because of the reasons such as, non-repetitiveness, and viewer's unpredictability [2]. Machine learning algorithms are now widely being used in the field of medical imaging to assist doctors in diagnosing and detecting this disease. These diagnostic tackles can benefit in initial recognition of melanoma and therefore, an active retrieval of the victims [3].

1.1 Background and Significance

Normal practice to detect melanoma skin cancer is to visualize lesion, followed by a biopsy. Identification of melanocytic lesions was done using like ABCD [4], 7-point checklist [5], and Menzies Method. ABCD explains the medical features of melanoma using labels A (standing for Asymmetry), B (Border irregularity), C (Color) and D (Diameter). The main criteria for Glascow7-pointchecklist are changes in shape, size and color, and the secondary criteria are inflammation, diameter, crusting and sensual changes. Finally, Menzies method uses positive and negative feature sets to classify the dermoscopic features of benign melanocytic lesions from melanoma. However, the disadvantage of this technique is lack of visualization of those morphological features which are not noticeable with the naked eye. Due to complexity of classification done by humans, as well as the variety of types of lesions and presence of confounding factors faced during study of dermoscopic images has become a vital research area to aid diagnosis [6]. In this paper, we aim to resolve the issues related to Melanoma detection applications, using vigorous image analysis algorithm.

1.2 Proposed Scheme

In proposed article, we have following precise objectives: 1) to implement segmentation technique followed by LBP for feature extraction. 2) To design an algorithm for extraction of morphological features and non-redundant textural features. 3) To present a method capable to classify melanocytic lesions from non-melanoma. Paper's order is as follows, Sect. 2 explains the methodology implemented in this work, followed by Sect. 2.5 which addresses dataset description. After that, we shall discuss our classification results in Sect. 3. Finally, Sect. 4 presents recommendations and conclusions.

2 Methodology

2.1 Dataset

The dataset which we have used in this thesis is known as PH2 dataset. It is a free of cost and publically used data base, built to do experiments, and compare the assessment of several systems. PH2 is composed of 200 images of Dermoscopy. Among which, 160 (80%) are of nevus and 40 (20%) images are of melanoma. This dataset was developed by group of people in Pedro Hispano Hospital, Matosinhos, Universidade do Porto and Técnico Lisboa [7].

2.2 Pre Processing

Keeping in view the importance of this step, we implemented it for the purpose of up gradation and improvement of image quality. Therefore, we implied DullRazor1 Software [8] available online for hair elimination. It is a great program to remove and eliminate hair from images which becomes obstacle in vision, using a pre-processing program. We applied this algorithm on images with hair as seen in the Fig. 1.

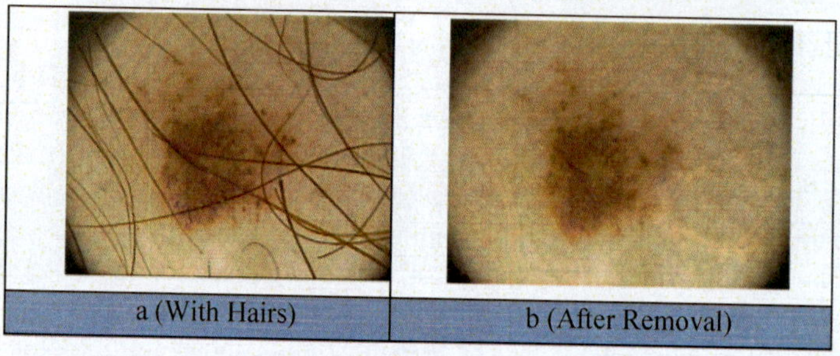

Fig. 1. Hair removals by Dull Razor software

2.3 Segmentation

Among the category of dermoscopic image investigation, there are mainly three categories of segmentation methods namely region based, edge based method and thresholding [9]. Due to various challenges, segmentation becomes tough to implement. We can observe and have a look at those challenges in PH2 dataset (Fig. 2).

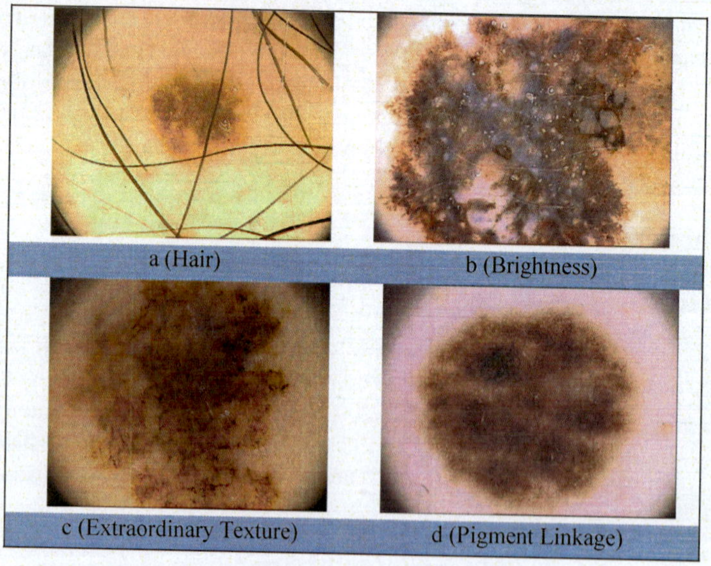

Fig. 2. Challenges in segmentation

Another category of segmentation methods which has got noteworthy importance in recent years is active contours [10]. Active contours have the advantage among other methods because even in edge based methods, active contours perform optimization on basis of local best within the region round the curve.

We intend to syndicate the techniques of region based, edge based, and thresholding characteristics for segmentation of lesion images by active contours. This proposed algorithm aims to maximize the Kullback-Leibler divergence between the lesions and skin and to minimize the mean shift vectors for the lesion and skin with help from the image edges. We designated to do so owing to the fact that, it is non-negative and if the subject distributions are same/equal, Kullback divergence becomes zero. This is a distinctive property of Kullback divergence than other methods like, Bhattacharya distance which becomes a number other than zero even when the distributions are same [11]. In this work, a novel approach based to maximize the distance between the background and distributions of the object is presented. This proposed energy functional is fixed as an external energy term in the previous work named distance regularized level sets evolution (DRLSE), for image segmentation.

Consider that Ω belongs to the image, I be a gray scale image such that I: $\Omega \rightarrow R2$. This Ω comprises of two subdomains, Ω+ (background) and Ω− (object). These two sub domains of Ω are separated by a function called level set φ: $\Omega \rightarrow R2$ which signifies the curve representing the boundary of ROI in the dermoscopic image. This level set function, also recognized as an active contour is developed to fit the borders of the things in the image. We can predict the order of functions by using variational framework. To be price, this function can be explained using the gradient flow which minimize a suitable cost function. Classically the cost function in variational level sets [17] is described as J (φ),

$$J(\phi) = \mu R(\phi) + \lambda L(\phi) + \alpha A(\phi) \tag{1}$$

In Eq. 1, R (φ) corresponds to regularization, A (φ) to the area L (φ) to the length in the sets. However, factors like, μ, λ and α are the ones which control the curve evolution. We can explain the above mentioned length and regularization terms as,

$$R(\phi) = \frac{1}{2}\int_{\Omega} (\nabla\phi - 1)^2 d\Omega \tag{2}$$

$$L(\phi) = \int_{\Omega} \delta(\phi)|\nabla\phi| d\Omega \tag{3}$$

These equations (Eq. 2) and (Eq. 3) have been modified from DRLSE method. A (φ) is new term named area for curve evolution on the basis of the external energy. Adaptive thresholding is used to select initial contour. Further, we employ Li's [12] applications for the regularization and length terms. Whereas the third term i.e. area is comprises of two vital variable terms, m (Ω+), m (Ω−). These variables stipulate the mean shift that persistently converges to their local maxima at the termination of the procedure. Four parameters are noticeable in the current method. i.e., λ, μ, α and ∇t (time step). This feature extraction system is modified form of DRLSE, which can be subjected to any further modification. In this work, these parameters are given following values, as $\alpha = 1.0$, $\lambda = 5.0$, $\nabla t = 5.0$ and $\mu = 0.04$. We can witness the segmentation using proposed scheme in Fig. 3.

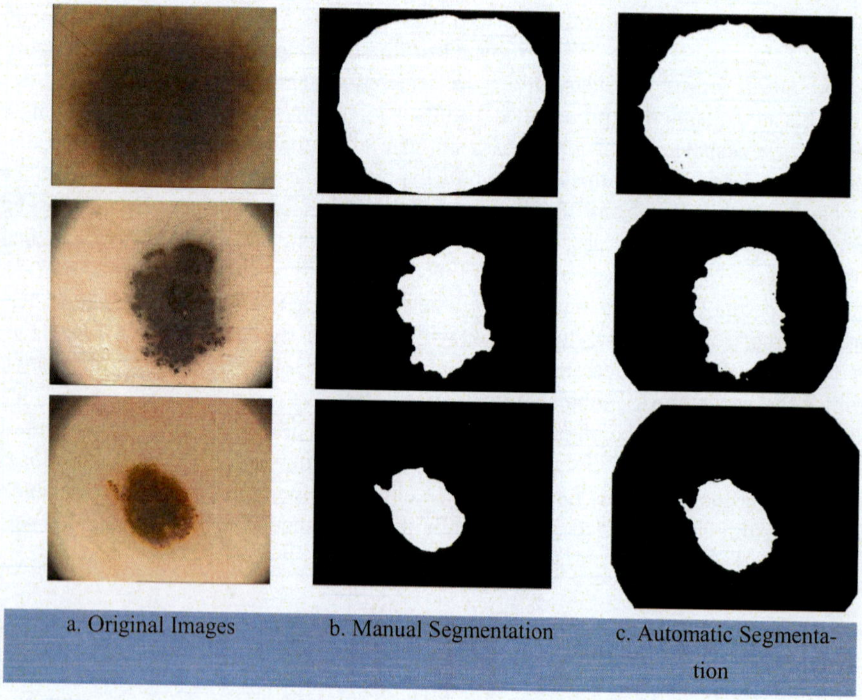

Fig. 3. Description of manual and automatic segmentation results

2.4 Features Extraction

Scientific findings show that the characteristics which differentiate melanoma from non-melanoma lies on peripheral area of lesion. Based on the scientific studies, we have presented a modification of LBP for detection of differential structures. The Local Binary Patterns (LBP), an invariant operator used for texture description. It is a great operator to know about the statistics of the texture within the lesion. In this paper, uniform LBP pattern is used for features extraction. Uniform LBP consists of maximum of two bit wise shifts from 0 to 1 and vice versa. Uniform LBP descriptor is easy to use and efficient to detect the most important characteristics such as spots, corners, vertical and horizontal edges, etc. (Fig. 4).

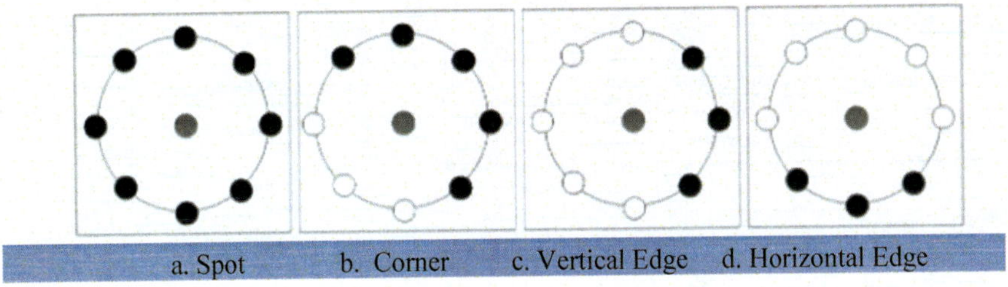

Fig. 4. Examples of structures using LBP

Generally, LBP is a good selection to study textural features deeply, but to analyze more about textural features; we need to focus on the contrast too. Therefore, the goal is to accomplish feature extraction using a combined histogram in which we make use of uniform LBP as well as contrast.

$$C(g_c) = \sum_{i=0}^{P-1} |g_p - g_c| \tag{4}$$

Here $C(g_c)$ signifies the contrast of LBP. We used various sizes of neighborhood values, LBPs are calculated by making circle of different radius. Like P = 8, R = 1, R = 1.5, R = 2. Among these values of R, we selected that value which gave stronger contrast because it contains most relevant information.

2.5 Classification and Recognition

Classification and identification is the concluding stage in the CAD system that provides us the end result about the tested image and categorize it to a certain class. In the current work, we have used SVM and KNN classification techniques. SVM linear kernel is implemented in this thesis. KNN is an unsupervised method of classification. In our case, we have implemented with K = 1. Weka is such a platform which has a collection of algorithms built in, related to machine learning. We used this version of Weka (https://www.cs.waikato.ac.nz/ml/weka/) data toolkit in our experiment.

3 Experimental Results

We have certain parameters as performance metrics like Accuracy, Sensitivity, and Specificity for computation of results. Classification results in Table 1 and Table 2 correspond to the performance measures of our method along with previous methods. By the term "Manual segmentation", we mean those images, whose segmentation was done by physicians and features extraction and classification done by us. However, with Automatic segmentation, we mean segmentation done by method proposed in Sect. 2 followed by the same features extraction and classification.

Table 1. Comparison of proposed method with state of art algorithms using SVM Classifier

Methods	Manual segmentation			Automatic segmentation		
	Sensitivity (SE) %	Specificity (SP) %	Accuracy (Acc.) %	Sensitivity (SE) %	Specificity (SP) %	Accuracy (Acc.) %
LBP (Without contrast)	88.3	71.4	86	86.8	64.7	82.5
Homogeneous Texture (HT)	64	86	81	79	63	79
Proposed method	89.2	84	90.1	89.3	75.4	88.2

Table 2. Comparison of proposed method with state of art algorithms using KNN Classifier

Methods	Manual segmentation			Automatic segmentation		
	Sensitivity (SE) %	Specificity (SP) %	Accuracy (Acc.) %	Sensitivity (SE) %	Specificity (SP) %	Accuracy (Acc.) %
LBP (Without contrast)	85.8	62.5	82.5	87.4	63.7	79.5
Homogeneous (HT)	69	71	79.5	78	20	66.5
Proposed method	88.4	74	87.6	88.2	70.25	86.3

Results reveal that the method homogenous texture demonstrates poorest performance among the rest methods. Traditional LBP (Without contrast) indicates comparatively low accuracy. We assume that it is due to ignorance of contrast factor during features extraction. Yet, our proposed model shows healthier performance in contrasting to the previous algorithms stated in the table.

4 Discussion and Conclusion

In this work, an automatic computer aided diagnosis (CAD) system is suggested for the purpose of illustration of lesions and detection of presence of melanomas using an optimized segmentation alongside LBP and contrast. The research of the offered procedure is carried out on both datasets, named manual and automated using performance measures of sensitivity, specificity and Accuracy. The top performance of the system was attained by manually segmented masks, i.e. SE = 89.2, SP = 84, Accuracy = 89.2. While, SE = 89.3, SP = 75.4, Accuracy = 88.2 results were shown on automated masks. On the basis of results, we observed a little less accuracy in automatic segmentation criteria, rather than manual. We can say that of course combination of LBP and contrast is a key point of identifying melanomas. Moreover, the overall precision accuracy of our suggested system is greater than old methods.

Although we have attained some decent classification accuracies, but there is always a room for improvement at every step of CAD system. For example, we used an active contours method based segmentation algorithm in this scheme, but sometimes it does not yield good segmentation output due to the fact that among many medical images, there is no specific or sharp transition between skin and lesion. In other words, it is difficult to detect the boundary. In such situation, some other segmentation methods might work well. So in upcoming future, it is essential to emphasize on such issues. There is a need to build color descriptors; they could yield much better accuracy. In future, we would like to syndicate LBP with drivable features to achieve best possible results. It would be inspiring to evaluate and investigate the performance of various permutations of segmentation and features extraction procedures on different datasets.

References

1. Torre, L.A., Bray, F., Siegel, R.L., Ferlay, J., Lortet-Tieulent, J., Jemel, A.: Global cancer statistics, 2012. CA Cancer J. Clin. **65**, 87–108 (2015)
2. Sadeghi, M.: Towards Prevention and Early Diagnosis of Skin Cancer: Computer Aided Analysis of Dermoscopy Images (2012)
3. Riaz, F., Naeem, S., Nawaz, R., Coimbra, M.: Active contours based segmentation and lesion periphery analysis for characterization of skin lesions in dermoscopy images. IEEE J. Biomed. Health Inform. **23**(2), 489–500 (2018)
4. Stolz, A.R., Cognetta, A.B.: ABCD rule of dermatoscopy: a new practical method for early recognition of malignant melanoma. Eur. J. Dermatol. **4**, 521–527 (1994)
5. Delfino, M., Argenziano, G., Fabbrocini, G., Carli, P., Giorgi, V.D., Sammarco, E.: Epiluminescence microscopy for the diagnosis of doubtful melanocytic skin lesions. Comparison of the ABCD rule of dermatoscopy and a new 7-point checklist based on pattern analysis. Arch. Dermatol. **134**, 1563–1570 (1998)
6. Fleming, M.G., Steger, C., Zhang, J., Gao, J., Cognetta, A.B., Pollak, I., Dyer, C.R.: Techniques for a structural analysis of dermatoscopic imagery. Comput. Med. Imaging Graph. **22**(5), 375–389 (1998)
7. Mendoncÿa, T., Ferreira, P.M., Marques, J., Marcÿal, A.R.S., Rozeira, J.: A dermoscopic image database for research and benchmarking. Accepted for Presentation in Proceedings of the PH2 IEEE EMBC (2013)
8. Lee, T., Ng, V., Gallagher, R., Coldman, A., McLean, D.: DullRazor: a software approach to hair removal from images. Comput. Biol. Med. **27**, 533–543 (1997). Dermweb, Dull Razor. https://www.dermweb.com/dull_razor/. Accessed 2020
9. Celebi, M.E., Wen, Q., Iyatomi, H., Shimizu, K., Zhou, H., Schaefer, G.: A state-of-the-art survey on lesion border detection in dermoscopy images. Dermoscopy Image Anal. **10**, 97–129 (2015)
10. Caselles, V., Kimmel, R., Sapiro, G., Bovik, A.: Geometric active contours for image segmentation. In: Handbook of Video and Image Processing, pp. 613–627 (2005)
11. Tabibian, S., Akbari, A., Nasersharif, B.: Speech enhancement using a wavelet thresholding method based on symmetric Kullback-Leibler divergence. Sig. Process. **106**, 184–197 (2015)
12. Li, C., Xu, C., Gui, C., Fox, M.D.: Distance regularized level set evolution and its application to image segmentation. IEEE Trans. Image Process. **19**(12), 3243–3254 (2010)

Cartoon Zero-Watermark Method Based on Edge Contourlet Feature and Visual Cryptography

Dongxing Li, De Li, and Hua Jin(✉)

College of Engineering, Yanbian University, Yanji 133002, Jilin, China
hjin@ybu.edu.cn

Abstract. By analyzing the characteristics of the cartoon images, we propose a new method by using edge feature. In this method, image normalization is applied to better resistance to geometric attacks firstly. Then the secondary contourlet transform are applied to extract the general contour and edge features effectively. Next, the non-negative matrix factorization is applied to obtain better feature information, after binary processing, the matrix which is basic obtained is used as the standard sequence of cartoon images. In order to make the algorithm proposed in this paper more secure, visual cryptography and other technologies are applied for the original watermarking image. Finally, the feature sequence extracted from the cartoon image and the watermark sequence representing copyright information are logically XOR operated to obtain the zero-watermark. Analysis of the test results of this paper can be seen that the cartoon zero-watermark algorithm is robust and can resist noise attacks, geometric transformation attacks, JPEG compression attacks.

Keywords: Zero-watermark · Image normalization · Contourlet transform · Visual cryptography

1 Introduction

It can be seen from the situation in the past few years, the copyright protection of cartoon images has attracted extensive attention, and digital watermarking technology can better protect the copyright of the authors of digital works. Article [1] uses traditional wavelet transform to extract image features. The feature extraction algorithm is based on the DWT-DCT domain, but its robustness is not good. When the image is subjected to geometric attack, noise attack, cropping attack or other attack methods, it will no longer be able to extract image features normally. Article [2] first performs DCT transform on the image to compress the image size, and then extracts the contour features of the image through Contourlet transform. The algorithm can resist cropping attacks, but not some other possible attacks against images.

The focus of this study is to take the cartoon as the main carrier, and a kind of zero-watermark method with strong robustness and high security is put forward. Not only does this method further broaden the research ideas of robust zero-watermark technology, but also has certain theoretical significance and academic value.

M. Tavana et al. (Eds.): IISA 2020, AISC 1304, pp. 984–990, 2021.
https://doi.org/10.1007/978-3-030-63784-2_119

2 Research on Related Technology

For the features extracted by the watermarking algorithm to better characterize the host image, multiple techniques need to be combined. This article is mainly based on the following two algorithms, which will be introduced in detail in this chapter.

2.1 Contourlet Transform

Contourlet transform [3, 4] is a kind of multiscale and multi-direction signal representation method which can't be separated. Contourlet transform can effectively capture the image's edge and texture feature accurately to different scales and different direction subbands.

The process of Contourlet transform is as follows. First, singular signals are captured by performing Laplace pyramid decomposition on the image. Secondly, directional filter bank is used to capture directional information of high frequency components. Laplace Pyramid transform downsamples the image to obtain a coarse-scale approximate image, and then different images make directional decomposition and capture high-frequency components to represent directional information of the image. Each level of Contourlet transform generates many subband frequencies of high and low components. Eventually, an iterative process is formed.

In this paper, the decomposition of images can be achieved through Contourlet transform, extract frequency with lower components, and compare with the wavelet transform. Experimental results show that the features extracted after Contourlet transform can better represent the edge and rough outline information of the host image.

2.2 Visual Cryptography

Visual cryptography [5, 6] is a new kind of image encryption technology. By designing secret sharing algorithm, thc image can be decomposed into different amount of shared parts by using the code, and each shared parts need preserving by introducing more participants. Once the original image needs restoring, all share parts must be superimposed together. The core algorithm of visual cryptography is to design secret shared basis matrix. Naor and Shamir [5] give a general solution through the experiment. And it is proved that if there are n share parts, the secret image can be recovered when the N shares are shared, and the N-1 overlay can't restore the original image, so the minimum pixel expansion degree is 2N-1.

In this paper, visual cryptography is applied to process the watermarking image, which can improve the security of the watermarking information more effectively.

3 The Zero-Watermark Algorithm Design

3.1 Zero-Watermark Construction Algorithm

The algorithm steps to generate zero watermark can be seen in Fig. 1. The main processing steps of this algorithm will be described in detail below.

Step1: Read the original carrier image 512 × 512, converted to grayscale image and deal with it based on gray the gray image; Step2: Set parameters to 512 × 512 of the normalized transformation. Then, select edge control points. Next, use image space transform to obtain a standard normalized image; Step3: Based on the standard image, the DCT of 2 × 2 block is carried out and the low frequency component is extracted; Step4: From the Contourlet transform, the frequency with lower components is carried out at the two stage and the sub-band coefficients with low frequency are extracted; Step5: The low frequency coefficients after Contourlet transformation are adopted with NMF [7] to obtain basis matrix (W) and the coefficient matrix (H). Finally, binarization is used for the basis matrix and the characteristic sequence (T) with the size of 64 × 64 is generated following; Step6: Read the watermark image with the size of 32 × 32 firstly. Then, chaotic scrambling is carried out for the image by Logistic mapping. Next, use visual cryptography to obtain the two shared parts Share1 & Share2 with the same size of 32 × 32. And the two shared parts are constructed into a secret sharing part matrix with the size of 64 × 64. Step7: Make (T) XOR with Share3 to obtain the zero-watermark. Last step, the result is registered into zero-watermark database.

3.2 The Zero-Watermark Extraction Algorithm

The zero watermark detection algorithm and the zero watermark generation algorithm are inverse. The following will introduce the zero watermark detection steps in detail.

Step1: Firstly, convert the original color image to gray image and make some attacks experiments. Then, carry out the next operation based on the attacked image; Step2: Make the image normalization and obtain the standard image; Step3: Based on the standard image, the DCT of 2 × 2 block is carried out and the low frequency component is extracted; Step4: From the Contourlet transform, the frequency with lower components is carried out at the two stage and the sub-band coefficients with low frequency are extracted; Step5: The low frequency coefficients after Contourlet transformation adopts NMF. Then, binarization is used for the basis matrix and the characteristic sequence (T') is generated; Step6: Read corresponding zero-watermark from the information database. Then, make it XOR with (T') in order to recover the secret sharing part; Step7: In accordance with the constructional rule of the secret sharing part matrix, recover the two original sharing parts Share1&Share2. Following, make them XOR with each other. Step8: The obtained image makes reverse chaotic scrambling and the watermark image would be recovered successful.

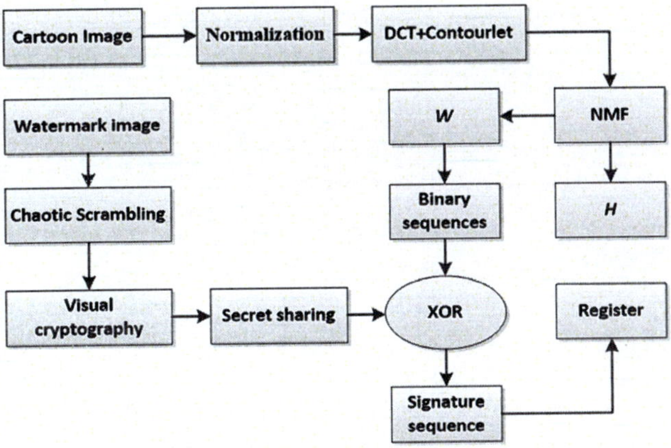

Fig. 1. Zero watermark generation step diagram

4 Experimental Results and Analysis

This article will continue to verify the effectiveness of the proposed animation zero watermark algorithm, the zero watermarking algorithm and scheme will be analyzed experimentally. This paper will use 24-bit true color images as the carrier for the experiment, the size is 512 × 512, as shown in Fig. 2(1). Use a binary image as the copyright image, with "YD" written on this image, the size is 32 × 32.

Experimental environment: operating system Windows7 ProfessionalX64, software: MATLAB2013a.

(1) (2) (3) (4)

Fig. 2. Images used in this experiment

4.1 Zero-Watermarking Uniqueness Verification

The uniqueness is to evaluate the distinguishing ability of two different cartoon images, that is, use the same algorithm to extract features from other images, and then perform the XOR operation with the zero watermark generated by the protected software, and the original copyright image cannot be obtained. Experimental analysis was carried out using Fig. 2(2), 2(3), 2(4) as comparative images. The NC value in Table 1 is small, and the BER value is large, indicating that the obtained image is very different from the copyright image, that is, the three comparison images are very different from the host image. Therefore, the features extracted by the algorithm can uniquely represent the iconic characteristics of the host.

Table 1. Zero watermark uniqueness test results

Image coding	1	2	3	4
NC	1.000	0.583	0.564	0.572
BER	0.00%	35.23%	36.67%	32.88%

4.2 Attack Experiments

This paper will continue to test the algorithm's ability to resist various common attacks, many attack operations with different types and different parameters were performed on the host image.

1) Noise attacks: To verify the method's robustness to against noise attacks, the paper makes some experiments. We can see the experimental results in Table 2. When the noise attacks are gradually enhancing and even affect the normal use of the image, it still can recover the watermark image well.

Table 2. Noise attacks test results

Noise attacks	Gauss noise attacks		Salt & pepper noise attacks	
Attack parameters	Variance 0.01	Variance 0.05	Density 0.01	Density 0.05
NC	0.992	0.983	0.992	0.982
BER	0.46%	1.39%	0.57%	1.59%

2) Geometric attacks: Rotation attack, zoom attack and crop attack are common geometric attack methods for images. In this article, we performed the above three geometric attack methods on the host image separately. In each attack method, we changed the attack parameters and conducted two sets of experiments to verify the pros and cons of the algorithm against common geometric attacks. We can see the experimental results in Table 3, where NC is larger and BER is smaller. From the final result, we can see that the algorithm proposed in this paper is very resistant to geometric attacks.

Table 3. Geometric attack test results

Geometric attacks	Rotation attacks		Scaling attacks		Cropping attacks	
Attack parameters	Rotation 30°	Rotation 60°	Scaling 1/2	Scaling 4	UL (100 × 100)	Center (200 × 200)
NC	0.987	0.982	0.992	0.992	0.931	0.923
BER	1.10%	0.97%	0.59%	0.63%	6.17%	7.85%

3) JPEG compression attack: Compress the host image. Set the compression parameters to 50, 40, 30, and 20. Table 4 records the results of this experiment, where NC is larger and BER is smaller. Experimental results show that after the JPEG compression attack is performed on the image, the features extracted by the algorithm can still authenticate the copyright.

Table 4. JPEG compression attacks test results

Compression facts	50	40	30	20
NC	0.989	0.991	0.996	1.000
BER	0.17%	0.12%	0.10%	0.00%

4.3 Algorithm Performance Comparison

In order to compare the advantages and disadvantages of the proposed algorithm with the existing related algorithms, after the above series of attacks on the host image, the zero watermark is constructed according to the algorithm proposed in this paper and the algorithm proposed in the Ref. [1], Ref. [2] and Ref. [8]. The results are recorded in Table 5. It can be seen that this algorithm can better resist Gaussian noise, salt & pepper noise, rotation, scaling, JPEG compression and other attack methods, and can resist on cropping attacks to some extent..

Table 5. Performance comparison

Attack type	NC comparison			
	Ref. [1]	Ref. [2]	Ref. [8]	Proposed
Gaussian noise (0.1)	0.913	0.987	0.984	0.979
Salt & pepper noise (0.1)	0.912	0.988	0.982	0.992
Rotation (30°)	0.936	0.953	0.921	0.979
Scaling (1/2)	0.922	0.986	0.968	0.996
Cropping (1/16)	0.957	0.977	0.922	0.931
JPEG compression (40)	0.935	0.991	0.935	0.991

5 Conclusion

Contourlet transform is performed on the host cartoon image, and then used normalization, DCT and NMF technologies, the extracted features can more accurately characterize the cartoon copyright. To improve the security of the method more effectively, while using chaotic scrambling, visual cryptography is applied. We can see from the results that the algorithm in this paper can better against noise attacks, geometric attacks, JPEG compression attacks and neighborhood filtering attacks than Ref. [1], Ref. [2] and Ref. [8], and the performance of resisting cropping attacks is acceptable.

Acknowledgments. This research project was supported by the Education Department of Jilin Province ("Thirteen Five" Scientific Planning Project, Grant No. JJKH20180898KJ), Jilin Education Science Planning Project ("Thirteen Five" Plan, Grant No. GH170043).

References

1. Amit, M., Neelesh, M.: Adaptive lossless medical image watermarking algorithm based on DCT & DWT. Procedia Comput. Sci. **78**, 88–94 (2016)
2. Shi, H.F.: Research on edge feature-based zero watermarking technology of comic and animation. Yanbian University, pp. 25–35 (2015)
3. Wang, Y.X., Wang, C.R.: A new digital image zero watermark algorithm based on contourlet transform. Acad. J. Yanshan Univ. **34**(6), 528–532 (2010)
4. Lou, O.J., Wang, Z.X.: Contourlet domain resisting geometric attacks algorithm research based on edge feature point template. J. Comput. **32**(2), 308–317 (2009)
5. Naor, M., Shamir, A.: Visual cryptography II: improving the contrast via the coverbase. Lecture Notes in Computer Science, pp. 197–202 (1996)
6. Qu, C.B., Yang, X.T., Yuan, Z.N.: Balanced multiwavelet visual password zero watermarking algorithm. Comput. Eng. **9**, 178–182 (2014)
7. Wan, H.N., Chen, P.: A digital watermarking algorithm based on non-negative matrix factorization and orthogonal projection transform. Trans. Beijing Inst. Technol. **31**(6), 727–731+744 (2011)
8. Naderahmadian, Y., Hosseini-Khayat, S.: Fast watermarking based on QR decomposition in wavelet domain. In: Intelligent Information Hiding and Multimedia Signal Processing, vol. 39, pp. 127–130 (2010)

Removing Cloud from Remote Sensing Digital Images Based on Anisotropic Kernel Function

Guohong Liang(✉) and Junqing Feng

Air Force Engineering University, Xi'an 710051, China
liangguohong321@163.com

Abstract. Kernel regression is a kind of nonparametric estimation method. Its remarkable advantage is that the regression model is completely driven by data and does not need specific signal form. In this paper, the kernel regression method is summarized comprehensively, the definition of kernel regression from low dimension to high dimension, some important useful properties of kernel regression are deduced.

Keywords: Regression · Kernel · Bilter kernel function · Anisotropic kernel function

1 Introduction

Regression analysis is the general term of a class of statistical techniques used to estimate the relationship between variables, including linear regression, nonlinear regression, logical regression, nonparametric regression, etc. Kernel regression method is a nonparametric estimation method. Its main advantages are: the regression model is completely driven by data, with strong adaptability and high robustness. The general form of regression model is as follows:

$$Y = z(T,\ \beta) \tag{1}$$

Where z is the regression function. It can also be expressed as:

$$Y_i = z(T_i) + N_i(0, \sigma^2)\,(i = 1, \cdots, n) \tag{2}$$

Equation (1) is written as conditional mathematical expectation:

$$E(Y|T = t_i) = z(T, \beta)$$

Kernel regression is a nonparametric regression method for estimating conditional mathematical expectation of variables. As a nonparametric method, kernel regression avoids the solution of unknown parameters in the regression model, and only through training data can learn the nonlinear functional relationship between input variables and target variables. It is widely used in signal processing, image processing, computer vision and other fields. The basic idea of the nuclear regression theory is to use the kernel regression interpolation functionLine non-parametric estimation. Literature

M. Tavana et al. (Eds.): IISA 2020, AISC 1304, pp. 991–995, 2021.
https://doi.org/10.1007/978-3-030-63784-2_120

detailed the theory of nuclear regressionIntroduction, it is widely used in image noise reduction [1, 2], Image fusion [3]. And imageInterpolation [4]. And so on, and achieved very significant denoising and interpolation effects.Takeda et al. [1] used the traditional kernel return method (Classic Kernel Based on the theory of Regression, CKR, a control kernel regression method is proposed(Steering Kernel Regression, SKR) and apply it to the image Denoising. Compared with CKR, SKR has a more significant noise reduction effect and betterGood image edge preservation effect and image interpolation effect [5, 6].

The original idea of kernel function regression is to use nonparametric method to estimate the probability density function in the case of discrete observation. Kernel function is widely used in the estimation of probability density function in statistics. It has the property that the integral result is 1 and symmetric about 0.

$$\int_{R^1} K(t) = 1, \int_{R^1} tK(t) = 1, \int_{R^1} t^2K(t) = C.$$

Common kernel functions include:

Quadratic kernel: $K(t) = \frac{15}{16}(1 - t^2)^2$.
Cubic Kernel: $K(t) = \frac{35}{32}(1 - t^2)^3$.
Cosine Kernel: $K(t) = \frac{\pi}{4}\cos\left(\frac{\pi}{2}t\right)$.
Gaussian Kernel: $K(t) = \frac{1}{\sqrt{2\pi}}\exp\left(-\frac{1}{2}t^2\right)$.

Given that T is in a neighborhood of T_i, according to the n-order Taylor expansion, we can obtain the following results:

$$\begin{aligned} z(T_i) &\approx z(T) + \{\nabla z(T)\}^T(T_i - T) + \frac{1}{2!}(T_i - T)^T\{Hz(T)\}^T(T_i - T) + \cdots \\ &= \beta_0 + \beta_1^T(T_i - T) + \beta_2^T\, vech\{(T_i - T)(T_i - T)^T\} + \cdots \end{aligned} \tag{3}$$

Define $vech(\cdot)$ as the semi-vectorized operator of the "lower triangle" part of a symmetric matrix, for example,

$$vech\left(\begin{bmatrix} a_{11} & a_{12} \\ a_{12} & a_{22} \end{bmatrix}\right) = [a_{11}\ a_{12}\ a_{22}]^T$$

According to the least square multiplication, the following problems are obtained by optimizing formula:

$$\min_{\{\beta_n\}_{n=0}^N} \sum_{i=1}^{n} \left[Y_i - \beta_0 - \beta_1(T_i - T) - \beta_2(T_i - T)^2 - \cdots - \beta_2(T_i - T)^N\right]\frac{1}{h}K\left(\frac{T_i - T}{h}\right)$$

Where $K(\cdot)$ is the kernel function, which takes the estimation point as the center, and is used to control the weight of each sampling point: the closer the point is to t, the greater the weight is. h is the smoothing parameter (scalar) that controls the scale of the kernel.

The form of kernel function is uncertain, and only needs to satisfy the symmetry about y-axis and take the maximum value at zero point.

In particular, when $N = 0$,

$$\widehat{z} = \frac{\sum_{i=1}^{N} K_h(T_i - T)y_i}{\sum_{i=1}^{N} K_h(T_i - T)}, K_H(T_i - T) = \frac{1}{\det(H_i)} K\left(\frac{T_i - T}{H_i}\right)$$

2 Anisotropic Kernel Function

In the interpolation model, the core idea of interpolation method is how to select kernel K. choosing different kernel determines different interpolation methods, as well as the accuracy and calculation amount of interpolation. The traditional kernel function only considers the distance between sampling points, while the two-sided kernel not only considers the distance between sampling points, but also effectively utilizes the gray value of the sampling points.

$$K_{Bilter} = (T_i - T, Y_i - Y) = K_{h_s}(T_i - T)K_{h_r}(Y_i - Y)$$

Where, K_{h_s} and K_{h_r} is the kernel function, and Gaussian function is generally used. h is the smoothing function of kernel function, but its kernel is not adaptive and the distance is still Euclidean distance in the usual sense, The anisotropic distance was proposed later:

$$\|t - t_i\|_A = (t - t_i)^T A(t - t_i)$$

Where, A is a symmetric matrix of 2×2. Replacc A with covariance matrix C.

$$\|t - t_i\|_A = (t - t_i)^T A(t - t_i) = (t - t_i)^T C_i(t - t_i)$$

A two-step method is proposed: the first step is to use the image gradient estimated by the classical kernel regression method as the initial value. Secondly, use this initial estimate to measure the dominant orientation of the image's local kernel, that is, use this initial estimate to adaptively "control" the local kernel, so that a shape similar to an elliptical contour spreading along the direction of the local edge structure can be formed. Kernel function. With these local adaptive kernels, the edges of the image can be effectively restored, making the details after denoising richer. The form of the adaptive core is as follows:

$$K_{adapt}(t_i - t, Y_i - Y) = K_{H_i^{steer}}(t_i - t)$$

Then the above distance is applied to the Gaussian kernel function to obtain an adaptive Steering kernel,

$$K(t - t_i) = \frac{\sqrt{\det_i(C_i)}}{2\pi h^2 \mu_i^2} \exp\left\{ -\frac{(t - t_i)^T C_i (t - t_i)}{2\pi h^2 \mu_i^2} \right\}$$

Definition of local covariance matrix C_i,

$$C_i = \begin{bmatrix} \sum_{X_j \in \omega_i} z_{x_1}(T_j) z_{x_1}(T_j) & \sum_{X_j \in \omega_i} z_{x_1}(T_j) z_{x_2}(T_j) \\ \sum_{X_j \in \omega_i} z_{x_1}(T_j) z_{x_1}(T_j) & \sum_{X_j \in \omega_i} z_{x_2}(T_j) z_{x_2}(T_j) \end{bmatrix}$$

Where, ω_i is the local window opened, $z_{t_1}(\cdot), z_{t_2}(\cdot)$ is the first derivative in the t_1, t_2 direction. The covariance matrix C_i is a symmetric matrix, Therefore, it can be orthogonal decomposed into:

$$C_i = \gamma_i U_{\rho_i} \Lambda_i U_{\rho_i}^T, U_{\rho_i} = \begin{bmatrix} \cos\rho_i & \sin\rho_i \\ -\sin\rho_i & \cos\rho_i \end{bmatrix}, \Lambda_i = \begin{bmatrix} \delta_i & 0 \\ 0 & \delta_i^{-1} \end{bmatrix}$$

Then the covariance matrix is determined by the parameters γ, ρ, δ, and the three parameters are selected according to the local gradient matrix C_i, Singular value decomposition of local gradient matrix:

$$C_i = \begin{bmatrix} \vdots & \vdots \\ z_{t_1}(T_j) & z_{t_2}(T_j) \\ \vdots & \vdots \end{bmatrix} = U_i S_i V_i^T$$

Where $S_i = \begin{bmatrix} s_1 & 0 \\ 0 & s_2 \end{bmatrix}$ represents the energy in the main direction and $V_i = [V_1, V_2] = \begin{bmatrix} * & v_1 \\ * & v_2 \end{bmatrix}$ determines the angle of the main direction, Define the elongation parameter $\delta_i : \delta_i = \frac{s_1 + \lambda}{s_2 + \lambda}, \lambda \geq 0, \lambda$ is a regularization parameter used to suppress noise and limit δ_i. Definition:

$$H_i^{steer} = h\mu_i C_i^{-\frac{1}{2}}$$

3 Experiments of Removing Cloud

In the experiment, three images are selected as grayscale images. Based on wavelet transform, multi-scale Retinex (k = 1), homomorphic filtering (n = 1), and JIANG Xing-fang method [7–9], the histogram of the obtained image is further considered. In the case of equalization, the calculation results are as follows.

	Brightness		Contrast		Production		Entropy	
	Normal	Equalizaiton	Normal	Equalizaiton	Normal	Equalizaiton	Normal	Equalizaiton
Original image	128.8	27.5	43.53	74.61	4260	7400	7.201	7.937
Wavelet	126.8	127.9	78.37	74.47	7720	7411	7.739	7.937
Retinex (k = 1)	171.2	128.0	58.47	75.59	7795	7526	7,227	7.844
High pass filter (n = 1)	128.6	127.5	127.5	33.30	5322	7396	7.523	7.954
Method of this article	128.5	127.5	127.6	74.59	6453	7489	7.324	7.675

The above table lists the comparison between the original image and the four images after enhancement and equalization to obtain the average brightness, contrast, product of brightness and contrast, and information entropy.

References

1. Takeda, H.: Kernel regression for image processing and reconstruction. IEEE Trans. Image Process. **16**(2), 349–366 (2007)
2. Wang, X., Wen, H., Luo, Y.: An image enhancement method based on robust estimation. Comput. Appl. **26**(7), 1611–1613 (2006)
3. Chen, H., Yang, Y.: Large-scale image based on normalized gradient phase correlation fast registration algorithm. Pattern Recogn. Artif. Intell **28**(8), 694–701 (2015)
4. Zhuo, G., Wu, B., Zhu, X.: An adaptive 3D kernel regression based remote sensing spatiotemporal fusion method. J. Wuhan Univ. (Information Science Edition) **43**(4), 563–570 (2018)
5. Shi, X., Xia, K., Wang, B., et al.: Remote sensing image using improved RPCALike the cloud algorithm. Comput. Eng. Des. **39**(6), 1653–1658 (2018)
6. Wu, F., Zhu, X., Xiangruxi, et al.: Remote sensing image based on dual-tree complex wavelet transformLike the cloud and fog system design. Appl. Optics **39**(1), 64–70 (2018)
7. Li, C., Deng, X., Zhao, H.: Remote sensing image thin cloud removal based on wavelet analysis research on division algorithm. Digital Technol. Appl. **6**, 137–139 (2017)
8. Lu, W., Wang, J., Cao, M.: Variable order dictionary learning AO-DL resource number three remote sensing image cloud removal. J. Surv. Map. **46**(5), 623–630 (2017)
9. Jiang, X.: Analysis of quality of removing cloud for monochromatic remote sensing imag. Leser Technol. **33**(3), 331–332 (2009)

Author Index

M. Tavana et al. (Eds.): IISA 2020, AISC 1304, pp. 997–1001, 2021.
https://doi.org/10.1007/978-3-030-63784-2